Progress in Clean Energy, Volume 2

Ibrahim Dincer • C. Ozgur Colpan
Onder Kizilkan • M. Akif Ezan
Editors

Progress in Clean Energy, Volume 2

Novel Systems and Applications

Springer

Editors
Ibrahim Dincer
Department of Mechanical Engineering
University of Ontario Institute of Technology
Oshawa, ON, Canada

Onder Kizilkan
Department of Energy Systems Engineering
Suleyman Demirel University
Isparta, Turkey

C. Ozgur Colpan
Department of Mechanical Engineering
Dokuz Eylul University
Izmir, Turkey

M. Akif Ezan
Department of Mechanical Engineering
Dokuz Eylul University
Izmir, Turkey

ISBN 978-3-319-17030-5 ISBN 978-3-319-17031-2 (eBook)
DOI 10.1007/978-3-319-17031-2

Library of Congress Control Number: 2015943850

Springer Cham Heidelberg New York Dordrecht London

Printed on acid-free paper

Springer International Publishing AG Switzerland is part of Springer Science+Business Media (www.springer.com)

Preface

Since energy is identified by many as one of humanity's most significant problems, clean energy solutions are recognized as the only option to achieve a more sustainable future. The prerequisite of this is to develop novel energy systems and implement them accordingly for practical applications.

The second volume of this edited book entitled *Progress in Clean Energy* focuses on novel systems and their applications and contains 76 uniquely selected papers out of the conference papers presented in the 13th International Conference on Clean Energy (ICCE-2014), which was held in Istanbul, Turkey, on June 8–12, 2014. This distinctive event was organized not only to bring all the researchers, scientists, policy makers, and engineers conducting research on the clean energy field together, but also to honor Dr. Veziroglu's 90th birthday in Istanbul, which is also his birthplace. This conference provided a forum for the exchange of latest technical information, the dissemination of the high-quality research results on the issues, the presentation of the new developments in the area of clean energy, and the debate and shaping of future directions and priorities for better environment, sustainable development, and energy security. The recent research findings in the clean energy topics including solar, wind, hydropower, nuclear, and hydrogen energy technologies, fuel cells, biomass and biofuels, clean fossil fuels, carbon sequestration and carbon tax, energy storage and energy conservation, environmental impact and remediation, and sustainable development and energy management were presented and discussed in this conference.

There is diverse coverage of clean energy topics in this volume, ranging from battery technologies to biomass and biofuels, from clean coal technologies to renewable energies, and many more to serve as a sustainable source of knowledge and information for researchers, scientists, engineers, practitioners, etc. The clean energy topics are linked to many areas, including energy and environment policies, energy conversion technologies, energy management and conservation, energy saving, energy security, renewable and sustainable energy technologies, emission reduction, sustainable development, pollution control and measures, policy development, global energy stability and sustainability, carbon tax, and waste management.

We hope that the second volume of this book will provide a unique source of clean energy systems and applications with a prime focus on novel energy systems and their implementation. We sincerely appreciate the help and assistance provided by various individuals and conference organizing committee members who deserve a clear acknowledgement.

Oshawa, ON, Canada — Ibrahim Dincer
Izmir, Turkey — C. Ozgur Colpan
Isparta, Turkey — Onder Kizilkan
Izmir, Turkey — M. Akif Ezan

Contents

Part I
Battery Technology

Chapter 1
Electrochemical Study of the Cobalt-Containing and Non-Cobalt-Containing AB_5 Alloys Used as Negative Electrodes in Nickel–Metal Hydride Batteries

Chokri Khaldi and Jilani Lamloumi

Abstract The thermodynamic parameters of $LaNi_{3.55}Mn_{0.4}Al_{0.3}Co_{0.75}$ and $LaNi_{3.55}Mn_{0.4}Al_{0.3}Fe_{0.75}$ alloys have been evaluated from the electrochemical isotherms determined by OCV method. A comparative study has been done between the parameter values deduced from the electrochemical OCV method and the solid–gas method. It was found that the difference can be explained by the electrochemical corrosion of these alloys in aqueous KOH electrolyte. The corrosion behavior of the $LaNi_{3.55}Mn_{0.4}Al_{0.3}Co_{0.75}$ and $LaNi_{3.55}Mn_{0.4}Al_{0.3}Fe_{0.75}$ electrodes after activation was investigated using the method of the potentiodynamic polarization. The evolution of corrosion current density and potential values as function of the state of charge (SOC), show that the total substitution of cobalt by iron leads to accentuate the corrosion process.

Keywords Metallic hydride • Nickel-metal hydride batteries • Electrochemical techniques • Electrochemical isotherm • Thermodynamic parameters • Corrosion parameters

1.1 Introduction

The commercial negative electrodes of nickel–metal hydride (Ni–MH) rechargeable batteries, especially a nominal composition $MmNi_{3.55}Mn_{0.4}Al_{0.3}Co_{0.75}$, has been found to offer the best compromise between a high hydrogen capacity and a good cycle life. Generally, these commercial alloys contain 10 wt% of cobalt, which constitutes about 40–50 % of the total cost of raw materials [1–6].

C. Khaldi (✉) • J. Lamloumi
Equipe des Hydrures Métalliques, Laboratoire de Mécanique, Matériaux et Procédés, Ecole Nationale Supérieure des Ingénieurs de Tunis, Université de Tunis, 5 Avenue Taha Hussein, 1008 Tunis, Tunisia
e-mail: chokri.khaldi@esstt.rnu.tn

I. Dincer et al. (eds.), *Progress in Clean Energy, Volume 2*,
DOI 10.1007/978-3-319-17031-2_1

The substitution of Co by other less-expensive elements or the complete elimination will significantly reduce the cost of hydrogen storage alloys. So low-cobalt or cobalt-free AB_5-type hydrogen storage alloys continues to be attractive.

Metal hydrides for battery applications are characterized in terms of their hydrogen storage capacity at moderate hydride stability, approximately constant equilibrium pressure during hydride phase transformation and reasonably fast sorption kinetics. The information about these characteristics is generally derived from pressure–composition–temperature (PCT) isotherms, which describe the dependence of the hydrogen equilibrium pressure on the amount of hydrogen absorbed or incorporated into the hydrogen storage material at fixed or at various temperatures [7, 8]. Thermodynamic properties such as the change in the enthalpy and the change in entropy can be calculated from PCT isotherms [9]. The PCT isotherms can also be used to identify different phases in metal hydride alloys [10]. Hydrogen concentration at various stages such as in the α-phase, during transition of α-phase to β-phase, and in the β-phase in equilibrium with the α-phase can also be obtained from PCT isotherms [11]. PCT diagrams can be constructed for metal hydride alloys in two ways, namely, a solid–gas PCT isotherm and an electrochemical PCT isotherm. In general, the plateau pressure for the metal hydride electrode at room temperature is lower than 1 atm, which calls for specially designed high-vacuum equipment as well as a long period for attaining hydrogen equilibrium pressure. If the plateau pressure is much lower than 1 atm, then it is impossible to attain the equilibrium pressure [12]. The electrochemical approach to construct PCT isotherms, therefore, is a simple and convenient alternative procedure. Despite the importance of the heat of hydride formation of metal hydrides for battery applications, sufficient studies have not appeared in the literature, especially by electrochemical methods.

In the previous works [13] we have shown that it is possible to obtain a satisfactory electrochemical stability of cobalt-free AB_5 hydrogen storage; In this chapter we present a detailed description of the thermodynamic properties of the commercial $LaNi_{3.55}Mn_{0.4}Al_{0.3}Co_{0.75}$ and the cobalt-free $LaNi_{3.55}Mn_{0.4}Al_{0.3}Fe_{0.75}$ negative electrodes. Solid–gas and electrochemical isotherms are determined. Solid–gas and electrochemical capacities are compared. The corrosion effect on the electrochemical properties of $LaNi_{3.55}Mn_{0.4}Al_{0.3}Co_{0.75}$ and $LaNi_{3.55}Mn_{0.4}Al_{0.3}Fe_{0.75}$ electrodes after activation is investigated.

1.2 Experimental

The $LaNi_{3.55}Mn_{0.4}Al_{0.3}Co_{0.75}$ and $LaNi_{3.55}Mn_{0.4}Al_{0.3}Fe_{0.75}$ alloys were prepared by induction melting of the pure elements followed by an appropriate annealing to ensure a good homogeneity. The results of the metallographic examination, the electron probe microanalysis (EPMA), and the structure characterization by X-ray diffraction (XRD) of these alloys are given in Table 1.1. The structural characterization shows that these alloys are indexed in the $CaCu_5$ hexagonal crystalline structure (P6/mmm space group). The $LaNi_{3.55}Mn_{0.4}Al_{0.3}Co_{0.75}$ and $LaNi_{3.55}Mn_{0.4}Al_{0.3}Fe_{0.75}$ ingot alloy were grounded mechanically and sieved (to less than 63 μm) in a glove box under an argon atmosphere [14].

Table 1.1 Annealing treatment conditions and characterization results of the $LaNi_{3.55}Mn_{0.4}Al_{0.3}Co_{0.75}$ and $LaNi_{3.55}Mn_{0.4}Al_{0.3}Fe_{0.75}$ alloys

Compounds	Annealing treatment condition	Electron probe microanalysis (EPMA)	X-ray diffraction (XRD) (cell parameters and volume)		
			a (Å)	c (Å)	V (Å^3)
$LaNi_5$	12 days, 1,000 °C	–	5.012	3.984	86.680
$LaNi_{3.55}Mn_{0.4}Al_{0.3}Co_{0.75}$	12 days, 1,000 °C	$LaNi_{3.52}Mn_{0.38}Al_{0.28}Co_{0.73}$	5.069	4.044	89.987
$LaNi_{3.55}Mn_{0.4}Al_{0.3}Fe_{0.75}$	12 days, 1,000 °C	$LaNi_{3.53}Mn_{0.38}Al_{0.28}Fe_{0.79}$	5.092	4.066	91.310

The "latex" technology has been used for the electrode preparation [15]. Ninety percent of the alloy powder was mixed with five percent of black carbon (to obtain a good conductivity) and 5 % of polytetrafluoroethylen (PTFE). Two pieces of 0.5 cm^2 of this latex have been pressed on each side of a nickel grid, playing the role of a current collector [16]. This assembly forms the negative electrode of Ni-MH battery. All the electrochemical measurements were performed at a room temperature in a conventional three electrodes open-air cell coupled directly with a Voltalab40 system (Radiometer Analytical) constituted by a Potentiostat–Galvanostat PGZ301. A nickel mesh and an Hg/HgO electrode have been used as a counter electrode and a reference electrode, respectively. The electrolyte consisted of KOH (1 M) solution, which is prepared using highest purity grade chemicals and triply distilled water, which is paddled by a continuous flow of argon.

The electrodes were cycled by the galvanostatically charging and discharging at a C/10 rate during a long cycling. The electrochemical isotherms were conducted in galvanostatic mode after activation by the method of open-circuit voltage (OCV) measurements during the discharge process. The electrodes are fully charged at C/30 regime for 45 h (50 % over charge). We realize several successive discharges of 5 % of state of discharge at D/30 rate during 1.5 h until the complete discharge electrodes. After each 5 % of state of discharge, the circuit is open to stabilize the potential (OCV).

The potentiodynamic polarization tests were carried out at a scan rate of 1 mVs^{-1}, between (−600 mV) and (−1,200 mV) vs. Hg/HgO at different state of charge (%SOC), to estimate the corrosion current density and potential of the electrodes in KOH solution.

1.3 Results and Discussion

1.3.1 Lifetime of the Electrodes

Figure 1.1 shows the discharge capacity versus the number of cycles of the negative electrodes consisting of $LaNi_{3.55}Mn_{0.4}Al_{0.3}Co_{0.75}$ and $LaNi_{3.55}Mn_{0.4}Al_{0.3}Co_{0.75}$ alloys. After few cycles, a maximum discharge capacity was obtained. The total

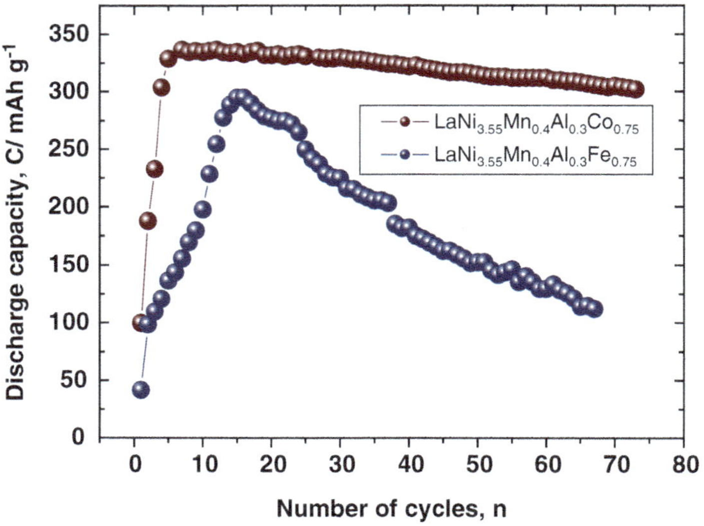

Fig. 1.1 Discharge capacity versus cycle number for the $LaNi_{3.55}Mn_{0.4}Al_{0.3}Co_{0.75}$ and $LaNi_{3.55}Mn_{0.4}Al_{0.3}Co_{0.75}$ electrodes in open test cells, at C/10 rate

cobalt substitution by iron leads to a decrease in the maximum capacity from 338 to 292 mAh g^{-1}, with a bad cycling held. Indeed, the discharge capacity loss after 60 cycles are about 8 % and 57 %, respectively, for $LaNi_{3.55}Mn_{0.4}Al_{0.3}Co_{0.75}$ and $LaNi_{3.55}Mn_{0.4}Al_{0.3}Co_{0.75}$ alloys.

1.3.2 Thermodynamic Parameters of the Alloys

The electrochemical isotherms of the $LaNi_{3.55}Mn_{0.4}Al_{0.3}Co_{0.75}$ cobalt containing and $LaNi_{3.55}Mn_{0.4}Al_{0.3}Fe_{0.75}$ non-cobalt-containing alloys were determined by a succession of discharges followed by the technique of open-circuit voltage (OCV). The electrodes are fully charged at C/30 regime for 45 h (50 % overloads). We realized many successive discharges of 5 % of state of discharge at a very low rate D/30 for 1.5 h, until the complete discharge electrodes. After each little discharge, the circuit is open to stabilize the potential.

The maximum discharge capacity is expressed in terms of the sum of successive time t_{di} associated with each small discharge of 1.5 h:

$$C_{max} = \frac{i_d}{m}\sum_i t_{di} \tag{1.1}$$

Where, i_d is the discharge current (mA) and m is the mass of alloy (g).

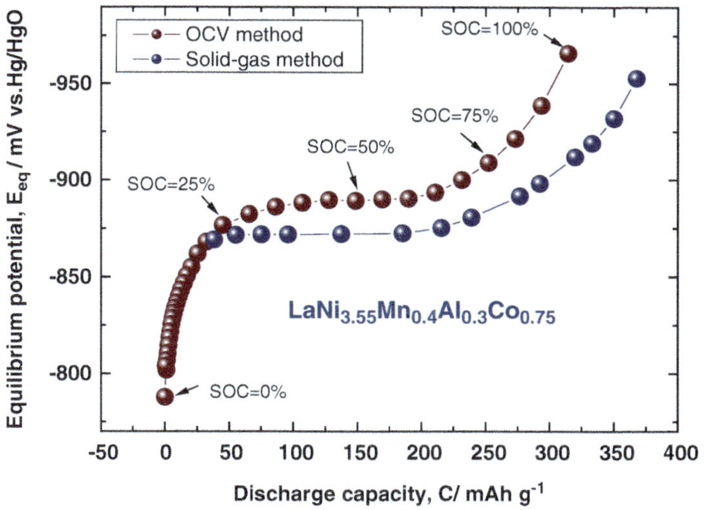

Fig. 1.2 Electrochemical isotherms determined by solid–gas and by a succession of discharges followed by the technique of open-circuit potential (OCV) of $LaNi_{3.55}Mn_{0.4}Al_{0.3}Co_{0.75}$ electrode, after activation

The electrochemical capacity of hydrogen remaining in the electrode during the jth discharge is:

$$C_j = C_{max} - \frac{i_d}{m}\sum_{k=1}^{k=j} t_{dk} \tag{1.2}$$

The jth equilibrium potential E_j associated with each discharge state is determined by the technique of open-circuit voltage (OCV). So we draw the electrochemical isotherms of $LaNi_{3.55}Mn_{0.4}Al_{0.3}Co_{0.75}$ and $LaNi_{3.55}Mn_{0.4}Al_{0.3}Co_{0.75}$ electrodes, respectively, shown in Figs. 1.2 and 1.3 by coupling points (E_j, C_j). These electrochemical OCV isotherms are compared with those determined by solid–gas isotherms [17], where the pressure (bar) and the hydrogen content (H mol^{-1}), in solid–gas, are converted into potential (mV) and electrochemical capacity (mAh g^{-1}) by the following equations:

$$E(\mathrm{mV}) = -0.926 - \frac{RT}{2F}\,\mathrm{Ln}\left(P_{H2}(\mathrm{bar})\right) \tag{1.3}$$

and

$$C\left(\mathrm{mAhg^{-1}}\right) = \frac{F}{3.6M} \times \left(\mathrm{Hmol^{-1}}\right) \tag{1.4}$$

Where, R is the ideal gas constant (8.32 J mol^{-1} K^{-1}), F is the Faraday constant (96,500 C mol^{-1}), T is the temperature (K), M is the molar mass of alloy (g) and × is the number of hydrogen atoms per alloy formula (H mol^{-1}).

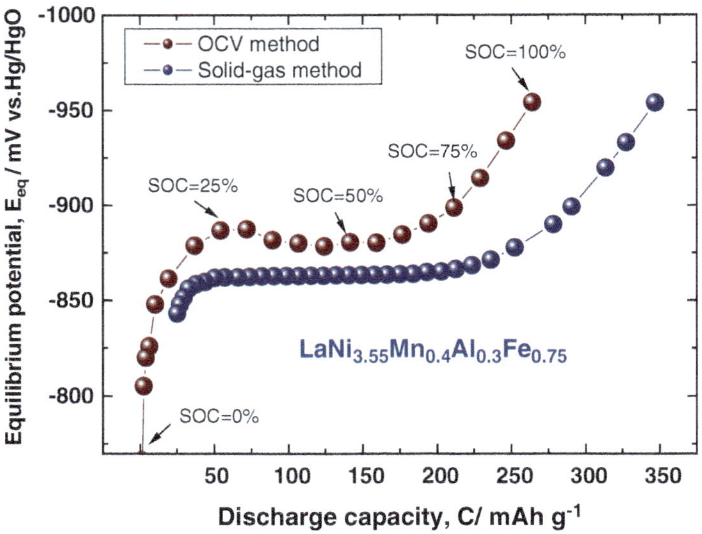

Fig. 1.3 Electrochemical isotherms determined by solid–gas and by a succession of discharges followed by the technique of open-circuit potential (OCV) of $LaNi_{3.55}Mn_{0.4}Al_{0.3}Fe_{0.75}$ electrode, after activation

Table 1.2 OCV and solid–gas thermodynamic parameters of the $LaNi_{3.55}Mn_{0.4}Al_{0.3}Co_{0.75}$ and $LaNi_{3.55}Mn_{0.4}Al_{0.3}Fe_{0.75}$ alloys

Compounds	Solid–gas				
	Equilibrium potential, E_{eq} (mV)	Equilibrium pression, P_{H2} (bar)	Solid–gas capacity, C_{sg} (H/mol)	Solid–gas capacity, C_{sg} (mAh/g)	
$LaNi_{3.55}Mn_{0.4}Al_{0.3}Co_{0.75}$	−874	0.015	5.78	368	
$LaNi_{3.55}Mn_{0.4}Al_{0.3}Fe_{0.75}$	−867	0.009	5.45	347	
Compounds	OCV				
	Equilibrium potential, E_{eq} (mV)	Equilibrium pression, P_{H2} (bar)	Electrochemical capacity, $C_{él}$ (mAh/g)	$\frac{\Delta P}{P}$ (%)	$\frac{\Delta C}{C}$ (%)
$LaNi_{3.55}Mn_{0.4}Al_{0.3}Co_{0.75}$	−877	0.020	315	25	17
$LaNi_{3.55}Mn_{0.4}Al_{0.3}Fe_{0.75}$	−878	0.022	265	60	31

All the electrochemical isotherms determined by succession of discharges followed by the technique of open-circuit voltage (OCV) are almost similar to the isotherms determined by solid–gas technique. Indeed, these methods are established under the equilibrium after the relaxation time long enough. Note that the electrochemical isotherms are shifted, to the capacity decrease, relative to those found by solid–gas. This discrepancy is due to the corrosion of these compounds in the KOH solution and will be studied in detail in the end of this chapter.

In Table 1.2, we have grouped the different values of the electrochemical capacity (C_{el} and C_{sg} expressed in mAh g^{-1}), equilibrium potential (E_{eq}) and the equilibrium pressure (P_{H2}) obtained by solid–gas and electrochemical OCV methods, of the $LaNi_{3.55}Mn_{0.4}Al_{0.3}Co_{0.75}$ and the $LaNi_{3.55}Mn_{0.4}Al_{0.3}Fe_{0.75}$ metal alloy electrodes and the uncertainty of measurements of the equilibrium pressure and the electrochemical capacity between the two methods ($\Delta P/P$ and $\Delta C/C$).

Although both methods, electrochemical method (OCV) and solid–gas method, are established at equilibrium state, the values of equilibrium pressure and discharge capacities, obtained by the both methods, are different. Note that this difference is smaller for the compound containing cobalt. This result is explained by the pronounced corrosion for the compounds rich in iron. Also we note that for the non-cobalt-containing alloy, more corrosive in KOH, the obtained electrochemical capacity is lower.

Indeed, such a layer of oxide or hydroxide once formed, its thickness increases with the cycling and therefore acts as a barrier to the diffusion of hydrogen in the volume of intermetallic compound in question.

1.3.3 Corrosion Parameters of the Alloys

The study of the corrosion behaviour of the $LaNi_{3.55}Mn_{0.4}Al_{0.3}Co_{0.75}$ and $LaNi_{3.55}Mn_{0.4}Al_{0.3}Fe_{0.75}$ electrodes after activation was performed by the technique of cyclic voltammetry at fast scan rate of 1 mV s^{-1} using the method of the potentiodynamic polarization. The 1 mV s^{-1} fast scan rate is used to study the electrochemical properties of metal hydrides [4, 5]. In order to estimate the anodic corrosion rate at the open-circuit potential (corrosion), the methods of Tafel extrapolation were applied. The complete Tafel equation for the anodic and cathodic reaction is [18]:

$$I = I_{\mathrm{corr}}\left\{\exp\left[\frac{0.0023}{b_{\mathrm{a}}}(E - E_{\mathrm{corr}})\right] - \exp\left[\frac{0.0023}{b_{\mathrm{c}}}(E - E_{\mathrm{corr}})\right]\right\} \tag{1.5}$$

Where b_a and b_c are the Tafel slopes, E_{corr} is the corrosion potential and I_{corr} is the corrosion current density. Equation (1.5) can therefore be used to calculate the corrosion rates, expressed as current densities and potentials using the software Origin.

Figs. 1.4 and 1.5 represent the evolution of the Tafel curves with the state of charge of the $LaNi_{3.55}Mn_{0.4}Al_{0.3}Co_{0.75}$ and $LaNi_{3.55}Mn_{0.4}Al_{0.3}Fe_{0.75}$ electrodes, respectively.

By increasing the state of charge, the Tafel curves shift to the negative direction, i.e., the corrosion potential is more negative for the larger state of charge.

In Figs. 1.6 and 1.7 we represent, respectively, the evolution of the corrosion current density and potential with the state of charge of the $LaNi_{3.55}Mn_{0.4}Al_{0.3}Co_{0.75}$ and $LaNi_{3.55}Mn_{0.4}Al_{0.3}Fe_{0.75}$ alloys.

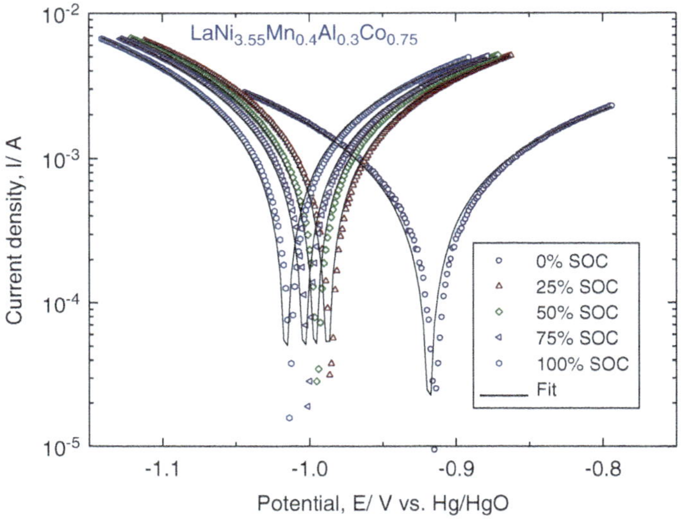

Fig. 1.4 Evolution of the Tafel curves as function of state of charge (SOC) of the $LaNi_{3.55}Mn_{0.4}Al_{0.3}Co_{0.75}$ electrode, at 1 mV s^{-1} potential scan rate after activation

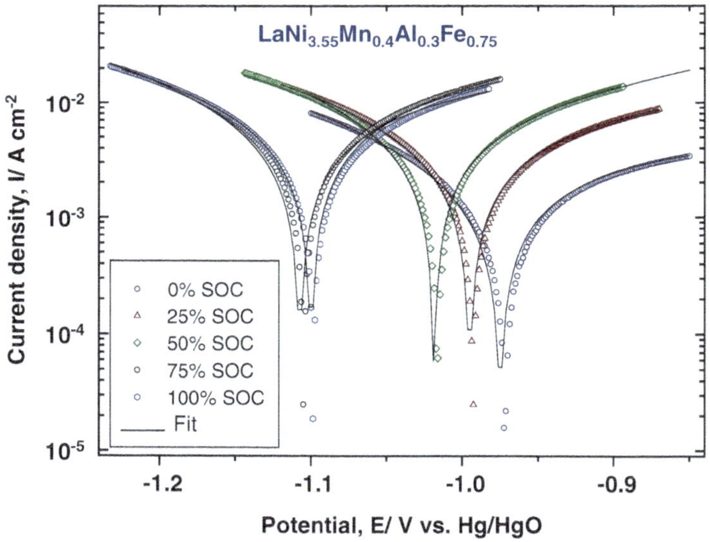

Fig. 1.5 Evolution of the Tafel curves as function of state of charge (SOC) of the $LaNi_{3.55}Mn_{0.4}Al_{0.3}Fe_{0.75}$ electrode, at 1 mV s^{-1} potential scan rate after activation

For the $LaNi_{3.55}Mn_{0.4}Al_{0.3}Co_{0.75}$ compound, rich in cobalt, the corrosion current density and the corrosion potential are practically constant independent of the state of charge beyond 25 % of state of charge, i.e., the elements La, Ni, Mn, Al, and Co are practically less corrosive. However, for the $LaNi_{3.55}Mn_{0.4}Al_{0.3}Fe_{0.75}$

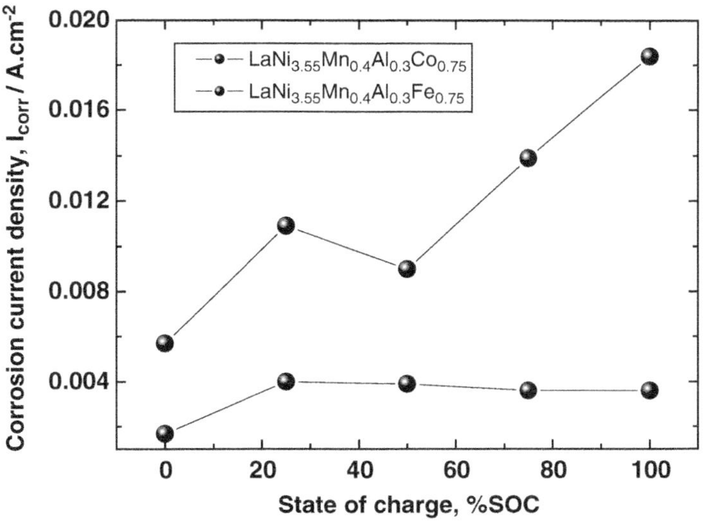

Fig. 1.6 Evolution of the corrosion current density as function of state of charge (SOC) of the $LaNi_{3.55}Mn_{0.4}Al_{0.3}Co_{0.75}$ and $LaNi_{3.55}Mn_{0.4}Al_{0.3}Fe_{0.75}$ electrodes after activation

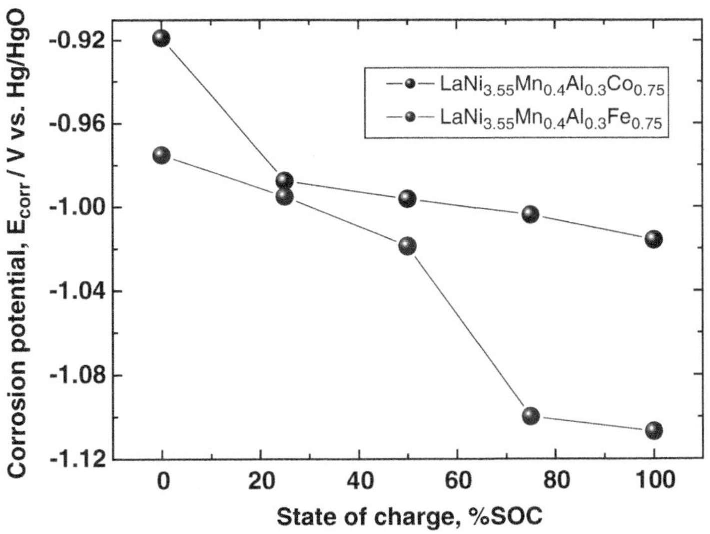

Fig. 1.7 Evolution of corrosion potential as function of state of charge (SOC) of the $LaNi_{3.55}Mn_{0.4}Al_{0.3}Co_{0.75}$ and $LaNi_{3.55}Mn_{0.4}Al_{0.3}Fe_{0.75}$ electrodes after activation

compound, rich in iron, the corrosion current density increases with increasing the state of charge and the corrosion potential shift towards the more negative values with increasing the state of charge. This result is explained by the pronounced corrosion for the compounds rich in iron.

1.3.4 Morphology Studies of the Alloys

The exploitation of the results obtained by different electrochemical techniques requires a morphological study of the electrode surface during cycling. This characterization is performed by scanning electron microscopy SEM. The decrepitation of intermetallic compounds during hydrogenation is an important feature because of its impact on their properties in particular the kinetics and aging. The micrographs obtained by scanning electron microscopy (SEM) allow us to determine the grain size and observe the surface morphology of the $LaNi_{3.55}Mn_{0.4}Al_{0.3}Co_{0.75}$ and $LaNi_{3.55}Mn_{0.4}Al_{0.3}Fe_{0.75}$ electrodes before and after cycling.

Figures 1.8 and 1.9 show, respectively, the SEM micrographs of the latex of the $LaNi_{3.55}Mn_{0.4}Al_{0.3}Co_{0.75}$ and $LaNi_{3.55}Mn_{0.4}Al_{0.3}Fe_{0.75}$ electrodes before and after cycling.

The morphology of the electrode surface before cycling differs from that of the electrode which has undergone electrochemical repeated cycles of charging and discharging. After cycling, the micrographs show the appearance of a high rate of cracks. Indeed, during cycling of the electrode, a phenomenon of decrepitation of the hydride is observed. The absorption of hydrogen in the network causes the fracture of the grains and hence the size grain of alloy decreases with the number of cycles. In addition, these micrographs show that the cracking of the electrode grain is remarkably observed for the alloy rich in iron. This is due to preferential oxidation of iron at the surface of alloys in aqueous KOH electrolyte.

The accentuated corrosion of iron-rich compound can be correlated to the low value of the electrochemical capacity and the highest value of corrosion current for this compound.

Fig. 1.8 SEMs photography of the $LaNi_{3.55}Mn_{0.4}Al_{0.3}Co_{0.75}$ electrode: (**a**) before activation and (**b**) after activation

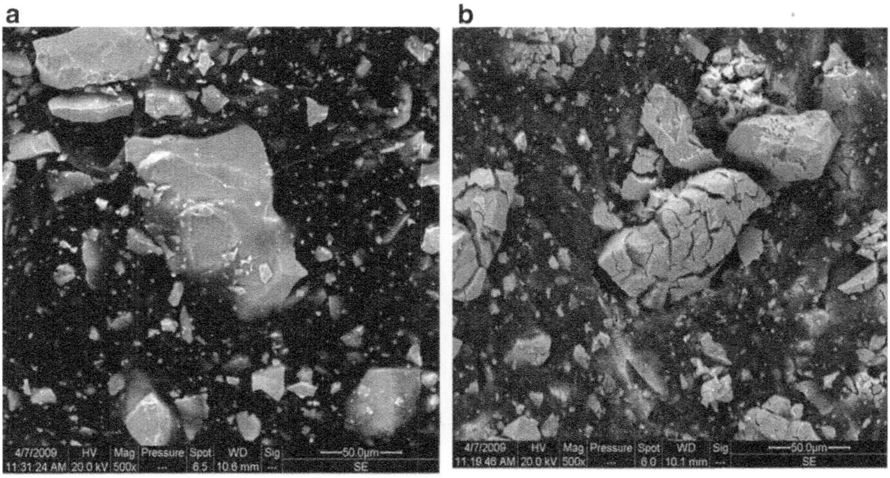

Fig. 1.9 SEMs photography of the $LaNi_{3.55}Mn_{0.4}Al_{0.3}Fe_{0.75}$ electrode: (**a**) before activation and (**b**) after activation

1.4 Conclusions

The aim of this work is to study the corrosion effect on the electrochemical properties of $LaNi_{3.55}Mn_{0.4}Al_{0.3}Co_{0.75}$ and $LaNi_{3.55}Mn_{0.4}Al_{0.3}Fe_{0.75}$ negative electrodes used in Ni-MH batteries. The following conclusions can be drawn up:

- For the $LaNi_{3.55}Mn_{0.4}Al_{0.3}Fe_{0.75}$ non-cobalt-containing alloy, rich in iron and more corrosive in KOH, the obtained electrochemical capacity is lower than that of the containing-cobalt alloy.
- For the $LaNi_{3.55}Mn_{0.4}Al_{0.3}Co_{0.75}$ cobalt containing alloy, rich in cobalt, the corrosion current density and the corrosion potential are practically constant independent of the state of charge beyond 25 % of SOC. However, for the $LaNi_{3.55}Mn_{0.4}Al_{0.3}Fe_{0.75}$ alloy, rich in iron, the corrosion current density increases with increasing the state of charge and the corrosion potential shift towards the more negative values with increasing the state of charge. This result is explained by the pronounced corrosion for the compounds rich in iron.

The accentuated corrosion of iron-rich alloy can be correlated to the low value of the electrochemical capacity and the highest value of corrosion current for this alloy.

Acknowledgement The authors would like to express their gratefulness towards Mr. M. Latroche and Mrs. A. Percheron-Guégan (LCMTR, CNRS, France) for having given them the opportunity to prepare the alloys in their laboratory.

References

1. Huang YX, Zhang H (2000) Characteristics of a low-cobalt AB -type hydrogen storage alloy obtained 5 by a gas-atomization processing. J Alloys Compd 305:76–81
2. Hu WK (1999) Studies on cobalt-free AB -type hydrogen storage alloys. J Alloys Compd 289:299–305
3. Hu WK (2000) Improvement in capacity of cobalt-free Mm-based hydrogen storage alloys with good cycling stability. J Alloys Compd 297:206–210
4. Wang YJ, Yuan HT, Wang G, Zhang YS (2002) Characteristics of novel Co-free $MlNi_{4.2-x}Cu_{0.5}Al_{0.3}Zn_x$ ($0 < x < 0.1$) hydrogen storage alloys. Solid State Ionics 148:509–512
5. Lei YQ, Zhang SK, Lu GL, Chen LX, Wang QD, Wu F (2002) Influence of the material processing on the electrochemical properties of cobalt-free $Ml(NiMnAlFe)_5$ alloy. J Alloys Compd 330–332:861–865
6. Tang R, Liu LQ, Liu YN, Yu G (2003) Study on the microstructure and the electrochemical properties of $Ml_{0.7}Mg_{0.2}Ni_{2.8}Co_{0.6}$ hydrogen storage alloy. Int J Hydrog Energy 28:815–819
7. Kleperis J, Wójcik G, Czerwinski A, Skowronski J, Kopczyk M, Beltowska-Brzezinska M (2001) Electrochemical behavior of metal hydrides. J Solid State Electrochem 5:229–249
8. Jain IP, Abu Dakka MIS (2002) Hydrogen absorption–desorption isotherms of $La_{(28.9)}Ni_{(67.55)}Si_{(3.55)}$. Int J Hydrog Energy 27:395–401
9. Züttel A (2004) Hydrogen storage methods. Naturwissenschaften 91:157–172
10. Schlapbach L, Anderson I, Burger J (1994) Hydrogen in metals. In: Buschow KHJ (ed) Electronic and magnetic properties of metals and ceramics part II. John Wiley, New York, NY, pp 271–331
11. Bliznakov S, Lefterova E, Bozukov L, Popov A, Andreev P (2004) Proceedings of the International workshop on advanced techniques for energy sources investigation and testing. Sofia, Bulgaria, September 4–9, 2004
12. Wang CS, Wang XH, Lei YQ, Chen CP, Wang QD (1997) A new method of determining the thermodynamic parameters of metal hydride electrode materials. Int J Hydrog Energy 22(12):1117–1124
13. Khaldi C, Mathlouthi H, Lamloumi J, Percheron-Guégan A (2004) Electrochemical study of cobalt-free AB5-type hydrogen storage alloys. Int J Hydrog Energy 29:307–311
14. Khaldi C, Mathlouthi H, Lamloumi J, Percheron-Guégan A (2004) Electrochemical impedance spectroscopy and constant potential discharge studies of LaNi3.55Mn0.4Al0.3Co0.75−xFex hydrides alloy electrodes. J Alloys Compd 384:249–253
15. Mathlouthi H, Lamloumi J, Latroche M, Percheron-Guégan A (1997) Study of poly-substituted intermetallic hydrides: electrochemical applications. Ann Chimie Sci Materiaux 22:241–244
16. Li CJ, Wang FR, Cheng WH, Li W, Zhao WT (2001) The influence of high-rate quenching on the cycle stability and the structure of the AB_5-type hydrogen storage alloys with different Co content. J Alloys Compd 315:218–223
17. Ayari M, Paul-Boncour V, Lamloumi J, Mathlouthi H, Percheron-Guégan A (2006) Study of the structural, thermodynamic and electrochemical properties of $LaNi_{3.55}Mn_{0.4}Al_{0.3}(Co_{1-x}Fe_x)_{0.75}$ ($0 \le x \le 1$) compounds used as negative electrode in Ni-MH batteries. J Alloys Compd 420:251–255
18. Zhu BH, Lindbergh G, Simonsson D (1999) Comparison of electrochemical and surface characterization methods for investigation of corrosion of bipolar plate materials in molten carbonate fuel cell: Part I. Electrochemical study. Corros Sci 41:1497–1513

Chapter 2
Nanocomposite Based on $Li_4Ti_5O_{12}$ Structures for High-Rate Li-Ion Battery Applications

Aslıhan Erdaş, Şeyma Özcan, Deniz Nalci, Mehmet Oğuz Güler, and Hatem Akbulut

Abstract Lithium titanate is synthesized from titanium isopropoxide and lithium nitrate solution via sol–gel processes. The obtained nanocrystalline lithium titanates were then subjected to electroless deposition in order to obtain Cu/$Li_4Ti_5O_{12}$ nanocomposite structures. The crystalline structure and morphological observation of the as-synthesized $Li_4Ti_5O_{12}$ are characterized by X-ray diffraction (XRD) and scanning electron microscopy, respectively. It is demonstrated that the electrochemical performance is significantly improved by the copper deposition onto lithium titanate structure. The copper-coated lithium titanate exhibits a stable capacity of 170 mAh g^{-1} at 1 C. Besides, the reversible capacity at 80 C remains over half of that at 1 C. The superior C-rate performance is associated with the copper/lithium titanate nanocomposite structure, facilitating lithium transportation ability during cycling.

Keywords $Li_4Ti_5O_{12}$ • Cu/$Li_4Ti_5O_{12}$ • Sol–gel • Electroless coating

2.1 Introduction

Lithium-ion (Li-ion) batteries are the now dominating chemistry within battery applications, having 60 % of worldwide sale values in portable batteries [1]. Li-ion chemistries have the highest energy density among the commercialized rechargeable batteries today which originate from their high working potential combined with their high capacities and low weight [1–3]. Yet their performances are not enough for some future demands, which include electric vehicles, power grid load leveling, and peak shaving, which together set a new list of demands on batteries in terms of lifetime, power, and energy density. Inspired by sustainable mobility, Li-ion battery research has focused toward improving the energy and power

A. Erdaş (✉) • Ş. Özcan • D. Nalci • M.O. Güler • H. Akbulut
Department of Metallurgical and Materials Engineering, Faculty of Engineering, Sakarya University, TR-54187 Serdivan, Sakarya, Turkey
e-mail: aslihan_erdas@hotmail.com

I. Dincer et al. (eds.), *Progress in Clean Energy, Volume 2*,
DOI 10.1007/978-3-319-17031-2_2

density, under the conditions of safety, long cycle life, and the use of low-cost materials. Therefore, spinel lithium titanate $Li_4Ti_5O_{12}$ (LTO), which is called zero-strain insertion material, has aroused researcher's attention recently, due to its potential application as anode material in high-rate lithium batteries. Spinel host has excellent structure stability and a very small volume change during cycling.

In a typical spinel LTO crystal structure, the 32e positions are occupied by oxygen atoms, while 5/6 of the 16d positions are occupied by Ti atoms, and the rest of the 16d positions are occupied by Li atoms, forming a highly stable $[Li_{1/3}Ti_{5/3}]_{16d}O_4$ framework. The tetrahedral (8a) sites are occupied by Li atoms, while the octahedral (16c) sites are empty, and the structure is denoted as $[Li]_{8a}[Li_{1/3}Ti_{5/3}]_{16d}O_4$. It is well known that the electronic structure of LTO spinels can be characterized by the empty Ti 3d states with band gap energy of 2–3 eV gives an insulating character to this material which is also reported by Yi et al. [4]. The calculated electrical conductivity of these materials is only 10^{-13} S cm^{-1} at room temperature. Therefore, low-rate capability due to the low electronic conductivity limits the practical applications, although the electrochemical stability of these electrodes is far beyond the graphite-based commercial electrodes.

Several methods have been proposed to improve these obstacles. Several researches are being carried out to develop the lattice structure of LTOs via reducing the particle size which can shorten the Li-ion diffusion path and provide extended contact area between the electrode and electrolyte, consequently improving Li-ion intercalation kinetics studied by Ladislav et al. [5], Borghols et al. [6], and Naiqing et al. [7]. Another effective way is also reported as the cation doping by incorporating cations of conventional, transition, and rare earth metals by Pierre et al. [8], Kiyoshi et al. [9], Biao et al. [10], and by Damien et al. [11] and anions of F by Shahua et al. [12] and Br by Yanling et al. [13] into the framework of LTO. Surface modifications with high-conductive phases, such as conductive oxides, metals, or carbon-based materials, were also reported by Shahua et al. [14], Yan et al. [15], Arumugam et al. [16], Cheng et al. [17], Zhimin et al. [18], Yonggang et al. [19], Hun et al. [20], Liang et al. [21] and Ilias et al. [22]. It has been previously reported that the surface modification with Ag significantly improved the electronic conductivity of the LTO and which in turn greatly improved the high-rate performance of the LTO by Zhimin et al. [18] and Michals et al. [23]. However, the cost of silver limited the practical applications of LTO-based composites in commercial batteries. On the other hand, copper can be a good candidate due to its low resistivity and high resistance to electron migration and being a cheap and convenient conductive agent when compared with Ag.

From the studies of Yi-R et al. [24, 25] and by Ting et al. [26], LTO synthesis is mostly based on solid-state synthesis using TiO_2 and Li_2CO_3 or LiOH. However, the obtained LTO particle morphologies are mostly in large particle size from submicron to micron range, and these particles contain impurity phases. Therefore, recent reports on LTO synthesis by Wei et al. [27], Meng et al. [28, 29], Zhenjiang et al. [30], Feixiang et al. [31], Daisuke et al. [32], Chien-Te et al. [33], Zhaoyin et al. [34], Jin et al. [35], and Milica et al. [36] clearly indicate that low-temperature sol–gel, spray drying, and hydrothermal methods are the most efficient procedures for high-purity nanocrystalline LTO structures. In this study, we aimed to produce spinel LTO anode materials with high electronic conductivity. LTO spinel structure is produced via facile sol–gel method, and the surfaces of the spinel LTO particles were then coated with Cu via electroless coating techniques. The effect of copper coating on the electrochemical properties is investigated.

2.2 Experimental Details

2.2.1 Synthesis of LTO Anode Electrodes

LTO powders were produced via sol–gel method using citric acid (Merck, purity: >99 %) as a chelating agent. 5 mmol of titanium isopropoxide, $Ti[OCH(CH_3)_2]_4$ (Sigma Aldrich, %99.999 trace metal basis), was dissolved in 150 mL of 2-propanol (Merck, Emsure) to obtain a saturated solution. 4 mmol of lithium nitrate, $LiNO_3$ (Merck, purity: 99.995 % Suprapur), was then added with mild stirring. A saturated aqueous citric acid ($C_6H_8O_7$, Merck, >99 %) dissolved in 150 mL of 2-propanol (Merck, Emsure) was then added at a molar ratio of 9 mmol. The pH of the final solution was kept at 7.0 by adding ammonium hydroxide (Alfa Aesar, %25 solution). The solution was then heated to 90 °C with vigorous stirring to remove the excess ammonia and water until a transparent gel was obtained. The gel was then kept for 12 h in air oven at 120 °C in order to obtain metal citrate precipitation. After the drying process, the precursors were decomposed at 450 °C for 4 h and 850 °C for 1 h in air in order to eliminate organic contents.

2.2.2 Electroless Cu Coating of LTO Anode Electrodes

Pristine LTO particles were used as the cores of the composite powders in the electroless plating. Before starting the deposition process, the surfaces of the anode active powders should have to be pretreated as follows: (1) the surfaces of the as-synthesized powders were cleaned by immersing powders into 1 M sodium hydroxide (Merck, Titripur) solution for 10 min. (2) As-cleaned LTO powders were then immersed into boiling hydrochloric acid (Merck, 1 N) for 15 min to coarsen their surfaces. The coarsened surfaces will also help to improve the adhesion between the LTO-based powders and copper particles. (3) Since the surfaces of LTO powders were nonconductive, an activation process is needed in order to improve the catalytic properties. One-step sensitization and activation process was performed in this study. The etched particles of LTO were immersed in a palladium activation solution consisting of palladium dichloride (Merck, $PdCl_2$) (0.3 g/L), stannous chloride (Merck, $SnCl_2.2H_2O$, >99.9 %) (16 g/L), sodium chloride (Merck, NaCl, >99.99 %) (150 g/l), and hydrochloric acid HCl (Merck, 1 N) (60 mL/L) and stirred strongly at room temperature for 15–20 min. After each pretreatment process, particles of LTO were repeatedly washed by using bidistilled water until pH is reached to 7. The pretreated powders were then dried after the activation step in an oven at 80 °C for 12 h.

2.2.3 Characterization

The crystal structures of the LTO and Cu/LTO anode materials were examined using an XRD spectrometer (XRD, Rigaku D/max 2200) with monochromated Cu Kα radiation at a scanning rate of 1° min^{-1} in the range of 10°–80°. Surface morphology was examined using SEM (Jeol 6060 LV).

2.2.4 Electrochemical Characterization

Galvanostatic charging and discharging tests were performed using CR2016 coin-type cells by using MTI BST8A battery tester system. The anode electrode was prepared by blending active material, carbon black (Alfa Aesar, >99.9 %), and polyvinylidene fluoride (Alfa Aesar, >99.9 %) (80:10:10) in *N*-methyl-2-pyrrolidone (Alfa Aesar, >99 %) with a weight ratio of 8:1:1. The slurry was casted onto an aluminum foil by the "doctor blade" technique. The weight of active material was ~1.786 mg cm^{-2}. Coin-type (CR2016) test cells were assembled in an argon-filled glove box (Mbraun, Labstar) using a microporous polypropylene film (Celgard 2400) as a separator, 1 M $LiPF_6$ (Sigma Aldrich, ≥99.99 %) in ethylene carbonate (Sigma Aldrich, 99 %) and dimethyl carbonate (Sigma Aldrich, ≥99 %) (EC/DMC, 1:1 vol) as electrolyte, and Li foil as the counter and reference electrodes. Constant current charge/discharge was performed at various rates within a voltage window of 1–2.5 V(versus Li^+/Li). Electrical impedance spectroscopy (EIS) experiments were carried out on a Gamry Reference 3000 advanced electrochemical system in the frequency range mainly from 1 MHz to 50 MHz at an amplitude of 10 mV. Before all electrochemical measurements, all samples were aged for 24 h at room temperature.

2.3 Results and Discussions

2.3.1 Structural and Morphological Observations

The X-ray diffraction patterns of the synthesized powders are shown in Fig. 2.1. All the sharp diffraction peaks can be indexed on the basis of a cubic spinel structure, $Li_4Ti_5O_{12}$ (JCPDS file no. 26-1198). The peaks at $2\theta = 18.4°$, 35.6°, 43.3°, 47.4°, 57.2°, 62.8°, 66.1°, 74.3°, and 79.3°, which correspond to (111), (311), (400), (331), (511), (440), (531), (533), and (444) planes of a face-centered cubic spinel LTO with a *Fd3m* space group, respectively. This suggests that high-purity LTO can be prepared by calcinating the sol–gel-derived precursors at 850 °C for only 1 h. No trace amount of impurity of Li phase can be detected, as shown in Fig. 2.1. In preliminary studies, we found that TiO_2 and Li_2TiO_3 phases would coexist with the main LTO phase when the sintering temperature was lower than 600 °C. It is also well known that when increasing the heat treatment temperature, TiO_2 or Li_2TiO_3 would disappear depending on the starting composition, Li rich or Ti rich. The sol–gel synthesis leads to samples with notably higher purity. Indeed, no Li impurities were detected in the X-ray diffraction pattern. The lattice parameters were calculated by the least-squares method using ten diffraction lines, and silicon was used as a reference to calibrate the peak position and intensity ratio. The crystal lattice parameter, *a*, was also calculated and was 8.359 Å for the samples, in good agreement with that obtained by Zhu et al. [3]. The crystal size of the LTO crystallites was evaluated by the Scherrer formula and found to be ca. 59 nm using the (111) reflection.

In order to determine the effects of copper coating on the crystal structure of LTO, the X-ray powder diffraction was carried out for uncoated and coated spinel materials and presented in Fig. 2.1b according to the JCPDS 04-0836. The intensities of the LTO peaks were reduced after coating the surfaces of the based spinels

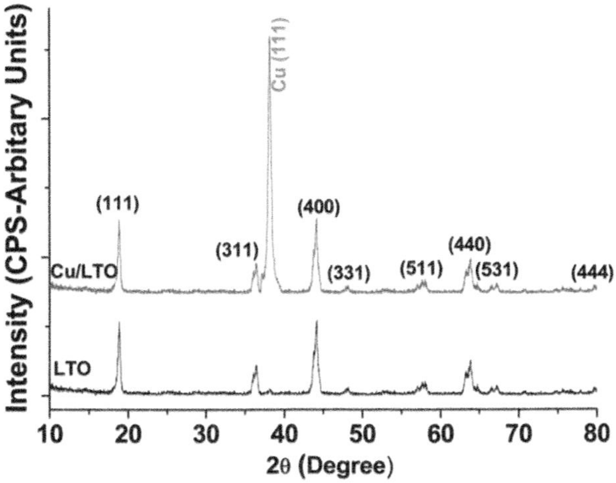

Fig. 2.1 XRD patterns of the (**a**) pristine LTO and (**b**) Cu-coated LTO spinel samples

uniformly with copper, which is the expected result. The lattice parameters were also calculated by the least-squares method for copper-coated LTO spinel samples and found to be 8.354 Å. The calculated lattice parameters show that there was almost no change in the lattice parameter for all samples which indicated that copper ions were not incorporated into the spinel structure but are just presented on the surfaces of LTO spinel samples. Hui et al. [37] reported that copper ion introduction to the spinel structure will lead to a significant change of the lattice parameter. The data obtained from the Fig. 2.1b also indicate that copper crystals largely grow in two preferential orientation (111) plane. Hui et al. [37] also reported that the peaks due to copper crystallites with (111) orientation are predominant. This is an important result knowing that stress-induced voiding is significantly inhibited in (111)-textured metals which is also reported by Zhaoyin et al. [34]. Furthermore, the (111)-textured Cu film has higher resistance to electromigration which suppresses grain boundary and interfacial diffusion of metal atoms previously reported by Jin et al. [35].

Surface morphologies of the LTO and Cu/LTO powders calcined at 850 °C are presented in Fig. 2.2. A typical SEM image in Fig. 2.2a shows that the sample has polyhedron morphology and a uniform particle size distribution. Higher magnification SEM image in the inset of Fig. 2.2a indicates that the polyhedron structures are composed of subgrains that have a particle size ranges from 60 to 90 nm, which is in agreement with the XRD result. Figure 2.2b shows the SEM images of

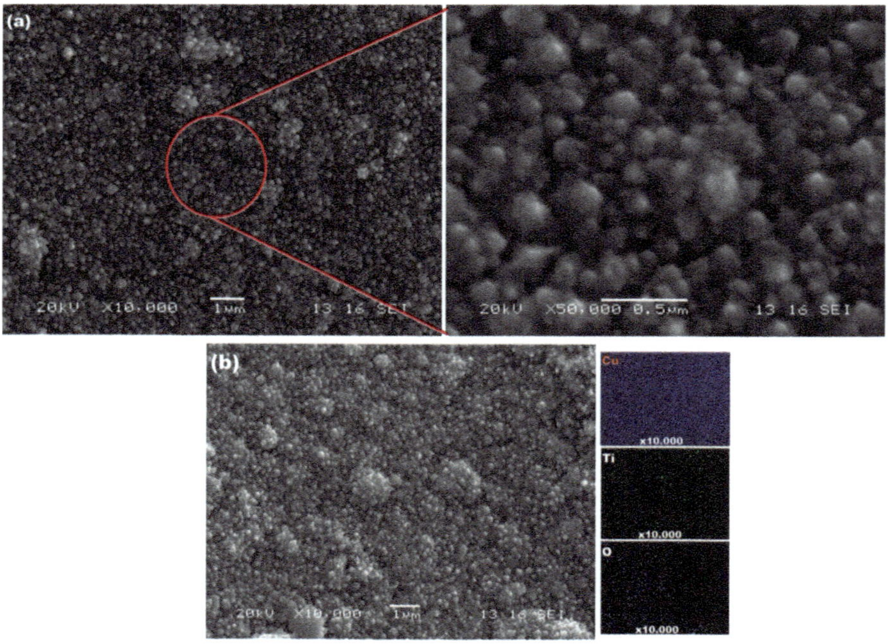

Fig. 2.2 SEM images of the (**a**) pristine LTO and (**b**) Cu-coated LTO with EDS Dot map analysis of spinel samples

Cu-coated as-synthesized LTO powders with EDS dot map analysis. No free copper particles and uncoated surface of pristine LTO particles are found in powders. From Fig 2.1a, b, it can also be concluded that surfaces of polyhedrons have not been changed after electroless deposition process from the SEM images. In addition, EDS dot map analysis also confirms that the surfaces of the LTO powders were coated uniformly with Cu.

2.3.2 Electrochemical Results

As shown in Fig. 2.3, only one pair of anodic/cathodic peak is observed in the CV curves of pristine LTO calcined at 850 °C. In the first cycle, the cathodic and anodic peaks are at 1.47 and 1.69 V, respectively. These peaks can be regarded as the signature of lithium insertion into and extraction from the spinel LTO framework, which is consistent with the previous studies by Daisuke et al. [32], Chien-Te et al. [33], Zhoyin et al. [34], and Jin et al. [35]. Upon completion of the 1st and 4th cycles, the CV curves still coincide well with that of the first cycles. The well-defined small potential interval between cathodic and anodic peaks and the good overlap of CV curves clearly indicate that the sample has high electrochemical reaction activity and reversibility.

Galvanostatic charge/discharge curves of pristine LTO and Cu/LTO as anode electrodes for Li-ion batteries are evaluated and given in Fig. 2.4. The electrode density and percentage of the active mass were adjusted to be similar for all samples

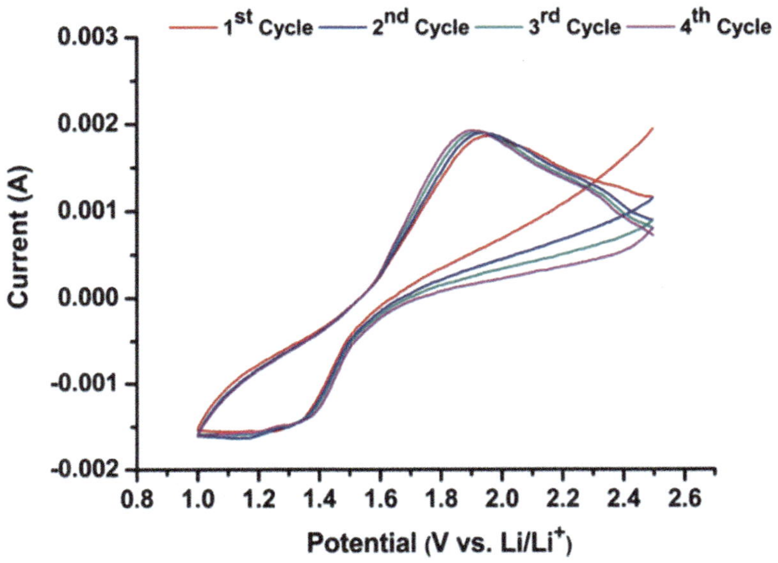

Fig. 2.3 Cycling voltammetry curves of the pristine LTO anode electrodes for four cycles

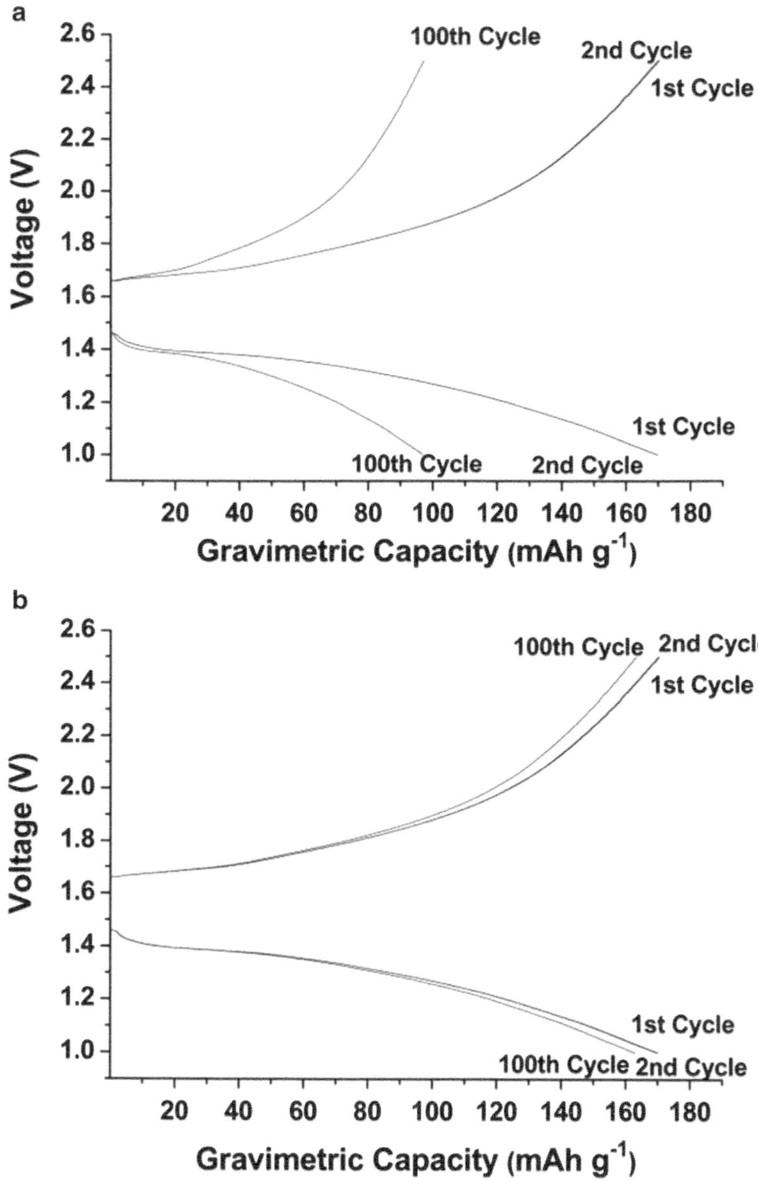

Fig. 2.4 Galvanostatic charge/discharge curves of the (**a**) pristine LTO and (**b**) Cu-coated LTO spinel samples

to compare the relationship between the particle size and morphology. The cells were cycled between 1 and 2.5 V at 1 C rate, and an initial capacity of 170 mAh g^{-1} was obtained for both samples which is closer to the theoretical discharge capacity of LTO spinels. Even at 1 C state of charge conditions, the difference between the

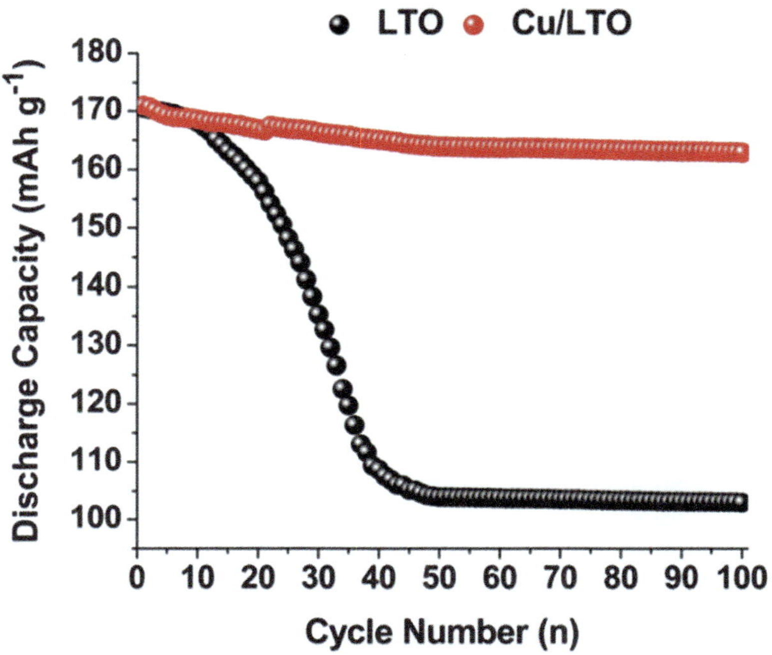

Fig. 2.5 Galvanostatic charge/discharge vs. cycling number curves of the pristine LTO and Cu-coated LTO spinel samples

capacity of the pristine LTO and Cu/LTO is still not so large. This is reasonable because Li^+ insertion/extraction is sufficient at this relatively low current rate. The Li grains in the inner part of the aggregated micron-sized particles are still electrochemically active due to the long diffusion time. However, with increasing the discharge–charge current rate, the difference between the lithium storage capacities of these two samples becomes evident. A specific capacity of 102 mAh g^{-1} and 163 mAh g^{-1} after 100 cycles for pristine LTO and Cu/LTO nanocomposite anode electrodes shows that the Cu coating on LTO samples significantly improved the electronic conductivity of LTO spinels.

Discharge capacity vs. cycle curves of the pristine LTO and Cu/LTO samples were given in Fig. 2.5. It can also be concluded from the figure that the Cu/LTO electrodes show higher lithium storage capability than that made from the pristine powders. Moreover, in addition to the lower capacity, the capacity–cycle profile of the pure sample is not as smooth as that of the Cu/LTO sample, namely, fluctuation of capacity on cycling occurred. Such a capacity fluctuation was also observed by other authors in large aggregated LTO electrode [10, 13]. This capacity fluctuation suggests that the severe aggregation makes the Li^+ insertion/extraction in individual LTO grains inhomogeneous in different cycles. On the contrary, very uniform electrochemical reaction may take place in the well-dispersed LTO nanocomposite electrode, which leads to a very smooth capacity–cycle profile with an average

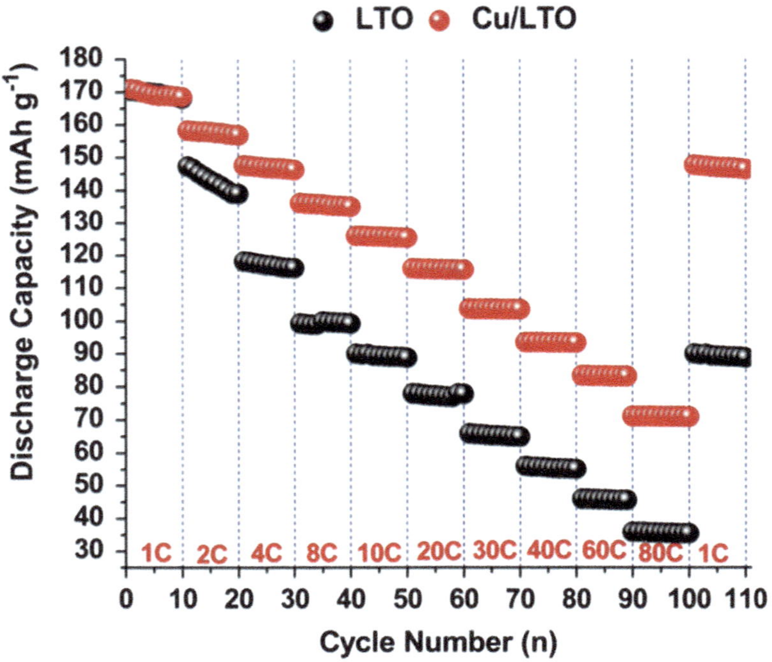

Fig. 2.6 Rate performance of pristine LTO and LTO/C composites over the voltage range of 1 and 2.5 V

capacity fading of 2 % per cycle within the first 50 cycles, and no capacity fading has been observed after 50 cycles. The electrochemical data obtained in this study indicates that the cycling performance of the pristine and Cu-coated LTO has been significantly improved. Based on the results from Fig. 2.5, it can be concluded that the Cu-coated LTO had the best cycling performance among pristine LTO.

Figure 2.6 shows the capacity vs. cycle number at various current densities for pristine LTO and Cu/LTO composites. The Cu/LTO samples show much better cycle stability and much better rate capability than pristine samples at the same discharge rate. Upon completion of the 100th charge/discharge cycles, the current density lowers to 1 C again, while the discharge capacity decreases for the first 1 C-rate cycle with a capacity loss of 11 % only for Cu/LTO composites while capacity loss of 47 % for pristine LTO samples. It can also be concluded from Fig. 2.6 that at a low discharge current rate of 1–10 C, the capacity of pristine LTO decays faster and that at a high discharge current rate of 10–80 C, the capacity of Cu/LTO becomes gentle. Figure 2.6 also demonstrates a remarkable rate performance.

Figure 2.7 shows the Nyquist plots of pristine and Cu/LTO composites. The EIS data were collected with a two-electrode coin cell after activation (i.e., after subjecting the coin cell to one charge/discharge cycle). The initial activation was aimed to suppress the Li–electrolyte interfacial resistance that arises from the passivating film formed on lithium metal in contact with the electrolyte. EIS results

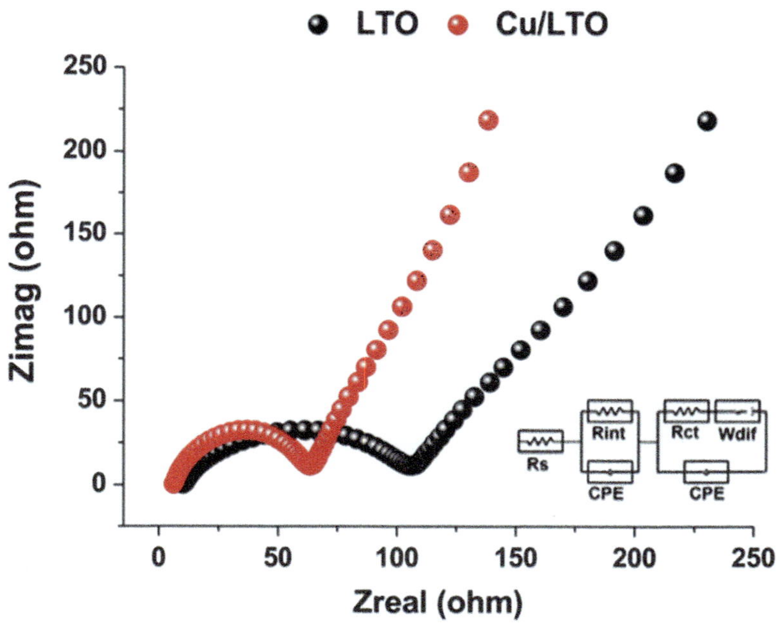

Fig. 2.7 EIS spectra of LTO and Cu/LTO composite electrodes with the frequency range of 10^{-2}–10^{6} Hz

were then fitted with the equivalent circuits inset in Fig. 2.7. On this circuit, R_s is the resistance associated with the electrolyte and cell component, R_{int} is the interface resistance of any film formation of the anode surface (first high-frequency semicircle), R_{ct} is the charge-transfer resistance of the electrode reaction with lithium ions (middle-frequency semicircle), and W_{dif} is the resistance of the lithium ion diffusion to the electrode (low-frequency semicircle). The Nyquist plots of Cu/LTO composites demonstrate unambiguously that the charge transfer of Cu/LTO composite is faster, which can be reflected by the diameter of a semicircle in the medium frequency region. The EIS plots comprised of three parts, and the high-frequency intercept at axis x represents the Ohmic resistance mainly contributed by electrolyte, while the depressed semicircle in the high-to-medium frequency range is normally the charge-transfer resistance and reflects the solid-state diffusion of Li^+ in the bulk of the active material and the constant phase-angle element involving double-layer capacitance. The inclined line in the lower frequency range is attributed to the Warburg impedance, which is associated with the lithium-ion diffusion through the LTO electrode. Apparently, the depressed semicircle of Cu/LTO composites is much smaller than pristine samples calcined, which means that the Cu/LTO composite has better electronic conductivity and ionic conductivities than other samples. Consequently, oxidation of electrolyte and degradation of electrolyte/electrode interface are suppressed effectively by Cu coating, which leads to decreased R_{ct} value and improved electrochemical performance.

2.4 Conclusions

In this study, LTO anode active electrodes were produced via a facile sol–gel process, and Cu/LTO composite electrodes were also prepared using electroless coating process. A pure spinel structure LTO can be obtained by calcination at 850 °C for 1 h. The intimate contact between Cu coating and LTO nanoparticles not only affords a highly conductive matrix for Li-ion insertion but also suppresses the agglomeration and growth of LTO nanoparticles after the calcination process. The confinement of LTO particles in Cu coating reduces lithium-ion and electron transport diffusion resistance. The capacity of Cu/LTO synthesized via electroless coating techniques is 170 mAh g^{-1} at 1 C rate, and it retains more than 79 and 42 % of its capacity at the discharge rate of 20 and 80 C. Irrespective of the rate used, Cu/LTO basically retain their initial capacity up to 100 cycles at 1 C state of charge conditions. Cu/LTO nanocomposite synthesized at 850 °C is a superior lithium storage material, and it has a promising application in power lithium-ion battery.

Acknowledgments This work is supported by the Scientific and Technological Research Council of Turkey (TUBITAK) under the contract number 111M021—Improving the Capacity of Li-Ion Batteries by Using New Semi-Conducting Metal Oxide Based Anodes. The authors thank the TUBITAK MAG workers for their financial support.

References

1. Yuan T, Wang K, Cai R, Ran R, Shao Z (2009) Cellulose-assisted combustion synthesis of $Li_4Ti_5O_{12}$ adopting anatase TiO_2 solid as raw material with high electrochemical performance. J Alloys Compd 477:665–672
2. Li J, Jin Y-L, Zhang X-G, Yang H (2007) Microwave solid-state synthesis of spinel Li4Ti5O12 nanocrystallites as anode material for lithium-ion batteries. Solid State Ionics 178:1590–1594
3. Zhu Y-R, Xie Y, Zhu R-S, Shu J, Jiang L-J, Qiao H-B, Yi T-F (2011) Kinetic study on LiFePO $_4$-positive electrode material of lithium-ion battery. Ionics 17:437–441
4. Yi T-F, Jiang L-J, Shu J, Yue C-B, Zhu R-S, Qiao H-B (2010) Recent development and application of Li4Ti5O 12 as anode material of lithium ion battery. J Phys Chem Solids 71:1236–1242
5. Ladislav K, Jan P, Timothy MS, Martin K, Markéta Z, Thierry D, Michael G (2003) Li Insertion into $Li_4Ti_5O_{12}$ (spinel): charge capability *vs.* particle size in thin-film electrodes. J Electrochem Soc 150:A1000–A1007
6. Borghols WJH, Wagemaker M, Lafont U, Kelder EM, Mulder FM (2009) Size effects in the $Li_{4+x}Ti_5O_{12}$ spinel. J Am Chem Soc 131:17786–17792
7. Naiqing Z, Zhimin L, Tongyong Y, Chenglong L, Zhijun W, Kening S (2011) Facile preparation of nanocrystalline $Li_4Ti_5O_{12}$ and its high electrochemical performance as anode material for lithium-ion batteries. Electrochem Commun 13:654–656
8. Pierre K, Aurélie G, Manfred W, Laurent A, Josette O-F, Jean CJ, Pierre EL (2003) Phase transition in the spinel Li4Ti5O12 induced by lithium insertion: influence of the substitutions Ti/V, Ti/Mn, Ti/Fe. J Power Sources 119–121:626–631
9. Kiyoshi K, Nao A, Kaoru D (2005) Three dimensionally ordered composite solid materials for all solid-state rechargeable lithium batteries. J Power Sources 146:86–89

10. Biao Z, Zhen DH, Sei WO, Jang KK (2011) Improved rate capability of carbon coated Li3.9Sn0.1Ti5O12 porous electrodes for Li-ion batteries. J Power Sources 196:10692–10697
11. Damien D, Ilias B, Jiwei M, Khalil A (2011) Template-assisted synthesis of high packing density $SrLi_2Ti_6O_{14}$ for use as anode in 2.7-V lithium-ion battery. J Power Sources 196:2871–2874
12. Shahua H, Zhaoyin W, Zhonghua G, Xiujian Z (2005) Preparation and cycling performance of Al^{3+} and F^- co-substituted compounds $Li_4Al_xTi_{5-x}F_yO_{12-y}$. Electrochim Acta 50:4057–4062
13. Yanling Q, Yudai H, Dianzeng J, Shu JB, Guo ZP (2009) Preparation and characterization of novel spinel Li4Ti5O12 – xBrx anode materials. Electrochim Acta 54:4772–4776
14. Shahua H, Zhaoyin W, Xiujian Z, Xuelin Y (2005) Research on Li4Ti5O12/CuxO composite anode materials for lithium-ion batteries. J Electrochem Soc 152:A1301–A1305
15. Yan YW, Yan JH, Qiong YL, Ji ZL, Yuan DC, Xiao YJ (2007) A new composite material Li4Ti5O12–SnO2 for lithium-ion batteries. Ionics 14:85–88
16. Arumugam S, Sukumaran G, Ramasamy T, Chandrasekaran N, Shanmuga P (2011) Novel $Li_4Ti_5O_{12}$/Sn nano-composites as anode material for lithium ion batteries. Mater Res Bull 46:492–500
17. Cheng CL, Qiu HL, Li BC, Tai HW (2012) A facile titanium glycolate precursor route to mesoporous Au/Li4Ti5O12 spheres for high-rate lithium-ion batteries. ACS Appl Mater Interfaces 4:1233–1238
18. Zhimin L, Naiqing Z, Zhijun W, Kening S (2012) Highly dispersed Ag nanoparticles ($<$10 nm) deposited on nanocrystalline $Li_4Ti_5O_{12}$ demonstrating high-rate charge/discharge capability for lithium-ion battery. J Power Sources 205:479–482
19. Yonggang W, Haimei L, Kaixue W, Hosono E, Yarong W, Haoshen Z (2009) Synthesis and electrochemical performance of nano-sized $Li_4Ti_5O_{12}$ with double surface modification of Ti (III) and carbon. J Mater Chem 19:6789–6795
20. Hun GJ, Junghoon K, Bruno S, Yang KS (2011) Micron-sized, carbon-coated Li4Ti5O12 as high power anode material for advanced lithium batteries. J Power Sources 196:7763–7766
21. Liang C, Jing Y, Guan NZ, Jia YL, Cong XW, Yong YX (2010) General synthesis of carbon-coated nanostructure Li4Ti5O12 as a high rate electrode material for Li-ion intercalation. J Mater Chem 20:595–602
22. Ilias B, Gary MK, Amine K (2011) Electrochemistry and safety of Li4Ti5O12 and graphite anodes paired with LiMn2O4 for hybrid electric vehicle Li-ion battery applications. J Power Sources 196:10344–10350
23. Michal K, Monika M, Bartosz H, Dominika Z, Krzysztof PK, Jacek BJ, Maria K, Ludwika L, Andrzej C (2014) $Li_4Ti_5O_{12}$ modified with Ag nanoparticles as an advanced anode material in lithium-ion batteries. J Power Sources 245:746–771
24. Yi RJ, Jenq GD (2012) Synthesis of entanglement structure in nanosized $Li_4Ti_5O_{12}$/multi-walled carbon nanotubes composite anode material for Li-ion batteries by ball-milling-assisted solid-state reaction. J Power Sources 198:294–297
25. Yi RJ, Jenq GD (2012) Electrochemical performance and low discharge cut-off voltage behavior of ruthenium doped Li4Ti5O12 with improved energy density. Electrochim Acta 63:9–15
26. Ting FY, Ying X, Qiuju W, Haiping L, Lijuan J, Mingfu Y, Rongsun Z (2012) High rate cycling performance of lanthanum-modified $Li_4Ti_5O_{12}$ anode materials for lithium-ion batteries. J Power Sources 214:220–226
27. Wei F, Xin QC, Peng JZ, Yu LM, Geping Y (2013) A facile strategy to prepare nano-crystalline $Li_4Ti_5O_{12}$/C anode material via polyvinyl alcohol as carbon source for high-rate rechargeable Li-ion batteries. Electrochim Acta 93:173–178
28. Meng LL, Yu HL, Shih CL, Jing MC, Jien WY, Han CS (2013) $Li_4Ti_5O_{12}$-coated graphite anode materials for lithium-ion batteries. Electrochim Acta 112:529–534
29. Meng LL, Yu HL, Shih CL, Jing MC, Jien WY, Han CS (2013) Li4Ti5O12-coated graphite as an anode material for lithium-ion batteries. Appl Surf Sci 258:5938–5942
30. Zhenjiang H, Zhixing W, Feixiang W, Huajun G, Xinhai L, Xunhui X (2012) Spherical $Li_4Ti_5O_{12}$ synthesized by spray drying from a different kind of solution. J Alloys Compd 540:39–45

31. Feixiang W, Zhixing W, Xinhai L, Huajun G, Peng Y, Xunhui X, Zhenjiang H, Qian Z (2012) Characterization of spherical-shaped $Li_4Ti_5O_{12}$ prepared by spray drying. Electrochim Acta 78:331–339
32. Daisuke Y, Yoshihiro K, Jung-Min K, Koichi U, Naoaki K, Naoto K, Yasushi I (2010) Spray-drying synthesized lithium-excess Li4 + xTi5 − xO12 − δ and its electrochemical property as negative electrode material for Li-ion batteries. Electrochim Acta 55:1872–1879
33. Chien-Te H, I-Ling C, Yun-Ru J, Jia-Yi L (2011) Synthesis of spinel lithium titanate anodes incorporated with rutile titania nanocrystallites by spray drying followed by calcination. Solid State Ionics 201:60–67
34. Zhaoyin W, Zhonghua G, Shahua H, Jianhua Y, Zuxiang L, Osamu Y (2005) Research on spray-dried lithium titanate as electrode materials for lithium ion batteries. J Power Sources 146:670–673
35. Jin YL, Xingcheng X, Drew H, Dongun L, Fathy H, Zhongwei C (2013) Hierarchical $Li_4Ti_5O_{12}$-TiO_2 composite microsphere consisting of nanocrystals for high power Li-ion batteries. Electrochim Acta 108:104–111
36. Milica V, Ivana S, Miodrag M, Slavko M, Nikola C (2013) Hydrothermal synthesis of $Li_4Ti_5O_{12}$/C nanostructured composites: morphology and electrochemical performance. Mater Res Bull 48:218–223
37. Hui W, Jianfeng J, Hongzhang S, Xing H, Hongwei S, Delin Y (2011) The preparation of Cu-coated Al_2O_3 composite powders by electroless plating. Ceram Int 37:2181–2184

Chapter 3
A Novel Battery System for Electric Vehicles

Turev Sarikurt and Abdulkadir Balikci

Abstract Battery electric vehicles are the optimum way of clean transportation without any pollutant emission. The power source needs to have both high power and high energy densities in electric vehicles. Lithium-ion is the most energy- and power-dense battery type, compared with the other types; however, none of the other commercial batteries provide both of these two features together. Therefore, mostly battery-supercapacitor hybrid systems including high-energy batteries and high power supercapacitors are used. Supercapacitors have higher power densities and longer life cycle than batteries; however, they are heavy and have longer charge and discharge times. Besides, following the development in lithium battery technologies, features of lithium batteries are improving while the prices decrease. Thus, a suitable battery can be used rather than a supercapacitor in order to maintain high power. In this study, a new structure named hybrid battery system is proposed. The system includes two different lithium batteries with different energy and power characteristics. One of these batteries has high energy density, while the other has high power density. In this paper, the proposed system is introduced in details, energy management of the system is performed, and the results are validated by real-time experiments by using Kokam SLPB55205130H and Altairnano 13 Ah automotive-grade batteries.

Keywords Hybrid battery system • Lithium-ion batteries • Electric vehicles

3.1 Introduction

While the need for energy is ever increasing worldwide, the amount of fossil fuel deposits is declining. Following the awareness about environmental pollution and stress caused by fossil fuel consumption as well, the renewable energy technologies have gained momentum. Moreover, due to recent legal restrictions on carbon-dioxide emission of vehicles with combustion engines, the use of electric vehicles is expected to become more common.

T. Sarikurt (✉) • A. Balikci
Department of Electronics Engineering, Gebze Institute Technology, TR-41400 Gebze, Kocaeli, Turkey
e-mail: turev@gtu.edu.tr

I. Dincer et al. (eds.), *Progress in Clean Energy, Volume 2*,
DOI 10.1007/978-3-319-17031-2_3

Electric vehicles can be classified under three headlines. These are: battery electric vehicles (BEVs) in which the motive mechanism is only electric motors. The major problems of BEVs are the short driving distance with single charge and the low life cycle. Additionally, these vehicles are insufficient in performance demanding conditions such as accelerating or when uphill climbing is needed. Consequently, hybrid electric vehicles (HEVs) and plug-in hybrid electric vehicles (PHEVs) are commonly preferred instead of BEVs. HEVs are vehicles that mostly use an internal combustion engine with an electric motor, and the energy sources of these vehicles are both fossil fuels and electricity. In PHEVs, the batteries for the electric motor can be charged from outside. Nevertheless, BEVs are the cleanest transportation alternative, but the disadvantages obstruct the spread of BEVs.

To overcome these disadvantages and to improve performance using supercapacitors with batteries are common strategies for BEVs [1]. Supercapacitors also have drawbacks when compared to batteries, such as low energy densities and high self-discharge rates, and they are more prone to voltage imbalances [2].

Lithium batteries are commonly preferred when compared with other rechargeable battery types because of their high energy densities [3]. These types of batteries are lighter than other secondary battery types. When compared with lead-acid, nickel-metal-hydrate, and nickel-cadmium batteries, lithium-ion-polymer batteries have higher open-circuit voltages and lower self-discharge rates [4]. Also, the development in lithium technology gives rise to new lithium-based battery structures and enables lithium batteries to compete with supercapacitors. In the near future, with the help of a suitable control strategy, lithium-based batteries would be able to take the place of supercapacitors in the electric vehicle area, in which their light structures and high energy densities are of great importance. Today, the commonly preferred lithium battery types are lithium-cobalt-oxide ($LiCoO_2$), lithium-manganese-dioxide or spinel ($LiMnO_2$), lithium-iron-phosphate ($LiFePO_4$), lithium-nickel-manganese-cobalt-oxide ($LiNiMnCoO_2$), lithium-nickel-cobalt-aluminum-oxide ($LiNiCoAlO_2$), and lithium-titanate ($Li_4Ti_5O_{12}$) [1].

In this study, the hybrid battery concept is introduced, where two batteries, one with high energy density and one with high power density, are being used together. Also, the feasibility of setting up a hybrid battery system and the handling of its energy management by a passive strategy is shown with experimental results.

In the following section, the hybrid battery concept is introduced, and the hybrid battery system is proposed. In the latter sections existing energy management systems are listed, passive hybridization techniques are detailed, and the experimental setup is described with details of the batteries being used. And in the final sections, experimental results were given, and the verification of the control strategy is discussed.

3.2 System Description

It is obvious that the best alternative transport is the BEV if the clean transportation notion is considered for environmental awareness. The major problems of the BEVs are the short driving distance and the low system durability. Besides, these vehicles are insufficient in performance demanding road types which can be illustrated as the paths in which uphill climbing or accelerating is needed. Because of these insufficiencies, hybrid systems are preferred. The most common pure electrical solution is the battery-supercapacitor hybrid system in which battery stacks and supercapacitors are used together. Supercapacitors have higher power densities and longer life cycles than batteries; however, they are heavy and have longer charge and discharge times. Also, supercapacitors have several disadvantages like high self-discharge rates and low energy densities [2, 5].

The hybrid battery concept which is introduced in this study consists of two different batteries instead of a battery-supercapacitor hybrid. One of these batteries has high energy density, and the other one has high power density. The battery with high power density will be employed in high-performance (hence high power) demanding conditions, and the battery with high energy density will help to increase the driving distance. In this study, a Kokam SLPB55205130H model $LiNiMnCo_4$ battery [6] is chosen as the battery with high energy density, and the Altairnano 13 Ah model $Li_4Ti_5O_{12}$ battery [7] is chosen as the one with high power density.

The general block diagram of a hybrid battery system can be seen in Fig. 3.1.

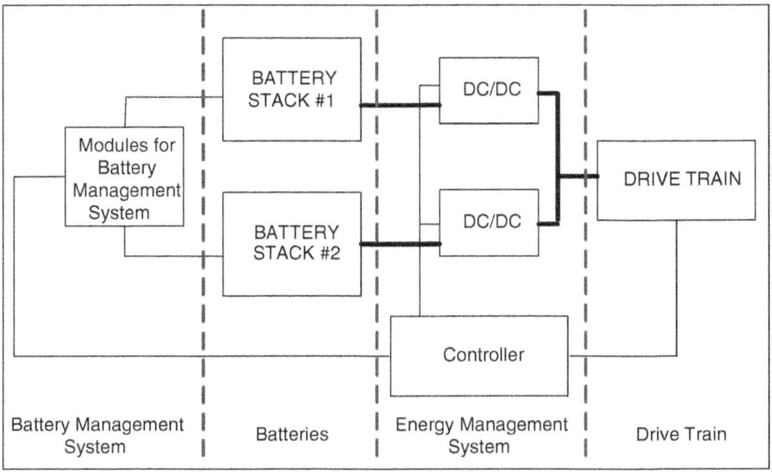

Fig. 3.1 Block diagram of a general hybrid battery system

3.3 Energy Management

Energy management systems (EMSs) are computer-aided tools which handle basic tasks in order to control and optimize the performance of the system. Generally in electric vehicles, EMSs take on the coordination of the energy supplied by several sources which exist in the vehicle and the recovery of the energy which is generated by regenerative braking system. This coordination can be achieved by the use of some factors such as driving cycles, charge and discharge characteristics, and energy storage parameters [8]. Basically EMSs decide when and which source to use.

In hybrid battery systems (HBSs), the task of EMS is to determine which battery to use and the amount of energy to draw depending on the driving and route conditions. In contrast with HBS, in HEVs, EMSs are mostly used to determine the ratio and frequency of the employment of internal combustion engine or electric motor. For instance, an internal combustion engine has low torque values in low speeds, while an electric motor can generate high torques. In this manner, a general strategy is to use the electric motor until the specified speed limit and then switching to the engine [9]. Gonder and Markel described three different strategies: a strategy where only the electric motor is employed, the electric motor used dominantly, and the internal combustion engine used dominantly [10].

Primarily, numerical methods are preferred in energy management of hybrid vehicles. Commonly, dynamic programming is used for the optimization in a particular time period. A-ECMS (Adaptive Equivalent Consumption Minimization Strategy) is used [11], but numerical analysis takes a long time. In another study, target routes are divided into a series of time segments and optimized separately [12]. By this approach also, the solution duration is shorter, but the solution itself is suboptimal. Also in [13], a simple method based on static optimization is proposed. Another approach to decrease the effort for calculations is based on Hamilton-Jacobi-Bellman equations and principles [14]. Some heuristic algorithms in which Boolean or fuzzy rules were employed are also used for the same purpose [15]. For this purpose, rule-based strategies are developed in [16], and fuzzy logic is used in [17]. In order to obtain a global and optimal solution, linear [18] and dynamic programming methods [19, 20] are studied. The drawback of these methods is that an online solution can't be obtained because the proposed strategies are performed for known driving cycles. Thereby in these approaches, the whole route information is assumed to be known. Nevertheless, solutions of these studies may be used as a benchmark while developing a different strategy or used while generating rules for a rule-based method.

To avoid this, [21] used an alternative driving cycle, which is a combination of several driving cycles instead of a single cycle. Although these methods use independent driving cycles, still it will be difficult to cover a real-life driving situation. In some studies, the current conditions of the vehicle and route are taken into consideration by EMS [22, 29]. Some methods in which the EMS predicts future conditions and events such as model predictive control [23] or global optimization [24] methods are also studied. Thus, a connection between the upcoming situations and EMS can be established. Estimation accuracy and estimation duration are the parameters to determine the success of the strategy. Estimation of conditions of the near future allows the EMS to perform optimization even in small time frames and gives good results [25]. For these approaches, the benefits of the control strategy directly depend on the accuracy of estimations. In some studies, the required parameters for optimization were obtained by using pattern recognition [26] collected continuously during the ride [22] or battery state of charge as a parameter [27].

The main motivations of energy management strategies for HEVs are decreasing the fuel consumption and/or reducing the carbon monoxide emissions. The objective is to use electric energy more efficiently. However in HBS, these motivations and objectives are invalid. Therefore, the studies about battery-supercapacitor hybrids are also examined.

In applications of battery-supercapacitor hybrids, the main motivation is reducing the usage frequency of battery thus reducing the number of charge-discharge cycles and increasing the battery life. The basic principle is supplying power demand by employing the supercapacitor while supplying energy demand by employing the battery [28]. With this approach, fast charge and discharge cycles and high power to be drawn from the battery can be limited. Besides, there are also ongoing studies about satisfying the energy demand of oscillating loads and optimizing the parameters and ratio of the supercapacitor.

Battery-supercapacitor hybrids can be divided into three categories as passive, semi-active, and active [29]. In passive topology, battery and supercapacitor are connected in parallel and directly coupled to the load. Although the semi-active topology is more successful than the passive topology, a dc/dc converter is needed. In active hybridization, two dc/dc converters are needed which makes this topology complicated [29, 30].

The advantage of the passive topology is its simple structure without a control circuit or power electronics. Thus, both the price and the volume of the system decrease, while durability and reliability increase. Also, the regenerative breaking is possible without spending extra effort. Major drawback of this topology is that the load sharing between the battery and the supercapacitor is totally uncontrolled.

General approach in passive hybridization is deciding the voltage and capacitance values of the capacitor by using a load profile of a specific pulse train [29–33]. Although this approach gives a good point of view about steady-state operation of the system, it is insufficient for transient process [30]. Generally in passive hybridization studies, it is shown that the driving distance increases when compared with a system without capacitors [34–36]. For example, Smith et al. hybridized a 600 mAh lithium battery with a 600 mF supercapacitor, used a load profile of 2 A, and proved that the driving distance is increased without giving any analytical explanation. Similar studies were also made [33, 37, 38]; however, Sikha et al., made assumptions by using chemical inner structure and simulated in time domain and by a pulse train. Michalczuk et al. [39] used a novel driving cycle which was generated in Warsaw including the effect of temperature.

In this study, a passive approach is undertaken. But in HBS, the nature of the system doesn't allow a totally uncontrolled operation unlike battery-supercapacitor hybrid. Thus, a choke inductance is also used in the system.

3.4 Experimental Setup

A totally uncontrolled operation is not possible while using two different types of batteries instead of a battery-supercapacitor hybrid. In passive battery-supercapacitor hybrids, the peak power demand is satisfied by the supercapacitor, and the battery is used in order to provide longer system operation. In HBS, the peak power demand is shared between both batteries although one of them has higher power density. To avoid this situation and to operate a system like a battery-supercapacitor hybrid, an additional passive power electronics element, a choke inductor, is added to the system. The choke inductor will be connected in series with the battery which has higher energy density to limit the energy consumption from that battery during the peak power phases of the route.

The circuit of the proposed system is demonstrated in Fig. 3.2.

The setup consists of a programmable DC power supply, Chroma 62050P; a DC electronic load, Chroma 63297; data acquisition cards, NI 9206 and NI 9213; and a system design software, LabView. The experimental environment is shown in Fig. 3.3. The most important part of the setup is of course the batteries. A Kokam SLPB55205130H model $LiNiMnCo_4$ battery which has higher energy density and Altairnano 13 Ah model $Li_4Ti_5O_{12}$ battery which has higher power density are used. The specifications of the batteries which were given by the manufacturer were listed in Table 3.1 [6, 7].

As seen from the table, Altairnano has higher power density, and the energy density of the Kokam is almost twice as that of the Altairnano. Batteries are used in

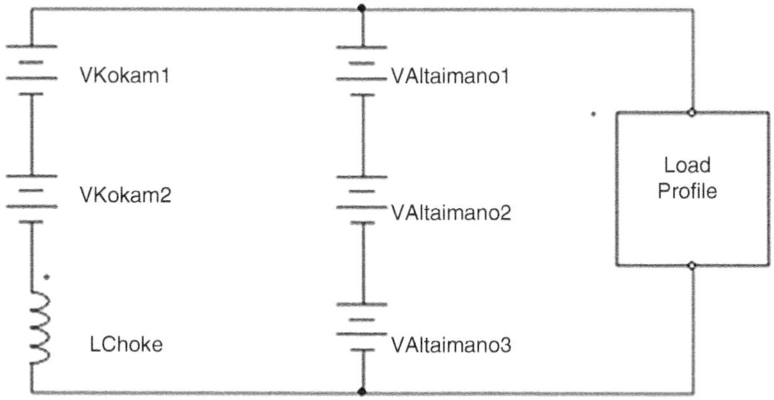

Fig. 3.2 Circuit diagram of HBS

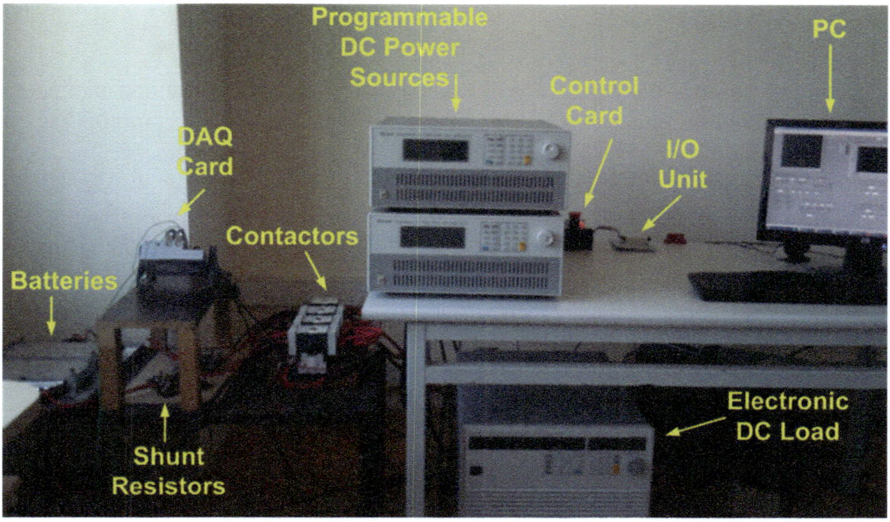

Fig. 3.3 Experimental environment

Table 3.1 Specifications of the batteries which were used in this study

Parameter	Battery model	
	Kokam SLPB55205130H	Altairnano 13 Ah
Structure	$LiNiMnCo_4$	$Li_4Ti_5O_{12}$
Nominal voltage	3.7 V	2.25 V
	Max: 4.2 V	Max: 2.8 V
	Min: 2.7 V	Min: 2.0 V
Nominal capacity	11.3 Ah	13 Ah
Max. continuous charge	33 A	130 A
Max. continuous discharge	88 A	130 A
Max. instant discharge	110 A	260 A
Weight	280 g	400 g
Energy	42 Wh	29 Wh
Max. power	380 W	670 W
	At 50 % SoC	At 50 % SoC
Energy density	150 Wh/Kg	73 Wh/Kg
Power density	1,357 W/Kg	1,675 W/Kg
Life	1,400 Cycles	16,000 Cycles

stacks to avoid voltage imbalances. Kokam and Altairnano have maximum voltages of 4.2 and 2.8 V, respectively. Thus, Kokam is used in a stack of two batteries connected in series, and Altairnano is used in a stack of three batteries. Hence, a bus voltage of 8.4 V is reached in both branches.

Because neither of the battery specifications given by the manufacturers is sufficient, an analytical explanation for the system couldn't be available for this stage of the study. Thus, a choke inductor of 3H is used.

3.5 Driving Cycles

Driving cycles are charts that show the change of velocity or consumed energy of a vehicle versus a certain route. Three basic driving cycles are commonly used; however, there are several driving cycles generated by different companies or different countries for different purposes. First of these cycles is the WLTP (Worldwide harmonized Light vehicles Test Procedure) which was produced by the United Nations European Commission of Economics (UNECE) with the help of the European Union (EU), Japan, and India. In WLTP, there are three different cycles for different power categories. It is commonly preferred among other cycles. Another cycle is NEDC (New European Driving Cycle). It is a cycle which is a combination of three repeats of ECE 15 Urban Driving Cycle and a single application of EUDC (Extra Urban Driving Cycle). ECE 15 and EUDC are produced by ECE in 1975 and 1990, respectively. It has a widespread use in the European region and the so-called MVEG (Motor Vehicle Emission Group Cycle). It lasts for 1180 s with an average speed of 33.6 km/h. It has criticisms that it cannot reflect an actual drive [40], leaving its place to WLTP. The last cycle is FTP 75 (Federal Test Procedure) and produced by EPA (US Environmental Protection Agency). It is commonly used in the USA, and vehicles are classified and labeled by this cycle. There are different variations of this cycle for urban or intercity uses or for different situations such as aggressive use or different weather conditions. A new cycle was generated for this study in order to observe both high power and high-energy situations. The cycle lasts for 360 s, and the maximum current drawn is 80 A, which was a level chosen for not exceeding the maximum continuous discharge currents of both batteries. The produced cycle is given in Fig. 3.4.

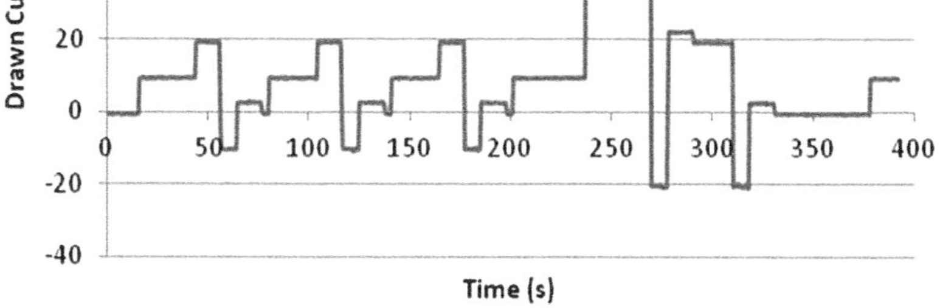

Fig. 3.4 The driving cycle used in experiments

3.6 Results and Discussion

An analytical explanation of the system couldn't be available because of the lack of information given by the manufacturers in battery specifications. Thus, a series of experiments were conducted in order to verify the feasibility of the HBS. Also a driving cycle which represents both high energy and high power demands was used as the load in experiments.

In the first experiment, two batteries were connected in parallel with each other and directly coupled to a load of the driving cycle. The results of the first experiment can be seen in Fig. 3.5 where the straight and dotted lines indicate the responses of batteries with high power and high energy densities, respectively.

In the next step, the experiment was repeated by connecting a 3 mH inductor in series with the battery with high energy density. The results of the experiment can be seen in Fig. 3.6 where the straight and dotted lines indicate the responses of batteries with high power and high energy densities, respectively.

Finally in Fig. 3.7, the comparison of the results of the battery with high energy density in both experiments is shown. The dotted and straight lines demonstrate the results with 3 mH inductor and without inductor, respectively.

From the figures, it can be seen that there is already a difference between the responses of the batteries. Also the parallel inductor has a significant impact on the result of the battery with higher energy level. Considering that the main expectation is limiting the drawn energy from the battery, it can be said that passive hybridization with an inductor with the correct value can be used in the control of HBS.

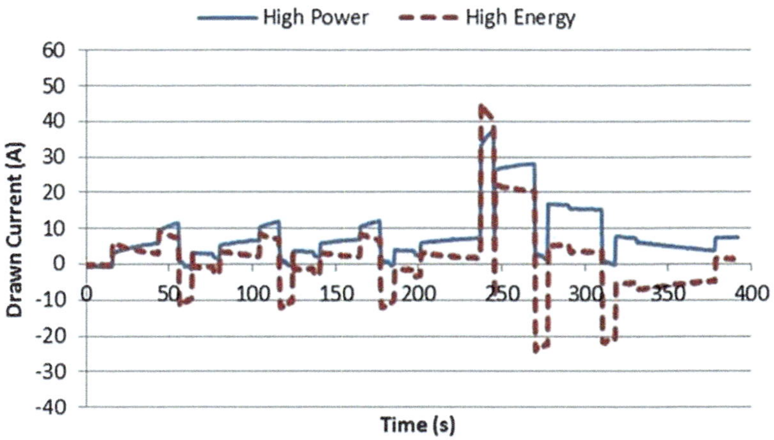

Fig. 3.5 The result of the experiment without the choke inductor

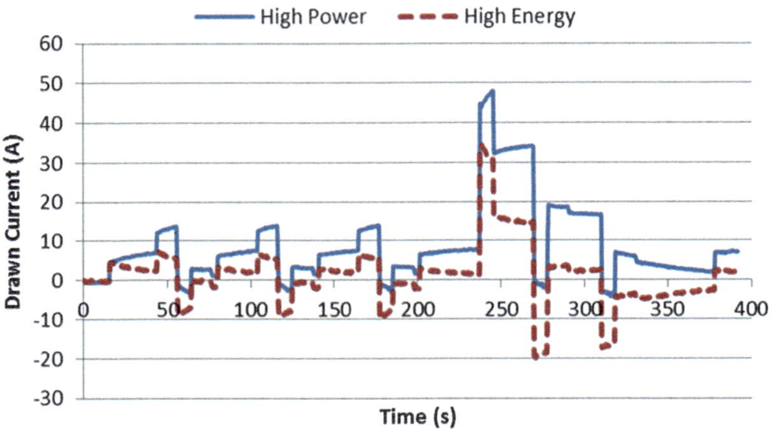

Fig. 3.6 The result of experiment with 3 mH inductor

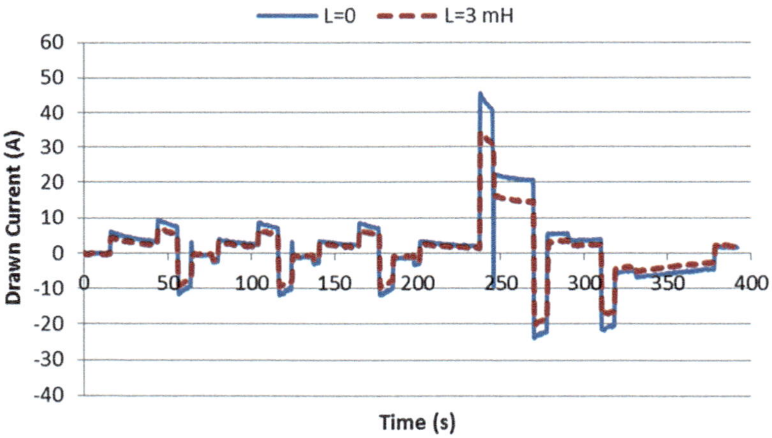

Fig. 3.7 Comparison of responses of battery with high energy density

3.7 Conclusions

In this study, a new hybridization concept for battery electric vehicles (BEVs) and hybrid battery system (HBS) is proposed. The energy management strategies for electric vehicles were listed. A passive hybridization approach for battery-supercapacitor hybrids is adapted to the HBS. Thus, a resulting passive control strategy with inductor is introduced, and the feasibility of this method was verified for HBS with experimental results. However, an analytical explanation couldn't be available.

The main aim is to employ the battery with higher power density in peak power demands and to limit the current drawn from the battery with higher energy density. From Fig. 3.7, it can be seen that the choke inductor limits the current drawn from the battery which has higher energy density. Future studies are directed to develop a battery model in order to gather the battery parameters, to perform an analytical representation and solution of the system, and to obtain ideal inductance to be connected in series with the battery with a higher energy density.

Acknowledgment This work was financially supported by the Scientific and Technological Research Council of Turkey (TÜBİTAK), project number 110E209.

References

1. Nazri G-A, Pistoia G (2008) Lithium batteries, science and technology. Springer, New York, NY
2. Obreja VVN (2008) On the performance of supercapacitors with electrodes based on carbon nanotubes and carbon activated material—a review. Physica E Low Dimens Syst Nanostruct 40(7):2596–2605
3. Köhler U, Kümpers J, Ullrich M (2002) High performance nickel-metal hydride and lithium-ion batteries. J Power Sources 105(1):139–144
4. Winter M, Brodd RJ (2004) What are batteries, fuel cells, and supercapacitors? Chem Rev 104 (10):4245–4269
5. Omar N, Coosemans T, Martin J, Salminen J, Kortschak B, Van Mierlo J, Van Den Bossche P, Sauvant-Moynot V, Volker H (2012) SuperLib Project: advanced dual-cell battery concept for battery electric vehicles. EVS 26 international battery, hybrid and fuel cell electric vehicle symposium, 2012, pp. 1–6
6. Kokam (2009) SLPB (superior lithium polymer battery) technical specification. pp 1–12
7. Altairnano (2011) 13 Amp hour cell, nano lithium-titanate battery cell specifications. pp 1–2
8. Yang Y, Hu T (2007) A new energy management system of directly-driven electric vehicle with electronic gearshift and regenerative braking. American Control Conference. pp 4419–4424
9. Sciarretta A, Guzzella L (2007) Control of hybrid electric vehicles. IEEE Control Syst Mag: 27 (2):60–70
10. Gonder J, Markel T (2007) Energy management strategies for plug-in hybrid electric vehicles distance. SAE World Congress, pp 1–5.
11. Musardo C, Rizzoni G, Guezennec Y, Staccia B (2005) A-ECMS: an adaptive algorithm for hybrid electric vehicle energy management. Eur J Control 11(4–5):509–524
12. Zhang R, Chen Y (2001) Control of hybrid dynamical systems for electric vehicles. American Control Conference. pp 1–6
13. Barsali S, Miulli C, Possenti A (2004) A control strategy to minimize fuel consumption of series hybrid electric vehicles. IEEE Trans Energ Convers 19(1):187–195
14. Delprat S, Lauber J, Guerra TM, Rimaux J (2004) Control of a parallel hybrid powertrain: optimal control. IEEE Trans Veh Technol 53(3):872–881
15. Guzzella L, Sciarretta A (2013) Electric and hybrid-electric propulsion systems, Vehicle Propulsion Systems. Springer, Berlin, pp 67–162
16. Baumann BM, Washington G, Glenn BC, Rizzoni G (2000) Mechatronic design and control of hybrid electric vehicles. IEEE ASME Trans Mechatron 5(1):58–72
17. Schouten NJ, Salman MA, Kheir NA (2002) Fuzzy logic control for parallel hybrid vehicles. IEEE Trans Control Syst Technol 10(3):460–468
18. Tate ED, Boyd SP (1998) Finding ultimate limits of performance for hybrid electric vehicles. SAE Trans 109:2884–2889

19. Hofman T, van Druten R (2004) Energy analysis of hybrid vehicle powertrains. IEEE international symposium of vehicular power propulsion. pp 1–8
20. Lin C, Peng H, Grizzle JW, Kang J-M (2003) Power management strategy for a parallel hybrid electric truck. IEEE Trans Control Syst Technol 11(6):839–849
21. Kolmanovsky I, Siverguina I, Lygoe B (2002) Optimization of powertrain operating policy for feasibility assessment and calibration: stochastic dynamic programming approach. Proceedings of the 2002 American Control Conference (IEEE Cat. No.CH37301) 2(i):1425–1430
22. Johnson V, Wipke K, Rausen D (2000) HEV control strategy for real-time optimization of fuel economy and emissions. SAE, Future Car Congress, 2000
23. Back M, Simons M, Kirschaum F, Krebs V (2002) Predictive control of drivetrains. World Congr. 15(1):1506
24. Koot M, Kessels JTBA, de Jager B, Heemels WPMH, van den Bosch PPJ, Steinbuch M (2005) Energy management strategies for vehicular electric power systems. IEEE Trans Veh Technol 54(3):771–782
25. Arsie I, Graziosi M, Pianese C, Rizzo G, Sorrentino M (2004) Optimization of supervisory control strategy for parallel hybrid vehicle with provisional load estimate. AVEC04, 2004, pp 483–488
26. Jeon S, Jo S, Park Y, Lee J (2002) Multi-mode driving control of a parallel hybrid electric vehicle using driving pattern recognition. J Dyn Syst Meas Control 124(1):141
27. Kleimaier A, Schroder D (2002) An approach for the online optimized control of a hybrid powertrain. Proceedings of the IEEE 7th international workshop on advanced motion control, 2002. pp 215–220
28. Burke A (2000) Ultracapacitors: why, how, and where is the technology. J Power Sources 91 (1):37–50
29. Kuperman A, Aharon I (2011) Battery-ultracapacitor hybrids for pulsed current loads: a review. Renew Sustain Energy Rev 15(2):981–992
30. Kuperman A, Aharon I, Kara A, Malki S (2011) A frequency domain approach to analyzing passive battery–ultracapacitor hybrids supplying periodic pulsed current loads. Energy Convers Manag 52(12):3433–3438
31. Penella MT, Member S, Gasulla M (2010) Runtime extension of low-power wireless sensor nodes using hybrid-storage units. IEEE Trans Instrum Meas 59(4):857–865
32. Dougal RA, Liu S, White RE (2002) Power and life extension of battery – ultracapacitor hybrids. IEEE Trans Compon Packag Techn 25(1):120–131
33. Zheng JP, Jow TR, Ding MS (2001) Hybrid power sources for pulsed current applications. IEEE Trans Aerosp Electron Syst 37(1):288–292
34. Jarvis LP, Atwater TB, Cygan PJ (1999) Fuel cell/electrochemical capacitor hybrid for intermittent high power applications. J Power Sources 79(1):60–63
35. Atwater TB, Cygan PJ, Leung FC (2000) Man portable power needs of the 21st century I. Applications for the dismounted soldier. II. Enhanced capabilities through the use of hybrid power sources. J Power Sources 91(1):27–36
36. Smith TA, Mars JP, Turner GA (2002) Using supercapacitors to improve battery performance. Power Electronics Specialists Conference PESC 1:124–128
37. Holland CE, Weidner JW, Dougal RA, White RE (2002) Experimental characterization of hybrid power systems under pulse current loads. J Power Sourc 109(1):32–37
38. Sikha G, White RE, Popov BN (2005) A mathematical model for a lithium-ion battery/ electrochemical capacitor hybrid system. J Electrochem Soc 152(8):A1682
39. Michalczuk M, Grzesiak LM, Ufnalski B (2012) A lithium battery and ultracapacitor hybrid energy source for an urban electric vehicle. Prz. Elektrotechniczny (Electrical Rev., no. 4), pp 158–162
40. Mock P, German J, Bandivadekar A, Riemersma I (2012) Discrepancies between type-approval and 'real-world' fuel- consumption and CO 2 values

Chapter 4
Cr- and V-Substituted $LiMn_2O_4$ Cathode Electrode Materials for High-Rate Battery Applications

Ahsen Akbulut Uludag, Aslıhan Erdaş, Şeyma Özcan, Deniz Nalci, Mehmet Oğuz Güler, Tuğrul Çetinkaya, Mehmet Uysal, and Hatem Akbulut

Abstract Spinel Cr- and V-substituted $LiMn_2O_4$ cathode materials were prepared by facile sol–gel process, which used lithium carbonate and manganese carbonate as starting materials and citric acid as a chelating agent. In order to increase electronic conductivity and prevent the Mn ion dissolution into the electrolyte, surfaces of the as-synthesized powders were coated with Cu via electroless deposition technique. The structure and physicochemical properties of the obtained Cr- and V-substituted $LiMn_2O_4$ powders were investigated by X-ray diffraction (XRD), scanning electron microscopy (SEM), galvanostatic charge discharge tests, and electrochemical impedance spectroscopy (EIS). The results have shown that the successful formation of Cr- and V-substituted $LiMn_2O_4$ product was highly dependent on its second-stage calcination temperature.

Keywords Li-ion batteries • $LiCr_{0.2}V_{0.4}Mn_{1.4}O_4$ • Sol–gel • Electroless Cu coating

4.1 Introduction

The excess of fossil fuels for energy releases a large amount of gas emissions such as CO_x, NO_x, and SO_x, resulting in severe environmental pollution and global warming. One way to solve the problem is to explore renewable clean energy such as solar, wind, and tidal energies. Another way is to use electric cars to replace internal combustion engine cars to reduce emissions and improve energy efficiency [1]. The lithium-ion (Li-ion) battery is a high-capacity rechargeable electrical

A.A. Uludag (✉)
Engineering Faculty, Department of Environmental Engineering, Sakarya University, Esentepe Campus, 54187 Sakarya, Turkey
e-mail: aakbulut@sakarya.edu.tr

A. Erdaş • Ş. Özcan • D. Nalci • M.O. Güler • T. Çetinkaya • M. Uysal • H. Akbulut
Engineering Faculty, Department of Metallurgical & Materials Engineering, Sakarya University, Esentepe Campus, 54187 Sakarya, Turkey

I. Dincer et al. (eds.), *Progress in Clean Energy, Volume 2*,
DOI 10.1007/978-3-319-17031-2_4

energy storage device with applications in portable electronics and growing applications in electric vehicles, military, and aerospace [2, 3]. Within the past decade, various cathode and anode materials have been studied, but these are within the boundaries of $LiCoO_2$, $LiNiO_2$, $LiMn_2O_4$, $LiFePO_4$, electrochemically active metals, and metal oxides [4]. At present, the main cathode material is $LiCoO_2$, which has high theoretical capacity and good cycling stability, but possesses high cost and toxicity of Co [5, 6]. Spinel $Li_{1-x}[Mn_2]O_4$ exhibiting a rapid Li extraction/insertion at 4 V (versus Li/Li^+) is considered as a promising potential cathode material for lithium-ion batteries used for EV, HEV, and PHEV due to its low cost, low toxicity, easy preparation, and high safety properties [7]. However, the major limitation of $LiMn_2O_4$ as an electrode material reported by Jayaprakash et al. [8] is that the reversible 4 V capacity tends to fade upon electrochemical cycling, due to factors such as (1) fracture of the particle surface due to local Jahn–Teller distortions at high rates of discharge, (2) dissolution of manganese from lithiated and delithiated spinels, (3) electrolyte oxidation on the spinel surface, (4) cation mixing or Li/Mn site exchange, (5) loss of oxygen from the spinel, and (6) structural failure in the two-phase reaction region.

Aklaloucha et al. [9] reported that stoichiometric spinel $LiMn_2O_4$ takes the cubic framework by close-stacked O^{2-} ions at the 32e sites and can be simply expressed as $[Li]_{8a}[Mn^{4+}Mn^{3+}]_{16d}\,[O_4]_{32e}$. The Li^+ ions occupy the tetrahedral 8a sites, and the Mn^{4+} and Mn^{3+} ions are located at the octahedral 16d sites, with an average valence of 3.5 for Mn. When $LiMn_2O_4$ is charge–discharge cycled in the 4 V (vs. Li^+/Li) region, Li^+ ions de-intercalate from and intercalate into $LixMn_2O_4$ over the x range $0 \leq x \leq 1$. However, Xu et al. presented that [10] when it is overdischarged at the 3 V level ($x > 1$), a structural transformation from cubic spinel phase ($LiMn_2O_4$) to tetragonal rock-salt phase ($Li_2Mn_2O_4$) takes place, i.e., the so-called Jahn–Teller distortion, and results in a fast capacity fading. The capacity fading of pure lithium manganese oxide is strongly influenced from the Jahn–Teller distortion occurring on the surface of the particles and manganese dissolution into the electrolyte, especially at elevated temperatures. To overcome these drawbacks, Sun et al. [11] reported two strategies were mainly pursued: (1) element substitution or oxygen excess to increase the oxidation state of Mn for suppressing the Jahn–Teller effect and (2) surface modification or coating for suppressing the dissolution of manganese into the electrolyte.

The first strategy of doping is known to be an effective route. It is reported by Deng et al. [12], Takashi et al. [13], and Yang et al. [14] that substitution of Mn ions with other metals such as Li, Mg, Al, Cr, Co, Ni, Zn, Mg, and Ga slows dissolution of Mn from the cathode into the electrolyte. Li et al. [15], Wang et al. [16], Liu et al. [17], Kakuda et al. [18], Wang et al. [19], and Kim et al. [20] have reported that the introduction of cations with a low oxidation number increases the oxidation state of Mn ions. Partially replacing Mn could increase the average valence of Mn in $LiMn_2O_4$ (decreasing the Mn^{3+} in $LiMn_2O_4$) and could depress the Jahn–Teller effect to a certain degree and hence could improve the cyclability of $LiMn_2O_4$ in nonaqueous electrolytes. This seems to prevent the generation of acids in the electrolyte or relieve the Jahn–Teller distortion, resulting in the reduction of Mn dissolution.

Because of the smaller ion radius or the stronger bond energy, Li [16], Co [21], Ni [22], and V [23] have been used to substitute the Mn^{3+} of $LiMn_2O_4$ to restrain the Jahn–Teller effect. It has been verified by many researchers that single metal element doping is one good way to improve the cycling performance of $LiMn_2O_4$. There has been much work on single metal element doping as also reported by Liu et al. [24]. However, it is thought that single ion substitution for the Mn^{3+} cannot resolve all the factors, which causes capacity loss. A synergistic effect has been reported by Haijun et al. [25] when using double cation or multiple cation substitution of Mn^{3+} to improve the cycling life of the $LiMn_2O_4$ cathode material. Dual metal element doping has also been found effective in improving the cycling performances of $LiMn_2O_4$ [26]. These include Li^+Ni [27], Li^+Al [28], Li^+Co [29], Li^+Cr [30], Li^+Mg [31], Co^+Gd [32], Cr^+V [33], and Ce^+Zn [34]. The functions of $LiMn_2O_4$ could also be enhanced by co-substitution [4, 5, 27], i.e., by one metal element and one nonmetal element, especially the F element [35].

Although the elemental substitution cathode materials still suffer from significant capacity decline at elevated temperature (50–60 °C), the second strategy to overcome the Jahn–Teller distortion is to coat $LiMn_2O_4$ with various protective layers. The surface modification technique has been reported as an effective approach to improve the cycling performance of $LiMn_2O_4$ by coating different conductive carbon materials, phosphates, amphoteric oxides, and composite oxides. It has reported by Ju et al. [36] that many efforts were performed to improve the cyclic performance via metallic oxide coating layers, which inhibit the Mn^{3+} dissolution of $LiMn_2O_4$, such as Al_2O_3, SiO_2, La_2O_3, TiO_2, ZrO_2, CeO_2, MgO, and $LiAlO_2$. The coating of electrical insulating materials such as alumina or magnesia improved the high-temperature cyclability, but degraded the rate capability, which is another important requirement for high-power-source applications as stated by Kim et al. [37]. Therefore, using a conductive metal coating over the surfaces of $LiMn_2O_4$ can be more beneficial for providing both high capacity and conductivity. It is well known that gold and silver are the metals which possess the lowest resistance; hence, they can be expected to enhance electron conduction of coated $LiMn_2O_4$ and then improve its electrochemical performance. Jarvis et al. [38] have reported that a nano-gold film-coated $LiMn_2O_4$ shows better capacity retention at room temperature than that of uncoated $LiMn_2O_4$, which is attributed to reduced contact area of electrode/electrolyte interface and suppressed dissolution of manganese during electrochemical cycling. Kim et al. [39] reported that the initial discharge capacity was decreased when the amount of Ag coating was increased. However, Ag-coated $LiMn_2O_4$ particles have shown very good capacity retention after 40 cycles. It was also reported by Yi et al. [40] that the silver-coated nanoparticle $LiMn_2O_4$ shows excellent cyclability at high-charging-rate galvanostatic conditions. It can be concluded that the surface modifications with conductive silver and gold nanoparticles will improve the cyclability of $LiMn_2O_4$-based cathode electrodes by enhancing the electron conduction due to the low electronic resistance of silver and gold.

In this study, it is aimed to investigate both the bication and conductive metal coating on the $LiMn_2O_4$ spinels. Specifically, this is the second ever attempt made to explore the electrochemical characteristics of pentavalent V dopant associated with the Cr^{3+} dopant in $LiMn_2O_4$ spinel structures. The first work was carried out by Ke et al. [41]. To perform this study, we have synthesized pristine $LiMn_2O_4$ and $LiV_{1.4}Cr_{0.2}Mn_{0.4}O_4$ cathode materials. The copper coating is aimed to increase the conductivity and prevent Mn dissolution into the electrolyte and also provide spontaneous doping agent in the case of Mn dissolution depending on the cyclic electrochemical process.

4.2 Experimental

4.2.1 Sample Preparation, Electroless Coating, and Characterization

$LiMn_2O_4$ and its substituted spinel $LiV_{1.4}Cr_{0.2}Mn_{0.4}O_4$ were prepared by a solution-based gel method from a stoichiometric mixture of lithium acetate [$Li(CH_3COO)\cdot 4H_2O$], manganese acetate [$Mn(CH_3\text{-}COO)_2\cdot 4H_2O$], and metal ion sources such as vanadium pentanedionate ($C_{15}H_{21}O_6V$) and chromium nitrate ($[Cr(H_2O)_6](NO_3)_3\cdot 3H_2O$). The mixture was dissolved in distilled water into which the solution of acetate and ethanol was added dropwise under continuous stirring. Thereafter, ammonium hydroxide [NH_3HO] was added slowly to the solution to control the pH at ~8. The prepared solution was heated to 70–80 °C to evaporate the water until a transparent gel was obtained. The resulting gel precursor was decomposed at 400 °C for 5 h in air to remove the organic contents. The heat-treated precursor was ground and calcined at 800 °C in air for 10 h to obtain the final spinel product.

Pristine and substituted spinel $LiV_{1.4}Cr_{0.2}Mn_{0.4}O_4$ were used as the cores of the composite powders in the electroless plating. Before starting the deposition process, the surfaces of the cathode active powders have been pretreated as follows: (1) The surfaces of the as-synthesized powders were cleaned by immersing powders into 1 M sodium hydroxide (Merck, Titripur) solution for 10 min. (2) The cleaned pristine and $LiV_{1.4}Cr_{0.2}Mn_{0.4}O_4$ powders were then immersed into boiling hydrochloric acid (1 N) for 15 min to coarsen their surfaces. The coarsened surfaces will increase the adhesion between the $LiMn_2O_4$-based powders and copper particles. (3) Since the surfaces of pristine and doped particles of $LiMn_2O_4$ were nonconductive, an activation process is needed in order to improve the catalytic properties. One-step sensitization and activation process was performed in this study. The etched pristine and $LiV_{1.4}Cr_{0.2}Mn_{0.4}O_4$ were immersed in a palladium activation solution consisting of palladium dichloride ($PdCl_2$) (0.3 g/L), stannous chloride ($SnCl_2\cdot 2H_2O$, >99.9 %) (16 g/L), sodium chloride (NaCl, >99.99 %) (150 g/L), and hydrochloric acid (HCl) (Merck, 1 N) (60 mL/L) and stirred strongly at room temperature for 15–20 min. After each pretreatment process, the pristine and $LiV_{1.4}Cr_{0.2}Mn_{0.4}O_4$ were repeatedly washed by using bidistilled water until pH is reached to 7. The pretreated powders were then dried after the activation step in an oven at 80 °C for 12 h.

The copper plating solution was composed of copper sulfate solution (28 g/L $CuSO_4 \cdot 5H_2O$ (>99.9 %), 30 g/L EDTA (Merck, Titriplex III), 14 g/L $C_4H_4O_6KNa \cdot 4H_2O$ (>99 %), and 10 mg/L $K_4[Fe(CN_6)] \cdot 3H_2O$ (Alfa Aesar, >98 %)) and formaldehyde solution (30 mL/L HCOOH (37 %), 100 mL/L CH_3OH (99.5 %), and 16 g/L NaOH). Plating was carried out at 30 °C in a container with ultrasonic bath (Bandelin Sonorex 10P). During the reaction process, system pH value was kept as a constant by adding NaOH (Merck, Titripur) solution until a pH value of 12 is obtained. A magnetic stirrer is necessary to make pristine and $LiV_{1.4}Cr_{0.2}Mn_{0.4}O_4$ particles suspend in the solution and make the hydrogen produced during plating easily escape. After plating, the powders were washed with distilled water and then dried in a vacuum oven at 80 °C for 12 h.

Powder X-ray diffraction (XRD) using D/Max-2200 Rigaku (Japan) with CuKα ($\lambda = 1.54178$ Å) monochromated radiation was used to identify the crystalline phase of the samples. XRD data were collected in the 2θ ranges from 10 to 80°. The particle size and morphology were investigated by scanning electron microscopy (SEM, JSM-6600LV).

4.2.2 *Electrochemical Measurements*

The positive electrodes were fabricated by intimately mixing the active material (80 wt%), conductive graphite (10 wt%), and binder (10 wt% polyvinylidene fluoride, PVDF, dissolved in *N*-methyl-2-pyrrolidone, NMP) to form a slurry. The mixed slurry was coated onto an aluminum current collector. The electrodes were dried under vacuum at 80 °C overnight and then punched out and weighed. The cells were assembled in a glove box under dry argon. The complete cell comprises a cathode, a Celgard (2300) separator, and a Li foil anode. 1 M $LiPF_6$ (battery grade) dissolved in a mixture of ethylene carbonate (EC, battery grade) and dimethyl carbonate (DMC, battery grade) (1:1 by volume) was used as the electrolyte. Charge–discharge performance of the cell was characterized galvanostatically at 1 C charge–discharge rate between 3.0 and 4.3 V (vs. Li/Li^+). The electrochemical impedance measurements were carried out by applying 100 kHz to 0.01 Hz frequency ranges with AC oscillation amplitude of 5 mV. In this study, the cells were discharged to a designated potential and then were kept in the open-circuit condition for 1 h before impedance tests were performed, so that the equilibrium of the cells could be ensured. Both of the electrochemical measurements were done using a Gamry Reference 3000 electrochemical workstation.

4.3 Results and Discussions

4.3.1 Structure and Morphology Characterization of Samples

The typical powder XRD patterns of pristine $LiMn_2O_4$ and substituted spinel $LiV_{1.4}Cr_{0.2}Mn_{0.4}O_4$ are shown in Fig. 4.1. The XRD patterns of the pristine and substituted $LiMn_2O_4$ particles present eight characteristic peaks at 18.6°, 36.4°, 38.3°, 44.3°, 48.5°, 58.6°, 64.2°, and 67.8° that correspond to crystal planes of (111), (311), (222), (400), (331), (511), (440), and (531), respectively. It can be concluded that the diffraction peaks of the three samples are in agreement with those from the JCPDS 35-0782 card, which corresponds to stoichiometric spinel $LiMn_2O_4$ with a cubic unit cell and an *Fd3m* space group; no other impurities such as Li_2MnO_3, $LiMnO_2$, or MnO_x were detected [26]. It was observed that no apparent changes are observed in the two patterns. Only the spinel phase is observed, and no additional peaks of chromium or vanadium oxides can be found which indicates that the modification does not change the spinel structure of $LiMn_2O_4$. But due to atom size effect, it can slightly change the lattice parameters [15, 16]. The lattice parameters were calculated by the least-squares method using ten diffraction lines, and silicon is used as a reference to calibrate the peak position and intensity ratio. The lattice parameters for the pristine and Cr–V-substituted spinels were calculated as 8.247 Å and 8.096 Å, respectively. After the modification, the lattice shrinkage is due to the smaller ionic radii of Cr^{2+} and V^{2+}, which replaced the Mn^{3+} at 16d sites. On the other hand, Cr^{2+} and V^{2+} substituting part Mn^{3+} can enhance the content of Mn^{4+} to keep the charge balance and Mn^{4+} with smaller ionic radii than that of Mn^{3+}. The lattice shrinkage increases the spinel structural stability of an active material [18], which is beneficial to the suppression of Jahn–Teller distortion.

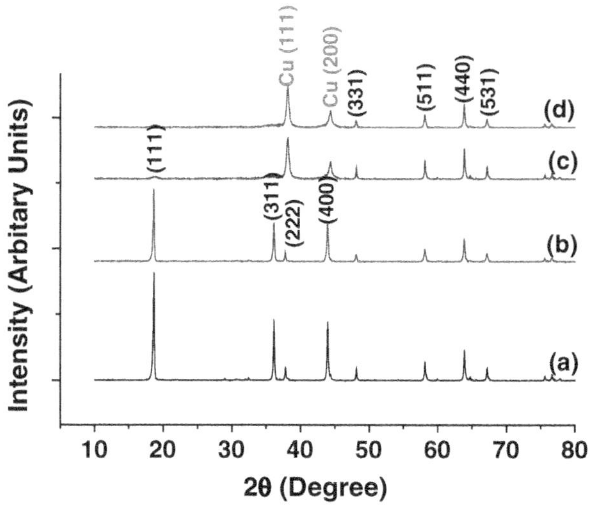

Fig. 4.1 XRD spectra of the (*a*) pristine $LiMn_2O_4$, (*b*) substituted spinel $LiV_{1.4}Cr_{0.2}Mn_{0.4}O_4$, (*c*) Cu/$LiMn_2O_4$, and (*d*) Cu/$LiV_{1.4}Cr_{0.2}Mn_{0.4}O_4$ cathode active electrodes

In order to determine the effects of copper coating on the crystal structure of $LiMn_2O_4$, XRD analyse were carried out for uncoated and coated spinel materials and presented in Fig. 4.1c, d according to the JCPDS 04-0836 and JCPDS 98-0094, respectively. Since the surfaces of the pristine and $LiV_{1.4}Cr_{0.2}Mn_{0.4}O_4$ cathode electrode materials were coated with copper, (111) and (221) planes for $LiMn_2O_4$ disappeared. However, no significant change is observed for the (551), (441), and (531) planes after coating with copper when compared with the uncoated pristine $LiV_{1.4}Cr_{0.2}Mn_{0.4}O_4$ spinels. However, the intensities of the peaks were reduced after coating the surfaces of the based spinels uniformly with copper, which was an expected result. The lattice parameters were also calculated by the least-squares method for copper-coated pristine and $LiV_{1.4}Cr_{0.2}Mn_{0.4}O_4$ spinel samples and found to be 8.244 Å and 8.101 Å, respectively. The calculated lattice parameters show almost no change in the lattice parameter for all samples, which indicated that copper ions were not incorporated into the spinel structure but were just presented on the surfaces of pristine and substituted spinel samples.

It is well known that copper ion introduction to the spinel structure will lead to significant change of the lattice parameter [31, 32]. The data obtained from Fig. 4.1c, d also indicate that copper crystals largely grow in two preferential orientations, (111) and (200). It has been reported that the peaks due to copper crystallites with (111) orientation are predominant. This is an important result knowing that stress-induced voiding is significantly inhibited in (111)-textured metals [33]. Furthermore, the (111)-textured Cu film has higher resistance to electromigration which suppresses grain boundary and interfacial diffusion of metal atoms [34].

The morphologies of the produced spinel cathode materials and their Cu-coated structures are presented in Figs. 4.2 and 4.3. The particle morphologies of the pure $LiMn_2O_4$ and $LiV_{1.4}Cr_{0.2}Mn_{0.4}O_4$ cathode electrode material powders are shown in Figs. 4.2a, b and 4.3a, b. As shown in Fig. 4.2a and the high-magnification image in Fig. 4.3a, the pristine $LiMn_2O_4$ powders annealed at 800 °C have a uniform crystalline shape and wide particle size distribution. The average size of particles is between 200 and 300 nm with homogeneous distribution. Additionally, the particles exhibit distinctive visible polyhedrons, which indicate good crystallinity. All of the as-prepared powders have nearly cubic structure morphology with narrow size distribution. From Figs. 4.2b and 4.3b, it can be concluded that the substituted samples show a slightly smaller particle size value and a more regular morphology than the pristine $LiMn_2O_4$. In addition, the particles of $LiV_{1.4}Cr_{0.2}Mn_{0.4}O_4$ samples are distributed more uniformly than those of $LiMn_2O_4$. Furthermore, the increase in agglomeration of the particles results in effective interparticle contact leading to good capacity and good cyclability. From Figs. 4.2b and 4.3b, it could be seen that the samples with Cr and V substitution appear to be more cohesive in nature, which consist of aggregates of regular crystallites with different crystallite sizes generally ranging from ca. 100 to 300 nm. The aforementioned features of $LiV_{1.4}Cr_{0.2}Mn_{0.4}O_4$ are very desirable for being employed as the electrode material to improve the electrochemical properties of spinel $LiMn_2O_4$ cathode materials for lithium rechargeable batteries. Figures 4.2c, 4.3c, 4.2d, and 4.3d show the SEM

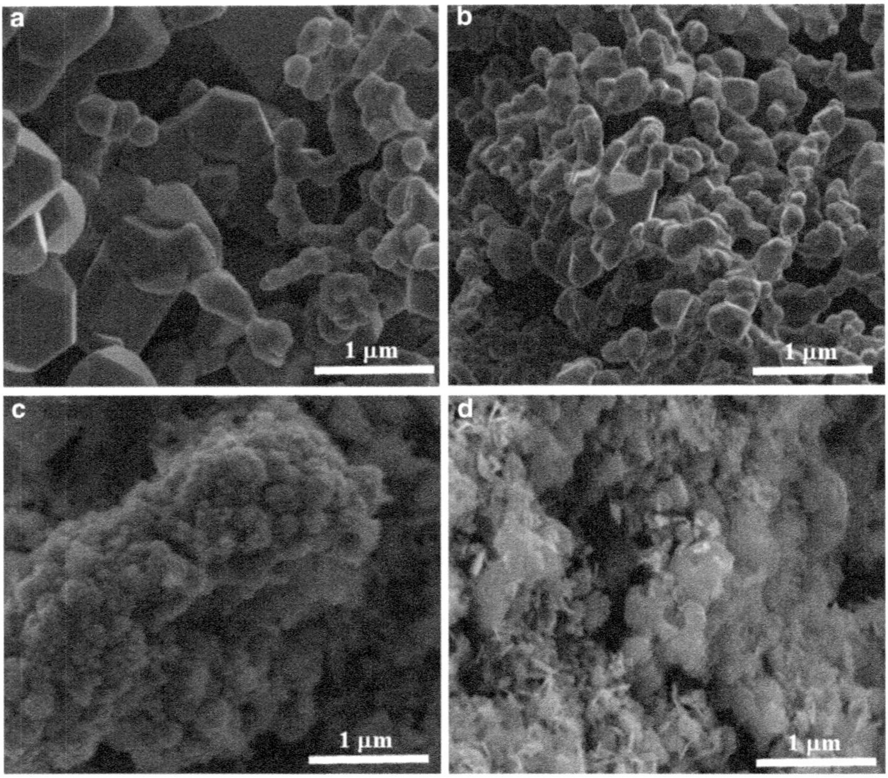

Fig. 4.2 SEM images of the (**a**) pristine $LiMn_2O_4$, (**b**) substituted spinel $LiV_{1.4}Cr_{0.2}Mn_{0.4}O_4$, (**c**) $Cu/LiMn_2O_4$, and (**d**) $Cu/LiV_{1.4}Cr_{0.2}Mn_{0.4}O_4$ cathode active electrodes

images of Cu-coated as-synthesized pristine $LiMn_2O_4$ and substituted $LiMn_2O_4$ powders. No free copper particles and uncoated surface of pristine $LiMn_2O_4$ and substituted $LiMn_2O_4$ particles are found on the powders. As also shown in the higher-resolution images in Figs. 4.2c, d and 4.3c, d, it is clear that the surface of cathode active particles is coated with copper particles of 30–60 nm. From Figs. 4.2c, 4.3c, 4.2d, and 4.3d, it can also be concluded that surfaces of truncated octahedron were damaged after the surface coarsening process. However, a tight, continuous, and void-free Cu coating is obtained after the deposition process. The island of coatings that nucleate starts spreading both in the vertical and horizontal directions [37].

4.3.2 *Electrochemical Properties of Samples*

The galvanostatic charge and discharge tests were performed by using coin-type CR2016 cells, and the results are shown in Fig. 4.4a–d for pristine, substituted

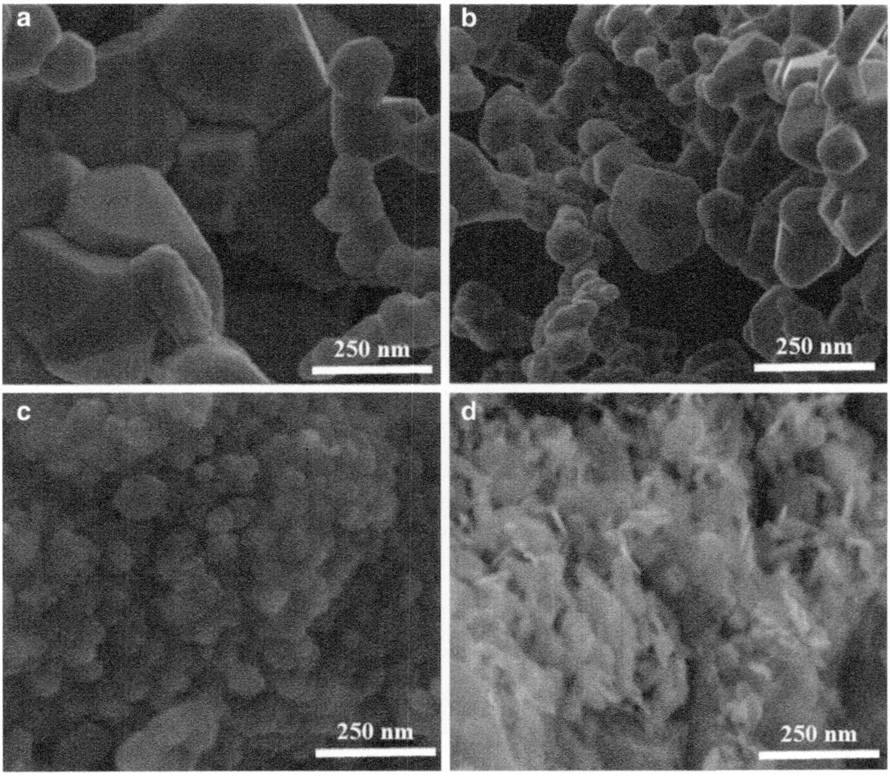

Fig. 4.3 High-magnification SEM images of the (**a**) pristine $LiMn_2O_4$, (**b**) substituted spinel $LiV_{1.4}Cr_{0.2}Mn_{0.4}O_4$, (**c**) Cu/$LiMn_2O_4$, and (**d**) Cu/$LiV_{1.4}Cr_{0.2}Mn_{0.4}O_4$ cathode active electrodes

spinel LiV1.4Cr0.2Mn0.4O4, Cu/LiMn2O4, and Cu/LiV1.4Cr0.2Mn0.4O4 cathode active electrodes, respectively. The cycling tests were performed between 3 and 4.3 V because of the electrolyte decomposition above 4.3 V [39]. The capacities obtained correspond to the oxidation of Mn^{3+} to Mn^{4+} and M^{3+}/M^{4+} (where M is Cr and V). The two-step de-intercalation/intercalation process in the case of Cr- and V-substituted $LiMn_2O_4$ may, for example, be represented as below:

$$\begin{aligned} &\mathrm{Li}\left[\mathrm{M}^{3+}y\mathrm{Mn}^{+3}\mathrm{Mn}^{4+}\right]\mathrm{O}_4 \\ &\quad \underset{4\mathrm{V}}{\leftrightarrow} 1-y\mathrm{Li}_y\left[\mathrm{M}^{3+}\mathrm{Mn}^{4+}\right]\mathrm{O}_4 + \left[(1-y)\mathrm{Li}^+ + (1-y)e^- \underset{4.8\mathrm{V}}{\leftrightarrow} \mathrm{M}^{4+}y\mathrm{Mn}^{4+}\right]\mathrm{O}_4 \\ &\quad + y\mathrm{Li}^+ + ye^- \end{aligned} \tag{4.1}$$

As can be concluded from Fig. 4.4a, the pristine spinel exhibited a specific capacity of 110 mA h g^{-1}. However, introducing Cr and V ions as a replacement for Mn^{3+}

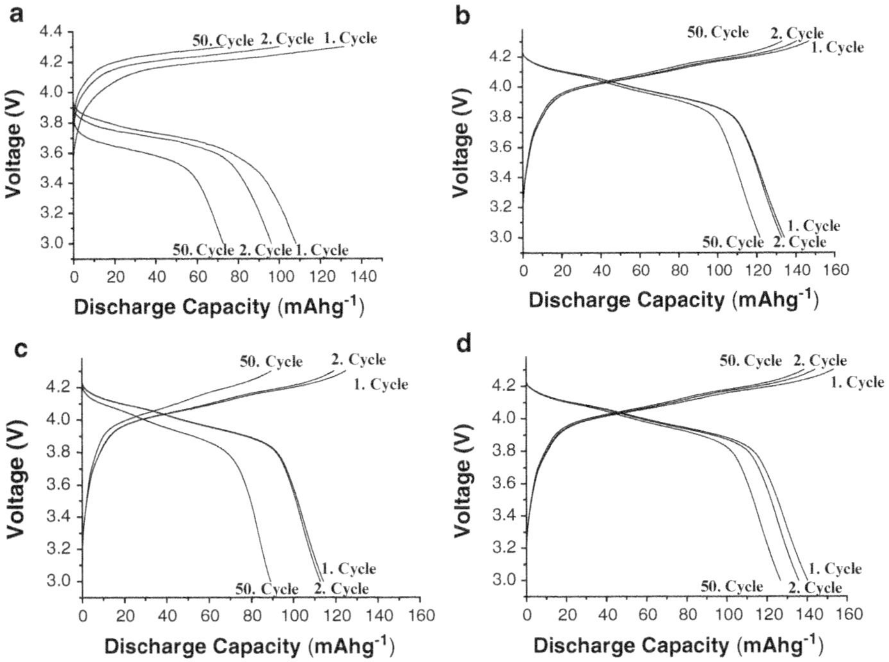

Fig. 4.4 Galvanostatic charge and discharge tests of the (**a**) pristine $LiMn_2O_4$, (**b**) substituted spinel $LiV_{1.4}Cr_{0.2}Mn_{0.4}O_4$, (**c**) Cu/$LiMn_2O_4$, and (**d**) Cu/$LiV_{1.4}Cr_{0.2}Mn_{0.4}O_4$ cathode active electrodes

ions gives a higher capacity as shown in Fig. 4.4b. Introducing Cr and V into the spinel structure, the oxidation of a similar amount of Mn^{3+} to the Mn^{4+} state leads to an increase in the average oxidation state of manganese [40, 41]. The diminished Mn^{3+} ion concentration causes a reduction in the unit cell volume of the spinel, which results in increased structural stability. The cyclic performances of Cu-coated spinels for the first, second, and 50th cycles are also shown in Fig. 4.4c, d. The initial charge curves of Cu-coated samples show a shorter plateau compared to the pristine sample, leading to lower initial charge capacity. This can be attributed to the strong mechanical bonding of Cu with the pristine samples, which suppresses the oxygen vacancy migration and reduces the activity of the evolved oxygen species, leading to less oxygen removal and less electrolyte oxidation during the initial charge process [42].

Table 4.1 and Fig. 4.5 summarize the cyclic performances of the four cathode electrodes at a constant rate of 1 C with a cutoff voltage of 3.0–4.3 V (versus Li/Li^+) at room temperature. As seen in Fig. 4.4c, the cycling performance of the Cr- and V-substituted spinel is significantly improved. The initial discharge capacity of the $LiV_{1.4}Cr_{0.2}Mn_{0.4}O_4$ cathode electrode sample is 134 mA h g^{-1}, and the total capacity loss is obtained as 9 % after 50 cycles. With the substitution of Cr and V into the spinel $LiMn_2O_4$ structure, the unit cell volume is decreased, which

Table 4.1 Total battery performance of $LiMn_2O_4$, $LiV_{1.4}Cr_{0.2}Mn_{0.4}O_4$, Cu-coated $LiMn_2O_4$, and Cu-coated $LiV_{1.4}Cr_{0.2}Mn_{0.4}O_4$ cathode active electrodes

Samples	Discharge capacity (mA h g^{-1})			Capacity fading (%)
	1st cycle	2nd cycle	50th cycle	
$LiMn_2O_4$	110	98	72	34
$LiV_{1.4}Cr_{0.2}Mn_{0.4}O_4$	134	132	122	9
Cu-coated $LiMn_2O_4$	115	103	89	22
Cu-coated $LiV_{1.4}Cr_{0.2}Mn_{0.4}O_4$	140	135	127	9

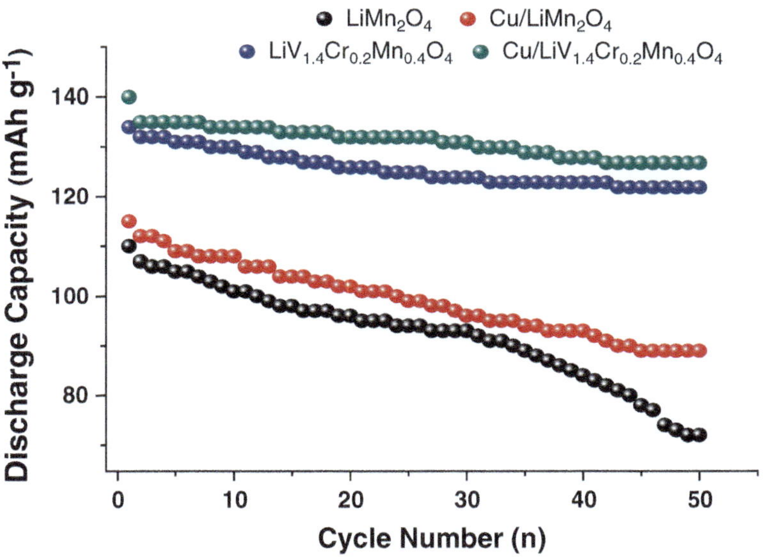

Fig. 4.5 Galvanostatic discharge vs. cycling number curves of the pristine $LiMn_2O_4$, $LiV_{1.4}Cr_{0.2}Mn_{0.4}O_4$, Cu-coated $LiMn_2O_4$, and Cu-coated $LiV_{1.4}Cr_{0.2}Mn_{0.4}O_4$ spinel cathode active electrodes

stabilizes the Li–Mn spinel structure notably. The content of Mn^{3+} within the spinel structure is decreased as a result of suppressed Jahn–Teller distortion effect. It is well known that during the lithiation and delithiation process, the cubic to tetragonal phase transition, often referred to as Jahn–Teller distortion, is accompanied by a 6.5 % increase in the unit cell volume, which damages the structural integrity of the electrode during charge–discharge cycling and results in rapid capacity fading [43, 44]. The results obtained from Fig. 4.4c, d clearly indicate that the Jahn–Teller distortion is suppressed by substituting Cr and V into the lattice structure of spinel $LiMn_2O_4$. In addition, the cycling properties of pristine electrode and Cr- and V-substituted $LiMn_2O_4$ cathode powders are increased after coating with Cu.

The discharge capacities of the pristine $LiMn_2O_4$, $LiV_{1.4}Cr_{0.2}Mn_{0.4}O_4$, Cu-coated $LiMn_2O_4$, and Cu-coated $LiV_{1.4}Cr_{0.2}Mn_{0.4}O_4$ cathode electrodes were measured as 110, 134, 115, and 140 mA h g^{-1} in the first cycle and are shown in Fig. 4.4. Moreover, the discharge capacities of the pristine $LiMn_2O_4$, $LiV_{1.4}Cr_{0.2}Mn_{0.4}O4$, Cu-coated $LiMn_2O_4$, and Cu-coated $LiMn_2O_4$ cathode electrodes were obtained as 72, 122, 89, and 127 mA h g^{-1} after 50 cycles. The capacity fadings of the pristine $LiMn_2O_4$, Cu-coated $LiMn_2O_4$, $LiV_{1.4}Cr_{0.2}Mn_{0.4}O_4$, and Cu-coated $LiV_{1.4}Cr_{0.2}Mn_{0.4}O_4$ cathode electrodes were 34, 22, 9, and 9 % of the initial capacity after 50 cycles. It has been reported earlier that the capacity loss caused by the dissolution of manganese accounted 23 % of the overall capacity loss [35, 45]. HF is generated during charging and discharging processes when $LiPF_6$-based electrolyte was used. Practically, it is not possible to prepare H_2O-free electrolyte containing $LiPF_6$ in organic solvent. Trace amount of water, which is less than 20 ppm, facilitates decomposition of $LiPF_6$-based salt, and therefore, HF is formed as a product according to the following reaction [46]:

$$LiPF_6 + H_2O \rightarrow LiF + POF_3 + 2HF \tag{4.2}$$

HF then corrodes the $LiMn_2O_4$ structure, and as a result, the spinel structure decomposes during the charging and discharging processes as shown in the reaction below:

$$2LiMn_2O_4 + 4H^+ \rightarrow 3\lambda\text{-}MnO_2 + Mn^{+2} + 2Li^+ + 2H_2O \tag{4.3}$$

Therefore, in this study, it was aimed to reduce the Mn dissolution by modifying the surfaces of pristine and $LiV_{1.4}Cr_{0.2}Mn_{0.4}O_4$ cathode electrodes with Cu via electroless deposition technique. The electrochemical data obtained in this study indicates that the cycling performance of the pristine and Cr- and V-substituted $LiV_{1.4}Cr_{0.2}Mn_{0.4}O_4$ has been significantly improved. Based on the results shown in Table 4.1, it can be concluded that the Cu-coated $LiV_{1.4}Cr_{0.2}Mn_{0.4}O_4$ has the best cycling performance among all four samples. The initial discharge capacity of the $LiV_{1.4}Cr_{0.2}Mn_{0.4}O_4$ sample is 140 mA h g^{-1}, and the total capacity loss is obtained as 9 % after 50 cycles.

The electrochemical impedance spectroscopies were performed to investigate the variation of electrode resistance of $LiMn_2O_4$ samples. The Nyquist plots of the $LiMn_2O_4$ samples after the first charge–discharge cycle are shown in Fig. 4.6. The perturbation potential was 5 mV with frequencies from 0.01 Hz to 100 kHz. EIS results were then fitted with the equivalent circuits inset in Fig. 4.5. On this circuit, R_s is the resistance associated with the electrolyte and cell component, R_{int} is the interface resistance of any film formation of the cathode surface (first high-frequency semicircle), R_{ct} is the charge–transfer resistance of the electrode reaction with lithium ions (middle-frequency semicircle), and W_{dif} is the resistance of the lithium ion diffusion to the electrode (low-frequency semicircle) [41]. The fitted results show obviously decreased R_{ct} after Cr and V substitution and Cu coating. The R_{ct} values obtained from the equivalent circuit for pristine $LiMn_2O_4$, $LiV_{1.4}Cr_{0.2}Mn_{0.4}O_4$, $Cu/LiMn_2O_4$, and $Cu/LiV_{1.4}Cr_{0.2}Mn_{0.4}O_4$ were 564 Ω, 455 Ω, 266 Ω, and 250 Ω, respectively. The results have shown that a thicker

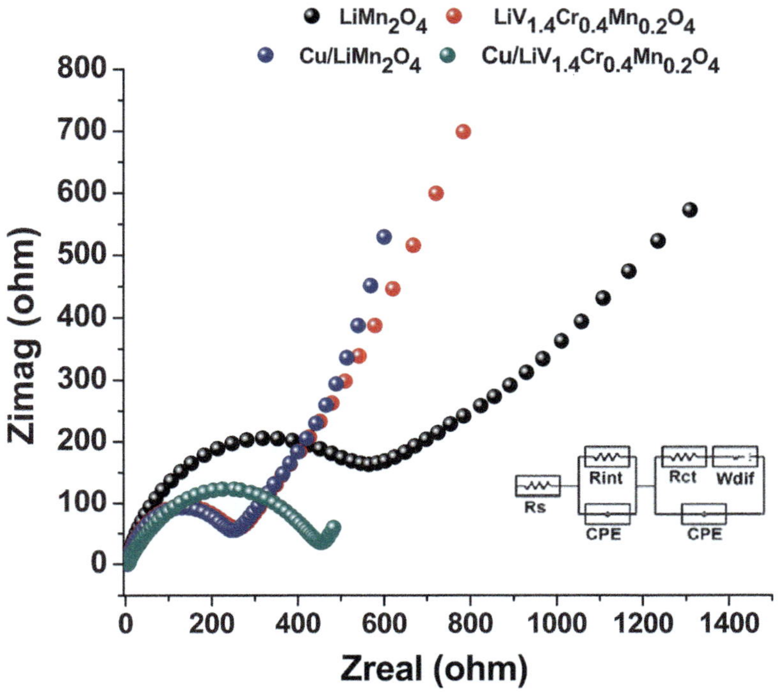

Fig. 4.6 EIS spectra of the pristine $LiMn_2O_4$, $LiV_{1.4}Cr_{0.2}Mn_{0.4}O_4$, Cu-coated $LiMn_2O_4$, and Cu-coated $LiV_{1.4}Cr_{0.2}Mn_{0.4}O_4$ spinel cathode active electrodes

surface layer was formed on the pure $LiMn_2O_4$ compared with the doped $LiCr_{0.1}Mn_{1.9}O_4$ electrode, which supports the proposal that the high self-discharge rate of the pure $LiMn_2O_4$ electrode is due to the decomposition of the electrolyte on the surface of the highly charged electrode. However, it is not possible to claim that this surface layer is robust or not. If it is not robust enough, a debris will form and enter into the electrolyte during subsequent charge–discharge cycles. In that way, the electrolyte will undergo continuous decomposition, causing a loss in capacity of the cell which will be detrimental to the operation of the battery. Furthermore, the charge transfer resistance of the Cu coating of pristine $LiMn_2O_4$ and $LiV_{1.4}Cr_{0.2}Mn_{0.4}O_4$ samples has significantly improved. Pristine $LiMn_2O_4$ shows rapid capacity fading, and one reason of this capacity fading is the Mn dissolution. The dissolved Mn migrates to the lithium anode through the electrolyte, and then a reduction occurs on the surfaces of the anode. Hence, it is believed that dissolved Mn and/or the Mn-containing complexes form a solid electrolyte interface on the surfaces of the lithium anode, and as a result, the charge transfer resistance increases. In the present study, oxidation of electrolyte and degradation of electrolyte/electrode interface are suppressed effectively by Cu coating, which leads to decreased R_{ct} value and improved electrochemical performance [35, 39–46].

4.4 Conclusions

In summary, spinel-based pristine $LiMn_2O_4$ and $LiV_{1.4}Cr_{0.2}Mn_{0.4}O_4$ powders have been synthesized via facile sol–gel method based on a two-stage calcination process, in which the temperature of the second stage plays a key role in the structure and physicochemical properties of the obtained $LiMn_2O_4$ products. In order to increase the electrochemical efficiencies of both pristine $LiMn_2O_4$ and $LiV_{1.4}Cr_{0.2}Mn_{0.4}O_4$ electrodes, the powders were coated with metallic Cu via electroless deposition technique. The XRD results of all synthesized cathode active electrodes were identified as a single phase of cubic spinel structure with the space group Fd3m. Compared with the other samples, the Cu-coated $LiV_{1.4}Cr_{0.2}Mn_{0.4}O_4$ powders are the most uniform, and the charge transfer resistance (R_{ct}) is the lowest, accompanying the best electrochemical activity and highest discharge capacity. The initial discharge specific capacity of this electrode is 140 mA h g^{-1}, and the discharge capacity retention remains 91 % after 50 cycles. The considerably large capacity may originate from the uniform particle size; the better rate capability and higher cycling performance may be related to the $LiV_{1.4}Cr_{0.2}Mn_{0.4}O_4$ shell layer formed, which is expected to enhance its electrical conductivity and to restrain the Jahn–Teller distortion and Mn dissolution during electrochemical reactions. The work opens up a facile and effective route for the synthesis of the shell/core Cr- and V-incorporated $LiMn_2O_4$ with excellent electrochemical performance.

Acknowledgment This work is supported by the Scientific and Technological Research Council of Turkey (TUBITAK) under the contract number 111M021—Improving the Capacity of Li-Ion Batteries by Using New Semi-Conducting Metal Oxide Based Anodes. The authors thank the TUBITAK MAG workers for their financial support.

References

1. Lilong X, Youlong X, Tao T, Jie S, Goodenough JB (2012) Excellent stability of spinel $LiMn_2O_4$-based composites for lithium ion batteries. J Mater Chem 22:24563–24568
2. Guler MO, Cevher O, Cetinkaya T, Tocoglu U, Akbulut H (2014) Nanocomposite anodes for lithium-ion batteries based on SnO_2 on multiwalled carbon nanotubes. Int J Energy Res 38:487–498
3. Chabot V, Farhad S, Chen Z, Fung AS, Yuland A, Hamdullahpur F (2013) Effect of electrode physical and chemical properties, on lithium-ion battery performance. Int J Energy Res 37:1723–1736
4. Kim MG, Cho J (2009) Reversible and high-capacity nanostructured electrode materials for Li-Ion batteries. Adv Funct Mater 19:1497–1514
5. Wei C, Deng J, Xi L, Huaiying Z, Wang Z, Chung CY, Yao Q, Rao G (2013) High power $LiMn_2O_4$ hollow microsphere cathode materials for Lithium Ion batteries. Int J Electrochem Sci 8:6775–6783
6. Liu Q, Wang S, Tan H, Yang Z, Zeng J (2013) Review preparation and doping mode of doped $LiMn_2O_4$ for Li-Ion batteries. Energies 6:1718–1730

7. Wang J, Xu Y, Wang J, Zhu J, Bai Y, Xiong L (2014) Study on capacitance evolving mechanism of polypyrrole during prolonged cycling. J Phys Chem B 118:1353–1362
8. Jayaprakash N, Kalaiselvi N, Gangulibabu CHD, Bhuvaneswari D (2010) A new class of Sol–gel derived $LiM_{1x}M_{2y}Mn_{22x2y}O_{3.8}F_{0.2}$ ($M_1 = Cr$, $M_2 = V$; $x = y = 0.2$) cathodes for lithium batteries. J Appl Electrochem 40:2193–2202
9. Aklaloucha M, Amarillaa JM, Rojasa RM, Saadoune I, Rojo JM (2008) Chromium doping as a new approach to improve the cycling performance at high temperature of 5 V $LiNi_{0.5}Mn_{1.5}O_4$-based positive electrode. J Power Sources 185:501–511
10. Xu W, Yuan A, Tian L, Wang Y (2011) Improved high-rate cyclability of sol–gel derived Cr-doped spinel $LiCr_yMn_{2-y}O_4$ in an aqueous electrolyte. J Appl Electrochem 41:453–460
11. Sun H, Chen Y, Xu C, Zhu D, Huang L (2012) Electrochemical performance of rare-earth doped $LiMn_2O_4$ spinel cathode materials for Li-ion rechargeable battery. J Solid State Electrochem 16:1247–1254
12. Deng B, Nakamura H, Zhang Q, Yoshio M, Xia Y (2004) Greatly improved elevated-temperature cycling behavior of $Li_{1+x}Mg_yMn_{2-x-y}O_{4+\delta}$ spinels with controlled oxygen stoichiometry. Electrochim Acta 49:1823–1830
13. Takahashi M, Yoshida T, Ichikawa A, Kitoh K, Katsukawa H, Zhang Q, Yoshio M (2006) Effect of oxygen deficiency reduction in Mg-doped Mn-spinel on its cell storage performance at high temperature. Electrochim Acta 51:5508–5514
14. Yang Y, Xie C, Ruffo R, Peng H, Kim D, Cui Y (2009) Single nanorod devices for battery diagnostics: a case study on $LiMn_2O_4$. Nano Lett 9:4109–4114
15. Li J, He X, Fan M, Wan C, Jiang C, Zhang S (2006) Capacity fading of $LiCr_{0.1}Mn_{1.9}O_4$/MPCF cells at elevated temperature. Ionics 12:153–157
16. Wang C, Lu S, Kan S, Pang J, Jin W, Zhang X (2009) Enhanced capacity retention of Co and Li doubly doped $LiMn_2O_4$. J Power Sources 189:607–610
17. Liu YJ, Guo HJ, Li XH, Wang ZX (2012) Study on the storage performance of manganese spinel battery. Adv Mater Res 347–353:1395–1398
18. Kakuda T, Uematsu K, Toda K, Sato M (2007) Electrochemical performance of Al-doped $LiMn_2O_4$ prepared by different methods in solid-state reaction. J Power Sources 167:499–503
19. Wang XQ, Tanaike O, Kodama M, Hatori H (2007) High rate capability of the Mg-doped Li–Mn–O spinel prepared via coprecipitated precursor. J Power Sources 168:282–287
20. Kim Y, Lim J, Kang S (2013) Investigation on the dissolution of Mn ions from $LiMn_2O_4$ cathode in the application of lithium ion batteries: first principle molecular orbital method. Int J Quantum Chem 113:148–154
21. Huang H, Wang C, Zhang WK, Gan YP, Kang L (2008) Electrochemical study on $LiCo_{1/6}Mn_{11/6}O_4$ as cathode material for lithium ion batteries at elevated temperature. J Power Sources 184:583–588
22. Shaik A, Rudramoorthy R, Neelakrishnan S, Varman KSR, Arjunan TV (2011) Evaluation of energy requirements for all-electric range of plug-in hybrid electric two-wheeler. Energy 36:1623–1629
23. Eftekhari A (2006) Bundled nanofibers of V-doped $LiMn_2O_4$ spinel. Solid State Commun 140:391–394
24. Liu J, Sun Z, Xie J, Chen H, Wu N, Wu B (2013) Study of electrochemical performances of multi-doped spinel $Li_{1.1}Mn_{1.85}Co_{0.075}Ni_{0.075}O_4$ at 4.3 and 5 V. Ionics 19:1867–1874
25. Haijun Y, Haoshen Z (2013) High-energy cathode materials (Li_2MnO_3–$LiMO_2$) for lithium-Ion batteries. J Phys Chem Lett 4:1268–1272
26. Tian J-K, Wan F-C, Battaglia VS, Zhang H-L (2014) Synthesis and electrochemical performance of nanosized multiple-doped $LiMn_2O_4$ prepared at low temperature for Li-ion battery. Int J Electrochem Sci 9:931–942
27. Lee K-S, Myung S-T, Bang HJ, Chung SJ, Sun Y-K (2007) Co-precipitation synthesis of spherical $Li_{1.05}M_{0.05}Mn_{1.9}O_4$ (M = Ni, Mg, Al) spinel and its application for lithium secondary battery cathode. Electrochim Acta 52:5201–5206
28. Myoung YS, Ik HK, Hye RP, Daniel RM (2011) Electrochemical properties of $LiCo_yMn_{2-y}O_4$ synthesized using a combustion method in a voltage range of 3.5–5.0 V. Ceram Int 37:2215–2220

29. Amarilla M, Petrov K, Picó F, Avdeev G, Rojo JM, Rojas RM (2009) Sucrose-aided combustion synthesis of nanosized $LiMn_{1.99-y}Li_yM_{0.01}O_4$ ($M = Al^{3+}$, Ni^{2+}, Cr^{3+}, Co^{3+}, $y = 0.01$ and 0.06) spinels: characterization and electrochemical behavior at 25 and at 55 °C in rechargeable lithium cells. J Power Sources 191:591–600
30. Changju C, Hongyeol P, Dongwook K, Jongsik K, Eun S, Jung KL (2013) A Li-ion battery using $LiMn_2O_4$ cathode and MnO_x/C anode. J Power Sources 244:214–221
31. Ouyang CY, Zeng XM, Sljivancanin Z, Baldereschi A (2010) Oxidation states of mn atoms at clean and Al_2O_3-covered $LiMn_2O_4$(001) surfaces. J Phys Chem C 114:4756–4759
32. Arumugam D, Kalaignan GP (2011) Electrochemical characterizations of surface modified $LiMn_2O_4$ cathode materials for high temperature lithium battery applications. Thin Solid Films 520:338–343
33. Jayaprakash N, Gangulibabu NK, Bhuvaneswari D (2011) Effect of mono- (Cr) and bication (Cr, V) substitution on $LiMn_2O_4$ spinel cathodes. J Solid State Electrochem 15:1243–1251
34. Thirunakaran R, Sivashanmugam A, Gopukumar S, Rajalakshmi R (2009) Cerium and zinc: dual-doped $LiMn_2O_4$ spinels as cathode material for use in lithium rechargeable batteries. J Power Sources 187:565–574
35. Seung TM, Ki SL, Yang KS, Hitoshi Y (2011) Development of high power lithium-ion batteries: layer $Li[Ni_{0.4}Co_{0.2}Mn_{0.4}]O_2$ and spinel $Li[Li_{0.1}Al_{0.05}Mn_{1.85}]O_4$. J Power Sources 196:7039–7043
36. Ju B, Wang X, Wu C, Wei Q, Yang X, Shu H, Bai Y (2014) Excellent cycling stability of spherical spinel $LiMn_2O_4$ by Y_2O_3 coating for lithium-ion batteries. J Solid State Electrochem 18:115–123
37. Kim J, Park SM, Roh KC, Lee JW (2013) Effect of manganese vanadate formed on the surface of spinel lithium manganese oxide cathode on high temperature cycle life performance. Bull Kr Chem Soc 34:2573–2576
38. Jarvis KA, Deng ZQ, Allard LF, Manthiram A, Ferreira PJ (2011) Atomic structure of a lithium-rich layered oxide material for lithium-ion batteries: evidence of a solid solution. Chem Mater 23:3614–3621
39. Kim JM, Lee G, Kim BH, Huh YS, Lee GW, Kim HJ (2012) Ultrasound-assisted synthesis of Li-rich mesoporous $LiMn_2O_4$ nanospheres for enhancing the electrochemical performance in Li-ion secondary batteries. Ultrason Sonochem 19:627–631
40. Yi T-F, Zhu Y-R, Zhu X-D, Shu J, Yue C-B, Zhou A-N (2009) A review of recent developments in the surface modification of $LiMn_2O_4$ as cathode material of power lithium-ion battery. Ionics 15:779–784
41. Ke D, Fei Y, Guorong H, Zhong DP, Yanbing C, Kwang SR (2013) Sodium additive to improve rate performance of $Li[Li_{0.2}Mn_{0.54}Ni_{0.13}Co_{0.13}]O_2$ material for Li-ion batteries. J Power Sources 244:29–34
42. Mohamed A, Jose MA, Ismael S, Jose MR (2011) $LiCr_{0.2}Ni_{0.4}Mn_{1.4}O_4$ spinels exhibiting huge rate capability at 25 and 55 °C: analysis of the effect of the particle size. J Power Sources 196:10222–10227
43. Yumei L, Zhenzhen L, Yongliang L, Caifeng C, Yi H, Xiaojing Y (2011) Preparation and electrochemical properties of Li-rich spinel-type lithium manganate coated $LiMn_2O_4$. Mater Res Bull 46:2450–2455
44. Rajive MT, Anil KKM, Anand PB, Jayalekshmi S (2011) Effect of annealing on structural and electrical properties of the Li–Mn–O thin films, prepared by high frequency RF magnetron sputtering. J Phys Chem Solid 72:1251–1255
45. Dario C, Petr N, Alexander W, Rudiger K (2011) Mixed bi-material electrodes based on $LiMn_2O_4$ and activated carbon for hybrid electrochemical energy storage devices. Electrochim Acta 56:8403–8411
46. Kedi Y, Jing S, Li Z, Yunfei L, Xiaoyan L, Yanxuan W (2012) Urea combustion synthesis of $LiNi_{0.5}Mn_{1.5}O_4$ as a cathode material for lithium ion batteries. Particuology 10:765–770

Part II
Biomass and Biofuels

Chapter 5
An Overview of Science Teachers' Knowledge of Bioenergy and the Need for Future Research: A Case from India

Pradipta Halder

Abstract The study aimed to explore science teachers' knowledge of bioenergy and their views regarding bioenergy education in schools. Data were collected from 28 science teachers from four schools based in New Delhi and Bengaluru in India. The results showed that although the teachers knew some of the facts of bioenergy, the majority did not appear to know that the use of bioenergy could release CO_2 into the atmosphere. Moreover, not all the science teachers were aware that bioenergy was included in the science syllabus of the tenth grade, and it indicated that bioenergy was not taught with much importance by the science teachers. The study recommends improving science teachers' knowledge of bioenergy including other renewable energy sources so that they can provide a high quality science education related to energy topics to their students.

Keywords Teacher • Knowledge • Bioenergy • India

5.1 Introduction

A significant growth in the modern bioenergy sector is expected due to strong policy support that it has been receiving from governments in both developing and developed countries. However, a transition from a society that is heavily dependent on fossil fuels to a society embracing renewable energy technologies is a great challenge. In this context, the emergence of an innovative, informed, and motivated young citizenry such as school students is paramount in all countries as they are the key to build a sustainable future. Nevertheless, schools all over the world are facing numerous challenges to come up with a curriculum that remains up to date and provide students with necessary knowledge and information of the latest scientific advancements in the field of energy science. In this backdrop, the role of school teachers becomes highly significant as Chedid [1] has suggested that "teachers are the primary link for preparing students for the future energy related jobs," whereas

P. Halder (✉)
School of Forest Sciences, University of Eastern Finland, Joensuu 80101, Finland
e-mail: pradipta.halder@uef.fi

I. Dincer et al. (eds.), *Progress in Clean Energy, Volume 2*,
DOI 10.1007/978-3-319-17031-2_5

Seraphin et al. [2] have emphasized that "teachers should have high level of knowledge and confidence for teaching energy science in schools."

India is one of the major emerging economies in the world, and the domestic energy consumption has increased rapidly in the country over the last decades. At present, around 95 % of India's commercial energy demand is met by fossil fuels, while about 70 % of the Indians rely on traditional biomass for cooking and heating, particularly in the rural areas of the country [3]. The country has a number of universities and research laboratories, which are engaged in advanced level bioenergy-related research and training. However, the status of bioenergy-related education in Indian schools under the broad theme of environmental education is unknown. Indian education system is the second largest in the world and there are 28 states and 7 union territories in the country that have different schooling systems and curricula. There is a National Council of Educational Research and Training (NCERT) under the Department of Education (Government of India), which is responsible for developing school curriculum, syllabus, and textbooks. Teaching *science* in schools starts at the *primary stage* in almost all schools in the country [4], and most recently, the NCERT has made changes in the *science* curriculum for schools and the revised curriculum includes energy-related themes including bioenergy in class X (tenth grade). Therefore, it is expected that science teachers should be knowledgeable of different energy sources and concepts including bioenergy and they will also be able to teach those topics to students effectively through curricular as well as extracurricular methods. However, there has not been any study conducted in India to explore teachers' knowledge of bioenergy and their views toward the possibility of bioenergy-related education in schools.

Previous studies from other countries have revealed that teachers have difficulties in teaching energy-related topics in schools as they do not have much knowledge in those topics. In this regard, Trumper [5] found several misconceptions about energy-related concepts among elementary school teachers in Israel, and Diakidoy and Iordanou [6] also found that the knowledge of energy as a scientific concept among teachers in Cyprus was not very satisfactory. Saglam-Arslan [7] has mentioned that since energy concepts are intrinsically complex and difficult to understand conceptually by many people, school teachers are no exception. In a recent study, Zyadin et al. [8] have found that the school teachers in Jordan have several misconceptions regarding bioenergy.

5.2 Aims of the Study

The study, based on the above discussions related to the challenges in energy education in schools, aimed to explore knowledge of bioenergy among science teachers in India as well as their views to the possibility of teaching bioenergy in schools. Therefore, the study expects to identify the challenges in bioenergy education in schools in India, and it provides some recommendations to overcome those challenges.

5.3 Methodology

A questionnaire-based survey was conducted among 28 science teachers in four schools based in New Delhi and Bengaluru—two of the largest cities in India. The first section of the questionnaire included questions related to the teachers' background (e.g., age, gender, years of teaching experience, and educational qualification). The second section consisted of questions related to the teachers' knowledge of bioenergy. The third section consisted of questions related to the teachers' views toward the possibility of teaching bioenergy in their schools. The four schools were randomly selected from the two cities, and after necessary permissions obtained from the school authorities, questionnaires were sent to them by post. A science teacher in each school was responsible for conducting the survey. The questionnaire items were selected through literature review and by consulting a few bioenergy experts. Among the 28 science teachers who participated in the survey, 82 % were female and the mean age of the respondents was 35 years (SD = 7.75). The average teaching experience of the respondents was 8 years (SD = 5.81). The majority (ca. 61 %) of the teachers had a master's degree, while about one-third of them had a bachelor's degree in one of the subjects such as *physics*, *chemistry*, *mathematics*, and *biology*.

5.4 Results

5.4.1 Science Teachers' Knowledge of Bioenergy

Science teachers' knowledge of bioenergy was measured by a four-item "true-false" scale (Table 5.1). The results showed that the majority of the science teachers knew that bioenergy was considered as a renewable source of energy. However, slightly above 50 % of the respondents only knew that bioenergy could be used as liquid fuel in motor vehicles, whereas one-fifth of them did not know about such a possibility and one-fifth did not respond to that statement. Furthermore, the majority of the respondents appeared to know that bioenergy could be used for electricity production. However, it appeared that the majority of the science teachers did not know that the use of bioenergy could release CO_2 into

Table 5.1 Knowledge of bioenergy among science teachers in India ($N = 28$). Adapted from Halder et al. [9]

Statements measuring bioenergy knowledge	True	False	No response
Bioenergy is a renewable source of energy	89 %	7 %	4 %
Bioenergy is used in liquid form in motor vehicles	54 %	21 %	25 %
Bioenergy is used for electricity production	82 %	–	18 %
Use of bioenergy does not release CO_2 into the atmosphere	61 %	18 %	21 %

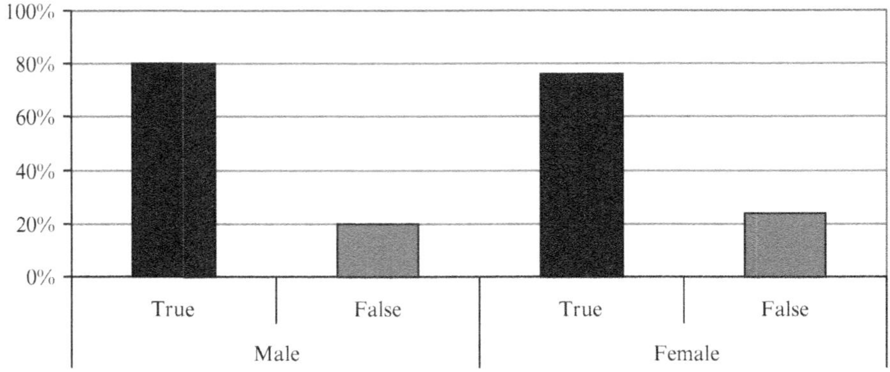

Fig. 5.1 Science teachers' responses to the statement "use of bioenergy does not release CO_2 into the atmosphere" ($N = 28$)

the atmosphere whereas only a few of them were able to recognize that fact correctly. Gender parity appeared between the science teachers related to their knowledge of bioenergy regarding the themes: renewability of bioenergy, use of bioenergy as liquid biofuels, and its use for producing electricity. Almost an equal percentage of the male and female science teachers appeared to be unaware of the issue concerning the release of CO_2 into the atmosphere from the use of bioenergy (Fig. 5.1).

5.4.2 *Science Teachers' Views Regarding the Possibility of Bioenergy Teaching in Schools*

The majority of the science teachers (ca. 86 %) agreed that their students should be aware of bioenergy. Two-thirds of the respondents confirmed that the school curriculum included bioenergy as a topic. Those who confirmed that the school curriculum contained bioenergy-related topic further informed that the topic was part of the biology, environmental science, social science, and geography subject syllabus. In terms of classes where bioenergy was taught in their schools, the respondents informed classes ranging from II to XI. All the science teachers informed that teaching bioenergy in schools would be possible and students could learn about bioenergy from school teachers and other sources such as newspapers, science books, science magazines, Internet, and TV.

5.5 Conclusions

The study investigated science teachers' knowledge of bioenergy and their views regarding bioenergy teaching from a survey in four schools, which were located in New Delhi and Bengaluru in India. The science teachers appeared to know that bioenergy was a renewable source of energy and it could be used for producing electricity. However, a sizeable proportion of them did not appear to know that bioenergy could be used as liquid fuels in motor vehicles. The majority of the science teachers were not also aware that the use of bioenergy could release CO_2 into the atmosphere. Therefore, it indicates that it is necessary to improve science teachers' knowledge of bioenergy so that they can teach their students this topic effectively. All the teachers agreed that their students should be aware of bioenergy and teaching such a subject in their schools could be possible. Nevertheless, significant discrepancies appeared in the teachers' responses concerning the class and subject where bioenergy was included as a topic in their schools. They suggested a variety of classes and subjects, and there were also discrepancies in the teachers' responses even within a same school. It indicates that even if bioenergy is part of a *science* syllabus in class X, it is not taught with much attention by the science teachers. In India, the number of full-time teachers teaching at the *secondary* schools is about 1.05 million and about 83–87 % of them are trained [10]. However, the learning environment in rural schools in India is severely affected by the absence of minimum basic facilities and most of them are often understaffed [11]. Therefore, an effective delivery of energy education is a challenging task in these schools since such education system demands that knowledgeable teachers are teaching in the designated science classes and are also motivating their students to become aware of energy and renewable energy concepts through various extracurricular activities. In this context, it is urgent to investigate the quality of overall teaching of energy topics in schools in India as it is directly related to the students' learning of energy science. The study was only limited to four schools in two large Indian cities, and therefore, future studies need to recruit more schools from both urban and rural areas in India to have a deeper understanding of the science teachers' knowledge and capacity related to the delivery of bioenergy education in schools.

Acknowledgments The author is thankful to the schools and teachers for their cooperation in the study. The author greatly acknowledges the funding support by the Kone Foundation (Helsinki, Finland) to carry out this research. Finally, the travel grant from the Kone Foundation for participating in the *International Conference on Clean Energy 2014*, Istanbul, Turkey, is also acknowledged by the author where the author presented this work.

References

1. Chedid LG (2005) Energy, society, and education, with emphasis on educational technology policy for K-12. J Sci Educ Technol 14:75–85
2. Seraphin KD, Philippoff J, Parisky A, Degnan K, Warren DP (2013) Teaching energy science as inquiry: reflections on professional development as a tool to build inquiry teaching skills for middle and high school teachers. J Sci Educ Technol 22:235–251
3. International Energy Agency (2011) Energy for all. Financing access for the poor. International Energy Agency, Paris, France

4. Yadav SK (2013) National study on ten year school curriculum implementation. National Council of Educational Research and Training, Department of Teacher Education, New Delhi, India
5. Trumper R (1997) The need for change in elementary school teacher training: the case of the energy concept as an example. Educ Res 39:157–174
6. Diakidoy IAN, Iordanou K (2003) Preservice teachers' and teachers' conceptions of energy and their ability to predict pupils' level of understanding. Eur J Psychol Educ 18:357–368
7. Saglam-Arslan A (2010) Cross-grade comparison of students' understanding of energy concepts. J Sci Educ Technol 19:303–313
8. Zyadin A, Puhakka A, Ahponen P, Pelkonen P (2014) Secondary school teachers' knowledge, perceptions, and attitudes toward renewable energy in Jordan. Renew Energy 62:341–348
9. Halder P, Havu-Nuutinen S, Pietarinen J, Zyadin A, Pelkonen P (2014) Subject knowledge and perceptions of bioenergy among school teachers in India: results from a survey. Resources 3:599–613
10. National Council of Educational Research and Training (2002) Seventh all India school educational survey – provisional statistics. National Council of Educational Research and Training, Department of Teacher Education, New Delhi, India
11. Cheney GR, Ruzzy BB, Muralidharan K (2005) A profile of the indian education system. Paper prepared for the New Commission on the Skills of the American Workforce. National Centre on Education and the Economy, USA

Chapter 6
Advances in Algae Harvesting and Extracting Technologies for Biodiesel Production

Farshad Khademi, İlhami Yıldız, Asena Cansu Yıldız, and Soheila Abachi

Abstract Bioenergy production by microalgal cell factories via different culturing, harvesting, extracting, and conversion methods offers a potentially viable approach towards renewable and sustainable energy sources. The approach, however, should be an integrated approach, and always be coupled with other benefits such as wastewater treatment, greenhouse gas sequestration, food and feed production, or biosynthesis of valuable biopharmaceutics. With a growth rate of up to 50-folds higher than crops, oleaginous eukaryotic microalgae and prokaryotic cyanobacteria are considered as the most promising versatile feedstock for the production of biofuels. These potent microscopic biomanufacturers are able to biosynthesize primary metabolites such as proteins, lipids, and carbohydrates which can be converted into different forms of bioenergy as well as valuable secondary metabolites such as food supplements, enzymes, polymers, toxins, and pigments. Depending on microbial species, cultivation, harvesting, and extraction technologies employed, microalgal feedstock can flexibly and continuously produce multiple biofuels such as biodiesel, bioethanol, biogas, and biohydrogen. However, a large-scale microalgae biomass production system cannot lead to a sustainable "greenenergy" production unless low-cost technologies are developed and employed in upstream and downstream processing. This article reviews the most-recent and relevant investigations in microalgal upstream and downstream bio-processing technologies highlighting potential cost-effective and feasible separation and extraction technologies.

F. Khademi • İ. Yıldız (✉)
Department of Engineering, Faculty of Agriculture, Dalhousie University, Truro-Bible Hill, NS, Canada B6L 3H9
e-mail: iyildiz@dal.ca

A.C. Yıldız
Department of Law, Queen Mary, University of London, Mile End Road, London, UK E1 4NS

S. Abachi
Department of Environmental Sciences, Faculty of Agriculture, Dalhousie University, Truro-Bible Hill, NS, Canada B6L 3H9

I. Dincer et al. (eds.), *Progress in Clean Energy, Volume 2*,
DOI 10.1007/978-3-319-17031-2_6

Keywords Microalgae harvesting • Oil extraction • Membrane filtration • Flocculation • Centrifugation • Flotation • Magnetic separation • Immobilization • Cell disruption • Solvent extraction • Supercritical fluid extraction • Ultrasonication

Nomenclature

ARA	Arachidonic acid
ATP	Adenosine triphosphate
DHA	Docosahexaenoic acid
EPA	Eicosapentaenoic acid
EPS	Extracellular polymeric substances
GHG	Greenhouse gas
SFE	Supercritical fluid extraction
TAGs	Triacylglycerols

6.1 Introduction

Climate change mitigation, fossil fuels dilemma, human and environmental health as the main challenges for humankind have inspired researchers around the world to look for alternative energy sources with the most economic advantages and minimal impact on the environment [1–11]. Fossil fuels currently supply almost 80 % of global energy demand, and microalgal biofuels seem to be the most sustainable option to secure future fuel supplies [12–15].

Microalgal production systems have considerable potential for not only biofuel production but also greenhouse gas (GHG) emissions reduction, food and feed production, wastewater treatment, and biosynthesis of valuable biopharmaceuticals or a combination of these processes [16–19]. For instance, these potent microscopic biomanufacturers are able to biosynthesize primary metabolites such as proteins, lipids, and carbohydrates, as well as valuable secondary metabolites such as food supplements, nutraceuticals such as antioxidants, antiatherogenic compounds, enzymes, polymers, toxins, biopharmaceuticals such as anti-inflammatory and anticancer compounds, photosynthetic carotenoids, and pigments such as astaxanthin, β-carotene, lutein, zeaxanthin, and some other precious supplements [20–26]. In their review study [27], researchers complied and reported the eicosapentaenoic acid (EPA) production in autotrophic and heterotrophic cultivation systems, docosahexaenoic acid (DHA), and arachidonic acid (ARA) productions in heterotrophic cultivation systems, β-Carotene production in *Dunaliella salina*, astaxanthin production in photoautotrophic and mixotrophic *Haematococcus pluvialis*, lutein production in photoautotrophic and heterotrophic cultivation systems, polysaccharide production in *Porphyridium* sp., phycoerythrin and phycocyanin productions in *Porphyridium cruentum* and *Spirulina platensis*, respectively. Our readers are referred to this review study for production details of specific components derived from different microalgae species in different cultivation systems and the original research literature.

Despite all the aforementioned advantages, the scaling up of microalgae-based processes has not been seamless; and therefore, employment of low-cost upstream or downstream technologies, genetic manipulation or microbial consortium engineering, and microbial screening for achieving higher yields and lower production costs have been the subject of many research and development activities globally.

Crop-based "first-generation" biofuel systems are already extensively used for the production of biofuels such as bioethanol, biodiesel, and biomethane, most notably in Brazil, USA, Southeast Asia, and Europe [28, 29]. However, one of the main obstacles in large-scale application of crop and non-crop biofuel systems is the non-sustainable or limited- to small- scale biodiesel feedstocks in these systems. Increasing pressure on food supplies and GHG emissions have resulted in growing concern and led to a heated "food and land versus fuel" debate [30, 31]. In addition to exploring other avenues, researchers have started working on microalgae-based fuel as a new generation of biofuel production systems which do not require arable land or clean water resources [32]. Microalgal biofuel systems have strengths and great potential in delivering clean and sustainably produced fuels for the future [33–35]. With a rapid growth rate of up to 50-folds greater than higher plants, oleaginous eukaryotic microalgae and prokaryotic cyanobacteria are considered as the most promising versatile feedstock for the production of new generation biofuels [36–40].

However, although microalgal biofuel systems have the potential to replace first- and second-generation feedstocks, and therefore drop the "food and land versus fuel" debate of the older versions of biofuel systems, more biofuel research and development in advanced technologies are required for microalgal biofuel systems to achieve economic viability. A large-scale microalgae-based biofuel production system cannot lead to a sustainable "greenenergy" production unless low-cost integrated technologies are developed and deployed in upstream and downstream processing. This work presents the recent investigations and advances in microalgal separation and extraction technologies.

6.2 Microalgae Harvesting and Biomass Concentration

Microalgal biodiesel production can be viable only if it is accompanied with a proper microalgal biorefinery approach. Harvesting is reflected as an expensive and challenging component of industrial production of microalgal biomass due to low cell density. Microalgal cultivation ends up with a dilute aqueous suspension including microalgae biomass, and depending on the applied cultivation method, its concentration ranges from 0.1 to 4 g dried microalgal biomass per liter culture medium. The microalgae biomass must be harvested, concentrated, and treated to release *triacylglycerols* (TAGs) which can then be utilized to produce biodiesel potentially at a reduced cost [41, 42]. There are a number of harvesting technologies currently available such as membrane filtration, flocculation, centrifugation, air flotation, centrifugation, ultrasound wave, magnetic separation, immobilization, and combined harvesting systems which are presented below.

6.2.1 Membrane Filtration

Membrane filtration is a simple method used for microalgae biomass removal and collection. Advantages such as simple and continuous operation, high separation efficiency, and requiring no chemicals in the process made membrane filtration a popular approach in biotechnological applications. Membrane filtration is able to shorten following steps of extraction, conversion, and refining [43]. When it comes to microalgae-based biofuel production, researchers stated that membrane filtration provides one more advantage by enabling the use of the recycled culture medium for the cultivation. In laboratory scale harvesting, some researchers [44] used membrane filtration for *B. braunii* and, while others [45] successfully filtered *Botryococcus braunii* through filter paper (0.2 μm pore size) to recover the algal cells grown in treated domestic wastewaters. Another study [42] noted that microfiltration is a cost effective method for small volumes of microalgae compared to centrifugation which also is discussed later in this article. Membrane fouling however is the critical restriction of membrane filtration and one big disadvantage of this system. For avoiding cell penetration into the filter, a reverse-flow vacuum filter is utilized to move the liquid upward across the membrane [46]. Researchers [43] indicated that significant reduction in the permeate fluidity caused by the attachment of organic polymers and the accumulation of the microalgae cells have reduced the interest of biofuel industry for applying this technique for microalgae biomass separation.

6.2.2 Flocculation

Due to the low energy requirement, flocculation seems to be an advantageous method of dewatering. In the course of flocculation, microalgae cells adhere to each other and form flocs which eventually settle and concentrate. Microalgae cells possess a negatively charged surface, which allows the separation from one another upon suspension. The negatively charged microalgae cells can be neutralized and destabilized with coagulants, such as polyvalent cations and cationic polymers [41]. And there are a number flocculation techniques which can be employed to aggregate microalgae cells for separation including autoflocculation, chemical flocculation, and electroflocculation (also known as electrolysis) [20, 21, 41].

6.2.2.1 Autoflocculation

Microalgal cells in the presence of calcium and magnesium at a high pH show aggregation. This phenomenon is called autoflocculation [47]. Ion-rich wastewater and saltwater can be the ideal culture medium for autoflocculation because, for an efficient flocculation of microalgae, the concentration of calcium, magnesium, and phosphorus ions in the medium should be sufficient. Some microalgae tend to precipitate out of solution naturally when the solution reaches fairly alkaline conditions in a phenomenon which is another form of autoflocculation. This concept originated from the natural changes in microalgae populations in sunlight while CO_2 is being absorbed for the process of photosynthesis, which also corresponds to a time when an increase in pH levels is observed. Investigators [48] suggest that the solution pH has to exceed 10.5 to induce high levels of precipitation of microalgae cells. However, reaching these pH levels by CO_2 uptake by photosynthesis only is unlikely. It has been reported that photosynthetic activity of most microalgae species ceases at pH levels beyond 9.0 [48]. However, instances where certain microalgae strains continue to uptake CO_2 until the pH reaches 10.0 and induce autoflocculation [49]. Experimentation of autoflocculation in microalgae strains offers some insight into the opportunity of employing this method for bulk harvesting. *Chlorella vulgaris* flocculated at 95 % efficiency at pH levels of 11.5–12.0 after adding sodium hydroxide [47]. Another study [48] reported removal rates for *Phaeodactylum tricornutum* and *Scenedesmus obliquus* as 85 and 70 % at pH levels of 10.5 and 11.3, respectively. Overall advantages of autoflocculation can be listed as the need for no binding chemicals, minimal energy requirement, potential to recycle culture medium with pH correction, and applicability to marine environments due to the reasons discussed above.

6.2.2.2 Chemical Flocculation

The surface charge of microalgae cells can be disrupted by the addition of alum, iron, lime, cellulose, salts, polyacrylamide polymers, surfactants, and chitosan, which would result in cell flocculation and eventually settling. In most cases, flocculation techniques are applied alone, but in some cases, can be used as a pre-concentration technique before employing other harvesting technologies. A number of different approaches have been used to flocculate microalgae cells to build heavy aggregates that are more suitable for separation. Researchers [46] reported that filtration using this method first resulted in 95 % recovery of the microalgae cells from the culture media. Researchers [50] indicated that microalgae flocculation was effective at salinity levels less than 5 g L^{-1}. In another study [51], it was noted that marine microalgae *Isochrysis galbana* and *Chlorella stigmatophora* require 5–10 times higher flocculent dosages than those required by freshwater microalgae. It was also observed that a mixture of two bacterial species of *Pseudomonas stutzeri* and *Bacillus cereus* were able to induce effective harvesting of the seawater microalgae, *Pleurochrysis carterae*. However, due to high ionic strength of seawater and apart from the availability of different types of flocculants, investigators [50, 51] concluded that the harvesting efficiency using flocculation considerably declined when dealing with saltwater microalgae. More research and development efforts seem to be necessary to improve flocculation for harvesting marine microalgae.

Investigators [52] reported a 100 % flocculation efficiency using chitosan in batch cultures of *Chlorella* sp. and *Thalassiosira nordenskoldii*. Extracellular Polymeric Substances (EPS) have been developed as environmentally friendly flocculants with more advantages for microalgal harvesting. EPS such as polysaccharides, functional proteins, and glycoproteins are termed as "bioflocculants" as they are synthesized by a variety of organisms, such as algae, bacteria, fungi, and actinomyces [53]. A recent study [54] reported γ-poly glutamic acid as a low-cost bioflocculant originated from *Bacillus subtilis*, and its efficiency for both freshwater and marine microalgae flocculation. In another study [55], it was shown that a bioflocculant produced by *Paenibacillus polymyxa* performed a successful harvesting of *Scenedesmus* sp. up to 95 %.

6.2.2.3 Electrocoagulation and Electroflocculation

In microalgae harvesting by electrocoagulation, polyvalent cations such as Al^{3+} and Fe^{2+}/Fe^{3+}, are dissolved by an anode. Electrodes are made up of either iron or aluminum, which sacrifice ions for the coagulation of microalgae cells [41]. In essence, the metal is providing the flocculant cation instead of adding a chemical flocculant. This means that the cost of electricity is replacing the cost of the chemicals, and the electrodes will need occasional replacement [41]. However, the same challenge exists with chemical flocculation where the harvested microalgae biomass is contaminated with the chemical flocculant. The amount of metal entering the solution as ions depends on the amount of electricity passing through the electrodes. These ions react with water molecules to form metal hydroxides which then in the next step bind to the microalgal cell surface, and destabilize the microalgal suspension through charge neutralization.

Particularly, when it comes to harvesting marine microalgae species, investigators [55, 56] noted that the electrolysis approach has a remarkable advantage compared to other harvesting technologies. Researchers [47] showed that the electrical energy input of the electrolytic harvesting approach for marine microalgae culture medium of marine microalgae has a high ionic strength that can be a major problem reducing or blocking the efficiency of other harvesting methods.

In microalgae harvesting by electroflocculation, the electrodes do not sacrifice the ions into solution, instead they attract the microalgae cells to the anode and neutralize the charges and form aggregates [21]. In an electroflocculation experiment, researchers [56] observed microalgae removal efficiency of 85–98 %. Another extensive study of electrocoagulation/flocculation explored various environmental factors and their impact upon the removal efficiency of their system. The system consistently removed 100 % of the cyanobacteria *Microcystis aeruginosa* [57].

Overall, using flocculants for microalgae separation has a number of advantages including high efficiency, low energy consumption, applicability to broad microalgae strains and pH spectrum, ability to treat large volumes, and minimal shear forces upon the cells. However, in chemical flocculation large concentration of flocculant is required to induce solid–liquid separation, which can result in the formation of a chemical sludge [58]. The process also requires a great deal of control over the solution pH to ensure optimum removal efficiency. Furthermore, the harvested microalgae biomass may be contaminated with the chemical flocculant which would not be acceptable for nutrition applications [21, 42]. However, organic flocculants can be used to address the chemical contamination of the harvested microalgae biomass. Chitosan is an example for organic flocculants derived from the chitin in crustacean shells [42], which is a biodegradable, nontoxic flocculant, which does not chemically contaminate the harvested microalgae biomass [21]. Researchers [41, 58] indicated that the optimal pH for maximal flocculation using chitosan is not difficult to achieve either. However, researchers [42] reported that the chitosan dosage required for the maximum coagulation efficiency is highly depended upon the microalgae strain.

6.2.3 Centrifugation

Centrifugal force can also be used for harvesting microalgae cells from dilute suspension. Centrifugation is commonly used either on its own or as a second step for further dewatering. High microalgae removal efficiency, improved concentration of microalgae biomass and shorter time required for harvesting are the main advantages of this technique. Particularly for high-value products, such as food or aquaculture applications, centrifugation is often recommended to recover high-quality microalgae without chemical and bacterial contamination of the raw product. Even milk separators are used to separate large volumes of microalgae cultures, and the authors of this article have used a milk separator successfully to separate microalgae biomass in their studies. In another study [59] investigators compared air flotation, drum filtration, and centrifugation techniques, and indicated that the most efficient technique for microalgae biomass recovery was centrifugation. Another study [60] noted removal efficiencies of 95–100 % using centrifugation. One more study [21] as well reported 80–90 % removal efficiencies from the growth media using centrifugation. Another research group [42] noted that the preferred method for microalgae biomass recovery is centrifugation. However, the energy requirement of centrifugation technique has a negative impact on the net energy and CO_2 balances in microalgae-based biodiesel production.

6.2.4 Flotation

In flotation, upward gas bubbles capture microalgae cells and bring them to the vacuole layer on top of the suspension. Different types of flotation techniques may be employed including air flotation, dissolved air flotation, and suspended air flotation. In general, the flotation efficiency is dependent on the size of the created bubble. Because of an increased surface area-to-volume ratio, researchers [61] stated that a smaller bubble size shows an extended durability and a better carrying capacity. Along with the bubble size, the surface hydrophobicity and the charge of the microalgae are also critical elements which control the interaction between the cells and the bubbles. Researchers highlighted the fact that in contrary to large bubbles, small bubbles which move slowly can better attach to microalgae cells and carry them to the top layer of the culture medium. It was also noted that the pre-accumulation of various microalgae species was shown to be effective to attain the mass required for effective flotation. In another study [62], it was reported that suspended air flotation and dissolved air flotation capture efficiencies of microalgae were approximately 77 and 85 %, respectively. In relatively an old study [63], researchers compared flotation, filtration and centrifugation and concluded that among these separation techniques centrifugation was the only economically feasible method for microalgae removal. Another study [64] on the other hand noted a biomass removal efficiency of 100 % by means of $FeCl_3$ flocculation and dissolved air flotation using *Isochrysis galbana*. And other researchers [65] concluded that microalgae removal is more efficient by flotation compared to sedimentation. They however noted that this technique is expensive and impractical because it requires a lot of energy for air compression.

6.2.5 Ultrasound

Among the available downstream processes in the production line of microalgal biodiesel, ultrasound is the only technique that can be used not only for harvesting microalgal biomass but also for the extraction of microalgae oil. High-frequency (MHz) ultrasound with low amplitude induces cell aggregation useful in cell harvesting. The microalgae cells agglomerate to the low-pressure nodes of ultrasound waves generated by high-frequency ultrasound waves. When the ultrasonic field is turned off, the cells settle by gravity. This dewatering technique offers continuous operation, is nonfouling and free of mechanical failures as there is no moving parts involved. On the other hand, ultrasound with a low frequency (KHz) and high amplitude induces cell break which is used in the oil extraction process [66].

In a microalgae harvesting study, a high resonance frequency acoustic wave was applied for continuous separation of *Monodus subterraneus* as a freshwater microalgae [66]. And it was reported that the efficiencies in this study were consistently higher than 90 % using ultrasound separation. Another study [67] reported that ultrasonic application improved microalgae settling; however it also changed the microalgae cell structure. And additionally, high energy consumption of this method was noted [46].

6.2.6 Magnetic Separation

Magnetic microalgae separation requires magnetic particles, an external magnetic field, and cationic polyelectrolytes to harvest microalgae cells since both the microalgae cells and magnetic particles have negatively charged surfaces in aqueous medium, cationic polyelectrolytes are needed as bridges between the magnetic particles on the microalgae cells [68]. These concept investigations date back to late seventies but have been seldom investigated until recently. The basic principle is to introduce magnetic nanoparticles into a microalgae solution and allow the agglomeration of the nanoparticles to the cells. The cells are attached to the magnetic nanoparticles first and then can be harvested from the aqueous solution using an external magnetic field. The specific adsorption mechanisms however have not been confirmed conclusively [69]. Thereafter, a simple magnet will attract the magnetically active microalgae cells to the bottom of the vessel for collection [69, 70]. It was also reported that *Botryococcus braunii* and *Chlorella ellipsoidea* were effectively separated from the growth media in a very short period of time (greater than 98 % of the microalgae in less than 3 min) [70]. Furthermore, they were able to recover approximately 95 % of the magnetic nanoparticles after separation.

Magnetic separation is an emerging technology as a potential microalgae separation technique, and not much data exist on the costs or energy consumption. We may expect to have minimal energy consumption, but the preparation of the magnetic nanoparticles may require high energy demand. However, if economically viable, magnetic separation would have potential advantages of rapid microalgae separation, neutral pH requirements, and having no microalgae strain specificity. Adverse effects upon cellular products (e.g., pigments and lipids) are not known; however, it was stated that it might be potentially important as the regeneration of the magnetic nanoparticles requires some solvent processing [70].

6.2.7 Immobilization

This technique uses an entrapment matrix such as alginate which immobilize microalgae cells way before the cultivation step begins next, where they grow and are indeed already embedded (harvested) by the entrapment matrix. Therefore, in the immobilization approach, there is actually no separate harvesting step which needs to come after cultivation step. The beads where microalgae grow into decline phase are detached through a simple filtering without the need for a large amount of energy input.

In one study [71], researchers successfully immobilized the freshwater microalgae *C. vulgaris* and grew in alginate beads. The authors indicated that this form of embedding beds is suitable for simplifying the overall separation process. In another study [72], it was suggested that the co-immobilization of microalgae with specific bacteria (plant-growth-promoting) could be a solution to improve the microalgae growth attained over an immobilization technology. Other researchers [43] on the other hand reported that pelletization by filamentous fungi, such as *Aspergillus niger*, could be an appropriate entrapment matrix to immobilize and grow the freshwater microalgae *C. vulgaris* either in a photoautotrophic system or under a heterotrophic growth condition. They suggested that immobilization by oleaginous filamentous fungi pelletization would be a good contribution to the improvement of the quality and quantity of oil produced from microalgae.

6.2.8 Combined Harvesting Techniques

Different harvesting techniques have their own advantages and disadvantages. Efficiency of one method can be increased if combined with another method; for instance, integrating electro-flocculation with dispersed-air flotation was used for harvesting *Botryococcus braunii* [73]. According to one study [74], flocculation in combination with flotation or sedimentation followed by centrifugation or filtration is the most energy and cost-efficient choice. Combined harvesting compared with others is a more viable technology for harvesting microalgae biomass.

6.3 Microalgae Oil Extraction

Harvesting microalgae is followed by oil extraction. Both harvesting microalgae biomass and oil extraction, as two consecutive steps of microalgae downstream processing, are energy-intensive processes. Prior to conversion of microalgae biomass to biodiesel, oil extraction must take place first [73]. The extracted oil is then converted into biodiesel [75]. Oil extraction is done by the physical processes, chemical processes in the form of solvent extractions, or a combination of the two. Overall, four common techniques are used for oil extraction, namely mechanical oil extraction, solvent extraction, supercritical fluid extraction, and ultrasonication [76]. Method used for extraction needs to be easily scalable, effective, and should not damage the extracted lipids [75]. Researchers [77] stated that not every lipid fraction is suitable for biodiesel production, and moreover sometimes nonlipid contents get extracted as well along with lipid contents. Therefore, the extraction process chosen should not only be lipid specific but also be selective towards desirable lipid fractions. Removing water beyond 10–30 wt% dry biomass is energy-intensive [77]; and therefore, if a lipid extraction methodology could be applied to a wet feedstock, it could save a lot of energy [78].

6.3.1 Cell Disruption Methods

Sometimes, pre-treatment of biomass may be required before oil extraction. Depending on the microalgae wall and on the product nature to be obtained either mechanical action-based (e.g., bead mills, and ultrasounds) or nonmechanical action-based (e.g., freezing, organic solvents, and osmotic shock) methods were reported to be used. Cell disruption efficiency for lipid extraction in microalgae differs from species to species, and also depends on the employed extraction methods which are discussed below. For further details, our readers are referred to the works of groups [79] and [80].

6.3.2 *Mechanical Oil Extraction*

Mechanical oil extraction techniques directly break cells through applying physical force, and they can universally be applied to any microbial biomass regardless of the species [81, 82]. For instance, the oil press is a very simple technique for oil extraction, and reported to extract 70–75 % of the oil from microalgae biomass [76, 83, 84]. It was stated that the microalgae biomass should be dried first for maximum extraction efficiency [84]. And it was noted that this method is still less efficient compared to other methods due to a long extraction time required [83].

6.3.3 *Solvent Extraction*

Microalgae oil extraction can also be realized via solvent extraction. In this technique, oil extraction from microalgae biomass is done by using solvents (benzene, cyclohexane, hexane, acetone, or chloroform) which break the microalgae cell walls, and then the oil in the solvent extract is separated via distillation [84]. The choice of solvent depends on the microalgae species chosen. Further, it was noted that the solvent should be inexpensive, nontoxic, volatile, nonpolar, and poor extractor of other nonlipid components of the cell [75]. Soxhlet extraction [80] and Bligh and Dyer's [85] methods are the two typically used techniques for extraction of lipids from microalgae biomass. The Soxhlet method uses hexane and the Bligh and Dyer's method uses mixture of chloroform and methanol as solvents to extract lipids [79, 80]. The other solvents include benzene and ether, but hexane has gained more popularity as a chemical for solvent extraction and it is also reported to be relatively inexpensive [76]. It was noted that the mixture of chloroform–methanol provided the highest extraction efficiency of microalgae lipids [46]. Recently ionic liquids have also been explored successfully for extraction of lipids. However, it was stated that the solvent extraction methods would not be practical on a large commercial scale application as the solvents used are not environment friendly chemicals, and as well do come with a high price tag [84, 86].

6.3.4 *Supercritical Fluid Extraction*

Supercritical fluid extraction (SFE) is one of the promising green technology methods, which is considered to have the potential to displace the traditional organic solvent lipid extraction methods. Fluids above their critical points are known as supercritical fluids. The diffusivity is enhanced and the viscosity of the fluid is decreased upon its critical point. Such properties allow fluids to diffuse easily through solid materials as opposed to non-super critical fluids [46]. A typical SFE unit consists of a feed pump for compression and transportation of liquid CO_2 to the extraction vessel, which is installed inside an oven module, and a heated micro-metering valve to depressurize incoming supercritical carbon dioxide (SC-CO_2). Once the oven is heated, the compressed CO_2 enters the heated oven in a supercritical state and extracts lipid from the microalgae. Once completely decompressed, CO_2 evaporates as gas to the ambient, and forces the extracted lipid to precipitate out and collect in the adjoining glass vial [78]. Researchers [87] reported that the pressure and temperature of SFE do not influence the yield of the extracted compounds, however influence the extraction rate. In another study, it was also reported that a temperature range of 45–55 °C and a pressure range of 400–700 bar had no impact on the extraction of bioactive lipids from the species *Nanochloropsis* [88]. SC-CO_2 has high solvating power and low toxicity. Intermediate diffusion/viscosity properties of the fluid lead to favorable mass transfer equilibrium, and this process produces solvent-free extract. Researchers [76] indicated that supercritical fluid is capable of extracting 100 % of oils; however, high infrastructure and operational cost associated with this process are the main disadvantages [78]. The SFE method for oil extraction is the most efficient of the methods discussed here due to its time efficiency, high selectivity, high-purity product generation, and no toxicity [46, 76, 84].

6.3.5 Ultrasonication

Ultrasonication is another method that can be used for oil extraction from microalgae. This method utilizes the cavitation effect caused by ultrasound in a liquid, a well-known technique for the cell disruption of microorganisms. When ultrasound is emitted to liquid media, small "vacant regions," called microbubbles, are generated as the liquid molecules are moved by the acoustic waves. If ultrasound is used with a sufficient intensity, the microbubbles are compressed to their minimum radii and implode, thereby producing heat, light (sonoluminescence), free radicals, and shockwaves, which can damage the cell envelopes of microorganisms [89, 90]. When the bubbles collapse and emit shockwaves that shatter the cell walls, the desired compounds are then released into the solution [84]. Using ultrasound extraction, one study [91] reported a 90 % extraction efficiency of fatty acids and pigments from *Scenedesmus obliquus*. Compared with other methods, ultrasonication seems to be the best cell disruption method for some microalgae species [92]. Researchers [93] conducted an in-depth study of ultrasonication utilizing simulations of microbubble sizes and microturbulence velocity depending on different extraction solvents and determined that ultrasonication could greatly increase the extraction yield from dry *Scenedesmus* sp. cells compared with the methods developed by Soxhlet and Bligh and Dyer. In ultrasound applications, high power consumption and scale up difficulty were noted by [94] and [95].

6.3.6 Chemical Methods

Microalgae cells can be disrupted by chemical means, such as treatments with acids, alkalis, and surfactants, which can degrade chemical linkages on the cell envelope, or osmotic pressure, which induces the "pop-out" of the cells. Main advantage is the lower energy consumption compared with the mechanical methods because it does not require a large amount of heat or electricity. Our readers are referred to the specific study [96] for further details.

6.3.7 Biological Methods

Biological methods refer to those methods which degrade the cell envelope using enzymes. Although there are other biological methods that use phages or autolysis [97, 98], most investigations of biological cell disruption utilize enzymes because they are commercially available and easily controlled biological materials. The advantages of enzymatic techniques are the mild reaction conditions and the high selectivity. An enzyme can selectively degrade a specific chemical linkage, whereas mechanical methods destroy almost every particle existing in the solution, and chemical methods sometimes induce side-reactions of the target products. The cell envelope of microalgae, such as *Chlorella*, has very resistant sporopollenin layers, but these can be degraded by a mixture of enzymes [99]. In this study, researchers incubated *Chlorella* sp. with a mixture of cellulase, hemicellulase, and pectinase for 90 h and found that 80 % of the cells were converted into cells in an osmotic labile state without rigid cell walls. Compared with mechanical methods, the enzymatic methods exhibited very competitive results [54]. If the enzymes are chosen carefully, enzymatic cell disruption is very effective. However, the critical downfall of this method is the high cost of the enzymes. There are two ways to reduce the cost of an enzymatic process: immobilization of the enzymes, and the combination of this process with other methods. Immobilized enzymes efficiently degrade the cell envelopes of *C. pyrenoidosa*, and they increase the lipid extraction yield by 75 %.

6.4 Concluding Remarks

Microalgae must be separated from the growth medium in order to produce biodiesel. We can conclude from the above review that there is no single best method of harvesting microalgae. There is not much valuable data about large scale separation of microalgae biomass, and the reported findings are derived from very different production conditions and scales. The identification of optimum harvesting technology depends on microalgae species, density, size, end product, and resource consumption. In this review, we also tried to cover various microalgae oil separation technologies. However, as it can be concluded from the review, identifying an ideal method is an extremely difficult task, and this task has not been accomplished yet. We can easily cite a few major problems associated with the extraction processes as well. First, reported findings are not economically feasible or applicable for large-scale microalgae oil extraction applications; second, as each investigation was performed under totally different conditions, the resource consumption and performance of each were quite different, so it is impossible to reasonably compare the different technologies that have been reported; and finally, various oil extraction techniques were reported to affect the transesterification or purification processes differently. And this suggests that the economic feasibility of an oil extraction technology needs to be assessed together with the post-extraction processes. In conclusion, more research and development is required to find out efficient and commercially viable, not individual, rather, integrated technologies for harvesting, extraction, and post-extraction processes.

References

1. DeCicco JM (2013) Biofuel's carbon balance: doubts, certainties and implications. Clim Chang 121:801–814
2. Fink R, Medved S (2013) Health impact assessment of liquid biofuel production. Int J Environ Health Res 23:66–75
3. Kumar S, Shrestha P, Salam PA (2013) A review of biofuel policies in the major biofuel producing countries of ASEAN: production, targets, policy drivers and impacts. Renew Sustain Energy Rev 26:822–836
4. Muldowney J, Mousney J, Kinsella L (2013) Agriculture in the climate change negotiations; ensuring that food production is not threatened. Animal 7:206–211
5. Rosch C, Skarka J (2008) European biofuel policy in a global context: trade-offs and strategies. Gaia 17:378–386
6. Schleifer P (2013) Orchestrating sustainability: the case of European Union biofuel governance. Regulation & Governance 7:533–546
7. Shafiee STE (2009) When will fossil fuel reserves be diminished? Energy Policy 37:181–189
8. Tainter JAPTW (2012) Drilling down the Gulf Oil debacle and our energy dilemma. Copernicus Books, New York, NY
9. ter Veld F (2012) Beyond the fossil fuel era: on the feasibility of sustainable electricity generation using biogas from microalgae. Energy Fuel 26:3882–3890
10. United Nations (2013) Framework convention on climate change. http://www.unfccc.int/. Accessed 21 March 2014
11. Williams TG (2010) Climate change negotiations the United Nations Framework Convention on Climate Change, the Copenhagen Accord and emissions reduction targets. Library of Parliament, Ottawa
12. Delrue F, Li-Beisson Y, Setier PA, Sahut C, Roubaud A (2013) Comparison of various microalgae liquid biofuel production pathways based on energetic, economic and environmental criteria. Bioresour Technol 136:205–212
13. Hennecke AM, Faist M, Reinhardt J, Junquera V, Neeft J (2013) Biofuel greenhouse gas calculations under the European Renewable Energy Directive – a comparison of the BioGrace tool vs. the tool of the Roundtable on Sustainable Biofuels. Appl Energy 102:55–62

14. Tiquia-Arashiro SM, Mormile M (2013) Sustainable technologies: bioenergy and biofuel from biowaste and biomass. Environ Technol 34:1637–1638
15. Yeboah A, Naanwaab C, Yeboah O, Owens J, Bynum J (2013) Economic feasibility of sustainable high oilseed-based biofuel production: the case for biodiesel in North Carolina. Int Food Agribusiness Manag Rev 16:41–65
16. Arbib Z, Ruiz J, Alvarez-Diaz P, Garrido-Perez C, Perales JA (2014) Capability of different microalgae species for phytoremediation processes: wastewater tertiary treatment, CO bio-fixation and low cost biofuels production. Water Res 49C:465–474
17. Hu G, Ji S, Yu Y, Wang S, Zhou G (2013) Organisms for biofuel production: natural bioresources and methodologies for improving their biosynthetic potentials. Adv Biochem Eng Biotechnol. doi:10.1007/10_2013_245
18. Vanthoor-Koopmans M, Wijffels RH, Barbosa MJ, Eppink MH (2013) Biorefinery of microalgae for food and fuel. Bioresour Technol 135:142–149
19. Vasudevan V, Stratton RW, Pearlson MN, Jersey GR, Beyene AG (2012) Environmental performance of algal biofuel technology options. Environ Sci Technol 46:2451–2459
20. Brennan L, Owende P (2010) Biofuels from microalgae – a review of technologies for production, processing, and extractions of biofuels and co-products. Renew Sustain Energy Rev 14:557–577
21. Chen CY, Yeh KL, Aisyah R, Lee DJ, Chang JS (2011) Cultivation, photobioreactor design and harvesting of microalgae for biodiesel production: a critical review. Bioresour Technol 102(1):71–81
22. Markou G, Nerantzis E (2013) Microalgae for high-value compounds and biofuels production: a review with focus on cultivation under stress conditions. Biotechnol Adv 31:1532–1542
23. Muramatsu Y, Sato H, Ninomiya M, Koketsu M (2012) Isolation of secondary metabolites from microalgae. Pharm Biol 50:595
24. Plaza M, Herrero M, Cifuentes A, Ibanez E (2009) Innovative natural functional ingredients from microalgae. J Agric Food Chem 57:7159–7170
25. Sahu A, Pancha I, Jain D, Paliwal C, Ghosh T (2013) Fatty acids as biomarkers of microalgae. Phytochemistry 89:53–58
26. Yildirim A, Demirel Z, Isleten-Hosoglu M, Akgun IH, Hatipoglu-Uslu S (2014) Carotenoid and fatty acid compositions of an indigenous *Ettlia texensis* isolate (*Chlorophyceae*) under phototrophic and mixotrophic conditions. Appl Biochem Biotechnol 172:1307–1319
27. Kermanshahi-Pour A, Zimmerman JB, Anastas PT (2013) Microalgae-derived chemicals: opportunity for an integrated chemical plant. In: Razeghifard R (ed) Natural and artificial photosynthesis: solar power as an energy source. John Wiley, New York, NY
28. Keeler BL, Krohn BJ, Nickerson TA, Hill JD (2013) U.S. federal agency models offer different visions for achieving Renewable Fuel Standard (RFS2) biofuel volumes. Environ Sci Technol 47:10095–10101
29. Li RP, di Virgilio N, Guan QF, Feng S, Richter GM (2013) Reviewing models of land availability and dynamics for biofuel crops in the United States and the European Union. Biofuels Bioprod Bioref 7:666–684
30. Farine DR, O'Connell DA, Raison RJ, May BM, O'Connor MH (2012) An assessment of biomass for bioelectricity and biofuel, and for greenhouse gas emission reduction in Australia. Glob Change Biol Bioenergy 4:148–175
31. Frank S, Bottcher H, Havlik P, Valin H, Mosnier A (2013) How effective are the sustainability criteria accompanying the European Union 2020 biofuel targets? Glob Change Biol Bioenergy 5:306–314
32. Franco ALC, Lobo IP, da Cruz RS, Teixeira CMLL, Neto JAD (2013) Biodiesel from microalgae: progress and challenges. Quim Nova 36:437–448
33. Gikonyo B (2013) Advances in biofuel production algae and aquatic plants. Apple Academic, Hoboken, NJ, p 393
34. Holma A, Koponen K, Antikainen R, Lardon L, Leskinen P (2013) Current limits of life cycle assessment framework in evaluating environmental sustainability – case of two evolving biofuel technologies. J Clean Prod 54:215–228
35. Oncel SS (2013) Microalgae for a macroenergy world. Renew Sustain Energy Rev 26:241–264

36. Alam F, Date A, Rasjidin R, Mobin S, Moria H (2012) Biofuel from algae – is it a viable alternative? International Energy Congress 49:221–227
37. Huesemann MH, Hausmann TS, Bartha R, Aksoy M, Weissman JC (2009) Biomass productivities in wild type and pigment mutant of *Cyclotella* sp. (Diatom). Appl Biochem Biotechnol 157:507–526
38. Rodolfi L, Zittelli GC, Bassi N, Padovani G, Biondi N (2009) Microalgae for oil: strain selection, induction of lipid synthesis and outdoor mass cultivation in a low-cost photobioreactor. Biotechnol Bioeng 102:100–112
39. Sheehan J, Dunahay T, Benemann J, Roessler P (1998) A look back at the U.S. Department of Energy's Aquatic Species Program: biodiesel from algae. NREL/TP-580-24190. Closeout report. U.S. Department of Energy, National Renewable Energy Laboratory, Golden, CO
40. Singh UB, Ahluwalia AS (2013) Microalgae: a promising tool for carbon sequestration. Mitig Adapt Strateg Glob Chang 18:73–95
41. Uduman N, Qi Y, Danquah MK, Forde GM, Hoadley A (2010) Dewatering of microalgal cultures: a major bottleneck to algae-based fuels. J Renew Sustain Energy 2:1–15
42. Grima EM, Belarbi E-H, Acien Fernandez F, Robles Medina A, Chisti Y (2003) Recovery of microalgal biomass and metabolites: process options and economics. Biotechnol Adv 20:491–515
43. Zhang JG, Hu B (2012) A novel method to harvest microalgae via co-culture of filamentous fungi to form cell pellets. Bioresour Technol 114:529–535
44. Tsukalhara K, Sawayama S (2005) Liquid fuel production using microalgae. J Jpn Petrol Inst 48(5):251–259
45. Sawayama S, Minowa T, Dote Y, Yokoyama S (1992) Growth of the hydrocarbon-rich microalga *Botryococcus braunii* in secondarily treated sewage. Appl Microbiol Biotechnol 38(1):135–138
46. Chen P, Min M, Chen Y, Wang L, Li Y, Chen Q, Wang C, Wan Y, Wang X, Cheng Y, Deng S, Hennessy K, Lin X, Liu Y, Wang Y, Martinez B, Ruan R (2009) Review of the biological and engineering aspects of algae to fuels approach. Int J Agr Biol Eng 2(4):1–30
47. Vandamme D, Foubert I, Fraeye I, Meesschaert B, Muylaert K (2012) Flocculation of *Chlorella vulgaris* induced by high pH: role of magnesium and calcium and practical implications. Bioresour Technol 105:114–119
48. Spilling K, Seppala J, Tamminen T (2011) Inducing autoflocculation in the diatom Phaeodactylum tricornutum through CO_2 regulation. J Appl Phycol 23:959–966
49. Blanchemain A, Grizeau D (1999) Increased production of eicosapentaenoic acid by *Skeletonema costatum* cells after decantation at low temperature. Biotechnol Tech 13:497–501
50. Bilanovic D, Shelef G (1988) Flocculation of microalgae with cationic polymers – effects of medium salinity. Biomass 17(1):65–76
51. Sukenik A, Beardall J, Kromkamp JC, Kopecky J, Masojidek J, van Bergeijk S, Gabai S, Shaham E, Yamshon A (2009) Photosynthetis performance of outdoor *Nannochloropsis* mass cultures under a wide range of environmental conditions. Aquat Microb Ecol. doi:10.3354/ame01309
52. Morales J, de la Noue J, Picard G (1985) Harvesting marine microalgae species by chitosan flocculation. Aquac Eng 4(4):257–270
53. Abd-El-Haleem DAM, Al-Thani RF, Al-Mokemy T, Al-Marii S, Hassan F (2008) Isolation and characterization of extracellular bioflocculants produced by bacteria isolated from qatari ecosystems. Pol J Microbiol 57:231–239
54. Zheng H, Yin J, Gao Z, Huang H, Ji X, Dou C (2011) Disruption of *Chlorella vulgaris* cells for the release of biodiesel-producing lipids: a comparison of grinding, ultrasonication, bead milling, enzymatic lysis, and microwaves. Appl Biochem Biotechnol 164:1215–1224
55. Kim DG, La HJ, Ahn CY, Park YH, Oh HM (2011) Harvest of *Scenedesmus* sp. with bioflocculant and reuse of culture medium for subsequent high-density cultures. Bioresour Technol 102(3):3163–3168

56. Poelman E, De Pauw N, Jeurissen B (1997) Potential of electrolytic flocculation for recovery of micro-algae. Resour Conserv Recycl 19:1–10
57. Gao S, Yang J, Tian J, Ma F, Tu G, Du M (2010) Electro-coagulation–flotation process for algae removal. J Hazard Mater 177:336–343
58. Shelef G, Sukenik A, Green M (1984) Microalgae harvesting and processing: a literature review. Solar Energy Research Institute, Golden, Colorado, United States
59. Sim TS, Goh A, Becker EW (1988) Comparison of centrifugation, dissolved air flotation and drum filtration techniques for harvesting sewage-grown algae. Biomass 16(1):51–62
60. Heasman M, Diemar J, O'Connor W, Sushames T, Foulkes L (2000) Development of extended shelf-life microalgae concentrate diets harvested by centrifugation for bivalve molluscs – a summary. Aquac Res 31:637–659
61. Zimmerman WB, Tesar V, Bandulasena HCM (2011) Towards energy efficient nanobubble generation with fluidic oscillation. Curr Opin Colloid Interface Sci 16(4):350–356
62. Wiley PE, Brenneman KJ, Jacobson AE (2009) Improved algal harvesting using suspended air flotation. Water Environ Res 81(7):702–708. doi:10.2175/106143009X407474
63. Golueke CG, Oswald W (1965) Harvesting and processing sewage-grown planktonic algae. J Water Pollut Contr Fed 37(4):471–498
64. Boussiba SE, Sandback G, Shelef Z, Cohen A, Vonshak AB-A, Arad S, Richmond A (1988) Outdoor cultivation of the marine microalga *Isochrysis galbana* in open reactors. Aquaculture 72:247–253
65. Chen YM, Liu JC, Ju YH (1998) Flotation removal of algae from water. Colloids Surf B: Biointerfaces 12:49–55
66. Bosma R, Spronsen WA, Tramper J, Wijffels RH (2003) Ultrasound, a new separation technique to harvest microalgae. J Appl Phycol 15:143–153
67. Zhang G, Wang B, Zhang P, Wang L, Wang H (2006) Removal of algae by sonication-coagulation. J Environ Sci Health A 41:1379–1390
68. Toh PY, Yeap SP, Kong LP, Ng BW, Chan DJC, Ahmad AL, Lim JK (2012) Magnetophoretic removal of microalgae from fishpond water: feasibility of high gradient and low gradient magnetic separation. Chem Eng J 211:22–30
69. Cerff M, Morweiser M, Dillschneider R, Michel A, Menzel K, Posten C (2012) Harvesting fresh water and marine algae by magnetic separation: screening of separation parameters and high gradient magnetic filtration. Bioresour Technol 118:289–295
70. Xu L, Guo C, Wang F, Zheng S, Liu C-Z (2011) A simple and rapid harvesting method for microalgae by in situ magnetic separation. Bioresour Technol 102:10047–10051
71. Lam MK, Lee KT (2012) Microalgae biofuels: a critical review of issues, problems and the way forward. Biotechnol Adv 30:673–690
72. Gonzalez LE, Bashan Y (2000) Increased growth of the microalga *Chlorella vulgaris* when coimmobilized and cocultured in alginate beads with the plant-growth-promoting bacterium *Azospirillum brasilense*. Appl Environ Microbiol 66:1527–1531
73. Li YG, Xu L, Huang YM, Wang F, Guo C, Liu CZ (2011) Microalgal biodiesel in China: opportunities and challenges. Appl Energy 88(10):3432–3437
74. Salim S, Bosma R, Vermue MH, Wijffels RH (2011) Harvesting of microalgae by bioflocculation. J Appl Phycol 23:849–855
75. Rawat I, Kumar RR, Mutanda T, Bux F (2011) Dual role of microalgae:phycoremediation of domestic wastewater and biomass production for sustainable biofuels production. Appl Energy 88:3411–3424
76. Demirbas A, Demirbas MF (2011) Importance of algae oil as a source of biodiesel. Energy Convers Manag 52:163–170
77. Amaro HM, Guedes AC, Malcata FX (2011) Advances and perspectives in using microalgae to produce biodiesel. Appl Energy 88(10):3402–3410
78. Halim R, Gladman B, Danquah MK, Webley PA (2011) Oil extraction from microalgae for biodiesel production. Bioresour Technol 102:178–185

79. Grimi N, Dubois A, Marchal L, Jubeau S, Lebovka NI, Vorobiev E (2014) Selective extraction from microalgae Nannochloropsis sp. using different methods of cell disruption. Bioresour Technol 153:254–259
80. Yoo G, Yoo Y, Kwon JH, Darpito C, Mishra SK, Pak K, Park MS, Im SG, Yang JW (2014) An effective, cost-efficient extraction method of biomass from wet microalgae with a functional polymeric membrane. Green Chem 16:312–319
81. Coustets M, Al-Karablieh N, Thomsen C, Teissie J (2013) Flow process for electroextraction of total proteins from microalgae. J Membr Biol 246:751–760
82. Ilavarasi A, Pandiaraj D, Mubarakali D, Ilyas MH, Thajuddin D (2012) Evaluation of efficient extraction methods for recovery of photosynthetic pigments from microalgae. Pak J Biol Sci 15:883–888
83. Popoola TOS, Yangomodou OD (2006) Extraction, properties and utilization potentials of Cassava seed oil. Biotechnology 5(1):38–41
84. Singh J, Gu S (2010) Commercialization potential of microalgae for biofuels production. Renew Sustain Energy Rev 14:2596–2610
85. Kanda H, Li P, Ikehara T, Yasumoto-Hirose M (2012) Lipids extracted from several species of natural blue-green microalgae by dimethyl ether: extraction yield and properties. Fuel 95:88–92
86. Demirbas MF (2011) Biofuels from algae for sustainable development. Appl Energy 88:3473–3480
87. Canela APRF, Rosa PTV, Marques MOM, Meireles MAA (2002) Supercritical fluid extraction of fatty acids and carotenoids from the microalgae. Ind Eng Chem Res 41:3012–3018
88. Andrich G, Nesti U, Venturi F, Zinnai A, Fiorentini R (2005) Supercritical fluid extraction of bioactive lipids from the microalga *Nannochloropsis* sp. J Lipid Sci Technol 107:381–386
89. Miller NJ, Sampson J, Candeias LP, Bramley PM, Rice-Evans CA (1996) Antioxidant activities of carotenes and xanthophylls. FEBS Lett 384:240–242
90. Miller I, Blunt J (2002) Evaluation of the structure of the polysaccharides from *Chondria macrocarpa* and *Ceramium rubrum* as determined by 13C NMR spectroscopy. Bot Mar 45:1–8
91. Wiltshire KH, Boersma M, Moller A, Buhtz H (2000) Extraction of pigments and fatty acids from the green alga *Scenedesmus obliquus* (Chlorophyceae). Aquat Ecol 34:119–126
92. Prabakaran P, Ravindran AD (2011) A comparative study on effective cell disruption methods for lipid extraction from microalgae. Lett Appl Microbiol 53:150–154
93. Ranjan A, Patil C, Moholkar VS (2010) Mechanistic assessment of microalgal lipid extraction. Ind Eng Chem Res 49:2979–2985
94. Martin PD (1993) Sonochemistry in industry progress and prospects. Chem Ind 7:233–236, ISSN: 0009-3068
95. Luque-Garcia JL, Luque De Castro MD (2003) Ultrasound: a powerful tool for leaching. Trends Analyt Chem 22:41–47, ISSN:0165-9936
96. Marcilla A, Catala L, Garcia-Quesada JC, Valdes FJ, Hernandez MR (2013) A review of thermochemical conversion of microalgae. Renew Sustain Energy Rev 27:11–19
97. Geciova J, Bury D, Jelen P (2002) Methods for disruption of microbial cells for potential use in the dairy industry-a review. Int Dairy J 12:541–553
98. Harrison ST (1991) Bacterial cell disruption: a key unit operation in the recovery of intracellular products. Biotechnol Adv 9:217–240
99. Braun T, Farag A (1975) Cellular and foamed plastics as separation media: a new geometrical form of the solid phase in analytical liquid-solid contact. Talanta 22:699–705

Chapter 7
Investigation of Photobioreactor Design for Biomass Production by Green Microalgae

Amel Ounnar, Fayrouz Kaidi, Lamia Benhabyles, and Majda Amina Aziza

Abstract In order to improve biological processes involved in biofuel production from microalgae, photobioreactors (PBR) should be adapted to control the production in optimal conditions. It should be first to study and control adapted reactors to model the entire process and facilitate the decision on the choice of raw material and design means, and finally to demonstrate the experimental feasibility of the process and allow biomass conversion for energy.

In this chapter, an overview on the role of PBR design over the microalgae growth efficiency is presented. Also, the study of the growth of *Chlorella sp.* microalgae, locally isolated from freshwater from southern Algeria, is followed inside a lab-scale PBR under controlled conditions. Some process parameters such as dry weight, and optical density were followed during the process. The obtained results show clearly the effectiveness of the closed-controlled reactor for a better growth of the *Chlorella sp.* microalgae.

Keywords Biofuel • Microalgae growth • *Chlorella sp.* • Air-lift photobioreactor

7.1 Introduction

Microalgal species are recently in the spotlight for biofuel production [1, 2]. They are also used as a biofertilizer, and source or potential source of high-value products such as polyunsaturated fatty acids or lipids [3, 4], natural colorants [5], biopolymers, and therapeutics [6]. In addition, a significant quantity of microalgal biomass is produced as essential aquaculture feed for shellfish and fish juveniles.

In fact, photosynthesis plays a fundamental role in all biofuel production. It drives the first step in the conversion of light to chemical energy and is therefore ultimately responsible for the production of the feedstocks required for all biofuel synthesis: protons and electrons (for biohydrogen), sugars and starch (for bioethanol), oils (for biodiesel), and biomass (for BTL products and biomethane).

A. Ounnar (✉) • F. Kaidi • L. Benhabyles • M.A. Aziza
Centre de Développement des Energies Renouvelables (CDER), BP. 62, Route de l'Observatoire - Bouzaréah, 16340 Alger, Algerie
e-mail: a.ounnar@cder.dz

I. Dincer et al. (eds.), *Progress in Clean Energy, Volume 2*,
DOI 10.1007/978-3-319-17031-2_7

Consequently, any increase in photosynthetic efficiency will also enhance the competitiveness of biofuel production.

In this way, it is of interest to find the most appropriate system to cultivate microalgae with very high productivities. It is essential to provide water, light energy, nutrients, carbon dioxide, as well as specific environmental conditions.

So, cultivation of microalgae can be done in open systems (lakes, ponds) and in controlled closed systems called photobioreactors (PBR). In order to improve biological processes from microalgae growth, PBR should be adapted to control the production in optimal conditions. It should be first to study and control adapted reactors to model the entire process and facilitate the decision on the choice of raw material and design means, and finally to demonstrate the experimental feasibility of the process and allow biomass conversion for energy.

Some photosynthetic microorganisms such as unicellular green microalgae (*Chlamydomonas reinhardtii*, *Chlorella fusca*, and *Scenedesmus obliquus*) are able to produce biofuel from light (artificial or solar energy) using water as electron and proton donor. In Algeria, the availability of algal resources that can be used as energy source makes the process of biofuel production interesting. Therefore, improving biological process involved in this technique led us to study air-lift PBR adapted to biofuel production control in optimal conditions.

In this work, an overview on the role of PBR design over the microalgae growth efficiency is presented. Also, the study of the growth of *Chlorella sp.* microalgae, locally isolated from freshwater from southern Algeria, is followed inside a lab-scale air-lift PBR under controlled conditions. Some process parameters such as dry weight, and optical density were followed during the process.

7.2 Photobioreactor Design

A PBR can be described as an enclosed, illuminated culture vessel designed for controlled biomass production. PBR refers to closed systems that are closed to the environment having no direct exchange of gases and contaminants with the environment [7].

When designing a PBR, some points should be taken into consideration [8]. The reactor must provide a uniform illumination of the culture surface and a fast mass transfer of CO_2 and O_2; its design must prevent or minimize the fouling of the reactor, particularly of its light-transmitting surfaces. Also, high rates of mass transfer must be attained by means that neither damage cultured cells nor suppress their growth.

The PBR which are used for the purpose of growing microalgae are tubular, flat panel, horizontal tubular, helical type, stirred tank, etc. These PBR have their own advantages and disadvantages.

7.2.1 Vertical Tubular Photobioreactor

It is made up of vertical tubing that is transparent in nature to allow light penetration. Sparger is attached at the bottom of the reactor which converts the sparged gas into tiny bubbles. Sparging with gas mixture provides overall mixing and mass transfer of CO_2 and also removes O_2 produced during photosynthesis [9]. Vertical tubular PBR can be divided into bubble column [10], torus shape reactor, and air-lift reactor based on their mode of liquid flow (Fig. 7.1).

7.2.2 Flat-Panel Photobioreactor

The flat-panel reactor has cuboidal shape with minimal light path (Fig. 7.2). It can be made from transparent materials like glass, plexiglas, and polycarbonate [13]. It is characterized by high surface area-to-volume ratio and open gas disengagement systems. Agitation is provided either by bubbling air from its one side through perforated tube or by rotating it mechanically through motor. Some works applied especially to the study of air-lift bioreactors showed the benefits of this type of reactor compared to conventional ones (low energy dissipation, controlled fluid circulation, good distribution of light, effective heat and light dissipation, etc.). Limiting parameters in the culture of microalgae are light, nutrients, temperature, and pH. Agitation is also an important parameter because it strongly influences the efficiency of gas-liquid transfer in the reactor. It serves to increase the conversion efficiency of light. A good mixing in the air-lift bioreactor could be obtained without causing too much shear force in the liquid phase, which could inhibit the growth of the algae. In addition, it was mentioned that the well-defined circulation pattern resulted in a better light utilization particularly for the system with high density of cells [14].

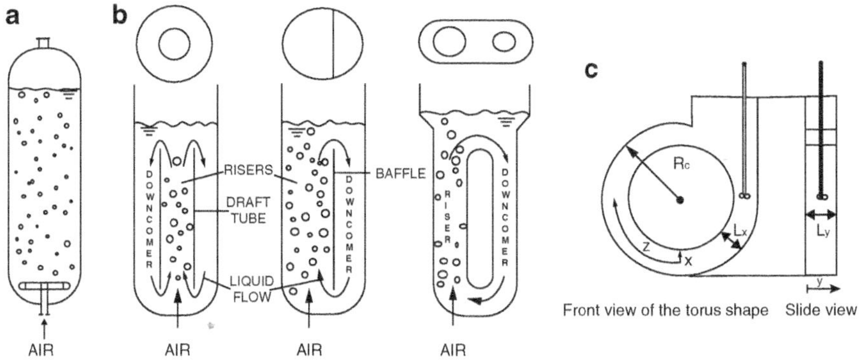

Fig. 7.1 Vertical tubular photobioreactor [11, 12]. (**a**) Bubble column, (**b**) air-lift reactor, (**c**) torus reactor

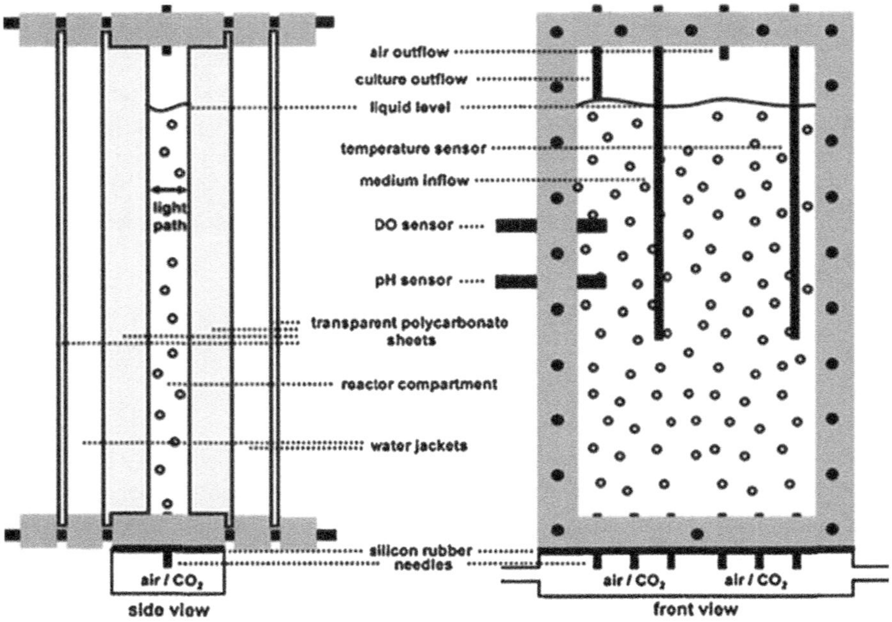

Fig. 7.2 Flat panel photobioreactor [7]

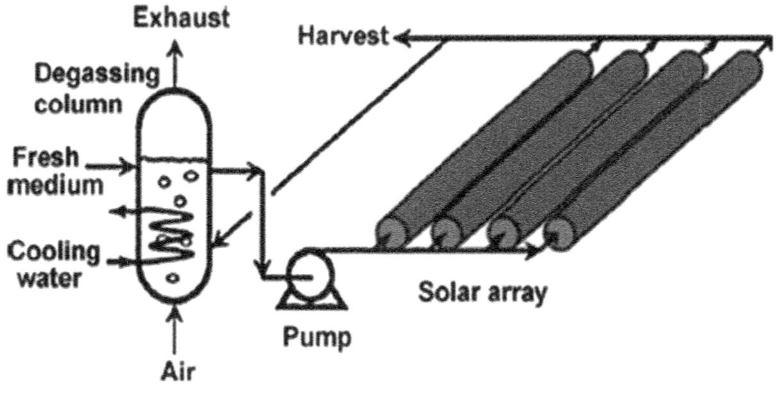

Fig. 7.3 Horizontal tubular photobioreactor [7]

7.2.3 Horizontal Tubular Photobioreactor

Horizontal tubular reactors are placed horizontally giving the design of parallel set of tubes, loop shape, inclined tubular shape, or horizontal tubular reactor (Fig. 7.3). Their shape gives advantage in outdoor culture for their orientation towards sunlight resulting in high light conversion efficiency.

7.2.4 Helical Type Photobioreactor

Helical type PBR consists of coiled transparent and flexible tube of small diameter with separate or attached degassing unit. A centrifugal pump is used to drive the culture through long tube to the degassing unit (Fig. 7.4). Travieso et al. experimented different algal strains with this system [15]. CO_2 gas mixture and culture medium can be circulated from either direction but injection from bottom gives better photosynthetic efficiency [16].

7.2.5 Stirred-Tank Photobioreactor

Stirred-tank reactor is most conventional where agitation is provided mechanically with the help of impeller of different sizes and shapes. Baffles are used in order to reduce vortex (Fig. 7.5). CO_2-enriched air is bubbled at the bottom to provide carbon source for the growth of algae. This type of bioreactor has been turned into PBR by illuminating it externally by fluorescent lamps or optical fibers but the main disadvantage of this system is low surface area-to-volume ratio which in turn decreases light-harvesting efficiency.

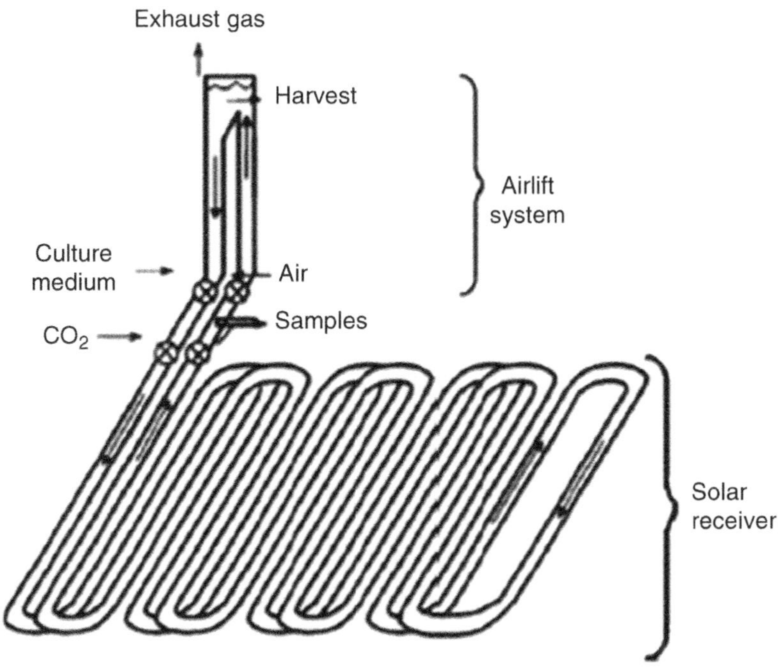

Fig. 7.4 Helical type photobioreactor [7]

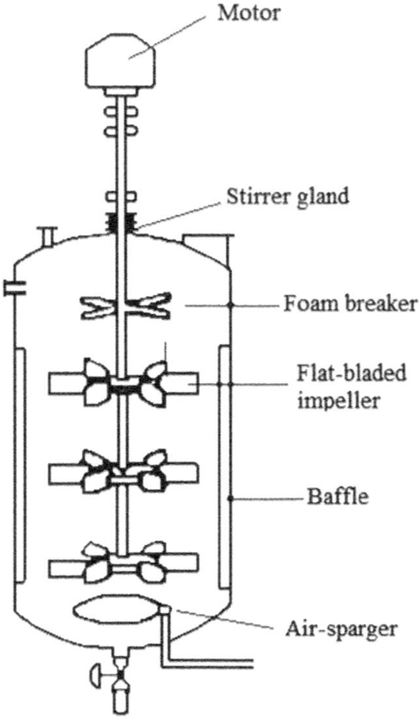

Fig. 7.5 Stirred-tank photobioreactor [7]

Despite their costs, PBR have several major advantages over open systems. They minimize contamination and allow axenic algal cultivation of monocultures; they offer better control over conditions such as pH, temperature, light, and CO_2 concentration. They also lead to less CO_2 loss, prevent water evaporation, and permit higher cell concentrations. In this work, the air-lift PBR is used to study the microalgae growth. It has characteristic advantage of creating circular mixing pattern where liquid culture passes continuously through dark and light phase giving flashing light effect to algal cells [13].

The microalgae cultivation in different types of PBR from locally identified microalgae was the subject of several research works in the Research Center of Renewable Energy Development (CDER) in Algeria [10, 17].

7.3 Experimental

7.3.1 Microalgae Strain

Microalgae specie used in this work is *Chlorella sp.* locally isolated from freshwater samples collected from southern Algeria (Fig. 7.1). The *chlorella* is a green microalga which has, most often, spherical or ellipsoidal shape. It is a solitary cell of maximum 20 μm and has a distinctive membrane and one or rarely two plasts [18]. It has a very fast reproduction. *Chlorella* is widespread in different locations such as freshwater, air, and soil [19]. For the purpose of determining favorable growth conditions for this microalga, it was cultured in a triacetate phosphate (TAP) medium containing (quantities in g L^{-1}) NH_4Cl, 0.4; $MgSO_4$, $7H_2O$, 0.1; $CaCl_2$, $2H_2O$, 0.05; K_2HPO_4, 0.108; KH_2PO_4, 0.056; Tris, 2.42; glacial acetic acid, 1 mL; and trace of metal. The pH of the medium was adjusted to 7.2.

Chlorella sp. was maintained as pure culture in 250 mL Erlenmeyer flasks containing 50 mL of the medium. The culture was kept at 25 °C under light intensity of 11.718 μmol m^{-2} s^{-1}. Every 3 weeks 0.5 mL of a culture was transferred to a new flask containing fresh medium.

7.3.2 Air-Lift Photobioreactor

The batch cultivation of the microalgae was performed by inoculating 100 mL starter culture into 700 mL of TAP medium. The culture was grown in a 1 L air-lift PBR made of borosilicate glass with an internal diameter of 12 cm (Fig. 7.6). It was equipped with an external loop in the bottom which allows introduction of air through a central sparger diffusing from riser (where gas is sparged) to downcomer.

In the external loop, riser and downcomer are separated physically by two different tubes. Mixing is done by gas bubbling through sparger in the riser tube without any physical agitation. Riser is similar to bubble column where sparged gas moves upward randomly and haphazardly. This decreases the density of the riser making the liquid to move upward. This upward movement is assisted by the gas hold up of riser.

Fig. 7.6 Experimental setup

Supply of air was provided from a compressor to the loop in order to allow liquid circulation. The aeration rate was controlled by the calibrated flow meter and was fixed along the experiment to a value of 286.4 mL min^{-1}. Light was supplied through ten fluorescent white lamps connected in parallel and equidistant at the side around the length of the reactor, which yielded approximately 113.67 μmol m^{-1} s^{-1} of light intensity for a continuous photoautotrophic cultivation (24-h/24-h light photoperiod). The temperature was kept at system condition and was measured in the range of 30 °C. The top of the reactor has holes for sampling and measuring pH through appropriate probes.

In order to study the growth kinetic of microalgae, samples of the culture were taken over the time of experiment and analyzed in a spectrophotometer UV-Visible Biomat 3 for absorbance determination.

7.4 Results and Discussion

7.4.1 *Growth of* Chlorella sp. *Microalgae Under Photoautotrophic Conditions*

As shown in Fig. 7.7, cultivation of microalgae in controlled conditions of light and air aeration inside the air-lift PBR is very efficient. It is clear that microalga concentration increases considerably in this type of reactor under continuous irradiation of 113.67 μmol m^{-1} s^{-1} of light intensity.

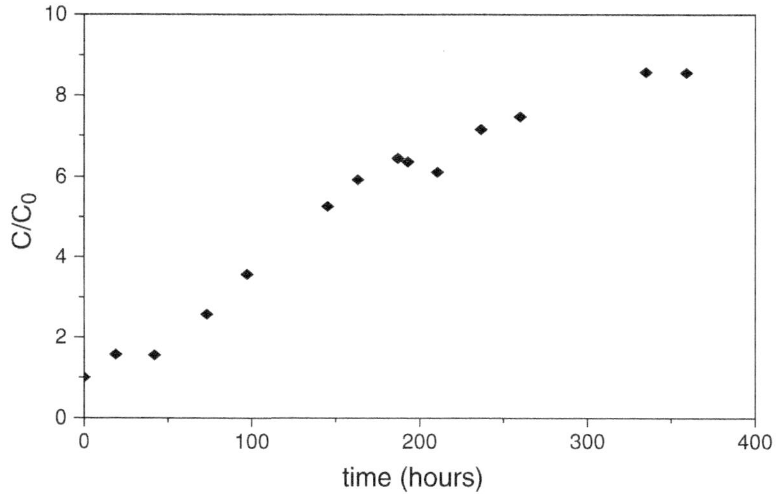

Fig. 7.7 Growth kinetic of *Chlorella sp.*

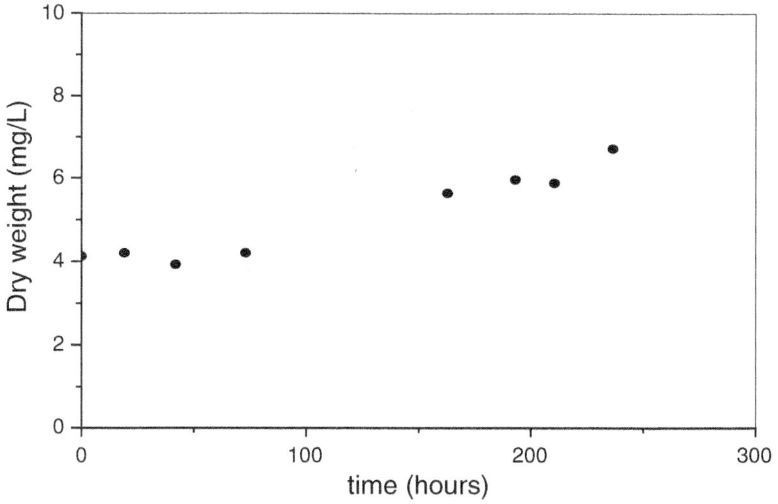

Fig. 7.8 Evolution of microalga dry-weight versus time

The growth of *Chlorella sp.* microalgae is characterized by a latency phase of about 48 h, and an acceleration phase of 100 h above the exponential phase. Thus, we observe a phase of slower growth after 300–350 h of culture.

7.4.2 Dry Weight Evolution

Figure 7.8 shows the evolution of the dry weight of microalga samples collected during the growing period. It is clear that the trend of this parameter reflects the growth of microalgae in the air-lift reactor. The high density of cells observed at the end of the experiment reflects the good mixing in the air-lift bioreactor obtained without causing too much shear force in the liquid phase.

7.5 Conclusion

The air-lift PBR are the most suitable for microalga growth because of the performance of mass transfer and present mixing phases. The operating parameters must be optimized for better system performance. In this work, the obtained results show clearly the effectiveness of this type of reactor for a better growth of the *Chlorella sp.* microalgae. This way of microalga cultivation for producing biofuels is promising in the field of renewable energy and the environment. It is necessary to increase research and development studies to promote the production technology of biofuels.

References

1. Chisti Y (2007) Biodiesel from microalgae. Biotechnol Adv 25(3):294–306
2. Brennan L, Owende P (2010) Biofuels from microalgae-A review of technologies for production, processing, and extractions of biofuels and co-products. Renew Sustain Energy Rev 14 (2):557–577
3. Belarbi EH, Molina E, Chisti Y (2000) RETRACTED: A process for high yield and scaleable recovery of high purity eicosapentaenoic acid esters from microalgae and fish oil. Process Biochem 35(9):951–969
4. Kanda H, Li P, Ikehara T, Yasumoto-Hirose M (2012) Lipids extracted from several species of natural blue–green microalgae by dimethyl ether: Extraction yield and properties. Fuel 95:88–92
5. Lorenz RT, Cysewski GR (2000) Commercial potential for Haematococcus microalgae as a natural source of astaxanthin. Trends Biotechnol 18(4):160–167
6. Borowitzka M (1999) Pharmaceuticals and agrochemicals from microalgae. Chemicals from microalgae. Taylor & Francis, London, pp 313–352
7. Singh RN, Sharma S (2012) Development of suitable photobioreactor for algae production-A review. Renew Sustain Energy Rev 16:2347–2353
8. Tsoglin LN, Gabel BV, Falkovich TN, Semenenko VE (1996) Closed photobioreactors for microalgal production. Russian J Plant Physiol 43(1):131–136
9. Kumar K, Dasgupta CN, Nayak B, Lindblad P, Das D (2011) Development of suitable photobioreactors for CO_2 sequestration addressing global warming using green algae and cyanobacteria. Bioresour Technol 102:4945–4953
10. Kaidi F, Rihani R, Ounnar A, Benhabyles L, Naceur MW (2012) Photobioreactor design for hydrogen production. Procedia Eng 33:492–498
11. Chisti Y, Moo-Young M (2001) Bioreactors encyclopedia of physical science and technology. Meyers RA (ed.) Academic Press: New York 247–271
12. Pruvost J, Pottier L, Legrand J (2006) Numerical investigation of hydrodynamic and mixing conditions in a torus photobioreactor. Chem Eng Sci 61(14):4476–4489
13. Barbosa MJ, Janssen M, Ham N, Tramper J, Wijffels RH (2003) Microalgae cultivation in air-lift reactors: modeling biomass yield and growth rate as a function of mixing frequency. Biotechnol Bioeng 82(2):170–179
14. Merchuk JC, Ronen M, Geris S, Arad S (1998) Light/dark cycles in the growth of the red microalga *Porphylidium sp.* Biotechnol Bioeng 59:705–713
15. Travieso L, Hall DO, Rao KK, Benitez F, Sanchez E, Borja R (2001) A helical tubular photobioreactor producing Spirulina in a semi continuous mode. Int Biodeterior Biodegradation 47:151–155
16. Morita M, Watanabe Y, Okawa T, Saiki H (2001) Photosynthetic productivity of conical helical tubular photobioreactors incorporating *Chlorella sp.* Under various culture medium flow conditions. Biotechnol Bioeng 74(2):137–144
17. Rihani R, Bensmaili A, Kaidi F, Belhamel M (2006) Etude de la distribution des pertes de charge des systèmes diphasiques dans un réacteur torique en pyrex. Revue des Energies Renouvelables 9(3):355–361
18. Barthet L (2003) Contribution à l'évaluation de l'impact sur les écosystèmes de la valorisation de résidus de procédés thermiques en BTP. Thesis, 217 p. Institut National des Sciences Appliquées de Lyon
19. Wu HL (2001) Identification of *Chlorella sp.* isolated using ribosomal DNA sequences. Bot Bull Acad Sinica 42:115–121

Chapter 8
Bioethanol Production from Molasses by Pervaporation Membrane Bioreactor

Filiz Ugur Nigiz and Nilufer Durmaz Hilmioglu

Abstract Bioethanol is a clean and green energy source that can be produced from fermented biomass. Environmental regulations require bioethanol to be blended with gasoline. Thus researchers and producers accelerate ethanol production from biobased sources and waste. There are many bioethanol production plants all over the world. Ethanol production steps consist of pretreatment, fermentation, and deep purification steps. The impurities in fuel ethanol should be below the limit values. Therefore, some specific purification techniques are required. The major production drawback is the complicated separation step that covers a very large portion of the production cost. Hence researchers place great emphasis on the separation step. For some time, the separation–fermentation hybrid system has garnered much attention owing to its economical and environmental features. The pervaporation membrane bioreactor (PVMBR) is a pervaporation-coupled fermentation hybrid process. In PVMBR, ethanol is selectively removed from the fermentation broth.

In this study bioethanol production from molasses by using *Saccharomyces cerevisiae* has been studied in PVMBR. Ethylene propylene diene monomer (EPDM)/Poly(dimethyl siloxane) (PDMS) blended membrane has been used as membrane material. The membrane structure has been investigated using a scanning electron microscope. Process efficiency has been determined as a function of ethanol productivity, total flux, and ethanol selectivity. Membrane reactor data have been compared with batch reactor data in order to evaluate system efficiency.

Keyword Pervaporation membrane bioreactor • Ethanol production • Clean energy

F.U. Nigiz (✉) • N.D. Hilmioglu
Chemical Engineering Department, Kocaeli University, Kocaeli 41380, Turkey
e-mail: filiz.ugur@kocaeli.edu.tr; niluferh@kocaeli.edu.tr

I. Dincer et al. (eds.), *Progress in Clean Energy, Volume 2*,
DOI 10.1007/978-3-319-17031-2_8

Nomenclature

Abbreviations

EPDM	Ethylene propylene diene monomer
PDMS	Poly(dimethyl siloxane)
PV	Pervaporation
PVMBR	Pervaporation membrane bioreactor

Symbols

α	Selectivity
A	Effective membrane area (m^2)
F_w	Weight percent of water in feed
J	Flux (kg/m^2 h)
P_w	Weight percent of water in permeate
W	Weight of permeated sample (g)

8.1 Introduction

Depletion of fossil fuel resources, the increase in hazardous gas emission in fuels, and increase in air pollution have expanded the investigations on clean fuel production. Bioethanol is a renewable and sustainable energy source that can be produced from a fermented biosource such as corn, algae, sugar cane, beet, basic lignocellulosic biomass, and agricultural waste [1, 2]. It produces lower amounts of carbon monoxide and dioxide. As it does not include sulfur- or nitrate-based chemicals it contributes far less to global warming [3]. Bioethanol is used as a fuel additive with gasoline in different ratios. In many countries, the use of bioethanol has become mandatory by law [4, 5]. Compared to the other biobased fuels, great progress has been achieved for the production of bioethanol. It is predicted that by 2020 bioethanol production capacity will have increased worldwide. However, process sensitivity and purification insufficiency restrict production efficiency.

Pervaporation (PV) is a promising energy-intensive and environmentally friendly membrane technology to separate azeotropic, thermally sensitive, and closely volatile liquid solvent. As opposed to other membrane separation processes such as microfiltration, ultrafiltration reverse osmosis, nonporous polymeric, inorganic or composite membranes are used [6–9]. The membrane is the "key device" of the system, therefore researchers have focused on appropriate membrane production. The separation character of the membrane is directly related to the efficiency of the PV. Membrane selectivity and flux are the main factors that determine membrane availability. A basic experimental pervaporation system is

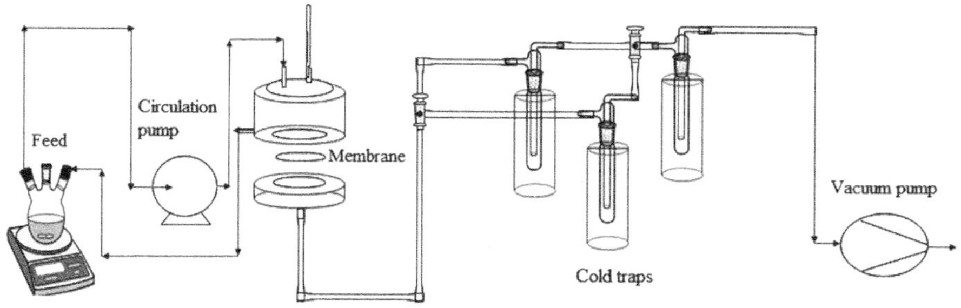

Fig. 8.1 Lab scale pervaporation unit

shown in Fig. 8.1. It consists of a feed tank, a circulation pump, membrane cell, cold traps, and vacuum pump.

Separation via nonporous membrane can be better understood by the solution–diffusion theory [10]. In contrast to a porous membrane, separation through a nonporous membrane depends on the membrane affinity to the selected component, solvent–solvent interaction, hydrogen bonding ability, operation conditions, solvent diffusivity, and vapor pressure difference as well as membrane structure [11, 12]. According to the solution diffusion model, separation in pervaporation occurs in three main steps. First, the selected component dissolves on the membrane surface according to the membrane–solvent affinity and bonding ability. Then the selected component diffuses through the intermolecular structure of the membrane according to the concentration gradient that is maintained by the pressure difference. Finally, the selected component desorbs to the downstream side of the membrane. Due to the supplied vacuum pressure, the liquid component converts to the vapor phase after desorption [13–18]. Then the vapor condenses into cold traps. Owing to the intermolecular separation ability of the nonporous membrane, PV is an effective selective separation technique for bioethanol dehydration [19, 20].

Either a hydrophobic membrane is preferred for dilute organic mixture separation or a hydrophilic membrane is used for organic mixture dehydration [20, 21]. Poly (dimethyl siloxane) is a kind of rubbery polymer that has mechanical and thermal resistance and is mostly used as a nonporous membrane material in PV for the separation of organic compounds from water. It is one of the rare commercial hydrophobic membrane materials [6, 22–25]. EPDM is also known as hydrophobic membrane material to separate organic compounds from water.

The pervaporation-aided reactor system has been known for several years for its process efficiency feature. In this system, a fermenter is supported by a pervaporative separation step [26, 27]. This system—which can be called a pervaporation membrane bioreactor (PVMBR)—has two different configurations [28, 29]. If a selective membrane is employed inside the fermenter and separation occurs simultaneously with fermentation, then it is called an internal PVMBR as seen in Fig. 8.2. If the pervaporation unit is settled outside the fermenter, it is called an external PVMBR as seen in Fig. 8.3. The internal PVMBR is more suitable

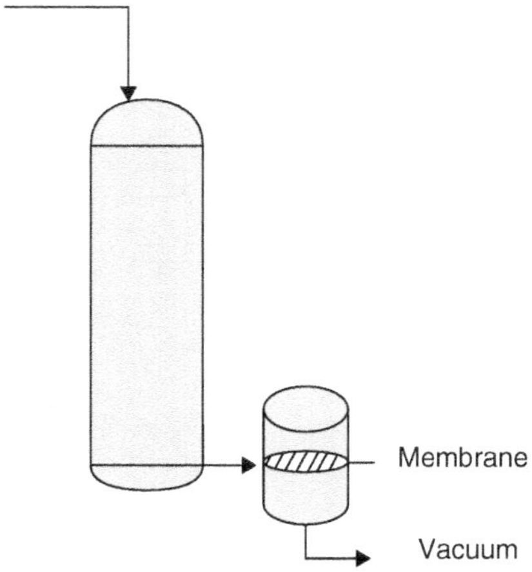

Fig. 8.2 External PVMBR

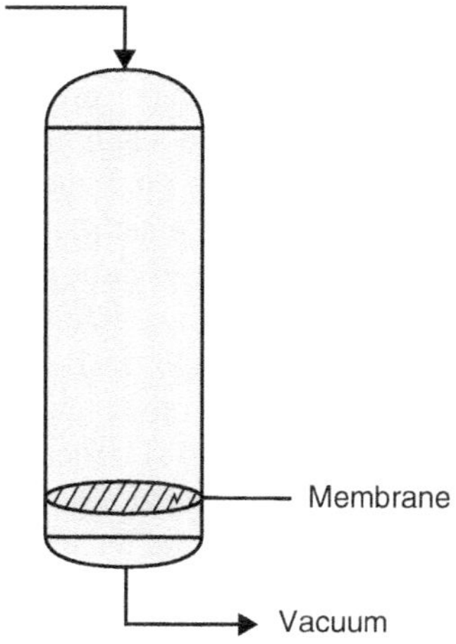

Fig. 8.3 Internal PVMBR

regarding process intensification rules, owing to less equipment usage, energy consumption, and fewer process step requirements.

This study focuses on ethanol productivity and the separation performance of a hydrophobic blended membrane. An internal PVMBR has been designed and used for ethanol fermentation. The performance of the system has been characterized by flux, selectivity, and ethanol concentration in the fermentation broth. Flux determines membrane productivity, selectivity defines the selective properties of EPDM/PDMS membrane, and the availability of the system is determined by ethanol production.

8.2 Experimental

8.2.1 *Material*

PDMS and EPDM have been kindly supplied by the Plastics and Rubber Processing Laboratory at Kocaeli University. Chloroform and benzeoil peroxide have been purchased from Merck Chemicals. Molasses and *Saccharomyces cerevisiae* have been kindly supplied by PAKMAYA, Kocaeli.

8.2.2 *Membrane Preparation*

A desired amount of PDMS and EPDM (PDMS:EPDM weight ratio was obtained as 2) were dissolved in chloroform and the solution was stirred for 7 h at 55 °C. Four weight percent of benzeoil peroxide (with respect to the total polymer weight) was added to the polymer solution and stirred for 2 h. After a homogeneous solution was obtained, the solution was poured onto a Teflon plate and dried overnight. Then the polymer film was cured in a vacuum oven at 185 °C for 6 h.

8.2.3 *Membrane Characterization*

The EPDM/PDMS structure was analyzed using a JEOL JSM-6335F field emission scanning electron microscope. Liquid nitrogen was used to break the membrane samples. The samples were coated with gold before the analysis.

8.2.4 *Molasses Solution Preparation*

Molasses was diluted to 10 Brix. The diluted solution was acidified to pH 4.0 with 10 N H_2SO_4 [30]. It was boiled about 30 min, decanted, and centrifuged. The clarified solution was obtained after the filtration.

8.2.5 *Batch Fermentation*

The fermentation experiment was carried out at room temperature for 7 h. Ten weight percent of yeast—with respect to the total sugar weight—was added to the molasses solution. The reaction rate was determined by the CO_2 volume and ethanol concentration per hour. The ethanol concentration was determined by gas chromatography. A basic batch reactor is shown in Fig. 8.4.

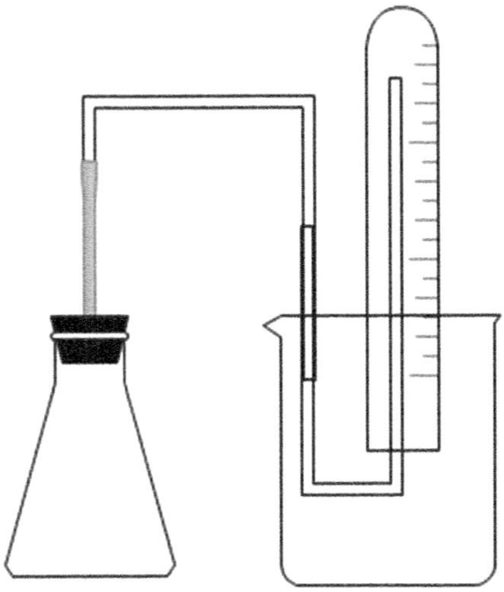

Fig. 8.4 Batch fermentation unit

8.2.6 *Pervaporation Membrane Bioreactor*

In PVMBR, fermentation was carried out for 7 h. The diluted molasses solution was started upstream of the membrane; 10 wt% of yeast (with respect to the total sugar weight—was added to the solution and the reaction was started 1 h later. The one-step fermentation–separation system can be seen in Fig. 8.5.

The effective membrane area was 28.26 cm^2, and the cell capacity was 500 ml. Upstream of the membrane was kept at atmospheric pressure and downstream of the membrane was kept at 30 mbar. As the fermentation was carried out, ethanol was continuously separated from the fermentation broth.

Flux (J) and selectivity (α) are calculated as seen in Eqs. (8.1) and (8.2);

$$J = W_{\mathrm{p}}/At \tag{8.1}$$

$$\alpha = (P_{\mathrm{w}}/P_{\mathrm{et}})/(F_{\mathrm{w}}/F_{\mathrm{et}}) \tag{8.2}$$

W_{p} represents the total permeate weight of the mixtures, A is the effective membrane area, and t is the time. P_{w} and F_{w} are the weight percent of water in permeate and feed, P_{et} and F_{et} are the weight percent of ethanol in permeate and feed [31].

8.2.7 *Analysis*

Ethanol and water concentrations were determined by using Agilent 7980 gas chromatography with a TCD detector. A HP-FFAP polyethylene glycol capillary column was used. Detector temperature was 280 °C and oven temperature was 220 °C.

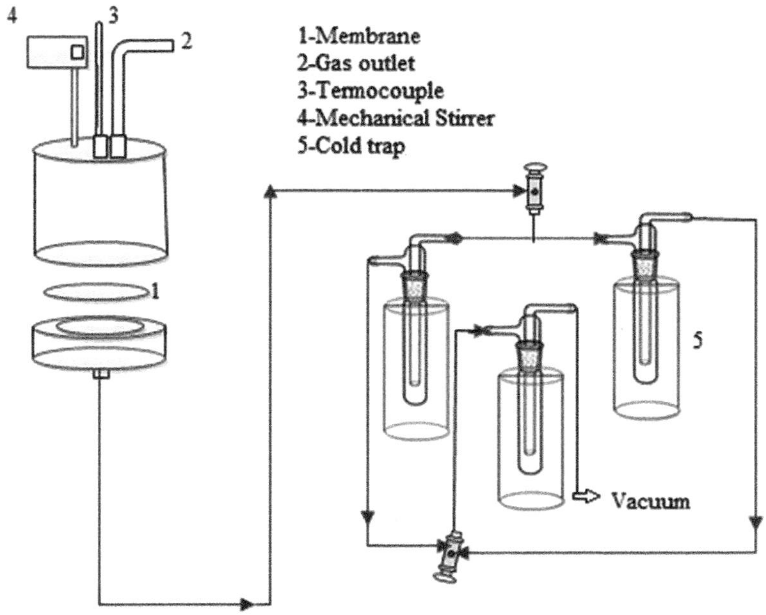

Fig. 8.5 Experimental PVMBR system

8.3 Results and Discussion

Results were discussed in terms of ethanol concentration in broth, ethanol selectivity, and total membrane flux values. Selectivity was calculated by means of permeate concentration.

8.3.1 Membrane Characterization

Membrane characterization is important to observe blended membrane miscibility and harmony. Although EPDM and PDMS have an amorphous chain structure, they have different chemical groups. Thus it is important to achieve full compatibility in order to obtain a nonporous pervaporation membrane. As seen in Fig. 8.6a, there is no polymeric hole or porosity on the top surface of the membrane. Figure 8.6b also confirms the nonporous structure.

In Fig. 8.6c, the EPDM and PDMS structures can clearly be seen separately. As shown in the micrograph, there is no contact-free region between two different polymers. Hence, it can be concluded that the PDMS and EPDM are fully compatible materials and it is possible to produce a nonporous pervaporation membrane with the blending procedure.

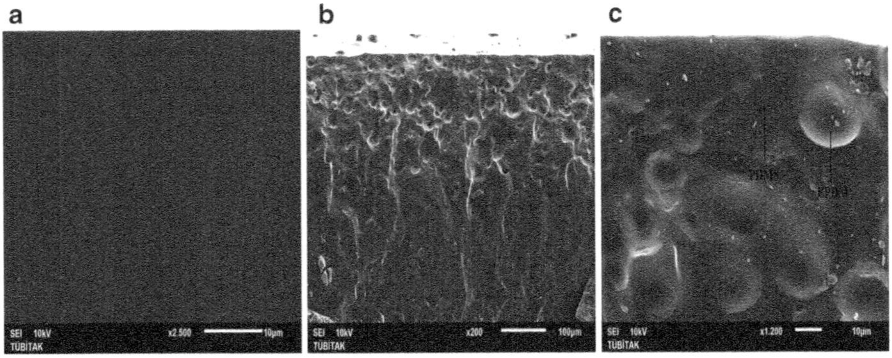

Fig. 8.6 SEM micrographs of blend membrane: (**a**) surface 2,500×; (**b**) cross-sectional 200×; (**c**) cross-sectional 1,200×

8.3.2 *Ethanol Productivity*

Ethanol productivity can be defined as the ethanol production rate per time and it can be basically determined by measuring the ethanol concentration in the fermentation broth. In this study for comparison of system efficiency, fermentation was carried out under the same conditions in the PVMBR and batch reactor. Also the concentration of the molasses solution was the same. Reactions were carried out at room temperature for 7 h. In internal PVMBR, a blended hydrophobic membrane was used to separate reactant–yeast and product media. Meanwhile, the membrane indirectly contributed to ethanol production and system efficiency.

In the PVMBR system, it is desired to separate ethanol selectively and increase ethanol productivity. As seen in Fig. 8.7, although the ethanol was selectively removed from the media, the ethanol concentration was higher than that of the batch reactor.

If the purification is taken into account in PVMBR, ethanol production is at least two times higher than with the batch reactor. As in PVMBR, a nearly steady-state condition was obtained in 5 h; in the batch reactor the ethanol concentration slightly increased. Continuous removal of alcohol was also profitable because it prevented the alcohol inhibition effect on yeast. Hence the yeast's stability could be preserved during the long operation period. In addition, product removal increased the substrate–yeast contact area.

8.3.3 *Total Flux Values of Blended Membrane*

Flux is the determinant factor for membrane productivity. Compared to the porous membrane, in nonporous membranes flux values can be lower due to the intermolecular passage rate. In the case of a membrane bioreactor—in which ethanol production is directly affected by membrane permeability—the flux value is the prominent factor. The operation temperature, pressure, solvent concentration, membrane thickness, and swelling tendency are the effective parameters for flux change.

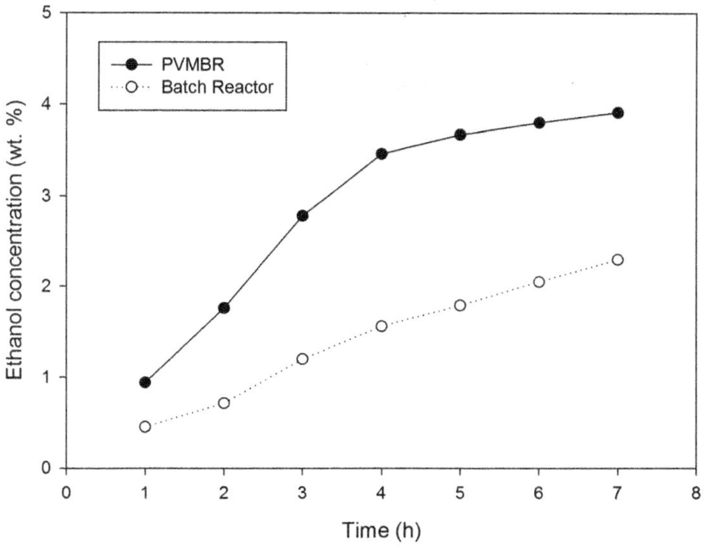

Fig. 8.7 Ethanol concentrations in fermentation broth

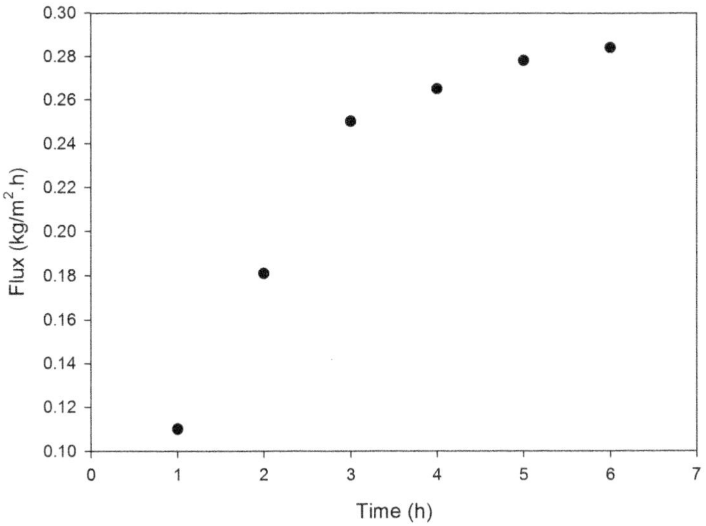

Fig. 8.8 Total flux value of blended membrane

Figure 8.8 represents the changing flux values with reaction time. In this study, effective parameters such as reaction pressure, temperature, and membrane thickness were kept stable. Hence the polymer structure and feed concentration are the main factors regarding flux change. As the ethanol concentration increased in the reaction media, flux was enhanced due to the increasing swelling degree of the polymer. As can be seen in Fig. 8.8, the flux shows the same increment trend with ethanol productivity.

Both PDMS and EPDM have amorphous flexible organic chain structures. It is well known that the hydrophobic polymers show a high affinity to organic solvent. As the organic content increases in the feed mixture, the organic polymer chain absorbs much solvent and the chain spaces expand. Swelling is an essential factor that contributes to the permeation of molecules. Before the PV experiments, membranes are mostly immersed in a feed solution for a while in order to increase membrane permeability. But there is a critical swelling limit. Excessive swelling of polymeric material can constrict the free chain spaces and block the passage of molecules. Therefore the flux decreases drastically. Permeability, productivity, and effectiveness of the membrane can suffer. In this study, flux value increases with the increasing ethanol contented as expected.

8.3.4 *Ethanol Selectivity of Blended Membrane*

Selectivity is the other determinant factor for membrane availability. Selectivity defines the selective character of the membrane and it directly contributes to system efficiency in a reactive separation system such as PVMBR. However, parameters that positively affect the flux values can show an adverse effect on selectivity. In this study the only effective variable is the ethanol concentration. Thereby, it is possible to consider that the selectivity change depends on the ethanol concentration. As can be seen in Fig. 8.9, selectivity values decreased with time.

As the swelling degree and free chain spaces increased, selectivity decreased. The factors contribute to the flux increment and adversely affect selectivity because expanded chain spaces prevent the selective separation ability. It is clearly known

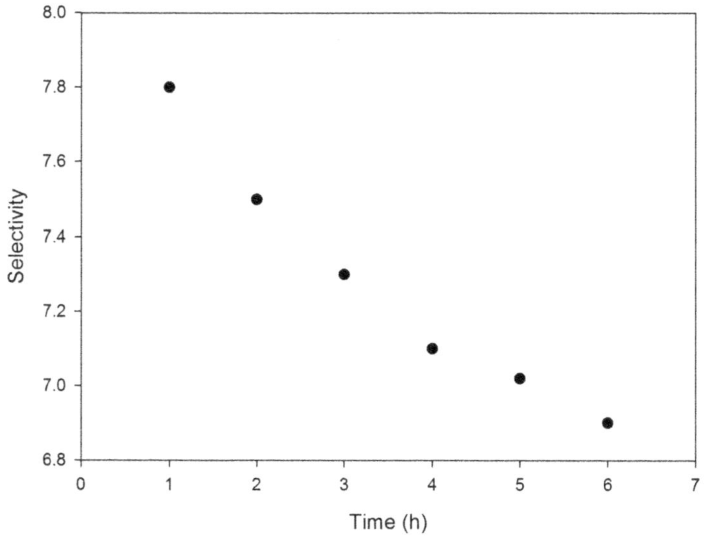

Fig. 8.9 Ethanol selectivity value of blended membrane

that the most important differentiation between PV and another membrane process is the selective separation ability of the nonporous membrane. If the chain spaces expand, separation turns into a filtration-type removal. Owing to the small kinetic diameter of water, the nonselective polymer structure allows uncontrolled passage of water molecules and selectivity decreases as can be understood from Fig. 8.9.

8.4 Conclusions

PVMBR is a promising technology to overcome process intensification problems. It is an energy-intensive, cost-effective, environmentally friendly process, and it provides significant advantages to the reaction system such as fewer process step requirements, less process stability, and less reaction time. Moreover; recovery of yeast and purification of ethanol in one step is possible with PVMBR. In this study, fermentation and ethanol purification were utilized in PVMBR and the membrane performances were evaluated as a function of ethanol production, flux, and selectivity. Compared to the batch reactor, a higher ethanol concentration was obtained in less reaction time. Also flux and selectivity values proved the availability of the system.

Acknowledgment This study was financially supported by the Scientific Research Projects Unit of Kocaeli University.

References

1. Wu FC, Huang SS, Shih IL (2014) Sequential hydrolysis of waste newspaper and bioethanol production from the hydrolysate. Bioresour Technol 167:159–168
2. Boateng CO, Lee KT (2014) Ultrasonic-assisted simultaneous saccharification and fermentation of pretreated oil palm fronds for sustainable bioethanol production. Fuel 119:285–291
3. Mojovic L, Pejin D, Rakin M, Pejin J, Nikolic S, Vukovic AD (2012) How to improve the economy of bioethanol production in Serbia. Renew Sustain Energy Rev 16:6040–6047
4. Cotana F, Cavalaglio G, Gelosia M, Nicolini A, Coccia V, Petrozzi A (2014) Production of bioethanol in a second generation prototype from pine wood chips. Energy Proc 45:42–51
5. Patrizi N, Pulselli FM, Morandi F, Bastianoni S (in press) Evaluation of the energy investment needed for bioethanol production in a biorefinery using residual resources and energy. J Cleaner Prod 96:549–556
6. Wei P, Cheng L, Zhang L, Xu XH, Chen H, Gao C (2014) A review of membrane technology for bioethanol production. Renew Sustain Energy Rev 30:388–400
7. Ramana KV, Ganesan K, Singh L (2006) Pervaporation performance of a composite bacterial cellulose membrane: dehydration of binary aqueous–organic mixtures. World J Microbiol Biotechnol 22:547–552
8. Roza M, Maus E (2006) Industrial experience with hybrid distillation – pervaporation or vapor permeation applications. Institution of Chemical Engineers, Rugby, pp 619–627

9. Sanchez OJ, Moncada JA, Cardona CA (2005) Modeling and simulation of ethanol dehydration by pervaporation and energy analysis of separation schemas for fuel ethanol production. Paper presented at 9th conference of Process integration, modeling and optimization
10. Shao P, Huang RYM (2007) Polymeric membrane pervaporation. J Membr Sci 287:162–179
11. Semenova S, Ohya H, Soontarapa K (1997) Hydrophilic membranes for pervaporation: an analytical review. Desalination 110:251–286
12. Trinh LTP, Cho EJ, Lee YJ, Bae H, Lee H (2013) Pervaporative separation of bioethanol produced from the fermentation of waste newspaper. J Ind Eng Chem 19:1910–1915
13. Baker RW (2004) Membrane technology and applications. Wiley, England
14. Wijmans JG, Baker RW (1995) The solution-diffusion model: a review. J Membr Sci 107:1–21
15. Lipnizki F, Field RW, Ten PK (1999) Pervaporation-based hybrid process: a review of process design, applications and economics. J Membr Sci 153:183–193
16. Vallieres C, Favre E (2004) Vacuum versus sweeping gas operation for binary mixtures separation by dense membrane processes. J Membr Sci 244:17–23
17. Wee SL, Tye CT, Bhatia S (2008) Membrane separation process—pervaporation through zeolite membrane. Sep Purif Technol 63:500–516
18. Hoof VV, Abeele LV, Buekenhoudt A, Dotremont C, Leysen R (2004) Economic comparison between azeotropic distillation and different hybrid systems combining distillation with pervaporation for the dehydration of isopropanol. Sep Purif Technol 37:33–49
19. Chovau S, Gaykawad S, Straathof AJJ, Van der Bruggen B (2011) Influence of fermentation by-products on the purification of ethanol from water using pervaporation. Bioresour Technol 102:1669–1674
20. Van der Bruggen B, Luis P (2014) Curr Opin Chem Eng 4:47–53
21. Niemistö J, Pasanen A, Hirvela K, Myllykoski L, Muurinen E, Keiski RL (2013) Pilot study of bioethanol dehydration with polyvinyl alcohol membranes. J Membr Sci 447:119–127
22. Vane LM, Namboodiri VV, Bowen TC (2008) Hydrophobic zeolite–silicone rubber mixed matrix membranes for ethanol–water separation: effect of zeolite and silicone component selection on pervaporation performance. J Membr Sci 308:230–241
23. Adnadjevid B, Jovanovid J, Gajinov S (1997) Effect of different physicochemical properties of hydrophobic zeolites on the pervaporation properties of PDMS-membranes. J Membr Sci 136:173–179
24. Liu G, Xiangli F, Wei W, Liu S, Jin W (2011) Improved performance of PDMS/ceramic composite pervaporation membranes by ZSM-5 homogeneously dispersed in PDMS via a surface graft/coating approach. Chem Eng J 174:495–503
25. Lu SY, Chiu CP, Huang HY (2000) Pervaporation of acetic acid/water mixtures through silicalite filled polydimethylsiloxane membranes. J Membr Sci 176:159–167
26. Fan S, Xiao Z, Tang X, Chen C, Zhang Y, Deng Q, Yao P, Li W (2014) Inhibition effect of secondary metabolites accumulated in a pervaporation membrane bioreactor on ethanol fermentation of Saccharomyces cerevisiae. Bioresour Technol 162:8–13
27. Fan S, Xiao Z, Zhang Y, Tang X, Chen C, Li W, Deng Q, Yao P (2014) Enhanced ethanol fermentation in a pervaporation membrane bioreactor with the convenient permeate vapor recovery. Bioresour Technol 155:229–234
28. Ikegami T, Yanagishita H, Kitamoto D, Negishi H, Haraya K, Sano T (2002) Concentration of fermented ethanol by pervaporation using silicalite membranes coated with silicone rubber. Desalination 149:49–54
29. Hyder MN, Huang RYM, Chen P (2009) Pervaporation dehydration of alcohol–water mixtures: optimization for permeate flux and selectivity by central composite rotatable design. J Membr Sci 326:343–353
30. Goksungur Y, Zorlu N (2001) Production of ethanol from beet molasses by Ca-Alginate immobilized yeast cells in a packed-bed bioreactor. Turk J Biol 25:265–275
31. Nunes SP, Peinemann KV (2006) Membrane technology in the chemical industry. Wiley, Germany

Chapter 9
Biofuel Production from Sunflower Oil and Determination of Fuel Properties

Imane Boumesbah, Zohra Hachaïchi-Sadouk, and Aida Cherifa Ahmia

Abstract The depletion of crude oil is central to the debate on energy, and the subject becomes even more pressing because of the rapid industrial development of emerging countries that hangs over demand. Growing emissions of combustion generated pollutants, and their increasing costs will make biomass sources more attractive. Currently, biodiesel is becoming popular as a more environment-friendly fuel, because it is a renewable, domestic resource with an environment-friendly emission profile, readily biodegradable and nontoxic.

The objective of our work is to produce biodiesel from a renewable and sustainable energy resource which is sunflower oil through transesterification process using alkaline catalyst and methanol, to optimize some parameters in the aim to obtain the best reaction yield, and to study some biodiesel's properties such as kinematic viscosity, density, cloud, and flash points; the biodiesel was analyzed by infrared spectroscopy and gas chromatography–mass spectrometry. The results of the analysis confirm that the synthesized biodiesel is a mixture of fatty acid methyl esters, a comparative study of biodiesel has been conducted versus standard biodiesel ASTM D6751, and the results obtained show good properties when compared to those of biodiesel's standard.

Keywords Biodiesel • FT-IR • Fuel properties • GC-MS • Transesterification • Sunflower oil

Nomenclature

T	Temperature, °C
Wt.	Weight, g

I. Boumesbah (✉) • Z. Hachaïchi-Sadouk • A.C. Ahmia
Laboratory of Applied Organic Chemistry, Faculty of Chemistry, University of Sciences and Technology Houari Boumediene, B.P. 32 El-Alia, 16111 Bab-Ezzouar, Algiers, Algeria
e-mail: boumesbah.i@gmail.com

I. Dincer et al. (eds.), *Progress in Clean Energy, Volume 2*,
DOI 10.1007/978-3-319-17031-2_9

Subscripts

ASTM The American Society for Testing and Material
FAME Fatty acid methyl esters
FT-IR Fourier transform infrared
GC-MS Gas chromatography–mass spectrometry
SFO Sunflower oil

9.1 Introduction

For some time, the depletion of crude oil reserves is central to the debate on energy, and the subject is becoming even more pressing due to the rapid industrial development of emerging countries, which weighs on the demand [1–3]. The tensions induced by the energy demand, combined with the consideration of the emissions of greenhouse, led to search alternative sources of petroleum-based fuel, including diesel and gasoline fuels [4–7]. Among the alternative possible sources, the biodiesel known also as fatty acid alkyl esters (FAAE) seems to be an interesting solution. Biodiesel is a renewable fuel that can be produced from a range of organic feedstock including fresh or waste vegetable oils, animal fats, and oilseed plants. It is also nontoxic and biodegradable, more compatible with the environment [8–10].

Biodiesel is traditionally produced by transesterification which is a technique of choice for its production. During this reaction, the triglycerides contained in vegetable oils react with a short-chain alcohol such as methanol or ethanol in the presence of a catalyst to obtain fatty acid methyl esters (FAME) or fatty acid ethyl esters (FAEE) [11, 12] according to the reaction scheme shown in Fig. 9.1.

Several parameters affect the transesterification reaction, such as the type and the amount of the catalyst, the alcohol to oil ratio, the reaction's time, the reaction's temperature, and the amount of the free fatty acids [13, 14].

The aim of this work is the synthesis of fatty acid methyl esters from sunflower oil (SFO). The optimization of the most important factors affecting the yield of the reaction such as time and temperature of the reaction is established. The biodiesel

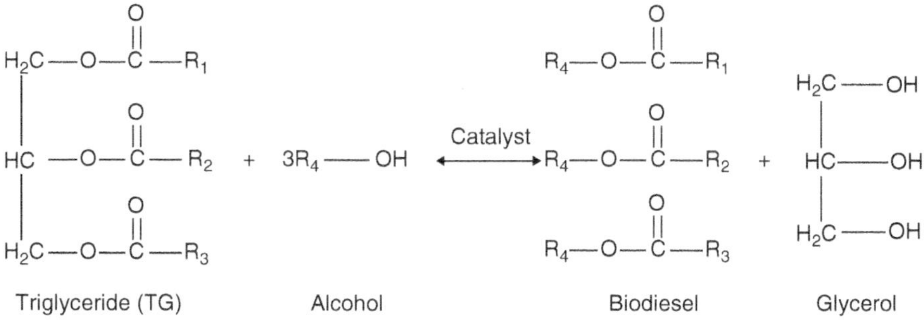

Fig. 9.1 Overall reaction of the transesterification of vegetable oils [11]

produced is analyzed using Fourier transform infrared and gas chromatography–mass spectroscopy, and the fuel properties of the biodiesel have been determined and discussed.

9.2 Experimental Part

In the first step, a solution of sodium methoxide was prepared from a required amount of methanol and sodium hydroxide. This solution was introduced into a dropping funnel and immediately added to the sunflower oil preheated to the desired temperature; vigorous stirring and a constant temperature were maintained throughout the duration of the experiment. When the reaction reached the preset reaction time, the reaction mixture was transferred into a separating funnel. The upper layer contains fatty acid methyl esters, residual alcohol, and catalyst, whereas the lower layer contains a mixture of glycerol, excess of alcohol, and catalyst. The bottom glycerol phase was removed and the fatty acid methyl esters layer was then purified.

The study was carried out at two different temperatures using methanol at various reaction times ranging from 30 to 360 min. The other factors such as molar ration and type and amount of catalyst were fixed as common parameters in all experiments.

9.3 Results and Discussion

9.3.1 Effect of Reaction Temperature

The experiments were carried out at two temperatures such as 25 and 50 °C with 1 wt% of NaOH, methanol/SFO molar ratio 6:1, and agitation speed 1,100 rpm. The FAME yield versus time at the different temperature is shown in Fig. 9.2. It was observed that it is better to work at 25 °C than at 50 °C.

9.3.2 Effect of Reaction Time

Figure 9.3 shows the effect of reaction time on the FAME yields. Transesterification experiments of sunflower oil are carried out at optimal temperature (25 °C) between 30 and 360 min.

Results obtained from the experiments reveal that fatty acid methyl esters yield increase with reaction time at the beginning to achieve the maximum yield at 60 min. Then the yield decreases slightly with increasing reaction time. This is in

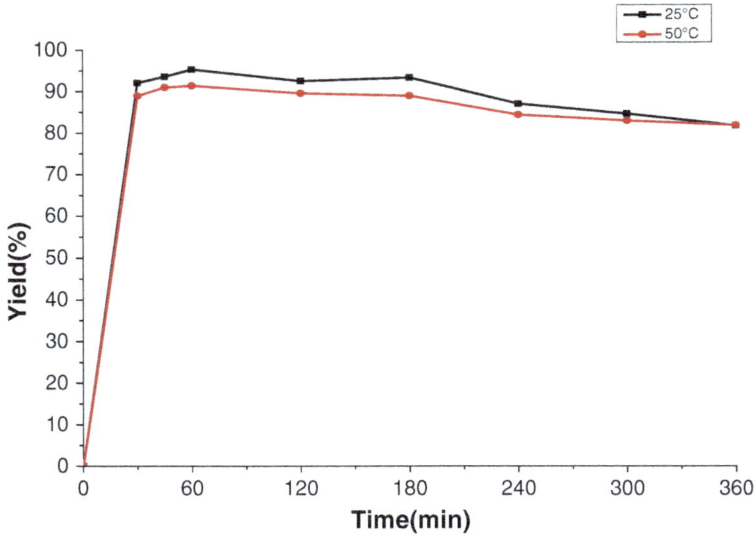

Fig. 9.2 Effect of reaction temperature on the yield of FAME

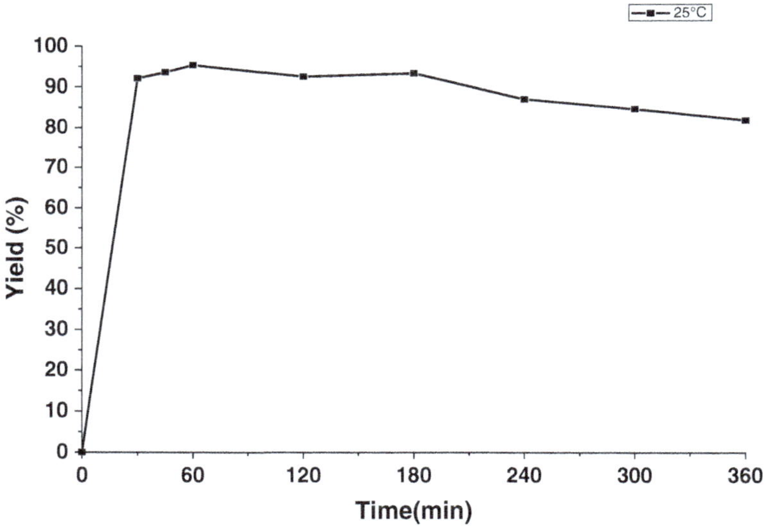

Fig. 9.3 Effect of time reaction on the yield of FAME

agreement with literature data, which shows that longer reaction time will lead to a reduction in the yield product due to the backward reaction of transesterification (hydrolysis), which tends to produce more fatty acids to form soap [15]. Due to reaching the maximum biodiesel yield at 60 min, it has been selected as the optimal reaction time for sunflower oil transesterification using methanol.

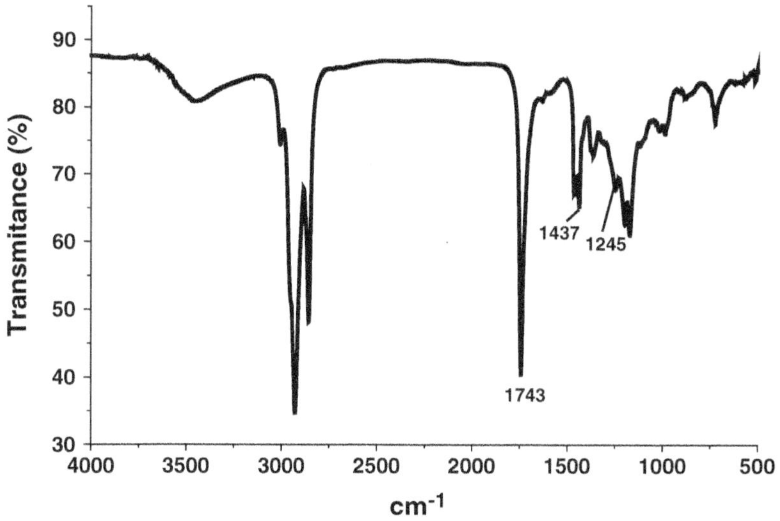

Fig. 9.4 FT-IR spectrum of FAME

9.3.3 IR Analysis

Infrared Fourier transform analysis of the principal product was performed. Spectrum in Fig. 9.4 shows that the bands of C=O and –C–O (ester function) appear at 1,743 cm^{-1} and 1,245 cm^{-1}, respectively, and that of O–CH_3 appears at 1,437 cm^{-1}; those results are in agreement with literature [16, 17].

9.3.4 GC-MS Analysis

Analysis by mass spectroscopy reveals the presence of the molecular ion peaks of all compounds. For example, the peak at *m/z* 270 present in the mass spectrum (Fig. 9.5) which correspond to $[C_{17}H_{34}O_2]^+$, $[C_{16}H_{31}O]^+$ fragment appears at *m/z* 239 ($[M\text{-}31]^+$) and represents the loss of methoxy group. The peak at *m/z* 74 is probably due to Mac Lafferty rearrangement.

9.4 Biodiesel Properties

Fuel properties of the biodiesel produced under optimum conditions including density, kinematic viscosity, flash point, and cloud point are studied and compared with the ASTM D6751 standards of biodiesel (Table 9.1) [18].

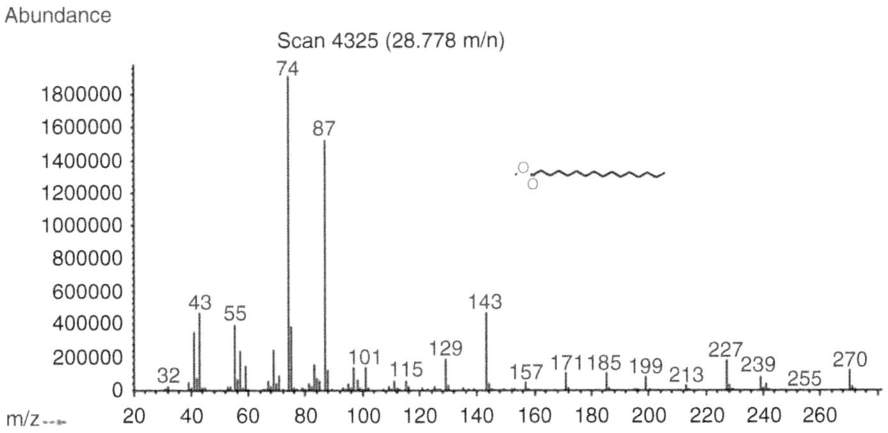

Fig. 9.5 Mass spectrum of methyl hexadecanoate

Table 9.1 Fuel properties of FAME produced from SFO under optimal conditions compared with the ASTM D6751 standards of biodiesel

Property	Unit	FAME	ASTM D6751 biodiesel
Density at 15 °C	g/cm^3	0.886	0.870–0.900
Kinematic viscosity at 40 °C	mm^2/s	4.51	1.9–6.0
Flash point	°C	>190	>130
Cloud point	°C	5	Not specified

The obtained results show that the kinematic viscosities and the density of the samples obtained at the optimal conditions are consistent and in the range of the standard of the biodiesel. We note also that the value of the flash point is higher than 190 °C; this result represents a benefit that reduces the risk of autoignition during storage and transport at high temperatures [19].

9.5 Conclusion

The results of the present study demonstrated that 1 h of reaction and a temperature of 25 °C (which can be considered as an ambient temperature and therefore does not require energy expenditure) present the optimum conditions for the production of biodiesel from sunflower oil and give a high yield (95.28 %). The results of analysis (IR and GC-MS) confirm that the structure of the produced biodiesel and the fuel properties are in good agreement with the standards of biodiesel. The transesterification of vegetable oil is an alternative for producing biodiesel, which could reduce pollution and protect the environment.

References

1. Ma F, Hanna MA (1999) Biodiesel production: a review. Bioresour Technol 70:1–15
2. Srivastava A, Prasad R (2000) Triglycerides-based diesel fuels. Renew Sustain Energy Rev 4(2):111–133
3. Dmytryshyn SL, Dalai AK, Chaudhari ST, Mishra HK, Reaney MJ (2004) Synthesis and characterization of vegetable oil derived esters: evaluation for their diesel additive properties. Bioresour Technol 92:55–64
4. Xu G, Wu GY (2003) The investigation of blending properties of biodiesel and No. 0 diesel fuel. J Jiangsu Polytech Univ 15:16–18
5. Andrade JE, Perez A, Sebastian PG, Eapen DA (2011) Review of bio-diesel production processes. Biomass Bioenergy 35:1008–1020
6. Apostolakou AA, Kookos IK, Marazioti C, Angelopoulos KC (2009) Techno-economic analysis of a biodiesel production process from vegetable oils. Fuel Process Technol 90:1023–1031
7. Leung DYC, Wu X, Leung MKH (2010) A review on biodiesel production using catalyzed transesterification. Appl Energy 87:1083–1095
8. Antolin G, Tinaut F, Briceno Y, Castano V, Perez C, Ramirez A (2002) Optimization of biodiesel production by sunflower oil transesterification. Bioresour Technol 83:111–114
9. Serio MD, Tesser R, Dimiccoli M, Cammarota F, Nastasi M, Santacesaria E (2005) Synthesis of biodiesel via homogeneous Lewis acid catalyst. J Mol Catal A Chem 239:111–115
10. Helwani Z, Othman MR, Aziz N, Fernando WJN, Kim J (2009) Technologies for production of biodiesel focusing on green catalytic techniques: a review. Fuel Process Technol 90:1502–1514
11. Alcantara R, Amores J, Canoira L, Fidalgo E, Franco MJ, Navarro A (2000) Catalytic production of biodiesel from soy-bean oil, used frying oil and tallow. Biomass Bioenergy 18:515–527
12. Hoang DH, Nguyen TD, Carmen S, Kenji O, Yasuaki M, Rokuro N (2008) Methanolysis of triolein by low frequency ultrasonic irradiation. Energy Convers Manag 49:276–280
13. Meher LC, Sagar DV, Naik SN (2006) Technical aspects of biodiesel production by transesterification: a review. Renew Sustain Energy Rev 10(3):248–268
14. Meher LC, Dharmagadda VSS, Naik SN (2006) Optimization of alkali-catalyzed transesterification of *Pongamia pinnata* oil for production of biodiesel. Bioresour Technol 97:1392–1397
15. Freedman B, Pryde EH, Mount TL (1984) Variables affecting the yields of fatty esters from transesterified vegetable oils. J Am Oil Chem Soc 61:1638–1643
16. Silverstein RM, Baster GC, Merill TC (2004) Identification spectrométriques des composés organiques, 1st edn. De Boeck Université, Bruxelles, Belgique
17. Stuart B (2004) Infrared spectroscopy: fundamentals and applications. Wiley, West Sussex
18. (1998) ASTM, American Society for Testing and Materials, D445, D93, D2500, D97
19. Durrett TP, Benning C, Ohlrogge J (2008) Plant triacylglycerols as feedstocks for the production of biofuels. Plant J 54:593–607

Chapter 10
Biodiesel Production from Non-comestible Oils

Aida Cherifa Ahmia, Fetta Danane, Rahma Bessah, and Imane Boumesbah

Abstract Increase in the price of petroleum, finite nature of fossil fuels, increasing concerns regarding environmental impact, especially related to greenhouse gas emissions, and health and safety considerations are driving the search for new energy sources. Consequently, recent years have seen the development of several sectors of renewable energy production. Among these energies, biofuels are a promising way to reduce the environmental degradation by increasing pollutants. Research on biofuel production is considerably focused on producing biodiesel from comestible oils. This production has been debated recently because of the use of fertile lands needed to produce biofuels and increased prices of staple foods making them more scarce which is dangerous to the food security of poor people around the world. Therefore, it is clever to produce biodiesel from vegetable species not intended for food. The purpose of this chapter is to consider the employment of non-comestible oils into biodiesel production, to present the various processes of oil extraction and oil conversion with emphasis on the transesterification, and, finally, to highlight several uses of biodiesel in the world.

Keywords Renewable energy • Biodiesel • Non-comestible oil • Transesterification

A.C. Ahmia (✉)
Division Bioénergie et Environnement, Centre de Développement des Energies Renouvelables, CDER, Algiers 16340, Algeria

Laboratoire de Chimie Organique Appliquée, Faculté de Chimie, Université des Sciences et de la Technologie Houari Boumediene, USTHB, Algiers 16111, Algeria
e-mail: ahmia.a.c@gmail.com

F. Danane • R. Bessah
Division Bioénergie et Environnement, Centre de Développement des Energies Renouvelables, CDER, Algiers 16340, Algeria

I. Boumesbah
Laboratoire de Chimie Organique Appliquée, Faculté de Chimie, Université des Sciences et de la Technologie Houari Boumediene, USTHB, Algiers 16111, Algeria

I. Dincer et al. (eds.), *Progress in Clean Energy, Volume 2*,
DOI 10.1007/978-3-319-17031-2_10

10.1 Introduction

Our society has an increased awareness that the combustion of fossil carbon was associated with an increase of the carbon dioxide concentration in the atmosphere, which is one reason for a drastic global warming causing massive environmental, economical, and social problems. Moreover, the most public attention has been paid to the sharp increases in energy costs, especially of crude oil prices [1]. These reasons, additionally with finite nature of fossil fuels and health and safety considerations, are driving the development towards new bio-based, renewable, and therefore environmental-friendly substitutes for fossil energies [2]. Renewable energies are considered important resources in many countries around the world. However, on a global scale, less than 15 % of the primary energy supply comes from renewable sources [3].

Biomass is the most common form of renewable energy and is, among the renewable forms of energy, the major source of the primary energy supply [4]. It accounts for about 10 % of the world's energy consumption and can be converted to other usable forms of energy like biofuels [5].

There is a great mindfulness in a diesel fuel substitution at the present all over the world with a clean, renewable fuel such as biodiesel, which has a lot of technical advantages over fossil fuels such as lower overall exhaust emission and toxicity, biodegradability, derivation from a renewable and domestic feedstock, negligible sulfur content, superior flash point, and higher combustion efficiency [6].

Developing countries have a comparative advantage for biodiesel production because of greater availability of land, favorable climatic conditions for agriculture, and lower labor costs [7]. In developed countries there is a growing trend towards employing modern technologies and efficient bioenergy conversion using a range of biofuels, which are becoming cost-wise competitive with fossil fuels.

Introduction and commercialization of biodiesel in many countries around the world have been accompanied by the development of standards to ensure high product quality and user confidence. Some biodiesel standards are ASTM D6751 (American Society for Testing and Materials) and EN 14214 (European norm) [8]. Biodiesel is defined by ASTM International as a fuel composed of monoalkyl esters of long-chain fatty acids derived from renewable vegetable oils or animal fats meeting the requirements of ASTM D6751 [9].

Biodiesel fuels are attracting increasing attention worldwide as a blending component or a direct replacement for diesel fuel in vehicle engines [10]. B5 used in Europe contains 5 % biodiesel (B100) and 95 % of petrodiesel. Biodiesel blends up to B20 can be used in nearly all diesel equipment and are compatible with most storage and distribution equipment. These low-level blends generally do not require any engine modifications.

In this context, the purpose of this chapter is to consider the employment of non-comestible oils into biodiesel production, to present the various processes of oil extraction and oil conversion with emphasis on the transesterification, and, finally, to highlight several uses of biodiesel in the world.

10.2 Employment of Non-comestible Oil

Although biodiesel has a lot of advantages related to petrodiesel, the high price of its production is the main barrier to its commercial use. Biodiesel price depends mainly on the cost of feedstocks, the price of which makes 70–95 % of the total biodiesel cost [11].

There are four main groups of feedstocks for biodiesel production, categorized into vegetable oils (edible and nonedible), animal fats, waste cooking oils, and algal oils. Most current research is considerably focused on producing biodiesel from comestible oils.

However, continuous and large-scale production of biodiesel from edible oils has recently been of great concern because they compete with food materials. Knowing that, nearly 60 % of humans in the world are malnourished [12]. Therefore, the use of cheap nonedible oils can be a way to improve the economy of biodiesel production and its commercial production at the industry scale [13].

Desirable characteristics of alternative non-comestible oil feedstocks for biodiesel production include adaptability to local growing conditions (rainfall, soil type, latitude, etc.), regional availability, high oil content, favorable fatty acid composition, compatibility with existing farm infrastructure, low agricultural inputs (water, fertilizer, pesticides), definable growth season, uniform seed maturation rates, potential markets for agricultural by-products, and ability to grow in agriculturally undesirable lands and/or in the off-season from conventional commodity crops.

Various oils extracted from seeds or kernels of nonedible crops are potential feedstocks for biodiesel production. The important nonedible oil plants are jatropha, karanja, tobacco, mahua, neem, rubber, sea mango, castor, and cotton. Of all these feedstocks, jatropha, karanja, mahua, and castor oils are the most often used in biodiesel synthesis [14].

Jatropha plant is one of the most promising potential oil sources for biodiesel production in Southeast Asia, Central and South America, and India. Today, it is the major feedstock for production of biodiesel in developing countries like India, where the annual production is about 15,000 t [15]. Jatropha can also be grown in Mediterranean countries. "Evaluation of the energy crop Jatropha curcas as a mean to promote renewable and sustainable energy for the Mediterranean region" (JatroMed) is a 4-year demonstration project involving five countries from the Mediterranean region: Greece (Agricultural University of Athens-Research Committee) which is the project coordinator, Italy (Consiglio per la Ricerca e Sperimentazione in Agricoltura), Egypt (City of Scientific Research and Technological Applications), Morocco (Centre de Developpement de la Region de Tensift), and Algeria (Centre de Dévelоppement des Energies Renouvelables). Eight Jatropha curcas L. genotypes collected over the world have been established in depressed rural areas of the North African partner countries in order to introduce this crop and the profits from its cultivation to local farmers and population. The main scope of JatroMed is to reinforce and upgrade the natural and socioeconomic conditions of the target areas and to give local populations the opportunity to produce sustainable energy to cover their needs [16].

Jatropha can grow almost anywhere, on waste, sandy, and saline soils, under different climatic conditions as well as under low or high rainfall and frost. Its cultivation is easy, without intensive care and minimal effort. Its healthy life cycle of 30–50 years eliminates the yearly replantation. Jatropha oil content varies depending on the types of species, but it is about 40–60 % in the seeds and 46–58 % in the kernels. Jatropha has comparable properties to diesel, such as calorific value and cetane number. It has a great potential as an alternative fuel since it does not require any modification of the engine [17]. The serious problem with jatropha oil is its toxicity to people and animals.

10.3 Extraction of Non-comestible Oil

There are three main methods that have been identified for the extraction of the oil, mechanical extraction, solvent extraction, and enzymatic extraction. Mechanical pressing and solvent extraction are the most commonly used methods for commercial oil extraction. Before the oil extraction takes place, seeds have to be dried. Seeds can be either dried in the oven (105 °C) or sundried (3 weeks). Mechanical expellers or presses can beefed with either whole seeds or kernels or a mix of both, but common practice is to use whole seeds. However, for chemical extraction only kernels are used as feed [18].

The technique of oil extraction using mechanical presses is the most conventional practice. In this type, either a manual ram press or an engine-driven screw press can be used. It has been found that engine-driven screw press can extract 68–80 % of the available oil while the ram presses only achieved 60–65 %. This broader range is due to the fact that seeds can be subjected to a different number of extractions through the expeller [18].

10.4 Conversion of Non-comestible Oil

Physicochemical properties are the main barrier that prevent the use of direct non-comestible oil in conventional diesel engines. Therefore, there are many techniques, methods, and processes that have been used recently to produce biodiesel from various nonedible feedstocks.

Pyrolysis is the thermal conversion of the organic matters in the absence of oxygen and in the presence of a catalyst. Thermal decomposition of triglycerides produces alkanes, alkenes, alkadienes, aromatics, and carboxylic acids. The liquid fractions of the thermally decomposed vegetable oils are likely to approach diesel fuels. The pyrolyzate had lower viscosity, flash point, and pour point than diesel fuel and equivalent calorific values [15].

Nonedible oil can be diluted with diesel to reduce the viscosity and improve the performance of the engine. This method does not require any chemical process. It has been reported that substitution of 100 % vegetable oil for diesel fuel is not practical. Therefore, blending of 20–25 % vegetable oil to diesel has been considered to give good results for diesel engine [15].

Microemulsification is defined as transparent, equilibrium thermodynamically stable colloidal dispersion of microstructure with diameter ranging from 100 to 1,000 Å. Microemulsion can be made of vegetable oils with an ester and dispersant (cosolvent), or of vegetable oils, and alcohol such as ethanol. Microemulsification has been considered as a reliable approach to solve the problem of the high viscosity of vegetable oils [15].

The reversible transesterification reactions are the most common method of converting triglycerides (TAG) from oils into biodiesel and the most promising solution of the high-viscosity oil problem.

10.5 Transesterification of Non-comestible Oil

Transesterification or alcoholysis is defined as the chemical reaction of alcohol with vegetable oils. This reaction has been widely used to reduce the viscosity of vegetable oil and conversion of the triglycerides into ester, as shown in Fig. 10.1. Then, the fuel produced has approximately the same property of petrodiesel and can be used in conventional diesel engines without any change in this last.

Transesterification can be non-catalyzed or catalyzed by an acid, a base, or an enzyme depending on the solubility of the chemical catalyst in the reaction mixture. Transesterification reaction can be homogeneously or heterogeneously catalyzed. These reactions can be accomplished as one-step (base or acid) or two-step (acid/base) processes, depending on the content of FFA.

The main factors affecting transesterification reaction and produced ester yield are the molar ratio of alcohol:oil, type of alcohol, type and amount of catalyst, reaction temperature, pressure and time, mixing intensity, as well as the contents of free fatty acids (FFA) and water in oils.

A further complicating factor of high FFA content is the production of water upon reaction with homogeneous base catalysts, as shown in Fig. 10.2.

In literature, we found that the most commonly used alcohols are ethanol and methanol, especially the latter with its low cost and its physical benefits (chains

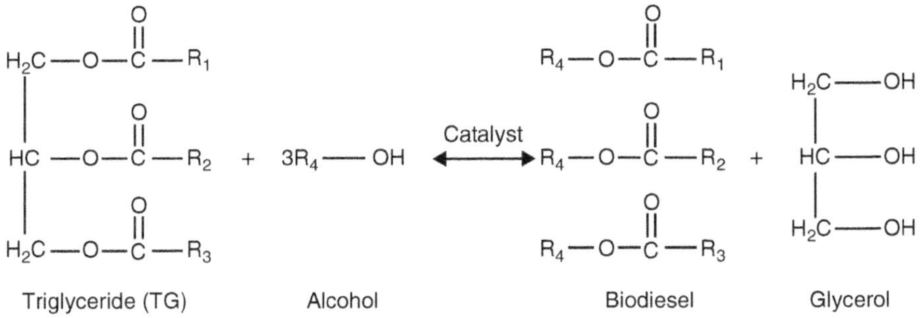

Fig. 10.1 Transesterification reaction of triglycerides with alcohol [16]

Fig. 10.2 Production of water upon reaction with homogeneous base catalysts

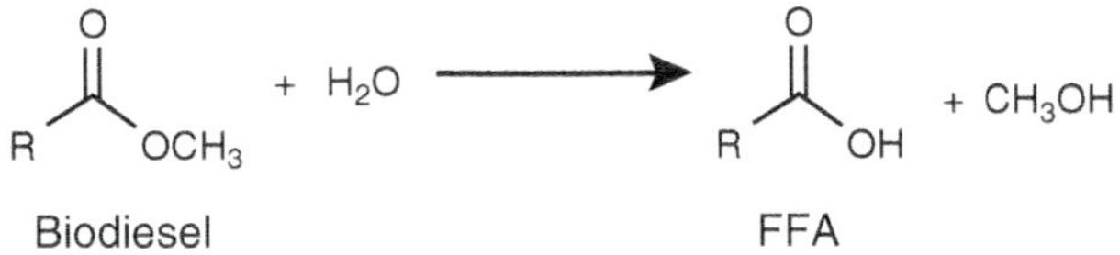

Fig. 10.3 Production of FFA by the hydrolysis of biodiesel

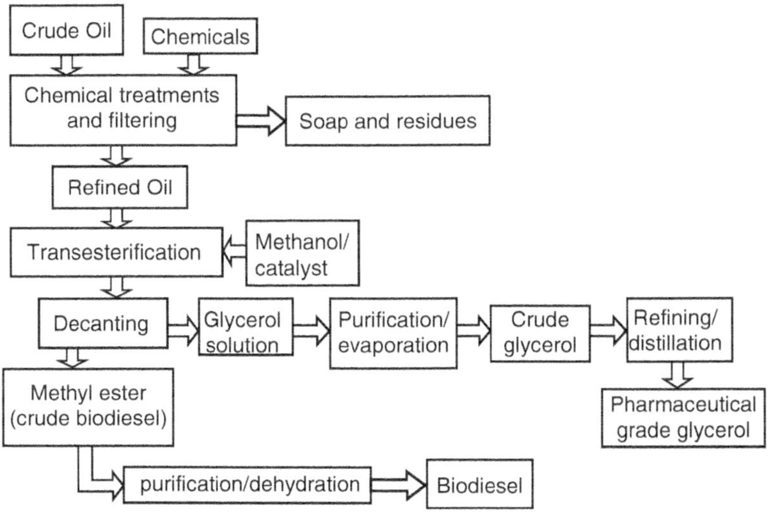

Fig. 10.4 Different steps of biodiesel production from non-comestible oils by transesterification

shorter and more polar alcohol). However, ethanol has the advantage to come from a renewable source by fermentation of sugar derived from sugarcane or beet. Biodiesel thus obtained is 100 % renewable.

Water is particularly problematic because in the presence of any remaining catalyst, it can participate in hydrolysis with biodiesel to produce additional FFA and methanol, as shown in Fig. 10.3.

The following diagram (Fig. 10.4) shows the different steps of biodiesel obtained from nonedible oils by transesterification process producing glycerol as coproduct.

The glycerol produced may be used as a chemical feedstock in the production of polyurethanes, polyesters, polyethers, and other materials. Glycerol may also be found in lubricants, wrapping and packaging materials, foods, drugs, cosmetics, and tobacco products. Applications and products that displace existing petroleum-derived materials or feedstocks are of particular interest. A recent significant advance is the development of a synthetic route to propylene glycol (1,2-propanediol) from glycerol, which represents a viable alternative to the classic petrochemical route from propylene [19]. Propylene glycol represents a replacement for the common toxic antifreeze component ethylene glycol.

10.6 Other Uses of Biodiesel Produced from Non-comestible Oil

Biodiesel has attracted considerable interest as an alternative bio-based diesel fuel for combustion in ignition engines. However, a number of additional applications have been developed or discovered for these versatile oleochemical materials. For instance, biodiesel may be used as a replacement for petroleum as a heating oil [20]. In the USA, blends of up to 5 % biodiesel in heating oils (B5 Bioheat) have recently been approved for inclusion in the ASTM heating oil standard, D396 [21]. Another combustion-related application of biodiesel is as an aviation fuel, although the relatively poor low-temperature properties of biodiesel restrict its use to low-altitude aircraft [22]. Moreover, the use of biodiesel in diesel-fueled marine engines to reduce environmental impact is another important application of this biodegradable and nontoxic fuel [23]. Because there are less harmful exhaust emissions from biodiesel than those from petrodiesel, the use of biodiesel to power underground mining equipment is another combustion-related application. Biodiesel may also be used as a fuel for generators and turbines for the generation of electricity [24]. Lastly, biodiesel in conjunction with certain surfactants can act as a contact herbicide to kill broadleaf weeds in turfgrass [25].

10.7 Conclusions

The use of edible oil in biodiesel production has an influence on the global imbalance to the market demand and the food supply by its high prices, the reduction of food sources, and the growth of commercial plant capacities. Thus, focus should be shifted to nonedible resources, which are not used in the human nutrition and could grow in the barren lands. This chapter considered the employment of non-comestible oils into biodiesel production, presented the various processes of oil extraction and oil conversion with emphasis on the transesterification, and highlighted other numerous uses of biodiesel.

References

1. Uthoff S, Bröker D, Steinbüchel A (2009) Current state and perspectives of producing biodiesel-like compounds by biotechnology. Microb Biotechnol 2(5):551–565
2. Sigar CP, Soni SL (2009) Performance and emission characteristics of vegetable oil as diesel fuel extender. Energy Source 31:139–148

3. Lund H (2007) Renewable energy strategies for sustainable development. Energy 32 (6):912–919
4. McKendry P (2002) Energy production from biomass Part 1 overview of biomass. Bioresour Technol 83(1):37–46
5. AnselmoFilho P, Badr O (2004) Biomass resources for energy in North-Eastern Brazil. Appl Energy 77(1):51–67
6. Bankovi, Ivana B, Olivera, Stamenkovi´c, S. Veljkovi´c, Vlada B (2012) Biodiesel production from non-edible plant oils. Renew Sust Energ Rev 16:3621–3647
7. Jull C, Renondo P. (2007) Recent trends in the law and policy of bioenergy production, promotion and use. Legislative study 95, FAO, Rome
8. Knothe G (2005) Dependence of biodiesel fuel properties on the structure of fatty acid alkyl esters. Fuel Process Technol 86:1059–1070
9. ASTM Standard specification for biodiesel fuel (B100) blend stock for distillate fuels. In: Annual Book of ASTM Standards, ASTM International, West Conshohocken, Method D6751-08, 2008
10. Demirbas A (2009) Progress and recent trends in biodiesel fuels. Energy Convers Manage 50:14–34
11. Balat M (2011) Potential alternatives to edible oils for biodiesel production—a review. Energy Convers Manage 52:1479–92
12. Pimentel D, Marklein A, Toth MA, Karpoff M, Paul GS, McCormack R et al (2009) Food versus biofuels: environmental and economic costs. Hum Ecol 37:1–12
13. Leung DYCW, Leung X, MKH (2010) A review on biodiesel production using catalyzed transesterification. Appl Energy 87:1083–1095
14. Ivana B, Olivera S, Vlada B (2012) Biodiesel production from non-edible plant oils. Renew Sust Energ Rev 16:3621–3647
15. Atabani A, Silitonga AS, Ong HC, Mahlia TMI, Masjuki HH, Fayaz H (2013) Non-edible vegetable oils: a critical evaluation of oil extraction, fatty acid compositions, biodiesel production, characteristics, engine performance and emissions production. Renew Sust Energ Rev 18:211–245
16. http://www.jatromed.aua.gr/index.php?option = com_content&view = article&id = 33&Item id = 18&lang = en
17. Jain S, Sharma MP (2010) Prospect of biodiesel from Jatropha in India: a review. Renew Sust Energ Rev 14:763–771
18. Achten WMJ, Verchot L, Franken YJ, Mathijs E, Singh VP, Aerts R et al (2008) Jatropha bio-diesel production and use. Biomass Bioenergy 32(12):1063–1084
19. Dasari MA, Kiatsimkul PP, Sutterlin WR, Suppes GJ (2005) Low pressure hydrogenolysis of glycerol to propylene glycol. Appl Catal A Gen 281:225–231
20. Mushrush G, Beal EJ, Spencer G, Wynne JH, Lloyd CL, Hughes JM, Walls CL, Hardy DR (2001) An environmentally benign soybean derived fuel as a blending stock or replacement for home heating oil. J Environ Sci Heal A 36:613–622
21. ASTM Standard specification for fuel oils (2008) In: Annual Book of ASTM Standards, ASTM International, West Conshohocken, Method D396-08b
22. Dunn RO (2001) Alternative jet fuels from vegetable oils. Trans ASAE 44:1751–1757
23. Nine RD, Clark NN, Mace BE, Morrison RW, Lowe PC, Remcho VT, McLaughlin LW (2000) Use of soy-derived fuel for environmental impact reduction in marine engine applications. Trans ASAE 43:1383–1391
24. Lin YC, Tsai CH, Yang CR, Jim Wu CH, Wu TY, Chang Chien GP (2008) Effects on aerosol size distribution of polycyclic aromatic hydrocarbons from the heavy-duty diesel generator fueled with feedstock palm-biodiesel blends. Atmos Environ 42:6679–6688
25. Vaughn SF, Holser RA (2007) Evaluation of biodiesels from several oilseed sources as environmentally friendly contact herbicides. Ind Crop Prod 26:63–68

Chapter 11
Development of a Bi-fuel SI Engine Model

K. Rezapour

Abstract Natural gas is a promising alternative fuel, with the potential to meet strict engine emission regulation, and is cheaper than other fuels in many countries. Use of natural gas as an automotive fuel may bring a reduction of environmental pollutants and reduce the economic costs of the transportation sector. As an intermediate step, and an alternative to dedicated CNG engines bi-fuel engines, powered by gasoline and compressed natural gas (CNG), provide many an opportunity. In support of the development of such engines and to aid analysis and improvement in this study, a four-stroke bi-fuel spark ignition (SI) engine model is developed. The engine model is based on the two-zone combustion model, and it has the ability to simulate turbulent combustion and compared to computational fluid dynamic (CFD) models it is computationally faster and efficient. The selective outputs are cylinder temperature and pressure, heat transfer, brake work, brake thermal and volumetric efficiency, brake torque, brake power (BP), brake-specific fuel consumption (BSFC), brake mean effective pressure (BMEP), concentration of CO_2, brake-specific CO (BSCO) and brake-specific NO_x ($BSNO_x$). In this research, the effect of engine speed, equivalence ratio and performance parameters using gasoline and CNG fuels are analysed. In addition, the model has been validated by experimental data using the results obtained from bi-fuel engine tests. Therefore, this engine model is capable for prediction, analysis and useful for optimisation of the engine performance parameters and minimisation of the emissions. In addition, in this chapter, a specific bi-fuel engine is studied and discussed that is used in the vast majority (almost are taxi). Therefore, the model and its results are significant.

Keywords Bi-fuel • Engine performance • Emissions • Engine modelling

K. Rezapour (✉)
Department of Mechanical Engineering, Faculty of Engineering, Islamic Azad University, Karaj Branch, Alborz, Iran
e-mail: rezapour@kiau.ac.ir

I. Dincer et al. (eds.), *Progress in Clean Energy, Volume 2*,
DOI 10.1007/978-3-319-17031-2_11

Nomenclature

A	Area exposed to heat transfer (m^2)
aBDC	After BDC
aTDC	After TDC
b	Bore of cylinder (m)
bBDC	Before BDC
bTDC	Before TDC
C_p	Specific heat at constant pressure (J kg^{-1} k^{-1})
C_b	Blow by coefficient (s^{-1})
E	Total energy (J)
EVO	Exhaust valve opening
H	Enthalpy (J)
h	Specific enthalpy (J kg^{-1})
h	Heat transfer coefficient (W m^{-2} K^{-1})
IVC	Inlet valve closing
m	Mass (kg)
P	Pressure (Pa)
PPM	Particle per million
Q	Heat transfer (J)
r	Compression ratio
R	Gas constant
RON	Research octane number
s	Specific entropy (J kg^{-1} K^{-1})
T	Temperature (K)
u_p	Engine piston speed (m/s)
V	Volume (m^3)
W	Work done (J)
WOT	Wide open throttle
x	Burnt mass fraction

Greek Letters

v	Specific volume (m^3 kg^{-1})
θ	Crank angle (°CA)
θ_0	Start of combustion (°CA)
$\Delta\theta$	Total combustion duration (°CA)
ω	Angular velocity (rad s^{-1})
ϕ	Equivalence ratio
φ_{ed}	Charge-up efficiency
γ_r	Mole fraction

11.1 Introduction

The Kyoto protocol called for reduction in greenhouse gas emissions between 2008 and 2012 to levels that are 5.2 % below 1990 levels in 38 industrialised countries [1]. Additionally, ever-increasing oil prices and security of supply issues are focussing attentions on alternative sources of motive energy. Consequently, vehicle manufacturers are focussing their interests on a diversity of engine technologies. This includes the development of engines that are capable of making use of alternative fuels such as CNG. CNG consists of 88 % methane and may be used in either CNG or liquefied gas forms in vehicle. CNG is cheaper and cleaner than gasoline but it reduces the engine brake power [2].

With regard to the climatic situation of some countries, and considering the existence of broad networks of gas distribution, natural gas can be a suitable alternative to conventional fuels. The bi-fuel vehicle in some countries, along with the implementation of strategies for the gasification of vehicles, has been identified, i.e., workshop conversion of vehicle in move (short-term approach), factory production of bi-fuel engines (midterm approach) and designing and producing base CNG engine (long-term approach) [2]. Therefore, developing bi-fuel engines (gasoline and CNG) in the short and midterm is a strategy for achieving this important aim. A major support for better achievement of this subject is applied studies for analysis and improvement of the engine performance.

Many studies and experimental works have been undertaken on CNG-fuelled engines; for example, Lapetz et al. [3] developed a Ford compressed natural gas bi-fuel truck. To ensure safety and control emissions, they modified the base vehicle's specification for conversion to operation of bi-fuel CNG. Flame speed in natural gas is lower than gasoline. For this reason, the duration of the total combustion is to extend compared with gasoline and diesel [4]. Zuo and Zhao [5] developed a QD model to analyse combustion process in SI pre-chamber natural gas engine. Evans and Blaszcsky [6] in their study characterising the performance and emissions of a bi-fuel Ricardo single-cylinder SI research engine showed a 12 % power and 5–50 % emission reduction when the engine is fuelled using natural gas. Further similar studies [7–9] have also been undertaken looking at CNG and related engine development. This chapter includes a model development and its experimental validation on a bi-fuel engine, as well as simulation results, discussion and conclusions.

11.2 Mathematical Model

The engine model developed herein is a quasi-dimensional, two-zone combustion model that solves the differential equations related to compression, combustion and expansion. Intake and exhaust processes computationally are calculated using an approximation method. In this model, the combustion chamber is divided into two zones including an unburned mixture (zone 1) and burned mixture (zone 2). The distance between the two zones is the flame front. The flame is propagated turbulently and expanded in the combustion chamber over a spherical flame front:

$$\frac{dE}{d\theta} = \frac{dQ}{d\theta} - \frac{dW}{d\theta} + \sum_{in} \dot{m}h - \sum_{out} \dot{m}h \tag{11.1}$$

$$\frac{dm}{d\theta} = \sum_{in} \frac{dm}{d\theta} - \sum_{out} \frac{dm}{d\theta} \tag{11.2}$$

Equation (11.1) can be written as

$$\frac{d(mu)}{d\theta} = \frac{dQ}{d\theta} - p\frac{dV}{d\theta} + \sum_{in} h\frac{dm}{d\theta} - \sum_{out} h\frac{dm}{d\theta} \tag{11.3}$$

The thermodynamic properties are provided using the relations proposed by the following expressions [10] that are curve-fitted to the tabulated JANAF thermochemical tables [11]:

$$\frac{C_{P,i}}{R} = U_{i1} + U_{i2}T + U_{i3}T^2 + U_{i4}T^3 + U_{i5}T^4 \tag{11.4}$$

$$\frac{h_i}{RT} = U_{i1} + \frac{U_{i2}}{2}T + \frac{U_{i3}}{3}T^2 + \frac{U_{i4}}{4}T^3 + \frac{U_{i5}}{5}T^4 + \frac{U_{i6}}{T} \tag{11.5}$$

$$\frac{s_i}{R} = U_{i1}\ell nT + U_{i2}T + \frac{U_{i3}}{2}T^2 + \frac{U_{i4}}{3}T^3 + \frac{U_{i5}}{4}T^4 + U_{i7} \tag{11.6}$$

where C_p is the specific heat measured at a constant pressure, h is the specific enthalpy and s is the specific entropy. The coefficients U_{i1} to U_{i7} are calculated over two different temperature ranges [11], (1) $300 < T < 1{,}000$ K and (2) $1{,}000 < T < 5{,}000$ K.

When modelling with a single fuel, the equivalence ratio can be written as [12]

$$\phi = \left(\frac{\text{Fuel}}{\text{Air}}\right)_{\text{Act.}} \Big/ \left(\frac{\text{Fuel}}{\text{Air}}\right)_{\text{St.}} \tag{11.7}$$

where subscript Act. denotes the actual and St. denotes the stoichiometric air/fuel ratios.

The mass in a control volume may be calculated [12]
for $\theta_{IVC} \geq \theta \geq -360°$ CA (intake)

$$m = \frac{V(\theta)}{\nu_u} \tag{11.8}$$

for $\theta_{\mathrm{EVO}} \geq \theta \geq \theta_{\mathrm{IVC}}$ (valve closed)

$$m = m_{\mathrm{IVC}} \exp[-C_b(\theta - \theta_{\mathrm{IVC}})/\omega] \tag{11.9}$$

for 360° CA $\geq \theta \geq \theta_{\mathrm{EVO}}$ (blow down and exhaust)

$$m = \frac{V(\theta)}{\nu_b} \tag{11.10}$$

where subscripts b and u denote the burned gas and unburned gas regions, respectively. The cylinder volume is known at any crank angle, with compression ratio r, volume at TDC V_c (clearance volume) and ε = stroke/2 × length of rod [12]:

$$V(\theta) = V_c\left\{1 + \frac{r-1}{2}\left\{1 - \cos\theta + \frac{1}{\varepsilon}\left[1 - \left(1 - \varepsilon^2 \sin^2\theta\right)^{0.5}\right]\right\}\right\} \tag{11.11}$$

Here, the combustion model is the two-zone model that divides the combustion chamber into unburned and burned zone. These zones are distinct by a turbulent flame front that it is solved numerically. Therefore, the combustion parameters such as burned mass fraction ($x = m/m_{\mathrm{b}}$) combined into the model consist of laminar and turbulent flame speed.

The adiabatic flame temperature is the maximum temperature that the products of combustion will reach in the limiting case of no heat loss to the surroundings during the combustion process. The adiabatic flame temperature reaches its maximum value when complete combustion happens with the theoretical value of air. Recalling the definition of enthalpy [13], this can be stated as

$$H_{\mathrm{react}}(T_i, p) = H_{\mathrm{prod}}(T_{\mathrm{ad}}, p)$$

where subscript react denotes reactants and prod denotes products, T_{ad} is the adiabatic flame temperature and T_i is the initial flame temperature.

The laminar flame with gasoline and CNG (methane) fuels, according to Metghalchi and Keck [14], is calculated as follows:

$$u_L = u_{L,0}\left(\frac{T_u}{T_0}\right)^{\alpha}\left(\frac{P}{P_0}\right)^{\beta}\left(1 - 2.0{x_b}^{0.77}\right) \tag{11.12}$$

in which P is the pressure and T_u is the unburned zone temperature. $T_0 = 298\,K$ and $P_0 = 1$(atm) are the reference temperature and pressure; α, β and $u_{L,0}$ are constants and x_b is the mole fraction of the residual gas in the unburned mixture. These constants are defined as follows for gasoline fuels:

$$\alpha = 2.18 - 0.8(\phi - 1)$$

$$\beta = -0.16 + 0.22(\phi - 1)$$

$$u_{L,0} = 0.305 - 0.549(\phi - 1.21)^2$$

The flame speed of the natural gas and air mixture has been calculated using the relations presented by Gu et al. [15]. This relation is

$$u_L = u_{L,0}\left(\frac{T_u}{T_0}\right)^{\gamma}\left(\frac{P_u}{P_0}\right)^{\kappa} \tag{11.13}$$

γ and κ depend on ϕ. They have determined the quantities with a non-significant error (0.014 %) for different quantities as shown below:

$$u_L = \begin{cases} 0.314\left(\frac{T_u}{T_0}\right)^{2.000}\left(\frac{P_u}{P_0}\right)^{-0.438} & \phi = 1.2 \\ 0.36\left(\frac{T_u}{T_0}\right)^{1.162}\left(\frac{P_u}{P_0}\right)^{-.0374} & \phi = 1.0 \\ 0.259\left(\frac{T_u}{T_0}\right)^{2.105}\left(\frac{P_u}{P_0}\right)^{-0.504} & \phi = 0.8 \end{cases} \tag{11.14}$$

There are different methods that may be used for the calculation of the turbulent flame speed. In this chapter, the Damkohler method [16] has been used to calculate the turbulent flame speed:

$$u_t = u' + u_L \tag{11.15}$$

$$u' = 0.75\overline{u_p}\left(1 - 0.5\frac{\theta - 360}{45}\right) \tag{11.16}$$

In the above relations, θ is the crank angle at the end of the compression stroke, which is equal to 360 degrees. In addition, u_p is the engine piston speed.

Burned mass amount during the combustion can be determined using the relation as follows [17]:

$$\frac{\mathrm{d}m_b}{\mathrm{d}\theta} = A_f.\rho_u.u_L.\left(1 + \frac{u_t}{u_L}\right)/6N \tag{11.17}$$

$$A_f = 4\pi R_f^2 \tag{11.18}$$

$$R_f = \left(\frac{3V_b}{4\pi}\right)^{\frac{1}{3}} \tag{11.19}$$

where N is the engine speed in rad/s, ρu is the unburned mass density (g/m^3), Af is the flame front area (m^2) and R_f is the radius of flame (m).

Under the atmospheric air composition assumption (79 %V nitrogen and 21 %V oxygen), and conditioned $\phi < 3$, the species including O, H, OH and NO are important due to dissociation [12]. Therefore, the combustion reaction becomes

$$\begin{aligned} &\varepsilon\phi C_\alpha H_\beta O_y N_\delta + 0.21O_2 + 0.79N_2 \rightarrow \\ &x_1CO_2 + x_2H_2O + x_3N_2 + x_4O_2 + x_5CO \\ &+x_6H_2 + x_7H + x_8O + x_9OH + x_{10}NO \end{aligned} \quad (11.20)$$

x_1 to x_{10} represent the products' mole fractions. Moreover, with two additional mole fractions in the products including N and Ar, which they [18] are made preparation content user specified air quality, and Depcik [19] improved the Olikara and Borman model. In terms of heat loss, heat transfer model is expressed [12]:

$$\frac{dQ}{d\theta} = \frac{-\dot{Q}_{\text{loss}}}{\omega} = \frac{-\dot{Q}_b - \dot{Q}_u}{\omega} \quad (11.21)$$

with

$$\dot{Q}_b = h \sum_{i=h,\,p,\,\ell} A_{bi}(T_b - T_{wi}) \quad (11.22)$$

$$\dot{Q}_u = h \sum_{i=h,\,p,\,\ell} A_{ui}(T_u - T_{wi}) \quad (11.23)$$

where A_{bi} and A_{ui} are the burned and unburned gas areas in heat transfer model in contact at temperature T_{wi} with the combustion chamber component, x is the mass fraction burned and subscripts h, p and ℓ denote the cylinder head, piston crown and linear, respectively. The following relations are [12]:

$$A_{bi} = A_i x^{0.5} \quad (11.24)$$

$$A_{ui} = A_i\left(1 - x^{0.5}\right) \quad (11.25)$$

Here, $A_i = A_h + A_\ell$, and are determined by $A_h = \frac{\pi b^2}{2}$ (hemispherical cylinder head), $A_p = \frac{\pi b^2}{4}$ (flat piston crown) and $A_\ell = \frac{4V(\theta)}{b}$ (linear surface area exposed to gases).

The heat transfer rate is calculated using the following equation from Woschni [20]:

$$\dot{Q} = A_w\left(c.b^{-0.2}p^{0.8}.T^{-0.55}.u^{0.8}\right).(T_w - T) \quad (11.26)$$

In this equation, the speed u is determined from

Table 11.1 Woschni's formula parameters [20]

Gas exchange	$c_1 = 6.18$	$c_2 = 0$
Compression	$c_1 = 2.28$	$c_2 = 0$
Combustion and expansion	$c_1 = 2.28$	$c_2 = 3.24\text{E-}3$

$$u = c_1 u_p + c_2 \frac{V.T_r}{P_r.V_r}(P - P_m) \tag{11.27}$$

where

$$u_p = 2LN \tag{11.28}$$

The parameters P_r, T_r and V_r are evaluated at any reference condition, such as inlet valve closure. In addition, A_w, P_m, L and N are cylinder wall area, motoring pressure, piston stroke and engine speed, respectively. The values for c_1 and c_2 suggested by Woschni are listed in Table 11.1.

By solving the equations of energy conversion for each stage, the pressure and temperature rate changes can be calculated [12]:

$$\frac{\mathrm{d}T_b}{\mathrm{d}\theta} = \frac{-h \sum\limits_{i=h,\,p,\,\ell} A_{ui}(T_b - T_{wi})}{m\omega c_{p_b}} + \frac{v_b}{C_{p_b}} \frac{\partial \ell n v_b}{\partial \ell n T_b} \frac{\mathrm{d}p}{\mathrm{d}\theta} + \frac{h_u - h_b}{x c_{p_b}} \left[\frac{\mathrm{d}x}{\mathrm{d}\theta} - \left(x - x^2\right) \frac{C_b}{\omega} \right] \tag{11.29}$$

$$\frac{\mathrm{d}T_u}{\mathrm{d}\theta} = \frac{-h \sum\limits_{i=h,\,p,\,\ell} A_{ui}(T_u - T_{wi})}{m\omega c_{p_u}(1 - x)} + \frac{v_u}{C_{p_u}} \frac{\partial \ell n v_u}{\partial \ell n T_u} \frac{\mathrm{d}p}{\mathrm{d}\theta} \tag{11.30}$$

$$\frac{\mathrm{d}p}{\mathrm{d}\theta} = \frac{A + B + C}{D + E} \tag{11.31}$$

where

$$A = \frac{1}{m}\left(\frac{\mathrm{d}V}{\mathrm{d}\theta} + \frac{VC_b}{\omega}\right) \tag{11.32}$$

$$B = \frac{h}{m\omega}\left[\frac{v_b}{c_{p_b}} \frac{\partial \ell n v_b}{\partial \ell n T_b} \frac{\sum_{i=h,\,p,\,\ell} A_{bi}(T_b - T_{wi})}{T_b} + \frac{v_u}{c_{p_u}} \frac{\partial \ell n v_u}{\partial \ell n T_u} \frac{\sum_{i=h,\,p,\,\ell} A_{ui}(T_u - T_{wi})}{T_u}\right] \tag{11.33}$$

$$C = -(v_b - v_u)\frac{dx}{d\theta} - v_b\frac{\partial \ell n v_b}{\partial \ell n T_b}\frac{h_u - h_b}{c_{p_b}T_b}\left[\frac{dx}{d\theta} - \left(x - x^2\right)\frac{C_b}{\omega}\right] \tag{11.34}$$

$$D = x\left[\frac{v_b^2}{c_{p_b}T_b}\left(\frac{\partial \ell n v_b}{\partial \ell n T_b}\right)^2 + \frac{v_b}{p}\frac{\partial \ell n v_b}{\partial \ell n p}\right] \tag{11.35}$$

$$E = (1 - x)\left[\frac{v_u^2}{c_{p_u}T_u}\left(\frac{\partial \ell n v_u}{\partial \ell n T_u}\right)^2 + \frac{v_u}{p}\frac{\partial \ell n v_u}{\partial \ell n p}\right] \tag{11.36}$$

Equations (11.29)–(11.36) are functions of θ, p, T_b and T_u and have been solved using a fourth-order Runge-Kutta solver. A detailed solution procedure of the quasi-dimensional combustion model is shown in Fig. 11.1.

Intake and exhaust processes computationally are calculated using an approximation method [21]. In this method, pressure loss is determined during the intake process by the Bernoulli equation for one-dimensional incompressible flow. In addition, intake pressure and temperature, exhaust pressure and temperature and volumetric efficiency are determined as

$$p_i = p_0 - \Delta p_i \tag{11.37}$$

$$T_i = (T_0 + \Delta T + \gamma_r T_e)/(1 + \gamma_r) \tag{11.38}$$

$$p_e = (1.05 \div 1.25)p_0 \tag{11.39}$$

$$T_e = T_b/(p_b/p_e)^{1/3} \tag{11.40}$$

$$\eta_v = \varphi_{\text{ed}}[r/(r-1)](p_i/p_0)[T_0/(T_0 + \Delta T + \gamma_r T_e)] \tag{11.41}$$

p_i, Δp_i, T_i, p_e, T_e, T_b, P_b, f, φ_{ed} and η_v are intake pressure, pressure loss (manifold), intake temperature, exhaust pressure and temperature, burned temperature and pressure, mole fraction, charge-up efficiency and volumetric efficiency, respectively.

Therefore, simulation and modelling of pressure, temperature, work and heat transfer are possible for the bi-fuel four-stroke SI engine running on gasoline and CNG fuels. The solution procedure of the quasi-dimensional combustion model is shown in Fig. 11.1.

The total friction work consists of three major components. These components are pumping work, rubbing friction work and accessory work. Data at WOT for several four-stroke cycle, four-cylinder SI engines, for providing total motored friction mean effective pressure (FMEP), as an engine speed function are adequately correlated by a relation as [22]

$$\text{FMEP(bar)} = 0.97 + 0.15\left(\frac{N}{1{,}000}\right) + 0.05\left(\frac{N}{1{,}000}\right) \tag{11.42}$$

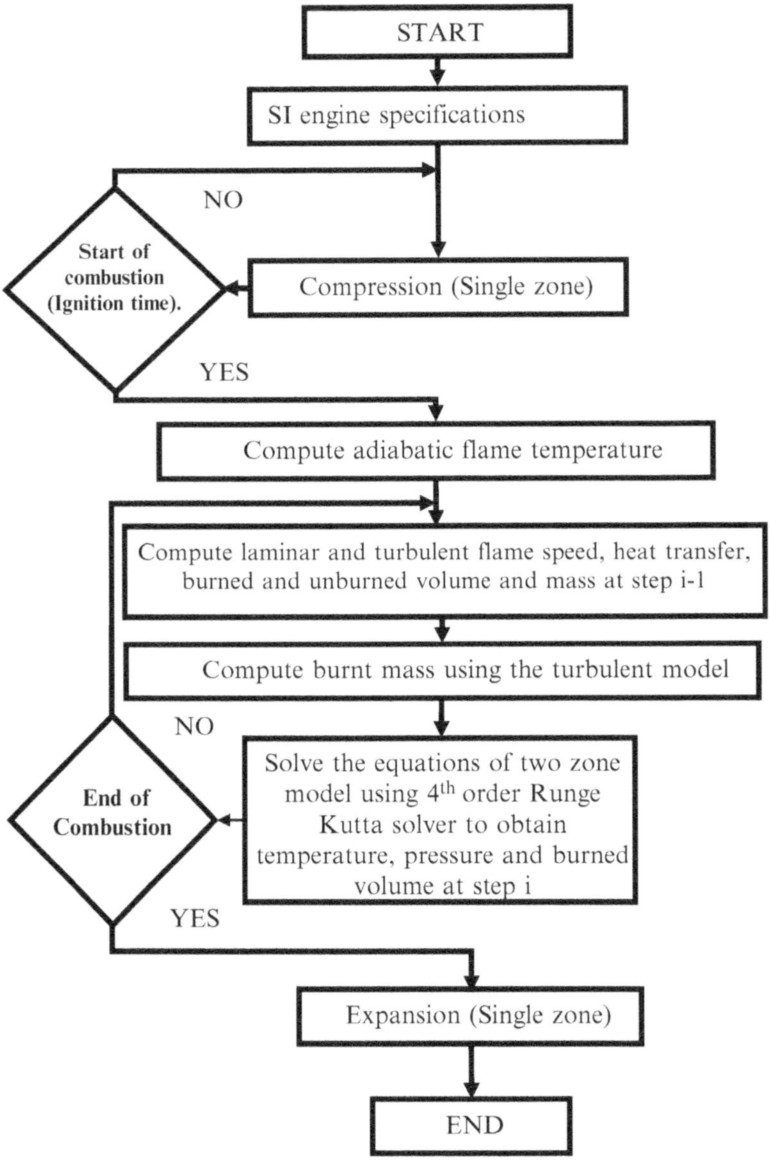

Fig. 11.1 Solution procedure of the quasi-dimensional combustion model

11.3 Model Validation

The model validation is undertaken through experimentation using the engine specified in Table 11.2. The engine is operated over its speed range, 1,500–6,000 r/min, at wide open throttle (WOT). Other experimental hardware include:

- Four-cylinder SI engine
- Eddy current dynamometer, Schenck WS230
- Exhaust gas analyser, Pierburg (HGA 400) 5 Gas Analyser
- Air fuel ratio device, Lambda Sensor Horiba (Mexa-700)
- CNG mass flow meters, Fisher Rosemount
- Gasoline mass flow metres, Maxmeter 214–410
- Cylinder Pressure device, Indimeter 619
- Data acquisition system, Ricardo
- Water and oil temperature control, Engine master software
- CNG storage

The engine and dynamometer specifications are listed in Tables 11.2 and 11.3. In addition, the layout of the test rig shows in Fig. 11.2. The test engine was a bi-fuel (gasoline and CNG) engine and prepared with an appropriate bi-fuelling system. Sensor applied for data gathering include an angle encoder, lambda, air mass flow metre, intake manifold, oil and fuel temperature and pressure, exhaust manifold, outlet water and oil thermocouples and intake manifold and oil pressure gauges. Data were gathered contemporaneously from the sensors and transfer to a data acquisition system. In addition, data for brake torque, brake power and exhaust gases was recorded, which included concentration of No_x, CO, CO_2, total-unburned hydrocarbon (THC) and O_2 in this study. In this model, CNG and gasoline have

Table 11.2 The engine specifications [23]

Engine type	Four-stroke, bi-fuel spark ignition
Fuel system	MPFI
Induction	Naturally aspirated
Number of cylinder	Four cylinder—In line
Bore (mm)	83
Stroke (mm)	81.4
Connecting rod (mm)	150.2
Displacement volume (cm^3)	1,761
Compression ratio	9.25
Maximum power	68.65 kW @ 6,000 rpm
Maximum torque	143 Nm @ 2,500 rpm
Inlet valve opening (IVO)	32° bTDC
Inlet valve closing (IVC)	64° aBDC
Exhaust valve opening (EVO)	59° bTDC
Exhaust valve closing (EVC)	17° aBDC

Table 11.3 Schenck dynamometer specifications

Dynamometer type	Schenck WS230
Maximum torque	750 (Nm)
Maximum speed	10,000 (rpm)
Maximum power	230 (kW)
Torsional spring	593 (Nm/rad × 1,000)
Weight	485 (kg)
Inertia	0.53 (kg/m^2)

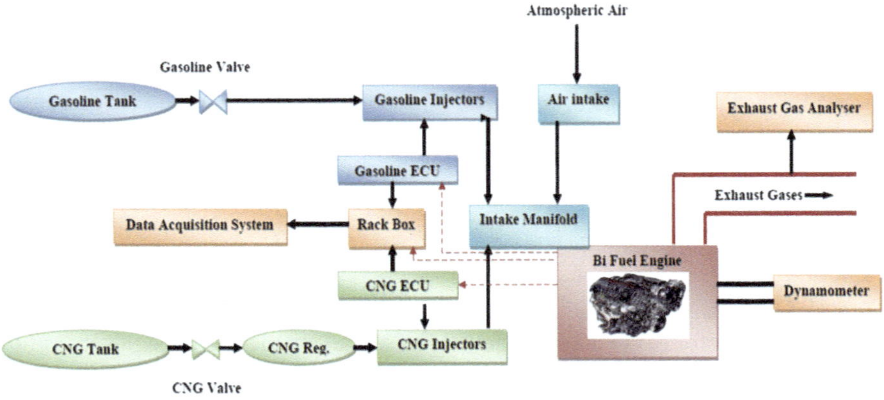

Fig. 11.2 A layout of test rig

been considered with composition of CH_4 and C_7H_{14} based on the properties and compositions of CNG and gasoline that are used in the tests [23], respectively.

For model validation, the experimental results are now compared. In running the model, the composition of CNG is taken as methane, CH_4 and gasoline, C_7H_{14} in accordance with the literature [23]. Model and experimental results such as brake power (BP), brake-specific CO (BSCO) and brake-specific NO_x ($BSNO_x$) are compared in Figs. 11.3, 11.4, 11.5, 11.6, 11.7 and 11.8. The results show good correspondence (with an average 8 % mean error). Therefore, the results support the fact that the model is valid for prediction of performance and emissions of the bi-fuel engine through the range tested.

11.4 Engine Thermodynamic Characteristics, Performance and Emissions

The validated model can be used to predict cylinder pressure, work done, heat transfer, brake thermal and volumetric efficiency, brake power (BP), brake mean effective pressure (BMEP), brake-specific fuel consumption (BSFC), equivalence

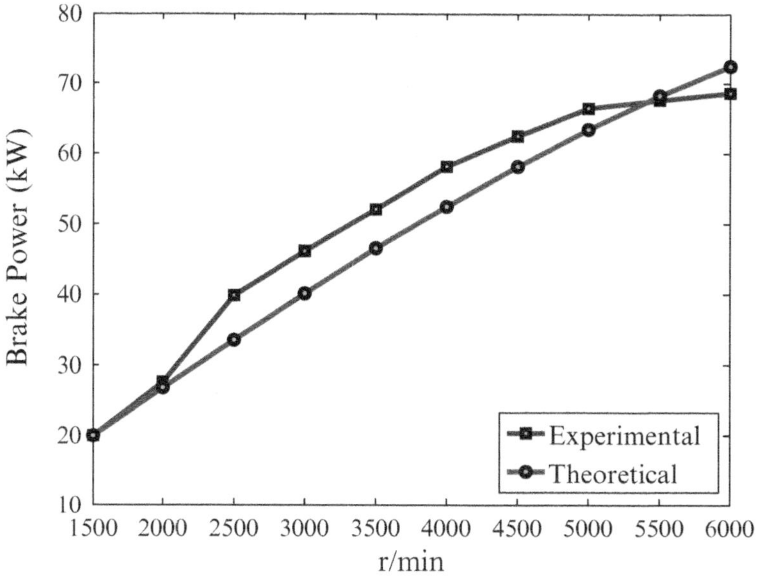

Fig. 11.3 A comparison of brake power results (gasoline)

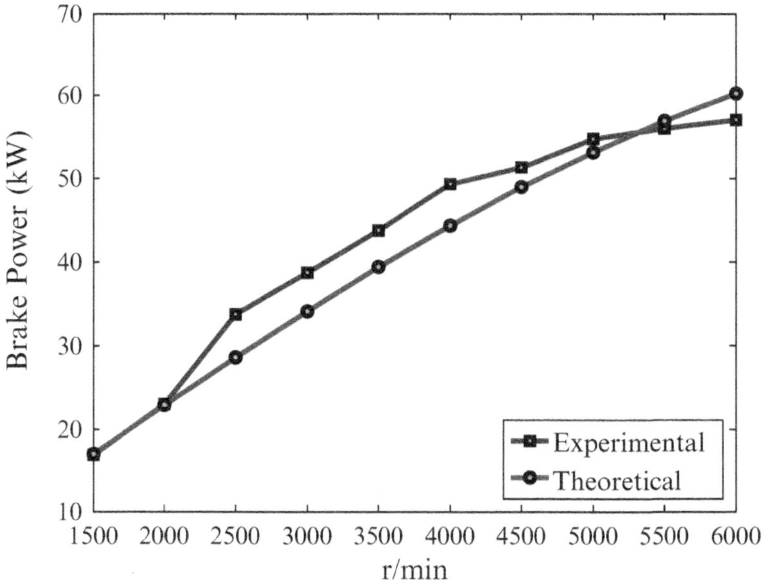

Fig. 11.4 A comparison of brake power results (CNG)

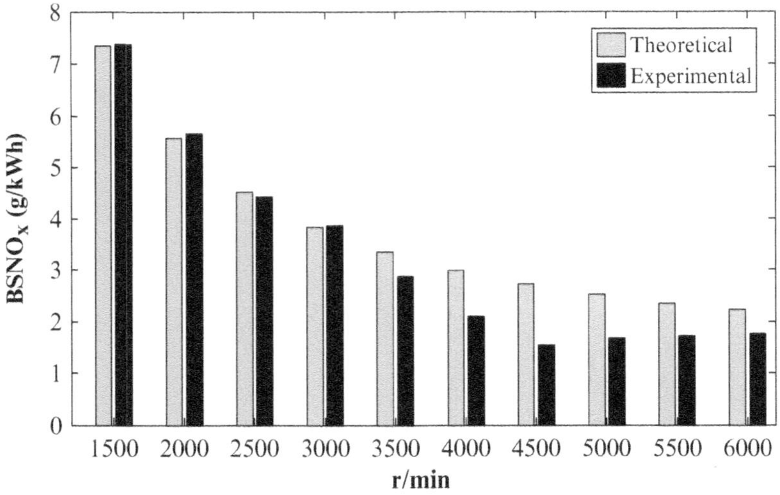

Fig. 11.5 Comparison of BSNO$_x$ results (gasoline)

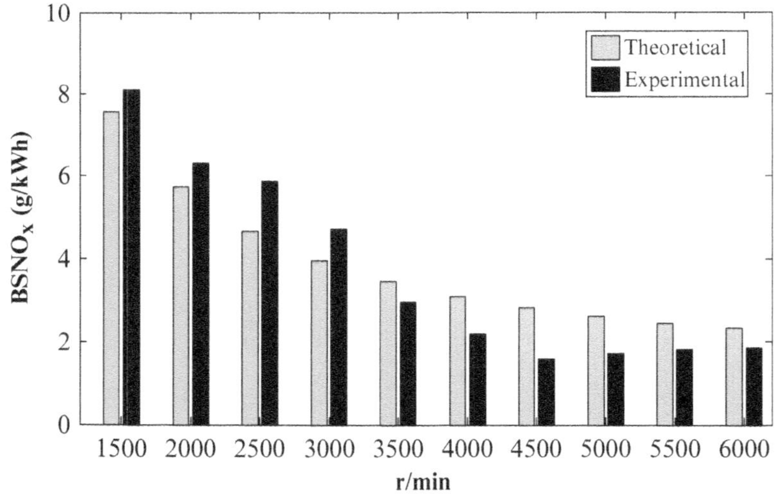

Fig. 11.6 Comparison of BSNO$_x$ results (CNG)

ratio and BSNO$_x$, BSCO and CO_2 concentration in exhaust gases. The engine performance and emissions for both fuels are now analysed and discussed.

In Figs. 11.9 and 11.10, cylinder pressure, work done for gasoline and CNG fuels as calculated by the validated model are shown. In these predictions $N = 3{,}000$ rpm and a spark timing of 25° bTDC is assumed. It is clear that cylinder pressure and work done for CNG engines are less than gasoline. In addition, the engine

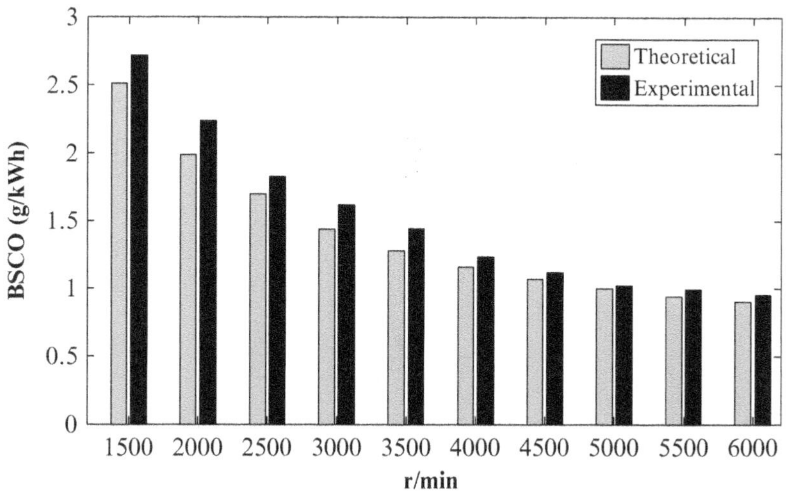

Fig. 11.7 Comparison of BSCO results (gasoline)

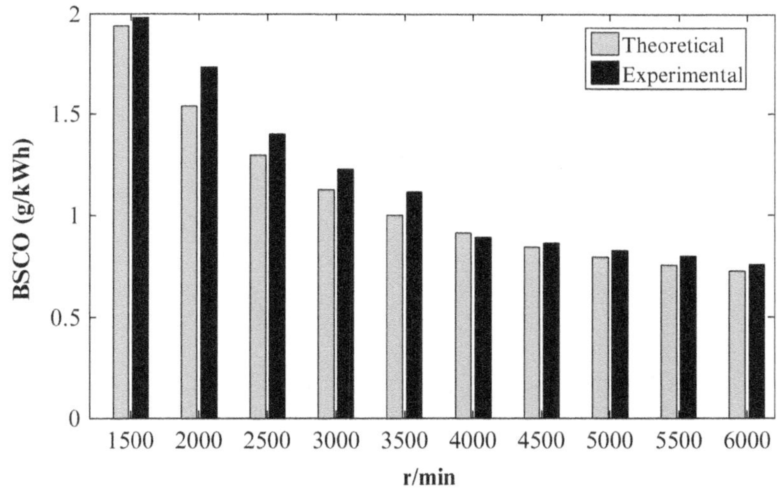

Fig. 11.8 Comparison of BSCO results (CNG)

performance in a specific engine has a high dependency on the physical condition inside the cylinder mixture.

The power produced in a specific engine has a high dependency on the physical condition of the cylinder mixture. Therefore, the volumetric efficiency performs one of the significant roles among the other engine parameters.

In Fig. 11.11, the calculated volumetric efficiency of the engine is shown at an engine speed for the gasoline and CNG fuels. Generally, the volumetric efficiency of a CNG engine is less (c.11 %) than gasoline.

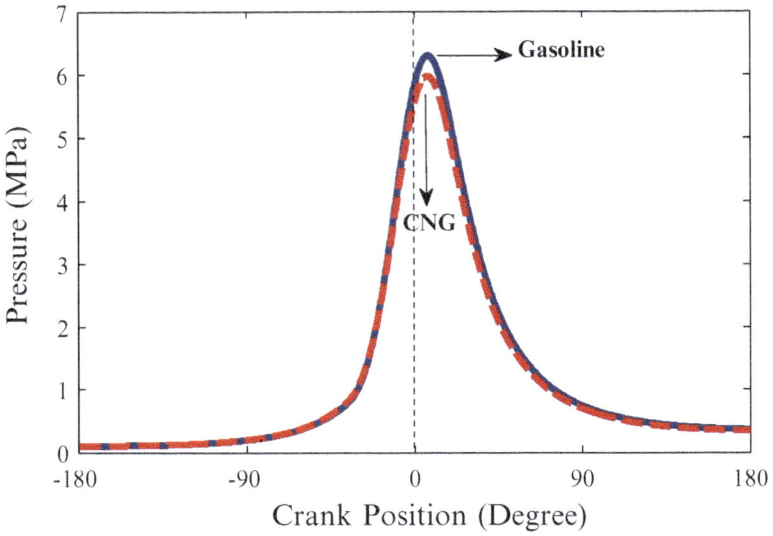

Fig. 11.9 Comparison of cylinder pressure for gasoline and CNG fuels in various crank positions

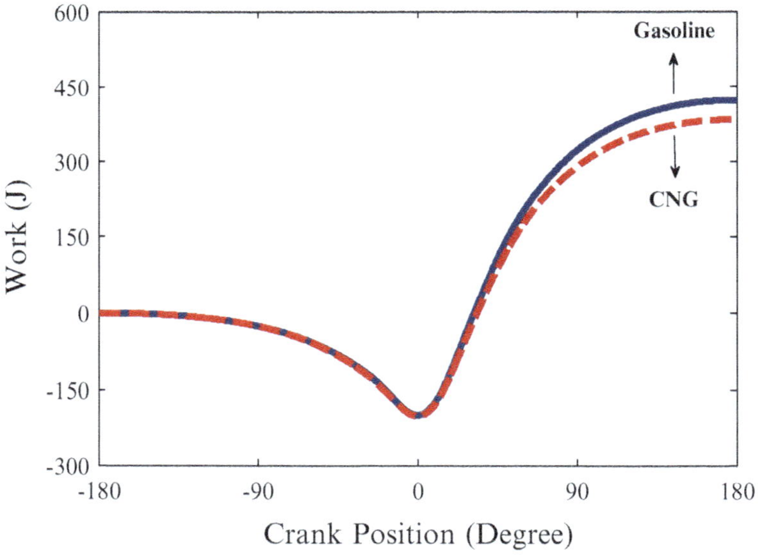

Fig. 11.10 Comparison of work done for gasoline and CNG fuels in various crank positions

This reduction in volumetric efficiency is due to two main reasons: Firstly, the vaporisation of gasoline produces a cooling effect on the intake charge. Therefore, the density of the charge is increased and the volumetric efficiency increases. Whereas with CNG, as it is already in the gaseous form at ambient vehicle

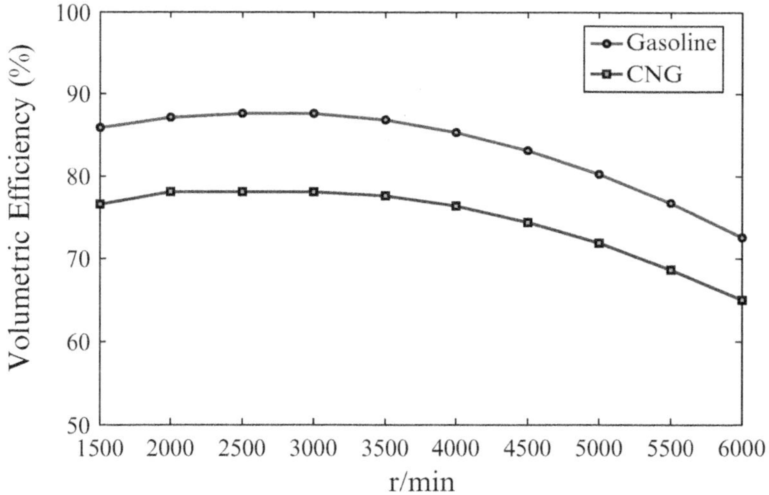

Fig. 11.11 Comparison of volumetric efficiency for gasoline and CNG fuels in various engine speeds

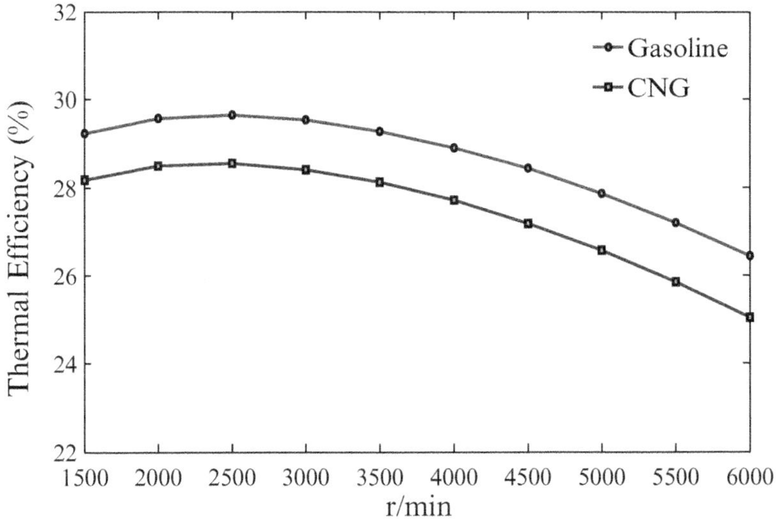

Fig. 11.12 Comparison of thermal efficiency for gasoline and CNG fuels in various engine speeds

temperatures cooling will not take place. Secondly, CNG fuel occupies a large volume in the inlet mixture; this displaces the oxygen available for combustion. These are the main reasons for a decrease in volumetric efficiency when the engine is CNG fuelled. Figure 11.12 shows that the brake thermal efficiency of a CNG engine is less (c. 4.5 %) than a gasoline-fuelled engine; hence for the CNG engine

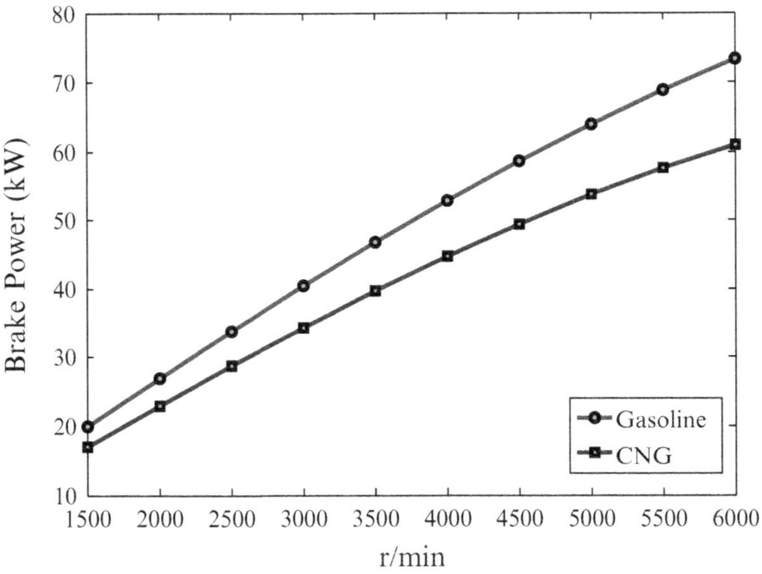

Fig. 11.13 Comparison of brake power for gasoline and CNG fuels in various engine speeds

the work produced is less even though the heating value of CNG fuel is greater than gasoline.

In Fig. 11.13, the comparative brake power (BP) of fuels is observed. As can be seen CNG produces less power (c. 15.5 %) when compared with gasoline. The reason is due to the lower volumetric efficiency of the engine when fuelled with natural gas. It should be noted that this engine has been designed for use with gasoline and not CNG. If the engine had been designed for CNG, it would have had a better performance. In order to alleviate this problem, it is possible to use turbo charging and redesign the intake manifold. Additionally the compression ratio of the engine may be increased because natural gas has a higher octane number compared with gasoline, and thus the knock limit is raised.

In Fig. 11.14, the predicted BMEP of CNG and gasoline fuels is compared. For naturally aspirated engines, the maximum BMEP is normally between 850 and 1,050 kPa [4]. As can be seen from the figure the engine BMEP when fuelled with CNG is less than gasoline by a maximum of 17 %. This reduction is due to two main reasons. Firstly, the flame speed of CNG is less than gasoline [4, 24, 25] for the same spark advance. The part of BMEP reduction happens with CNG operation that is due to longer ignition delay and lower flame speed of CNG. Therefore, the combustion should be started earlier with respect to top dead centre (TDC), and there is greater negative work done on the piston before TDC compared to gasoline. In addition, the remainder of the BMEP reduction is due to the displacement of air by CNG fuelled when the engine is gasoline base designed. Secondly, the

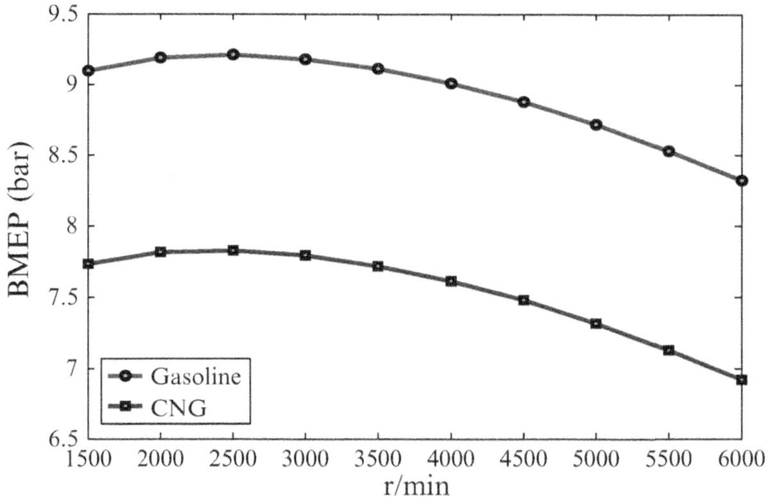

Fig. 11.14 Comparison of BMEP for gasoline and CNG fuels in various engine speeds

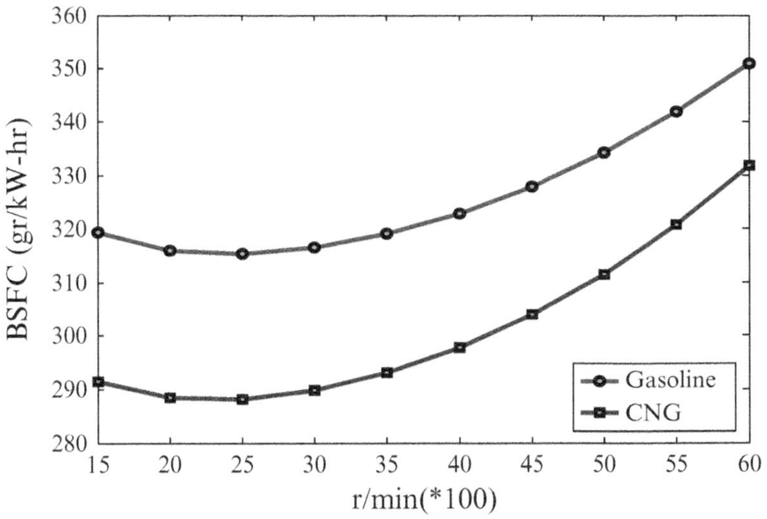

Fig. 11.15 Comparison of BSFC for gasoline and CNG fuels in various engine speeds

volumetric efficiency that plays one of the most important roles in CNG engine is less than gasoline. For these reasons, the BMEP of CNG engine is less than gasoline.

The BSFC for the fuels under study is compared in Fig. 11.15. It is obvious that the BSFC for CNG engine is less than (c. 9 %) gasoline. The main reason is the greater natural gas heating value compared to gasoline.

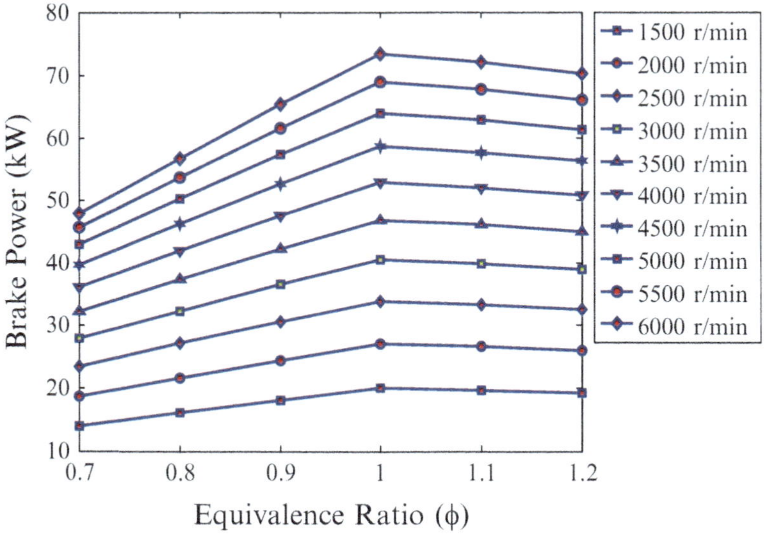

Fig. 11.16 Comparison of BP variations in various equivalence ratios (gasoline)

The equivalence ratio (ϕ) has an important effect on engine performance and emissions. Fig. 11.16 shows this effect, and it shows that brake power changes from 14 to 72 kW over the range of ϕ and speeds of the engine. Also, ϕ has a significant effect on the rate of NO_x emissions. The point of maximum NO_x emission occurs for all engine speeds at near $\phi = 0.8$; leaning or enriching the mixture from this point decreases NO_x emission rate (Fig. 11.17). However, the model predicts that CO emission is low when the mixture is lean ($\phi < 0.8$), and after $\phi > 0.8$, CO emission increases (Fig. 11.18).

Figure 11.19 shows $BSNO_x$ emissions for both CNG and gasoline fuels. It is clear that more $BSNO_x$ is created by CNG fuel than gasoline. The reactions that lead to the NO_x formation take place mainly at high temperatures. As mentioned earlier, the effect of cooling at the time of evaporation does not occur for CNG. Consequently, the initial temperature of CNG air/fuel charge at the start of combustion will be greater than gasoline.

This will lead to the increase of the maximum temperature in cylinder and finally produce more NO_x. On the other hands, with regard to the fact that the flame speed of CNG is less than gasoline, there will be a need to have a greater spark advance as compared to gasoline. The greater spark advance will increase the maximum temperature and pressure inside the cylinder. Three-way catalytic converters are used in vehicle emission control system and can be used to treat NO_x reduction specifically with the CNG operation ($0.91 < \phi < 0.95$). In addition, natural gas contains very little sulphur oxide rate (10 PPM) and for this reason has the lowest destructive effect to catalytic converters compared with gasoline [2].

In Figs. 11.20 and 11.21, the concentration of CO_2 and BSCO in exhaust gases may be observed. The amount of CO_2 in hydrocarbon combustion is proportional to

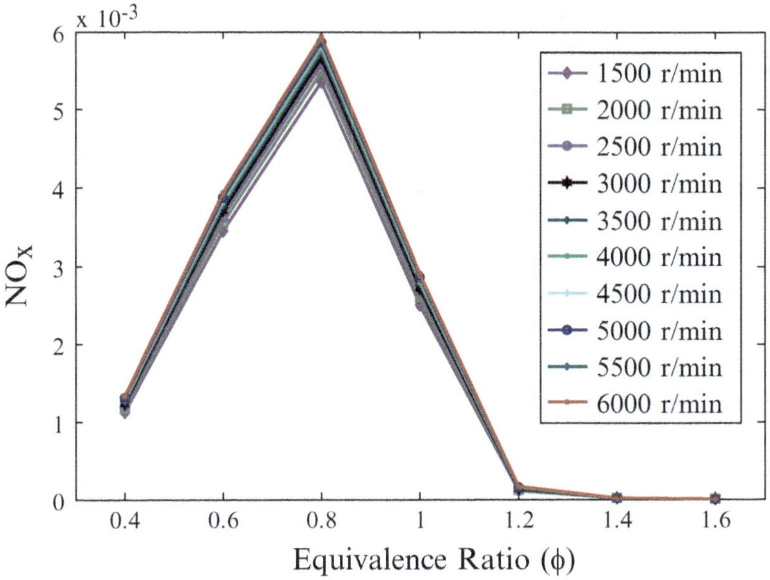

Fig. 11.17 Comparison of NO_x mole fraction in various equivalence ratios (gasoline)

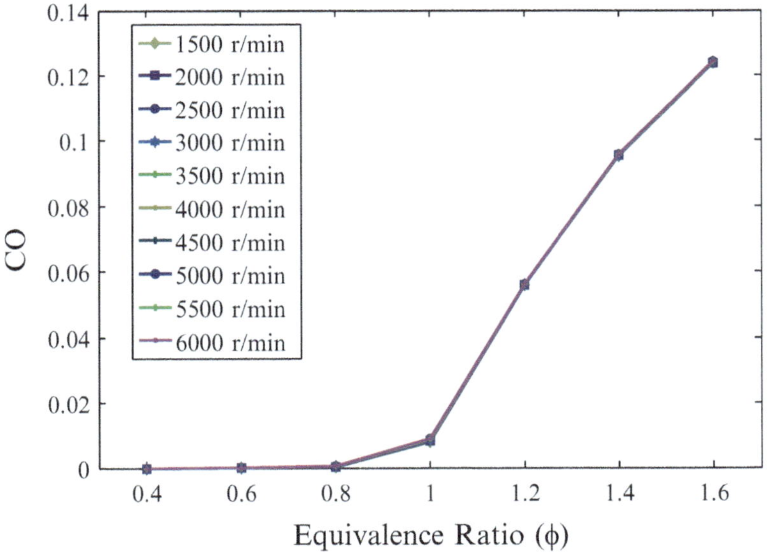

Fig. 11.18 Comparison of CO mole fraction in various equivalence ratios (gasoline)

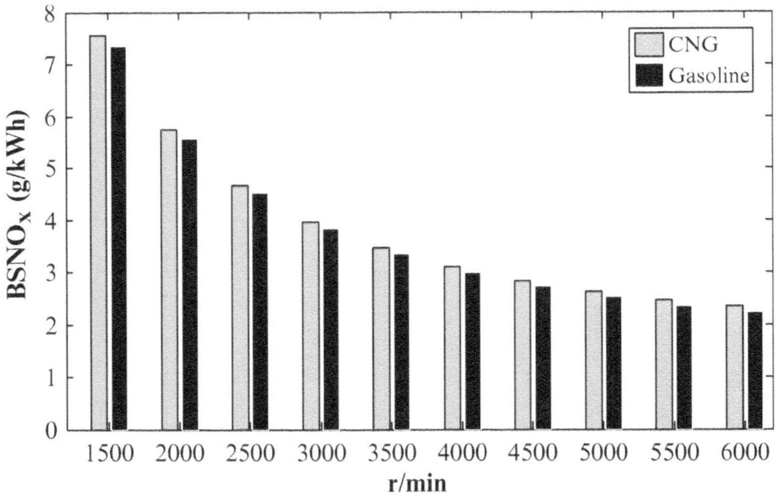

Fig. 11.19 Comparison of $BSNO_x$ for gasoline and CNG fuels in various engine speeds

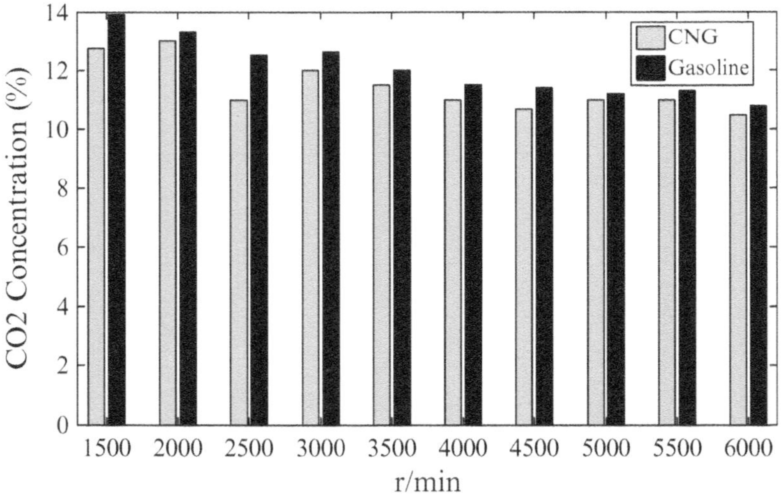

Fig. 11.20 Comparison of CO_2 concentration for gasoline and CNG fuels in various engine speeds

the carbon-to-hydrogen ratio. The main component of natural gas is methane, which has the lowest carbon-to-hydrogen ratio (C/H ratio) compared to other hydrocarbons. Therefore, the CO_2 produced in CNG combustion is less than gasoline (Fig. 11.20).

The corresponding of CO_2, C/H ratio of fuel has affecting the production of CO, for this main reason CO produced in CNG combustion less than gasoline

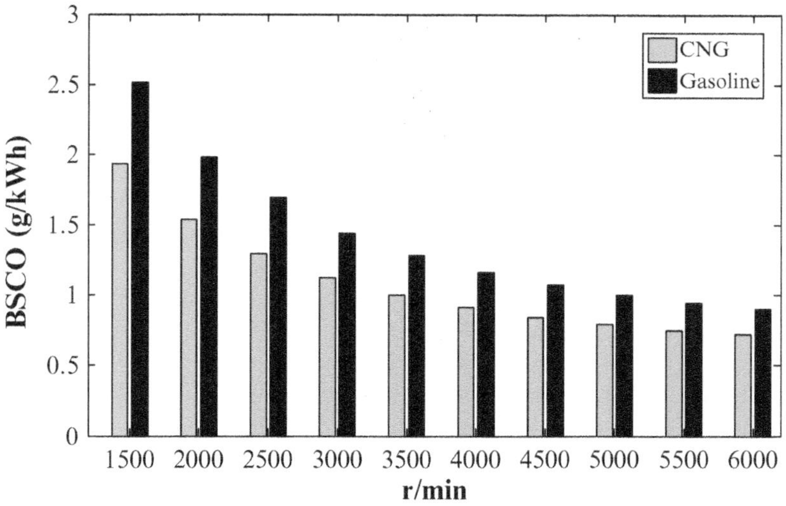

Fig. 11.21 Comparison of BSCO for gasoline and CNG fuels in various engine speeds

(Fig. 11.21). In addition, flame quenching at the walls of the cylinder and the wall oil film deposits are additional sources of CO.

Finally, as a significant result, this case study has shown that average rate reductions of CO_2 and CO for CNG engine compared to gasoline are about 29 g/km and 8 g/km [2, 23, 25], respectively. With the assumption of mean travel through the distance of each vehicle about 30,000 km (that almost are taxi), annual rate reduction of CO_2 and CO for each CNG engine will be about 860 kg/year and 240 kg/year, respectively, compared to gasoline engine.

11.5 Conclusions

A quasi-dimensional thermodynamic model of bi-fuel (CNG and gasoline) spark ignition engine was developed. It was able to simulate turbulent combustion and compared to CFD it is computationally faster and efficient. The results of the model were compared to experimental data and the validity of the model was confirmed. Therefore, this model was capable of prediction, analysis and useful for optimisation of the engine performance parameters.

In this chapter CNG was presented as an alternative fuel; it has advantages and disadvantages compared to gasoline when the engine is gasoline base designed (midterm approach).

Natural gas has smaller C/H ratio in comparison to gasoline and for this main reason it produces lower amounts of CO_2 and CO. These emissions reduction are significant annually when the vehicles specifically used such as taxi in the big city.

CNG fuel decreases volumetric efficiency, increases temperature of combustion, and finally produces more $BSNO_x$ when compared to gasoline. However, three-way catalytic converter is a part of vehicle emission control system and can treat NO_x reduction specifically with the CNG operation ($0.91 < \phi < 0.95$). Moreover, natural gas in this study contains very little sulphur oxide and for this reason has a lower destructive effect upon catalytic converters as compared to gasoline. In addition, it is cheaper than gasoline and therefore it is economic fuel. The BSFC of an engine fuelled with CNG is less than gasoline fuelled and the main reason is the greater heating value of natural gas compared to gasoline.

The volumetric efficiency plays the most important role between the other engine parameters; that is, the decreasing of volumetric efficiency in CNG will decrease the BMEP and finally decrease the work done. Therefore, the thermal efficiency of a CNG-fuelled engine is less than gasoline. Using CNG will decrease brake power (BP) in gasoline base engine designed. In order to remove this problem it is possible to use turbo charger, redesign intake manifold and increase the compression ratio.

In order to obtain an engine with less pollution, better performance and the result of this chapter, engines should be designed specifically for each type of fuel. Therefore, in the bi-fuel engine, the optimality of the performance parameters should be sacrificed.

References

1. Cho HM, He BQ (2007) Spark ignition natural gas engines-a review. J Energy Conversion Manage 48-608-618
2. Rezapour K, Ebrahimi KM (2007) The necessity of using CNG fuel and gasification of vehicle in Iran. Proceeding of 3rd international energy, exergy and environment symposium (IEEES3)
3. Lapetz J, McCarthny D, Greenfield N, Czapski R, Geftos T, Rosson J (1996) Development of the Ford QVM CNG Bi-Fuel 4.9L F-Series Pickup Truck. Society of Automotive Engineering. Paper No. 960850
4. Duan SY (1996) Using natural gas in engines: laboratory experience with the use of natural gas fuel in IC engine. IMechE Seminar Publication 39–46
5. Zuo C, Zhao KA (1997) Quasi-dimensional model of si natural gas engines with pre-chamber. Society of Automotive Engineers. Paper No. 972994
6. Evan RL, Blaszczyk JA (1997) Comparative Study of The Performance and Exhaust Emissions of a Spark Ignition Engine Fuelled by Natural Gas and Gasoline. Proc. Inst. Mech. Engrs., D00295
7. Sun X, Lutz A, Vermiglio E, Arold M, Wiedmann T (1998) The development of the GM 2.2 CNG Bi-fuel passenger cars. Society of Automotive Engineers. Paper No.982445
8. Valpato O, Theunissen F, Mazara R (2005) Engine management for multi fuel plus compressed natural gas vehicles. Society of Automotive Engineers. Paper Series 2005-01-3777
9. Alsam MU, Masjuki HH, Kalam MA, Abdessalam H, Mahlia TMI, Amalina MA (2006) An experimental investigation of CNG as an alternative fuel for a retrofitted gasoline vehicle. J Fuel Res Fuel 85:717
10. Gordon S, McBride BJ (1971) Computer program for calculation of complex chemical equilibrium composition, rocket performance, incident and reflected shocks, and Chapman Jouguet detonations. NASA publication, SP-273

11. JANAF Thermo-Chemical Tables (1971) United States National Bureau of Standard Publication, NSRDS-NBS 37
12. Ferguson CR (1968) Internal combustion engines, applied thermo-sciences. John Wiley and Sons, New York
13. Cengel YA (2005) Thermodynamic an engineering approach. Fifth Edition, McGraw Hill, New York
14. Methghalchi M, Keck JC (1980) Laminar burning velocity of propane air mixture at high temperature and pressure. Combustion Flame 38:143–154
15. Gu XJ, Haq MZ, Lawes M, Woolley R (2000) Laminar burning velocity and Markstein lengths of methane-air mixtures. Combustion and Flame, 121 (pp. 41–58)
16. Verhelst S (2005) A study of the combustion in hydrogen-fuelled internal combustion engines. PhD thesis, Ghent University, Gent, Belgium
17. Boulouchos K, Steiner T, Dimopoulos P (1994) Investigation of flame speed models for flame growth during premixed engine combustion. Society of Automotive Engineers. Paper No. 940476
18. Olikara C, Borman GL (1975) Calculating properties of equilibrium combustion products with some application to I.C. engines. Society of Automotive Engineers. Paper No. 750468
19. Depcik C (2000) Open-ended thermodynamic cycle simulation. M.S. Thesis, University of Michigan, Ann Arbor
20. Woschni G (1967) A universally applicable equation for the instantaneous heat transfer coefficient in the internal combustion engine. Society of Automotive Engineers. Paper No.670931
21. Bayraktar H, Durgun O (2003) Mathematical modelling of spark ignition engine cycles. Energy Sources 25:651–666
22. Ferguson CR, Kirkpatrick AT (2001) Internal combustion engines, applied thermo science, 2nd edn. John Wiley & Sons Inc, New York
23. Rezapour K, Ebrahimi MK, Wood AS, Nikranjbar A (2010) Simulation and modelling Bi fuel engine for the improving of performance. Society of Automotive Engineers, doi: 10.4271/2010-01-2034
24. Heywood JB (1998) Internal combustion engines fundamentals. McGraw Hill, New York
25. Rezapour K (2011) Exergy based si engine model optimisation. PhD thesis, University of Bradford, The United Kingdom

Chapter 12
Gasification of Biomass for Hydrogen and Power Production: Efficiency and Environmental Assessment

Rami Salah El-Emam, Ibrahim Dincer, and Salah H. El-Emam

Abstract This study focuses on efficiency and environmental impact assessments of steam biomass gasification and gasification-solid oxide fuel cell (SOFC) integrated system for power and hydrogen production. Biomass is considered as the main energy source of the system, and it is changed into gaseous components through gasification process where steam is used as a drying and gasification medium. The gasifier is proposed to be integrated with SOFC unit for power production after extracting part of the produced syngas for hydrogen production. Thermodynamic model is introduced to analyze the gasification process and the electrochemical process through the SOFC. The effect of steam biomass ratio on the system performance is investigated. Environmental assessment is performed based on the carbon dioxide produced from the system with respect to the generated useful products.

Keywords Biomass • Gasification • Fluidized bed • SOFC • Hydrogen • Efficiency

R.S. El-Emam (✉)
Clean Energy Research Laboratory (CERL), Faculty of Engineering and Applied Science, University of Ontario Institute of Technology, 2000 Simcoe Street North, Oshawa, ON, Canada L1H 7K4

Faculty of Engineering, Mansoura University, Mansoura, Egypt
e-mail: rami.elemam@uoit.ca

I. Dincer
Clean Energy Research Laboratory (CERL), Faculty of Engineering and Applied Science, University of Ontario Institute of Technology, 2000 Simcoe Street North, Oshawa, ON, Canada L1H 7K4
e-mail: Ibrahim.Dincer@uoit.ca

S.H. El-Emam
Faculty of Engineering, Mansoura University, Mansoura, Egypt
e-mail: selemam45@yahoo.com

I. Dincer et al. (eds.), *Progress in Clean Energy, Volume 2*,
DOI 10.1007/978-3-319-17031-2_12

12.1 Introduction

One of the most promising alternatives for combined heat and power (CHP) cogeneration is integrating gasification and solid oxide fuel cell (SOFC) technologies. This alternative becomes more forthcoming and appealing with using the indigenous biomass fuel which makes it more environmentally benign. Biomass gasification is also considered a promising and sustainable source of hydrogen-rich synthetic gas used for hydrogen production [1].

Different studies in the literature investigated the biomass gasification. Schuster et al. [2] performed an extensive parametric study on a dual-steam gasifier for CHP. They studied the effect of the gasification temperature, fluidization agent, and water content on the performance of the system. Cohce et al. [3] also performed an efficiency assessment for biomass gasification process for hydrogen production. They presented a simplified model for energy and exergy analyses considering chemical equilibrium. SOFC, among the different types of fuel cells, is the most suitable type to be integrated with biomass gasifiers. High operating temperature of SOFC, relative insensitivity tolerance to fuel contaminants, and high-energy conversion efficiency make SOFC the suitable candidate to be integrated with biomass gasification technology and being fed with the produced syngas for clean energy production. Athanasiou et al. [4] performed energy analysis and optimization of integrated gasification and SOFC system. Wongchanapai et al. [5] investigated the performance of a small-scale integrated biomass gasification and SOFC system. They studied the effect of different gasifier and fuel cell operating parameters on the system performance.

In this chapter, a performance assessment of biomass gasification process and integrated biomass gasification and direct internal reforming SOFC are investigated based on energy and exergy analyses. Steam is used as a drying and gasification medium in this study. Parametric studies are undertaken to investigate the effect of changing the operating parameters on the performance of the gasifier, fuel cell, and final output of the system. An environmental impact assessment of the proposed system is carried out parametrically.

12.2 Integrated System Description

The schematic diagram depicted in Fig. 12.1 shows the integrated biomass gasification-SOFC system. In the proposed system, superheated steam is used as a drying medium for the fed biomass. The dry biomass leaves the dryer at 10–30 % of moisture content. The gasification process in this study is proposed to be steam gasification. Steam is proposed to be supplied by the HRSG to the dryer and the gasifier at the same pressure and temperature. The syngas produced from the gasification process is cleaned and then cooled to the SOFC inlet temperature through a heat exchanger for heating the air fed to the SOFC cathode. Part of the syngas is directed to the hydrogen production section after it is cooled down, heating up the SOFC air to its operating temperature. The SOFC proposed in this

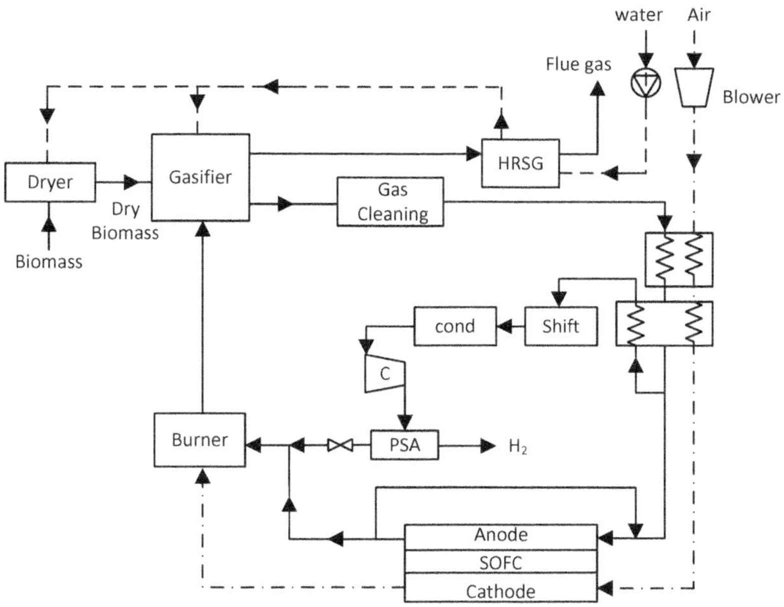

Fig. 12.1 Schematic of the proposed integrated system

system is modeled as DIR-SOFC. The percentage of syngas directed to the fuel cell is controlled by the electric power required. Complete combustion is considered in the burner and adiabatic flame temperature is estimated and adapted for the combustion process.

12.3 System Analysis

In this section, energy, exergy, chemical equilibrium, kinetics of gasification, and SOFC analyses, for the main sections of the system in Fig. 12.1, are illustrated. Energy and concentration equations of the gasifier are also introduced to find the axial distribution of the gasifier temperature. Mass, energy, and exergy balances are applied to each component of the proposed system:

$$\dot{Q}_{cv} - \dot{W}_{cv} = \sum\nolimits_{\text{in}} \dot{m}_i h_i - \sum\nolimits_{\text{out}} \dot{m}_i h_i$$

$$0 = \sum \left(1 - \frac{T_o}{T}\right)\dot{Q}_{cv} - \dot{W}_{cv} + \sum_{\text{in}} \dot{m}_i \text{ex}_i - \sum_{\text{out}} \dot{m}_i \text{ex}_i - \dot{E}x_{d_i}$$

The exergy efficiency in this study is defined for the different system components, through the analyses, as the exergy associated with the useful products of the system with respect to the exergy associated with the fuel input stream:

$$\eta_{\text{sys}}^{\text{ex}} = \frac{\dot{E}x_{\text{p}}}{\dot{E}x_{\text{f}}}$$

For the integrated system, the exergy of the useful outputs is calculated considering the following: exergy of the produced hydrogen, net work out of the system considering system auxiliary power, and exergy of extra heat of the exhaust gas from the burner that passes through the heat recovery steam generator. The extra heat is considered to produce process steam at the same gasification steam condition.

12.3.1 Modeling of Gasification Process

In this study, steam is considered as a drying medium for the biomass fuel before it is fed to the gasifier. Biomass leaves the dryer at 10–30 % of moisture content. In this section, the subscripts db, b, m, s and p refer to dry biomass, wet biomass, moisture, steam, and product, respectively. Below are energy and exergy balance equations applied on the modeled dryer; the energy balance equation, considering the heat carried by the drying steam, can be written as follows:

$$\dot{m}_{db,\text{in}}h_{db,\text{in}} + \dot{m}_{m,\text{in}}h_{m,\text{in}} + \dot{m}_{s,\text{in}}h_{s,\text{in}} = \dot{m}_{db,\text{out}}h_{db,\text{out}} + \dot{m}_{m,\text{out}}h_{m,\text{out}} + \dot{m}_{s,\text{out}}h_{s,\text{out}}$$

And the exergy balance equation can be written as follows:

$$\begin{aligned}\dot{m}_{db,\text{in}}ex_{db,\text{in}} + \dot{m}_{m,\text{in}}ex_{m,\text{in}} + \dot{m}_{s,\text{in}}ex_{s,\text{in}} &= \dot{m}_{db,\text{out}}ex_{db,\text{out}} + \dot{m}_{m,\text{out}}ex_{m,\text{out}} \\ &\quad + \dot{m}_{s,\text{out}}ex_{s,\text{out}} + \dot{E}x_{d,\text{dryer}}\end{aligned}$$

Steady-state equilibrium gasification model is considered for predicting the rich hydrogen product gas from the steam gasification process. The general gasification reaction for a general biomass fuel can be written as follows:

$$\begin{aligned}&C_aH_bO_cN_d + SC \cdot H_2O + ER \cdot (O_2 + 3.76N_2) \\ &\quad \rightarrow \alpha_{CO}CO + \alpha_{CO_2}CO_2 + \alpha_{H_2}H_2 + \alpha_{CH_4}CH_4 + \alpha_{H_2O}H_2O + \alpha_{N_2}N_2 + \alpha_C C\end{aligned}$$

Steam is controlled by the steam-to-carbon SC ratio. This ratio can be described in the form of steam-to-biomass ratio, S/B, as it will be referred to in this study. Air would be considered as a provider of the heat required for the gasification process and it is controlled by the air equivalence ratio, ER.

Three gasification equilibrium reactions are to be considered for carbon conversion into gaseous components: the endothermic Boudouard reaction that requires high temperature, the endothermic water-gas reaction, and the exothermic hydrogasification reaction [6]:

$$CO_2 + C \leftrightarrow 2CO$$
$$C + H_2O \leftrightarrow CO + H_2$$
$$C + 2H_2 \leftrightarrow CH_4$$

The following two equilibrium reactions, water-gas shift reaction and methane formation, are coupled if the gasification occurs at a temperature where zero solid carbon is left unconverted at the gasification process [7]:

$$CO + H_2O \leftrightarrow CO_2 + H_2$$
$$CO + 3H_2 \leftrightarrow CH_4 + H_2O$$

The energy balance equation for the gasifier can be written in the molar form as follows:

$$\dot{n}_{b,\mathrm{in}}\left(\overline{h}_b^f + \overline{h}_b - \overline{h}_b^o\right) + \dot{n}_{s,\mathrm{in}}\left(\overline{h}_s^f + \overline{h}_s - \overline{h}_s^o\right) + Q_g = \sum_1^n \dot{n}_i\left(\overline{h}_i^f + \overline{h}_i - \overline{h}_i^o\right)_p$$

The exergy destruction in the gasifier can be calculated from the exergy balance equation as follows:

$$\dot{E}x_{b,\mathrm{in}} + \dot{E}x_{s,\mathrm{in}} = \dot{E}x_p + \dot{E}x_{d,\mathrm{gasifier}}$$

The gasification energy efficiency is calculated here as follows:

$$\eta_{\mathrm{G}}^{\mathrm{en}} = \frac{\mathrm{LHV}_p \cdot \dot{m}_p}{\mathrm{LHV}_b \cdot \dot{m}_b + \dot{Q}_{st}}$$

The exergy efficiency of the gasification process is determined considering the total exergy of the biomass fuel and the calculated total exergy of the produced gas, considering chemical and physical exergy values:

$$\eta_{\mathrm{G}}^{\mathrm{ex}} = \frac{\dot{E}x_p}{\dot{E}x_b + \dot{E}x_{st}}$$

Also the gasification process was assessed by defining the energy and exergy efficiencies of hydrogen production considering the hydrogen yield from the gasification process:

$$\eta_{\mathrm{H_2}}^{\mathrm{en}} = \frac{\mathrm{LHV}_{\mathrm{H_2}} \cdot \dot{m}_{\mathrm{H_2}}}{\mathrm{LHV}_b \cdot \dot{m}_b + \dot{Q}_{st}}$$
$$\eta_{\mathrm{H_2}}^{\mathrm{ex}} = \frac{\dot{E}x_{\mathrm{H_2}}}{\dot{E}x_b + \dot{E}x_{st}}$$

The chemical and physical exergy terms are considered in calculating the total exergy of the produced hydrogen and the input exergy flows into the system. Considering an atmospheric bubbling fluidized bed, the temperature distribution through the bed height is modeled through the gasification process. The model is designed considering three different zones: bed zone, which is considered to be isothermal zone, splashing zone, and freeboard zone [8]. The model used the following temperature distribution equation to predict the temperature profile over the bed height [8–10], neglecting the effect of radiation heat transfer:

$$\frac{dT_g}{dz} = \frac{1}{N_g \cdot c_{p,g}} \left[A_{cs} \left(\sum R_i \, \mathrm{HR}_i \right) - \left(\frac{6\, M_s\, A_{cs} h_s \left(T_g - T_s \right)}{d_s \cdot \rho_s} \right) - Q_w \right]$$

where $\frac{dT_g}{dz}$ is the temperature distribution of the gases through the gasifier height, and N_g and $c_{p,g}$ are the total number of moles and the specific heat of the gases at every section through the gasifier height. The first term in the right-hand side of the equation represents the heat of reaction which is calculated based on reaction rates, R (kmol/m^3 s), and heat of reaction HR (J/kmol) for the gas species at the corresponding height. A_{cs} is the cross-sectional area of the bed. The second term represents the local heat transfer rate between ejected particles and the gas species (J/m s), and the third term is the local heat loss through the gasifier wall. Ejected sand particles are assumed to be of uniform temperature, T_s. For sand particles, d_s, ρ_s and h_s are the diameter, density, and heat transfer coefficient through the used sand particles. M_s in the previous equation is defined as the mass flux of ejected sand (kg/m^2 s) and is determined as follows:

$$M_s = \beta_M \rho_s (U_f - U_{f\text{-}m})(1 - \varepsilon_{f,m}) \exp[-C_M (z - z_{eb})]$$

Here, β_M and C_M are constants; for the current model their values are taken as listed in Table 12.2, as adapted from the literature [10, 11], U_f is the fluidization velocity, and $U_{f.m}$ is the minimum fluidization velocities. $\varepsilon_{f,m}$ is the void fraction at minimum fluidization. All hydrodynamics parameters are calculated from the correlations and equations presented in different references in the literature [10–15]. z_{eb} is the extended bed height. The upper limit of splashing zone is considered to be bounded by ejected sand flux of 0.001. However, there is no distinct limit considered between the splashing and freeboard zones [10].

The model is designed to analyze the axial temperature and gas species' concentrations through the fluidized bed height for combustion, and steam or air/steam gasification of biomass fuels. The gas species' concentrations are modeled over the dense bed zone considering the bubbling and emulsion phases [8]. Table 12.1 shows the specification of the considered biomass fuel for the current study and Table 12.2 shows the main parameters that are used as a base case in the system analyses.

12.3.2 SOFC Modeling

A direct internal reforming SOFC is considered for the proposed system. The steam reforming and shifting reactions along with the electrochemical reaction occur simultaneously at the cell [16]:

Table 12.1 Specification of the biomass fuel

Beech chip specifications [2]	
Ultimate analysis wt.%	
Carbon	48.26 %
Hydrogen	5.82 %
Oxygen	45.67 %
Nitrogen	0.22 %
Sulfur	0.03 %
Lower heating value	12,430 kJ/kg

Table 12.2 Gasification operating parameters

Gasifier input data	[Base case]
Gasification temperature	800 °C
Temperature of steam	350 °C
Pressure of the gasifier	1 bar
Water content of fed biomass	55 %
Water content at the gasifier	20 %
Steam-to-biomass ratio, S/B	0.7
Fluidization velocity	1 m/s
Extended bed height	70 cm
Bed material	Silica sand
Constant in sand mass flux, β_M	0.1
Constant in sand mass flux, C_M	18

$$CH_4 + H_2O \leftrightarrow CO + 3H_2$$

For the shifting reaction:

$$CO + H_2O \leftrightarrow CO_2 + H_2$$

The net of the fuel cell is given by

$$H_2 + \frac{1}{2}O_2 \rightarrow H_2O$$

The applied SOFC model is illustrated in previous work [16, 17]. The equilibrium constants for the two governing reactions of the SOFC can be calculated as function of the molar concentrations as follows:

$$K_r = \frac{P_{H_2}^3 P_{CO}}{P_{CH_4} P_{H_2O}} = \frac{y_{eq,\ H_2}^3 \cdot y_{eq,\ CO}}{y_{eq,\ CH_4} \cdot y_{eq,\ H_2O}} x \left(\frac{P}{P_o}\right)^2$$

$$K_s = \frac{P_{H_2} P_{CO_2}}{P_{CO} P_{H_2O}} = \frac{y_{eq,\ H_2} \cdot y_{eq,\ CO_2}}{y_{eq,CO} \cdot y_{eq,\ H_2O}}.$$

where P_i represents the partial pressure values for the respective component and the equilibrium constants can be represented in the form of y_i which is the molar fraction as shown in the same equations. The value of K is to be calculated from the following correlation:

$$\log K = aT^4 + bT^3 + cT^2 + dT + e$$

The exit gas composition can be determined by solving the chemical equilibrium constant equations with the equations of gas mixture composition equilibrium and the equation relating electric current and inlet hydrogen flow rate which is function of recirculation ratio and fuel utilization ratio. The value of the constants in the previous equation, the equilibrium gas mixture composition equations, and the electric current-inlet flow rate relation are illustrated previously in El-Emam et al. [17]. The SOFC power output is calculated as follows:

$$W_{el} = i \cdot V \cdot A \cdot N_{\text{stack}} N_{\text{element}}$$

where the cell voltage, V, is to be calculated by subtracting the different polarizations affecting the cell from the Nernst open-circuit voltage defined as

$$E = \frac{\Delta G^o}{2F} - \frac{R \cdot T}{2 \cdot F} \ln\left(\frac{P_{H_2O}}{P_{H_2} \cdot P_{O_2}{}^{1/2}}\right)$$

The ohmic polarization in this study is represented as a function of the dimensions and the specific electrical resistance of the different cell components as follows:

$$V_{\text{ohm}} = i \cdot \sum \rho_k L_k$$

The values of L_k for the different components are listed in Table 12.3. The values of the specific electrical resistance, ρ_k, are calculated as described in Wongchanapai et al. [18].

The activation polarization is to be calculated based on the following formula which is derived from reforming the Butler-Volmer equation:

$$V_{\text{act}} = \frac{R \cdot T}{F} \cdot \sinh^{-1}\left(\frac{i}{2 \cdot i_o^{\text{a}}}\right) + \frac{R \cdot T}{F} \cdot \sinh^{-1}\left(\frac{i}{2 \cdot i_o^{\text{c}}}\right)$$

where i_o^{a} and i_o^{c} are the exchange current densities for the anode and cathode, respectively. They can be calculated from the semiempirical correlations below [19]:

$$i_o^{\text{a}} = \gamma_{\text{a}} \cdot \exp\left(-\frac{E_{\text{act}}^{\text{a}}}{R \cdot T}\right) \cdot \left(\frac{P_{H_2}}{P_o}\right) \left(\frac{P_{H_2O}}{P_o}\right)$$

$$i_o^{\text{c}} = \gamma_{\text{c}} \cdot \exp\left(-\frac{E_{\text{act}}^{\text{c}}}{R \cdot T}\right) \cdot \left(\frac{P_{O_2}}{P_o}\right)^{1/4}$$

Table 12.3 SOFC model input data

Anode-supported SOFC parameters	[Base case]
Desired output	500 kW
Fuel cell operating temperature	750 °C
Fuel cell operating pressure	1 bar
Electrolyte thickness, L_e	0.04 mm [16]
Anode thickness, L_a	0.75 mm [16]
Cathode thickness, L_c	0.05 mm [16]
Operating current density	0.4 A/cm^2
Anode effective gas diffusion factor, D^a_{eff}	3.5×10^{-5} m^2/s [23]
Cathode effective gas diffusion factor, D^c_{eff}	7.3×10^{-6} m^2/s [23]
Anode pre-exponential factor, γ_a	5.7×10^7 A/m^2 [23]
Cathode pre-exponential factor, γ_c	7.0×10^9 A/m^2 [23]
Anode activation energy, E^a_{act}	140 kJ/mol [18]
Cathode activation energy, E^c_{act}	137 kJ/mol [18]
Fuel utilization	83 %
Recirculation ratio	20 %

The pre-exponential factors γ_a and γ_c and the values of the activation energies E^a_{act} and E^c_{act} are given in Table 12.3. The concentration polarization which is caused by the mass transport limitations of the reactants and products is calculated as a function of the limiting current density at the anode and cathode:

$$V_{con} = -\frac{R \cdot T}{2 \cdot F} \ln\left(1 - \frac{i}{i_s^a}\right) + \frac{R \cdot T}{2 \cdot F} \ln\left(1 + \frac{P^{ex}_{H_2} \cdot i}{P^{ex}_{H_2O} \cdot i_s^a}\right) - \frac{R \cdot T}{4 \cdot F} \ln\left(1 - \frac{i}{i_s^c}\right)$$

The limiting current densities i_s^a and i_s^c for the anode and cathode are functions of temperature, hydrogen and oxygen partial pressure at the cell exit conditions, effective gas diffusion factors, and thicknesses of anode and cathode. The values of these parameters considered for the current model are given in Table 12.3.

12.3.3 Greenhouse Gas Emissions

In this study, the environmental assessment is performed considering the amount of carbon dioxide produced per unit energy of the useful product. In steam gasification process for hydrogen production, the energy of the produced hydrogen is considered. The carbon dioxide emissions are also calculated for the integrated system considering the energy content of the system useful output.

12.4 Results and Discussion

In this section, the results of thermodynamics performance assessment and environmental impact evaluation of the steam gasification integrated system are presented. The effect of the operating conditions of gasification on the performance of the proposed system and carbon dioxide emissions are investigated.

The gasifier energy analysis for axial temperature distribution is validated with combustion measurements of rice straw in fluidized bed combustor with height of 3 m and bed temperature of 850 °C with fluidization velocity of 0.5 m/s and 1.2 excess air. The results showed good agreement with the developed code. Figure 12.2 shows the results of the energy analysis through the bed height. Steam gasification is considered in this case; however, it is compared with the case of combustion of the biomass fuel at the same bed temperature. Steam gasification process is performed at the base case condition and the combustion is modeled at 1.2 excess air. As shown in the figure, the temperature profile exhibits an overheating of 41 °C above the bed temperature compared with 140 °C for the combustion case. The increase in the temperature starts at the splashing zone where most of the heat gets recovered by the ejected sand particles. The temperature increases much more through the freeboard zone and the effect of the ejected particles on the gas temperature becomes less. This overheating phenomenon is reported in the literature [20–22].

The gasification performance is investigated and parametric studies are performed to show the effect of the steam biomass ratio on the hydrogen yeild. It is measured in terms of energy and exergy efficiencies of the produced hydrogen as defined in the analysis section. The hydrogen efficiency is found to decrease with the increase in the gasification temperature; efficiency increases with the steam-to-biomass ratio until it reaches a maximum at the steam biomass ratio of 2.2 with

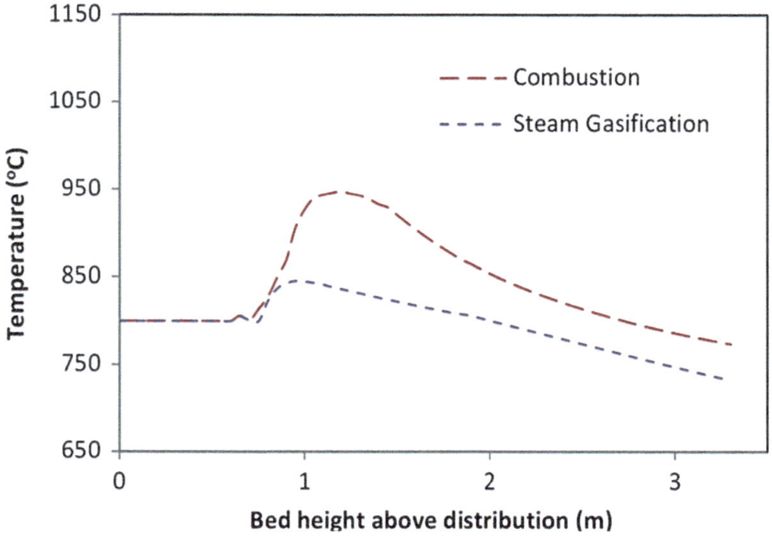

Fig. 12.2 Axial temperature profile through the gasifier

maximum efficiency of 56.46 %, 53.50 %, and 51.48 % at 750 °C, 800 °C, and 850 °C, respectively. Exergy efficiency of hydrogen from gasification has same trend of energy efficiency. However, the maximum values of the exergy efficiency occur at lower values of steam-to-biomass ratios at 0.670 S/B, 0.677 S/B, and 0.69 S/B ratios for 750 °C, 800 °C, and 850 °C, with efficiency values of 36.76 %, 35.92 %, and 35.12 %, respectively.

The results shown in Figs. 12.3 and 12.4 show the effect of steam-to-biomass ratio on molar percentage of gas species produced from the gasification process.

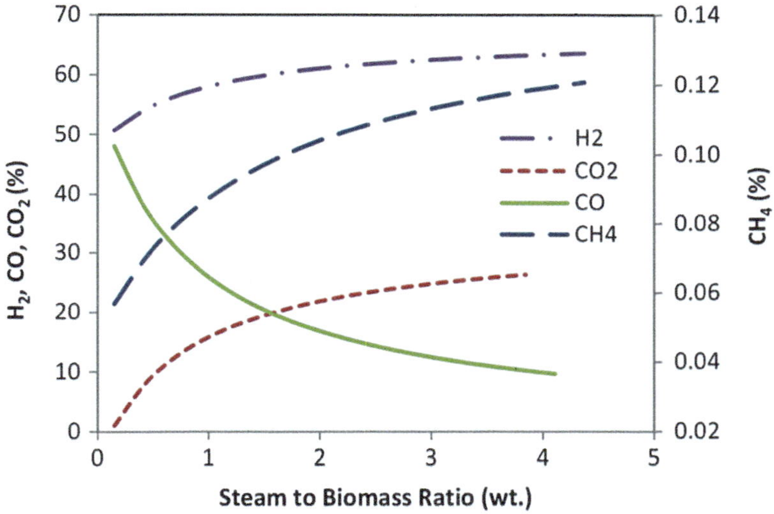

Fig. 12.3 Effect of S/B ratio on product gas species concentrations (dry basis)

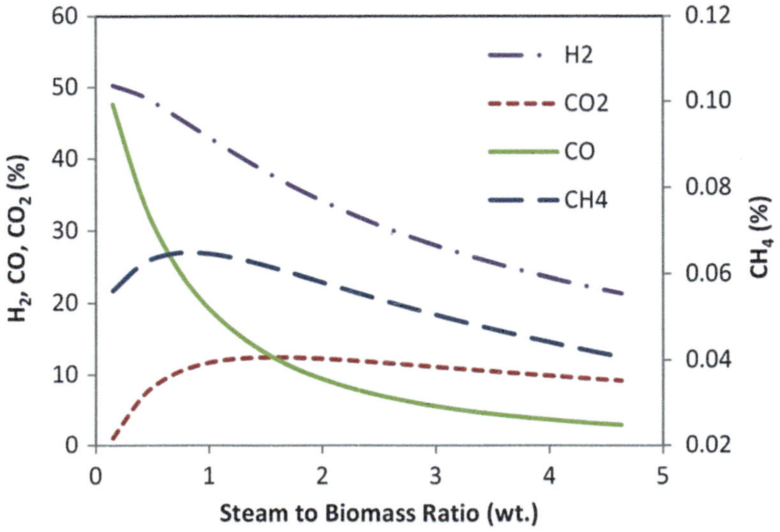

Fig. 12.4 Effect of S/B ratio on product gas species concentrations, wet basis

Table 12.4 Integrated system performance under different S/B ratios

S/B	0.7	1.2	2
Gasification product syngas composition % vol dry			
H_2	56.86	59.15	61.26
CO_2	13.47	18.16	22.24
CO	29.58	22.58	16.39
CH_4	0.083	0.093	0.105
Syngas thermal value, kJ/kmol			
LHV_{syngas}	176,408	141,157	106,928
Efficiency, %			
$\eta_{H_2}^{en}$	47.10	50.09	51.5
$\eta_{H_2}^{ex}$	38.54	37.16	35.02
η_{Gasif}^{en}	75.97	72.72	67.93
η_{Gasif}^{ex}	58.36	52.20	45.11
η_{SOFC}^{en}	46.90	47.15	47.85
η_{SOFC}^{ex}	36.73	37.45	39.12
η_{Sys}^{en}	54.74	55.31	55.92
η_{Sys}^{ex}	45.21	45.92	47.42

In Fig. 12.3, the values are represented with respect to dry condition, and Fig. 12.4 represents the species concentrations considering wet gas basis. It is noticed that the increase of steam–biomass ratio shifts the product gas equilibrium towards the production of hydrogen. However, this results in reduction in the lower heating value of the produced gas.

The SOFC model presented in this study is a direct internal reforming cell. The model is validated with the results reported by Colpan et al. [28]. The results with respect to the integrated system are concluded in studying the system efficiency as shown in Table 12.4. The effect of changing the steam-to-biomass ratio on the system performance and the heating value of the produced gas are shown.

Figure 12.5 shows the effect of changing the gasification temperature over the integrated system energy and exergy efficiency values. The energy efficiency increases from 55.49 to 58.01 % with changing the gasification temperature from 700 to 800 °C. Exergy efficiency also increases from 45.64 to 46.89 % over the same temperature range. This is considering the less amount of extra heat utilized in producing process steam where most of the heat of the combustion gases is directed to heat the gasifier. The produced process steam is calculated as 356 kg/h, 230.4 kg/h, and 165.5 kg/h of steam at 350 °C and 3 bar, at 700 °C, 750 °C, and 800 °C of gasification temperature, respectively.

The two cases of considering steam gasification for hydrogen production and the integrated system are compared on the bases on energy and exergy efficiency values as shown in Fig. 12.6. Greenhouse gas emissions are also shown represented as carbon dioxide emissions per unit of produced useful energy. The base cases are considered for the shown results. Carbon dioxide emissions for steam gasification process are calculated as 180 t of carbon dioxide per TJ of produced hydrogen energy. The results show also the energy and exergy efficiency values as 47.1 and 38.53 %. For the integrated system, the exergy and energy efficiencies increased

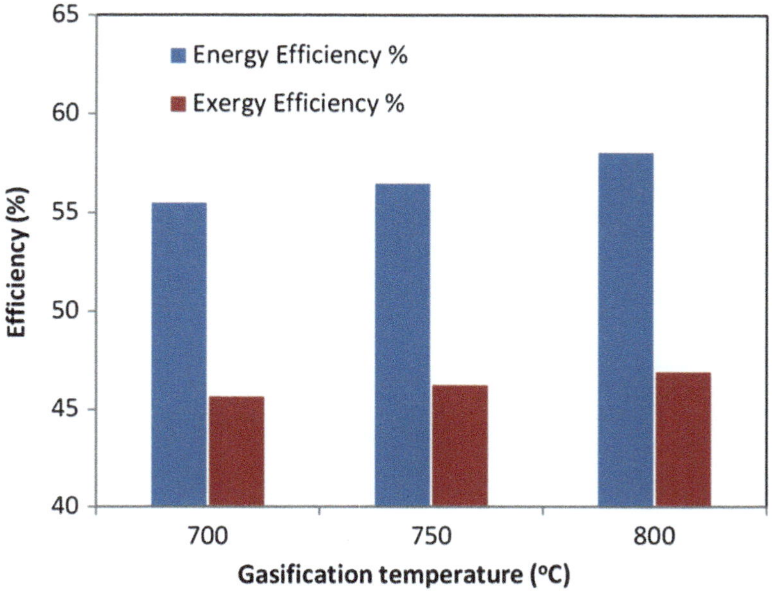

Fig. 12.5 System performance assessment for the integrated system at different gasification temperatures

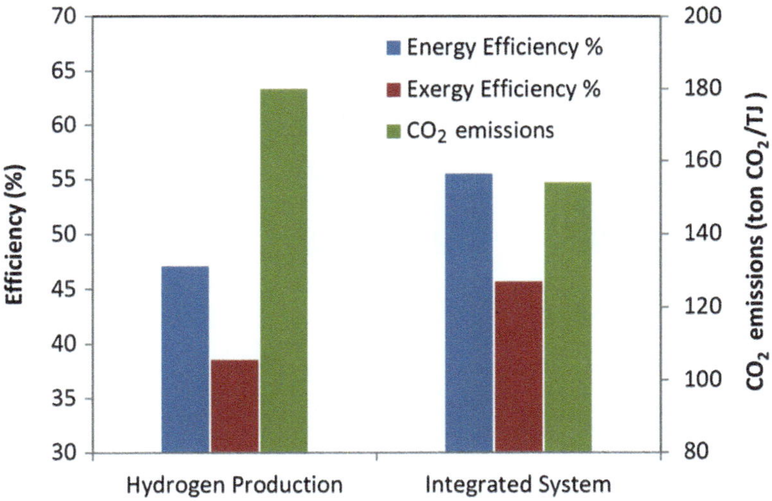

Fig. 12.6 Efficiency and greenhouse gas emissions for hydrogen production and integrated system options

compared with steam gasification without integrating the fuel cell. This is because of the produced power and the extra heat utilized to produce the process steam. However, the integrated system shows an increase of 8.39 % and 7.1 % in the energy and exergy efficiencies, and also lower carbon dioxide emission is achieved.

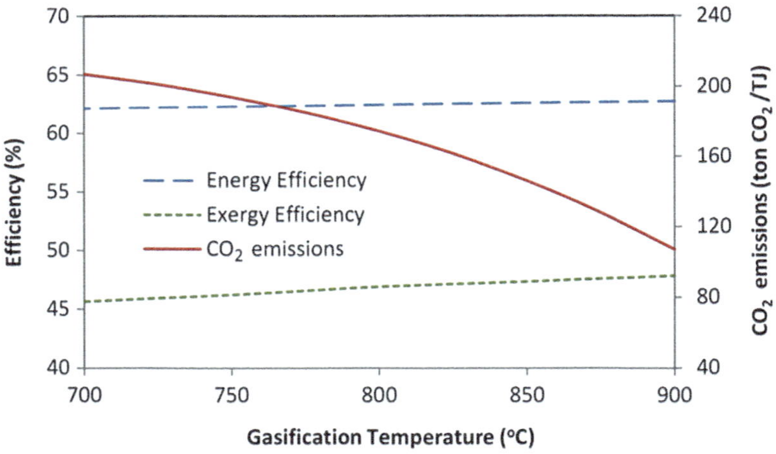

Fig. 12.7 Integrated system performance at different gasification temperature values

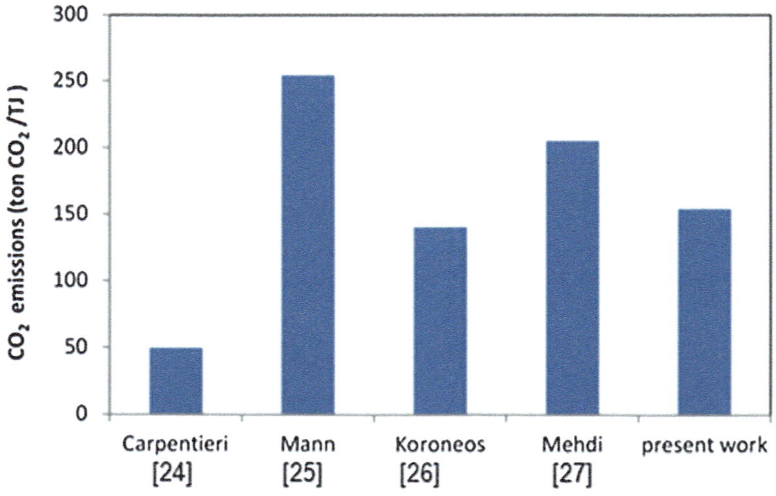

Fig. 12.8 Comparison of CO_2 emissions of the present system with the results reported in the literature

The results in Fig. 12.7 show the effect of varying the gasification temperature on the overall performance of the integrated system. The increase of gasification temperature causes an enhancement in the system energy and exergy efficiencies, which is accompanied with reduction in the carbon dioxide emissions per unit of energy produced. Figure 12.8 shows a comparison of the carbon dioxide emissions of the integrated system proposed in the current study with results reported in the literature. The following studies are considered for the comparison: Carpentieri et al. [24]

studied integrated biomass gasification system with carbon dioxide removal and the emissions are calculated as 49.4 t CO_2 per TJ of produced energy for this system. Mann and Spath [25] reported a higher value of emissions, 254 t CO_2 per TJ for the conventional integrated biomass gasification combined cycle. Koroneos et al. [26] investigated biomass gasification plant for hydrogen production and reported 140 t CO_2 per unit energy produced. The results of Hosseini et al. [27] show emission of 205 t CO_2/TJ of produced energy for integrated biomass gasification, SOFC, and gas turbine system.

12.5 Conclusions

In this chapter, an efficiency evaluation of steam gasification for hydrogen production is performed considering beech chips as biomass fuel. The gasifier is modeled considering bubbling fluidized bed gasifier and axial energy analysis is performed, and the axial temperature distribution is presented. Direct internal reforming SOFC model is introduced and integrated with the proposed gasifier model to investigate the efficiency of the integrated system. Its efficiency is investigated at different operating parameters. The results presented show the following:

- Steam-to-biomass ratio has a great effect on gasification performance and the produced gas species concentrations.
- Optimum steam-to-biomass ratio for maximum hydrogen exergy efficiency for the base case is 0.677.
- The effect of gasification temperature on the hydrogen efficiency is significant where higher energy and exergy efficiency values are achieved at lower gasification temperature values.
- Integration of biomass gasification and SOFC system is one of the promising alternatives in integrated systems. The efficiency of the integrated system is studied under different operating steam-to-biomass ratios considering power and hydrogen production.
- The greenhouse gas emissions decrease with the increase of the system exergy efficiency.

References

1. Toonssen R, Woudstra N, Verkooijen AHM (2008) Exergy analysis of hydrogen production plants based on biomass gasification. Int J Hydrog Energy 33(15):4074–4082
2. Schuster G et al (2001) Biomass steam gasification—an extensive parametric modeling study. Bioresour Technol 77:71–79
3. Cohce MK, Rosen MA, Dincer I (2011) Efficiency evaluation of a biomass gasification-based hydrogen production. Int J Hydrog Energy 36(17):11388–11398
4. Athanasiou C et al (2007) From biomass to electricity through integrated gasification/SOFC system-optimization and energy balance. Int J Hydrog Energy 32(3):337–342
5. Wongchanapai S et al (2012) Performance evaluation of an integrated small-scale SOFC-biomass gasification power generation system. J Power Sources 216:314–322

6. Prins MJ, Ptasinski KJ, Janssen FJJG (2003) Thermodynamics of gas-char reactions: first and second law analysis. Chem Eng Sci 58(3–6):1003–1011
7. Karamarkovic R, Karamarkovic V (2010) Energy and exergy analysis of biomass gasification at different temperatures. Energy 35(2):537–549
8. Basu P (2006) Combustion and gasification in fluidized bed. CRC Press, Taylor & Francis, Florida, USA
9. Mostoufi N, Cui H, Chaouki J (2001) A comparison of two- and single-phase models for fluidized-bed reactors. Ind Eng Chem Res 40(23):5526–5532
10. El-Emam RS, Dincer I (2015) Thermal modeling of fluidized bed gasification of rice straw. In: Dincer I et al (eds) Progress in clean energy, vol I. Springer International Publishing, Switzerland, DOI 10.1007/978-3-319-16709-1_4
11. El-Emam RS, Dincer I, El-Emam SH (2014) Efficiency and environmental assessment of biomass gasification for hydrogen and power production. In global conference on global warming (GCGW-14), Beijing, China
12. Chirone R et al. (1999) Fluidized bed combustion of high-volatile solid fuels: an assessment of char attrition and volatile matter segregation. In the 15th international conference on FBC, ASME
13. Davidson JF, Harrison O (1963) Fluidized particles. Cambridge Univ Press, New York
14. Gómez-Barea A, Leckner B (2010) Modeling of biomass gasification in fluidized bed. Prog Energy Combust Sci 36(4):444–509
15. Kunii D, Levenspiel O (1991) Fluidization engineering. Elsevier, Stoneham, MA
16. El-Emam RS, Dincer I (2015) Thermal modeling and efficiency assessment of an integrated biomass gasification and solid oxide fuel cell system. Int J Hydrogen Energy. doi:10.1016/j.ijhydene.2015.02.061
17. El-Emam RS, Dincer I, Naterer GF (2012) Energy and exergy analyses of an integrated SOFC and coal gasification system. Int J Hydrog Energy 37(2):1689–1697
18. Wongchanapai S et al (2012) Selection of suitable operating conditions for planar anode-supported direct-internal-reforming solid-oxide fuel cell. J Power Sources 204:14–24
19. Campanari S, Iora P (2004) Definition and sensitivity analysis of a finite volume SOFC model for a tubular cell geometry. J Power Sources 132(1–2):113–126
20. Dounit S, Hemati M, Andreux R (2008) Modelling and experimental validation of a fluidized-bed reactor freeboard region: Application to natural gas combustion. Chem Eng J 140 (1–3):457–465
21. Guedea I et al (2013) On the modeling of oxy-coal combustion in a fluidized bed. Chem Eng J 228:179–191
22. El-Emam RS, Okasha FM, El-Emam SH (2014) Clean combustion of low quality fuel in fluidized bed combustor. In: Dincer I et al (eds) Progress in sustainable energy technologies. Creating sustainable development, vol II. Springer International Publishing, Switzerland, pp 531–546
23. Costamagna P, Honegger K (1998) Modeling of Solid Oxide Heat Exchanger Integrated Stacks and Simulation at High Fuel utilization. J Electrochem Soc 145(11):3995–4007
24. Carpentieri M, Corti A, Lombardi L (2005) Life cycle assessment (LCA) of an integrated biomass gasification combined cycle (IBGCC) with CO2 removal. Energy Convers Manag 46 (11–12):1790–1808
25. Mann MK, Spath PL (1997) Life cycle assessment of a biomass gasification combined-cycle power system. National Renewable Energy Laboratory, Golden, CO
26. Koroneos C, Dompros A, Roumbas G (2008) Hydrogen production via biomass gasification—A life cycle assessment approach. Chem Eng Process Process Intensif 47(8):1261–1268
27. Hosseini M, Dincer I, Rosen MA (2013) Integrated renewable energy-based systems for reduced greenhouse gas emissions in causes, impacts and solutions to global warming. In: Dincer I et al (eds) Causes, impacts and solutions to global warming. Springer, New York, pp 779–802
28. Colpan C, Dincer I, Hamdullahpur F (2007) Thermodynamic modeling of direct internal reforming solid oxide fuel cells operating with syngas. Int J Hydrogen Energy 32(7):787–795

Part III
Clean Coal Technologies

Chapter 13
Influence of Oxygen Concentration and Equivalence Ratio on MSW Oxygen-Enriched Gasification Syngas Compositions

Haoran Yuan, Tao Lu, Dandan Zhao, Yazhuo Wang, and Noriyuki Kobayashi

Abstract Oxygen-enriched gasification for the treatment of municipal solid waste (MSW) is proposed in this chapter; mechanism analysis and thermodynamic calculations results show the advantages of lower heat loss than air gasification. The effects of oxygen concentration and equivalence ratio (ER) on gas products compositions were investigated. It was found that, during the ER range of 0.23–0.29, CO, H_2, and CO_2 content in combustible gas continuously increased, and CH_4 content continuously decreased with the oxygen concentration in the gasification agent rising from 20 % to 100 %. Furthermore, when the oxygen concentration was 20 % or 40 %, the CO_2 content in combustible gas grew constantly, and the CH_4 content gradually decreased within the ER span of 0.23–0.29, however, CO and H_2 contents increased then decreased when ER exceeded 0.27. In addition, when the oxygen concentration was 80 % or 100 %, the CO_2 content grew, CO and CH_4 content decreased gradually, and H_2 content decreased when ER exceeded 0.25 and 0.27, respectively. Therefore, gas product quality could be improved through increasing oxygen concentration within a certain ER range.

Keywords MSW • Gasification • Oxygen-enriched

13.1 Introduction

Solid waste was divided into general solid waste and hazardous solid waste. Municipal solid waste was one of general solid waste including solid waste, semi-solid, or liquid waste emanating from human activities. Landfill, partial recovery, partial fermentation, and heat disposal are several typical

H. Yuan • T. Lu • D. Zhao • Y. Wang (✉)
Guangzhou Institute of Energy Conversion, Chinese Academy of Sciences, Guangzhou 510640, China
e-mail: wangyz@ms.giec.ac.cn

N. Kobayashi
EcoTopia Science Institute, Nagoya University, Nagoya 464-8603, Japan

I. Dincer et al. (eds.), *Progress in Clean Energy, Volume 2*,
DOI 10.1007/978-3-319-17031-2_13

methods used for MSW treatment. The heat disposal technique was divided into incineration and pyrolysis; the former was applied widely in developed countries due to the many advantages of MSW heat disposal, such as high efficiency, large volume reduction rate, and part of the energy available without wasting land resources [1–4]; the latter is a new technique, which obtained rapid development in recent years.

Some valuable theories and methods regarding the MSW heat disposal technique have been achieved through extensive studies. Guo et al. [5] divided the complex combustible waste materials into fiber and polymer substances, and established a mathematical model of MSW pyrolysis and combustion, then illustrated the mechanism showing why pyrolysis and combustion inhibited dioxin generation. Galvagno et al. [6] and Dai et al. [7] investigated the pyrolysis mechanism of used tires and proposed the pyrolysis conditions. Li et al. [8] proposed heavy metals migration rules in the combustion process based on the analysis result of heavy metals in MSW fly ash. Buekens and Huang [9] investigated a dioxin formation mechanism to proposed control methods during the MSW combustion process. In addition, Wang et al. [10] and McLachlan [11] studied the combustion mechanism and features of refuse-derived fuel and agricultural waste, respectively. However, these studies were mainly focused on the MSW combustion process or a single-component pyrolysis process. Recently, MSW gasification has drawn much attention ascribed to many advantages, such as converting solid waste with different sizes, physical properties, and heating values into biogases with a wide range of applications, high energy recovery efficiency, and low pollution. A series of research results has been obtained through abundant theoretical and experimental research on MSW gasification [12–14]. However, some critical questions need to be solved before this technique is widely applied. The oxygen-enriched technique proposed by Horne and Steinberg [15] possessed many advantages such as sufficient combustion, substantial emissions reduction, high efficiency, and so on. The oxygen-enriched technique has currently been used in power stations, magnetic fluid power technology, fuel cells, IGCC, and combined energy production systems. The technique also was used in MSW fluidized bed combustion [16, 17].

Some drawbacks still exist in common gasification or combustion, such as high nitrogen content that resulted in low heat value and a large amount of smoke. The common oxygen-enriched combustion technique cannot fundamentally solve these issues of controlling solid-phase (nonhomogeneous) combustion and restraining secondary pollution. The oxygen-enriched gasification technique may be an alternative choice for MSW treatment, however, the MSW oxygen-enriched gasification mechanism is still not clear enough, hence the MSW oxygen-enriched gasification experiment was conducted and the mechanism analyzed in this study.

13.2 Experiment

13.2.1 Experimental System

The main device used in experimental studies was a small downdraft gasifier with a handling capacity of 5 kg/h. The experimental system is shown in Fig. 13.1. MSW oxygen-enriched gasification experiments were carried out at different oxygen concentrations of the gasification agent and the equivalence ratio (ER), and the gas products were also analyzed with a gas chromatograph.

13.2.2 Experimental Materials

MSW collected from an urban waste recycle bin in Nanling Anhui province was used as raw material in all the experiments. After squeeze and drying pretreatment, partly dried MSW (moisture percent = 10 %) was loaded into the oxygen-enriched pyrolysis system. The chosen MSW ($\rho = 520$ kg/m^3) was a mixture of irregularly

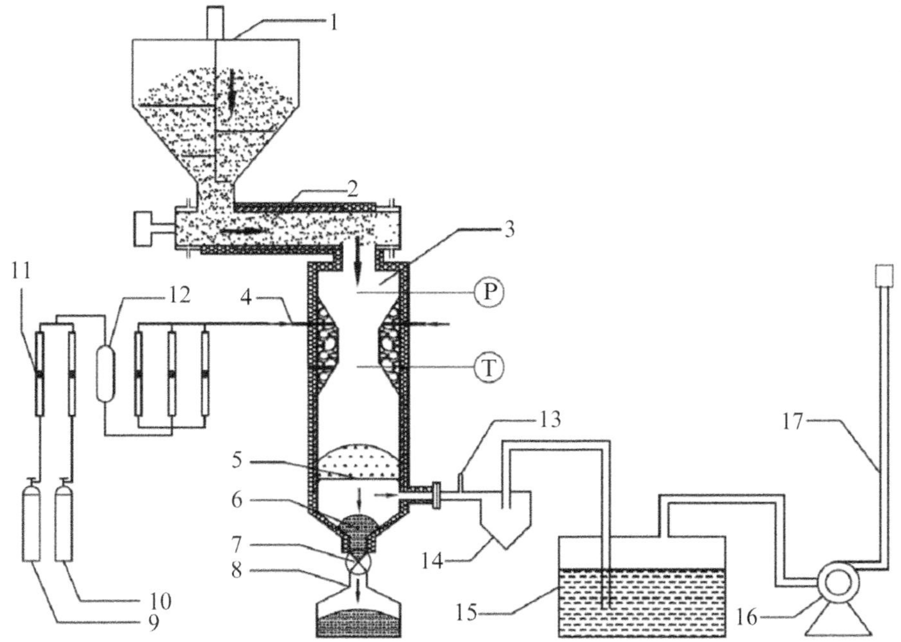

Fig. 13.1 Schematic diagram of the MSW experimental system. (*1*) Silo, (*2*) Anger feeder, (*3*) Gasifier main body, (*4*) Gasification furnace, (*5*) Grate, (*6*) Ashes, (*7*) Ash valve, (*8*) Ash bucket, (*9*) Canned oxygen, (*10*) Canned nitrogen, (*11*) Gas flow meter, (*12*) Mixing tank, (*13*) Sampling orifice, (*14*) Ash dischargers, (*15*) Water washing chamber, (*16*) Roots blower, (*17*) Outlet pipe, *T* temperature measuring point, *P* pressure measuring point

Table 13.1 Elemental analysis of experimental materials

Elements (%)	Ash free basic (%)	Dry basis (%)	As-fired basis (%)
C	43.79	29.25	13.06
H	5.87	3.92	1.75
N	0.60	0.40	0.18
S	0.39	0.26	0.12
O	21.53	14.38	6.42

shaped substances, whose main components were plastic, paper, cloth, vegetation, white plastic, and food waste. The elemental analysis results of MSW are shown in Table 13.1.

13.2.3 Experimental Process

First, ignite charcoal to heat the gasifier, then load the MSW feedstock with vaguer feeder into the gasifier body to form the MSW layer, while adjusting and keeping the oxygen concentration and equivalence ratio at the desired values for 30 min through modifying the oxygen and nitrogen valves; then sample syngas through the sampling orifice, and analyze its composition through a gas chromatograph.

13.3 Mechanism of MSW Oxygen-Enriched Gasification

13.3.1 MSW Oxygen-Enriched Gasification Reaction

The C, H, and O elements, volatile matter, and fixed carbon contents were high in MSW, and N and S element contents were low in MSW, therefore, MSW was suitable for gasification. The whole process of MSW oxygen-enriched gasification could be divided into four steps: drying, pyrolysis, reduction, and oxidation, and the chemical reaction steps shown below [12]:

$$C(s) + O_2 \rightarrow CO_2 \tag{13.1}$$

$$C(s) + CO_2 \rightarrow 2CO \tag{13.2}$$

$$C(s) + H_2O \rightarrow CO + H_2 \tag{13.3}$$

$$CO + H_2O \rightarrow CO_2 + H_2 \tag{13.4}$$

$$C(s) + 2H_2 \rightarrow CH_4 \tag{13.5}$$

$$CH_4 + H_2O \rightarrow CO + H_2 \tag{13.6}$$

The MSW feedstock lost water (constitutional water) in the drying zone, then cracked into some small molecules in the pyrolysis zone, in which CH_4 was the main component. The water–gas reaction and CO_2 reduction reaction mainly occurred in the reducing zone. The oxidation reaction of these small molecules and carbon with oxygen occurred in the oxide zone, and emitted heat to sustain gasification. Therefore, the produced gases were mainly composed of CO, CO_2, H_2, CnHm, and so on [18]. In the dry region, MSW was dried and dehydrated, then cracked further into macromolecules; the heat to dry and dehydrate MSW derived from the gasifier walls with high temperature and high-temperature gas produced from other reaction regions [14]. In the pyrolysis region, the pyrolysis reaction includes cracking and dehydrogenation reactions (main reaction Eqs. 13.7–13.9), and the energy required for the pyrolysis process is mainly provided by oxidation of the available partial fuel.

$$C_2H_6 \xrightarrow{\Delta} C_2H_4 + H_2 \tag{13.7}$$

$$C_2H_4 \xrightarrow{\Delta} CH_4 + C \tag{13.8}$$

$$CH_4 \xrightarrow{\Delta} C + 2H_2 \tag{13.9}$$

In the pyrolysis process, the chemical bonds of organic compounds in MSW were broken into small molecules and solid residues. Polymer molecules were accelerated at high temperature, and their chemical bonds then broken to produce free radicals. These highly active free radicals can react with the surrounding various types of molecules or radicals to form more free radicals, causing polymer degradation and cross-linking. The chemical bonds with a low binding energy can be easily broken when heated, and the corresponding element or group was the structural weakness, which became an active point during the pyrolysis process. Pyrolysis reactions eventually transformed MSW into combustible gas (CnHm, H_2, CO, et al.), tar, solid carbon, and ash.

13.3.2 Theoretical Calculations of MSW Oxygen-Enriched Gasification

The theoretical temperature of MSW oxygen-enriched pyrolysis was obtained according to the enthalpy of gasification products. However, the actual furnace temperature was lower than the theoretical temperature. The air contained more than 79 % nitrogen, which resulted in more inert gas being brought into the furnace, then a decrease in furnace temperature, and a great amount of heat loss with the discharge of excessive smoke. The oxygen-enriched gas contained little inert gas; hence the oxygen-enriched pyrolysis technique overcame the above shortcoming. The energy input to the reactor was energy carried by fuel, oxidants, and steam. The calculation equation is shown below:

$$t = \left(Q_d^y + C_r t_r + C_k t_k L_a\right)/C_y V_y \tag{13.10}$$

where Q_d^y is the fuel calorific, C_r is the heat capacity of fuel, C_k is the heat capacity of oxidant, C_y is the heat capacity of gasification products, t is the reaction temperature, t_r and t_k are the temperature of fuel and oxidant, respectively, V_y is the volume of smoke generated by gasification, and L_a is the oxidant for unit of fuel consumption.

It can be seen from Eq. (13.10) that the smoke generated from oxygen-enriched gasification is much smaller; correspondingly, the reaction temperature is greater. Therefore, oxygen-enriched gasification has better energy-saving efficiency. The heat loss during reactions was calculated according to the following formula.

$$(I_{py} - \alpha_{py} I_{ik})(100 - q_4)/Q_{dw} \tag{13.11}$$

where I_{py} is the enthalpy of gasification products, α_{py} is the excess air coefficient at the exit, I_{ik} is the enthalpy of cold air, q_4 is the loss of incomplete gasification, and Q_{dw} is the fuel calorific.

It can be seen from Eq. (13.11) that enthalpy of gasification products contained enthalpy of the inert N_2 which caused massive heat loss. Oxygen-enriched gas contained a very small fraction of N_2; therefore the heat loss of oxygen-enriched gasification was much lower than that of air gasification.

13.3.3 Thermodynamics of MSW Oxygen-Enriched Gasification

The equilibrium constant method and Gibbs minimum free energy method were usually used to evaluate the complex chemical equilibria in thermodynamics. The Gibbs minimum free energy method was used in this study. Under the isothermal isobaric conditions, the condition was regarded as equilibrium when the system's Gibbs free energy reached minimum. When conducting the oxygen-enriched gasification experiment, oxygen-enriched gas and MSW were fed into the experimental device under a certain condition (certain temperature and pressure) for complex chemical reactions. The device was considered to be an ideal closed reaction system. When the system reached chemical equilibrium, the Gibbs free energy of the entire system reached minimum [19, 20], which could be calculated according to the formula shown below:

$$\mu_i = \left[\frac{\partial G}{\partial n_i}\right]_{n_i \neq n_j} \tag{13.12}$$

where μ_i is the chemical potential energy of component (i), G is the Gibbs free energy, and n_i is the amount of substance for component (i). With the transformation of Eq. (13.12), the Gibbs minimum free energy could be expressed as

$$\mathrm{Min}(G) = \mathrm{Min}\left(\sum G_i\right) = \sum (n_i \mu_i) \tag{13.13}$$

The parameters of various substances in the calculation were determined according to a relative thermodynamic data reference manual [21]. Chemical reactions during the pyrolysis process were calculated according to Eqs. (13.1)–(13.6). The pressure was set as 1.013 × 105 Pa ascribed to the actual equipment pressure. The reaction temperature could be calculated in accordance with the first law of thermodynamics [22]:

$$Q_{\mathrm{system}} = H_{\mathrm{P}} - H_{\mathrm{R}} \tag{13.14}$$

where H_{P} is the enthalpy of the resultant, and H_{R} is the enthalpy of the reactant. Given the condition that only MSW released heat during the reaction process, and the enthalpy of other substances just changed in various final temperatures, the reaction temperature was thus regarded as a constant, and the reaction temperature could be calculated from the low heat value of MSW.

The effects of the equivalence ratio (ER, the ratio of input oxygen to the oxygen required for complete combustion of the substances) and oxygen concentration on the composition of gas products were investigated in this study. Oxygen concentrations of 20 %, 40 %, 60 %, 80 %, and 100 % were chosen and 0.20, 0.23, 0.25, 0.27, 0.29, and 0.38 were selected as ERs, respectively. The results are shown in Figs. 13.2 and 13.3.

As shown in Fig. 13.2, ER had a significant effect on gas composition under the pure oxygen condition. With ER rising from 0.20 to 0.38, both CO and CH_4

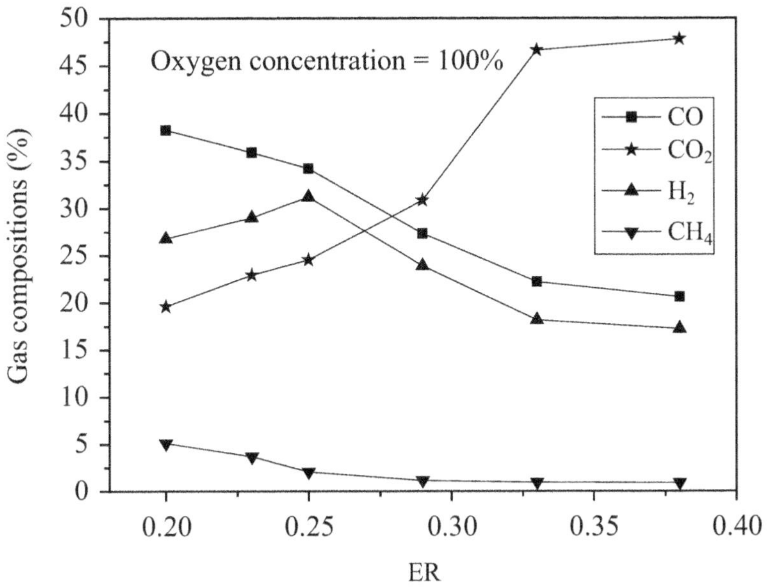

Fig. 13.2 Effect of ER on gas product compositions

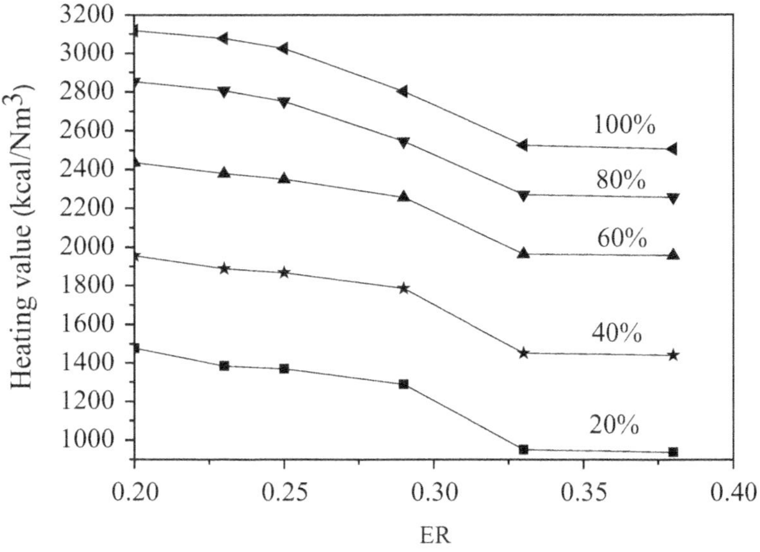

Fig. 13.3 Effect of oxygen concentration on gas product heat values

contents decreased, whereas H_2 content increased up to peak when ER was 0.25, then decreased. Different from other gas compositions, CO_2 content rose slowly during the ER span of 0.20 and 0.29, then rose rapidly during the ER span of 0.29 and 0.33, and finally rose slowly again. Clearly, the oxygen was very scarce when ER was below 0.25, therefore, the main gas compositions were derived via MSW cracking (this was the pyrolysis stage), and with ER rising, oxygen concentration increased, thus the main gas compositions were derived from the reaction of oxygen and MSW (the combustion stage).

As shown in Fig. 13.3, the effect of ER on gas product heat value at various oxygen concentrations was similar. With ER rising from 0.20 to 0.38, there was adequate oxygen to induce more MSW combustion, hence the CO_2 content in the gas product increased, and gas product heat value decreased; especially during the ER span of 0.28 and 0.33, the CO_2 content in the gas product increased dramatically, hence the heat value decreased sharply. Figure 13.3 also illustrates that the lower the oxygen concentration, the more intensive the effect of ER on gas product heat value. Because of the lower oxygen concentration, and the higher nitrogen concentration, the great amount of nitrogen diluted the combustible gas composition, hence the gas product heat value was very low at the lower oxygen concentration (20 % and 40 %).

According to the results of theoretical calculations, the gas product compositions could be modified through controlling ER, and 0.20 and 0.30 may be a proper ER span for MSW gasification. A proper oxygen concentration also needed to be chosen, because that increasing oxygen concentration may raise the gas product heat value, however, the cost and energy waste also increased with the rise in oxygen concentration.

13.4 Results and Discussion

13.4.1 Effect of Oxygen Concentration on Gas Product Compositions Under Different ER

The theoretical calculation results also indicated that the main reaction was combustion when ER exceeded 0.30, and the gasification reaction was restrained when ER was too small, hence 0.20–0.29 was chosen as the ER in the gasification experiment study. The experiment results indicated that increasing oxygen concentration may then increase combustible gas content to improve gas product quality. As demonstrated in Fig. 13.4a–c, the results showed that when ERs were 0.23, 0.25, and 0.27, the gas composition content in gas products followed the order of $CO > H_2 > CO_2 > CH_4$, and the content of CO, H_2, and CO_2 increased with oxygen concentration rising from 20 % to 100 %. With the increase of oxygen concentration in the gasifying agent, the nitrogen concentration decreased, and the gasification reaction temperature also increased, which resulted in the increase of CO, CO_2, and H_2 contents, and CH_4 cracked into C and H_2 at high temperature, hence CH_4

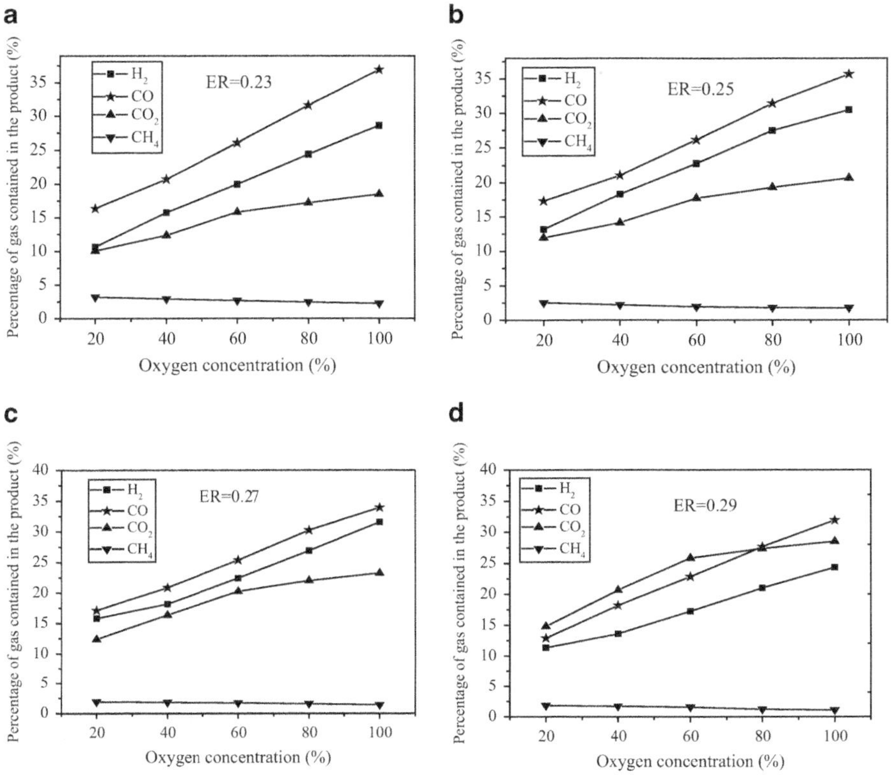

Fig. 13.4 Effects of oxygen concentration on gas product compositions

content decreased with the oxygen concentration rising from 20 % to 100 %. In contrast to these results (shown in Fig. 13.4d), the gas composition content in gas products followed the order of $CO_2 > CO > H_2 > CH_4$ when ER was 0.29 during the oxygen concentration span of 20 % and 60 %; furthermore, the gas composition contents in gas products followed the order of $CO > CO_2 > H_2 > CH_4$, because the increase of CO_2 content was very slow when oxygen concentration rose from 60 % to 100 %.

13.4.2 Effect of ER on Composition of Gas Products Under Same Oxygen Concentration

As shown in Fig. 13.5a, b, when oxygen concentration was 20 % or 40 %, the organic matter decomposition and combustion contributed CO_2 content ongoing growth during the ER span of 0.23–0.29, and the increase of CO and H_2 contents during the ER span of 0.23–0.27, and then a decrease during the ER span of 0.27–0.29. The cracking of CH_4 resulted in its content gradually decreasing with ER rising from 0.23 to 0.29. As shown in Fig. 13.5c, d, when oxygen concentration was

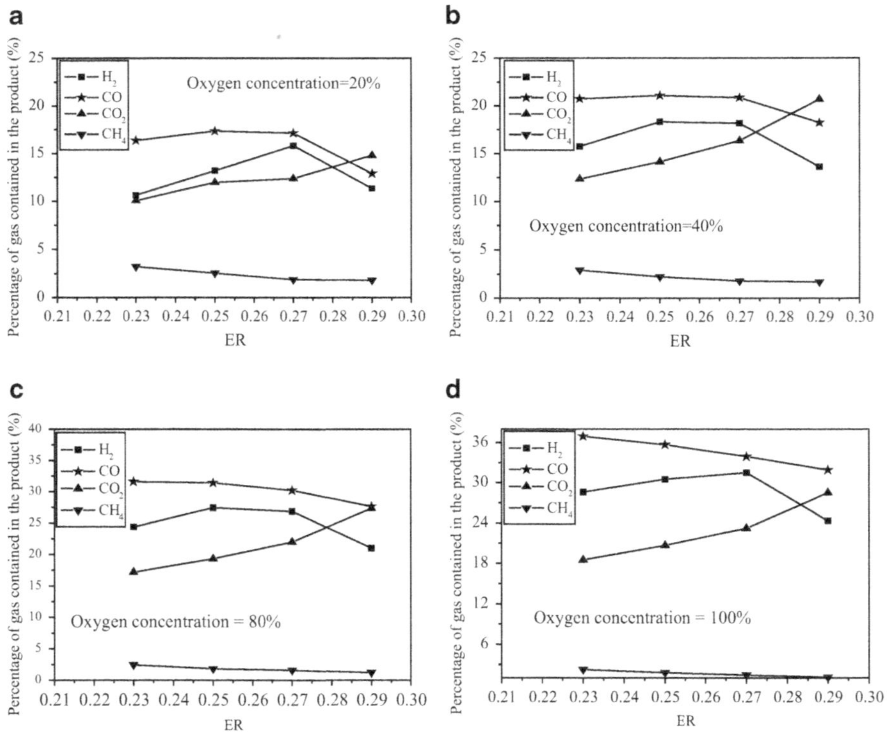

Fig. 13.5 Effect of ER on gas product compositions

80 % or 100 %, CO_2 content grew and CH_4 content decreased gradually with ER rising from 0.23 to 0.29; however, the abundant oxygen contributed to the conversion of pyrolysis to combustion, hence the CO content continuously decreased with the ER rising from 0.23 to 0.29, and H_2 content decreased when the ER exceeded 0.25 and 0.27, at oxygen concentrations of 80 % and 100 %, respectively.

13.5 Conclusions

Oxygen-enriched gasification was an alternative technique to dispose of MSW, which converted MSW into high heat value combustible gas. The oxygen concentration and ER had a critical effect on gas composition and heat value. Increasing the oxygen concentration may increase combustible gas content to improve gas product quality. During the ER span of 0.23–0.29, CO, H_2, and CO_2 contents in combustible gas continuously increased, and CH_4 content continuously decreased with oxygen concentration in the gasification agent rising from 20 % to 100 %. When the oxygen concentration was 20 % or 40 %, CO_2 content in the combustible gas grew constantly, and CH_4 content gradually decreased within the ER span of 0.23–0.29; however, CO and H_2 contents increased then decreased when ER exceeded 0.27. When the oxygen concentration was 80 % or 100 %, CO_2 content grew, CO and CH_4 content decreased gradually, and H_2 content decreased when ER exceeded 0.25 and 0.27, respectively.

References

1. Cheng HF, Hu YA (2010) Municipal solid waste (MSW) as a renewable source of energy: Current and future practices in China. Bioresour Technol 101:3816–3824
2. McKay G (2002) Dioxin characterization, formation and minimization during municipal solid waste (MSW) incineration: review. Chem Eng J 86:343–368
3. Ruth LA (1998) Energy from municipal solid waste: a comparison with coal combustion technology. Prog Energy Combust Sci 24:545–564
4. Lisk DJ (1988) Environmental implications of incineration of municipal solid waste and ash disposal. Sci Total Environ 74:39–66
5. Guo XF, Yang XL, Li HB, Wu CZ, Chen Y (2001) Release of hydrogen chloride from combustibles in municipal solid waste. Environ Sci Technol 35:2001–2005
6. Galvagno S, Casu S, Casabianca T, Calabrese A, Cornacchia G (2002) Pyrolysis process for the treatment of scrap tyres: preliminary experimental results. Waste Manage 22:917–923
7. Dai XW, Yin XL, Wu CZ, Zhang WN, Chen Y (2001) Pyrolysis of waste tires in a circulating fluidized-bed reactor. Energy 26:385–399
8. Li JX, Yan JH, Jin YQ, Ni MJ, Cen KF (2004) Characteristic analysis of heavy metals in MSW fly ash. J Zhejiang Univ (Eng Sci) 38:48–52
9. Buekens A, Huang H (1998) Comparative evaluation of techniques for controlling the formation and emission of chlorinated dioxins/furans in municipal waste incineration. J Hazard Mater 62:1–33

10. Wang Z, Huang H, Li H, Wu C, Chen Y, Li B (2002) Pyrolysis and combustion of refuse-derived fuels in a spouting-moving bed reactor. Energy Fuel 16:136–142
11. McLachlan MS (1995) Accumulation of PCCD/F in an agricultural food chain. Organohalogen Comp 26:105–108
12. Yan JH, Zhu HM, Jiang XG (2008) Pyrolysis and gasification characteristic of multi-component medical waste. J Zhejiang Univ (Eng Sci) 42:885–889
13. Liu K, Ma XQ, Xiao HM (2010) Experimental and kinetic modeling of oxygen-enriched air combustion of paper mill sludge. Waste Manage 30:1206–1211
14. Bao XJ, Cai JJ, Luo GQ, Qi LJ (2003) The new regenerating technology of municipal refuse pyrolysis. Metallurgy Energy 22:44–48
15. Horn FL, Steinberg M (1982) Control of carbon dioxide emissions from a power plant (and use in enhanced oil recovery). Fuel 61:415–422
16. Liang ZY, Ma XQ (2010) Mathematical modeling of MSW combustion and SNCR in a full-scale municipal incinerator and effects of grate speed and oxygen-enriched atmospheres on operating conditions. Waste Manage 30:2520–2529
17. Frey HH, Peters B, Hunsinger H, Vehlow J (2003) Characterization of municipal solid waste combustion in a grate furnace. Waste Manage 23:689–701
18. Zhang Y, Tao H (2002) Waste treatment technology and projects. Chemical Industry Press, Bei Jing, p 106
19. Flemming F (1995) Trace elements from coal combustion. Technical University of Denmark, Copenhagen
20. Xu ZH, Wang LS (1987) Inorganic thermochemical database. Bei Jing: Science Press. Barin, I. 1995. Thermochemical Data of Pure Substances, New York: VCH
21. Malcolm W, Chase Jr. (1998) NIST-JANAF Thermochemical tables, Washington DC: American chemical society; wood-bury New York; American Institute of Physics for the Nation-al Institute of Standards and Technology
22. Potter MC, Somerton CW (1998) Schanms's outline of theory and problems of thermodynamics for engineers. New York: McGraw-Hill Inc

Chapter 14
Oxy-fuel Combustion for Carbon Capture and Sequestration (CCS) from a Coal/Biomass Power Plant: Experimental and Simulation Studies

Nelia Jurado, Hamidreza G. Darabkhani, Edward J. Anthony, and John E. Oakey

Abstract Oxy-fuel combustion is a promising and relatively new technology to facilitate CO_2 capture and sequestration (CCS) for power plants utilising hydrocarbon fuels. In this research experimental oxy-combustion trials and simulation are carried out by firing pulverised coal and biomass and co-firing a mixture of them in a 100 kW retrofitted oxy-combustor at Cranfield University. The parent fuels are coal (Daw Mill) and biomass cereal co-product (CCP) and experimental work was done for 100 % coal (w/w), 100 % biomass (w/w) and a blend of coal 50 % (w/w) and biomass 50 % (w/w). The recirculation flue gas (RFG) rate was set at 52 % of the total flue gas. The maximum percentage of CO_2 observed was 56.7 % wet basis (73.6 % on a dry basis) when 100 % Daw Mill coal was fired. Major and minor emission species and gas temperature profiles were obtained and analysed for different fuel mixtures. A drop in the maximum temperature of more than 200 K was observed when changing the fuel from 100 % Daw Mill coal to 100 % cereal co-product biomass. Deposits formed on the ash deposition probes were also collected and analysed using the environmental scanning electron microscopy (ESEM) with energy-dispersive X-ray (EDX) technique. The high sulphur, potassium and chlorine contents detected in the ash generated using 100 % cereal co-product biomass are expected to increase the corrosion potential of these deposits. In addition, a rate-based simulation model has been developed using Aspen Plus® and experimentally validated. It is concluded that the model provides an adequate prediction for the gas composition of the flue gas.

Keywords Oxy-fuel combustion • Carbon capture and sequestration (CCS) • Co-firing coal and biomass • Process modelling

---

N. Jurado • H.G. Darabkhani (✉) • E.J. Anthony • J.E. Oakey
Centre for Combustion and CCS, School of Energy, Environment and Agrifood (EEA), Cranfield University, Cranfield, Bedfordshire MK43 0AL, UK
e-mail: h.g.darabkhani@cranfield.ac.uk

I. Dincer et al. (eds.), *Progress in Clean Energy, Volume 2*,
DOI 10.1007/978-3-319-17031-2_14

List of Abbreviations

CCP	Cereal co-product
CCS	CO_2 capture and sequestration
db	Dry basis
EDX	Energy-dispersive X-ray
ESEM	Environmental scanning electron microscopy
FB	Fluidised bed
FTIR	Fourier transform infrared
NO	Nitric oxide
NO_x	Nitrogen oxides
PF	Pulverised fuel
RFG	Recirculation flue gas
SO_x	Sulphur oxide
w/w	Weight ratio
wb	Wet basis

14.1 Introduction

There are three major carbon capture technologies that could reduce emissions from fossil fuel power plants: pre-combustion capture, post-combustion capture, and oxy-fuel combustion capture. Oxy-fuel combustion has several features which makes it a very attractive technology for implementation in both existing air-firing and new power plants. These features include generation of a flue gas with a high percentage of CO_2 which is more easily captured, reduction in the size of the flue gas conditioning equipment, no additional space required for the post-combustion capture equipment, and also reduced environmental impacts compared with the other carbon capture technologies [1].

The main disadvantage of oxy-fuel combustion technology is the efficiency loss and cost associated with generating high-purity O_2 by cryogenic separation. Nonetheless, oxy-firing remains an extremely promising option as it requires minimal modification of existing coal-fired plants, and for coal firing at least it can already be considered to be near commercial technology. This is, however, not the case for co-firing with biomass. Such studies that have been done on co-firing of coal and biomass have noted that the ignition temperature decreases as the proportion of biomass in the fuel increases [2, 3]. It is found that the fuel burnout improves when oxy-firing blends of coal and biomass [2, 4, 5]. Recycling the flue gas to the oxy-combustor decreases or returns the nitric oxide (NO) content and sulphur oxide (SO_x) emissions (sulphur is retained in the ashes) per energy unit of fuel combusted. It has been suggested that boiler corrosion problems are likely to appear as the acid gases increase in the flue gases, as compared with the air-firing case [6, 3]. It has been also pointed out that some problems are likely to appear during the utilisation of the cement or concrete fabricated with the fly ashes generated in the oxy-combustion process, due to the higher content of SO_x retained by them [3]. It is, therefore, clear that while much is known, more research is required on a wide range of biomass fuels co-fired with coal in oxy-fuel systems. Ideally, experimental programs should be supported by process modelling, so the results of such tests can be generalised and used to provide data on scale-up.

Typical commercial packages used to support experimental studies or model full-scale systems include Fluent, Chemkin Pro, Aspen Plus or Hysys®, gPROMS® or Thermoflex®. However, prior to the use of this software to study combustion processes, such as conventional combustion, fluidised circulating bed combustion and oxy-firing, rigorous research was undertaken on the characterisation of char combustion [7, 8]. Researchers have investigated the oxy-firing process using Aspen Plus® [9], and circulating fluidised bed using coal as fuel in air-firing conditions [10, 11], both using coal as fuel in oxy-firing assuming equilibrium conditions. However, to date few authors have performed simulations on oxy-firing for co-firing using Aspen Plus®. The work endeavours to fill this gap, by providing such an experimental study using a 100 kW unit, together with a simulation of co-firing coal and biomass blends. The rate-based simulation model was designed to predict the gas composition and temperatures reached in the oxy-combustion mode, although with the simplifying assumptions proposed [12].

14.2 Experimental Approach

Air-firing of the fossil fuels results in relatively low concentration of CO_2 in flue gases which make the capture of CO_2 difficult and expensive. Oxy-firing combustion is a novel method of using enriched oxygen for coal/biomass combustion with recycled flue gases (RFG) to control the adiabatic flame temperature and to increase the CO_2 concentration of the off-gases up to 60–70 % in oxy-firing mode (compared to air-fired mode, around 12–14 %). This new technology is being applied at Cranfield University to retrofit an existing 100 kWth air-firing down-fired multi-fuel combustor to the oxy-firing mode. The retrofitting process of this combustor consisted of several phases and during that the following modifications and measurement systems have been implemented on the rig (see Fig. 14.1):

- Installation of flue gas recirculation line (recycle mode)
- Installation of the primary and secondary lines of the oxygen feedings
- Fitting an axial swirler in the air inlet port of the burner to improve mixing
- Installation of a gas-tight fan for recirculation of the flue gases
- Employing a gas-tight and CO_2 purged fuel feeder facility
- Water and SO_x removal facilities (in general acidic species)
- Gas analyser, thermocouples for measurement of the process environment
- Collection and analysis of the ash deposits

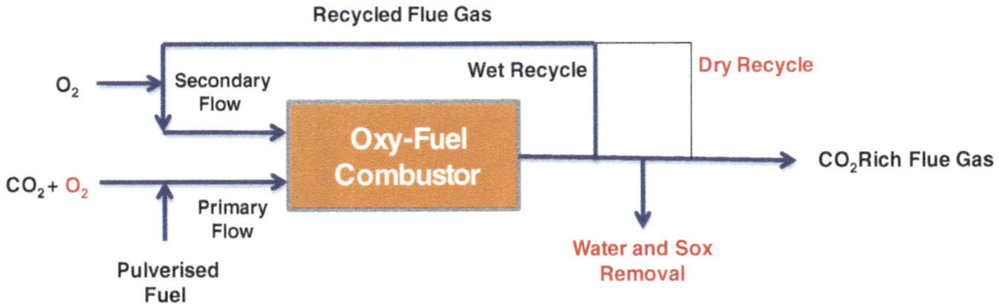

Fig. 14.1 Simple flow diagram of Cranfield University oxy-fuel combustor (*red*: recent modifications)

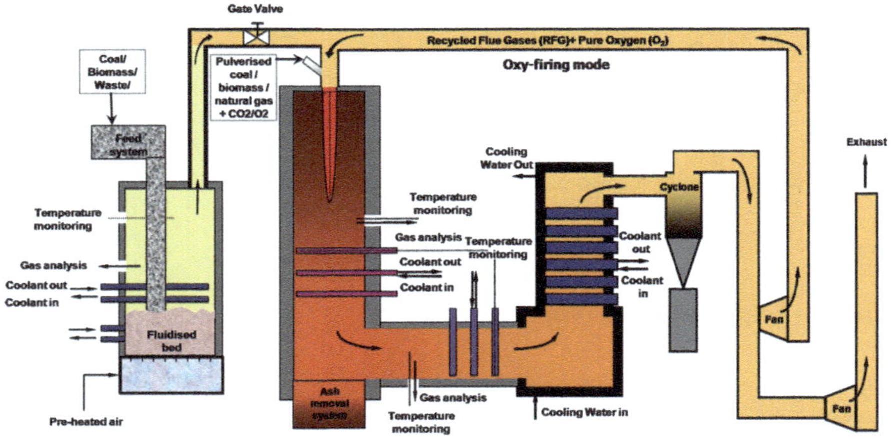

Fig. 14.2 Diagram of multi-fuel combustion rig at Cranfield University including the RFG pipe for the oxy-firing mode

Figure 14.2 illustrates a schematic of the multi-fuel combustor rig at Cranfield University. This rig, which has been retrofitted for the oxy-fuel combustion experiments, comprises both a fluidised bed (50 kWth) and a down-fired pulverised fuel combustor (100 kWth). For the oxy-combustion tests here, only the pulverised fuel (PF) combustor was used, with the gate valve, placed between the fluidised bed (FB) and the PF, closed to isolate the fluidised bed section of the reactor.

The pulverised fuel is fed to the oxy-combustor at a constant rate using a fuel feeder provided by a metering screw, a vibratory tray and a venturi eductor. Additionally, the feeder has three purge points located at the main storage hopper, at the feeding hopper and at the end of the vibratory tray feeder. CO_2 is injected through these purge points to keep a positive pressure inside the pulverised fuel feeding system so as to avoid air ingress once the pulverised fuel reaches the venturi eductor; it is fluidised and conveyed to the burner by a stream of pure CO_2 gas coming from CO_2 cylinders. Prior to this stream being fed to the burner, the primary O_2, coming from O_2 cylinders, is also injected into the primary stream.

The combustor is equipped with a down-fired burner, provided with a pilot flame port and a flame detector. The pulverised fuel is combusted in the vertical zone of the combustor (3.7 m height). The combustor has a square cross section with sides of 650 mm and thermal isolation with thickness of 175 mm. The vertical zone has four view ports and ports where sensors are used to measure wall temperatures (six type K thermocouples). The data acquisition system is a Pico Logger Unit-TC-08. More thermocouples are installed along the rig monitoring: inlet and outlet temperature of the water used in the refrigeration system, at the heat exchanger section, temperature after the cyclone and six more thermocouples to measure the temperature along the recirculation line. These sensors are sampled using the data logger, with 24 sampling ports distributed along the vertical and horizontal sections of the chamber. Three of these ports, placed at the bottom part of the vertical section, are used by the deposition probes. To collect the deposits, it was necessary to allow the chamber to cool down for 18–24 h, and then the ash deposited on the probes was sampled and analysed using ESEM/EDX. An online high-resolution multi-component Fourier transform infrared (FTIR) gas analyser is connected to one of the aforementioned sampling ports located at the vertical section of the chamber to measure the composition of the oxy-combustion gas.

The exhaust gas goes through two water-cooled heat exchangers, one in the horizontal and one in the vertical section, before exiting the oxy-combustor. Finally, after leaving the chamber, the gas enters a cyclone where the suspended particles are removed. Then part of the gas is recirculated to the combustion chamber and the rest is sent to the stack. The pipelines that convey the recycled flue gas are thermally isolated with a trace heating system to avoid a temperature drop below the acid dew point of the flue gas. The secondary O_2 is injected into this stream prior to feeding it to the oxy-combustor.

The PF oxy-combustor at Cranfield University is based on a retrofit to the existing air-firing combustor. The retrofitting process involved the installation of the recirculation line including setting up the recirculation fan, the O_2 injection to the recycle flue gas (secondary O_2) and trace heating wiring to prevent condensation in the flue gas recirculation lines. Additionally, a line to supply the CO_2 was installed to convey the pulverised fuel from the hopper to the burner itself. The second phase covered the design and fitting of an axial swirler in the air inlet port of the burner; the implementation of a gas-tight fan to recirculate the flue gas; the primary O_2 injection, added to the stream conveying the fuel to the combustor; installation of the gas-tight fuel-feed hopper and the design and replacement of a new CO_2 supply line, to respond to the design requirements of the new fuel feeder.

For these tests 100 % coal (w/w), 100 % biomass (w/w) and blend of coal 50 % (w/w) and biomass 50 % (w/w) were used. The coal used was Daw Mill and the biomass was cereal co-product (CCP), the analyses of which are shown in Table 14.1.

The experiments have been carried out with recirculation of 52 % (v/v) flue gas to the oxy-combustor. Calculations have been carried out for the cases in which the

Table 14.1 Analysis of Daw Mill and cereal co-product (CCP)

Properties	Daw Mill	CCP
CV, kJ/kg (as received)		
– Gross	25,260	17,610
– Net	24,107	16,340
Proximate analysis (% (w/w))		
– Moisture	4.6	8.1
– Fixed carbon	62.78	18.39
– Volatile matter	32.8	77.04
– Ash	4.4	4.57
Ultimate analysis (% (w/w))		
– C	77.7	47.22
– H	4.5	6.46
– N	1.2	3
– Cl	0.2	0.18
– S	1.3	0.17
– O	10.9	38.4
Sulphur analysis (% (w/w))		
– Pyritic	0.2	0.06
– Sulphate	0	0.05
– Organic	1.1	0.06

O_2 would be fed to the process at a 5 %, 10 % and 21 % (v/v) excess. To ensure safe conditions for the oxygen injection, the oxygen percentage in the primary stream must not exceed 21 % (v/v). For the secondary stream, which mainly comprises the recycled flue gas, the upper limit for the oxygen content is 28 %. These limits are selected to ensure that the oxygen leaving the combustor varies between 3 and 5 % (v/v). The operating pressure in the combustor must be below atmospheric pressure due to design conditions. However, to minimise the air ingress into the chamber, the pressure during the experimental tests was kept as close as possible to atmospheric pressure.

14.3 Simulation Approach

Limited publications are available giving the results of oxy-combustion of blends of coal and biomass considering the kinetics for the char combustion, and this chapter seeks to explore these issues further. In the model suggested here, a number of reactors have been defined to recreate as faithfully as possible the experimental process. The main effort has been focused on providing a mathematical model of the reactor where the char particles are oxy-fired. Furthermore, an equation has been proposed to provide the kinetics in the oxy-combustor. Additional reactors have been included in the model to simulate the gas treatments (particle removal and desulphurisation) before the recirculation of the flue gas to the combustor or its proceeding to the next stage of the carbon capture process (CO_2 purification and compression).

The implementation of the model presented in this chapter has been carried out using Aspen Plus® V7.3 and has taken place in two stages. A model based on the equilibrium of the reactions was first defined. The aim of this stage was to perform an initial analysis of the resulting mass and energy balances, providing an assessment of the contribution of different inputs including: percentage of oxygen excess; percentage of recycled flue gas (RFG); heat loss in the RFG stream and air ingress into the process. The equilibrium model was developed taking into account the published works [10, 11], as well as Aspen Plus® tutorials/guidelines. In the second stage, a kinetic model was developed. The additional output of this model provides a more realistic approach to the process than the equilibrium one. The criterion adopted here is to ensure the same heat transfer as in the air-firing case [13]. However, this chapter focuses on the explanation of the kinetic simulations.

The oxy-combustor is simulated using five reactors: the first one serves to convert the fuel, which is regarded as a nonconventional solid, to a conventional one; then there is a reactor to perform the combustion of the volatile species of the fuel; a reactor to simulate the combustion of the char; a reactor where the thermal nitrogen oxides (NO_x) are generated and, lastly, a reactor to simulate the generation of NO_x from the N of the fuel. A variation of the previous case study was considered, in which an additional block was implemented to simulate the partial condensation of H_2O vapour during recirculation of the flue gas to the oxy-combustor. This set of simulations was necessary so as to have a kinetic model with operating conditions as similar as possible to the experimental tests performed in the pilot plant. The box plot for the rate-based case is presented in Fig. 14.3.

In the next stage a rate-based model with partial condensation will be developed for dry recirculation modelling. To define the amount of H_2O vapour that condenses in this reactor, the vapour pressure of the H_2O at the minimum temperature reached in the flue gas will be considered. This value established the maximum concentration in the gaseous phase that the H_2O vapour could have. Thus, if the H_2O vapour concentration is higher, it will condense.

The methodology adopted to compare results from simulations and experimental data was, first, to set up operating conditions as similar as possible to the experiments, and, second, to perform the simulations comparing the results generated by the model in Aspen Plus® regarding gas composition and temperatures reached.

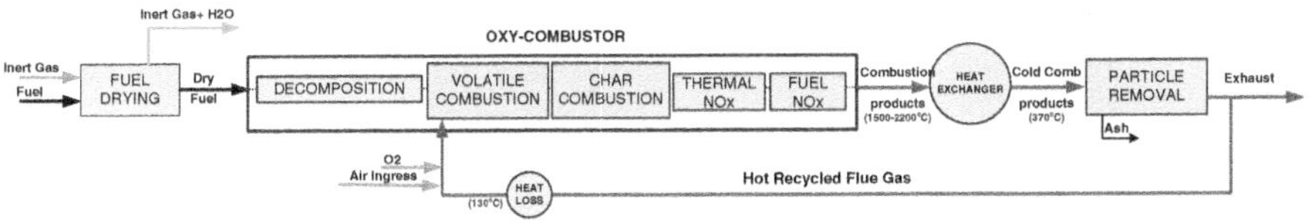

Fig. 14.3 Box plot of the rate-based model

For the first step (establish similar operating conditions), it was necessary to find out the amount of air ingress into the process which occurred during the experiments. This was accomplished using mass balances based on the experimental data. In addition air ingress was varied in the model, until analogous values generated by the simulations to the empirical gas composition were obtained. Although the air ingress is an unknown variable that needed to be characterised, other operating parameters were changed in the simulations to find out the combination that best represented the experimental conditions during the tests. These parameters were percentage of RFG, excess of oxygen fed and amount of water vapour condensed.

14.4 Results and Discussion

14.4.1 Experimental Results

Major emission species (CO_2, O_2, H_2O) and minor emission species (SO_2, CO, NO, NO_2, N_2O, HCl, CH_4) and gas temperature profiles were obtained and analysed for different fuel mixtures. Figure 14.4 shows the typical emission profiles of gaseous species using different fuels (100 % Daw Mill coal, 100 % CCP biomass and 50 % Daw Mill coal-50 % CCP biomass).

As expected, the highest percentage of CO_2 in the flue gas is achieved when 100 % Daw Mill is burned. A summary of the maximum values for CO_2 and average values for H_2O, O_2, SO_2 and CO achieved is given in Table 14.2. These data are in qualitative agreement with typical gas compositions obtained in different facilities by other authors [14–16] as shown in Table 14.2. The quantitative effect of the amount of air ingress on decreasing of the CO_2 concentration follows the trend suggested by [14]. In the graph proposed by these authors an increment in the air ingress into the process of 1 % causes a drop in the CO_2 content of around 3.5 %. The main points where the air leakage takes place are believed to be in the recirculation fan and along the oxy-combustor. In reality, the same type of issues would be faced in retrofitting an existing power plant, so it is necessary to explore the effect of such leakage. In addition, such data are relevant to hybrid concepts, where oxy-fuel is combined with amine scrubbing [17].

The content of H_2O vapour measured in the flue gas increases when the content of biomass in the fuel as expected is higher. Problems related to the performance of the FTIR analyser were observed during the test where 100 % CCP was used and in the future a condenser will be placed between the cyclone and the recirculation fan as conditioning equipment for the RFG.

The measured SO_2 concentration follows the trend of decreasing when the content of biomass in the fuel is higher. This was expected given that the biomass used here has one-eighth the sulphur present in the coal used. However, while the case where co-firing 50 % coal-50 % biomass shows a content of SO_2 in the flue gas in agreement (same order of magnitude) with the theoretical prediction (1,217 ppm

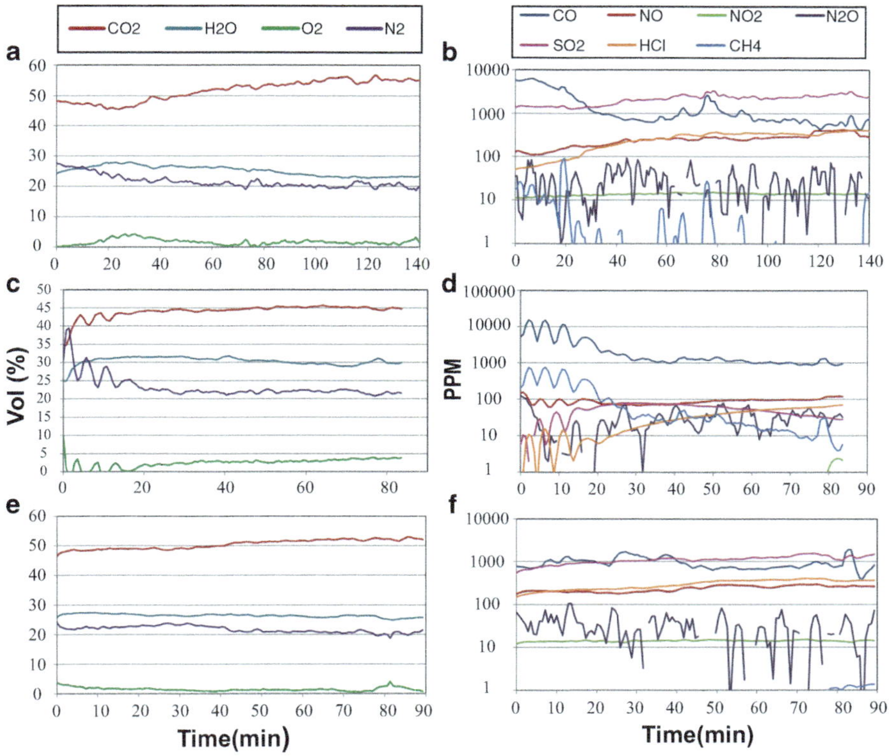

Fig. 14.4 Gas composition of the flue gases from the experimental work. (**a**) Main species 100 % Daw Mill. (**b**) Minor species 100 % Daw Mill. (**c**) Main species 100 % CCP. (**d**) Minor species 100 % CCP. (**e**) Main species 50 % Daw Mill-50 % CCP. (**f**) Minor species 50% Daw Mill-50 % CCP

measured against 1,450 ppm predicted), that is not the case for the test where 100 % of biomass was oxy-fired. The decrease in the SO_2 content when burning 100 % cereal co-product is higher than expected (47 ppm measured against 320 ppm predicted). The most likely reason is that the FTIR analyser results were affected by the very high concentration of H_2O vapour. An additional explanation for the SO_2 content not following the expected trend may be that part of the SO_2 was oxidised to SO_3 and condensed together with the H_2O vapour as H_2SO_4. This may also be a consequence of the temperature dropping below the acid dew point in a particular location of the RFG pipeline.

The CO presence is very high (around 10,000 ppm) during the initial stage of the tests using either 100 % Daw Mill or 100 % cereal co-product. But once the steady operation regime is reached, it drops down to around 850 ppm, on average (800 ppm for 100 % Daw Mill coal; 1,000 ppm for 100 % cereal co-product; 700 ppm for 50 % coal-50 % biomass). It is possible that part of the explanation for this result is that the gaseous species are sampled inside the combustor and are

Table 14.2 Experimental values for gas compositions with comparisons from the literature [14–16]

	Cranfield University						[14]		[15]		[16]	
	100 % Daw Mill coal		100 % CCP biomass		50 % Daw Mill-50 % CCP blend		100 % Coal (10 % air ingress)		Blends of coal and biomass		100 % coal	
	wb	db	wb	db	wb	db	wb	db	wb	db	wb	db
CO_2 (%)	56.7	73.6	45.7	65.3	53	72	62.4	–	55–65	96	68	90
H_2O(%)	23	–	30	–	26.4		4.2	–	–	–	20	–
O_2(%)	1.5	–	3	–	1.5		3	–	–	–	–	–
SO_2 (ppm)	2,000	–	100	–	1,500		–	–	–	–	–	–
CO (ppm)	1,000	–	1,200	–	900		–	–	–	–	–	–

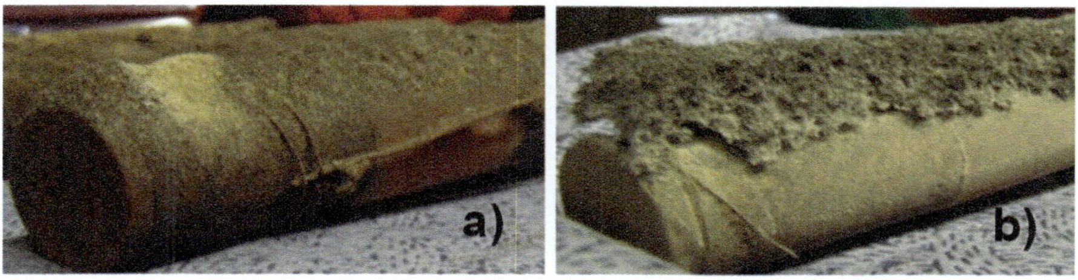

Fig. 14.5 Close-up view of the deposits formed from oxy-firing (**a**) Daw Mill:CCP(50:50 %,wt), (**b**) CCP(100 %,wt)

relatively close to the flame. As such the equilibrium of the exothermic reaction, $CO + 0.5O_2 \leftrightarrow CO_2$, would be shifted towards the generation of CO instead of CO_2, due to the high temperatures and elevated CO_2 [18]. The maximum flue gas temperatures reached during the experimental tests were achieved when oxy-firing 100 % Daw Mill coal (1,100 °C). The temperatures decrease as the content of biomass in the fuel increases, due to the lower heat value of this fuel. The HCl and NO contents decrease in the flue gas when there is more biomass in the fuel.

In terms of ash sampling data, the appearance of the deposits generated using different types of fuel indicates that when using 100 % CCP the structure of the deposit is more fibrous and porous than when oxy-firing 100 % Daw Mill coal or the blend of 50 % Daw Mill-50 % CCP (see Fig. 14.5). Additionally, from the SEM images the larger size of the particles of the 100 % CCP case is observed.

Figure 14.6 shows the elemental concentration (%wt) of the ash generated during the oxy-combustion of different fuels (100 % Daw Mill coal, 100 % cereal co-product biomass and 50 % Daw Mill coal-50 % cereal co-product biomass).

For elements such as K, Mg, Fe and Ti, the trend followed agrees with the previous analysis. It is worth mentioning that K and Mg increase with the percentage of biomass, and Fe and Ti rise with the percentage of coal. Other elements (O, Si) do not show a clear pattern and Ca content increases as the share of biomass increases, showing the opposite behaviour to that expected. The S content in ash increases with the share of biomass in the fuel fired, contrary to expectations given the elemental analysis of the parent fuels shown in Table 14.1 (the S content in coal is eight times higher than in biomass). This may be due to the higher content of K in deposits generated from 100 % CCP, as higher K in the fuel means more K_2SO_4 is likely to form. In addition, the lower temperatures experienced in the reactor may possibly have increased the conversion from SO_2 to SO_3 compared to cases using coal or co-firing [19]. However, the resulting trend for the S content in ash is not very different from the one shown by Stanger and Wall [20], where coals with low S content (0.24 %, db) generated ash with similar SO_3 content to the ash from oxy-firing higher S-content coals (0.88 %, db).

Cl content was only detected in the deposits generated when oxy-firing 100 % CCP biomass. This, together with the higher concentration of HCl in the flue gas

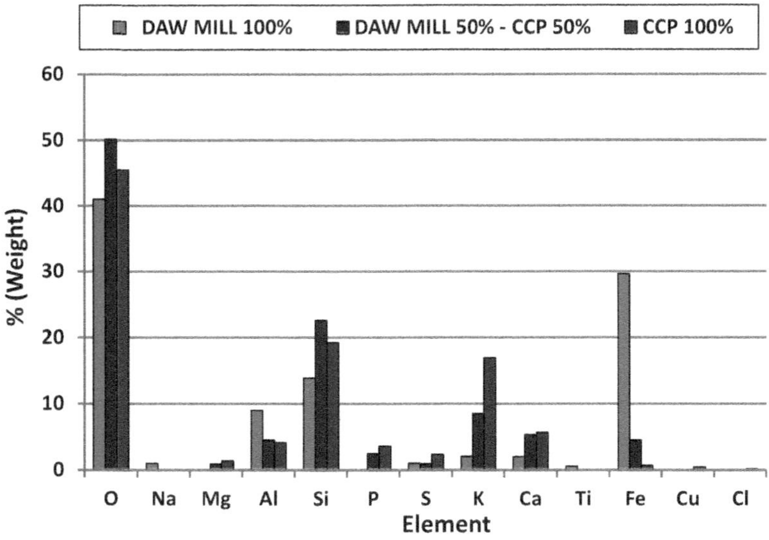

Fig. 14.6 Elemental concentrations of the top ash deposits from oxy-firing different blends of the fuels

generated by 100 % coal and 50 % coal-50 % biomass (also reported by Khodier and Simms [21], indicates that Cl tends to be in the gaseous phase and not retained in the ash deposit when there is a low share of biomass in the fuel.

14.4.2 Simulation Results (with Comparison with the Experimental Data)

To carry out the comparison between the simulation and the experimental data, key parameters such as amount of air ingress into the process, % RFG and % O_2 excess in the combustion stream have been varied within a range to recreate the same operating conditions as in the experimental tests. The experimental data used to validate the rate-based simulation model were generated at the retrofitted Cranfield University oxy-combustor. Each of the parameters varied for the simulations was changed while keeping a fixed value for the rest. The first phase of the rate-based simulations was focused on the study of the effects when the percentage of air ingress varied (2 %, 10 % and 18 % of the total flow rate fed to the oxy-combustor), and the selection of the amount of air ingress that better represented the experimental conditions during the operation of the pilot plant. Figure 14.7 shows the comparison between the experimental data and the results from the intermediate rate-based simulation model (box plot shown in Fig. 14.3).

For these simulations 55 % RFG was used, and two different excess O_2 feed values (0 and 5 %). The CO_2 percentage is better predicted by the simulations that

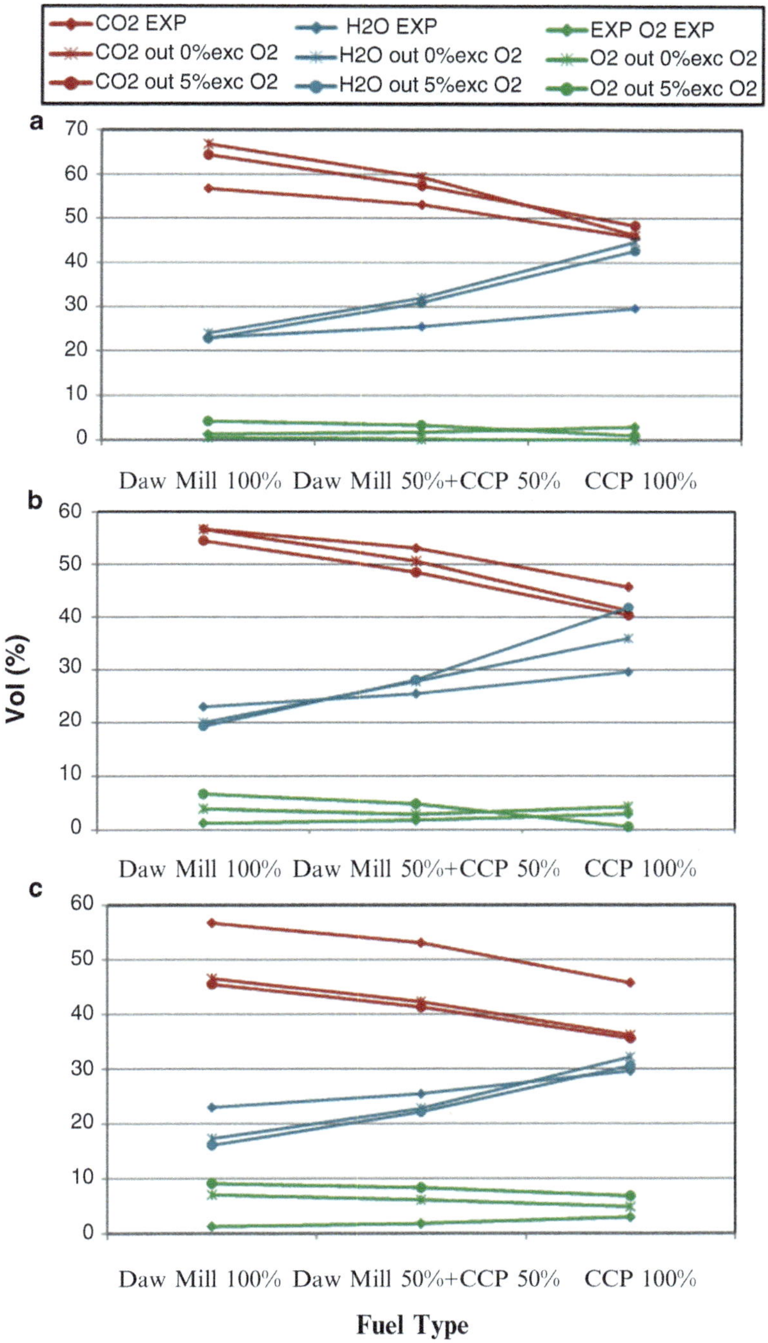

Fig. 14.7 Comparison between experimental results and rate-based simulation data burning different blends of fuel considering different percentages of air ingress. (**a**) 2 % Air ingress. (**b**) 10 % Air ingress. (**c**) 18 % Air ingress

used 2 and 10 % of air ingress. For the simulation case for 18 % of air ingress, the flue gas appears to be more diluted than in the experimental cases. Regarding the H_2O vapour content, the best estimation for the 18 % air ingress case is obtained using 100 % CCP as fuel. For the O_2 percentage at the exit of the oxy-combustor, the simulations using 2 and 10 % of air ingress give a good approximation in comparison with the experimental data. Consequently, supplementary simulations to improve the rate-based model will be done using 10 % air ingress.

The temperature estimation given by the model is in good agreement with the measurements of Smart et al. [6], for a 500 MWth test rig. Likewise, these calculations were checked using the FACT program, allowing for flame dissociation, and gave very similar results to the ASPEN calculation (T-Aspen:1,940 °C against T-FACT:1,944 °C). These results are generated using 100 % Daw Mill coal as fuel, feeding 5 % of excess of oxygen over the stoichiometric and recirculating 55 % of the flue gas to the oxy-combustor block. It should be noted that given that temperatures are at or below 2,000 K, dissociation has been ignored in the ASPEN simulation.

14.5 Conclusions

This chapter summarises the retrofitting experience and the experimental results in a 100 kW oxy-combustor located at Cranfield University in the UK, as well as the rate-based simulation results using Aspen Plus®. The fuel blends used for this study are 100 % Daw Mill coal, 50 % Daw Mill coal-50 % cereal co-product (biomass) and 100 % cereal co-product. The performance of the combustor was explored, keeping recirculation of the flue gas around 52 %, and the maximum percentage of CO_2 achieved was 56.7 %, wb (73.6 %, db) when 100 % Daw Mill coal was fired. The high content of H_2O vapour in the flue gas when firing 100 % cereal co-product biomass caused operational complications. Thus a water condenser will be installed as part of the flue gas conditioning equipment with the aim of avoiding these difficulties in forthcoming tests.

The ash deposition was evaluated taking into account different fuels and it appears that similar sulphate contents existed in deposits generated using 100 % Daw Mill coal and 100 % cereal co-product, even though the biomass has much lower S content in its elemental analysis. The higher K content in the deposits generated by oxy-firing of 100 % cereal co-product biomass was also noticeable. Consequently, operational problems, related to material corrosion, are likely to occur. In general the oxy-fuel environment appears to be more "corrosive" in comparison to the air-firing environment due to higher levels of K_2SO_4 formation in the deposits as well as the higher SO_x content in the flue gas generated in the oxy-combustion process. Also higher concentrations of K and Cl in the deposits generated by 100 % cereal co-product biomass suggest the formation of KCl, which would contribute to more corrosion problems as it is a highly fouling compound. Further corrosion research is required to study this effect in more detail.

A kinetic simulation model was developed and validated to determine the amount of air ingress in comparison with the experimental data. It was estimated that the air ingress into the reactor is 10 % of the total gas fed to the combustor. Simulation results using different values for % RFG and % O_2 excess fed to the oxy-combustor are achieved. In the second stage a rate-based model with partial condensation in the RFG is under development.

Acknowledgments The authors would like to thank the UK Engineering and Physical Sciences Research Council (EPSRC) and EON to the Oxy-Cap UK consortium for their financial support. The authors also acknowledge Dr. Jinsheng Wang from Canmet Energy for his help with FACT simulations.

This chapter is an augmented version of a paper presented at the International Conference on Clean Energy 2014 (ICCE-2014), in June 2014, Istanbul, Turkey.

References

1. Corsten M, Ramírez A, Shen L, Koornneef J, Faaij A (2013) Environmental impact assessment of CCS chains—lessons learned and limitations from LCA literature. Int J Greenhouse Gas Control 13:59–71
2. Arias B, Pevida C, Rubiera F, Pis JJ (2008) Effect of biomass blending on coal ignition and burnout during oxy-fuel combustion. Fuel 87(12):2753–2759
3. Toftegaard MB, Brix J, Jensen PA, Glarborg P, Jensen AD (2010) Oxy-fuel combustion of solid fuels. Prog Energy Combust Sci 36(5):581–625
4. Smart JP, Patel R, Riley GS (2010) Oxy-fuel combustion of coal and biomass, the effect on radiative and convective heat transfer and burnout. Combustion Flame 157(12):2230–2240
5. Borrego AG, Garavaglia L, Kalkreuth WD (2009) Characteristics of high heating rate biomass chars prepared under N_2 and CO_2 atmospheres. Int J Coal Geol 77(3–4):409–415
6. Smart J, Lu G, Yan Y, Riley G (2010) Characterisation of an oxy-coal flame through digital imaging. Combustion Flame 157(6):1132–1139
7. Field MA (1969) Rate of combustion of size-graded fractions of char from a low-rank coal between 1 200°K and 2 000°K. Combustion Flame 13(3):237–252
8. Murphy JJ, Shaddix CR (2006) Combustion kinetics of coal chars in oxygen-enriched environments. Combustion Flame 144(4):710–729
9. Sotudeh-Gharebaagh R, Legros R, Chaouki J, Paris J (1998) Simulation of circulating fluidized bed reactors using ASPEN PLUS. Fuel 77(4):327–337
10. Xiong J, Zhao H, Chen M, Zheng C (2011) Simulation Study of an 800 MWe Oxy-combustion Pulverized-Coal-Fired Power Plant. Energy Fuel 15:2405
11. Hu Y, Yan J (2012) Characterization of flue gas in oxy-coal combustion processes for CO_2 capture. Appl Energy 90(1):113–121
12. Edge P, Gharebaghi M, Irons R, Porter R, Porter RTJ, Pourkashanian M, Smith D, Stephenson P, Williams A (2011) Combustion modelling opportunities and challenges for oxy-coal carbon capture technology. Chem Eng Res Des 89(9):1470–1493
13. Wall T, Liu Y, Spero C, Elliott L, Khare S, Rathnam R, Zeenathal F, Moghtaderi B, Buhre B, Sheng C, Gupta R, Yamada T, Makino K, Yu J (2009) An overview on oxyfuel coal combustion—State of the art research and technology development. Chem Eng Res Des 87 (8):1003–1016
14. Steinmetz C, Bergins C, Weckes P, Dieter K (2011) Oxyfuel power plant design: retrofit options for different fuels. 2nd Oxyfuel Combustion Conference, Vol. 1, 12–16 September, Queensland, Australia, IEAGHG, UK

15. Haykiri-Acma H, Turan AZ, Yaman S, Kucukbayrak S (2010) Controlling the excess heat from oxy-combustion of coal by blending with biomass. Fuel Process Technol 91 (11):1569–1575
16. Tai Z, Zhaohui L, Xiaohong H, Jingzhang L, Dingbang W, Chuguang Z (2011) O_2/RFG Coal Combustion Results on a 300 kW Pilot Scale Facility. Second Oxyfuel Combustion Conference, 12-16 September, Queensland, Australia, IEAGHG, UK
17. Huang Y, Wang M, Stephenson P, Rezvani S, McIlveen-Wright D, Minchener A, Hewitt N, Dave A, Fleche A (2012) Hybrid coal-fired power plants with CO_2 capture: A technical and economic evaluation based on computational simulations. Fuel 101:244–253
18. Chen L, Yong SZ, Ghoniem AF (2012) Oxy-fuel combustion of pulverized coal: Characterization, fundamentals, stabilization and CFD modeling. Prog Energy Combust Sci 38 (2):156–214
19. Ahn J, Okerlund R, Fry A, Eddings EG (2011) Sulfur trioxide formation during oxy-coal combustion. Int J Greenhouse Gas Control 5(Suppl 1):S127–S135
20. Stanger R, Wall T (2011) Sulphur impacts during pulverised coal combustion in oxy-fuel technology for carbon capture and storage. Prog Energy Combust Sci 37(1):69–88
21. Khodier A, Simms N (2010) Investigation of gaseous emissions and ash deposition in a pilot-scale PF combustor co-firing cereal co-product biomass with coal. Conference on Renewable Energies and Power Quality, 2010. Conference on Renewable Energies and Power Quality, 2010

Part IV
Energy Storage

Chapter 15
Numerical Simulation of Wallboards Constructively Incorporated with Different PCM Content Solutions for Passive Cooling in Southern of Algeria

R. Ghedamsi, N. Settou, N. Saifi, S. Rahmouni, B. Dokkar, and B. Recioui

Abstract Buildings in Algeria represent 41.4 % of the total energy consumption. Heating and cooling represent the largest component of energy use. Opportunities to reduce the heating and cooling demand in buildings therefore represent a potentially significant part of the future energy and CO_2 emission savings. The heating and cooling loads of buildings are most of the time due to heat transfer through building envelope. From the point of view of energy saving, the most effective way to reduce these loads is to carry out thermo isolation with building envelope, between others, by using phase change materials (PCMs). The effectiveness of wall protection systems depends on several parameters such as orientation, size, and their thermal operation, with respect to the climate. The objective of this work is to study the thermal behavior of plaster wall containing three different types of PCM (eicosane, $Na_2CO_3 \cdot 10H_2O$, and $CaCl_2 \cdot 6H_2O$) with variable concentrations to highlight the importance of quantity and quality of PCM under the climatic conditions in Ouargla city (at south of Algeria). The results indicate that $CaCl_2 \cdot 6H_2O$ gives good performances compared with other PCM. The increase in PCM percentage makes it possible to increase thermal inertia.

Keywords Buildings • PCM • Energy saving • Thermal inertia

R. Ghedamsi (✉) • N. Settou • N. Saifi • S. Rahmouni • B. Dokkar • B. Recioui
Fac. des sciences appliquées, Lab. Promotion et valorisation des ressources sahariennes, Univ Ouargla, Ouargla 30 000, Algeria
e-mail: rghedamsi@gmail.com

I. Dincer et al. (eds.), *Progress in Clean Energy, Volume 2*,
DOI 10.1007/978-3-319-17031-2_15

Nomenclature

T	Temperature, °C
c	Specific heat, J/kg °C
k	Thermal conductivity, W/m °C
l_f	Fraction of melted PCM
T_0	Initial temperature of PCM melting point, °C
T_{end}	Final temperature of PCM melting point, °C
k	Conductivity of the materials, W/m °C
k_i	Conductivity of the building materials excepting the PCM, W/m °C
w_i	Weight fraction of the building materials excepting the PCM
w_{PCM}	The weight fraction of the PCM
k_{PCM}^{liq}	Conductivity of the liquid PCM, W/m °C
k_{PCM}^{sol}	Conductivity of the solid PCM, W/m °C
$\dot{Q}$	Entering heat flux, W/m^2
T_{ini}	Initial temperature of the plaster block, °C
T_{ini}^{cork}	Initial temperature of the cork, °C
k_{Cork}	Cork thermal conductivity, W/m °C
h_c	Convection heat transfer coefficient, W/m^2 °C
T_e	External wall temperature, °C
T_∞	Environmental temperature, °C
T_{max}	Maximum interior surface temperature, °C
T_{min}	Minimal interior surface temperature, °C

Greek Letters

δ	Width of the plaster block, m
ΔT	Time of dephasing, h
ρ	Density, kg/m^3
ρ_i	Density of the building materials excepting the PCM
ρ_{PCM}^{liq}	Density of the liquid PCM
ρ_{PCM}^{sol}	Density of the solid PCM

15.1 Introduction

Energy is a crucial factor for the social and economic development of societies. Today, energy consumption contributes to pollution, environmental deterioration, and global greenhouse emissions. Increases in energy consumption are driven by population growth (it is estimated that Algeria population will be 47 million by 2050) [1] and economic development that tends to increase energy use per capita. The building sector (residential and tertiary) is the largest energy consumer. Algeria is one of the major energy consumers and its contribution toward global energy consumption was 41.4 % in 2012 [2]. Most of this consumption is due to heating and air-conditioning systems which ensure an interior temperature compatible with comfort condition. Energy conservation of residential buildings has become an important part of national energy strategies and will continue growing in importance in future. The most effective way of reducing this energy is achieved by applying thermal insulation to the building envelope. The use of phase change materials (PCMs) in the walls can cure this problem. Since antiquity, the building cooling thanks to PCM energy thermal storage is a natural used process. In 1940s, Telkes studied the use of sodium sulfate decahydrate ($Na_2SO_4 \cdot 10H_2O$) to store solar energy and using this energy to heat buildings at night and cloudy days. In the beginning, its work did not cause much interest until the energy crisis (end of 1970s and beginning of 1980s) [3]. The use of PCM in buildings is the subject of very thorough studies in order to limit energy consumption. The first building boards containing PCM are carried out in Mexico by Wright and Balcomb. In 1970s, they built "passive" houses with systems called direct profits (construction materials with integrating paraffins in the concretes) [4]. In the same year, in France, it was made the first panel plaster-paraffin to improve thin wall inertia. However, inflammability and seepage of this paraffin on these two materials slow down the development of the integration of the PCM in building sector. The installation of "microencapsulation" technique which eliminates these two problems can restart the renewal of these materials in buildings from 2000 [5]. Many researchers studied the thermal performance of wall with PCM (floor, ceiling, and vertical wall) by experiments and simulations for building constructors in order to limit buildings' energy consumption and carried out thermal comfort. For example, Athienitis et al. carried out experimental and numerical simulation study in a full-scale outdoor test room with PCM gypsum board as inside wall lining. The PCM gypsum board used contained about 25 % by weight proportion of butyl stearate. It was shown that utilization of the PCM gypsum board may reduce the maximum room temperature by about 4 °C during the day and can reduce the heating load at night significantly [5]. Shilei et al. executed a comparative study between ordinary room phase change wall rooms. The PCM was manufactured by impregnation of capric acid and lauric acid (fatty acid based) mixture in the proportion of 82:18 % having freezing and melting temperatures of 19.14 and 20.39 °C. They observed that the room which integrates the PCM gives a good performance in terms of maintenance of heat and thermal comfort during the winter. The effects of room heat loss were at least reduced in winter. Moreover, the electric rate of power consumption for the heating was appreciably reduced [6]. Borreguero et al. examined thermal characterization and heat storage capacity of distinct PCMs. They have simulated and tested various PCM contents and reported that thermal conductivity of gypsum was autonomous on contents of wallboard microcapsules. Besides, they concluded that the higher the content of PCM in wallboard, the higher would be the heat storage capacity. In addition, gypsum block containing 5 % microcapsules would reduce the thickness of gypsum by 8.5 % without changing its insulating effects [7]. Darkwa et al. studied laminated and randomly mixed PCM drywalls. The analysis showed that the laminated PCM drywall performed thermally better. The laminated drywall sample also 3 % deviation of the average experimental result from the numerical about 55 % of the phase change process as against 48 % for the randomly distributed drywall sample. The laminated board sample also released about 27 %, more latent heat than the randomly distributed type at the optimum time of 90 min, thus validating previous simulation study [8]. Necib et al. studied the thermal behavior of a brick filled with PCM to increase the thermal inertia of buildings walls. This study relates to a clay brick with 20 cm as wall thickness; its hole cells were filled by paraffin and the thermocouples were judiciously fixed. This breadboard construction is limited thus to the analysis of heat transfer phenomena inside brick. The analysis of experimental data showed that PCM integration is advantageous to create a thermal comfort. The results of simulation show an operation similar to that revealed by measurements of the temperature variation in the tests on the breadboard construction [9]. The objective of this work is to study the thermal behavior of plaster wall containing three different PCM types (eicosane, $Na_2CO_3 \cdot 10H_2O$, and $CaCl_2 \cdot 6H_2O$) with variable concentrations to highlight the importance of quantity and quality of PCM under the climatic conditions in Ouargla city (at south of Algeria).

15.2 Model Description

Figure 15.1 shows the physical model which consists of a hollow metallic box of aluminum, through which water constantly flows by means of a pump from a thermostatic bath at the desired temperature. This water allowed for the manipulation of the temperature on the aluminum cell. Dimensions of the aluminum cell

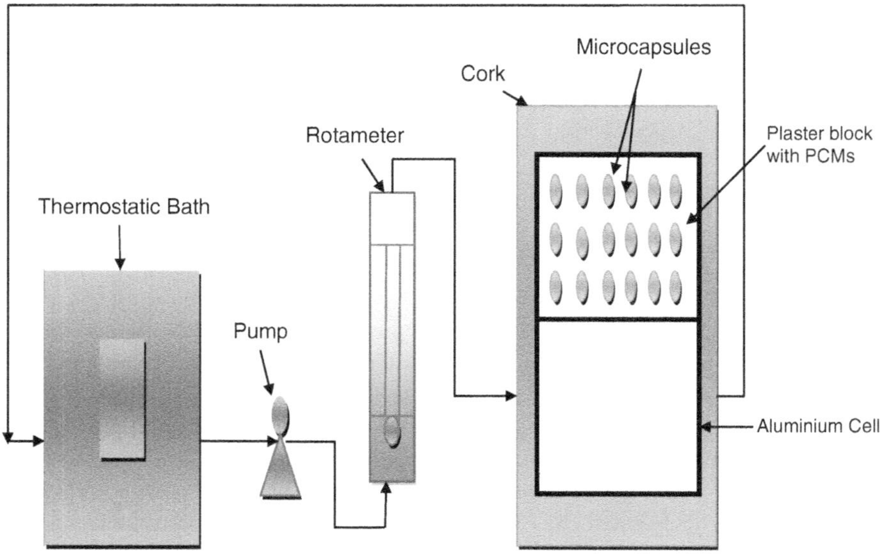

Fig. 15.1 Physical model for studying the thermal behavior of walls

Table 15.1 Properties of the materials [7, 11–13]

Material	k [W/m.°C]	ρ_{PCM}^{sol} [kg/m^3]	ρ_{PCM}^{liq} [kg/m^3]	Cp [kJ/kg °C]	T_F (°C)	L_f[kJ/kg]
Polystyrene	0.08	1,050	–	1.3	–	–
Cork	0.06	272	–	3.2	–	–
Plaster	0.2	900	–	0.9	–	–
Eicosane	0.15	856	778	2.21	37	247
$Na_2CO_3.10H_2O$	0.544	1,458	1,485	2.16	32	252
$CaCl_2.6H_2O$	0.54	1,802	1,562	2.16	29	190.8

were 10 × 6 × 3 cm with a wall thickness of 1.0 mm. The cell was incorporated in an internal diffuser plate to improve the liquid distribution avoiding the formation of preferential flow paths. The plaster wall containing three different types of PCM (eicosane, $Na_2CO_3 \cdot 10H_2O$, and $CaCl_2 \cdot 6H_2O$) with the same dimensions of the aluminum cell was placed on the upper surface of the aluminum cell and further insulated with 2 cm cork board thickness. All the wallboard characterizations were carried out by applying a thermostatic bath. The obtained bath temperatures are at the same level of periodic ambient temperature in Ouargla city. Maximal and minimal temperature used to determine the periodic ambient temperature are obtained by averaging measurements recorded in meteorological data over the years 2000–2010 in the month of July [10]. The properties of materials used in the physical model are given in Table 15.1.

15.3 Mathematical Model

In this work, we study the thermal behavior of plaster wall containing three different types of PCM (eicosane, $Na_2CO_3 \cdot 10H_2O$, and $CaCl_2 \cdot 6H_2O$) with variable concentrations to highlight the importance of quantity and quality of PCM under the climatic conditions of Ouargla. The mathematical model is formulated using the following assumptions [7]:

1. Microcapsules are homogeneous and uniformly distributed inside the wall.
2. The rate of encapsulation equals to 50 %.
3. The total volume of the wall is constant.
4. The contribution of the PCM latent heat to the storage capacity of the wall is established by the variation of the apparent specific heat capacity with the temperature.
5. Thermal conductivities and densities of the building materials are independent of temperature but different for solid and liquid phases. Thus, the PCM density and conductivity are function of its melted fraction.
6. The heat transfer through the wall is one dimensional.

7. The variation of the thermo-physical properties with the temperature of the insulating material is negligible due to the smaller range of temperature in which it is working.

The mathematical model is based on Fourier equation of heat conduction. Several types of walls will be studied by associating various materials: plaster and PCMs. The numerical simulation will be carried out by solving the equation of heat conduction through these walls with suitable boundary conditions. The transient one-dimensional heat conduction equation may be written as

$$\frac{\partial(\rho \cdot c \cdot T)}{\partial t} = \frac{\partial}{\partial x}\left(k \cdot \frac{\partial T}{\partial x}\right) \tag{15.1}$$

where x and t are the space and time coordinates, respectively. T is the temperature, and ρ, c, and k are the density, the specific heat, and the thermal conductivity, respectively.

The wall density and conductivity at any temperature can be obtained as function of the initial densities and conductivities of the original materials, respectively. They depend on the fraction of melted PCM (l_{f}) at each temperature. The fraction of melted PCM can be obtained as [7]

$$T \leq T_0; \quad l_{\mathrm{f}} = 0 \tag{15.2}$$

$$T_0 < T \leq T_{\mathrm{end}}; \quad l_{\mathrm{f}} = \frac{\int_{T_0}^{T} C_{\mathrm{P}}^{\mathrm{eff}}\, dt}{\int_{T_0}^{T_{\mathrm{end}}} C_{\mathrm{P}}^{\mathrm{eff}}\, dt} \tag{15.3}$$

$$T > T_{\mathrm{end}} \quad l_{\mathrm{f}} = 1 \tag{15.4}$$

where T_0 and T_{end} are the initial and final temperature of PCM melting point, respectively. Thus, if the total volume is assumed constant, the wallboard density and conductivity at any temperature can be obtained as follows [7]:

$$\rho = \sum_{i=1}^{N} \rho_i w_i + w_{\mathrm{PCM}}\left(l_{\mathrm{f}}\rho_{\mathrm{PCM}}^{\mathrm{liq}} + (1 - l_{\mathrm{f}})\rho_{\mathrm{PCM}}^{\mathrm{sol}}\right) \tag{15.5}$$

$$k = \sum_{i=1}^{N} k_i w_i + w_{\mathrm{PCM}}\left(l_{\mathrm{f}}k_{\mathrm{PCM}}^{\mathrm{liq}} + (1 - l_{\mathrm{f}})k_{\mathrm{PCM}}^{\mathrm{sol}}\right) \tag{15.6}$$

where ρ_i, k_i, and w_i are the density, conductivity, and weight fraction of the building materials excepting the PCM; w_{PCM} is the weight fraction of the PCM; $\rho_{\mathrm{PCM}}^{\mathrm{liq}}$ and $k_{\mathrm{PCM}}^{\mathrm{liq}}$ are the density and the conductivity of the liquid PCM; and $\rho_{\mathrm{PCM}}^{\mathrm{sol}}$ and $k_{\mathrm{PCM}}^{\mathrm{sol}}$ are their respective solid-phase density and conductivity. N is the number of wall building materials excepting the PCM and i is a counter. Finally, l_f is the melted fraction of PCM.

To solve Eq. (15.1), it is necessary to specify initial conditions and two boundary conditions. The initial conditions and the boundary conditions are given as follows [7]:

(a) For the wall zone

$$t = 0; \quad T = T_{\text{ini}} \tag{15.7}$$

$$x = 0; \quad k\frac{dT}{dx}\Big|_{x=0} = \dot{Q} \tag{15.8}$$

$$x = \delta; \quad k\frac{dT}{dx}\Big|_{x=\delta^-} = k_{\text{Cork}} \frac{dT}{dx}\Big|_{x=\delta^+} \tag{15.9}$$

where $\dot{Q}$ is the entering heat flux, T_{ini} is the initial temperature of the plaster block, k_{Cork} is the cork thermal conductivity, and δ is the width of the gypsum block.
(b) For the insulated zone, Eq. (15.1) becomes

$$\frac{\partial T}{\partial t} = \alpha_{\text{cork}} \frac{\partial^2 T}{\partial x^2} \tag{15.10}$$

$$t = 0; \quad T = T_{\text{ini}}^{\text{cork}} \tag{15.11}$$

$$x = \delta + \delta_{\text{cork}}; \quad k_{\text{cork}}\frac{dT}{dx}\Big|_{x=\delta+\delta_{\text{cork}}} = h_c\,(T - T_\infty) \tag{15.12}$$

where k_{cork} and α_{cork} are the conductivity and thermal diffusivity of the cork, respectively; h_c is the convection heat-transfer coefficient. In this study, h_c is taken as 10 W/m$^{2\circ}$C; T_∞ is the environmental temperature and δ_{cork} is the width of the insulation slab. Both T_∞ and T_i^{cork} are supposed constant and equal (23 °C).

Figure 15.2 shows a typical heat flux curve when a cool plaster block is placed on a hot flat plate.

The system above of the equations was discretized by the finite difference method. By applying this numerical method, we can transform the partial differential Eqs. (15.1) and (15.10) to an ordinary equation system.

15.4 Results and Discussion

15.4.1 Validation of the Numerical Results

The first steep consists in validating our numerical results by comparing them with experimental results presented by Borreguero et al., 2010. Figure 15.3 shows the change of the temperature of wall external surface. We observe a good agreement between the two curves.

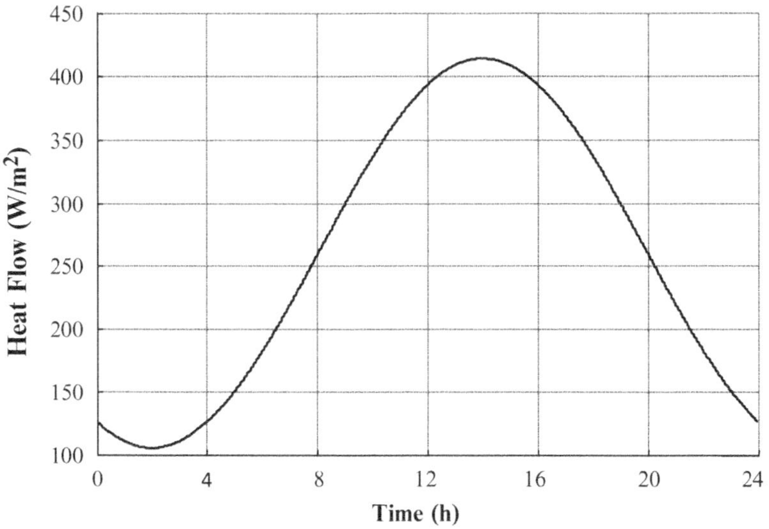

Fig. 15.2 Inlet heat flow of the plaster blocks as a function of time

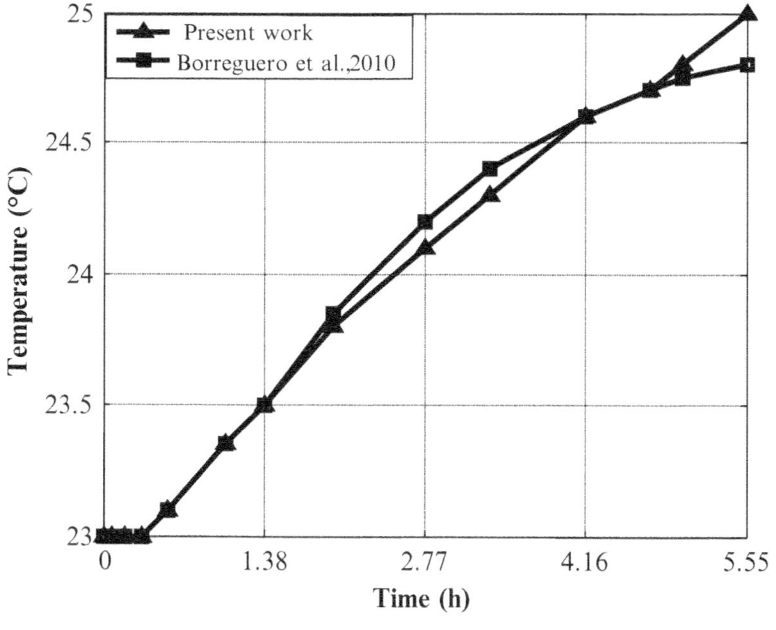

Fig. 15.3 Comparison of present work with Borreguero et al.'s work

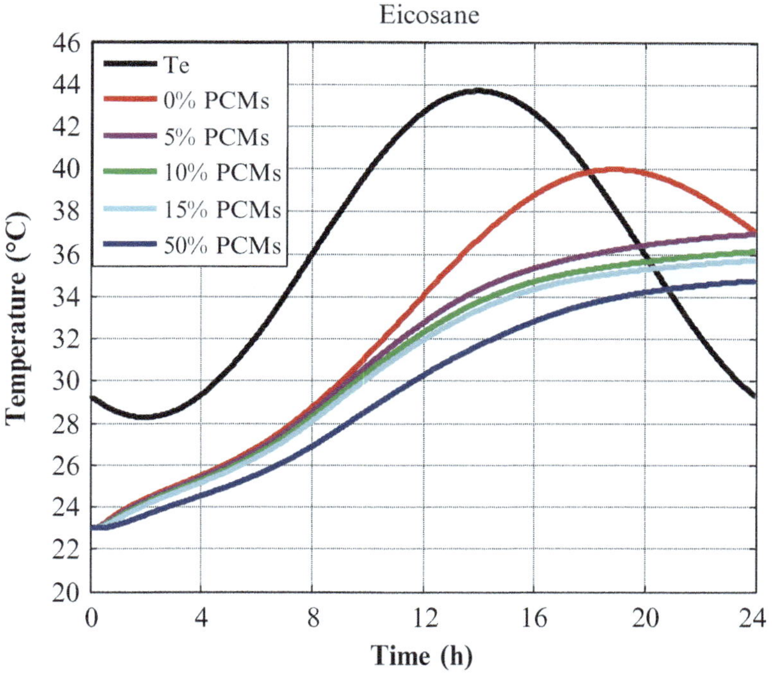

Fig. 15.4 Comparison of the evolutions of the interior temperature for different concentrations (case: eicosane)

15.4.2 Interior Surface Temperature According to Concentration and Type of PCM

A wall plaster containing three different types of PCMs (eicosane, $Na_2CO_3 \cdot 10H_2O$, and $CaCl_2 \cdot 6H_2O$) with variable concentrations (0 %, 5 %, 10 %, 15 %, and 50 %) was studied to highlight the importance of the quantity and the quality of PCM.

The following Figs. 15.4, 15.5, and 15.6 present the variation in the temperature on the interior wall surface of mixture plasters/PCMs with different concentrations of PCMs.

Storage by the plaster is storage by significant heat. The efficiency of the plaster on the attenuation of the amplitude of the oscillations is well noticed. There is practically dephasing of 5 h.

It is noticed, moreover, that the stability of the temperature of interior surface is related to the rate (the increase in the quantity of the PCM has a positive effect) and to the type of the PCM contained in the wall (the nature of the PCM, in particular, the melting point of the PCM). The optimal value of the melting point depends on the interior temperature, which also depends on the outside temperature and thermal resistance of the wall.

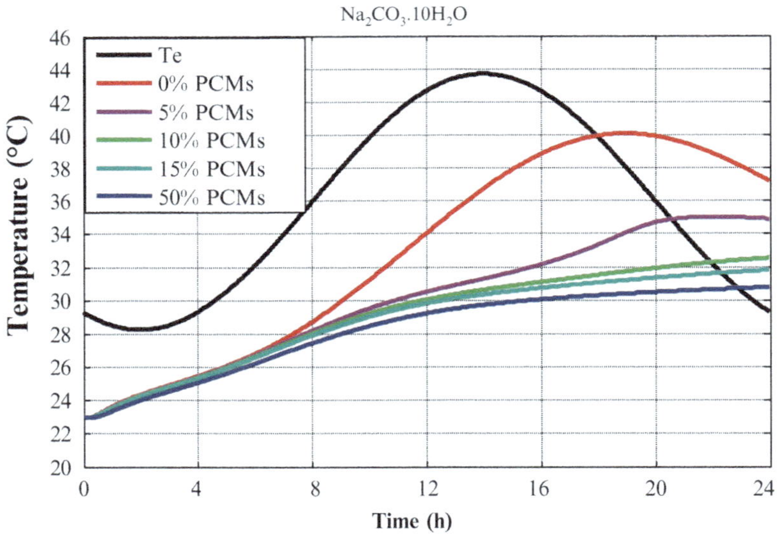

Fig. 15.5 Comparison of the evolutions of the interior temperature for different concentrations (case: $Na_2CO_3.10H_2O$)

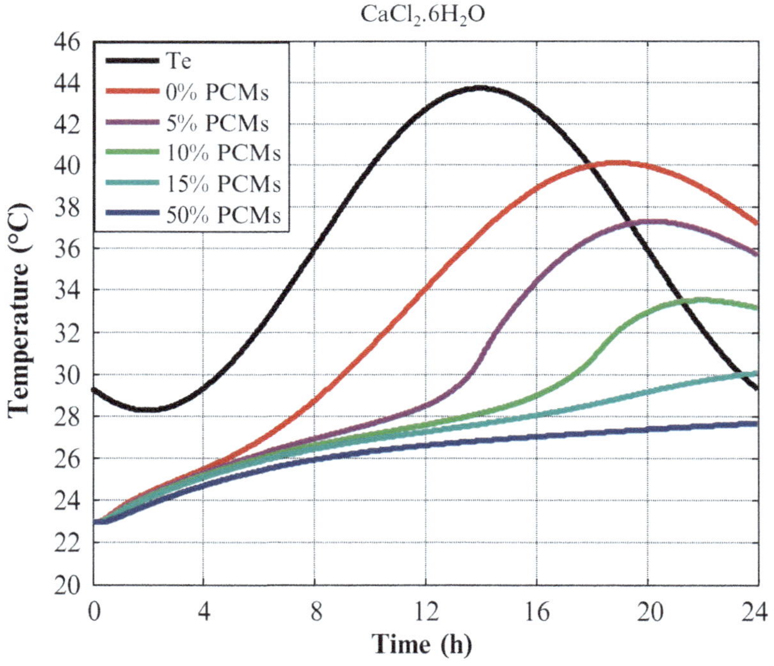

Fig. 15.6 Comparison of the evolutions of the interior temperature for different concentrations (case: $CaCl_2.6H_2O$)

Table 15.2 Variation of the maximum and minimal temperatures on interior surface according to the concentration (case: eicosane)

Concentration	0 %	5 %	10 %	15 %	50 %
$T_{max}(°C)$	40.12	37	36.15	35.75	34.77
$T_{min}(°C)$	23	23	23	23	23
ΔT	17.12	14	13.15	12.75	11.77

Table 15.3 Variation of the maximum and minimal temperatures on interior surface according to the concentration (case: $Na_2CO_3 \cdot 10H_2O$)

Concentration	0 %	5 %	10 %	15 %	50 %
T_{max}(°C)	40.12	34.91	32.57	31.85	30.8
T_{min}(°C)	23	23	23	23	23
ΔT	17.12	11.91	9.57	8.85	7.8

Table 15.4 Variation of the maximum and minimal temperatures on interior surface according to the concentration (case: $CaCl_2.6H_2O$)

Concentration	0 %	5 %	10 %	15 %	50 %
T_{max}(°C)	40.12	37.33	33.57	30.1	27.7
T_{min}(°C)	23	23	23	23	23
ΔT	17.12	14.33	10.57	7.1	4.7

The comparison between the maximum value and the minimal value of the temperature of interior surface also as the difference for each concentration is presented in Tables 15.2, 15.3, and 15.4:

According to the results presented, one notices that the difference between the maximum and minimal temperatures decreases with the increase in the concentration of PCM.

15.4.3 Comparison of Simulations for Different PCMs

We will compare in what follows the profiles of temperatures interior and external of the walls for three different PCMs. The three studied mixtures are, respectively, (eicosane/plaster), ($CaCl_2 \cdot 6H_2O$/plaster), and ($Na_2CO_3 \cdot 10H_2O$/plaster), and this for various concentrations of PCMs (10 %, 15 %, and 50 %) (Figs. 15.7, 15.8, and 15.9).

The variations in temperature decrease as long as the concentrations of PCM in the mixtures increase. This report is valid for the types of PCM used. The mixture plaster/$CaCl_2 \cdot 6H_2O$ in the three cases of figure shows the best characteristics with respect to the thermo isolation. Dephasing between the maximum temperature without PCM and the maximum temperature with PCM appeared for various concentrations. With regard to eicosane and $NaCO_3 \cdot 10H_2O$, breadth dephases residence constant equal to 7 h and remains insensitive with the concentration used. For $CaCl_2 \cdot 6H_2O$, dephasing is function of the concentration and reaches at 7 h a maximum starting from a concentration of 15 %.

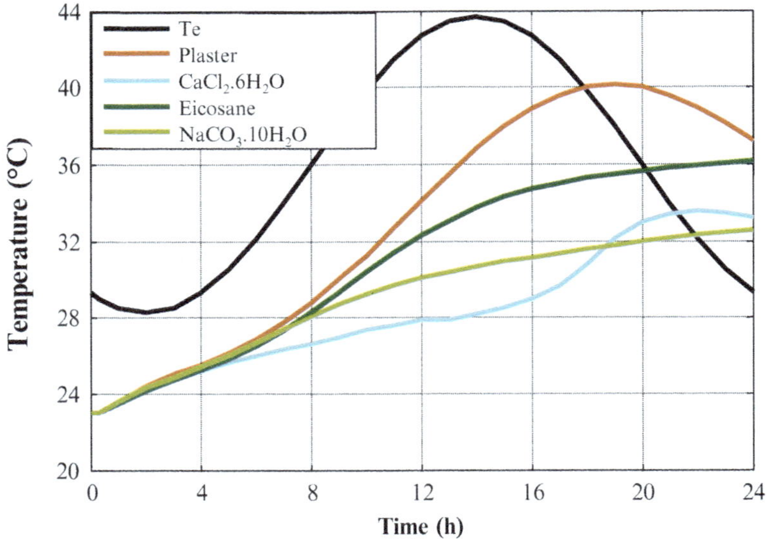

Fig. 15.7 Comparison of the variation in the interior temperature for three walls of identical concentrations (case 10 %)

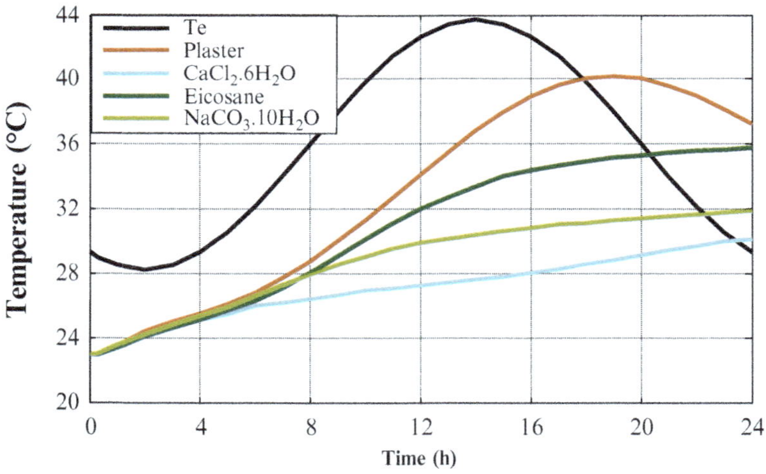

Fig. 15.8 Comparison of the variation in the interior temperature for three walls of identical concentrations (case 15 %)

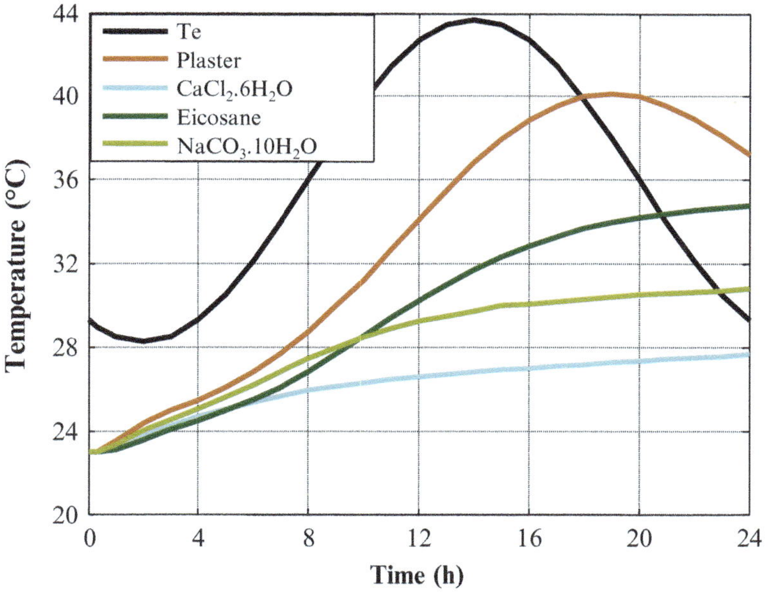

Fig. 15.9 Comparison of the variation in the interior temperature for three walls of identical concentrations (case 50 %)

From results obtained, it can be said that $CaCl_2 \cdot 6H_2O$ having a melting point equalizes at the temperature (29 °C) and carries out best thermal comfort according to the selected environment of Ouargla.

15.5 Conclusion

In this work, we studied the thermal behavior of a wall containing PCM to carry out the thermal and economic energy comfort in the buildings at summer. For that, numerical simulation was carried out with Matlab7 software on various wall types. We studied the thermal behavior of plaster wall containing three different types of PCMs (eicosane, $Na_2CO_3 \cdot 10H_2O$, and $CaCl_2 \cdot 6H_2O$) with variable concentrations (0 %, 5 %, 10 %,15 %, and 50 %) to highlight the importance of the quantity and the quality of PCM.The results obtained highlighted the following:

- $CaCl_2 \cdot 6H_2O$ gives good performances compared with other PCMs studied. The increase in percentage of PCM makes it possible to increase thermal inertia.
- Dephasing between the maximum temperature without PCM and the maximum temperature with PCM appeared for various concentrations. For $CaCl_2 \cdot 6H_2O$, dephasing is function of the concentration and reaches at 7 h a maximum starting from a concentration of 15 %.

References

1. www.dii-eumena.com. Accessed on January 10, 2014.
2. www.sonatrach.com. Accessed on January 15, 2014.
3. Zhu N, Ma Z, Wang S (2009) Dynamic characteristics and energy performance of buildings using phase change materials: A review. Energy Convers Manag 50:3169–3318
4. Riedere P, Catarina O (2007) Comparaison international bâtiment et énergie, C7- Stockage de chaleur, ADEME-PUCA-CSTB C151
5. Athienitis AK, Liu C, Hawes D, Banu D, Feldman D (1997) Investigation of the thermal performance of a passive solar test-room with wall latent heat storage. Build Environ 32:405–410
6. Shilei L, Neng Z, Guohui F (2006) Impact of phase change wall room on indoor thermal environment in winter. Energy Build 38:18–24
7. Borreguero AM, Sanchez ML, Valverde JL, Carmona M, Rodiguez JF (2010) Thermal testing and numerical simulation of gypsum wallboards incorporated with different PCMs content. Appl Energy 88:930–937
8. Darkwa K, Kim J-S (2005) Dynamics of energy storage in phase change drywall systems. Int J Energy Res 47: 335–343
9. Necib H, Settou N, Saifi N, Damene D (2013) Brick Experimental and numerical study of a usual brick filled with PCM to improve the thermal inertia of buildings, Terragreen 13 international conference, Beirut, Lebanon, 15-17 February
10. Office National de la Météorologie ONM (2011) Wilaya de Ouargla, Edition
11. Ahmad M (2004) Nouveaux composants actifs pour la gestion énergétique de l'enveloppe légère des bâtiments, Thèse de Doctorat: Mécanique-Energétique, Université Joseph Fourier, Grenoble
12. Zalba B, Marin JM, Cabeza LF, Mehling H (2003) Review on thermal energy storage with phase change: materials, heat transfer analysis and applications. Appl Therm Eng 23:251–283
13. Alawadhi EM (2008) Thermal analysis of a building brick containing phase change material. Energy Build 40:351–357

Chapter 16
The Energy Conservation Potential of Using Phase-Change Materials as Thermal Mass Material for Air Source Heat Pump-Driven Underfloor Heating System in a Building

Ming Jun Huang and Neil Hewitt

Abstract Improved energy efficiency in buildings is the key element to reduce the greenhouse gas emissions while contributing to energy security. Underfloor heating is a more efficient and economical method for home heating with improved thermal comfort than any other heating methods. Due to the low-temperature heating source requirement the underfloor heating is widely accepted as the most efficient form of heating. Heat pumps are energy-efficient equipment to provide low-temperature heat source which is suitable for underfloor heating applications. Phase-change materials (PCMs) are attractive for use in thermal energy store for underfloor heating applications due to their high-energy storage density over a small temperature range, therefore allowing the air source heat pump to operate during winter warmer afternoon ambient air conditions or in an electricity tariff management mode. A numerical simulation model has been validated and used to analyse the thermal performance of PCM-layered underfloor heating under different heating modes. Different layouts of the underfloor heating pipes with PCMs as floor mass material were analysed for realistic diurnal temperature boundary conditions and temperature distribution was predicted.

16.1 Introduction

Buildings account for more than 40 % of the total energy consumption and in cold climate the heat loss to the ground might be responsible for up to one-third or even a half of total heat losses. Underfloor heating, in which the hot water pipes are built in the floor as the heating medium, is a more efficient and economical method for

M.J. Huang (✉) • N. Hewitt
Centre for Sustainable Technologies, School of Built Environment,
University of Ulster, Newtownabbey, Co. Antrim, N. Ireland BT37 0QB, UK
e-mail: m.huang@ulster.ac.uk; nj.hewitt@ulster.ac.uk

I. Dincer et al. (eds.), *Progress in Clean Energy, Volume 2*,
DOI 10.1007/978-3-319-17031-2_16

home heating with improved thermal comfort than any other heating methods [1–4]. Uniform temperature distribution and low-temperature heating source requirements make the underfloor heating widely accepted as the most efficient form of heating across the mainland Europe and many countries in the world [2, 5]. Unlike the conventional radiator heating where the supplied hot water temperature needs to be above 70 °C through burning fossil fuel, the supplied hot water can be less than 50 °C for the underfloor heating system. Heat pumps are energy-efficient equipment to provide low-temperature heat source for residential applications. They have some impact on the environment as they need electricity to run, but the heat they extract from the ground, air, or water is constantly being renewed naturally. Air source heat pumps (ASHPs) are easy to install, and will be capable to provide low-temperature heat for underfloor heating to keep the floor surface temperature to around 27 °C. However when the ambient temperature is low the efficiency of the ASHPs will reduce and there is frost formed on the surface of outdoor heat exchanger; this will reduce the efficiency of the ASHP further. Defrost for ASHPs has to be carried even when the ambient temperature is above 5 °C. Therefore operating the ASHPs during daytime for the underfloor heat store and releasing the stored heat for later night use will achieve the benefit of using ASHPs for underfloor heating and avoid dropping on ASHP efficient due to defrost for long time continuing operation. In addition, when there is no defrosting required in night-time, operating the ASHPs for underfloor heating during the off-peak time can save the energy and the stored heat in floor can be released for early morning use to reduce the electrical demands on electricity distribution network. The shift of electrical consumption from peak period to off-peak periods will provide significant economic benefit and balance the electricity demanding. The issues are how to efficiently store enough heat from ASHP to the floor and discharge it for later use with low cost.

Electrical space heating with heat storage for underfloor heating has been well investigated for many years [6, 7]. Heavy thermal mass materials are often used to store the thermal energy for electrical floor heating system [7]. The increased temperature may cause the heat loss during the storing period and also the quantity of thermal mass will increase the cost. Phase-change materials (PCMs) are attractive for use in thermal energy store for underfloor heating applications due to their high-energy storage density over a small temperature range, therefore allowing the ASHP to operate during winter warmer afternoon ambient air conditions or in an electricity tariff management mode. This energy is then stored in the floor material for evening heat release, with the added benefit of delaying the ASHP start until after the peak electricity demand time of early evening and keeping the thermal comfort with narrow temperature swing.

Floor cover materials and structure are important in the design of underfloor heating systems with thermal storage. Improved information regarding the dynamic response of a structure should lead to a more successful holistic low-energy design. Jin et al. [5] have developed a finite-volume model to study the impacts of water velocity on the performance of the underfloor cooling system. The effect of water velocity on the performance of the underfloor cooling is not significant. De Monte [8] studied the transient response of the multilayered composite conducting slab using one-dimensional simulation method. Lu and Tervola [9] derived an analytical solution for composite slab with period temperature changes. Zhang et al. [10] developed a simplified model to calculate the performance of radiant floor with cooling and heating capacity.

Ling and Poon [11] reviewed the means of PCM incorporation in concrete and the properties of concrete. Three predominant methods are immersion, impregnation, and mixing. They found out from the literature review that organic PCM, and particularly paraffin wax, seems to be one of the most suitable latent heat storage materials that can be used in concrete, despite the low thermal conductivity, leakage, and flammability properties. Concrete materials incorporating PCMs can provide a more uniform temperature distribution which ensures that temperatures conductive to human thermal comfort in the buildings' internal space can be maintained. The improvement of the thermal storage of PCM concrete may make it more widely used in construction and building applications. Some investigators have studied the improvement of the thermal conductivity of PCM by mixing with a particle metal or carbon-containing substance that has a high thermal conductivity [12–14]. Kim and Drzal [15] have presented the process of preparing exfoliated graphite nanoplatelets (xGnP) as high-thermal-conductive PCMs for heat storage. Jeon et al. [16] studied the performance of three composite PCMs (hexadecane, octadecane and paraffin wax) loading xGnP and found that for the paraffin loaded with xGnP the melting temperature was maintained, latent heat was decreased less than 1.2 % and thermal conductivity can increase 28 %. Therefore octadecane, hexadecane and paraffin used as PCMs can be considered as energy-saving building materials for residential buildings. Shape-stabilised PCM plates have attracted the thermal storage application due to ease of handling to integrate to buildings and avoidance of leakage. A new shape-stabilised polymer composite PCM containing 85 % of paraffin, with a latent heat of 110 kJ/kg and melting point at about 27 °C, has been inserted to the hollow of lightweight concrete floor panel to increase the thermal inertia [17]. The experimental results show that the PCM can store a great quality of thermal energy and shift the peaks of load. Therefore in this chapter the shape-stabilised polymer composite PCM will be used as thermal mass for underfloor heating application.

A two-dimensional finite-volume numerical simulation model has been validated and used to examine the thermal performance of PCM-augmented underfloor heating system for a heat pump energy management application. The three most significant thermal characteristics of a floor heating are the heat capacity, U value and location of insulation and mass layers. These parameters as well as others characterise the building itself by affecting heating/cooling loads and floor surface temperatures. In this chapter, the floors with different types of PCM were heated by ASHP through pipes with different spacings. Those PCMs are available commercially with different melting temperatures. Different layouts of the underfloor heating pipes with PCM as floor mass material were analysed for realistic diurnal temperature boundary conditions and the heat transfer and temperature distribution were calculated.

16.2 Description of Numerical Model

The two-dimensional finite-volume model will be validated and used to predict the effect of the design and operating parameters, i.e. the pipe spacing, filling layer thickness and pipe water temperature, on the floor heat storage and heat release. The temperature distribution has been analysed with the effect of heating and heating period.

16.2.1 Governing Equations for Underfloor Heating System with PCM Layer

A good understanding of the fundamental heat transfer processes involved is essential for accurately predicting the thermal performance of the PCM floor heating system driven by HP and for avoiding costly system design mistakes. The simulation model used in this work is modified from a two-dimensional temperature-based finite-volume-based conjugated heat transfer numerical model which was used for predicting the temperature rise in BIPVs in a PV/PCM system which was validated by theoretical analysis and experimental results [14]. This modified model for underfloor heating will be validated by the experimental data from other researchers' published work on underfloor heating purposes. The developed model can be used to predict the thermal performance of transient temperature distribution, discharge and heat flux within a two-dimensional region in the floor heating system with PCM as thermal mass material for different supplied water temperatures, heat supply, floor structure, hot water pipe spacing, floor layer materials and ambient boundary conditions. The modified numerical model can be used for the PCMs with different transient temperatures. The following assumptions are made:

1. The heat conduction in the PCM-combined underfloor heating system is two dimensional and the end sides at the floor system edges are adiabatic.
2. The PCM is homogeneous and isotropic.
3. The convection effect in the molten PCM is neglected.
4. The thermophysical properties of the floor materials are constant except the specific heat capacity and density of the PCM during phase changes.
5. The room ambient conditions are assumed as constant and there is no solar radiation gain.
6. The interfacial resistances between the pipe and filling material are negligible.
7. Since the tube is very thin and its thermal conductivity is very high, the thermal resistance of the tube can be neglected. Thus, the model assumes that the heat transfer fluid directly contacts the solid and the tube thickness can be assumed to be zero.
8. Advection, not conduction, dominates the heat transfer in the fluid. Therefore, thermal conduction in the axial direction in the fluid is negligible.

The energy equation for the floor:

$$\rho_S c_S \frac{\partial T}{\partial t} + \nabla \cdot (k_S \nabla T) = 0 \tag{16.1}$$

where the same equations hold good for all the layers of floor material and PCM by incorporating suitable k, ρ, and C_p values. S is the heat flux from the pipe. The instantaneous continuity of heat flux and temperature at the interfaces of pipes with concrete layer or PCM layer are preserved.

In the upper and down surface boundary, where the floor system is exposed to indoor environment and ground, the boundary conditions are

$$k\frac{\partial T}{\partial x}\bigg|_{x=0} = S_{\text{rad}} + h_0(T_{\text{amb, indoor}} - T_{x=0}) \tag{16.2}$$

$$k\frac{\partial T}{\partial x}\bigg|_{x=L} = h_L(T_{x=L} - T_{\text{amb, ground}}) \tag{16.3}$$

where h_0 and h_L are the heat transfer coefficients from the upper and under surfaces of the floor system to the surroundings. S_{rad} is the heat received by the floor system on the upper surface from the incident solar energy which is ignored in this work. The boundary conditions with the interfaces of pipes with floor-filling materials are

$$T(x_{\text{pipe}}, t) = T_w \quad (x = \text{pipe position}, t = \text{during heating}) \tag{16.4}$$

$$S(x_{\text{pipe}}, t) = S_{\text{w}} \quad (x = \text{pipe position}, t = \text{during heating}) \tag{16.5}$$

$$k\frac{\partial T}{\partial x}\bigg|_{x=\text{pipe interface}} = S_w \tag{16.6}$$

where T_{w} is the average pipe water temperature and S_{w} is the heat flux supplied by water pipe to the floor heating system.

The phase change occurs at a set temperature. In the case of constant specific heat capacities for each phase, the temperature field can be defined as

$$T = \begin{cases} E/c_S & & T < T_{\text{m}}\,(\text{solid line}) \\ T_{\text{m}} & 0 < E < H, & T > T_{\text{m}}\,(\text{melt zone}) \\ T_{\text{m}} + (E - H)/c_L & E \geq H, & T > T_{\text{m}}\,(\text{liquid phase}) \end{cases} \tag{16.7}$$

The latent heat value E of the PCM in the melt zone is modelled as high-sensible heat value in each time step and accumulated with time. With time elapsed when the accumulated heat is larger than the specified latent heat H of PCM the PCM is changed to liquid phase.

The governing equation along with the boundary conditions is discretised using the full-implicit finite volume method. The space grid size is 1 mm and the time step is variable.

16.2.2 Physical Model Description

A scheme of the potential floor structure is shown in Fig. 16.1. The potential material properties used in the underfloor structure are shown in Table 16.1. The spacing between two pipes in 150, 200 and 300 mm is studied. The external diameter of the heating pipe is 20 mm. The average temperature of the supply and return water is used as the calculated water temperature in this study. Standard concrete and timber joist underfloor heating without PCM and with PCM are studied.

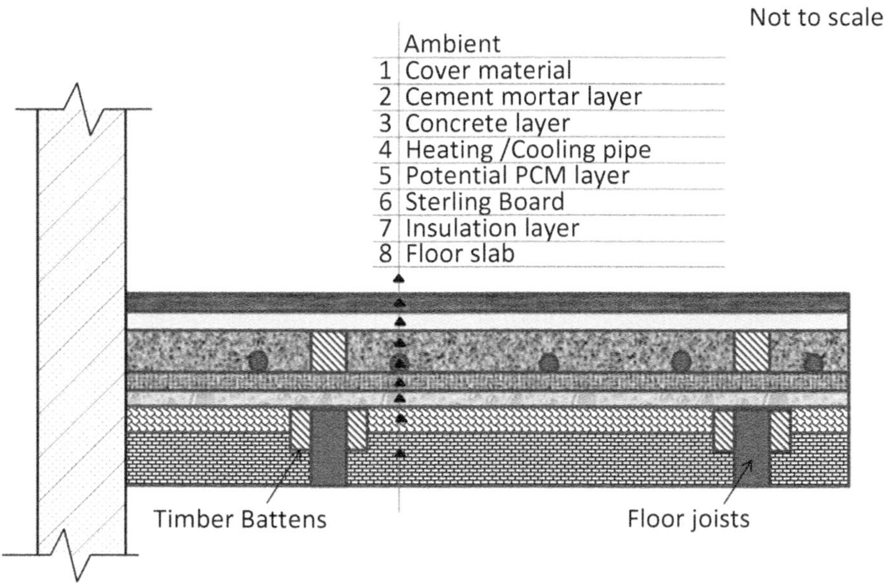

Fig. 16.1 Schematic diagram of the example of experimented underfloor structure with potential layers with joists

Table 16.1 Thermo-physical properties of floor materials [18]

Structure layers	Thermal conductivity ($Wm^{-1} K^{-1}$)	Thickness (mm)	Density ($kg\ m^{-3}$)	Specific thermal conductivity ($J\ Kg^{-1} K^{-1}$)
Wood layer	0.14	10	650	1200
Tile layer	1.1	10	1900	1050
Cement mortar layer	1.51	10	2300	920
Concrete layer	1.28	40	2400	
PPR pipe	0.22	20		325
Insulation layer	0.027	20	30	2000
Soil	2		1500	1350
Sterling board	0.13		800	1700

16.3 Validation of the Underfloor Heating/Cooling Model

16.3.1 Validation of Model in Steady-State Cooling Process

In order to validate the underfloor heating/cooling model, the numerical simulation results are compared with the experimental measurements for cooling in steady state. Experimental measurements on underfloor cooling were carried by Jin et al. [5] in the conditions of ambient temperature in the range of 24.7–27.2 °C and water temperature from 10.8 to 20.2 °C. The main structure of the floor is similar with the schematic diagram presented in Fig. 16.1 and the material information is listed in Table 16.2. The cooling is provided by parallel water pipes embedded in the filling layers. The spacing between two pipes is 150 mm. The numerical simulations have been carried out under the same conditions. Table 16.3 shows four cases of working conditions in steady heat transfer with different average supply water temperatures in pipes along with the floor surface temperatures from experimental measurements [5] and the simulation results. The absolute discrepancies between experimental results and simulation prediction on the floor surface temperature are within 0.34 °C and relative error is within 2.6 %. A good agreement between the numerical simulation and experimental measurement has been achieved.

Figure 16.2 shows one of the four cases on predicting the thermal isotherms for the cross section of underfloor with cooling water 16.8 °C-air 24.7 °C after 900 min to the steady process. It can be seen that in the steady state the temperature distribution on the floor surface is relatively uniform, although the temperature variation around the pipes is intense.

Table 16.2 The thermal conductivity and thickness of floor layers for cooling test

Structure layers	Thermal conductivity ($Wm^{-1}K^{-1}$)	Thickness (mm)
Wood layer	0.14	10
Cement mortar layer	0.93	10
Concrete layer	1.28	40
Insulating layer	0.035	20

Table 16.3 Comparison of measured [5] and numerical predicted floor surface temperature

		Floor surface temperature			
Air temperature (°C)	Pipe water temperature (°C)	Measured (°C)	Simulated (°C)	Discrepancies (°C)	Relative error (%)
27.2	10.8	20.7	20.75	0.05	0.2 %
25.7	13.8	21.1	20.93	0.17	0.8 %
24.7	16.8	21.8	21.46	0.34	1.6 %
25.8	20.2	23.8	23.18	0.62	2.6 %

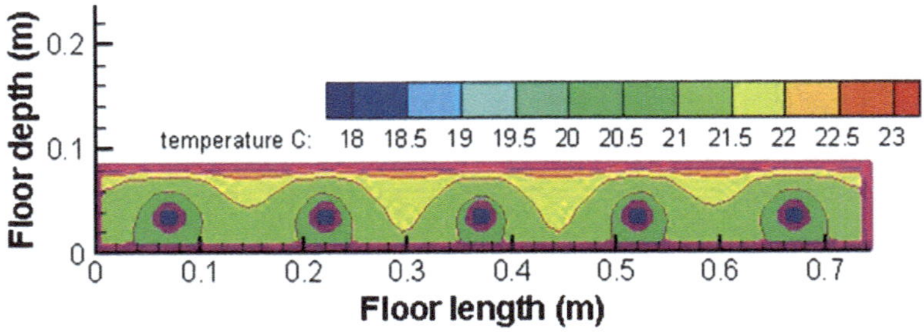

Fig. 16.2 Predicted thermal isotherms for the cross section of underfloor with cooling after 900 min to the steady process with water 16.8 °C-air 24.7 °C

16.3.2 Validation of Model in Dynamic Heating and Cooling Processes

The numerical model has been validated through the comparisons between the measured and predicted temperature for the preheating and heat release processes in the dynamic period.

The floor heat storage in the preheating and heat release processes in slab floor has been experimentally investigated by Wang et al. [18]. The main structure of the measured floor consisted of face brick layer of 5 mm ceramic tile, 50 mm filling layers of crushed stone and mortar and 50 mm insulation layer of polystyrene cystosepiment. The heating is provided by 30 °C parallel water pipes embedded in the filling layers. The 20 mm diameter pipe is in spacing of 300 mm. The measurements were recorded every 10 min. The indoor air temperature is at around 17.5 °C. The measured floor surface temperatures from Wang et al. [18] have been used to validate the developed numerical simulation model. The predicted floor surface temperatures are compared with the measured temperatures under the same operation conditions (Fig. 16.3). For the heating to be at steady state, the mean measured floor surface temperature is 22.8 °C while the predicted average temperature is 22.95 °C; the difference is 0.6 %. The average temperature differences between the measured and predicted during the dynamic heating and discharge period are 0.12 °C and 0.14 °C, respectively, and equivalent to 0.6 % and 0.7 %. More detailed comparison can be seen in Fig. 16.4. The predicted results have a good agreement with the measurements in the dynamic heating and heat releasing processes towards steady states.

The effect of wood joists has been seen through the heating and heat discharge processes. Figure 16.5 presents the predicted thermal isotherms for the cross section of underfloor structure with heating to the steady process after 812 min. The temperature at the floor surface towards uniform and the joists increases the heat transfer through the floor layers. The increased heat transfer by the joists can be seen clear in Fig. 16.6 with the heat discharge after 15 h to a steady state. The heat loss through the underneath can be observed in the isotherm figure. Insulation plays an important factor for underfloor heating.

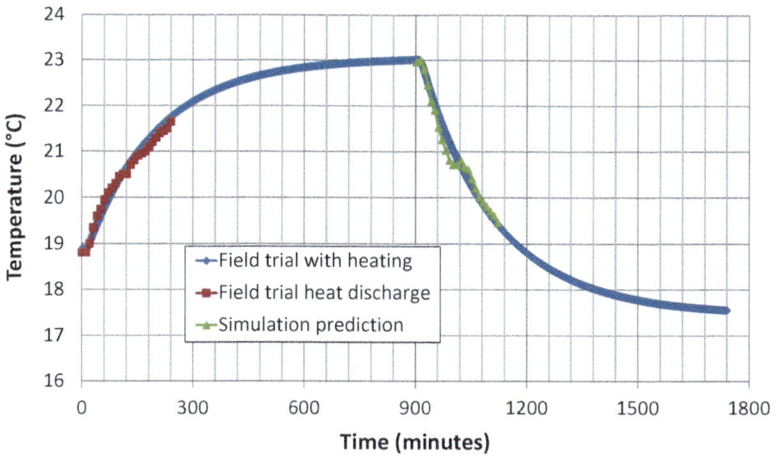

Fig. 16.3 Average measured and predicted temperatures at the surface of floor with whole process with heating and intermittent to a steady state

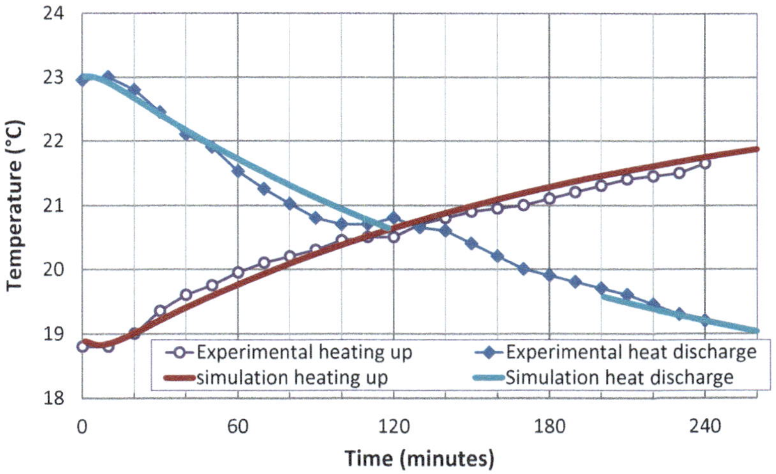

Fig. 16.4 Average measured and predicted temperatures at the floor surface with heating and intermittent period

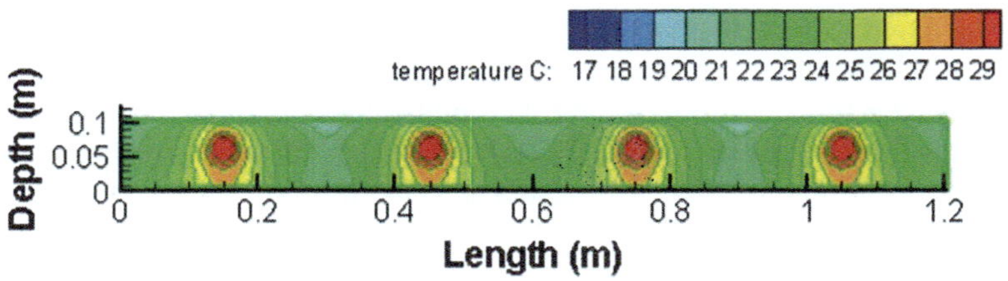

Fig. 16.5 Predicted thermal isotherms for the cross section of underfloor with heating after 812 min to the steady-state process

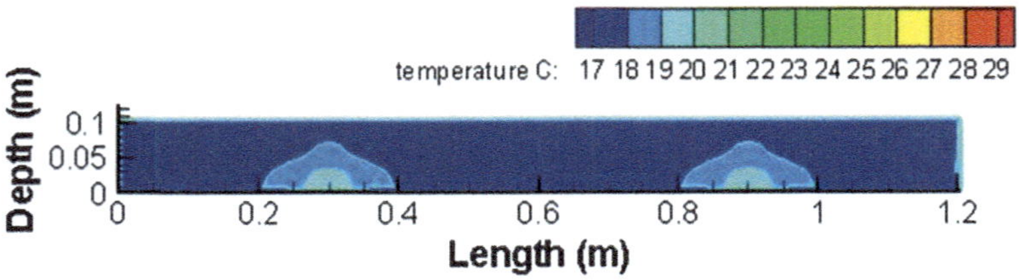

Fig. 16.6 Predicted thermal isotherms for the cross section of underfloor with intermittent period after 1,720 min to the steady process

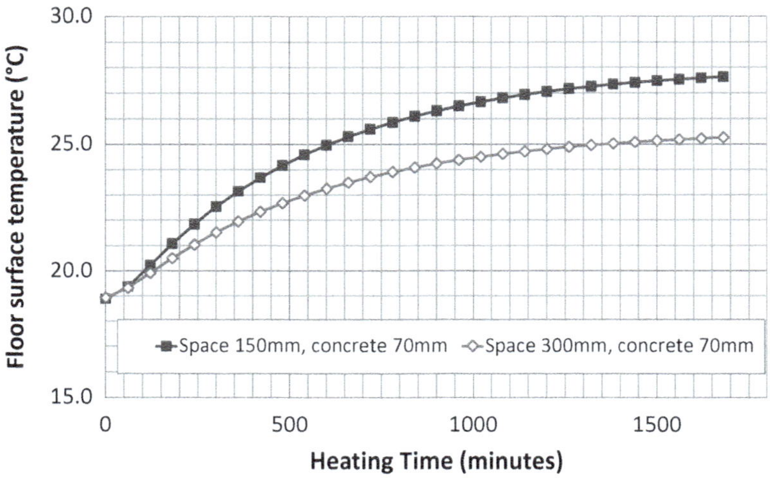

Fig. 16.7 Predicted floor surface temperature evolution with 40 °C water pipe heating and the pipe spacing at 150 and 300 mm with concrete layer of 70 mm

16.3.3 Validation of Numerical Model Using Other Developed Underfloor Heating Models

Zhang et al. [10] have developed and validated a simplified model for radiant floor surface temperature calculation. Two simulation results for heating process from their published paper have been compared with the current simulation predictions under the same operation conditions. The predicted floor is structured with 70 mm of gravel concrete (as concrete layer), 25 mm of cement mortar (as towelling layer) and 25 mm of granite (as surface layer). A 20 mm diameter pipe is used to provide 40 °C hot water for underfloor heating. The ambient and ground temperatures are at 20 °C and 16 °C, respectively. The predicted dynamic heating processes for underfloor from initial 18.5 °C to steady state for pipe spacing of 150 and 300 mm are presented in Fig. 16.7. The thermal mass inertia can be observed. Increasing the pipe spacing will increase the time to research the thermal steady

Table 16.4 Predicted floor surface temperatures from the validated model and the results from Zhang et al. [10]

Pipe water temperature (°C)	Pipe spacing (mm)	Floor surface heat flux (W m^{-2})	Floor surface temperature [Zhang, et al.] (°C)	Predicted floor surface temperature (°C)	Diff (%)
40	300	59.4	25.4	25.3	0.4 %
40	150	84.7	27.7	27.6	0.4 %

state for the floor while lowering the floor surface temperature. The predicted steady-state results are compared with the work from Zhang et al. [10] in Table 16.4. The difference between the results from Zhang et al. [10] and the current model is less than 0.4 % for the steady state. The agreement between the predicted results with the other researchers' results has been achieved.

16.4 Results and Discussions

Due to the complexity of heat flows between a structure and the surrounding ground a good understanding on the dynamic performance insight is necessary. In this chapter the thermal performances of the underfloor heating with different structure and thermal mass are analysed using the validated model.

16.4.1 Effect of Pipe Spacing on the Underfloor Heating

The thermal performance of the underfloor heating with pipe space at 200 and 300 mm is studied. The supplied hot water temperature by the pipe is 40 °C; the heat flux from pipe is same for all of the simulations as 360 W m^{-2}. The concrete thickness is 70 mm and the insulation layer is 70 mm. Figure 16.8 presents the predicted thermal isotherms for the cross section of underfloor with pipe spacing at 200 and 300 mm at different times during heating process. The insulation layer at the underneath of the floor can resist the heat loss. The underfloor system with 200 mm pipe space can charge the floor mass more efficiently and quickly raise the floor surface temperature to a steady state with more uniform distribution than the 300 mm.

16.4.2 Effect of Concrete Layer Thickness with Pipe Spacing on the Underfloor Heating

The thermal mass plays an important factor on the underfloor heating. Concrete layer inside of the underfloor heating with thickness of 30 and 70 mm has been studied for pipe spacing of 150, 200 and 300 mm. Figure 16.9 shows the floor surface

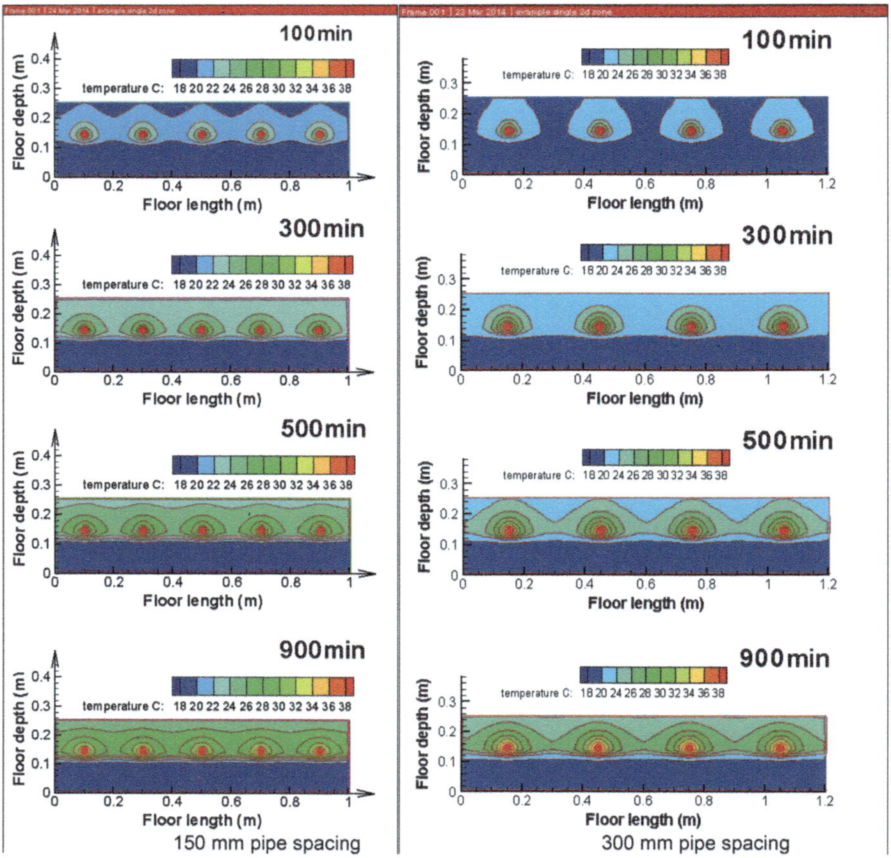

Fig. 16.8 Predicted thermal isotherms for the cross section of underfloor with pipe spacing at 200 and 300 mm at different times during heating process

temperature comparisons for the heating pipe spacing as 150 and 300 mm and concrete layers as 30 and 70 mm. The effect of pipe spacing and concrete thickness is significant. The thickness of the concrete filling around the pipe will affect the floor surface temperature. The 30 mm concrete filling is more efficient to reach the steady stage with higher floor surface temperature than the 70 mm concrete filling floor. The difference of the floor surface temperature for pipe spacing at 150 mm and 300 mm is around 2.5 °C for the both concrete layer thickness. For the same pipe spacing, although the temperatures in the steady state are approached with different concrete layer thicknesses, the floor surface temperature rising rate is higher for the 30 mm than the 70 mm concrete filling layer.

Figure 16.10 presents the predicted thermal isotherms for the cross section of underfloor concrete layer of 30 and 70 mm with pipe spacing at 200 mm at different times during heating process. The underfloor system with 30 mm concrete layer can charge the floor mass more efficiently and quickly raise the floor surface

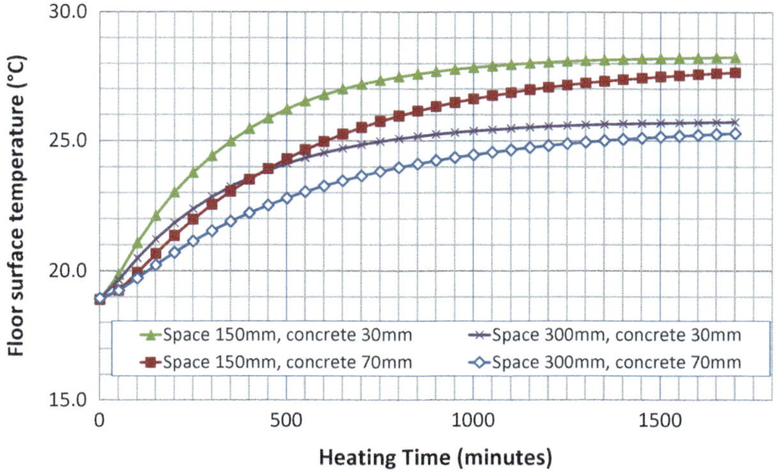

Fig. 16.9 Floor surface temperature comparisons for the heating pipe spacing as 150 and 300 mm with concrete filling layers as 30 and 70 mm

temperature to a steady state with more uniform distribution than the 70 mm concrete layer of underfloor system. Same conclusion can be achieved from Fig. 16.11 on the temperature profile of the floor surface temperature comparison between the two pipe spacing cases.

For the residential application, the 30 mm concrete layer with 150 mm pipe spacing will be more efficient for the convention heating purpose. The following numerical simulation will study the thermal performance of integrating PCM layer to this floor structure.

16.4.3 Effect of PCM Layer on Underfloor Heating with Consistence Heating and Off Heating

The effect of underfloor heating will depend on the thermal inertia of the selected material and structure of the floor. PCM can absorb a large amount of energy during the phase change, and is therefore widely investigated for thermal storage for underfloor heating. The main barriers for the PCM to be implied in the residential are the low thermal conductivity which will affect the thermal inertia of floor and the leakage for some organic PCMs. The heat store rate in PCM is as important as the heat discharge rate for underfloor heating. The thermal performance of the underfloor heating with the same floor structure as presented in Sect. 16.4.2 and the 30 mm concrete layer will be replaced by PCM for comparison in Fig. 16.12. The floor has been preheated for 29 h to reach to a steady state and then heating switched off for heat discharge to steady state.

The floor surface temperature for the PCM layer underfloor system is lower than the concrete layer floor system for the heating process, but the PCM floor can keep

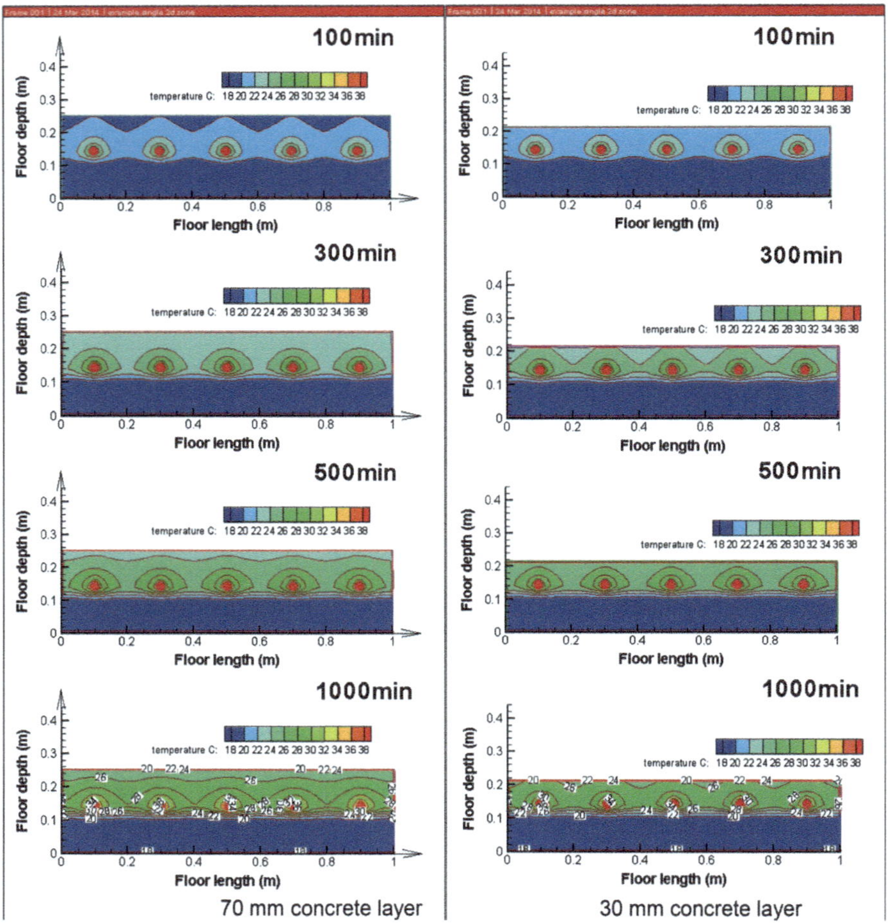

Fig. 16.10 Predicted thermal isotherms for the cross section of underfloor with 30 and 70 mm concrete filling and 200 mm pipe spacing with 40 °C water at different times during heating process

the surface temperature higher than the concrete floor during heat discharge stage. That means that the PCM has high thermal mass and makes the floor surface temperature stable during the heat charge and discharge.

16.4.4 Effect of PCM Layer on Underfloor Heating with Consistence Heating and Off Heating for Daily Operation

The thermal performance of the underfloor system with daily heating and off heating routine has been analysed. In the PCM-layered underfloor heating system, the 30 mm cement mortar and 30 mm concrete will be replaced by 30 mm PCM

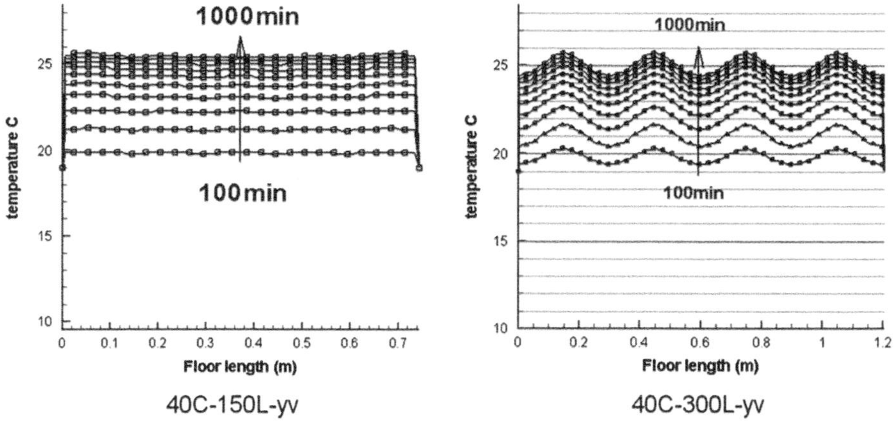

Fig. 16.11 Floor surface temperature profile comparison between the 150 and 300 mm pipe spacing

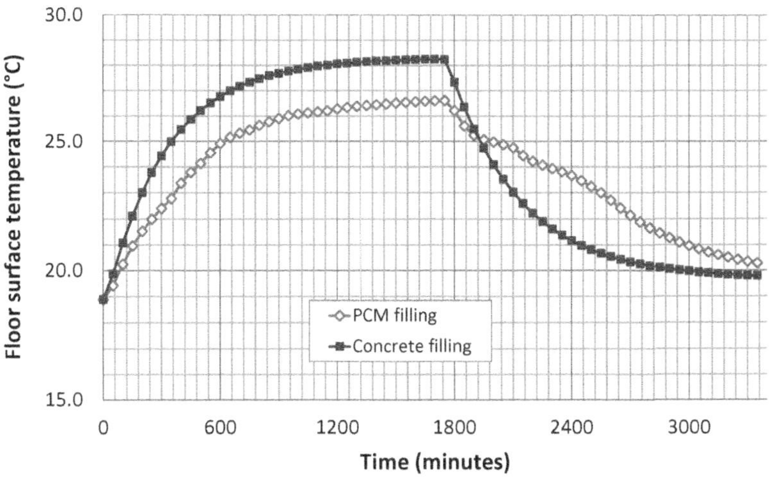

Fig. 16.12 Floor surface temperature comparison between the 30 mm concrete and PCM fillings

layer. Figure 16.13 shows the floor surface temperature comparison with concrete and PCM layer for underfloor heating with 14-h heating and 10-h off heating in the daily operating routine. In order to reduce the impact from the initial and boundary conditions, the system has been preheated for 29 h to reach a steady state before carrying on the heating schedule for 3 days. The data from the second day routine have been analysed to reduce the impact from thermal inertia. From the comparison it can be seen that in 81 % of the daily time the floor surface temperature is higher for the PCM layer floor than the concrete floor at an average of 1 °C. The heat flux

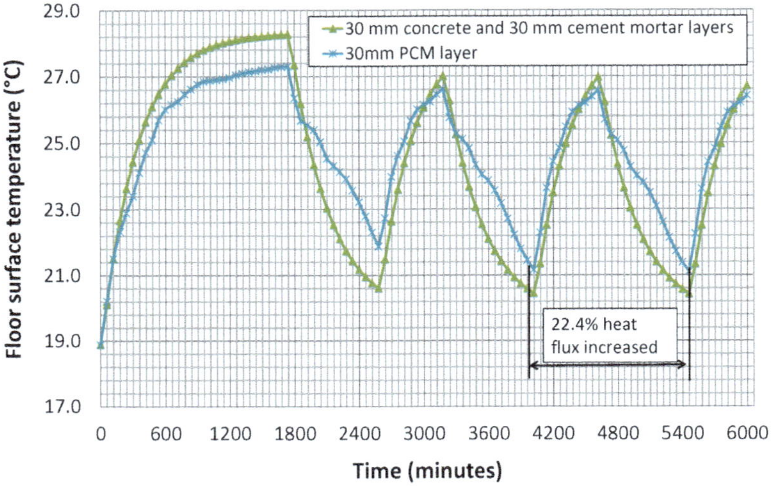

Fig. 16.13 Floor surface temperature comparison with concrete and PCM layer for underfloor heating with 14-h heating and 10-h off heating in daily routine

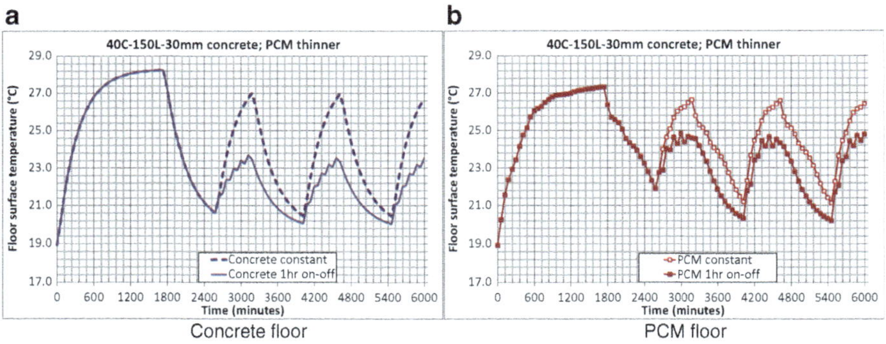

Fig. 16.14 Floor surface temperature comparisons for concrete and PCM floors with heating modes of 10-h continuous and 1-h on-off heating

from the floor surface with PCM layer is 22.4 % higher than the original concrete floor during the 24-h operation. It is only in 19 % of the 24-h time period that the floor surface temperature for the concrete layer is higher than the PCM layer floor with the average higher temperature less than 0.3 °C. The benefit of using PCM layer floor is clear with reduced floor material and steady desirable floor surface temperature and heat supply.

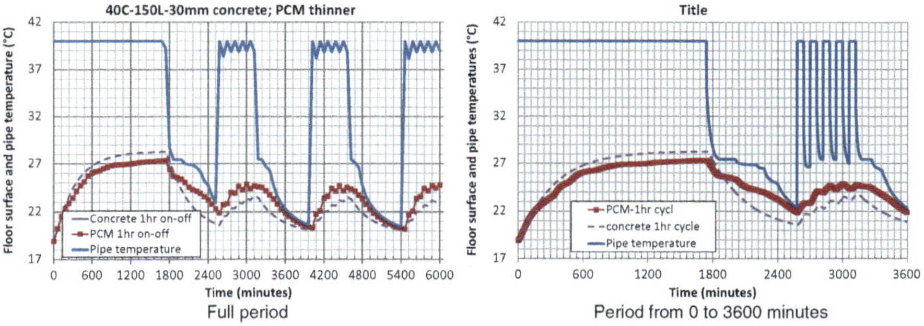

Fig. 16.15 Floor surface temperature comparisons with on-off heating for the concrete and PCM-layered floors

16.4.5 *Effect of PCM-Layered Underfloor Heating System with On-Off ASHP Heat Supply for Daily Operation*

The ASHP is normally operated in the on-off cycle control. In order to integrate the ASHP to the underfloor heating system, the underfloor thermal performance with ASHP on-off cycle heat supplied has been studied. The ASHP will supply heat in the mode of on-off cycle in each 1 h. Floor surface temperature profiles have been compared for the PCM floors with the daily continuing heating modes of 10-h continuous and ASHP 1-h on-off heating. The underfloor heating system performance has been compared between the PCM floor and concrete floor under the two heating modes as well. The floor heating system has been preheated for about 29 h to a steady state before carrying on with the different heating modes. Figure 16.14a shows the floor surface temperatures for the concrete floor in the heating modes of 14-h continuous heating and ASHP on-off heating in each hour. Figure 16.14b shows the floor surface temperatures for the PCM floor in the heating modes of 14-h continuous heating and ASHP on-off heating in each hour. The floor surface temperature with the ASHP on-off heating mode is much lower than the 14-h continuous heating mode for both of the PCM and concrete floor due to lack of heat supply. However the surface temperatures are more stable with the ASHP on-off heating mode. The floor surface temperature fluctuation in the PCM-layered floor is moderate compared with the concrete floor for both of the heating modes. The PCM floor can achieve a better performance than the concrete floor for the ASHP on-off heating mode due to the high thermal energy level stored in the preheating process (Fig. 16.15). The PCM-layered floor can achieve more stable floor surface temperature for both of the heating modes than the concrete floor system.

16.5 Conclusions

The use of underfloor heating coupled with energy-efficient heat sources, ASHP systems, will decrease the carbon footprint of building. The heat source from ASHP will be stored in the floor materials and released in the off heating period. Depending on ambient conditions, a PCM underfloor material with ASHP system may enable the indoor environment to operate near thermal comfort range efficiently and it is believed that it will reduce the electricity cost for a residential home. A transient numerical model for predicting the thermal behaviour has been used to optimise the PCM underfloor heating system design. PCM floor can be used to stabilise the room temperature. The simulation results have also indicated that the thermal characteristics of the PCM and the configuration of the PCM floor can result in advantageous control of the heating operation with reduced insulation material cost. In the future, the thermal performance of the PCM floor can be studied through combining with the ASHP's different defrost strategies and innovative PCM structures.

References

1. Lin K, Zhang Y, Xu X, Di H, Yang R, Qin P (2005) Experimental study of under-floor electric heating system with shape-stabilized PCM plates. Energy Build 37:215–220
2. Song D, Kim T, Song S, Hwang S, Leigh SB (2008) Performance evaluation of a radiant floor cooling system integrated with dehumidified ventilation. Appl Therm Eng 28:1299–1311
3. Niu JL, Zhang LZ, Zuo HG (2002) Energy savings potential of chilled ceiling combined with desiccant cooling in hot and humid climates. Energy Build 34:487–495
4. Olesen BW (1997) Possibilities and limitations of radiant floor cooling. ASHRAE Trans 103 (1):42–48
5. Xing J, Xiaosong Z, Yajun L, Rongquan C (2010) Numerical simulation of radiant floor cooling system: The effects of thermal resistance of pipe and water velocity on the performance. J Build Environ 45:2545–2552
6. Farid M, Chen X (1999) Domestic electrical space heating with heat storage, Proceedings of the Institution of Mechanical Engineers Part A. J Power Energy 213:83–92
7. Athienitis AK, Chen TY (1993) Experimental and theoretical investigation of floor heating with thermal storage. ASHRAE Trans 99(1):1049–1057
8. De Monte F (2000) Transient heat conduction in a one-dimensional composite slab. A 'natural' analytic approach. Int J Heat Mass Transf 43(19):3607–3619
9. Lu X, Tervola P (2005) Transient heat conduction in the composite slab-analytical method. J Phys A: Mathematical General 38:81–96
10. Zhang L, Liu XH, Jiang Y (2012) Simplified calculation for cooling/heating capacity, surface temperature distribution of radiant floor. Energy Build 55:397–404
11. Tung-Chai L, Chi-Sun P (2013) Use of phase change materials for thermal energy storage in concrete: An overview. J Construct Build Mater 46:55–62
12. Cabeza LF, Mehling H, Hiebler S, Ziegler F (2002) Heat transfer enhancement in water when used as PCM in thermal energy storage. Appl Therm Eng 22:1141–1151
13. Cui Y, Liu C, Hu S, Yu X (2011) The experimental exploration of carbon nanofiber ad carbon nanotube additives on thermal behaviour of phase change materials. Sol Energy Mater Sol Cells 95:1208–1212

14. Huang MJ, Eames PC, Hewitt NJ (2006) The application of a validated numerical model to predict the energy conservation potential of using phase change materials in the fabric of a building. Sol Energy Mater Sol Cells 90(13):1951–1960
15. Sumin K, Lawrence DT (2009) High latent heat storage and high thermal conductive phase change materials using exfoliated graphite nanoplatelets. Sol Energy Mater Sol Cells 93 (1):136–142
16. Jeon J, Jeong SG, Lee JH, Seo J, Kim S (2012) High thermal performance composite PCMs loading xGnP for application to building using radiant floor heating system. Sol Energy Mater Sol Cells 101:51–56
17. Karim L, Barbeon F, Gegout P, Bontemps A, Royon L (2014) New phase-change material components for thermal management of the light weight envelope of buildings. Energy Build 68(Part B):703–706
18. Dengjia W, Yanfeng L, Yingying W, Jiaping L (2014) Numerical and experimental analysis of floor heat storage and release during an intermittent in-slab floor heating process. Appl Therm Eng 62:398–406

Chapter 17
Microcapsulation and Macrocapsulation of Phase Change Materials by Emulsion Co-polymerization Method

Murat Unal, Yeliz Konuklu, and Halime O. Paksoy

Abstract In this study, decanoic acid suitable for thermal energy storage applications was microencapsulated with poly(styrene-co-ethyl acrylate) by emulsion co-polymerization method. Chemical structures, morphological characteristics, and thermal properties of microcapsules and macrocapsules were determined using Fourier Transfer Infrared Spectroscopy (FT-IR), Scanning Electron Microscopy (SEM), and Differential Scanning Calorimeters (DSC) respectively. The microPCMs and macroPCMs were synthesized successfully and the encapsulation ratio was about up to 65.5 %. As a result, the as-prepared microcapsules show good potentials for thermal energy storage and could be used in many applications.

Keywords Phase change material (PCM) • Decanoic (capric) acid • Co-polymerization and microcapsulation and macrocapsulation

17.1 Introduction

Generally, renewable energies are discontinuous and seasonal due to its nature that is dependent on the meteorological conditions of the location. Similar troubles also occur in heat recovery systems where the waste heat availability and utilization periods are different. Therefore, energy storage plays important roles in energy conversion to reduce the mismatch between supply and demand and improve its utilization [1]. Lately, phase change materials (PCMs) have been shown increased interest due to increased energy consumption [2]. Phase change materials (PCMs) are substances that store energy as latent heat. Recently, many thermal energy storage applications [3–6] have benefited from using PCMs such as salt hydrates,

M. Unal (✉) • H.O. Paksoy
Faculty of Science and Arts, Department of Chemistry, Cukurova University, Adana, Turkey
e-mail: murat_unal33@hotmail.com

Y. Konuklu
Department of Food Processing, Bor Vocational School, Nigde University, Nigde, Turkey

I. Dincer et al. (eds.), *Progress in Clean Energy, Volume 2*,
DOI 10.1007/978-3-319-17031-2_17

paraffins, fatty acids, and their mixtures (e.g., high-chain fatty acid esters of higher alcohols). Latent heat storage capacity significantly affects the energy saving that is obtained by these applications. PCMs that are used for these applications should have a suitable phase change application, a high enthalpy near the temperature of use, a high density, should be stable and compatible with the container materials, should be nontoxic, non-polluting, cheap, and abundant in order to ensure the highest efficiency [6, 7].

To the best of our knowledge, no study has been reported on the preparation of leakage-free, thermally stable decanoic acid microcapsules. Decanoic acid is a fatty acid that is essentially environmentally friendly since it is obtained from vegetable and animal oils. Decanoic acid can be derived from coconut oil, which is abundant in tropical countries [8]. We present here an efficient and simple method to synthesize decanoic acid microcapsules with poly(styrene-co-ethyl acrylate) in emulsion polymerization technique. The properties of microPCMS and macroPCMs have been characterized by scanning electron microscopy (SEM), differential scanning calorimetry (DSC), particle size analyzer, thermal cycle, and Fourier transform infrared (FTIR) spectra analysis.

17.2 Experimental Part

17.2.1 Materials

Decanoic acid (Capric acid) (Meck, Germany) was used as core material; styrene (Sigma-Aldrich, USA) and ethyl acrylate (Merck, Germany) were used as wall materials; ethyleneglycoldimethacrylate (Merck, Germany) was used as cross-linker; sodium thiosulfate (Merck, Germany) and tertbutylhydroperoxide solution (70 %) (Merck, Germany) were used as initiators; and ferrous sulfate (Panreac, Spanish), ammonium persulfate and tertbutylhydroperoxide solution (70 %) were obtained from Merck, Germany. Triton X-100 (Merck, Germany) was used as emulsifier.

17.2.2 Preparation of the Microcapsules with Emulsion Co-polymerization

All the polymerizations were carried out in a five-neck reactor equipped with reflux condenser and mechanical stirrer. Total distilled water (120 ml), capric acid (decanoic acid) (48 g), and Triton X-100 (1.6 g) were added into a 250 ml five-necked reactor, and the reaction mixture was stirred at 45 °C. This mixture was prepared at 45 °C because decanoic acid (DA) is in liquid phase at this temperature. After 10 min, total Ethyl acrylate (14 ml) and styrene (14 ml), cross-linker (8 ml), freshly prepared $FeSO_4 \cdot 7H_2O$ solution (1 ml) (prepared by mixing 0.15 g $FeSO_4 \cdot 7H_2O$ with 100 ml distilled water) and ammonium persulfate (0.25 g) were added to this mixture. The resultant mixture was stirred to form an emulsion at 1,000 rpm for 30 min by mechanical stirrer. Finally, $Na_2S_2O_7$ (0.25 g) and tertbutylhydroperoxide solution (70 %) (1 ml) were added and heated to 90–95 °C under nitrogen atmosphere. During reaction, the resultant mixture was stirred at 1,000 rpm. The mixture was stirred for 3 h. After completion of 3 h reaction time, all reaction mass was dried in an oven at 45 °C. The dried samples were weighed and then characterized using FT-IR, DSC, particle size analyzer, and SEM.

17.2.3 *Characterization of the Microcapsules*

The spectroscopic analyses of microcapsules were obtained by using Fourier transform Infrared spectrophotometer (FT-IR, PerkinElmer FTIR spectrometer) (wave-numbers 350–4,000/cm) at room temperature.

The morphology of the decanoic acid with poly(styrene-co-ethyl acrylate) microcapsules was obtained by using a scanning electronic microscope (SEM, Carl Zeiss AG-EVO® 40 Series) and appraised using a particle size analyzer (ZetasizerNano ZS90). SEM analysis samples were coated with a layer of gold in vacuum conditions.

Thermal properties of microencapsulated PCMs were determined using differential scanning calorimeter (DSC, PerkinElmer DSC 4000) at a heating and cooling rate of 10 °C/min between −30 °C and 60 °C under a constant stream of nitrogen at a flow rate of 10 ml/min. An average 5 mg of sample weight was used.

17.3 Results and Discussion

17.3.1 *Morphology Properties of Microcapsulation and Macrocapsulation*

In this study, co-polymer wall material was used in the microcapsulation and macroencapsulation studies of decanoic acid. The morphologies of the microcapsules and macrocapsules obtained in different mixing cases could be examined on the SEM images shown in Fig. 17.1. Moreover, it was showed that mixing speed influences size encapsulation [9]. As shown in Fig. 17.1a, b, when it was mixed at 1,000 rpm and 500 rpm, spherical structure and regular microcapsules and macrocapsules were obtained with a smooth surface. Furthermore, as seen in Fig. 17.1b, the macrocapsule surface was obtained from thousands of microcapsulations. Figure 17.2 images the particle size and distribution of the microPCM.

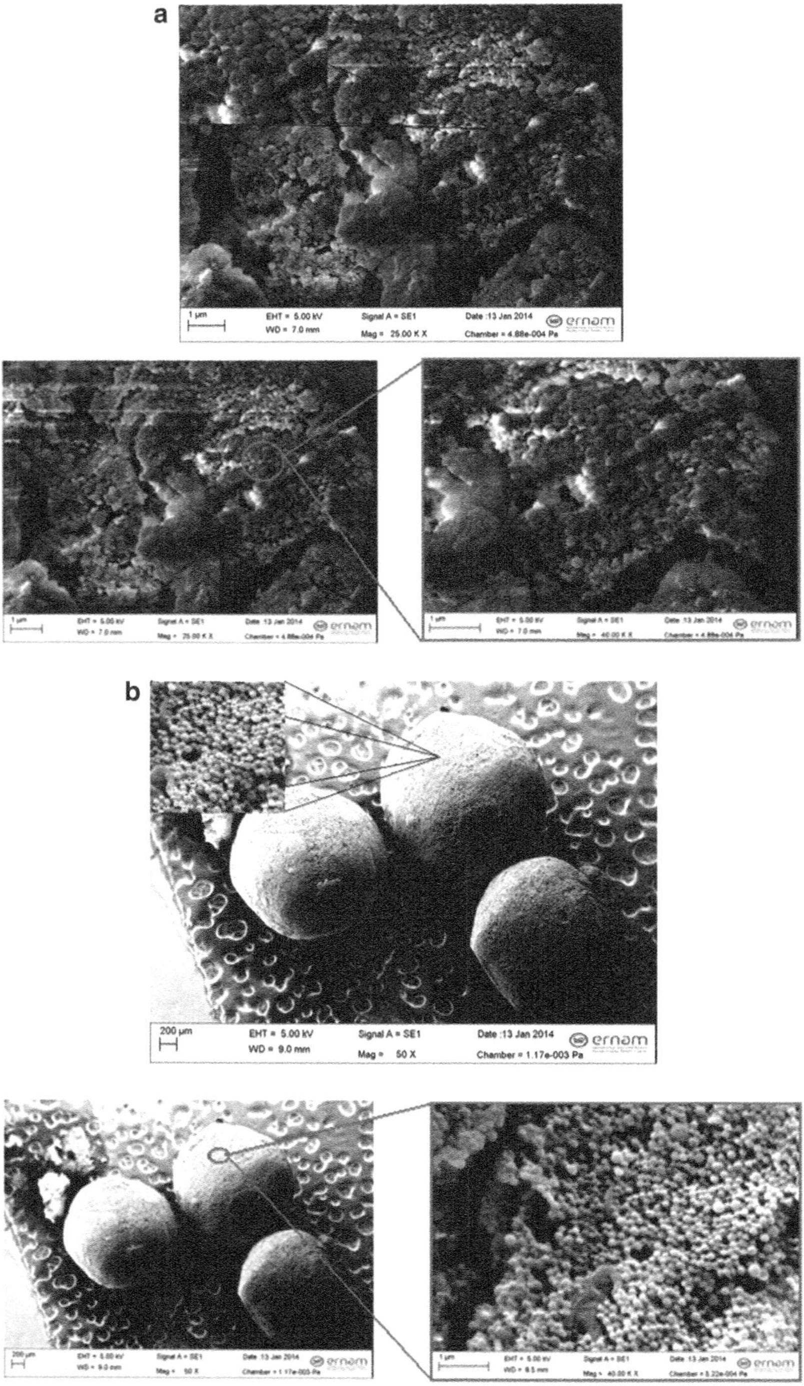

Fig. 17.1 SEM images of the microcapsules and macrocapsules synthesized at different mixing speeds with decanoic acid (**a**) microPCM, (**b**) macroPCM

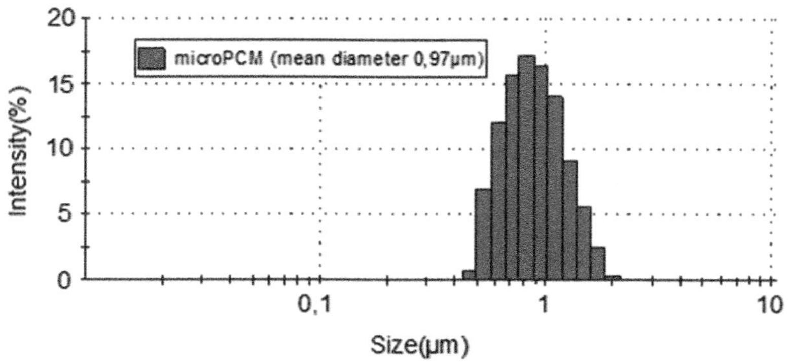

Fig. 17.2 Particle size distribution of microcapsules

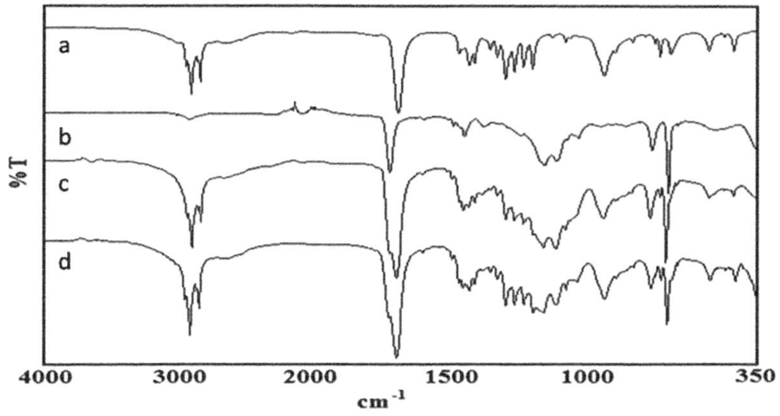

Fig. 17.3 Microcapsulated and macrocapsulated FT-IR spectra (**a**) decanoic acid, (**b**) poly(styrene-co-ethylacrylate), (**c**) microPCM, (**d**) macroPCM

The particle size ranges from 0.46 to 1.99 μm. The average particle size is measured as 0.97 μm. Also macroPCM was seen to have an average particle size of 1.5 mm with SEM (Fig. 17.1).

17.3.2 FT-IR Analysis

FT-IR spectra of decanoic acid, poly(styrene-co-ethylacrylate), and the microcapsulated and macrocapsulated PCM are showed in Fig. 17.3. As seen in the core material spectrum, –C–H stretching in the aliphatic groups was observed at 2,850–2,900/cm and the stretching was strongly seen at microcapsulated and macrocapsulated spectrums. Characteristic carbonyl –C=O stretches were observed at 1,700/cm in core material and poly(styrene-co-ethylacrylate) spectrums, and co-polymer special stretching phenyl ring oop–C–H bending for styrene was observed at 700/cm and the stretching was showed in microcapsulated and macrocapsulated PCM spectrums. These evidences indicate that decanoic acid encapsulated successfully at both micro and macro levels.

17.3.3 Thermal Properties Measurement and Performance

The phase change conducts of pure decanoic acid and the microcapsules and macrocapsules prepared by using styrene and ethyl acrylate were researched by DSC, and their thermograms are showed in Fig. 17.4. The heat storage capacity of

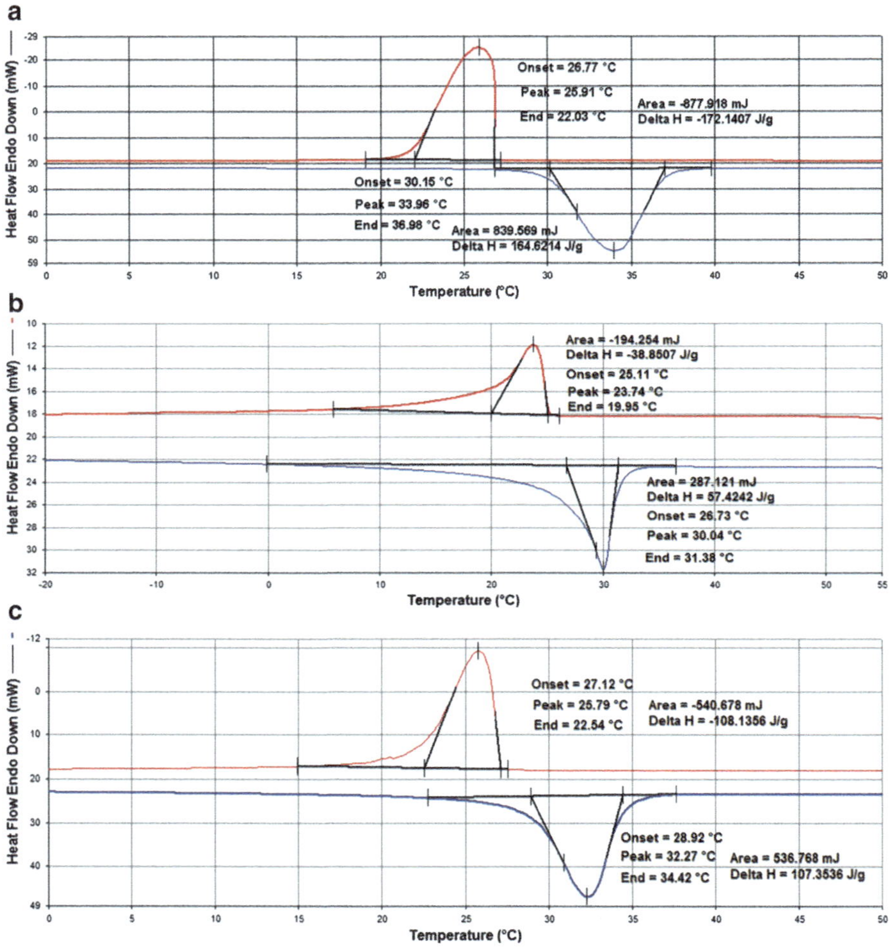

Fig. 17.4 DSC thermograms of (**a**) decanoic acid, (**b**) microcapsulation, (**c**) macrocapsulation

decanoic acid is about 164.62 J/g described as the latent heat storage capacity of endothermic peak (melting), while endothermic peak (melting) temperature is 30.15 °C. The endothermic peak (melting) and exothermic peak (freezing) properties of all the samples are epitomized in Table 17.1 in conditions of the sample abbreviations. Encapsulation ratio, encapsulation efficiency, and thermal storage capability are very important in the characterization of the capsules and calculated using the following equation [10]:

$$\text{micro-macroencapsulation ratio }\% = \frac{\Delta H_{\mathrm{m}}, \text{micro} - \text{macroPCM}}{\Delta H_{\mathrm{m}}, \text{PCM}} \times 100 \quad (17.1)$$

$$\text{micro-macroencapsulation efficiency }\% = \frac{\Delta H_{\mathrm{m}}, \text{micro} - \text{macroPCM} + \Delta H_{\mathrm{c}}, \text{micro} - \text{macroPCM}}{\Delta H_{\mathrm{m}}, \text{PCM} + \Delta H_{\mathrm{c}}, \text{PCM}} \times 100 \quad (17.2)$$

$$\text{Thermal storage capability }\% = \frac{\text{micro} - \text{macroencapsulation efficiency }\%}{\text{micro} - \text{macroencapsulation ratio }\%} \times 100 \quad (17.3)$$

The calculated results of encapsulation ratio, encapsulation efficiency, and thermal storage capability are showed in Table 17.1. Microencapsulated and macroencapsulated decanoic acid (DA) can effectively store the latent heat through phase change and microencapsulation and macroencapsulation ratio reached to 34.88 and 65.21 %, respectively.

17.3.4 Thermal Reliability of Microcapsules and Macrocapsules

There should be no significant change in thermal properties after the repeated heating and cooling. Thermal cycling test was conducted to determine the thermal reliability of decanoic acid microcapsules and macrocapsules. Figure 17.5 shows the DSC thermograms of core material, microcapsules and macrocapsules uncycling and after 500, 1,000 cycles. Table 17.2 shows the melting temperature value of the core material and the microcapsules and macrocapsules has changed by (+)0.31 °C, (−)3.12 °C, (−)0.11 °C and their freezing temperature value changed by (+)1.48 °C, (+)4.11 °C, (−)2.81 °C, respectively, after repeated 1,000 thermal cycles. On the other hand, the latent heats of fusion and crystallization after repeated 1,000 thermal cycles changed by (−)4.39 J/g, (−)7.89 J/g, and (−) 8.37 J/g, respectively. According to these results there was no significant temperature or enthalpy change of the core material and the microcapsules and macrocapsules after 1,000 thermal cycles.

Table 17.1 Thermal properties and mixing speeds of decanoic acid and microcapsules and macrocapsules

Sample abbreviations	Mixing speed (rpm/min)	Melting process				Freezing process				Microcapsulation–macrocapsulation ratio (%)	Microcapsulation–macrocapsulation efficiency (%)	Thermal storage capability
		T_{om} (°C)	T_{pm} (°C)	T_{em} (°C)	H_m (J/g)	T_{oc} (°C)	T_{pc} (°C)	T_{ec} (°C)	H_c (J/g)			
DA	–	30.15	33.96	36.98	164.62	26.77	25.91	22.03	−172.14	–	–	–
microPCM	1,000	29.73	30.04	31.38	57.42	21.11	23.74	19.95	−38.85	34.88	28.59	81.96
macroPCM	500	28.92	32.27	34.12	107.35	27.12	25.79	22.54	−108.14	65.21	63.99	98.13

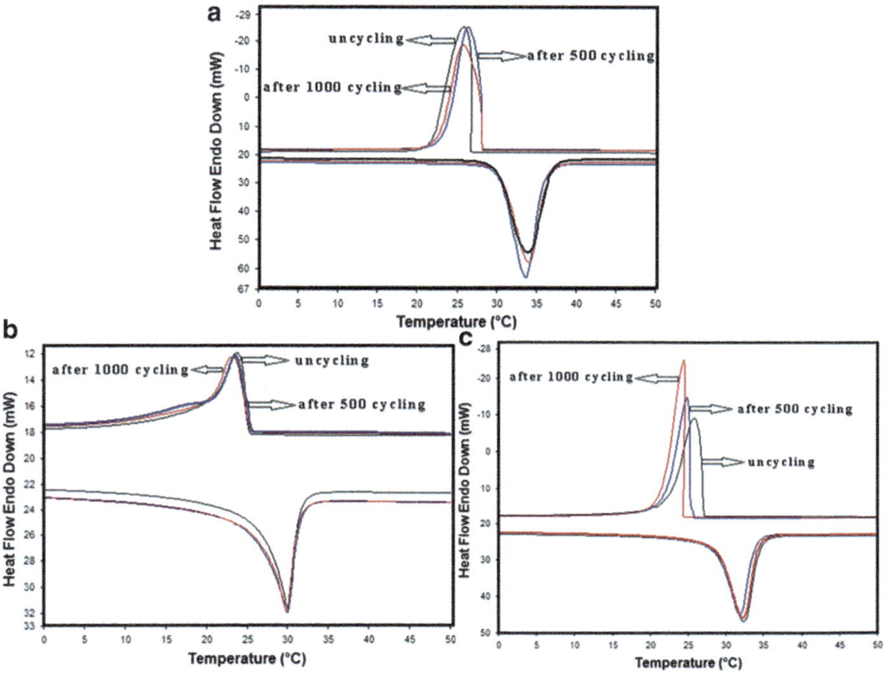

Fig. 17.5 DSC thermogram of (**a**) core material and (**b**) microcapsules (**c**) macrocapsules uncycling and after 500, 1,000 thermal cycles

Table 17.2 Thermal properties of microcapsules and macrocapsules after 1,000 cycles

Sample abbreviations	Number of thermal cycle	Melting process		Freezing process	
		T_{om} (°C)	H_m (J/g)	T_{oc} (°C)	H_c (J/g)
DA	Uncycle	30.15	164.62	26.77	−172.14
DA	1,000	30.46	159.73	28.25	−165.96
microPCM	Uncycle	29.73	57.42	21.11	−38.85
microPCM	1,000	26.61	49.53	25.22	−30.82
macroPCM	Uncycle	28.92	107.35	27.12	−108.14
macroPCM	1,000	28.81	98.98	24.31	−99.7

17.4 Conclusion

In this study, decanoic acid suitable for thermal energy storage applications was microencapsulated with poly(styrene-co-ethyl acrylate) by emulsion co-polymerization method. SEM results proved that decanoic acid was encapsulated by the polymer shell. FTIR results verified that the capsule core material kept its chemical structure after microencapsulation and macroencapsulation with polymer shell. It has been observed that the diameter and the melting heat in two experiments vary with the mixing speed. As shown by DSC thermograms, the latent heats of melting and freezing of macrocapsules were identified to be 107.35 and −108.14 J/g, respectively. Thermal cycling results demonstrated that the macrocapsules have quite well thermal reliability after 1,000 cycles.

Acknowledgments We would like to thank The Scientific and Technical Research Council of Turkey (TUBITAK) (The Project Code: TUBITAK 111M614) for their financial support for this study. And we also would like to thank Mrs. Fatma Aydin for SEM analysis for this study.

References

1. Zhang T, Wang Y, Shi H, Yang W (2012) Fabrication and performances of new kind microencapsulated phase change material based on stearic core and polycarbonate shell. Energ Convers Manag 64:1–7
2. Sarı A, Alkan C, Karaipekli A (2010) Preparation, characterization and thermal properties of PMMA/n-heptadecane microcapsules as novel solid–liquid microPCM for thermal energy storage. Appl Energ 87:1529–1534
3. Konuklu Y, Paksoy H (2009) Phase change material sandwich panels for managing solar gain in buildings. J Sol Eng 131:042012-1–2-7
4. Beyhan B, Paksoy H, Das_gan Y (2013) Root zone temperature control with thermal energy storage in phase change materials for soilless greenhouse applications. Energ Convers Manag 74:446–453
5. Aydın AA, Okutan H (2013) Polyurethanerigid foam composite sincorporated with fatty acid ester-based phase change material. Energ Convers Manag 68:74–81
6. Konuklu Y, Paksoy OH, Unal M, Konuklu S (2014) Microencapsulation of a fatty acid with Poly(melamine–urea–formaldehyde). Energ Convers Manag 80:382–390
7. Zalba B, Marin JMB, Cabeza L, Mehling H (2003) Review on thermal energy storage with phase change materials, heat transfer analysis and applications. Appl Ther Eng 23:251–83
8. Dimaano MNR, Watanabe T (2002) The capric lauric acid and pentadecane combination as phase change material for cooling applications. Appl Ther Eng 22:365–377
9. Özonur Y, Mazman M, Paksoy ÖH, Evliya H (2006) Microencapsulation of coco fatty acid mixture for thermal energy storage with phase change material. Int J Energ Res 30:741–749
10. Yu S, Wang X, Wu D (2014) Microencapsulation of n-octadecane phase change material with calcium carbonate shell for enhancement of thermal conductivity and serving durability: synthesis, microstructure and performance evaluation. Appl Energ 114:632–643

Chapter 18
Investigation of Heat Storage Performance of a Solar Pond with Potassium Chloride

Mehmet Karakilcik, Ismail Bozkurt, Ilker Balkaya, and Ibrahim Dincer

Abstract This chapter concerns an experimental investigation of heat storage performance of a solar pond saturated with potassium chloride. The solar pond consists of potassium chloride water zones. The heat storage zone (HSZ) is formed as saturated brine with potassium chloride to collect and storage reaching the solar radiation. The gradient zone (GZ) is called non-convective zone (NCZ) with various density layers prepared with potassium chloride brine decreasing from HSZ to upper convective zone (UCZ). The layers consist of five different concentrations with a thickness of 10 cm each. These layers form a brine gradient to prevent heat transfer by convection from HSZ and brine layers to UCZ. The brine gradient layers act as an insulator between HSZ and UCZ. UCZ is a clean water layer. Solar radiation is especially absorbed by saturated brine zone through UCZ and NCZ. The mass capacity of the HSZ is approximately 430 kg. The measurements of the temperatures and densities of the layers are obtained by using thermocouples and hydrometers from August to November. The exergy efficiency of saturated potassium chloride brine is defined in terms of heat storage capacity of saturated brine and average representative solar energy. As a result, the maximum and minimum exergy efficiencies of the HSZ are obtained as 25.33 % in August and 9.77 % in November, respectively.

Keywords Energy • Exergy • Efficiency • Potassium chloride solar pond

---

M. Karakilcik (✉) • I. Balkaya
Department of Physics, Faculty of Sciences and Letters, University of Cukurova, Adana 01330, Turkey
e-mail: kkilcik@cu.edu.tr

I. Bozkurt
Department of Mechanical Engineering, Faculty of Engineering, University of Adiyaman, Adiyaman 02040, Turkey

I. Dincer
Clean Energy Research Laboratory (CERL), Faculty of Engineering and Applied Science, University of Ontario Institute of Technology (UOIT), 2000 Simcoe Street, North Oshawa, ON, Canada L1H 7K4

I. Dincer et al. (eds.), *Progress in Clean Energy, Volume 2*,
DOI 10.1007/978-3-319-17031-2_18

Nomenclature

A	Surface area, m^2
C	Specific heat, J/kg K
E	Total solar energy reaching to the pond, MJ/m^2
Ex	Exergy, J
F	Absorbed energy fraction at a region of δ-thickness
h	Solar radiation ratio
HSZ	Heat storage zone
m	Mass, kg
NCZ	Non-convective zone
S	Entropy, J/K
s	Salinity
T	Temperature, °C
UCZ	Upper convective zone
V	Volume, m^3

Greek Letters

δ	Thickness where long-wave solar energy is absorbed, m
β	Incident beam entering rate into water
θ	Angle
ρ	Density, kg/m^3
ψ	Exergy efficiency

Subscripts

d	Destruction
g	Gain
i	Incident
l	Loss
r	Refraction
rec	Recovered
side	Side wall
surr	Surrounding
sys	System

18.1 Introduction

Solar energy is the most important source of renewable energy in Turkey. There are many potential applications of solar energy, including solar ponds. A solar pond is an integral device for collecting and storing solar energy; by virtue of having built in thermal energy storage, it can be used irrespective of time and season [1]. Recently, there has been increasing interest in environmentally solar energy. In this regard, solar ponds receive some increasing attention for implementation. Bozkurt and Karakilcik [2] investigated heat storage performance of integrated solar pond and collector system. The integrated solar pond efficiencies were calculated experimentally and theoretically according to the number of collectors. Karakilcik et al. [3] presented an experimental investigation of the exergetic performance of a solar pond integrated with solar collectors. The energy efficiencies are compared with the corresponding exergy efficiencies. Karakilcik et al. [4] investigated an experimental investigation of energy distribution, energy efficiency, and ratios of the energy efficiency with respect to shading effect on each zone of a small rectangular solar pond. Bozkurt et al. [5] studied the effect of the transparent covers (glass, polycarbonate, and mica) on the small cylindrical solar pond performance. Sodium chloride salt has been used in most of these studies. In few studies, however, magnesium chloride salt has been used [e.g., 6–8]. The performance of the solar ponds is changed with its construction parameters. So, many parameters such as dimensions, materials (e.g., NaCl, MgCl) for concentration and transmission of the layers, inner and outer insulations, collection and storing capacity, material climatic conditions, and location of the pond effect on the performance. Therefore, many studied are conducted on the better parameters in order to improve the performance of the solar pond. Hence, some experimental and theoretical studies dealing with solar ponds for different salty waters were done [9].

In this study, a new experimental work is carried out with a small potassium chloride solar pond. The exergy analysis of potassium chloride solar pond has not yet been studied anywhere. Therefore, this study is the first work in the area dealing with the investigation of exergetic performance analysis of saturated potassium chloride brine solar pond. For this purpose, the small solar pond was filled with potassium chloride salty water in order to build the HSZ and layers from bottom to up. Some thermophysical properties of potassium chloride salt water (e.g., thermal conductivity, concentration, and heat capacity) are investigated and calculated to use in performing exergy analysis.

18.2 Experimental Apparatus and Procedure

A solar pond generally consists of three zones. The surface zone is called as upper convective zone (UCZ). UCZ is the freshwater layer at the top of the pond. The middle zone is called as non-convective zone (NCZ). NCZ is composed of

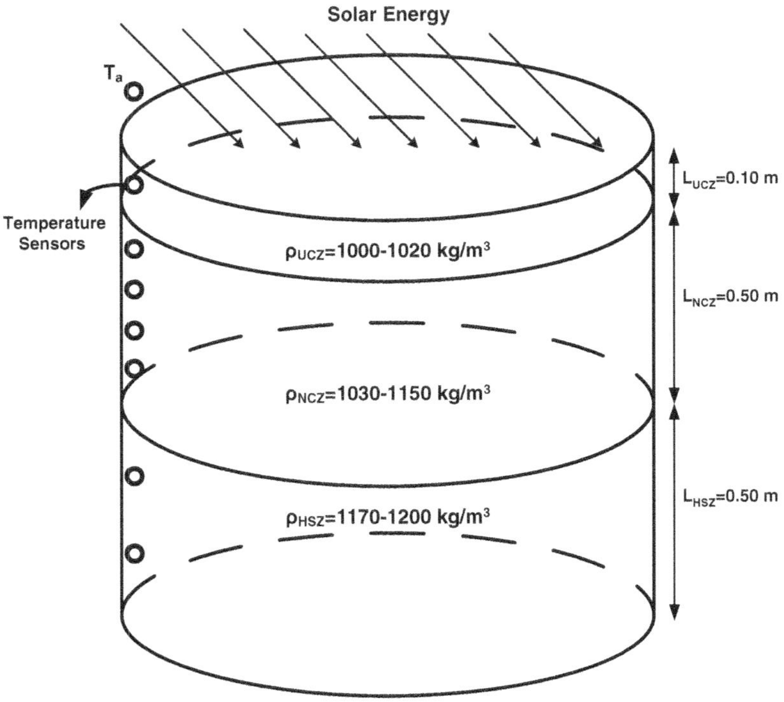

Fig. 18.1 A schematic representation of the potassium chloride solar pond

salty water layers whose brine density gradually increases toward the bottom of the pond. This zone plays a key role in the solar pond because this zone constitutes a transparent insulating layer to prevent convection heat losses. Solar radiation that reaches the bottom of the pond is absorbed by heat storage zone (HSZ) which is composed of salty water with highest density. An experimental model solar pond with the area of 0.72 m^2 and a depth of 1.10 m was built in Cukurova University in Adana, Turkey (i.e., 35°18′ E longitude, 37°05′ N latitude). Figure 18.1 shows a schematic representation of the experimental solar pond system by using potassium chloride salt.

The pond's bottom and side wall was insulated by using 0.10 m thickness glass wool. The pond temperature was measured at 7 points, starting from the bottom, at 0.25, 0.40, 0.55, 0.65, 0.75, 0.85, and 1.05 m heights by using thermocouples with an accuracy of about ± 1 °C. The density distributions are also measured and analyzed by taking samples from the same point of the temperature sensors. The pond was filled in August 2013 and worked. The thicknesses of the UCZ, NCZ, and HSZ are 0.10 m, 0.50 m, and 0.50 m, respectively. The range of salt density in the zones is 1,000–1,020 kg/m^3 in UCZ, 1,030–1,150 kg/m^3 in NCZ, and 1,170–1,200 kg/m^3 in HSZ.

18.3 Exergy Analysis

Exergy analysis is a method that uses the conservation of mass and conservation of energy principles together with the second law of thermodynamics for the design and analysis of systems and processes [10]. The exergy analysis of solar thermal energy systems is usually used to determine exergy efficiencies and identify and quantify exergy destructions so that directions for improved efficiency can be determined. One of these thermal energy systems is solar pond. It describes solar pond system and their components (e.g., size and construction material and its thermophysics and thermodynamic properties of the pond) and discusses the use of exergy analysis to assess and improve solar pond systems. Exergy methods provide a physical basis for understanding, refining, and predicting the variations in solar pond's zone behavior.

This study describes a method of exergy analyzing the quality work of the solar ponds on the basis of exergy dynamics that is related to the efficiency with which a system performance itself, and if steady, maintains its complexity. Once the exergy ratio has been calculated, it would then be possible to understand what the performance of the system is. Thus the exergy ratio can indicate the role of the performance in time. Furthermore, valuations of the exergy ratio at different times can give further indications on the behavior of the system. If the value of the ratio is increasing the system will move toward forms with lower thermodynamic efficiency, due to environmental factors like reference temperature and climatic changes. In contrast, if it is decreasing, the system is moving toward an efficient use of the available resources [11].

In this study, we focus on to improve the capacity of HSZ by using saturated brine with potassium chloride because of useful heat energy to be stored in this zone of the pond.

The exergy balance equation for HSZ of the pond can be written as

$$\begin{aligned}\Delta \mathrm{Ex}_{\mathrm{stored}} &= \mathrm{Ex}_{\mathrm{rec,NCZ}} - (\mathrm{Ex}_{\mathrm{d,HSZ}} + \mathrm{Ex}_{\mathrm{l,HSZ}} + \mathrm{Ex}_{\mathrm{side,HSZ}} + \mathrm{Ex}_{\mathrm{down,HSZ}}) \\ &= \beta \mathrm{Ex}_{\mathrm{solar}} A_{\mathrm{HSZ}}[(1-F)h(x-\delta)] - \\ &\left\{\begin{array}{c} T_0(\Delta S_{\mathrm{net,HSZ}}) + m_{\mathrm{HSZ}} C_{p,\mathrm{HSZ}}\left[(T_{\mathrm{HSZ}} - T_{\mathrm{m,NCZ}}) - T_0 \ln\left(\dfrac{T_{\mathrm{HSZ}}}{T_{\mathrm{m,NCZ}}}\right)\right] \\ + m_{\mathrm{HSZ}} C_{p,\mathrm{HSZ}}\left[(T_{\mathrm{HSZ}} - T_{\mathrm{side,HSZ}}) - T_0 \ln\left(\dfrac{T_{\mathrm{HSZ}}}{T_{\mathrm{side,HSZ}}}\right)\right] \\ + m_{\mathrm{HSZ}} C_{p,\mathrm{HSZ}}\left[(T_{\mathrm{HSZ}} - T_{\mathrm{down,HSZ}}) - T_0 \ln\left(\dfrac{T_{\mathrm{HSZ}}}{T_{\mathrm{down,HSZ}}}\right)\right] \end{array}\right\}\end{aligned} \tag{18.1}$$

where $\Delta \mathrm{Ex}_{\mathrm{stored}}$ is the exergy stored in HSZ, $\mathrm{Ex}_{\mathrm{rec,NCZ}}$ is the recovered exergy from NCZ to HSZ, $\mathrm{Ex}_{\mathrm{d,HSZ}}$ is the exergy destruction in HSZ, $\mathrm{Ex}_{\mathrm{l,HSZ}}$ is the exergy loss from HSZ to NCZ, and $\mathrm{Ex}_{\mathrm{side,HSZ}}$ is the exergy loss through side walls. $\mathrm{Ex}_{\mathrm{down,HSZ}}$

is the exergy loss through bottom wall, and ΔEx_{stored} is the exergy stored in HSZ. β is the fraction of the incident solar radiation that actually enters the pond and is given by Hawlader [12] as follows:

$$\beta = 1 - 0.6\left[\frac{\sin\theta_i - \sin\theta_r}{\sin\theta_i + \sin\theta_r}\right]^2 - 0.4\left[\frac{\tan\theta_i - \tan\theta_r}{\tan\theta_i + \tan\theta_r}\right]^2 \tag{18.2}$$

Here, θ_i and θ_r are the incidence and refraction angles. h represents the ratio of the solar energy reaching the depth in the layer I to the total solar incident falling on to the surface of the pond and is given by Bryant and Colbeck [13] as

$$h = 0.727 - 0.056\ln\left[\frac{(x_I - \delta)}{\cos\theta_r}\right] \tag{18.3}$$

where x_I is the thickness of the layer, and δ thickness of the layer in the UCZ where long-wave solar energy is absorbed. The exergy of solar radiation can be expressed by Petela [14]:

$$Ex_{solar} = E_{net}\left\{1 - \frac{4T_0}{3T} + \frac{1}{3}\left(\frac{T_0}{T}\right)^4\right\} \tag{18.4}$$

where E_{net} is the net incident solar radiation reaching the surface of HSZ. As seen in Eq. (18.1), $\Delta S_{net,HSZ}$ is the net entropy change for HSZ of the pond which is defined as $\Delta S_{net,HSZ} = \Delta S_{sys} + \Delta S_{surr}$. The exergy destruction within HSZ can be written as

$$Ex_{d,HSZ} = T_0\left[m_{HSZ}C_{p,HSZ}\ln\left(\frac{T_{HSZ}}{T_0}\right) - \left(\frac{Q_{g,HSZ}}{T_{HSZ}} + \frac{Q_{side,HSZ}}{T_0}\right) + \left(\frac{Q_{down}}{T_0}\right)\right] \tag{18.5}$$

where $m_{HSZ} = \rho_{HSZ}V_{HSZ}$ is the mass of potassium chloride salty water in HSZ, ρ_{HSZ} is the averaged density of HSZ, and V_{HSZ} is the volume of HSZ. T_0 is the reference air temperature; T_{HSZ} is the temperature of HSZ. $Q_{g,HSZ}$, $Q_{side,HSZ}$, and Q_{down} are the gain and heat losses from side and bottom wall, respectively. $C_{p,HSZ}$ is the specific heat capacity of potassium chloride salt water in HSZ. The specific heat capacity is calculated by using an empirical equation [15]:

$$C_{p,HSZ} = (-0.0044s_{HSZ} + 4.1569)10^3 \tag{18.6}$$

where s_{HSZ} is the salinity of potassium chloride salty water. The density difference at low temperature takes place approximately in linear relationship between the density and salinity. We use an empirical correlation as given below to determine the salinity of the saturated brine zones:

$$s_{\mathrm{HSZ}} = \frac{(\rho_{\mathrm{HSZ}} - 998.24)}{0.756} \tag{18.7}$$

The exergy efficiency is written for HSZ as follows:

$$\psi_{\mathrm{HSZ}} = \frac{\Delta \mathrm{Ex}_{\mathrm{stored}}}{\mathrm{Ex}_{\mathrm{r,NCZ}}} = 1 - \frac{(\mathrm{Ex}_{\mathrm{d,HSZ}} + \mathrm{Ex}_{\mathrm{l,HSZ}} + \mathrm{Ex}_{\mathrm{side,HSZ}} + \mathrm{Ex}_{\mathrm{down,HSZ}})}{\mathrm{Ex}_{\mathrm{r,NCZ}}} \tag{18.8}$$

18.4 Results and Discussion

The performance of the solar pond depends not only on thermal energy flows (e.g., heat losses and/or heat gains in the zones), but also on incoming solar radiation (e.g., accounting for reflection, transmission, and absorption) and insulating by better insulator material, both inner gradient layer and side walls [4]. Moreover, properties of the saturated brine with potassium chloride can affect the performance of small solar ponds (Fig. 18.1). In this study, we present the results of the experimental calculations for exergy efficiencies of HSZ of an experimental potassium chloride solar pond. These results show how thermophysical properties are crucial for absorbing and storing capacity of the performance effect of HSZ of the pond.

Here we now present the results of the exergy efficiency of potassium chloride solar pond. It has been demonstrated that the stability of the salt density distributions in a solar pond is of great significance. Figure 18.2 shows the variation of the experimental salty solution densities with height from bottom to surface of the pond, throughout the 4 months. As seen in Fig. 18.2, the density distribution is kept approximately.

The average experimental temperature distributions are shown in Fig. 18.3. As seen here, the average temperature of HSZ is observed to be a maximum of 53.21 °C in August, and a minimum of 27.17 °C in November. The temperature of the solar pond increases toward the bottom like the density distribution. The average temperature of UCZ closes to the ambient temperature because this zone is surface zone.

Figure 18.4 shows both average energy and exergy distribution during 4 months. As seen in Fig. 18.4, the minimum energy and exergy content were observed as 298.11 $\mathrm{MJ/m^2}$ and 279.03 $\mathrm{MJ/m^2}$ in November, respectively. The maximum energy and exergy content were observed as 629.95 $\mathrm{MJ/m^2}$ and 587.71 $\mathrm{MJ/m^2}$ in August, respectively.

Figure 18.5 shows the variation of the exergy input, exergy stored, and destruction and losses taking place in the HSZ of potassium chloride solar pond during 4 months. As obviously seen here, the exergy inputs are equivalent to the summation of exergy stored and exergy destruction and losses. The exergy stored in HSZ appears to be maximum as 27.08 MJ in August and minimum as 5.02 MJ in November.

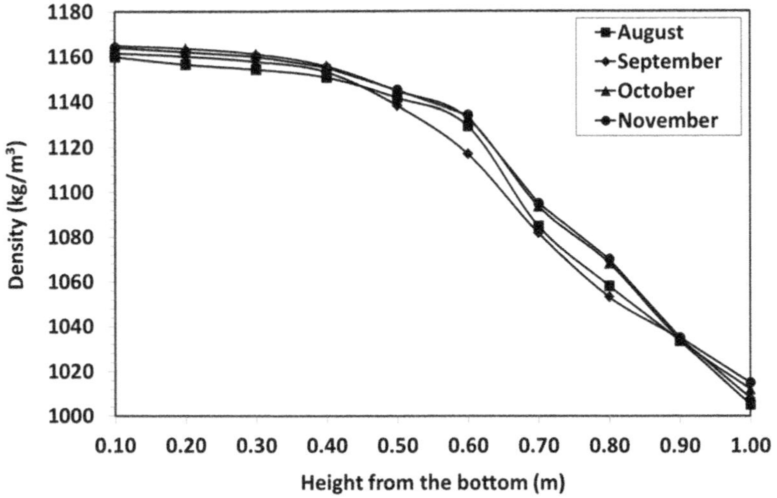

Fig. 18.2 Density distributions of potassium chloride in the solar pond

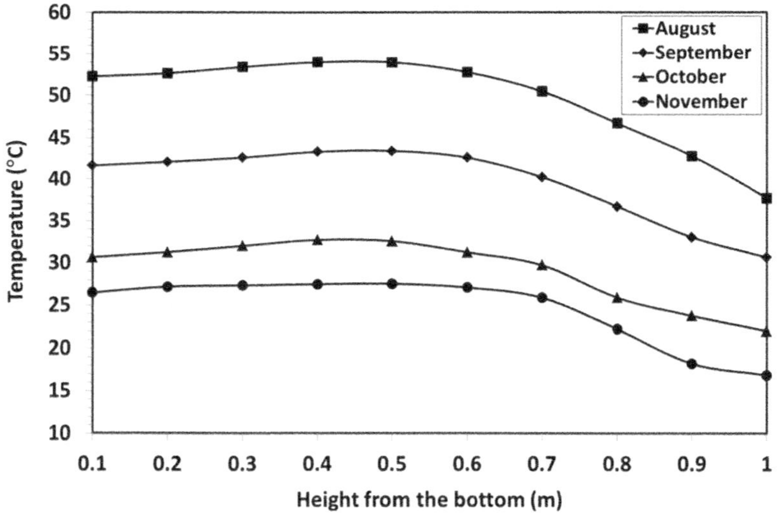

Fig. 18.3 Temperature distributions of potassium chloride in the solar pond

The exergy efficiency variations are given in Fig. 18.6 and the maximum and the minimum exergy efficiencies of HSZ are observed in August as 25.33 % and 9.77 % in November, respectively. The exergy analysis takes into account the true magnitudes of the destructions and losses and these should be minimized for performance improvement of the solar pond.

The results have been found for the potassium chloride solar pond for August, September, October, and November. The exergy efficiency is determined for HSZ

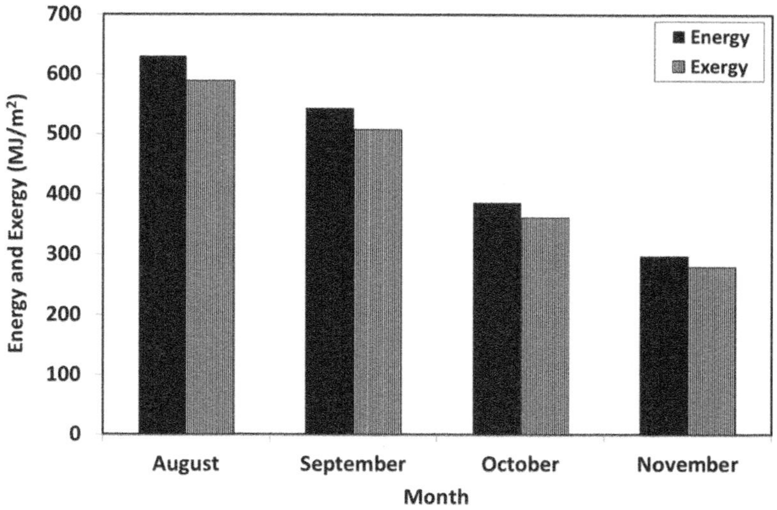

Fig. 18.4 Energy and exergy variations of the solar energy in Adana, Turkey

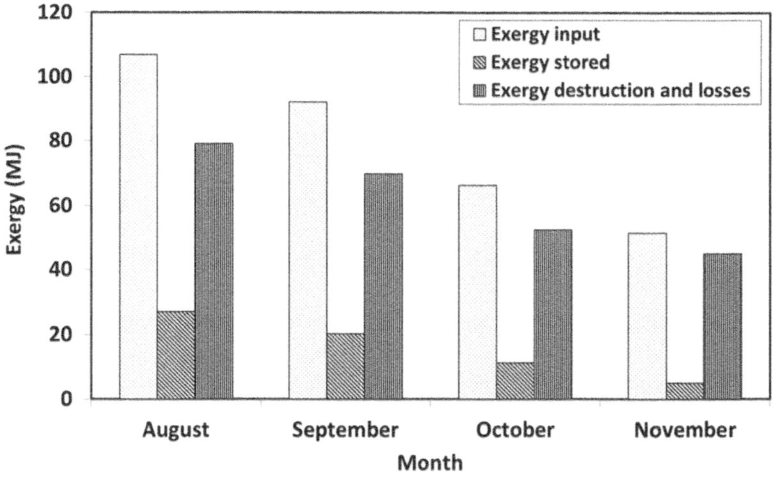

Fig. 18.5 Variations of the exergy input, storage, destruction, and losses of HSZ of the solar pond

of the pond. As seen in Fig. 18.6, the maximum exergetic efficiencies are obtained in August, and the minimum efficiencies in November although the greatest amount of thermal energy has been stored in August than in November. This is because of the decreasing incident radiation to per square meter of the pond's surface and the fact that it is responsible for the largest heat losses from HSZ by decreasing the environment temperature in November.

The temperature distribution profiles (in Fig. 18.3) for the inner zones are usually different, causing the zone's exergy efficiencies to also differ. The temperature

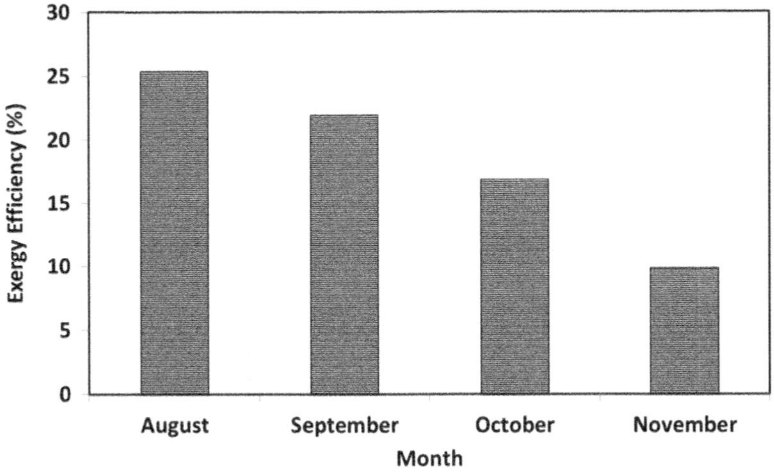

Fig. 18.6 Exergy efficiencies of HSZ in the solar pond

distributions, thus, have an important effect on the performance of the pond. The energy efficiency of the pond is decreased because of thermal energy losses due to heat transfer from the surface of the pond to air. The gradient zone's efficiency consequently has a greater effect on the performance of the HSZ. Most of the energy is stored in the HSZ of the pond. As seen in Fig. 18.5, HSZ has higher capacity to store thermal energy since this zone is insulated by gradient zone and insulated side walls. This zone plays a key role in the improvement of storage performance of the solar pond. As a result, the inner regions of the pond store more thermal energy in August than in November due to the considerable temperature differences between the zones. Not only heat losses affect the exergy efficiencies of the HSZ but also shading areas in the zones by side walls affect the storage performance. We thus suggest that heat storage, heat losses, shading areas, and solar radiation absorption should be carefully considered when determining the thermal performance of solar ponds. In the future studies, if the tilt angle of the side walls and better thermodynamic properties of the salty water of the solar ponds are selected appropriately, it can obtain much more efficiency by eliminating the shading effect of the walls and increasing the heat capacity and protecting the heat energy in the HSZ.

18.5 Conclusions

In this study, we carried out experiment in potassium chloride solar pond to see its effect on the exergy efficiency of the gradient zone in order to demonstrate the effect of saturated brine HSZ on the thermal performance of the solar pond. The temperature of each zone depends on the incident radiation, concentration, conduction and zone thicknesses, shaded surface area of the zones, and heat losses from upper surface area, bottom wall, and side walls. So, to increase the performance of the pond, the zone thicknesses should be modified to achieve higher efficiency and stability of the pond. Through careful design and thermodynamic parameter modifications, pond performance can be maintained even if the incoming solar radiation reaching and storing in the HSZ. Experimental data are used to determine the efficiency for HSZ for a real insulated solar pond. Several parameters of the pond having influences on the thermal performance are discussed. It is shown that the introduction of the other two zones (upper and non-convective zones) provides many conveniences in calculating the storage efficiency in the HSZ, and in determining the relations between heat loads and a best operating state. Therefore, the exergy efficiency of the HSZ of a solar pond is an important parameter in practical applications. Exergy efficiencies of the HSZ decreased from August to November due to the decreasing ratio of solar radiation reaching the pond. Also, while environmental temperature decreases surrounding the pond, the heat losses increase from HSZ to the air. It is important to determine the true magnitudes of these losses for performance improvement studies of HSZ. The results demonstrate the effect of the potassium gradient layers (gradient zone) as an insulation zone and materials and ambient temperature on the efficiency of the pond. It is important that exergy is a potential to help achieve better efficiency of the pond.

Acknowledgement The authors are thankful to University of Cukurova for financial support of this work (Grant No. FEF2012YL12).

References

1. Karakilcik M, Kıymaç K, Dincer I (2006) Experimental and theoretical distributions in a solar pond. Int J Heat Mass Transfer 49:825–835
2. Bozkurt I, Karakilcik M (2012) The daily performance of a solar pond integrated with solar collectors. Sol Energ 86:1611–1620
3. Karakilcik M, Bozkurt I, Dincer I (2013) Dynamic exergetic performance assessment of an integrated solar pond. Int J Exergy 12:70–86
4. Karakilcik M, Dincer I, Bozkurt I, Atiz A (2013) Performance assessment of a solar pond with and without shading effect. Energ Convers Manag 65:98–107
5. Bozkurt I, Atiz A, Karakilcik M, Dincer I (2014) Transparent covers effect on the performance of a cylindrical solar pond. Int J Green Energ 11:404–416
6. Subhakar D, Murthy SS (1991) Experiments on a magnesium chloride saturated solar pond. Renew Energ 5–6:655–660
7. Subhakar D, Murthy SS (1993) Saturated solar ponds: 2. Parametric studies. Sol Energ 50:307–319
8. Subhakar D, Murthy SS (1994) Saturated solar ponds: 3. Experimental verification. Sol Energ 53:469–472
9. Kurt H, Ozkaymak M, Binark AK (2006) Experimental and numerical analysis of sodium-carbonate salt gradient solar pond performance under simulated solar radiation. Appl Energ 83:324–342
10. Dincer I (2002) The role of exergy in energy policy making. Energ Pol 30:137–149

11. Dincer I, Rosen MA (2013) Exergy: energy, environment and sustainable development, 2nd edn. Elsevier Ltd., Oxford, UK. ISBN 978-0-08-097089-9
12. Hawlader MNA (1980) The influence of the extinction coefficient on the effectiveness of solar ponds. Sol Energ 25:461–464
13. Bryant HC, Colbeck I (1977) A solar pond for London. Sol Energ 19:321–322
14. Petala R (2003) Exergy of undiluted thermal radiations. Sol Energ 74:469–488
15. Sun H, Feistel R, Koch M, Markoe A (2008) New equations for density, entropy, heat capacity, and potential temperature of a saline thermal fluid. Deep Sea Res Oceanogr Res Paper 55:1304–1310

Chapter 19
Performance Comparison of Sodium and Magnesium Chloride-Saturated Solar Ponds

Ismail Bozkurt, Sibel Deniz, Mehmet Karakilcik, and Ibrahim Dincer

Abstract This chapter deals with an experimental investigation of energy efficiency of sodium and magnesium chloride-saturated solar ponds. The solar pond systems are filled with varying-density sodium and magnesium chloride water in order to form gradient layers. A solar pond generally consists of three zones. The density of the zones increases toward the bottom. Solar radiation is absorbed by salty water and the temperature rises. The high-temperature salty water at the bottom of the solar pond remains denser than less salty water above it. Thus, the convective heat losses are prevented by gradient layers. The temperature distributions of the solar pond are obtained by using thermocouples from August to November. The density of the layers was also measured and analyzed by taking samples from the same point of the temperature sensors. The efficiencies of the solar pond are defined in terms of temperatures as the average representative solar energy. As a result, the maximum energy efficiencies of the heat storage zone for the sodium and magnesium chloride-saturated solar ponds are found as 25.41 % and 27.40 % for August, respectively.

Keywords Solar energy • Solar pond • Sodium and magnesium chloride layers

I. Bozkurt
Department of Mechanical Engineering, Faculty of Engineering, University of Adiyaman, Adiyaman 02040, Turkey

S. Deniz • M. Karakilcik (✉)
Department of Physics, Faculty of Sciences and Letters, University of Cukurova, Adana 01330, Turkey
e-mail: kkilcik@cu.edu.tr

I. Dincer
Faculty of Engineering and Applied Science, University of Ontario Institute of Technology (UOIT), 2000 Simcoe Street, North Oshawa, ON, Canada L1H 7K4

I. Dincer et al. (eds.), *Progress in Clean Energy, Volume 2*,
DOI 10.1007/978-3-319-17031-2_19

Nomenclature

A	Surface area, m^2
E	Total solar energy reaching to the pond, MJ/m^2
F	Absorbed energy fraction at a region of δ-thickness
h	Solar radiation ratio
HSZ	Heat storage zone
k	Thermal conductivity, J/m °C
L	Thickness of the inner zones, m
NCZ	Non-convective zone
Q	Heat, J
r	Inner radius, m
T	Temperature, °C
UCZ	Upper convective zone

Greek Letters

η	Energy efficiency
δ	Thickness where long-wave solar energy is absorbed, m
β	Incident beam entering rate into water
θ	Angle

Subscripts

a	Ambient
i	Incident
r	Refraction
sw	Side wall

19.1 Introduction

Turkey is an energy-importing country; more than half of the energy requirement has been supplied by imports. On the other hand, Turkey has young population and growing energy demand per person [1]. The renewable energy sources should be used in Turkey to get rid of dependence on foreign energy. Solar energy is accepted as a key alternative energy source for the future. Therefore, solar energy is considered for satisfying significant part of the energy demand in Turkey [2]. Solar ponds appear to have significant potential for solar energy storage as thermal energy.

Recently, there has been increasing interest in solar pond's performance. Karakilcik et al. [3] investigated the thermal performance of an insulated salt gradient solar pond. Bozkurt and Karakilcik [4] studied heat storage performance investigation of integrated solar pond and collector system. Karakilcik et al. [5] presented an experimental investigation of energy distribution, energy efficiency, and ratios of the energy efficiency with respect to shading effect on each zone of a small rectangular solar pond. Bozkurt et al. [6] studied the effect of the transparent covers (glass, polycarbonate, and mica) on the small cylindrical solar pond performance. Sodium chloride salt has been used in most of these studies. However, magnesium chloride salt was used in few solar ponds [e.g., 7–9].

This work deals with experimental investigation of energy efficiency of sodium and magnesium chloride-saturated solar ponds and comparison of the energy efficiencies for heat storage zones (HSZ) of the solar ponds.

19.2 System Description

An artificially constructed pond in which significant temperature rises are caused to occur in the lower regions by preventing convection is called "solar pond" [10]. Solar ponds are composed of three zones. The bottom zone of the solar pond is filled by using saturated salty water. This zone is called as HSZ and useful heat energy is stored in here. The middle zone of the pond is non-convective zone (NCZ). NCZ is composed of salty water layers whose brine density gradually increases toward the surface of the solar pond. The density gradient in NCZ is very important to keep the stored energy in HSZ. The surface zone of the pond is upper convective zone (UCZ). UCZ is the freshwater layer at the top of the pond. This zone is fed with freshwater in order to maintain its density close to the density of freshwater and to replenish the lost water due to evaporation [4].

In this study, we constructed sodium and magnesium chloride-saturated solar ponds to determine the energy efficiency. Figure 19.1 shows a schematic representation of the solar pond.

The solar ponds have a radius of 0.48 m and a depth of 1.10 m; the bottom and the side walls of the pond were plated with the PVC sheets in 5 mm thickness, and in between with a glass wool of 10 cm thickness as outer surface insulating layer. The solar pond's inner zones possess different density levels of salty water. Salty waters with different density are prepared in the plastic tanks in order to form the solar pond. The thicknesses of the UCZ, NCZ, and HSZ are 0.1 m, 0.50 m, and 0.50 m, respectively. The range of salt gradient in the inner zones is such that the densities are 1,000–1,020 kg/m^3 in UCZ, 1,020–1,160 kg/m^3 in NCZ, and 1,150–1,180 kg/m^3 in HSZ.

19.3 Energy Analysis

The temperature distributions in the solar pond and the incident radiation reaching on the surface of the pond were determined, experimentally. To calculate the heat storage efficiency of the solar pond, the energy balance equations were written.

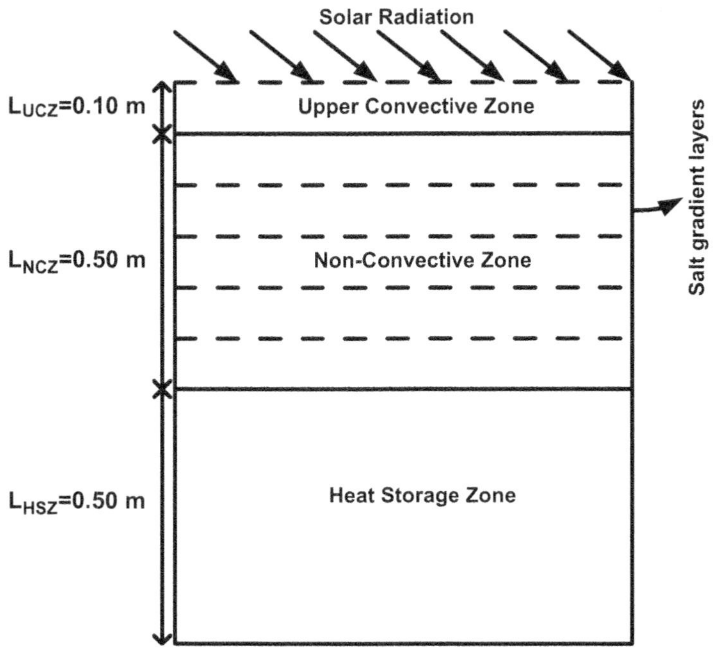

Fig. 19.1 Schematic representation of the solar pond

In this study, we focus on the heat fluxes in HSZ because the useful heat energy is stored in HSZ. The energy balance equation of HSZ is written as follows:

$$Q_{\text{stored}} = Q_{\text{solar, HSZ}} - Q_{\text{loss, HSZ}} \tag{19.1}$$

where Q_{stored} is stored heat energy in the HSZ of the solar pond, $Q_{\text{solar,HSZ}}$ is the amount of net solar energy which is absorbed in the zone, and $Q_{\text{loss,HSZ}}$ is the heat loss from the HSZ to outside of the pond.

$Q_{\text{solar,HSZ}}$ is defined as

$$Q_{\text{solar, HSZ}} = \beta \text{EA}_{\text{HSZ}}[(1 - F)h(x - \delta)] \tag{19.2}$$

where E is the total solar energy reaching the solar pond surface, A_{HSZ} is the area of the HSZ which is subjected to solar insolation, F is the fraction of energy absorbed at a region of δ-thickness, h is the solar radiation ratio, and x is the thickness of the UCZ. β is the fraction of the incident solar radiation that actually enters the pond and is given by Hawlader [11] as follows:

$$\beta = 1 - 0.6\left[\frac{\sin\theta_{\text{i}} - \sin\theta_{\text{r}}}{\sin\theta_{\text{i}} + \sin\theta_{\text{r}}}\right]^2 - 0.4\left[\frac{\tan\theta_{\text{i}} - \tan\theta_{\text{r}}}{\tan\theta_{\text{i}} + \tan\theta_{\text{r}}}\right]^2 \tag{19.3}$$

Here, θ_i and θ_r are the incidence and refraction angles. h represents the ratio of the solar energy reaching the depth in the layer I to the total solar incident falling on to the surface of the pond and is given by Bryant and Colbeck [12] as

$$h = 0.727 - 0.056\ln\left[\frac{(x_I - \delta)}{\cos\theta_r}\right] \quad (19.4)$$

where x_I is the thickness of the layer, and δ thickness of the layer in the UCZ where long-wave solar energy is absorbed. $Q_{\mathrm{loss,HSZ}}$ is defined as

$$\begin{aligned} Q_{\mathrm{loss,HSZ}} &= Q_{\mathrm{up}} + Q_{\mathrm{side}} + Q_{\mathrm{down}} \\ &= \frac{k_s A}{\Delta x_{\mathrm{HSZ-NCZ}}}(T_{\mathrm{HSZ}} - T_{\mathrm{NCZ}}) + \frac{k_{\mathrm{sw}} 2\pi r L_{\mathrm{HSZ}}}{\Delta x_{\mathrm{side}}}(T_{\mathrm{HSZ}} - Ta) + \frac{k_{\mathrm{sw}} A}{\Delta x_{\mathrm{down}}}(T_{\mathrm{down}} - T_a) \end{aligned} \quad (19.5)$$

where Q_{down} is the total heat loss to the down wall from HSZ, and Q_{up} is the heat loss from HSZ to the above zone. Q_{side} is the total heat loss to the side walls of the solar pond. A is the surface area of the solar pond, T_a is the ambient air temperature, k_{sw} is the thermal conductivity of the side and bottom walls, k_s is the thermal conductivity of the salty water, L_{HSZ} is the thickness of the HSZ (m), r is the inner radius of the cylindrical solar pond, Δx_{down} is the thickness of the down wall, Δx_{side} is the thickness of the side wall, and $\Delta x_{\mathrm{HSZ\text{-}NCZ}}$ is the thickness of the HSZ's middle point and the NCZ's middle point. The energy efficiency of the solar pond can be defined as follows:

Substituting Eq. (19.5) in Eq. 19.1 for the HSZ yields the following expression for the energy efficiencies:

$$\eta = \frac{Q_{\mathrm{stored}}}{Q_{\mathrm{solar,HSZ}}} = 1 - \frac{\{Q_{\mathrm{down}} + Q_{\mathrm{up}} + Q_{\mathrm{side}}\}}{Q_{\mathrm{solar,HSZ}}} \quad (19.6)$$

Substituting equations for each parameter in Eq. 19.6 provides us with the following energy efficiency:

$$\eta_{\mathrm{HSZ}} = 1 - \frac{\left\{\frac{k_s A}{\Delta x_{\mathrm{HSZ-NCZ}}}(T_{\mathrm{HSZ}} - T_{\mathrm{NCZ}}) + \frac{k_{\mathrm{sw}} 2\pi r L_{\mathrm{HSZ}}}{\Delta x_{\mathrm{side}}}(T_{\mathrm{HSZ}} - T_a) + \frac{k_{\mathrm{sw}} A}{\Delta x_{\mathrm{down}}}(T_{\mathrm{down}} - T_a)\right\}}{\beta \mathrm{EA}_{\mathrm{HSZ}}[(1 - F)h(x - \delta)]} \quad (19.7)$$

19.4 Results and Discussion

In this work, we present the results of the energy efficiency for both sodium and magnesium chloride-saturated solar ponds. The stability of salt density distribution has a great significance to protect convection heat losses from the HSZ of the solar pond. Figure 19.2 shows the density distribution of the sodium and magnesium chloride solar pond. As seen in Fig. 19.2, there are little differences between these

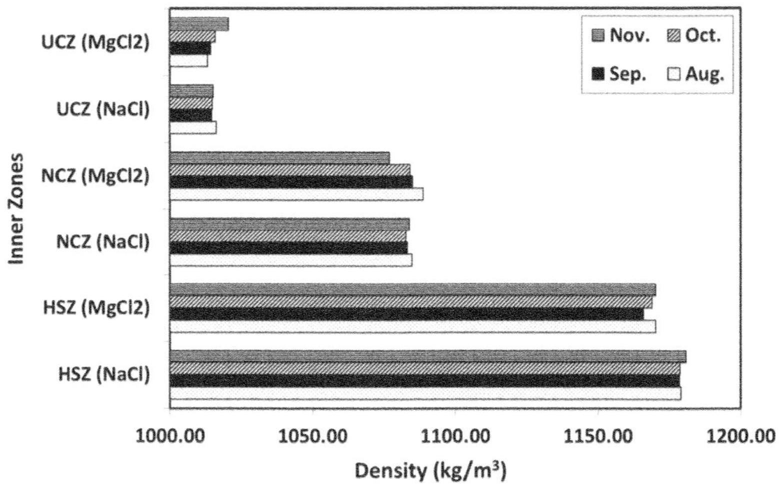

Fig. 19.2 The density distributions of the sodium and magnesium chloride solar ponds

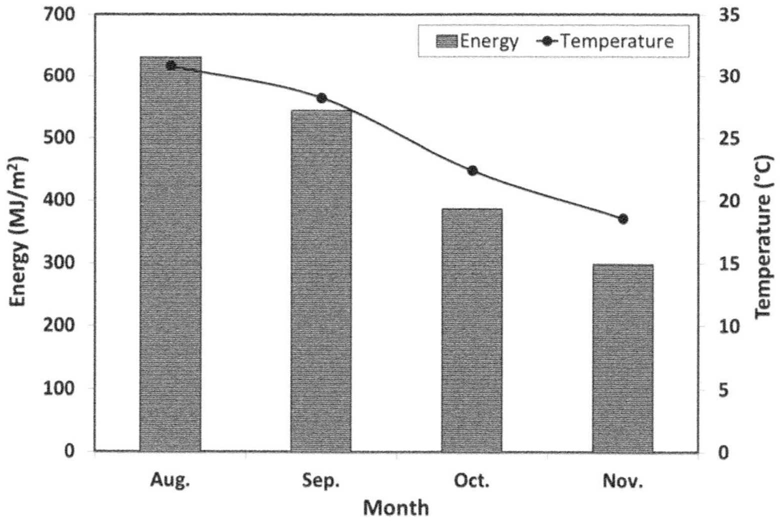

Fig. 19.3 The variations of the solar energy and air temperature distributions

density distributions. The density distributions are kept approximately stable by using the salt gradient protection system in order to inject high-density saline water at the top layer of the HSZ. Figure 19.3 shows the variations of the solar energy and air temperature in Adana, Turkey.

As seen in Fig. 19.3, the maximum and minimum solar energy are 630.00 MJ/m^2 in August and 298.12 MJ/m^2 in November during 4 months, and the maximum and minimum average air temperature are 30.80 °C in August and 18.56 °C in November during 4 months.

Sodium and magnesium chloride-saturated solar ponds were used to store the heat energy. The experimental temperature measurements were taken from the solar ponds on an hourly basis. Figure 19.4 shows average experimental temperature distributions measured inside the solar ponds during 4 months. As shown in Fig. 19.4 the maximum average temperatures of HSZ are 41.87 °C and 52.42 °C in August for sodium and magnesium chloride-saturated solar ponds, respectively. Figures 19.5 and 19.6 show the variation of the energy input, stored, and losses from August to November for sodium and magnesium chloride-saturated solar ponds.

The energy distributions were calculated by using temperature distributions of the solar pond and the reference air temperature. As seen in Figs. 19.5 and 19.6, the

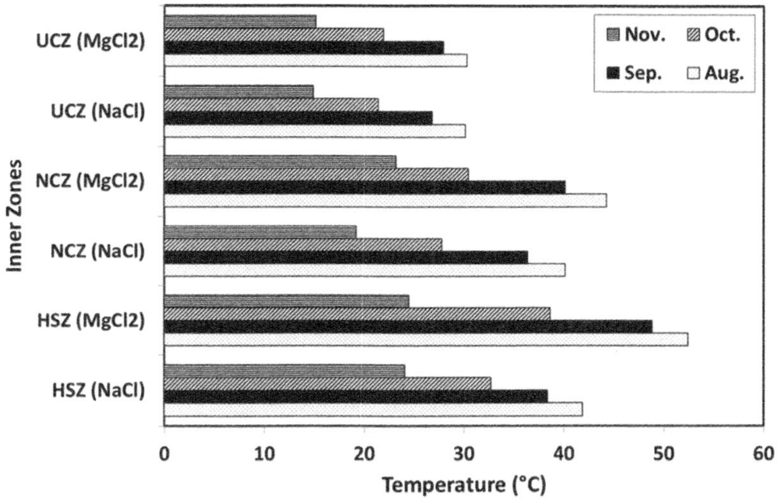

Fig. 19.4 The average experimental temperature distributions of the inner zones

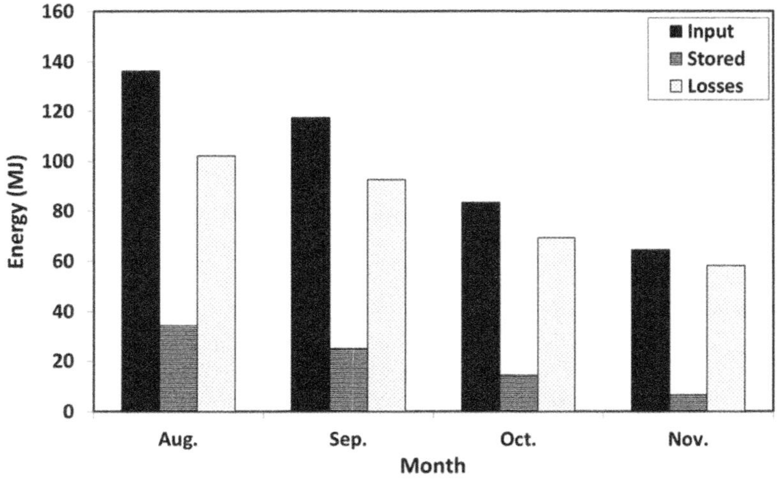

Fig. 19.5 Variations of the energy input, stored, and losses for sodium chloride-saturated pond

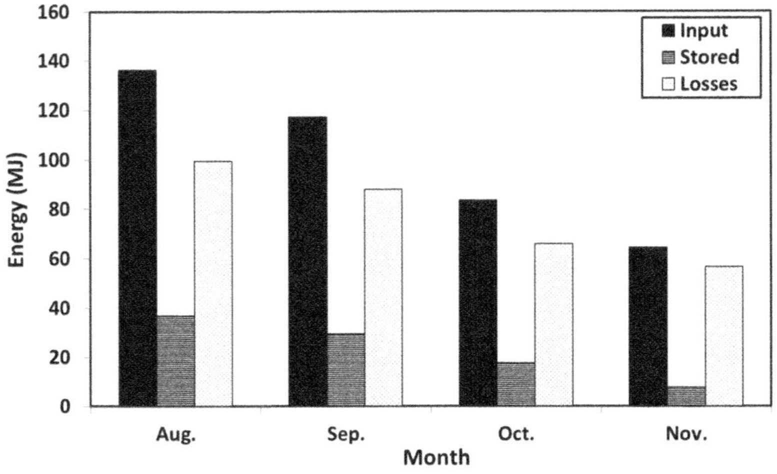

Fig. 19.6 Variations of the energy input, stored, and losses for magnesium chloride-saturated ponds

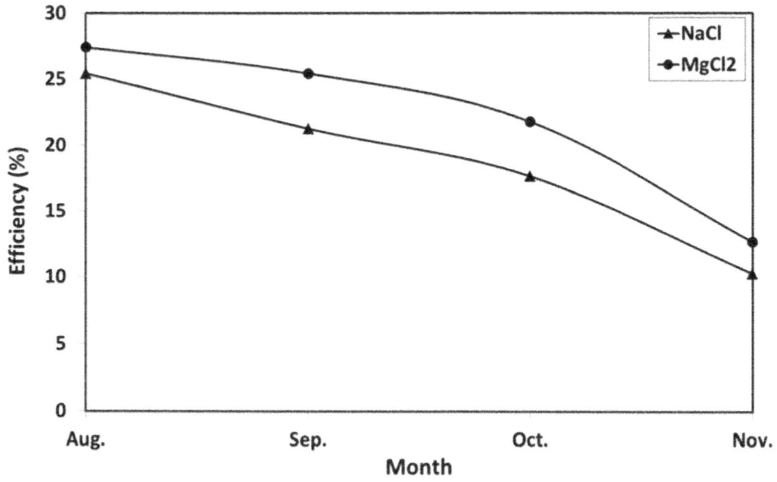

Fig. 19.7 The efficiencies of the sodium and magnesium chloride-saturated ponds

energy stored appears to be maximum at 34.02 MJ and 36.74 MJ in August and minimum at 6.44 MJ and 7.73 MJ in November for sodium and magnesium chloride-saturated solar ponds, respectively.

The efficiencies are dependent on the heat losses, and the temperatures of the salty water and air. The energy losses in the zones were calculated. The energy efficiencies were determined for the solar ponds during the months August, September, October, and November. Figure 19.7 shows the efficiencies of the sodium and magnesium chloride-saturated ponds.

As seen in Fig. 19.7, the maximum and minimum efficiencies of the solar pond were seen to occur in August and November. The highest efficiencies were observed to be 25.41 % and 27.40 % for sodium and magnesium chloride-saturated solar ponds, respectively.

19.5 Conclusion

In this chapter, we have studied both sodium and magnesium chloride-saturated solar ponds' performances. The energy efficiencies are determined for the HSZ of the solar ponds by using the experimental data. The efficiencies are dependent on the heat losses, temperatures of the inner zones and ambient air, incoming solar radiation, and turbidity of the salty water. By comparison, magnesium chloride-saturated solar pond stores more heat energy than sodium chloride-saturated solar pond.

Acknowledgements The authors are thankful to University of Cukurova for financial support of this work (Grant No. FEF2010BAP5 and FEF2012YL13).

References

1. Canyurt OE, Ozturk HK (2008) Application of genetic algorithm (GA) technique on demand estimation of fossil fuels in Turkey. Energ Pol 36:2562–2569
2. Bulut H, Buyukalaca O (2007) Simple model for the generation of daily global solar-radiation data in Turkey. Appl Energ 84:477–491
3. Karakilcik M, Kıymaç K, Dincer I (2006) Experimental and theoretical distributions in a solar pond. Int J Heat Mass Transfer 49:825–835
4. Bozkurt I, Karakilcik M (2012) The daily performance of a solar pond integrated with solar collectors. Sol Energ 86:1611–1620
5. Karakilcik M, Dincer I, Bozkurt I, Atiz A (2013) Performance assessment of a solar pond with and without shading effect. Energ Convers Manag 65:98–107
6. Bozkurt I, Atiz A, Karakilcik M, Dincer I (2014) Transparent covers effect on the performance of a cylindrical solar pond. Int J Green Energ 11:404–416
7. Subhakar D, Murthy SS (1991) Experiments on a magnesium chloride saturated solar pond. Renew Energ 5–6:655–660
8. Subhakar D, Murthy SS (1993) Saturated solar ponds: 2. Parametric studies. Sol Energ 50:307–319
9. Subhakar D, Murthy SS (1994) Saturated solar ponds: 3. Experimental verification. Sol Energ 53:469–472
10. Ranjan KR, Kaushik SC (2014) Thermodynamic and economic feasibility of solar ponds for various thermal applications: a comprehensive review. Renew Sustain Energ Rev 32:123–139
11. Hawlader MNA (1980) The influence of the extinction coefficient on the effectiveness of solar ponds. Sol Energ 25:461–464
12. Bryant HC, Colbeck I (1977) A solar pond for London. Sol Energ 19:321–322

Chapter 20
Investigation of Effect of Using Evacuated Tube Solar Collector on Solar Pond Performance

Ayhan Atiz, Ismail Bozkurt, Mehmet Karakilcik, and Ibrahim Dincer

Abstract The present study deals with the performance investigation of a solar pond integrated with an evacuated tube solar collector system. The experimental cylindrical solar pond system (with a radius of 0.80 m and a depth of 1.65 m) with an evacuated tube solar collector was built in Cukurova University in Adana, Turkey. The solar pond was filled with salty water of various densities to form three salty water zones (e.g. upper convective zone, non-convective zone and heat storage zone). Heat energy was collected by solar pond and evacuated tube solar collectors. The heat collected by collector was transferred to the heat storage zone of the solar pond with a heat exchanger system. Several temperature sensors connected to data acquisition system were placed vertically inside of the solar pond and inlet and outlet of the heat exchanger in the storage zone of the pond. The studies were performed using an evacuated tube solar collector integrated with the solar pond.

Keywords Solar energy • Solar pond • Evacuated solar collector • Heat transfer

A. Atiz • M. Karakilcik (✉)
Department of Physics, Faculty of Sciences and Letters, University of Cukurova, Adana 01330, Turkey
e-mail: kkilcik@cu.edu.tr

I. Bozkurt
Department of Mechanical Engineering, Faculty of Engineering, University of Adiyaman, Adiyaman 02040, Turkey

I. Dincer
Faculty of Engineering and Applied Science, University of Ontario Institute of Technology (UOIT), 2000 Simcoe Street, North Oshawa, ON, Canada L1H 7K4

I. Dincer et al. (eds.), *Progress in Clean Energy, Volume 2*,
DOI 10.1007/978-3-319-17031-2_20

Nomenclature

A	Surface area (m^2)
C	Specific heat capacity (J/kg °C)
D	Declination angle
I	Total solar energy reaching to the pond (J)
F	Absorbed energy fraction at a region of δ-thickness
G	Solar constant
h	Solar radiation ratio
HSZ	Heat storage zone
k	Thermal conductivity (J/m °C h)
k_t	Clearness index
$\dot{m}$	Mass flow (kg/s)
n	Number of the day
NCZ	Non-convective zone
Q	Heat (J)
R	Reflectivity
S	Salinity (g/kg)
T	Temperature (°C)
UCZ	Upper convective zone

Greek Letters

δ	Thickness where long-wave solar energy is absorbed (m)
β	Incident beam entering rate into water
γ	Latitude angle
α	Tilt angle
w	Hour angle
θ	Incident angle
χ	Beam radiation tilt factor
ρ	Density (kg/m^3)
Δx	Thickness of horizontal layers (m)

Subscripts

bm	Beam
bt	Bottom
coll	Collector
d	Declination
dif	Diffuse
exc	Exchanger
g	Ground

i	Incident
in	Inner
out	Outer
r	Reflection
s	Salinity
sc	Solar constant
sw	Salty water
swall	Side wall
st	Heat stored
u	Useful
up	Just above the zone
z	Zenith

20.1 Introduction

Turkey's renewable energy potential appears to be enormous with the high potential of solar, hydro, biomass, geothermal and wind and offers an exceptional opportunity for implementation of renewable energy-based systems for better environment and sustainability.

During the past decade, the studies of the solar energy applications have steadily been increasing. Some of these applications especially thermal power-generating systems are flat-plate solar and evacuated tube collectors and integrated with the solar ponds. The most effective systems of these systems are the systems integrated with each other to overcome disadvantages of individual systems [1]. The integrated thermal energy applications, including solar ponds, offer better efficiency and better cost-effectiveness.

Many studies (e.g. 2–7) have been done to investigate the performance of the solar ponds. Date et al. [8] suggested a new model to predict transient behaviour of a solar pond. The transient thermal performance of the solar pond with heat extraction was determined. In recent years, lots of studies have been carried out about the performance investigation of the evacuated tube solar collectors [9–12]. Sharma and Diaz [13] investigated the thermal analysis of an evacuated solar collector based on mini channels by developing an appropriate heat transfer model of the radiative and convective thermal exchange. Ma et al. [14] determined the thermal performance of the individual evacuated glass tubular solar collector based on the energy balance for the U-tube glass evacuated tube solar collector. Karakilcik et al. [15] determined the dynamic exergetic efficiencies of the solar pond integrated with flat-plate collector system. Bozkurt et al. [16] compared energy efficiency assessment of the integrated and non-integrated solar ponds.

In this chapter, our aim was to investigate the effect of using evacuated tube solar collector on the performance of an experimental solar pond. For this purpose, an integrated solar pond system was constructed and tested at Cukurova University in Adana, Turkey (with the coordinates of 35°18′ E longitude, 37°05′ N latitude).

Some properties (such as temperature distributions) of the integrated system were continuously measured and recorded by data acquisition system. A heat exchanger system was used to transfer the heat collected by evacuated tube solar collectors.

20.2 System Description

An integrated system, which consists of a solar pond with a radius of 0.80 m and a depth of 1.65 m and evacuated tube solar collector, was built in Cukurova University in Adana, Turkey (Fig. 20.1). A heat exchanger system was placed in the heat storage zone of the solar pond. Solar energy was collected by both solar pond and evacuated tube solar collector. Therefore, much more heat was captured and store in the heat storage zone of the solar pond for a long time. Solar pond consists of three different zones from the bottom to the top of the pond, namely, heat storage

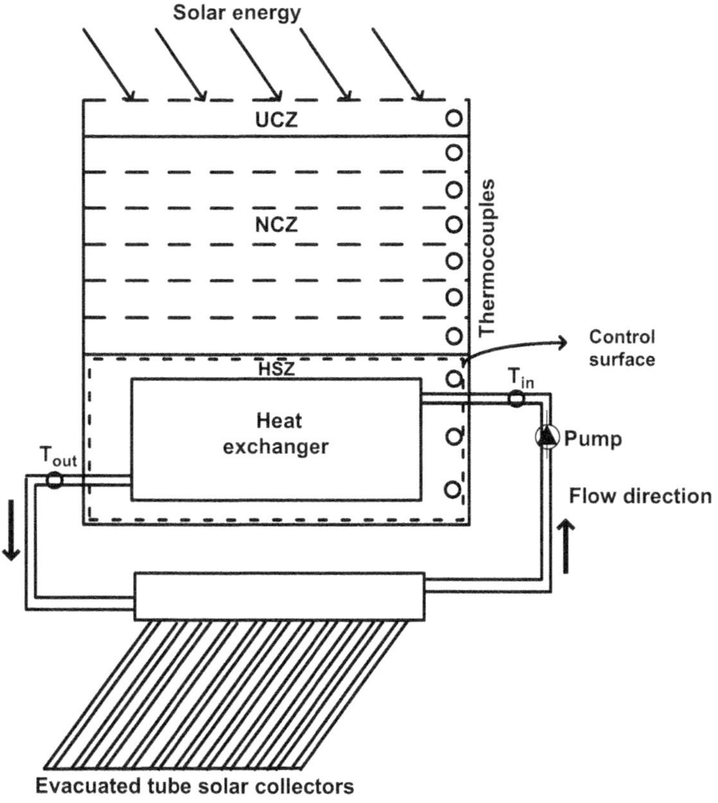

Fig. 20.1 A schematic diagram of the integrated solar pond system

zone (HSZ), non-convective zone (NCZ) and upper convective zone (UCZ). First, salty water solutions were prepared by using a mixer for desired concentrations. Then, such three zones were created. The density of the solar pond increased through to the bottom of the solar pond. The thicknesses of HSZ, NCZ and UCZ were 0.60, 0.90 and 0.15 m, respectively. The range of the salt gradient in the inner zones was such that the density ranges were 1,000–1,030 kg/m^3 in UCZ, 1,030–1,150 kg/m^3 in NCZ and 1,180–1,200 kg/m^3 in HSZ.

The experimental temperature distributions were measured by using temperature sensors as they were placed into the inner zones and inlet-outlet of the heat exchanger. Here after, the temperature distribution profiles of these regions at any time were experimentally obtained by a data acquisition system which was connected to a computer for data recording, monitoring and processing. The measurement system had 16 temperature sensors. Several temperature sensors were placed into the inner zones, starting from the bottom to the upper layer at 0.10, 0.30, 0.50, 0.70, 0.90, 1.10, 1.30, 1.50 and 1.60 m heights. Two sensors were placed to the inlet and outlet pipes of the collector in order to measure the temperature distributions. As seen in Fig. 20.1, the circulation pump was connected to the inlet pipe of the exchange in the storage zone to measure the mass flow rate from the collector to storage thermal energy in the solar pond. Heat captured by evacuated tube solar collectors was transferred to the exchanger by using a pump and hence to the hot water which was circulated through. The experiments help to improve the performance of the integrated system.

20.3 Energy Analysis

The thermal performance of the solar pond integrated with evacuated collector was determined by using energy balance equation of the system. It is thermodynamically very important to identify the boundary of the system. The heat energy collected by solar pond and evacuated tube collector was stored in the heat storage zone of the solar pond. Therefore, the system was considered as a control volume, and the control surface of the system is taken as the volume surface of HSZ (as shown in Fig. 20.1). First, the mass balance equation of the system is written as

$$\sum \dot{m}_{\mathrm{in,exc}} = \sum \dot{m}_{\mathrm{out,exc}} \tag{20.1}$$

where $\dot{m}_{\mathrm{in,exc}}$ is the mass flow rate of the inlet of the exchanger and $\dot{m}_{\mathrm{out,exc}}$ is the mass flow rate of the outlet of the exchanger. The heat exchanger in steady operation requires that the sum of the inlet mass flow rates equals the sum of the outlet mass flow rates, so the total amount of mass contained within the system does not change with time. Heat exchangers typically involve no work interactions and negligible kinetic and potential energy changes [17]. During the experiments, the mass flow rate of the pump is 0.035 kg/s. Therefore, the energy balance equation of the system is written as follows, depending on the control volume:

$$\sum Q_{st} = \sum Q_{in,HSZ} + \sum Q_{u,coll} - \sum Q_{loss,HSZ} = I\beta A\{(1-F)h(x-\delta)\} + \dot{m}C(T_{in,exc} - T_{out,exc}) - \left\{ k_{swall}A_{swall}\frac{(T_{HSZ} - T_{air})}{\Delta x_{swall}} + k_{swall}A_{bt}\frac{(T_{HSZ} - T_{air})}{\Delta x_{bt}} + k_{sw}A_{sw}\frac{(T_{HSZ} - T_{NCZ})}{\Delta x_{HSZ\text{-}NCZ}} \right\} \quad (20.2)$$

where Q_{st} is the stored energy in HSZ, $Q_{in,HSZ}$ is the reaching solar energy from the surface of the solar pond to HSZ, $Q_{u,coll}$ is the useful heat energy from the evacuated tube solar collector and $Q_{loss,HSZ}$ is the heat losses from the above zone, side wall and bottom wall to surrounding air of the solar pond. I is the total solar radiation reaching the solar pond surface, A_{swall} is the side wall area of the HSZ, F is the fraction of energy absorbed at a region of δ-thickness, h is the solar radiation ratio, A is the surface area of the solar pond, T_{air} is the air temperature, k_{swall} is the thermal conductivity of the side and bottom walls, k_{sw} is the thermal conductivity of the salty water, Δx_{bt} is the thickness of the bottom wall, Δx_{swall} is the thickness of the side wall, $\Delta x_{HSZ\text{-}NCZ}$ is the thickness between HSZ's middle point and NCZ's middle point and β is the fraction of the incident solar radiation which enters the pond and is given by Hawlader [18]:

$$\beta = 1 - 0.6\left[\frac{\sin\theta_i - \sin\theta_r}{\sin\theta_i + \sin\theta_r}\right]^2 - 0.4\left[\frac{\tan\theta_i - \tan\theta_r}{\tan\theta_i + \tan\theta_r}\right]^2 \quad (20.3)$$

where θ_i and θ_r are the incidence and refraction angles. h represents the ratio of the solar energy reaching the depth in the layer to the total solar incident falling onto the surface of the pond and is given by Bryant and Colbeck [19] as

$$h = 0.727 - 0.056\ln\left[\frac{(x-\delta)}{\cos\theta_r}\right] \quad (20.4)$$

where x is the thickness of the layer and δ is the thickness of the layer in the UCZ where long-wave solar energy is absorbed. The heat capacity is calculated by the empirical equation as follows [20]:

$$C_p = (-0.0044S + 4.1569)\ 1000 \quad (20.5)$$

where C_p is heat capacity (J/kg °C) and S is salinity (g/kg). The density difference at low temperature takes place approximately in linear relationship between density and salinity. We use an empirical correlation, as given below, to calculate the salinity of the zones [20]:

$$S = \frac{(\rho - 998.24)}{0.756} \tag{20.6}$$

The hourly extraterrestrial radiation on a horizontal surface is calculated as [21]

$$I_0 = \frac{86400\mathrm{G_{sc}}}{\pi}\left[1 + 0.033\cos\left(\frac{360}{365}n\right)\right]\left[\cos\gamma\cos D_\mathrm{d}\sin w + \left(\frac{\pi w}{180}\right)\sin\gamma\sin D_\mathrm{d}\right] \tag{20.7}$$

where G_{sc} is the solar constant, n is the number of day and γ is the latitude angle which is calculated using solar elevation angle. It is positive (+) for northern hemisphere and negative (–) for southern hemisphere. w is the sunset hour in degrees and is given as

$$w = \cos^{-1}\left[-\tan 360\frac{284 + n}{365}\right] \tag{20.8}$$

Here, D_d is the declination angle as given by Cooper [22]:

$$D_\mathrm{d} = 23.25\sin\left[360\frac{284 + n}{365}\right] \tag{20.9}$$

The amount of the reaching solar radiation on a tilted evacuated solar collector surface is given as

$$I_\mathrm{col} = I_\mathrm{bm}\chi + \frac{I_\mathrm{dif}(1 + \cos\alpha)}{2} + R_\mathrm{g}I\,(1 - \cos\alpha)/2 \tag{20.10}$$

where I_{col} is the total radiation reaching a given aperture area of collector (W/m^2), I_{bm} is the beam radiation, χ is the tilt factor for south-facing tilted surface, I_{dif} is the diffuse radiation, R_g is the ground reflectivity (given as 0.4 for December), I_{dif} is the direct solar radiation which is the sum of the diffuse and beam radiation [23] and α is the tilt angle which is zero for horizontal surface. The beam radiation tilt factor is calculated by using the following equation:

$$\chi = \frac{\cos\theta}{\cos\theta_\mathrm{z}} = \frac{\sin(\gamma - \alpha)\sin(D_\mathrm{d}) + \cos(\gamma - \alpha)\cos(D_\mathrm{d})\cos(w)}{\sin(\gamma)\sin(D_\mathrm{d}) + \cos(\gamma)\cos(D_\mathrm{d})\cos(w)} \tag{20.11}$$

where θ is the incident angle and θ_z is the zenith angle. The diffuse radiation was determined by using Erbs et al. [24] correlation. $\alpha = 0$ for horizontal surface and

Eq. 20.10 can be rewritten for horizontal surface. So, I_{col} is equal to the total solar radiation reaching on the solar pond as

$$I = I_{bm} + I_{dif} \tag{20.12}$$

where I is the measurement of the total solar radiation on a horizontal surface as

$$\frac{I_{dif}}{I} = \begin{cases} 1 - 0.09k_t & 0 \le k_t \le 0.22 \\ 0.9511 - 0.1604k_t + 4.388k_t^2 - 16.638k_t^3 + 12.336k_t^4 & 0.22 < k_t \le 0.8 \\ 0.165 & k_t > 0.8 \end{cases} \tag{20.13}$$

where k_t is the index of the clearness and it can be defined as $k_t = \frac{I}{I_0}$.

20.4 Results and Discussion

The integrated solar pond system was completed in December 2013. After that, it was tested under actual solar conditions. The behaviour of the solar pond was investigated after integrating it with an evacuated tube solar collector. The solar energy collected during the day by an evacuated tube collector was transferred to HSZ. The density distribution was measured at the initial and final of the experiment. Figure 20.2 shows the variations of the experimental density distribution for the inner zones of the integrated solar pond system. The solar pond consists of density gradient layers. This gradient is very important to prevent convection heat losses.

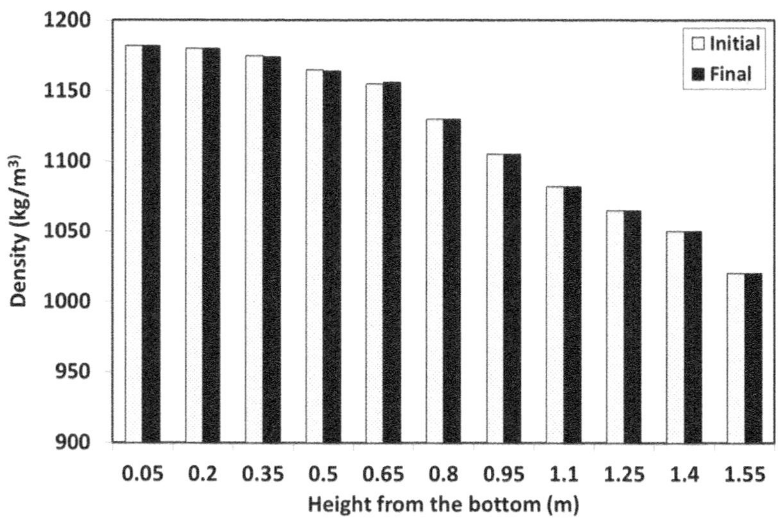

Fig. 20.2 Variations of the experimental obtained densities

As seen in Fig. 20.2, despite the rapid temperature increase by using evacuated tube solar collector, the salty water gradient of the solar pond was kept stable. The slight change of the density was observed, especially in the upper parts of HSZ. Figure 20.3 shows the monthly total global solar radiation and air temperature distribution in Adana, Turkey. As seen in Fig. 20.3, the maximum and minimum solar radiations are 713.91 MJ/m^2 in June and 218.48 MJ/m^2 in January. But the maximum and minimum air temperatures are 30.80 °C in August and 11.30 °C in January, respectively.

Figure 20.4 shows the solar energy distribution reaching the evacuated tube solar collector's surface during the experiments. As seen in Fig. 20.4, the minimum and

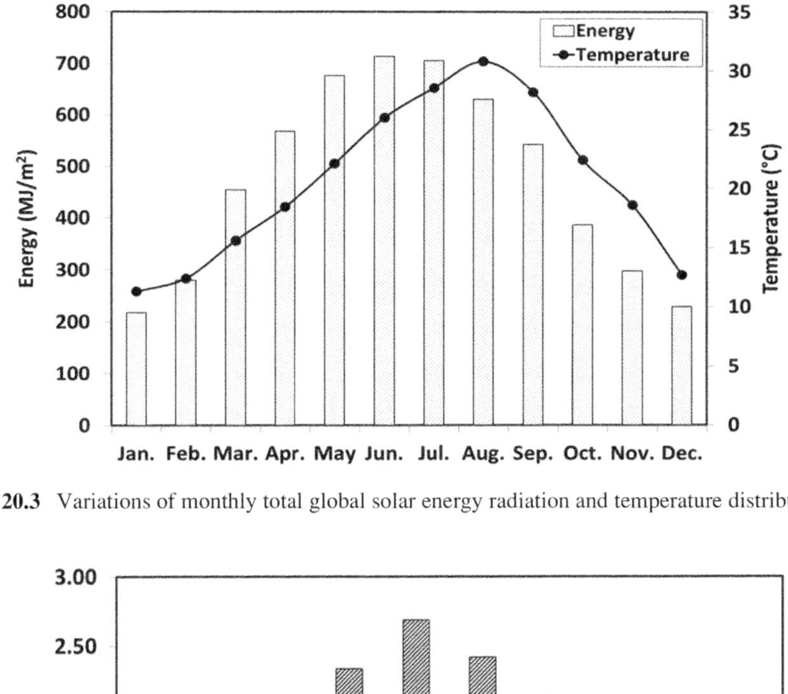

Fig. 20.3 Variations of monthly total global solar energy radiation and temperature distributions

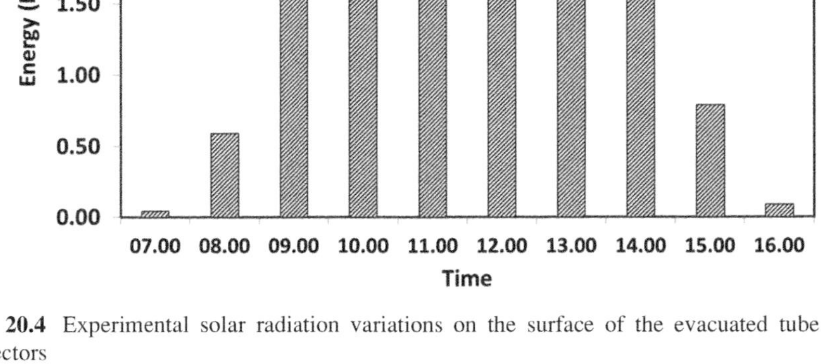

Fig. 20.4 Experimental solar radiation variations on the surface of the evacuated tube solar collectors

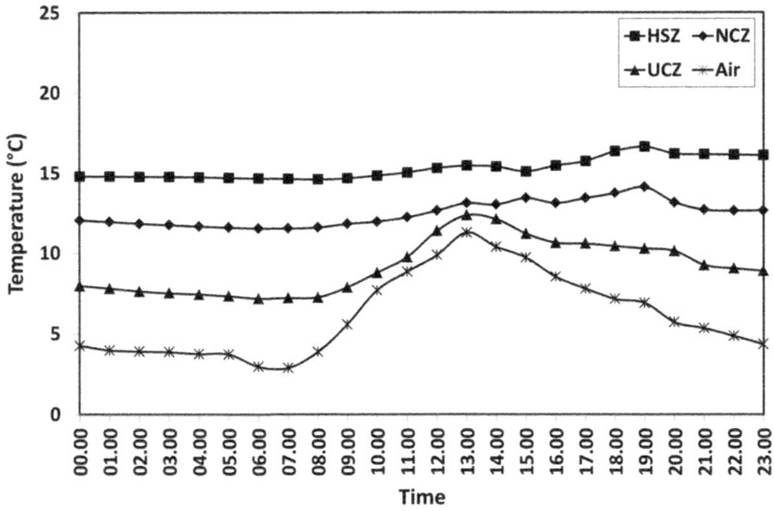

Fig. 20.5 Average temperature distributions in the zones of the solar pond on December 17, 2013

maximum reaching energy of the collector surface are 0.04 MJ/m^2 at 07.00 a.m. and 2.69 MJ/m^2 at 11.00 a.m., respectively. The aperture area of each collector is 2.22 m^2. The total energy contents, coming to the collector surface, are calculated by using the energy distribution and the aperture area.

The experimental temperature measurements were taken from inner zones of the solar pond and inlet-outlet of the heat exchanger by using thermocouples in December. The temperature of the solar pond was low at first. The evacuated tube solar collector was used to increase the temperature of the solar pond. Figure 20.5 shows the temperature distributions of the integrated system by using an evacuated tube solar collector. As seen in Fig. 20.5, the temperature of HSZ is increased by 1.52 °C in a day. The temperature changes of the UCZ and NCZ are less because the heat exchanger was placed in HSZ. Figure 20.6 shows the inlet and outlet temperatures of the heat exchanger. As seen in Fig. 20.6, the heat is transferred to the solar pond until 18.00 because the inlet temperature of the heat exchanger is close to the outlet temperature due to the sunset and because the evacuated tube collector is active during the sunrise.

The inner zones of the solar pond consist of different density salty water layers. Heat capacity and salinity of the inner zones were calculated by using Eqs. 20.5 and 20.6. Figure 20.7 shows the heat capacity and salinity of the solar pond. As seen in Fig. 20.7, the heat capacity and salinity appear to be inversely proportional.

The solar energy was measured at the location of the system. The reaching energy of the surface of the solar pond and evacuated tube collector was calculated. Figure 20.8 shows the energy contents of the integrated system. As seen in

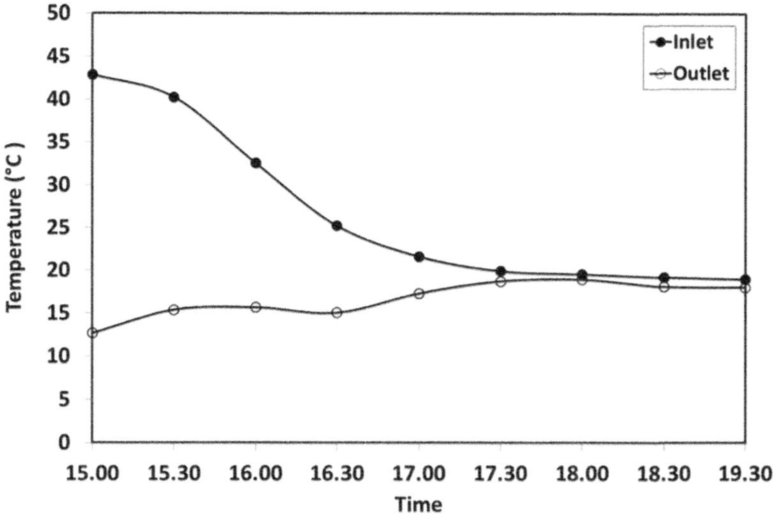

Fig. 20.6 Variations of inlet and outlet temperatures of the heat exchanger

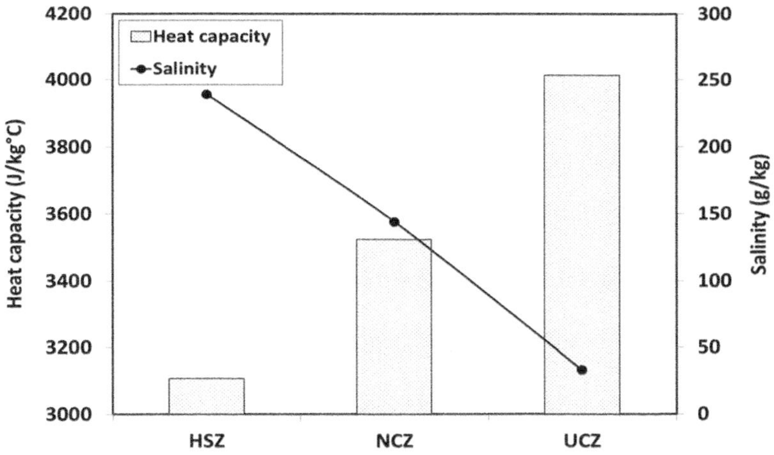

Fig. 20.7 Heat capacity and salinity values of the solar pond

Fig. 20.8, the stored and lost heat energy in the integrated solar pond system during the day is 4.89 and 31.64 MJ for using an evacuated tube collector, respectively. Furthermore, the reaching energy on the surface of the solar pond and collector is 16.28 and 20.25 MJ, respectively.

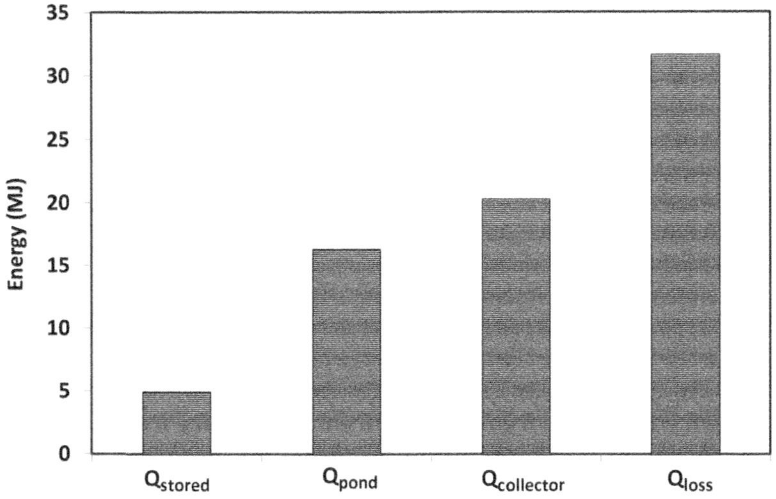

Fig. 20.8 Energy contents of the integrated solar pond system

20.5 Conclusions

The experiments were conducted in order to demonstrate the effect of the evacuated tube solar collector on solar pond performance. The experiments were carried during the sunny days in December which was the coldest of the year. The temperature of the heat storage zone was significantly increased within a day by using an evacuated tube solar collector. As a result, the heat storage performance of the solar pond was increased and also no disturbed the structure of the solar pond. A significant deterioration in the structure of the layers of the pond was not observed.

Acknowledgements The authors are thankful to Cukurova University for the financial support provided for the present work (Grant No. FEF2012D20).

References

1. Bozkurt I, Karakilcik M (2012) The daily performance of a solar pond integrated with solar collectors. Sol Energ 86:1611–1620
2. Karakilcik M, Kiymac K, Dincer I (2006) Experimental and theoretical temperature distributions in a solar pond. Int J Heat Mass Transfer 49:825–835
3. Karakilcik M, Dincer I, Rosen MA (2006) Performance investigation of a solar pond. Appl Therm Eng 26:727–735
4. Karakilcik M, Dincer I, Bozkurt I, Atiz A (2013) Performance assessment of a solar pond with and without shading effect. Energ Convers Manag 65:98–107

5. Dehghan AA, Movahedi A, Mazidi M (2013) Experimental investigation of energy and exergy performance of square and circular solar ponds. Sol Energ 97:273–284
6. Bozkurt I, Atiz A, Karakilcik M, Dincer I (2014) An investigation of the effect of transparent covers on the performance of cylindrical solar ponds. Int J Green Energ 11:404–416
7. Suárez F, Ruskowitz JA, Childress AE, Tyler SW (2014) Understanding the expected performance of large-scale solar ponds from laboratory-scale observations and numerical modeling. Appl Energ 117:1–10
8. Date A, Yaakob Y, Date A, Krishnapillai S, Akbarzadeh A (2013) Heat extraction from non-convective and lower convective zones of the solar pond: a transient study. Sol Energ 97:517–528
9. Kroll JA, Ziegler F (2011) The use of ground heat storages and evacuated tube solar collectors for meeting the annual heating demand of family-sized houses. Sol Energ 85:2611–2621
10. Nkwetta DN, Smyth M, Zacharopoulos A, Hyde T (2012) Optical evaluation and analysis of an internal low-concentrated evacuated tube heat pipe solar collector for powering solar air-conditioning systems. Renew Energ 39:65–70
11. Zhang M, Miao L, Kang YP, Tanemura S, Fisher CAJ, Xu G, Li CX, Fan GZ (2013) Efficient, low-cost solar thermoelectric cogenerators comprising evacuated tubular solar collectors and thermoelectric modules. Appl Energ 109:51–59
12. Liang R, Zhang J, Zhao L, Ma L (2014) Research on the universal model of filled-type evacuated tube with U-tube in uniform boundary condition. Appl Therm Eng 63:362–369
13. Sharma N, Diaz G (2011) Performance model of a novel evacuated-tube solar collector based on minichannels. Sol Energ 85:881–890
14. Ma L, Lu Z, Zhang J, Liang R (2010) Thermal performance analysis of the glass evacuated tube solar collector with U-tube. Build Environ 45:1959–1967
15. Karakilcik M, Bozkurt I, Dincer I (2013) Dynamic exergetic performance assessment of an integrated solar pond. Int J Exergy 12:70–85
16. Bozkurt I, Karakilcik M, Dincer I (2014) Energy efficiencies assessment of the integrated and nonintegrated solar ponds. Int J Low-Carbon Tech 9:45–51
17. Cengel YA, Boles MA (2011) Thermodynamics, an engineering approach, 7th edn. McGraw Hill, Boston, ISBN: 978-0-07-736674-2
18. Hawlader MNA (1980) The influence of the extinction coefficient on the effectiveness of solar ponds. Sol Energ 25:461–464
19. Bryant HC, Colbeck I (1977) A solar pond for London. Sol Energ 19:321–322
20. Sun H, Feistel R, Koch M, Markoe A (2008) New equations for density, entropy, heat capacity, and potential temperature of a saline thermal fluid. Deep-Sea Res I 55:1304–1310
21. Kalogirou SA (2009) Solar energy engineering: processes and systems. Elsevier, London
22. Cooper PL (1969) The absorption of solar radiation in solar stills. Sol Energ 12:333–346
23. Duffie JA, Beckman WA (eds) (2013) Solar engineering of thermal processes. Wiley, New York
24. Erbs DG, Klein SA, Duffie JA (1982) Estimation of the diffuse radiation fraction for hourly, daily, and monthly-average global radiation. Sol Energ 28:293–302

Part V
Geothermal Energy

Chapter 21
Numerical Modeling of Vertical Earth Pipe Cooling System for Hot and Humid Subtropical Climate

S.F. Ahmed, M.M.K. Khan, M.T.O. Amanullah, M.G. Rasul, and N.M.S. Hassan

Abstract Energy crisis is one of the major problems facing the progress of human society. There are several energy-efficient technologies that can be applied to save energy and make a sustainable environment. Passive air cooling of earth pipe cooling technology is one of them to reduce the energy consumption for hot and humid subtropical climates. The technology works with a long buried pipe with one end for intake air and the other end for providing air cooled by soil to the desired space such as residential, agricultural, or industrial buildings. It can be an attractive economical alternative to conventional cooling since there are no compressors or any customary mechanical unit. This chapter reports the performance of a vertical earth pipe cooling system for a hot and humid subtropical climatic zone in Queensland, Australia. A series of buried pipes were installed in vertical arrangement in order to increase earth pipe cooling performance. To measure the performance of the system, a numerical model was developed and simulated using the CFD software Fluent in ANSYS 15.0. Data were collected from two modeled rooms built from two shipping containers and installed at the Sustainable Precinct at Central Queensland University, Rockhampton, Australia. The impact of air temperature and velocity on room cooling performance has also been assessed. A temperature reduction of 1.82 °C was observed in the room connected to the vertical earth pipe cooling system, which will save the energy cost for thermal cooling in buildings.

Keywords Earth pipe cooling • Air temperature • Relative humidity • Subtropical climate

S.F. Ahmed (✉) • M.M.K. Khan • M.G. Rasul • N.M.S. Hassan
School of Engineering and Technology, Central Queensland University, Rockhampton Campus, North Rockhampton, QLD 4702, Australia
e-mail: s.f.ahmed@cqu.edu.au

M.T.O. Amanullah
School of Engineering, Deakin University, Burwood, VIC 3220, Australia

I. Dincer et al. (eds.), *Progress in Clean Energy, Volume 2*,
DOI 10.1007/978-3-319-17031-2_21

Nomenclature

k	Turbulence kinetic energy, m^2s^{-2}
ε	Dissipation rate for turbulence kinetic energy, m^2s^{-3}
u_i, u_j	Fluctuating velocity components, ms^{-1}
x_i	Components of length, m
p	Pressure, Pa
υ	Molecular kinetic fluid viscosity, m^2s^{-1}
μ_{eff}	Effective viscosity $(\mu + \mu_t)$
μ	Viscosity of the fluid, kg m^{-1} s^{-1}
ρ	Density of the fluid, kg m^{-3}
t	Time, s
G_k	Turbulence kinetic energy generation due to the mean velocity gradients, kg m^{-1} s^{-2}
G_b	Turbulence kinetic energy generation due to buoyancy, kg m^{-1} s^{-2}
Y_M	Contribution of the fluctuating dilatation in compressible turbulent flow to the overall dissipation rate, kg m^{-1} s^{-2}
α_k	Inverse effective Prandtl numbers for k, kg m^{-1} s^{-2}
α_ε	Inverse effective Prandtl numbers for ε, kg m^{-1} s^{-2}
R_ε	Gas-law constant for ε (8.31447×10^3 J $Kgmol^{-1}$ K^{-1})
S_K	Mean strain rates for k, m^2s^{-2}
S_ε	Mean strain rates for ε, m^2s^{-2}
$C_{1\varepsilon}$	Turbulence model constant
$C_{2\varepsilon}$	Turbulence model constant
$C_{3\varepsilon}$	Turbulence model constant
k_{eff}	Effective conductivity, W m^{-1} K^{-1}
k_t	Thermal conductivity for turbulent flow
J_j	Component of diffusion flux, m^{-2} s^{-1}
T	Temperature, K
h	Enthalpy, J kg^{-1}
S_h	Total entropy, JK^{-1}

21.1 Introduction

Energy plays an important role in the economic and social development of human society. Energy consumption in various forms is gradually rising all over the world. Income and population growth are the main drivers behind this growth of energy demand. China and India are leading in both world energy demand and economic growth [1]. Energy consumption in both the nations as a share of total energy use has expanded substantially. They accounted together for around 10 % of total world energy consumption in 1990 and about 24 % in 2010 [2]. Their combined energy use is predicted to rise 34 % of estimated total world energy consumption in 2040. Moreover, China became the largest energy consumer of the world recently and it is

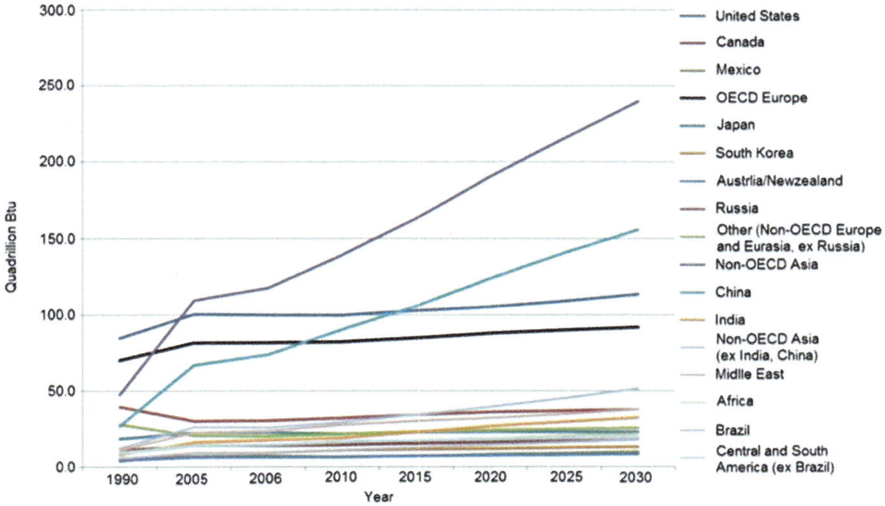

Fig. 21.1 Total world primary energy consumption by region [5]

projected to consume more than twice the extent that the United States (previous largest energy consumer) will consume in 2040 as shown in Fig. 21.1.

By 2030, the world population is estimated to reach 8.3 billion, which indicates 1.3 billion more people will need energy as the world income in 2030 is projected to be nearly the double of the 2011 level [3]. The energy consumption worldwide is expected to rise 1.6 % per annum from 2011 to 2030, which will add 36 % more to the total world energy consumption by 2030 [4]. The buildings are responsible for more than 40 % of global energy use in the major sectors of industry, agriculture, commercial, and residential where Australian buildings use up to one-third of their electricity on air conditioning. People use this energy for cooling and heating their residences, operating their appliances, heating water, and for lighting. Apart from these residential uses, energy is also used for agricultural, commercial, and industrial purposes.

The energy consumption for the Australian residential sector in 1990 was around 299 petajoules (PJ) and that by 2008 had increased to about 402 PJ and is projected to increase to 467 PJ by 2020 under the current trends [6]. This signifies 56 % increase in energy consumption in this sector over the period from 1990 to 2020. The contribution of electricity to this residential energy consumption is predicted to increase from 46 % of total energy consumption in 1990 to 53 % in 2020 as shown in Fig. 21.2. Gas consumption is also expected to increase from 30 % in 1990 to 37 % in 2020 while wood is predicted to decrease from 21 to 8 % over the same period [6]. LPG use in the residential sector remains unchanged and is expected to contribute to 2 % in 2020.

Residential electricity consumption grew 1.5 % per annum at a compound rate between 2008 and 2011 [7]. Energy consumption in this sector has been less responsive to electricity price increases since 2008 than before 2008. Per capita

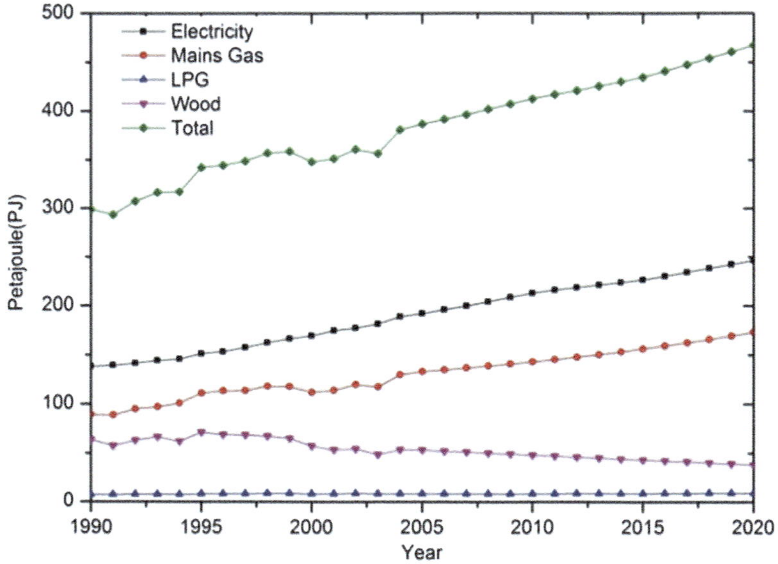

Fig. 21.2 Total energy consumption of Australia by fuel. (Data source: Department of Environment, Water, Heritage and the Arts)

residential energy consumption was nearly static between 2003 and 2012, suggesting that higher consumption due to population growth rather than higher spending per capita. This residential energy consumption varies state to state in Australia. Victoria, New South Wales (NSW) and Queensland show steady growth of residential energy consumption over the period of 1990–2020 as shown in Fig. 21.3. This energy consumption is primarily due to extensive use of gas for space heating and cooling.

Australian buildings use 38 % of their total energy consumption on cooling and heating purposes [8]. Moreover, the average annual temperature in Australia has increased by 0.9 °C since 1950 [9]. Projections from Commonwealth Scientific and Industrial Research Organization (CSIRO) indicate an increase of average annual temperature by a further 0.6–1.5 °C by 2030 and 1–5.0 °C by 2070, depending upon the levels of global emissions [10]. According to this projection, more energy will be required for survival and achieving thermal comfort. It is essential to save energy today in Australian residential sector as well as in the world for the world economy and environment.

Australia is the eighteenth leading energy consumer in the world and fourteenth on a per person basis [11]. The energy demand in Australia is predicted to increase rapidly in future. Net energy consumption of Australia was increased by 1.8 % average rate per annum over the 10 years from 1999–2000 to 2009–2010, compared with 2.3 % over the previous 10 years [11]. Domestic energy consumption growth

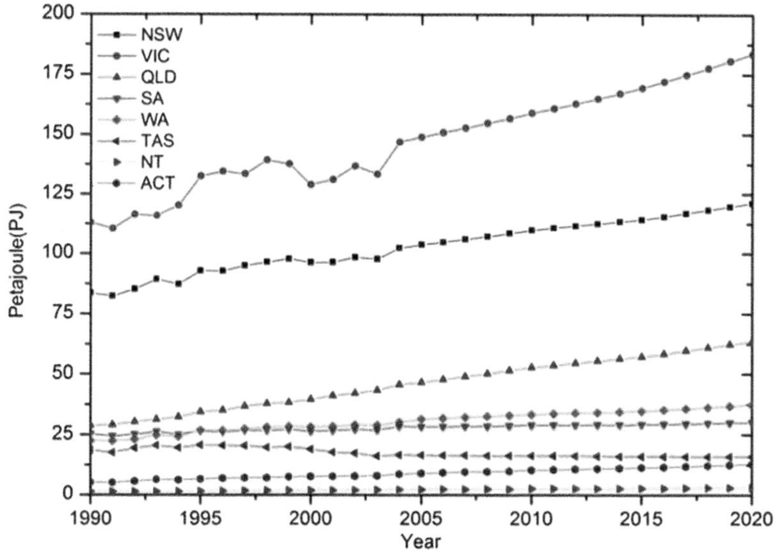

Fig. 21.3 Total residential energy consumption by state. (Data source: Department of Environment, Water, Heritage and the Arts)

was at a slower rate than the energy generation over the past 20 years. The residential energy consumption trend per person demonstrates a consistent but modest growth from 17 gigajoules (GJ) in 1990 to 20 GJ in 2020. So, it is essential to save energy in the Australian residential sector as most of the energy is used in this sector. The energy consumption in a building can be reduced by utilizing low-energy or passive cooling techniques. Passive air cooling (PAC) system is seen as an attractive option to reduce the energy consumption in the buildings for all subtropical zones.

The passive air cooling system provides cooling in passive process without using customary mechanical units such as fan, compressor etc. in regions where cooling is a dominant problem. It has heat transferring capacity from a building to different natural heat sinks [12]. Passive air cooling system is based on the interaction of the building envelope and its surroundings. It minimizes heat gain from outer environment and facilitates heat loss to the cooling natural sources, for example, cooling breezes, air movement, earth coupling, and evaporation. The most suitable passive cooling strategy selection procedure was discussed for a hot and humid subtropical climate in Queensland, Australia [13]. The result showed that the passive cooling strategy of natural ventilation would be suitable for Rockhampton, Brisbane, Mackay, and Townsville, whereas high thermal mass would be appropriate for Mackay and Townsville.

Four different passive cooling techniques namely evaporative cooling, natural ventilation, high thermal mass, and high thermal mass with night ventilation were discussed by Brown [14]. All of these passive cooling techniques depend on daily temperature and relative humidity changes. Among these passive cooling techniques, the thermal mass effect with night ventilation was calculated for hot and humid climate of Israel in summer [15]. The result showed that a temperature reduction of 3–6 °C can be achieved using this technique in a heavy constructed building without running any air conditioning unit. A simple design tool was also found to predict the indoor temperature reduction which depends on the thermal mass amount, night ventilation rate, and the temperature swing of the place between day and night. To reduce the energy consumption an air-cooled chiller system in an office building was analyzed using passive cooling [16]. The impact of economizer usages and pre-cooling was simulated and analyzed in this study to measure the annual energy demand savings. It was found that the pre-cooling system can save 115 kW/m^2/month of total cooling energy demand.

Three passive cooling strategies, namely natural ventilation, radiant cooling, and indirect evaporative cooling were tested with shading and nonshading system to determine their applicability in warm climates [17]. A smart controller was utilized to control an operable shading framework which was tested inside and outside the test cell's window. The radiant cooling used in this investigation was tested with different configurations of mass and insulation. The temperature distribution and swing inside the cell were affected by the amount of solar radiation, position of the mass inside the space and the wall's insulation level. Dobson [18] developed a thermal model for a night sky radiation cooling system in Namibia. It was inferred that this thermal model can be utilized as a design tool for measuring night sky radiation cooling framework.

Earth pipe cooling is one of the passive air cooling methods where underground spaces provide several additional benefits including safeguard from dust, noise, storms, radiation, and restricted air infiltration. The technology utilizes the earth as a heat sink to cool a room in hot climates, where the ambient air passes through the pipe buried underground and exchanges excess heat to the earth by convection. It is the lowest costly method for cooling the interior spaces of a building with the most minimal ecological effect. This technology offers a great potential for energy saving for hot and humid climate like Northern Queensland, Australia since it can supplement the air-conditioning load of many homes with virtually no negative environmental impact.

In some cases, the earth pipe cooling system is assisted with a heat pump as a heat exchanger located within the buried pipe. It can increase cooling capacity and hence improve the coefficient of performance. Applications of this technique were discussed by Fanciotti and Scudo [19] and Saini [20]. The cooling performance of earth pipe cooling system was investigated using two different piping systems for a subtropical climate in Queensland, Australia [21–23]. The major difference between the summer cooling and the winter preheating potential using buried pipe systems was examined by Hollmuller and Lachal under Central European climate from an economic point of view [24]. It was concluded that the air preheating with buried pipes remains more expensive than with fuel in all cases.

The infinite thermal capacity of earth has made it a very useful heat sink for building cooling or even heating purposes. This fact has been studied and proven by several researchers [25]. The rationale behind this is due to the daily and seasonal temperature variation, which is greatly reduced in the earth with growing depths up to a certain depth where the soil temperature remains steady during the year [26]. Many researchers found that the resulting pipe outlet temperature decreases with decreasing pipe diameter, increasing pipe length, growing depths up to 4 m and decreasing mass flow rate through the pipe [27, 28].

Low-energy cooling technique using earth became increasingly popular in Europe and America after the oil crisis in 1973 [26]. The underground pipes concept can be followed back several centuries [29]. Most of these studies concentrated mainly on modeling and analysis of PAC systems with some experimental studies for some selected climatic conditions. Since this passive cooling technology has not been practiced in Australia, this study is very timely and important for Australian economy and environment. The aim of this research is to investigate the earth pipe cooling performance using a series of buried pipes laid in vertical arrangement in the ground.

21.2 Earth Pipe Cooling Technology

Earth pipe cooling technology involves a long pipe with one end for air intake and the other end for supplying cool air cooled by the soil to the building as shown in Fig. 21.4. The pipe is covered underground at an optimum depth that could provide

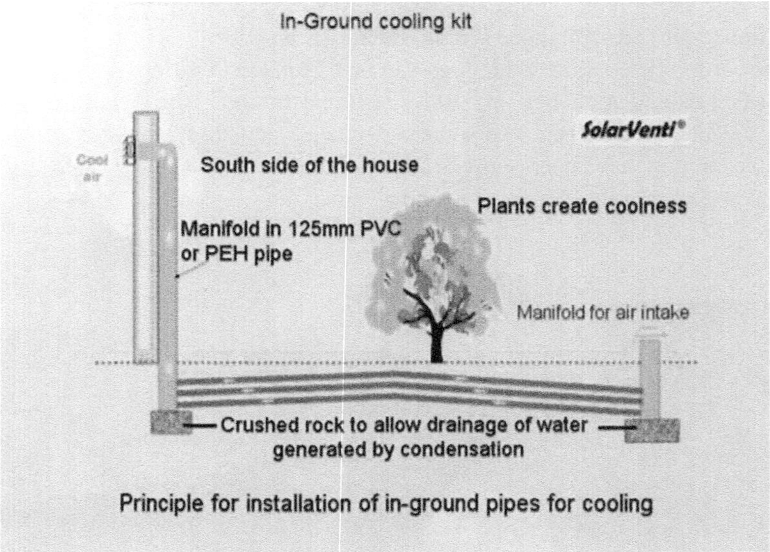

Fig. 21.4 Earth pipe cooling installation system [30]

most effective results. This technology utilizes the ground as a heat sink to cool a desired space in hot humid climates. A blower is fitted to suck intake air into the room of the building. An adequate supply of cool air is provided through the pipe end connected to the building.

There are two main strategies for earth pipe cooling: direct and indirect earth contact [31]. Direct earth contact involves total or partial building envelope placing in contact with the ground surface directly, and the indirect earth contact involves an earth-to-air heat exchanger system through which air is circulated between interior and exterior of the building and then is brought into the building [32]. The direct earth-to-building contact ground cooling is a low-maintenance passive cooling strategy with minimal heat gains and solar exposure. To use this advantage, underground houses were built in southern Tunisia and eastern Spain to protect occupants from their hot and arid climates and large houses were excavated in northern China to cope with severe winter climate [26]. Despite the advantages of the direct-to-building contact ground cooling, it also creates environmental problem such as indoor condensation and poor indoor air quality [31]. Moreover, in some places such as desert and semi-desert countries, large excavation is not suitable due to their geological condition. This study investigates the indirect earth contact ground cooling system.

Earth pipe cooling system can be organized in two ways: open loop and closed loop as shown in Fig. 21.5. Air which comes through the buried pipes into the room in an open-loop system and passes through a ventilation system was investigated by Goswami and Biseli [33]. It provides ventilation while optimistically cooling interior of a house [34]. In a closed-loop system, air is constantly re-circulated from the buried pipes underground to the room. The system is more effective than the open-loop system. This system does not exchange air with the air outside the house. The latter pipe arrangement can also reduce the tunnel length since the conditioned air is re-circulated within the buried pipe [33].

The earth pipe cooling technology can be organized in different piping systems such as horizontal trenches or borings, vertical borings, slinky coil, or even in surface water body. There are some advantages and disadvantages using these systems and they have been summarized in Table 21.1.

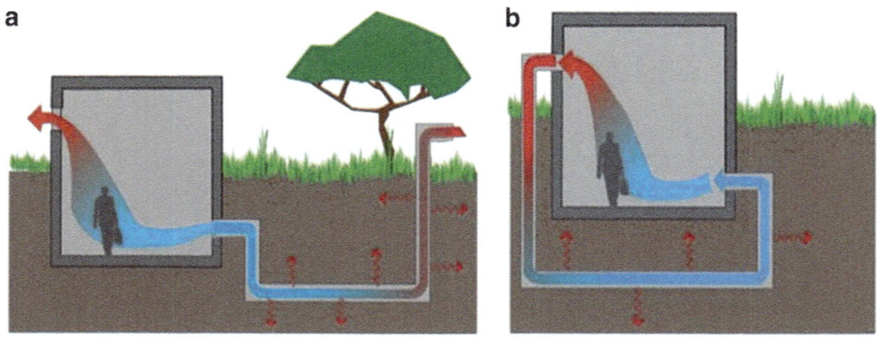

Fig. 21.5 Earth pipe cooling strategies (**a**) Open loop; and (**b**) Closed loop

Table 21.1 Comparison between closed- and open-loop system [35]

System	Advantages	Disadvantages
Closed loop	• Can be installed almost anywhere • Fewer maintenance issues • Less temperamental	• Earth is not as good a conductor • Less energy per foot of loop
Open loop	• More energy per foot of loop • Water is a better conductor of energy than earth	• More maintenance issues • Could have local environment risks • Requires lake or well nearby • More complex

21.3 Experimental Design and Measurement

Two shipping containers, each of dimension 5.63 × 2.14 × 2.26 m were installed and fitted in the Sustainable Precinct at Central Queensland University, Rockhampton, Australia for the experimental measurement. One was connected to the vertical earth pipe cooling system as shown in Fig. 21.6 and the other was without any earth pipe cooling system. For excess rain water runoff, the containers were set with a 3° pitch from North to South.

An exhaust fan (blower) was set into a manifold to suck intake air through another manifold as shown in Fig. 21.6. The blower sucks intake air (from pipe inlet) which is cooled as it passes through a series of buried pipes and finally into the room. Air conditioner was set in all the modeled rooms to cool the interior space in combination with the PAC systems. All the air conditioner was set at 24 °C to achieve optimal comfort and energy savings.

An excavation of dimension 8.1 × 1 m was made for installing the buried pipes in vertical arrangement. Excavation maintained a slope of 0.2 m down the length of the excavation to enable moisture to fall towards the sand and gravel drainage pit. The sloped base ensures the flow of water to a low point, where a gravel rubble drain enables water runoff and/or provides a point from which excess water can be mechanically extracted by pump.

The vertical pipe installation system shown in Fig. 21.7 consists of two simple Polyvinyl Chloride (PVC) pipes of outside diameter 0.125 m, also known as manifold. PVC pipes were set at the end points outside the container. The manifold contains 20 holes of 21 mm diameter each to accept 20 tubes of 20 mm each. 20 corrugated PVC tubes of 8 m each in length with diameter 21 mm were connected with the manifold. These corrugated PVC tubes with a wall thickness of 1 mm were pressed (friction fitting) into the manifold in five rows, i.e., each row contains four corrugated PVC tubes. Each row is separated from one another by 100 mm and each tube in each row was separated from its neighbor by approximately 20 mm.

Fig. 21.6 Container connected to vertical earth pipe cooling system

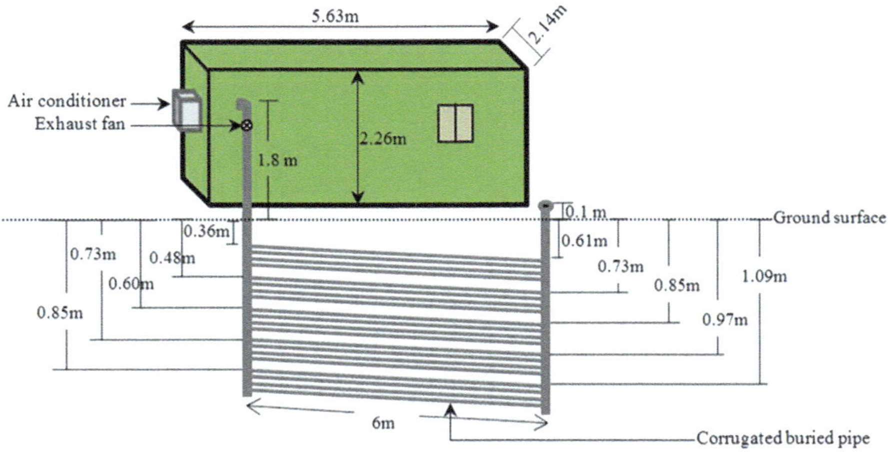

Fig. 21.7 Vertical installation of earth pipe cooling system

21.3.1 Shading

Small trees and plants were used as a garden shown in Fig. 21.8 to provide shade to the soil which covered the pipes buried underground. This was intended to reduce the amount of solar energy absorbed by the soil surface. An auto-irrigation system was installed to water the plants.

21.3.2 Data Collection

Through a suite of experimental design, experimental tests, and field investigation, the cooling performance was assessed against various parameters. Air temperature

Fig. 21.8 Garden closed to the container with irrigation system

and air velocity were measured to investigate their impact on cooling performance of the system. Average indoor air temperature and velocity, pipe inlet average temperature and velocity data were collected using the measuring tools from both the containers. Data were collected for the months of July, August, and September, 2013 from 12:00 PM to 5:00 PM. The data for this time period was taken into consideration for this study as the maximum heat loads occurred in this period during a day.

21.4 Model Description

The vertical earth pipe cooling system involves the heat transfer problem where the earth transfers the heat to the buried pipes underground. Computational fluid dynamics (CFD) is a well known and powerful tool to study heat transfer process. CFD codes are designed around the numerical algorithms that can tackle fluid flow problems. It solves partial differential equations numerically governing air flow and heat transfer in a discretized form. A CFD model, namely RNG $k - \varepsilon$ turbulence model was used for the heat transfer problem of VEPC system. It is more accurate and reliable for a wider class of turbulent flows in industrial flow and heat transfer simulations. Turbulence model was selected for the thermal modeling as the flow passes through the buried pipes was turbulent. The model satisfies certain mathematical constraints on the Reynolds stresses and consistent with the physics of turbulent flows. The heat transfer problem of vertical earth pipe cooling system was solved numerically by using the CFD code "Fluent in ANSYS 15.0", which uses the finite volume method for discretization of the computational domain.

21.4.1 Modeling Equation

The RNG $k - \varepsilon$ model is derived from Navier-Stokes equations. The Navier-Stokes equation of motion and the transport equations for the RNG $k - \varepsilon$ model are given by [36]

$$\frac{\partial u_i}{\partial t} + u_j \frac{\partial u_i}{\partial x_j} = -\frac{1}{\rho}\frac{\partial p}{\partial x_i} + \upsilon \frac{\partial^2 u_i}{\partial x_j \partial x_j} \tag{21.1}$$

and

$$\frac{\partial(\rho k)}{\partial t} + \frac{\partial(\rho k u_i)}{\partial x_i} = \frac{\partial}{\partial x_j}\left(\alpha_{\mathrm{k}}\mu_{\mathrm{eff}}\frac{\partial k}{\partial x_j}\right) + G_{\mathrm{k}} + G_{\mathrm{b}} - \rho\varepsilon - Y_{\mathrm{M}} + S_{\mathrm{K}} \tag{21.2}$$

$$\frac{\partial(\rho \varepsilon)}{\partial t} + \frac{\partial(\rho \varepsilon u_i)}{\partial x_i} = \frac{\partial}{\partial x_j}\left(\alpha_{\varepsilon}\mu_{\mathrm{eff}}\frac{\partial \varepsilon}{\partial x_j}\right) + C_{1\varepsilon}\frac{\varepsilon}{k}\left(G_{\mathrm{k}} + C_{3\varepsilon}G_{\mathrm{b}}\right) - C_{2\varepsilon}\rho\frac{\varepsilon^2}{k} - R_{\varepsilon} + S_{\varepsilon} \tag{21.3}$$

where, u_i, u_j are the fluctuating velocity components (m s^{-1}); x_i are the components of length (m); p is the pressure (Pa), $\upsilon = \mu/\rho$ is the molecular kinetic fluid viscosity (m^{2}s^{-1}); μ_{eff} is the effective viscosity $(\mu + \mu_{\mathrm{t}})$, where μ is the fluid viscosity (kgm^{-1} s^{-1}); ρ is the fluid density (kg m^{-3}); t is the time (s); k is the kinetic energy for turbulent flow (m^{2}s^{-2}); ε is the energy dissipation rate (m^{2}s^{-3}); G_{k} is the turbulence kinetic energy generation due to the mean velocity gradients (kg m^{-1} s^{-2}); G_{b} is the turbulence kinetic energy generation due to buoyancy (kg m^{-1} s^{-2}); Y_{M} represents the contribution of the fluctuating dilatation in compressible turbulent flow to the overall dissipation rate (kg m^{-1} s^{-2}); α_{k} and α_{ε} are the inverse effective Prandtl numbers for k and ε respectively (kg m^{-1} s^{-2}); R_{ε} is the gas-law constant for ε (8.31447×10^3 J Kgmol^{-1} K^{-1}); user-defined source terms S_{K} and S_{ε} are the mean strain rates for k and ε respectively (m^{2}s^{-2}); $C_{1\varepsilon}$, $C_{2\varepsilon}$ and $C_{3\varepsilon}$ are constants.

Energy equation for this heat transfer problem is solved all over the whole domain and is given by

$$\frac{\partial(\rho E)}{\partial t} + \nabla\cdot\left(\vec{u}(\rho E + P)\right) = \nabla\cdot\left(k_{\mathrm{eff}}\nabla T - \sum_j h_j \vec{J}_j + \left(\bar{\bar{\tau}}_{\mathrm{eff}}.\vec{u}\right)\right) + S_{\mathrm{h}} \tag{21.4}$$

Where $k_{\mathrm{eff}} = (k + k_{\mathrm{t}})$ is the effective conductivity (W m^{-1} K^{-1}), where k_{t} is the thermal conductivity for turbulent flow, J_j is the component of diffusion flux (m^{-2} s^{-1}), T is the temperature (K), $\nabla\cdot(k_{\mathrm{eff}}\nabla T)$ is the heat transfer due to convection, h is the enthalpy (J kg^{-1}), $\nabla\cdot\left(\sum_j h_j \vec{J}_j\right)$ species diffusion, $\nabla\cdot\left(\bar{\bar{\tau}}_{\mathrm{eff}}\cdot\vec{v}\right)$ is the heat transfer due to viscous diffusion and S_{h} is the total entropy (JK^{-1}).

21.4.2 Geometry of the Model

A 2D geometry for this model was created in DesignModeller which is shown in Fig. 21.9. Geometry consists of a pipe with diameter 0.125 m connected to one of the containers with dimension of 5.63 × 2.26 m.

21.4.3 Mesh Generation

A typical mesh is presented for an earth pipe cooling system in Fig. 21.10. This mesh was generated using DesignModeler in ANSYS 13.0. A study was carried out to check the effect of the grid variation and to establish the optimum mesh size which ensures the consistent results for every mesh size. The mesh

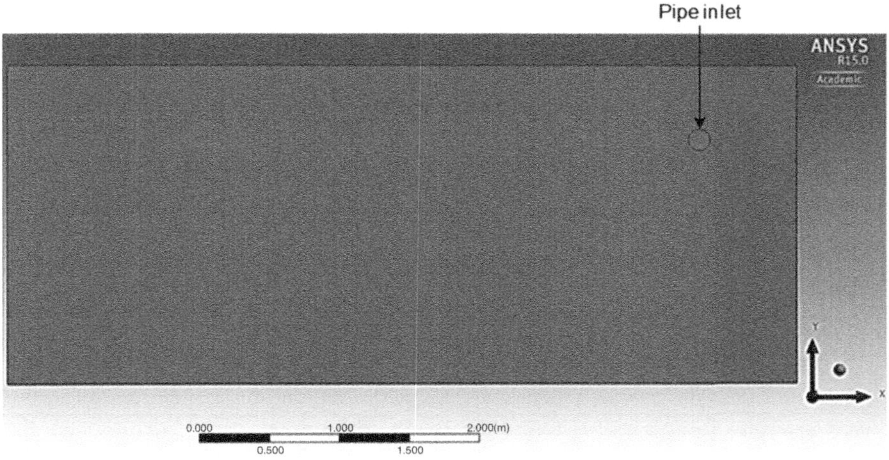

Fig. 21.9 Model geometry in DesignModeler

Fig. 21.10 A typical 2D mesh

produces a high number of cells in a Cartesian layout in the far field to provide the accurate fluid flow results. An optimum number of cells, namely 124,785 were created in the meshing to get an accurate result for this study as a large number of cells may result in long solver runs and a small number of cells may lead to inaccurate results [37]. Cell zone condition for the surface body was defined as fluid. All the zones used in the geometry were defined such as pipe inlet, pipe wall, and room wall. No slip boundary conditions were applied on the pipe walls and room walls.

21.4.4 Solver Approach

A two dimensional pressure-based-segregated solver was used for the modeling of vertical earth pipe cooling system. The solver uses an algorithm where the governing equations are solved consecutively. Several iterations of the solution loop were performed before obtaining a converged solution, since the governing equations are nonlinear. The pressure-velocity coupling scheme of pressure implicit with splitting operators (PISO) was adopted for numerical calculations as it is more suitable for steady-state schemes on highly skewed meshes. PISO is based on the higher degree of the approximate relation between the velocity and pressure corrections [38]. It allows for a rapid rate convergence without a substantial loss of solution accuracy and stability [39, 40]. Pressure was discretized with a PRESTO scheme in view of its strong convergence ability. The PRESTO scheme is also available for all meshes such as tetrahedral, triangular, hexahedral, quadrilateral, and hybrid meshes.

Spatial discretization of second-order upwind scheme was used for turbulent dissipation rate, turbulent kinetic energy and momentum as the second-order discretization of the viscous terms is always accurate in Fluent. Moreover, the differencing scheme of second-order upwind was utilized to overcome numerical dispersion. At the end of the solver iteration, the moment coefficient, lift and drag are calculated and stored to make convergence history. The standard initialization in the entire domain used in this study allows setting the initial values of the flow variables and initializing the solution with these values.

21.5 Results and Discussion

Experimental results were obtained through a series of experiments and measurements. The measuring tool of Reed Vane Anemometer was used for measuring air velocity and HOBO Pendant Temp was used for measuring air temperature and relative humidity. The tests were conducted while an air-conditioner of power input

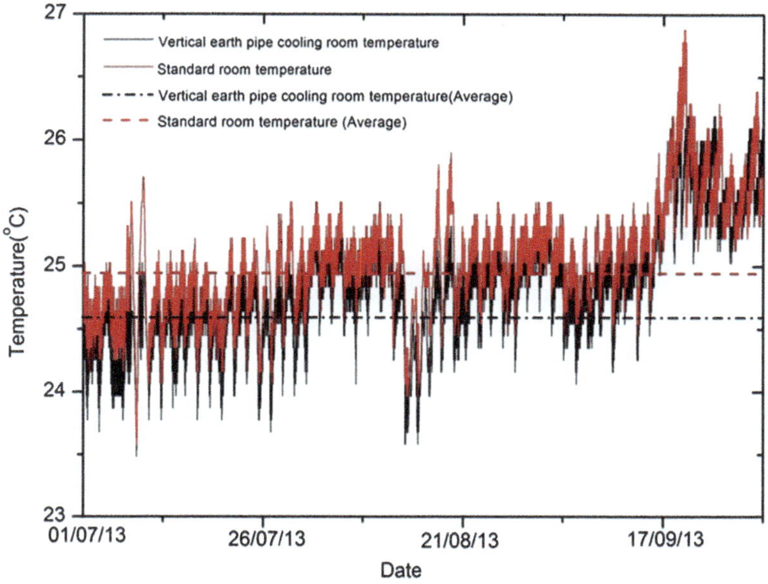

Fig. 21.11 Temperature profile for vertical earth pipe cooling system and standard room

1.01 kW and capacity output 5.70 kW set in both vertical earth pipe cooling and standard room (not connected to any technology). All the measurements were taken from July to September, 2013.

The HOBO Pendant Temps were set at different points inside the rooms to measure indoor air temperature and relative humidity. The indoor room temperature profile at the mid points of both vertical earth pipe cooling and standard room are shown in Fig. 21.11. The figure illustrates that the indoor room temperature varies from 23.48 to 26.39 °C for vertical earth pipe cooling system and 23.54 to 26.88 °C for standard room. The average room temperature for vertical earth pipe cooling system and standard room was 24.61 and 24.87 °C respectively. The highest temperature difference of 1.16 °C occurs on 22nd July, 2013 at 12:25 PM.

Figure 21.11 also shows that the indoor temperatures are increasing gradually for both vertical earth pipe cooling and the standard room. In Australia, the winter which ends on 31 August and the spring that starts on 1 September are the main reasons behind this. Temperature was fallen on 15 and 16 August, 2013 due to the rainfall. In winter the temperature of Rockhampton ranges from 23 to 26 °C in daytime pick hours and in spring it ranges from 24 to 27 °C.

One HOBO Pendant Temp logger was also set outside the rooms to measure the outdoor air temperature. The maximum temperature of 41.47 °C and minimum temperature of 25.87 °C was recorded outside the rooms. The temperature difference between outdoor and indoor is comparatively small in winter but substantially varies in summer due to geographical location of Rockhampton, Australia. Rockhampton lies within the southeast trade wind belt, too far south to experience regular North West monsoonal influence, and too far to gain much benefit from higher latitude cold fronts [41].

An average air velocity of 3.40 m/s and air temperature of 21.01 °C (294.16 K) were measured at the pipe inlet of the modeled room connected to the vertical earth pipe cooling system. In many cases, inadequate supply of outdoor air causes the poor indoor air quality. Indoor air velocities below 0.25 m/s do not produce substantial distraction even in tasks requiring substantial attention. Relative humidity of the air has a noticeable impact in the thermal comfort. It is still an essential variable to measure, especially in hot climates. The relative humidity of the rooms was measured to assess its impact inside the rooms, which is summarized in Table 21.2. Figure 21.12 shows the indoor relative humidity that varies from 55.19 to 83.46 % for vertical earth pipe cooling and 48.18 to 78.99 % for standard room.

Relative humidity levels underneath 20 % can result discomfort through drying of the mucous membranes and skin, whereas relative humidity above 70 % may lead to the development of condensation on surfaces and within the interior equipment and building structures. Average indoor relative humidity of around 68 % was observed for vertical earth pipe cooling system, which is acceptable in context of thermal comfort for the measured temperature of 24.61 °C. It was also found that the relative humidity of vertical earth pipe cooling system is higher than the standard room. Air came through the inlet to the room of the vertical earth pipe cooling system increases this relative humidity.

The measurement carried for the vertical earth pipe cooling system was compared with ASHRAE standard 55-2010 using adaptive method in the center for the built environment (CBE) thermal comfort tool. It was found that the measured average temperature and relative humidity complies with the ASHRAE standard 55-2010 [42], Brown's bioclimatic chart [14] and Givoni's chart [12, 43].

Performance of the vertical earth pipe cooling system was also calculated numerically at different iterations. The solution was convergent at the iteration of 25,478. Temperature and velocity profiles are shown in Figs. 21.13 and 21.14 respectively at the iterations of 500 and 25, 478. The flow and thermal variables for the boundary and cell zone conditions were set on the boundaries of the model. Average air velocity of 3.40 m/s and air temperature of 21.01 °C (294.16 K) were set at the pipe inlet of the boundary conditions. Average maximum room air temperature of 24.87 °C (298.02 K) collected from the standard container was put as the indoor room temperature of the model. The simulation result was obtained under some boundary conditions which are summarized in Table 21.3.

The predicted indoor room temperature of 23.05 °C (296.2 K) was found in the modeled room of vertical earth pipe cooling system. This predicted temperature shown in Fig. 21.13d is 1.82 °C less than the measured standard room temperature of 24.87 °C (298.02 K). Cool air came through the inlet to the room is the main reason behind this temperature reduction. The air comes through the series of buried underground pipes that was cooled by the soil in summer due to heat convection.

Table 21.2 Air temperature and relative humidity for all the rooms

Modeled rooms	Indoor temperature			Outdoor temperature			Relative humidity		
	Min (°C)	Max (°C)	Avg (°C)	Min (°C)	Max (°C)	Avg (°C)	Min (%)	Max (%)	Avg (%)
VEPC	23.48	26.39	24.61	25.87	41.47	31.10	55.19	83.46	67.95
Standard room	23.54	26.88	24.87	–	–	–	48.18	78.99	58.58

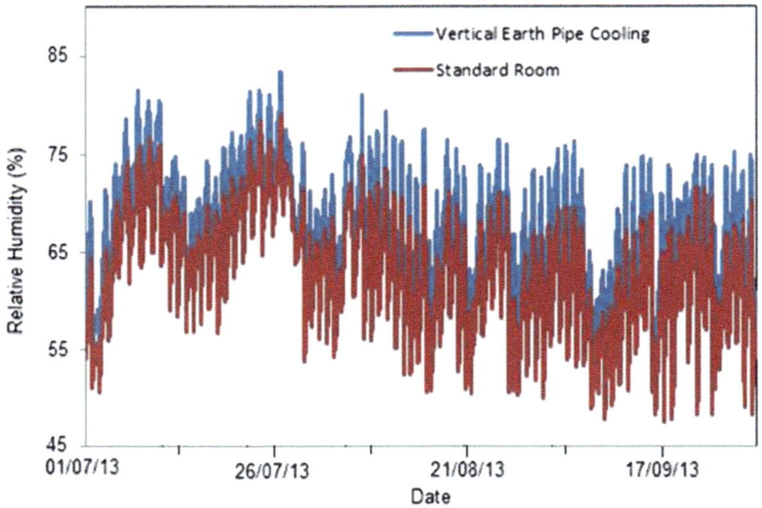

Fig. 21.12 Indoor relative humidity for both vertical earth pipe cooling and standard room

The results obtained in this study were also validated inside the room by comparing the numerical data with the experimental data. The comparison between experimental and numerical results is summarized in Table 21.4. The numerical simulation results for air velocity and temperature showed some deviations from the experimental results because of the air pressure drop inside the pipe. However, the overall simulated results are in very good agreement with the corresponding experiments and lie within 0.76–6.34 % limits.

21.6 Conclusions

Earth pipe cooling system is well known as an energy-efficient technology which uses geothermal energy to reduce cooling needs in buildings. Recent progresses on this technology can help to enhance the quality of indoor environment of the buildings. The main purpose of this study was to develop a numerical model for vertical earth pipe cooling system to measure the thermal performance of the system. The numerical model was developed for a subtropical climate in Queensland, Australia. RNG turbulence model was used to develop the model. Impact of air temperature, air velocity, and relative humidity was assessed on the room cooling performance using simulation. Inlet air temperature and velocity were set at the pipe inlet to predict their effects on the room temperature. Simulation result shows the minimum temperature reduction of 1.82 °C in the vertical earth pipe cooling room. The developed model was validated inside the room by comparing the numerical results with the experimental results. A good agreement was obtained between these two results. Further investigation is being undertaken in this study.

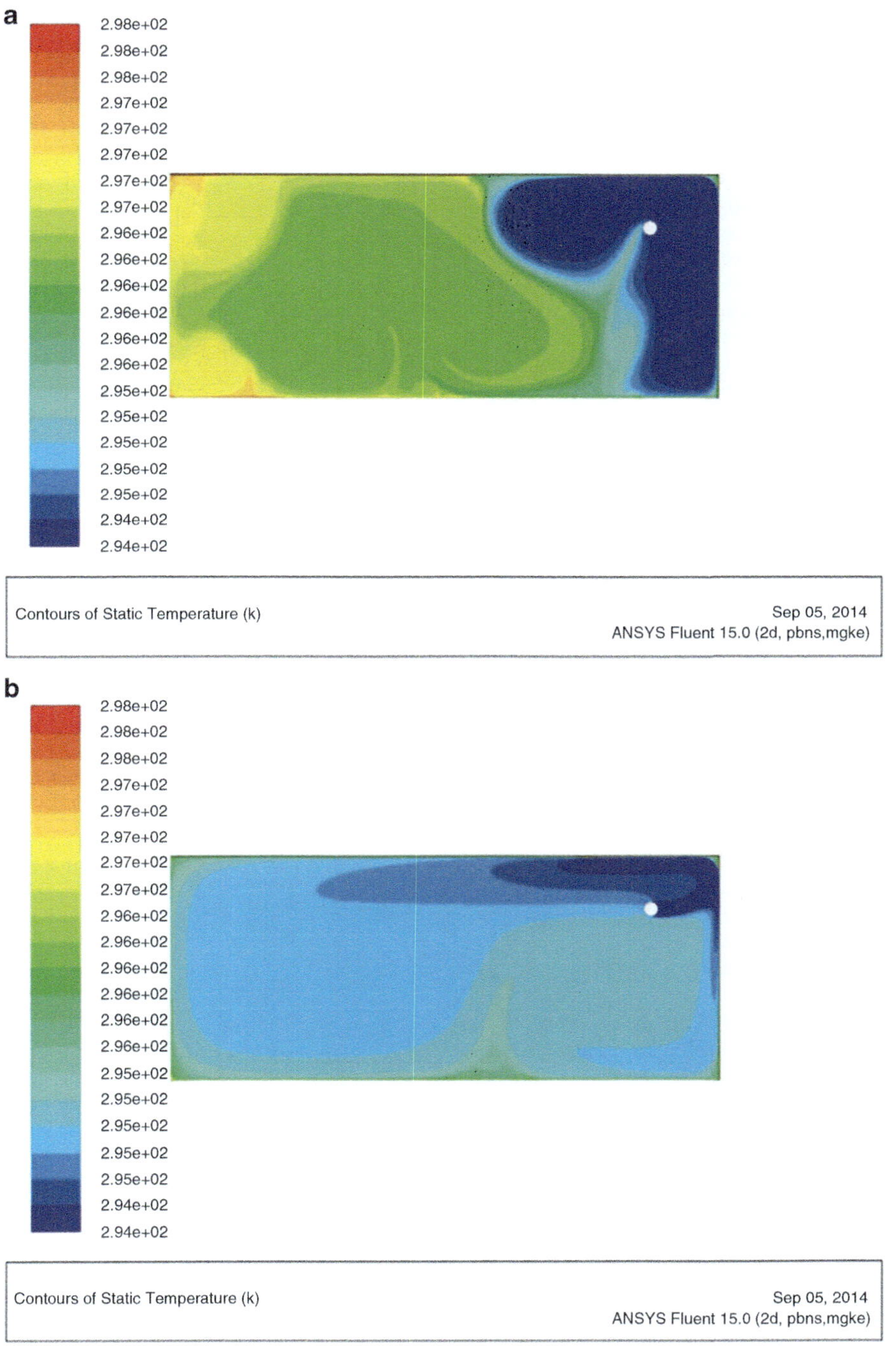

Fig. 21.13 Temperature profile (**a**) at the iteration of 500; and (**b**) at the iteration of 25,478

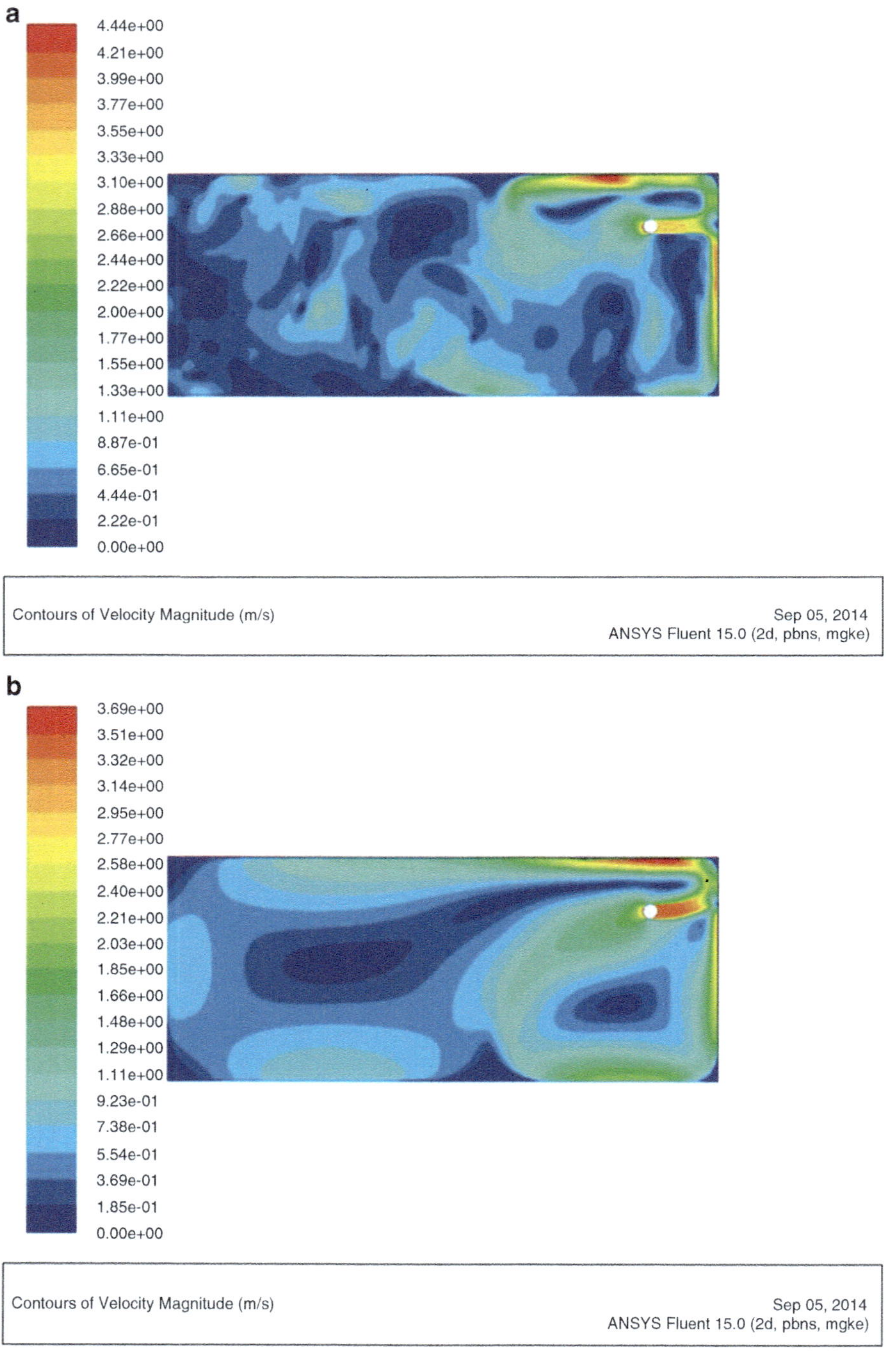

Fig. 21.14 Velocity profile (**a**) at the iteration of 500; and (**b**) at the iteration of 25,478

Table 21.3 Parameters used in boundary conditions of the model

Parameters	Value
Pipe inlet velocity	3.40 m/s
Air temperature at pipe inlet	21.01 °C
Indoor room temperature	24.87 °C
Air thermal conductivity	0.024 W/m-K
Air density	1.204 kg/m^3
Specific heat of air	1006.43 J/kg-K
Air viscosity	1.850387e-05 kg/m-s

Table 21.4 Comparison between experimental and numerical results

Data	Experimental	Numerical	Average differences (%)
Average air velocity at pipe inlet (m/s)	3.40	3.54	4.12
Average air temperature at pipe inlet (°C)	21.01	20.85	0.76
Average room air temperature of vertical earth pipe cooling system (°C)	24.61	23.05	6.34

Acknowledgements The authors gratefully acknowledge the support provided by Ergon Energy, Australia, for supplying PAC system and the installation cost of the measuring tools. S.F. Ahmed is grateful to Central Queensland University, Australia, for an international postgraduate research award (IPRA) to support this study.

References

1. Ahmed SF, Khan MMK, Rasul MG, Amanullah MTO, Hassan NMS (2014) Comparison of earth pipe cooling performance between two different Piping Systems. Energy Procedia 61:1897–1901.
2. U.S. Energy Information Administration (2013) International Energy Outlook, 2013. http://www.eia.gov/forecasts/ieo/pdf/0484%282013%29.pdf. Accessed 25 Mar 2014
3. (2013) BP Energy outlook 2030. London, UK
4. (2013) Securing our energy future: Chinese and Australian perspectives. http://www.griffith.edu.au/__data/assets/pdf_file/0008/524780/2013-Second-Track-Dialogue-Background-Paper.pdf. Accessed 1 Sept 2014
5. (2009) World energy consumption, 1990–2030. http://rainforests.mongabay.com/energy/energy.html. Accessed 4 Apr 2014
6. Department of the Environment, Water, Heritage and the Arts 1921298146 (2008) Energy use in the Australian residential sector 1986–2020. http://ee.ret.gov.au/sites/default/files/documents/04_2013/energy-use-australian-residential-sector-1986-2020-part1.pdfAccessed 4 Apr 2014
7. (2013) Analysis of electricity consumption, electricity generation emissions intensity and economy-wide emissions. http://climatechangeauthority.gov.au/node/155. Accessed 4 Apr 2014
8. Heating and cooling. http://www.originenergy.com.au/2673/Heating-and-cooling. Accessed 9 Apr 2014

9. Shahiduzzaman M (2012) Energy consumption, economic growth and CO_2 emissions in Australia: the potential for energy conservation. PhD Thesis, University of Southern Queensland
10. Commonwealth Scientific and Industrial Research Organization (CSIRO) (2009) The science of tackling climate change. www.csiro.au. Accessed 2 Apr 2014
11. (2012) Energy in Australia. http://www.bree.gov.au/documents/publications/energy-in-aust/energy-in-australia-2012.pdf. Accessed 15 Sept 2014
12. Givoni B (1994) Passive low energy cooling of buildings. John Wiley, New York, NY
13. Ahmed SF, Khan MMK, Amanullah MTO, Rasul MG (2014) Selection of suitable passive cooling strategy for a subtropical climate. Int J Mech Mater Eng 9:1–11
14. DeKay M, Brown G (2013) Sun, wind, and light: architectural design strategies. John Wiley, New York, NY
15. Shaviv E, Yezioro A, Capeluto IG (2001) Thermal mass and night ventilation as passive cooling design strategy. Renew Energy 24:445–452
16. Chowdhury AA, Rasul MG, Khan MMK (2009) Modelling and analysis of air-cooled reciprocating chiller and demand energy savings using passive cooling. Appl Therm Eng 29:1825–1830
17. La Roche P (2004) Passive cooling strategies for buildings in hot climates with specific application to Venezuela. PhD Thesis, University of California, Los Angeles
18. Dobson R (2005) Thermal modelling of a night sky radiation cooling system. J Energy South Afr 16:21
19. Fanciotti A, Scudo G (1981) An evaluation of the architectural design of 16th century buildings in Italy. In: International passive and hybrid cooling conference, Miami Beach, 1981, pp 179–194
20. Saini BS (1973) Building environment: an illustrated analysis of problems in hot dry lands. Angus and Robertson, Sydney
21. Ahmed SF, Khan MMK, Amanullah MTO, Rasul MG, Hassan NMS (2013) Thermal performance analysis of earth pipe cooling system for subtropical climate. In: 12th international conference on sustainable energy technologies, Hong Kong, 2013, pp 1795–1803
22. Ahmed SF, Khan MMK, Amanullah MTO, Rasul MG, Hassan NMS (2014) Performance analysis of vertical earth pipe cooling system for subtropical climate. In: International conference on clean energy, Istanbul, Turkey, 2014, pp 691–700
23. Ahmed SF, Khan MMK, Rasul MG, Amanullah MTO, Hassan NMS (2014) Comparison of earth pipe cooling performance between two different piping systems. Energy Procedia 61:1897–1901
24. Hollmuller P, Lachal B (2001) Cooling and preheating with buried pipe systems: monitoring, simulation and economic aspects. Energ Build 33:509–518
25. Mihalakakou G, Santamouris M, Asimakopoulos D, Tselepidaki I (1995) Parametric prediction of the buried pipes cooling potential for passive cooling applications. Sol Energy 55:163–173
26. Krarti M, Kreider JF (1996) Analytical model for heat transfer in an underground air tunnel. Energy Convers Manag 37:1561–1574
27. Santamouris M, Mihalakakou G, Balaras C, Argiriou A, Asimakopoulos D, Vallindras M (1995) Use of buried pipes for energy conservation in cooling of agricultural greenhouses. Sol Energy 55:111–124
28. Ghosal M, Tiwari G (2006) Modeling and parametric studies for thermal performance of an earth to air heat exchanger integrated with a greenhouse. Energy Convers Manag 47:1779–1798
29. Mihalakakou G, Santamouris M, Asimakopoulos D (1992) Modelling the earth temperature using multiyear measurements. Energ Build 19:1–9
30. Dwyer G (2011) An overview of passive air cooling
31. Jacovides C, Mihalakakou G, Santamouris M, Lewis J (1996) On the ground temperature profile for passive cooling applications in buildings. Sol Energy 57:167–175

32. Mihalakakou G, Santamouris M, Lewis J, Asimakopoulos D (1997) On the application of the energy balance equation to predict ground temperature profiles. Sol Energy 60:181–190
33. Goswami D, Biseli K (1993) Use of underground air tunnels for heating and cooling agricultural and residential buildings. Fact Sheet EES 78:
34. Zaki AK, Amjad AM, Almssad A (2007) Cooling by underground earth tubes. In: 2nd PALENC conference and 28th AIVC conference on building low energy cooling and advanced ventilation technologies in the 21st century, Crete island, Greece, 2007
35. Johnston S (2011) Using the Earth's heat. http://www.homepower.com/articles/home-efficiency/project-profiles/using-earths-heat. Accessed 29 Jan 2014
36. Fluent A (2010) ANSYS FLUENT theory guide: Version 13.0. Ansys Inc., Canonsburg
37. (2014) ANSYS Meshing. http://www.idac.co.uk/products/products/meshing.htmAccessed 3 Sept 2014
38. Bukhari SJ, Siddiqui MK (2007) Characteristics of air and water velocity fields during natural convection. Heat Mass Transf 43:415–425
39. Hassan NMS, Khan MMK, Rasul MG (2010) A modelling and experimental study of the bubble trajectory in a non-Newtonian crystal suspension. Fluid Dyn Res 42:065502
40. Akhtar A, Pareek VK, Tadé MO (2007) CFD simulations for continuous flow of bubbles through gas-liquid columns: application of VOF method. Chem Prod Process Model 2:1–19
41. (2013) Rockhampton area, climate of Rockhampton. http://www.bom.gov.au/qld/rockhampton/climate.shtml. Accessed 3 Sept 2014
42. ASHRAE A (2010) Standard 55-2010: thermal environmental conditions for human occupancy. ASHRAE, Atlanta USA
43. Givoni B (1992) Comfort, climate analysis and building design guidelines. Energ Build 18:11–23

Chapter 22
Determination of Geothermal Fields at Kızılcahamam (Ankara) Using Vertical Electrical Sounding (VES) and Spontaneous Potential (SP) Methods

Hatice Karakilcik

Abstract Geophysical exploration of geothermal resources deals with measurements on the physical properties of the Earth with two methods. First, the vertical electrical sounding (VES) method which is useful in determining the depth of overburden and depth, structure, and resistivity of flat-lying sedimentary beds and possibly of the basement also if it is not too deep. Furthermore the search for geothermal reservoirs normally involves resistivity surveying, and it is also employed routinely in groundwater exploration, which is of increasing worldwide importance. Second, spontaneous potential (SP) method is applied for determining the possible faults and finding the places with and without the liquid flow in the study area. At the SP graphic is analyzed that the natural voltage in a few measuring points varied between +28 mV and −48 mV values. At the natural potential, the plus (+) and minus (−) transitions indicate faulting zone. In this work, VES and SP methods at Kızılcahamam (Ankara) geothermal area were used to delineate location of aquifer zones and site wells or estimate properties of the system. Schlumberger electrode arrays were used in the VES measurements at 60 point in the study area. Schlumberger soundings at the study area have been carried out to try and estimate the bedrock resistivity at different depths. The measured quantity at the study area was called apparent resistivity. Interpreting the resistivity data consists of two steps: first, a physical interpretation of the measured data, resulting in a physical model, and second, a geological interpretation of the resulting physical model. The resistivity structure at study area is dominated by two coherent low resistivity layers that underlie most of the field. The shallower layer ($<10\,\Omega$m) lies within 300 m depth, while the deeper one is at about 900 m depth. Resistivity of rocks depends on porosity, saturation, content of clay, and resistivity of pore water. As a result of all these studies, geothermal anomalies having low resistivity values at the five different VES points have been identified. For geothermal drilling of the suggested VES, points constitute the appropriate fields.

H. Karakilcik (✉)
Department of Geology, Faculty of Engineering and Architecture, University of Cukurova, Balcali, Adana 01330, Turkey
e-mail: hkilcik@cu.edu.tr

I. Dincer et al. (eds.), *Progress in Clean Energy, Volume 2*,
DOI 10.1007/978-3-319-17031-2_22

Keywords Geothermal exploration • Vertical electrical sounding (VES) • Spontaneous potential (SP) • Geothermal anomaly

22.1 Introduction

The production of geothermal energy is increasingly growing worldwide. In Turkey, there are 17 geothermal fields discovered by General Directorate of Mineral Research and Exploration of Turkey (MTA), which are suitable for geothermal power production. When all the fields are developed, the total installed capacity will reach to about 600 MW of electricity. The first geothermal electricity production in Turkey started in Kızıldere geothermal field by MTA in 1974 as a pilot plant with an installed capacity of 0.5 MW_e. Afterward in 1984, the Kızıldere Geothermal Power Plant was installed by Turkish Electricity Establishment (TEK), renamed as (TEAS) with a total installation capacity of 20.4 MW_e; today, its installed capacity is 15 MW_e. This power plant generates an average of 12–15 MW_e electricity (nine wells are productive). MTA, while continuing to prospecting deeper fluid circulating geothermal fields, plans to identify sites in the future years where hot rocks having insufficient porosity and permeability with high temperatures exist and, by determining their physical and chemical parameters, plans to perform mapping and inventory studies beginning from the year 2009 [1].

During the early stages of geothermal exploration, most methods use data from surface manifestations except geophysical methods that aim at locating and describing subsurface structures that relate to geothermal potentials in the study area. In addition to the ability of geophysical methods to delineate subsurface structures, the methods are cost effective compared to drilling costs. Cost effectiveness and risk minimization make geophysical surveys a vital technique prior to a drilling program in the development of geothermal resources. Detailed geophysical exploration resolves geophysical anomalies and narrows down the targets. Various geophysical methods are used for geothermal exploration purposes. Geophysical methods in geothermal exploration are classified into direct and structural methods [2]. Direct methods include electrical and thermal methods. Direct methods are based on physical parameters that relate directly to the presence of hot water or geothermal fluids. Structural methods, including gravity, magnetic, and seismic methods, are used to delineate geological structures such as intrusive bodies or tectonic lineaments that might control the flow of the fluid.

The most popular and useful methods in estimating the depth to the reservoirs and locating the drilling targets in the world are electrical resistivity methods. Electrical methods have been used to identify and delineate geothermal resources in Turkey. Resistivity that is among the most easily measurable from the surface is the physical parameter characterizing the porous crust. As resistivity is highly dependent on temperature, resistivity methods are very useful in finding and delineating geothermal systems. The first electricity resistivity studies at Kızılcahamam geothermal area were carried out by Hacısalihoğlu [3]. This area has a low temperature which is approximately 80 °C. The geothermal field is located 70 km away from Ankara, Turkey. The produced geothermal water from this geothermal field is used to heat more than 2,500 houses.

Self-potential (SP) surveys have been carried out on a number of geothermal areas in the world. Self-potential (SP) anomalies are generated by the flow of fluids, heat, and ions in the Earth's interior. In geothermal areas, hot fluid circulation takes place along faults, and SP investigations have been used to help locate and delineate the features associated with geothermal fluid motions [4]. In this study, we determine the subsurface structure around Kızılcahamam hot spring with SP and VES surveys. The aims of the survey are to identify the location of aquifers and delineate the fracture pattern and a possible heat source that is making the hot springs alive. We report the results of SP and VES data and discuss the mechanism of the observed SP and VES changes

22.2 Geothermal Technologies for Exploration and Resource Evaluation

As a development risk (bringing about unfavorable result) is not negligible on steam field development, phased progress of steam field development is usually adopted. One typical process on steam field development is shown in Fig. 22.1. Development process is composed of four stages as follows:

- First stage: exploration stage
- Second stage: feasibility study stage
- Third stage: construction stage
- Fourth stage: operation and maintenance stage

The aim of the first stage is to identify chemical and physical property of geothermal resource and to estimate resource capacity (optimal output to keep sustainable operation). This is subdivided into three phases as follows:

- Phase 1, regional exploration phase: to select promising area (or areas)
- Phase 2, detailed exploration phase: to select drilling targets
- Phase 3, well drilling and resource evaluation phase: to identify chemical and physical property of targeting geothermal reservoir

A conceptual model of geothermal resources must be required to estimate at the end of each phase and to draw up strategy for the development. There are many kinds of geothermal exploration technology contributing to estimate the model, though some kinds of exploration technology must be applied to estimate sufficient model for geothermal development. And comprehensive analysis will be required, based on obtained results from applied geothermal explorations. Basic design of a future geothermal power plant is drawn up, based on estimated optimum output and steam quality clarified by production tests. After that, feasibility of the aiming geothermal project is examined, based on the basic design of the plant and a scenario of steam field development proposed by the reservoir simulation [5].

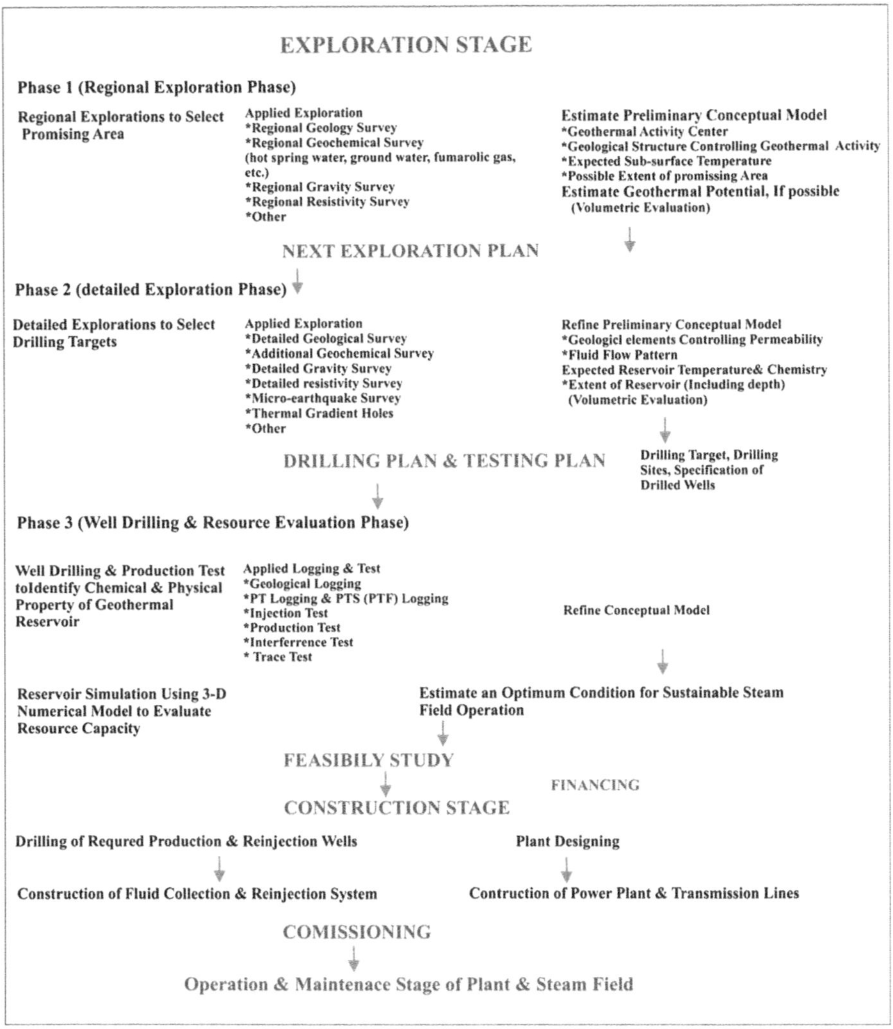

Fig. 22.1 General geothermal power development flow (modified from Akasako [5])

22.3 Results and Discussion

22.3.1 Geologic and Hydrogeologic Setting of the Field

The investigation areas are highlands with deep valleys. Koca stream flowing in the direction of N–S is formed by the union of Perçin and Sey streams. Kızılcahamam geothermal field is located within the Tertiary Galatian Volcanic Complex which consists of lava flows, tuffs, and agglomerates intercalated with lacustrine sediments (Fig. 22.2). The basement beneath the complex consists of Paleozoic schist

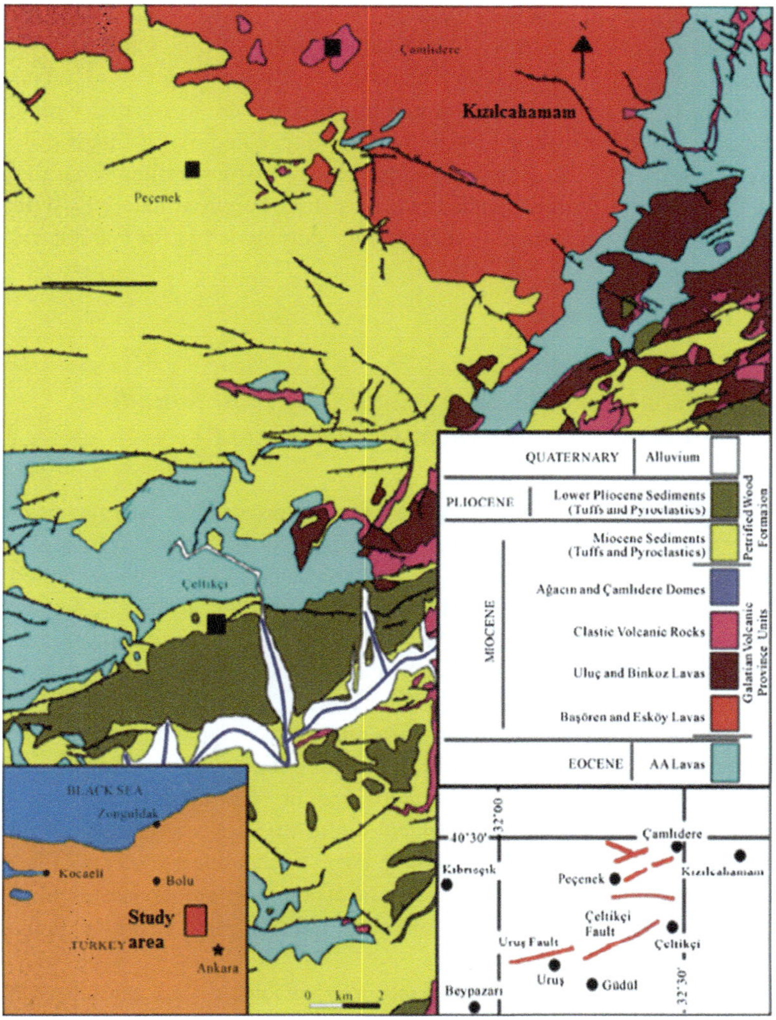

Fig. 22.2 Geological map of study area (modified from Keskinsezer and Beyhan [9])

and Permo-Triassic limestones. The Mesozoic, development mainly in Lower Cretaceous limestone and Upper Cretaceous flysch facies, comes over the Paleozoic basement and is overlain by the Galatian Volcanic Complex. The volcanic activity which built up this complex is reported to have started at the end of the Upper Cretaceous but reached its climax during the Miocene. Quaternary alluvium uncomfortably covers the Galatian Volcanic Complex. The complex is cut across by faults which strike dominantly in ENE–WSW direction. The volcanic activity seems to have started with the lava flows of dacitic–andesitic composition, followed by and locally intercalated with the pyroclastic products of (tuffs and agglomerates), and ended with andesitic–basaltic lava flows [6].

The Kızılcahamam area is drained by the Koca stream which flows southward. Groundwater is recharged through the fractures in the volcanics and emerges, at the elevation of about 1000 m, as both hot and cold springs through these fractures, tuff contacts, and fault zones which, on a regional scale, are associated with an E–W trending graben, the presence of which has been revealed by gravimetric as well as geological studies [7, 8]. Being accompanied by various antithetic and synthetic faults, the master fault in Kızılcahamam area strikes approximately E–W direction in conformity with the general trends of faults cutting across the Galatian Volcanic Complex.

22.3.2 Geophysical Data: Vertical Electrical Sounding (VES) and Spontaneous Potential (SP)

Combining the results of VES data and SP surveys provides an integrated evaluation of a geothermal system by detecting geothermal zones in a very efficient manner. In particular, it is believed that the VES and SP anomalies can be generated by the same geological source (i.e., a zone with a flow of fluid, heat, and ions is the SP source). Vertical changes in subsurface electrical properties are measured using the VES method. At this method, an electrical current is introduced directly into the ground through a pair of electrodes. The resulting voltage difference is measured between another pair of electrodes. The subsurface apparent resistivity is then calculated. Resistivity is the reciprocal of conductivity. Thus, measuring resistivity provides information on ground conductivity. The VES method measures vertical changes in the resistivity of the geological strata. VES are used to laterally trace clay layers and, in conjunction with borehole data, to characterize electrically distinct layers. In the field, a series of resistivity measurements are made at Schlumberger electrode spacing centered on a common point. Sampling depth is increased by increasing electrode spacing.

Integrated geophysical surveys using vertical electrical sounding (VES) and self-potential (SP) measurements are carried out to characterize the geothermal area around a hot spring in Kızılcahamam district. The study was performed to detect and identify the location of aquifers and delineate the fracture pattern and possible heating source. For interpreting the data set of the 60 VES soundings, Schlumberger (electrode) array and shallow DC resistivity measurements were used. Used electrode spacing ranges from AB/2 = 100 m up to 2000 m in successive steps. Interpretation of resistivity data by using model curves has furnished significant information related to the setting and extensions of geothermal fields. The locations of the VES sites were considerably restricted by logistical difficulties. The presence of narrow valleys and topography prevented a wider coverage. The maximum AB/2 spacing of the Schlumberger array ranged from 15 to 1,000 m. The separation of the current electrodes was arranged as 3, 4, 6, 8, 10, 12, 16, 20, 24, 30, 40, 50, 60, 80, 100, 150, 200, 250, 300, 350, 400, 500, 600, 800, 1,000,

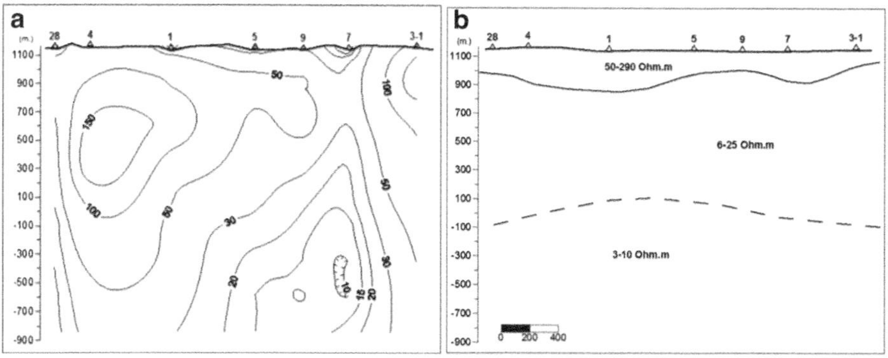

Fig. 22.3 Profile CC′. (**a**) Kızılcahamam, apparent resistivity cross section. (**b**) Kızılcahamam, geoelectric section based on the interpretation of the vertical electrical soundings

1,200, 1,400, 1,600, 1,800, and 2,000 m. The potential electrode separation was arranged as 1, 10, 20, and 40 m. The increase of the potential electrode separation MN allowed that readings from the same current electrode spread AB with the previous and expanded MN were taken. The results of resistivity surveying carried out in Kızılcahamam region along profile CC′ are shown in Fig. 22.3a. Interpretation of VES data at study area suggests a three- to four-layered structure in the study area (Fig. 22.3b). Resistivity survey near the hot spring suggests that weathered and fractured formations constitute the main aquifer system. Interpreted results of all soundings show three- to four-layered formations. Resistivity soundings 28, 4, 1, 5, 9, 7, and 3-1 lie approximately along the line CC′ (Fig. 22.3b). A resistivity cross section along line CC′ is prepared using these sounding results and shown in Fig. 22.3b. The subsurface resistivity cross sections along line CC′ show maximum depth of a relatively low resistive zone exactly below the hot spring. This low resistivity zone can be interpreted as a fracture zone filled with hot water. Receding from the hot spring, the thickness of the low resistive zone reduces.

Water absorption is very dependent on the type of rock. The rocks at study area are water saturated. Fluid flow and hydrologic cycle processes in geothermal systems originated from presence of water seeping into the ground through fractured rocks. The presence of fault structures at study area can create a wide fracture zone that can make groundwater contact with hot rock easier. Major faults may be connected to deep zones of high-temperature bodies. Then, these fractures tap heat from the crust by the injection of cold water and the discharge of hot water to form geothermal reservoirs.

Recommended VES curve for mechanical drilling at study area is shown in Fig. 22.4. Depth of drilling is 1,000 ± 200 m. Geothermal fields often exhibit highly disparate and diverse characteristics, both between different fields and, indeed, within the same field. Wells are typically targeted to reach a particular area within a geothermal reservoir since variations in geology, the properties of resource fluids, subsurface temperatures and pressures, as well as the presence of fractured formations can significantly affect the quality and quantity of production.

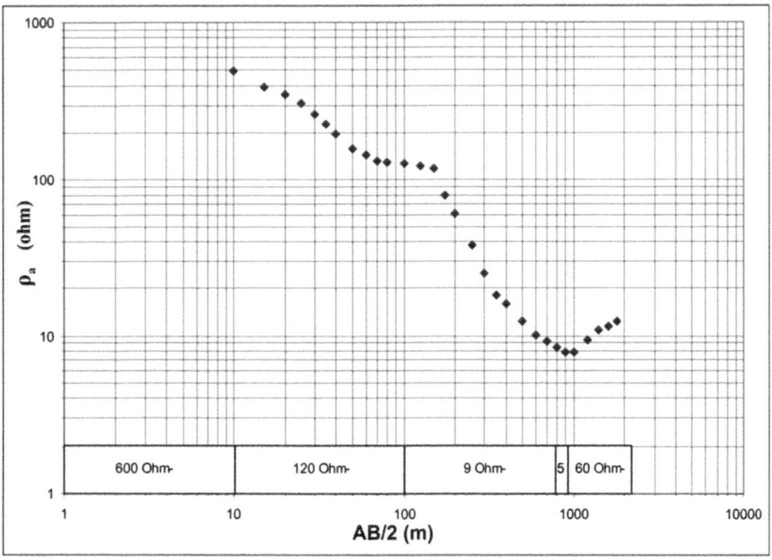

Fig. 22.4 Recommended VES curve for mechanical drilling at study area (depth of drilling 1,000 ± 200 m)

There are several significant surface manifestations of geothermal phenomena at the study area, for example, hot springs, upwelling hot waters, and rich mineralization created by the hydrothermal alteration or deposition. Thermal waters issuing through the tertiary-aged volcanics in the Kızılcahamam geothermal area are all alkali–bicarbonate waters with temperatures ranging from 28 to 86 °C. As seen in Fig. 22.5, the stratigraphic columns determined from KHD-1 well are displayed. The size and depth of casings of this well are also reported in this figure to indicate the completion designs. In the Kızılcahamam geothermal field, Middle–Upper Miocene volcanics, Pliocene sediments, and Quaternary alluviums exist. The volcanics called as a group name of Kızılcahamam volcanites are separated into four formations. Starting from the bottom, Akyarlar formation is the main unit and consists of tuff. The Çakal formation is composed of pyroclastics, while the Gevrekdoruk formation is formed by andesitic, and the Taslica formation is formed by basaltic lavas [10].

Geothermal fluids encountered in Turkey can be classified chemically as 95 % incrusting, and two to three geothermal fields have highly corrosive geothermal fluids. In three of the 140 geothermal fields, geothermal fluid containing total dissolved solids (TDS) exceeds 5,000 ppm.

Self-potential (SP) method or "spontaneous polarization" method is based on the surface measurement of natural potentials resulting from electrochemical reactions in the subsurface. Two nonpolarizing electrodes (trailing and leading), a voltmeter, and a long connecting wire were utilized for the SP survey. The measurements were made with copper–copper sulfate nonpolarizing electrodes and a high impedance voltmeter. SP survey is carried out exactly over the hot spring along profile P1

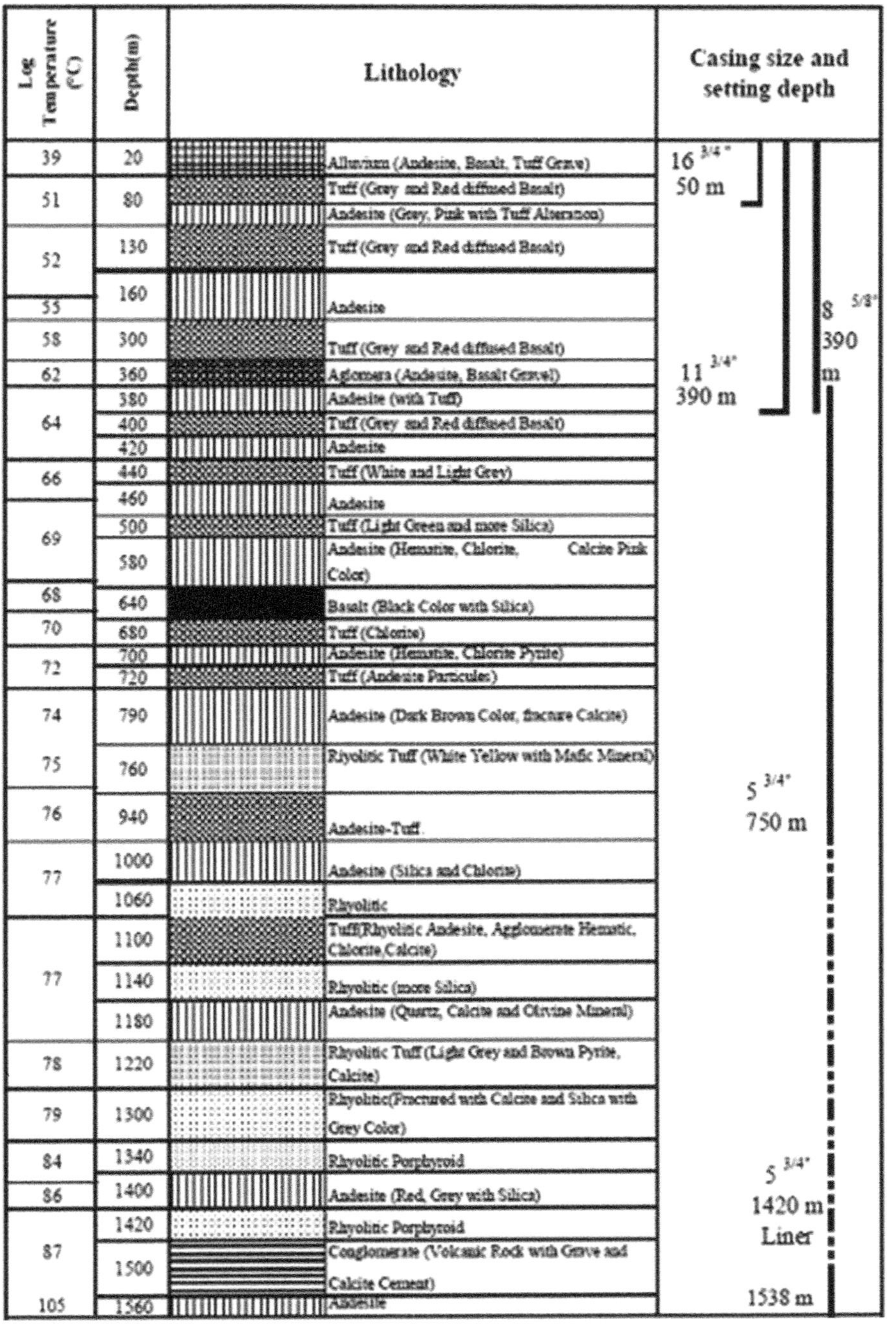

Fig. 22.5 Stratigraphic column of well KHD-1, Kızılcahamam geothermal field (Kaya [6])

(Fig. 22.1). Groundwater is thought to be a common factor responsible for SP anomaly at study area. Potentials are generated by the flow of water and by water reacting as an electrolyte and as a solvent of different minerals. The method is not widely used as the interpretation of the data which is difficult to quantify, but it still can give important information. The most important applications are for reconnaissance studies, mapping boundaries, and tracing faults. The anomalies may relate to temperature difference creating thermoelectrical potential, or fluid flow of hot or cold water, but may also be influenced by conductive mineral deposits or chemical variations in rock or pore fluid.

Fluid injection and/or circulation within geothermal reservoirs can produce surface self-potential (SP) anomalies of several mV that are correlated in space and in time to reservoir fluid flow. In addition, the sign of these anomalies depends on the direction of the flow, i.e., negative when fluid is injected and positive when fluid is produced [11]. SP anomalies are often interpreted qualitatively by profile shape, amplitude, polarity (+ or −), and contour pattern. More concentrated areas of high temperature at shallow depth, such as thermal fluids in a fault zone, could give rise to anomalies of greater amplitude. As seen in Fig. 22.6, the polarization effects in sequence are observed along the on E–W direction. The eastern side of the study area shows a marked difference compared to the rest of the study area. The eastern side of P1 profile exhibits a negative SP values.

A broad positive self-potential anomaly is observed over the geothermal field, and it is interpretable in terms of the circulation of the thermal waters. The polarity and waveform of the observed anomalies vary, with positive, negative, bipolar, and multipolar anomalies having been reported from different areas at study area.

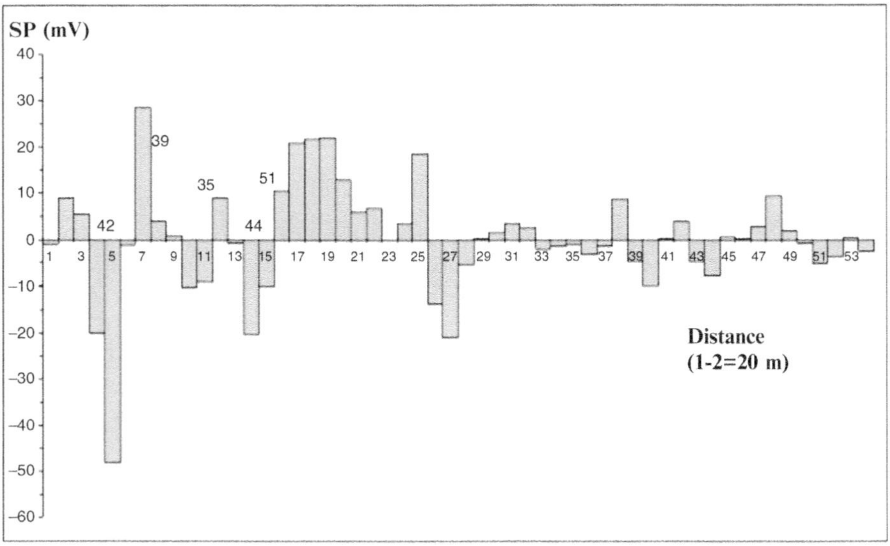

Fig. 22.6 Graph of derivative values on profile P1. The SP values in P1 profile are between −48 and 28 mV. The decreasing SP value in this line is due to fluids in the fracture zone

Steep potential gradients often are seen over faults which are thought to act as conduits for thermal fluids. The SP anomaly at study area is related to a thermal anomaly and to an anomalous under groundwater. These SP signals are usually attributed to electrokinetic processes as water circulates through fractures within the reservoir.

22.4 Conclusions

Geophysical resistivity techniques are based on the response of the Earth to the flow of electrical current. The VES method is effective for geoelectrical characterization of contaminated zones, allowing future geochemical study with an optimized well location and drilling depths. The low resistivity values indicate the presence of geothermal environment. In addition to this, the tectonic structure in the region and the high seismic activity provide important evidence about the existence of geothermal potential in the region. The SP method, employed with extra fieldwork precautions, can provide valuable information about a geothermal system when used as the main reconnaissance technique along with other geophysical data to correlate with the SP model. Small positive SP anomaly suggests possible lateral movement of hot water. Then, a more site-specific survey that includes audiomagnetotellurics (AMT) and transient electromagnetic (TEM) method, depending on the depth of exploration, should be done.

Acknowledgment The author is thankful to Geophysical Engineer Ahmet Lezgi that supports the achievement of VES and SP data at the study area.

References

1. Dagistan H, Dogdu N, Karadaglar M (2010) Geothermal explorations and investigations by MTA in Turkey. In: Proceedings World Geothermal Congress, Bali, Indonesia, 25–29 Apr 2010
2. Hersir GP, Björnsson A (1991) Geophysical exploration for geothermal resources, principles and applications. UNU-G.T.P, Iceland, report 15. pp 94
3. Hacısalihoğlu I (1999) Ankara-Çamlıdere-Peçenek-Bolu-Gerede-Aktaş-Salur-Dereköy field, geothermal energy explorations. Geophysic survey report, General Directorate of Mineral Research and Exploration of Turkey (MTA), Ankara, Turkey
4. Corwin RF (1990) The self-potential method for environmental and engineering applications. In Ward SW (ed) Geotechnical and Environmental Geophysics I: 127–145
5. Akasako H (2011) General process of geothermal power development (geothermal technologies for exploration and resource evaluation). West Japan Engineering Consultants, Inc., Compiled by: Fujii k, pp1–48
6. Kaya T (2005) Characterization of Kizilcahamam geothermal field by tracer testing. MSc. Thesis, Graduate School of Natural and Applied Sciences of Middle East Technical University, Ankara, Turkey

7. Kurtman F, Samilgil E (1976) Geothermal energy possibilities, their exploration and evaluation in Turkey. In: Proc. 2nd U.N. symposium on the development and geothermal resources, San Francisco, CA, pp 447–457
8. Simsek S, Okandan E (1990) Geothermal energy development in Turkey. Geothermal Resources Council Trans 14:257–266
9. Keskinsezer A, Beyhan G (2013) Geothermal modeling of Kizilcahamam-Çamlidere area using 3D imaging technique. Geomaterials 3:126–131
10. Toker M, Durak S (1990) Kızılcahamam KHD-1 ve MTA-1 wells, Test report. General Directorate of Mineral Research and Exploration of Turkey (MTA), Ankara, Turkey
11. Darnet M, Maineult A, Marquis G (2004) On the origins of self-potential (SP) anomalies induced by water injections into geothermal reservoirs. Geophys Res Lett 31: L19609, DOI:10.1029/2004GL020922

Part VI
Green Buildings

Chapter 23
The Sustainable Potential of Digital Fabrication Process and De-standardisation of Architectural Products

Rossella Siani

Abstract Digital fabrication processes consist in a parametric designing and in a product manufacturing by computer numerical control (CNC) machines.

At the design stage, the architectural object can reach a great grip with the environmental and cultural context, because the parameters measured by the environment become the engine of the virtual model. In this way a change of external parameters results in a change of the project.

In this chapter the possibility to optimise, through this method of project generation, is investigated. Parametric design increases the performance of the architecture in terms of structural, thermal and functional needs, with a reduction in resources, material and energy, ensuring greater sustainability.

In the production stage, the use of CNC machines ensures the realisation of the objects, designed with parametric methods, that, for intrinsic reasons, cannot be standardised and thus they are unique.

This feature of digital fabrication could radically change the industrialised society.

23.1 Introduction

Nowadays, in the automation of information era, the mass production of objects gradually loses importance in favour of the operating cycle, controlled by a set of logically arranged and automated instructions that can be preprocessed or are instantly able to be processed within the organisation: the organisation takes, therefore, the upper hand.

The technology of power automatism and automation of leading control, ordered according to a plot governed by automatic processing of information, tends to free itself from the slavery of the industrial series, defeating the importance already

R. Siani (✉)
DIARC, Dipartimento Architettura, Università degli Studi di Napoli, Federico II, Naples, Italy
e-mail: arch.rossellasiani@gmail.com

I. Dincer et al. (eds.), *Progress in Clean Energy, Volume 2*,
DOI 10.1007/978-3-319-17031-2_23

attributed to the "quantity" factor so as to provide the product with intrinsic values rather than with accessory values.

The electronic progresses, applied to automation, transform the notion of standard, as formal milestone of objectual series, into those systems of informative stimuli, able to give birth to continuously different objects, manufactured in many or few copies, or even, into a unique piece [2].

This is the prophetic thought of Giuseppe Ciribini, which in the mid-1980s describes, with efficiency and precision, a phenomenon that will emerge in the decades to come: what we now call the digital fabrication.

Nearly 30 years later, Achim Menges argues: the opportunities of contemporary computer numerical control (CNC) fabrication technologies have ... superseded the paradigms of serialisation and repetition that were predominant in the twentieth-century architecture [7].

Digital fabrication is a complex process, which has great potential in the context of sustainable development. In this chapter the phases, tools, properties and some experiments are analysed.

23.2 Digital Fabrication Phases

Digital fabrication is divided into two phases: the first one is the designing phase, and takes place with digital software; the second one is the manufacturing phase, which transforms the virtual project into a real object through the use of CNC machines or robots.

Both phases evolved in recent decades and promise a steady growth in the coming years.

The first tools for the design are limited to representation and are based on a computer-aided design (CAD) system, widely used today.

It is known that CAD is used in computer systems to assist in the creation, modification, analysis or optimisation of a design [8].

CAD software has the advantage of increasing the productivity of the designer and improving the quality of design and communications through documentation, and is quickly creating a database for manufacturing [8].

CAD output is often in the form of electronic files for print, machining or other manufacturing operations.

In recent years, CAD tools have been joined by computational design (CD) tools, based on the theory of computational complexity. The CD software enables the processing of the input data through a complex method; it uses algorithms and advanced computational techniques not for drawing shapes, but for creating formal possibilities. It deals not with producing a solution, but with generating a family of possible outcomes. This software is not only a support for the representation, but it is also a real way to design: it is parametric design.

The field of parametric design is generating geometry from the definition of a family of initial parameters and designing the formal relations they keep each other.

The use of variables and algorithms generate a hierarchy of mathematical and geometric relations that allow to generate a certain design, but also to explore the whole range of possible solutions that the variability of the initial parameters may create.

The benefits of this process are immediate. It is a huge quality leap for users, since tools do not constrain them anymore; now it will be the users themselves who design their own tools. On the other hand, parametric design is fundamental in order to minimise the effort needed to create and test design modification. The automated process eliminates tedious repetitive tasks, complicated calculations, and possibility of human error, and can generate huge shifts in the outcomes with slight variations of the original parameters. It is the difference between using the "Cube" command 1,000 times, entering center point and dimensions and customising the design of a "Group of Variable Height Cubes" that command out of our own predefined variability rules [16].

The generative tools of parametric design are expressed in the field of virtuality; and only in the second manufacturing stage, when a real model is created, the process can be completed.

CNC machines and robots are tools able to translate a virtual project into a real object.

The CNC machines, in the most basic versions, are available from the middle of the twentieth century, but only after the creation of a universal language, the G-code, these mechanical instruments had a direct dialogue with the software tools.

The CNC machines can realise both projects defined by CAD software and computational design software.

The application field of these machines is wide, and in recent years several research groups had drawn their attention to them. It is easy to preview that, in a short time, the performances of these tools will be greatly improved.

But what are the CNC machines?

In modern CNC systems, end-to-end component design is highly automated using CAD or computational design and computer-aided manufacturing (CAM) software.

These software produce a computer file that is interpreted to extract the commands needed to operate a particular machine via a post-processor, and then loaded into the CNC machines for production [17].

CNC machines can be divided into several groups, depending on the technology used to transform the virtual file into the actual product: cutting, subtraction or adduction.

Lamination process, which is based on cutting, is divided into two categories:

- The simple cutting (by blade or laser) of material panels and the various parts of the element, which will then be manually assembled by the operator.
- Laminated object manufacturing (LOM): Layers of adhesive-coated paper, plastic or metal laminates are successively glued together and cut to shape with a knife or laser cutter. Machining or drilling, after printing, can additionally modify objects, printed by this technique. Typical layer resolution for this

process is defined by the material feedstock and usually ranges in thickness from one to a few sheets of copy paper.

The subtraction process in CNC machines is based on the removal of matter from a solid block of material. CNC machines usually employed for these technologies are the milling machine and the lathe:

- The milling machine is a machine tool used for the machining of complex shapes in metal parts or other materials.
- The lathe is a machine tool used for the machining of a work piece placed in rotation.

The CNC machines that work with the method of adduction of the materials are divided into two groups, depending on the technique of processing the material: in the first case part from solid material moulded by melting, and in the second case, part of a liquid material transformed via solidification.

The first group, which is based on fusion, uses different techniques:

- Selective laser sintering (SLS): SLS uses lasers as its power source to sinter powdered material, binding them together so as to create a solid structure. The main materials used with this technique are thermoplastics, sand, metals and glass.
- Selective laser melting (SLM) is similar to SLS, the difference being that the materials are not only sintered after having put together but are also able to achieve a full melt.
- Fused deposition modelling (FDM) works on an "additive" principle by stratifying down material in layers; a plastic filament or metal wire is unwound from a coil and supplies material to produce a part. They use mainly thermoplastics.
- Laser deposition technology (LDT) is a process in which metal powder is injected into the focused beam of a high-power laser under tightly controlled atmospheric conditions. The focused laser beam melts the surface of the target material and generates a small molten pool of base material.
- Syringe extrusion works by extruding a fluid material. The syringe is connected to a carefully controlled air compressor. As the printer begins its print, the pressure in the syringe is manually raised, and continuously extrudes material until the compressor is shut off. This technology uses different materials that can be rendered fluid, such as the clay and earth, but also chocolate.

In the second case of adduction, there is only a methodology.

Stereo lithography apparatus (SLA) production separates one layer per time by curing a photo-reactive resin with a UV laser or another similar power source. The term "stereo lithography" was coined in 1986 by Charles (Chuck) W. Hull, who patented it as a method and apparatus for making solid objects by successively "printing" thin layers of an ultraviolet curable material, one on top of the other. Hull's patent described a concentrated beam of ultraviolet light focused onto the surface of a vat filled with liquid photopolymer. The light beam draws the object onto the surface of the liquid layer by layer, and using polymerisation or cross-linking to create a solid.

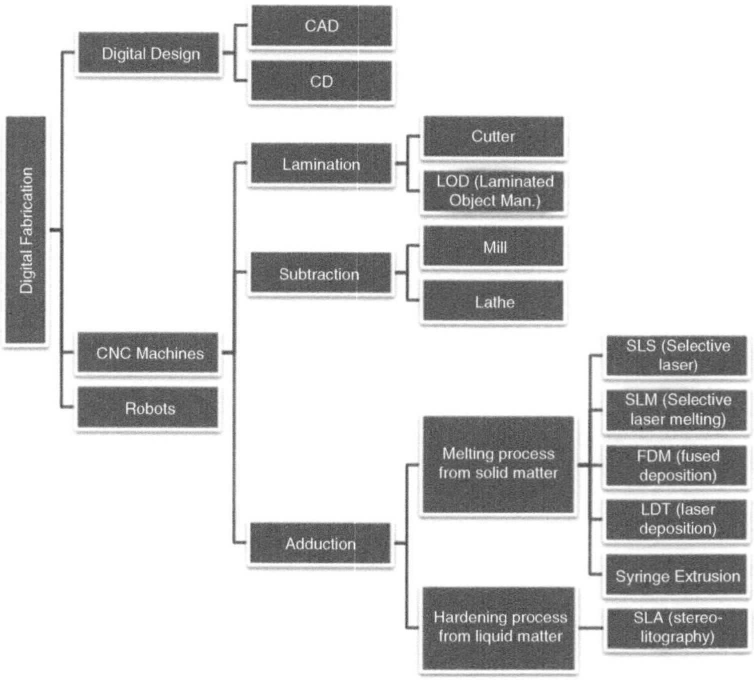

Fig. 23.1 Digital fabrication tools

Of all the CNC machines, those attracting more interest from the media are based on the process of adduction, commonly defined as 3D printer. They are the most recent, and also the most promising: easy to use and, in the basic models, easy to accomplish.

Digital fabrication takes advantage also by using robots. In this case their use is so varied that cannot be summarised. The design and programming of the robot is very wide and the only limits are the human imagination and technological feasibility (Fig. 23.1).

The materials used with these techniques are several. In cutting, we found mainly traditional materials such as wood, cardboard and aluminium foil. In the processes of adduction, the predominant materials have synthetic origin, made up with different types of plastic, but can also be metallic, or mixtures (powders of stone, wood, etc.) with a resin that glue all together.

The combination of these two phases produces different effects of considerable interest.

The first is the reciprocity between design and object: you can have a constant cross-reference between the virtual object and the real one.

In Computational Design Institute of Stuttgart Achim Menges defines and tests this property.

> The robot, being at the intersection of the binary and discrete world of the computer and material world in which exists, represents the interface between design computation and physical materialization. This opens up the opportunity non only to question the presumed unidirectional of the flow of information from file to factory, but rather to investigate the possibilities for reversing or short-circuiting this flow by informing the digital design tools themselves with the procedural logic of fabrication and the physical characteristics of material systems [7].

It is just the response of the materials that have attracted the attention of Menges in this study.

23.3 Digital Fabrication Properties and Some Experiments

> Structural design and form-finding process, allowing for the generation of design variation through adaptation to changing system-external and system-internal constraints, have to be closely intertwined. Using finite element (FE) analysis, the structural behaviour of the system can been simulated and continuously driven by the evolution of the design; in parallel, the load-bearing capacity of physical prototypes can be tested and the result thrown into the simulation, which in turn informed the global topology of the system. Through this integrative computational design process based on the reciprocal effects of fabrication, form, structures, material and performance the nominal material thickness could be minimised [7].

The material used in the testing by Menges is wood, but the logic of this process can be adopted for different materials.

The consequence of the material optimisation is directly related to the use of a lower quantity of material and, while still achieving the same performance, it follows the saving of resources. This is one of the first features that make this method a tool for sustainable development.

Till now in architecture the shape is expensive, while the material employment is cheap. With this new system the situation is reversed: the shape will be cheap and the material use expensive.

The material, for its intrinsic characteristics, becomes part of the parameters to consider in the realisation of the objects. Toni Kotnik and Michael Weinstock of EmTech programme at the Architecture Association (AA) in London gives another example of this trail of experimentation.

Even in this case, thanks to an interactive design process based on digital fabrication, that integrates material, form and force, digital fabrication has the potential to unfold a new generative logic of shape finding [5].

The ability to control the shape of the object using exogenous and endogenous parameters involves the monitoring of performance and the subsequent performance optimisation.

The stages of analysis and design are placed side by side; thus the project is processed and analysed at the same time, shaping the results on unique result.

The performances may be structural that are the most studied in the generation of optimised forms. But if the parameter is the energy, a project can be developed to generate devices for energy savings, increasing the sustainable potential of digital fabrication process.

But how is it possible to design architecture with energy parameters?

An example may be to provide a passive system, capable of interacting with the surrounding environment, modulating the action of climatic factors. Architecture can be "made", taking into account the thermal comfort and indoor humidity: so as to allow a less energy use to adjust the internal comfort. This process is possible by the control of environmental factors such as sunlight, ventilation and precipitation.

While achieving the same comfort performances, you will consume less power, expressing energy savings.

A good example of this theme is a study project for the envelope of Piraeus Tower in Athens (MSc Dissertation of Ioannis Douridas) that is derived from an interactive algorithmic procedure with the environment. "Various lines of research within the Emergent Technologies and Design context examined the way that building and components may be utilised to contribute in effective way to environmental modulation" [4] (Fig. 23.2).

MArch Dissertation of Juan Subercaseaux shows an other interesting study about a second skin: "the overall geometry of the grid shell was subsequently elaborated through a series of environmental performance criteria within context-specific defined ranges of environmental condition, including the insulation of rainwater, ventilation and thermal modulation, as well as controlled sunlight exposure of the interior" [4] (Fig. 23.3).

Intents of performance optimisation, the research work at MIT by Neri Oxman, investigates the new design approach to digital fabrication that offers the potentialities to programme physical matter. In this research, form generation is driven by maximal performance with minimal resources through local material property variation [9].

With the use of a single material it is possible to satisfy different requirements (structural, insulating, thermal, etc.) varying the composition of the material itself. Let us take as example an external wall; it is traditionally composed by several layers, with different materials; each of them has a specific characteristic: flow, insulation and waterproofing. With the innovation proposed by Neri Oxman it is possible to design a stratigraphic variability of density (Fig. 23.4), elasticity or other characteristics within a single material, and thus to obtain different performances in the different layers of the product, without the need of adopting a plurality of materials [9].

This experimentation can be applied to many products that now require stratification with more materials.

The main benefits are the increasing of performances and the reduction of material, two essential aspects for the sustainability, but there is more: the use of a single material makes it more effective because the step of recycling or disposal will be cheaper at the end of product life. The recyclability increases the sustainability of the product.

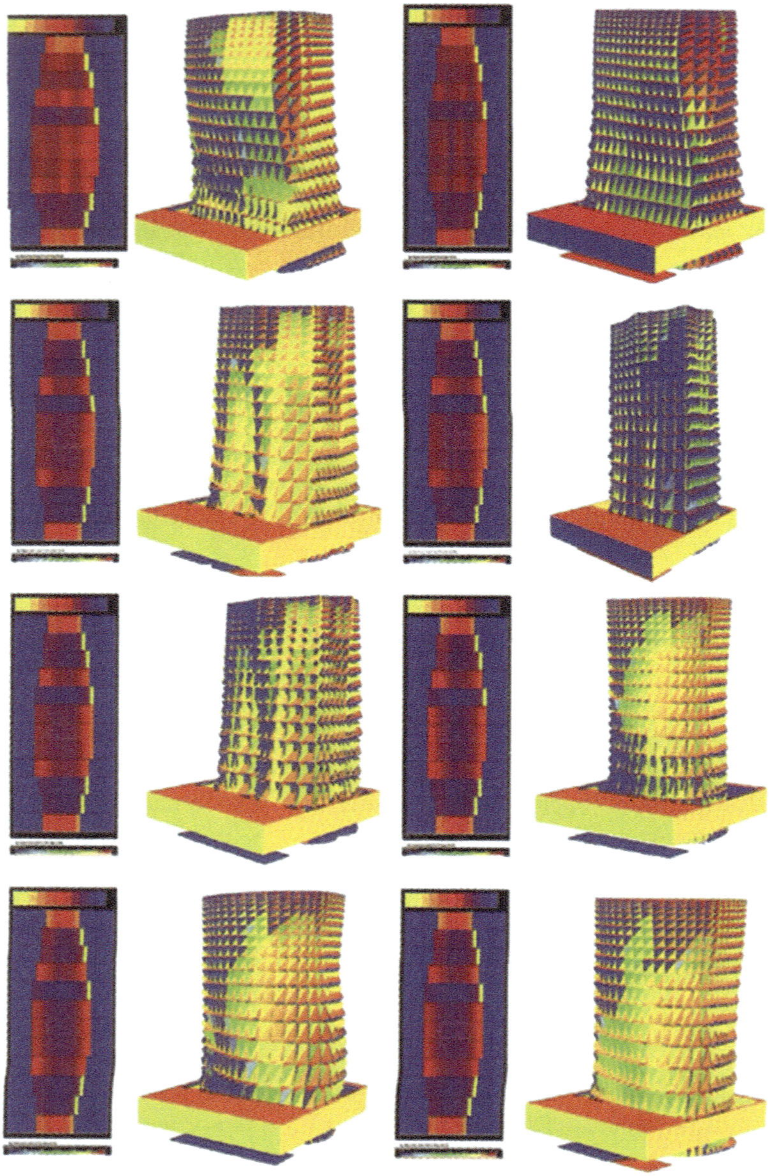

Fig. 23.2 Multiple generation of the Pireaus Tower's envelope is computationally evolved and analysed in relation to a number of environmental fitness criteria in order to explore the self-shading capacity of both the global envelope shape and local articulation of the skin components. MSc Dissertation of Ioannis Douridas. October 2005

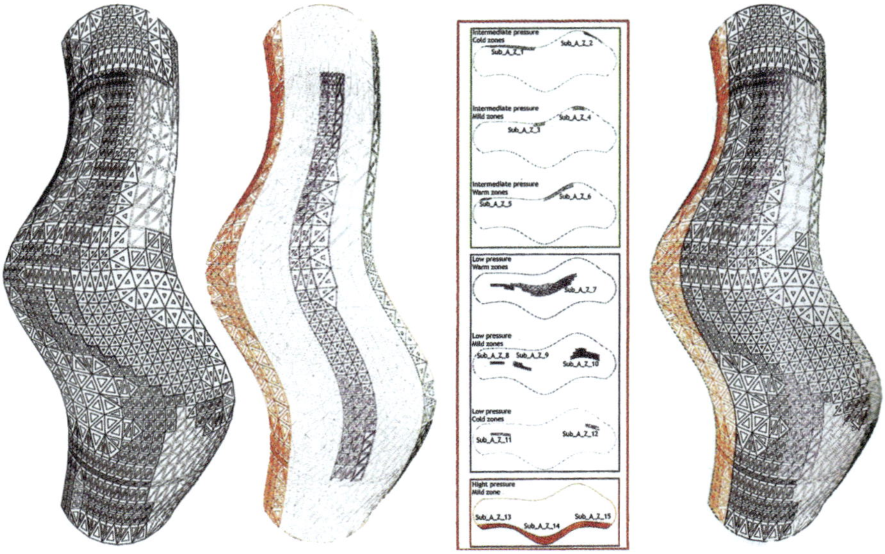

Fig. 23.3 Final tessellation and distribution of frames over the double-curved surface of the building envelope (*left*); air pressure zones and water-collection zones indicated on the envelope (*centre*); articulation of enveloper informed by the combined input parameters of light-exposure and air-pressure zones (*right*). MArch Dissertation of Juan Subercaseaux. February 2006

Fig. 23.4 Material tests for a concrete extruder head capable of dynamic density and aggregate control. Mechanical foaming techniques are automated using a six-axis robotic arm to produce lightweight, floating concrete structure with programmable porosity. Steven Keating and Timothy Cook, Variable-Density Printing, Mediated Matter Group, MIT Media Lab, Cambridge, Massachusetts, 2011

23.4 Accessible Test

An interesting aspect of digital fabrication process is the accessibility, both of information and instruments, in particular regarding the economic aspect.

Due to the low cost of the process, from design to construction, an emergent system of delocalised testing is dawning.

We are witnessing the birth of Fab Lab, fabrication laboratory [3], "physical locations where these new models of process and product are realised, are the places where it is possible to revolutionise the foundations of the goods' exchange process principles and ideas in order to meet the global sustainability and solidarity" [11].

Digital fabrication field is constantly and rapidly evolving, so part of the activities by Fab Lab is dedicated to research and test.

The experiments are no longer preserve of research centres and universities, but it is fragmented. Experimenting, but mainly making: these are the key words that move these small organisations of digital fabrication.

If you do something and you share it on the Web you are a Maker [1], so even without organising Fab Lab, the process of digital manufacturing is spreading more and more widely: just with the enthusiasm of a single person.

Digital fabrication teaches people to make and to repair too. "This new relationship between production and use seems to be a reminiscent of the pre-industrial culture, but instead of taking life from the local community, it tends to form itself a community, even global, which gathers around the single object. It tends to reduce, as it was for the craft, the changes of the finished product ..." [11].

The information and the basic tools for the design and production are available on the Web.

The consequence of this phenomenon, in different aspects and even in the context of sustainability, is not to be foreseen. However, the multiplication of experimentation and research creates a competitive environment and a greater number of possible solutions; therefore it is desirable to promote sustainability, or at least to raise its different aspects.

Hereinafter two projects are shown: Cocoon Pavilion and Cocoon Evo Pavillion.

Cocoon Pavilion (Fig. 23.5) is a project born from the collaboration of Medaarch, Mediterranean FabLab, Co-de-iT and CRTS Cartotecnica, made by corrugated cardboard, and it was the result of a workshop organised by the Mediterranean FabLab [13]. The project gave the opportunity to investigate the processes that lead from design to manufacturing of architecture, until their physical realisation. The most interesting note is that the entire process is self-financed; all the partners have contributed to an aspect. It is a clear example of an experiment by small groups of individuals.

The second project is an evolution of the first. Cocoon Evo Pavillion (Fig. 23.6) is an installation presented at the Maker Faire Rome 2013 designed by Mediterranean FabLab, Co-de-iT and PicernoCerasoLab and realised thanks to the machines and professionalism of Tekla company [15].

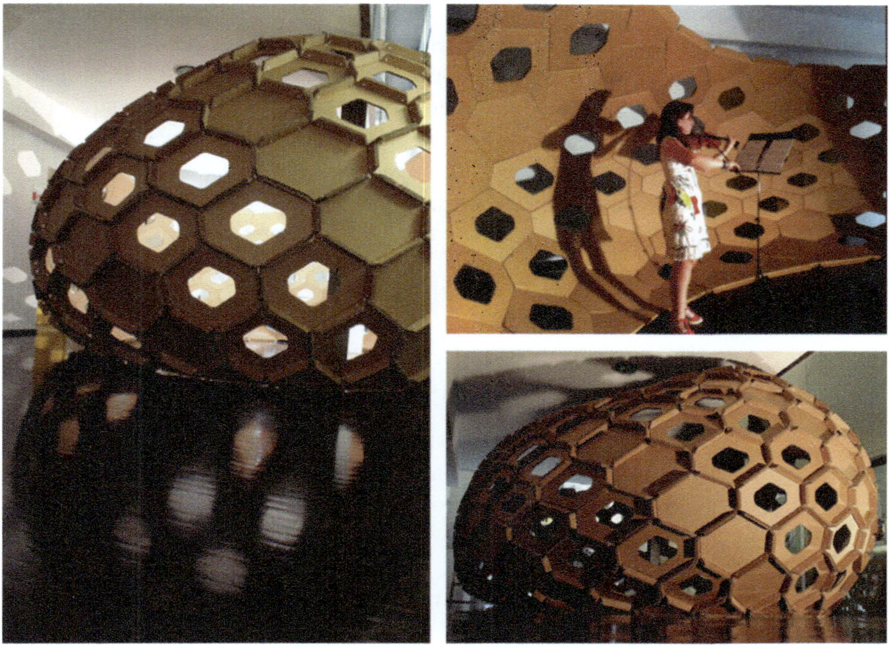

Fig. 23.5 Cocoon Pavilion is made by corrugated cardboard. It is born from the collaboration of Medaarch, Mediterranean FabLab, Co-de-iT and CRTS Cartotecnica

Fig. 23.6 Cocoon Evo Pavillion is an installation presented at the Maker Faire Rome 2013. Realised by Mediterranean FabLab, Co-de-iT, PicernoCerasoLab and Tekla srl, ALURAME

A new era is emerging.

In a first step, an increase in the creation of objects, dictated by the ability of having customisable products, seems nothing more than another tool of consumerism and is clearly at odds with the logic of sustainability. In fact many of the products, made using these processes, are in plastic material, fact that can brings to an increasing of pollution and wastes, in the realisation phase.

But one of the most desirable solutions, aimed at overcoming these problems, is the experimentation on materials with a low environmental impact and easy recyclability.

23.5 Relocation and Customisation Production

By the spread of small production companies, based on digital fabrication process, it will be possible to have access to a delocalised and custom manufacturing.

The products will have less need to travel from one continent to another, simply by sending projects via e-mail.

One of the solutions on which various experiments are starting concerns the possibility to create architecture with the parametric fabrication using local materials, as sand, earth or clay and so on. Such an approach, in addition to being considerably ecological, is definitely cheaper than traditional techniques.

The use of such a process will allow best performances, lower energy resources and lesser amount of material than traditional methods, and it will therefore be preferable from the sustainability point of view.

All analysed aspects consider the digital fabrication as a moderate factor of sustainable development. It is true that the optimisation of materials affects energy performances; it is also true that the relocation of production leads to the decreasing of transports, but in contrast, the growing flourish of objects of each type and material does not contribute to the increasing pollution and wastes.

So far, the balance of advantages and disadvantages seems to be equal.

The sustainability factor of digital manufacturing is not in the direct consequences, but in the indirect ones.

This manufacturing process makes it possible to create unique objects at no additional cost as much as create thousands and thus undermines economies of scale. The spread of digital manufacturing could have a profound impact on the world, as it was the advent of the industrial era. Just as nobody could have predicted the impact of the steam engine in 1750—or the printing press in 1450, or the transistor in 1950—it is impossible to predict the long-term impact of these new technologies. And probably it will subvert every field it will touch.

With this enthusiasm The Economist welcomes the advent of digital fabrication, with a focus on 3D printers [14].

So the principal quality of this process is the potentialities to compete with the economy of scale and replace it.

23.6 Conclusion

The current system is based on the industrial mass production.

Mass production is the production of large amounts of standardised products, and is an established and effective practice, because its inherent logic allows lowering the cost of unique product increasing the number of production.

The low cost of the product ensures spread to all sections of society. The spread of standardised products has expanded exponentially, as these products are more economically affordable compared to handmade products. Standardised objects invade today all fields of goods production. The standardised product fulfills the needs of the consumer universally and not particularly. Today the needs and performances have to undergo an adaptation, providing approximate results. The standardised products then provide standardised performances; they are neither adaptable nor flexible, and sometimes perform only in a minimal way as per the required needs.

This process lowers the quality of the industrial products, which are preferred to the handmade one for their cheapness. The market thus supplies the development of mass production. The production of goods increases, and the market becomes saturated and is no longer able to absorb that particular product. To deal with this gap there are several strategies in the system that maintain a high level of sales of that particular product.

A possible way to increase the number of consumers can be that of operating a constant market extension (to other territories, to other social groups).

Another possibility lies in increasing the need for that product.

When consumers are led to the need for replacing some goods they already own, we talk about planned obsolescence [10]. This process is triggered by the production of goods subject to a rapid decay of functionality and is realised by means of appropriate devices introduced in the production phase (use of poor-quality materials, planning of repair costs higher than those of purchase, etc.). Product's life is reduced to a shorter time of effetive technological possibilities, so as to push for purchases of replacement. Detailed further acceleration of planned obsolescence of the goods is made by the advertising of new models which have irrelevant changes in functional terms, but great ones in substance and form. This is particularly relevant in the advanced technology industries that operate in oligopolistic market structures, in which the positioning of companies depends on the ability to differentiate technological products which are homogeneous: in these market conditions, planned obsolescence is realised through sophisticated marketing strategies with which it is possible to create fashions and trends, aimed to elicit consumer needs to replace products [12].

These logics involve a higher economic use of resources and energy and a greater production of waste. As demands increase the profit rises and the system is repeated, growing to the infinity in a finite environment, threatening the balance of the ecosystem.

Just in opposition to this indiscriminate growth, different ideologies and strategies are born, including the need to adopt a "happy degrowth" [6] and a sustainable development.

If the digital fabrication is proposed as an alternative to mass production, which is mainly responsible for the increase of the ecological footprint of the industrialised lifestyle, it has the inevitable implications of sustainability.

Digital fabrication creates a customised product, which fulfills the demands with increasing accuracy, and varies the performance, with a return on investment-based materials and energy savings. The difference of cost between digital fabrication and standardised products is offset by the higher quality of performances. Finally customised digital fabrication products are cheaper than handmade ones.

However, these factors may not be sufficient to induce consumers to prefer customised products to those standardised.

Perhaps the crisis of Western consumption can be a contributory factor to a new model.

A decrease in sales, which is typical of a crisis, requires a reduction in production, which leads to an increase in costs of each product in the long term.

If the crisis will be enough strong to topple the mass production, maybe a minimum gap will result between the costs of standardised and digital fabrication products.

In short, until there are no tools to compete with the low costs of the mass production, it is not conceivable to implement a comprehensive sustainable development. Digital fabrication phenomenon seems to get characteristics suitable to perform this task.

In this case, there are many changes that industrialised society will have to face, including substantial changes in habits and consciences.

References

1. Anderson C (2012) 2013 Makers. Il ritorno dei produttori. Rizzoli, Milano
2. Ciribini G (1984) Tecnologia e progetto. Celid, Torino
3. Gershenfeld N (2012) How to make almost anything. The digital fabrication revolution. In: Foreign affair, vol 91(6), November/December 2012
4. Hensel M (2010) Material system and environmental dynamics feedback. In: Hensel M, Menges A, Weinstock M. Emergent technologies and design, towards a biological paradigm for architecture. Routledge 2010, ISBN 10 0-415-49343-9
5. Kotnik T, Weinstock M (2012) Material, form and force: material computation. AD architecture design, ISBN 978-0-470-97330-1
6. Latouche S (2007) Petit traité de la décroissance sereine,. Tr. it.: 2008. Breve trattato sulla decrescita serena, Torino, Bollati Boringhieri, ISBN 978-88-339-1869-3
7. Menges A, Shwinn T (2012) Manufacturing reciprocities: material computation. AD architecture design, ISBN 978-0-470-97330-1
8. Narayan KL (2008) Computer aided design and manufacturing. Prentice Hall of India, New Delhi. ISBN 812033342X

9. Oxman N (2012) Programming matter. In: Material computation. AD architecture design, p 90–95, ISBN 978-0-470-97330-1
10. Packard V (1957) The hidden persuaders, trad. Fruttero C (1968) I persuasori occulti. Torino: Einaudi, Milano. 1989—I persuasori occulti rivisitati negli anni ottanta. In: Il Saggiatore. Einaudi
11. Pone S, Colabella S (2014) Maker. In: Op. Cit. Selezione della critica d'arte contemporanea n. 149—gen 2014
12. http://www.treccani.it/enciclopedia/obsolescenza-programmata/
13. http://www.medaarch.com/2128-cocoon-a-temporary-pavilion/
14. http://www.economist.com/node/18114327. The Economist. 10 feb 2011
15. http://www.co-de-it.com/wordpress/cocoon-evo-pavilion.html
16. http://www.parametriccamp.com/
17. http://en.wikipedia.org/wiki/Cnc

Chapter 24
A Building Energy Performance Model and Advisor System

Ali Palizban and Mehrzad Tabatabaian

Abstract The *AFRESH* home is an *a*ffordable, *f*lexible, *r*esilient, *e*nergy efficient, *s*ustainable, and *h*ealthy home designed and built to integrate different technologies that help improve energy efficiency. The home is located at BCIT (British Columbia Institute of Technology, Burnaby, BC, WWW.bcit.ca) campus and is constructed with energy efficient, nontoxic, and environmentally friendly materials and features: photovoltaic (solar roof panel), wind turbine, fuel cell, energy storage, and distributed energy generation. The home is also fitted with a heat-recovery and geo-exchange HVAC system. This article explains the characteristics of AFRESH home under study and its energy performance model, followed by a description of the actual design and implementation of the home energy management system. Results of the research and the analysis of the results are also included in the paper.

24.1 Introduction

In collaboration with BCIT's School of Energy research teams, Group for Advanced Information Technology (GAIT), and Products and Process Applied Research Team (PART), we are developing methods to monitor, simulate, and estimate the power consumption, generation, and storage of the AFRESH. Using this data we will simulate common home appliance usage scenarios in order to predict daily electric energy consumption costs. For this purpose a prototype Home Energy Advisor System is designed [1]. For the first prototype, a limited amount of automated control is considered; instead, there is a greater emphasis on human control to verify the simulation. Ideally the prototype would be used to provide homeowners with sustainable and practical power consumption habits.

Appliances used in the home are intelligent and can communicate via a home area network with a load control system. The AFRESH home (Fig. 24.1) is a collaboration of green technologies that help the homeowners make informed

A. Palizban • M. Tabatabaian (✉)
School of Energy, British Columbia Institute of Technology (BCIT), Burnaby, BC, Canada
e-mail: Ali_Palizban@bcit.ca; Mehrzad_Tabatabaian@bcit.ca

I. Dincer et al. (eds.), *Progress in Clean Energy, Volume 2*,
DOI 10.1007/978-3-319-17031-2_24

Fig. 24.1 AFRESH home in BCIT Burnaby campus

decisions to decrease their energy consumption. The AFRESH home energy performance was modeled using the eQuest (Quick Energy Simulation Tool), version 3.64. The model inputs were based upon the information gathered during a site review and conversations with the building operator on the building systems and typical use. In general, the energy performance model focused on the building's total energy usage. Energy production systems were not included in the model; the performance of these systems was calculated separately and can be subtracted from the total energy usage to evaluate the building's "net-energy" performance [2].

24.2 Modeling Project Scope

Beginning in January 2013, a project was undertaken to investigate the energy performance of AFRESH home, located at the BCIT Burnaby campus. The objectives of the project are summarized as follows:

- Develop an understanding of the AFRESH home, including the building enclosure assemblies, mechanical systems, and power generation systems.
- Perform energy performance modeling based on the current configuration and usage of the AFRESH home.
- Validate the performance model using actual building data.
- Identify opportunities for improvements in building design (dependent upon modeling results and available data).
- Identify opportunities for modeling improvement.
- Model AFRESH home assuming residential occupancy.
- Provide recommendations for using AFRESH home designs in residential houses.

24.3 Building History

Originally designed as "Home 2000," AFRESH home was featured at the 2001 BC Home Show to demonstrate best practices in affordable, sustainable residential construction and energy systems. Following the Home Show, AFRESH home was relocated to BCIT.

In 2008, retrofits were conducted to allow for research into a Distributed Power Connections study by BC Hydro. This study intended to "transform the AFRESH Home to become a net energy producer and to study the way that energy migrates to and from the grid." Building retrofits included installing a new externally mounted photovoltaic panel array on the south building face, incorporating a natural gas fuel cell and a geo-exchange heating system, and integrating an energy management system (including wireless thermostats and smart meters). The controls for these systems are housed in a mechanical room addition located in the southeast corner of the building.

In 2011, a vertical-axis wind turbine, on the northwest corner of the property, and new south-facing photovoltaic panels were installed as part of a project from the GAIT research group. The purpose of this project is to research integration issues of renewable energy sources into grid-connected and off-grid scenarios.

24.4 Building Description

24.4.1 Structure, Enclosure, and Systems

ARESH home is a three-story, wood-framed house located at the BCIT Burnaby campus. The lower two floors are designed as living space, while the upper floor is intended to be an attic. The house has an interior floor area of about 185 m^2 (approximately 2,000 ft^2). There is a second floor balcony on the south elevation that the building drawings indicate to be an interior storage area.

There is a single-level addition on the west elevation that houses the majority of the mechanical equipment. This space is insulated but only heated by the mechanical equipment it houses.

The windows are vinyl-framed with awning and casement operable unit. The double-glazed insulated glazing units (IGUs) have metal spacers with low-e coatings.

The building has sloped metal roofs with claristory windows on the west elevation. There are single- glazed skylights on the east elevation with building-integrated photovoltaic panels. The building is serviced by multi-zone mechanical systems. A ground-source heat pump, consisting of 4 loops buried on the north side of the property, provides the heat for space heating and hot water. There are heat-recovery ventilation systems on each floor to capture heat from the exhaust air to heat the fresh makeup air.

There is an air-conditioning unit on the west side of the property; from our discussions with the building operator, it's unclear what the system's frequency of use is. The building's interior environment is controlled automatically by a wireless thermostat system. Building users are also able to control the system manually via the online GridPoint energy management system (EMS).

The building's electrical systems are sub-metered per device through the online GridPoint system. However, the summation of the device loads breakdown is less than the total annual consumption recorded both through EMS and through BCIT's data, indicating that some devices are consuming power but are not connected to the sub-metering monitoring system. The difference in energy usage each month was between 500 and 800 kWh, for a total of 7,000 kWh over the year.

24.4.2 Appliances

The building is equipped with LED lights and ENERGY STAR (efficient) appliances. The building contains a refrigerator, stove and range, microwave, clothes washer and dryer, dishwasher, and television, although many of these devices are seldom used by the current commercial occupants.

24.4.3 *Power Generation Systems*

The building has several alternative power generation systems; however, some of them are not currently being utilized. For clarity, the systems are categorized as either "active" (currently in use) or "inactive."

24.5 Active Power Generation Systems

24.5.1 *Photovoltaic Panels*

There are building-integrated photovoltaic (BIPV) panels incorporated into the east skylights, with 1 cm by 1.6 cm glass spacers to allow daylight into the attic. The capacity of this system is approximately 2 kW. These skylights were originally intended to be south facing; however, when the house was relocated to BCIT, the site geometry required orienting the building such that the panels were east facing.

There are externally mounted photovoltaic panels on the south-facing roof. The capacity of the six-panel array is approximately 1.2 kW. The south-facing panels were installed to take advantage of the great solar insolation and were tilted at an optimum angle for the local latitude.

24.5.2 *Ground-Source Heat Pump*

There is a "geo-exchange" ground-source heat pump system, discussed in the previous section. This system is reported to have a 3:1 production ratio (COP = 3), producing three units of heat energy for every one unit of electrical energy consumed. Although capable of providing heating and cooling, the system is currently only being used for space heating and hot water.

24.6 Power Generation Systems

24.6.1 *Natural Gas Fuel Cell*

The building is equipped with a natural gas fired fuel cell, intended to be used to reduce the electric grid loads during peak hours. However, the building operators have had difficulties maintaining and operating the fuel cell and it is not currently being used. Information on its performance was not made available to us for this project.

24.6.2 Vertical-Axis Wind Turbine

A 50 ft tall vertical-axis wind turbine is located in the northwest corner of the property. The turbine has the capacity to produce a maximum of 5 kW at a peak wind speed of 12 m/s and can generate power at a reduced efficiency at wind speeds as low as 2 m/s. The wind turbine is part of a research project by the GAIT research group and is not currently being used.

24.6.3 Actual Power Consumption

The building's energy consumption data was obtained from the online GridPoint monitoring system and compared to BCIT's microgrid energy management system data available. The total average annual energy purchased from the grid is about 15,850kWh. In addition, AFRESH home produced about 1,330 kWh from the photovoltaic panels in 2012. Figure 24.2 shows yearly energy generation consumption of AFRESH home.

24.7 Building Performance Modeling

The AFRESH home was modeled using the eQuest (Quick Energy Simulation Tool), version 3.64. The model inputs were based upon the information gathered during a site review and conversations with the building operator on the building systems and typical use.

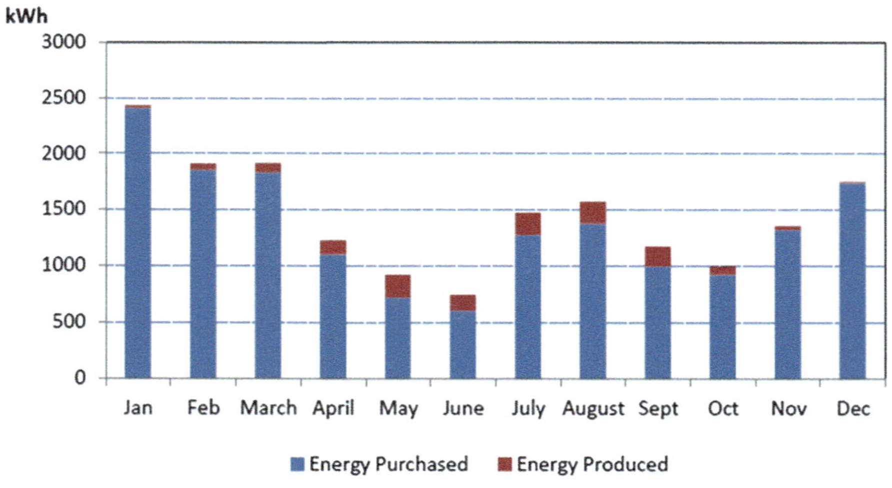

Fig. 24.2 Energy generated and consumed in AFRESH

In general, the energy performance model focused on the building's total energy usage. Energy production systems were not included in the model; the performance of these systems was calculated separately and can be subtracted from the total energy usage to evaluate the building's "net-energy" performance.

24.7.1 Modeling Approach

The modeling approach was to capture the essential elements relevant to energy performance. The following criteria were used to produce an effective energy model:

- The model should accurately represent heat load paths.
- The model should provide an estimate of interior heating and cooling loads.
- The model should provide the analyst with a clear understanding of the building's behavior under heating and cooling loads.
- The model should be relatively easy to modify so it can be used to investigate a building's sensitivity to changes in various input parameters.

These criteria were adopted from the course notes [3] for effective structural modeling and modified for energy modeling.

24.7.2 Model Design

Following the adopted modeling principles, a two-story, rectangular building shell was created in eQuest to represent AFRESH home. The actual building footprint consists of two offset rectangles, as illustrated in Fig. 24.3, which may introduce solar shading effects not encountered in the model.

Further, the attic floor plan is altered from the lower floors. However, the effects of the staggered floor plans were considered to be negligible on the load analysis in this case. A simplified building floor area was created with an equivalent footprint size to the original building and the mechanical room was not considered. The isometric representation of the energy model is shown in Fig. 24.4.

24.7.3 Loads

AFRESH home is currently being used as an office space, so the energy model was first designed to reflect office loads, to allow for calibration with the actual performance data. The building's actual occupancy scheduling was inconsistent and not accurately recorded, so the model inputs assumed that the building was occupied from 8 am to 5 pm on weekdays and closed weekends and holidays.

Fig. 24.3 Actual building footprint

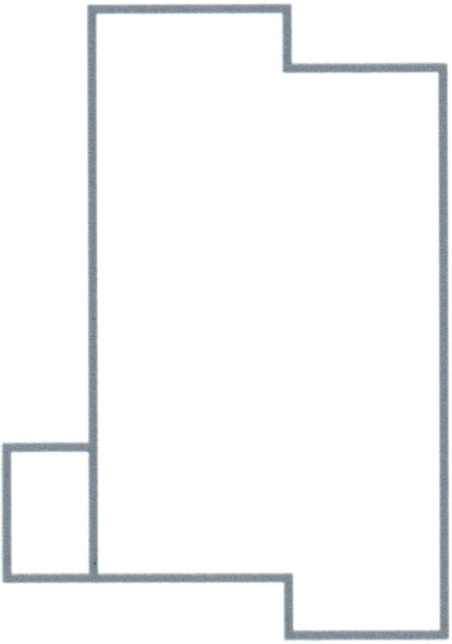

Fig. 24.4 The isometric representation of the energy model

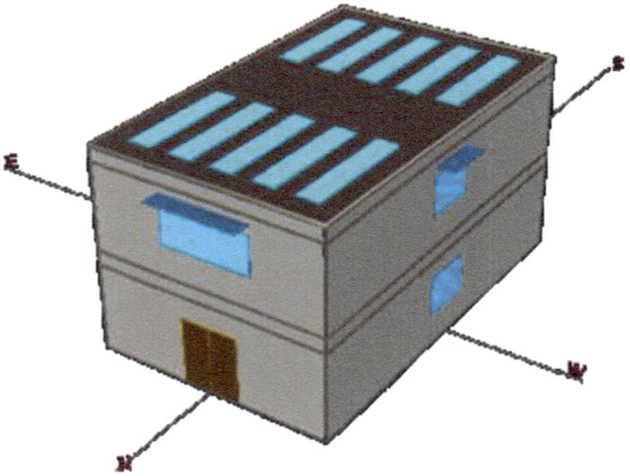

These values were based on typical energy modeling schedules and assumed constant throughout the year.

Internal loads were dependent on the building's occupancy type, based on the default values in eQuest and modified to reflect the actual occupant density of the home.

24.7.4 Model Calibration

The building performance model was calibrated using measured energy data, so changes in the model can reflect expected changes in actual building performance. Model calibration is used to produce a model that accurately predicts energy consumption

For this project, the level of calibration was limited to whole building performance, due to a lack of resolution in energy usage data and unknowns in the mechanical system design and interior loads.

Interestingly, the occupant densities and occupant schedules had a more significant effect on the whole building energy performance than mechanical system adjustments.

The model was well calibrated, and the final predicted energy usage closely matched the actual energy usage. The modeled usage is compared to the actual energy usage in Fig. 24.5.

24.7.5 Modeling Result and Suggestions

Strategies for improving the power generation of the photovoltaic system and wind turbine were evaluated to identify opportunities for reducing the building's grid energy usage. These strategies are discussed below.

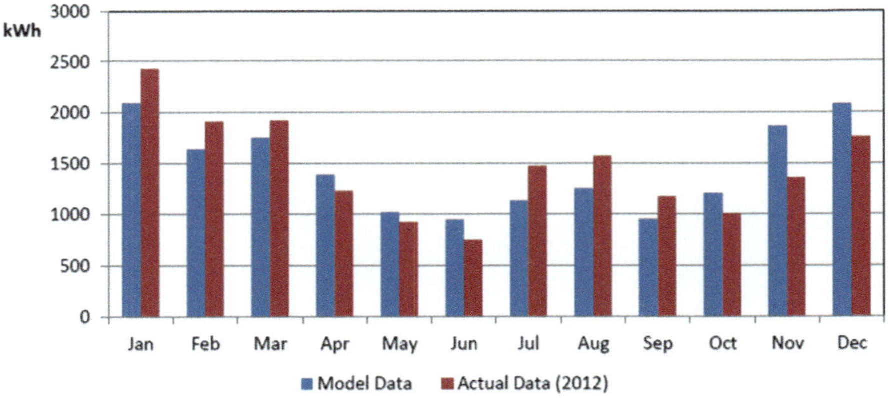

Fig. 24.5 Actual data vs. calibrated performance model outputs

24.8 Photovoltaic Improvements

The current configuration of the photovoltaic panels (2 kW east-facing BIPV and 1.2 kW south-facing externally mounted PV) was evaluated to assess the potential energy generation currently possible. The results are summarized in Table 24.1.

There is a large discrepancy between the potential energy generation and the actual generation of the solar panels, the cause of which was not immediately apparent. The historical energy production of the panels was reviewed. The energy production of the photovoltaic panels has been decreasing regularly every year, likely due to dust/debris build up on the panels. However, the total energy production in 2010 was only about 1,750 kWh, which is still significantly less than the energy production potential summarized in Table 24.1.

The photovoltaic energy production capacity was evaluated to consider the effects of orienting the BIPV system to the south. In this example, additional externally mounted PVs were still included to further increase the energy production. The results are summarized in Tables 24.1 and 24.2.

The energy modeling part of this project consisted of an energy performance analysis of the AFRESH home on BCIT Burnaby campus, summarizing the existing building systems, developing a calibrated whole building energy performance model based on the building's current usage as an office space, estimating the building's energy performance when used as a single-family house, and reviewing the building's current alternative energy production and identifying strategies for improvements. Based on the obtained results, AFRESH home achieves its goal of being an affordable, high-performing home. However, there is currently a considerable amount of "untapped" power generation potential that could be utilized to further reduce the home's reliance on grid energy.

Table 24.1 Summary of current and optimum configuration of AFRESH PVs

	Current configuration-potential energy output	Optimum configuration
2 kW BIPV	1,300 kWh (east facing)	2000 (south facing)
1.2 kW externally mounted PV	1,200 kWh (south facing)	856 (east facing)
Total	2,500 kWh	2,856 kWh

Table 24.2 Summary of potential and actual energy output of AFRESH PVs

	Current configuration-potential energy output (kWh)	Current configuration-actual energy output
2 kW BIPV (east facing)	1,300	1,330 kWh (combined, 2012)
1.2 kW externally mounted PV (south facing)	1,200	
Total	2,500	1,330 kWh

AFRESH home design principles can be affordably implemented in typical residential house construction to improve the stock of residential buildings and reduce the grid energy demand.

24.8.1 The Low Cost Home Energy Advisor

The Low Cost Power Advisor (LCPA) Simulation [1] was designed to provide the user with a tool to effectively determine optimum strategies for appliance power consumption. It does this by integrating what is known about the power consumption of various appliances and when they are being used. This can be done for as many appliances as you want, making it possible to simulate multiple use case environments. Consumption power is then converted into a cost which is dependent on the price scheme being implemented by the electricity provider. An analysis on this cost can help answer questions such as:

- How much will I save if I hold off activating the dishwasher until off peak hours?
- What will be the annual return for an installation of a 1 kW solar panel?
- What control can be implemented by appliance manufacturers to reduce the electric bills for the appliance users?
- What control can be implemented by appliance manufacturers to reduce power consumption during peak loads for electricity providers?
- Which control strategy should you implement in a fixed tariff vs. real-time based power pricing strategy?

The appliance model may either be analytical or numerical. Implementing an analytical model requires a good understanding on what impacts the appliances power consumption/generation. The analytical model may require some advanced math which may or may not be supported by the simulation. This may not always be viable due to a lack of theory and data available for the given appliance. In numerical model we can take the numerical data we collect, through monitoring the appliance, and map it onto its corresponding simulation counterpart.

Whether we want to verify the analytical model or input numerical models, we needed to acquire appliance power consumption data from the AFRESH home as shown in Fig. 24.6. The data we used was taken from:

- The Energy Storage System
- Electric Energy Meter
- Appliance specifications
- Energy consumption averages
- PLC data acquisition system

A block diagram of the LCPA is shown in Fig. 24.7. The system measures the power consumption data using sensor, takes feedback from the homeowner (user), and suggests optimum appliance adjustment to the user.

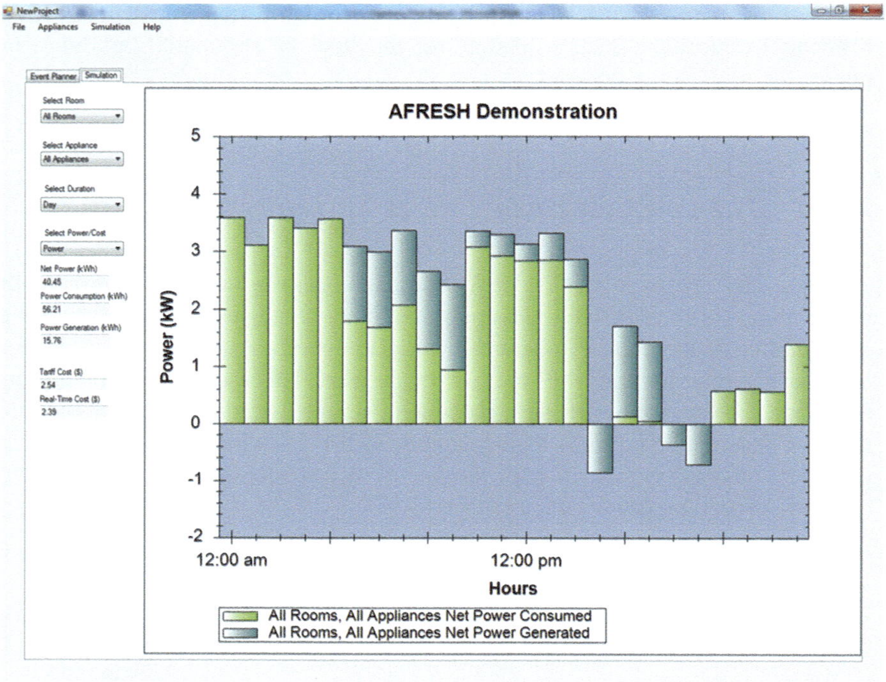

Fig. 24.6 LCPA user interface tab

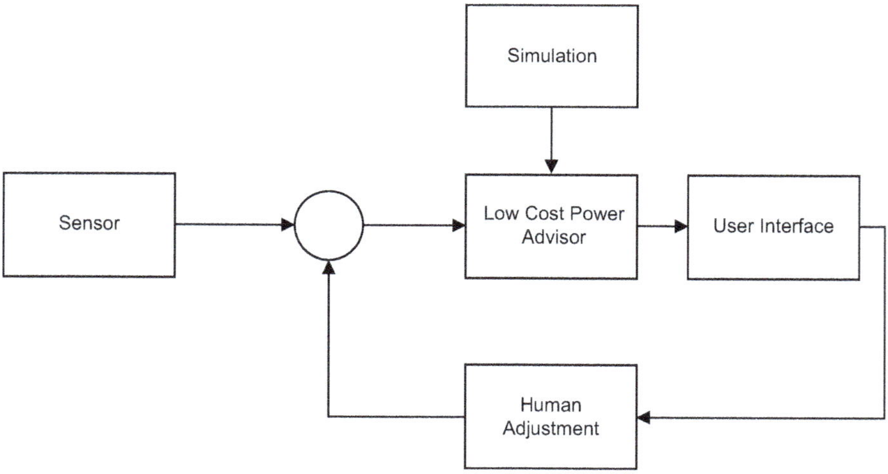

Fig. 24.7 Overall high-level design flow chart

24.8.2 *Application of the Home Energy Management System*

The Prototype Home Energy Managements system with the simulation software provided the user with the ability to evaluate control implementations and appliance usage scenarios. The simulation software was able to evaluate:

- Real-time pricing scenarios
- Tariff pricing scenarios
- Appliance control strategies
- Power and cost savings

Overall the Low Cost Power Advisor was successful in providing a proof of concept solution to reduce homeowner power consumption following British Columbia Hydroelectric Company guidelines [4]. The LCPA provides the AFRESH home owner insight of:

- Current power consumption
- Detailed consumption list of individual appliances
- Appliance record of total and high-tariff period activations with associated extra cost
- User settable consumption alarm set points
- Consumption trend

24.9 Conclusion

This paper discussed the development of energy performance analysis model of the AFRESH home on BCIT Burnaby campus, summarizing the existing building systems, developing a calibrated whole building energy performance model based on the building's current usage as an office space, estimating the building's energy performance when used as a single-family house, reviewing the building's current alternative energy production, and identifying strategies for improvements. AFRESH home achieves its goal of being an affordable, high-performing home. However, there is currently a considerable amount of "untapped" power generation potential that could be utilized to further reduce the home's reliance on grid energy.

In addition to energy analysis, a Low Cost Power Advisor was built and tested on the house. The system was successful in providing a proof of concept solution to reduce homeowner power consumption. LCPA is a prototype solution for a consumer that clearly and simply demonstrates the energy savings that can be achieved using the simulator by adjusting power-consuming habits or implementing automated control with the appliance. It will be necessary to have this tool for consumers to reliably make positive long-term decisions regarding their energy-consuming habits.

References

1. Palizban HA, Jappy A, Morrissey C, Tigor W (2011) "A prototype Home Energy Management System". International conference on condition monitoring, diagnosis and maintenance - CMDM 2011
2. Project report by Nichole Wapple and Michael Meslin (students) for BSCI9130; a Master-level course taught and supervised by M. Tabatabaian. April 2013 (with permission)
3. Course notes of Robert Schubak (2012) for effective structural modeling and modified for energy modelling
4. BC Hydro (2010) New Act powers B.C. forward with clean energy and jobs, April 28, 2010. https://www.bchydro.com/news/press_centre/news_releases/2010/new_act_powers_bc_forward.html
5. British Columbia Institute of Technology, Burnaby, BC, Canada. WWW.bcit.ca

Chapter 25
Shading Performance on Terraced House Facade Designs in Malaysia

Ahmad Sanusi Hassan, Yasser Arab, and Mohammed Salem Obaid Bakhlah

Abstract This study analyses shading performance on front house facades in Malaysia designed with early, modern, postmodern and neo-minimalist architectural style. For the case study, four front facades of the terraced houses are selected. The reason for the selection of terraced houses as the case studies is that terraced houses are the most popular house type built in urban areas in this country. The Early Modern Terraced House Style was built in 1950s to 1970s, and Modern Terraced House Style was popularly built later in 1980s to 1990s. Postmodern Terraced House Style was commonly integrated in the building design in 2000s while the Neo-Minimalist Terraced House Style has been integrated in the house design since 2010. The selected case studies are located in Petaling Jaya, Putrajaya and Shah Alam, the new towns of Kuala Lumpur, the capital city of Malaysia. The SunTool software is used in the survey to calculate shading percentages on the front house facades. The survey will be conducted at a position when the sun path is perpendicular to the house facade. The study finds that the shading performance is improving over time. Recessed wall, balcony, attached roof and roof overhang are commonly used in the house facade design with car porch on the ground facade level.

Keywords Shading • Facade • Terraced house • Architectural styles

Nomenclature

S_{WA}	Shaded window area (S_{WA})
S_{WH}	Shaded window height (S_{WH})
W_W	Window width (W_W)
S_{OA}	Shaded opaque area (S_{OA})
S_H	Shaded opaque height (S_{OH})
F_L	Facade length (F_L)

A.S. Hassan (✉) • Y. Arab • M.S.O. Bakhlah
School of Housing, Building and Planning, Universiti Sains Malaysia, 11800 USM, Penang, Malaysia
e-mail: sanusi@usm.my

I. Dincer et al. (eds.), *Progress in Clean Energy, Volume 2*,
DOI 10.1007/978-3-319-17031-2_25

E_{OA}	Exposed opaque area (E_{OA})
T_{OA}	Total opaque area (T_{OA})
F_H	Facade height (F_H)
T_{WA}	Total window (glazing) area (T_{WA})
W_H	Window height (W_H)
$W_N\&_T$	Window number and type ($W_N\&_T$)
E_{WA}	Exposed window area (E_{WA})

25.1 Introduction

This study discusses results of shading area on house facades of selected terraced houses designed with different architectural styles built from 1950s to present. Terraced house types are selected for the case studies because they are the most popular house types built in urban area in Malaysia representing more than 43 % of the total house units in 2000 [1]. The finding of this study is able to guide the architects with information on awareness to design terraced house facades with excellent sun shading elements. Design faults and lack of consciousness about the importance of shading elements by the architects when designing the house facade are among the major reasons of this poor design. With poor design, the house facade will be exposed to direct sunlight. The benefit of this study is the providing of empirical findings and contributions which lead to design recommendations on terraced house facade design as one of the important considerations. In a tropical region like Malaysia, excellent facade design to avoid intensity of solar radiation is necessary [2]. House facade exposed to direct sunlight causes problem of solar radiation. The sun energy will reradiate the heat from outside wall transmitted to the interior of the house [3, 4]. It generates extra heat gains inside the house which causes warm temperature to the indoor area. As a result, this creates uncomfortable thermal condition to the occupants. The objectives of this study are:

- To measure the level of shading performance on four different architectural styles of front facade terraced houses in Malaysia
- To identify the types of shading elements which provide excellent shade on the house facade

25.2 The Case Studies

The case studies consist of a survey on front facades of terraced houses built in four different periods which typify atypical design in Malaysia, namely the early modern style in 1950s, modern style in 1980s, postmodern style in 2000s and neo-minimalist style in 2010s. All the selected houses are located in new towns near Kuala Lumpur. The good examples of the early and modern terraced houses are in Petaling Jaya which is the first Garden City new town built in Malaysia.

The case studies of the postmodern style are selected in Putrajaya which is the latest new town and currently functioned as an administrative city for the federal government [5, 6]. Finally the selected case studies of neo-minimalist style are located in one of the newly garden housing estates in the existing new town of Shah Alam. The summary of design and addresses of selected terraced houses for the case studies are as follows:

1. The Early Terraced Houses (Fig. 25.1) were either built by the British authority slightly before the country's independence or local authority to house the government officers and city population from 1950 to 1970 [5]. The style typifies model village concept with simplified cottage style based on terraced house design and its site planning by Raymond Unwin and Barry Parker who built the first Garden City new town, Letchworth, in 1903 near London, England [7].
   (a) Case Study A: No. 2, 6/30 Street, Section PJS6, 46000, Petaling Jaya.
   (b) Case Study B: No. 9, 3/57D Street, Section PJS3, 46000, Petaling Jaya.
2. Modern Terraced House Style (Fig. 25.2) typifies simple geometric design influenced from modern architecture during Industrial Age [8, 9] with reference from a development of simplified cottage style in the Early Modern Terraced House Style.
   (a) Case Study A: No. 17, SS1/34 Street, Section SS1, 47300 Petaling Jaya.
   (b) Case Study B: No. 25, SS2/43 Street, Section SS2, 47300 Petaling Jaya.

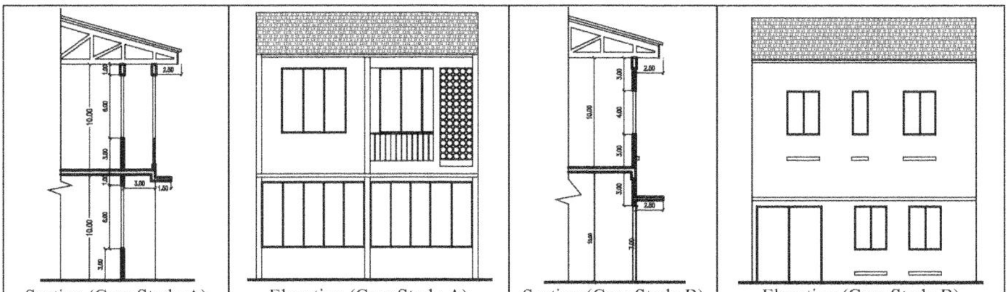

Section (Case Study A) | Elevation (Case Study A) | Section (Case Study B) | Elevation (Case Study B)

Fig. 25.1 Section and elevation in the Case Study A and B of the early terraced house facade

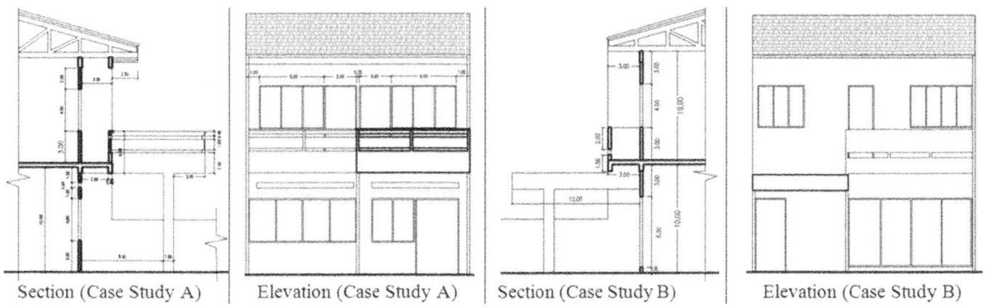

Section (Case Study A) | Elevation (Case Study A) | Section (Case Study B) | Elevation (Case Study B)

Fig. 25.2 Section and elevation in the Case Study A and B of the modern terraced house facade

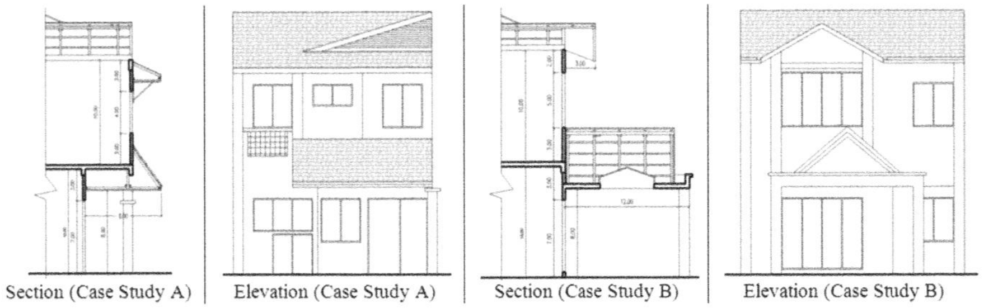

Fig. 25.3 Section and elevation in the Case Study A and B of the postmodern terraced house facade

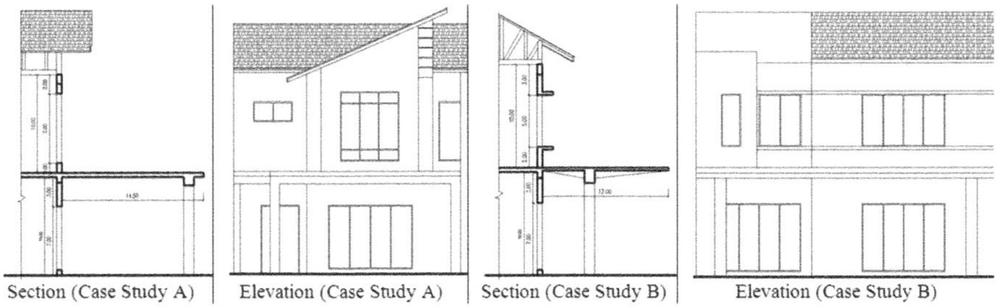

Fig. 25.4 Section and elevation in the Case Study A and B of the neo-minimalist terraced house facade

3. Postmodern Terraced House Style (Fig. 25.3) has architectural mixture of modern style with Palladian Villa, Mediterranean or Traditional Malay style. The design has complex geometric elements blending of modern, colonial and traditional styles with colourful paints on the house facade [10]. Its design deviates from regular and simple composition of modern architectural style which emphasises an expression of architectural simplicity with white colour scheme [11].

   (a) Case Study A: No. 12, Jalan P9D3 Street, Presint 9, 62250 Putrajaya.
   (b) Case Study B: No. 25, Jalan P16D2 Street, Presint 16, 62150 Putrajaya.

4. Neo-Minimalist Terraced House (Fig. 25.4) is an architectural style which typifies a style after 2010s with integration of modern design with complex simplified geometry with white and grey tones of colour facade. It has also known as "neo-geometric" or "neo-geo" art's concept [12].

   (a) Case Study A: Arabella Type A, Section 13, Shah Alam (under construction).
   (b) Case Study B: Marbella Type A, Section 13, Shah Alam (under construction).

25.3 Material and Methods

25.3.1 Computer Simulation

The SunTool software will be used in the survey to calculate the percentage of shading area on the selected front house facades. The survey will be conducted at a position when the sun path is perpendicular to the house facade during morning (east) and evening (west) session in each of the case studies. The reason having perpendicular orientation is that terraced houses are mass-produced house type built at a position of various orientations. In this study, the day time at which the sun path perpendicular to the front house facade will be used in order to generate the results when the house facade has been perpendicularly exposed to direct sunlight. By having this method, the survey will be able to do comparative analysis identifying the effectiveness of shading design on the house facades [13, 14]. Limitation of this survey is that the position of the sun path changes over time. In order to get perpendicular angle of the sun (sun path's azimuth) to the east (90°) and west (270°), the data were calculated using the SunTool software. Time and date when the sun paths were perpendicular to the house facade are illustrated in Table 25.1 and Fig. 25.5. The other limitation is that there are certain times and dates that the sun path's azimuths were not possible to have been perfectly at 90° [15, 16]. In these cases, the closest azimuths nearest to 90° will be used when the simulations are made from 7 a.m. to 6 p.m. (daytime hours) as in Table 25.1.

All data such as the location, facade orientation (east or west), time and date will be keyed in the solar position calculator in the SunTool programme (Fig. 25.6) in order to get the correct position before the simulation is made. Later, dimensions of the house facade which are the depth of exterior shading devices, floor height, wall width and sill height will be keyed in the SunTool programme. With these solar position and dimensions of the house facade, the programme will be able to generate in its drawing section to show the sun beam and shade of the house facade which provide the results of the percentage of shading area [17].

25.3.2 Calculating Shading Area on House Facade

In order to calculate shading area, the house facades will be divided into two main areas namely opaque and glazing area. The amount of shading area of opaque and glazing elements will be analysed by using the SunTool programme. The shading area will be converted to percentage of the total facade area for comparative analysis. The amount of shading area for the window (glazing), opaque and total (window and opaque) area as indicated in Fig. 25.7 will be calculated based on the following formula [18]:

Table 25.1 Time, date and azimuth of the sun when the simulations were generated

Orientation	Time	Date	Azimuth	Orientation	Time	Date (2013)	Azimuth
East 90°	7 a.m.	23 March	90°	West 270°	1 p.m.	16 September	90.5°
	8 a.m.	25 March	90°		2 p.m.	29 March	89.8°
	9 a.m.	27 March	89.8°		3 p.m.	18 September	89.8°
	10 a.m.	28 March	90.1°		4 p.m.	26 March	89.9°
	11 a.m.	29 March	90°		5 p.m.	24 March	89.9°
	12 p.m.	29 March	92.2°		6 p.m.	22 March	89.9°

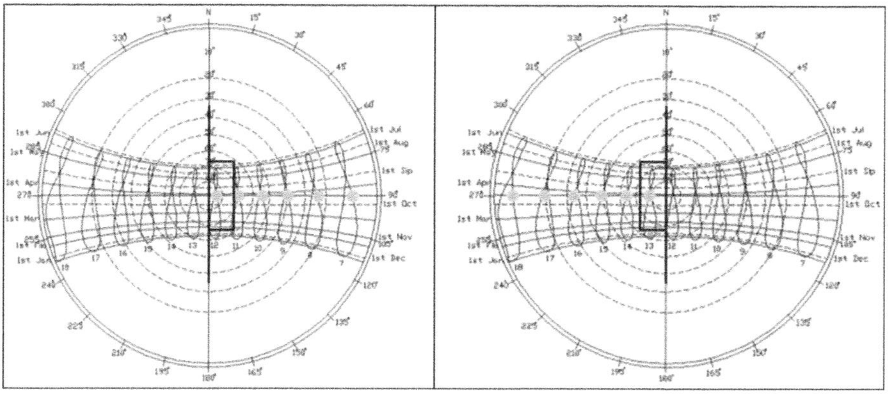

Fig. 25.5 Sun path diagrams show the position of the sun perpendicular to the house facade from 7 a.m. to 12 p.m. at east orientation (*left*) and from 1 p.m. to 6 p.m. at west orientation (*right*) during daytime hours. *Source*: SunTool software

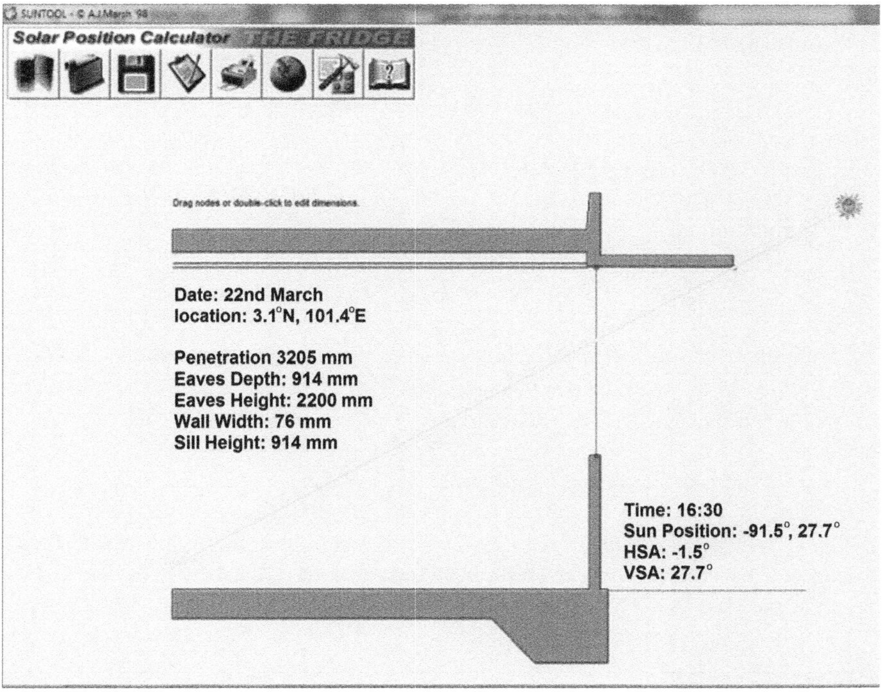

Fig. 25.6 The extent of sunlight penetration through window section. *Source*: SunTool software

1. Window (glazing) area of the house facade under shade is measured as shaded window area (S_{WA}) which is the area defined as shaded window height (S_{WH}) by window width (G_W). The equation is

$$S_{WA} = S_{WH} \times W_W$$

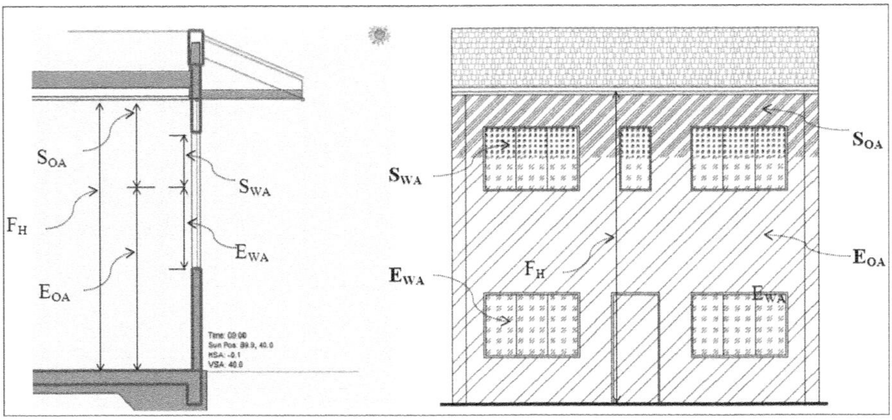

Fig. 25.7 Illustration of the house facade and section in calculation of shading area

2. Opaque (wall) area under shade is measured as shaded opaque area (S_{OA}) which is the area defined as shaded opaque height (S_{OH}) by length of the facade (F_L) minus shaded window area (S_{WA}). The equation is

$$S_{OA} = S_H \times F_L - S_{WA}$$

3. Opaque (wall) area exposed to the sunlight is defined as exposed opaque area (E_{OA}) which is total opaque area (T_{OA}) minus shaded opaque area (S_{OA}). The equation is

$$E_{OA} = T_{OA} - S_{OA}$$

4. Total opaque (wall) area of the house facade is defined as total opaque area (T_{OA}) which is the area defined as facade height (F_H) by facade length (F_L) minus total window area (T_{WA}). The equation is

$$T_{OA} = F_H \times F_L - T_{WA}$$

5. Total window (glazing) area (T_{WA}) is all window area of the house facade defined as window height (W_H) by window width (W_W) by window number and type ($W_{N\,\&T}$). The equation is

$$T_{WA} = W_H \times W_W \times W_{N\&T}$$

6. Exposed window (glazing) area exposed to the sunlight is defined as exposed window area (E_{WA}) which is total window area (T_{WA}) minus shaded window area (S_{WA}). The equation is

$$E_{WA} = T_{WA} - S_{WA}$$

25.4 Results and Discussion

The analysis compares results of shading area on the front house facade at the ground floor level (storey 1) and first floor level (storey 2) for each of the case studies. The scales of measurement are divided into four categories namely Category 1 from 0 to 25 % as very weak, Category 2 from 26 to 50 % as weak, Category 3 from 50 to 75 % as good and Category 4 above 75 % as excellent [17]. Table 25.2 shows percentages of the shading area of the house facades in the Case Studies from results of the computer simulation, which will be used in the analysis.

25.4.1 Early Terraced House Style

Tables 25.3 and 25.4 and Fig. 25.8 show the percentages of shading area for the two selected case studies of Early Terraced House Style. The results illustrate that in the Case Study A, shading area of storey 1 at 7:00 a.m. and 8:00 a.m. was under Category 1 in the scales of measurement with 9.36 % and 17.08 %, respectively. At 9:00 and 10:00 a.m., the shading areas were classified under Category 2 with 26.95 % and 43.26 % respectively. At 12:00 p.m., the facade had an excellent shading performance with 100 % shading area while at 1:00 p.m. its shading area slightly declined to 78.94 %. The results show that at 2:00 p.m., the shading area dropped to Category 3 with 64.06 %. Finally in the last 3 h, shading area was under Category 1 from 23.24 % at 4:00 p.m., 12.55 % at 5:00 p.m. to 3.54 % at 6:00 p.m. For the Case Study B, its shading area of storey 1 fell under Category 1 for the first 2 h at 7:00 and 8:00 a.m. before it had a gradual increase to Category 2 at 9:00 a.m. and Category 3 at 10:00 a.m. Shading area at 11:00 a.m. was under Category 4 with 84.82 %. At 12:00 and 1:00 p.m., shading area accounted 100 % before it had a steady decline to Category 3 with 73.98 % at 2:00 p.m., Category 2 with 40.67 % at 3:00 p.m. and Category 1 at 4:00 p.m., 5:00 p.m. and 6:00 p.m. with 24.66 %,

Table 25.2 Total facade area (percentages of the shading area of the house facades in the Case Studies from results of the computer simulation, which will be used in the analysis)

Cases		Total area (m^2)	Window (m^2)	Opaque (m^2)	Storey 1		Storey 2	
					Window (m^2)	Opaque (m^2)	Window (m^2)	Opaque (m^2)
1950s	Case A	38.5	16.7	21.7	11.1	8.1	5.6	13.7
	Case B	35.6	4.5	21.8	2.2	14.9	2.2	16.3
1980s	Case A	36.1	10.6	25.5	6.1	13.3	4.5	12.3
	Case B	40.6	11.1	29.5	7.2	13	3.9	16.4
2000s	Case A	37.2	6.9	30.3	3.2	15.4	3.7	14.9
	Case B	40.9	11.5	29.4	6.3	14.1	5.2	15.2
2010s	Case A	47.3	11.1	36.2	5.2	20.4	5.9	15.8
	Case B	53.5	16.1	37.4	9.7	17.1	6.4	20.3

Table 25.3 Shading percentages of the Case Study A (Early Terraced House Style)

Shading area								
		Storey 1		Storey 2		Total		
Facade	Time	Window %	Opaque %	Window %	Opaque %	Storey 1 %	Storey 2 %	Total shade
East 90	7:00 a.m.	3.41	15.31	0	12.81	9.36	6.40	7.88
	8:00 a.m.	17.08	17.08	12.6	28.54	17.08	20.57	18.82
	9:00 a.m.	36.16	17.74	36.03	46.03	26.95	41.03	33.99
	10:00 a.m.	64.75	19.78	66.5	76.02	42.26	71.26	56.76
	11:00 a.m.	82.16	54.67	100	77.86	68.41	88.93	78.67
West 270	12:00 p.m.	100	100	100	100	100	100	100
	1:00 p.m.	100	57.87	100	100	78.93	100	89.46
	2:00 p.m.	73.75	54.38	83.5	72.95	64.0658	78.22	71.14
	3:00 p.m.	48.08	18.15	50.66	56.98	33.11	53.82	43.47
	4:00 p.m.	29	17.49	34.06	27.09	23.24	30.57	26.91
	5:00 p.m.	12.25	16.91	13.8	11.31	14.58	12.55	13.56
	6:00 p.m.	0.33	13.67	0	7.09	7.01	3.54	5.27

Table 25.4 Shading percentages of the Case Study B (Early Terraced House Style)

Shade area 1950s case B								
		Storey 1		Storey 2		Total		
Facade	Time	Window %	Opaque %	Window %	Opaque %	Storey 1 %	Storey 2 %	Total shade %
East 90	7:00 a.m.	0	5.76	0	5.27	2.88	2.63	2.75
	8:00 a.m.	16.75	12.47	0	13.94	14.61	6.97	10.79
	9:00 a.m.	40	20.9	0	25.18	30.45	12.59	21.52
	10:00 a.m.	76.75	34.22	20	39.89	55.48	29.94	42.71
	11:00 a.m.	100	69.65	100	65.2	84.82	82.6	83.71
West 270	12:00 p.m.	100	100	100	100	100	100	100
	1:00 p.m.	100	100	100	100	100	100	100
	2:00 p.m.	94.5	53.47	73.12	52.52	73.98	62.82	68.4
	3:00 p.m.	55	26.34	0	31.98	40.67	15.99	28.33
	4:00 p.m.	31.5	17.82	0	20.96	24.66	10.48	17.57
	5:00 p.m.	10.75	10.3	0	11.01	10.52	5.5	8.01
	6:00 p.m.	0	3.2	0	2.92	1.6	1.46	1.53

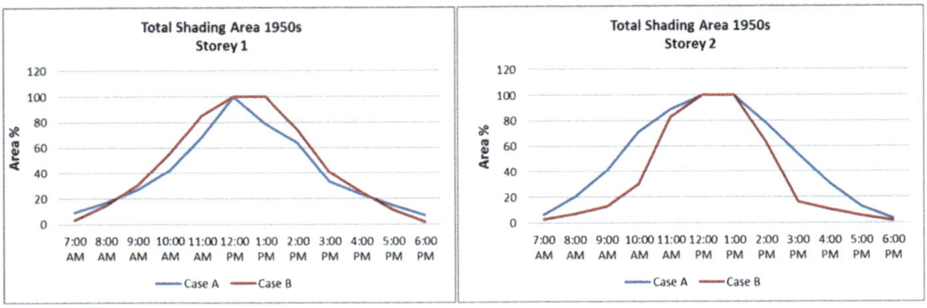

Fig. 25.8 Shading percentages of the Case Study A and B, the Early Terraced House Style. Storey 1 (*left line chart*), storey 2 (*right line chart*)

10.52 % and 1.6 %, respectively. Storey 1 in the Case Study B had slightly better overall shading performance (Fig. 25.7) than that in the Case Study A.

For storey 2 in the Case Study A, it had shading area at 7:00 a.m. and 8:00 a.m. under Category 1 with 6.4 % and 20.57 %, respectively. It had a gradual incline from Category 2 at 9:00 a.m. with 41.03 % to Category 3 at 10:00 a.m. with 71.26 %. From 11:00 a.m. to 2:00 p.m. the shading areas were under Category 4 with 88.93 % at 11.00 a.m. and 78.22 % at 2:00 p.m., and 100 % from 12:00 to 1:00 p.m. The shading percentages had a gradual decline at the last 4-h daytime from Category 3 at 3:00 p.m. with 53.82 % and Category 2 at 4:00 p.m. with 30.57 % to Category 1 at 5:00 p.m. and 6:00 p.m. with 12.55 % and 3.54 %, respectively. Compared to storey 2 in the Case Study A, storey 2 in the Case Study B had low percentages of shading area. It had lower percentages of shading area than Case A with less than 25 % at the first 3 h and last 4 h during the daytime. Its shading percentages were under Category 1 from 7:00 a.m. (2.63 %) to 9:00 a.m. (12.59 %) during morning hours and from 3:00 p.m. (15.99 %) to 6:00 p.m. (1.46 %) during evening hours. At 10:00 a.m., the shading area was under Category 2 with 29.94 %; however, at 11.00 p.m. it had a dramatic increase to Category 4 with 82.6 and 100 % at 12:00 to 1:00 p.m. At 2.00 p.m., the shading area dropped to Category 3 with 62.82 %, and after that, its percentage had a sudden decline to 15.99 % (Category 1) at 3:00 p.m. The reason for this sudden incline and drop factor is that the Case Study B does not have a recessed wall design on its facade only fitted with roof overhang as a shading element. Storey 2 in the Case Study A has good recessed wall design with a balcony as sun shading element which contributes better shading result in contrast to storey 2 in the Case Study B.

25.4.2 Modern Terraced House Style

The results are shown in Tables 25.5 and 25.6 and Fig. 25.9 which shows that both storey 1 and 2 of the house facade in the Case Study A had higher shading percentages than the Case Study B most of the time except during afternoon hours.

Table 25.5 Shading percentages of the Case Study A (Modern Terraced House Style)

Shade area 1980s case A								
		Storey 1		Storey 2		Total		
Facade	Time	Window %	Opaque %	Window %	Opaque %	Storey 1 %	Storey 2 %	Total shade %
East 90	7:00 a.m.	7.57	23.47	0	21.75	15.52	10.87	13.2
	8:00 a.m.	34.60	53.45	12.7	37.01	44.03	24.85	34.44
	9:00 a.m.	43.93	91.45	65.31	43.62	67.69	54.46	61.08
	10:00 a.m.	71.36	92.12	100	73.18	81.74	86.59	84.16
	11:00 a.m.	100	92	100	100	96	100	98
West 270	12:00 p.m.	100	97.96	100	100	98.98	100	99.49
	1:00 p.m.	100	97.96	100	100	98.98	100	99.49
	2:00 p.m.	100	91.82	100	100	95.91	100	97.95
	3:00 p.m.	55	91.57	95.93	47.03	73.28	71.48	72.38
	4:00 p.m.	39.39	89.33	44.97	40.71	64.36	42.84	53.60
	5:00 p.m.	23.03	43.84	0	34.16	33.43	17.08	25.26
	6:00 p.m.	3.78	15.31	0	21.43	9.54	10.71	10.13

Table 25.6 Shading percentages of the Case Study B (Modern Terraced House Style)

Shade area 1980s case B		Storey 1		Storey 2		Total		
Facade	Time	Window %	Opaque %	Window %	Opaque %	Storey 1 %	Storey 2 %	Total shade %
East 90	7:00 a.m.	0	23.97	0	21	11.98	10.5	11.24
	8:00 a.m.	0	54.58	0	29.41	27.29	14.7	20.99
	9:00 a.m.	2.76	93.38	0	36.83	48.07	18.41	33.24
	10:00 a.m.	30.61	94.07	34.25	44.95	62.34	39.6	50.97
	11:00 a.m.	93.84	93.95	100	78.42	93.89	89.21	91.55
West 270	12:00 p.m.	100	100	100	100	100	100	100
	1:00 p.m.	100	100	100	100	100	100	100
	2:00 p.m.	62.92	93.76	85.5	57.69	78.34	71.59	74.97
	3:00 p.m.	13.84	93.51	7.25	39.96	53.68	23.6	38.64
	4:00 p.m.	0	91.22	0	25.99	45.61	12.99	29.3
	5:00 p.m.	0	44.77	0	27.63	22.38	13.81	18.1
	6:00 p.m.	0	15.63	0	19.66	7.81	9.83	8.82

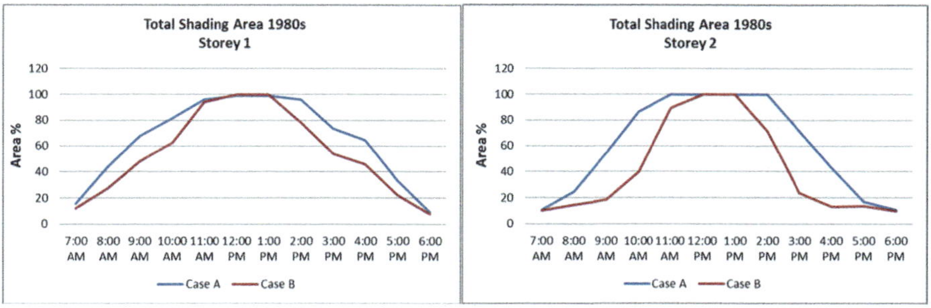

Fig. 25.9 Shading percentages of the Case Study A and B, the modern terraced house style. Storey 1 (*left line chart*), storey 2 (*right line chart*)

Storey 1 in the Case Study A had shading area started with Category 1 at 7:00 a.m. with 15.52 %. Shading area of storey 1 in the Case Study A was under Category 2 with 44.03 % at 8:00 a.m. and inclined to Category 3 at 9:00 a.m. with 67.69 %. From 11:00 a.m. to 2:00 p.m., the percentages were under Category 4 with almost 100 % shade of the front facade. The percentages had a gradual decrease from 3:00 to 6:00 p.m. with 73.28 % (Category 3), 64.36 (Category 3) and 33.43 % (Category 2) to 9.54 % (Category 1). Storey 1 of the Case Study B had shading area started with 11.98 %. The percentage later had a steady increase to 27.29 % at 8:00 a.m., 48.07 % at 9:00 a.m. and 62.34 % at 10:00 a.m. The facade had excellent shade with 93.89 % at 11:00 a.m., followed by 100 % at 12:00 and 1:00 p.m. After 1:00 p.m., the shading area had gradually dwindled to 78.34 % at 2:00 p.m., 53.68 % (Category 3) at 3:00 p. m., 45.61 % (Category 2) at 4:00 p.m., 22.38 % at 5:00 p.m. and 7.81 % (Category 1) at 6:00 p.m.

Storey 2 had slightly lower shading performance than storey 1 in both of the case studies. Shading area on a house facade of storey 2 in the Case Study A was 10.87 % (Category 1) at 8:00 p.m. It had a steady increase to 24.85 % at 9:00 a.m. and 54.46 % at 10:00 a.m. From 10:00 a.m. to 2:00 p.m., the facade had excellent shade (Category 4) with 86.59 % at 10:00 a.m. and 100 % from 11:00 a.m. to 2:00 p.m. After that, the shading area had dwindled to 71.48 % (Category 3) at 3:00 p.m., 42.84 % (Category 2) at 4:00 p.m., 17.08 % (Category 1) at 5:00 p.m. and 10.71 % at 6:00 p.m. Storey 2 in the Case Study B had shading percentage with 10.5 % at 7:00 a.m. and had a gradual incline with 14.7 % at 8:00 a.m., 18.41 % (Category 1) at 9:00 a.m. and 39.6 % (Category 2) at 10:00 a.m. Shading area had a dramatic increase at 11:00 a.m. with 89.21 %. Storey 2 had excellent shading area with 100 % at 12:00 and 1:00 p.m. but its shading area had dwindled to 71.59 % (Category 3) at 2:00 p.m. and sudden drop to 23.6 % (Category 2) at 3:00 p.m. Later, it had a steady decrease to 12.99 % (Category 1) at 4:00 p.m., 13.81 % at 5:00 p.m. and 9.83 % at 6:00 p.m. Unlike the Case Study A, storey 2 in the Case Study B had a sudden incline at 10:00 a.m. and decline at 3:00 p.m., about 50 % difference because its overhang roof has the same width (3 ft) as its cantilevered balcony.

25.4.3 Postmodern Terraced House Style

Tables 25.7 and 25.8 and Fig. 25.10 show the results of shading area in percentages in the Case Study A and B. The analysis finds that both the case studies of Postmodern Terraced House Style had the smallest number of shading percentages with 100 % compared to other terraced house styles. Front facade of storey 1 in the Case Study B had higher overall shading percentages than that in the Case Study A. Storey 1 in the Case Study A had shading area with 15.56 % at 7:00 a.m. followed by a gradual increase of shading percentage with 36.29 % at 8:00 a.m., 56.10 % at 9:00 a.m., 64.36 % at 10:00 a.m. and 67.89 % at 11:00 a.m. Shading area at 12:00 p.m. and 1:00 p.m. fell under Category 4 with 87.45 % and 80.09 %, respectively. However, shading percentage had gradually dwindled to 66.18 % at 2:00 p.m., 62.59 % (Category 3) at 3:00 p.m., 48.10 % (Category 2) at 4:00 p.m., 29.17 % at 5:00 p.m. and 11.35 % (Category 1) at 6:00 p.m.. Storey 1 in the Case Study B had shading area under Category 2 with 26.46 % at 7:00 a.m. and Category 3 with 57.84 % at 8:00 a.m. It had excellent shading percentage under Category 4 from 10:00 a.m. to 3:00 a.m. At 4:00 p.m., the shading area had a steady decrease from 74.49 % (Category 3) at 4:00 p.m. and 47.23 % (Category 2) at 5:00 p.m. to 17.78 % (Category 1) at 6:00 p.m.

The analysis also finds that front house facade of storey 2 in the Case Study A had better overall shading performance than that in the Case Study B. The shading percentage in the Case Study A was 24.72 % at 7:00 a.m., 38.06 % at 8:00 a.m. and 63.67 % at 9:00 a.m.. The shading percentage was excellent (Category 4) from 10.00 a.m. to 2:00 p.m. From 3:00 to 6:00 p.m., it had a gradual decrease from 72.43 % (Category 3), 50 and 33.35 % (Category 2) to 22.69 % (Category 1). Storey 2 in the Case Study B had shading area started with 9.39 % at 7:00 a.m., 12.6 % at 8:00 a.m., 20.76 % at 9:00 a.m. and 42.71 % at 10:00 a.m. The percentage had a sudden incline at 11:00 a.m. with 87.81 % (Category 4), and was 100 % at 12:00 and 1:00 p.m. At 2:00 p.m., the shading area declined to 61.75 % (Category 3) but at 3:00 p.m., it had a sudden drop to 28.47 % (Category 2). For the last 3 h of the daytime, the percentage had declined to 17.37, 11.66 and 7.74 %.

25.4.4 Neo-Minimalist Terraced House Style

Tables 25.9 and 25.10 and Fig. 25.11 show the results of shading percentages of Neo-Minimalist Terraced House Style. Both the case studies had excellent (Category 4) shading percentages on storey 1 front house facade from 9:00 a.m. to 4:00 p.m. They have wide car porch which gives shade to the facade of storey 1. Storey 1 house facade in the Case Study B had slightly higher overall shading percentages than that in the Case Study A. In the Case Study A, storey 1 had a shading area with 16.5 % at 7:00 a.m., followed by 59.72 % (Category 3) at 8:00 a.m. Similar pattern occurred in the evening hours with 43.89 % at 5:00 p.m. before it

Table 25.7 Shading percentages of the Case Study A (Postmodern Terraced House Style)

Shade area 2000s case A								
		Storey 1		Storey 2		Total		
Facade	Time	Window %	Opaque %	Window %	Opaque %	Storey 1 %	Storey 2 %	Total shade %
East 90	7:00 a.m.	4.7	26.43	0.6	48.84	15.56	24.72	20.14
	8:00 a.m.	34.7	37.87	19.05	57.08	36.29	38.06	37.17
	9:00 a.m.	56.23	55.96	65.1	62.25	56.1	63.67	59.89
	10:00 a.m.	64.7	64.02	84.575	77	64.36	80.79	72.57
	11:00 a.m.	64.7	71.08	100	89.22	67.89	94.61	81.25
West 270	12:00 p.m.	91.17	83.73	100	93.75	87.45	96.87	92.16
	1:00 p.m.	83.67	76.5	100	93.75	80.09	96.87	88.48
	2:00 p.m.	64.7	67.66	100	88.55	66.18	94.27	80.23
	3:00 p.m.	60.35	64.83	78.25	66.61	62.59	72.43	67.51
	4:00 p.m.	52.35	43.85	41.175	58.83	48.1	50	49.05
	5:00 p.m.	24.05	34.28	12.225	54.48	29.17	33.35	31.26
	6:00 p.m.	0	22.71	0	45.39	11.35	22.69	17.02

Table 25.8 Shading percentages of the Case Study B (Postmodern Terraced House Style)

Shade area 2000s case B								
		Storey 1		Storey 2		Total		
Facade	Time	Window %	Opaque %	Window %	Opaque %	Storey 1 %	Storey 2 %	Total shade %
East 90	7:00 a.m.	14.82	38.10	0	18.79	26.46	9.39	17.93
	8:00 a.m.	62.47	53.22	0	25.21	57.84	12.6	35.22
	9:00 a.m.	89.23	68.13	4.91	36.62	78.68	20.76	49.72
	10:00 a.m.	97.26	69.32	34.71	50.7	83.29	42.71	63
	11:00 a.m.	100	77.63	99.46	76.17	88.81	87.81	88.31
West 270	12:00 p.m.	100	100	100	100	100	100	100
	1:00 p.m.	100	87.96	100	100	93.98	100	96.99
	2:00 p.m.	100	71.6	64.35	59.14	85.8	61.75	73.77
	3:00 p.m.	92.32	68.59	14.01	42.93	80.45	28.47	54.46
	4:00 p.m.	86.32	62.66	6.5	28.24	74.49	17.37	45.93
	5:00 p.m.	45.97	48.49	1.07	22.26	47.23	11.66	29.45
	6:00 p.m.	1.98	33.59	0	15.48	17.78	7.74	12.76

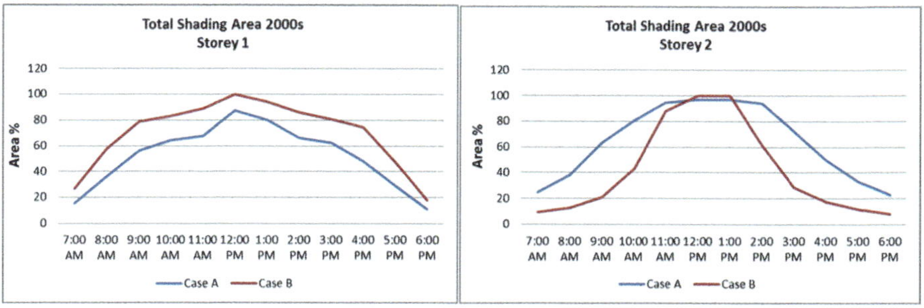

Fig. 25.10 Shading percentages of the Case Study A and B, postmodern terraced house style. Storey 1 (*left line chart*), storey 2 (*right line chart*)

declined to 10.5 % at 6:00 p.m. In the Case Study B, storey 1 had a shading percentage of 24.77 %. In contrast to storey 1 in the Case Study A, the shading percentage was excellent (Category 1) with 81.43 % at 8:00 a.m. At 5:00 p.m., it was 63.4 %, and had suddenly dwindled to 13.2 % at 6:00 p.m.

Storey 2 in both the Case Study A and B had lower shading percentages than storey 1. It has poor shading designed only with roof overhangs. Shading percentage of storey 2 in the Case Study A started with 1.85 % at 7:00 a.m. It had a gradual increase to 5.27 % at 8:00 a.m., 10.8 % at 9:00 a.m. and 30.2 % at 10:00 a.m. From 11:00 a.m. to 1:00 p.m., the shading area had excellent percentages (Category 4). The percentages had a steady decline from 40.85 % (Category 2) at 2:00 p.m., 15.27 % (Category 1) at 3:00 p.m., 8.42 % at 4:00 p.m. and 4.12 % at 5:00 p.m. to 2.02 % at 6:00 p.m. Storey 2 in the Case Study B had shading area with 5.88 % at 7:00 a.m., 12.99 % at 8:00 a.m., 24.25 % (Category 1) at 9:00 a.m., 38.05 % at 10:00 a.m. and 71.53 % (Category 3) at 11:00 a.m. It had excellent shading percentage at 12:00 and 1:00 p.m. with slightly above 75 %. Starting at 2:00 p.m., it had gradually dwindled from 49.25 % (Category 2) to 25.36 % at 3:00 p.m., 18.62 % (Category 1) at 4:00 p.m., 10.7 % at 5:00 p.m. and 4.19 % at 6:00 p.m.

Figure 25.12 shows the overall results on shading percentages of all terraced house styles. Storey 1 had better overall shading performance on front house facade than storey 2. Integration of car porch in the design of front house facade created excellent shading area in Modern, Postmodern and Neo-Minimalist Terraced House Style. Only Early Modern Terraced House Style does not have front facade design with a car porch. Neo-Minimalist Terraced House Style had the best shading percentage on storey 1 followed by Postmodern and Modern Terraced House Styles because it had the widest car porch design covering its ground facade level (storey 1). Postmodern and Modern Terraced House Styles do not have wide cover with a car porch but they have integrated a car porch with recessed wall design on the ground level of the house facade. For storey 2, Postmodern Terraced House Style had the best design with shading elements followed by Modern Terraced House Style and Early Modern Terraced House Style. Neo-Minimalist Terraced House Style had the worst shading performance on front facade of Storey

Table 25.9 Shading percentages of the Case Study B (Neo-Minimalist Terraced House Style)

Shade area 2010s case A								
		Storey 1		Storey 2		Total		
Facade	Time	Window %	Opaque %	Window %	Opaque %	Storey 1 %	Storey 2 %	Total shade %
East 90	7:00 a.m.	2.14	30.86	0	3.702	16.5	1.85	9.17
	8:00 a.m.	63.07	56.37	0	10.54	59.72	5.27	32.49
	9:00 a.m.	100	81.9	0	21.6	90.95	10.8	50.87
	10:00 a.m.	100	94.64	22.01	38.39	97.32	30.2	63.76
	11:00 a.m.	100	100	82.81	82.84	100	82.83	91.41
West 270	12:00 p.m.	100	100	100	100	100	100	100
	1:00 p.m.	100	100	100	97.74	100	98.87	99.43
	2:00 p.m.	100	100	34.31	47.38	100	40.85	70.42
	3:00 p.m.	100	100	3.045	27.50	100	15.27	57.63
	4:00 p.m.	100	77.38	0	16.84	88.69	8.42	48.55
	5:00 p.m.	38.15	49.64	0	8.25	43.89	4.12	24.01
	6:00 p.m.	0	21.01	0	2.02	10.5	2.02	6.26

Table 25.10 Shading percentages of the Case Study B (Neo-Minimalist Terraced House Style)

Shade area 2010s case B								
		Storey 1		Storey 2		Total		
Facade	Time	Window %	Opaque %	Window %	Opaque %	Storey 1 %	Storey 2 %	Total shade %
East 90	7:00 a.m.	4.61	44.93	0.98	10.77	24.77	5.88	15.32
	8:00 a.m.	81.76	81.10	2.72	23.26	81.43	12.99	87.93
	9:00 a.m.	100	100	5.39	43.12	100	24.25	112.12
	10:00 a.m.	100	100	26.26	49.85	100	38.05	119.02
	11:00 a.m.	100	100	76.4	66.66	100	71.53	135.76
West 270	12:00 p.m.	100	100	86.95	66.66	100	76.81	138.4
	1:00 p.m.	100	100	86.95	66.66	100	76.81	138.4
	2:00 p.m.	100	100	43.13	55.38	100	49.25	124.62
	3:00 p.m.	100	100	6.26	44.46	100	25.36	112.68
	4:00 p.m.	100	100	4.05	33.2	100	18.62	109.31
	5:00 p.m.	57.69	69.12	2.14	19.26	63.4	10.7	68.76
	6:00 p.m.	0	26.4	0.52	7.86	13.2	4.19	15.29

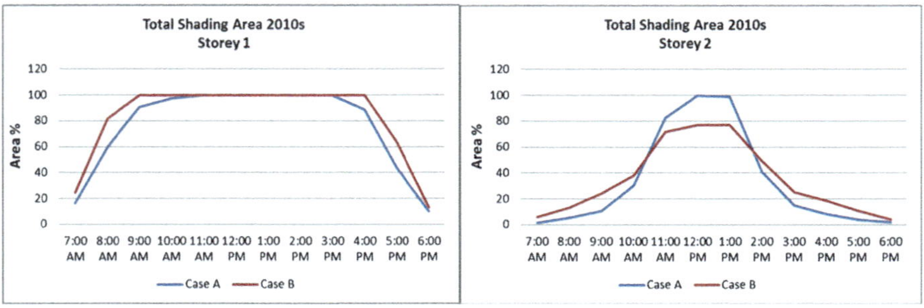

Fig. 25.11 Shading percentages of the Case Study A and B, neo-minimalist terraced house style. Storey 1 (*left line chart*), storey 2 (*right line chart*)

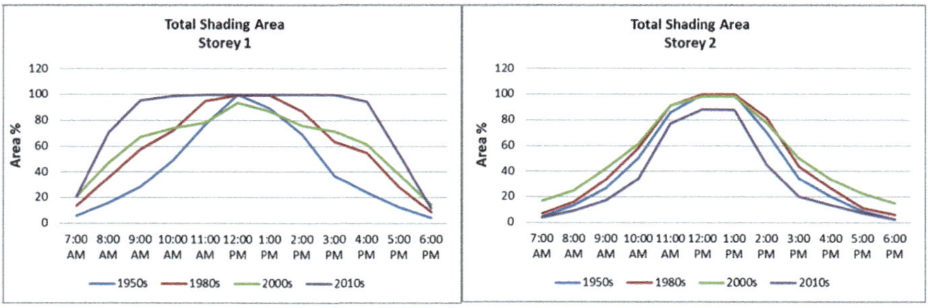

Fig. 25.12 Shading percentages of the Case Study A and B in all front facades of the terraced house styles. Storey 1 (*left*), storey 2 (*right*)

2 because it had only roof overhang as a part of its facade design. Postmodern, Early Modern and Modern Terraced House Styles had a front facade design integrated with roof overhang, attached roof, recessed wall and balcony.

25.5 Conclusion

The study finds that shading performance of the front house facades is poor in the early morning and late evening in all case studies of the terraced house styles except Neo-Minimalist Terraced House Style because the facades were exposed to direct sunlight due to the low angle of the sun position in the sky during these times [19]. However, this study does not concern to early morning hours before 10:00 a.m. due to cool morning temperature below 28 °C [3]. Having exposed to direct sunlight before 10:00 a.m. is permissible in a tropical country like Malaysia as it is morning sunlight. Having exposed human skin to UV radiation from morning sunlight is necessary formation for vitamin D [20]. The most concern hours are from 11:00 a.m. to 4:00 a.m., the time when the front facades are exposed to warm and harsh sunlight. This study finds that all facades in general are well designed with shading devices. Most facades had shading percentages under Category 3 and 4 during these hours. Most facades had 100 % shading at 12:00 to 2:00 p.m.. Car porch, recessed wall, balcony, attached roof and roof overhang are atypical sun shading elements integrated in the front facade design in Malaysia. These devices are very effective sun shading elements preventing solar radiations on the house facade. The study finds that shading performance is improving over time on the ground facade level (storey 1) of the terraced house design, except the facade on the first floor level (storey 2) of Neo-Minimalist Terraced House Style.

Acknowledgement The authors would like to express appreciation for the financial support under Research University Grant by Universiti Sains Malaysia [Grant Number = 1001/PPBGN/816200].

References

1. Department of Statistics Malaysia (2003) General report of the housing census. Department of Statistics Malaysia, Putrajaya
2. Bakhlah MSO, Hassan AS (2012) The study of air temperature when the sun path direction to Ka'abah: with a case study of Al-Malik Khalid Mosque, Malaysia. Int Trans J Eng Manag Appl Sci Technol 3(2):185–202
3. Hassan AS, Ramli M (2010) Natural ventilation of indoor air temperature: A case study of the Traditional Malay House in Penang. Am J Eng Appl Sci 3(3):521–528
4. Feriadi H, Nyuk HW (2004) Thermal comfort for naturally ventilated apartments in Indonesia. Energ Build 36:614–626
5. Hassan AS (2005) Konsep Rekabentuk Bandar di Semenanjung Malaysia: Kuala Lumpur dan Bandar-Bandar di Sekitarnya (Urban Design Concepts in Peninsular Malaysia: Kuala Lumpur and Its Surrounding Towns). Universiti Sains Malaysia, Penang
6. Hassan AS (1999) Putra Jaya: the direction of Malaysian new town, Proceedings of the 5th International Congress of Asian Planning Schools Association (APSA). Seoul: Seoul National University, p 165–175
7. Hall P (1988) Cities of tomorrow. Basil Blackwell, Oxford
8. Hassan AS (2004) Issues in sustainable development of architecture in Malaysia. Universiti Sains Malaysia, Penang
9. Trachtenberg M, Hyman I (1986) Architecture: from prehistory to post-modernism/the western tradition. Harry N. Abrams, New York, NY
10. Klotz H (1998) History of post-modern architecture. MIT Press, Cambridge
11. Curtis WJR (1996) Modern architecture since 1900, 3rd edn. Phaidon, London
12. Foster H (1994) What's Neo about the Neo-Avant-Garde? The Duchamp Effect. 70: 5–32. http://www.jstor.org/stable/779051. Accessed 15 Dec 2013
13. Mazloomi M, Hassan AS, Bagherpour PM, Ismail MR (2010) Influence of geometry and orientation on flank insolation of streets in an arid climate city. Am J Eng Appl Sci 3(3):540–544
14. Hassan AS, Bakhlah MSO (2013) Shading analysis on front facade of modern terraced house type in Petaling Jaya. Malaysia Procedia Soc Behav Sci 91:13–27
15. Hassan AS, Arab Y (2013) The essence of design with light: single pendentive dome mosque in Turkey and Bosnia Herzegovina during winter solstice. In: Hassan AS, Omer S (eds) From

Anatolia to Bosnia: perspectives on pendentive dome mosque architecture. Universiti Sains Malaysia Press, Penang
16. Arab Y, Hassan AS (2012) Daylighting analysis of pendentive dome's mosque design during summer solstice with case studies in Istanbul, Turkey. Int Trans J Eng Manag Appl Sci Technol 3(2):167–183
17. Hassan AS, Bakhlah MSO (2013) Façade analysis on sunlight penetration and shading in post modern terraced houses, Malaysia. J Selcuk Univ Nat Appl Sci 1:310–320
18. Hassan AS, Bakhlah MSO (2013) Shading and extent of sunlight penetration on house facades of the early terraced house type in Petaling Jaya Old Town, Malaysia. Int Trans J Eng Manag Appl Sci Technol 4(3):191–206
19. Arab Y, Hassan AS (2012) Lighting analysis of single pendentive dome mosque design in Sarajevo and Istanbul during summer solstice. Arab World Geographer 15(2):163–179
20. Landry M, Breton P (2009) Daylight simulation in Autodesk 3ds Max Design 2009: advanced concepts. Autodesk Inc., San Rafael

Chapter 26
Employing Second Matter from Agricultural Sector in Architecture: A Comparison Between the Italian and Romanian Situation

Dora Francese, Ana-Maria Dabija, Ovidiu-Horaţiu Teleche, and Nicolina Mastrangelo

Abstract This paper deals with the complex issue of wastes and makes an effort to synthesize possible solutions by reuse applications in the building sector.

Architecture as means of transformation of the environment has lately been responsible for a number of impacts, but at the same time it can be considered as a tool for trying to reduce the amount of waste through recycling and reusing by-products and wastes from various sectors.

The paper discusses the employment of wastes in architecture and the necessary change of route from linear to circular processes. Some case studies are shown to compare the situation in the building sectors in Italy and Romania.

The paradigms of the new sustainable society are defined in opposition to those of the last century, and an important input and incentive is noticed toward the promotion of a new market of wastes.

The Romanian architecture of the past had widely employed the green and natural products of the countryside, either from wastes or from specifically growth species. Some examples of local Romanian architecture of the past are shown, so as to illustrate the construction of good quality buildings with waste systems. Then the present situation is shown, both in Romania and in Italy, where wastes from the agricultural sector are recycled, reprocessed, and employed in contemporary architecture, with great amount of sustainability and health benefits for users and for the environment.

Keywords Agricultural wastes • Straw roof • Natural insulations • *Posidonia* • Hemp

D. Francese (✉) • N. Mastrangelo
Cittam-Interdepartmental research Center for the study of Traditional Techniques in the Mediterranean Area, University "Federico II" Naples, Latilla Palace, via Tarsia 31, Naples, Italy
e-mail: francese@unina.it; linamastrangelo@hotmail.com

A.-M. Dabija • O.-H. Teleche
"Ion Mincu" University of Architecture and Urbanism, Bucharest, Romania
e-mail: am.dabija@yahoo.com; ovidiu.teleche@yahoo.com

I. Dincer et al. (eds.), *Progress in Clean Energy, Volume 2*,
DOI 10.1007/978-3-319-17031-2_26

26.1 Introduction. The Need of a New Paradigm for Production of Goods in Architecture

One of the paradigms of this millennium can be synthesized in Pliny Fisk's sentence: "Disassembly and reassembly must be the wave of the future" [1]. Possible links between the waste issue on our planet and the role of architecture within the process of reconstruction for a participatory future take the step from the awareness of the fact that, on one hand, architecture and land transformation should abandon the idea of a consumerist and commercialized society (which had led to the application of bad practices) and that, on the other hand, a sustainable management of the artificiality can contribute to reduce the excessive mechanization without renounce to its benefits and to any new discovery [2].

In fact, the twentieth century had focused on the technological progress, and a part (unfortunately too reduced) of humanity had taken benefits from this progress. Paolo Soleri had, in fact, identified six principles, corresponding to an equal number of man's benefits (these principles, according to Soleri's words, are the increase of life length to 100 years except for African countries where the medium is 40, the exploration of space, both never-ending and microscopic nano-technologies, an abundant society which had won over hunger and had created the super-production hazard but 1/3 of humanity still suffer from hunger and thirst, the revolution of quality, and last but not least the authentic globalization, where by means of informatization everybody is citizen of the world).

In the twenty-first century, instead, it is possible to ascertain the contrast created between Soleri's principles and the new culture of sustainable development: growth and technological progress should have provided welfare for all humanity, who instead suffers from the collateral effects of this progress, demonstrating at latest disrespect toward the planet.

"Numeurous attempts to reshape the vague materialism in a reflected balance, where the production-consumption value set against an elegant and illuminated trinity, which works on the base of knowledge, apprehension and transcendence ..." [3], had led to reduce the damaging effects of artificiality, by means of employing non-adulterated materials, or in a state closer to their "natural" condition, so as to minimize the toxicity effects by reducing or completely avoiding the high-tech production methods which consume high amounts of energy and release polluting substances.

One of the possible routes to follow is to determine, for each stage of the building process, which actions can be carried out for respecting the nature's cyclical course, for assembling and disassembling the matter in order to make it always pure and usable, and to match the principles of sustainable development.

Currently, as far as secondhand materials are concerned, it can be said that the great amount of wastes produced by construction activities represents a big portion of the trash issue, for it has been calculated as one of the most affecting sectors (in 1997 the construction and demolition sector (C&D) produced 20.397 t × 1,000/ year, i.e., 23 % of the total amount of wastes; in 2004 it produced 46.458, i.e., 33 %;

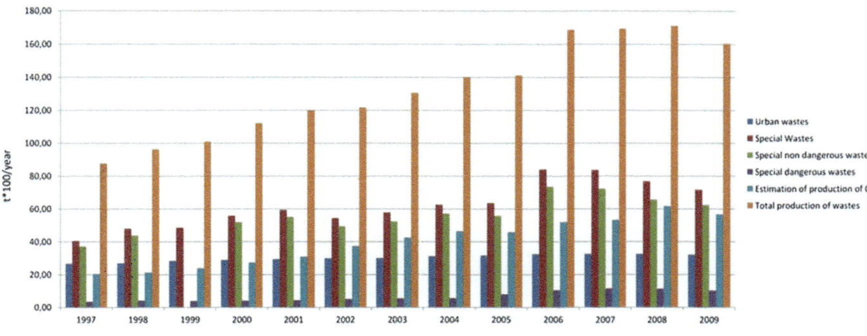

Fig. 26.1 Amount of waste produced in Italy per sectors (drawn by the authors)

and in 2009, it produced 56.681, i.e., 35 %). In Italy, the whole production is increasing; in fact urban wastes in 1997 were superior to the C&D ones (construction and demolition wastes from the building sector), while in 2009 they were inferior: the trend toward improvement is due to the reason that the situation is inverted and the special (dangerous or not) wastes are increasing (Fig. 26.1). This can depend on the fact that people have become more conscious about the danger arising from some materials but also on the fact that more and more artificial and chemical products have been lately employed in the building sector.

This high employment of artificiality reminds us that sustainable architecture requires the revision of processes from linear to circular and that these processes should be managed by natural elements, such as water, air, fire, and earth. In fact, as it has happened with energy, which already had been established as a precious resource and for which question new measures have been searched and applied in order to reduce the use of nonrenewable sources, concern has been extended to the renovation and renewability of other resources such as water, materials, soil, etc. Therefore, now is the time that architecture takes account of this scenario and re-considers the design parameters.

Moreover, since the gap between matter and energy is nowadays scientifically proved being bridged by phase changes, it can be underlined that a number of protocols and methods for assessing the impacts would refer to the concept of life cycle, which includes the idea of reusing for new goals.

Waste is in any case a design error which should be avoided by preventing upstream trash production and rethinking about waste design. It should be the producers' and sellers' task to support the disposal costs, instead of the buyers'. The future is going in the direction where finally the transition stage from waste to "secondary prime matter" should occur directly at a local level, with already tested technologies both in Italy and in the rest of Europe.

If we remember that matter is something that has a shape and a mass, whereas energy is everything that has the capacity to produce labor, the product is instead a term which should be redefined and never set against the words waste or rejected object.

In fact, the word "product" comes from Latin and means "result of the production, born and carried out for the profit, the advantage of someone"; if production was once meant for the benefit of people, now with the linear production systems, only the first segment of the life cycle of a product is really leading to an advantage for users, whereas the remaining 99 % of its life cycle is identified with waste, often a trouble in terms of financial as well as environmental targets (the environmental cost of wastes is due to the pollution in the compartments of water, of air, and of soil).

If "naturality" is understood as the path that reduces the impacts of the extraction, the consumption, the emission, and the conversion of gas, liquid, or solid resources that a product requires during its life cycle, then reduced manufacturing processes that consume less resources and produce less rejection in the environment have a high "naturality" level, while in the opposite case, its "artificiality" level would be high. The maximum level of a prime matter's naturality, i.e., at the pure state, is called "climax," or maximum potentiality condition, while on the other hand, when a product had assumed the "maximum-level transformation" [4], is completely artificial and thus has a low level of naturality.

The new sustainable approach for living and transforming human habitat creates the need of inverting the route by considering any existing object, even wastes, as a potential product, as long as it can be reinserted into the natural cycle.

The question of recycling regards mainly two great fields of action, one defined by the actual waste produced by the building sector and the other regarding the wastes produced by other human activities and that can be re-employed in the production of building materials.

Two different employments can be shown of the wastes from the building sector, the first being the use of substances and products in the same shape, dimension, and properties ("as it is") within a new structure and the second being its reintroduction in the cycle of production in special factories in order to become a new different system, with different shape, volume, characters, and performances (Fig. 26.2).

By recycling all the physical, cultural, and economic components of the environment, The waste can be again considered as a resource rather than a restriction.

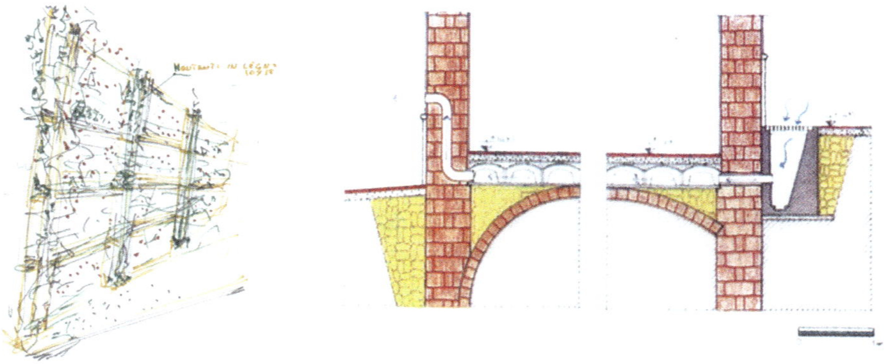

Fig. 26.2 Employing vegetable wastes as *green* wall (*left*) or as "Igloo system" in the foundation (*right*) (drawn by the authors)

It is known that "the technique is not an intrinsic fact, it is not an external mask of a spiritual internal fact, but it is vice versa the concrete synthesis of nature and art, the completion of shapes suggested by nature into natural-technical tools, where only the art masterpiece takes shape" [5]; architecture also, as a high form of art to serve man's benefit, employs natural-technical means, through which the designer's consciousness can lead to ecological or non-ecological solutions. In fact, the techniques are needed for transforming the society from linear to circular and for making architecture comprehensive of how the trash and the waste can become "new materials" for the construction: with this technique we can make a first step toward the circular processes of production and life. It can be then possible to say—with the words of the Nobel Prize winner Rita Levi-Montalcini—that "the crisis of the modern society, its life and thought stile, our production, wasting and consuming system are no longer compatible with people's and nature's rights": we can interact with these modalities by thinking and acting in a different way, starting from architecture as space constructor. And if "space" can be, on one hand, considered as possession taking, it should also be a tool for dialogue, reciprocity, responsibility, care, and agreement. Solving the duality between man and environment has had various alternative occurrences in history, by intervening with sensitivity and consciousness in the circularity of the world-subject interaction and by updating and maintaining alive our knowledge of the deep links between man and earth: we should not exclude a priori all the components that we call "wastes," only because we are not yet able to put them back into a circular logic.

26.2 The Use of Agricultural Wastes in the Traditional Romanian Architecture

One feature that can be recognized in all traditional building cultures throughout the world is the use of the force of nature, from concept to finalized product. From the choice of site to building orientation, from the geometry of the house to building materials, everything is carried out taking into consideration the local conditions and using the natural environment as a leverage for the building. Within this concept frame—mostly implied and transmitted from one generation of vernacular builders (merely peasants) to another, rather than expressed in writing—the use of secondary products resulting from agriculture in the process of building materials is a natural approach.

Straw and bale, as they are or mixed with other "ingredients," were used for roofs, bricks, plaster, and walls. It is a perennial principle that has to do mainly with common sense, comfort, and cost-effective building (reduced to common sense in the country) techniques and not with geography or history. Therefore, the same theory applies from China to Scandinavia for instance, whenever similar conditions and materials occur. However, in a contemporary society, it became clear that a return to the basic principles is necessary, as the use of means and materials tends to exhausts the planet of resources while suffocating it with waste.

In traditional Romanian architecture, the use of mud, clay, and some type of waste is common: straws represent "waste" in agriculture as the wheat is the product, thatch is "waste," and muck is obviously "waste." Still, these products have commonly been used as primary ingredients in the field of construction.

Reeds are plants that grow in marshes—which represent a great part of the Danube Delta—and are used both as food for animals that populate these marshes and as building material.

The most common use of reeds is for the sloped roofs of traditional houses in this part of Romania. As it is an organic material that deteriorates when exposed to natural environmental agents, in order to avoid leaking, thatch roofs have an impressive thickness, around 35–40 cm, thus accomplishing a good thermal protection of the space beneath the envelope as well.

As an example, we present a fisherman's house in Jurilovca, the Danube Delta, built at the end of the nineteenth century, which is now an exhibit at the Peasant Museum in Bucharest (Fig. 26.3).

Apart from the thatched roof, one can admire the beautiful blue color of the gates and the white-blue and red shutters that come into contrast with the white chalked walls.

Fig. 26.3 House from Jurilovca, Dobrogea, 1898. Peasant Museum, Bucharest. Photo Ana-Maria Dabija (photo by the authors)

Fig. 26.4 Cottage in the Romanian Plain. Peasant Museum, Bucharest. Photo Ana-Maria Dabija (photo by the authors)

With quite a different approach come the buried cottages in the plains of Romania: the use of straw roofs—thick to avoid leakage of the roofs but also to keep the warmth inside—is combined with the concept of half-buried dwellings, in the attempt of preserving the warmth/coolness of the inner space.

In Romania, in the territory between the Carpathian Mountains and the Danube, this seems to have been the most common building system. A "contemporary" concept, to consider the current need to preserve energy, building when the is under the ground, beyond the freezing depth where the temperature of the earth is around 10–12 °C, by using less wood for heating the rooms.

Winds are rough in the Romanian Plain (where these cottages were built) and the straw roofs, thick as it can be seen in Fig. 26.4, had to withstand the rain and snow as well as the wind.

Another reason of building in the ground was to dissimulate the houses and hide them from the attacks carried out by migratory populations.

In the central and northern part of Romania, straw roofs are common as well. Their high slopes prevent water and snow from remaining on the roof while, through thickness, the warmth is preserved in the space beneath (Fig. 26.5).

Another building material that is specific to the traditional Romanian architecture is the adobe ("chirpici" in Romanian). Made from clay, straw, and horse muck, it has the form (and more or less dimension) of a brick. It is dried in the sun. Due to their structure—clay and straw—the adobes are thermal insulating products. Some say that their behavior in seismic areas are fair, provided the buildings have foundations

Fig. 26.5 Dwellings from Sălciua, 1815, Peasant Museum, Bucharest. Photo Ana-Maria Dabija (photo by the authors)

or basements made with more resistant materials. The adobe itself allows for some deformations before breaking.

If a (wooden) frame is provided, with transversal rods, the mixture of clay and straw represents a fair weatherproof sealing. This system ("paianta" in Romanian language) is also traditional in the vernacular Romanian architecture and has recently been re-vitalized within the frame of eco-architecture.

Under seismic conditions, this type of buildings had a fair behavior, provided they were built on stone foundation (sometimes with a wooden beam over it). As the buildings are rather small, with also small openings, the fair behavior is almost natural, even in countries with seismic hazard.

Another use of straw in traditional buildings—in Romania and elsewhere—is in plasters. The use of clay with sand and chopped straw (as reinforcement) was a cheap and healthy solution, as everything that uses natural materials and product.

26.3 The Use of Agricultural Wastes in Contemporary Romanian Architecture

Although natural construction materials are highly appreciated for ecological, aesthetic, and health-related reasons, their use is often restrained by their limited availability or higher costs as compared to industrial materials. For this reason, there is an increasing focus on low-cost and renewable natural materials that can be used in construction.

As an agricultural by-product, straw is produced yearly and it is particularly cheap. And the packing of hay or straw into bales enlarged the possibilities of using these materials in contemporary civilized architecture.

Today in Romania there is an increasingly high interest in straw-bale construction, with over twenty such houses being built recently all over the country. The interest in this ecological and low-cost material made communities pursue the development of construction know-how for positive and reliable results.

The architects that are using straw-bale methods are generally open to sharing their experience around dedicated public events, which have been organized since a few years (one of the first seminars on straw-bale construction was held in Bucharest by Stefano Soldati in May 2007).

The interest of the public in this construction technique is also high. Moreover, a young but vibrant no-profit organization is largely dedicated to this field of knowledge in an attempt to find a solution for a "cheaper, healthier, and warmer home" for the people of Romania [6].

Currently, a new low-cost experimental house is being developed by this organization that is planning to develop and publish a straw-bale construction guide adapted to the local conditions in Romania.

The highest vulnerability during the lifetime of such a building comes from exposure to water and moisture (from capillarity, condensation, vapor diffusion, direct exposure to rain, roof leakages, or interior inundation).

The vapor diffusion process requires particular attention and must consider the inter-seasonal reversal of the interior-exterior vapor flow.

Typically, a vapor retarder is only a layer on the interior face of the roof insulation [7], allowing the walls to breathe. The interior layer of clay or lime plaster acts as a buffer system for internal humidity variation by its ability to absorb excessive air moisture and then compensate latter dryness. However, the hydrothermal performance of this system requires long-term observation.

Although the high thermal resistance of the straw-bale walls is a plus, their reduced thermal inertia can come as a downside if they are used for permanent dwellings.

In this case, the thermal inertia of the interior space can be insured by the interior walls, which can be made of brick (case study 1) (Fig. 26.6).

Sometimes these internal masonry walls are also structurally employed, but most of the straw-bale houses built so far in Romania are only supported by the wood-frame structural system.

Another vulnerability of the straw-bale construction systems comes from rodents, but this can be prevented by carefully removing the remaining seeds from the straw, before packing the bales. Careful detailing is also necessary in order to avoid any hidden un-plastered area of the straw-bale walls.

Additionally, there are some things that the architect should consider regarding straw-bale construction. Some of the construction works may require on-site experience, trial, and failure, so it is very important for the architect to be on the construction

Fig. 26.6 Application of hemp products in Italy [12]. New construction: Veghini House, 120 m^2; Location: Loria (TV); Client: Damiano Veghini; Architect: Arch. D. Favero, FAST Studio; Year: September 2013; Materials: supporting structure in steel and roof insulation panels of hemp and lime (20 cm). Stratigraphy wall perimeter (from inside to outside): (1) panel in magnetite, (2) lime thickness of 14 cm, and (3) outer wall bio-brick 15 cm

site. For the plaster works, for example, which are very important, even the most renowned specialists say that there is no formula, so trial is often necessary.

Moreover, the development of know-how in this field depends on the communication of the project results and the construction methods that were used. From start to end, for successful results, communication appears to be critical.

Basically, straw-bale construction only determines the structure of the walls, allowing for various architectural possibilities. However, unlike other natural construction materials, the use of straw-bales is more likely to lead the architects to follow a specific natural and traditional character in all other aspects of the project.

Stone foundations, wood structure and finishing, hay, reed, or wood shingle roofs are often used in the construction of straw-bale houses, although the straw-bales themselves are never visible nor exposed. In fact, it is the taste for natural construction or rustic architecture that appears to lead to promote of straw-bales, rather than the economic or ecological benefits of the material itself.

Using straw-bales as a construction material can partly reduce the costs of single-family or holiday homes in Romania. The widespread availability of raw materials should also allow for low transportation costs.

However, as long as the straw-bales are only used as a nonstructural infill, the economic benefits are limited. For this reason, load-bearing straw-bale wall systems may deserve more attention in the future. The greatest advantage of using straw-bales is the fact that the production of this material has an insignificant environmental impact and at the same time it allows for healthier homes.

Case study	Location	Floor area	Structural system	Construction period	Costs	Architecture
Single-family house	Cârligu Mare, Glodeanu-Siliştea, Buzău	70 m^2	Wood framing, internal masonry	2012–2014	200 EUR/m^2 (estimative only, many materials were donated by sponsors)	"Eco-Habitat" organization, Cătălina Grigore and Florin Grosu [13]
Holiday house	Lupşanu, Călăraşi	120 m^2	Double wood-framing walls	April–August 2012	330 EUR/m^2 (including long-distance transportation costs for stone and reed)	Iulian Ungureanu
Holiday house	Valea Nucului, Buzău		Wood framing	May–October 2012	540 EUR/m^2	"Earth Safe Design," Grzegorz Gorski

[8–10].

26.4 The Use of Wastes from the Agricultural Sector as Construction Material in Italy

In the field of renewable sources, the process of innovation involved the use of biomass in the Italian building sector. According to the National Action Plan for Renewable Energies, by 2020 Italy must cover 44 % of energy consumptions with biomasses. So, the question is whether and how the construction industry can benefit from their interaction with biomass. The state of the art of building materials based on residual origin or dedicated crops brings new innovations in sustainable, biodegradable, and recyclable products. The sustainability of the production chains associated with biomass allows to make assessments considering not only the environment but also the effects on economic and social aspects, essential for starting up new perspectives. Analyzing the characteristics of local renewable resources in Italy shows that hemp became a dedicated crop, largely redeveloped in Campania. The attention that researchers, professionals, farmers, and local authorities have placed to this natural fiber is mainly owing to the various uses and to the benefits that the production chain of hemp can bring to the economy. The profits and the incomes for farm could increase, while various organizations could promote subconsequent renewal to the community that is the basis of bio-regionalism strategies. The added value in the construction industry can be found in different life cycle stages of the product. The preproduction and production, stages do not require high energy. During the use phase, this biocompatible and eco-friendly material does not release toxic substances for human health and has the added value of capturing CO_2. Today, the greatest supporter of hemp cultivation is the Assocanapa association [11]. Since 1998, Assocanapa, in accordance with the provisions of the standard by the Ministry of Agriculture, promotes the cultivation of *Cannabis sativa* as long as they have a THC value of less than 0.02. Taking a look at the historical background, it is clear that Italy has been an important producer of hemp fiber, second only to Russia. The historical data of production, in particular the data from 1958, shows that the most important producer of "canapicola" in Italy is the Campania region. According to Assocanapa, in 2011 the cultivation of cannabis averaged 120 q/ha/year. From a structural point of view, the plant consists of a fibrous outer part, which is used in the building industry to make panels, felts and insulation, and an inner, ligneous part called "canapulo"; the latter mixed with lime can be operated by means of mechanical pumps or it can be used to make blocks of bio-bricks to be used in the guise of traditional bricks. The cellulose contained in the inner part is a raw material for recyclable bio-plastics. From the seeds of the hemp, it is possible to extract oils used in paints or glues.

Products like CAF can be obtained, or panels that cannot be recycled and are made up with polyester, hemp, crushed potato starch, or natural glues. The hemp mixed with cement can create an "insulating concrete," i.e., a cement with good insulating qualities, great lightness, and good mechanical properties (Fig. 26.7).

Another interesting material source is *Posidonia oceanica*, a marine phanerogamae with high environmental value, whose result attests the sustainability of the habitat sites. While the European community protects the meadows of *Posidonia* as a priority habitat, as real factories of oxygen, and as ideal places for countless ecological niches, many coastal disposals of *Posidonia* beaches are real environmental and economic troubles. A key feature of *Posidonia* is the high

Fig. 26.7 Application of hemp products in the internal design in Italy [13]. Reconstruction of the prototype shown at Made Expo 2012. Location: San Possidonio (MO). Year: October 2013. Materials: supporting structure in wood. Stratigraphy wall perimeter (from inside to outside): (1) panel mag and (2) outer wall bio-brick

production of biomass (between 2 and 10 t/ha per year). *Posidonia* cyclically loses its leaves that are deposited on the coasts. A strip of grassland of 1 km in width per meter of coast can also produce up to 125 kg biomass per year (in terms of dry matter). The management of *Posidonia* opens a complex argument. According to legislation, there are three options for the management of beached *Posidonia* which are, in order of priority, on-site maintenance, moving/storage accumulation, and removal/transfer to landfill. Legislative Decree 205/2010 confirms what said by Decree (January 22) 2009 which indicates that green phanerogamae (including *Posidonia*) and marine algae should be safe guarded. From the regulatory point of view, it would seem that the only way designed in Italy to deal with the disposal of biomass of *Posidonia* is the application of compostable material. The LIFE project PRIME [14] (*Posidonia* residues integrated management for eco-sustainability) carried out some test conducted in Puglia region and compared the use of alga with the rest of Europe, where the innovative building materials such as sea grass are common practice. In the island of Læs (a small island situated in the sea of Kattegat, between the Baltic Sea and the North Sea), insulating buildings with sea grass is part of the local identity; in Germany, the raw material imported from Tunisia is studied and tested as thermal insulating product. The Fraunhofer-Institut für Chemische Technologie ICT Pfinztal [15], in collaboration with several companies, including Neptutherm is conducting tests for the feasibility of this biomass in the construction industry (Figs. 26.7, 26.8, and 26.9).

Fig. 26.8 Application of *Posidonia* in walls [16]

Fig. 26.9 Application of *Posidonia* in roofs [17]

In addition to secondary materials, they are experimenting on the possibility of obtaining a panel of *Posidonia* as well as thermoplastic materials that can be used with 3D printers.

The impact of the construction industry on the exploitation of resources is very high; the designers' approach in the choice of building materials must be oriented toward sustainable materials. The decision to test biomass for sustainable and biocompatible materials allows to reduce CO_2 emissions and therefore complies with the objectives set by the European "Horizon 2020" strategy as well as is useful in addressing the economic crisis. The conversion of the traditional sector of agriculture toward new scenarios of growth and development can have positive impact in the social sphere and on the welfare of citizens. In order to promote sustainable development, biomass is a chance for investments. The "short chain" production process can activate local economies and business skills of farmers, who can prize the earning capacity of their land by giving a new value to waste. In this perspective, the trash becomes a resource, while recycling, reuse, and composting become the main roads along.

26.5 Conclusion

The comparison between the Romanian and the Italian situation regarding the use of agricultural by-products in architecture has shown how a more natural and environmentally friendly approach to the production processes could reduce the impacts and the amount of waste objects in the present society. In Romania, traditional architecture had widely employed agricultural products by creating at the same time an interesting style and good bioclimatic answers. Nowadays, the innovation in the market, both in Italy and in Romania, has started to address the reuse and recycling issues, thanks to the employment of by-products and wastes from the agricultural sector by means of sustainable design and technologies.

The trend is obviously increasing, as elevated architecture has started to reconsider what were once some of the most conventional construction materials specific to commonplace architecture. In the future, successful "short chain" applications of biomass resources in the construction industry will depend largely on the architects' skills of using these "weak" materials in a durable way. This requires simple but sound technical detailing and a stronger engagement on the construction site. The short-term development of practical solutions that meet present-day standards depends on the professionals' availability to communicate and share experience and result

References

1. Pliny Fisk http://www.cmpbs.org
2. Dorfles G (1968) Artificio e natura, Einaudi, Torino, p 22
3. Lefevre La rose de sable, in "le carré bleu" n 1, 2009
4. Abrami G (1987) Progettazione ambientale. Clup, Milano
5. Paci E (1957) "L'architettura e il mondo". In Lampugnani-Nigri,Relazioni e significati, Milano, p 155
6. Ciprian Anghel S (2010) Monumente de arhitectură tradiţională în muzeul Astra. Sibiu: Editura "ASTRA MUSEUM"
7. Dan A (2011) Nou în România. "Bordeie" moderne, din baloţi de paie şi lut. In: EVZ, November, 7. http://www.evz.ro/detalii/stiri/nou-in-romania-bordeie-moderne-din-baloti-de-paie-si-lut-952806.html. Accessed 13 Feb 2014
8. http://eco-habitat.ro/despre-noi. Accessed 13 Feb 2014
9. Pierre B (2012) Straw-bale construction seminar held in Bucharest, Nov. 22. Recording. http://arhitectura-1906.ro/2013/11/inregistrarea-prezentarii-case-din-baloti-de-paie. Accessed 13 Feb 2014
10. Ion D (2012) De ce case din baloţi de paie în România?. In: Romania Liberă, May, 25. http://www.romanialibera.ro/timpul-liber/casa-mea/de-ce-case-din-baloti-de-paie-in-romania-265112.html. Accessed 13 Feb 2014
11. http://www.assocanapa.org
12. http://www.equilibrium-bioedilizia.it/it/mappa-cantieri
13. http://www.equilibrium-bioedilizia.it/it/cantieri/san-possidonio-mo-cantiere-remade-equilibrium
14. P.R.I.M.E. "Posidonia residues integrated management for ecosustainability" (LIFE09 ENV/IT/000061), http://www.lifeprime.eu/wp/
15. Fraunhofer-Institut für Chemische Technologie ICT (2013). http://www.fraunhofer.de/en/press/research-news/2013/march/seaweed-under-the-roof.html
16. http://www.neptutherm.de/phpwcms/index.php?innenwand-1
17. http://www.neptutherm.com/index.php?generalsanierung-rheinstr-61-stutensee-friedrichstal#

Chapter 27
Comparison of Energy Reduction Potential of an Adobe House Under Different Climatic Conditions in India

Basharat Jamil and M. Jamil Ahmad

Abstract The present work is an evaluation of energy reduction (heating and cooling) potential of a dome-shaped adobe house. The energy potential has been evaluated on the basis of energy balance assuming quasi-steady state condition. This has further been analyzed by incorporating the effect of passive energy technologies namely ventilation and Earth–air heat exchanger (EAHE). The analysis for room air temperatures and quantification of energy potential has been carried out for three different locations in India. These locations come under different climatic conditions viz. Bangalore (*Moderate climate*), Jodhpur (*Hot climate*), and Srinagar (*Cold climate*). Annual heating potential for winter was found maximum for the case of cold climate (Srinagar) of the order of 43.96×10^7 kW h while it was estimated minimum for hot climate (Jodhpur) as 16.47×10^7 kW h. Annual cooling potential for summer was found maximum for hot climate (Jodhpur) 28.14×10^7 kW h while it was estimated minimum for moderate climate (Bangalore) as 8.87×10^7 kW h. Thus, a comparison of energy reduction potentials for three locations falling under different climates in India has been made. The performance of dome-shaped house was found to be satisfactory both in winters and summers for the locations considered under different climates. The temperature of room air with an EAHE was found varying due to low isothermal mass of the building. Adobe house was found to temperate the room air temperature, significantly reducing temperature fluctuations within the building.

Keywords Adobe house • Climate • Earth–air heat exchanger • Heating/cooling potential • Ventilation

---

B. Jamil (✉) • M.J. Ahmad
Department of Mechanical Engineering, Zakir Husain College of Engineering & Technology, Aligarh Muslim University, Aligarh 202002, Uttar Pradesh, India
e-mail: basharat1986@gmail.com; jamil.amu@gmail.com

I. Dincer et al. (eds.), *Progress in Clean Energy, Volume 2*,
DOI 10.1007/978-3-319-17031-2_27

Nomenclature

A	Area of wall/roof surface (m^2)
C_a	Specific heat of room air (J/kg K)
F_R	Heat removal factor for EAHE from underground earth's surface
h_{ca}	Convective heat transfer coefficient between surface and ambient (W/m^2 K)
h_o	Outside heat transfer coefficient of wall/roof surface (W/m^2 K)
h_i	Inside heat transfer coefficient of wall/roof surface (W/m^2 K)
h_{ra}	Radiative heat transfer coefficient between surface and ambient (W/m^2 K)
I	Solar radiation on surface (W/m^2)
K	Thermal conductivity of material (W/m K)
L	Length of PVC pipe (m)
l	Thickness of material (m)
M_a	Mass of room air (kg)
m_a	Mass flow rate of air through the Earth–air heat exchanger tube (kg/s)
N	Number of air change per hour (h^{-1})
Q	Heating/Cooling Potential (J)
Q_{Door}	Heat gain through door (W)
Q_{EAHE}/Q_u	Rate of thermal energy gained by air through Earth–air heat exchanger (W)
Q_{floor}	Heat loss into the ground (W)
Q_{Gain}	Heat gain into the room (W)
Q_{Loss}	Heat loss from room air to outside ambient air (W)
Q_{Roof}	Heat gain through roof surfaces (W)
$Q_{Ventilation}$	Heat loss from room air to outside ambient air by ventilation (W)
Q_{Wall}	Heat gain through wall surfaces (W)
Q_{Window}	Heat gain through window (W)
R	Radius of dome (m)
r	Radius of pipe (m)
Δt	Time interval (h)
T_a	Ambient air temperature (°C)
T_r	Air temperature of mud-house (°C)
T_{r0}	Initial room air temperature (°C)
T_{sol}	Sol–air temperature of the Sun-exposed surfaces (°C)
T_0	Underground temperature (°C)
U	Overall heat transfer coefficient of wall/roof/ground structure (W/m^2 K)
V	Velocity of air through the Earth–air heat exchanger (m/s)
v_a	Room air volume of mud-house (m^3)

Greek Symbols

α	Absorptivity of wall/roof/door surface
β	Slope of wall/roof surface with respect to horizontal
ε	Emissivity of roof surface
ρ	Reflection coefficient of ground
ρ_a	Density (kg/m^3)
τ	Transmissivity of glass

27.1 Introduction

Buildings are a major source of greenhouse gases especially in developing countries like India due to high requirement of maintaining favorable indoor comfort conditions [1]. Due to their high consumption of energy, buildings stand the third largest consumers of energy after industry and agriculture. Buildings annually consume more than 20 % of electricity used in India [2]. The optimal design of building and related systems can minimize operation costs while maintaining desirable indoor air quality and thermal comfort.

A building is always influenced by the climatic conditions since it is in direct contact with its environment [3]. Therefore, the extent of reduction in energy use strongly varies from climate of a location to another. It is important in this context to be able to analyze how buildings will respond to different climates, and assess the likely changes in energy use. To reduce energy consumption and increase energy efficiency, several strategies are necessary [4]. To evaluate the indoor air temperatures and energy potential, thermal energy performance of the building and type of climate the building exists in is significant [5]. A significant gap remains in the quantitative assessment of energy reduction potential in buildings, particularly from the point of different climatic conditions. Also, the literature lacks to identify the potential of energy efficiency measures for different climates.

Field experiences have shown that adobe buildings are environmentally friendly in comparison to conventional buildings [6]. Also, the dome-shaped roofs due to thermal stratification and reduced local radiant flux on rounded surface create comfortable feeling at floor level [7].

The objective here is to investigate energy use in a dome-shaped adobe building in different climates. Assessment of passive energy measures (ventilation and Earth–air heat exchanger) has also been done Ventilation through the air flow renews the air and doesn't let the inside wall or roof surface to warm-up. While an EAHE continuously temperatures the room air by exchanging heat with the Earth. In this regard, building energy performance is crucial to ascertain energy use in buildings and is the basis to make any decision for enhancing energy efficiency.

Table 27.1 Description of representative locations [8]

Location	Latitude	Longitude	Elevation (m)	Climate
Bangalore	12°58′	77°35′E	921	Moderate
Jodhpur	26°18′	73°01′E	224	Hot
Srinagar	31°06′	74°50′E	1,586	Cold

The goal of this study is to answer three main questions that are essential to understand the impact of climatic parameters on energy usage in building:

1. How sensitive are buildings to different climatic conditions?
2. How much energy efficiency measures (like ventilation and Earth–air heat exchanger) influence energy reduction potential?
3. Can adobe buildings provide thermal comfort with low energy investments?

The answers to these questions constitute a significant contribution to the field of building energy assessment where different organizations could benefit from the results obtained to improve buildings' energy performance combined with the climatic data.

In the present case, for the purpose of analysis three locations from India have been considered which happens to fall under different climate type as shown in Table 27.1.

27.2 Methodology

27.2.1 Description of Dome-Shaped Adobe House

An adobe house has been designed for serving mid-day meal in a government school. The layout of the house (shown in Fig. 27.1) has been adopted for study and analysis from Energy Research Applications (ERA), New Delhi [9]. There are two parts of the house, kitchen cum serving and store room. Only kitchen cum serving part has been studied for heating/ cooling potential, because occupants are not supposed to stay in store room. Orientation and dimensions of windows, door, roof, and floor are shown in Fig. 27.1. The design parameters used in thermal modeling of the adobe house and other related systems are provided in the Appendix.

27.2.2 Collection of Meteorological Data

The solar radiation and ambient air temperature data have been collected for the period of 1991–2001 from India Meteorology Department (IMD) Pune, India for the locations considered within India. Climate of each location has been classified

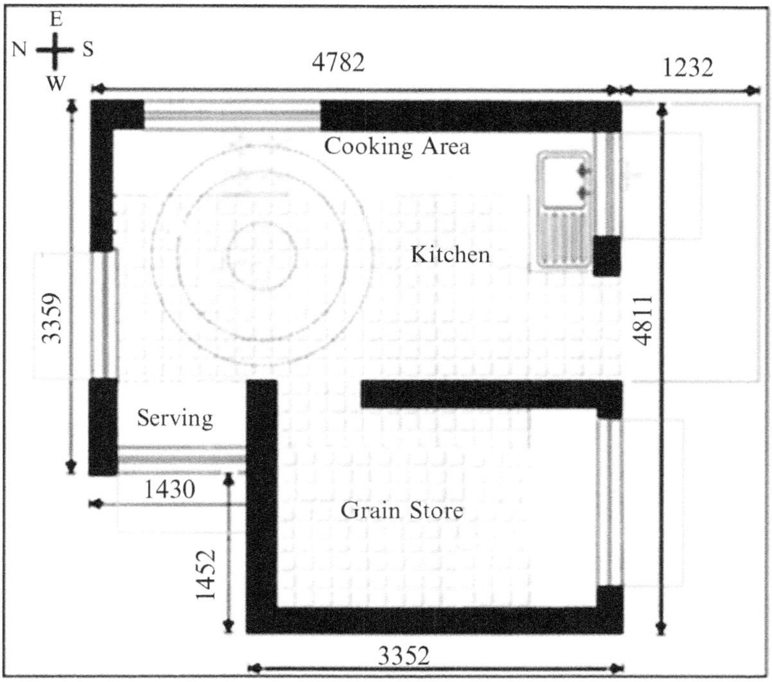

Fig. 27.1 Top view of adobe house (kitchen plan = 24.0 m^2, kitchen/serving = 13.2 m^2, store = 7.6 m^2, wash = 3.1 m^2)

into four weather conditions [10] depending upon sunshine hours and ratio of diffuse to global radiation as follows:

(a) *Clear day* (*blue sky*): If diffuse radiation is less than or equal to 25 % of global radiation and sunshine hour is more than or equal to 9 h.
(b) *Hazy day* (*fully*): If diffuse radiation is less than 50 % or more than 25 % of global radiation and sunshine hour is between 7 and 9 h.
(c) *Hazy and cloudy* (*partially*): If diffuse radiation is less than 75 % or more than 50 % of global radiation and sunshine hour is between 5 and 7 h.
(d) *Cloudy day* (*fully*): If diffuse radiation is more than 75 % of global radiation and sunshine hour is less than 5 h.

27.2.3 Assumptions

Following assumptions have been made while writing energy balance equation for the dome-shaped adobe house.

1. The heat transfer through roof and walls occurs in one direction along the thickness.

2. There is quasi-steady state heat transfer across the roof and wall.
3. The wall and roof structures are made of homogeneous material layers.
4. The ambient and room air temperatures are assumed constant for 1 h.
5. Average solar intensity for 1 h has been considered.
6. The values of parameters like air change per hour and inside (h_i) and outside (h_o) convective heat transfer coefficients are assumed constant.
7. All thermal properties of building materials e.g. thermal conductivity and specific heat are assumed constant.

27.2.4 Energy Modeling of Adobe House

Based on the assumptions made the simple energy balance for room air can be written as follows:

$$M_a C_a \frac{dT_r}{dt} = \sum Q_{gain} - \sum Q_{loss} \tag{27.1}$$

where, M_a is isothermal mass (kg), C_a is specific heat of room air (J/kg °C), T_r is room temperature (°C), t is time (s), Q_{gain} is rate of thermal energy gained by air (J/s), Q_{loss} is rate of thermal energy lost by air (J/s).

$$\sum Q_{gain} = Q_{wall} + Q_{roof} + Q_{window} + Q_{door} + Q_{floor} \tag{27.2}$$

where, Q_{wall} is rate of thermal energy gained by air through wall (J/s), Q_{roof} is rate of thermal energy gained by air through roof (J/s), Q_{window} is rate of thermal energy gained by air through window (J/s), Q_{door} is rate of thermal energy gained by air through door (J/s), Q_{floor} is rate of thermal energy gained by air through floor (J/s).

$$\sum Q_{loss} = Q_{ventilation} \tag{27.3}$$

where, $Q_{ventilation}$ is rate of thermal energy gained by air through ventilation (J/s).

The expressions for rate of heat gain and loss from different building components for quasi-steady state heat transfer analysis is given below.

The rate of heat gain through wall is:

$$Q_{wall} = (UA)_{wall}(T_{sol,wall} - T_r) \tag{27.4}$$

where, U is overall heat transfer coefficient, A is area of the wall (m^2), $T_{sol,wall}$ is sol-air temperature of wall (°C), T_r is room temperature (°C).

$$(UA)_{wall} = \left[\frac{1}{h_0} + \frac{L_1}{K_1} + \frac{L_2}{K_2} + \cdots + \frac{1}{h_i}\right] A_{wall} \tag{27.5}$$

h_o is outside convective heat transfer coefficient (W/m^2 °C), L is thickness of the layer (m), K is thermal conductivity (W/m °C), h_i is inside convective heat transfer coefficient (W/m^2 °C).

The expression of sol–air temperature on any inclined wall/roof surface can be written as

$$T_{sol} = \left[\frac{\alpha I}{h_o} + T_a - \frac{\varepsilon \Delta R}{h_o}\right] \quad (27.6)$$

where, T_{sol} is the sol–air temperature of bare roof surface (°C), T_a is ambient temperature (°C), I is solar radiation falling on different surfaces of dome-shaped house (W/m^2) and $h_o = h_{ra} + h_{ca}$

where, the terms h_{ra} and h_{ca} are radiative and convective heat transfer coefficients between surface and ambient, respectively. ΔR is the long wavelength radiation exchange between surface and sky; its values for different surface orientation are as follows.

$$\Delta R = \begin{cases} 60\frac{W}{m^2}, & \text{for horizontal surface} \\ 0, & \text{for vertical surface} \\ \left[\frac{\cos\beta}{\sin\beta} \times 60\right]\frac{W}{m^2}, & \text{for surface inclined at angle } \beta \end{cases} \quad (27.7)$$

The rate of heat gain through roof is

$$Q_{roof} = (UA)_{roof}(T_{sol,roof} - T_r) \quad (27.8)$$

The rate of heat gain through window is

$$Q_{window} = A_{window}\tau I + (UA)_{window}(T_{sol,window} - T_r) \quad (27.9)$$

The rate of heat gain through door is

$$Q_{door} = (UA)_{door}(T_{sol,door} - T_r) \quad (27.10)$$

The rate of heat gain/loss through ground is

$$Q_{floor} = (UA)_{floor}(T_r - T_0) \quad (27.11)$$

where, T_0 is base temperature (°C), $T_0 = 25$°C.

The equation for rate of heat loss/gain due to room air ventilation to ambient air can be expressed as follows:

$$Q_{ventilation} = \frac{\rho_a v_a C_a N(T_r - T_a)}{3600} = 0.33\ N v_a(T_r - T_a) \quad (27.12)$$

where, N is number of air change per hour, ρ_a is density of air; v_a is volume of air in the room.

When the building room air is integrated with recirculation type Earth–air heat exchanger (EAHE), the rate of heat gain in winter (or loss in summer) is given below;

$$Q_{\mathrm{EAHE}} = F_{\mathrm{R}} m_{\mathrm{a}} C_{\mathrm{a}} (T_0 - T_{\mathrm{r}}) \tag{27.13}$$

where, $T_0 = 25°\mathrm{C}$, Q_{EAHE} is rate of thermal energy gained by air through Earth–air heat exchanger (J/s), m_a is mass flow rate of air (kg/s).

$$F_{\mathrm{R}} = 1 - \exp\left(-\frac{2\pi r h_{\mathrm{ca}}}{m_{\mathrm{a}} C_{\mathrm{a}}} L_{\mathrm{pipe}}\right) \tag{27.14}$$

h_{ca} is convective heat transfer coefficient between surface and ambient (W/m^2 °C), L_{pipe} is length of PVC pipe (m), r is radius of PVC pipe (m).

27.2.5 Heat Balance Equation

Based on the above described Eqs. (27.1)–(27.14), the heat balance equation for room of the house is written as follows:

$$\begin{aligned} M_{\mathrm{a}} C_{\mathrm{a}} \frac{dT_{\mathrm{r}}}{dt} = {} & \sum\nolimits_i^4 (UA)_{\mathrm{wall},i} (T_{\mathrm{sol,wall},i} - T_{\mathrm{r}}) + \sum\nolimits_j^4 (UA)_{\mathrm{window},j} (T_{\mathrm{sol,window},j} - T_r) \\ & + (UA)_{\mathrm{door}} (T_{\mathrm{sol,door}} - T_r) + (UA)_{\mathrm{roof}} (T_{\mathrm{sol,roof}} - T_{\mathrm{r}}) + (UA)_{\mathrm{floor}} (T_{\mathrm{r}} - T_0) \\ & - 0.33 N v_{\mathrm{a}} (T_{\mathrm{r}} - T_{\mathrm{a}}) + F_{\mathrm{R}} m_{\mathrm{a}} C_{\mathrm{a}} (T_0 - T_{\mathrm{r}}) \end{aligned} \tag{27.15}$$

or,

$$\frac{dT_{\mathrm{r}}}{dt} = f(T_{\mathrm{r}}, g(t)) \quad \text{or} \quad \frac{dT_{\mathrm{r}}}{dt} = aT_{\mathrm{r}} + g(t) \tag{27.16}$$

where,

$$\begin{aligned} a = -\frac{1}{M_{\mathrm{a}} C_{\mathrm{a}}} \Bigg[& \sum_i^4 (UA)_{\mathrm{wall},i} + \sum_j^4 (UA)_{\mathrm{window},j} + (UA)_{\mathrm{door}} + (UA)_{\mathrm{roof}} + (UA)_{\mathrm{floor}} \\ & - 0.33 N v_{\mathrm{a}} + F_{\mathrm{r}} m_{\mathrm{a}} C_{\mathrm{a}} \Bigg] \end{aligned} \tag{27.17a}$$

and,

$$g(t) = \frac{1}{M_a C_a}\left[\sum_i^4 (UA)_{\text{wall},i} T_{\text{sol,wall},i} + \sum_j^4 (UA)_{\text{window},j} T_{\text{sol,window},j} + (UA)_{\text{door}} T_{\text{sol,door}} + (UA)_{\text{roof}} T_{\text{sol,roof}} + (UA)_{\text{floor}} T_{\text{sol,floor}} - 0.33 N v_a T_a + F_r m_a C_a T_0\right] \tag{27.17b}$$

The constant "a" is the coefficients of room air temperature. The term $g(t)$ represents the function of time "t" in above and comprises of the time-dependent terms like sol–air surface temperature (T_{sol}), ambient air temperature (T_a) and solar radiation on surface (I). The exact solution of the above first-order linear differential Eq. (27.16) is:

$$T_r = \left[\frac{\overline{g(t)}}{a}\right][1 - \exp(-at)] + T_{r0}\exp(-at) \tag{27.18}$$

where, (t) is average of $g(t)$ over time interval 0 to t.

The values of room air temperature (T_r) for each month of the year were obtained from Eq. (27.18) by the simulation in MATLAB computational environment. Based on the values of room air temperature, the heating/cooling potential of the adobe building can be evaluated.

27.2.6 Energy Reduction Potential

Monthly energy (heating and cooling) reduction potential obtained for an adobe house is obtained as:

$$Q = \sum m_a C_a (T_r - T_a)\Delta t \tag{27.19}$$

where, Q is heating or cooling potential (J), m_a is mass flow rate of air (kg/s).

27.3 Results and Discussion

27.3.1 Energy Reduction Potential for Moderate Climate

The room air temperatures for the moderate climate of Bangalore have been computed using Eq. (27.18). The resulting hourly variation of room air temperature during the month of January with and without Earth–air heat exchanger is shown in Fig. 27.2. The variation of atmospheric temperature lies between 13.0 and 25.0 °C as shown in Fig. 27.2. The room air temperatures are in the range of 20.1–28.8 °C without Earth–air heat exchanger and 24.6–28.3 °C with Earth–air heat exchanger.

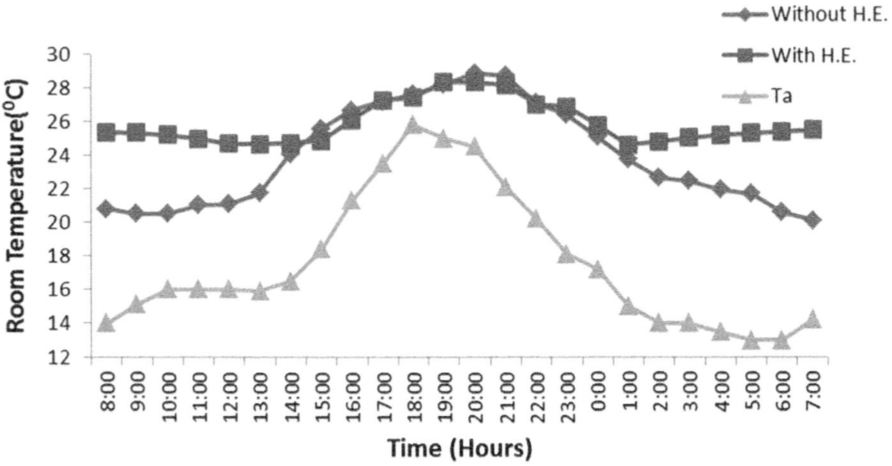

Fig. 27.2 Variation of room temperature with time in January for Bangalore

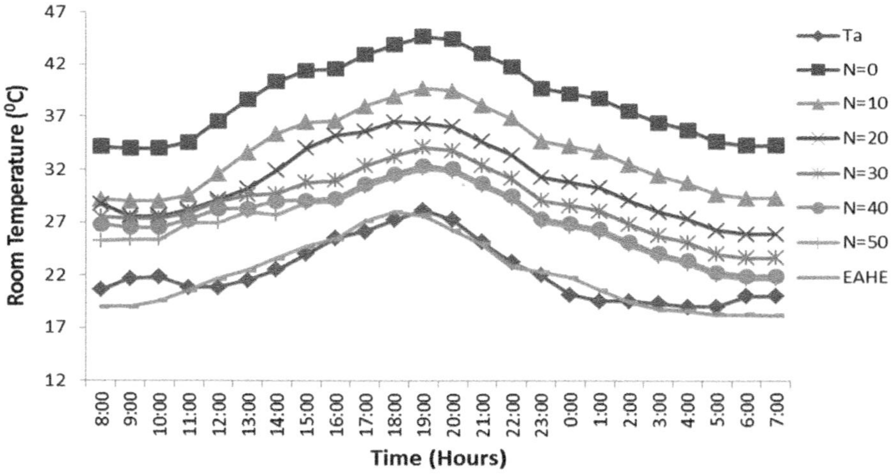

Fig. 27.3 Variation of room temperature with time in June for Bangalore

Difference between atmospheric temperature and room temperature lies between 1.8 and 8.7 °C for the case of no ventilation (i.e., $N = 0$).

Here, it can be observed that although, the use of Earth–air heat exchanger does not substantially improve the heating of the house but, it is beneficial to use it during the early day hours and late night hours to improve the temperature of the room air and thus provide thermal comfort to the occupants. During the rest of hours, temperature is itself maintained in the comfort zone by the adobe house.

The hourly variation of room temperature at different rate of ventilation (or number of air change per hour, N), in the month of June at Bangalore is shown in Fig. 27.3. The atmospheric temperature variation from 19.0 to 28.1 °C has also been shown. The variation of room temperature when the room is integrated with an Earth–air heat exchanger is also depicted.

At $N=0$, room temperature is in the range of 33.9–44.6 °C. At $N=10$, room temperature falls considerably and lies in the range 28.9–39.6 °C. At $N=20$, room temperature again falls and varies from 25.8 to 36.4 °C. At $N=30$, room temperature falls in the range of 23.6–34.0 °C. At $N=40$, room temperature varies from 21.9 to 32.3 °C. At $N=50$, room temperature varies from 21.6 to 32.0 °C. With the Earth–air heat exchanger (length of PVC pipe=78 m, air velocity=3 m/s) integrated to the room, the hourly room temperature varies from 18.1 to 27.9 °C. Maximum difference between atmospheric temperature and room temperature is 13.5 °C if there is no ventilation ($N=0$). With ventilation at $N=10$, 20, and 30, maximum temperature difference falls to 5.0, 4.2, and 3.4 °C, respectively. At $N=40$ and 50, maximum temperature difference is 2.2 and 1.6 °C respectively, i.e., no significant reduction in room temperature is further observed. Therefore ventilation is optimized at $N=30$. The use of ventilation is recommended to lower the temperature in such climate.

Use of Earth–air heat exchanger improves the cooling of the room air. With Earth–air heat exchanger, the room temperature closely follows the atmospheric temperature. The maximum difference in atmospheric temperature and room temperature (with EAHE) is found to be 1.8 °C. Thus the use of Earth–air heat exchanger in summer for the moderate climate of Bangalore is recommended.

Using Eq. (27.19), heating/cooling potential for four weather types of dome-shaped house was computed for each month. It is observed that for the climate of Bangalore heating requirement occurs in the 5 months of January, February, October, November, and December.

Figure 27.4 shows the heating potential during winter for the four weather types (a–d). In February, "a" type weather yield the maximum heating potential. While for rest of the winter months i.e. January, October, November, and December "b"

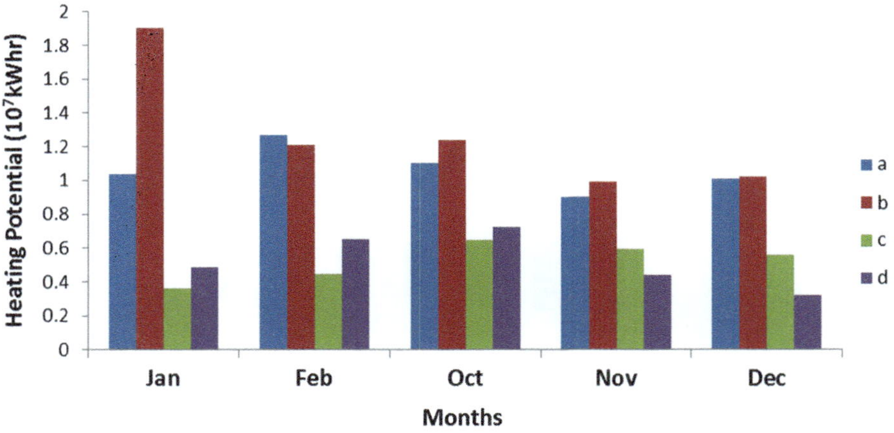

Fig. 27.4 Heating potential in winter at Bangalore ($N=0$)

type weather dominates and produces maximum heating potential owing to the maximum number of days in a month falling in that category.

Cooling potential for the 7 months of March to September for four weather types of dome-shaped house were also estimated and is shown in Fig. 27.5. For March, April, June, and September type "a" weather yields maximum cooling potential. For the months of May, July, and August "b" type weather dominates and hence produces maximum cooling potential.

Total monthly heating/cooling potential of dome-shaped house were estimated and shown in Fig. 27.6. During winter, the month of January shows maximum heating potential of 3.78×10^7 kW h and during summer May shows maximum cooling potential of 1.45×10^7 kW h.

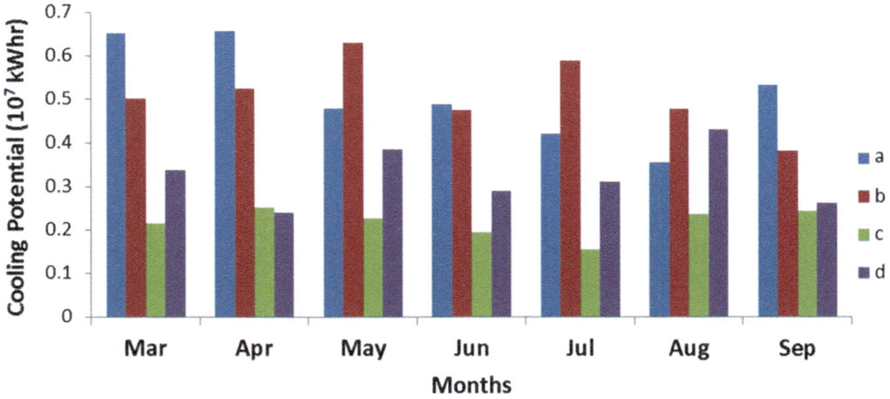

Fig. 27.5 Cooling potential in summer at Bangalore ($N = 30$)

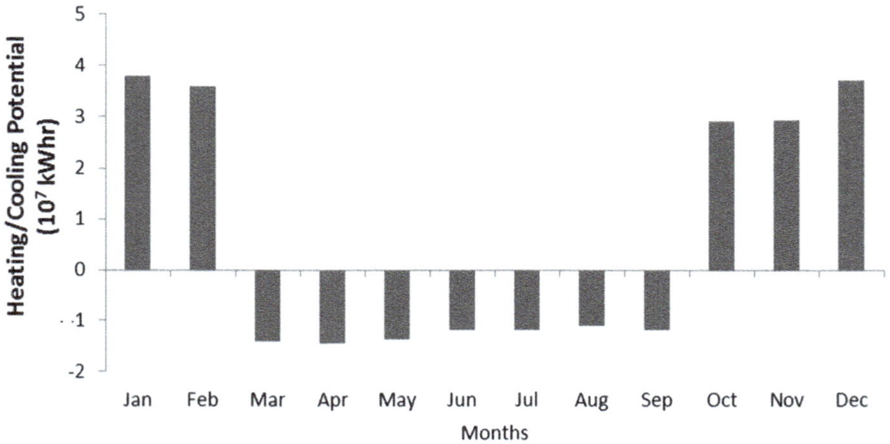

Fig. 27.6 Monthly heating/cooling potential at Bangalore

The magnitude of monthly heating potentials for the months of winter and monthly cooling potential for summer directly depends on the difference between room temperature and atmospheric temperature. The magnitude of monthly heating potential for winter is thus larger than that of monthly cooling potential due to the larger difference in room temperature and ambient temperature in winter.

The performance of dome-shaped house is, thus, found satisfactory in both summer and winter for the moderate climate of Bangalore. Hence, the adobe house under study can provide natural thermal comfort in moderate climate.

27.3.2 *Energy Reduction Potential for Hot Climate*

The variation of room air temperature for a typical day for 24 h has been computed using Eq. (27.18). Figure 27.7 shows the resulting hourly variation of room temperature for the month of January with and without Earth–air heat exchanger at Jodhpur. Variation of atmospheric temperature (10.5–24.2 °C) with time has also been shown in Fig. 27.7.

The room air temperature is in the range of 22.9–37.2 °C without Earth–air heat exchanger and 21.0–31.2 °C with Earth–air heat exchanger. Difference between atmospheric temperature and room temperature lies between 10.9 and 18.1 °C for the case of no ventilation (i.e., $N=0$). Earth–air heat exchanger does not improve heating of the room air. This is because the difference in base temperature T_0 and room temperature at $N=0$ is very less.

The hourly variation of room temperature at different rate of ventilation (or number of air change per hour, N), in the month of June at Jodhpur is shown

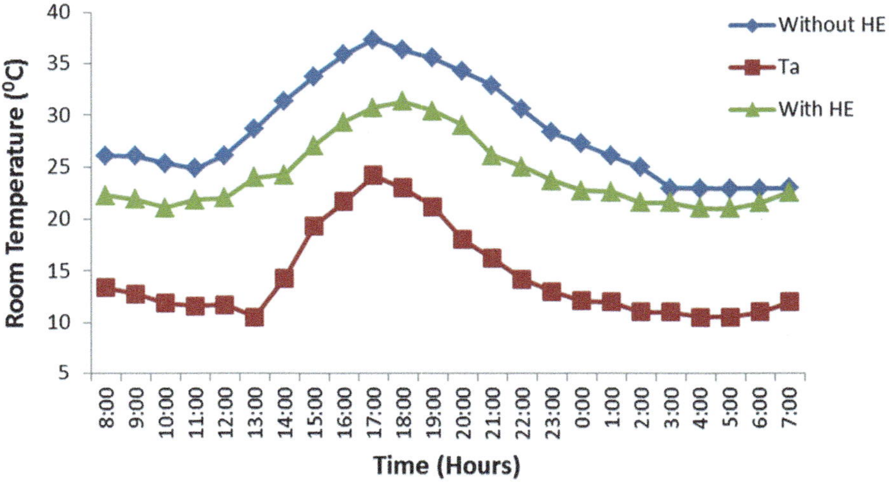

Fig. 27.7 Variation of room temperature with time in January for Jodhpur

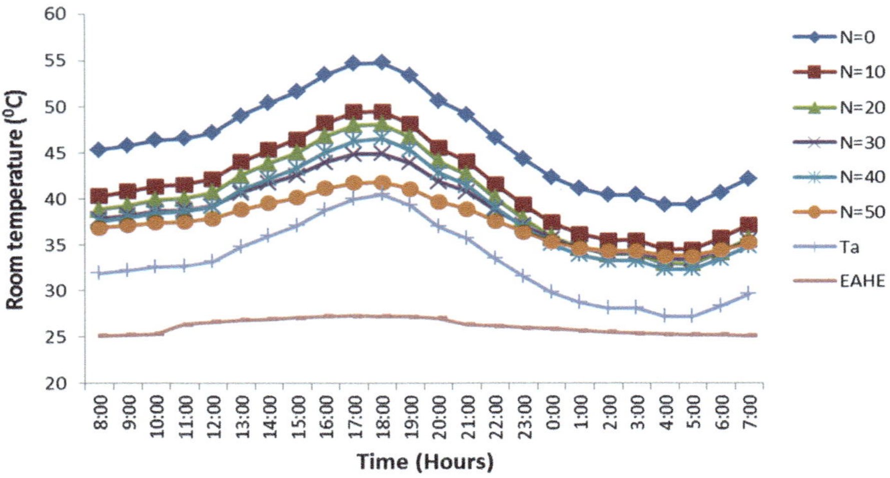

Fig. 27.8 Variation of room temperature with time in June for Jodhpur

in Fig. 27.8. Hourly variation of atmospheric temperature (27.2–38.8 °C) for 24 h has also been shown.

Effect of Earth–air heat exchanger on room temperature has also been shown in Fig. 27.8. At $N = 0$, room temperature is very high in the range of 39.3–54.7 °C. At $N = 10$, room temperature falls considerably and lies in the range 34.4–48.2 °C. At $N = 20$, room temperature again falls and varies from 32.9 to 46.8 °C. At $N = 30$, room temperature again fall and varies from 32.2 to 45.1 °C. At $N = 40$, room temperature varies from 33.3 to 43.9 °C. At $N = 50$, room temperature varies from 33.7 to 41.0 °C. With the Earth–air heat exchanger (length of PVC pipe = 78 m, air velocity = 3 m/s) integrated to the room, the hourly room temperature falls by a significant amount and varies from 25.0 to 27.1 °C. Maximum difference between atmospheric temperature and room temperature is 14.6 °C if there is no ventilation ($N = 0$). With ventilation at $N = 10$, 20, and 30, maximum temperature difference falls to 9.4, 8.0, 6.1 and 4.6 °C, respectively. At $N = 50$ maximum temperature difference is less than 2.8 °C, i.e., no significant reduction in room temperature is further observed. Therefore ventilation is optimized at $N = 40$. Earth–air heat exchanger improves cooling of the room air. With Earth–air heat exchanger, the room temperature becomes lower than atmospheric temperature. The maximum difference in atmospheric temperature and room temperature (with EAHE) is found to be 11.6 °C.

Using Eq. (27.19), heating/cooling potential for four weather types of dome-shaped house was computed for each month. It was observed that for the climate of Jodhpur heating requirement occurs in the 4 months of January, February, November and December. Figure 27.9 shows the heating potential during winter for the four weather types (a–d). For all the 4 months "b" type weather yields maximum heating potential owing to the maximum number of days in a month

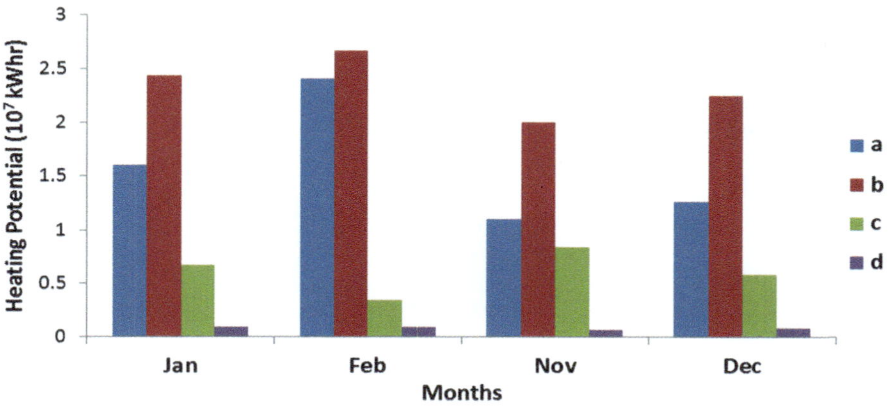

Fig. 27.9 Heating Potential in winter at Jodhpur ($N = 0$)

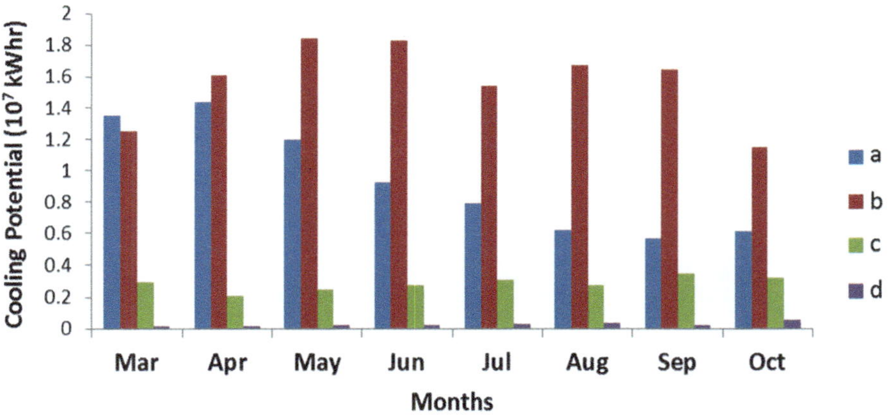

Fig. 27.10 Cooling Potential in summer at Jodhpur ($N = 40$)

falling in that category. Heating potential of type "b" is then followed by "a," "c," and "d," for each month of winter.

Cooling potential for the months of March to October for four weather types of dome-shaped house was also estimated and is shown in Fig. 27.10. In March, type "a" weather yields maximum cooling potential followed by "b," "c" and "d." While for the rest of months (April, May, June, July, August, September, and October), "b" type weather dominates and hence produces maximum cooling potential.

Total monthly heating/cooling potential of dome-shaped house was estimated and is shown in Fig. 27.11. During winter, the month of February shows maximum heating potential of 5.5×10^7 kW h and during summer May shows maximum cooling potential of 4.11×10^7 kW h. The magnitude of monthly heating potentials for the

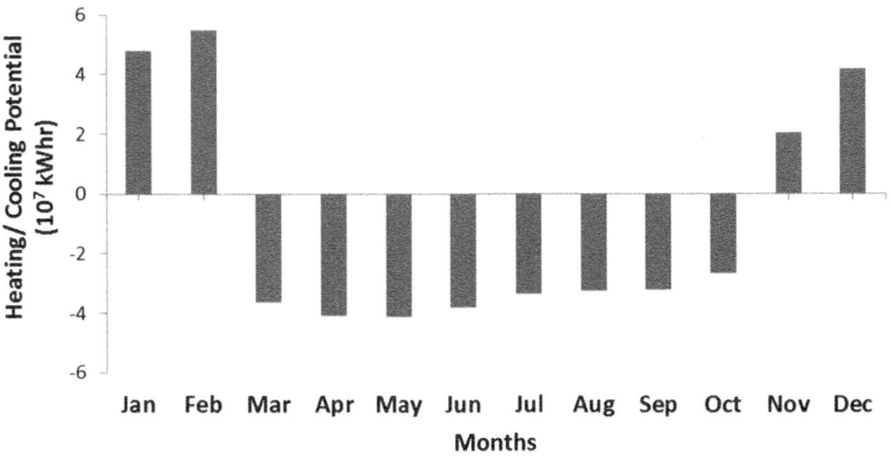

Fig. 27.11 Monthly heating/cooling potential at Jodhpur

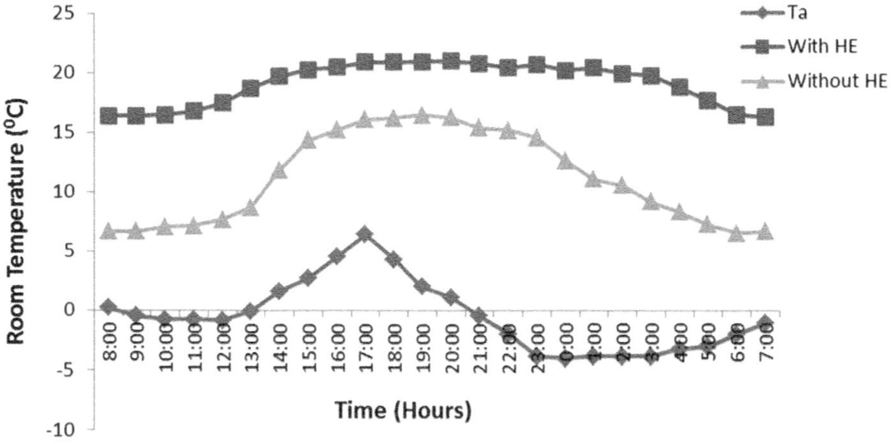

Fig. 27.12 Variation of room temperature with time in January for Srinagar

months of winter and monthly cooling potential for summer directly depends on the difference between room temperature and atmospheric temperature (in January, the maximum temperature difference is 13.0 °C and for June it is 12.1 °C).

27.3.3 Energy Reduction Potential for Cold Climate

The room air temperature for the cold climate of Srinagar has been computed by the use of Eq. (27.18) as for the previous cases. Figure 27.12 depicts the resulting hourly variation of room temperature during the month of January with and without Earth–air heat exchanger.

The atmospheric temperature dips below zero degrees centigrade during the night. It lies between −4.0 and 6.4 °C as shown in Fig. 27.12. The room air temperature is in the range of 6.5–16.2 °C without Earth–air heat exchanger and 16.2–20.9 °C with Earth–air heat exchanger. Difference between atmospheric temperature and room temperature lies between 6.4 and 18.3 °C for the case of no ventilation (i.e., $N = 0$).

The temperature of the room air is below the comfort level. In this case, it can be observed that Earth–air heat exchanger substantially improves the room air temperature. Therefore, it is beneficial to use an Earth–air heat exchanger to heat the room air in winters for the cold climate of Srinagar.

The hourly variation of room temperature at different rate of ventilation (or number of air change per hour, N), in the month of June at New Delhi is shown in Fig. 27.13. The Atmospheric temperature variation from 19.0 to 30.3 °C has also been shown in Fig. 27.13. The variation of room temperature when the room is integrated with an Earth–air heat exchanger has also been shown.

At $N = 0$, room temperature is in the range of 28.3–34.5 °C. At $N = 10$, room temperatures lies in the range 27.5–31.9 °C. At $N = 20$, room temperature again falls and varies from 25.8 to 30.1 °C. At $N = 30$, room temperature falls in the range of 24.3–29.7 °C. At $N = 40$, room temperature varies from 24.7 to 28.6 °C. At $N = 50$, room temperature varies from 24.9 to 28.1 °C. With the Earth–air heat exchanger integrated to the room, the hourly room temperature varies from 22.8 to 25.5 °C. Maximum difference between atmospheric temperature and room temperature is 11.5 °C if there is no ventilation ($N = 0$). With ventilation at $N = 10$ and 20 maximum temperature difference falls to 9.5 and 6.8 °C, respectively. At $N = 30$, 40, and 50 maximum temperature difference is 2.8, 2.3, and 2.0 °C,

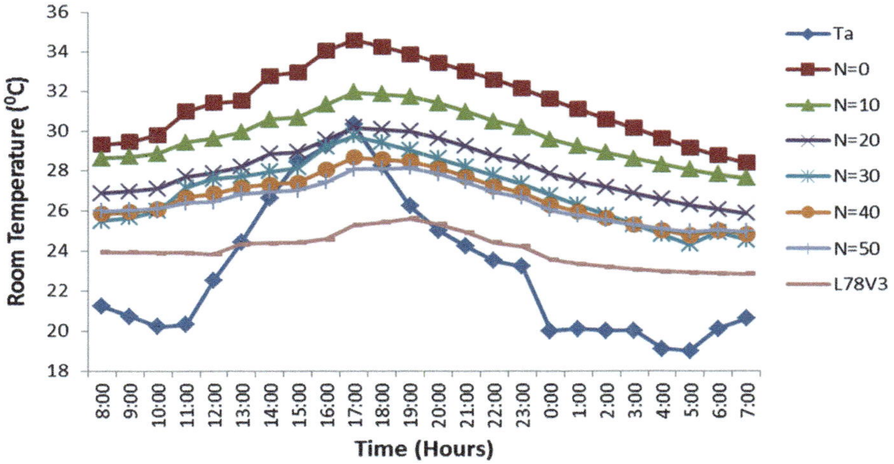

Fig. 27.13 Variation of room temperature with time in June for Srinagar

respectively, i.e., no significant reduction in room temperature is further observed. Therefore ventilation is optimized at $N = 20$.

Earth–air heat exchanger in this case does not substantially improve cooling of the room air. With Earth–air heat exchanger, the room temperature remains around the atmospheric temperature. Although, the maximum difference in atmospheric temperature and room temperature (with EAHE) is found during the day and is of the order of 3.6 °C. But, during the early day and late night hours, the room temperature with Earth–air heat exchanger is above the ambient temperature. Thus, the use of Earth–air heat exchanger in summer for the cold climate of Srinagar is not recommended.

Heating/cooling potential for four weather types of dome-shaped house was computed for each month. It was observed that for the climate of Srinagar heating requirement occurs in the 7 months of September to March. Figure 27.14 shows the heating potential during winter for the four weather types (a–d). In December, "a" type weather yields the maximum heating potential. While for rest of the winter months "b" type weather dominates and produces maximum heating.

Cooling potential for the 5 months of April to August for four weather types of dome-shaped house was also estimated and is shown in Fig. 27.15.

For May and June type "a" weather yields maximum cooling potential. For the months of April, July, and August "b" type weather dominates and hence produces maximum cooling potential.

Total monthly heating/cooling potential of dome-shaped house was estimated and is shown in Fig. 27.16. During winter, the month of January shows maximum heating potential of 12.56×10^7 kW h and during summer May shows maximum cooling potential of 1.74×10^7 kW h.

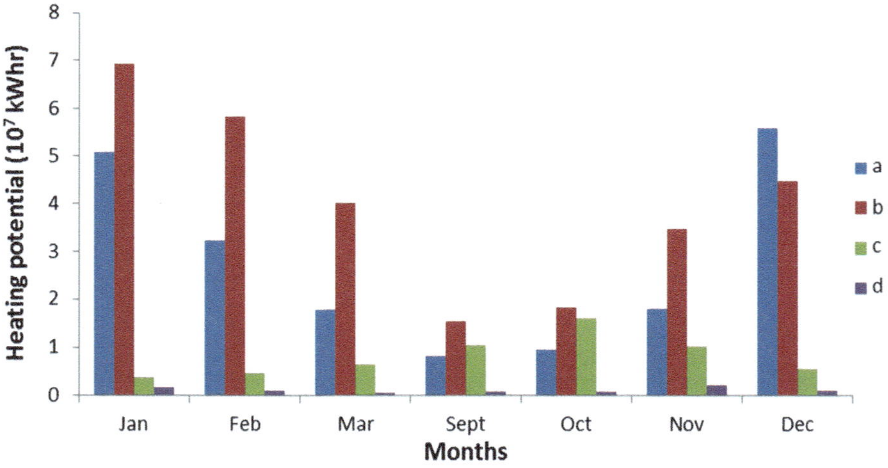

Fig. 27.14 Heating potential in winters at Srinagar ($N = 0$)

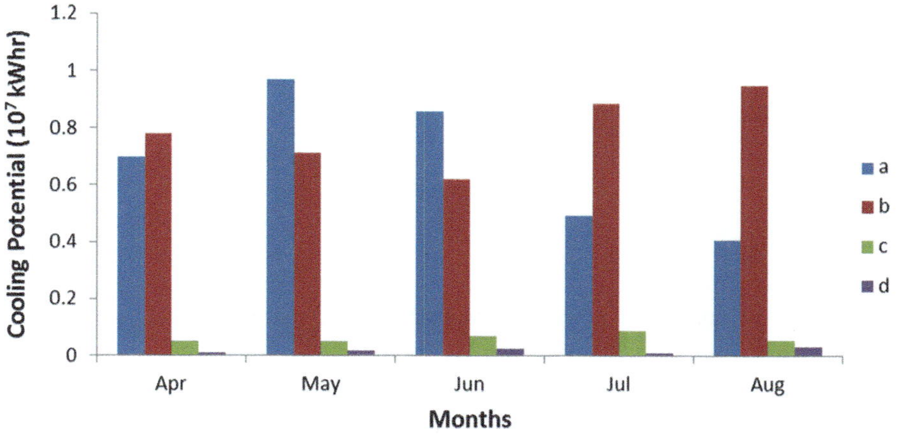

Fig. 27.15 Cooling potential in summers at Srinagar ($N = 20$)

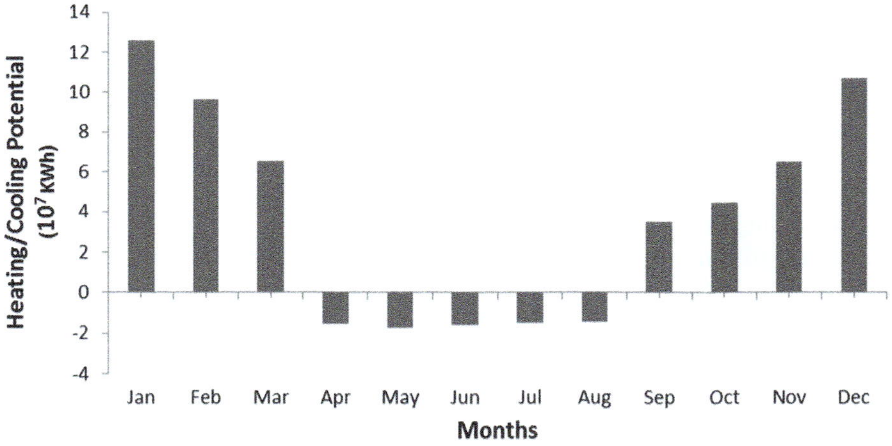

Fig. 27.16 Monthly heating/cooling potential at Srinagar

Table 27.2 Annual heating/cooling potential (10^7 kW h) for different locations

Climate type	Moderate	Hot	Cold
Location	Bangalore	Jodhpur	Srinagar
Annual heating potential	16.89	16.47	43.96
Annual cooling potential	8.87	28.14	9.77

The performance of dome-shaped house is, thus, found satisfactory in both winter and summer for the cold climate of Srinagar. Hence, the adobe house under study can provide natural thermal comfort in cold climatic locations.

Table 27.2 summarize the results obtained. Annual heating potential for winter was found maximum for the case of cold climate (Srinagar) of the order of

43.96×10^7 kW h while it was estimated minimum for hot and dry climate (Jodhpur) as 16.47×10^7 kW h. Annual cooling potential for summer was found maximum for hot and dry climate (Jodhpur) 28.14×10^7 kW h while it was estimated minimum for moderate climate (Bangalore) as 8.87×10^7 kW h.

27.4 Conclusions

The following conclusions can be made from the present work:

1. In the moderate climatic zone (Bangalore), cooling period is larger than the heating period. During winter, incorporation of EAHE improves the heating of the room air. During summer, ventilation is found to be a good option for cooling. Ventilation is optimized at $N = 30$ and the room air temperature is in the range of 23.6–34.0 °C. Earth–air heat exchanger improves the cooling of room air and brings the room air temperature close to comfort condition and is recommended for use in summer.
2. In the hot and dry climatic zone (Jodhpur) Earth–air heat exchanger could not improve the heating of room air and is not recommended for use in winter. While during summer, the EAHE maintains the room temperature in the range of 25.0–27.1 °C when the atmospheric temperature is in the range of 27.2–38.8 °C. Ventilation for Jodhpur, optimized at $N = 40$, is also found to be a good option for cooling in summer.
3. In the cold climatic zone (Srinagar) the room air temperature is quite low in winter due to low atmospheric temperature of below zero degrees centigrade. By the use of Earth–air heat exchanger, the room air temperature rises and lies in the range 16.2–20.9 °C. Thus, the use of EAHE during winters in cold climate is highly recommended. During summer, ventilation (optimized at $N = 20$) is found to be good options for cooling the house. The use of EAHE could not improve the room air temperature and hence is not recommended for use in summer.
4. Annual heating potential for winter was found maximum for cold climate (Srinagar) of the order of 43.96×10^7 kW h while it was estimated minimum for hot and dry climate (Jodhpur) as 16.47×10^7 kW h. Annual cooling potential for summer was found maximum for hot and dry climate (Jodhpur) 28.14×10^7 kW h while it was estimated minimum for moderate climate (Bangalore) as 8.87×10^7 kW h. The temperature of room with Earth–air heat exchanger was found varying in all the cases due to low isothermal mass of the building. Also, adobe house was found to temperate the room air temperature, thus significantly reducing temperature fluctuations.
5. The performance of dome-shaped house was found satisfactory in both winter and summer for the locations considered in various climatic zones. The temperature of room with Earth–air heat exchanger was found varying in all the cases due to low isothermal mass of the building. Also, adobe house was found to temperate the room air temperature, thus significantly reducing temperature fluctuations.

Acknowledgment The authors would like to thank the Indian Meteorological Department (IMD), Pune, for providing the climatic data for the locations considered.

Appendix: Design Parameters

Table 27.3 Design parameters for thermal modeling

Parameters for house	Value
h_o	9.50 W/m^2 K
h_i	2.80 W/m^2 K
K value of soil	0.44 W/m K
K value of brick	0.84 W/m K
K value of bamboo	0.17 W/m K
K value of Khapra	1.28 W/m-K
K value of wood (door)	0.14 W/m K
Air density	1.20 kg/m^3
Emissivity of roof	0.90
Absorptivity of roof	0.40
Absorptivity of wall	0.40
Absorptivity of door	0.60
Transmissivity of glass	0.90
Air change per hour	10–50
Volume of room	58 m^3
Floor area of room	16 m^2

Table 27.4 Design parameters of EAHE

EAHE parameters	Value
Depth of PVC pipe	1.5 m
Radius of PVC pipe (r)	0.3 m
Length of PVC pipe (l)	78 m
Air velocity at outlet (V)	3.0 m/s
h_{ca} inside pipe of EAHE	5.7 W/m^2
Specific heat of air	1,006 J/kg K

Table 27.5 Construction details of roof structure of the house

Roof material layers (inside to outside)	Thickness (mm)
Bamboo	4
Mud	4
Khapra	5

Table 27.6 *U*-value of building components

Building components	U (W/m^2 K)
Wall (brick)	1.36
Door (wood)	1.48
Window (glass)	2.15
Roof	2.00
Ground	0.44

References

1. BEE (2007) Annual Report 2006-07. Bureau of Energy Efficiency. http://beeindia.in/about_bee/documents/annualreports/2006-07E.pdf
2. International Energy Agency, IEA (2008a) India energy efficiency policies and measures: energy conservation building code, www.iea.org/textbase/pm/?mode=pm&id=4162&action=detail. Accessed from October 2008
3. Nik VM, Kalagasidis AS, Kjellström E (2012) Statistical methods for assessing and analysing the building performance in respect to the future climate. Build Environ 53:107–118
4. Koronakis P, Kalligeris Y, Santamouris M (1989) The contribution of appropriate energy design strategies and appraisal of a tourist residence complex in Crete, Second European Conference on Architecture, Paris, France, 2008
5. Lam JC, Wan KKW, Tsang CL, Yang L (2008) Building energy efficiency in different climates. Energ Convers Manag 49:2354–2366
6. Shukla A, Tiwari GN, Sodha MS (2009) Embodied energy analysis of adobe house. Renew Energ 34:755–761
7. Sanjay CP (2008) Passive cooling techniques of buildings: past and present – a review. ARISER 4:137–146
8. Tiwari GN (2002) Solar energy fundamentals, design, modeling and applications. Narosa Publishing House, India
9. Ahmad MJ, Tiwari GN, Singh AK, Sharma M, Singh HN (2010) Heating/cooling potential and carbon credits earned for dome shaped house. Energ Environ 1(1):133–148
10. Ahmad MJ, Tiwari GN (2008) Estimation of hourly global solar radiation for composite climate. Open Environ Sci 2:34–38

Chapter 28
The Proper Utilization of Passive Solar Energy in Residential Buildings, Northern Cyprus

Seyedeh Ayeh Mirrezaei, Harun Sevinç, and Ahadollah Azami

Abstract In the past, humans have always tried to have a relationship with the environment in order to create comfortable conditions for their lives. Over time, different cultures of the world created the architectural styles and techniques to adapt to climatic conditions. Unfortunately, building adaptation to the climate was gradually ignored with the arrival of the industrial revolution. Followed by high-energy consumption, environmental pollution and an energy crisis emerged. Thus there has been a revised consideration by architects motivated toward climatic design which is an effective approach to reduce the energy consumption in buildings. Undoubtedly, one of the important issues in compatible design with climate in mind is the proper utilization of solar energy. In this context, attention to the form and orientation of the building can significantly reduce energy loss especially in countries that enjoy solar energy such as Cyprus. This chapter focuses on the evaluation of appropriate orientation and form of dwellings regarding energy efficiency in Northern Cyprus based on a problem-solving method. The main purpose of this work is reducing building heat in summer and gaining heat in winter in order to achieve a favorable residential space. Ignoring climatic design can lead to harmful effects on human comfort.

Keywords Solar energy • Form • Orientation • Energy efficiency • Climatic design

28.1 Introduction

Today, building adaptation to climate has unique significance due to high-energy consumption, shortage of fossil fuels, environmental pollution, and human comfort. Climatic design is an effective way to achieve comfortable conditions and energy efficiency in buildings. Different characteristics of each climatic region have a great influence in shaping architecture. In this context, solar architecture plays a

S.A. Mirrezaei (✉) • H. Sevinç • A. Azami
Department of Architecture, Eastern Mediterranean University, Gazimagusa, North Cyprus via Mersin 10, Turkey
e-mail: ayeh.m373@gmail.com; harun.sevinc@emu.edu.tr; ahadollah.azami@emu.edu.tr

I. Dincer et al. (eds.), *Progress in Clean Energy, Volume 2*,
DOI 10.1007/978-3-319-17031-2_28

particular role in building design. The combination of architecture and solar energy can create convenient space for our living as well as being cost effective and affordable, therefore optimum utilization of solar energy is one of the effective factors for providing indoor comfort in residential buildings. In order to achieve this goal it is necessary to focus and apply the form and orientation of dwellings integrated with solar architecture in regard to energy efficiency. In this research, there is an attempt to present solutions for the problem of high-energy consumption of residential buildings and settlements in Northern Cyprus that have been evaluated in the terms of form and orientation. Mesan Apartment was selected as a case study in the city of Famagusta. Cyprus, due to geographical location and climate, can achieve energy savings by taking advantage of natural energy such as solar. However, current construction could not properly find ways to use this energy which is accessible and renewable, so the question comes to mind of why the ideal orientations and forms in contemporary dwellings are neglected regarding the appropriate utilization of solar energy and how we can achieve energy efficiency. The first step in solar design is based on analyzing the location and climate of Cyprus then compliance of the building with the climate. In the following, form and function of space should be evaluated in architectural design.

The methodology of this chapter is applied through problem solving and followed for the aim of this research via observation, questionnaires, and interviews. The case study research is a type of qualitative and quantitative research. Observation has been done to show the relationship between the building and sun and apperceive the best form and orientation for the building. The questionnaires designed for the people account for different ages and genders living in Mesan Apartment. The content of questions was based on location, orientation, shape, and plan of apartment relationship to the sun, cost of energy, use of passive solar system, and the tendency of energy efficiency measurements for the house. The outcomes revealed a significant dissatisfaction of spatial performance in two dimensions toward enjoying the sun. Also energy consumption accelerated enormously especially in winter and summer which led to an increase in energy prices. At the end, interviews with professional people have been done in order to deduce the best solution to the problem. The purpose of this chapter is based on heating the interior space in winter and cooling it in summer by attention to passive solar energy. As a result, disregarding building compatibility with the climate can have the greatest negative impact on the trend of energy use. Thus by proper form design, dwelling, and orientation in a proportionate and reasonable manner we may take advantage of the free resource of solar energy in our lives.

28.2 Literature Review

For energy efficiency one of the important characters in efficient building design is understanding the regional climate. Thus, compatible design with nature is one approach to providing comfortable conditions for the inhabitants. In this context, consideration of solar radiation has a significant impact on the process of suitable design of buildings. In order to create pleasant spaces for our habitat, it is necessary to pay attention to the relationship between the sun and architectural features of the building so we can take advantage of passive solar energy. In this regard, orientation, shape, and space performance of a building are the main architectural features in order to capture or lose solar energy during the different seasons [1].

28.2.1 Orientation

Proper building orientation is one of the most important features in architecture in order to reflect natural light and bring solar rays and fresh air to indoor spaces. Thus design seeks to improve user well-being and quality of spaces by considering the relationship between the path of the sun and the building [2]. As seen in Fig. 28.1, the best orientation of the building toward the sun is the southern side because of the sun's position toward the building. To utilize solar radiation in the building, it is necessary to situate it on the ground using the largest amount of solar radiation. Therefore the site should be identified and determined in order to find the proper location having a vast view to the southern side. Thus, the important factor about the building location is the low angle of solar radiation and latitude of the site (Fig. 28.1).

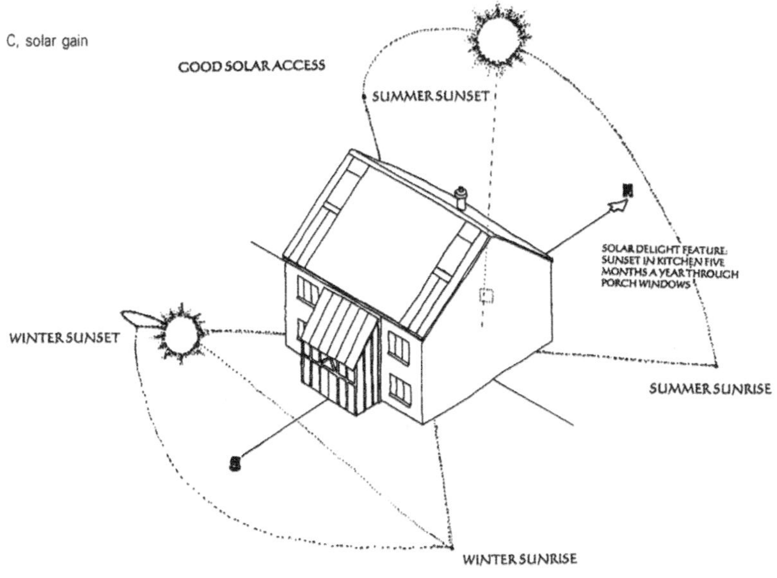

Fig. 28.1 Sun direction in summer and winter (Roaf [4])

28.2.2 Form

Consideration of climate, environment, and time (day, night, and seasons) is a particular part of designing the appropriate form in order to discover the amount of heating and cooling for creating comfortable spaces [3]. In this context, the southern facade, roof, and the number of windows should be considered in the building form. In the process of designing the plan these items should be considered:

- East–west elongation of the plan, in order to utilize maximum solar radiation.
- Allocation of main spaces on the south side and nonsignificant spaces on the west and east sides.
- Placement of spaces: for example, the kitchen which produces heat is better situated in the center of the plan in order to avoid heat loss.
- Arrangement of the plan due to penetration of solar radiation in interior spaces.
- Design of solar greenhouse on the southern side and near the main spaces.
- Attention paid to the path of the sun in indoor spaces [4].

Thus, efficient performance of spaces arises from an appropriate form of plan. The form of building plays an important role in the preservation of energy. The best building form should provide more absorption of energy and minimum heat loss in winter and maximal sorption in summer. Shaping the form is based on the demands of our shelters.

28.2.3 Passive Solar Design in Buildings

Passive solar design consists of direct and indirect gain systems. In the direct type, the roof, walls, and floor are integrated parts of the building for adsorbing the sun during the day and then transferring it to internal spaces at night. In this issue, windows play an important role in order to bring solar radiation directly into interior spaces. In an indirect gain system, solar radiation penetrates into interior spaces through intermediaries. A solar greenhouse (sunspace), the Trombe wall, atrium, the water wall, Thermo-syphon, and roof ponds are the main components of an indirect gain system. The performance of these elements causes the interior spaces to be heated in winter and cooled in summer. For instance, solar greenhouses capture solar energy on sunny days and then by natural convection come into indoor spaces in winter. By opening the windows of a sunspace, natural ventilation happens during the summer for cooling the spaces. In addition, it creates a pleasant space for users. As a result, designing the proper form in dwellings can allow us to bring passive solar energy into our shelters [5].

28.3 Field Study

Famagusta is a small coastal city, located in Northern Cyprus with a Mediterranean temperature. The north latitude of Famagusta is 35.13° and the east longitude is 33.95°. The height difference from sea level is 1 m. The population of the city is about 42,500 [5]. The maximum temperature of Cyprus reaches 41 °C in the summer and then in winter reaches 5 °C. Therefore, Cyprus has hot summers and cold winters [5] (Fig. 28.2).

The city of Famagusta was selected because it is in the best position to use solar energy. This city has an international university with many students, necessitating more residential buildings for their students. Therefore much new construction was built. Unfortunately, climatic design was ignored in these buildings. Disregarding this factor led to the emergence of the essential problem of high-energy consumption. Because of Cyprus' climate, paying attention to solar energy is an effective answer in this realm.

28.3.1 Data Collection and Methodology

The methodology of this chapter is based on problem solving along with both qualitative and quantitative analyzing. Data gathering was executed through the use

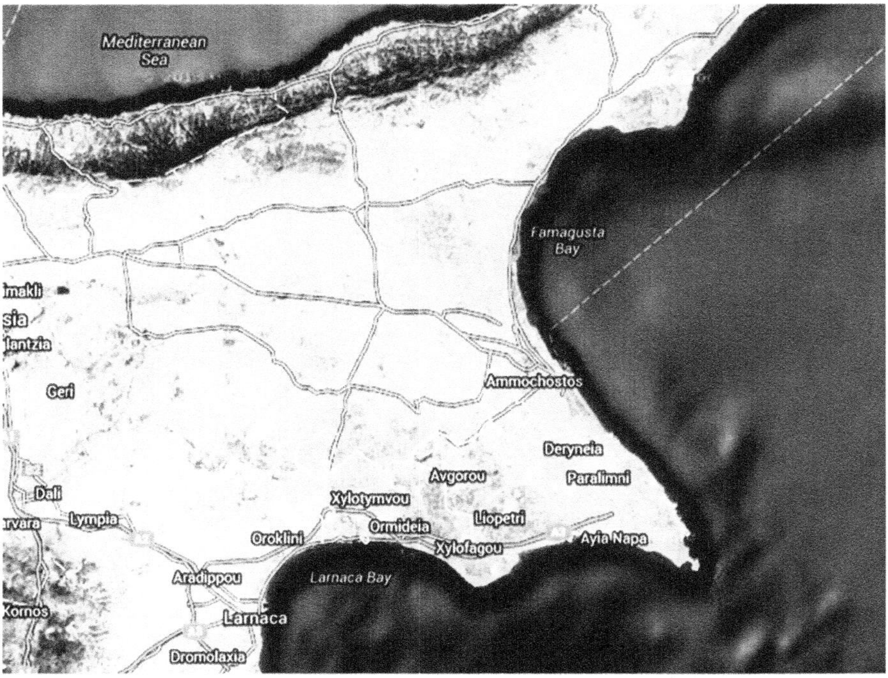

Fig. 28.2 Famagusta, Cyprus (18 Dec 2013. [7])

of observation, interviews, and questionnaires. The quantitative analysis of residential buildings was applied in order to improve the dwellings for their users. In this context, different questions were drawn in various fields such as satisfaction of daylight and sunlight, sources of energy for cooling or heating, arrangement of space performance, importance of windows, and also general questions about the financial investment for using solar energy. The qualitative analysis was conducted via observation and interview. During the observation, pictures for the case study were taken at different times of the morning, noon, and afternoon. Thus, the movement of the sun toward this building was evaluated. In order to comprehend how people provided comfortable conditions in their interior spaces and could reduce energy consumption by using passive solar energy in the future, the interview was done with regard to proficiency.

28.3.2 Interview with Local People

The questionnaires were conducted with the residents in theMesan Apartment in order to analyze the form and plan of the building regarding the sun. Over 45 users participated in this research. In the pie chart shown in Fig. 28.3 a large majority of inhabitants [87.7 %] complained about daylight. They have to use electricity for lighting their rooms. Only a few were happy: they liked dark spaces and avoided incandescents in their spaces during the day (Fig. 28.3).

The pie chart in Fig. 28.4 shows that 89 % of residents were disappointed in the amount of sunlight in heating their spaces during cold days, and only a handful of them [11 %] were satisfied. They explained that they weren't at home during the day so the sunlight was never important for them.

The chart in Fig. 28.5 expresses the various types of sources for domestic heating and cooling. As seen, electricity was allocated the highest level in both cooling and heating, whereas solar energy was in the lowest position.

Figure 28.6 indicates the space performance arrangement in the plan of Mesan Apartment. According to this chart, the majority of user dissatisfaction was about

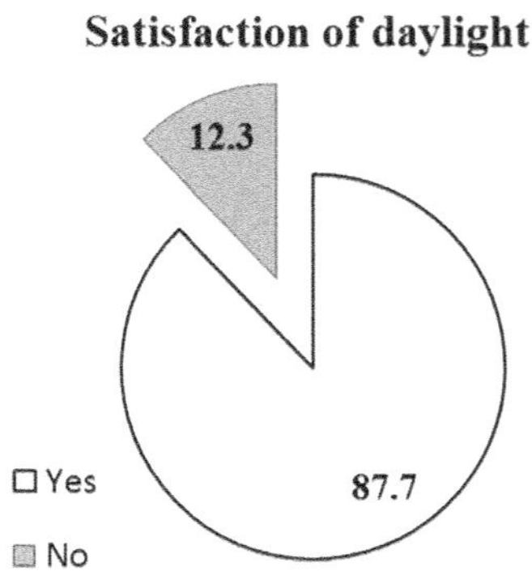

Fig. 28.3 Satisfaction of daylight (drawn by author)

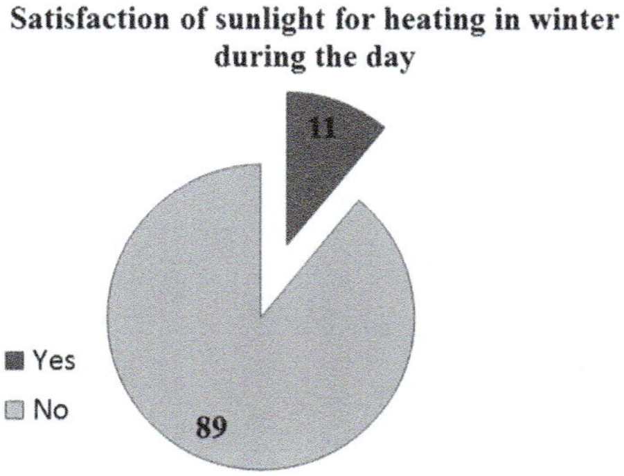

Fig. 28.4 Satisfaction of sunlight for heating in winter (drawn by author)

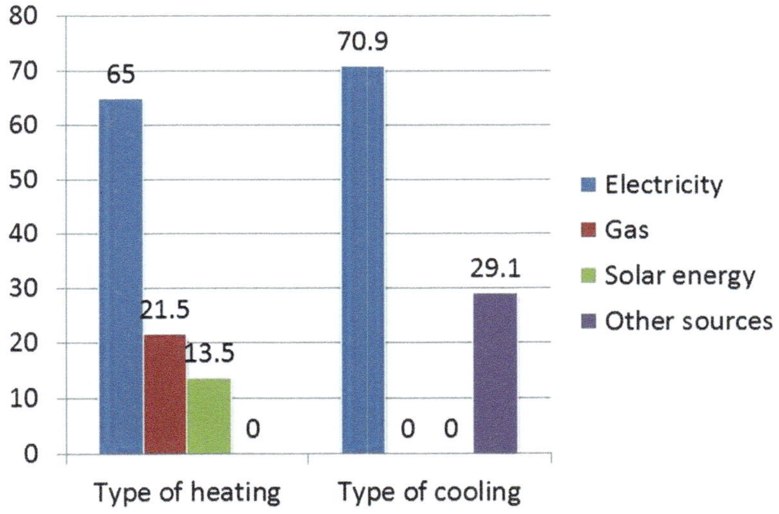

Fig. 28.5 Prototypes of cooling and heating

the situation of the living room, master bedroom, and other bedrooms. They explained that they could not benefit from solar radiation especially in winter for heating and lighting their rooms. The interior spaces were very cold and also gloomy, so they had to use electricity or gas for heating and lighting. On the other side, the position of the kitchen was the best one. They pointed out that this part was the heated section of the building in winter because it enjoyed the sun. They believed that these spaces should be shifted. The other three graphs reflect the importance of windows from various aspects (size, numbers, and orientation) in order to absorb the sunlight. As expected, the majority of occupants were dissatisfied with the number and size of windows because the apartment had small-sized insufficient windows on the southern side. Almost all of them showed their agreement that the windows should be located on the southern side in order to provide

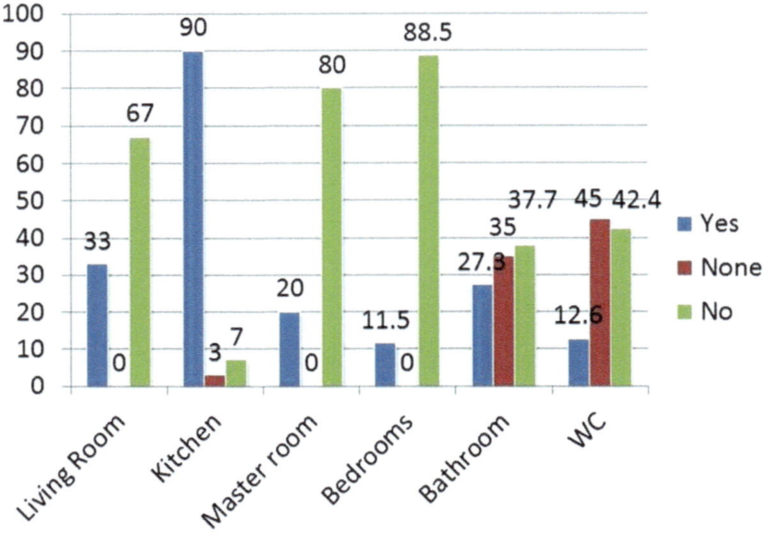

Fig. 28.6 Space performance arrangement satisfactory (drawn by author)

Fig. 28.7 Size of windows

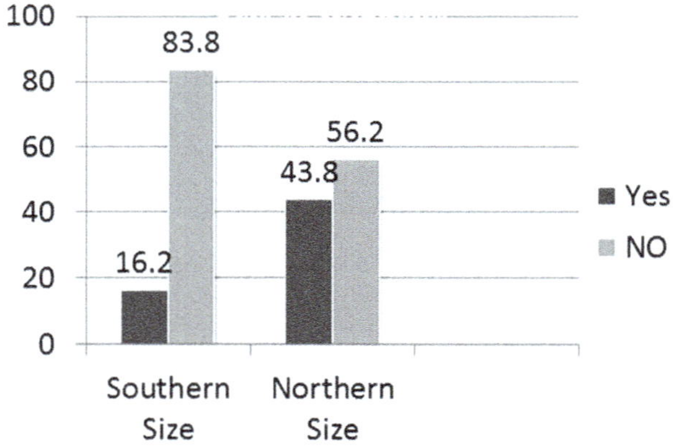

sufficient sunlight. Due to the cold wind in winter, people were dissatisfied with the large number of with huge-size northern windows (Figs. 28.7, 28.8, and 28.9).

This question was drawn from different people in the city, not only those in the mentioned apartment. It was a general question to determine the importance of using solar energy. The pie chart illustrates the tendency of people to make a financial investment in order to benefit from solar energy in their flats. It is apparent from the charts that more than half the people disagreed (64.9 %). Approximately, 21 % of them agreed and a very low percentage (13.5 %) didn't have any idea. The most surprising indicator was that they knew the advantages of using solar energy in

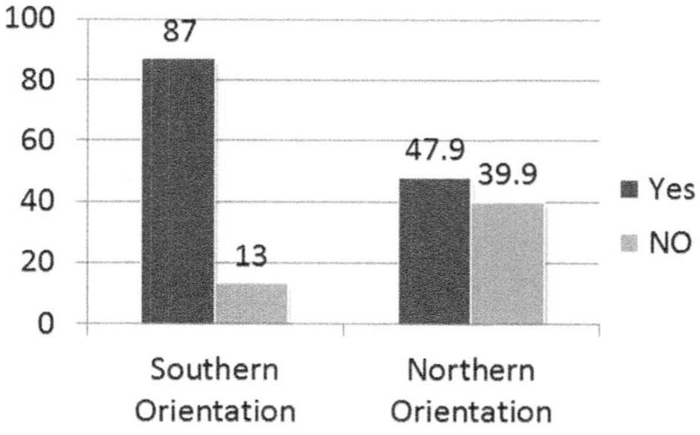

Fig. 28.8 Orientation of windows (drawn by author)

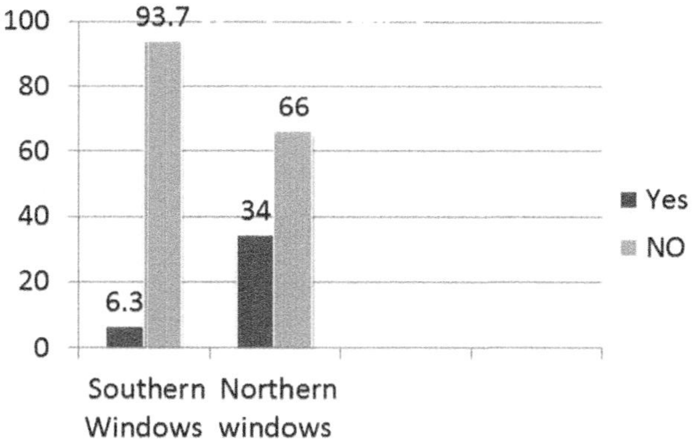

Fig. 28.9 Number of windows

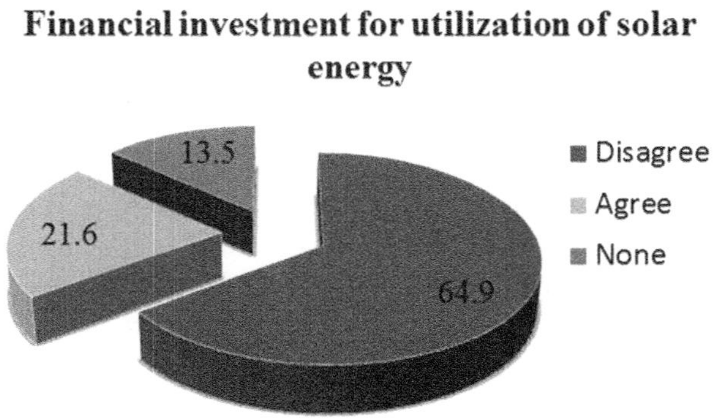

Fig. 28.10 Financial investment for solar energy utilization (drawn by author)

the near future and also about high energy consumption. Despite this fact, they did not show a tendency to invest money for reconstruction of their houses in order to benefit from energy efficiency. They were of the opinion, "Who will give us the money," even those who were financially capable (Fig. 28.10).

28.3.3 Observation

The pictures from Mesan Apartment were taken with the intention of evaluating the relationship between the sun and this building at different times and examining the best form and orientation of the building toward the optimal utilization of solar energy.

- Sunrise will be from the southeast and sunset will be from the southwest at this time of the year. Thus the relationship between the building and position of the sun is very important. According to these pictures, the southern walls receive the largest amount of heat and light from the sun. Therefore openings on the south can make a significant contribution to winter heating. Unfortunately, the southern side of this building has remarkably limited openings for enjoying the sun especially in winter. Moreover, these windows do not have any shading system for sun protection in summer. Instead of a southern wall, the northern walls are arranged with many windows. The height and size of the southern windows are not proportional because solar radiation comes directly to interior spaces and solar glare disturbs human comfort. Southwest walls can enjoy the sun especially in winter but you cannot see the windows in this part of building (Figs. 28.11, 28.12, and 28.13).
- The building form should be designed to have minimal absorption in summer and maximum achievement of solar energy in winter, but we cannot see this feature in this case study.
- A semicircular form of wall is very effective for absorbing the sunlight but in this case, the design is applied only on the balcony of the eastern wall so it is not useful.
- Due to high humidity in the summer, bilateral airflow is particularly important in order to comfort people in interior spaces. However, windows and ventilation are not in appropriate locations for airflow.
- The roof is an effective building component in solar architecture because of design integration. This building has a flat roof that has not been used to utilize solar energy properly.
- On the second floor, the northern side of the envelope incorporates a large balcony that faces the street without any buffer zone. It is not efficient for all of the seasons. People can use it only in summer despite vehicle sound pollution. On other floors there are small narrow balconies on the northern sides that do not have a profitable performance. Other balconies are situated on the eastern side which faces the Mediterranean wind. Thus, they are not beneficial for autumn and winter.
- In general, the planning system was badly designed. The quality of plan performance is very poor in this building. Because the living room and bedroom are located in the north, it is not possible to take advantage of solar radiation. The shortage of daylight causes people to turn on the lights at 4 p.m. In winter, interior spaces of houses are very cold. Only the kitchen and bathrooms are situated on the southern side (Figs. 28.14 and 28.15).

Fig. 28.11 Photos, Mesan Apartment, at 12:00 p.m.

As seen in Fig. 28.16, the southern side does not have efficient windows. Southern windows are very important in order to add their particular contribution for heating internal spaces in winter. Unfortunately, this building cannot enjoy this feature.

28.3.4 Interview

At the end, interviews with professional masters categorized the executable solution for the problem of the case study. Assistant Professor Dr. Harun Sevinç and Ahadollah Azami (solar architects) from the architecture faculty at EMU University

Fig. 28.12 Photos, Mesan Apartment, at 4:00 p.m. (taken by author)

participated in this interview. In this context, attention to the climate of each region during the design stage is an efficient approach to achieving energy efficiency. The position and orientation of the building toward the sun on the site, proper performance spaces in the plan, a suitable form of building, efficient openings, and use of passive and active solar are some factors in solar design. Unfortunately, current construction shows the opposite because architects do not have sufficient knowledge about state organization related to climate effect. It seems that they try to get money rather than quality of their design. As a result, a lack of balance between internal and external temperature, shortage of human comfort in the house, high energy consumption, the energy crisis, and environmental pollution are happening.

Fig. 28.13 Photos, Mesan Apartment, at 8:00 a.m. (taken by author)

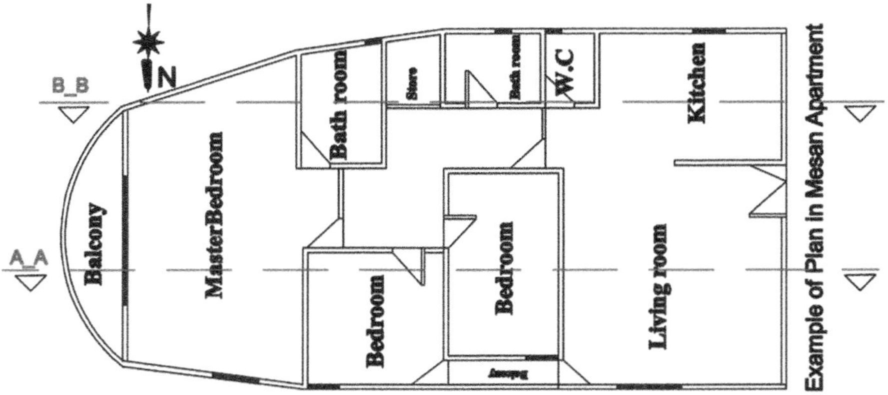

Fig. 28.14 Plan in Mesan Apartment (drawn by author)

Fig. 28.15 Location of case study, Google map (31 Dec 2013. [8])

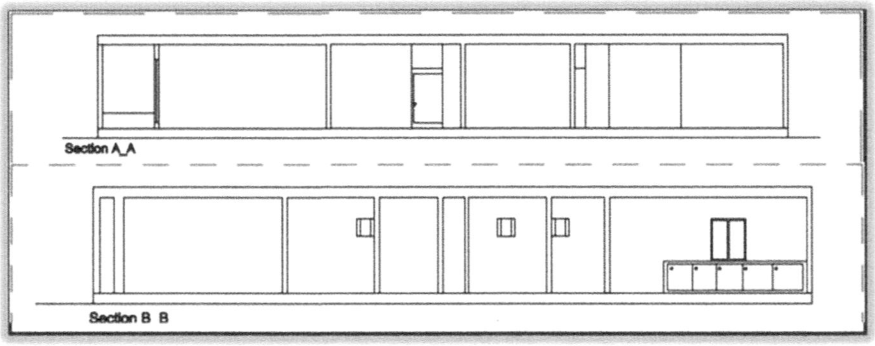

Fig. 28.16 Mesan Apartment section (drawn by author)

28.3.5 Data Evaluation

Throughout the data collection, it was revealed that dissatisfaction of Mesan's residents included the orientation and plan arrangement of their flat, the high cost of energy, high energy consumption, and shortage of solar radiation in internal spaces. In addition, the most surprising part of the questionnaires implied that people were reluctant to make a financial investment for energy efficiency despite awareness of the advantages of solar energy. During the observation, the relationship between the sun and the building was determined by taking photos at different times. According to the images, inattention to this relationship was proved. This present study has set up an evaluation of the form and orientation of the dwelling regarding energy efficiency and the importance of the relationship between the buildings and the sun. By regarding the outcomes of observations, questionnaires, and interviews, these facts revealed:

- Disregard of coordination of the building with the environment in order to create comfortable conditions for humans in interior spaces.
- Shortage of solar radiation in internal spaces because of inappropriate orientation and form of apartment.
- Inefficient and limited windows in southern walls for enjoying the sun.
- Insufficient daylight in indoor spaces.
- High energy consumption during the cold winter and hot summer.
- Improper arrangement of spaces in the plan of houses.
- Unwillingness of people for reconstruction of their houses in order to use solar energy because of financial issues.
- Architects did not have sufficient knowledge about building compatible with climate in the process of designing both form and plan.

28.3.6 Recommendation

According to the findings mentioned, there are various and effective solutions for improving the quality of residential buildings in order to utilize solar energy, reduce energy consumption, and provide comfortable conditions in interior spaces in different seasons. See Fig. 28.17.

Due to the climate of Cyprus, these solutions were offered:

- Considering the design principles of the building with climate and nature with the goal of reducing energy consumption.
- A possible east–west elongation in the plan.
- Having appropriate width for designing the plan in order to access better airflow and ventilation.
- Main areas requiring more heating should be set in the south side of the plan. It is better to track the sun's movement in designing the plan.
- Designing more windows with appropriate size on the southern side.
- Using a passive solar system such as windows with direct gain and proper sunshade, Trombe wall, roof ponds, and green house.
- Covering the external surface with greenery to modulate temperature and decrease temperature fluctuation.
- Designing a suitable form of building in order to integrate a passive solar system.

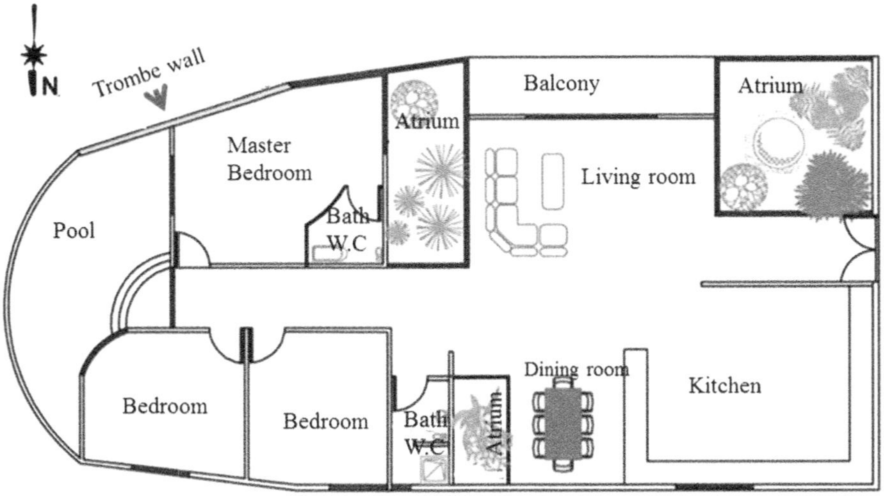

Fig. 28.17 Recommended plan for Mesan Apartment (drawn by author)

- The best range of adsorbing solar radiation is between the 15° West and 30° east toward the south.
- Using efficient shading systems for protection from direct solar radiation (Fig. 28.17).

28.4 Conclusion

Buildings play a key role as climatic modifiers, therefore adaptation of buildings to climate is an effective step for reducing energy consumption. This chapter established that the form and orientation of shelters have a great impact on the reduction of energy consumption and achieve positive results toward the utilization of solar energy. Providing comfortable and pleasant conditions in internal spaces is strongly based on the relationship between the climate and demands of buildings. As a particular prototype, the sun is a significant climate factor and can have both positive and negative effects in a building. By giving consideration to the proper design of form and paying attention to the orientation of the building toward the sun, it is possible to use the eternal energy of the sun. Thus it follows that we will have a decline in energy consumption, environmental pollution, and the energy crisis. Due to the climatic conditions on Cyprus, more utilization of solar energy power and a new approach to solar energy in creating the European Union building in Nicosia for a solar plan or solar energy can be suitable in the near future. In addition, cooling and thermal storage systems are being designed into new and existing buildings during construction. Although some arrangements have been accomplished it is not enough and we need more development. The development of solar energy usage depends on people's beliefs and cultures in society. As an example, people should be taught and be familiar with the advantages of using solar energy and how to access it. In this trend, government plays an important role in encouraging people toward the optimal utilization of energy. Awareness of people and allocating subsidization can be effective in motivating people to use it.

Cyprus has a high number of sunny days per year and in the case of obtaining an amount of solar energy has suitable conditions, but unfortunately, it does not pay attention to such an immense source of energy. If climatic characteristics of each city are not taken into account we may have big problems such as a negative impact on energy utilization in buildings. Thus, it is better to find new technology designs in order to provide solar energy rather than the consumption of other fuels. Compatible building design can reduce energy dissipation and create a desirable effect on human life. In addition, Cyprus can achieve high independence in the field of importing fuels by the widespread usage of solar energy, thus the government sought to include solar energy in future programs in order to move dramatically into a new era of clean and renewable energy. I hope that this research creates a positive step toward reducing energy consumption and the energy crisis, and motivates architects to revise their designs in order to provide affordable conditions for residents.

References

1. Galloway T (2004) Solar House, a guide for the solar designer. Architectural press An Imprint of Elsevier Limited Linacre House, New York, NY, pp 17, 18
2. Kasmaee M (2003) Climate and architecture. Isfahan, Khak Publications, New York, NY. pp 305–311
3. Phillips D (2004) Daylighting: natural light in architecture. Architectural Press An imprint of Elsevier Linacre House, New York, NY, p 25
4. Roaf S (2001) Eco house – a design guide. Architectural Press An imprint of Butterworth-Heinemann Linacre House p17, New York, NY
5. URL1: Famagusta Maps - road map, satellite view, street view and GPS navigation (2011) http://www.maps-streetview.com/Cyprus/Famagusta/. Accessed 18 Dec 2013
6. URL2: Climatemps.com (2012) http://www.famagusta.climatemps.com/january.php. Accessed 18 Dec 2013
7. URL3: Famagusta satellite view & Famagusta Street view (2011) http://www.maps-streetview.com/Cyprus/Famagusta/satelliteview.php. Accessed 18 Dec 2013
8. URL4: Google Maps (2013) https://maps.google.com. Accessed 31 Dec 2013

Chapter 29
A Study on Thermo-Physical Properties of Building Materials: Concrete of Vegetable Fiber

Drifa Belkharchouche and Abla Chaker

Abstract The objective of this work is to make our contribution to the development of local resources, namely vegetable fibers (grignons of olives) of low cost and from a renewable source, and integrate it in a rational way in the field of construction. To date, these fibers remained undeveloped except for some traditional uses. A judicious choice of proportion additions (fiber) and technical implementation are considered. Particular attention is paid to the thermal characteristics (comfort and insulating properties of the habitat) and mechanical resistance which is a decisive criterion for the choice of a material in the construction. We determine the thermal conductivity of the materials studied with an experimental apparatus that allows us to make measurements of the thermal properties under actual operating conditions (temperature and humidity). Results show a significant reduction in the density and thermal conductivity with increasing dosage of vegetable fibers in the concrete. Thus, improving of the thermal insulation capacity is confirmed. The compressive strength decreases with increasing dosage of plant fibers.

Keywords Porous materials • Concrete of vegetable fiber • Thermo-physical properties • Box method

Nomenclature

ΔT	Temperature difference between indoor and outdoor environments (°C)
λ	Apparent thermal conductivity (W/m °C)
q	Heat flux (W)
S	Surface of the sample (m^2)
e	Thickness of the sample (m)
C	Heat loss coefficient (W/°C)

D. Belkharchouche (✉) • A. Chaker
Energy Physics Laboratory, University Constantine 1, Constantine, Algeria
e-mail: belkharchouchehayet@yahoo.fr

I. Dincer et al. (eds.), *Progress in Clean Energy, Volume 2*,
DOI 10.1007/978-3-319-17031-2_29

W Water content (%)
V Voltage across the heater (V)
R Electrical resistance (Ω)

29.1 Introduction

Concrete is a heterogeneous composite of the resulting intimate mixture of cement, aggregates, water, and small amounts of adjuvant. These constituents are dosed so as to obtain, after curing, a solid product whose mechanical properties can be very superior to those of natural rocks. Insofar as cement is a hydraulic binder made from natural minerals, concrete can be considered as an artificial rock, but its fragility and density remain a handicap of its thermal and mechanical behavior, which is why it is reinforced with steel bars showing the tensile stresses. But instead of reinforcing concrete, and especially the thin parts which pose problems of size and protection of reinforcement, we can consider incorporating fibers throughout the material. Indeed different types of fibers are used as reinforcement of various materials to increase their mechanical strength and to improve their stability and thermal insulation. Before the development of composite materials, concrete is not the exception; researchers and fiber producers have thought of incorporating fibers as reinforcing materials in cement matrices [1]. Since then, the use of fibers in concrete has become a practice increasingly common and applications are developed through their proven ability by experience. The study allowed us to find new lightweight building materials and the valorisation of plant fibers and of low cost from renewable sources; such as grignons of olives, currently burned or used as animal feed.

29.2 Materials

After a deep investigation of the various existing local materials in Algeria and taking into account their availability and their cost, our choice is fixed on vegetable fibers to lighten the concrete. The proportions of additions are studied subsequently to develop new materials that can be insulating-carriers at competitive prices. Also thermophysical and mechanical characterization of these materials is carried out to confirm these choices. Natural fibers selected for this study are: The grignons of olives.

29.3 Experimental Determination of the Thermal Conductivity

The technique used to measure the thermal conductivity is called "box method" Fig. 29.1. It presents the advantage of a very simple implementation and the measurement precision is comparable to that obtained by conventional methods (hot wire, etc.) [2]. The box method allows to measure thermal conductivity of materials tested in permanent regime by realizing an energy balance of the system. The measuring principle is based on achieving a permanent unidirectional heat flow through the sample, supposed to be homogeneous and without internal generation of heat, by creating a temperature gradient between its two faces. Indeed, the sample of parallelepiped form is placed between two enclosures, one of which is heated and the other is cooled, in such a way that the lateral flows are negligible. Once the permanent regime is established, the thermal conductivity is given by the Fourier law:

$$\lambda_a = e \cdot q/S \cdot \Delta T = e/S \cdot \Delta T\left(V^2/R + C\Delta T'\right) \tag{29.1}$$

For measuring the moisture influence on the thermal conductivity, the used variable is the volumetric water content W. This is done by practicing successive weightings of the material from the partial saturation state until the dry state. Thus, the first wet measure is obtained after immersion of the material, for a few days, in water until its mass remains constant for 24 h (mass variation <0.1 %), which corresponds to the partial saturation state. For the intermediate wet measures, the first two drying operations are done in the ambient air with the measure of the corresponding values of thermal conductivity. Then, the drying is continued in a ventilated and regulated oven at 60 °C until the dry state. This is for obtaining certain continuity of the curves which give the variation of λ and a as function of volumetric water content.

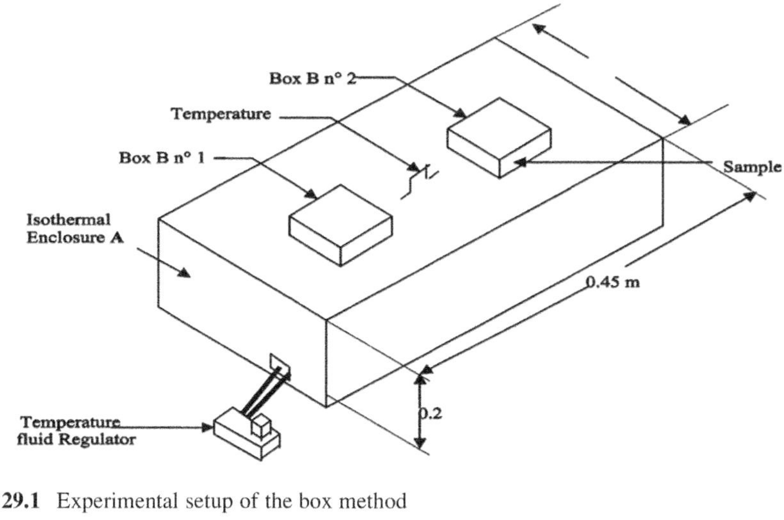

Fig. 29.1 Experimental setup of the box method

29.3.1 Preparation of Specimens

The homogenized mixture is then introduced into parallelepiped molds; two dimensions of molds were made according to the type of tests to be realized. As regards the thermal aspects, the molds of dimension ($27 \times 27 \times 6$ cm^3) were used. For mechanical strength, the molds ($16 \times 4 \times 4$ cm^3) were used. These different sizes are related to measuring devices (thermal box, mechanical press) whose dimensions are imposed. The use of the same manufacturing process, regardless of the mold, allows working on the same material. The specimens are preserved before and after turning out into the room test at a controlled temperature and humidity (Ta = 20 °C, RH = 60 %). These conditions correspond to a standard climate and allow reproducing real conditions of the use of the material [3].

29.3.2 Formulations

The method of composition of the classic concrete used is the one of BARON-LESAGE and GORISSE [4] for a ratio E/C (water on cement) minimally given; these methods aim at the optimization of the ratio S/G (sand on gravel) to obtain better workability [5].

- The weight of the fresh concrete is 2,350 kg/m^3.
- Good workability is obtained for a concrete having an S/G ratio of between 0.58 and 0.89.

To improve the thermal properties of concrete vegetable fibers (grignons of olives) were incorporated at rates ranging from 0 to 2 % (0, 1, 2 %). The sample thickness is 6 cm.

The aggregates used in the different mixtures were prepared from two sizes 5/8 and 8/15, sand (quarries) and cement portland CPJ45 Hamma Bouziane.

Water is the main component of agents that can degrade the material; we conducted a study on the influence of humidity on the thermo-physical properties of the material (Table 29.1).

Table 29.1 Mix design proportion

Materials	Mix 1 (0 %)	Mix 2 (1 %)	Mix 3 (2 %)
Cement (kg/m^3)	400	400	400
Aggregate (kg/m^3)	1,000	982	973
Sand (kg/m^3)	750	750	750
Water (kg/m^3)	240	270	290
Vegetal fiber (kg/m^3)	0	18	29

29.4 Results and Discussion

The observation of Fig. 29.2 shows that the thermal conductivity increases with the apparent density. It is evident that the critical factor in this growth is that the porosity of the material decreases as the material is dense.

Furthermore, the observation of these curves shows that the thermal conductivity of the concrete in the saturated state is higher than that of the concrete in the dry state. This is explained by the fact that by adding water to the dry concrete, we replace a component of low thermal conductivity (thermal conductivity of air is approximately 0.026 W/m °C) by forming a higher conductivity (the thermal conductivity of water equal to 0.6 W/m °C), 30 times higher than that of the dry air. This results in an increase of the overall thermal conductivity of the material [6, 7].

Figure 29.3 shows the shape of the experimental curves expressing the variation of the apparent thermal conductivity as a function of water content for different fiber percentages; measurement of the average temperature is between 16 and 25 °C. These graphs highlight the heat-insulating power and water absorption of these composite materials. We also note that the apparent thermal conductivity increases significantly with the volumetric water content without having a linear variation. This has been observed on other materials such as plaster [8]. The incorporation of fibers in the concrete, during its mixing, improves its thermal insulation power [9].

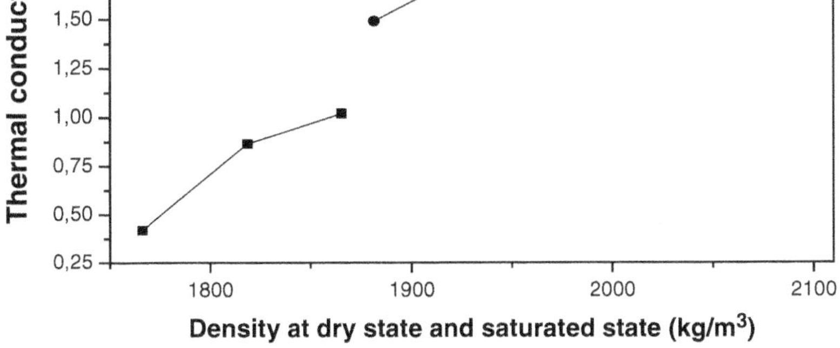

Fig. 29.2 Influence of density in the dry state and saturated state on the thermal conductivity

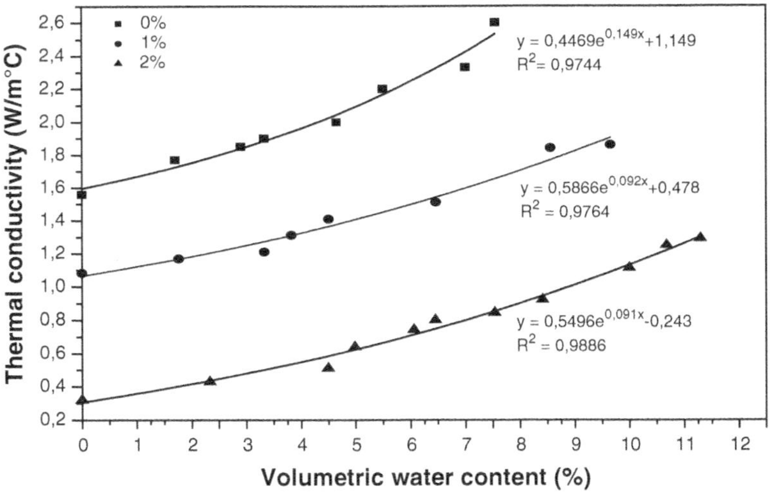

Fig. 29.3 Variation of thermal conductivity as a function of water content for different percentages of fiber

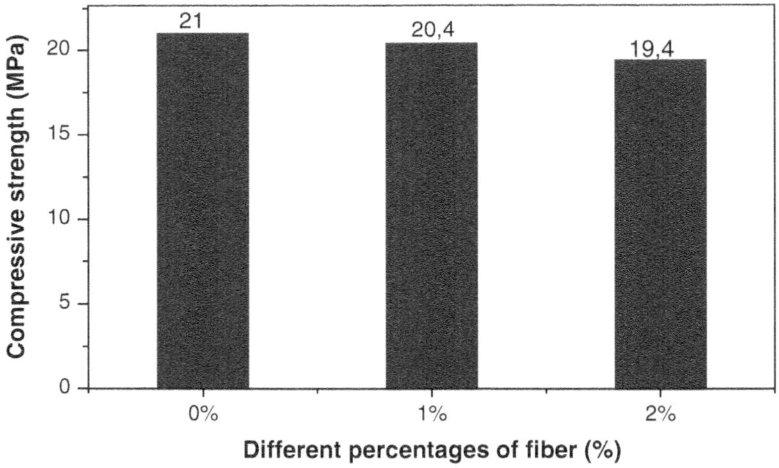

Fig. 29.4 Variation of the compression strength as a function of different percentages of fiber

29.4.1 Compressive Strength

The mechanical compression strength after 28 days is obtained by crushing test pieces with a hydraulic press. In Fig. 29.4, we have represented the variation of the compressive strength as a function of different percentages of fiber. Then, in Fig. 29.5 is shown the variation of the compressive strength as a function of density. It is clear that the gain of the light is accompanied by loss of mechanical strength compression, due to the decrease of the compactness of the material.

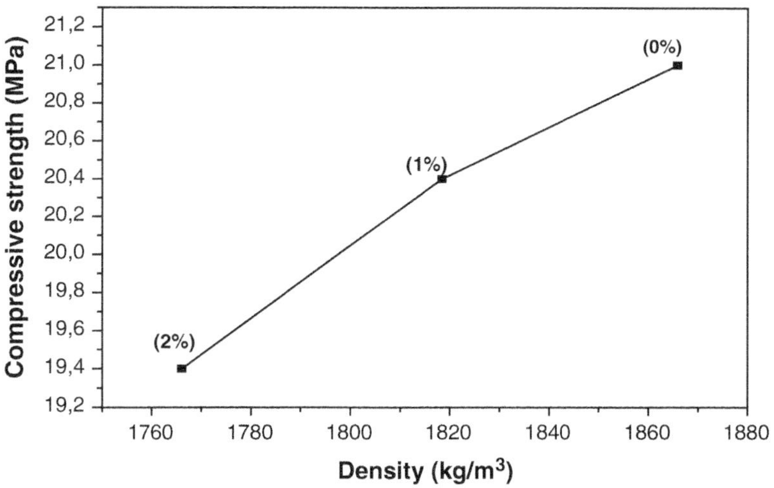

Fig. 29.5 Variation of the compression strength as a function of density

29.5 Conclusion

All the results obtained in these thermal and mechanical tests allow to note that adding fiber seems slightly improve the thermal and mechanical performance. It appears in particular a decrease in thermal conductivity with increasing fiber content, on the moisture content, we have shown a very significant increase in the thermal conductivity at low water contents and then a slower increase to values average and again more rapidly to levels close to the saturation state. Nevertheless to refine our judgment proportions of other fibers, most important and specific provisions of the fibers in the material should be considered and studied.

The thermal properties of the composite material are not alone a selection criterion, we must also take into account the state of the material, its availability and its mechanical bending of the effect of other additions to the thermal and mechanical behavior of materials developed. A more general study on its thermomechanical properties is in progress.

References

1. Khedari J, Watsanasathaporn P, Hirunlabh J (2005) Development of fibre-based soil–cement block with low thermal conductivity. Cement Concr Compos 27(1):111–116
2. Al-Ajlam SA (2006) Measurements of thermal properties of insulation materials by using transient plane source technique. Appl Therm Eng 26:2184–2191
3. Kriker A, Bali A (2008) Durability of date palm fibres and their use as reinforcement in hot dry climates. Cement Concr Compos 30:639–648

4. Baron J, Lesage R (1976) La composition du béton hydraulique, du laboratoire au chantier. Rapport de recherche LCPC No. 64
5. Rossi P. Les bétons de fibres métalliques, presses de l'école nationale des Ponts et chaussées, ISBN2-85978-2923
6. Taoukil D et al (2013) Moisture content influence on the thermal conductivity and diffusivity of wood concrete composite". Construct Build Mater 48:104–115
7. Bal H, Jannot Y (2013) Measurement and modelisation of the thermal conductivity of a wet composite porous medium: laterite based bricks with millet waste additive. Construct Build Mater 41:586–593
8. Cherki A-b, Remy B (2014) Experimental thermal properties characterization of insulating cork-gypsum composite. Construct Build Mater 54:202–209
9. Bal H, Jannot Y (2012) Water content dependence of the porosity, density and thermal capacity of laterite based bricks with millet waste additive. Construct Build Mater 31:144–150

Chapter 30
Theoretical Investigation on a Building-Integrated PV/T (BiPVT) System for Electrical and Thermal Energy Saving Case Study: Integrated Solar Village of Bou Saada

Oualid Sotehi, Abla Chaker, Mostefa Lamine Benamra, and Esma Ramoul

Abstract The aim of this work is to study and estimate, using TRNSYS software, the production and performance of hybrid water PV/T collectors and to assess the impact of integrating solar devices on energy and environmental balances of existing building.

The results show a positive influence when integrating a BiPVT in the building on the economic and environmental balance. In addition to the electric production, heat productivity obtained is important, the solar fraction is satisfactory and the efficiency is high. The installation can supply the entire demand of hot water of the building and all the heating requirements during the entire period of using heater. The electrical production covers all consummations of solar pumps and regulation devices. This production also allows covering the cooling booster for air conditioning during months requiring air conditioning, which makes the installation completely autonomous during all months of the year. A zero-energy building is then obtained.

Keywords BiPVT • Thermal energy • Electric production • Solar fraction • Efficiency

---

O. Sotehi (✉) • E. Ramoul
Department of Project Management, Faculty of Architecture and Urban Planning, Constantine 3 University, Constantine 25000, Algeria
e-mail: sotehioualid@gmail.com

A. Chaker
Energetic physics laboratory, Constantine 1 University, Constantine 2500, Algeria

M.L. Benamra
Department of Architecture, Mohamed Khider University, Biskra 07000, Algeria

I. Dincer et al. (eds.), *Progress in Clean Energy, Volume 2*,
DOI 10.1007/978-3-319-17031-2_30

Nomenclature

Abbreviations and Acronyms

f	Solar fraction (%)
COP	Coefficient of performance
BiPV	Building integrated photovoltaic
BiPVT	Building integrated photovoltaic thermal
PV	Photovoltaic
PV/T	Photovoltaic thermal

Greek letters

τ	Economy rate (%)

Subscripts

Economy	Economy

30.1 Introduction

During the preceding 30 years, the energy consumption increased significantly. According to statistics, the world consumption of primary energy has reached 89.5 % leading to an increase in CO_2 emissions of 79 % in the period 1973–2006 [1].

The building sector is the main responsible of the high-energy consumption. In 2005, statistics showed that the building sector in Algeria consumed 41 % of the national energy production, adding 9 % for the construction, which makes this sector the largest consumer of the energy. This consumption generates 25 % of the national CO_2 emission [2]. These statistics are likely to rise over the recent years due to the growing demand for comfort in the building sector.

In this context, resorting to thermal and electrical solar energy in the building sector is a real challenge for environmental, economic and social issues.

Algeria offers one of the most significant solar fields in the world, with more than 3,000 h of sunshine per year. The daily energy received per horizontal surface unit over most of the national territory is estimated at about 5 KWh, approximately 1,700 kWh/m^2/year in the northern regions and 2,263 kWh/m^2/year in the southern ones. However, the market for solar equipment in Algeria remains relatively modest till now as their distribution is constrained.

According to Cengel, the energy required for a building heating is estimated at 40 % of the building consumption. The domestic hot water, lighting, air-conditioning and cooling freezing, respectively, represent 17 %, 7 % and 19 % of the total energy consumption of a building [3]. Building-integrated photovoltaic (BiPV) technology has been one major field of photovoltaic (PV) applications. The technology provides a cohesive design and sustainable solution for the build environment [4].

In addition to electric production, the hybrid photovoltaic-thermal (PV/T) has two possible applications: heating the air [5–7] or heating water [8–10].

In this way, we have developed a theoretical study concerning the production and performance of hybrid solar water PV/T integrated in an individual house. This installation ensures simultaneously three functions, i.e. supplying domestic hot water applications, feeding a coil integrated in the floor with hot water for space heating and supplying an absorption system for space cooling. In addition to the thermal energy, electrical energy produced by the photovoltaic cells is used to offset the power consumption of auxiliary systems (pumps, fans, regulators . . .).

30.2 Technical Considerations

30.2.1 Solar Circuit

The solar loop consists of a series of hybrid solar collectors providing thermal energy to a storage tank through an internal heat exchanger located in the lower region of the tank. The thermal energy supplied to the solar storage tank can also come from an auxiliary hydraulic heating system, via a heat exchanger located in the upper area of the tank. The photovoltaic cells are connected to a converter and an electrical storage battery. Figure 30.1 shows such a system.

Characteristics of the solar system are shown in Table 30.1.

The slope of solar collectors is 35°, facing south with a total area of 33 m^2. The characteristics of the storage tank, pumps, booster and thermal absorption chiller are shown in Table 30.2.

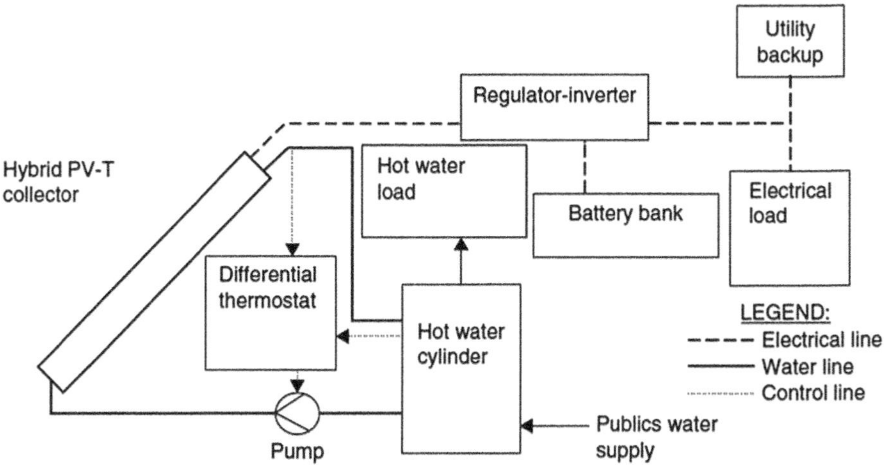

Fig. 30.1 Descriptive diagram: A solar hybrid PV/T water system

Table 30.1 Hybrid PV/T water collector characteristics

Characteristics	Value	Unit
Absorber plate thickness	0.0005	m
Thermal conductivity of the absorber	1,386	KJ/h.m.k
Number of tubes	10	–
Tube diameter	0.01	m
Bond width	0.01	m
Bond thickness	0.001	m
Bond thermal conductivity	1,386	KJ/h.m.k
Resistance of substrate material	0.01	h.m^2.k/Kj
Resistance of back material	3	h.m^2.k/Kj
Fluid-specific heat	4.190	KJ/Kg.k
Reflectance	0.15	Fraction
Emissivity	0.9	Fraction
PV cell reference temperature	20	°C
PV cell reference radiation	3,600	KJ/h.m^2
PV efficiency at reference condition	0.12	Fraction
Efficiency modifier—temperature	−0.005	1/°C
Efficiency modifier—radiation	0.000025	h.m^2/KJ

Table 30.2 Solar tank, pumps and absorption machine characteristics

Designation	Characteristics	Value	Unit
Solar tank	Tank volume	2,310	Litres
	Tank loss coefficient	1.93	KJ/hm°C
	Auxiliary (gas)	24.379	KW
Primary pump	Maximum power	41.853	W
	Nominal flow rate	2,623.50	Kg /h
Secondary pump	Maximum power	42.294	W
	Nominal flow rate	2,436.14	Kg/h
Pump connection	Maximum power	25.00	W
	Nominal flow rate	150	Kg/h
Pump generator	Maximum power	51.142	W
	Nominal flow rate	2,454.82	Kg/h
Cooling circuit pump	Maximum power	63.395	W
	Nominal flow rate	3,651.55	Kg /h
Evaporator pump	Maximum power	41.667	W
	Nominal flow rate	1,718.38	Kg /h
Absorption machine	Cooling power	10.00	KW
	COP nominal	0.7	–

30.2.2 Prototype Study

The situation of the village is chosen in a semi-arid area, with an intermediate climate between the high plateaus and that of the Sahara. The region is located on latitude 35.4°N, the sun is high in the summer and the angle exceeds 70°. From early April to mid-September, the facade protection in summer is easy with solar shading, and provides no shadow in winter. In midseason, we opt for a protection with leafy trees or movable protections.

In winter, additional collector surfaces are advisable. Movable greenhouses, incorporated into a house, offer a suitable solution.

The buffer zones increase insulation and reduce the space heating in winter or the space cooling in summer. Ventilation is ensured through appropriate openings. Natural convective phenomena are used to distribute heat during the day in winter, and the coolness of the night during the summer.

The prototype shown in the figure is designed as R + 0 and is composed of four rooms, a kitchen and a toilet block (Fig. 30.2).

The two main facades are facing south and north, whereas East and West ones are completely blind (Fig. 30.3).

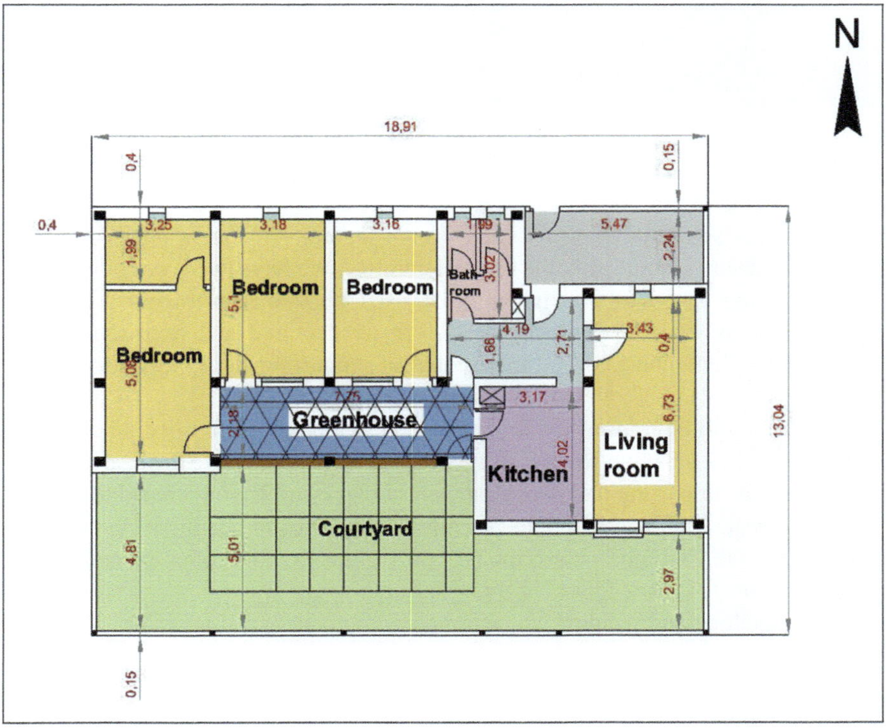

Fig. 30.2 Prototype study

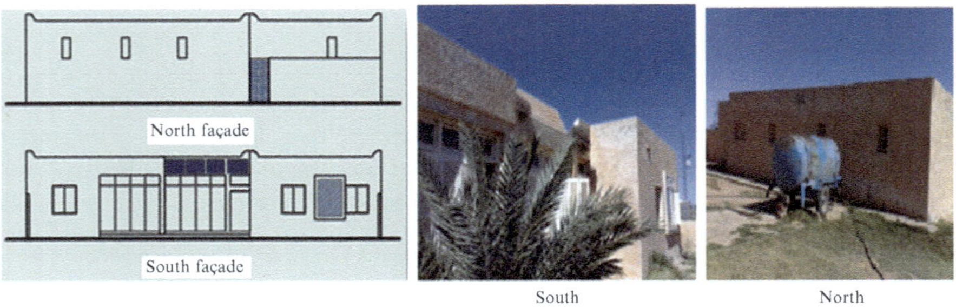

Fig. 30.3 North and south facades

Fig. 30.4 The greenhouse with the ventilation openings for the rooms

A passive solar system is adopted. Room layout, orientation and use of local materials ensure significantly the thermal comfort. We have two types of operation:

The living spaces of the house often used by family members are of type sun-space-mass-space, where the heat is firstly captured and imprisoned in the space called "greenhouse", and then stored in supporting stone walls as well as in floors (Fig. 30.4).

The heat will be transmitted to the room with a phase shift schedule coinciding with nightfall. The living room is of type sun-mass-space, where the heat is directly stored in the supporting wall stone painted in black and having coated glass witch traps the heat. High and low slots provide ventilation and heating of the air in the room; this wall "Trombe" stores the heat during the day to be released and emitted, with a lag time, in the night [11] (Fig. 30.5).

The envelope and window characteristics are given in Table 30.3.

Fig. 30.5 View to outside of the Trombe wall

Table 30.3 Prototype construction characteristics

Wall/window or door		Composition	Thickness (cm)	Surface (m^2)
South facade	External walls	Stone	40	20.15
	External walls	Stone	40	11.60
	Sun-exposed windows	Single glazing	–	3.6
	Windows with shadow	Single glazing	–	4.2
	External doors	Wood	–	2.25
East facade	External walls	Parpaing	40	14.70
	External walls	Parpaing with Liege insulation (5 cm)	40	4.75
North face	External walls	Parpaing	40	21.30
West facade	External walls	Parpaing with Liege insulation (5 cm)	40	28.20
	Windows with shadow	Single glazing	–	4.2
	External doors	Wood	–	2.25

30.3 Performance Evaluation of the Installation

To evaluate the performance of the studied solar installation, certain criteria are used [12]:

30.3.1 Solar Fraction

The solar fraction f is used to estimate the proportion of energy needs covered by solar energy. It is defined by the following equation:

$$f = \frac{\text{Useful solar energy}}{\text{Energy requirements}} \times 100\ \% \tag{30.1}$$

30.3.2 Economy Rate

The economy rate is defined by the following relationship:

$$\tau_{\text{economy}} = \left(1 - \frac{\text{Booster energy with solar}}{\text{Booster energy without solar}}\right) \times 100\ \% \tag{30.2}$$

30.3.3 Environmental Criteria

The amount of CO_2 released into the atmosphere is estimated depending on the type of fuel used. In our case, the facility operates with natural gas, so there is 0.28 kg of CO_2 emitted for 1 kWh of fuel consumed by the booster.

30.4 Results and Discussion

30.4.1 Energy Requirements and Solar Production

30.4.1.1 Domestic Hot Water

Figure 30.6 shows the energy requirements for domestic hot water. These needs are estimated for a family of five members. According to the standard ASHRAE, we have 40 l per person per day. The temperature of domestic hot water is 45 °C. The temperature at the outlet of the storage tank to the consumption is 60 °C to eliminate any risk of contamination with the Legionella bacteria.

According to the figure, we note that these needs vary during the year. We observe an increase in winter period, while a slight decrease in summer.

Figure 30.7 shows the solar fraction covered by the solar installation. We find that the coverage of needs varies between 84 and 100 %. The installation covers all domestic hot water requirements in the period that extends from February to October, while for the remaining months the solar fraction is high and satisfactory from the point of view of efficiency. In this period, the hot water needs increase and the incident energy is less important than the rest of the year. A quantity of the latter

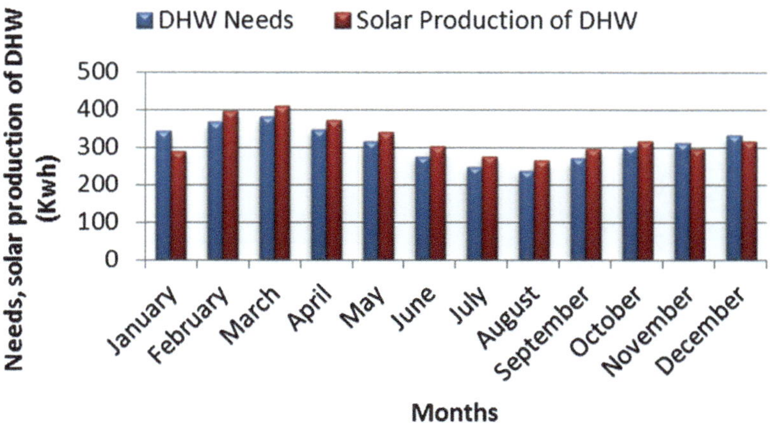

Fig. 30.6 DHW needs and facility production of the DHW

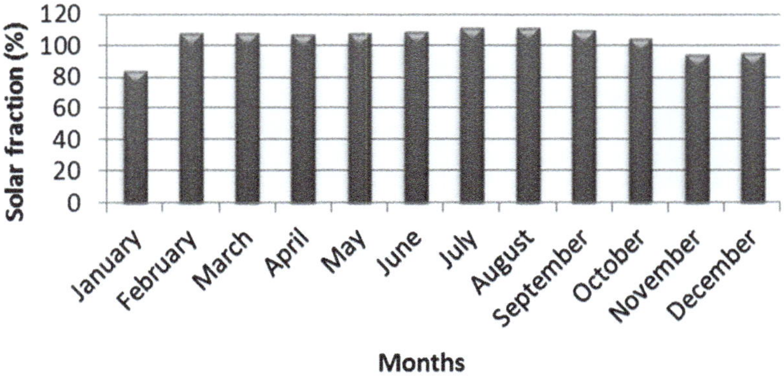

Fig. 30.7 Solar fraction of DHW requirements

is converted into electricity, adding losses in each organ of the solar installation especially in the two exchangers that separate the solar collector and the tank, as the collector circuit is filled with an antifreeze and the exchanger of domestic hot water, which separate the tank from the consumption circuit to limit the risk of Legionella.

30.4.1.2 Solar Heating

Figure 30.8 shows the need for heating and producing thermal energy for heating through the solar system. We note that the needs are considerable for the two months of December and January. Slight heating demand is noticed for the months of November and February while for the remaining months of the year, the heating demand is zero. These results can be explained by the fact that the architecture of

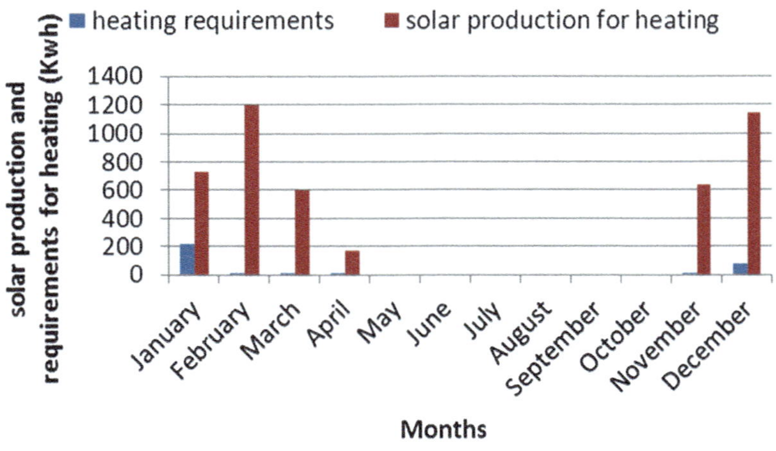

Fig. 30.8 Heating requirements and facility production in heating

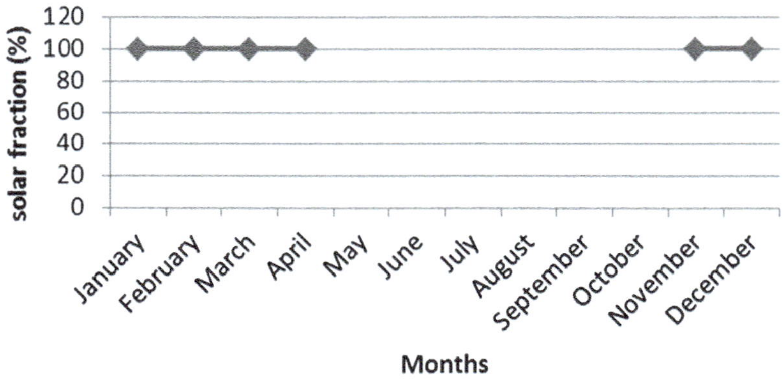

Fig. 30.9 Solar fraction for heating needs

the prototype promotes the use of passive solar energy. The solar production of thermal energy for heating exceeds the heating needs.

According to Fig. 30.9, we may notice that this system covers all the heating needs for all months of the year.

30.4.1.3 Solar Air Conditioning

In Fig. 30.10, we may notice that the requirements for air conditioning spread the period from February to October. Requirements become significant from May to September. The solar production represents high values than the requirements except for the period that ranges from May to September.

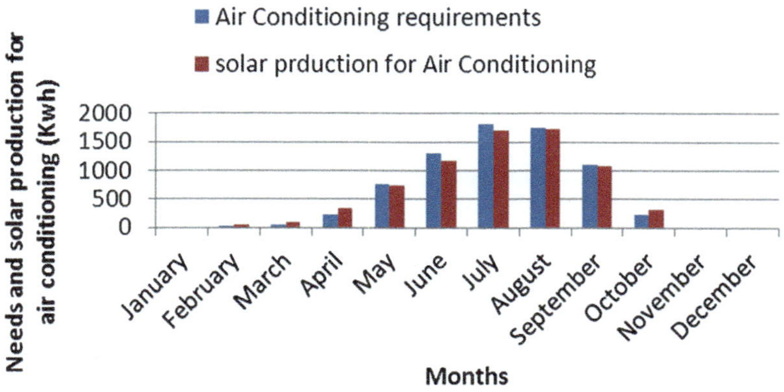

Fig. 30.10 Air conditioning needs and solar production of air conditioning

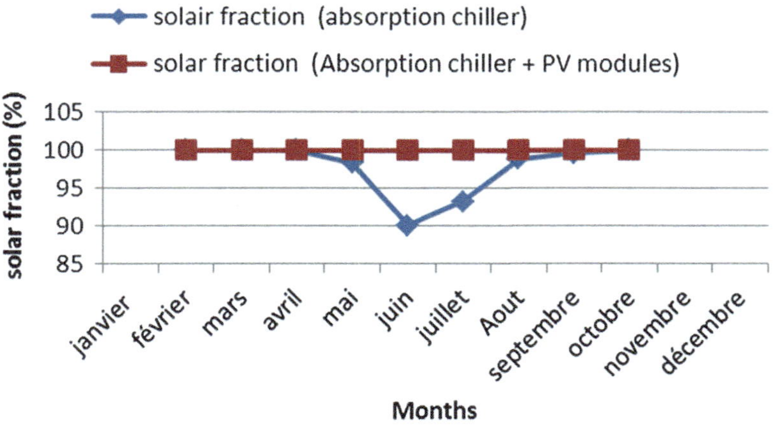

Fig. 30.11 Solar fraction of air conditioning needs

Figure 30.11 shows that this system has a high coverage rate of needs going from 91 to 100 %. The electricity production of PV modules can however cover cold booster, making the installation completely autonomous.

30.4.1.4 Electrical Production of PV Modules and Auxiliary Device Consumption

Figure 30.12 shows the electricity production of PV modules and the rest of the electricity produced after covering booster of the absorption chiller. The electrical consumption of the pumps is also represented. We may notice that the electricity production of PV modules can largely cover the electricity consumption of pumps for the two cases.

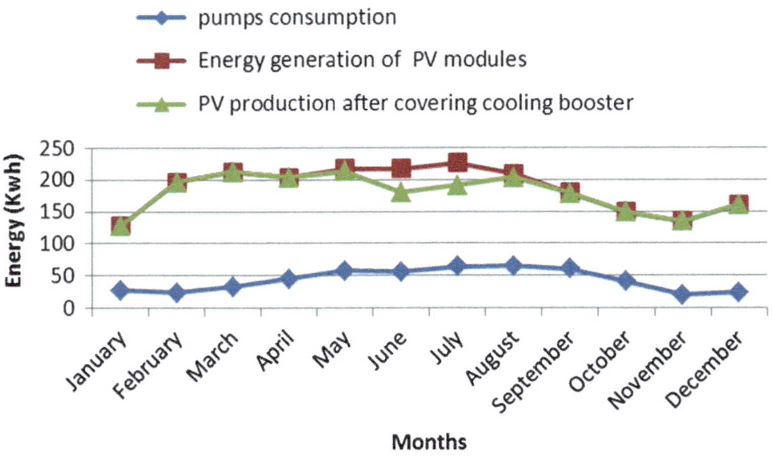

Fig. 30.12 Consumption of auxiliaries and solar production of electricity

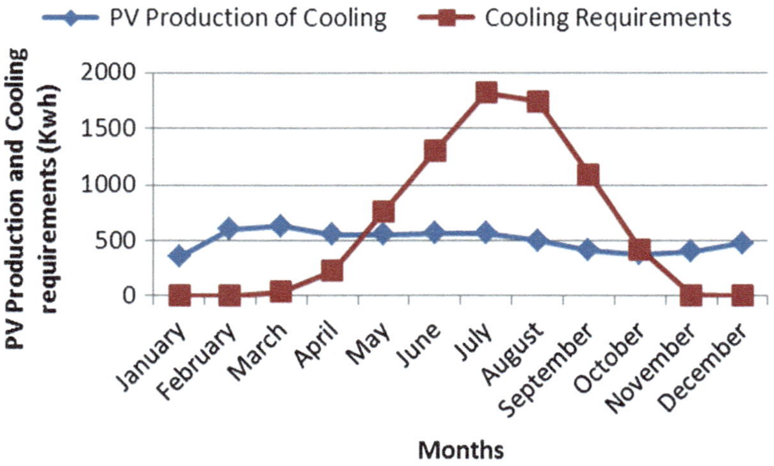

Fig. 30.13 PV solar production of refrigeration and air conditioning requirements

Figure 30.13 illustrates the electricity production of PV modules converted into cold in the case of using a mechanical compression machine instead of the thermal compression machine (absorption machine). We note that the chilling requirements largely exceed the cold production of PV modules in period that extends from May to September. The system can cover the needs for March and April months.

Figure 30.14 shows PV module solar fraction of cooling needs. We note that the coverage of needs ranges between 30 and 100 %. The summer period is characterized by a strong cooling demand; the installation may cover only 30–40 % of these needs.

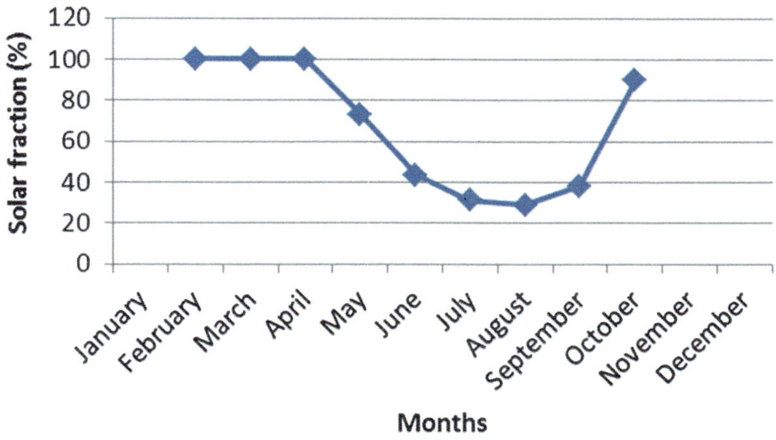

Fig. 30.14 Coverage of cooling needs by PV panels

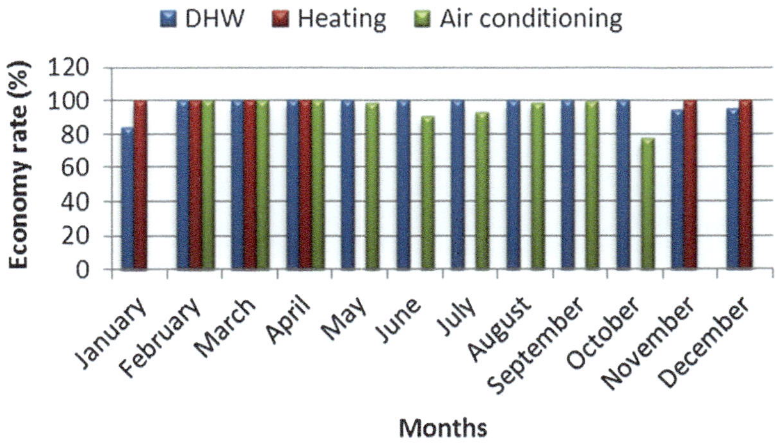

Fig. 30.15 Economy rate achieved by the solar installation

30.4.2 Rate of Economy

Figure 30.15 shows the economy rate to achieve by this installation. The values vary between 84 and 100 % for domestic hot water. High rates of economies are achieved going from 84 to 95 % for the period spanning from November to January where the demands for domestic hot water are high, while the production is less important and the heating demands are high during this period. For the period that runs between February and October, the economy rate achieved is

100 %; the installation is fully efficient during this period. The economy rate for heating requirements is 100 % for the entire heating period; this is due to the particular architecture of the prototype which significantly reduces the heat loss on one side and the use of techniques that contributed to the passive heating of the prototype, thus reducing heating needs.

The economy rate achieved by the installation for air conditioning requirements is represented in the same figure. We note that the values obtained period from May to October. The architecture of the prototype promotes passive cooling and the insulation used in walls reduces the heat gain significantly.

30.4.3 Environmental Balance

Figure 30.16 illustrates the amount of CO_2 avoided by the facility for each month. This amount depends essentially on the used booster. In our case, the booster works with natural gas, so there is 0.28 kg per kWh (natural gas is less polluting). The highest values are found in the summer when the need to cover is high. The facility covers a large part of air conditioning requirements, which increases the amount of CO_2 avoided. The annual amount of CO_2 avoided by the installation is 3,242 Kg.

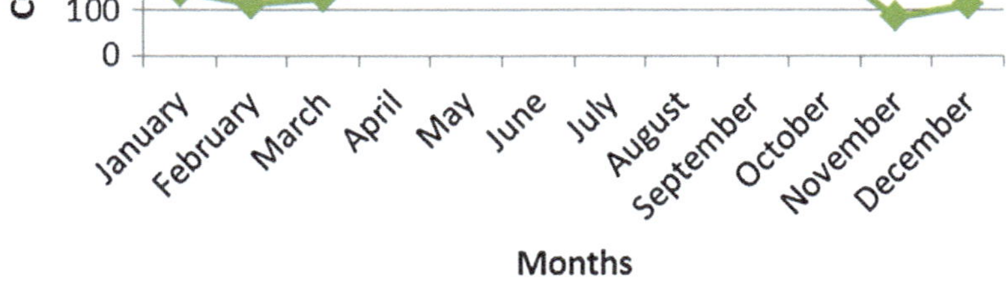

Fig. 30.16 Environmental balance of the facility

30.5 Conclusion

The main objective of this work is to study the performance of hybrid solar collector PV/T water, and to assess the impact of the integration of solar devices on the energy and environmental balances of existing buildings. The results clearly show the importance of using these facilities to cover the energy requirements of buildings. Our interest has focused on the thermal and electrical performances of the installation.

These results led us to a number of observations:

- The installation covers thermal needs of domestic hot water during the months of the year.
- The installation covers all the heating requirements during the entire period of using heater; wall insulation and the use of passive techniques such as Trombe wall can significantly reduce the thermal requirements.
- The rate of recovery of air conditioning requirements is high when using an absorption chiller.
- The electrical energy produced by the PV modules can cover the consumption of pumps and regulators. This production also allows covering the cooling booster for air conditioning during the months requiring air conditioning, which makes the installation completely autonomous in all months of the year. While if we use the electrical energy for the direct production of cooling without using absorption chiller, PV modules cover half the needs of air conditioning.
- The rate of economy achieved is very high and the amount of CO_2 avoided is important.
- According to the results, a net zero-energy building is obtained by the use of this type of installation. The passive architecture of the prototype can significantly reduce heating and cooling requirements, associated with a solar installation allows reducing the use of fossil energy to zero, which allows achieving energy security and protecting the environment.

References

1. Ricardo R, Braun P (2009) Energetic contribution potential of building-integrated photovoltaic on airports in warm climates. Sol Energy 83:1923–1931
2. Sotehi Oualid, Abla Chaker (2012) Comportement Energétique des capteurs hybrides PV/T à eau intégrés dans les bâtiments algériens (Energetic behavior of hybrid water collectors PV/T integrated in Algerian buildings). Paper presented at the first Seminar on Energy, Alger, 25–27 Mar 2012
3. Basant A, Tiwari GN (2010) Life cycle cost assessment of building integrated photovoltaic thermal (BIPVT) systems. Energy Build 42:1472–1481
4. Chow TT et al (2003) Building integrated photovoltaic and thermal applications in subtropical hotel building. Appl Therm Eng 23:2035–2049
5. Hegazy AA (2000) Comparative study of the performance of four photovoltaic/thermal solar air collectors. Energ Convers Manag 41:861–881

6. Infield D, Mei L, Eicker U (2004) Thermal performance estimation of ventilated PV facades. Sol Energy 76:93–98
7. Tripanagnostopoulos Y, Nousia TH, Souliotis M, Yianoulis P (2002) Hybrid photovoltaic/thermal solar system. Sol Energy 72:217–234
8. Zondag HA, de Vries DW, van Helden WGJ, van Zolengen RJC, Steenhoven AA (2002) The thermal and electrical yield of a PV thermal collector. Sol Energy 72:113–128
9. Kalogirou SA (2001) Use of TRYNSYS for modeling and simulation of a hybrid PV – thermal solar system for Cyprus. Renew Energy 23:247–260
10. Garg HP, Agarwall RK, Joshi JC (1994) Experimental study on a hybrid photovoltaic thermal solar water heater and its performance prediction. Energy Convers Manag 35:621–633
11. Boudiaf M, Michel C (1983) Etude thermique des trois prototypes du village solaire intégré. Édition centre de recherche d'architecture et d'urbanisme. Algeria
12. Johannes Kévyn (2009) Optimisation des installations solaires collectives d'eau chaude sanitaire: application des techniques« des faibles débits » et « du stockage divisé. Ph.D. thesis. Université de Savoie. http://tel.archivesouvertes.fr/docs/00/36/76/30/PDF/these.pdf

Part VII
Hydrogen Energy and Fuel Cells

Chapter 31
Water Is the Ultimate Source of Hydrogen Energy: Scientific Citations and Quotations

Hussein K. Abdel-Aal and Nejat Veziroglu

Abstract From the ancient times, mankind has always been aware that all life on earth depends upon water, the principal ingredient of all living cells. Its importance in forming the creation in all of its aspects in general and the living creatures in particular, as well as human kind, animals, and plants, is evident to all of us. It is highly important to know that water has played a major role for human kind. When the Ionian philosopher Thales of Miletus (624–545 BC) replaced the Gods with Natural Laws as the force governing all phenomena; he made *water* the central element in his theory.

This chapter addresses the issue that water, besides its indispensable usage by human beings in drinking and in everyday life, is the key element in providing life with energy, in the form of *hydrogen*.

The main aim of this work is to shed some light on the relationship between hydrogen, water, and energy. A review of some scientific unique physical and powerful properties of water is presented. The high specific heat, caused by hydrogen bonding, is behind the resistance to temperature change. Its availability for mankind made it the standard of the thermometric scale. *Next*, the work presents to the community of scientists some scientific citations and quotations that support our statement that water is the ultimate source of hydrogen energy.

Newton, an avid alchemist set forth in De Natura Acidorium, views that all substances can be reduced to water. *Eventually, hydrogen, one of the components of water, can be derived from it.* What is most amazing is that all of these citations, scientific statements, and notions had been proven to be in perfect agreement with science and our modern-day scientific discoveries that were not known to man 1,500 years ago.

This understanding of the use of water in providing energy in the form of *hydrogen* adds a new dimension to our scientific thinking that life on earth depends totally upon water, the principal ingredient of living cells.

H.K. Abdel-Aal (✉)
NRC, Cairo, Egypt
e-mail: habdelaal@link.net

N. Veziroglu
IAHE, Miami, FL, USA

I. Dincer et al. (eds.), *Progress in Clean Energy, Volume 2*,
DOI 10.1007/978-3-319-17031-2_31

Keywords Hydrogen • Water resources • Water and energy • Hydrogen and energy

31.1 Introduction

It is rather interesting to report the analogy that shows that our bodies are approximately 2/3 water, the earth is approximately 2/3 water, and the universe is again approximately 2/3 water. The majority of everything around us and in us beyond what we can see is water.

A quick look at the chemical equation for the formation of water tells us more:

$$2H_2 + O_2 = 2H_2O \tag{31.1}$$

This formula for water carries with it a very significant piece of information [1, 2]. Water is two parts hydrogen; the most abundant element in the universe and one part oxygen; and the third abundant element on earth. Hydrogen however is considered the central part of water. It constitutes 2/3 of water, with ratio 1:1, hydrogen:water.

Hence, we can establish the amazing formula of the 2/3. In other words, we have a plentiful source of hydrogen equivalent to the (2/3) times (2/3) times (quantity of available water on earth).

31.2 Hydrogen and Water

In the equation of water formation, as shown by Eq. (31.1), the following question is raised: What about the energy of formation? The formation of water from its elements produces, in addition to water, a tremendous amount of energy, 572 kJ per 2 mol of water.

We therefore should rewrite Eq. (31.1):

$$2H_2 + O_2 = 2H_2O + \text{Energy} \tag{31.2}$$

This is what we refer to as an exothermic reaction. It is also recognized as a combustion reaction, where hydrogen is combusted with oxygen producing water [3, 4].

A question is raised again: If we reverse the reaction given above in the other direction, can we generate hydrogen from water by adding energy?

$$\text{Energy} + \text{water} = \text{Hydrogen} + \text{oxygen} \tag{31.3}$$

Equation (31.3) represents an endothermic reaction, where water is used as a fuel to produce hydrogen and oxygen.

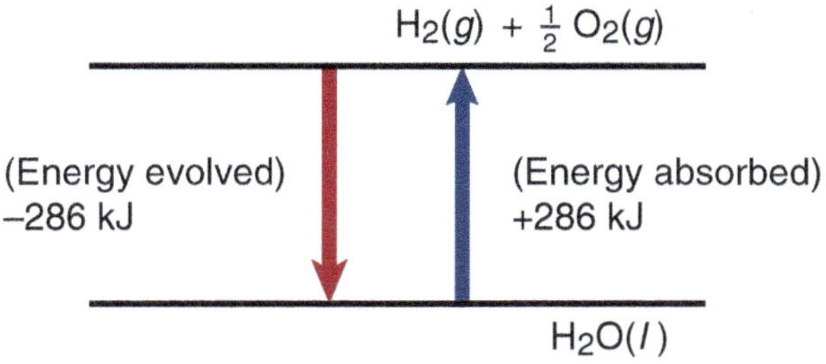

Fig. 31.1 Energy involved in the formation of water from H_2 and O_2. *Source*: Chemistry in Context Wm C Brown Publishers, Dubuque Iowa, 2nd edition, A project of the American Chemical Society, ed: A. Truman Schwartz et al., 1997, Chapter 5 The Wonder of Water

The next crucial question comes: Is it feasible to use energy to break water to form hydrogen, which eventually will be used as a fuel to combine with oxygen to give energy back? In fact, because of the laws of thermodynamics, you cannot **breakeven** in this exchange of energy. However, the answer could be reached by using sustainable energy sources (renewable energy sources), solar energy and wind energy, to split water-releasing hydrogen (Fig. 31.1).

31.3 Water Resources

Water covers 71 % of the earth's surface, and is vital for all known forms of life. Almost, 97 % of the water on the earth is salt water and only 3 % is freshwater; slightly over two-thirds of this is frozen in glaciers and polar ice caps. The remaining unfrozen freshwater is found mainly as groundwater, with only a small fraction present above ground or in the air.

Water on earth moves continually through the hydrological cycle of evaporation and transpiration (evapotranspiration), condensation, precipitation, and runoff, usually reaching the sea. Evaporation and transpiration contribute to the precipitation over land.

The world's water resources include the entire range of natural waters that occurs on earth in all different phases: vapor, liquid, and solid.

Distribution of water on earth is presented next in Table 31.1.

Of these, the most available resources for use are the waters of the oceans, rivers, and lakes. Other available waters include groundwater and deep subsurface water.

In addition, it should be observed that about:

- 97.2 % of the total earth's water occurs in oceans and inland seas
- 2.2 % in ice caps and glaciers
- 0.6 % is liquid freshwater

Table 31.1 Distribution of the earth's water

	Volume (1,000 km^3)	Annual loss (1,000 km^3)	Renewal period (years)
Entire earth	1,460,000	520 by evap.	2,800
In oceans	1,370,000	449 by evap.	3,100
In earth's crust:	60,000	13 underground	4,600
(a) Zone of exchange	4,000	13 underground	300
(b) Lakes	750	–	–
(c) Glaciers AND snowfield	29,000	1.8 runoff	16,000
(d) Soil/subsoil moisture	65	65 evap.	1
(e) Atmospheric vapor	14	520 precipitation	0.03
(f) River	1.2	36 runoff	0.04

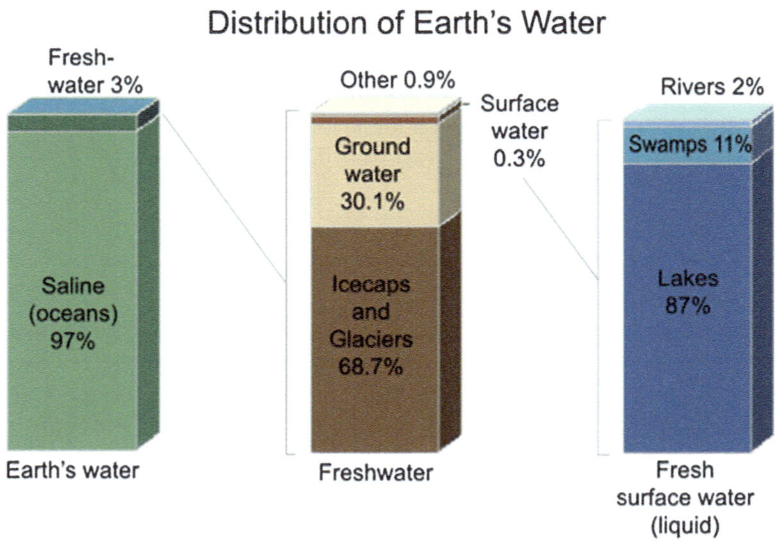

Fig. 31.2 Earth's water distribution. *Source*: Wikipedia, the free encyclopedia [5]

However, most of the liquid freshwater occurs as groundwater (97.74 %) and the balance accounts for lake water, surface soil water, and river water, as illustrated next in Fig. 31.2.

31.4 Some Unique Physical Properties

All three phases of water, solid, liquid, and gas, form vital links to life. Liquid water accounts for two-thirds of our body weight. To stay healthy, we need about one liter of water each day. Water helps blood and its components transport oxygen and

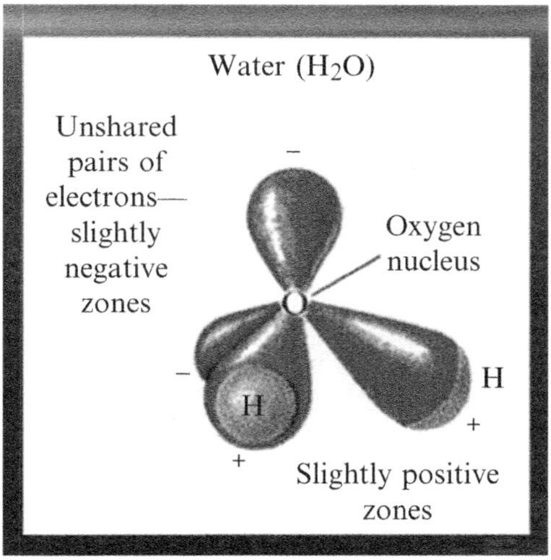

Fig. 31.3 Structure of water molecule

nutrients and remove waste products through our circulatory system. Each time we exhale, water vapor leaves our bodies. The following physical properties are significant to be presented [6, 7]:

First—Water is an extraordinary substance, anomalous in nearly all of its physical and chemical properties compared to other familiar single-chemical compounds.

Second—A quick look at the chemical equation for the formation of water tells us more:

$$2H_2 + O_2 = 2H_2O$$

It takes two molecules of the diatomic hydrogen gas combined with one molecule of the diatomic oxygen gas to produce two molecules of water. The result is a V-shaped, triangular molecule, as shown in Fig. 31.3.

While water molecules are electrically neutral, the oxygen atom holds a small negative charge and the two hydrogen atoms hold small positive charges. It is believed that this unusual electrical balancing, which is known as polarity, gives water some of its remarkable and unique properties. The high specific heat is caused by hydrogen bonding.

Third—Because so much energy is required to break the hydrogen bonds, water resists temperature change. This is important because a number of biological processes can occur only within a narrow temperature change.

31.5 The Hydrologic Cycle

The predominance of water on our globe is manifested by the fact that about 75 % of its surface is covered with water. The global water balance is controlled by what is called "hydrologic cycle."

The principal factor governing this cycle is solar energy. Evaporation from land and oceans takes place accordingly. The cycle reflects the relationship between the heat and water balances of both land and sea.

31.6 Water and Energy

31.6.1 Scientific Citations

Ancient men sensed the importance of *water* and *sun* and made them the central elements in their life. This is explained by the following:

First, for water, the Greeks considered the Titans Oceans and Tethys the parents of creation. The Ionian philosopher/scientist Thales substituted natural forces for Gods for being the causes of natural phenomena. However, he retained *water* as the central principle or "element" of his cosmology. His successors, in particular Aristotle, added other primal elements. These are *fire, earth, and air*.

For many centuries the four elements, including *water*, formed the basis of *alchemical* beliefs.

Newton, an avid alchemist, set forth in De Natura Acidorium, views that "all substances can be reduced to water." Eventually, *hydrogen*, one of the components of water, was considered to be the primal element, largely because it was found to be the lightest of all elements. This hypothesis was not abandoned until the nineteenth century.

Second, for sun, the pharaoh Amenhotep (Akhenaten), his wife queen Nefertiti, and their children honored the solar disc "Aten"—with its life-giving rays—as the only God (1350 BC), as shown in Fig. 31.4.

31.6.2 The Breakthrough

When *water* provides *heat*, then our daily living needs of both *food* and *energy* are realized from a *single* source: that is *water* [8–10]. This is established, of course, by using the *sun*, which represents the single most massive influx of energy reaching the earth. Hydrogen production, by splitting the water molecule using solar energy, is a scientific fact. When used as energy source, hydrogen combines with oxygen to give energy plus water vapor, which is recycled back to the atmosphere.

Water has been used over the years to provide hydroelectric energy in many parts of the universe. Now, it is time to utilize water as the ultimate source of hydrogen.

Taking the solar system, by itself, it is known that the maximum intensity of sunspots corresponds with the maximum intensity of magnetic storms on the earth. The universal law of gravitation seems to bind all mass and hold it together.

Fig. 31.4 Akhenaton honoring the sun

Physical facts, on the other hand, point out that planets were thrown off (clove) from the vast quantities of diffused nebular matter, of which the central condensed core is the *sun*.

It is God's well-ordered-providence that made from water every living thing to bring life on earth through *WATER*. About 72 % of the surface of our globe is covered with water; and it has been estimated *if* the inequalities on the surface were all leveled, then the whole earth's surface would be under water. This explains the fact the existence of firm mountains on earth is a source of security to life in this regard.

In addition, our latest knowledge of biological science proves that life began in water. The constitution of protoplasm is made up of about 85 % water.

It can be stated that *water* which is made up of hydrogen and oxygen provides mankind with life through the following:

- As liquid to provide our daily need of food, drink, etc.
- As a source of energy in the form of *HYDROGEN*
- As a source of oxygen as well

31.7 Conclusions

The relationship between solar energy and water was demonstrated, through the "hydrologic cycle." Most important, the link between water and hydrogen was explained by the hydrogen-water cycle, which offers different options to produce hydrogen from water using renewable energy sources, which are solar derived.

Citations and interpretations that may imply that water is indispensable for human beings in providing energy in the form of hydrogen are presented. In other words, when water provides ***heat***, in the form of hydrogen, our daily living needs of both food and energy are realized and maintained from a single source, that is, ***water***. This is realized by using the sun, which represents the single most massive influx of energy reaching the earth.

This understanding of the use of water in providing energy in the form of *hydrogen* adds a new dimension to our scientific thinking that life on earth depends totally upon water, the principal ingredient of living cells.

References

1. Truman Schwartz A et al. (1997) Chemistry in context, 2nd ed. A project of the American Chemical Society, Chapter 5. The wonder of water. Wm C Brown Publishers, Dubuque, Iowa
2. (1982) American encyclopedia, vol. 28
3. Abdel-Aal HK, Fahim MA, Gashgari MA (1974) Utilization of concentrated solar energy as a source of heat in the production of hydrogen. COMPLES, 2 Sept 1974
4. Abdel-Aal HK, Al-Somait F (1978) Solar energy prospects in Saudi Arabia. Energy Commun 4:3
5. WIKIPEDIA, the free encyclopedia
6. Abdel-Aal HK (1986) Solar hydrogen energy systems: present state of the art. Solar energy prospects in the Arab World. Pergamon Press, Oxford. pp 348–353
7. Abdel-Aal HK (1986) Water: feed stock for hydrogen production. Hydrogen energy progress VI, vol. 1. Proceedings of the 6th World Hydrogen Energy Conference, Vienna, Austria. pp. 31–36
8. Abdel-Aal HK, Mohammed MA (1989) Storage of solar energy in the form of hydrogen: a pioneering experiment in Egypt. Energy Sources 11:95–103
9. Abdel-Aal HK (1992) Storage and transport of solar energy on a massive scale: the hydrogen option. Int J Hydrogen Energy 17(11):875–882
10. Abdel-Aal HK et al (2005) A new approach to utilize hydrogen as a safe fuel. Int J Hydrogen Energy 30(13/14):1511–1514

Chapter 32
Twin-Head Platform

Oğuzcan Tokar and Şule Kapkin

Abstract In this work, we discuss the design of a platform that will utilize the huge potential of hydrogen sulfide in the deep waters of the Black Sea. It will be different from an ordinary oil platform as it will generate energy from hydrogen sulfide instead. The construction of this platform is foreseen as a Green Project. While the platform is cultivating energy out of hydrogen sulfide, it will also be purposed to contribute to the ecosystem at the same time. Once it completes its function, the platform is planned to be fully recycled and recovered.

Intended for the energy generation in the deep waters of the Black Sea, this platform is modelled as a multistage system, which consists of two major parts: (1) Sea water is drawn and processed to generate H_2, and (2) the resulting water is purified and released back to the nature. The H_2S concentrations are significantly high in 1,500 m and deeper in the Black Sea. Such information, as well as research data and the geographical conditions, is taken into consideration in the design of the platform.

The hydrogen-based energy potential in the Black Sea needs to be added to existing energy sources. Otherwise, the sensitive ecological balance will begin to shift for the worse. Initially, the sea life will be contaminated and become hazardous for human consumption and eventually, it will cease to exist altogether. This platform project is a precursor in generating energy while improving the nature at the same time.

Keywords Black Sea • Platform • Hydrogen sulfide

32.1 Introduction

Black Sea is an elliptical basin with a surface area of 423,000 km^2 and a volume of 534,000 km^3 [1]. The existence of H_2S was discovered by the Russian Geographic Society more than 100 years ago [2]. H_2S layer does not have any living creatures except bacteria. These anaerobic bacteria produce 10,000 t of H_2S per day under

O. Tokar • Ş. Kapkin (✉)
Department of Mechanical Engineering, Faculty of Engineering, Istanbul University, 34320 Avcilar, Istanbul, Turkey
e-mail: oguzcantokar@hotmail.com; skapkin@istanbul.edu.tr

I. Dincer et al. (eds.), *Progress in Clean Energy, Volume 2*,
DOI 10.1007/978-3-319-17031-2_32

current conditions. The total amount of this accumulation is reported to be approximately 4.6×10^9 t all across the Black Sea. Detected to start around 150 m in depth, the layer of hydrogen sulfide reaches a concentration of 9.5 mg/l in around 1,500 m [3]. It is also detected to become steadily closer to the surface. The sea reaches a depth of 100–120 m in 20–25 km from Turkish shores. In consideration of these data, the generation of hydrogen from H_2S will contribute to both people and environment. This energy potential in the Black Sea is a major candidate for future hydrogen energy systems.

Assuming that 100 % yield will be obtained, around 270,000,000 t of hydrogen is estimated to be generated from 4,587,000,000 of H_2S in the Black Sea. With this amount of hydrogen, 8,970,000 GWh of electrical energy and 38.3 million TJ of thermal energy can be produced [4]. This amount will meet the energy demand while also preventing the potential harm to the environment.

32.2 Platform on the Black Sea

A feasibility study was conducted before the construction of the platform. This study focused on the determination of the production method, as well as the assessment of the geographic location and existing reserves. Various manufacturing methods are known to be used in the generation of hydrogen from H_2S, including thermal, photochemical, electrochemical, plasma, and thermochemical ones. Given the geographical conditions of the Black Sea, the thermochemical platform was determined to be the most appropriate method. The generation of hydrogen sulfide in the Black Sea is affected by the enzymes and environmental factors. In addition, the existence of different metals, relations among the components, temperature, salt content, and the amount of oxygen in the environment are also important factors influencing the generation.

Moreover, the geographical location of the Black Sea affects the establishment of the hydrogen production process. The mountainous coast of the Black Sea is dense with green spaces and very rich in oxygen. A terrestrial facility to be built in this region should be located with the protection of green spaces in mind.

The second stage is the obtainment of pure H_2S from the Black Sea. In this stage, a great number of suggested processes were examined, including the construction of a manufacturing facility to draw and treat the seawater [2].

32.2.1 *Why Platform?*

What makes the platform significant may be explained by three major issues:

- *Portable*: Depending on the H_2S concentration in the sea, the platform can be relocated to serve other regions, when necessary.

- *It is affordable*: The costliest stage of the current process is the pumping of water from 1,500 m to a terrestrial platform. 1 t of hydrogen requires 10^9 m^3 of water and large indoor tanks.
- *Recycling*: Process the sea provide the balance when the fuel supply, coast guard, marine re-contamination blocked intermittently to make measurements on the Black Sea, an over-center acting as a research and investigation, the continuation of the form can be listed.

32.2.2 Hydrogen Production in the Platform

The industrial extraction pilot plant (IEPP) on the platform was designed using the Henry's law. As is known, the Henry's law states that the solubility of a gas in a liquid is directly proportional to the partial pressure of the gas above the liquid. This law is used in the extraction of hydrogen from the snow [5].

The H_2S separation process was designed based on this law. In the subsequent stages of the process, abbreviation (H_2S SP) will be used for the H_2S separation process.

The concentration of H_2S gas varies by depth in the Black Sea and reaches around 10 ppm in 1,500 ft. Accordingly, a two-stage system was designed. In the first stage, pure H_2S gas will be extracted from the Black Sea and transferred to the surface. In the second stage, H_2S gas will be separated in hydrogen and sulfur and stored in tanks.

32.3 Section of the Platform

32.3.1 Section I: H_2SSP (H_2S Separate Plant) Working Principle and Design

With the H_2SSP system, it is possible to extract pure H_2S. The system is suitable for both land and sea. Pure H_2S gas is transmitted from the platform to the work site with four motors via heat-insulated abutting pipes of up to 1,500 m long (Figs. 32.1, 32.2, and 32.3).

First, seawater is drawn with motor pumps. It is filtered into the main line and then carried into capillary lines with electric heaters, which distribute it into the sprayer.

While seawater is going through electric heaters, its H_2S content converts into the gas phase. Coming out from the sprayer, H_2S gas is directed to the platform and transferred into the storage tanks. Meanwhile, water is accumulated in the chamber located under the sprayer. When it reaches a particular level, water is treated through a second filter and returned back to sea, completing its cycle as a significant ecologic contribution.

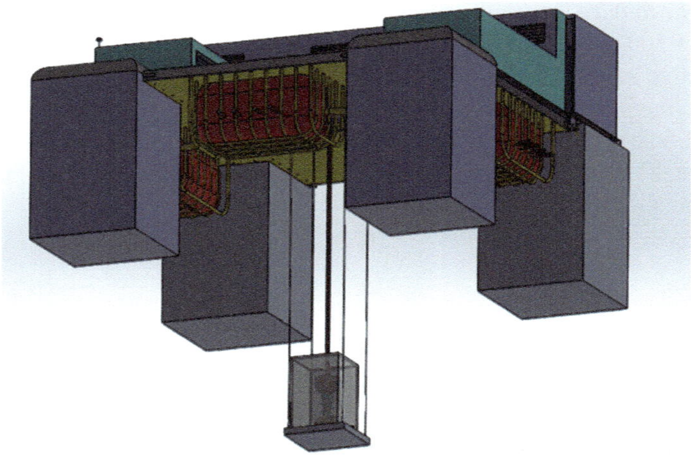

Fig. 32.1 Installation of H_2SSP

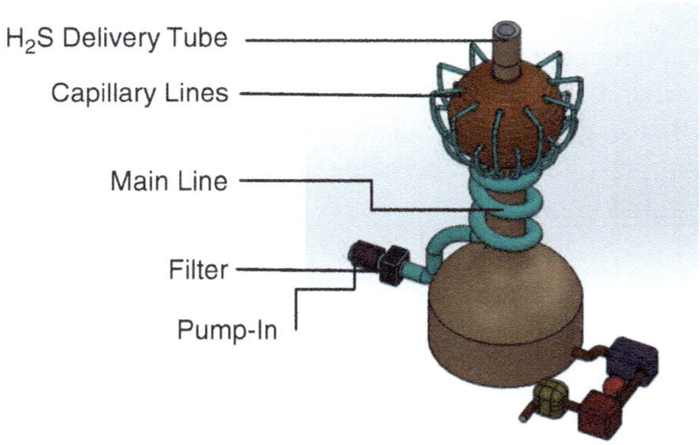

Fig. 32.2 General view of H_2SSP

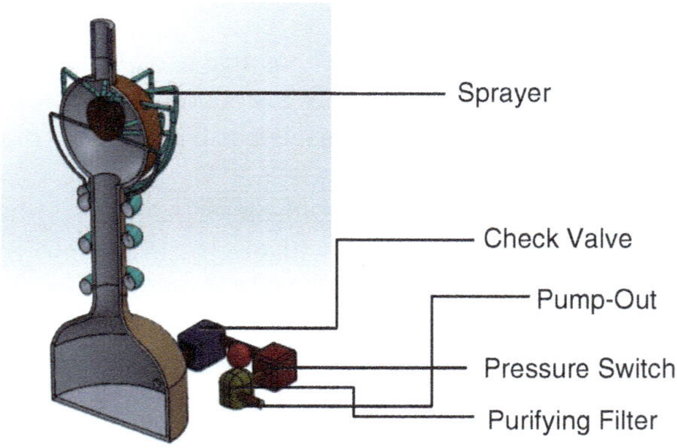

Fig. 32.3 Cross-sectional view of H_2SSP

Fig. 32.4 Connection between H_2SSP and the tank

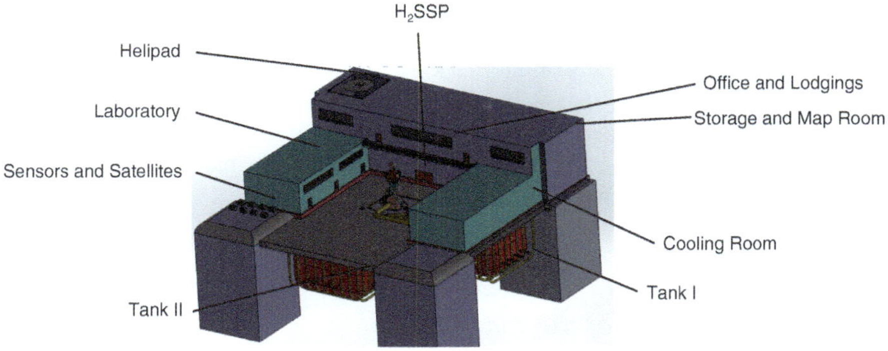

Fig. 32.5 Sections of the platform

32.3.2 Section II: Connection of the H_2SSP with the Tank

As shown in Fig. 32.4 the junction between the platform and holding the cylindrical H_2S SP pipelines is one of the most important parts in the platform. Which provide gas flow between the tank portions H_2S SP also shown in Fig. 32.5.

32.3.3 Section III: Superstructure of the Platform

The mechanism of the superstructure is designed as to ensure the maintenance of H_2SSP when necessary. In addition to the main production plant, offices, lodgings, laboratories, as well as cooling, storage, and map rooms were laid out. A helipad was also included for emergencies. Figure 32.5 shows the sections of the platform.

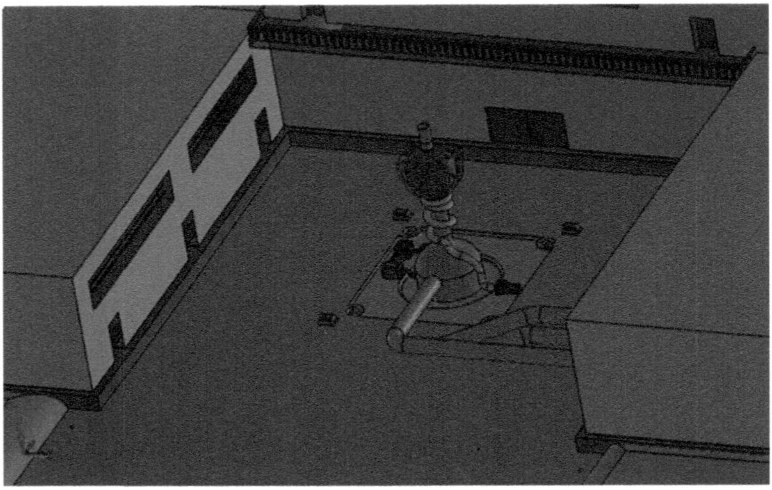

Fig. 32.6 Positioning of the H_2SSP on the platform

Located in the deep waters of the Black Sea, the H_2SSP will be removed once it has cleaned the necessary region. To do this, the H_2SSP is first disconnected from the platform and then pulled up with four motors, stopping every 50 m to remove connecting pipes. Finally, the H_2SSP is positioned on the platform as shown in Fig. 32.6 to rest until the next operation point.

32.3.4 Flow Chart of the Platform

The platform is shown in the flow chart of Fig. 32.7. Provided below is the process of operating on the platform:

1. Before positioning platform, H_2S ratio is determined by examining the dense regions.
2. The benchmark results were evaluated together with the platform can be positioned regions, the platform is positioned in the sea.
3. Located on the platform is installed in position H_2SSP sea. H_2SSP continuous data transfer between the platform makes up the various sans (H_2S density, the density of the sea, sea temperature, flow direction and intensity, images, data retrieval, etc.) that are recorded through the data.
4. Alignment provided by H_2SSP decomposes H_2S gas from seawater.
5. At the end of segregation in the platform pure H_2S gas is transferred to Tank 1. H_2S' dissociated seawater from the sea is passed through the filter.
6. Tank 1 H_2S gas in stepped or on demand, to thermochemical treatment is directed.

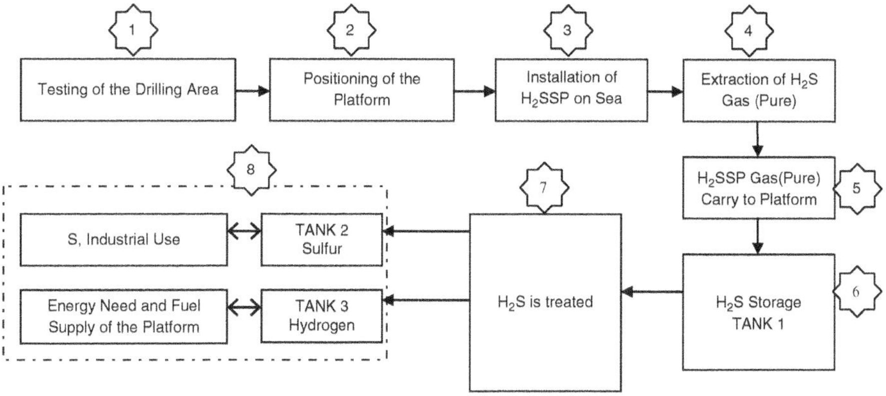

Fig. 32.7 Flow chart of the platform

7. Liquid H_2S subjected to thermochemical treatment, sulfur, and hydrogen are obtained from H_2S. Sulfur goes to Tank 2 and hydrogen goes to Tank 1.
8. On three sides of the platform was placed three tanks for storage purposes. In this tank, the sulfur in one of the other hydrogen, is storage of hydrogen sulfide in the other. Its sulfur H_2S and hydrogen are passed to respective tanks. Hydrogen platform is to meet the energy used to refuel at the same time. The remaining sulfur is sent ashore as the use of the industry.

32.4 Results and Discussion

At the designing platform is paid attention to there basic matters that balance ecology, use to be alternative way and produce energy. The Henry law is watched out in the H_2SSP's working principle.

The platform's cruise facilities and process are to be positioned at sea, and alternative evaluation facilities are brought into use. Cruise facility laboratories, coastal and marine area control, and meteorological services such as refueling points can be used for many purposes and are designed to evaluate the recycling together.

The navigational facilities of the platform can be used for various services at sea. Also, depending on the concentration of H_2S in the Black Sea, the position of the platform can be changed. With the development of H_2SSP'n, except using H_2S has a high potential energy contained in the sea bottom detected, H_2SO_4, HSR, S^- [6] will allow to the use.

If facility is on land:

- To pump water to the surface will spend a lot of energy.
- The same amount of water would have to be pumped in sprayer.

- A large closed tank on the beach would be needed.
- Water will be pumped to the surface was needed to stations.
- 10^9 m^3 of water should be achieved when 1 t of hydrogen is required.
- If H_2SSP on the land, both financially and advantage as it will not again be recovered. The advantages of this system being at sea are the following:
- No need to pump the water on surface.
- Surface will do the pumping, so stations will not be required. H_2S gas may be extracted with a suitable pump.
- Prototype, to be converted to use for different purposes and be able to move.
- Water tank will not be needed.
- At the same time the system will be placed in the water and will be purified through filters.
- Recycling spaced data in records that can be taken from the sea will be able to be intervened whenever necessary.

In this case, H_2SSP that the sea is more suitable to our design. If a system is to be produced, the gains at the cost of recycling are also important. In this system to pump seawater were calculated to shore. A per year cost is $28,34,165 [7]. This loss will be further increased when other possibilities are considered. A provision of this platform with the recycling of spent budget will be able to recover in time. Above all, in the Black Sea while H_2S is not cleaned in the future not only to live in the sea around the disaster which cannot be compensated for many years in the region compared to the experienced and the emergence of environmental pollution will cause irreparable.

32.5 Conclusions

As a result of research conducted in this study, effects of pollutants in the Black Sea and that are close to the surface have been demonstrated that increasingly deteriorate chemical balance. This platform is designed to ensure the improvement of the nature and meet the energy requirements. Energy gain is beneficial for humanity. In terms of cleanliness aims to bring balance to nature. Floating is a process and after the task according to the different needs with the provision of recycling is environmentally sensitive. Be economically profitable that enables the creation of platform. Day by day this condition can cause ecological disasters that cannot be ignored. Otherwise, the balance of nature will be disrupted and people will be affected directly or indirectly. It has to be solved as soon as possible.

Acknowledgement Thanks to Dr. T.N. Veziroglu for his encouragement and support throughout this study.

This work is supported by Research Fund of the Istanbul University, Project number UDP-44245, Istanbul, Turkey.

References

1. Gedik A, Dinçel A (1975) Chaın Oseonografi Gemisi ile Yapılan Karadeniz'deki Bilimsel Araştırmalar Hakkında Rapor, Mineral Research and Exploration (MTA), Derleme no: 5433
2. Apenizde M, Diasamidze R, Kordzakhia G, Jomidava R, Tsitskishvili M (2013) Complex investigation of ecological state of the black sea and actions for its protection, Chapter 1. In: Veziroğlu A, Tsitskishvili M (eds) Black Sea energy resource development and hydrogen energy problems, NATO science for peace and security series C: environmental security. Springer Science + Business Media, Dordrecht. doi:10.1007/978-94-007-6152-0_1
3. Baykaraa SZ, Figena EH, Kaleb A, Nejat Veziroglu T (2007) Hydrogen from hydrogen sulphide in Black Sea. Int J Hydrog Energy 32:1246–1250
4. Midillia A, Ay M, Ayfer K, Nejat Veziroglu T (2006) A parametric investigation of hydrogen energy potential based on H2S in Black Sea deep waters. Int J Hydrog Energy 32 (2007):117–124
5. Naman SA, Veziroğlu A (2013) The low cost hydrogen production from hydrogen sulfide in Black Sea, Chapter 9. In: Veziroğlu A, Tsitskishvili M (eds) Black Sea energy resource development and hydrogen energy problems, NATO science for peace and security series C: environmental security. Springer Science + Business Media, Dordrecht. doi:10.1007/978-94-007-6152-0_9
6. Turgay I (1969) Woods Hole Oseonografi Rnstitüsü Tarafından Karadeniz'de Yapılan Araştırma Raporu, Mineral Research and Exploration (MTA), Derleme no: 4268
7. Çekirge HM, Al-Waked R (2013) Feasibility of H_2S production from black sea waters, Chapter 8. In: Veziroğlu A, Tsitskishvili M (eds) Black Sea energy resource development and hydrogen energy problems, NATO science for peace and security series C: environmental security. Springer Science + Business Media, Dordrecht. doi:10.1007/978-94-007-6152-0_8

Chapter 33
Coke-Resistant Catalysts for Methane Steam Reforming in the Presence of Higher Hydrocarbons

Sofia D. Angeli, Fotis G. Pilitsis, and Angeliki A. Lemonidou

Abstract The catalytic performance on steam reforming and carbonaceous deposits was examined over $Ni/La/CeO_2$-ZrO_2 and $Rh/La/CeO_2$-ZrO_2 catalysts, prepared via wet impregnation. The catalytic performance tests were conducted using mixtures of C2–C4 alkanes and CH_4 as steam reforming feedstock for 4 h at the temperature range of 400–550 °C and for 10 h at 500 °C, with $S/C = 3$ and $GHSV = 70{,}000\ h^{-1}$. Analysis on the used catalysts was carried out via temperature-programmed oxidation (TPO). Both catalysts showed high activity towards the alkane feedstock achieving product distribution close to that predicted by thermodynamics with no significant deactivation during the experiments. TPO of the used catalysts after 10 h on stream showed increased carbon depositions compared to the pure methane reforming but still at very low levels, even though the concentration of the second alkane used was much higher than the real concentration in natural gas.

Keywords Methane steam reforming • Natural gas • Coke • Ni catalyst • Rh catalyst • Alkane

33.1 Introduction

Steam reforming of natural gas is one of the most significant technologies for the production of hydrogen. The conventional process consists of a steam-reforming reactor operating at temperature over 800 °C and high pressure, a WGS reactor, and a separator of the products and non-reacted gases. The high temperature and pressure of operation and the requirements for steam generation result in high-energy consumption, high capital and operation cost, as well as high GHG emissions.

S.D. Angeli (✉) • F.G. Pilitsis • A.A. Lemonidou
Department of Chemical Engineering, Aristotle University of Thessaloniki, GR-54124 Thessaloniki, Greece
e-mail: saggeli@auth.gr

I. Dincer et al. (eds.), *Progress in Clean Energy, Volume 2*,
DOI 10.1007/978-3-319-17031-2_33

Aiming at the intensification, a low-temperature steam reforming process is proposed at temperatures from 400 to 550 °C. Although the conversion rate and the hydrogen yield are considerably lower than the conventional process, there are significant advantages. Low reaction temperature decreases the operation cost, consumption of energy, and GHG emissions. Furthermore, the temperature range of operation is favorable for WGS reaction; thus both reactions can take place in a single reactor, lowering the capital cost of the process. The thermodynamic limitations in methane conversion due to low operation temperature can be surpassed by concomitant separation of a product such as hydrogen using H_2-selective membranes (membrane reactors) or CO_2 using CO_2 sorbents (sorption-enhanced steam reforming process).

The key issue for the successful realization of a low-temperature steam reforming process is the development of catalysts which are able to activate methane at low temperature and drive its conversion towards the equilibrium values at short contact times [1] and in addition to be resistant to deactivation factor such as carbon formation [2–4].

Steam reforming of natural gas is described through the reactions (33.1)–(33.3):

$$CH_4 + H_2O \leftrightarrow CO + 3H_2 \tag{33.1}$$

$$C_nH_m + nH_2O \leftrightarrow nCO + (n + 1/2m)\ H_2 \tag{33.2}$$

$$CO + H_2O \leftrightarrow CO_2 + H_2 \tag{33.3}$$

where C_nH_m represents hydrocarbons higher than methane, such as ethane, propane, and butane which are found as secondary components of natural gas. The presence of non-methane hydrocarbons can significantly affect the formation of surface carbon species and thus the effectiveness of the catalyst. The formation of carbon on the catalyst under steam reforming conditions may take place through cracking of hydrocarbons (Eqs. 33.4 and 33.5) and/or the Boudouard reaction (Eq. 33.6):

$$CH_4 \leftrightarrow C + 2H_2 \tag{33.4}$$

$$C_nH_m \leftrightarrow nC + m/2\ H_2 \tag{33.5}$$

$$2CO \leftrightarrow C + CO_2 \tag{33.6}$$

Catalysts containing CeO_2 or CeO_2-ZrO_2 mixed oxides show promise in the reforming of methane at low temperature, due to their redox properties, thus enhancing the steam reforming and water gas shift reactions as well as preventing the accumulation of coke [5, 6]. Our recent studies [7, 8] in low-temperature steam reforming demonstrated that ceria-zirconia support when doped with lanthana can result in the minimization of coking thanks to the unique property of ceria to provide active oxygen for the oxidation of carbonaceous deposits. Moreover, lanthana provides thermal stabilization, thus preventing possible sintering during the preparation and distress of the catalyst [9].

In this study, the effect of the presence of C2 to C4 alkanes in the feedstock in coke accumulation during low-temperature steam reforming of methane is investigated over Ni and Rh catalysts supported on lanthanum-doped ceria-zirconia mixed oxide. The amount of carbon deposited is qualitatively and quantitatively characterized by temperature-programmed oxidation (TPO). Preliminary evaluation of the deposited carbon was conducted after steam reforming of mixtures of methane with ethane, propane, or butane at 400–550 °C. The effect of the addition of non-methane hydrocarbons was investigated after isothermal steam reforming of mixed feedstock at 500 °C for 10 h on stream.

33.2 Experimental

33.2.1 Catalyst Preparation

The catalysts were prepared through the wet impregnation method. Ceria- and lanthana-doped zirconium oxide (78 % ZrO_2, 17 % CeO_2, and 5 % La_2O_3) provided by Mel Chemicals was used as the support. The pelleted support was crushed and sieved to attain particle size in the range of 250–355 μm. $Ni(NO_3)_2{\cdot}6H_2O$ (Merck) was the precursor for nickel (10 wt%) and $RhCl_3 \cdot xH_2O$ was the precursor for rhodium (1 wt%). The aqueous solution of the metal precursor was mixed with the support particles and stirred for 1 h at 70 °C. The solvent was removed with evaporation under mild vacuum conditions and the sample was dried overnight at 110 °C. Subsequently, the material was calcined in air flow at 800 °C for 5 h. The catalytic material is referred to as M(x)CeZrLa, where M stands for Ni or Rh and x is the percentage (wt %) of the active metal.

33.2.2 Catalyst Characterization

The surface area of the prepared materials was measured by N_2 adsorption at 77 K, using multipoint BET analysis method with an Autosorb-1 Quantachrome flow apparatus. The samples were dehydrated in vacuum at 250 °C overnight, before surface area measurements. X-ray diffraction patterns were obtained using a Siemens D500 diffractometer, with Cu Kα radiation.

Temperature-programmed reduction (TPR) experiments were performed in a gas flow system using a U-tube reactor connected online with a quadrupole mass analyzer (Omnistar). The catalyst sample (100 mg) was placed in the reactor and pretreated for 0.5 h at 250 °C at a rate of 10 °C/min in a 10 % H_2/He flow (50 mL/min). The mass numbers (m/z) 2 and 18 were used for H_2 and H_2O monitoring, respectively.

33.2.3 Catalytic Tests and Characterization of Carbon Depositions

The catalytic evaluation experiments were performed at atmospheric pressure in a laboratory unit equipped with a mass flow-controlled system for gases admission, a fixed bed quartz reactor, and an online gas chromatograph. A UFLC pump (Shimadzu) was used for the feeding of water to the reactor through a preheated line. The fixed bed reactor was heated electrically by a tubular furnace, with three independently controlled temperature zones. The temperature in the middle of the catalytic bed was measured with a coaxial thermocouple. The hot gases exiting the reactor were cooled to condense the unreacted steam. The gas-phase products were analyzed with an online gas chromatograph (Agilent Technologies 7890A) equipped with a TCD. To separate the products, two columns were used: PoraPlotQ for CO_2, C_2H_6, C_3H_8, and C_4H_{10} and MS 5A for H_2, CO, and CH_4. In order to investigate the effect of the presence of C_n alkanes ($1 < n < 4$) in the feedstock on the formation of carbon during steam reforming, two sets of experiments were conducted. The coking tendency was first evaluated by feeding mixtures of methane with C2–C4 alkanes at temperatures from 400 to 550 °C, atmospheric pressure, S/C molar ratio of 3, and gas hourly space velocity (GHSV) of 70,000 h^{-1}. The duration of preliminary screening tests was ca. 4 h; 1 h in each temperature, so as to obtain equilibration of the experimental conditions. The mixtures used as feed were 20 vol.% C_2H_6/CH_4, 10 vol.% C_3H_6/CH_4, and 2 vol.% n-C_4H_{10}/CH_4. The selected concentrations of higher alkanes in methane were higher than those in natural gas, in order to attain more severe conditions. The second set of experiments (also referred to as "coking" tests) includes steam reforming of the mixtures of hydrocarbons under isothermal conditions (500 °C) at atmospheric pressure and S/C ratio of 3. The duration of these tests was 10 h. Tests using only methane as feedstock were also carried as reference. Catalytic results of steam reforming of mixtures of hydrocarbons are expressed in terms of outlet gas composition.

Coke depositions on the used catalysts were qualitatively and quantitatively characterized in a flow system equipped with MS detector via TPO. The used catalysts were pretreated at 250 °C for 0.5 h in He flow and allowed to cool. The samples were then heated in the rate of 10 °C/min to 850 °C in a flow of 50 mL/min of oxidation gas (20 % O_2/He). Quantification analysis of the carbonaceous species was realized by comparing the peak area of experimentally produced CO_2 to that of known amount of CO_2 pulsed.

33.3 Results and Discussion

33.3.1 Synthesis and Characterization

The characteristic properties of the materials are shown in Table 33.1. The impregnation metal on the support mildly affects the specific surface area which decreases from 54.9 m^2/g to 37.2 m^2/g for the nickel and to 41.6 m^2/g for rhodium catalyst.

Table 33.1 Characteristic properties of the catalysts

Sample	Metal loading (%)	Support composition (wt. %)	Specific surface area (m^2/g)
Ni(10)CeZrLa	Ni-10	78 % ZrO_2, 17 % CeO_2, 5 % La_2O_3	37.2
Rh(1)CeZrLa	Rh-1		41.6
CeZrLa	–		54.9

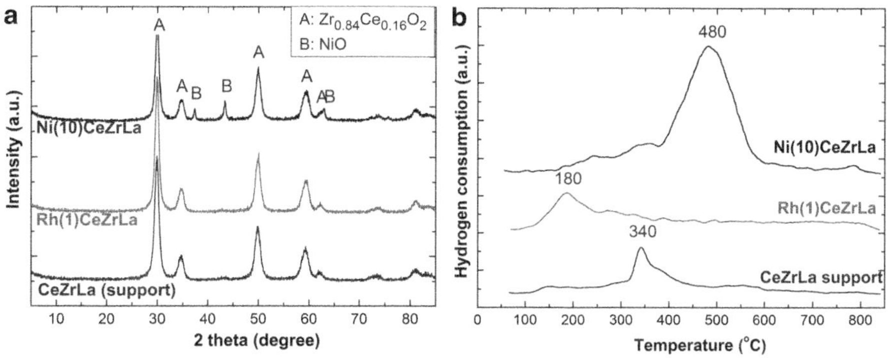

Fig. 33.1 (**a**) X-ray diffraction patterns of the prepared catalysts and the support. (**b**) TPR profiles of Ni(10)CeZrLa, Rh(1)CeZrLa, and the support

The diffraction patterns are shown in Fig. 33.1. Both materials show characteristic peaks of the $Zr_{0.84}Ce_{0.16}O_2$ phase (support), but no peaks are shown for La_2O_3, implying that the dopant is finely dispersed or it is in an amorphous state. On the Ni-based catalyst, the NiO peaks are revealed, but no relevant peaks of Rh_2O_3 can be seen for the Rh-based catalyst due to the low content of the metal.

The TPR profiles of the support and the two catalysts are shown in Fig. 33.2. The main peak observed at the TPR profile of the Ni-based catalyst at 380–550 °C is attributed to the reduction of NiO, while the shoulder appeared at lower temperature (300–380 °C) to the partial reduction of the support. For the Rh-based catalyst, the main peak at 110–250 °C is the result of the reduction of Rh_2O_3, and the shoulder at 250–350 °C, of partial reduction of the support. Lower reduction temperature of the support in the presence of Rh is observed, due to hydrogen activation by the metal at lower temperature and consequent migration to the support favoring reduction of the surface, the so-called H_2 spillover effect [10]. It should be emphasized that in low-temperature operating systems, the ability of the catalyst to be reduced in the same temperature range offers significant advantage since no further heating of the reactor is required.

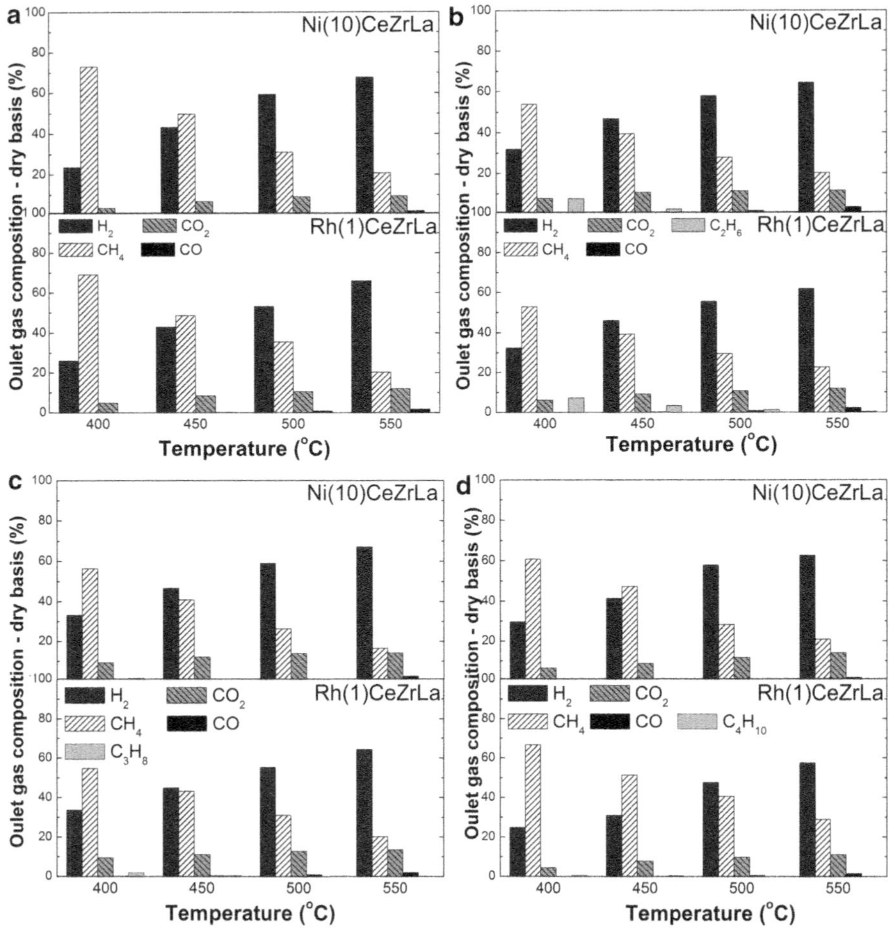

Fig. 33.2 Product distribution in steam reforming of methane and mixtures of hydrocarbons as a function of temperature over Ni(10)CeZrLa and Rh(1)CeZrLa. (**a**) CH_4 feed. (**b**) 20 % C_2H_6/CH_4 feed. (**c**) 10 % C_3H_8/CH_4 feed. (**d**) 2 % C_4H_{10}/CH_4 feed (conditions: $P = 1$ bar, S/C = 3, GHSV = 70,000 h^{-1})

33.3.2 Steam Reforming of Mixtures of Alkanes at 400–550 °C

Preliminary investigation of the coking tendency during steam reforming of contaminated methane feed was realized at the temperature range of 400–550 °C. Pure methane feed was also used as reference. The composition of the outlet dry gas over Ni(10)CeZrLa and Rh(1)CeZrLa is presented in Fig. 33.2.

The product distribution during steam reforming of pure methane over the two catalysts is very similar (Fig. 33.2a) probably due to synergistic effects of metal intrinsic activity, metal loading, and dispersion. The H_2 content increases with the increase in reaction temperature reaching almost 70 % at 550 °C, while CH_4 content decreases to a minimum of ca. 20 % at 550 °C, which corresponds to a conversion of 35 %. It should be emphasized, however, that according to thermodynamics, a maximum conversion of only 60 % can be achieved at such low temperature [11] due to the endothermic character of the reforming reaction. Considering also the high GHSV applied and the fact that no diluent was used (inert gas or inert solid in the catalyst bed), the two catalysts show promising activity in the low-temperature steam reforming. Furthermore, the two catalysts show very high activity in WGS reaction, since very low concentrations of CO were detected in all the temperature range (<3 %).

The addition of the second alkane in methane mildly affected the outlet concentration of H_2 compared to pure methane reforming. Low content of unreacted ethane was detected at 400 and 450 °C when C1/C2 mixture was used as feed and also traces of unreacted propane in the case of C1/C3 feed. No butane was detected in the outlet dry gas in steam reforming of C1/C4 mixture. It should also be noted that no formation of unsaturated hydrocarbons was observed (e.g., ethylene or propylene) in any of the tests, which could significantly affect the formation of carbonaceous species.

The TPO profiles of the catalysts used in non-isothermal steam reforming tests are shown in Fig. 33.3 and the quantification results, expressed as content of carbon

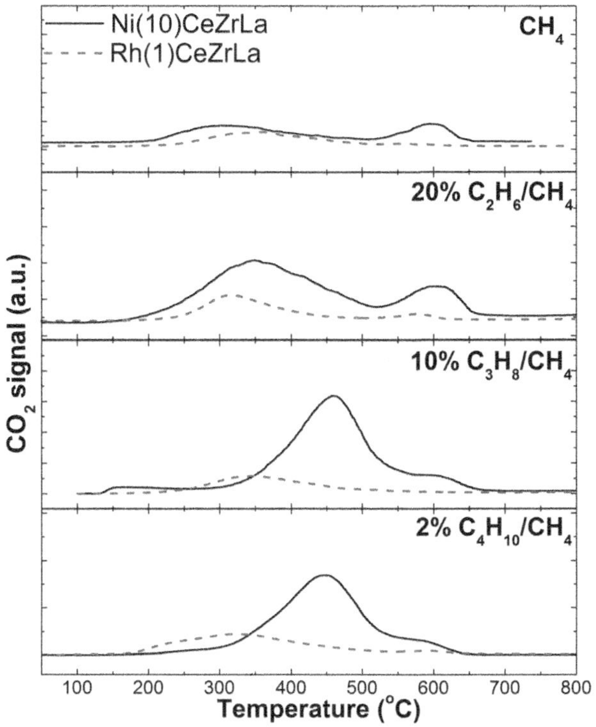

Fig. 33.3 TPO profiles of Ni(10)CeZrLa and Rh(1) CeZrLa after steam reforming of C1–C4 mixtures at 400–550 °C

Table 33.2 Carbon content after steam reforming of mixtures of methane and C2–C4 hydrocarbons at 400–550 °C.

	Carbonaceous deposits on used catalysts (wt%)			
Catalyst	CH_4	20 % vol. C_2H_6/CH_4	10 % vol. C_3H_8/CH_4	2 % vol. C_4H_{10}/CH_4
Ni(10)CeZrLa	0.006	0.440	0.294	0.119
Rh(1)CeZrLa	0.003	0.011	0.005	0.008

in the used catalyst on weight basis, are reported in Table 33.2. Steam reforming of pure methane at 400–550 °C resulted in almost no carbon formation which was 0.006 wt% for Ni(10)CeZrLa and 0.003 wt% for Rh(1)CeZrLa. In the case of Rh(1) CeZrLa, the amount of carbon detected after all screening tests was very low and was not affected by the different feed types. In the case of Ni(10)CeZrLa, the addition of ethane in the feed had the strongest impact on the amount of carbon formed which was increased to 0.44 wt%. However, according to the TPO profile, the main type of carbon formed was oxidized in lower temperature than that formed in the presence of propane and butane (Fig. 33.3).

33.3.3 Isothermal Steam Reforming of Methane in the Presence of Higher Alkanes: "Coking Tests"

To further investigate the resistance towards carbon formation, the two catalysts were tested in steam reforming of methane and mixtures of methane and C2–C4 hydrocarbons at 500 °C and S/C ratio of 3 for 10 h on stream. The catalytic results in terms of hydrogen content in the dry effluent stream as a function of time on stream during steam reforming of CH_4 and 20 % vol. C_2H_6/CH_4 are presented in Fig. 33.4. The hydrogen content remained constant in 10 h of operation. The two catalysts achieve 85–87 % of the equilibrium predicted value in both methane and ethane/methane mixture steam reforming. The respective catalytic results of 10 % C_3H_8/CH_4 and 2 % vol. C_4H_{10}/CH_4 are not presented here; however, the characterization of used samples is included.

No deactivation was observed in 10 h of reaction by both catalysts in steam reforming of mixtures of methane and C3-C4 hydrocarbons. Almost no content of the hydrocarbons higher than methane was observed and no unsaturated compounds such as ethylene or propylene were produced in any of the tests.

The investigation on the carbon formation was realized after steam reforming of mixtures of methane and higher alkanes for 10 h. Even though modifications on the nature of the carbon species may occur with time on stream our previous studies at the low-temperature reforming showed that both the amount and the nature of carbon formed on these catalysts remained stable after 5, 50, and 90 h on stream [8].

Qualitative and quantitative analyses of the carbonaceous deposits have been realized through TPO. The quantitative results of carbon accumulation are

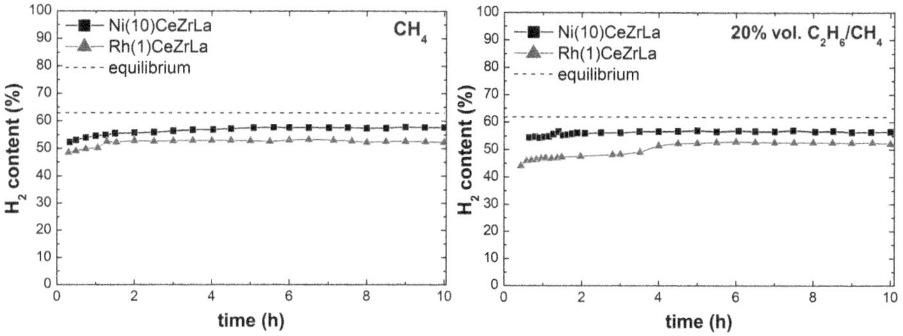

Fig. 33.4 Time on stream H_2 content during steam reforming of CH_4 and 20 % vol. C_2H_6/CH_4 over Ni(10)CeZrLa and Rh(1)CeZrLa (conditions: $T = 500$ °C, S/C = 3, GHSV = 70,000 h^{-1})

Table 33.3 Carbon content detected after steam reforming of mixtures of methane and C2–C4 hydrocarbons at 500 °C for 10 h

	Carbonaceous deposits on used catalysts (wt%)			
Catalyst	CH_4	20 % vol. C_2H_6/CH_4	10 % vol. C_3H_8/CH_4	2 % vol. C_4H_{10}/CH_4
Ni(10)CeZrLa	0.095	1.303	1.246	0.407
Rh(1)CeZrLa	0.004	0.010	0.015	0.027

presented in Table 33.3. In agreement with the results described in the previous section, Rh(1)CeZrLa showed nearly no carbon accumulation, which was not affected by the presence of higher alkanes. This finding, in conjunction with the high activity and selectivity in reforming reactions, makes Rh(1)CeZrLa a highly promising catalyst for low-temperature steam reforming of natural gas.

The presence of higher alkanes affected the accumulation of carbonaceous species on the surface of Ni(10)CeZrLa, which was however maintained in very low level. The addition of ethane in the reaction mixture resulted in an increase in the carbon content by an order of magnitude from 0.095 to 1.303 wt%. Similar effect was observed for the addition of propane, as the carbon content was 1.246 wt%. Taking into account that propane shows higher reactivity than ethane, one would expect carbon accumulation to be higher. Reforming of 2 % vol. C_4H_{10}/CH_4 resulted in a carbon content of 0.407 wt%, which was higher than methane but significantly lower than that of ethane and propane mixtures. This might be the result of the low concentration of butane, even though it is much higher than the one of natural gas.

The TPO patterns of the catalysts tested for 10 h in reforming of mixtures of hydrocarbons are presented in Fig. 33.5. The signal of produced CO_2 during oxidation of surface carbon on Rh(1)CeZrLa is of low intensity in all feedstock tested and it has been amplified to be better presented. A weak peak is observed in the low temperature range (200–350 °C), indicating high reactivity of the carbonaceous species, even when butane/methane mixture is used as reforming feedstock.

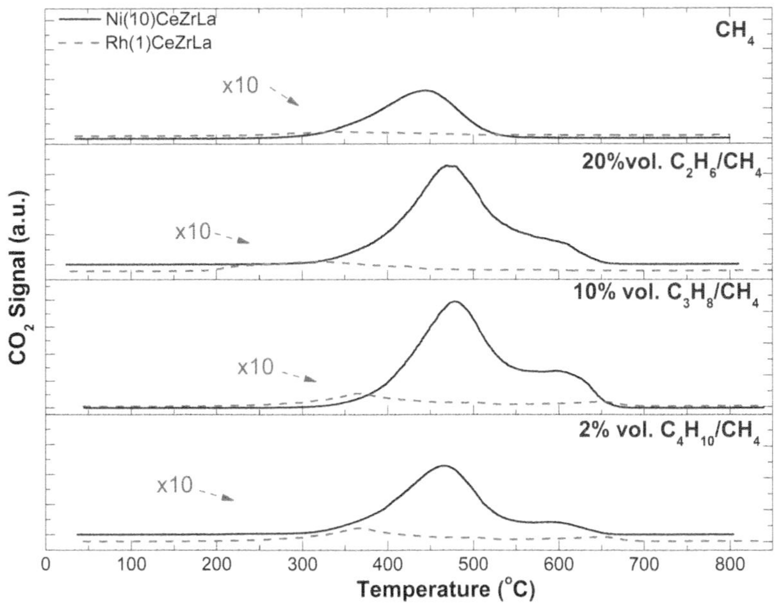

Fig. 33.5 TPO profiles after steam reforming of mixtures of methane and C2–C4 hydrocarbons

The TPO profile of Ni(10)CeZrLa of reforming of methane shows a wide peak in the temperature range of 250–600 °C, with peak maximum at ca. 550 °C.

Addition of ethane results in the appearance of a shoulder at ca. 600 °C, probably due to formation of filamentous carbon. This peak is further enhanced by the addition of propane, indicating that the increase in the carbon content (Table 33.3) is due to the formation of filamentous carbon [12]. It should also be noted that the relative intensity of the high temperature peak is higher in steam reforming of propane/methane mixture than in ethane/methane feed, even though the total amount of carbon detected was similar (Table 33.3). Moreover, when 20 % vol. C_2H_6/CH_4 is introduced, the C/H ratio of the feed is same as in the case of using 10 % vol. C_3H_8/CH_4. Thus, the amount of carbonaceous species on Ni(10)CeZrLa seems to be affected only by the C/H ratio of the feed and not by the C-number of the alkane. However, the latter has stronger influence on the nature of carbonaceous species.

Reforming of butane/methane resulted in a profile similar to that of ethane/methane, with a wide peak at 300–550 °C and a shoulder at ca. 600 °C due to filamentous carbon formation. Thus, it can be concluded that the addition of higher hydrocarbons in methane is responsible for the formation of filamentous carbon, which seems not to be formed in the reforming of methane.

33.4 Conclusions

The effect of the presence of alkanes higher than methane in terms of carbon formation was investigated in steam reforming of natural gas at low-temperature conditions (400–550 °C) over Ni and Rh catalysts supported on La/CeO_2-ZrO_2 (Ni(10)CeZrLa and Rh(1)CeZrLa). Both catalysts showed high activity towards the alkane feedstock with no apparent deactivation during the experiments.

Preliminary investigation was realized through steam reforming of mixtures of methane and ethane/propane/butane at 400–550 °C followed by TPO. Catalytic results showed high reactivity in steam reforming at low temperature over Ni(10)CeZrLa and Rh(1)CeZrLa. TPO analysis showed that almost no carbon was accumulated on Rh(1)CeZrLa independently of the reforming feedstock, while increase in the C-chain number of the feed affected both the amount and the nature of carbon detected on the used Ni(10)CeZrLa.

Further investigation on carbon formation was conducted by characterizing the catalysts used in steam reforming of mixtures of methane and ethane, propane, and butane for 10 h on stream at 500 °C. Both catalysts showed no deactivation or formation of olefins during these tests and were proved to be coke resistant, making these catalysts highly promising for steam reforming of natural gas at low temperature. The carbonaceous deposits on the spent Rh(1)CeZrLa were extremely low and highly reactive since they were oxidized in the temperature of 350 °C in all the types of feedstock tested. The presence of higher hydrocarbons was found to increase the accumulation of carbonaceous species on the surface of Ni(10)CeZrLa. Increase in the C-number of the additive alkane had almost no influence on the total amount of carbon formed (C/H feed ratio = constant) but favored the formation of filamentous carbon.

Acknowledgements The research leading to the results reported in this chapter has received funding from the European Union's Seventh Framework Programme (FP7/2007-2013) for the Fuel Cells and Hydrogen Joint Technology Initiative, for CoMETHy project under grant agreement no. 279075.

References

1. Halabi MH, De Croon MHJM, Van Der Schaaf J, Cobden PD, Schouten JC (2010) Intrinsic kinetics of low temperature catalytic methane-steam reforming and water-gas shift over Rh/$Ce_{\alpha}Zr_{1-\alpha}O_2$ catalyst. Appl Catal A Gen 389:80–91
2. Aasberg-Petersen K, Dybkjær I, Ovesen CV, Schjødt NC, Sehested J, Thomsen SG (2011) Natural gas to synthesis gas—catalysts and catalytic processes. J Nat Gas Sci Eng 3:423–459
3. Bartholomew CH (1982) Carbon deposition in steam reforming and methanation. Catal Rev 24:67–112
4. Sehested J (2006) Four challenges for nickel steam-reforming catalysts. Catal Today 111:103–110

5. Halabi MH, De Croon MHJM, Van Der Schaaf J, Cobden PD, Schouten JC (2010) Low temperature catalytic methane steam reforming over ceria-zirconia supported rhodium. Appl Catal A Gen 389:68–79
6. Purnomo A, Gallardo S, Abella L, Salim C, Hinode H (2008) Effect of ceria loading on the carbon formation during low temperature methane steam reforming over a Ni/CeO_2/ZrO_2 catalyst. React Kinet Catal Lett 95:213–220
7. Lemonidou AA, Vagia EC, Lercher JA (2013) Acetic acid reforming over Rh supported on La_2O_3/CeO_2-ZrO_2: Catalytic performance and reaction pathway analysis. ACS Catal 3:1919–1928
8. Angeli SD, Monteleone G, Giaconia A, Lemonidou AA (2013) Low temperature methane steam reforming: catalytic activity and coke deposition study. Chem Eng Trans 35:1201–1206
9. Teixeira ACSC, Giudici R (1999) Deactivation of steam reforming catalysts by sintering: experiments and simulation. Chem Eng Sci 54:3609–3618
10. Wang JA, Lopez T, Bokhimi X, Novaro O (2005) Phase composition, reducibility and catalytic activity of Rh/zirconia and Rh/zirconia-ceria catalysts. J Mol Catal A Chem 239:249–256
11. Angeli SD, Pilitsis FG, Lemonidou AA (2014) Methane steam reforming at low temperature: effect of light alkanes' presence on coke formation. Catal Today 242:119–128
12. Hardiman KM, Cooper CG, Adesina AA, Lange R (2006) Post-mortem characterization of coke-induced deactivated alumina-supported Co-Ni catalysts. Chem Eng Sci 61:2565–2573

Chapter 34
Ultrasonic Spray Coating Technique for High-Performance PEM Fuel Cell Electrode Manufacturing

Serdar Erkan and Inci Eroglu

Abstract The fabrication technique of membrane electrode assembly (MEA) significantly affects the performance and durability of proton exchange membrane (PEM) fuel cell. In the present study, ultrasonic spray coating technique has been used for manufacturing different sizes of gas diffusion electrodes up to 400 cm^2. The catalyst ink was composed of 70 wt% catalyst and 30 wt% Nafion in a solution of 1:7 water to 2-propanol. The platinum loading of the electrode has been adjusted to 0.4 mg Pt/cm^2. Commercial Pt/C catalysts have been used having different platinum content. Catalyst ink has been sprayed onto the gas diffusion layer using an ultrasonic spray coating instrument operating at 48 kHz. Coating process was fully computer controlled, and it was programmed according to the coating area, spray flow rate, and multiple layer coating. The performance of electrodes has been tested in a single PEM fuel cell. The power densities reached with the new method were 0.53, 0.74, 0.77, and 0.88 W/cm^2 for 20, 40, 50, 70 % Pt/C catalyst by keeping the platinum loading constant, respectively. The power density increased 2.5 times compared to "spraying of catalyst ink with air pressure atomizing spray gun". Uniform thin film has been achieved with ultrasonic coating method. Multiple layer formation retained the porosity of the electrode, impeded flooding on the electrode surface during the coating process, and enhanced transport of reactant during the fuel cell operation. Therefore, high-performance, reproducible, and large area electrodes could be manufactured by ultrasonic spray coating technique.

34.1 Introduction

The membrane electrode assembly (MEA) is the most important part of a proton exchange membrane (PEM) fuel cell stack. Its components (catalyst, membrane, gas diffusion layer) and preparation technique significantly affect the performance of the

S. Erkan • I. Eroglu (✉)
Department of Chemical Engineering, Middle East Technical University, Ankara, Turkey
e-mail: ieroglu@metu.edu.tr

I. Dincer et al. (eds.), *Progress in Clean Energy, Volume 2*,
DOI 10.1007/978-3-319-17031-2_34

fuel cell. A previously developed technique has been applied for many years by Middle East Technical University Fuel Cell Research Center (FCRC) group [1–4]. This technique is spraying of catalyst ink with air pressure atomizing spray gun (air brush) onto the micro porous layer coated side of the gas diffusion layer. The success of this manufacturing technique strongly depends upon the skill of the operator, since the operation is manual. Therefore, the membrane electrode assemblies produced by this method may not be identical. Besides, this method is also not applicable for producing larger active area and large number of membrane electrode assemblies for stacks, besides considerable amount of material is lost during manufacturing.

In order to improve the performance and enable consistent quality membrane electrode assemblies a new method has been implemented. Hand operated spray gun method has been improved by integrating computer automated tools. The experience showed that uniform coating of the electrodes cannot be achieved for large electrode areas with hand controlled spraying. Moreover, electrode structure depends on the operator of the sprayer. The new method developed here is also a spraying method but spraying mechanism is different. An ultrasonic nozzle is used for the atomization of the ink. When liquid is added to a resonating nozzle, waves are formed on the atomizing surface. Increasing the power causes the wave peaks to get so high that droplets fall off the tips of the wave. The uniform thin film coating achieved with ultrasonic retains the porosity of the electrode and does not impede the transport of reactants due to flooding of the electrode surface. The nozzle is attached to a computer controlled and programmable XYZ motion device. Therefore, high-performance and reproducible electrodes can be produced.

34.2 Experimental

34.2.1 Membrane Electrode Assembly Preparation by Spray Gun Method

The preparation technique of MEA can affect the performance of a PEM fuel cell seriously. The techniques shown in Fig. 34.1 can be applied for MEA preparation. In the first technique, a pneumatic spray gun was used for coating the catalyst onto the treated hydrophobic layer of the gas diffusion media. Once a single ultra thin layer was coated, the surface was dried by a hot air gun set at 80 °C or infrared heating lamp. This coating-drying cycle was repeated many times. The catalyst coated gas diffusion layer (GDL) was weighted in order to determine if the desired catalyst load was reached or not. When the desired load was reached for anode and cathode, the catalyst coated GDLs were hot pressed onto the both sides of the Nafion membrane. The second technique involved the use of Teflon sheet as an electrode transfer medium. First, catalyst ink was sprayed as described in Method 1, and the electrode was transferred to Nafion membrane by hot pressing the electrode coated Teflon sheet to the membrane. The third method was the application of spray method directly to the Nafion membrane placed on a heated vacuum table kept at 80 °C.

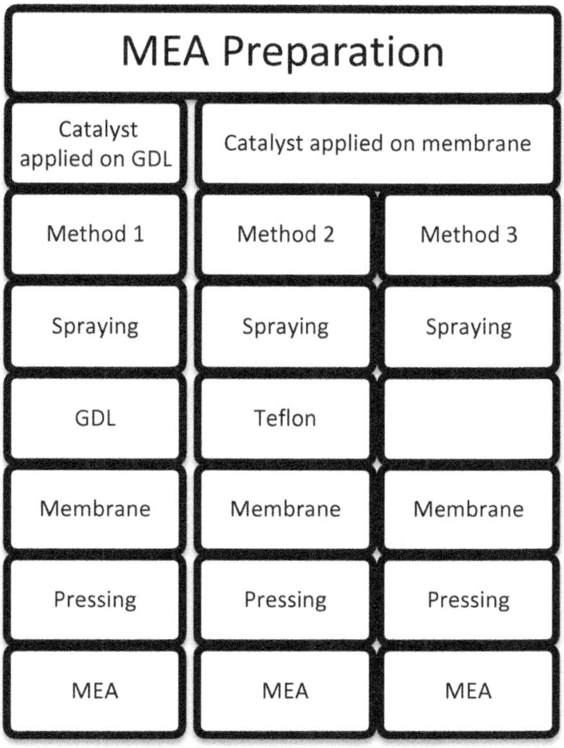

Fig. 34.1 MEA preparation techniques

In the present work, MEAs were prepared by spraying catalyst ink onto the GDLs (GDL 30 BC from SGL Carbon, Germany) via a spray gun. Nafion® loading was 30 wt% on dry basis in catalyst ink. The catalyst ink was sprayed until the desired catalyst loading (0.4 mg Pt/cm^2) for both anode and cathode sides were achieved. After spraying the catalyst ink onto the GDL a five-layer MEA was prepared by pressing these GDLs onto the membrane at 130 °C, 172 N/cm^2 (250 psi) for 3 min. Nafion® 212 which has a nominal thickness of 50 μm was used as the proton exchange membrane [1, 2].

34.2.2 Ultrasonic Spraying MEA Preparation Procedure

The pictures of the coating machine (SonoTek, USA) and the coating process are shown in Fig. 34.2. The coating was based on the ultrasonic atomization of the catalyst ink material onto a support layer which was either membrane or GDL. Coating process was fully computer controlled and could be programmed according to the coating area. Catalyst ink flow rate, spray head speed, spray width, stop times between the adjacent coatings can be varied. These parameters and their selected

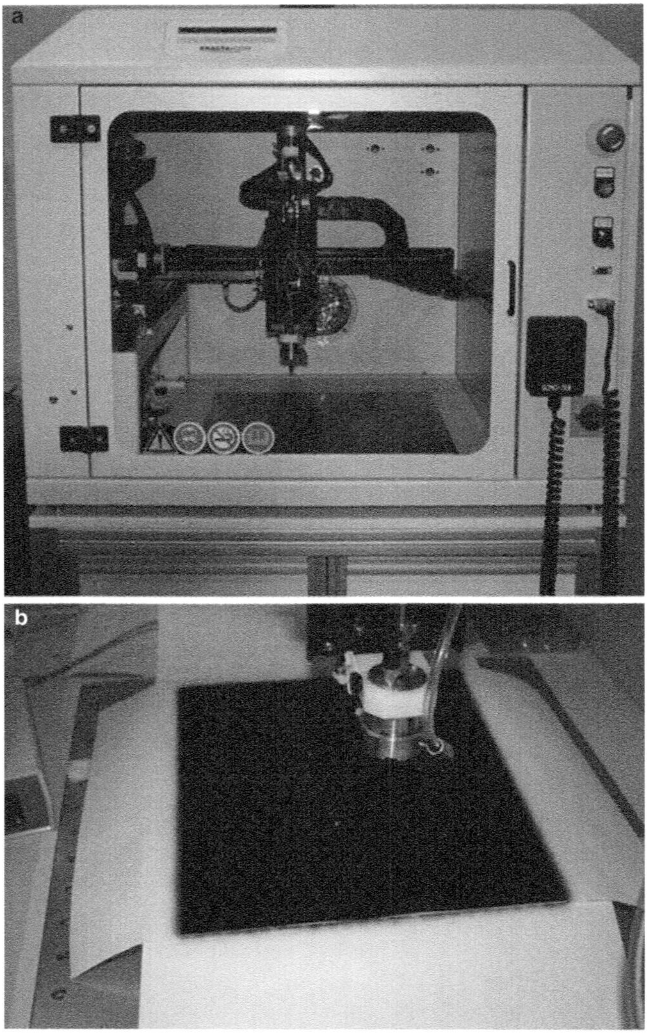

Fig. 34.2 (**a**) A view of the ultrasonic coating machine (**b**) coating process

values are listed in Table 34.1. The newly developed method is called "ultrasonic spray coating method".

The catalyst ink composition was exactly the same as the ink used in spray gun method as reported in a previous study [1]. It was composed of a catalyst having 20, 40, 50, or 70 % platinum on carbon (Pt/C), Nafion, and 1:7 ratio of water and 2-propanol. The prepared catalyst ink was mixed in ultrasonic mixer at least 15 min prior to filling the syringe pump of the coating machine. Catalyst loading could be changed by changing the number of passes and it was verified by gravimetric method. The number of passes was 88, 110, or 132 times. The ink was sprayed either onto membrane or GDL.

Table 34.1 Ultrasonic machine parameters during electrode coating

Spray head frequency	48 kHz
Catalyst ink flow rate	0.3 ml/min
Spray head speed	20 mm/s
Spray width	5 mm
Brake time between cycles (to allow drying of the layer)	5 s

34.2.3 PEM Fuel Cell Performance Tests

A single cell PEM fuel cell (Electrochem, FC05-01 SP REF) with hardware made of graphite with serpentine channels with a 5 cm^2 active area was used in the experiments. Performances of fabricated MEAs were measured via PEM fuel cell test station (Erdes, FCTS600). The current and voltage of the cell were monitored and logged throughout the operation of the cell by the fuel cell test software supplied with test station. The test cell was tightened with a torque of 1.7 N m applied to each bolt. The hydrogen and oxygen flows were set to 0.1 slpm. The cell temperature was controlled to be stable at 70 °C. In order to supply the reactant gases at 100 % relative humidity, the humidifier and heated transfer lines were kept at 70 °C. The cell was operated at 0.5 V until it reached to steady state. Then, starting with the open circuit voltage (OCV) value, the current–voltage data was logged by changing the load.

34.3 Results and Discussion

34.3.1 Performance of PEM Fuel Cell of MEAs Prepared with Hand Spray Coating

34.3.1.1 The Effect of Gas Diffusion Media

Figure 34.3 illustrates the comparison of the polarization curves obtained with fuel cells assembled with membrane electrode assemblies manufactured by hand spraying on carbon paper or carbon cloth GDL having 20 % Pt/C catalyst. The highest power density reached with MEA prepared with the carbon paper GDL was 0.39 W/cm^2, whereas it was 0.28 W/cm^2 for the MEA prepared with carbon cloth GDL at the same operating conditions. The cell built with the carbon paper GDL had a resistance of 0.51 Ω whereas the resistance of the cell assembled with carbon cloth GDL was 0.60 Ω. Because of the superior performance, carbon paper was selected as GDL material for the succeeding studies.

34.3.1.2 The Effect of Gasket Thickness on the Performance of PEM Fuel Cell

Figure 34.4 illustrates the effect of gasket thickness on the performance of fuel cell assembled with the MEAs prepared on carbon paper GDL. It is clearly seen from the

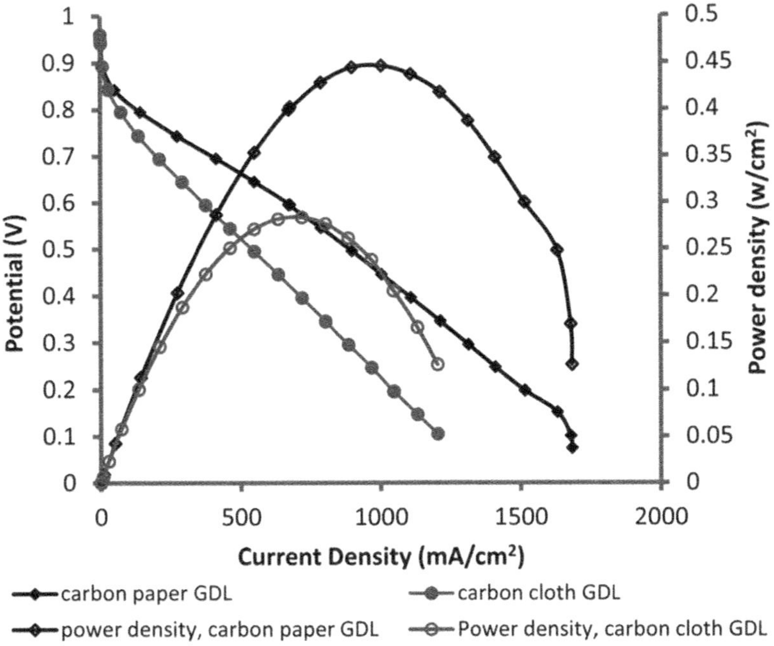

Fig. 34.3 Polarization curves for the fuel cells assembled with the MEAs prepared with 20 % Pt/C catalyst by hand spraying method on carbon paper or carbon cloth GDL

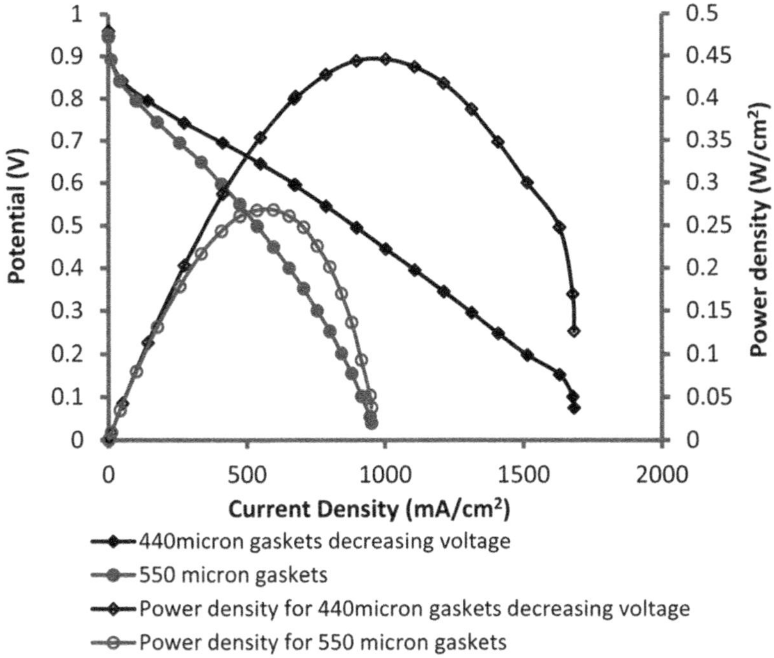

Fig. 34.4 The effect of gasket thickness on the performance of serpentine channeled fuel cell with GDL 30BC carbon paper GDL operated at 70 °C in stable conditions

figure that by an increase in gasket thickness from 440 to 550 μm the cell resistance increased from 0.51 to 0.94 Ω. In consequence, the maximum power decreased from 0.45 W/cm^2 to 0.28 W/cm^2. The smaller gasket thickness decreased the contact resistance; therefore it has been selected for further performance tests in PEMFC.

34.3.2 Performance of Newly Developed MEAs

Ultrasonic coating machine was used to prepare MEAs having 5 cm^2 active areas in order to determine the performance of the newly developed MEA production method. Two different techniques were tried. In first method, the catalyst ink directly coated to the membrane to form three layers MEA first and the GDLs were pressed to this three layer MEA to form a five layer MEA. In the second method, the catalyst ink was coated on the GDLs to form gas diffusion electrodes (GDEs) and these GDEs were pressed both side of the membrane to form five layers MEA.

The comparison of the performances of the MEAs prepared by membrane or GDL coating is shown in Fig. 34.5. The same amount of platinum loading

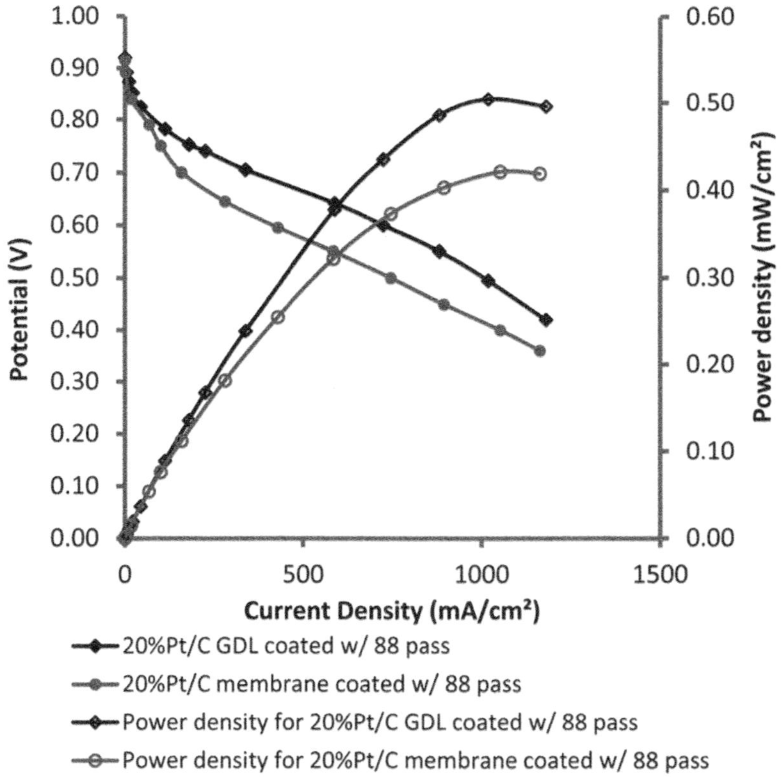

Fig. 34.5 Comparison of the polarization curves of the MEAs produced by membrane coating or GDL coating

(0.32 mg Pt/cm^2) was applied with 88 pass to both MEAs. Membrane coated MEA exhibited the highest power at 0.4 V and the corresponding power was 0.42 W/cm^2. Whereas, the MEA produced by GDL coating showed the highest power at 0.5 V and the corresponding power was 0.5 W/cm^2. The performance of the later was 20 % higher. It should be noted that higher voltages are always preferred in PEMFC operations since the efficiency of fuel cell utilization is higher. Both cells had the same resistance which was 0.32 Ω, but GDL coating lead to faster kinetics. These results revealed that GDL coating is more effective than the membrane coated MEA in this new technique.

The performances of the GDL coated MEAs prepared by 88 pass (0.32 mg Pt/cm^2) or 110 pass (0.4 mg Pt/cm^2) are compared in Fig. 34.6. Insignificant increase in the power density was observed if the Pt loading had been increased from 0.32 to 0.4 mg Pt/cm^2. However, the characteristics of the both cells were very similar as it can be seen in the current-potential (IV) curves. These results designate that Pt load might be reduced by the new MEA preparation method.

Different catalysts were used during catalyst ink preparation. The basic differences in them were their platinum percent on Pt/C catalyst. The catalyst load had been kept constant on the gas diffusion electrodes for both anode and cathode which was 0.4 mg Pt/cm^2. All the other conditions were kept the same. The effect of using different catalysts on the performance of the fuel cells having MEAs prepared with ultrasonic coated electrodes is shown in Fig. 34.7. The performance of the fuel cell with the MEA prepared using 40 % Pt/C catalyst was incredibly high. The highest power reached with this cell is 0.74 W/cm^2 at 0.45 V and the performance is higher than the commercial MEAs which were

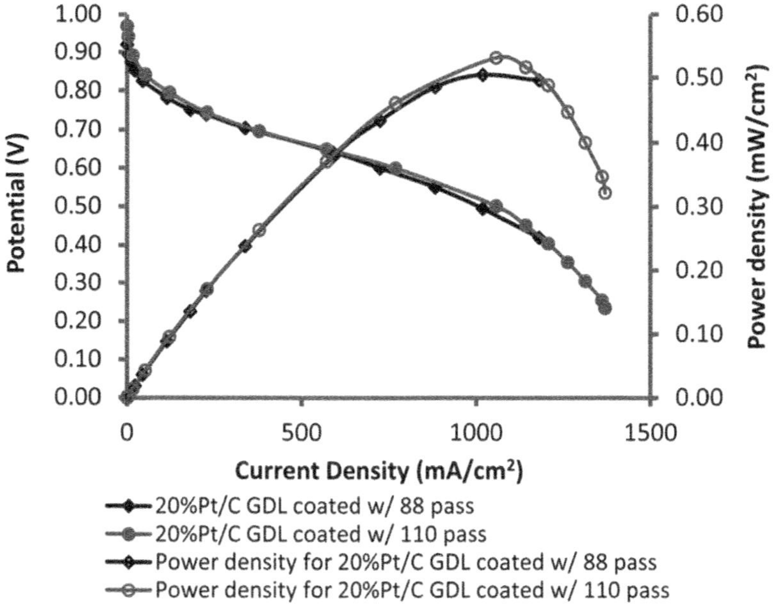

Fig. 34.6 Comparison of the polarization curves of the GDL coated MEAs produced by 88 (0.32 mg Pt/cm^2) pass or 110 pass (0.4 mg Pt/cm^2) with the same catalyst ink

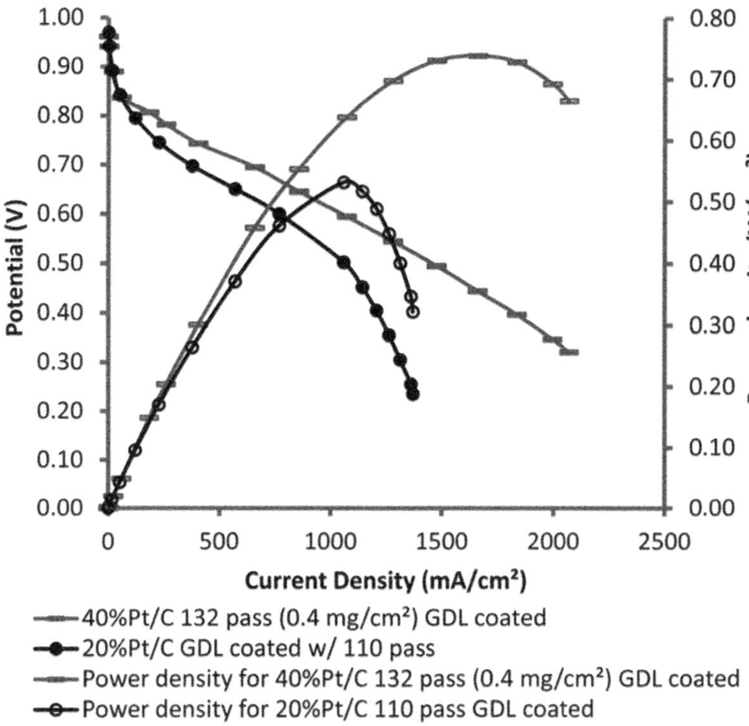

Fig. 34.7 The effect of the catalyst Pt content on the performance of the fuel cells having MEAs prepared with ultrasonic coated electrodes

the highest performance MEAs ever tested in FCRC laboratory, previously. The comparison of the performances of the MEAs produced in this work with the commercial MEA is shown in Fig. 34.8. Even with the 20 % Pt containing catalyst newly developed MEAs presented in this study have a better performance than the commercial MEA (OMG).

The comparison of the performances of the MEAs prepared by 40, 50 or 70 % Pt/C catalysts coated onto GDL is shown in Fig. 34.9. The same amount (0.4 mg Pt/cm^2) of platinum loading was applied with 110 pass to MEAs. The MEAs prepared with 40, 50, and 70 % Pt/C catalysts has the highest power at 0.45, 0.45, and 0.5 V, respectively. The corresponding power densities were 0.74, 0.77, and 0.88 W/cm^2, respectively. The resistances calculated from the linear part of the polarization curve were 0.24, 0.22, and 0.19 Ω for the fuel cells assembled with MEAs prepared with 40, 50 and 70 % Pt/C catalysts, respectively. Figure 34.9 revealed that increase in the Pt weight percentages on carbon support resulted in an increase on the PEM fuel cell performance. These results confirmed our previous findings with hollow core mesoporous shell carbon supported Pt electrocatalysts with high Pt loading [5]. The increase in the Pt amount over the carbon support enhances the contact of the catalyst with the electrolyte and also it decreases the contact resistance.

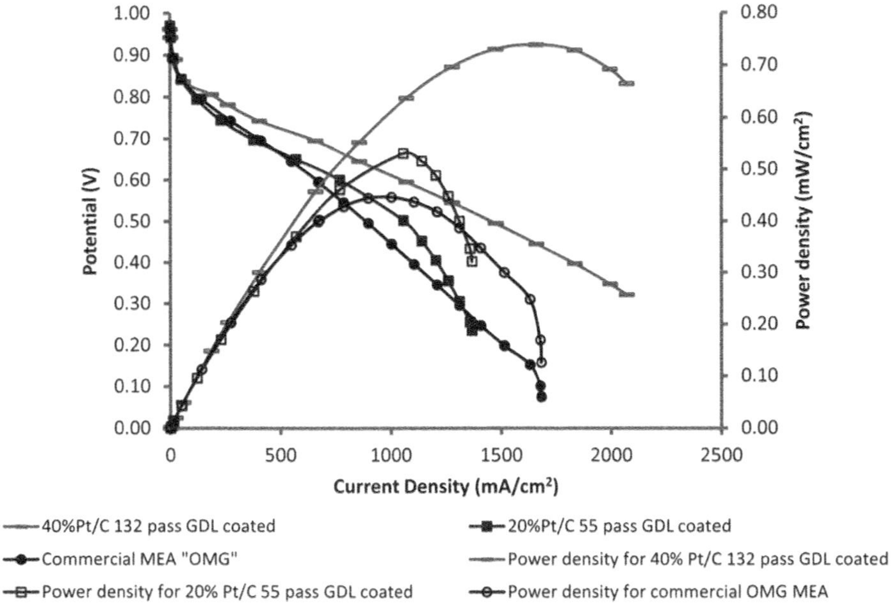

Fig. 34.8 The comparison of the performances of the MEAs produced with 20 and 40 % Pt/C catalysts with a commercial MEA's performance

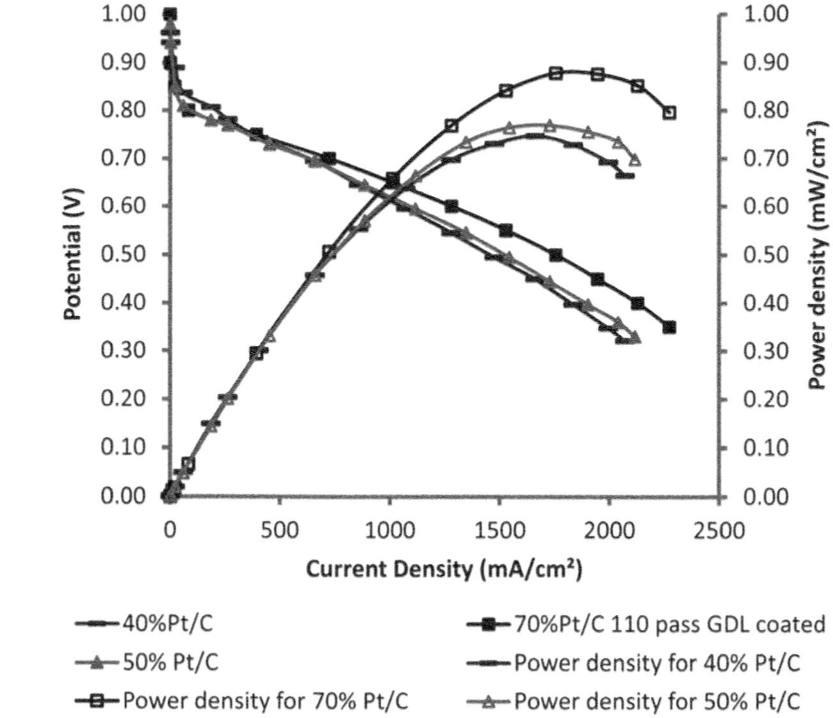

Fig. 34.9 Comparison of the polarization curves of the MEAs produced 40, 50, or 70 % Pt/C catalysts

The power or current density of a PEM fuel cell is affected by many different parameters other than the MEA. The preparation method and materials of the MEA also differ in the published studies. There is no straightforward comparison with the literature. However, it is known that 1,000 mA/cm^2 current density is a typical value for a single PEM fuel cell operating at 0.6 V potential. Qi and Kaufman [6] achieved power densities as high as 0.72 W/cm^2 at cell temperature of 75 °C, for a cathode with a Pt loading of 0.12 mg/cm^2.

Endoo et al. [7] studied the identification of the key variables for the MEA preparation that affects the PEM fuel cell performance. The maximum performance achieved was 393 mA/cm^2 (Pt loading: 1 mg Pt/cm^2) and 566 mA/cm^2 for the in-house prepared and commercial MEAs, respectively. Thanasilp and Hunsom [8] evaluated the effect of MEA fabrication techniques on the performance of Pt-Pd/C catalyst used in PEM fuel cell. The maximum power density achieved was 0.21 W/cm^2 at 0.6 V cell potential with catalyst loading of 0.15 mg/cm^2.

The ultrasonic-sprayed MEA were tested and compared with hand-sprayed with spray gun MEAs with Nafion/$TiSiO_4$ membrane [9]. It was found that the GDEs prepared by the ultrasonic-spray method exhibited better performances compared to those prepared by the spray technique. The maximum power densities obtained at 75 °C for ultrasonic coating and spray gun were 0.803 and 0.298 W/cm^2, respectively. The high performance of the MEA prepared by using ultrasonic coating is due to the homogenous catalyst distribution on the gas diffusion electrode [10].

The new method is also highly recommended for PEMFC fuel cell stacks' MEAs manufacturing [11].

34.4 Conclusions

The membrane electrode assembly is an important part of a PEM fuel cell. Its components (catalyst, membrane, gas diffusion layer) and preparation technique significantly affect the performance of the fuel cell. A new technique called "ultrasonic spray coating technique" was applied to increase the MEA performance. The power densities reached with the new method were 0.53, 0.74, 0.77, and 0.88 W/cm^2 for 20, 40, 50, 70 % Pt/C catalyst by keeping 0.4 mg Pt/cm^2 loading constant, respectively. The power density increase with the new technique was 267 % compared to "spraying of catalyst ink with air pressure atomizing spray gun". Advantages of the new technique can be listed as follows:

1. Reproducible MEAs can be produced.
2. Performance can be guaranteed.
3. Larger surface area MEAs can be produced up to 400 cm^2.
4. Uniform distribution of the catalyst can be achieved.
5. Electrodes with complex patterns or different designs can be produced.

6. Catalyst loss can be minimized.
7. Parameters of coating can be optimized.
8. Pt loading can be reduced.

Acknowledgement This study was supported by Turkish Boron Institute (BOREN) with project 2009.Ç0219.

References

1. Bayraceken A, Erkan S, Turker L, Eroglu I (2008) Effects of membrane electrode assembly components on proton exchange membrane fuel cell performance. Int J Hydrogen Energ 33:165–170
2. Sengul E, Erkan S, Bac N, Eroglu I (2009) Effect of gas diffusion layer characteristics and addition of pore-forming agents on the performance of polymer electrolyte membrane fuel cells. Chem Eng Comm 196:161–170
3. Fıçıcılar B, Bayrakçeken A, Eroglu I (2010) Pt incorporated hollow core mesoporous shell carbon nanocomposite catalyst for proton exchange membrane fuel cells. Int J Hydrogen Energ 35:9924–9933
4. Devrim Y, Erkan S, Bac N, Eroglu I (2013) Nafion/titanium silicon oxide nanocomposite membranes for PEM fuel cells. Int J Energ Res 37:435–442
5. Güvenatam B, Fıçıcılar B, Bayrakçeken A, Eroglu I (2012) Hollow core mesoporous shell carbon supported Pt electrocatalyst with high Pt loading for PEMFCs. Int J Hydrogen Energ 37:1865–1874
6. Qi Z, Kaufman A (2003) Low Pt loading high performance cathodes for PEM fuel cells. J Power Sources 113:37–43
7. Endoo S, Pruksathorn K, Piumsomboon P (2010) Identification of the key variables in membrane electrode preparation for PEM fuel cells by a factorial design. Renew Energy 35:807–813
8. Thanasilp S, Hunsom M (2010) Effect of MEA fabrication techniques on the cell performance of Pt- Pd/C electrocatalyst for oxygen reduction in PEM fuel cell. Fuel 89:3847–3852
9. Devrim Y, Erkan S, Bac N, Eroglu I (2012) Improvement of PEMFC performance with Nafion/inorganic nanocomposite membrane electrode assembly prepared by ultrasonic coating technique. Int J Hydrogen Energ 37:16748–16758
10. Millingtona B, Whippleb V, Pollet BG (2011) A novel method for preparing proton exchange membrane fuel cell electrodes by the ultrasonic-spray technique. J Power Sources 196:8500–8508
11. Erkan S (2011) Development of 100 W Portable fuel cell system working with sodium borohydride. Dissertation, Middle East Technical University

Chapter 35
Role of Sulfur to Acquire Hydrogen from the Black Sea

Sevgi Güneş Durak and Şule Kapkin

Abstract In Turkey, the Black Sea, which features both hydrogen sulfur and wave energy, can be considered as one of the important alternative energy sources. Research in areas of about 200 m in depth has proved that there is a quite intense amount of H_2S. Even at the depths of 30–40 m, the presence of the gas has been detected. It is possible to obtain hydrogen and sulfur from H_2S by using a converter or by means of other methods. Sulfur is a very valuable gas in economic terms. Quantity of sulfur-containing resources decreases day by day in our country. The Black Sea is rich in hydrogen and sulfur. The aim of this work is to develop new methods for acquiring hydrogen and sulfur from H_2S, which is handled as energy source and raw material.

Therefore, the development of a new method has been attempted to obtain solid sulfur from sea bottom. Thus, in addition to obtaining new sources for energy, we will be able to prevent the formation of anaerobic conditions which will result in improvement in the environment, and will eliminate the risk of poisoning that is predicted to happen in the next 20–30 years.

Keywords Hydrogen • Hydrogen sulfur • Sulfur • Sulfur bacteria • Black Sea

35.1 Introduction

Kızılırmak, Yeşilırmak, Sakarya, Çoruh, Kuban, Don, Dinyeper Rivers, and Tuna River, which go around whole Eastern and Middle Europe, pour into the Black Sea. As the amount of organic substances in these rivers is much more than the maximum amount of substances decomposable by bacteria, bacteria obtain oxygen from sulfide ions, a component of the sea water, rather than using dissolved oxygen in

S. Güneş Durak
Environmental Engineering Department, Faculty of Civil Engineering,
Yildiz Technical University, Davutpasa Campus, 34220 Esenler, Istanbul, Turkey
e-mail: sgdurak@yildiz.edu.tr

Ş. Kapkin (✉)
Department of Mechanical Engineering, Faculty of Engineering,
Istanbul University, 34320 Avcilar, Istanbul, Turkey
e-mail: skapkin@istanbul.edu.tr

I. Dincer et al. (eds.), *Progress in Clean Energy, Volume 2*,
DOI 10.1007/978-3-319-17031-2_35

the seawater. This process releases a very dangerous gas—hydrogen sulfide—which disables organic life under 200 m depth [1].

In the closed basins, fresh seawater entering the basin subsides via evaporation. This heavy water cannot reach open seas. So normal salty water is always on the top and the very salty water is at the bottom.

This oxygen-free environment constitutes 87 % of the Black Sea. Therefore, there is no chance of survival of aerobic organisms in this portion of the Black Sea. Domestic and industrial waste from rivers flowing into the Black Sea is responsible for the impoverishment of its flora and fauna. Although the Black Sea is the largest hydrogen sulfide reserve in the world, it is not the only sea accumulating H_2S. There are some oxygen-free areas in the bottom of the Baltic Sea and in some of the Norwegian fjords. H_2S sometimes rises to the surface off the coasts of Peru, creating a disaster called "El Nino." This disaster kills the entire ecosystem, destroying fishery in that region and blackening the bottom of boats (Callao painter). Since the Black Sea is an inland sea, a similar incident may cause a huge disaster.

There are a lot of studies to realize hydrogen technology, economics, environmental impact, special system applications, and hydrogen energy status around the world at the end of the twentieth century [2, 3]. And there must be more studies to create awareness for sulfur of the Black Sea.

If studies on the use of the H_2S reserves in the bottom of the Black Sea are not carried out, there will be a disaster because of the H_2S accumulation.

A study titled "Methane research in the Black Sea" was conducted by the General Directorate of Mineral Research & Exploration [4]. Samples were collected from 11 different stations in the Black Sea for a month. The study concluded that the water of the Black Sea cannot contain methane because of its high salt amount. However, salinity does not affect H_2S negatively. Methane- and CO_2-producing bacteria begin forming organic matter. Excessive salinity does not seem to affect the organic matter, and the existing salt content, combined with H_2S, enters into a chemical reaction, causing H_2S to convert into sulfate. Therefore, this study is the evidence of the existing H_2S reserves in the Black Sea.

The Scripps Institution of Oceanography (University of California) collected samples from three locations from May 20, 1975, to June 11, 1975. Drillings were made in 624.5, 1,073, and 503 m. Samples were evaluated by sedimentologists, geochemists, and paleontologists. Studies concluded that the Black Sea had had four glacial stages in the last million years and consequently, the water level descended to 100–150 m, causing the Black Sea to become a big lake. Drillings revealed that the organic mud which consists 75 % of the deposition in the lake floor was diatomaceous and turbid mud. All three wells abounded with methane and ethane gases formed from organic substances due to the pressure and temperature. In addition, the Black Sea water contains plenty of hydrogen sulfide from 100 to 2,200 m depth. This substance causes the death of all living organisms, except anaerobic bacteria [5].

In the Black Sea, the hydrogen-sulfide concentration increases regularly from the anoxic interface (at 1,000 m depth) to the seafloor. Near the seafloor, 2,000 m depth, it attains maximum values of about 400 μM. Hydrogen sulfide and dissolved oxygen concentrations are shown in Fig. 35.1 [6, 7].

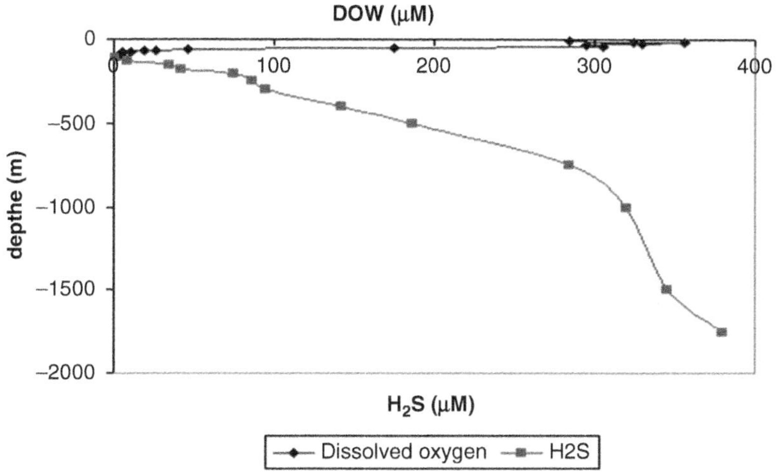

Fig. 35.1 Hydrogen sulfide and dissolved oxygen concentrations in the Black Sea, obtained from the data from March to April 1995 cruise of R/V Bilim [6, 7]

In another study by the Woods Hole Oceanography Institute, total magnetic field was compared between the Black Sea and the Mediterranean Sea. It was determined that, devoid of dissolved oxygen but rich in dissolved sulfide species (H_2S, HS, S-) as from about 250 m in onshore areas and 150 m in offshore areas, the Black Sea was the world's largest anoxic system [8].

This study evaluates the H_2S potential in the Black Sea in terms of sulfur. A sustainable method is developed to enable the production of sulfur with minimal environmental damage.

35.2 Production of Sulfur from H_2S

This study prefers the Claus process to obtain sulfur from hydrogen sulfide in terms of efficiency.

$$1/3H_2S + 1/2O_2 \rightarrow 1/3SO_2 + 1/3H_2O \quad (35.1)$$

$$\underline{2/3H_2S + 1/3SO_2 \rightarrow S + 2/3H_2O} \quad (35.2)$$

$$H_2S + 1/2O_2 \rightarrow S + H_2O + Q \quad (35.3)$$

Recovery of sulfur from H_2S by Claus process and the progressive processing of sulfur are shown in Fig. 35.2.

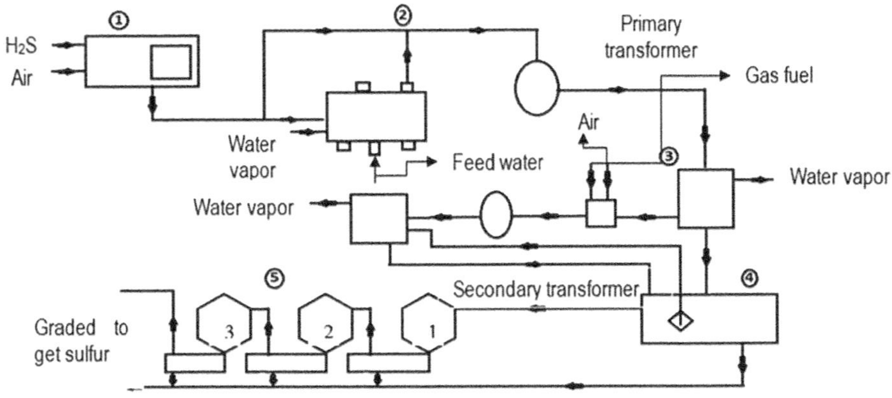

Fig. 35.2 Claus process and processing of sulfur

In the thermal step, H_2S oxidizes partially with air. This process occurs in the furnace at high temperatures (1,000–1,400 °C). In the furnace, sulfur is formed but reaction is not performed completely and thus SO_2 is released. In the catalytic step, hydrogen sulfide reacts and consequently, more sulfur can be obtained due to low operating temperatures.

The resulting concentrated H_2S and oxygen are dehumidified and then combusted in the furnace to release SO_2 for the production of sulfuric acid. However, H_2S can also be converted into "recovered elemental sulfur" through a variety of the original Claus process, which leads to the ever-increasing amount of sulfur recovered in this way. According to the modified process, reactions occur in two steps.

$$H_2S(g) + 3/2O_2(g) \rightarrow SO_2(g) + H_2O(g) \quad \Delta H = -123.9\,\text{kcal} \tag{35.4}$$

$$SO_2(g) + 2H_2(g) \xrightarrow{Fe_2O_3} 3S(l) + 2H_2O \quad \Delta H = -34.2\,\text{kcal} \tag{35.5}$$

In the first step, one third of H_2S is burned and converted into SO_2, which is then reacted with the remaining hydrogen sulfide. In modern Claus furnaces, heat of reaction is used to produce vapor. The second step is carried out very effectively with catalysts. In general, bauxite is used as catalyst. In plants with two transformers, sulfur recovery can be approximately 90–93 %.

In Fig. 35.2, Section 5 the resulting liquid from the Claus process is arranged, via additional processing of sulfur, for industrial use. These three sections are applied to serve the industrial use of sulfur directly. In the first section, liquid sulfur recovered via the Claus process is stored to ensure its use in liquid form. In the second section, liquid sulfur is evaporated to enable the precipitation of solid sulfur, which is sent to the third section where it is processed in powder and granules. Thus, efficient industrial use of sulfur is ensured without damaging the nature.

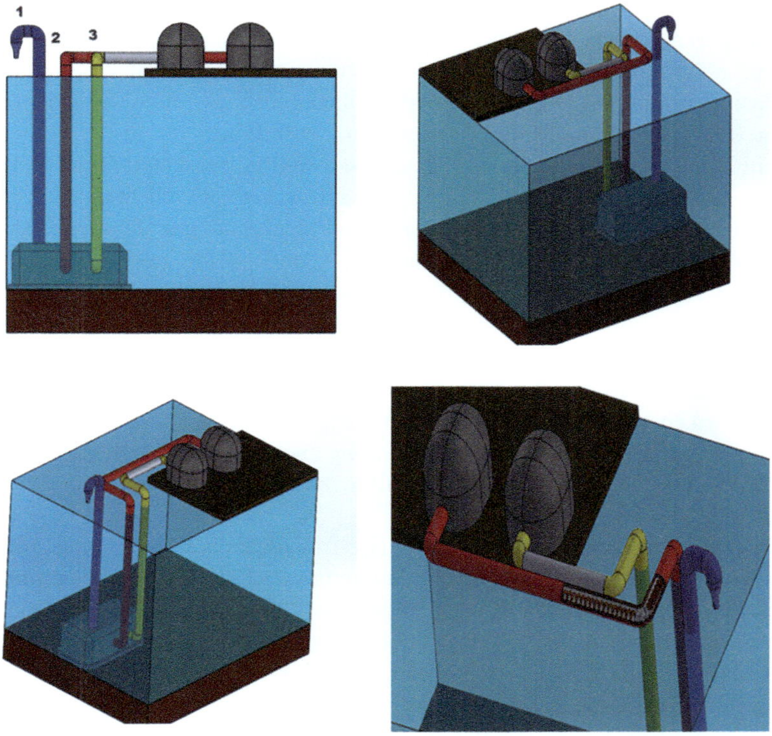

Fig. 35.3 Diagram of the system, which takes solid sulfur from the seabed and converts it into molten sulfur

On the other hand, precipitated solid sulfur is found in the seabed of the Black Sea. Therefore, sulfur should be removed to prevent the formation of an anaerobic environment. To do this, a unit with a certain volume was lowered into areas with the highest content of sulfur to drain and dehydrate the sulfur-containing mass with a predefined pressure, as shown in Fig. 35.3. There are three lines connected to this unit. The first line (1) is responsible for drainage. The second line (2) provides superheated water, which will melt the sulfur depending on the temperature of the seawater. Then, the third line (3) raises the molten sulfur to the surface. The molten sulfur recovered via the Claus process is adapted to liquid sulfur and thus made available for industrial use.

35.3 Results and Discussion

As is known, the adverse effects of fossil fuels such as coal, oil, and natural gas are not limited to their immediate environment but also spread over the atmosphere. This pollution has caused and continues to cause climate change and deterioration of the ecological system all around the world. Sulfur dioxide (SO_2) that mixes into the air is transformed in sulfuric acid, which is a serious threat to human health. Cancer is the most common health risk resulting from this process. In addition to increasing the amount of carbon dioxide in the atmosphere, fossil fuels also release sulfur. This leads to multifaceted atmospheric pollution.

Sulfur and sulfuric acid are used in almost everything essential in our daily lives. On the other hand, the industrial use of sulfur includes chemical, agricultural fertilizer, rubber, tire, paint, paper, iron and steel, and petroleum industries, as well as production of pesticides, gunpowder, and paving stones. Production of sulfuric acid accounts for 88 % of the consumption of sulfur.

Sulfur recovered via the Claus cycle should undergo an additional process to be converted in liquid form. Only then can we prevent its mixture into air. We can thus minimize the environmental damage caused by the industrial use and production lines. When energy is recovered from the Black Sea, the resulting sulfur should also be considered. Simultaneous project planning and processing will ensure economical production and industrial use of sulfur, as well as prevention of its possible damage to the environment.

This study focuses on the recovery of H_2S from the Black Sea and the production of sulfur via the Claus process. In addition, removal of sulfur from the bottom of the Black Sea through a newly developed process will also prevent the formation of an anaerobic environment. Claus process enables the recovery and storage of elemental sulfur in liquid form. A certain portion is evaporated to be converted into solid sulfur. Then, solid sulfur is converted into powder and granules suitable for industrial use. Sulfur recovered from the seabed is taken into a special unit where it is first drained and dehydrated and then superheated to obtain molten sulfur. Thus, both economic and ecological benefits will be provided.

35.4 Conclusions

Huge environmental catastrophes or irrecoverable environmental pollution will be possible unless equilibrium is preserved in the Black Sea and improvement efforts are carried immediately. The first effects of pollution will be seen in surface waters, and with increasingly deteriorating chemical equilibrium, hydrogen and sulfur ions in deep waters will be much likelier to pass into the flammable, explosive gas phase.

As it is currently known that the danger persists, it should be considered a silent bomb with an unpredictable time of explosion. As a precaution, the existing energy must be recovered and the sea must be desulfurized. Under this light, the seemingly negative scenario will end up in a positive and productive manner.

Acknowledgment Thanks to Dr. T. N. Veziroglu for his encouragement and support throughout this study.

This work is supported by Research Fund of the Istanbul University, project number UDP-44245, Istanbul, Turkey

References

1. URL. tr.wikipedia.org/wiki/Karadeniz. Accessed on August 18, 2013
2. Momirlan M, Veziroglu TN (2002) Current status of hydrogen energy. Renew Sustain Energy Rev 6:141–179
3. Veziroglu TN (2003) Hydrogen energy system: a permanent solution to global problems
4. Karababa I (1964) Investigation methane gas in the Black Sea. Turkey General Directorate of Mineral Research and Exploration. Compilation number: 3484
5. Senalp M (1975) Investigation reports of Glomar Challenger Ship in the Black Sea. Turkey General Directorate of Mineral Research and Exploration. Compilation number: 5430
6. Haklidir M, Kapkin S (2005) Black Sea, A hydrogen source. IHEC 16/13–15 June 2005 – Istanbul, Turkey
7. Haklidir M, Tut FS, Kapkin S (2006) Possibilities of production and storage of hydrogen in the Black Sea. WHEC 16/13–16 June 2006 – Lyon, France
8. Turgay I (1969) Investigation report of Black Sea by Woods Hole Oceanography Institute. Turkey General Directorate of Mineral Research and Exploration. Compilation number: 3490

Chapter 36
Hydrogen Generation and Storage from Sodium Borohydride

Valentina G. Minkina, Stanislav I. Shabunya, and Vladimir I. Kalinin

Abstract Hydrogen generator with circulation scheme that generates high-purity hydrogen from the solution of sodium borohydride has been developed. In circulation scheme working solution is repeatedly passed through catalytic unit in reactor; the velocity of solution flow and mass transfer processes are more intensive and the entire catalyst is working under approximately the same conditions. Alumina-supported Pt-, Pd-, and Ni-granulated materials have been used as model catalysts during performance testing of the hydrogen generator. Instead of water-alkaline solution it is proposed to use sodium metaborate solution saturated at room temperature as a solvent and stabilizer to prepare working solutions. Optimal parameters of working process (working solution composition, temperature, pressure) were defined. Technical specifications of the generator allow achieving hydrogen performance up to 3 Nm^3/h and higher.

Keywords Sodium borohydride • Sodium metaborate • Hydrogen generator • Circulation scheme • Catalyst

36.1 Introduction

Aqueous solutions of alkali metal borohydrides can be used to produce high-purity hydrogen. Regarding the mass percent of hydrogen and the safety of storage, sodium borohydride $NaBH_4$ can be considered as one of the most promising compounds [1, 2]. $NaBH_4$ does not detonate at contact with water, as opposed to lithium borohydride which reacts without the ignition, and in contrast to potassium borohydride has a high content of hydrogen. Contrary to traditional kinds of fuel it is a renewable energy source and safer, as it is not combustible and nonflammable. Usage of $NaBH_4$ as fuel has additional advantages. These solutions are nontoxic and possess high mass energy and high volume density [3, 4]. In the reaction of hydrolysis hydrogen is the only one gaseous product that can be easily separated.

V.G. Minkina (✉) • S.I. Shabunya • V.I. Kalinin
Heat and Mass Transfer Institute, National Academy of Sciences of Belarus,
P. Brovka Street 15, 220072 Minsk, Belarus
e-mail: minkina@dnp.itmo.by

I. Dincer et al. (eds.), *Progress in Clean Energy, Volume 2*,
DOI 10.1007/978-3-319-17031-2_36

Sodium borohydride is environmentally friendly since the products of the reaction of hydrolysis are mostly water and sodium metaborate $NaBO_2$.

The cost and environmental issues are of primary importance for hydrogen energy. The cost of $NaBH_4$ is high enough. Obviously, the area of application for hydrogen storage will be limited by possibility of sodium borohydride recycling. Only when the energy-efficient ways will be found to convert end product—sodium metaborate—into sodium borohydride, we can discuss the large-scale applications such as energy or transportation. The works in this direction are in progress [5–8], and we hope that this problem will be solved in the near future.

Current study is focused on developing flowing-type hydrogen generators among which the choice between single-pass and circulation rector is the most important. To the best of our knowledge most of the studies are focused on the single-pass reactor [9–14]. In the single-pass scheme, the front part of catalytic unit works with a concentrated solution, the output—with the lean one.

In our proposed circulating scheme solution repeatedly passes through the catalytic unit; the velocity of solution flow and mass transfer processes are more intensive and the entire catalyst works under approximately the same conditions.

This chapter discusses the operation principle of such a hydrogen generator and the results of its testing with the use of different catalysts.

36.2 Experimental

36.2.1 Materials

The commercial sodium borohydride (JSC Aviabor, granules powder 98 wt% $NaBH_4$) is used. Three samples of the granulated catalysts, alumina-supported Pt, Pd (Sigma Aldrich Co. LLC), and Ni (Liaoning Haitai Sci-Tech Development Co., Ltd.), have been used. These granular catalysts are traditionally used in the processes of partial oxidation of hydrocarbons [15–17]. Though their performance has not been specifically optimized for the hydrolysis of $NaBH_4$, these catalysts demonstrated significant acceleration of the hydrolysis reaction and catalytic activity that was suitable for the study of the hydrogen generator capabilities.

Table 36.1 shows the characteristics of these catalysts deposited on alumina carrier. Surface area is measured by BET technique using analyzer ASAP 2020 V3.04 H.

36.2.2 Hydrogen Generator

The scheme and photograph of our hydrogen generator are shown in Fig. 36.1. Inside the reactor there is catalytic unit. Casing of catalytic unit is a cylinder 80 mm

Table 36.1 Characteristics of used alumina-supported catalysts Pt, Pd, and Ni

Catalyst	Content (wt%)	Shape	Dimensions (mm)	Surface area (m^2/g)
Pt	0.5	Cylinder	$\Delta\varnothing$ 3.2, h = 3.6	82.3
Pd	1.0	Cylinder	$\varnothing$ 3.2, h = 3.7	86.1
Ni	20–30	Cylinder	$\varnothing$ 2.0, h = 3.0	89.0

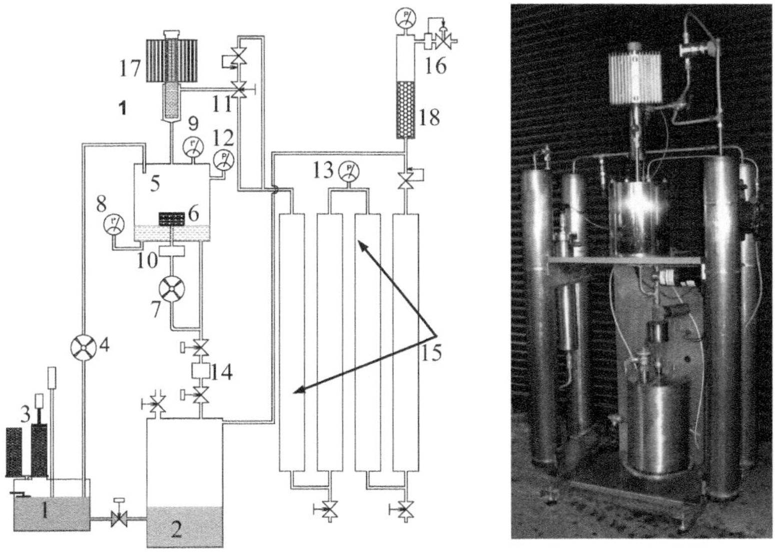

Fig. 36.1 Scheme and photograph of hydrogen generator. 1—fuel tank mixer, 2—waste tank containing $NaBO_2$ solution, 3—system supplying dry $NaBH_4$, 4—pump supplying initial solution to reactor, 5—reactor, 6—catalytic unit, 7—pump to circulate working solution through catalytic unit, 8, 9—platinum Pt100 sensor, 10—electric heater for initial solution preheating, 11—valve to disconnect reactor and receivers, 12, 13—sensors measuring pressure in reactor and in the receiver, respectively, 14—dosing vessel, 15—high-pressure receiver, 16—flow controller of hydrogen, 17—dephlegmator, 18—low-pressure receiver containing silica gel

in diameter and 60 mm in height. Catalytic unit contains 0.2 L of catalyst granules. In order to prepare working solution, $NaBO_2$ solution is supplied from waste tank to fuel tank mixer. The bottom of waste tank is removable, which allows removing accumulated solid sodium metaborate. The aim of precipitation is the recycling of solution having lowered $NaBO_2$ concentration in duty cycle, to maximum use water.

A series of experiments on $NaBH_4$ hydrolysis is conducted with working solutions having molar ratio of $H_2O/NaBH_4 = 6$ and $H_2O/NaBH_4 = 7$ and with weigh $NaBH_4$ concentrations in solution equal to 26 % and 23 %, respectively. In case of 26 wt% $NaBH_4$ solution a very viscous supersaturated sodium metaborate is formed after complete hydrolysis. Even when draining it from hot reactor, this solution begins to thicken and converts into a solid mass. This state of working solution may hinder its removal from reactor, and is dangerous because of the high probability of crystallization in a bypass channel. In case of 23 % wt% $NaBH_4$ solution after complete hydrolysis there is formation of $NaBO_2$ solution which maintains consistency at draining from hot reactor, while at room temperature this solution is quickly converted into a solid crystalline hydrate form of sodium metaborate.

Thus, the solution of the final product ($NaBO_2$) should not be supersaturated at reactor outlet and must contain a minimum amount of water. High solubility of $NaBH_4$ does not impose additional restrictions when considering the loss of water.

If hydrogen generator start-up occurs, when the reactor and working solution are cold, then there is temporary preheating of the whole system by an electric heater. After heating the working solution in the reactor electric heater is turned off, and further it is not required for operation of the generator. The temperature of solution inside the reactor is measured by platinum Pt100 sensor.

The circulation of working solution in reactor occurs up to complete hydrolysis of $NaBH_4$. Several cycles of filling and emptying the dosing vessel must be repeated to ensure complete discharge of consumed solution from reactor. Such a phased removal of waste solution from the reactor allows controlling the amount of hot $NaBO_2$ solution in the reactor. Hot solution remaining in reactor plays the role of the thermal buffer, which preheats fed fresh $NaBH_4$ solution mixing with it.

In the case of concentrated solution, it is inconvenient to use total volumes of the working solution for circulation, due to two significant factors. First, the final product can be obtained in the form of a liquid only at temperatures above 70 °C [18]; that is it is necessary to maintain all solution at this temperature. Second, the flow rate of the generated hydrogen constantly decreases with consumption of $NaBH_4$ and in the final stage it will be difficult to maintain a nominal flow rate. As alternative to continuous circulation, periodic circulation of solution can be considered, when a portion of working solution is used for circulation, until a completion of the hydrolysis process. The solution is then drained into waste tank, and the circulation is resumed with the next portion of fresh solution. In this operation mode, the generation of hydrogen generation is not constant during the cycle: fast at the beginning of the cycle and slow at the end. Fluctuations in hydrogen consumption will cause corresponding pressure fluctuations in the receiver, and the amount of hydrogen supplied to the fuel cell is controlled by flow controller.

36.3 Results and Discussion

36.3.1 The Operating Temperature

The temperature of the rector is one of the most important parameters. Increase of the temperature greatly accelerates hydrolysis of $NaBH_4$ and thereby reduces the requirements for the volume and efficiency of the catalyst. On the other hand a high

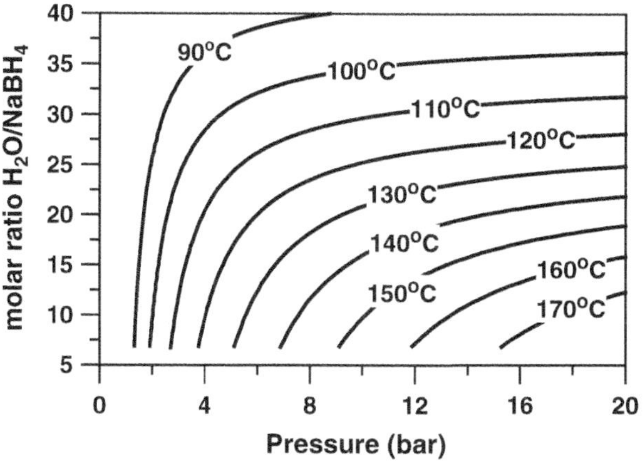

Fig. 36.2 Molar ratio $H_2O/NaBH_4$ changes versus pressure for the complete hydrolysis in adiabatic conditions

temperature of sodium metaborate solution at reactor outlet reduces the probability of sodium metaborate formation in solid phase. The higher is the temperature, the higher concentration of $NaBO_2$ is permissible, and therefore, the initial solution of $NaBH_4$ having a higher concentration can also be used. The reaction of hydrolysis is accompanied by heat release:

$$NaBH_4 + 2H_2O = 4H_2 + NaBO_2 + 212\ \text{kJ/mol}.$$

Since the reaction of hydrolysis is exothermic, the released heat can be used to self-heat the reactor. The values of specific heat of $NaBH_4$, water, $NaBO_2$, and hydrogen are 86.54 J/mol/K, 75.3 J/mol/K, 34.85 J/mol/K, and 28.83 J/mol/K, respectively [19]. The results of the corresponding temperature changes for $NaBO_2$ solution (ΔT) formed as a result of complete hydrolysis of the working solution at different concentration of $NaBH_4$ in adiabatic conditions are presented in Fig. 36.2. The curves in Fig. 36.2 give an idea of the temperature regime in the catalytic unit depending on the solution concentration and operating pressure.

It can be seen that at the molar ratios of $H_2O/NaBH_4 \leq 28$, the temperature changes $\Delta T > 100$ °C which requires cooling of the reactor in the case of single-pass approach or cooling of the solution for a circulation scheme. Considering the differences between the single-pass and circulation schemes, it seems easier to cool down the circulating solution than the whole reactor itself.

36.3.2 Storage of Working Solutions

All hydrogen generators known from the literature operate with working solutions consisting of $NaBH_4$, H_2O, and NaOH. The increase of solution pH by addition of alkali strongly inhibits hydrolysis, but in order to achieve low rates of hydrolysis at 30 °C too high alkali concentration would be required. Currently, the amount of alkaline is comparable to the amount of $NaBH_4$ in solution [20, 21] and can reach up to 10 wt% which has a negative influence on the rate of catalytic processes in general.

Suspension is an intermediate storage variant between solution and solid forms. Suspensions have a small advantage in stability compared with solutions. Both suspensions and solutions require substantial stabilizing additives of alkali in case of temperature rise. Both suspensions and $NaBH_4$ solid forms have difficulty with dosing and require dilution with water before being fed into reactor with a catalytic unit.

Based on the results of conducted studies [22] we concluded that the best way to store $NaBH_4$ is in anhydrous dry form; working solution has to be prepared before the usage and stability characteristics of such a solution are chosen taking into account the terms and way of consumption of this solution. Granulated $NaBH_4$ can be stored for a long time in closed containers. In the case of $NaBH_4$ pellet usage the problems of technology to prepare stable compositions are resolved.

In case of an anhydrous $NaBH_4$ supply, it is necessary to either have additional vessel for water or use sodium metaborate solution as a solvent. In this regard, the stability of $NaBH_4$ solutions is investigated, where the product of hydrolysis—$NaBO_2$—is used as a solvent.

To understand the solubility of $NaBO_2$ at different temperatures, Fig. 36.3 was created based on the data reported earlier [18]. It can be seen that at 25 °C solubility of $NaBO_2$ corresponds to a molar ratio of $H_2O/NaBO_2 = 13$. It was taken into account when working solutions of $NaBH_4$ with a given molar ratio of $H_2O/NaBH_4$ were prepared. Hydrolysis of the $NaBH_4$ solutions was tested in a sealed reactor. The degree of $NaBH_4$ decomposition versus time was calculated from the pressure change in the reactor that is proportional to the amount of the hydrogen formed in the reaction [23]. The curves in Fig. 36.4 show the degree of hydrolysis of $NaBH_4$ in solution of $NaBO_2$ saturated at 25 °C.

After an hour at 25 °C and molar ratio of $H_2O/NaBH_4$ in the range of 6–12 the decomposition of $NaBH_4$ was 0.5–0.8 %. If to consider these solutions for

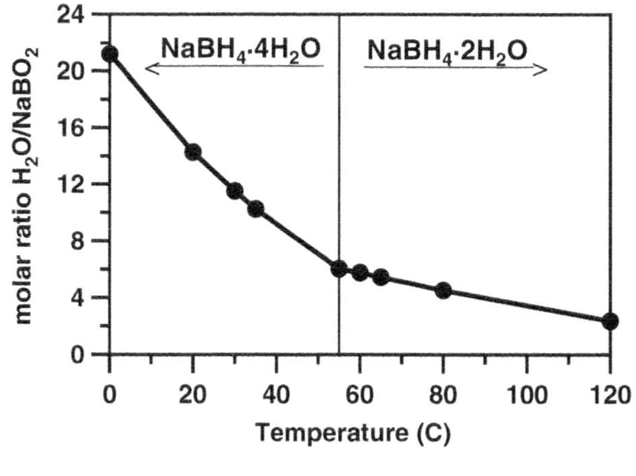

Fig. 36.3 Solubility curve of sodium metaborate

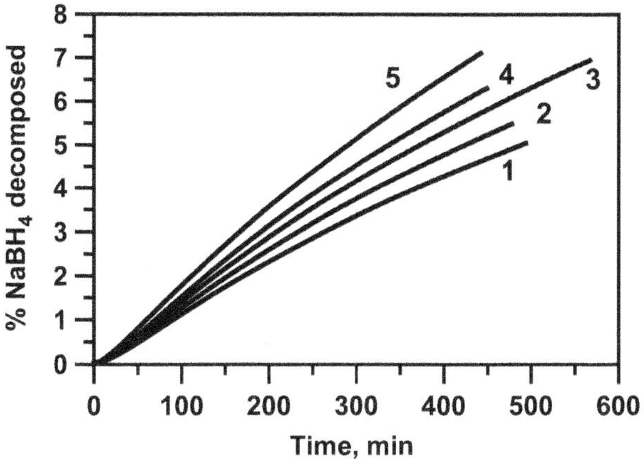

Fig. 36.4 Curves of $NaBH_4$ hydrolysis in solution of $NaBO_2$ saturated at 25 °C. $H_2O/NaBH_4 = 6$ (1), 7 (2), 8 (3), 9 (4), 12 (5)

long-term storage, the rates of hydrolysis are rather high at 25 °C. In case of short-term storage (e.g., few minutes), during which a new portion of working solution is prepared, such a loss is not significant. The effectiveness of such a stabilization is lower than in case of alkali solution, but if the storage interval is not too long, then this option is acceptable. Namely this stabilization is used in the hydrogen generator. In this case, hot solution of concentrated $NaBO_2$ is collected in waste tank, where at cooling sodium metaborate is partially deposited as a solid precipitate; above there is a solution having concentration close to equilibrium. This solution saturated at room temperature is used as a solvent for sodium borohydride.

36.3.3 Tests of Hydrogen Generator

At carrying out the tests we stated the task to demonstrate an opportunity of steady work of hydrogen generator at various hydrogen flow rates. The hydrogen generator is tested to identify the $NaBH_4$ concentration for preparation of working solution, choice of a suitable temperature regime reactor, and level of working pressure in reactor.

In tests with the catalyst 0.5 % Pt/Al_2O_3 the first portion of solution was prepared of three moles of $NaBH_4$ and 700 ml of $NaBO_2$ solution saturated at room temperature, which corresponds to a molar ratio $H_2O/NaBH_4 = 12$. The heater was not switched on because we wanted to find out the possibility of self-heating of working mixture. As the heating occurred, it resulted in the acceleration of hydrolysis in reactor. The test was performed using three portions of $NaBH_4$ solution. The values of temperature and pressure in reactor, pressure in receiver, and hydrogen flow rate are measured during the experiment. The indications of thermocouple 9, pressure sensors 12 and 13, and flow controller 16 are shown in Fig. 36.5.

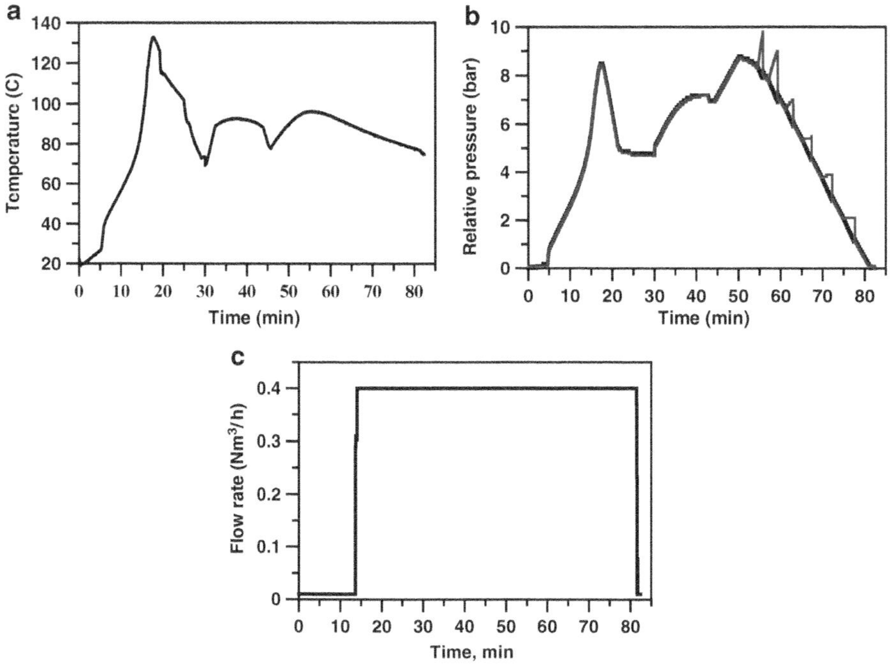

Fig. 36.5 Evolution of solution temperature inside reactor (**a**), pressure inside reactor (*magenta curve*) and receiver (**b**), and hydrogen production (**c**). $NaBH_4/H_2O = 12$, 0.5 % Pt/Al_2O_3 (color figure online)

The complete hydrolysis of $NaBH_4$ is controlled by the valve 11, and pressure sensors 12 and 13 (see Fig. 36.1). In operating mode the valve 11 is opened, and hydrogen from reactor is supplied to receiver. At that pressure value in reactor and receiver is the same. If the valve 11 is closed, the pressure sensor 12 registers pressure in reactor, which now is not dependent on other elements of installation. If pressure rises, it means the continuation of hydrolysis process. If pressure does not increase, the process of $NaBH_4$ hydrolysis is completed (Fig. 36.5b).

When supplying the next portion of cold solution (200 ml with a molar ratio of $H_2O/NaBH_4 = 12$) to reactor, the temperature therein has dropped below 100 °C and not rising above 100 °C until the end of the experiment (Fig. 36.5a). The necessity of solution heating does not appear. The hydrogen flow rate by flow controller equal to 0.4 Nm^3/h demonstrates stability of system operation for more than an hour (Fig. 36.5c).

In tests with the catalyst 1.0 % Pd/Al_2O_3 the first portion of solution was prepared of 5 moles of $NaBH_4$ and 700 ml of $NaBO_2$ solution saturated at room temperature, which corresponds to a molar ratio of $H_2O/NaBH_4 = 7$. Because of the high concentration of $NaBH_4$ self-heating of the working solution was expected. Therefore in the beginning of the experiment the heater was not switched on. But the solution temperature is decreased, and the pressure in reactor is not

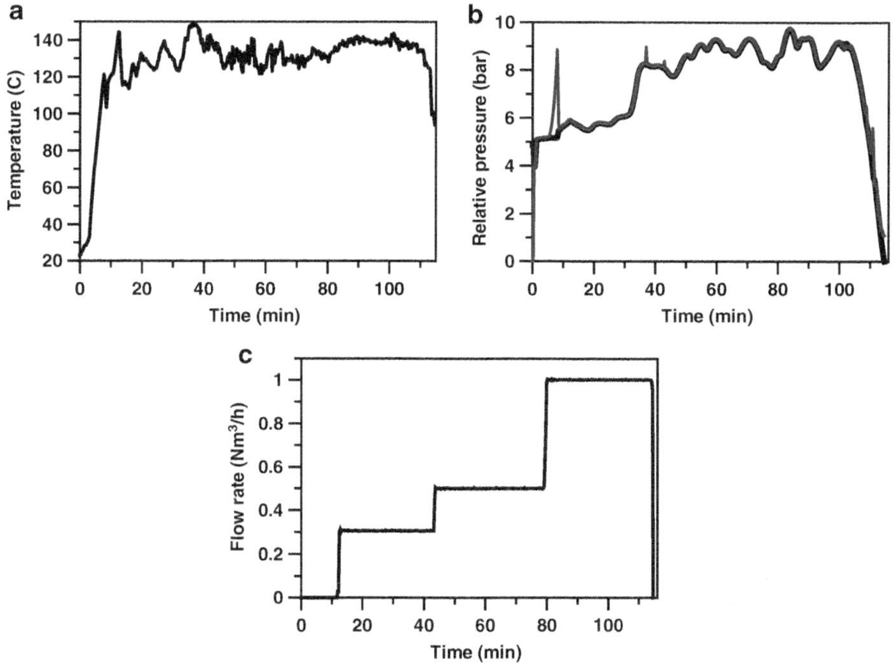

Fig. 36.6 Evolution of solution temperature inside reactor (**a**), pressure inside reactor (*magenta curve*) and receiver (**b**), production of hydrogen (**c**). $NaBH_4/H_2O = 7$, 1 % Pd/Al_2O_3 (color figure online)

increased. Therefore the heater is switched on, to warm up a working solution to the temperature of 80 °C (Fig. 36.5a).

After start-up procedure (loading and preheating of a solution) at time instant, when pressure has achieved the value equal to 6 bar, the removal of hydrogen having flow rate equal to 0.3 Nm^3/h was started. Such a flow rate was kept approximately for 0.5 h. Pressure graph (Fig. 36.6b) demonstrates the time instants of fresh solution supply resulting in pressure growth. During this regime changes the rate of working solution supply trying to minimize pressure fluctuations close to 6 bars. When we were convinced that it can be achieved, we increased pressure (having increased the supply of working solution), and stabilization of pressure at a level equal to 8 bars was easily achieved. Further the transition to regime with the flow rate of hydrogen equal to 0.5 Nm^3/h is done. Such a regime was maintained approximately for 40 min. At this productivity the value of pressure changed with small amplitude close to 9 bars (Fig. 36.6b).

With increase of productivity the amplitude of fluctuations will grow, as supply of fresh solution is forcedly discrete. It is clear that having increased pressure it is possible to smooth such fluctuations. However the limit of working pressure in our installation is equal to 10 bars and consequently the further running of this experiment was close to this limit. After the transition from 0.5 Nm^3/h productivity to the value equal to 1 Nm^3/h the amplitude of pressure fluctuations has grown, and the temperature was stabilized at a level of 140 °C—all elements of installation have heated up (Fig. 36.6a). Having convinced that installation works steadily and is controllable, we stopped granule powder supply. In total 12 moles of $NaBH_4$ has been used during this experiment. Increasing the hydrogen flow rate by flow controller from 0.3 to 1.0 Nm^3/h demonstrates stability of system operation (Fig. 36.6c).

In tests with the Ni/Al_2O_3 catalyst the first portion of solution was prepared of 5 moles of $NaBH_4$ and 700 ml of $NaBO_2$ solution saturated at room temperature, which corresponds to a molar ratio of $H_2O/NaBH_4 = 7$.

Based on the experience accumulated during previous experiments, the heater automatically turns off when the temperature inside the reactor becomes equal to 80 °C. The system is heated up to 150 °C (Fig. 36.7a) due to self-heating. After complete hydrolysis of the first portion of $NaBH_4$ 5 more portions of the working solution was applied to 200 ml with a molar ratio of $H_2O/NaBH_4 = 7$. The hydrogen flow rate by flow controller equal to 1.5 Nm^3/h demonstrates stability of system operation for about an hour (Fig. 36.7b).

The tests have demonstrated the serviceability of installation, and during experiments certain correlations were observed between such working parameters of the process as concentration of sodium borohydride in solution supplied to reactor; supplying rate of this solution; volume of solution in the reactor; temperature and pressure in the reactor; and productivity of hydrogen generator.

Analyzing the experiment results it has been found that stability of reactor work rises with increase of pressure and temperature in reactor. As we have accepted the lower level of working temperature is equal to 120 °C, and the lower level of pressure is equal to 7–8 bars. The necessity to have significant volume of preheated solution in reactor and prepare working solution with enough high concentration of $NaBH_4$ is also emphasized. The solution having molar ratio of $H_2O/NaBH_4 = 12$ (15 wt% $NaBH_4$) is accepted as the upper limit.

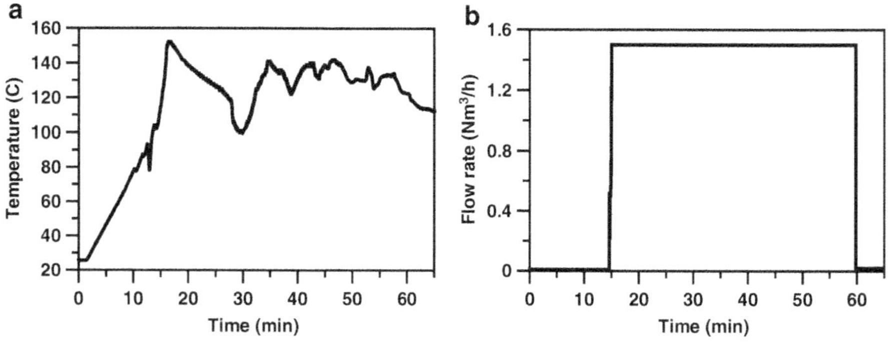

Fig. 36.7 Evolution of solution temperature inside reactor (**a**), production of hydrogen (**b**) $H_2O/NaBH_4 = 7$, Ni/Al_2O_3

After completing the experiments and emptying hydrogen receiver the yield of hydrogen is 97–98.5 %. $NaBH_4$ losses (about 3 %) may be associated with the presence of residual $NaBH_4$ in waste tank when waste solution is partially drained during experiment. Chromatographic analysis of hydrogen gas produced by hydrolysis of $NaBH_4$ demonstrates that the main reaction product is hydrogen. The significant impurities of other components are not found; only traces of solvents used in the technology of sodium borohydride production are detected: 1.6×10^{-4} vol.% C_6H_5OH and 1.8×10^{-4} vol.% CH_3OH.

Conducted tests have demonstrated the operation of generator in steady-state regime, when hydrogen production is in the range of 0.3–1.5 Nm^3/h. Generating 1.5 Nm^3/h of hydrogen is equivalent to 2.5 kW. Our hydrogen generator is used to power a 1.2 kW PEM fuel cell (Nexa™ power module).

36.4 Conclusions

We have developed a generator that generates high-purity hydrogen from aqueous solution of $NaBH_4$. Flow-type reactor is used and the circulation scheme is proposed. Circulation scheme has the following advantages: there is no need to complete hydrolysis for a single pass; heterogeneity, if it is at reactor outlet, is averaged during flow in the tubing; the ability to use a less effective catalyst; the volume of the catalytic unit is less; thermal regime of catalyst is more uniform; surface of catalyst is efficiently washed with working solution; it is easier to control the final degree of hydrolysis. Also there is an additional opportunity to control the process by regulating the rate of solution pumping. Thus, it is possible to assume that if the mass transfer processes in the reactor of the hydrogen generator are optimized, relatively cheap catalysts can be used instead of highly effective but expensive ones. The presence of heater in installation is a positive thing because it allows starting up of hydrogen generator even under very cold conditions. Instead of water-alkaline solution it is proposed to use sodium metaborate solution saturated at room temperature as a solvent and stabilizer to prepare working solutions. Circulation scheme provides stable operation of the generator with high-performance hydrogen. Technical specifications of the generator allow achieving hydrogen performance up to 3 Nm^3/h and higher.

References

1. Schlesinger HI, Brown HC, Finholt AE, Gilbreath JR, Hoekstra HR, Hyde EK (1953) Sodium borohydride, its hydrolysis and its use as a reducing agent and in the generation of hydrogen. J Am Chem Soc 75(1):215–226
2. Amendola SC, Sharp-Goldman SL, Janjua MS, Kelly MT, Petillo PJ, Binder M (2000) An ultrasafe hydrogen generator: aqueous, alkaline borohydride solutions and Ru catalyst. J Power Sources 85(2):186–189

3. Kojima Y, Suzuki K, Fukumoto K, Sasaki M, Yamamoto T, Kawai Y, Hayashi H (2002) Hydrogen generation using sodium borohydride solution and metal catalyst coated on metal oxide. Int J Hydrog Energy 27(10):1029–1034
4. Liu BH, Li ZP, Suda S (2009) Solid sodium borohydride as a hydrogen source for fuel cells. J Alloys Compd 468(1):493–498
5. Kojima Y, Haga T (2003) Recycling process of sodium metaborate to sodium borohydride. Int J Hydrog Energy 28(9):989–993
6. Park EH, Jeong SU, Jung UH, Kim SH, Lee J, Nam SW, Lim TH, Park YJ, Yu YH (2007) Recycling of sodium metaborate to borax. Int J Hydrog Energy 32(14):2982–2987
7. Minkina V, Barral K (2007) Preparation of an alkaline alcoholate and its implementation for the regeneration of sodium borohydride from sodium metaborate. European Patent No. 1787952 A1, 2007
8. Ved AS, Miley GH, Seetaraman TS (2010) Recycling sodium metaborate to sodium borohydride using wind-solar energy system for direct borohydride fuel cell. In: Proceedings of the ASME 2010 eighth international fuel cell science, engineering and technology conference fuelcell2010, June 14–16, 2010, New York, USA, vol 1. pp 139–141
9. Kojima Y, Suzuki K-I, Fukumoto K, Kawai Y, Kimbara M, Nakanishi H, Matsumoto S (2004) Development of 10 kW-scale hydrogen generator using chemical hydride. J Power Sources 125(1):22–26
10. Rusta-Sellehy A, Frank D, Rady-Pentek R (2005) Chemical hydride hydrogen generation system and an energy system incorporating the same. US Patent No. 6946104 B2, 2005
11. Mohring RM, Strizki M (2006) System for hydrogen generation. US Patent No. 7083657 B2, 2006
12. Zhang J, Zheng Y, Gore JP, Mudawar I, Fisher TS (2007) 1 kWe sodium borohydride hydrogen generation system: part II: reactor modeling. J Power Sources 170(1):150–159
13. Shurtleff K, Ladd E, Patton J (2010) System for generating hydrogen from a chemical hydride. US Patent No. 7651542 B2, 2010
14. Muir SS, Yao X (2011) Progress in sodium borohydride as a hydrogen storage material: development of hydrolysis catalysts and reaction systems. Int J Hydrog Energy 36(10):5983–5997
15. Tang S-B, Qiu F-L, Liu S-J (1996) Combined partial oxidation and carbon dioxide reforming of methane process to synthesis gas. J Nat Gas Chem 8(3):272–277
16. Tsang SC, Claridge JB, Green MLH (1995) Recent advances in the conversion of methane to synthesis gas. Catal Today 23(1):3–15
17. Beretta A, Forzatti P (2004) Partial oxidation of light paraffins to synthesis gas in short contact-time reactors. Chem Eng J 99(3):219–226
18. Adams RM, Siedel AR (1964). In: Adams RM (ed) Boron, metalloboron compounds and boranes. Interscience Publisher, New York, pp 380–390
19. Charles D Hodgman MS (ed) (1961–1962) Handbook of chemistry and physics. Chemical Rubber Publishing Co, Cleveland, OH
20. Dai HB, Liang Y, Wang P, Yao XD, Rufford T, Lu M, Cheng HM (2008) High-performance cobalt–tungsten–boron catalyst supported on Ni foam for hydrogen generation from alkaline sodium borohydride solution. Int J Hydrog Energy 33(16):4405–4412
21. Liang Y, Dai HB, Ma LP, Wang P, Cheng H-M (2010) Hydrogen generation from sodium borohydride solution using a ruthenium supported on graphite catalyst. Int J Hydrog Energy 35(7):3023–3028
22. Minkina VG, Shabunya SI, Kalinin VI, Martynenko VV, Smirnova AL (2008) Long-term stability of sodium borohydride for hydrogen generation. Int J Hydrog Energy 33(20):5629–5635
23. Minkina VG, Shabunya SI, Kalinin VI, Martynenko VV, Smirnova AL (2012) Stability of alkaline aqueous solutions of sodium borohydride. Int J Hydrog Energy 37(4):3313–3318

Chapter 37
System Design and Optimal Management Strategy for Photovoltaic Energy and Hydrogen Production

Hammou Tebibel and Siffedine Labed

Abstract Currently, hydrogen is used in a wide variety of applications mainly in the industrial sector. However, hydrogen as an energy carrier and chemical element can be obtained by several methods such as reforming and gasification of fossil fuel or electrolysis of water. Hydrogen from renewable energy sources is a clean and sustainable way to satisfy the world demand of this element. As hydrogen is mostly used in the industry where electrical energy is needed to run the various production machines, design of stand-alone photovoltaic system for both clean electrical energy and hydrogen production is proposed in this chapter as an environmentally friendly and zero-emission solution. This system, which constituted principally with a photovoltaic array, electrolyser, compressor, fuel cell and hydrogen tank storage, has a DC bus bar, AC bus bar and hydrogen piping. The latter connects in the same time the electrolyser to the hydrogen demand and storage tank, and the storage tank to the hydrogen demand and fuel cell.

The mismatch between the intermittent solar irradiation and the time-varying of both electrical energy and hydrogen demands impacts the system performances and increases the system cost. In this context, optimal energy and hydrogen management strategy with a hysteresis band scheme is investigated to achieve high system efficiency with rational cost. Case study is carried out to verify the relevance and effectiveness of the proposed management strategy. Simulation results show that the use of the proposed stand-alone photovoltaic system design with the associate electrical energy and hydrogen management strategy provides a good solution for green energy and hydrogen production with high energy efficiency and low system cost.

H. Tebibel (✉)
Centre de Développement des Energies Renouvelables, CDER, 16340 Algiers, Algeria

Laboratoire de dispositifs de communication et de conversion photovoltaïque (LDCCP), Ecole Nationale Polytechnique, 10 Avenue Hassen Badi, El Harrach, 16200 Algiers, Algeria
e-mail: h.tebibel@cder.dz

S. Labed
Centre de Développement des Energies Renouvelables, CDER, 16340 Algiers, Algeria

I. Dincer et al. (eds.), *Progress in Clean Energy, Volume 2*,
DOI 10.1007/978-3-319-17031-2_37

Keywords Stand-alone photovoltaic system • Energy and hydrogen clean production • System design • Management strategy

Nomenclature

$m_{h_2,elec}$	Electrolyser hydrogen production, kg
$m_{h_2,user}$	Hydrogen end-user demand, kg
η_{elec}	Electrolyser efficiency, %
η_{fc}	Fuel cell efficiency, %
η_{inv}	DC/AC converter efficiency, %
η_{mppt}	Maximum power point tracker efficiency, %
P_a	PV array rated power, kWp
P_{elec}	Electrolyser rated power, kW
$P_{elec,max}$	Electrolyser maximum power input, kW
$P_{elec,min}$	Electrolyser minimum power input, kW
P_{fc}	Fuel cell rated power, kW
P_{load}	Load demand, kW
$P_{pv,o}$	PV array output power, kW
SoC_b	Battery state of charge, %
$SoC_{b,int}$	Battery initial state of charge, %
$SoC_{b,max}$	Battery maximum threshold state of charge, %
$SoC_{b,min}$	Battery minimum threshold state of charge, %
SoC_{ht}	Hydrogen storage tank state of charge, %
$SoC_{ht,int}$	Hydrogen storage tank initial state of charge, %
$SoC_{ht,max}$	Hydrogen storage tank minimum threshold state of charge, %
V_{ht}	Hydrogen storage tank capacity, kg

37.1 Introduction

Hydrogen is the most present element in the universe and the main component of the stars. Nevertheless, this element is less present in independent form and with a large quantity in the earth, it is produced to meet industrial needs. Hydrogen is used in a wide variety of applications: ammonia and fertiliser manufacture, poly-silicon production, margarine manufacture, cooling of power plant and so on [1]. However, hydrogen as an energy carrier and chemical element can be obtained mainly by reforming and gasification of the fossil fuel or by water electrolysis. The yearly world production of hydrogen is estimated at 50 million tons. Unfortunately, more than 90 % of this amount is produced by thermal process known as steam methane reforming. Although this technique ensures the world's needs of hydrogen, it is very pollutant and consumes considerable amounts of natural gas. The electrolytic production of hydrogen via renewable energy sources is positioned to become the preferred method due to the inevitable price increase of natural gas and harmful environmental impacts. Moreover, photovoltaic (PV) electricity source is considered as one of the most promising options for the production of clean hydrogen since the PV plant prices are more competitive.

In the past decades, many investigations are carried out on the stand-alone photovoltaic systems with and without hydrogen storage around the world [2–11], with proposed modelling, optimal sizing and energy management strategy models. At the best author's knowledge, no study is conducted on the stand-alone photovoltaic energy and hydrogen production system.

This chapter presents a design of stand-alone photovoltaic system for both the energy and hydrogen clean production. The mismatch between the intermittent solar irradiation and the time-varying of both electrical energy and hydrogen demands is addressed through the development of an optimal energy and hydrogen management strategy.

37.2 System Description

Figure 37.1 shows the configuration of the stand-alone photovoltaic energy and hydrogen production system (SAPV-E-H_2) which consists of a PV array associated to maximum power point tracker device (MPPT) for solar energy conversion, a lead-acid battery bank for electricity energy buffer, an advanced alkaline electrolyser with a DC/AC converter for hydrogen production, a compressor for compressing the produced hydrogen at high pressure, a pressurised storage tank for high-density hydrogen storage, gas regulators for regulating the pressure of hydrogen intended to the H_2 user or to regenerate the electricity and a polymer exchange membrane fuel cell (PEMFC) with DC/AC converter for the electrical energy generation.

During the daylight hours, PV array provides electrical energy to meet the AC load consumption and run the advanced alkaline electrolyser. Lead-acid battery absorbs excess energy to cover subsequently energy deficit and night consumption. Electrolyser uses clean energy from PV array to produce hydrogen that is directly used in this form or stored in pressurised tank. Stored hydrogen gas can be used to cover the deficit at the H_2 user level or via fuel cell to regenerate electrical energy in the case of power shortage.

37.3 System Objective

The present stand-alone photovoltaic energy and hydrogen production system with a proposed design aims to provide sufficient and reliable energy and hydrogen to the end user at any meteorological conditions. Also, system sustainability imposes that the final energy and hydrogen residue stored in both battery and tank, and after at least 1-year continuous operation, must exceed its initial value. Clean production

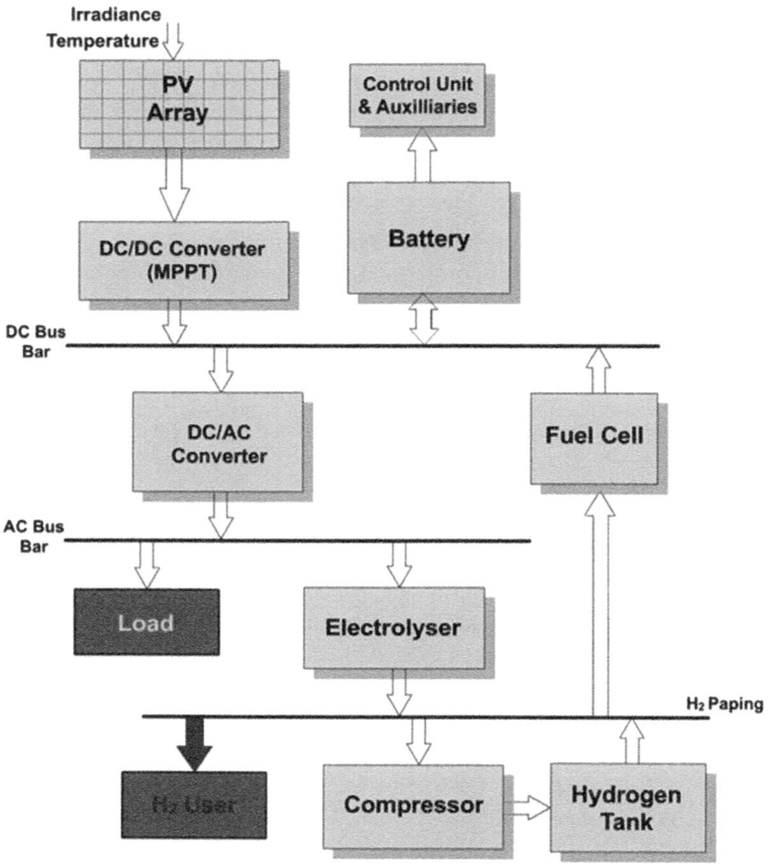

Fig. 37.1 Schematic diagram of the proposed stand-alone photovoltaic energy and hydrogen production system

means that all the energy and hydrogen required by the end user come from the solar irradiation through the photovoltaic array. However, energy and hydrogen management strategy enhances significantly the performances of the current stand-alone photovoltaic energy and hydrogen production system.

37.4 Optimal Management Strategy

In addition to the traditional key decision parameters considered in the literature [12, 13] such as battery state of charge SoC_b at the minimum and maximum level ($SoC_{b,min}$, $SoC_{b,max}$), hydrogen tank state of charge at the maximum level $SoC_{ht,max}$ and electrolyser minimum and maximum safety power input ($P_{elec,min}$, $P_{elec,max}$), proposed energy and hydrogen management strategy is based on the key decision parameters showed in Fig. 37.2. Basically, if the instantaneous electrical

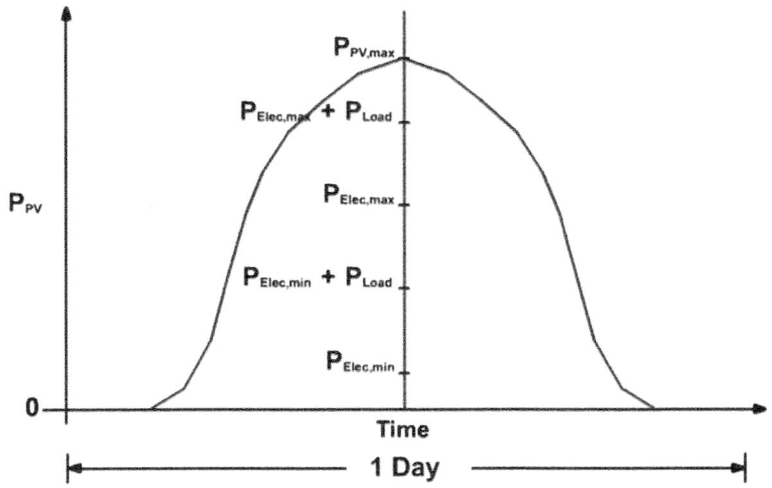

Fig. 37.2 Diagram of the management strategy power level

power $P_{pv,o}$ generated by the PV array is zero, the necessary power (P_{load}) and hydrogen ($m_{h_2,user}$) to satisfy the end user demand are provided by the lead-acid battery or fuel cell and hydrogen tank, and if $P_{pv,o} > 0$ then the surplus energy is used to charge the lead-acid accumulator, or in the electrolyser for the hydrogen production. In detailed way, the optimal energy and hydrogen management strategy can be described in the following scenarios chronologically encountered in the system operation:

1. In the case of $P_{pv,o} = 0$ and (1) if $SoC_b > SoC_{b,min}$ then the lead-acid battery and hydrogen tank provide the required energy and hydrogen; (2) if $SoC_b \leq SoC_{b,min}$ then the hydrogen tank provides the required energy to the load through the fuel cell and (3) if $SoC_b \leq SoC_{b,min}$ and $SoC_{ht} \leq 0$ then for the system safety the load and H_2 user should be disconnected.
2. In the case of $P_{pv,o} > 0$ and $P_{pv,o} - P_{load} < P_{elec,min}$ and (1) if $P_{pv,o} < P_{load}$ and $SoC_b > SoC_{b,min}$ then the battery covers the energy deficit and hydrogen tank ensures the total H_2 user consumption; (2) if $P_{pv,o} < P_{load}$, $SoC_b \leq SoC_{b,min}$ and $SoC_{ht} > 0$ then the hydrogen tank ensures H_2 user consumption and covers also the energy deficit through the fuel cell; (3) if $P_{pv,o} < P_{load}$, $SoC_b \leq SoC_{b,min}$ and $SoC_{ht} \leq 0$ then for the system safety the load and H_2 user should be disconnected with the use of photovoltaic produced energy to charge battery; (4) if $P_{pv,o} = P_{load}$ then the photovoltaic produced energy is directly sent to the load while the hydrogen tank ensures the total H_2 user consumption and (5) if $P_{pv,o} > P_{load}$ and $SoC_b < SoC_{b,max}$ then the PV array energy is directly sent to the load and hydrogen tank ensures the total H_2 user consumption while the lead-acid battery absorbs the energy surplus.
3. In the case of $P_{elec,min} < P_{pv,o} - P_{load} < P_{elec,max}$, then the PV array ensures directly the total load consumption while the energy surplus is used in the

electrolyser to produce hydrogen and (1) if $m_{h_2,elec} < m_{h_2,user}$ then the hydrogen tank corves the hydrogen deficit; (2) if $m_{h_2,elec} = m_{h_2,user}$ then the electrolyser feeds directly and totally the H_2 user; (3) if $m_{h_2,elec} > m_{h_2,user}$ then the electrolyser feeds directly and totally the H_2 user and the hydrogen surplus is stored in the hydrogen tank and (4) if $m_{h_2,elec} > m_{h_2,user}$, $SoC_{ht} \geq SoC_{ht,max}$ and $SoC_b < SoC_{b,max}$ then in this case the surplus energy from the load demand and the required energy for the electrolyser to ensure the H_2 user consumption are used to charge lead-acid battery.

4. In the case of $P_{pv,o} - P_{load} > P_{elec,max}$ and if $SoC_b < SoC_{b,max}$ then the same management logic of the previous condition (3) is applied with the use of the energy surplus to charge battery.

37.5 Case Study

The main objective of the proposed SAPV-E-H_2 system design is to provide sufficient and reliable energy and hydrogen to the end user at any meteorological conditions. Furthermore, the energy and hydrogen management strategy described below allows to achieve high energy efficiency with rational system costs. To verify the relevance of the proposed management strategy, Algiers (36.8 °N) is chosen as the location test for setting up a SAPV-E-H_2 system. As shown in Fig. 37.3 that illustrates the hourly average solar irradiation over 1 typical year for Algiers site, the maximal hourly yearly value and daily average yearly value of the recorded solar irradiation are 994.7 Wh/m^2 and 4.552 kWh/m^2, respectively [14]. In the simulation, the MPPT converter efficiency is $\eta_{mppt} = 95\%$, the DC/AC converter efficiency is $\eta_{inv} = 95\%$, the alkaline electrolyser efficiency is $\eta_{elec} = 70\%$ and the PEMFC efficiency is $\eta_{fc} = 60\%$. However, the current SAPV-E-H_2 system must provide to the end user daily and all along the year amounts of energy and hydrogen chosen as 298.8 kWh and 12 kg, respectively. In addition, these quantities are used within 24 h of the day without interruption and in an equitable way. The main characteristics and parameters of the stand-alone photovoltaic energy and hydrogen production system used in the simulation are summarised in Table 37.1.

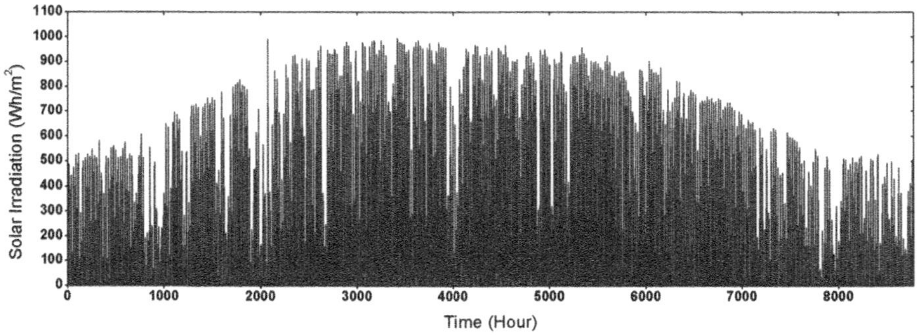

Fig. 37.3 Hourly solar irradiation over 1 typical year

Table 37.1 Stand-alone photovoltaic energy and hydrogen production system characteristics and parameters

Parameters	Values	Dimensions
P_a	266.04	kWp
P_{fc}	13.15	kW
P_{elec}	180	kW
$P_{elec,max}$	180	kW
$P_{elec,min}$	45	kW
V_{ht}	2,350	kg
$SoC_{b,int}$	50	%
$SoC_{b,max}$	95	%
$SoC_{b,min}$	30	%
$SoC_{ht,int}$	25	%

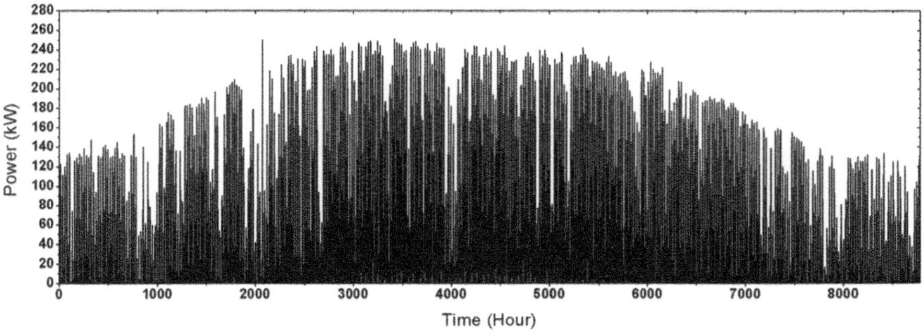

Fig. 37.4 Hourly photovoltaic array power output over 1 typical year

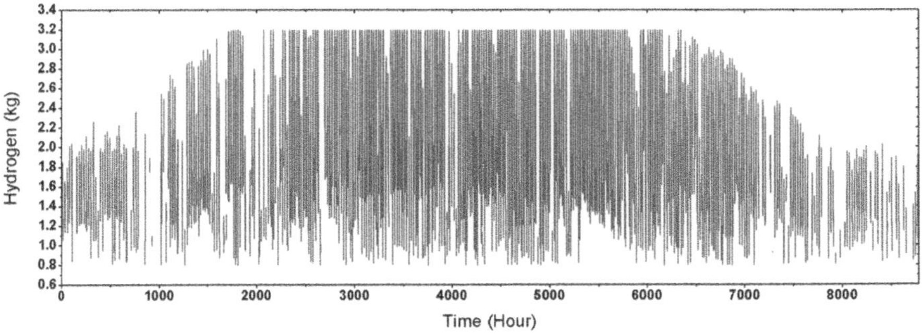

Fig. 37.5 Hourly electrolyser hydrogen production over 1 typical year

Figure 37.4 presents the hourly photovoltaic array power output $P_{pv,o}$ over 1 typical year. The maximal photovoltaic array power output is 251.4 kW and the daily average yearly energy output is 1,151.27 kWh. The main part of this energy is delivered daily to the electrolyser to produce hydrogen. Moreover and as Fig. 37.5 shows, the highest value of the hourly produced hydrogen is 3.19 kg. This quantity of hydrogen is provided by the electrolyser using the maximal instantaneous allowed electrolyser input power ($P_{elec,max} = 180$ kWh). The lowest electrolyser

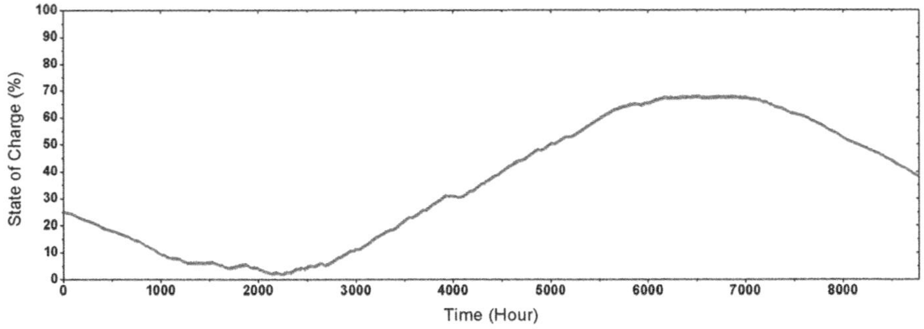

Fig. 37.6 Hourly hydrogen tank state of charge over 1 typical year

hydrogen production value is 0.8 kg and the daily average yearly value is about 15.25 kg. The latter value is clearly higher than the daily hydrogen consumption, and this is because the surplus of the hydrogen quantity is used as a way of long-term energy storage.

Figure 37.6 illustrates the hourly hydrogen tank state of charge evolution during the simulation period. This figure shows that due to the low irradiation level at the beginning of the year, the hydrogen tank covers the hydrogen deficit which results in a continuous decrease in the state of charge until it reaches the minimum threshold with 1.9 %. After a fluctuation period around this state of charge level, the hydrogen tank starts charging. This situation lasts for a long period about 6 months to reach the maximal threshold of the hydrogen tank state of charge with 68.05 %. A further period of fluctuation around the maximum threshold is undertaken before starting the last phase of decreasing.

This phase that represents another period of low irradiation is terminated at the hydrogen tank state of charge level of 38.4 %. Note that during the two periods of decreasing hydrogen tank state of charge, the hydrogen tank feeds not only the H_2 user but also the load through the fuel cell.

Through Fig. 37.7 which illustrates the hourly lead-acid battery state of charge evolution during the simulation period, the period of 1 year can be divided into two main phases. The first is the one that has lasted the longest and is characterised by a discharge battery state with state of charge fluctuating around the minimum level ($SoC_{b,min} = 30\ \%$) considered to protect the battery against deep discharge. This phase lasts for approximately 4 months of the beginning of the year and 4 months of the end of the year. However, during this phase of charge and discharge battery and fuel cell ensure the load consumption. The other phase is characterised by a steady increase of the battery state of charge until it reaches the maximum threshold with 85.6 % followed also by a steady decrease of the battery state of charge.

Figure 37.8 presents the hourly fuel cell hydrogen input over 1 typical year. The fuel cell that provides energy to the load in the case of low battery state of charge has operated for 1,424 h consuming more than 788 kg of hydrogen restored from the hydrogen tank. As the fuel cell is considered as a complementary element

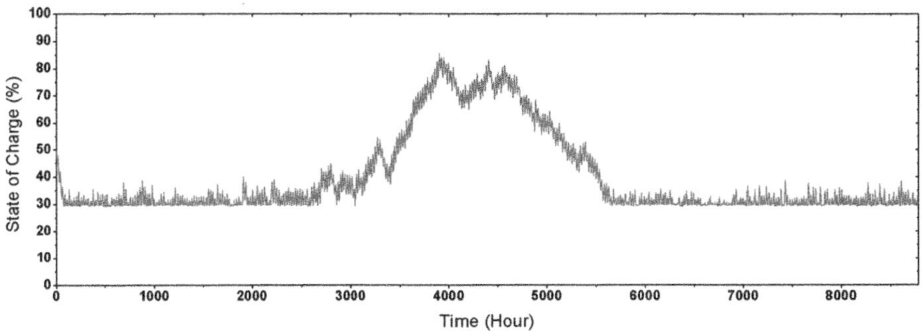

Fig. 37.7 Hourly battery state of charge over 1 typical year

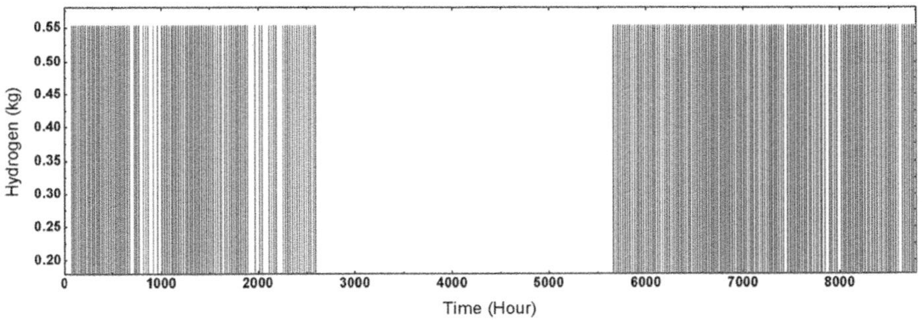

Fig. 37.8 Hourly fuel cell hydrogen input over 1 typical year

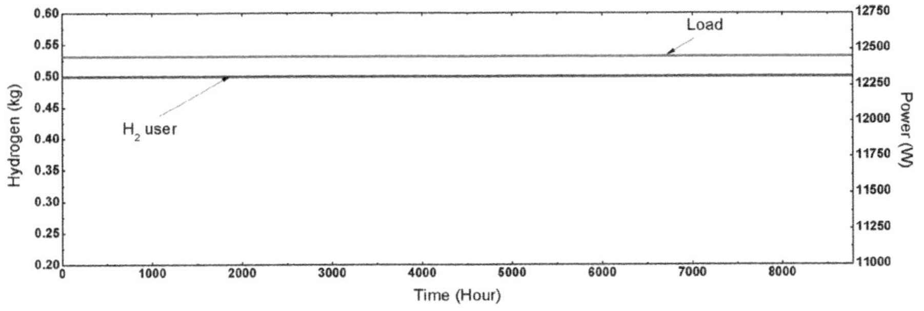

Fig. 37.9 Hourly power and hydrogen delivered to the end user over 1 typical year

to the battery, it supports the load during some hours of the night after that the battery state of charge decreases below 30 %. However, the fuel cell is shut down during the entire period of high irradiation.

Finally, Fig. 37.9 demonstrates and proves at the same time the relevance of the proposed energy and hydrogen management strategy through the continues feed of the end user with the hydrogen and energy over 1 year.

37.6 Conclusions

This study provides a solution to the greenhouse gas emissions and the energy insecurity clearly noticed through the quick depletion of the natural fossil fuel. Clean and sustainable energy and hydrogen production to satisfy the world needs are very possible using the proposed stand-alone photovoltaic energy and hydrogen production system. The system is designed to meet at the same time the load with sufficient clean electrical energy, store the energy surplus in the accumulator and produce hydrogen via water electrolyser. Also, the system design allows the electrolyser to ensure the required hydrogen of the H_2 user with or without the hydrogen tank support, and store the surplus of hydrogen in the hydrogen tank. Fuel cell is used to support the load in case of low battery state of charge.

The other part of the presented work consists of an energy and hydrogen management strategy. The relevance of the latter is tested during 1-year continuous operation. The results show that the proposed management strategy allows a continuous PV array power production, uninterrupted supply of the end user, a complementary role between the battery and fuel cell, battery protection against deep discharge and overcharge and safe and secure electrolyser operation within an appropriate power input interval. In another level, this management strategy guarantees a high energy efficiency with low system cost.

References

1. Rajeshwar K, McConnell R, Licht L (2008) Solar hydrogen generation: toward a renewable energy future. Springer, New York, NY
2. Ulleberg Ø (2004) The importance of control strategies in PV-hydrogen systems. Sol Energ 76:323–329
3. Hatti M, Meharrar A, Tioursi M (2011) Power management strategy in the alternative energy photovoltaic/PEM fuel cell hybrid system. Renew Sustain Energy Rev 15:504–510
4. Tesfahunegn SG, Ulleberg Ø, Vie PJS, Undeland TM (2011) Optimal shifting of photovoltaic and load fluctuations from fuel cell and electrolyzer to lead acid battery in a photovoltaic/hydrogen standalone power system for improved performance and life time. J Power Sources 196:10401–10414
5. Zhou K, Ferreira JA, de Haan SWH (2008) Optimal energy management strategy and system sizing method for standalone photovoltaic-hydrogen systems. Int J Hydrog Energ 33:477–489
6. Ganguly A, Misra D, Ghosh S (2010) Modeling and analysis of solar photovoltaic-electrolyzer-fuel cell hybrid power system integrated with a floriculture greenhouse. Energ Build 42:2032–2043
7. Zhang L, Huang AQ (2011) Model-based fault detection of hybrid fuel cell and photovoltaic direct current power sources. J Power Sources 196:5197–5204
8. Li CH, Zhu XJ, Cao GY, Sui S, Hu MR (2009) Dynamic modeling and sizing optimization of stand-alone photovoltaic power systems using hybrid energy storage technology. Renew Energy 34:815–826
9. Hwang JJ, Lai KL, Wu W, Chang WR (2009) Dynamic modeling of a photovoltaic hydrogen fuel cell hybrid system. Int J Hydrog Energ 34:9531–9542

10. Cetin E, Yilanci A, Oner Y, Colak M, Kasikci I, Ozturk HK (2009) Electrical analysis of a hybrid photovoltaic-hydrogen/fuel cell energy system in Denizli, Turkey. Energ Build 41:975–981
11. Tebibel H, Labed S (2013) Performance results and analysis of self-regulated PV system in Algerian Sahara. Renew Energy 60:691–700
12. Dimitris I, Voutetakis S, Seferlis P, Stergiopoulos F, Papadopoulou S, Elmasides C (2008) The effect of the hysteresis band on power management strategies in a stand-alone power system. Energy 33:1537–1550
13. Dimitris I, Voutetakis S, Seferlis P, Stergiopoulos F, Elmasides C (2009) Power management strategies for a stand-alone power system using renewable energy sources and hydrogen storage. Int J Hydrog Energ 34:7081–7095
14. Tebibel H, Labed S (2014) Design and sizing of stand-alone photovoltaic hydrogen system for HCNG production. Int J Hydrog Energ 39:3625–3636

Chapter 38
Production of Synthesis Gases from Catalytic Steam Reforming of Ethanol and Propane Processes

Murat Ağbaba, Menderes Levent, and Yusuf Şahin

Abstract In this study, steam reforming of ethanol and propane was purposed for production of hydrogen and synthesis gases. In the first stage of this project, ethanol steam reforming reaction studies were carried out at different experimental conditions for production of hydrogen and synthesis gases. In the second stage of this work, propane steam reforming reaction studies at different experimental conditions were performed. At the beginning, a commercial BASF catalyst at different experimental conditions was used for production of synthesis gases, and also, a catalyst with different metal compositions and supports was prepared in department research laboratory and then, it was used for production of synthesis gases under different experimental conditions.

Compositions of synthesis gases over commercial BASF catalyst and compositions of synthesis gases over laboratory-prepared catalyst were compared on the basis of synthesis and hydrogen yields. The determined exit gas compositions for both catalyst studies were identical about hydrogen and synthesis gas yields. However, laboratory-prepared catalyst has longer life, but it has slightly lower activity than commercial BASF catalyst at 300–700 °C. In both ethanol and propane steam reforming reaction measurements, higher hydrogen ratios were obtained at the reactor exit. Hydrogen and synthesis gas compositions are increasing with catalyst amounts at the reactor exit, but other effluent gas compositions were determined in lower ratios with respect to exit hydrogen percentages.

This paper is a revised version of a paper entitled 'Production of Synthesis Gases from Catalytic Steam Reforming of Ethanol and Propane Processes' presented at 13th International Conference on Clean Energy(ICCE2014), June 8-12,2014, Istanbul, Turkey.

M. Ağbaba • Y. Şahin
Department of Physics, Faculty of Science, Atatürk University, 25240 Erzurum, Turkey
e-mail: muratagbaba@mynet.com.tr; ysahin@atauni.edu.tr

M. Levent (✉)
Department of Chemical Engineering, Faculty of Engineering,
Uşak University, 64100 Uşak, Turkey
e-mail: menderes.levent@usak.edu.tr

I. Dincer et al. (eds.), *Progress in Clean Energy, Volume 2*,
DOI 10.1007/978-3-319-17031-2_38

For analysis of kinetics data, similar complex reaction mechanisms were considered for both ethanol and propane steam reforming processes. Ethanol and propane steam reforming reactions are considered to have first-order reaction mechanism. Computed kinetics parameters and related graphs have realized this fact, clearly. According to obtained activation energies related to ethanol and propane steam reforming reactions, we may say that both reactions are chemical reactions controlled over both catalysts with lower diameters and lower catalyst quantities.

Keywords Synthesis gases • Steam reforming of ethanol • Steam reforming of propane • First order reaction • Kinetic parameters • Preparation of catalyst • Space time

38.1 Introduction

38.1.1 Ethanol Steam Reforming Studies

A thermodynamic analysis of hydrogen production by steam reforming of ethanol via response reactions was studied, recently [1]. A simple algorithm for deriving a unique set of response reactions has been purposed. They have concluded that ethanol was unstable with respect to a mixture of H_2O, H_2, CH_4, CO and CO_2. Galvita and co-workers have been purposed synthesis gas production by steam reforming of ethanol in a two-layer fixed-bed catalytic reactor [2]. Klouz and co-workers were focused at optimizing an ethanol-reforming process over a Ni/Cu catalyst to produce a hydrogen-rich stream in order to feed a solid polymer fuel cell [3]. In this study, an overall reaction scheme was purposed by researchers as a function of the temperature domain.

Production of hydrogen for fuel cells by steam reforming of ethanol over supported noble metal catalysts was studied by Liguras and co-workers [4]. They found that 5 % Ru/Al_2O_3 catalyst is able to completely convert ethanol with selectivities toward hydrogen above 95 %. Comas et al. have investigated bioethanol steam reforming on Ni/Al_2O_3 catalyst. They concluded that higher water/ethanol ratio (6:1) and higher temperature (773 K) promote hydrogen production with 91 % of selectivity [5]. Akande et al. have investigated synthesis, characterisation and performance evaluation of Ni/Al_2O_3 catalysts for reforming of crude ethanol for hydrogen production. They found an 85 % of crude ethanol conversion over 15 % Ni-loading catalysts [6].

Steam reforming of ethanol over CuO/CeO_2 catalyst was studied by Nishiguchi and co-workers at 260 °C. With main product of acetone, 2 mol of hydrogen was produced from 1 mol of ethanol at 380 °C. The addition of MgO to CuO/CeO_2 resulted in the production of hydrogen at lower temperatures [7]. Hydrogen production by ethanol reforming over NiZnAl catalysts was studied, previously [8]. The catalyst has been prepared by the citrate sol-gel method. The nickel amount was varied from 0 to 25 wt%. The complete ethanol conversion was achieved at 500 and 600 °C. Zhang and co-workers studied hydrogen production from steam reforming of ethanol and glycerol over ceria-supported metal catalysts. At higher temperatures, all the ethanol and the intermediate compounds, like acetaldehyde and acetone, were completely converted into hydrogen [9].

Vizcaino et al. studied ethanol steam reforming on Ni/Al_2O_3 catalysts with effect of Mg addition. Their obtained results showed that Mg addition increases the catalytic activity [10]. Steam and autothermal reforming of bioethanol over MgO- and CeO_2Ni-supported catalysts were studied by Frusteri and co-workers. They obtained very high H_2 selectivity (>98 %) [11].

38.1.2 Propane Steam Reforming Studies

Catalytic autothermal reforming of methane and propane over supported metal catalysts was studied by Ayabe et al. They observed little carbon deposition for the autothermal reforming of methane; however, they reported a large amount of carbon deposition for the propane autothermal reforming. Propane always gave rise to carbon deposition even in the region expected from the equilibrium to be deposition free [12]. Propene versus propane steam reforming for hydrogen production over Pd-based and Ni-based catalysts was studied by Resini et al. They concluded that propane steam reforming over the Pd catalyst is definitely worst than over Ni catalyst. They found similar conversions for both propane and propene over Ni-$NiAl_2O_4$ catalyst [13].

Catalytic steam reforming of ethane and propane over CeO_2-doped Ni/Al_2O_3 at solid oxide fuel cell conditions (700–900 °C) was studied by Laosiripojana et al. They concluded that 14 % CeO_2-doped Ni/Al_2O_3 catalyst was a good candidate for the reforming of ethane and propane at 900 °C temperature [14]. Deactivation characteristics of lanthanide-promoted sol-gel Ni/Al_2O_3 catalysts in propane steam reforming were studied by Natesakhawat and co-workers. They concluded that the presence of lanthanide elements (i.e. La, Ce and Yb) was found to significantly enhance catalytic activity and stability. The cerium promotion and excess steam improved catalyst activity [15].

Pino et al. studied the performance of Pt/CeO_2 catalyst for propane oxidative steam reforming. The catalytic activity, as a function of O_2/C_3H_8, H_2O/C_3H_8 molar ratios in the feed, reaction temperature and gas hourly space velocity, was evaluated [16]. Ceramic microreactors for on-site hydrogen production from high-temperature steam reforming of propane were studied by Christian et al. They concluded that reaction order with respect to propane is 0.50 which indicates that rate-limiting step in the steam reforming of propane is the dissociative adsorption of propane on the Ru catalyst. At 900 °C and 1.095 of S/C, 99 % conversion of C_3H_8 was achieved with a H_2 selectivity of 74 % [17].

38.2 Experimental

Experimental system was made essentially from four sections (Fig. 38.1). These parts are the following: (1) Flow system: It contains four rotameters and each one has a 0–1,000 ml/min of measurement capacity. (2) Steam system can be run in 0–4 bars operational pressure conditions; it has 0–6 ml/min water feeding capacity of one peristaltic pump which was made of nickel-chromia alloy steel. The steam system has a 1.5 m in length vertical pipe and it consists of connectors to reactor and pump, steam safety valve and one pressure gauge. Peristaltic pump is in the bottom of the system. Water in here is pumped into the oven in reverse direction and it flows by evaporating into the reactor from top of the oven. (3) Two fixed-bed reactors with higher temperature profile of chromia-nickel alloy: Each one has 2 cm external and 1.5 cm internal diameters and each reactor is 93 cm and 83.5 cm in length, respectively [18–21].

Tubular oven has a cubic shape of external part and it was completely insulated against any temperature losses. The oven has a 5 cm internal diameter and 80 cm length. The ceramic pipe was completely rolled by a resistance wire material. This oven was covered with fire cement refractor. It can be heated up with 380 V industrial electrical voltage and can be heated up to 1,200 °C temperature. (4) Gas

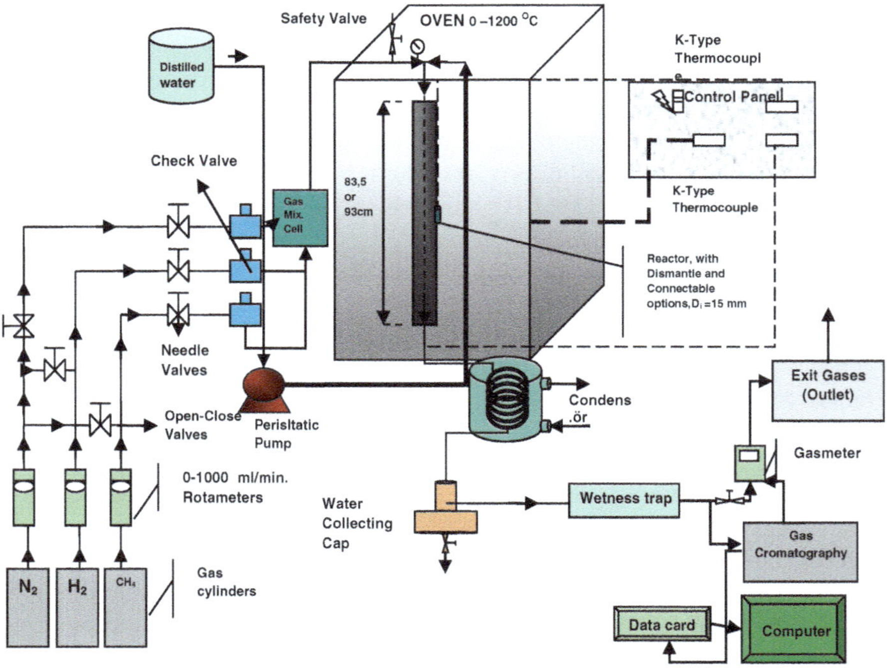

Fig. 38.1 Flow diagram of experimental system (this experimental setup was designed by M. Levent for two completed projects (TÜBİTAK-MAG No: 106 M162 and AÜ BAP No: 2004/101)) [18–21]

analysis system is made up of two parts (units): (a) Gas mixing system, to be able to make up various gas mixtures. This system contains four rotameters; each rotameter has a capacity of 0–1,000 ml/min. Each of these rotameters was calibrated separately with pure H_2, CH_4, CO and CO_2 gases [18–21]. (b) Mini GC (Model 2100 from Perichrom, Paris, France) gas analysis equipment has a 15 kg weight and a cubic shape equipment. There was a local control panel in front of the equipment. There was a TCD detector in the equipment. There were two packed columns in the equipment in which one of them was Porapak Q and the other one was molecular sieve column. The configuration of the equipment was adjusted to do the analysis of exit gases in hydrocarbon-reforming process. There are two automatic valves next to the equipment. One of the valves was commutation and the other one was the equilibrium valve [18–21].

Before starting to experimental studies, 1–5 g catalysts were loaded into the reactor. In order to prevent any carbon deposition on catalyst surface, the catalyst was conditioned at 500 °C with 100 ml/min H_2 flow for 6 h. Thus, NiO in the catalyst was reduced to metallic nickel. Therefore, the activity of catalyst was provided. At the beginning of the experiment, 200 ml/min N_2 gas was fed to the system. When system temperature has reached to 250 °C, nitrogen and hydrogen flows were cut off. In order to prevent catalyst surface from any coke formation, the conditioning experiments were carried out for known time intervals every day. The catalyst deformation related to the transmitted water droplets to the catalyst bed was prevented, by heating up the EtOH/water mixtures to higher temperatures. After reduction of the catalyst with H_2, in order to prevent the catalyst from any carbon deposition, the experiments were conducted, continuously, every day [18–21].

After preparation of the catalyst for experiments, 0.5–1 ml/min EtOH/water mixtures were fed to the system. Temperature was raised gradually and against different temperatures the gas compositions were analyzed at the exit of the system. The system has an option of to be able to analyse all the gases in the present study. Helium and nitrogen were used as carriers for GC. The exit voltages of GC analysis equipment were read continuously with an ADC card. The exit voltages were continuously observed in a monitor of Pentium-IV computer by aid of a drawing program which was developed by Dizge analytical co., Ankara, Turkey, recently. All output peaks related to each experiment were stored in the computer. Then, peak areas of each gas in unknown samples were determined by GC and recorded in computer. By comparison of unknown sample peaks with pure gas samples, the composition of each component in unknown samples was determined [18–21].

The inlet pressure of carrier gas to GC was 2.1 bars and the temperature of GC oven for analysis was 80 °C. When doing analysis with helium, detector current was 150 mA and accuracy was adjusted to 10. When analysing gas mixtures with nitrogen, detector current was adjusted to 60 mA and accuracy (gain) was adjusted as 10. GC accuracy balances were changed between 1 and 10. The value of 10 was the most accurate value of balance and 1 was the least accuracy value of gain [18–21].

38.3 Results and Discussions

38.3.1 Ethanol Steam Reforming Results

Exit H_2 percentages at 1/8 EtOH/H_2O ratio against quantity of commercial BASF catalyst are given in Fig. 38.2. As seen in this graph, when the quantity of commercial catalyst was increased from 0 to 6 g, then hydrogen percentages in exit gas composition were also increased for different temperature values (between 300 and 700 °C). At temperatures of 500–700 °C, H_2 percentages were found around 66–70 % for catalyst quantity of 1 g. When catalyst quantities were increased, H_2 percentages were also gradually increased to 75 %. At lower temperatures (300–400 °C) and lower catalyst quantities (1–2 g), H_2 percentages in the exit gas composition were found around 15–20 %. When catalyst ratios were increased from 2 to 4 g, H_2 percentages were also parabolically increased to 60–65 % at 300–400 °C and further increments in catalyst quantities up to 6 g resulted in some small increments in H_2 percentages and composition of H_2 reached 66 % at 400 °C [20, 21].

Exit H_2 percentages against different quantities of laboratory-prepared catalyst are given in Fig. 38.3. As seen on graph, H_2 percentages are increasing with increments in catalyst quantities. At lower catalyst quantities (1–2 g) and lower temperatures (300–400 °C), hydrogen compositions are around 15–20 % and at higher temperatures over lower catalyst quantities, hydrogen composition is around 50–60 %. While catalyst quantities were increased from 2 to 6 g and temperatures were increased from 400 to 700 °C, then hydrogen composition was increased up to 65 %. When we compared Figs. 38.2 and 38.3, hydrogen percentages over commercial BASF catalyst are slightly higher than hydrogen percentages over laboratory-prepared catalyst.

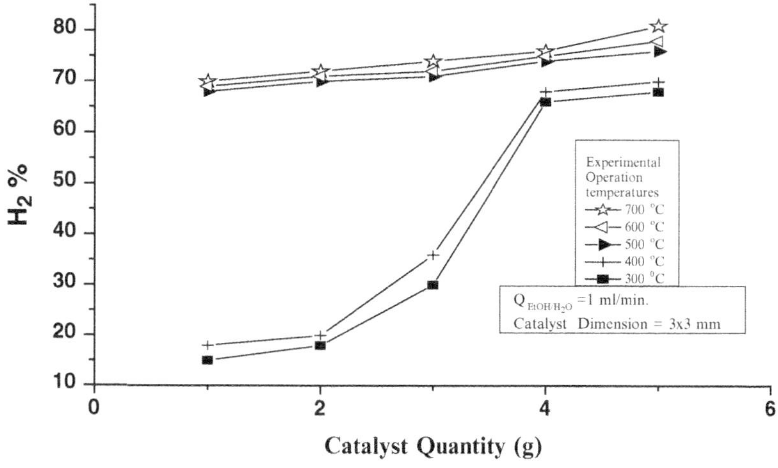

Fig. 38.2 Exit H_2 percentages for 1/8 EtOH/H_2O values against different quantities of commercial BASF catalyst [20, 21]

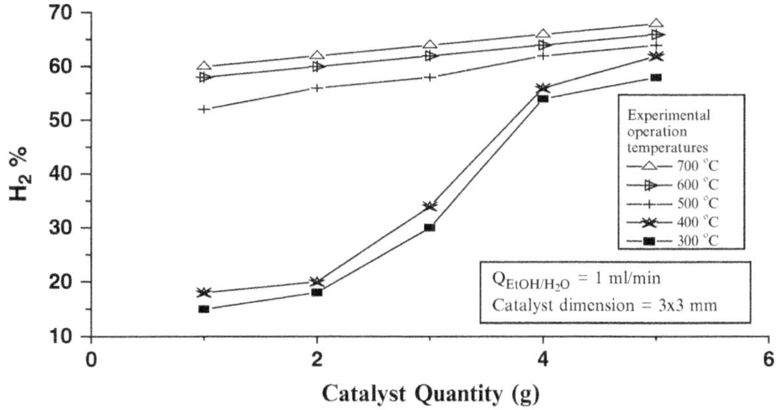

Fig. 38.3 Exit H_2 percentages of reactor against different quantities of laboratory-prepared catalyst at 1/8 EtOH/H_2O ratio [20, 21]

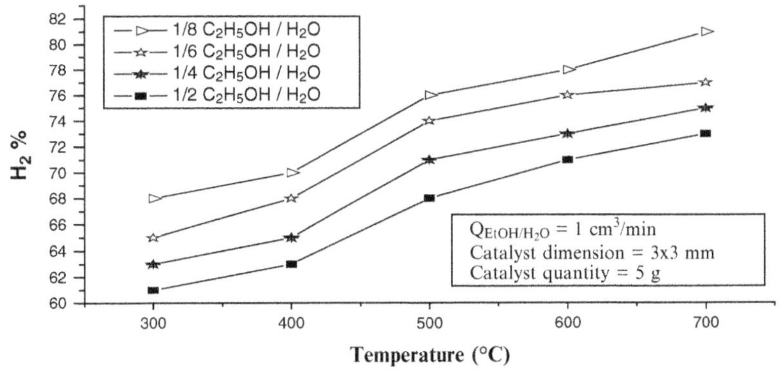

Fig. 38.4 Exit H_2 percentages for different EtOH/H_2O ratios against different temperatures between 300 and 700 °C with 5 g of commercial BASF catalyst [20, 21]

H_2 percentages against different temperatures and different ethanol and water ratios are given in Fig. 38.4. H_2 percentages are linearly increased with temperatures. When EtOH/H_2O ratios were 1/2 to 1/8 at 300 °C, then H_2 percentages were 61–68 % over 5 g commercial BASF catalyst. When EtOH/H_2O ratios were 1/2 to 1/8 at 700 °C, then H_2 percentages were changed from 68 to 81 % at the exit of the reactor. As a result, all H_2 percentages are linearly proportional with temperature and inversely proportional with decrements in EtOH/H_2O ratios.

Peak areas of ethanol steam reforming process against different temperatures are given in Figs. 38.5 and 38.6. As seen in both graphs, peak areas of H_2 are higher than other gases for commercial BASF catalyst and laboratory-prepared catalyst. Peak areas of H_2 slightly increase with temperature between 3,500 and 4,800 mVs in both catalysts. Peak areas of CO_2 are variable between 500 and 1,000 mVs for both catalysts and peak areas of CO_2 decrease from 1,000 to 500 mVs between

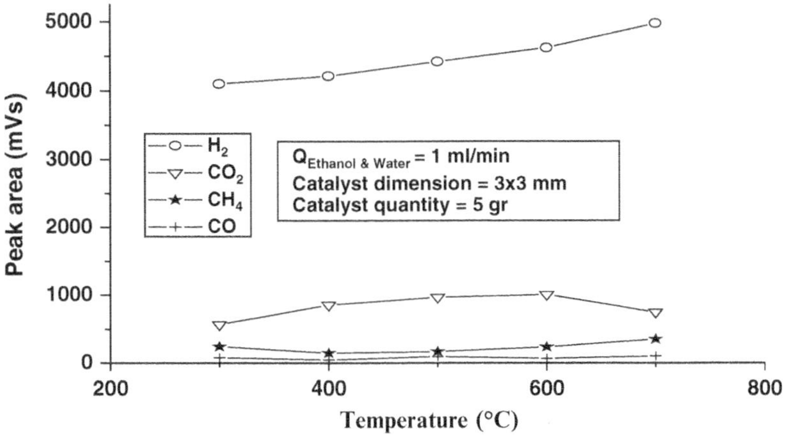

Fig. 38.5 Peak areas against different temperatures at 1/8 EtOH/H_2O ratio over 5 g of commercial BASF catalyst [20, 21]

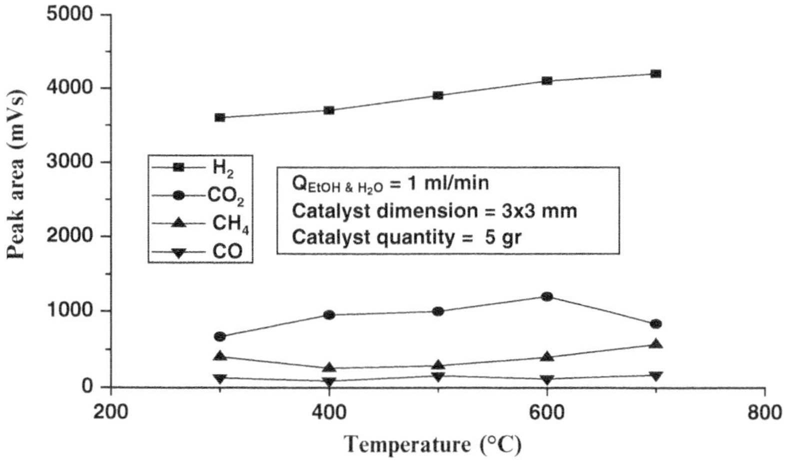

Fig. 38.6 Peak areas against different temperatures at 1/8 EtOH/H_2O ratio over 5 g of laboratory-prepared catalyst [20, 21]

600 and 700 °C temperatures. Peak areas of CH_4 and CO gases are smaller than H_2 and CO_2 and peak areas of CH_4 and CO are in the range of 100–400 mVs. This means that percentages of CH_4 and CO are extremely lower than percentages of H_2 and CO_2. While compositions of CH_4 and CO are lower for both catalysts, composition of CO_2 is identical in both catalysts. H_2 percentages for commercial BASF catalyst are slightly higher than H_2 percentages of laboratory-prepared catalyst.

Ethanol conversions against space time (W/F_{A0}) at different temperatures are given in Fig. 38.7. Ethanol conversion of 700 °C is approximately linear and is higher than ethanol conversion of 500 and 600 °C temperatures. Ethanol

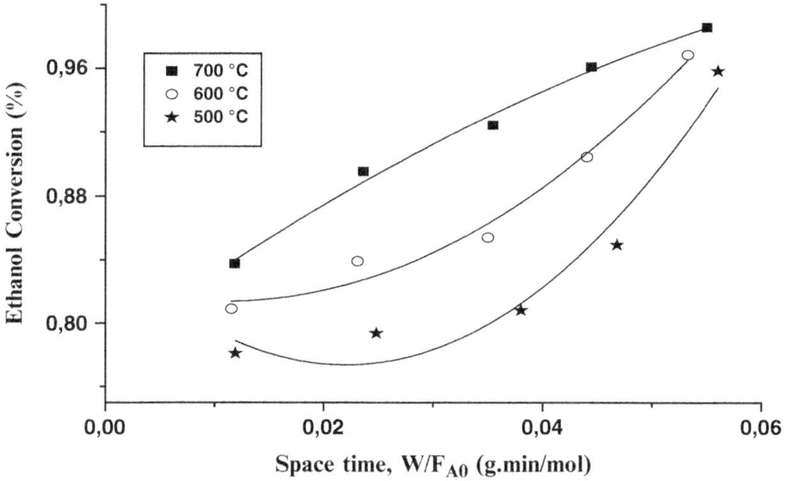

Fig. 38.7 Ethanol conversions against different space times (W/F_{A0}) of the reactor at different operational temperatures and 1/8 $EtOH/H_2O$ ratio over commercial BASF catalyst [20, 21]

conversions at temperatures of 500 and 600 °C increase exponentially in the conversion range of 0.70–0.94. Ethanol conversion at 700 °C rises from 0.84 to 0.98 value from 0.01 to 0.055 space time values. As a result, ethanol conversions increase in both catalysts while space times are increased.

38.3.2 Propane Steam Reforming Results

H_2 percentages of propane feeds against different catalyst quantities at different temperatures of commercial BASF and laboratory-prepared catalysts are presented in Figs. 38.8 and 38.9. Hydrogen percentages at the exit gas compositions in temperature range of 500–700 °C are higher than H_2 percentages of 300 and 400 °C temperatures. H_2 percentages at temperature range of 500–700 °C slightly increase with increments in catalyst quantities. However, H_2 percentages of 300 and 400 °C are lower and slightly increase up to 3 g of catalyst loads; further increments in catalyst quantities significantly increase hydrogen percentages at the exit of the reactor up to 60–65 %. Hydrogen percentages of propane steam reforming process over commercial BASF catalyst are slightly higher (5 %) than hydrogen percentages of laboratory-prepared catalyst.

Hydrogen percentages at different propane feed flow rates (50–400 ml/min) against various temperatures are given in Fig. 38.10. When temperature increases in the range of 300–700 °C, hydrogen percentages in the exit gas composition increase linearly from 62 % up to 81 %. H_2 percentages in exit gas composition at lower propane feed flow rates are always higher than the H_2 percentages of higher propane feed flow rates. Higher feed flow rates cause lower space times (W/F_{A0}),

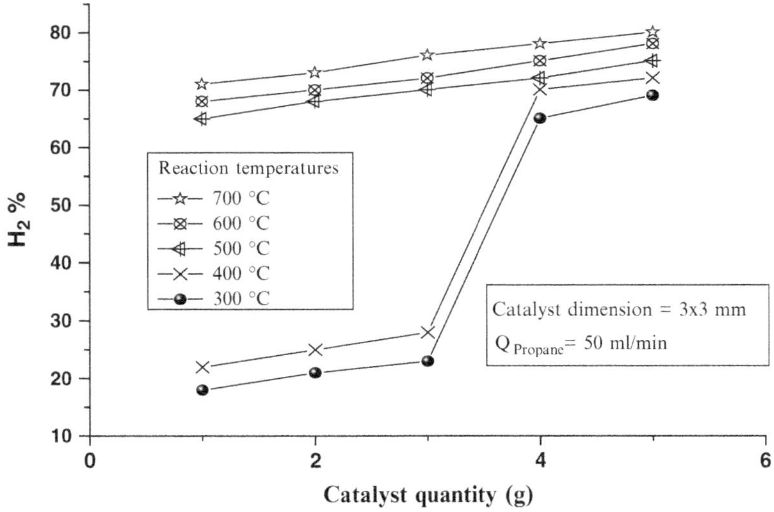

Fig. 38.8 H_2 percentages against different quantities of commercial BASF catalyst at different temperatures [21]

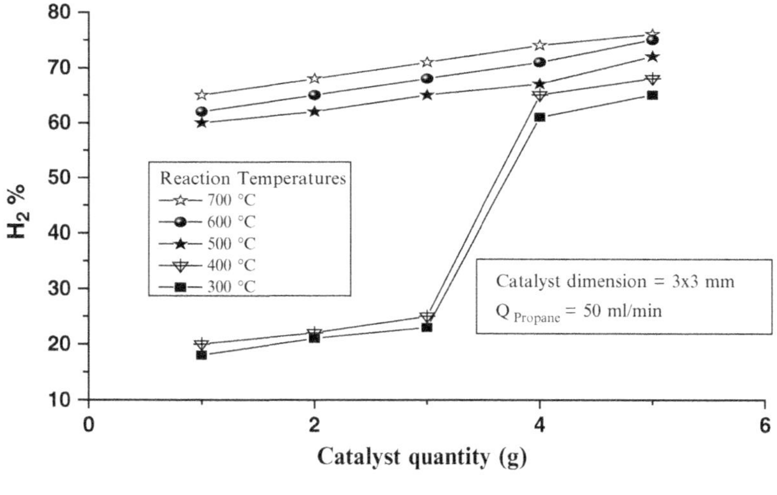

Fig. 38.9 H_2 percentages against different quantities of laboratory-prepared catalyst at different temperatures [21]

and result in lower hydrogen percentages over both commercial BASF and laboratory-prepared catalyst at the exit of the reactor.

Peak areas of propane steam reforming process against different temperatures over commercial BASF and laboratory-prepared catalysts are given in Figs. 38.11 and 38.12. Peak areas of H_2 at different temperatures are higher than peak areas of other gases (CO_2, CH_4 and CO) for both commercial BASF and laboratory-

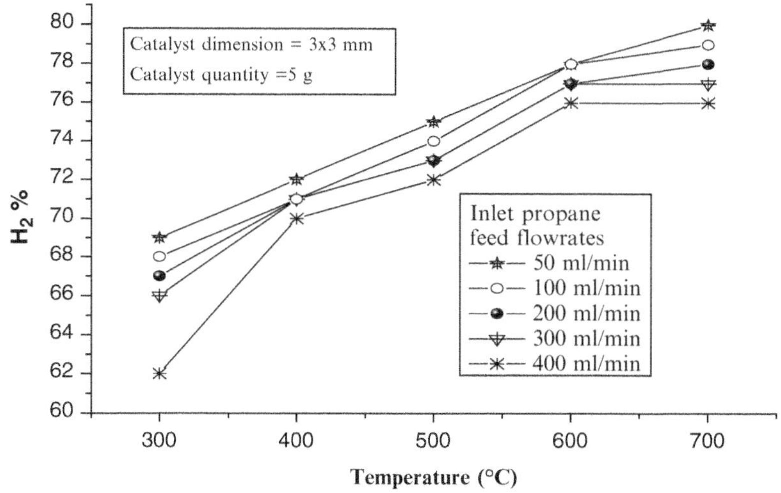

Fig. 38.10 H_2 percentages against different temperatures for various propane gas flow rates over commercial BASF catalyst [21]

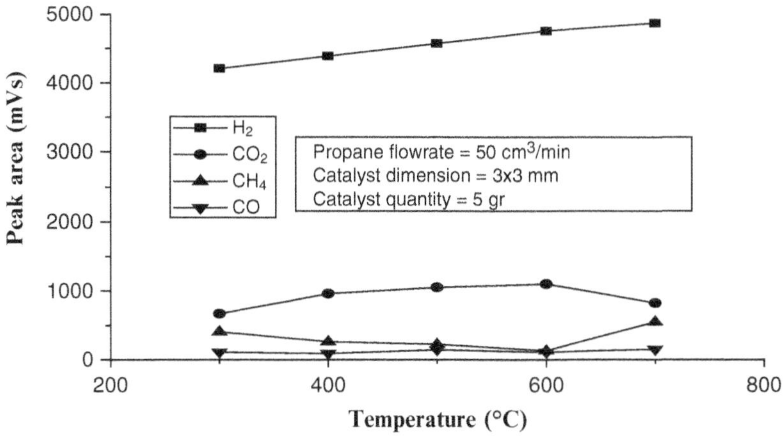

Fig. 38.11 Peak areas of all exit gases against different temperatures at 50 ml/min of propane feed flow rate over commercial BASF catalyst [21]

prepared catalysts. Peak areas of hydrogen over commercial BASF catalyst are higher than peak areas of laboratory-prepared catalyst. Peak areas of CO_2 level off around 1,000 up to temperature of 600 °C, then peak areas of CO_2 slightly reduce and peak areas of CH_4 slightly increase from 500 to 1,000 mVs. Hydrogen yields from both commercial BASF and laboratory-prepared catalysts are significantly higher; however, hydrogen yields of commercial BASF catalyst were higher (approximately, 5 %).

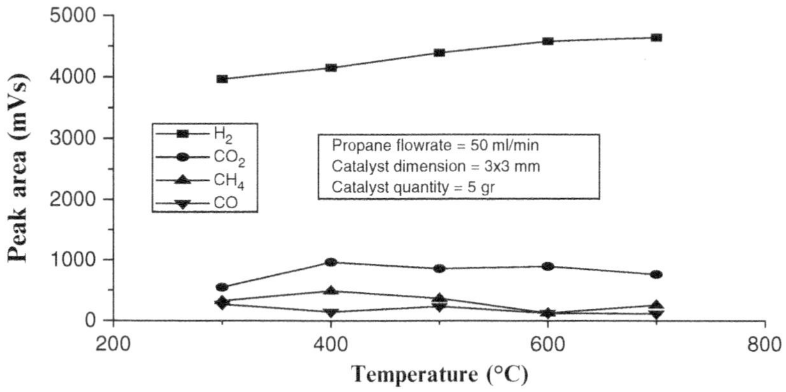

Fig. 38.12 Peak areas against different temperatures over 5 g of laboratory-prepared catalyst at 50 ml/min propane feed flow rate [21]

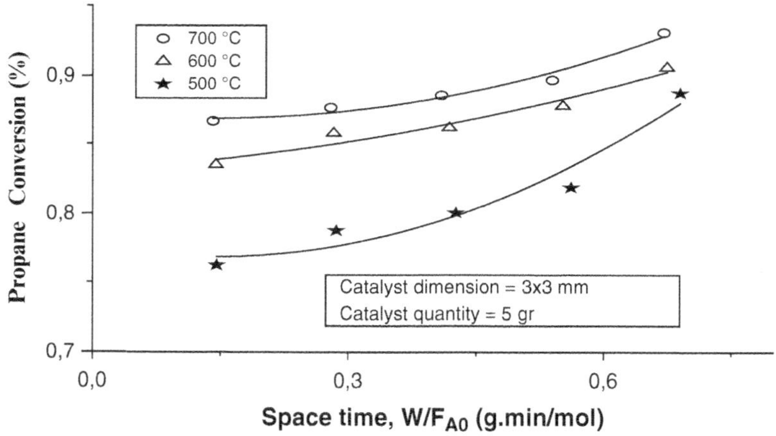

Fig. 38.13 Propane conversions against different space times (W/F_{A0}) at different temperatures and 50 ml/min of propane feed flow rate over commercial BASF catalyst [21]

Propane conversions at different temperatures and different space times (W/F_{A0}) are given in Fig. 38.13. Propane conversions are increasing exponentially with space times. Propane conversions of higher temperatures and higher space times are always higher than propane conversions of lower temperatures and lower space times. Propane conversion at 700 °C and 0.68 space time (W/F_{A0}) reaches approximately up to 0.95. However, maximum conversion value of 500 °C at 0.68 space time is approximately equal to 0.88 value.

38.4 Conclusions

So far, ethanol and propane steam reforming studies were realised under different experimental conditions. Hydrogen percentages over different quantities of commercial BASF and laboratory-prepared catalysts are determined at different temperatures and different ethanol and propane feed flow rates. Higher hydrogen percentages up to 85 % were found over both commercial BASF and laboratory-prepared catalysts. But, obtained H_2 percentages over commercial BASF catalysts were slightly higher (5 %) than H_2 percentages over laboratory-prepared catalyst. Similar H_2 percentages were obtained for both ethanol steam and propane steam reforming reactions between 300 and 700 °C. When catalyst quantities were increased, hydrogen percentages of both reactions over commercial BASF and laboratory-prepared catalyst were also increased in exit gas composition of both ethanol steam and propane steam reforming reactions.

Peak areas of hydrogen with temperature were found always higher than peak areas of other gases (CO, CO_2 and CH_4). Conversions of ethanol and propane were increased with space time (W/F_{A0}). Higher conversion values (98 % and 95 %, respectively) of ethanol and propane were obtained at higher temperatures and higher space times. Slightly lower conversion values of ethanol and propane were obtained at 600 °C with same space time values. On the basis of kinetic analysis, reactions of ethanol and propane with steam over commercial BASF and laboratory-prepared catalysts are fit to a first-order reaction model and because of those linear plots of reaction velocity constants, we may say that diffusion effects of catalyst on reaction rates are quite small and ethanol steam and propane steam reforming reactions were controlled by chemical reaction over catalyst surfaces.

During steam reforming of ethanol, less carbon deposition was detected on surfaces of both commercial BASF and laboratory-prepared catalysts. The activity of commercial BASF catalyst was found slightly higher than that of laboratory-prepared catalyst. Carbon deposition quantities on catalyst surface were increased in case of propane steam reforming measurements. The catalyst life (time on stream) for both commercial BASF and laboratory-prepared catalysts was decreased approximately, from 1 month to 1 week when propane was used as reactor feed.

Acknowledgements Authors wish to thank Chemical Engineering Department of Atatürk University, for the given opportunity to work in their laboratory during experimental measurements. Authors also thank to Turkish Scientific Research Council (TÜBİTAK) and Atatürk University for financial supports during project studies.

References

1. Fishtik I, Alexander A, Datta R, Geana D (2000) A thermodynamics analysis of hydrogen production by steam reforming of ethanol via response reactions. Int J Hydrog Energ 25:31–45
2. Galvita VV, Semin GL, Belyaev VD, Semikolenov VA, Tsiakaras P, Sobyanin VA (2001) Synthesis gas production by steam reforming of ethanol. Appl Catal Gen 220:123–127

3. Klouz V, Fierro V, Denton P, Katz H, Lisse JP, Bouvot-Mauduit S, Mirodatos C (2002) Ethanol reforming for hydrogen production in a hybrid electric vehicle: process optimisation. J Power Sources 105:26–34
4. Liguras DK, Kondarides DI, Verykios XE (2003) Production of hydrogen for fuel cells by steam reforming of ethanol over supported noble metal catalysts. Appl Catal Environ 43:345–354
5. Comas J, Marino F, Laborde M, Amadeo N (2004) Bio-ethanol steam reforming on Ni/Al_2O_3 catalyst. Chem Eng J 98:61–68
6. Akande AJ, Idem RO, Dalai AK (2005) Synthesis, characterization and performance evaluation of Ni/Al_2O_3 catalysts for reforming of crude ethanol for hydrogen production. Appl Catal Gen 287:159–175
7. Nishiguchi T, Matsumoto T, Kanai H, Utani K, Matsumura Y, Shen WJ, Imamura S (2005) Catalytic steam reforming of ethanol to produce hydrogen and acetone. Appl Catal Gen 279:273–277
8. Barroso MN, Gomez MF, Arrua LA, Abello MC (2006) Hydrogen production by ethanol reforming over NiZnAl catalysts. Appl Catal Gen 304:116–123
9. Zhang B, Tang X, Li Y, Xu Y, Shen W (2007) Hydrogen production from steam reforming of ethanol and glycerol over ceria-supported metal catalysts. Int J Hydrog Energ 32:2367–2373
10. Vizcaino AJ, Arena P, Baronetti G, Carrero A, Calles JA, Laborde MA, Amadeo N (2008) Steam reforming on Ni/Al_2O_3 catalysts: effect of Mg addition. Int J Hydrog Energ 33:3489–3492
11. Frusteri F, Freni S, Chiodo V, Donato S, Bonura G, Cavallaro S (2006) Steam and auto-thermal reforming of bio-ethanol over MgO and CeO_2Ni supported catalysts. Int J Hydrog Energ 31:2193–2199
12. Ayabe S, Omoto H, Utaka T, Kikuchi R, Sasaki K, Teraoka Y, Eguchi K (2003) Catalytic autothermal reforming of methane and propane over supported metal catalysis. Appl Catal Gen 241:261–269
13. Reisini C, Delgado MCH, Arrighi L, Alemany LJ, Marazza R, Busca G (2005) Propane versus propane steam reforming for hydrogen production over Pd-based and Ni-bas catalysts. Cataly Commun 6:441–445
14. Laossiripojana N, Sangtongkitcharoen W, Assabumrungrat S (2006) Catalytic steam reforming of ethane and propane over CeO_2-doped Ni/Al_2O_3 at SOFC temperature: improvement of resistance toward carbon formation by the redox property of doping CeO_2. Fuel 85:323–332
15. Natesakhawat S, Watson RB, Wang X, Ozkan US (2005) Deactivation characteristics of lanthanide-promoted sol-gel Ni/Al_2O_3 catalysts in propane steam reforming. J Catal 234:496–508
16. Pino L, Vita A, Cipiti F, Lagana M, Recupero V (2006) Performance of Pt/CeO_2 catalyst for propane oxidative steam reforming. Appl Catal Gen 306:68–77
17. Christian MM, Kenis PJA (2006) Ceramic microreactors for on-site hydrogen production from high temperature steam reforming of propane. Lab Chip 6:1328–1337
18. Levent M, Küçük Ö, Çalban T 3rd. Hydrogen production by hydrocarbon reforming process, final report of TÜBİTAK-MAG project and final report of BAP project of Atatürk University, Sept 2007 and May 2008
19. Levent M, Küçük Ö, Çalban T. Production of hydrogen by catalytic coal-steam gasification process, 7th national clean energy symposium (UTES'2008), Istanbul, Turkey, 17–18 Dec 2008
20. Levent M, Ağbaba M, Şahin Y. Hydrogen production by C_2H_5OH-steam reforming process in a fixed-bed reactor with commercial BASF catalyst, 5th int. advanced technologies symposium (IATS'09), Karabük, Turkey, 13–15 May 2009
21. Ağbaba, M (2013) Production of hydrogen from catalytic steam reforming of ethanol and propane processes, PhD thesis, Atatürk University, Erzurum, Turkey

Chapter 39
Clean Energy Technology Development: Hydrogen Production by *Escherichia coli* During Glycerol Fermentation

Karen Trchounian and Armen Trchounian

Abstract Hydrogen (H_2) is accepted as a clean, effective, and renewable energy source; the biotechnology of its production is intensively developed. Glycerol can serve as a cheap carbon source to produce H_2 and the other biofuel by *Escherichia coli* during mixed acid fermentation. Data on metabolic pathways of glycerol fermentation, hydrogenase enzymes responsible for H_2 production, and dependence of H_2 production on pH and other external factors during glycerol fermentation are summarized; some novel findings are presented. Metabolic engineering to enhance H_2 yield from glycerol has resulted in effective strains. The mixed carbon (glycerol and glucose) fermentation is a novel approach to improve H_2 production and to enlarge carbon sources containing wastes used: glycerol added to glucose-containing medium is shown to increase H_2 production. Taken together these are of significance for improving H_2 production biotechnology as clean energy technology.

Keywords Clean energy technology • H_2 • Glycerol fermentation • Mixed carbon • *Escherichia coli*

39.1 Glycerol Fermentation by *Escherichia coli*

Glycerol has been discovered to be fermented by *Escherichia coli*; and hydrogen gas (H_2) is detected among the end products during glycerol fermentation at slightly acidic pH [1]. This is a very interesting and intriguing phenomenon applicable in

K. Trchounian (✉)
Department of Microbiology, Plants and Microbes Biotechnology, Yerevan State University, 1 A. Manoukian Street, 0025 Yerevan, Armenia

Department of Biophysics, Faculty of Biology, Yerevan State University, 1 A. Manoukian Street, 0025 Yerevan, Armenia
e-mail: k.trchounian@ysu.am

A. Trchounian
Department of Microbiology, Plants and Microbes Biotechnology, Yerevan State University, 1 A. Manoukian Street, 0025 Yerevan, Armenia

I. Dincer et al. (eds.), *Progress in Clean Energy, Volume 2*,
DOI 10.1007/978-3-319-17031-2_39

clean energy production technology because H_2 is a clean, effective, and renewable energy source and glycerol is a very cheap carbon source for fermentation and H_2 production, compared to sugars and other sources [2–4]. Moreover, the growing production of biodiesel results in the increase of the yield of by-product—glycerol. Glycerol is a main residue (about 10 w/w%) of biodiesel production and has increased from 0.5 million m^3 in 2005 to 2.0 million m^3 in 2010 [5]. As a result, glycerol is changing from a chemical commodity to a chemical waste. The effective and economic utilization of these large quantities of waste glycerol is one of the main challenges for the biodiesel industry. The conversion of glycerol to value-added chemicals could promote the development of the biodiesel industry in return. The high degree of reduction of carbon atoms in glycerol confers the ability to produce H_2 at higher yields when compared with sugars [3]. Interestingly, fermentative H_2 production is more efficient than H_2 production by photosynthetic pathways [6]. But H_2 produced by multiple [Ni-Fe]-containing hydrogenases (Hyd) of *E. coli* has negative impact on the cell growth and glycerol fermentation.

The glycerol fermentation pathways, the mechanisms of Hyd enzyme biosynthesis, determination of their specific activity, and operation mechanisms during glycerol fermentation at different environmental conditions, especially pH, would be of great importance.

Glycerol could permeate into the bacterial cell by the facilitator protein complex, GlpF [7]; glycerol fermentation pathways within the cell are relatively simple interlinked biochemical reactions known from glycolysis but unlike glycolytic pathways they have a unique coupled reaction on the level of phosphoenolpyruvate (PEP) conversion into pyruvate generating a cycle pathway [4, 8–11]. These pathways of glycerol fermentation can be presented as shown; at the stage of PEP some intermediates may be used for succinate formation, whereas all other end products, including formic acid (HCOOH), are formed from pyruvate (PYP) [12–14] (Fig. 39.1). The pathways for glycerol fermentation are not clear yet although succinic, acetic, and formic acids and ethanol, as well as H_2 are shown to be produced but lower acetic and less lactic acids could be ended from glycerol [15]. Moreover, additional ATP could be obtained (see Fig. 39.1) which has significant bioenergetic advantage.

39.2 Hydrogen Production by *Escherichia coli* During Glycerol Fermentation and Hydrogenases

To produce H_2 *E. coli* has the ability to synthesize different membrane-associated Hyd enzymes [14, 16, 17]. However, these enzymes are reversible (not only produce but also oxidize H_2 [14, 18–22]) and therefore H_2 cycling is suggested. The latter is a novel phenomenon in the bioenergetics of mixed acid fermentation and can maintain a proton-motive force [17]. It is also important that genes coding these Hyd enzymes and participating in their maturation are known, and

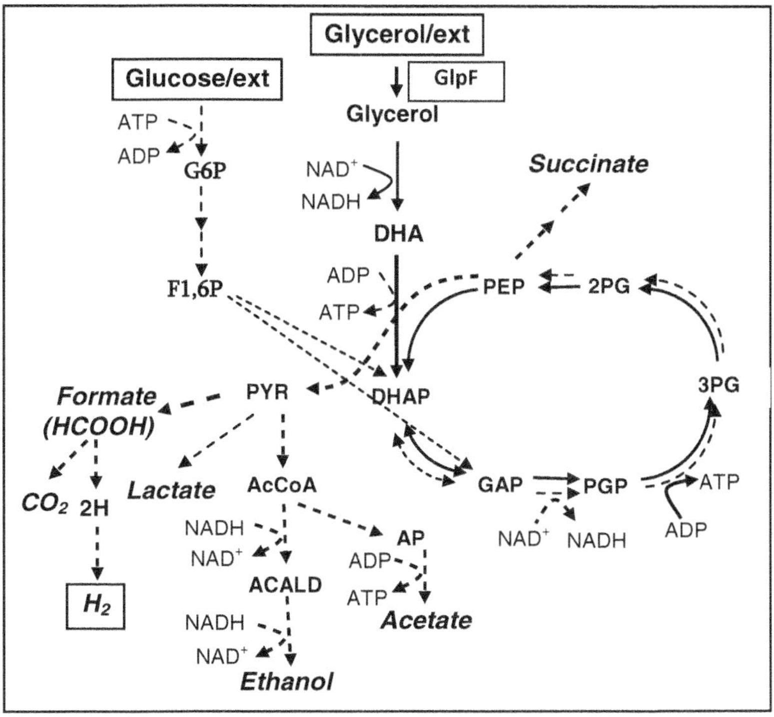

Fig. 39.1 Combined putative pathways of glycerol and glucose fermentation in *E. coli*. The pathways are adapted from [4, 8, 12–16]. *Linear arrows* indicate pathways only for glycerol fermentation, *broken arrows* indicate pathways only for glucose fermentation, and *solid broken arrows* indicate pathways for both glucose and glycerol fermentation. The products are formatted as *italics*. *2PG* 2-phosphoglycerate, *3PG* 3-phosphoglycerate, *AcCoA* acetyl-coenzyme A, *ADP* adenosine diphosphate, *ATP* adenosine triphosphate, *DHA* dihydroxyacetone, *DHAP* dihydroxyacetone phosphate, *GAP* glyceraldehyde-3-phosphate, *NADH* dihydrodiphosphopyridine nucleotide, *NAD*$^+$ diphosphopyridine nucleotide, *PGP* 1,3-diphosphate glycerate, *G6P* glucose 6 phosphate, *F1,6P* fructose-1,6 diphosphate, *AP* acetyl-phosphate. For the others, see the text

mechanisms of their expression and regulation are complex and have not been understood well [12, 18, 23–29]. However, manipulating with these genes is an effective tool to regulate H_2 production. The problem is to determine Hyd enzymes responsible for H_2 production during glycerol fermentation and to reveal their dependence on external and other factors.

The principal findings on H_2 production shown with *E. coli* during glycerol fermentation are the following as summarized in Table 39.1. The major changes during glycerol fermentation as a comparison data upon glucose fermentation are presented hereafter, in Table 39.2.

Indeed, during glycerol fermentation

- *E. coli* wild-type and mutant strains with defects in different genes coding Hyd enzymes and its maturation have different Hyd enzymatic activity and H_2

Table 39.1 H_2 production by *E. coli* Hyd enzymes during glycerol fermentation at different pHs

Characteristics or properties	H_2 production[a] pH 7.5	pH 5.5
Responsible Hyd enzymes	Hyd-2 is major, and Hyd-1 is less	Hyd-3
Recycling H_2	+[b]	+
Inhibition by DCCD	–	+
Requirement of the F_0F_1-ATPase	+	+
Coupled generation of proton-motive force	+	+
Intracellular pH	7.0	5.7
Sensitivity to osmotic stress	Hyd-3 and Hyd-4	Hyd-1, Hyd-3, and Hyd-4

[a]For references, see the text
[b]The signs "+" or "–" represent the presence or the absence of the characteristics, respectively

Table 39.2 H_2 production by *E. coli* Hyd enzymes during glucose fermentation at different pHs

Characteristics or properties	H_2 production[a] pH 7.5	pH 5.5
Responsible Hyd enzymes	Hyd-4	Hyd-3
Stimulation by external formate	Hyd-3	Hyd-3
Recycling H_2	+[b]	+
Inhibition by DCCD	+	+
Requirement of the F_0F_1-ATPase	+	+
Coupled generation of proton-motive force	+	+
Intracellular pH	7.5	6.1

[a]For references, see the text
[b]The signs "+" or "–" represent the presence or the absence of the characteristics, respectively

production rate depending on pH. Hyd-2 mostly and Hyd-1 partially are responsible for H_2 production at slightly alkaline pH (pH 7.5); Hyd-3 and Hyd-4 can work in H_2-oxidizing mode [19, 30]. This is confirmed by the other group [31] and by the findings with *E. coli* Hyd-1 and Hyd-3 or Hyd-2 and Hyd-3 double mutants [32]. In the stationary growth phase and in the absence of Hyd-3, Hyd-2 can be forced to evolve hydrogen at pH 7.5 [32]. Thus, the major contribution of Hyd-2 to H_2 production during glycerol fermentation at pH 7.5 results from changed metabolism altered intracellular pH and surprisingly influenced on proton reduction.

- *E. coli* wild-type and *atp* mutants lacked the proton F_0F_1-ATPase and have also different Hyd enzymatic activity and H_2 production rate depending on pH [33]. Hyd activity was inferred by native page electrophoresis to be dependent on the active F_0F_1-ATPase during glycerol, as well as glucose fermentation, especially at extreme pHs [21]. This might be resulted from the link between Hyd enzymes and F_0F_1-ATPase, physiological role of which is in maintaining a proton-motive force [17, 34].

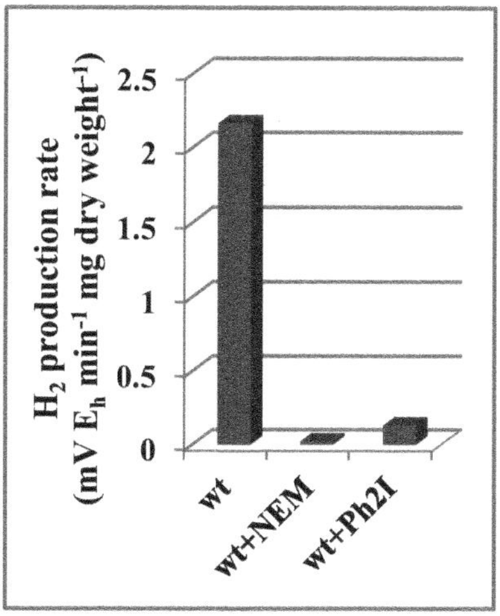

Fig. 39.2 H_2 production rate by *E. coli* wild-type (wt) strain (BW25113) is inhibited by *N*-ethylmaleimide (NEM) and diphenylene iodonium (Ph_2I). Bacteria were grown in peptone medium containing 20 g L^{-1} peptone, 15 g L^{-1} K_2HPO_4, 1.08 g L^{-1} KH_2PO_4, and 5 g L^{-1} NaCl, pH 7.5; glycerol was added in the concentration of 10 g L^{-1}. The assay medium was 150 mM Tris–phosphate buffer, pH 7.5, containing 0.4 mM $MgSO_4$, 1 mM NaCl, and 1 mM KCl. Glycerol was added in the concentration of 10 g L^{-1}. The dry weight was determined as described [39]. Average means are presented; error bars (not shown) are within 3 %. For the others, see the text

- H_2 production is inhibited by *N*,*N*′-dicyclohexylcarbodiimide (DCCD), an inhibitor for the F_0F_1-ATPase [34]: DCCD inhibited H_2 production at acidic (pH 5.5) but not slightly alkaline (pH 7.5) medium. It should be noted that DCCD inhibition was reversed during glucose fermentation [21, 30].
- Proton-motive force (Δp) is generated with its chemical (pH gradient, ΔpH) and electrical components (membrane potential, $\Delta\varphi$) at different pHs. The intracellular pH (pH_{in}) and, hence, proton-motive force were lower at pH 7.5 compared with those during glucose fermentation [35]. Hyd impact on overall proton-motive force generation is disclosed; H^+ pumping by Hyd is suggested. The lower Δp might result in change of working direction of Hyd.
- H_2 production is sensitive to osmotic stress provided by high concentration of sucrose: Hyd-3 and Hyd-4 are sensitive to osmotic stress but at different pHs, whereas Hyd-1 is osmosensitive at low pH [36].
- H_2 production is completely inhibited by 0.5 mM *N*-ethylmaleimide (NEM) (Fig. 39.2), inhibitor of glycerol kinase [37], which catalyzes the formation of GP and then DHAP (see Fig. 39.1). However, NEM has no effect on membrane potential (not shown) suggesting different mechanisms for NEM action on glycerol metabolism and H_2 production by *E. coli*. Moreover, H_2 production is very sensitive to diphenylene iodonium (Ph_2I), inhibitor of hydrogenases [38]: even 1 nM Ph_2I completely inhibits H_2 production (see Fig. 39.2). This result points out the absolute role of Hyd enzymes in H_2 production during glycerol fermentation.

For comparison, during glucose fermentation

- *E. coli* Hyd-4 is major for H_2 production at slightly alkaline pH, whereas Hyd-2 works in H_2 uptake direction; no activity of Hyd-1 is detected [40]. The activity of Hyd-4 depends on glucose concentration: it could be observed at low concentration of glucose [41].
- DCCD inhibited H_2 production in a pH-dependent manner; the inhibition was reversed compared to glycerol fermentation. Total inhibition of H_2 production was observed at slightly alkaline pH suggesting the major role of F_0F_1 at these conditions [14].
- Δp generated during glucose fermentation is different compared to glycerol fermentation. Especially (pH_{in}) differs in the cells grown at pH 7.5 and pH 5.5 [14].

Importantly, these data have been obtained by using electrochemical determination of H_2 with a pair of oxidation-reduction (redox) titanium-silicate and platinum electrodes as described elsewhere [10, 16, 19, 21, 22, 39, 40]. In contrast to platinum electrode, titanium-silicate electrode is not sensitive to H_2 and oxygen; therefore this pair of electrodes allows detecting exclusively H_2 during fermentation under anaerobic conditions (in the absence of oxygen). Various controls have ruled out interference by the other fermentation products (see Fig. 39.1). This approach is close to the method with Clark-type electrode employed by different groups [38, 42–45]: a correlation between redox potential measuring redox electrode reading difference and H_2 production was shown.

Moreover, the results obtained have been confirmed by chemical (with using permanganate in sulfur acid [46, 47]) and other (for instance, with Durham tubes) methods.

The progress with glycerol fermentation pathways (see Fig. 39.1) and H_2 production from glycerol has been achieved by metabolic engineering of *E. coli* [3, 4] obtaining an improved strain with 20-fold increased H_2 production in glycerol medium compared with precursor one [48] and creating a new strain with fivefold enhanced H_2 yield since the old one for H_2 production from glucose was not suitable [49]. Interestingly, the strains grow faster on glycerol, so they achieve a reasonable anaerobic growth rate [48] which is important for biotechnology.

39.3 Hydrogen Production by *Escherichia coli* During Mixed Carbon (Glycerol and Glucose) Fermentation

The novel approach would be H_2 production detected during mixed carbon source (glycerol and glucose) fermentation:

- Glycerol can be fermented in the presence of sugars (glucose) at slightly alkaline pH and DCCD inhibits H_2 production in a pH-dependent manner; especially in *fhlA* mutant (*fhlA* gene is coding for transcriptional activator of *hyc* operon [50]) total inhibition of H_2 production was observed suggesting direct relationship of FhlA and F_0F_1 [29, 51].

- In addition, Hyd-4 activity mainly and Hyd-2 activity to some extent to produce H_2 are revealed at low pH. Hyd-1, besides Hyd-2, has H_2-oxidizing activity at slightly alkaline pH, compared to glucose-only fermentation where Hyd-2 only was responsible for H_2 uptake [29, 51].
- Glycerol added into the medium with glucose significantly (two- and more fold) increased H_2 production [39]; this was observed at slightly alkaline pH (Table 39.3). Thus, glucose and glycerol concentrations might have significant role in H_2 production rate change.

Table 39.3 Changes in H_2 production by *E. coli* wild-type strain (BW25113) upon glycerol added into the peptone medium with glucose[a]

	H_2 production rate (mV E_h min^{-1} mg dry weight^{-1})			
			0.1 % glucose + 0.5 % glycerol	
pH	0.1 % glucose	0.5 % glycerol	Glucose assay	Glycerol assay
7.5	4.5	1.8	8.3	4.7

[a] H_2 production by the cells grown on mixed carbon sources (glucose and glycerol) was assayed with adding either glucose (glucose assay) or glycerol (glycerol assay) in the same concentrations as during the growth. For the others, see the legends to Fig. 39.2

To understand the role of Hyd in cell physiology in the presence of other carbon sources than combination of glucose with glycerol, glycerol and formate were investigated when added together. Different Hyd-defective mutants were investigated but one of the interesting results was obtained with *hycE* mutant (*hycE* gene coding large subunit of Hyd-3) [14]. Indeed, during growth of wild type on glycerol and formate H_2 production rate, when glycerol was added upon the assays, was the same compared to glycerol-only fermentation at both pHs (pH 7.5 and pH 6.5) (Fig. 39.3). At pH 7.5 when formate was supplemented upon the assays, H_2 production rate was ~1.5-fold higher than at pH 6.5. Moreover, investigation of *hycE* mutant revealed novel unidentified role of Hyd-3. When in the assays glycerol was supplemented at both pHs, H_2 production was the same as in wild type. These findings are of significance as it has been accepted that *hycE* is a negative control and up to nowadays under all conditions tested no H_2 gas production was detected by different groups [14, 18, 25]. Therefore, above idea is confirmed but only in the case when formate is added upon the assays; Hyd-3 becomes major and responsible for H_2 production as formate stimulates the *fhlA* transcriptional activator of *hyc* operon which encodes Hyd-3. But in the case of glycerol assays it seems that different Hyd enzymes are responsible for H_2 production or each Hyd compensates the other one for maintaining H_2 cycling and, thus, Δp generation.

The mixed carbon fermentation is a new phenomenon; its pathways are complex but particularly clear as represented (see Fig. 39.1), requiring a detailed study. However, this approach might be used in biotechnological applications for

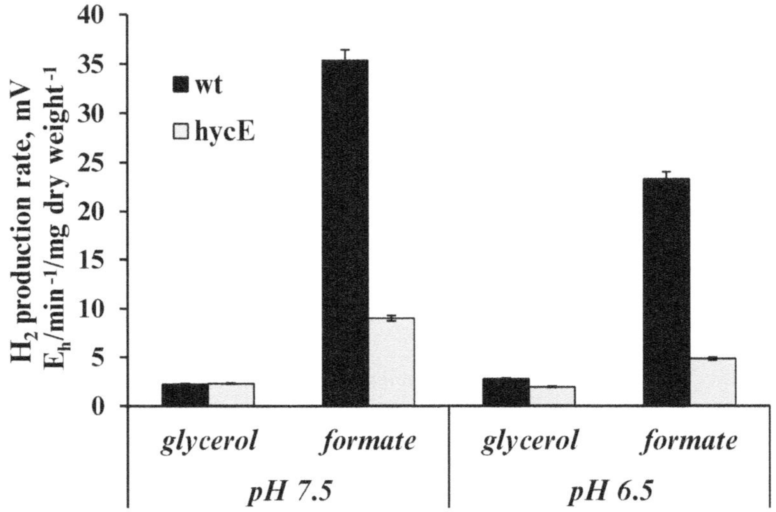

Fig. 39.3 H_2 production rate by *E. coli* wild-type and *hycE* mutant (lacking Hyd-3 large subunit) strains grown in the presence of glycerol and 10 mM formate at different pHs. Glycerol—in the assays with added glycerol, formate—in the assays with added formate. For the others, see the legends to Fig. 39.2

pre-cultivation of bacterial cells, regulation of Hyd enzymes activity towards enhancing H_2 production, and utilization of different carbon sources containing industrial, agricultural, and kitchen wastes as well as water wastes.

39.4 Concluding Remarks

The important remark is that Hyd enzymes are reversible depending on bacterial growth phase, pH, and fermentation substrates and possibly metabolic pathways [19, 20, 30, 31, 51, 52]. The other remark is with some link or cross talk between Hyd enzymes forming H_2 cycling suggested but the mechanisms underlying how this is controlled are still not clearly understood [17].

There are many unclear problems with glycerol fermentation, metabolic pathways and end products, Hyd enzyme activity and reversibility, and their dependence on pH, osmotic stress, and other external factors. Besides, H^+ and H_2 cycling are important to regulate H_2 production by bacteria [17]. All the findings together are of significance for development of H_2 production biotechnology as production of clean energy using glycerol as a carbon source for fermentation.

References

1. Dharmadi Y, Murarka A, Gonzalez R (2006) Anaerobic fermentation of glycerol by *Escherichia coli*: a new platform for metabolic engineering. Biotechnol Bioeng 94:821–829
2. Khanna S, Goyal A, Moholkar VS (2012) Microbial conversion of glycerol: present status and future prospects. Crit Rev Biotechnol 32:232–265
3. Clomburg JM, Gonzalez R (2013) Anaerobic fermentation of glycerol: a platform for renewable fuels and chemicals. Trends Biotechnol 31:20–28
4. Trchounian A (2015) Mechanisms for hydrogen production by different bacteria during mixed-acid and photo-fermentation and perspectives of hydrogen production biotechnology. Crit Rev Biotechnol 35(1):103–113
5. Yen HW, Hu IC, Chen CY, Ho SH, Lee DJ, Chang JS (2013) Microalgae based biorefinery—from biofuels to natural products. Bioresour Technol 135:166–174
6. Hallenbeck PC (2009) Fermentative hydrogen production: principles, progress, and prognosis. Int J Hydrog Energ 34:7379–7389
7. Fu D, Libso A, Stroud R (2002) The structure of GlpF, a glycerol conducting channel. Novartis Found Symp 245:51–61
8. Cintolesi A, Comburg JM, Rigou V, Zygourakis K, Gonzalez R (2011) Quantitative analysis of the fermentative metabolism of glycerol in *Escherichia coli*. Biotechnol Bioeng 109:187–198
9. Ganesh I, Ravikumar S, Hong SH (2012) Metabolically engineered *Escherichia coli* as a tool for the production of bioenergy and biochemicals from glycerol. Biotechnol Bioproc Eng 17:671–678
10. Poladyan A, Avagyan A, Vassilian A, Trchounian A (2013) Oxidative and reductive routes of glycerol and glucose fermentation by *Escherichia coli* batch cultures and their regulation by oxidizing and reducing reagents at different pHs. Curr Microbiol 66:49–55
11. Kim K, Kim SK, Park YC, Seo JH (2014) Enhanced production of 3-hydroxypropionic acid from glycerol by modulation of glycerol metabolism in recombinant *Escherichia coli*. Bioresour Technol 156:170–175
12. Bock A, Sawers G (2006) Fermentation. In: Neidhardt FG (ed) *Escherichia coli* and *Salmonella*. Cellular and molecular biology. ASM Press, Washington DC. http://www.ecosal.org
13. Booth IR (2006) Glycerol and methylglyoxal metabolism. In: Neidhardt FG (ed) EcoSal—*Escherichia coli* and *Salmonella*. Cellular and molecular biology. ASM Press, Washington DC. http://www.ecosal.org
14. Trchounian K, Poladyan A, Vassilian A, Trchounian A (2012) Multiple and reversible hydrogenases for hydrogen production by *Escherichia coli*: dependence on fermentation substrate, pH and F_OF_1-ATPase. Crit Rev Biochem Mol Biol 47:236–249
15. Murarka A, Dharmadi Y, Yazdani SS, Gonzalez R (2008) Fermentative utilization of glycerol by *Escherichia coli* and its implications for the production of fuels and chemicals. Appl Environ Microbiol 74:1124–1135
16. Bagramyan K, Trchounian A (2003) Structure and functioning of formate hydrogen lyase, key enzyme of mixed-acid fermentation. Biochemistry (Mosc) 68:1159–1170
17. Trchounian A, Sawers RG (2014) Novel insights into the bioenergetics of mixed-acid fermentation: can hydrogen and proton cycles combine to help maintain a proton motive force? IUBMB Life 66:1–7
18. Redwood MD, Mikheenko IP, Sargent F, Macaskie LE (2008) Dissecting the roles of *Escherichia coli* hydrogenases in biohydrogen production. FEMS Microbiol Lett 278:48–55
19. Trchounian K, Trchounian A (2009) Hydrogenase 2 is most and hydrogenase 1 is less responsible for H_2 production by *Escherichia coli* under glycerol fermentation at neutral and slightly alkaline pH. Int J Hydrog Energ 34:8839–8845
20. Lukey MJ, Parkin A, Roessler MM, Murphy BJ, Harmer J, Palmer T, Sargent F, Armstrong FA (2010) How *Escherichia coli* is equipped to oxidize hydrogen under different redox conditions. J Biol Chem 285:3928–3938

21. Trchounian K, Pinske C, Sawers RG, Trchounian A (2011) Dependence on the F_0F_1-ATP synthase for the activities of the hydrogen-oxidizing hydrogenases 1 and 2 during glucose and glycerol fermentation at high and low pH in *Escherichia coli*. J Bioenerg Biomembr 43:645–650
22. Poladyan A, Trchounian K, Sawers G, Trchounian A (2013) Hydrogen-oxidizing hydrogenases 1 and 2 of *Escherichia coli* regulate the onset of hydrogen evolution and ATPase activity, respectively, during glucose fermentation at alkaline pH. FEMS Microbiol Lett 348:143–148
23. Menon NK, Robbins J, Wendt JC, Shanmugan KT, Przybyla AE (1991) Mutational analysis and characterization of the *Escherichia coli hya* operon, which encodes [NiFe] hydrogenase 1. J Bacteriol 173:4851–4861
24. Menon NK, Chatelus CY, Dervartanian M, Wendt JC, Shanmugam KT, Peck HD, Przybyla AE (1994) Cloning, sequencing, and mutational analysis of the *hyb* operon encoding *Escherichia coli* hydrogenase 2. J Bacteriol 176:4416–4423
25. Sauter M, Bohm R, Bock A (1992) Mutational analysis of the operon (*hyc*) determining hydrogenase 3 formation in *Escherichia coli*. Mol Microbiol 6:1523–1532
26. Andrews SC, Berks BC, Mcclay J, Ambler A, Quail MA, Golby P, Guest JR (1997) A 12-cistron *Escherichia coli* operon (*hyf*) encoding a putative proton-translocating formate hydrogenlyase system. Microbiology 143:3633–3647
27. Richard DJ, Sawers G, Sargent F, McWalter L, Boxer DH (1999) Transcriptional regulation in response to oxygen and nitrate of the operons encoding the [NiFe] hydrogenases 1 and 2 of *Escherichia coli*. Microbiology 145:2903–2912
28. Hube M, Blokesch M, Bock A (2002) Network of hydrogenase maturation in *Escherichia coli*: role of accessory proteins HypA and HybF. J Bacteriol 184:3879–3885
29. Trchounian K (2012) Transcriptional control of hydrogen production during mixed carbon fermentation by hydrogenases 4 (*hyf*) and 3 (*hyc*) in *Escherichia coli*. Gene 506:156–160
30. Trchounian K, Sanchez-Torres V, Wood TK, Trchounian A (2011) *Escherichia coli* hydrogenase activity and H_2 production under glycerol fermentation at a low pH. Int J Hydrog Energ 36:4323–4331
31. Sanchez-Torres V, Yusoff MYM, Nakano C, Maeda M, Ogawa HI, Wood TK (2013) Influence of *Escherichia coli* hydrogenases on hydrogen fermentation from glycerol. Int J Hydrog Energ 38:3905–3912
32. Trchounian K, Soboh B, Sawers RG, Trchounian A (2013) Contribution of hydrogenase 2 to stationary phase H_2 production by *Escherichia coli* during fermentation of glycerol. Cell Biochem Biophys 66:103–108
33. Blbulyan S, Avagyan A, Poladyan A, Trchounian A (2011) Role of *Escherichia coli* different hydrogenases in H^+ efflux and the F_OF_1-ATPase activity during glycerol fermentation at different pH. Biosci Rep 31:179–184
34. Trchounian A (2004) *Escherichia coli* proton-translocating F_0F_1-ATP synthase and its association with solute secondary transporters and/or enzymes of anaerobic oxidation-reduction under fermentation. Biochem Biophys Res Commun 315:1051–1057
35. Trchounian K, Blbulyan S, Trchounian A (2013) Hydrogenase activity and proton-motive force generation by *Escherichia coli* during glycerol fermentation. J Bioenerg Biomembr 45:253–260
36. Trchounian K, Trchounian A (2013) *Escherichia coli* multiple [Ni-Fe]-hydrogenases are sensitive to osmotic stress during glycerol fermentation but at different pHs. FEBS Lett 587:3562–3566
37. Pettigrew DW (1986) Inactivation of *Escherichia coli* glycerol kinase by 5,5′-dithiobis (2-nitrobenzoic acid) and N-ethylmaleimide: evidence for nucleotide regulatory binding sites. Biochemistry 25:4711–4718
38. Hakobyan L, Gabrielyan L, Trchounian A (2012) Relationship of proton motive force and the F_0F_1-ATPase with bio-hydrogen production activity of *Rhodobacter sphaeroides*: effects of diphenylene iodonium, hydrogenase inhibitor, and its solvent dimethylsulphoxide. J Bioenerg Biomembr 44:495–502

39. Trchounian K, Sargsyan H, Trchounian A (2014) Hydrogen production by *Escherichia coli* depends on glucose concentration and its combination with glycerol at different pHs. Int J Hydrog Energ 39:6419–6423
40. Bagramyan K, Mnatsakanyan N, Poladyan A, Vassilian A, Trchounian A (2002) The roles of hydrogenases 3 and 4, and the F_0F_1-ATPase, in H_2 production by *Escherichia coli* at alkaline and acidic pH. FEBS Lett 516:172–178
41. Trchounian K, Trchounian A (2014) Hydrogen producing activity by *Escherichia coli* hydrogenase 4 (*hyf*) depends on glucose concentration. Int J Hydrog Energ 39:16914–16918
42. Fernandez VM (1983) An electrochemical cell for reduction of biochemical: its application to the study of th effect of pH and redox potential on the activity of hydrogenases. Anal Biochem 130:54–59
43. Eltsova ZA, Vasilieva LG, Tsygankov AA (2010) Hydrogen production by recombinant strains of *Rhodobacter sphaeroides* using a modified photosynthetic apparatus. Appl Biochem Microbiol 46:487–491
44. Noguchi K, Riggins DP, Eldahan KC, Kitko RD, Slonczewski JL (2010) Hydrogenase-3 contributes to anaerobic acid resistance of *Escherichia coli*. PLoS One 5:e10132
45. Piskarev IM, Ushkanov VA, Aristova NA, Likhachev PP, Myslivets TS (2010) Establishment of the redox potential of water saturated with hydrogen. Biophysics 55:13–17
46. Bagramyan KA, Martirosov SM (1989) Formation of an ion transport supercomplex in *Escherichia coli*. An experimental model of direct transduction of energy. FEBS Lett 249:149–152
47. Maeda T, Wood TK (2008) Formate detection by potassium permanganate for enhanced hydrogen production in *Escherichia coli*. Int J Hydrog Energ 33:2409–2412
48. Hu H, Wood TK (2010) An evolved *Escherichia coli* strain for producing hydrogen and ethanol from glycerol. Biochem Biophys Res Commun 391:1033–1038
49. Tran KT, Maeda M, Wood TK (2014) Metabolic engineering of *Escherichia coli* to enhance hydrogen production from glycerol. Appl Microbiol Biotechnol 98:4757–4770
50. Self W, Shanmugam KT (2000) Isolation and characterization of mutated FhlA proteins which activate transcription of the *hyc* operon (formate hydrogenlyase) of *Escherichia coli* in the absence of molybdate. FEMS Microbiol Lett 184:47–52
51. Trchounian K, Trchounian A (2013) *Escherichia coli* hydrogenase 4 (*hyf*) and hydrogenase 2 (*hyb*) contribution in H_2 production during mixed carbon (glucose and glycerol) fermentation at pH 7.5 and pH 5.5. Int J Hydrog Energ 38:3919–3927
52. Maeda T, Sanchez-Torres V, Wood TK (2007) *Escherichia coli* hydrogenase 3 is a reversible enzyme possessing hydrogen uptake and synthesis activities. Appl Microbiol Biotechnol 76:1036–1042

Chapter 40
Safety Rules and Measures to Be Taken Where Hydrogen Gas Is Stored

Fevzi Bedir, Muhammet Kayfeci, and Umran Elmas

Abstract Hydrogen is much safer than other energy sources. Hydrogen is nontoxic, not easily flammable and the products of oxygen water released during energy production are harmless to the natural environment. For pressurized tanks, the higher the pressure of the gas is, the smaller the tank volume needed, and the higher the operating cost. Safety regulations on high-pressure gas storage (including the compression process), utility, and transportation are strict, and complying with the regulations costs more than the hydrogen generators and storage equipment. Cryogenically cooling hydrogen into a liquid state is a well-established technology and considered the comparative benchmark for storing hydrogen. But the technology requires substantial energy to liquify the hydrogen, including continual "boil off" of hydrogen during storage and it requires very costly storage tanks and handling. Chemical or physically combined storage of hydrogen in other materials has potential advantages over other storage methods. Intensive research has been done on metal hydrides recently for improvement of hydrogenation properties. Metal-hydride–based chemical hydride and rechargeable hydride technologies offer storage efficiency and storage safety while providing the cost-saving advantage of being able to use the existing fossil fuel infrastructure to deliver and store a pumpable and nonexplosive hydride slurry as future hydrogen fuels. In this study we compare hydrogen storage methods, and show their advantages and disadvantages.

Keywords Hydrogen • Metal hydride • Hydrogen storage • Safety rules

F. Bedir
Department of Mechanical Engineering, Suleyman Demirel University, 32260 Isparta, Turkey

M. Kayfeci
Faculty of Technology, Karabuk University, 78060 Karabuk, Turkey

U. Elmas (✉)
The Scientific and Technological Research Council of Turkey (TUBITAK), 06100 Ankara, Turkey
e-mail: umran.elmas@tubitak.gov.tr

I. Dincer et al. (eds.), *Progress in Clean Energy, Volume 2*,
DOI 10.1007/978-3-319-17031-2_40

40.1 Introduction

Energy, which is one of the basic elements of economic development, is an indispensable requirement of humanity. Hydrogen, which can be stored, which is transportable, and which has high calorific value has been highlighted as a primary energy source because energy sources have limited reserves and existing fuels have some negative effects on environmental concerns. Hydrogen has attracted the attention of researchers since the 1970s as it has many advantageous properties regarding the absorption/desorption process in metal alloys. In recent years many studies have been released for the development of metal-hydride–based heating and cooling systems that are environmentally friendly technologies. They do not contain environmentally harmful products such as CFS and they act in reducing CO_2 emissions. During the use of hydrogen, which is preferred in waste heat or solar energy applications because of attractive performance data and environmental concerns, and is coming into wider use in industry, there are security rules and safety measures to be taken.

Hydrogen is found naturally in hydrogen-rich compounds. It cannot be extracted like natural gas or oil, but needs to be released by applying energy. Hydrogen is not a primary source of energy, but is an energy carrier that can be produced from a wide range of resources using various energy conversion processes [1, 2].

Technologies for hydrogen production from fossil fuels include steam reforming of natural gas, partial oxidation of hydrocarbons, and coal gasification. However, these technologies depend on fossil fuels and emit harmful greenhouse gases to the atmosphere. Hydrogen can also be produced by water electrolysis and this method is particularly suitable for use in conjunction with renewable wind and solar energy. Hydrogen production costs of water electrolysis depend a great deal on electricity prices.

The manner of hydrogen production and distribution mix needs to be clarified regarding some questions, which are electricity sources (solar, wind, nuclear, fossil, etc.), feedstock (natural gas, oil, coal, biomass, etc.), technologies (gasification, reforming, electrolysis, etc.), and distribution options (cryogenic trucks, pipelines etc.). These are relevant for the supply of hydrogen and their respective shares in the two scenarios. First, primary energy demand, CO_2 emissions and costs need to be evaluated. Second, additional evaluation criteria such as the security of supply, environmental issues, employment impacts, competitiveness, wealth creation, and the quality of life as well as aspects of feedstock availability should be evaluated, as shown in Fig. 40.1 [3, 4].

Hydrogen will play a crucial role for long-term seasonal storage on electricity grids relying mainly on renewable energy. Hydrogen storage is currently the only method with a technical potential for single-energy storage systems [5, 6].

The gravimetric energy storage density of hydrogen is excellent. One kilogram carries about 33 kWh of energy. Being the chemical element with the lowest density, the volumetric storage density of hydrogen is a huge problem though. Under ambient conditions 1 L of gaseous hydrogen stores only about 3 Wh of energy. In existing technical applications hydrogen is therefore either stored in its gaseous state under very high pressures up to 700 bar (called compressed gaseous

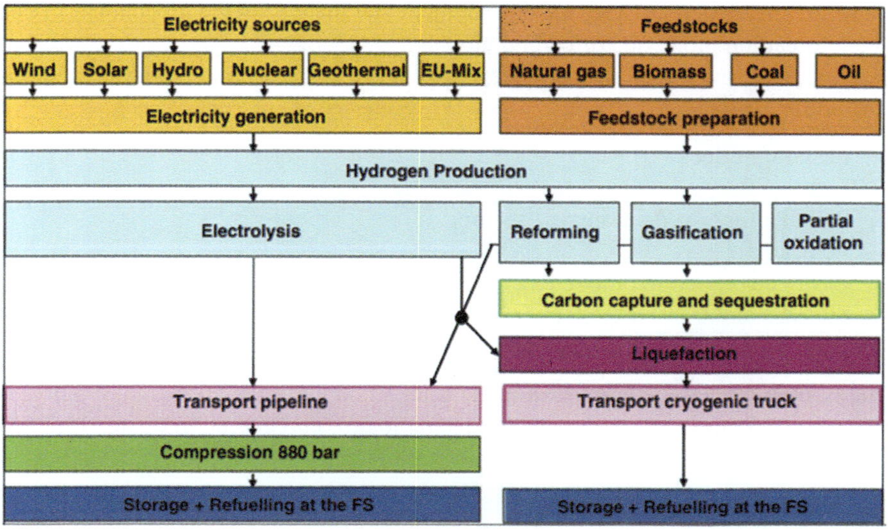

Fig. 40.1 Subsystems of hydrogen pathways

hydrogen—CGH_2) or in its liquid state which requires temperatures below 253 °C (called liquid hydrogen—LH_2) [7, 8].

Hydrogen can also be stored by metal hydride alloys. Hydrogen storage alloys are often crushed into powder packed in a reaction tank as the hydrogen source. This method utilizing storage alloys has many advantages in hydrogen storage compared to high-pressure hydrogen storage or liquid hydrogen storage, such as being more secure, and having a higher density of volumetric hydrogen storage, higher purity of hydrogen, and longer life. For these reasons, hydrogen storage alloys have great competitiveness as hydrogen storage media in the fuel cell area and some of the specific application areas, such as submarines and electric forklifts [9, 10].

40.2 Hydrogen Energy Systems and Applications

Figure 40.1 illustrates a cascade active magnetic refrigerator, which consists of four main processes. Hydrogen, which has the simplest configuration consisting of only a proton and an electron, constitutes more than 90 % of all atoms in the universe and more than three-quarters of the total mass. Hydrogen is colorless, odorless, non-metallic, tasteless under standard temperature and pressure, highly flammable, 14.4 times lighter than air, and is a completely nontoxic gas. It is the lightest element. It is found in compounds in nature such as water.

Hydrogen is clean and easy to use in all areas that require heat and explosion energy. The only product released to the atmosphere in the energy systems where hydrogen is used as a fuel is water and/or water vapor. It is not possible to get any harmful chemicals polluting the environment and increasing the greenhouse effect such as carbon monoxide or carbon dioxide.

Hydrogen, which is an ideal fuel and energy carrier, has several advantages including:

- Easy to produce
- Safer
- Suitable fuel for the transport sector
- Ecofriendly
- Easily converted to other energy forms
- Storable and transportable over long distances

The applications where hydrogen is used and tested as a fuel cell include: cars, buses, motorcycles, bicycles, golf carts, space technologies, airplanes, locomotives, ships, submarines, energy production technologies, co-generation plants, back-up power supplies, portable power sources, and others.

40.2.1 Applications Where Hydrogen Is Used

Hydrogen has been gaining importance in terms of environmental problems caused by the burning of fossil fuels and the limited reserves of natural resources such as oil and natural gas. Hydrogen is not a natural fuel, but a synthetic fuel produced from various primary energy sources including water, fossil fuels, and biomass. Hydrogen production from agricultural waste is also possible. Hydrogen is an energy carrier. It is used as a fuel in land and air transport, directly for generating heat energy by means of fuel cells or indirectly for generating electricity as a fuel in gas or steam turbines. Hydrogen has also been used in aerospace and aviation. Railway technology by diesel–electric systems is also suitable for the substitution of hydrogen. Alternative systems employing hydrogen as fuel aim not to increase the weight of vehicles and not to affect negatively the power of the engine.

Hydrogen, about 40 million tons of which are produced each year worldwide, is used in the applications below:

- Oil refineries
- Ammonia synthesis
- Methanol production
- Vegetable and animal oil sector
- Cooling of generator windings at power plants
- In meteorology as lifter gas
- In welding as a protective medium
- In the glass industry for melting and cutting quartz
- In the chemical industry
- In the steel industry at the annealing furnace, for cooling and preventing oxidation
- In low-temperature production, also called cryogenics

Table 40.1 Methods of hydrogen production

Methods	Process	Source	Energy
Thermal	Steam Conversion	Natural Gas	High-Temperature Steam
	Thermochemical water decomposition	Water	High-temperature nuclear reactors obtained
	Gasification	Coal, biomass	High temperature and pressure of water vapor and oxygen
	Pyrolysis	Biomass	Medium temperature water vapor
Electro-chemical	Electrolysis	Water	Electricity (wind, solar, nuclear)
	Electrolysis	Water	Electricity (coal, natural gas)
	Photoelectro-chemical	Water	Sunlight
Biological	Photobiological	Water and algae bacteria	Sunlight
	Anaerobic digestion	Biomass	High heat
	Fermentative micro-organisms	Biomass	High heat

- In metallurgy for heat treatment, bright annealing, brazing coating
- In the electronics industry to prevent oxidation, as a gas conduit, as an ambient gas, in the semiconductor works
- Monitoring nuclear parts in bubble chambers

40.2.2 Hydrogen Production

Hydrogen can be produced by thermal, electrochemical, and biological methods as listed in Table 40.1.

40.3 Widespread Use of Hydrogen Energy and Developments in the World

As technology develops, the need for energy increases and the energy gap between need and energy resources is growing by the day. Current energy systems are not sustainable. Alternative solutions are needed to close this gap. Hydrogen energy is outstanding as the ideal choice among all these solutions because of environmental friendliness, its efficiency, and sustainability.

Some examples of why hydrogen energy is widely used in the world are given below:

- Electricity is generated by fuel cell technology by hydrogen.
- Apollo and the Space Shuttle fuel cells used in space missions have proven their roles.

- Hydrogen filling stations, hydrogen cars, fuel-cell buses; a hydrogen-powered fishing boat fleet has been used in Iceland.
- Iceland is scheduled to use hydrogen energy exclusively by 2030.
- Mercedes-Chrysler Company has been testing 30 hydrogen-powered public transport buses by 70 people around 10 European capitals in all kinds of weather and terrain conditions.
- It has been testing a new bus powered by a fuel mixing 15–20 % hydrogen and 80–85 % of natural gas in Montreal since 1993.
- The British Columbia Hydrogen Highway Project has been studied in Canada.
- The Hydrogen Highway where hydrogen stations are placed in California are planned.
- A clean public transport network for Europe called CUTE (Clean Urban Transport for Europe) was established.
- HYDRA, which is the world's first fuel-cell–powered and –approved passenger ship, was built in Hamburg.
- Boeing has completed a manned aircraft project powered only by fuel cells and lightweight batteries.
- The electricity needs of 40,000 kW for the city of Tokyo have been supplied by hydrogen energy systems.

Hydrogen storage tanks are the main components to be employed in hydrogen storage systems. Two examples of important hydrogen storage systems as commissioned are in France and in Japan. These systems both employ hydrogen storage tanks. The systems in France and in Japan are detailed below.

40.3.1 Greenergy Box Settlement on HELION Site in France

For the first time in France, the CE-certified hydrogen-based energy storage system, the Greenergy Box system, has been developed by HELION-AREVA and is being deployed in several French public buildings (Fig. 40.2). The Greenergy Box is an integrated modular system that can offer power from 50 to 500 kW with a storage capacity from 0.2 to 2 MW h. The safety measures of the complete hydrogen chain are recapitulated into several parts, that is, leak suppression and control, prevention of formation of flammable or overoxygenated atmospheres, suppression/reduction of ignition sources, protection against overpressure, emergency and safety shutdown, and security of the installation.

The Greenergy Box and storage tanks are installed outside within a controlled area surrounded by fences, walls, or safety barriers and only accessible to authorized persons. Authorized personnel are trained and know the relevant emergency procedures [11].

Fig. 40.2 Greenergy Box™ settlement on HELION site

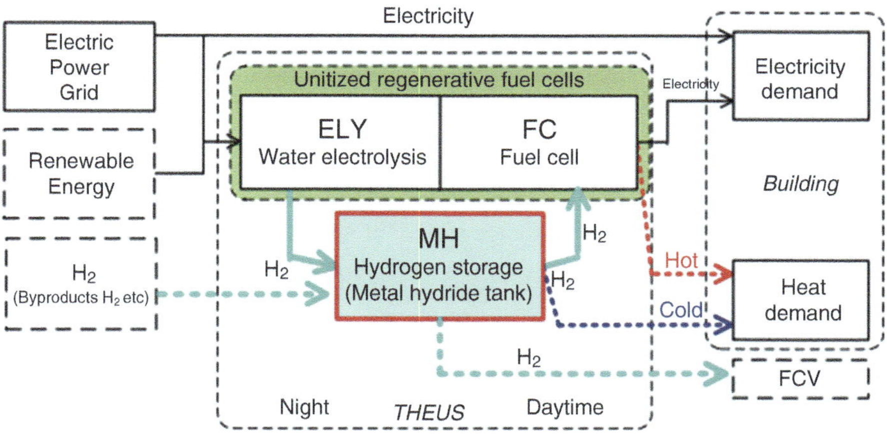

Fig. 40.3 Temperature distribution along the regenerator during hot to cold process for three different magnetic materials

40.3.2 Totalized Hydrogen Energy Utilization System (THEUS) in Japan

Variability in electricity demand due to air conditioning of commercial buildings necessitates installation of onsite energy storage systems. To meet that demand the energy storage system of THEUS, which stands for Totalized Hydrogen Energy Utilization System was proposed. THEUS has the potential to be installed onsite at commercial buildings in Japan and consists of a fuel cell, water electrolyzer, hydrogen storage tank, and their auxiliary machinery (Fig. 40.3). This system uses a water electrolyzer to store electricity energy via hydrogen at night and uses fuel cells to generate power during the day. The system also utilizes the cold heat of reaction heat during the hydrogen desorption process for air conditioning.

The basic operation of THEUS is as follows. At night, hydrogen is produced by water electrolysis using low-cost nighttime electric power with low CO_2 emission and stores the hydrogen in metal hydride tanks. During the day, electric power is generated by the fuel cell using the stored hydrogen to meet the electric power demand of the building. The chilled water and hot water generated in these processes are also utilized. Electric power from renewable energy, such as photovoltaic and wind power, can also be used for hydrogen production. In the future, THEUS can supply energy to fuel cell vehicles (FCV).

THEUS optimizes the integration of energy storage and polygeneration of electric power, cold source, heat source, and hydrogen to meet energy demand, balances load leveling, and stores renewable energy such as solar energy and wind energy. THEUS can bridge the gap between renewable energy and energy demand. Use of renewable energy sources becomes easier when THEUS is installed because fluctuations in electrical power loads can be managed.

Hydrogen storage tanks with metal hydrides are one of the important elements of THEUS (Fig. 40.4). The goal of this tank is to recover the cold heat of the endothermic process for air conditioning and thus improve the efficiency of THEUS. The storage tank contains 50 kg of metal hydride. Absorption and desorption of 5,400 NL of hydrogen was successfully attained when the absorption rate was 10 NL/min and the desorption rate was 6.9 NL/min. A 24-h cycle experiment emulating hydrogen generation at night and power generation during the day revealed that the system achieved a ratio of recovered thermal energy to the entire reaction heat of the hydrogen storage system of 43.2 % without heat loss [12, 13].

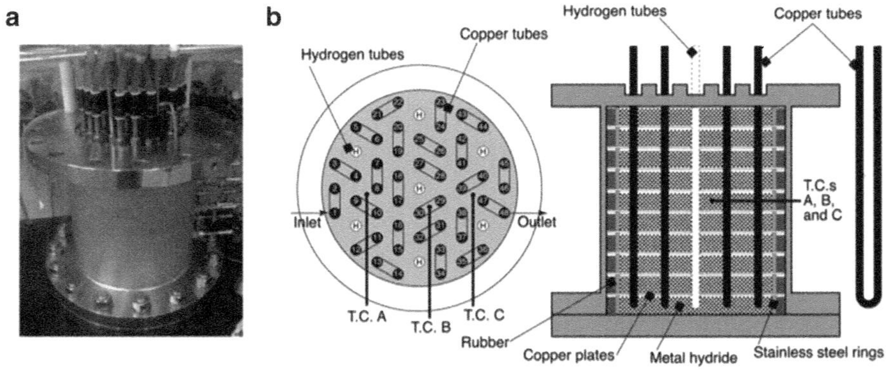

Fig. 40.4 Cooling power as function of fluid mass flow rate for different magnetic materials

40.4 Security in the Use of Hydrogen

Hydrogen is much safer than other energy sources. Hydrogen is nontoxic, not easily flammable, and the products of oxygen water released during energy production are harmless to the natural environment.

However, currently the biggest obstacle to the implementation of hydrogen energy is the hydrogen transport problem. As hydrogen is a light gas, when it is moved as compressed in a high-pressure hydrogen tank, it occupies only 3% of the weight of the hydrogen tank. On the other hand, the danger in the use of hydrogen is its being highly flammable when oxygen is mixed with air. Because it is odorless and burns by very faint flame its presence cannot be easily noticed. Hydrogen can easily leak and penetrate into the tank and pipes as it has small molecules. When it is ignited hydrogen leakage creates a narrow flame rising upward, thus harm to the environment is much less compared to other gas and liquid fuels.

There are a number of measures to be taken for safety in the applications where hydrogen energy is employed. Due to its very low viscosity hydrogen leaks at high speed in defective tanks. That's why liquid hydrogen is stored in double-wall insulated and air-free containers filled with perlite. Security precautions must be taken such as good ventilation to prevent any accumulation of hydrogen gas in case of leakage, keeping a proper distance between hydrogen and oxygen tubes, and employing hydrogen sensors for warning of the possible leakage of the gas [14, 15].

It should be noted that polysulfide rubber used as an insulating material in liquid hydrogen tanks caused the American space shuttle "Challenger" catastrophe in 1986.

40.5 Safety Rules and Measures to Be Taken Where Hydrogen Gas Is Applied

40.5.1 Use of Liquid Hydrogen

The hydrogen system assembly for a hydrogen gas vehicle is given below.

1. Tank
2. Automatic shut-off valve
3. Check valve or nonreturn valve
4. Flexible fuel line
5. Hydrogen filter
6. Heat exchanger
7. Mechanical or automatic valve
8. Pressure regulator
9. Pressure relief valve
10. Pressure, temperature, and flow sensors
11. Fuel tank refiller connection or tank
12. Hydrogen leak detection sensors

40.5.2 Applicable Test Procedures for Liquid Hydrogen Tanks

(a) Explosion test: The objective is to show that the hydrogen tank is OK at high pressure above its burst pressure.
(b) Combustion test: The aim is to show no explosion by the fire protection system when it is tested in fire conditions.
(c) Maximum filling level test: The purpose is to show that the overfilling prevention system is working adequately and the hydrogen level never leads to the opening of pressure relief devices during the filling process.
(d) Pressure test: The aim is to show that the hydrogen tank withstands the specified high-pressure level. For this, the tank is kept under pressure for a specified period. The tank should show no visible signs of leakage or permanent damage after the test.
(e) Sealing test: The objective is to determine no signs of leakage under the specified conditions. For this, the tank is kept at the rated working pressure. The tank should be free of cracks, pores, signs of leakage, or other similar defects.

The applicable test procedures for the hydrogen assembly rather than liquid hydrogen tanks are as shown in Table 40.2.

(a) Pressure Test: The aim is to show that the hydrogen assembly withstands a higher pressure than the operating pressure. The assembly should show no visible signs of leaking, deformation, tears, or cracks when hydrogen pressure is increased up to a certain level.
(b) External leakage test: The objective is to show that there is no external leakage as well as no signs of pores.
(c) Endurance test: The aim is to show that hydrogen components can work reliably and continuously. The test comprises a certain test cycle for hydrogen components at certain temperature and pressure conditions. The test cycle refers to the normal operation of the hydrogen components.
(d) Working test: The objective is to show that hydrogen components can work reliably.
(e) Corrosion resistance test: The objective is to show the hydrogen components can withstand corrosion. For this, chemicals are brought into contact with the hydrogen assembly.
(f) Dry heat resistance test: The objective is to demonstrate that nonmetallic hydrogen components can withstand high temperatures. For this, the components are exposed to air at a maximum operating temperature.
(g) Ozone aging test: The objective is to demonstrate that nonmetallic hydrogen components can withstand ageing due to ozone. For this, the components are exposed to air whose ozone density is high.
(h) Temperature cycling test: The objective is to show that hydrogen components can withstand high temperature variations. For this, the parts are subjected to

Table 40.2 The applicable test procedures for the hydrogen assembly

Hydrogen Assembly	Type of Test										
	Pressure Test	External Leakage Test	Endurance Test	Working Test	Corrosion Resistance Test	Dry Heat Resistance Test	Ozone Aging Test	Temperature Cycling Test	Pressure Cycling Test	Hydrogen Compatibility Test	Slot Leakage Test
Pressure relief devices	✔	✔		✔	✔			✔		✔	
Valves	✔	✔	✔		✔	✔	✔	✔		✔	✔
Heat exchanger	✔✔	✔			✔	✔	✔	✔		✔	
Fuel tank refiller connection or tank	✔	✔	✔		✔	✔	✔	✔		✔	✔
Pressure regulators	✔	✔	✔		✔	✔	✔	✔		✔	✔
Sensors	✔	✔			✔	✔	✔	✔		✔	
Flexible fuel lines	✔	✔			✔	✔	✔	✔	✔	✔	

temperature cycles at certain times from minimum to maximum operating temperature.

(i) Pressure cycle test: The aim is to show that hydrogen components can withstand high pressure variations. For this, the components are subjected to pressure variation from atmospheric pressure to a maximum allowable working pressure (MAWP) and then back to atmospheric pressure again in a short time.

(j) Hydrogen compatibility test: The aim is to demonstrate that metallic hydrogen components (i.e., cylinders and valves) are not sensitive to hydrogen embrittling. It should be avoided to begin with, as well as the spread of fatigue cracks and of conditions that can lead to local fatigue in the structure which is subjected to frequent load cycles.

(k) Slot leakage test: The aim is to show that the hydrogen assembly placed in the hydrogen system has no leakage.

40.5.3 Use of Compressed Gaseous Hydrogen

The applicable test procedures for the hydrogen assembly rather than gaseous hydrogen tanks are as shown in Table 40.3.

1. Material tests.
   (a) Hydrogen compatibility test.
   (b) Ageing test: The objective is to check if the nonmetallic materials used in the assembly rely on ageing. Any cracks in the test samples are not allowed.
   (c) Ozone compatibility test: The objective is to check the elastomeric material in the assembly to discover whether it is compatible with hydrogen-related ozone exposure. Any cracks in the test samples are not allowed.
2. External leakage test.
3. Endurance test.
4. Corrosion resistance test.
5. Pressure cycle test: Hydrogen-related components should show no visible signs of deformation or extrusion and must meet the requirements of internal and external leakage tests.
6. Internal leakage test: The aim is to show that the hydrogen components are free of internal leakage. For this, the parts of hydrogen are held in different pressure and temperature conditions and leakage is observed. The hydrogen assembly should be free from blisters and there should be no leakage as specified.

40.5.4 Placement of Hydrogen Components and Systems

1. The hydrogen system should be protected against damage and against all types of corrosion, and should be insulated from heat sources.

Table 40.3 The applicable test procedures for the hydrogen assembly

Hydrogen Assembly	Type of Test					
	Material Tests	Corrosion Resistance Test	Endurance Test	Pressure Cycling Test	Internal Leakage Test	External Leakage Test
Pressure relief device	✓	✓	✓	✓	✓	✓
Automatic valves	✓	✓	✓	✓	✓	✓
Mechanical valves	✓	✓	✓	✓	✓	✓
Nonreturn valves	✓	✓	✓	✓	✓	✓
Pressure relief valves	✓	✓	✓	✓	✓	✓
Heat exchangers	✓	✓		✓		✓
Fuel tank refiller connection or tank	✓	✓	✓	✓	✓	✓
Pressure regulators	✓	✓	✓	✓	✓	✓
Sensors for hydrogen systems	✓	✓	✓	✓		✓
Flexible fuel lines	✓	✓	✓	✓		✓
Fittings	✓	✓	✓	✓		✓
Hydrogen filters	✓	✓		✓		✓
Removable storage system connectors	✓	✓	✓	✓		✓

2. The hydrogen tank can only be removed in order to be replaced with another hydrogen tank for the purpose of refueling or maintenance. For internal combustion engines, the tank should not be placed in the engine compartment.
3. Precautions should be taken to make sure there is no hydrogen leakage during filling and refilling of the vehicle and that the removable hydrogen storage system is safely dismantled.
4. Fuel-filling connections must be secure against misalignment and be free from dirt and water. Fuel tank refiller connections must be connected with a nonreturn valve. If the fuel tank filler connection is not directly mounted to the tank, it must be secured by a valve mounted directly on the tank or by a nonreturn valve inside the tank.
5. The hydrogen tank must be installed and fixed in such a way that the safety-related components can absorb securely for acceleration when the tanks are full.
6. The hydrogen fuel supply lines must be secured by an automatic shut-off valve directly mounted on the tank. The valve must be closed in case of leakage or a

failure in the hydrogen system. When the drive system is switched off, the fuel supply from the tank to the drive system must be closed until the system is on.

7. In case of an accident, the automatic shut-off valve in the tank must cut the gas flow.
8. Hydrogen components, including the protective materials of the assembly, should not protrude beyond the external line of the vehicle or the protective structure. This issue is not applicable for any part of the assembly that is adequately protected and which is placed inside the protective structure.
9. The hydrogen system is placed so as to protect against damage caused by moving parts of the vehicles, bumps, stone ejections, loading/unloading of the vehicle, or displacement of cargo.
10. Parts of hydrogen must not be placed near the internal combustion engine exhaust or any other heat source unless it is well protected against heat.
11. The passenger compartment, air conditioning or heating system, and all places where it is possible that any leakage or accumulation of hydrogen might occur should be designed such that hydrogen is not drawn into the vehicle.
12. In the case of an accident, the functioning of a pressure relief device and combined drain system should be provided. The removal system of the pressure relief device must be adequately protected against dirt and water.
13. The passenger compartment must be separate from the vehicle's hydrogen system to avoid the accumulation of hydrogen. Fuel leakage from the tank or accessories to the passenger compartment must be avoided.
14. The passenger or luggage compartment, or the parts of the unventilated hydrogen assembly where hydrogen might bleed should be secured by a gas-tight slot or by the equivalent solution.
15. To prevent sparks. hydrogen-employing electrical devices must be insulated such that no electric currents will pass through hydrogen-containing parts. Metallic parts of the hydrogen system must always be grounded to the vehicle system during the electrical maintenance period.
16. Labels or other means of identification should be used to point out that an ambulance or life-saving car is powered by liquid or compressed-gaseous hydrogen.

40.6 Conclusions

In order to guarantee the security of not only the storage system but also the public, safety around the tank system is extremely important. The security-related precautions must always be obeyed. For example, in these systems welded connections are preferred and are used wherever practical to minimize potential leak sources. The number of joints and fitted connections should be minimized. Hydrogen and oxygen leaks should also be detected by pressure differences during standby phases. If a tank or a portion of pipe loses pressure during a standby stage, it potentially means that there is a leak. Before commissioning, hydraulic and leak tests should be completed as required by the related pressure equipment directives. Regular inspections and preventive maintenance programs should be planned and organized to ensure the maximum safety level. In particular, leak tests on pressure regulators, valves, pipes, joints, connections, and so on should be realized regularly. Regular visual inspections should be organized to check the level of corrosion.

References

1. Yang FS, Meng XY, Deng JQ, Wang YQ, Zhang ZX (2009) Identifying heat and mass transfer characteristics of metal hydride reactor during adsorption—improved formulation and further discussion about parameter analysis. Int J Hydrog Energy 34:1852–1861
2. Lee M, Kim KJ, Hopkins RR, Gawlik K (2009) Thermal conductivity measurements of copper-coated metal hydrides ($LaNi_5$, $Ca_{0.6}Mm_{0.4}Ni_5$, and $LaNi_{4.75}Al_{0.25}$) for use in metal hydride hydrogen compression systems. Int J Hydrog Energy 34:3185–3190
3. Ozarslan A (2012) Large-scale hydrogen energy storage in salt caverns. Int J Hydrog Energy 37:14265–14277
4. Wietschela M, Hasenauer U, de Groot A (2006) Development of European hydrogen infrastructure scenarios—CO_2 reduction potential and infrastructure investment. Energy Policy 34:1284–1298
5. Andrews J, Shabani B (2012) Re-envisioning the role of hydrogen in a sustainable energy economy. Int J Hydrog Energy 37:1184–1203
6. Freni A, Cipiti F, Cacciola G (2009) Finite element-based simulation of a metal hydride-based hydrogen storage tank. Int J Hydrog Energy 34:8574–8582
7. Teichmann D, Arlt W, Wasserscheid P (2012) Liquid Organic Hydrogen Carriers as an efficient vector for the transport and storage of renewable energy. Int J Hydrog Energy 37:18118–18132
8. Alnıak MO (2009) Hidrojen Enerjisi Hakkında Düşünceler. Bilge Adamlar Stratejik Araştırmalar Merkezi (BİLGESAM), Turkey
9. Duana W, Junlin D, Wang Z, Niu Y, Huang T, Li Z, Chaohui P, Zhu W (2013) Strain variation on the reaction tank of high hydrogen content during hydrogen absorption-desorption cycles. Int J Hydrog Energy 38:2347–2351
10. Yang FS, Wang GX, Zhang ZX, Meng XY, Rudolph V (2010) Design of the metal hydride reactors—a review on the key technical issues. Int J Hydrog Energy 35:3832–3840
11. Verbecke F, Vesy B (2013) Safety strategy for the first deployment of a hydrogen-based green public building in France. Int J Hydrog Energy 38:8053–8060
12. Maeda T, Nishida K, Tange M, Takahashi T, Nakano A, Ito H, Hasegawa Y, Masuda M, Kawakami Y (2011) Numerical simulation of the hydrogen storage with reaction heat recovery using metal hydride in the totalized hydrogen energy utilization system. Int J Hydrog Energy 36:10845–10854
13. Tangea M, Maeda T, Nakano A, Ito H, Kawakami Y, Masuda M, Takahashi T (2011) Experimental study of hydrogen storage with reaction heat recovery using metal hydride in a totalized hydrogen energy utilization system. Int J Hydrog Energy 36:11767–11776
14. T.C. Resmi Gazete, Başbakanlık Mevzuatı Geliştirme ve Yayın Genel Müdürlüğü
15. T.C. Çalışma ve Sosyal Güvenlik Bakanlığı, İş Sağlığı ve Güvenliği Enstitüsü Müdürlüğü İSG Dökümanları

Chapter 41
Inlet Methane Temperature Effect at a Planar SOFC Thermal Field Under Direct Internal Reforming Condition

Hafsia Abdenebi, Bariza Zitouni, Hocine Ben Moussa, Djamel Haddad, Hadda Zitouni, and Youcef Sahli

Abstract In this work, an Anode Supported Planar Solid Oxide Fuel Cell (ASP_SOFC) is applied. The thermal fields are shown for a standard SOFC: yttria stabilized zirconia for the electrolyte, nickel/zirconia cermet for the anode, and doped lanthanum manganite (LSM) for the cathode. It is operating under direct internal reforming condition of methane gas.

The present paper's purpose is the thermal field visualization of an ASP_SOFC. The temperature fields are discussed under the influence of heat sources caused by the internal reforming reactions occurring at the SOFC anode side. The reforming reactions in the anode in this case are described by the endothermic steam reforming reaction, the exothermic water-gas shift reaction, and the endothermic overall chemical reaction. The SOFC thermal behavior is influenced by several parameters such as mole fractions, pressure and temperature, etc. In this work we focus on two parameters: the inlet methane temperature and pressure.

This study requires coupling conservation equations; mass, energy, and species. To accomplish this study, it is necessary to calculate velocities. The latter is governed by Darcy's law. The thermal fields are studied by a two-dimensional numerical simulation in the plane perpendicular to the methane flow. The method adopted for solving numerically a complex system of equations is the finite

H. Abdenebi (✉) • D. Haddad
Institute of Hygiene and Security, University Hadj lakhder, Batna, Algeria
e-mail: abdenebihafsa@gmail.com; H2SOFC@gmail.com; H2SOFC@gmail.com

B. Zitouni
Faculty of Technology, University Hadj lakhder, Batna, Algeria
e-mail: H2SOFC@gmail.com

H. Ben Moussa • H. Zitouni
Department of Mechanical, University Hadj lakhder, Batna, Algeria
e-mail: H2SOFC@gmail.com; H2SOFC@gmail.com

Y. Sahli
Research Unit Renewable Energies in the Middle of Sahara URERMS,
Centre for Development of Renewable Energy CDER 01000, Adrar, Algeria
e-mail: H2SOFC@gmail.com

I. Dincer et al. (eds.), *Progress in Clean Energy, Volume 2*,
DOI 10.1007/978-3-319-17031-2_41

difference method. The thermal fields as results of this study are obtained by developing a program in FORTRAN language and the Tecplot software, respectively.

The results show the thermal fields together with the application of the parameters studied in the literature. Methane and air inlet conditions are methane temperature values of 1,173 K or $T = 1,273$ K and pressure values of 1 bar, 2 bar, or 3 bar. The inlet fuel is considered as a gas mixture of H_2, CH_4, CO, H_2O, and CO_2. Fuel and air inlet compositions are: CH_4, 0.29; H_2, 0.09; CO_2, 0.01; CO, 0.01; H_2O, 0.6; O_2, 0.21; and N_2, 0.79.

The analysis of the thermal fields is based on the source term introduced in the energy equation. The influence of the heat source is shown by the decrease or increase in the temperature.

Keywords SOFC • Direct internal reforming • Thermal fields • Heat sources

Nomenclature

C_p	Fluid specific heat at constant pressure, J/kg K
D_{eff}	Effective diffusion coefficients, m^2/s
E	Component thickness, μm
F	Faraday constant, C/mole
I	Current density, A/m^2
P	Pressure, Pa
R	Perfect gas constant, J/mole K
S	Heat source, W/m^3
T	Temperature, K
U	Velocity, m/s
X	Species mass fraction $(j = CH_4, H_2, H_2O, CO, CO_2)$
ΔH	Enthalpy variation, J/mole

Greek Letters

ρ	Density, kg/m^3
Γ	Diffusion coefficient, m^2/s
Φ	General variable
K	Permeability, m^2
ε	Porosity, %
λ	Thermal conductivity, W/m K
μ	Viscosity, kg/m s

Subscripts

an	Anode
cat	Cathode
ele	Electrolyte
r	Steam reforming
s	Shift
T	Thermal
Ref	Reforming
sol	Solid
max	Maximum
min	Minimum

41.1 Introduction

Fossil fuels are always widely used in the world. They respond about 80 % of world energy demand today, but its continued use causes two main problems: The first problem is that they are limited in amount and sooner or later will be depleted. According to the estimates by petroleum companies, the production of the most conveniently utilizable fossil fuels, petroleum and natural gas, will peak sometime between the years 2015 and 2020 and then begin to decrease. This means that there will be a gap between demand and production of fluid fuels, beginning around 2015. The second problem is that fossil fuels are causing serious environmental problems, such as global warming, climate changes, melting of ice caps, rising sea levels, acid rains, pollution, ozone layer depletion, oil spills, forest and agricultural land damage caused by surface mining of coal, and so on.

Early in the 1970s hydrogen energy system had been proposed as a solution for these two interconnected global problems. One of its unique properties is that through electrochemical processes, it can be converted to electricity in fuel cells [1].

In this work, we are interested with the solid oxide fuel cell (SOFC). It operates at high temperatures. The high operating temperature (700–1,000 °C) of solid oxide fuel cells (SOFCs) has a number of consequences, the most important of which is the possibility of running the cells directly on practical hydrocarbon fuels without the need for a complex and expensive external fuel reformer that is necessary for low-temperature fuel cells [2]. The most common fuel, particularly for stationary applications, is natural gas. It varies in composition, but its main compound is methane.

To achieve our objectives, a comprehensive study is conducted on the various works on SOFC fueled by methane. Cheekatamarla et al. [3] presented an SOFC that combines a catalyst layer with a conventional anode, allowing internal reforming via partial oxidation of fuels such as methane, propane, butane, biomass gas, etc., without coking and yielding stable power output. Cosimo Guerra et al. [4] investigated the catalytic properties of Ni/YSZ anodes as electrodes of solid oxide fuel cells to be operated under direct dry reforming of methane. In reference to Ozgur Colpan et al. [5], a direct internal reforming SOFC operating with syngas has been modeled thermodynamically. They considered the recirculation of the anode exit gas stream to obtain valid gas mixtures containing different gas compositions, and they also investigated the effect of the recirculation ratio and fuel utilization ratio. Analysis of chemical and electrochemical reactions and thermo fluid flow in methane-feed internal reforming SOFCs under gas concentrations and temperature effect was studied by 3D numerical simulations [6]. Thinh X. Ho et al. [7] investigated, using a detailed numerical model, the performance of a planar cathode-supported solid oxide fuel cell (SOFC) with composite electrodes fed by methane. Vakouftsi et al. [8] presented a three-dimensional CFD simulation of a planar SOFC fueled by modeled biogas/steam mixtures. The simulation considered species distribution of gas, current densities, potentials, and temperature gradients and confirmed that equimolar CH_4/CO_2 biogas leads to improved performance, while minimal steam addition can prevent carbon deposition.

The objective of the present work is to research the effects of heat sources of reforming reactions on the thermal fields. Therefore, a two-dimensional numerical simulation of the basic equations, continuity equation, mass equation, heat equation, and flow equation, are studied. The temperature distribution is shown for a complete ASP_SOFC (two interconnects, an anode, an electrolyte, and cathode).

41.2 Physical Model

A planar unit cell at anode supported is shown in Fig. 41.1. It is composed of two interconnects, an anode, an electrolyte, and a cathode. Figure 41.1 presents boundary conditions and heat sources taken in this paper. The fuel in the anode is a mixture of methane (CH_4), hydrogen (H_2), steam (H_2O), carbon monoxide (CO),

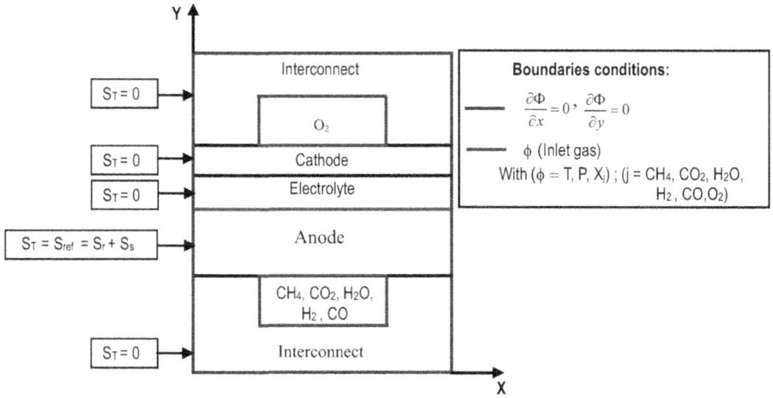

Fig. 41.1 Physical model, heat sources, and boundary conditions

and carbon dioxide (CO_2) whence the appearance of the direct reforming phenomenon. Consequently electricity production occurs in two steps successively: first hydrogen production by the reforming reactions and second the electricity production by electrochemical reactions. The reforming reactions of direct internal reforming (DIR) are: the steam reforming reaction (41.1), an endothermic process, requires energy, and favored by high temperatures; and the water-gas shift reaction (41.2), a weak exothermic reaction; and the endothermic overall reaction.

The reforming reactions are:

$$\text{Anode} : -\text{Steam reforming reaction} : \quad CH_4 + H_2O \rightarrow 3H_2 + CO \tag{41.1}$$

$$-\text{Water-gas shift reaction} : \quad CO + H_2O \rightarrow H_2 + CO_2 \tag{41.2}$$

$$-\text{Overall reforming reaction} : \quad CH_4 + H_2O \rightarrow 4H_2 + CO_2 \tag{41.3}$$

The electrochemical reactions are:

$$\text{Anode} : H_2 + O^{2-} \rightarrow H_2O + 2e^- \tag{41.4}$$

$$CO + O^{2-} \rightarrow CO_2 + 2e^- \tag{41.5}$$

$$\text{Cathode} : 0.5\,O_2 + 2e^- \rightarrow O^{2-} \tag{41.6}$$

The overall electrochemical reaction (41.7 and 41.8) is the sum of half reactions in the cathode (41.6) and the anode (41.4 and 41.5) corresponding to the exothermic reactions of water formation (41.7) and carbon dioxide (41.8).

$$\text{Anode} : \text{Hydrogen oxidation} : \quad H_2 + 1/2O_2 \rightarrow H_2O \tag{41.7}$$

$$\text{Carbon monoxide oxidation} : CO + 1/2O_2 \rightarrow CO_2 \tag{41.8}$$

41.3 Mathematical Model

Thermal field study of an anode-supported planar SOFC is shown in a perpendicular plane to the gas flow direction (Fig. 41.1). This study requires coupling conservation equations; mass, energy, and species. For the momentum equation in the porous electrodes, the flow is modeled using Darcy's law:

$$U = -\frac{K}{\varepsilon\mu}\,\text{grad}(P) \tag{41.9}$$

The generalized form of conservation equations, mass, energy, and species, is written as follows:

$$\mathrm{div}(\varepsilon\rho U\Phi) = \mathrm{div}\Gamma_{\Phi}\mathrm{grad}(P) + S_{\Phi} \tag{41.10}$$

where Φ is a generalized variable, Γ_{Φ} is the diffusion coefficient, and S_{Φ} is the source term.

The mass conservation equation, $\Phi = 1$, $\Gamma_{\Phi} = 1$, and $S_{\Phi} = 0$ and the species conservation equation Φ, Γ_{Φ} and S_{Φ} are replaced by the mass fraction of each species Y_j (with j = (CH_4, CO, CO_2, H_2, O_2, H_2O)), $(\rho\ D)_j$ and S_{Y_j} respectively. For the conservation energy equation, $\Phi = T$, $\Gamma_{\Phi} = \lambda_{(\mathrm{eff,i})}$ (with i = (ele, an, cat, int)), and $S_{\Phi} = S_T$.

The fuel used is a gas mixture of five constituents: CH_4, H_2, CO_2, CO, and H_2O; therefore, the fuel density can be defined as follows [9, 10]:

$$\rho_{\mathrm{fuel}} = \frac{P_{\mathrm{fuel}}(M_{CH_4}X_{CH_4} + M_{H_2}X_{H_2} + M_{CO_2}X_{CO_2} + M_{CO}X_{CO} + M_{CO}X_{CO} + M_{H_2O}X_{H_2O})}{RT} \tag{41.11}$$

For the fuel specific heat expression is given by the following relationship [11]:

$$C_{p_j} = \int_{T_1}^{T_2} \sum X_j C_{p_j} dT \tag{41.12}$$

where C_{p_j} is the specific heat of species j, and the effective thermal conductivity λ_{eff} can be calculated by the following formula [12, 13]:

$$\lambda_{\mathrm{fuel}} = \varepsilon\lambda_{\mathrm{fuel}} + (1-\varepsilon)\lambda_{\mathrm{sol}} \tag{41.13}$$

The fuel dynamic viscosity is calculated with the following relationship used by [10]:

$$\mu_{\mathrm{fuel}} = \sum_1^5 \mu_j X_j = \mu_{CH_4}X_{CH_4} + \mu_{H_2}X_{H_2} + \mu_{CO_2}X_{CO_2} + \mu_{CO}X_{CO} + \mu_{H_2O}X_{H_2O} \tag{41.14}$$

41.3.1 Source Terms

41.3.1.1 Mass Source

Within the anode and the anode/electrolyte interface, the mass source of species conservation equation S_{Y_j} results from the reforming reactions and the electrochemical reactions. In this work, we are only interested in the mass source due to the reforming reactions, the steam reforming reaction $S_{r,j}$, water gas shift reaction

Table 41.1 Mass sourcs expressions [13, 14]

	Species				
Source terms	CH_4	H_2	CO_2	CO	H_2O
$S_{r,j}$	$-R_r M_{CH4}$	$3R_r M_{H2}$	0	$R_r M_{CO}$	$-R_r M_{H_2O}$
$S_{s,j}$	0	$R_s M_{H_2}$	$R_r M_{CO_2}$	$-R_r M_{H_2O}$	$-R_s M_{H_2O}$
$S_{ref} = S_{r,j} + S_{s,j}$	$-R_r M_{CH_4}$	$(3R_r + R_s) M_{H_2}$	$R_r M_{CO_2}$	$(R_r - R_S) M_{CO}$	$(R_r + R_s) M_{H_2O}$

$S_{s,j}$ and the overall reforming reaction. Expressions of different mass sources due to the reforming reactions are shown in Table 41.1.

At the cathode, the mass source is the source resulting from the consumption of oxygen (O_2) reaction (41.6):

$$S_{cat,O_2} = -\frac{M_{O_2}}{2F} i \tag{41.15}$$

where M_{O_2}, i, and F are the oxygen molar mass, the current density, and the Faraday number respectively.

41.3.1.2 Heat Source

In this paper, the thermal source taken into account is the heat source due to the reforming reactions: heat sources due to the steam reforming reaction S_r, which is an endothermic process, the water-gas shift reaction S_s which is an exothermic reaction, and the endothermic overall reaction $S_{T,an}$. In the anode, heat source is described by the following equations [7, 15, 16]:

$$S_{T,an} = S_r + S_s \tag{41.16}$$

where

$$S_r = -R_r \Delta H_r \tag{41.17}$$

$$S_s = -R_s \Delta H_s \tag{41.18}$$

R and ΔH are the reaction rate and enthalpy change of the reforming reactions [15, 17, 18].

The reaction rate expressions are given as follows:

$$R_r = K_r^+ P_{CH_4} P_{H_2O} - K_r^- P_{CO} (P_{H_2})^3 \tag{41.19}$$

$$R_s = K_s^+ P_{CO} P_{H_2O} - K_s^- P_{CO_2} P_{H_2} \tag{41.20}$$

Thus, the rate constants of reactions 'k^+' 'k^-'and the equilibrium constant are used:

$$K_r^+ = 2,169\exp(-225,103/RT) \quad (41.21)$$

$$K_s^+ = 0.0183\exp(-103,842/RT) \quad (41.22)$$

The coefficients K_r^- and K_s^- are determined based on the following equilibrium constants for both reactions:

$$K_{er} = \frac{K_r^+}{K_s^-} \quad (41.23)$$

$$K_{es} = \frac{K_r^+}{K_r^-} \quad (41.24)$$

And equilibrium constants, K_{er} and K_{es}, can be calculated by the following empirical equations:

$$K_{er} = 1.0267 \times 10^{10}\exp(-0.2413Z^4 + 0.3665Z^3 + 0.5810Z^2 - 27.134Z + 3.2770) \quad (41.25)$$

$$K_{es} = \exp(-0.2935Z^3 + 0.6351Z^2 + 4.1788Z + 0.3169) \quad (41.26)$$

Z: is a variable depending on the temperature [14, 19].

$$Z = 1,000/T - 1 \quad (41.27)$$

And for the partial pressure P_j of each species j is equal to its mole fraction X_j by the total pressure P_{fuel} [10, 20].

$$P_j = X_j P_{fuel} \quad (41.28)$$

P_{fuel}: The gas mixture pressure at the channels inlet.

The enthalpies change of reforming reactions, steam reforming reaction (41.1), and water-gas shift reaction (41.2) are given by the following relations [15]:

$$\Delta H_r = 192.220 + 0.0541T - 2.062 \times 10^{-5}T^2 \quad (41.29)$$

$$\Delta H_s = -44.034 + 0.00847T - 5.819 \times 10^{-7}T^2 \quad (41.30)$$

The gas physical properties of fuel and air: the thermal conductivity, the specific heat, and the dynamic viscosity of species CH_4, H_2, CO, CO_2, H_2O, and O_2 are polynomial temperatures of order 4 (CFD Fluent). They are expressed by the following relations:

$$\lambda_j = a + bT + cT^2 + dT^3 + eT^4 \tag{41.31}$$

$$C_{p_j} = a + bT + cT^2 + dT^3 + eT^4 \tag{41.32}$$

$$\mu_j = a + bT + cT^2 + dT^3 + eT^4 \tag{41.33}$$

Or *a*, *b*, *c*, *d*, and *e* are empirical constants, different for each gas as shown in the Tables 41.2, 41.3, and 41.4. Data of physical properties of species and physical properties of the cell components are illustrated in Tables 41.5 and 41.6.

Table 41.2 Thermal conductivity coefficients of gases (CFD Fluent)

Gas	*a*	*b*	*c*	*d*	*e*
CH_4	0.0332	0	0	0	0
H_2	0.08269088	0.0003561521	1.072322e-08	−4.40516e-12	3.263276e-16
CO_2	0.0145	0	0	0	0
CO	0.025	0	0	0	0
H_2O	−0.007967996	6.881332e-05	4.49046e-08	−9.099937e-12	6.173314e-16
O_2	0.003921754	8.081213e-05	−1.354094e-08	2.220444e-12	−1.416139e-16

Table 41.3 Specific heat coefficients of gases (CFD Fluent)

Gas	*a*	*b*	*c*	*d*	*e*
CH_4	872.4671	5.305473	−0.002008295	3.516646e-07	−2.33391e-11
H_2	12,337.53	2.887275	−0.000232356	−3.807379e-08	6.527742e-12
CO_2	841.3765	0.5932393	−0.0002415168	4.522728e-08	−3.153134e-12
CO	897.9305	0.428231	−0.0001671392	3.023444e-08	−2.05137e-12
H_2O	1,233.234	1.410523	−0.0004029141	5.542772e-08	−2.949824e-12
O_2	811.1803	0.4282316	−0.0001750725	3.757596e-08	−2.973548e-12

Table 41.4 Dynamic viscosity coefficients of gases (CFD Fluent)

Gases	*a*	*b*	*c*	*d*	*e*
CH_4	1.087e-05	0	0	0	0
H_2	3.773319e-06	1.932317e-08	−3.231968e-12	4.917018e-16	−3.024337e-20
CO_2	1.37e-05	0	0	0	0
CO	0.025	0	0	0	0
H_2O	−4.418944e-06	4.687638e-08	−5.389431e-12	3.202856e-16	4.919179e-22
O_2	7.879426e-06	4.924946e-08	−9.851545e-12	1.527411e-15	−9.425674e-20

Table 41.5 Physical properties of species

Physical properties	Species						References
	CH_4	H_2O	H_2	CO_2	CO	O_2	
X_j (%)	0.29	0.6	0.09	0.01	0.01	–	[21]
M_j (g/mol)	16	18	2	44	28	32	
D_j (cm^2/s)	$D_j = 0.364(T\|273)^{1.5}$					$D_{O_2} = 0.181(T\|273)^{1.5}$	[22]

Table 41.6 Physical properties of the solid components

Parameters	Anode	Electrolyte	Cathode	Interconnects	References
K (m^2)	10–12	10–18	10–12	–	[23]
ε (%)	0.4	–	0.3	–	[24]
λ (W/m K)	2	2	2	2	[23]
e (m)	0.2e-3	0.05e-3	0.05e-3	0.3e-3	[15]

41.4 Results and Discussion

In this study, a DIR_SOFC operating with methane gas and air has been studied through a two-dimensional numerical simulation. The results present the gas operating parameters', temperature and pressure, effect at an ASP_SOFC. The heat sources effect due to the reforming reactions: the endothermic steam reforming reaction, the exothermic water-gas shift reaction, and the endothermic overall reforming reaction are also being investigated. The values of gas temperature and pressure for this numerical study are 1,173 K, 1,273 K, 1 bar, 2 bar, and 3 bar, respectively. Figures 41.2, 41.3, and 41.4 present thermal fields of an ASP_SOFC under the effect of the parameters listed above.

41.4.1 SOFC Thermal Fields

41.4.1.1 Gas Temperature Effect on the Thermal Fields at Different Gas Pressures

The thermal fields of an ASP_SOFC operating under the values of 1 bar and 1,173 K of gas pressure and temperature, respectively, without or with heat source effect, are invariant. We notice that we have heat transfer channels, with warmer environment, to other components of the SOFC (Fig. 41.2a–d). Under these operating conditions, the reforming reactions' heat sources have no effect. So the reforming phenomenon did not appear: $\Delta T = 0\,\text{K}$. At the same pressure but at a higher temperature, more than the previous case, with gas temperature equaling to 1,273 K, a change in the thermal fields shape appeared. The minimum temperature is located in the anode far from gas channel. The effect of heat source due to the

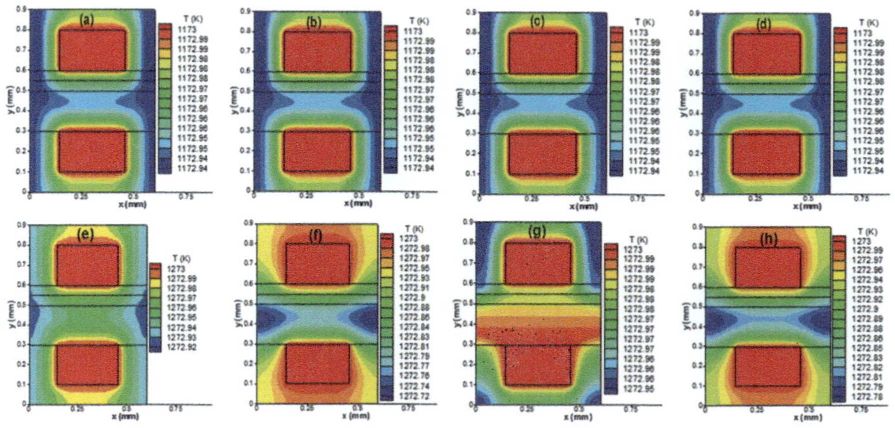

Fig. 41.2 Thermal fields, $P = 1$ bar, $T = 1{,}173$ K (**a–d**) and $T = 1{,}273$ K (**e–h**), without heat source (**a**, **e**), heat source of steam reaction (**b**, **f**), heat source of water-gas shift reaction (**c**, **g**), heat source of overall reaction (**d**, **h**)

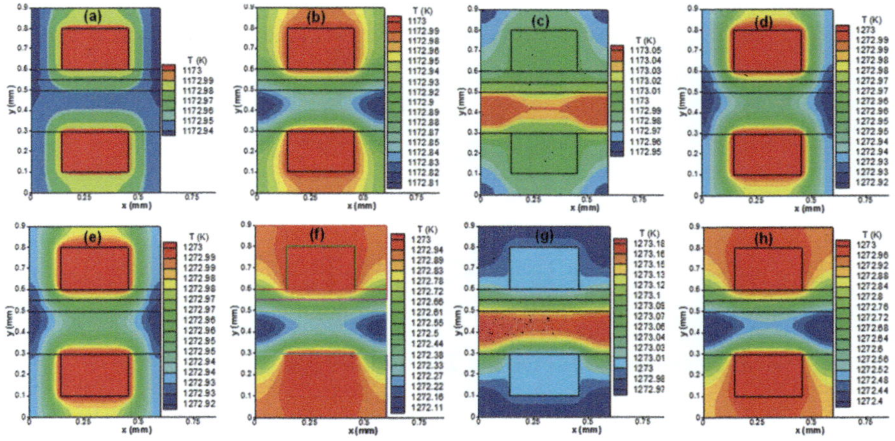

Fig. 41.3 Thermal fields, $P = 2$ bar, $T = 1{,}173$ K (**a–d**) and T = 1,273 K (**e–h**), without heat source (**a**, **e**), heat source of steam reaction (**b**, **f**), heat source of water-gas shift reaction (**c**, **g**), heat source of overall reaction (**d**, **h**)

steam reforming and the overall chemical reaction is shown by a decrease in the temperature: $\Delta T = -0.2\,\text{K}$ and $\Delta T = -0.14\,\text{K}$ correspondingly (Fig. 41.2f, h). The effect of heat source due to the water-gas shift reaction is shown by an increase in the temperature in the anode: $\Delta T = 0.03\,\text{K}$ (Fig. 41.2g). When the SOFC operates at gas pressure equal to 2 bar or 3 bars, the effect of the reforming reaction heat source is at gas temperature equal to 1,173 K. The energy consumption, a diminution in the temperature, is the steam reforming and overall reforming reactions' effect (Figs. 41.3 and 41.4: b, d, f, h). The heat source effect of water-gas shift reaction is shown by energy production (Figs. 41.3 and 41.4: c, g).

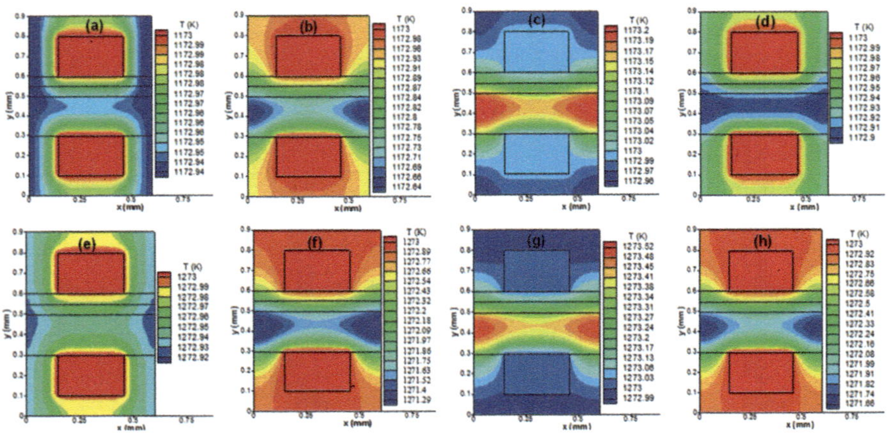

Fig. 41.4 Thermal fields for $P = 3$ bar, $T = 1{,}173$ K (**a**–**d**) and $T = 1{,}273$ K (**e**–**h**), without heat source (**a**, **e**), heat source of steam reaction (**b**, **f**), heat source of water-gas shift reaction (**c**, **g**), heat source of overall reaction (**d**, **h**)

Table 41.7 Temperatures; $T_{\max}$, $T_{\min}$, and the temperature variation; ΔT at $P = 1$ bar

	Temperatures					
	$T = 1{,}173$ K			$T = 1{,}273$ K		
Heat sources	$T_{\max}$	$T_{\min}$	ΔT	$T_{\max}$	$T_{\min}$	ΔT
$S_{\mathrm{T,an}} = 0_{\mathrm{r}}$	1,173	1,172.94	–	1,273	1,272.92	–
$S_{\mathrm{T,an}} = S_{\mathrm{r}}$	1,173	1,172.94	0	1,273	1,272.72	−0.2
$S_{\mathrm{T,an}} = S_{\mathrm{s}}$	1,173	1,172.94	0	1,273	1,272.95	0.03
$S_{\mathrm{T,an}} = S_{\mathrm{ref}} = S_{\mathrm{r}} + S_{\mathrm{s}}$	1,173	1,172.94	0	1,273	1,272.78	−0.14

41.4.1.2 Gas Pressure Effect on the Thermal Fields

The variety in the pressure values of the SOFC serves us to understand the DIR phenomenon. With gas pressure values of 1 bar, 2 bar, and 3 bar, the thermal fields without any heat source effect are unchanging (Figs. 41.2, 41.3, and 41.4: a, e). For gas pressure and temperature of 1 bar and 1,173 K, respectively, the thermal fields, in the absence or presence of a reforming reaction heat source, are constant (Fig. 41.2a–d) which signified that the DIR phenomenon is not done at low pressure. A weak appearance of the heat source effect is at $P = 1$ bar and $T = 1{,}273$ K. A temperature diminution is indicated under the heat source effect of the endothermic steam reforming and overall reforming reactions (Fig. 41.2f, h and Figs. 41.3 and 41.4: b, d, f, h). The effect of exothermic water-gas shift reaction is temperature elevation (Fig. 41.2 g, and Figs. 41.3, 41.4 c,g) and consumption of energy is caused by variations in temperature.

Tables 41.7, 41.8, and 41.9 show DIR_SOFC temperatures. The maximum and the minimum temperature ($T_{\max}$, $T_{\min}$) are calculated in the absence or presence of

Table 41.8 Temperatures; T_{max}, T_{min}, and the temperature variation; ΔT at $P = 2$ bar

	Temperatures					
	$T = 1{,}173$ K			$T = 1{,}273$ K		
Heat sources	T_{max}	T_{min}	ΔT	T_{max}	T_{min}	ΔT
$S_{T,an} = 0$	1,173	1,172.94	–	1,273	1,272.92	–
$S_{T,an} = S_r$	1,173	1,172.81	−0.13	1,273	1,272.11	−0.81
$S_{T,an} = S_s$	1,173.05	1,172.95	0.11	1,273.18	1,272.97	0.26
$S_{T,an} = S_{ref} = S_r + S_s$	1,173	1,172.92	−0.02	1,273	1,272.4	−0.52

Table 41.9 Temperatures; T_{max}, T_{min}, and the temperature variation; ΔT at $P = 3$ bar

	Temperatures					
	$T = 1{,}173$ K			$T = 1{,}273$ K		
Heat sources	T_{max}	T_{min}	ΔT	T_{max}	T_{min}	ΔT
$S_{T,an} = 0$	1,173	1,172.94	–	1,273	1,272.92	–
$S_{T,an} = S_r$	1,173	1,172.64	−0.3	1,273	1,271.29	−1.63
$S_{T,an} = S_s$	1,173.2	1,172.96	0.26	1,273.52	1,272.99	0.6
$S_{T,an} = S_{ref} = S_r + S_s$	1,173	1,172.9	−0.04	1,273	1,271.66	−1.26

heat source effect (S_r, S_s, and $S_r + S_s$) for three values of gas pressure: 1 bar, 2 bar, and 3 bar. The energy production/consumption is indicated by the temperature variation (ΔT). The last one (ΔT) is the difference between the DIR_SOFC temperature value (T_{min}) under heat source influence and the DIR_SOFC temperature value (T_{min}) without heat source effect. For a gas pressure of 1 bar, the temperature variation is null, $\Delta T = 0\,\text{K}$, but at 3 bar, the temperature variation (ΔT) is at maximum: $\Delta T = -1.63\,\text{K}$, $\Delta T = 0.60\,\text{K}$, and $\Delta T = -1.26\,\text{K}$ under heat source of steam reforming reaction, water-gas shift reaction, and overall reforming reaction, respectively.

For validation of the results, the numerical simulation results are compared to our previous results. First, the results of previous studies [25–28] show thermal fields of an SOFC fed by hydrogen as fuel and air. The analysis is done to determine the effects of several parameters such as SOFC configuration, inlet gas temperature and pressure, and different heat sources. The latter is due to the Joule effect, activation and concentration over voltage, and the exothermic chemical reaction. Second [29], we presented an SOFC supplied with methane and air under molar fraction effect. However, the present work shows the thermal fields of a DIR_SOFC in the presence of the effects of the heat sources due to the endothermic steam reforming reaction, the exothermic water-gas shift reaction, and the endothermic overall reforming reaction. The electrochemical reaction influences are ignored and the thermal field analysis is done for different values of gas temperature and pressure.

41.5 Conclusions

The work developed may be summarized as follows: temperature distributions of a DIR_SOFC are investigated. The details of these properties are studied by a FORTRAN programming language two dimensionally using the finite difference method. The numerical simulation used conservation equations: mass, energy, and species. For porous electrodes (anode and cathode), velocities are calculated by Darcy's law. The thermal fields' analysis consists of several steps. First, the effect of inlet gas temperature and pressure is considered. Second, the effect of heat sources result of reforming reactions is taken into account. These are the endothermic steam reaction, the exothermic water gas shift reaction, and the endothermic overall reforming reaction. Results show that the effect of heat sources is clear at inlet gas temperature and pressure of 1,273 K and 3 bar, respectively. Based on the results found, it can be concluded that the reforming phenomenon appeared at high gas temperatures and pressures.

References

1. Barbir F (2005) PEM fuel cells: theory and practice
2. Singhal SC, Kendall K (2003) High temperature solid oxide fuel cells: fundamentals, design and applications. Elsevier Ltd., Oxford, UK
3. Cheekatamarla PK, Finnerty CM, Cai J (2008) Internal reforming of hydrocarbon fuels in tubular solid oxide fuel cells. Int J Green Energ 33:1853–1858
4. Guerra C, Lanzini A, Leone P, Santarelli M, Brandon NP (2014) Optimization of dry reforming of methane over Ni/YSZ anodes for solid oxide fuel cells. J Power Sources 245:154–163
5. Ozgur Colpana C, Dincerb I, Hamdullahpur F (2007) Thermodynamic modeling of direct internal reforming solid oxide fuel cells operating with syngas. Int J Hydrog Energy 32:787–795
6. Park J, Li P, Bae J (2012) Analysis of chemical, electrochemical reactions and thermo-fluid flow in methane-feed internal reforming SOFCs: part II-temperature effect. Int J Green Energ 37:8532–8555
7. Ho TX, Kosinski P, Hoffmann AC, Vik A (2009) Modeling of transport, chemical and electrochemical phenomena in a cathode-supported SOFC. Chem Eng Sci 64:3000–3009
8. Vakouftsi E, Marnellos GE, Athanasiou C, Coutelieris F (2011) CFD modeling of a biogas fuelled SOFC. Solid State Ion 192:458–463
9. Huang X, Kromp A (2013) Cooling of a diesel reformate fuelled solid oxide fuel cell by internal reforming of methane: a modelling study. Chin J Chem Eng 21:324–331
10. Mounir H (2010) Caractérisation, Modélisation thermo fluide et électrochimique, simulation numérique et étude des performances des nouvelles piles à combustible types IP-SOFC (Integrated-Planar Solid Oxide Fuel Cell). Thèse de Doctorat, Univ. Rabat
11. Bertin M, Faroux J-P, Renault J (1981) Thermodynamique, 3rd edn. Dunod, Paris
12. Jiang TL, Chen M-H (2009) Thermal-stress analyses of an operating planar solid oxide fuel cell with the bonded compliant seal design. Int J Green Energ 34:8223–8234

13. Pramuanjaroenkij A, Zhou XY, Kakac S (2010) Numerical analysis of indirect internal reforming with self-sustained electrochemical promotion catalysts. Int J Green Energ 35:6482–6489
14. Iwai H, Yamamoto Y, Saito M, Yoshida H (2010) Numerical simulation of intermediate-temperature direct-internal-reforming planar solid oxide fuel cell. Energy 36:2225–2234
15. Wang Q, Li L, Wang C (2009) Numerical study of thermoelectric characteristics of a planar solid oxide fuel cell with direct internal reforming of methane. J Power Sources 186:399–407
16. Kang Y-W, Li J, Cao G-Y, Heng-Yong T, Li J, Yang J (2009) A reduced 1D dynamic model of a planar direct internal reforming solid oxide fuel cell for system research. J Power Sources 188:170–176
17. Brus G, Szmyd JS (2008) Numerical modelling of radiative heat transfer in an internal indirect reforming-type SOFC. J Power Sources 181:8–16
18. Zinovik I, Poulikakos D (2009) Modeling the temperature field in the reforming anode of a button-shaped solid oxide fuel cell. Electrochim Acta 54:6234–6243
19. Arpornwichanop A, Patcharavorachot Y, Assabumrungrat S (2010) Analysis of a proton conducting SOFC with direct internal reforming. Chem Eng Sci 65:581–589
20. Xi H, Sun J, Tsourapas V (2007) A control oriented low order dynamic model for planar SOFC using minimum Gibbs free energy method. J Power Sources 165:253–266
21. Li J, Kang Y-w, Cao G-y, Zhu X-j, Tu H-y, Li J (2008) Numerical simulation of a direct internal reforming solid oxide fuel cell using computational fluid dynamics method. J Zhejiang Univ Sci A 9:961–969
22. Ferguson JR, Fiard JM, Herbi R (1996) Three-dimensional numerical simulation for various geometries of solid oxide fuel cells. J Power Sources 58:109–122
23. Kang Y, Li J, Cao G, Tu H, Li J, Yang J (2009) One-dimensional dynamic modeling and simulation of a planar direct internal reforming solid oxide fuel cell. Chin J Chem Eng 17:304–317
24. Ho TX, Kosinski P, Hoffmann AC, Vik A (2010) Effects of heat sources on the performance of a planar solid oxide fuel cell. Int J Green Energ 35:4276–4284
25. Abdenebi H, Zitouni B, Haddad D, Ben Moussa H, Andreadis GM, Abdessemed S (2011) SOFC fuel cell heat production: analysis. Energy Procedia 6:643–650
26. Zitouni B, Andreadis GM, Ben Moussa H, Abdenebi H, Haddad D, Zeroual M (2011) Two-dimensional numerical study of temperature field in an anode supported planar SOFC:effect of the chemical reaction. Int J Hydrog Energy 36:4228–4235
27. Haddad D, Abdenebi H, Zitouni B, Ben Moussa H, Oulmi K (2013) Thermal field in SOFC fed by hydrogen: inlet gases temperature effect. Int J Hydrog Energy 38:8575–8583
28. Oulmi K, Zitouni B, Ben Moussa H, Abdenebi H, Andreadis GM (2001) Total polarization effect on the location of maximum temperature value in planar SOFC. Int J Hydrog Energy 346:4236–4243
29. Abdenebi H, Zitouni B, Ben Moussa H, Haddad D (2014) Thermal field in SOFC fed by CH4: molar fractions effect. J Assoc Arab Univ Basic Appl Sci 17:82–89
30. CFD Fluent 6.2.16

Chapter 42
Prospects and Analysis of Hydrogen Production from Renewable Electricity Sources in Algeria

S. Rahmouni, N. Settou, B. Negrou, N. Chennouf, and R. Ghedamsi

Abstract In response to the global concerns for reducing the greenhouse gas emissions and ensure the energy security supply, renewable energy-based hydrogen systems appear to be an interesting solution. The purpose of this study is on techno-economic analysis of massive hydrogen production system through water electrolysis using different renewable energy sources. Numerical simulations are carried out to study the performance of hydrogen production system. The potential of all electrical sources is analyzed and assessed. The sizing of a hydrogen production system and annual energy requirements are determined. Correct methodology is demonstrated for a case study in Ouargla region. An approximate cost analysis, which included a total investment estimate, was performed. The levelized cost of hydrogen production was also calculated for comparison purposes. Finally, the results are discussed.

Keywords Renewable energy sources • Alkaline electrolysis • Hydrogen production costs • Techno-economic study • Numerical modeling

Nomenclature

A	Area, m^2
C	Cost, \$
CF	Capacity factor
C	Specific heat capacity, kJ/kg K
E	Electric energy, kWh
G	Solar irradiation, kWh/m^2
g	Acceleration of gravity, m^2/s

S. Rahmouni (✉) • N. Settou • B. Negrou • N. Chennouf • R. Ghedamsi
Fac. Des sciences appliquées, Lab. Promotion et valorisation des ressources sahariennes, Universite Kasdi Merbah, Ouargla 30000, Algeria
e-mail: soumiarahmouni@gmail.com

I. Dincer et al. (eds.), *Progress in Clean Energy, Volume 2*,
DOI 10.1007/978-3-319-17031-2_42

H	Height, m
h	Well depth, m
I	Investment, \$
k	Theoretical specific energy, kWh/kg
m	Annual hydrogen production capacity, kg/year
O&M	Annual operation and maintenance cost, \$/year
P	Puissance, MW
Q	Volumetric flow rate, m^3/h
t	Running hours, h
T_a	Ambient temperature, °C

Greek Letters

η	Efficiency, %
Δ	Difference

Subscripts

BOS	Balance of system
c	Chimney
coll	Collector
elec	Electrolyzer
f	Fluid
Geo	Geothermal carbon dioxide system
h	Horizontal
i	Tilted
PCU	Power conversion unit
PV	Photovoltaic
RE	Renewable energy
SCPP	Solar chimney power plant
t	Turbine
th	Theoretical
u	Unit

42.1 Introduction

Total world consumption of energy has increased even more rapidly. The International Energy Agency forecasts that the global demand for energy is expected to increase 53 % by 2035 [1]. Conventional energy sources such as oil, gas, and coal are the main resources for world energy supply (cover roughly 80 %). Fossil fuels are easy to exploit to generate energy because they only require a simple direct combustion. However, their use leads to increased undesirable production of greenhouse gas, contributing to global warming. More importantly, fossil fuel resources are finite as the reserve depletion times for oil, gas, and coal are approximately 35, 37, and 107 years, respectively [2]. Today, the greatest challenge is how to manage and to meet the rising demand in sustainable, economical, secure, and responsible ways, including dealing with greenhouse gas emissions. The transition to renewable energy-based hydrogen systems appears to be an interesting solution and provides an opportunity to address the challenges [3, 4]. Renewable hydrogen is a promising energy carrier for the future energy supply; it benefits as an environmentally friendly, versatile, and efficient fuel. Hydrogen may be produced from renewable energy sources through a variety of pathways and methods [5–11]. Attention around the world has focused on the promise of hydrogen economy. Several projects have been realized, and numerous further hydrogen energy systems are being planned for stationary applications [12]. Algeria plays a key role in world energy markets as a leading producer and exporter of natural gas. Algeria's energy mix in 2010 was almost exclusively based on fossil fuels, especially natural gas (93 %) [13]. However, the country has enormous renewable energy potentialities in solar as well as in geothermal and wind energy [14]. The transition of energy from traditional fossil fuel-based economy towards a sustainable hydrogen economy in the next years could fundamentally transform Algeria's energy policy, creating opportunities to increase energy security through the use of a variety of renewable energy sources for hydrogen production while reducing environmental impacts and a lever for economic and social development, particularly through the establishment of wealth- and job-creating industries [15, 16]. The purpose of this study is on techno-economic analysis of massive hydrogen production by water electrolysis process using different renewable energy sources. This study focuses on three renewable electrical sources: solar PV, solar chimney power plant, and geothermal energy. The organization of the paper is as follows: first, the technical potential of selected renewable electricity sources was analyzed and assessed. Then, the sizing of the main components of the hydrogen production process, especially electrolysis, and annual energy requirements are determined. Correct methodology is demonstrated for a case study in Ouargla region. An approximate cost analysis, which included a total investment estimate, was performed. The levelized cost of hydrogen production was also calculated for comparison purposes. Finally, the results are discussed.

42.2 Renewable Resources Potential Assessment in Algeria

Due to their substantial hydrocarbon reserves, Algeria is a key oil and gas exporting country. According to OPEC, the proven crude oil reserves rose to 12.2 billion barrels, whereas natural gas reserves were at 4.5 billion standard cubic meters [17].

Algeria also possesses significant renewable energy resources; Sahara's total surface could capture enough solar energy to meet the electricity needs of the entire world [18]. Growing the renewable energy systems is perceived as a solution to:

- Protect the climate and the wider environment from impacts of fossil fuels use
- Bring its energy system out of dependence on oil
- Diversify its resources
- As an opportunity to support economic growth and energetic transition in Algeria

The national potential for renewable energy is strongly dominated by solar energy, which exceeds 6 billion GWh/year. The assessed potential is more than 13.9 TWh/year for solar photovoltaic and 169,440 TWh/year for solar thermal. The insulation time over the 95 % of the territory is between 2,000 and 3,900 h/year with a mean value of 5 kWh/m^2 [14]. The potential for wind energy generation in Algeria depends on the availability of the wind resource that varies with location, where the best wind energy potential is located in the southwestern region and is estimated at 35 TWh/year [19]. According to [20], the region of Adrar receives the most wind potential in the country, where its velocity is higher than 6 m/s at 10 m aboveground. The geothermal resources in Algeria are of low enthalpy type [21]. Most of these geothermal resources are located in the northeastern region of the country. The highest temperatures recorded were 68 °C for the western area, 80 °C for the central area, and 98 °C for the eastern area. In the south, the thermal springs have a mean temperature of 50 °C. The Central Sahara has a mean gradient of 21 °C km^{-1}, the two other regions have geothermal gradients of 32 °C km^{-1} and 26 °C km^{-1} for the Western Sahara and northern Algeria, respectively [22]. The biomass potentially in Algeria offers great promises. Forest reserves are estimated to be 37 Mtoe, which is being utilized at a 10 % recovery rate. In addition, roughly 1, 33 Mtoe per year of agricultural and urban wastes are available (365 kg per Algerian as urban wastes) [23]. The Algerian government takes measures to increase the deployment of renewable energy technologies by launching an ambitious program to develop renewable energies in March 2011. The renewable energy program (REP) provides to install 22 GW (between 2011 and 2030) which represents 40 % of total energy consumption from renewable source by 2030. According to the REP, the target for 2017 is to obtain 5 % of the total energy from nonfossil fuels. In the long run, by 2030, Algeria expects to reach 20 % overall renewable coverage, of which 37 % of national electricity production is generated by solar energy and 3 % by wind [24]. Table 42.1 below shows a breakdown of the targeted production of each individual clean energy resource.

Table 42.1 Algeria's renewable energy targets in the medium and long term

Resource	2013 actual	2020 target	2030 target
Solar PV energy	–	800 MW_p	200 MW_p
Solar thermal energy	25 MW	1,200 MW	1,100 MW
Wind energy	10 MW	20 MW (2015)	1,700 MW

Beyond its largest renewable energy sources, Algeria has huge reserves of groundwater hydraulics resources in which it has the ambition to develop with international partners many projects for development of the massive renewable hydrogen production [16]. Transition to renewable hydrogen as a major fuel for the next decades would have several positive consequences for Algeria, its partners, and the world community in general. In this context, Algeria, in partnership with Euro-Maghrebian countries, launched a Mediterranean Hydrogen Solar (MedHySol) demonstration project at the First Workshop on Hydrogen Energy Vector from a Renewable Origin, in 2005 [15]. The goal of the MedHySol project is to improve the system efficiency of producing hydrogen from solar energy in quantities large enough (10–100 kW) and at costs low enough by implementing the most effective and less expensive technologies to pilot great projects (1–1,000 MW).

42.2.1 Solar Photovoltaic Energy Conversion

Solar energy refers to energy that comes from the Sun's radiation. Photovoltaic (PV) devices generate electricity directly from sunlight via an electronic process that occurs naturally in certain types of material, called semiconductors. Electrons in these materials are freed by solar energy and can be induced to travel through an electrical circuit, powering electrical devices. To determine the PV electricity generation potential for a particular site, it is important to assess the average total solar energy received over the year (Eq. 42.1):

$$E_{\mathrm{PV}} = \eta_{\mathrm{PV}}(T_a, G_i) A_{\mathrm{PV}} G_i \tag{42.1}$$

where E_{PV} [kWh] is the electric energy output of a PV panels unit, η_{PV} is the module efficiency, A_{PV} [m^2] is the PV panel area, T_a [°C] is the ambient temperature, and G_i [kWh/m^2] is the solar irradiation incident on the till plan.

PV arrays must be placed to receive the most sunlight. There are several ways to mount the panels fixed, with adjustable tilt angles, manual tracking, passive tracking, and active trackers. Fixed mounts are the least costly and lowest energy producing mounting systems. Two categories of PV cells are used in most of today's commercial PV modules: c-Si and thin film [25]. The c-Si category, called first-generation PV, includes monocrystalline and polycrystalline PV cells, which are the most efficient of the mainstream PV technologies (15–25 %). The thin-film category and multi-junction PV cell, called second-generation PV, includes PV cells that produce electricity via extremely thin layers of semiconductor material (10–20 %). Various emerging technologies (include dye-sensitized, organic PV cells and quantum dots), known as third-generation PV, could become viable commercial options in the future (5–10 %). All types of PV systems are widely used today in a variety of applications.

42.2.2 Solar Chimney Energy Conversion

A solar updraft tower power plant, also called solar chimney or solar tower, offers a method for a large-scale generation of electricity from solar energy and is particularly suitable in deserts and sun-rich lands. This technology is several decades old, and it has been verified in many research and works [26]. The solar chimney power plant (SCPP) is a simple solar thermal power technology, which includes three main components, namely, solar collector, chimney, and power conversion unit (PCU), e.g., turbine generators. In the collector, the solar energy will be transformed into heat energy. The chimney converts the generated heat energy into kinetic energy, which will be transformed into electric energy by using a combination of a wind turbine and a generator [27]. The power output of a solar chimney depends on parameters such as the climatic conditions, the solar radiation intensity, and ambient temperature and structural dimensions of the system, and includes the height and radius of both the chimney and collector. The maximum electrical power from the solar chimney is obtained by multiplying the total pressure difference by wind turbine efficiency that contains blade, transmission, and generator efficiency (Eq. 42.2):

$$E_{\mathrm{SCPP,max}} = G_h A_{\mathrm{coll}} \frac{2}{3} \frac{g H_c}{C_P T_a} \eta_{\mathrm{coll}} \eta_t \tag{42.2}$$

where E_{SCPP} [kWh] is the maximum electric energy output from the solar chimney, G_h [kWh/m^2] is the solar irradiation on horizontal plan, A_{coll} [m^2] is the solar collector area, H_c [m] is the solar chimney height, g [m^2/s] is the acceleration of gravity, C_{P} [kJ/kg K] is the specific heat capacity of air, η_{coll} is the solar collector efficiency, and η_{t} is the turbine efficiency.

42.2.3 Geothermal Energy Conversion

Geothermal power generation is environmentally friendly, uses renewable resources, and provides reliable energy. There are many types of geothermal power plants depending on the state of the fluid (steam or water) and its temperature [28]. Combining the geothermal energy extraction and CO_2 sequestration to produce clean, renewable, and carbon-negative electricity becomes an attractive solution. This method of geothermal energy production is twice as efficient as traditional geothermal power systems because it uses carbon dioxide in place of groundwater as the working fluid. The system pumps CO_2 into geothermal reservoirs as depleted gas and oil fields or saline aquifers through an injection well, where the carbon dioxide is trapped in an existing geologic formation, heated, and put under high pressures. The high-temperature CO_2 then rises through a recovery well to drive a turbine that is connected to a power generator. After passing through the turbine, the carbon dioxide is reinserted into the cycle, forming a closed-loop system [29]. The amount of energy that can be recovered from the reservoir equals the amount of energy currently present in the reservoir; the following equation is used to calculate the total energy being extracted per year [30]:

$$E_{\mathrm{Geo}} = Q_v t_f \Delta T C_{P,f} \tag{42.3}$$

where E_{Geo} [kWh] is energy extracted per year, Q_v [m^3/h] is the volumetric flow rate, t_f [h] is the equivalent running hours, ΔT is the temperature difference between injected carbon dioxide and initial reservoir temperature, and $C_{P,f}$ [kJ/m^{3}K] is the volume-related heat capacity working fluid.

42.3 Renewable Electrolysis Hydrogen Production

The production of hydrogen via electrolysis is, in principle, very simple. It is a process by which electricity is passed through two electrodes to decompose water into its gaseous components. The overall reaction is:

$$H_2O + \text{Energy} \Rightarrow H_2 + 1/2\ O_2 \tag{42.4}$$

Figure 42.1 shows a schematic diagram of hydrogen production system that uses renewable electrical source. Three different configurations are presented, for each one; the system mainly consists of two subsystems: eclectic power source and industrial electrolyzer.

PV panels, SCPP, and geothermal carbon dioxide system are utilized to produce electricity. This electricity is used to operate the electrolyzer in which water splits into oxygen and hydrogen. To maximize the energy transfer between the subsystems, direct connection is more viable [31]. For this study, an industrial alkaline water electrolyzer (supplied by Hydrogenics), with its robust, reliable, and highly efficient operation, is used. At full capacity, the electrolyzer can generate up to 60 Nm^3 of hydrogen per hour, with a power consumption of 5.2 kWh/Nm^3 and high quality. Energy efficiency is defined as the higher heating value (HHV) of hydrogen divided by the energy consumed by the electrolysis system per kilogram of hydrogen produced; the energy efficiency of the electrolysis process is 68 % [32].

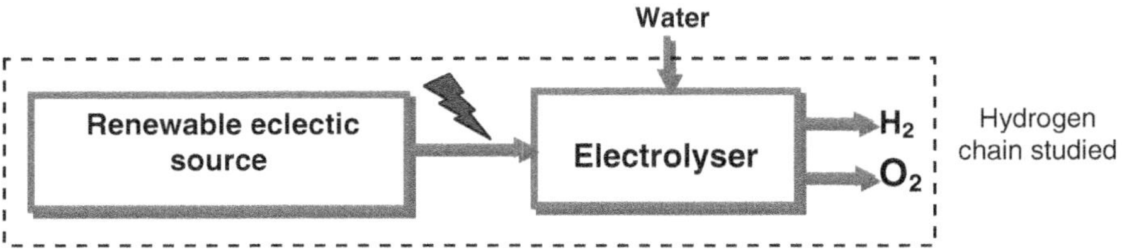

Fig. 42.1 Schematic diagram of hydrogen production system

42.3.1 Energy Requirements for Electrolysis

The energy required for the water electrolysis process is calculated from multiplication of the theoretical specific energy required by the electrolyzer at a given temperature (in kWh/kg) and the annual hydrogen production capacities (in kg/year) divided by the energy efficiency of the electrolyzer. The annual amount of electricity necessary for producing 10, 250, and 500 tons of hydrogen per year was calculated. For each electrolyzer, a specific system energy requirement used to determine the number of cells constituting a well-sized electrolyzer was established. Table 42.2 below illustrates the energy, power requirements, and number of cells for electrolysis systems.

The 10,000 kg/year unit would require 0.85 GWh annually. The 250 t/year unit would require 2.42 MW of power and the 500 t/year unit, which was the largest system, would require 4.85 MW of power, or 42 GWh annually. The cell stack consists of a series of interconnected, circular electrolysis cells, each containing two electrodes located on either side of an advanced patented inorganic ion-exchange membrane. The estimated AC power consumption for one cell stack is 78 kW [32]. An installation of 500 t/year size would require at least 62 of electrolyzer cell stacks.

42.3.2 Sizing of the Generating Station of Electricity

Electricity generation from renewable sources is highly variable through climatic conditions, seasonal changes, and the resource's geographical locations. Since any energy conversion is associated with a loss of energy. First, the technical potential is estimated by applying the electrolyzer operation efficiency and the energy losses of the electrolyzer (by Eq. 42.5), then the sizing of the energy supply unit can be determined by knowing the climatic data:

$$E_{RE} = \frac{E_{H_2}}{\eta_{elec}\eta} \tag{42.5}$$

where E_{RE} [kWh] is the renewable energy production, E_{H_2} [kWh] is the hydrogen energy produced, η_{elec} is the efficiency of the electrolysis process, and η is in additional the efficiency coefficient included to take into account the energy losses in the electrolyzer.

Table 42.2 Energy requirements and number of cells for electrolysis systems

m_{H_2} (t/year)	10	250	500
E_{el} (GWh/year)	0.85	21	42
P_{el} (MW)	0.01	2.42	4.85
Number of cell stacks	2	31	62

42.3.3 Sizing of the Photovoltaic Panels

PV-based hydrogen production is mainly associated with an electrolyzer unit that utilizes DC electricity produced by the PV panels to electrolyze water into hydrogen and oxygen [33]. The design of the photovoltaic system returns to determine the necessary area of the PV generator to adopt sufficient energy supply system to meet the needs of the used electrolyzer at any time. Once known the overall solar irradiation incident on the tilt plan of the generator and the electrical energy needs necessary to the electrolyzer, the corresponding surface of the PV array can be calculated. Assuming a PV system can provide 149 W per square meter of PV array [34], and the PV array has a capacity factor of 37.5 %, an installation of 10 t/year size would require at least 2,731 m^2 of PV cells. A station providing annually 250 tons and 500 tons of hydrogen would require a PV system of approximately 6 and 13 ha, respectively. For this, the surface of the photovoltaic generator is increased by increasing the capacity of hydrogen production.

42.3.4 Sizing of the Solar Chimney Power Plant

The sizing of the SCPP consists in determining the volume of the cylinder with a base the size of the collector and a height equal to that of the chimney (see Fig. 42.2).

Two parameters have much importance in the design of the solar chimney: the meteorological data such as incident solar radiation on a horizontal collector and the ambient temperature, and the specific consumption of the used electrolyzer. Considering efficiencies of the turbine and collector which are, respectively, 0.8 and 0.7 [35], the volume of the solar chimney can be calculated. The following

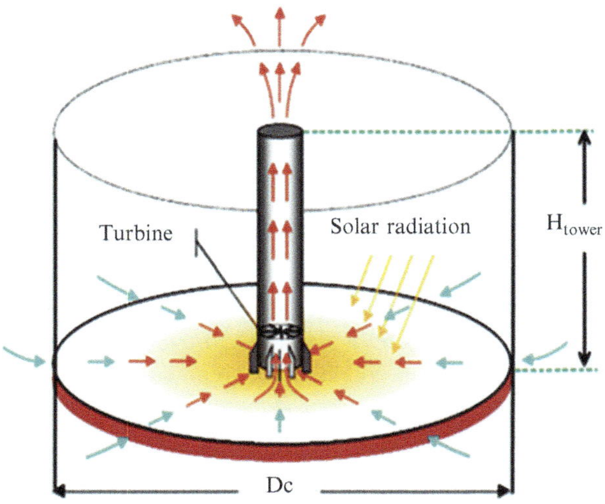

Fig. 42.2 Volume of solar chimney [27]

Table 42.3 Solar tower sizing

m_{H_2} (t/year)	10	250	500
$H_t \times A_{coll}$ (m^3)	4.7×10^7	1.17×10^9	2.35×10^9

Table 42.4 Collector diameter variation with the tower height

	H_t (m)			
	150	350	550	750
$H_t \times A_{coll}$ (m^3)	D_c (m)			
4.7×10^7	632	414	330	283
1.17×10^9	3,161	2,069	1,651	1,413
2.35×10^9	4,470	2,926	2,334	1,999

Table 42.5 Sizing of geothermal power plant

m_{H_2} (t/year)	10	250	500
P_{Geo} (MW$_e$)	0.01	2.42	4.85
m_{CO_2} (kg/s)	5	60	100
h_{Well} (m)	400	3,500	4,000

table illustrates the variation of volume according to three hydrogen production capacities retained (Table 42.3).

For each calculated volume, several configurations can be achieved while varying the height of the chimney to find the corresponding diameter. Table 42.4 summarizes all the geometries. We increment of 200 m height for new iteration.

For a chimney of a volume equal to 4.7×10^7 m^3, we obtain two geometrical proposals of the chimney: the first proposal is a chimney with a height equal to 150 m and a diameter of 632 m; second is for a 350 m height which gives us a diameter of 414 m. The same interpretation for the other volumes of chimney are presented in the table. By increasing the height of the chimney, the diameter of the collector is reduced. However, for the configurations in which the diameter of collector is less than the tower height, therefore we can share the overall volume in a finite number N of volumes and which correspond to N solar chimneys of realizable sizes.

42.3.5 Sizing of the Geothermal Carbon Dioxide System

Geothermal-based hydrogen production via electrolysis offers potential advantages over other sources for a growing hydrogen economy, and it provides reliable energy supply in an environmentally benign manner [36, 37]. The design of a geothermal power plant for hydrogen production is a relatively complex process involving many parameters. In this study, these parameters are the mass flow of the carbon dioxide CO_2 and the well depth. The results, presented in Table 42.5, are obtained by using for each hydrogen production capacity a turbine which a nominal operating point is given for the sizing mass flow.

The results show that for any delivered geothermic power, only one injecting well is able to satisfy the energy demand, by increasing each time the CO_2 flow injected and the depth of well. For a power of 4.85 MWe (equivalent to a production capacity of 500 t/year), we need to use a CO_2 mass flow of 100 kg/s and a depth of 4,000 m.

42.4 Renewable Hydrogen Production Potential in Ouargla Region

Ouargla region (latitude 31°N, longitude 5°24′E) is located in Sahara desert in southern Algeria. It benefits of a hot desert climate, with very hot, dry summers, mild winters, and very little precipitation throughout the year. The air temperature varies between 12.7 °C in January (the coldest month) and 38.05 °C in July (the hottest month), with an annual average value of 25.37 °C [38]. The Ouargla region has an important solar potential; the evaluation of the incident solar energy on an inclination of 0° (horizontal level) and an inclination equal to the latitude of the site (31°) is presented in Fig. 42.3. It is noted that the solar irradiation received per year on a tilted plan is in the range of 2,400 kWh/m^2 and equal to 2,071 kWh/m^2 for a horizontal plan.

The methodology adapted to estimate the hydrogen production potential by water electrolysis of each renewable energy source is thus different due to the varied nature of the source and due to the technological limitations of each source.

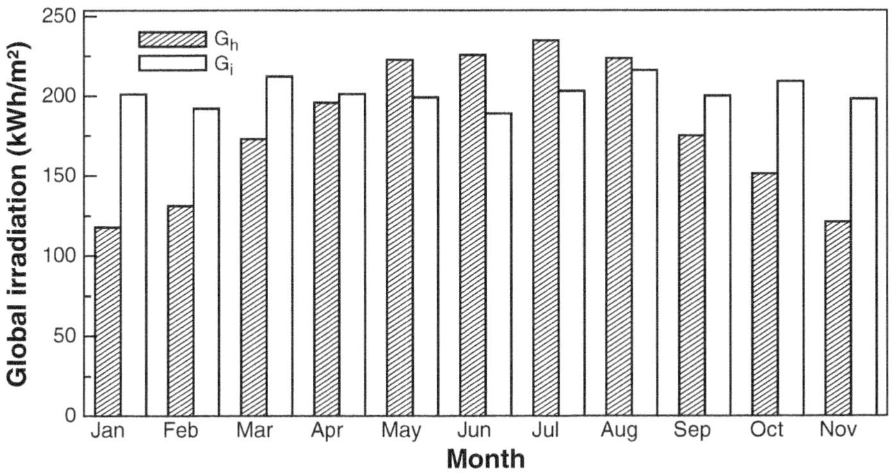

Fig. 42.3 Monthly average variations of the solar irradiation received at horizontal and a tilt angle equal to the latitude (31°) of the location

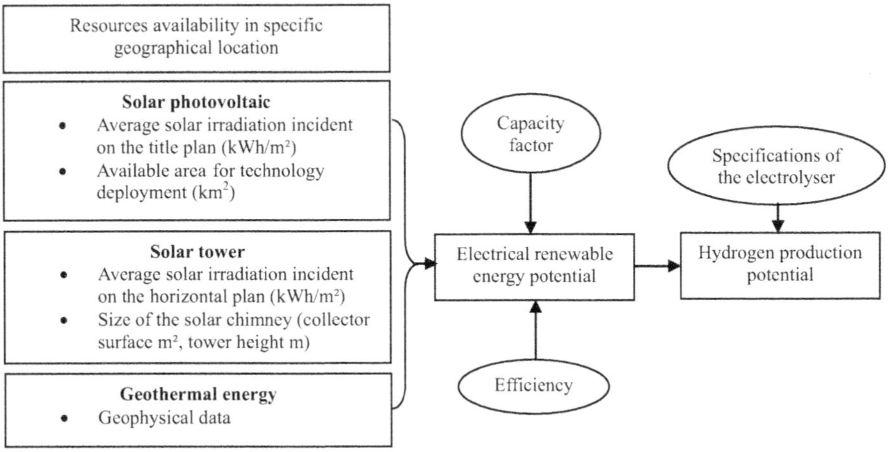

Fig. 42.4 Flowchart representation for the evaluation of the hydrogen production potential

However, the general methodology used for the assessment of the hydrogen production potential from different renewable energy resources is shown in Fig. 42.4.

In order to determine the most favorable renewable electric supply unit, we considered that the quantity of hydrogen produced would be equal to 250 t/year. As regards the electrolyzer, it operates at an efficiency of 68 %, and it has electrolysis process losses of 0.9. For this study, PV module type Sunmodule SW 250 poly of 250W_p is adopted. It is a standard 60-cell panel, using polycrystalline silicon, with an average efficiency of 14.91 % [32]. It is assumed that the PV module area is 69,000 m^2, which are tilted at the Ouargla region latitude, a 31° slope to the panels with a northern orientation. The hydrogen production is analyzed in which the solar chimney power plant is used; the height and the diameter of the chimney are 434 m and 25 m, respectively, and the diameter of the solar collector is 1,858 m. The geophysical data for geothermal power plant is given by [39]. The CO_2 is injected into reservoir/aquifer through one horizontal well up to a depth of 3,500 m in 20 m thick carboniferous sandstone of 10 mD permeability and 15 % porosity (Fig. 42.5).

It can be seen that the electrolytic hydrogen production from geothermal system is constant throughout the year, while that from the PV system has a very low monthly variation due mainly by the fixed tilt PV modules all year-round. SCPP-based hydrogen production has a strong variation of the monthly production; it doubles (20 t) for the period from January to April and September to December, with more 40 t for the summer period. For an exploitation view point, the ideal system for hydrogen production is characterized by a stable monthly production, and this allows us to make economies as for the design of operating system (tank, connection, control, etc.). The electrolyzer subsystem consists of a number of stacks operate at permanent regime chosen by the base of the nominal point of operation, which will allow us greater energy efficiency. Both geothermal and photovoltaic systems are characterized by a stabilization of production year-round.

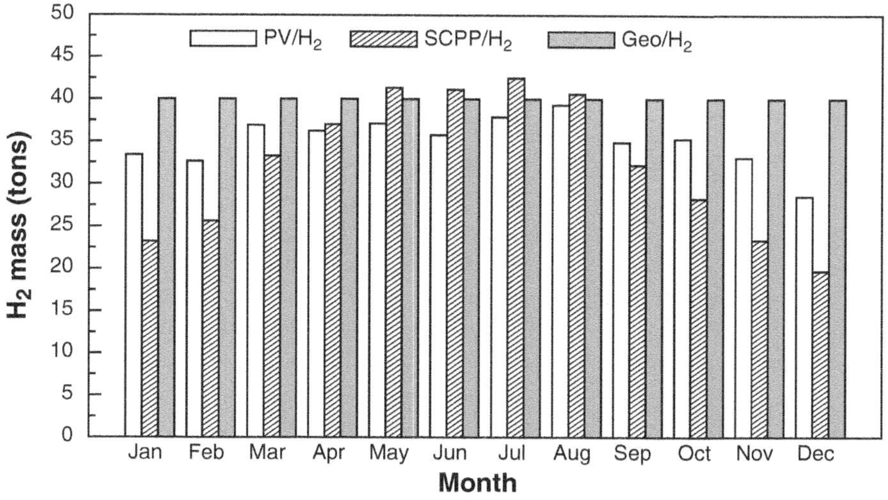

Fig. 42.5 Monthly hydrogen production

42.5 Hydrogen Production Cost

Once the technical hydrogen production potential using renewable electric sources is determined, the economic analysis can be carried out, which includes both costs and benefits of the system. The economic assessment is evaluated using the levelized cost of hydrogen production method (LCA) [40]. The LCA equation is an evaluation of the total life cycle cost and total lifetime hydrogen production. It allows renewable technologies to be compared when different scales of operation, investment, or operating time periods exist.

42.5.1 Estimation of Renewable Electricity Source Investment

In this section, we present the economic modeling of the electrical process equipments, which allow us to estimate the initial investment and operating costs. Overall system costs are greatly influenced by installation and subsystem costs.

The total investment in a solar PV system is the capital cost of the PV power plant, I_0^{PV} [\$] (Eq. 42.6), which include the PV panel costs, C_{PV} [\$] and the cost of the balance of system (BOS), C_{BOS} [\$] plus operation and maintenance (O&M) cost [41].

$$I_0^{\mathrm{PV}} = C_{\mathrm{PV}} + C_{\mathrm{BOS}} \tag{42.6}$$

PV panels have experienced remarkable cost reductions in the last years due to increased supply competition [25]. The annual growth rate for global PV cell production was 111 % from 2009 to 2010, with global cell production reaching 23.9 GW by the end of 2010. Average module prices reached all time lows despite robust demand and tight raw material supplies. In 2010, the average module price ranged from 1.64 $\$/W_p$ to 2.87 $\$/W_p$. In general, the BOS and labor costs represent 50 % of the PV system cost [42]. The operation and maintenance (O&M) of a PV power plant is relatively straightforward because there are few moving parts and no cooling systems. The fixed O&M costs are often expressed as a percentage of investment costs and are reported in the range of 1.5–5 %.

Total capital cost of SCPP includes the initial investments (Eq. 42.7), on the collector, chimney, and a power conversion unit (PCU) plus the O&M costs.

$$I_0^{\mathrm{SCPP}} = C_{\mathrm{Coll}} + C_t + C_{\mathrm{PCU}} + \mathrm{outhers} \tag{42.7}$$

where I_0^{SCPP} [$] is the initial investment of SCPP, C_{Coll} [$] is the collector construction cost, C_t [$] is the chimney construction cost, C_{PCU} [$] is the PCU cost, and outhers is the engineering cost.

The collector roof consists of glass (or plastic) and support structure. The support structure consists of steel columns supporting a support matrix from which the glass is suspended [43]. The cost of the collector includes the costs of materials, transportation costs, as well as construction costs. The orientation price for a collector roof made of single glass amounts to 6.0–9.0 €/m^2 [44]. The cost of the chimney is a function of the volume of material used, the specific material cost, the construction cost, and transport cost. The price of the chimney, which is made of normal or high performance concrete, amounts to 100–125 €/m^2 [45]. The cost analysis of power conversion system (turbine, generator, etc.) is most complex, as we have very little information regarding costs. The PCU cost increases with the decrease of the nominal power of turbine [46]. The cost of PCU is therefore calculated according to Eq. (42.8). The annual operation and maintenance costs amount to 4–5 % of the total investment.

$$C_{\mathrm{PCU}} = 2.513 \times P_{\mathrm{MW}}{}^{-0.259} \tag{42.8}$$

The total capital cost of geothermal power plant with CO_2 sequestration can be divided into the costs of capture and compression, pipeline transport, injection, and storage of carbon dioxide (CO_2) and the cost of the energy conversion system. The cost of transportation includes the capital cost for the pipeline, and the cost of storage includes the cost of the injection system including the injection wells and platforms. The cost of the capture depends on the types of combustion systems that have been proposed for capturing CO_2; the cost range from 15 to 75 $\$/tCO_2$ net captured from a coal- or gas-fired power plant and from 25 to 115 $\$/tCO_2$ net captured from other industrial sources. The injection equipment costs include supply wells, plants, distribution lines, headers, and electrical services. The O&M

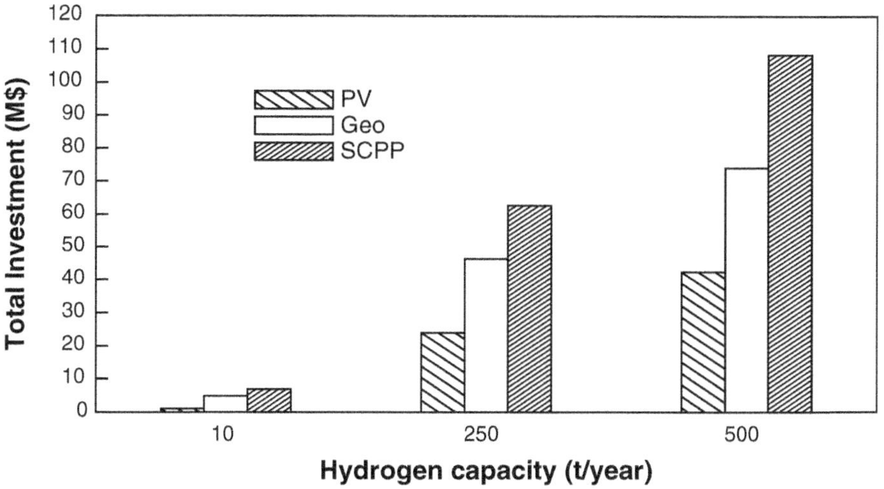

Fig. 42.6 The total capital cost of renewable electricity sources for three different hydrogen production capacities

costs include normal daily expenses and surface and subsurface maintenance costs. The annual operating cost of the sequestration project can be approximated at 5–10 % of the total capital cost [47, 48].

The cost of the energy conversion system is estimated from [30]. Equation (42.9) has been formed to calculate the price per kW_e, which is formed using turbines up to 20 MW. While this data contains only prices for gas turbines, it is assumed that carbon dioxide turbines have similar costs.

$$C_{PCU} = 0.0004\,kW_e + 0.4137 \tag{42.9}$$

The total project investment of a plant producing 10, 250, and 500 t/year was evaluated for the three renewable electrical sources. The cost breakdown for each source is presented in Fig. 42.6.

It can be seen that the investment cost of PV panels is much lower than the geothermal system costs and also lower than SCPP costs due to the remarkable unit cost reductions. However, the total capital cost of the SCPP is much higher than the two other sources for three hydrogen production capacities. The high total investment, a major part of the SCPP cost, is due to high investments in chimney construction cost. In particular, the chimney column share is about 54 % the SCPP cost. Moreover, the high impact of the chimney cost can be noted by the high cost of material construction. The total investment of geothermal system, including all the contributions abovementioned, was estimated to be at 4, 46, and 74 M$ for 10, 250, and 500 hydrogen production capacity.

42.5.2 Electrolysis Cost

The economic modeling of the electrolyzer is presented in many previous works, in which the investment includes three main costs, namely, capital, replacement, and O&M costs. The total investment of an electrolyzer depends on the size of the hydrogen production facility. The electrolyzer capital cost is determined by the required maximum hydrogen production rate, the effective electrolyzer efficiency, and the estimated specific capital cost per kW_e at nominal production [49]:

$$C_{elec} = C_{elec,u} \times \frac{m_{H_2} k_{el,th}}{8,760\,\mathrm{CF}\,\eta_{elec}} \tag{42.10}$$

where C_{elec} [\$] is the capital cost of the electrolyzer, $C_{elec,u}$ [\$/kWe] is the unit cost of the electrolyzer, $k_{el,th}$ [kWh/kg] is the theoretical specific energy required by the electrolyzer, and CF is the capacity factor.

The reference case considers an electrolyzer unit cost of 368 \$/kW_e, which corresponds to target values established by [25]. We assume that the replacement costs and annual operating costs equal to 25 % and 2 % of the first cells investment.

42.5.3 Electrolytic Hydrogen Cost

The cost of electrolytic hydrogen is obviously a key factor that will markedly influence the role a given renewable technology will play for future hydrogen production. The selected technical and economic parameters used for economic evaluation of hydrogen production system by water electrolysis are the following:

- Electricity consumption: 57.85 kWh/kg
- Investment cost: 368 \$/kW_e
- Operation life of the electrolyzer: 7 years
- Project lifetime: 30 years
- Discount rate: 6 %

Figure 42.7 shows the results from the analysis of hydrogen production costs. PV panels have experienced remarkable cost reductions in the last years; then, hydrogen produced from water electrolysis driven with PV electricity is the cheapest (about 7 \$/kgH_2) than hydrogen from other sources. Hydrogen from geothermal carbon dioxide and water electrolysis could be obtained at a price of 15 \$/kgH_2 for a small hydrogen capacity production and at a price of 5 \$/kgH_2 for a large capacity. The hydrogen-based solar chimney production cost will decrease to 12 \$/kgH_2 as the system capacity increases from 10 t/year to 500 t/year.

The hydrogen production cost breakdown for a medium hydrogen capacity production (250 t/year) shows that the electric energy cost accounts for 90 % of the total project investment of both SCPP and geothermal-based hydrogen station,

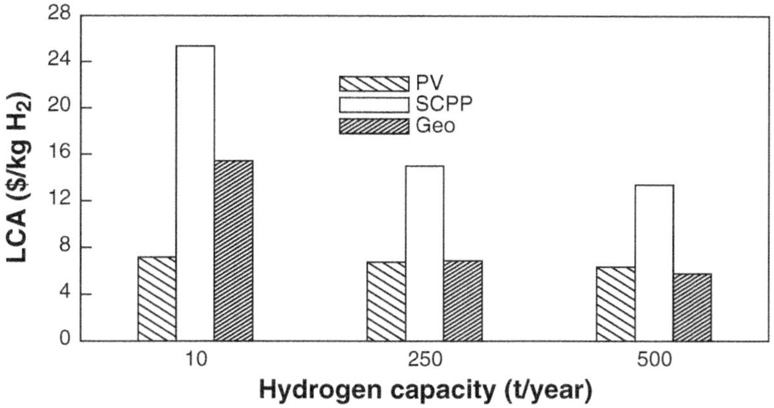

Fig. 42.7 Levelized cost of hydrogen production

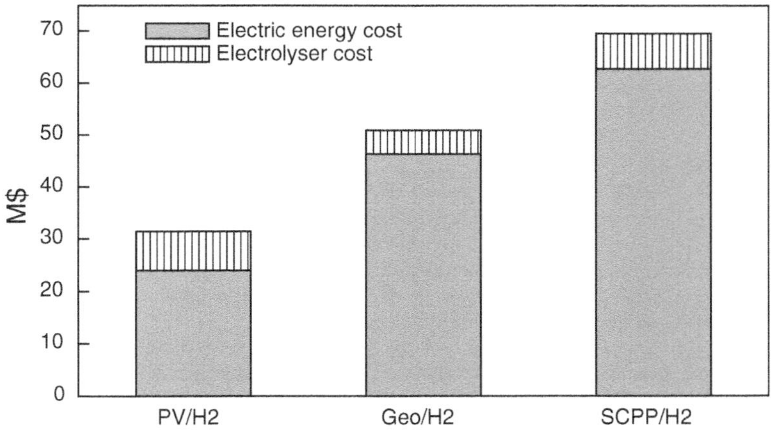

Fig. 42.8 Hydrogen production cost breakdown

while the capital cost of the electrolyzer represent around 10 % of the total cost (Fig. 42.8). However, for the PV-based hydrogen station, the electric cost contributes for more than 70 % of the hydrogen production cost. From these results, the strong impact of the electricity production costs is confirmed, followed by the impact of electrolyzer expenses and investment.

42.6 Conclusion

Technical and economic analysis of hydrogen production by water electrolysis using three different renewable sources was conducted in this paper. The economic feasibility of renewable hydrogen projects is increasingly being evaluated using the levelized cost of hydrogen (LCA) in order to compare the renewable hydrogen production cost of the different renewable technologies. The results of the evaluation of the technical hydrogen production potential showed a very high potential of hydrogen in the Ouargla region, where the application has been considered. The results of a cost analysis have indicated that the electric costs contribute significantly to the total project investment and have a major impact on the produced hydrogen price via water electrolysis process. In conclusion, an even stronger development of renewable energies would be required in the future to satisfy both electricity and hydrogen production and to contribute an opportunity to support economic growth and energy transition in Algeria.

References

1. International Energy Agency (IEA) (2013) CO_2 emissions from fuel combustion highlights
2. Shafiee S, Topal E (2009) When will fossil fuel reserves be diminished? Energy Policy 37:181–189
3. Barbir F (2009) Transition to renewable energy systems with hydrogen as an energy carrier. Energy 34:308–312
4. Abbasi T, Abbasi SA (2011) 'Renewable' hydrogen: prospects and challenges. Renew Sustain Energy Rev 15:3034–3040
5. Miltner A, Wukovits W, Pröll T, Friedl A (2010) Renewable hydrogen production: a technical evaluation based on process simulation. J Clean Prod 18:S51–S62
6. Kleijn R, van der Voet E (2010) Resource constraints in a hydrogen economy based on renewable energy sources: an exploration. Renew Sustain Energy Rev 14:2784–2795
7. Bicakova O, Straka P (2012) Review: production of hydrogen from renewable resources and its effectiveness. Int J Hydrog Energy 37:11563–11578
8. Dincer I (2012) Green methods for hydrogen production. Int J Hydrog Energy 37:1954–1971
9. Caliskan H, Dincer I, Hepbasli A (2013) Exergoeconomic and environmental impact analyses of a renewable energy based hydrogen production system. Int J Hydrog Energy 38:6104–6111
10. Escobar B, Hernandez J, Barbosa R, Verde-Gomez Y (2013) Analytical model as a tool for the sizing of a hydrogen production system based on renewable energy: The Mexican Caribbean as a case of study. Int J Hydrog Energy 38:12562–12569
11. Acar C, Dincer I (2014) Review: comparative assessment of hydrogen production methods from renewable and non-renewable sources. Int J Hydrog Energy 39:1–12
12. Gahleitner G (2013) Review: hydrogen from renewable electricity: an international review of power-to-gas pilot plants for stationary applications. Int J Hydrog Energy 38:2039–2061
13. Sonatrach (2011) Annual report. www.sonatrach.com
14. Boudghene Stambouli A, Khiat Z, Flazi S, Kitamura Y (2012) A review on the renewable energy development in Algeria: current perspective, energy scenario and sustainability issues. Renew Sustain Energy Rev 16:4445–4460
15. Mahmah B, Harouadi F, Benmoussa H, Chader S, Belhamel M, M'Raoui A, Abdeladim K, Cherigui AN, Etievant C (2009) MedHySol: future federator project of massive production of solar hydrogen. Int J Hydrog Energy 34:4922–4933
16. Boudries R, Dizene R (2008) Potentialities of hydrogen production in Algeria. Int J Hydrog Energy 33:4476–4487
17. OPEC (2013) Annual statistical bulletin
18. Boudghene Stambouli A (2011) Promotion of renewable energies in Algeria: strategies and perspectives. Renew Sustain Energy Rev 15:1169–1181

19. Boudia SM, Benmansour A, Ghellai N, Benmedjahed M, Tabet Hellal MA (2013) Temporal assessment of wind energy resource at four locations in Algerian Sahara. Energ Convers Manag 76:654–664
20. Diaf S, Notton G (2013) Evaluation of electricity generation and energy cost of wind energy conversion systems in southern Algeria. Renew Sustain Energy Rev 23:379–390
21. Kedaid FZ (2007) Database on the geothermal resources of Algeria. Geothermics 36:265–275
22. Saibi H (2009) Geothermal resources in Algeria. Renew Sustain Energy Rev 13:2544–2552
23. Ministry of Energy and Mines (MEM). www.mem-algeria.org
24. Renewable Energy and Energy Efficiency Program (2011) Ministry of Energy and Mines
25. Department of Energy (2011) 2010 solar technologies market report. National Renewable Energy Laboratory
26. Zhou X, Wang F, Ochieng RM (2010) A review of solar chimney power technology. Renew Sustain Energy Rev 14:2315–2338
27. Schlaich J, Bergermann R, Schiel W, Weinrebe G (2005) Design of commercial solar updraft tower systems-utilization of solar induced convective flows for power generation. J Sol Energ Eng 127:117–24
28. Gallup DL (2009) Production engineering in geothermal technology: a review. Geothermics 38:326–334
29. Randolph JB, Saar MO (2011) Coupling carbon dioxide sequestration with geothermal energy capture in naturally permeable, porous geologic formations: implications for CO_2 sequestration. Energy Procedia 4:2206–2213
30. Janse D (2010) Combining geothermal energy with CO_2 storage. Master thesis report, Utrecht University
31. Clarke RE, Giddey S, Ciacchi FT, Badwal SPS, Paulb B, Andrews J (2009) Direct coupling of an electrolyser to a solar PV system for generating hydrogen. Int J Hydrog Energy 34:2531–2542
32. HySTAT® Hydrogen Generators. www.hydrogenics.com
33. Joshi AS, Dincer I, Reddy BV (2011) Solar hydrogen production: a comparative performance assessment. Int J Hydrog Energy 36:11246–11257
34. SolarWorld. SW 250 poly/Version 2.0 and 2.5 Frame
35. Negrou B, Settou N, Chennouf N, Dokkar B (2011) Valuation and development of the solar hydrogen production. Int J Hydrog Energy 36:4110–4116
36. Belta MT, Dincer I, Hepbasli A (2009) Potential methods for geothermal-based hydrogen production. Int J Hydrogen Energy. doi:10.1016/j.ijhydene.2009.09.040
37. AlZaharani AA, Dincer I, Naterer GF (2013) Performance evaluation of a geothermal based integrated system for power, hydrogen and heat generation. Int J Hydrog Energy 38:14505–14511
38. Office National de la Météorologie (2011)
39. Kechiched R, Haddane A, Foufou A, Ameur Ziameche O (2013) Caracterisation petrophysique des reservoirs à l'aide de l'analyse statistique: application sur les données de la Zone 17-Champ de Hassi Messaoud-Algérie. The International conference on electronics & oil: from theory to applications, Ouargla, Algeria
40. Sigurvinsson J, Mansilla C, Lovera P, Werkoff F (2007) Can high temperature steam electrolysis function with geothermal heat? Int J Hydrog Energy 32:1174–1182
41. Boudries R, Dizene R (2011) Prospects of solar hydrogen production in the Adrar region. Renew Energy 36:2872–2877
42. Branker K, Pathak MJM, Pearce JM (2011) A review of solar photovoltaic levelized cost of electricity. Renew Sustain Energy Rev 15:4470–4482
43. Zhou X, Yang J, Wang F, Xiao B (2009) Economic analysis of power generation from floating solar chimney power plant. Renew Sustain Energy Rev 13:736–749
44. Nizetic S, Ninic N, Klarin B (2008) Analysis and feasibility of implementing solar chimney power plants in the Mediterranean region. Energy 33:1680–1690

45. Fluri TP, Pretorius JP, Van Dyk C, Von Backstrom TW, Kroger DG, Van Zijl GPAG (2009) Cost analysis of solar chimney power plants. Solar Energy 83:246–256
46. Le Sellin M (2006) Etude d'opportunité pour une filière de cheminée solaire à la Réunion. Ecole Centrale Marseille
47. McCollum DL, Ogden JM (2006) Techno-economic models for carbon dioxide compression, transport, and storage & correlations for estimating carbon dioxide density and viscosity. University of California
48. Shafeen A, Croiset E, Douglas PL, Chatzis I (2004) CO_2 sequestration in Ontario, Canada. Part II: cost estimation. Energ Convers Manag 45:3207–3217
49. Prince-Richard S (1996) A techno-economic analysis of decentralized electrolytic hydrogen production for fuel cell vehicles. Master thesis

Chapter 43
Electrochemical Evaluation of High Rate $Cu/LiCr_{2/3}V_{1/6}Mn_{1/6}O_2$ Nanocomposite Structures as Cathode Electrodes for Li-Ion Batteries

Şeyma Özcan, Aslıhan Erdaş, Deniz Nalci, M.Oğuz Güler, and Hatem Akbulut

Abstract In this study, a strategy used to design high-capacity (>200 mAh g^{-1}), Cu-coated $LiMO_2$ (M = Mn, Cr, V) electrodes for lithium-ion batteries is discussed. The advantages of the $LiCr_{2/3}V_{1/6}Mn_{1/6}O_2$ component and its influence on the structural stability and electrochemical properties of these layered electrodes are highlighted. In this study, $LiCr_{2/3}V_{1/6}Mn_{1/6}O_2$ cathode active materials were prepared via facile sol–gel method and were then coated with Cu via electroless deposition techniques. Structural, chemical and electrochemical properties of $LiCr_{2/3}V_{1/6}Mn_{1/6}O_2$ electrodes are considered. The phase purity and the structural characterisation of the synthesised cathode active electrodes were analysed by X-ray powder diffraction (XRD) using a Rigaku D/Max 2200 system and Cu Kα radiation. Surface morphology of the synthesised cathode electrodes was analysed via SEM tests. The electrochemical tests have shown that higher initial discharge capacities are obtained after first cycle. A total discharge capacity of 206 mAh g^{-1} and 194 mAh g^{-1} was obtained after first discharge cycle for $LiCr_{2/3}V_{1/6}Mn_{1/6}O_2$ and $Cu/LiCr_{2/3}V_{1/6}Mn_{1/6}O_2$ samples, respectively. Best results were obtained for $Cu/LiCr_{2/3}V_{1/6}Mn_{1/6}O_2$ electrodes, and a total capacity retention of 88 % was obtained after 50 cycles. From the experimental data, it is concluded that both Cr and V substitution and surface modifying with copper is an effective way to improve the cyclability of lithium-ion batteries for commercial applications.

Keywords Li-ion batteries • $LiCr_{2/3}V_{1/6}Mn_{1/6}O_2$ • Sol–gel • Electroless Cu coating

---

Ş. Özcan (✉) • A. Erdaş • D. Nalci • M.O. Güler • H. Akbulut
Department of Metallurgical and Materials Engineering, Faculty of Engineering, Sakarya University, 54187 Serdivan, Sakarya, Turkey
e-mail: ozcanseyma19@gmail.com

I. Dincer et al. (eds.), *Progress in Clean Energy, Volume 2*,
DOI 10.1007/978-3-319-17031-2_43

43.1 Introduction

To globally solve the problem of an environmentally clean and secure distribution of energy is a major task for scientist worldwide [1, 2]. The use of clean energy will have a direct positive influence on, for example, the emission of greenhouse gases and our future climate. Today the rechargeable battery with the highest energy density is the Li-ion or Li-polymer battery. Commercially this battery is used for mobile phones, laptops and camcorders. The development of new batteries of this type has to go towards larger batteries for hybrid vehicles or towards further miniaturisation for smaller electronic devices where "battery on a chip" is one solution [3–6].

Both these routes will, however, require basic materials research within chemistry and physics; new materials must be synthesised (some materials used today are too expensive for upscaling), and the complex chemistry occurring both in bulk electrode materials and at the interface between electrolyte and electrode must be better understood. Among different electrode materials, layered transition metal oxides (LMO) are the most successful cathode material in application at present time, but their further development is severely hindered due to the intrinsic safety limitation and the cost reported by several research groups such as Ramesh et al. [7], Jung-Min et al. [8], Pang et al. [9], Young-Il et al. [10] and Hongmei et al. [11]. After charging/discharging processes, Li_xCoO_2 decompose at elevated temperatures (almost at 60 °C) to form Li_yCoO_2, Co_3O_4, CoO and O_2 (disproportionation). In addition, the solvents in the electrolytes will also oxidise Li_xCoO_2 structure at elevated temperatures. A recent study by Yi et al. [12] has shown that doping with ions such as nickel has significantly improved the cathode activity during cycling processes. The redox activity takes place mainly in the nickel centre, and the $Ni^{2+}/Ni^{3+}/Ni^{4+}$ are believed to be involved during the lithium extraction/insertion reactions.

Many studies were reported that the amount of doping element drastically changes the rate capability and diffusion coefficient. Moreover, the cathode electrodes prepared particularly with nickel or cobalt requires relatively higher temperatures (more than 900 °C) and a long annealing time as stated by Qun et al. [13]. The aim of this study is to prepare $LiCr_{2/3}V_{1/6}Mn_{1/6}O_2$ type of rhombohedral layered oxide which will be suitable as a positive electrode material for Li-ion batteries. In this study, $LiCr_{2/3}V_{1/6}Mn_{1/6}O_2$ layered oxide structure was prepared via facile sol–gel method. Since the aim is to increase the electrochemical efficiencies, the surfaces of the electrodes were coated with copper via electroless deposition techniques in order to increase the electronic conductivities.

43.2 Experimental Details

43.2.1 Synthesis of $LiCr_{2/3}V_{1/6}Mn_{1/6}O_2$ Cathode Electrodes

$LiCr_{2/3}V_{1/6}Mn_{1/6}O_2$ samples were prepared using appropriate quantities of lithium acetate dehydrate, $LiCH_3COO{\cdot}2H_2O$, $Cr_3(OH)_2(OOCCH_3)_7$ (Alfa Aesar, 99.9 %), manganese acetate tetrahydrate, $Mn(CH_3COO)_2{\cdot}4H_2O$ (Alfa Aesar, 99.9 %) and vanadium (IV) oxide bis(2,4 pentanedionate, $C_{10}H_{14}O_5V$ (Alfa Aesar, 99.9 %)). Saturated solutions of these materials were prepared separately in distilled water and mixed with saturated solution of citric acid. The pH of the resultant solution mixture was maintained at about 9.0 by adding ammonium hydroxide solution. Viscous gel was obtained by stirring the solution maintained at a temperature of 80–90 °C. The gel was heated in air to 450 °C for 5 h inside a furnace and then cooled in it. The precursor thus obtained was ground and calcined at 850 °C for 18 h in air followed by cooling to room temperature.

43.2.2 Electroless Cu Coating of $LiCr_{2/3}V_{1/6}Mn_{1/6}O_2$ Cathode Electrodes

$LiCr_{2/3}V_{1/6}Mn_{1/6}O_2$ powders were used as the cores of the composite powders in the electroless plating. Before starting the deposition process, the surfaces of the cathode active powders should have to be pretreated as follows: (1) the surfaces of the as-synthesised powders were cleaned by immersing powders into 1 M sodium hydroxide (Merck, Titripur) solution for 10 min. (2) The cleaned $LiCr_{2/3}V_{1/6}Mn_{1/6}O_2$ powders were then immersed into boiling hydrochloric acid (Merck, 1 N) for 15 min to coarsen their surfaces. The coarsened surfaces will also help to improve the adhesion between the $LiCr_{2/3}V_{1/6}Mn_{1/6}O_2$-based powders and copper particles. (3) Since the surfaces of $LiCr_{2/3}V_{1/6}Mn_{1/6}O_2$ powders were nonconductive, an activation process is needed in order to improve the catalytic properties. One-step sensitisation and activation process was performed in this study. The etched particles of $LiCr_{2/3}V_{1/6}Mn_{1/6}O_2$ were immersed in a palladium activation solution consisting of palladium dichloride (Merck, $PdCl_2$) (0.3 g/L), stannous chloride (Merck, $SnCl_2{\cdot}2H_2O$, >99.9 %) (16 g/L), sodium chloride (Merck, NaCl, >99.99 %) (150 g/L) and hydrochloric acid HCl (Merck, 1 N) (60 mL/L) and stirred strongly at room temperature for 15–20 min. After each pretreatment process, particles of $LiCr_{2/3}V_{1/6}Mn_{1/6}O_2$ were repeatedly washed by using bidistilled water until pH is reached to 7. The pretreated powders were then dried after the activation step in an oven at 80 °C for 12 h.

43.2.3 Characterisation

XRD patterns of the calcined powders were obtained by X-ray diffractometer (XRD, Rigaku D/Max 2200) using Copper Kα radiation of wavelength 1. The scanning was made over $2\theta = 10–80°$, with a step size and dwell time of 0.1 and 5 s, respectively. Microstructural observations of the powder were made using SEM (Jeol 6060 LV). CR2016 coin-type cells were assembled in an Ar-filled glove box for the electrochemical measurements. The active materials, conductive graphite, and PVDF with a weight ratio of 80:10:10 were mixed and pressed into a film onto aluminium foil followed by drying at 140 °C for 4 h. The anode was lithium foil and separated by a Celgard 2400 microporous membrane. The applied electrolyte is 1.0 M $LiPF_6$ dissolved in EC/DMC (1:1 in volume). The charging/discharging measurements were carried out on a battery cycler by using the fabricated 2016 coin cells, and the electrochemistry impedance spectra (EIS) were recorded on a Gamry Reference 3000 electrochemical workstation by using a three-electrode cell with active material film as the cathode and lithium metal as both anode and reference electrodes. Cyclic voltammogram (CV) tests were carried on a MTI BST8A electrochemical workstation by CR2016 button cells.

43.3 Results and Discussions

43.3.1 Structural and Morphological Observations

XRD patterns of all the three powdered samples are shown in Fig. 43.1, which shows that the XRD patterns of $LiCr_{2/3}V_{1/6}Mn_{1/6}O_2$ are quite similar to that of single phase

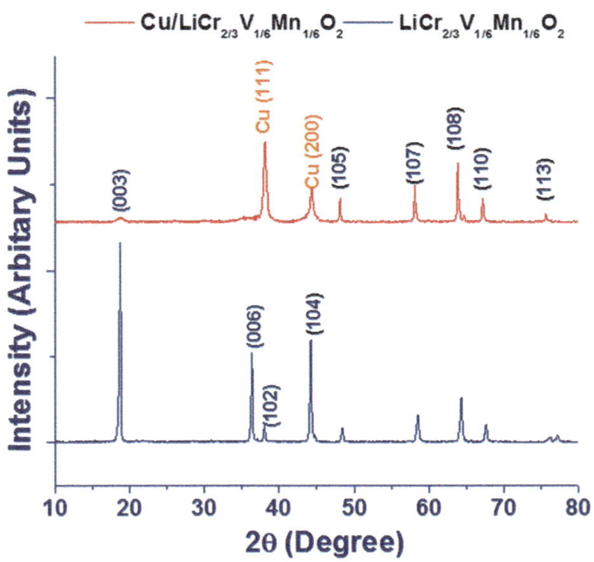

Fig. 43.1 XRD patterns of the (**a**) pristine $LiCr_{2/3}V_{1/6}Mn_{1/6}O_2$ and (**b**) Cu-coated $LiCr_{2/3}V_{1/6}Mn_{1/6}O_2$ samples

$LiCoO_2$ and $LiNiO_2$. The materials have single phase layered rhombohedral structure with space group of *R3m*. The obtained peaks also reveal that the synthesised material consists of close packed oxygen planes in ABCABC stacking sequence. The results also showed that lithium and transition metal ions were octahedrally coordinated with oxygen and formed alternative layers. No any other impurity phases were detected by XRD analysis, and the narrowness of the diffraction lines indicates good crystallinity. Rietveld refinement analysis was also performed by using the Jade package and showed that different type of cations was homogeneously distributed over the layered structure. The hexagonal unit cell parameters a and c were also calculated and found to be 2.813 Å and 14.218 Å, which are very similar to the values reported by Ramesh et al. [7] and Jung-Min et al. [8].

The results have shown that well-ordered layered structure of the $LiCr_{2/3}V_{1/6}Mn_{1/6}O_2$ is produced. Cu(111) and Cu(200) planes were obtained after electroless deposition process showing that $Cu/LiCr_{2/3}V_{1/6}Mn_{1/6}O_2$ has been successfully synthesized. The intensities of the $LiCr_{2/3}V_{1/6}Mn_{1/6}O_2$ peaks were reduced after coating the surface of the spinels uniformly with copper which is an expected result. The calculated lattice parameters show that there is almost no change in the lattice parameter for all samples which indicated that copper ions were not incorporated into the spinel structure but were just presented on the surfaces of $LiCr_{2/3}V_{1/6}Mn_{1/6}O_2$ samples. From the literature studies of Ouyang et al. [14] and Arumugam [15], it is well known that introduction of copper ions into spinel or layered structured materials will lead to significant change of the lattice parameter. It can also be concluded from the XRD spectra; copper crystals largely grow in two preferential orientation (111) planes. Copper growing preferentially over the (111) plane during the electroless deposition process plane will also inhibit the stress-induced voiding as also stated by Jayaprakash et al. [16]. In addition, the (111)-textured Cu film has higher resistance to electro-migration which suppresses grain boundary and interfacial diffusion of metal atoms as also indicated by Thirunakaran et al. [17].

Surface morphologies of the $LiCr_{2/3}V_{1/6}Mn_{1/6}O_2$ and $Cu/LiCr_{2/3}V_{1/6}Mn_{1/6}O_2$ powders calcined at 850 °C are presented in Fig. 43.2. A typical SEM image in Fig. 43.2a shows that the sample has agglomerated morphology and a nonuniform particle size distribution. Figure 43.2b shows the SEM images of Cu-coated as-synthesised $LiCr_{2/3}V_{1/6}Mn_{1/6}O_2$ powders with EDS dot map analysis. No free copper particles and uncoated surface of pristine $LiCr_{2/3}V_{1/6}Mn_{1/6}O_2$ particles are found in powders. In addition, EDS dot map analysis also confirms that surfaces of the $LiCr_{2/3}V_{1/6}Mn_{1/6}O_2$ powders were coated uniformly with Cu.

43.3.2 Electrochemical Results

Galvanostatic charge/discharge curves of pristine $LiCr_{2/3}V_{1/6}Mn_{1/6}O_2$ and $Cu/LiCr_{2/3}V_{1/6}Mn_{1/6}O_2$ as positive electrodes for Li-ion batteries are evaluated and given in Fig. 43.3. The discharge capacity for the first cycle to the next cycle

Fig. 43.2 SEM images of the (**a**) $LiCr_{2/3}V_{1/6}Mn_{1/6}O_2$ and (**b**) Cu-coated $LiCr_{2/3}V_{1/6}Mn_{1/6}O_2$ with EDS dot map analysis of samples

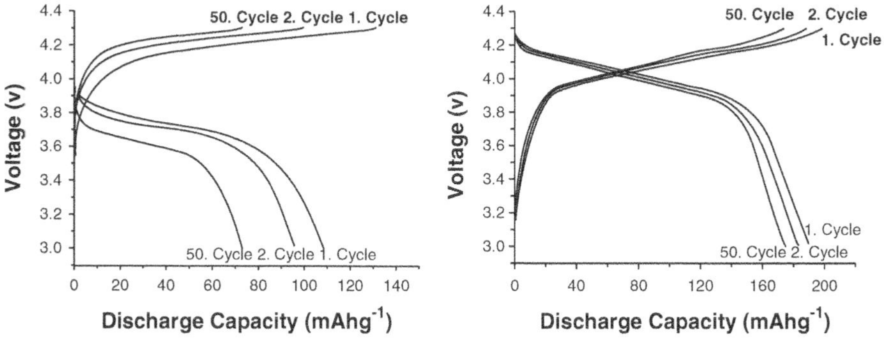

Fig. 43.3 Galvanostatic charge/discharge curves of the (**a**) $LiCr_{2/3}V_{1/6}Mn_{1/6}O_2$ and (**b**) $Cu/LiCr_{2/3}V_{1/6}Mn_{1/6}O_2$ samples

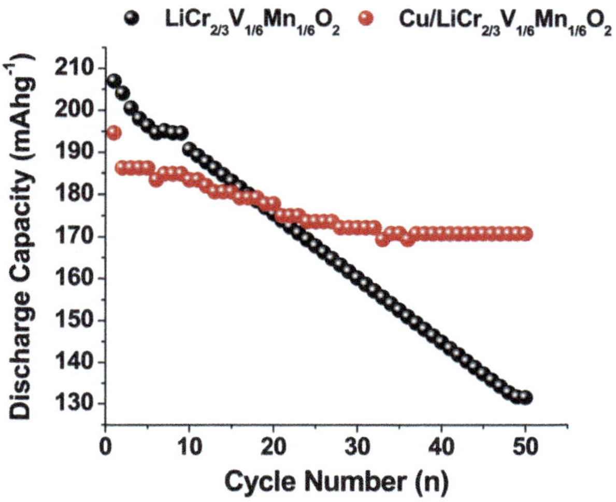

Fig. 43.4 Galvanostatic charge/discharge vs. cycling number curves of the $LiCr_{2/3}V_{1/6}Mn_{1/6}O_2$ and Cu-coated $LiCr_{2/3}V_{1/6}Mn_{1/6}O_2$ samples

decreases from 206.13 mAh g^{-1} to 204.10 mAh g^{-1} for Cu/$LiCr_{2/3}V_{1/6}Mn_{1/6}O_2$ and from 194.62 mAh g^{-1} to 186.22 mAh g^{-1}, which indicates that during the first charge of sample, a proportion of Li inserted to anode may not be reinserted back to cathode during the subsequent discharge because of irreversible internal chemical reactions. It is well known that irreversible capacity of $LiCr_{2/3}V_{1/6}Mn_{1/6}O_2$ is required to be reduced for a full scale commercial exploitation of such cathode material. The irreversible capacities of the $LiCr_{2/3}V_{1/6}Mn_{1/6}O_2$ and Cu/$LiCr_{2/3}V_{1/6}Mn_{1/6}O_2$ are 75 mAh g^{-1} and 24 mAh g^{-1} at a constant 1C charging and discharging conditions, respectively. All these results show that very stable cycling regime and remarkable cyclic performances of $LiCr_{2/3}V_{1/6}Mn_{1/6}O_2$ are reached.

Figure 43.4 shows the evolution of the capacity vs. the cycle number during the first 50 cycles between 3.0 and 4.3 V at 1C rate. Except for the first cycle, the capacity retention is excellent even for the faster rate for both samples. As also can be concluded from Fig. 43.1, the capacity retention is equal to 96.2 % for the first cycle; it becomes close to 97 % in the following cycles and the total capacity retention of 69.1 % is obtained after 50 cycles. However, the capacity retention for the Cu-coated sample was 99.1 % for the first cycle, and it becomes close to 99.5 % in the following cycles. Total capacity retention of 88 % after 50 cycles is obtained indicating that the Cu coating has significantly improved the electrochemical properties. These cycling performances are probably related to the weak amount of the Cr and V in the lithium plane. This enhances the lithium diffusion in the interslab plane even at relatively high regime.

The rate capability was tested by varying the discharge rate between 4.3 and 3 V, as shown in Fig. 43.5. The capacity fading is greater at 10 C discharge than at 1 C discharge. It is clear that the cycle performance and high rate discharge ability of the $LiCr_{2/3}V_{1/6}Mn_{1/6}O_2$ sample are poor. The maximum discharge capacity and cycle stability of cathode materials can be greatly improved by electroless Cu coating. The increment in the rate capability may be attributed to the Cu coating suppressing the

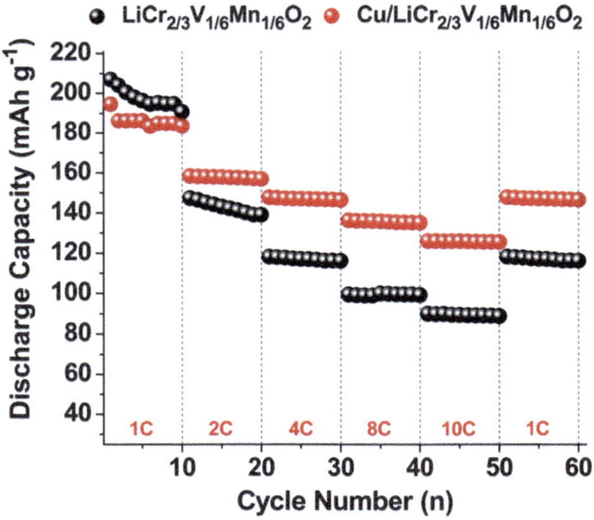

Fig. 43.5 Rate capability of the (**a**) pristine $LiCr_{2/3}V_{1/6}Mn_{1/6}O_2$ and (**b**) Cu-coated $LiCr_{2/3}V_{1/6}Mn_{1/6}O_2$ samples

Jahn–Teller distortion during the intercalation processes as also stated by Kanamori et al. [18]. Although the reversible capacity decreases slightly with increasing the discharge rate, the obtained discharge capacities are very good even for faster regimes. The Cu-coated electrode delivers a discharge capacity of 183, 156, 146, 132 and 125 mAh g^{-1}, when the battery is discharged at 1C, 2C, 4C, 8C and 10C, respectively. Also, it was observed that there is a shift of the discharge curve with the increase of the current cycling, as a result of the increase in the internal resistance and a high ohmic drop which cause the lowering of the operating potential of the electrochemical cell as also experienced by Dahn et al. [19]. These good rate performances are certainly related to the high homogeneity, the high purity and the higher crystalline degree of the studied electrode material as stated by Dahbi et al. [20].

Figure 43.6 shows the Nyquist plots of pristine and $Cu/LiCr_{2/3}V_{1/6}Mn_{1/6}O_2$ composites. The EIS data were collected with a two-electrode coin cell after activation (i.e. after subjecting the coin cell to one charge–discharge cycle). The initial activation was aimed to suppress the Li–electrolyte interfacial resistance that arises from the passivating film formed on lithium metal in contact with the electrolyte. EIS results were then fitted with the equivalent circuits inset in Fig. 43.6. On this circuit, Rs is the resistance associated with the electrolyte and cell component, Rint is the interface resistance of any film formation of the anode surface (first high frequency semicircle), Rct is the charge-transfer resistance of the electrode reaction with lithium ions (middle frequency semicircle) and Wdif is the resistance of the lithium-ion diffusion to the electrode (low frequency semicircle). It is noted that the AC impedance response of the cell forms a broad semicircle and a line to the real axis in the lowest frequency range. The semicircle in the high frequency is mainly related to the complex reaction process at the electrolyte/cathode interface including resistance of SEI film formed on the surface of $LiCr_{2/3}V_{1/6}Mn_{1/6}O_2$ particles, particle-to-particle contact resistance, charge-transfer resistance, and corresponding capacitances. The

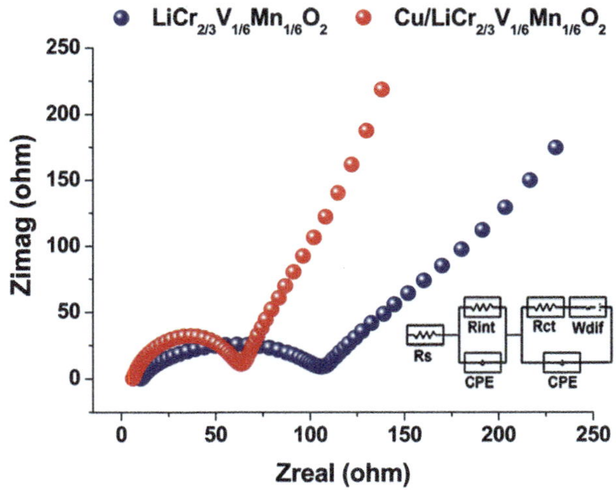

Fig. 43.6 EIS spectra of $LiCr_{2/3}V_{1/6}Mn_{1/6}O_2$ and $Cu/LiCr_{2/3}V_{1/6}Mn_{1/6}O_2$ composite electrodes with the frequency range of 10^{-2} to 10^6 Hz

inclined line in the lower frequency is attributed to the Warburg impedance, which is associated with lithium-ion diffusion of $LiCr_{2/3}V_{1/6}Mn_{1/6}O_2$ electrode. As can be seen from Fig. 43.6, the resistances are 107 Ω and 63 Ω for $LiCr_{2/3}V_{1/6}Mn_{1/6}O_2$ and $Cu/LiCr_{2/3}V_{1/6}Mn_{1/6}O_2$ electrodes, respectively. The results are clearly indicating that the oxidation of electrolyte and degradation of electrolyte/electrode interface are suppressed effectively by Cu coating, which leads to decreased Rct value and improved electrochemical performance reported by several research groups such as Kim et al. [21], Yi et al. [22], Ke et al. [23], Mohamed et al. [24], Yumei et al. [25] and Rajive et al. [26].

43.4 Conclusions

$LiCr_{2/3}V_{1/6}Mn_{1/6}O_2$ with excellent homogeneity and crystallinity has been synthesised by the sol–gel method. The optimum condition for the synthesis is calcination at 850 °C for 1 h. Structural characterisation by method showed that the single phase $LiCoO_2$ and $LiNiO_2$ is obtained. Galvanostatic charge–discharge cycling at 1C rate revealed better capacity retention and cyclability of the studied $LiCr_{2/3}V_{1/6}Mn_{1/6}O_2$ in comparison with the Cu-coated $LiCr_{2/3}V_{1/6}Mn_{1/6}O_2$ samples. The results have shown that $LiCr_{2/3}V_{1/6}Mn_{1/6}O_2$-based electrodes can easily be prepared via facile sol–gel methods and the electrochemical efficiencies can be enhanced via electroless copper coating techniques.

Acknowledgement This work is supported by the Scientific and Technological Research Council of Turkey (TUBITAK) under the contract number 111 M021—Improving the Capacity of Li-Ion Batteries by Using New Semi-conducting Metal Oxide-Based Anodes. The authors thank the TUBITAK MAG workers for their financial support.

References

1. Zu-Fei H, Han-Zhuang Z, Chun-Zhong W, Deng-Pan W, Xing M, Xing M, Gang C (2009) First-principles investigation on extraction of lithium ion from monoclinic $LiMnO_2$. Solid State Sci 11:271–274
2. Zu-Fei H, Chun-Zhong W, Xing M, Yuan S, Gang C (2008) Competition between ferromagnetic and antiferromagnetic interaction in monoclinic $LiMnO_2$. Comput Mater Sci 42:504–509
3. Young-Il J, Douglas W, Douglas Moorehead W, Yet-Ming C (2002) Synthesis of the monoclinic and orthorhombic phases of $LiMnO_2$ in oxidizing atmosphere. Solid State Ion 149:201–207
4. Zu-Fei H, Chun-Zhong W, Xing M, Deng-Pan W, Gang C (2006) Effects of Al-doping on the stabilization of monoclinic $LiMnO_2$. J Solid State Chem 179:1602–1609
5. Young-Il J, Yet-Ming C (2000) Stability of the monoclinic and orthorhombic phases of $LiMnO_2$ with temperature, oxygen partial pressure, and Al doping. Solid State Ion 130:53–59
6. Zu-Fei H, Xing M, Chun-Zhong W, Yuan S, Gang C (2006) First-principles calculations on the Jahn–Teller distortion in layered $LiMnO_2$. J Power Sources 158:1394–1400
7. Ramesh C, Hirofumi K, Yang-Soo K, Yoshitaka M, Kenta O (2001) Synthesis of layered-type hydrous manganese oxides from monoclinic-type $LiMnO_2$. J Solid State Chem 160:69–76
8. Jung-Min K, Hoon-Taek C (2003) Electrochemical characteristics of orthorhombic $LiMnO_2$ with different degrees of stacking faults. J Power Sources 115:125–130
9. Pang WK, Lee JY, Wei YS, Wu SH (2013) Preparation and characterization of Cr-doped $LiMnO_2$ cathode materials by Pechini's method for lithium ion batteries. Mater Chem Phys 139:241–246
10. Young-Il J, Chou FC, Yet-Ming C (1999) Magnetic properties of monoclinic phase $LiAl_{0.05}Mn_{0.95}O_2$. J Phys Chem Solid 60:1763–1771
11. Hongmei J, Gang Y, Xiaowei M, Anqin H (2010) Efficient microwave hydrothermal synthesis of nanocrystalline orthorhombic $LiMnO_2$ cathodes for lithium batteries. Electrochim Acta 55:3392–3397
12. Yi H, Ronghua L, Xiaokun D, Lilong J, Mingdeng W (2010) Hydrothermal synthesis and electrochemical properties of orthorhombic $LiMnO_2$ nanoplates. J Alloys Compd 492:601–604
13. Qun L, Dali M, Chengkang C, Fuqiang H (2007) Phase conversion and morphology evolution during hydrothermal preparation of orthorhombic $LiMnO_2$ nanorods for lithium ion battery application. J Power Sources 173:538–544
14. Ouyang CY, Zeng XM, Sljivancanin Z, Baldereschi A (2010) Oxidation states of mn atoms at clean and Al_2O_3-covered $LiMn_2O_4$(001) surfaces. J Phys Chem C 114:4756–4759
15. Arumugam D, Kalaignan GP (2011) Electrochemical characterizations of surface modified $LiMn_2O_4$ cathode materials for high temperature lithium battery applications. Thin Solid Films 520:338–343
16. Jayaprakash N, Gangulibabu NK, Bhuvaneswari D (2011) Effect of mono- (Cr) and bication (Cr, V) substitution on $LiMn_2O_4$ spinel cathodes. J Solid State Electrochem 15:1243–1251
17. Thirunakaran R, Sivashanmugam A, Gopukumar S, Rajalakshmi R (2009) Cerium and zinc: dual-doped $LiMn_2O_4$ spinels as cathode material for use in lithium rechargeable batteries. J Power Sources 187:565–574
18. Kanamori J (1959) Superexchange interaction and symmetry properties of electron orbitals. J Phys Chem Solid 10:87–98
19. Dahn JR, Fuller EW (1994) Thermal stability of $LixCoO_2$, $LixNiO_2$ and λ-MnO_2 and consequences for the safety of Li-ion cells. Solid State Ion 69:265–270
20. Dahbi M, Saadoune I, Gustaffson T, Edstrom K (2011) Effect of manganese on the structural and thermal stability of $Li_{0.3}Ni_{0.7-y}Co_{0.3-y}Mn_{2y}O_2$ electrode materials ($y = 0$ and 0.05). Solid State Ion 203:37–41

21. Kim JM, Lee G, Kim BH, Huh YS, Lee GW, Kim HJ (2012) Ultrasound-assisted synthesis of Li-rich mesoporous $LiMn_2O_4$ nanospheres for enhancing the electrochemical performance in Li-ion secondary batteries. Ultrason Sonochem 19:627–631
22. Yi T-F, Zhu Y-R, Zhu X-D, Shu J, Yue C-B, Zhou A-N (2009) A review of recent developments in the surface modification of $LiMn_2O_4$ as cathode material of power lithium-ion battery. Ionics 15:779–784
23. Ke D, Fei Y, Guorong H, Zhong DP, Yanbing C, Kwang SR (2013) Sodium additive to improve rate performance of $Li[Li_{0.2}Mn_{0.54}Ni_{0.13}Co_{0.13}]O_2$ material for Li-ion batteries. J Power Sources 244:29–34
24. Mohamed A, Jose MA, Ismael S, Jose MR (2011) $LiCr_{0.2}Ni_{0.4}Mn_{1.4}O_4$ spinels exhibiting huge rate capability at 25 and 55 °C: analysis of the effect of the particle size. J Power Sources 196:10222–10227
25. Yumei L, Zhenzhen L, Yongliang L, Caifeng C, Yi H, Xiaojing Y (2011) Preparation and electrochemical properties of Li-rich spinel-type lithium manganate coated $LiMn_2O_4$. Mater Res Bull 46:2450–2455
26. Rajive MT, Anil KKM, Anand PB, Jayalekshmi S (2011) Effect of annealing on structural and electrical properties of the Li–Mn–O thin films, prepared by high frequency RF magnetron sputtering. J Phys Chem Solid 72:1251–1255

Chapter 44
Hydrogen Production from Nuclear Energy: Comparative Cost Assessment

Rami Salah El-Emam and Ibrahim Dincer

Abstract Hydrogen is one of the most promising alternatives for sustainable energy solutions to meet the world energy demands. Recently, there has been great interest in nuclear hydrogen production. Hence, economic evaluation of the nuclear hydrogen production process has become a crucial issue for better implementation opportunities of the production technologies and the nuclear power plants that can be integrated in the process. This chapter presents the Hydrogen Economic Evaluation Programme (HEEP) which is developed by the International Atomic Energy Association (IAEA). Five different cases of nuclear hydrogen production using different technologies of reactors and production are presented. The cost of hydrogen production is studied for these cases, including several scenarios with integrating storage of the produced hydrogen as compressed gas and/or transportation through pipelines to investigate the effect of these parameters on the hydrogen cost.

Keywords Hydrogen • Nuclear power • Production cost • HEEP • IAEA

44.1 Introduction

Most of the world energy demand is still covered by conventional fossil fuels which are the major contributors, with negative impact, to the global warming [1]. Besides the environmental unbalance caused by using fossil fuels that deplete the atmospheric oxygen, they also cause geopolitical conflicts and limit the access of next

---

R.S. El-Emam (✉)
Clean Energy Research Laboratory (CERL), Faculty of Engineering and Applied Science, University of Ontario Institute of Technology, 2000 Simcoe Street North, Oshawa, ON, Canada L1H 7K4

Faculty of Engineering, Mansoura University, Mansoura, Egypt
e-mail: rami.elemam@uoit.ca

I. Dincer
Clean Energy Research Laboratory (CERL), Faculty of Engineering and Applied Science, University of Ontario Institute of Technology, 2000 Simcoe Street North, Oshawa, ON, Canada L1H 7K4
e-mail: Ibrahim.Dincer@uoit.ca

I. Dincer et al. (eds.), *Progress in Clean Energy, Volume 2*,
DOI 10.1007/978-3-319-17031-2_44

generations to a clean environment [2]. In order to achieve the commitments of Kyoto protocol with limiting greenhouse gases emissions, developments of clean mechanisms and technologies as well as potential options with more environmentally benign fuels and fuel production methods have has been considered [1, 3].

A shift to renewable and nuclear energy options to meet world's growing energy demands is one of the best options. It also helps in limiting carbon dioxide emissions to the environment. However, the high cost of energy produced from the relatively immature renewable technologies gives more credit to nuclear-based options [2, 4]. Nuclear energy, besides heat and power cogeneration, is also one of the most promising candidates for hydrogen production at a lower cost [5]. Hydrogen is taken into consideration as a promising alternative fuel to be integrated or to replace the current fossil fuels. It is believed by decision makers and researchers that it is a unique energy carrier for building a carbon-free economy.

Hydrogen economy was initially considered in late 1960s, and the interest in delving into hydrogen technology increased with the oil crisis at the first half of 1970s. Figure 44.1 shows the possible routes for producing hydrogen from water using nuclear heat and power. This figure shows the water splitting technologies that generate non-carbon based hydrogen. It also shows the carbon-based hydrogen generation methods utilizing fossil fuels through gasification and reforming technologies using nuclear heat. The generation IV nuclear reactors provide better opportunity for nuclear hydrogen production path as it co-produce high-temperature process heat and electric power [6].

Nuclear hydrogen generation through the conventional water electrolysis is driven by the electric power from the nuclear power plant. Electrolysis is a relatively mature technology, and achieved an industrial scale since it was developed in 1800s where hydrogen was utilized in fertilization industry. There was a great interest in a larger scale production through electrolysis which diminished with the appearance of the fossil fuel-based compotators; gasification and steam methane reforming, early in the twentieth century. Electrolysis then came back to the scene after the 1970s oil crisis.

High-temperature steam electrolysis (HTSE) is a more efficient alternative when integrated with nuclear energy. HTSE requires less electric power compared with conventional low temperature water electrolysis where the rest of the required energy is substituted by thermal energy. This provides an advantage of being a relatively cheaper option. The French Atomic Energy Commission (CEA) has a well-developed program to investigate the use of HTSE among other thermochemical options utilizing nuclear heat and power [7].

Thermochemical water splitting cycles provide a very promising path for green nuclear hydrogen production. As indicated in Fig. 44.1, heat is the only energy supplied to thermochemical cycles. Water is consumed as it decomposes into oxygen and hydrogen streams through a closed sequence of chemical reaction. All chemicals involved in the reactions of any thermochemical cycle are completely recycled. The interest in thermochemical cycles for nuclear hydrogen production started early 1960s. It is suitable for large-scale production of hydrogen. Sulfur Iodine (S-I) is a pure three-step thermochemical cycle that requires heat at temperature higher than 800 °C and up to 900 °C or 1,000 °C to operate. It can operate at efficiency of up to 56 % [8]. Figure 44.2 shows a schematic of the cycle

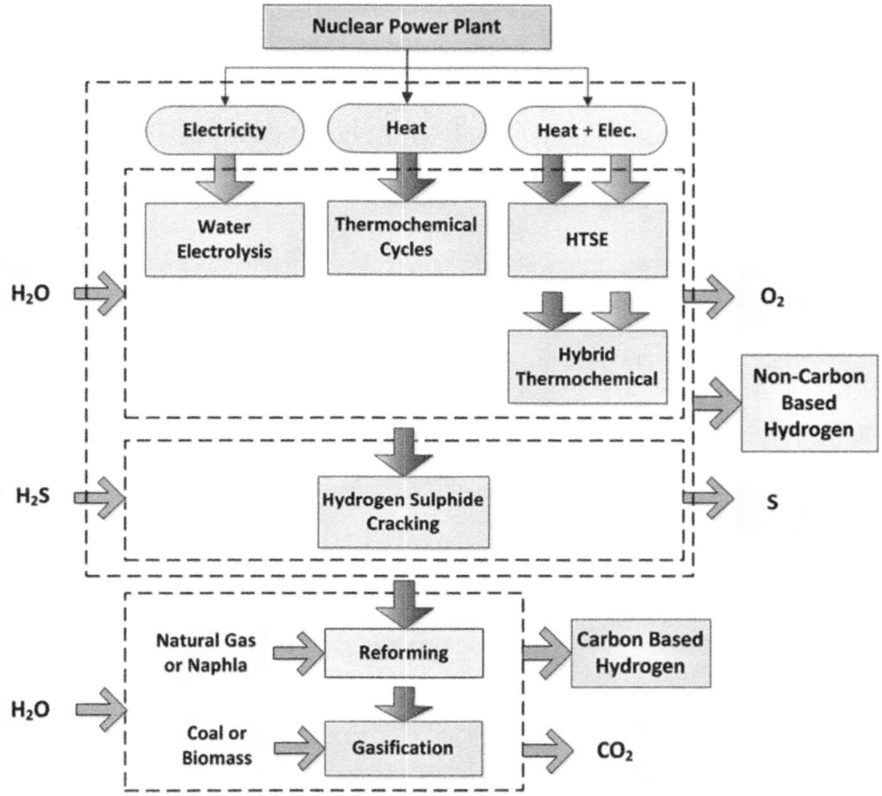

Fig. 44.1 Production methods of nuclear hydrogen production

Fig. 44.2 Schematic of Sulfur Iodine (S-I) thermochemical cycle

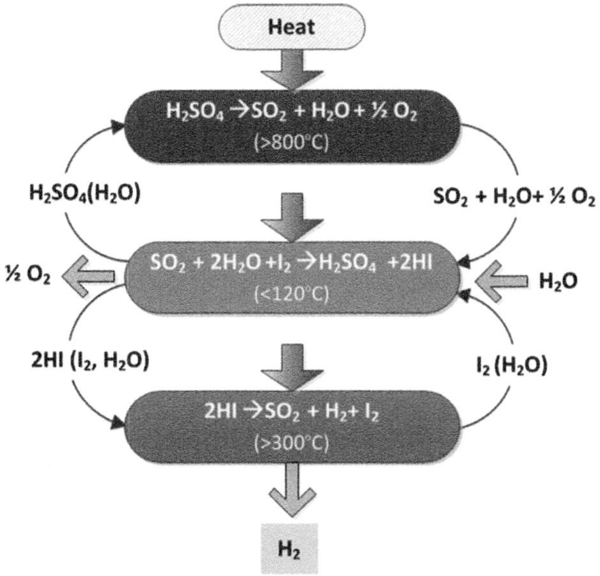

with its three chemical reactions. The first chemical reaction in this cycle is the Bunsen reaction that produces hydriodic acid, HI, and sulfuric acid, H_2SO_4, under relatively low temperature. HI acid gets into the second reaction which is high-temperature decomposition reaction producing hydrogen. The third reaction is high-temperature decomposition of H_2SO_4 produced from Bunsen reaction. This reaction produces SO_2, H_2O, and O_2 that are utilized in the solid catalyst [9].

In hybrid thermochemical cycles technology, water is split using thermal and chemical energies with conjunction with another form of energy. Electro-thermo-chemical hybrid cycles utilize electric power in electrochemical reaction step with thermal energy to accomplish the generation of hydrogen process. There are other types of hybrid cycles using photochemical or radiochemical reactions.

Hybrid sulfur cycle (HyS) is one of the promising hybrid electro-thermochemical cycles for nuclear hydrogen production. It is under development since 1975 when the cycle was patented by Westinghouse. It is a two-step water splitting process. In this cycle, electrochemical reaction of SO_2 and H_2O producing hydrogen is one step, and the other step is a thermochemical decomposition shared with the conventional S-I cycle. Figure 44.3 shows a schematic of the two processes of HyS cycle. A modified cycle is developed in Japan where the maximum temperature range of the cycle is reduced from 800 °C to 500–700 °C [10]. Another promising hybrid cycle is the Copper-Chlorine (Cu-Cl) cycle which was initially developed in 1970s. It is a medium-temperature cycle operating at around 550 °C in three to five thermochemical and electrochemical steps in different configurations. The efficiency of this cycle is calculated at about 40 % [11]. Cu-Cl cycle is studied and examined by Argonne National Laboratory (ANL), Atomic Energy of Canada Ltd. (AECL) and the French Atomic Energy Commission (CEA) with collaboration with different research institutes. Figure 44.4 shows a schematic of a five-step Cu-Cl thermochemical cycle.

There are several other thermochemical and hybrid thermochemical cycles under research for efficiently integrating nuclear heat and power for hydrogen production. Figure 44.5 shows a three-step Magnesium-Chlorine (Mg-Cl) cycle which includes

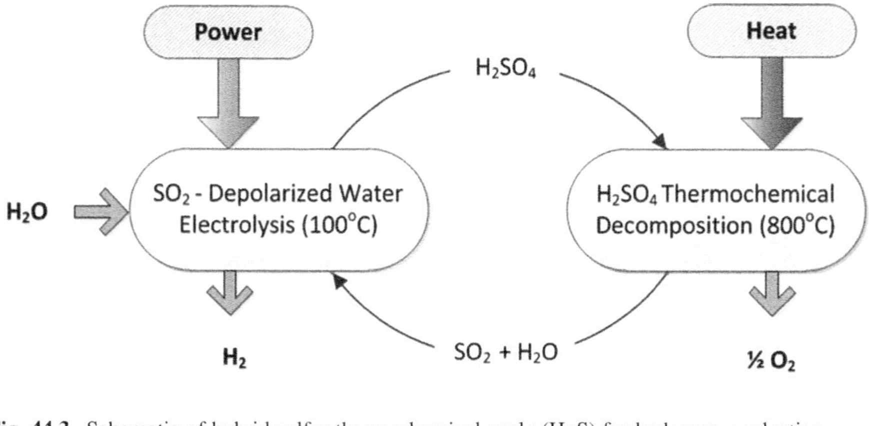

Fig. 44.3 Schematic of hybrid sulfur thermochemical cycle (HyS) for hydrogen production

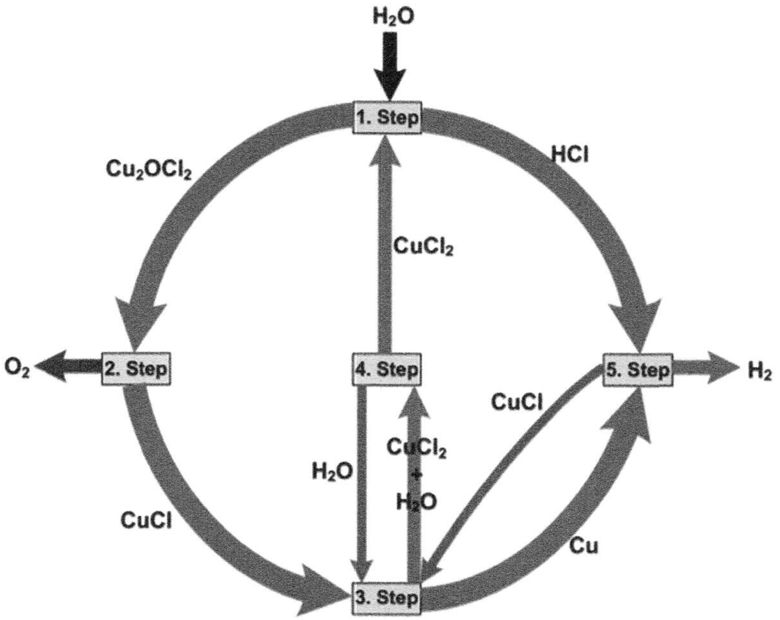

Fig. 44.4 The five-steps Cu-Cl thermochemical cycle [adapted from 13]

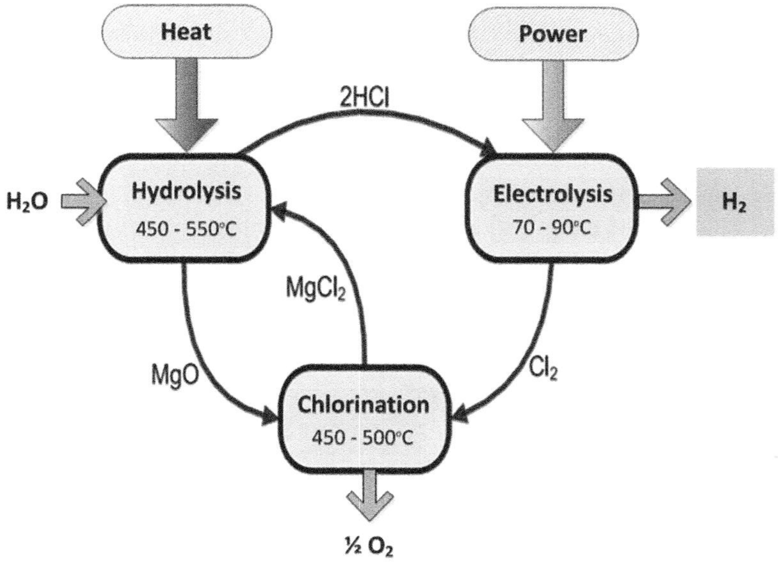

Fig. 44.5 Mg-Cl thermochemical cycle, modified from [adapted from 3]

hybrid thermochemical-electrochemical process for water splitting. This cycle is convenient for integration with nuclear power as it operates at about 500 °C. The efficiency of Mg-Cl cycle was reported as 37.4 % by Ozcan and Dincer [3].

Economics of nuclear hydrogen production is one of the most important factors that should be considered among national and international energy policies, environmental consideration and resources availability. The cost is highly affected by the technology of hydrogen generation and the pathway through the process from nuclear infrastructure to storage technology. In this study, a comparative cost assessment of nuclear hydrogen production is performed considering different nuclear power plants and hydrogen production technologies. The study also considers several scenarios of storage and transportation for better understanding of their effect on the costs of hydrogen production. The assessment is performed using HEEP software which is developed by the International Atomic Energy Agency (IAEA). The following sections cover the description of HEEP software structure, controlling parameters and formulation as well as case studies and their results for comparative cost assessment.

44.2 HEEP for Hydrogen Production Cost Assessment

The Hydrogen Economic Evaluation Programme (HEEP) is developed as a specific software by the International Association of Atomic Energy in collaboration with the Bhabha Atomic Research Centre (BARC) in India [12].

This software is utilized to estimate the cost of hydrogen generation from S-I, HTSE, water electrolysis, and other promising thermochemical and hybrid technologies combined with high-temperature nuclear reactors. Considering storage and transportation for sure has an impact on the hydrogen production cost. In HEEP, different scenarios of storage for compressed gas, liquefaction and metal hydride, and transportation for pipelines and vehicles are integrated to facilitate the evaluation of the effects of design, duration, quantity, and delivery condition, on the cost estimates. Figure 44.6 shows the different scenarios of storage and

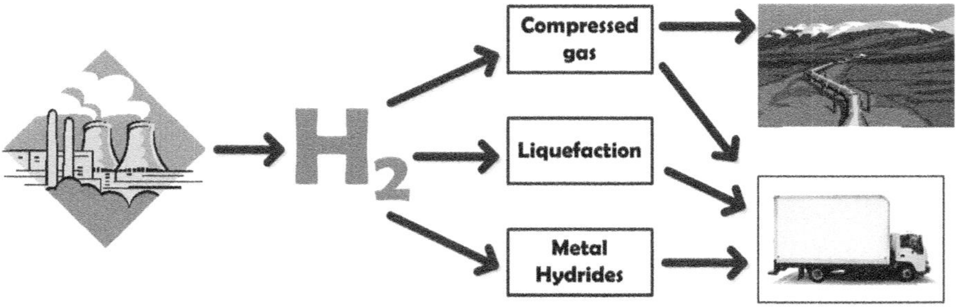

Fig. 44.6 Scenarios of nuclear hydrogen production to delivery in HEEP [adapted 14]

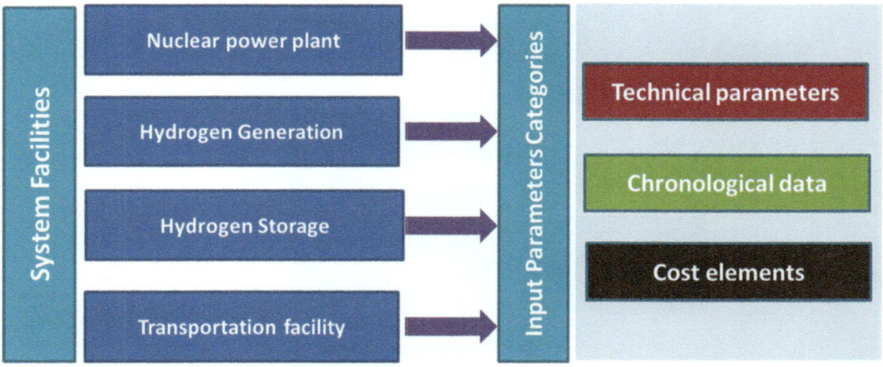

Fig. 44.7 System information categorization as defined for HEEP

Table 44.1 Technical features of HEEP

Nuclear plant	Hydrogen plant	Hydrogen storage plant	Transportation
• Reactor type • Thermal rating • Thermal power for hydrogen plant • Electricity rating • Plant efficiency • Unit numbers • Unit capacity • Availability factor	• Production technology • Production rate • Plant location • Heat consumption • Electricity required • Number of units • Capacity of the unit • Availability factor • Auxiliary power	• Storage option (compression, liquefaction, metal hydride) • Storage capacity • Electricity requirement • Cooling required • Compression power	• Transportation option (pipeline, vehicle) • Distance • Vehicle capacity • Speed of vehicle • Trips preparation • Delivery pressure • Pipes friction

transportation options for hydrogen production, from nuclear power plant to delivery of the product. The software is a single-window with friendly easy use interface. It is programmed in three modules: pre-processing module for data input programmed in Microsoft Visual Basic, execution module where the levelized cost calculations are performed and this module is programmed in Fortran, and the third module is the post-processing module for processing the results in graphs and producing the final report and it is programmed in Visual Basic [7].

The input parameters and variables required for HEEP are classified in three different categories for each of the four processes based on the activated ones in the studied scenario. Figure 44.7 shows a simple chart showing the parameters categories; technical, chronological, and cost elements data. For the technical parameters, Table 44.1 presents a list of the most important parameters to be fed to HEEP. There is a library with different cases provided with the software and some data would be used as default if not available as input parameters. The chronological data, the second set of input parameters, represent the period of different activities through

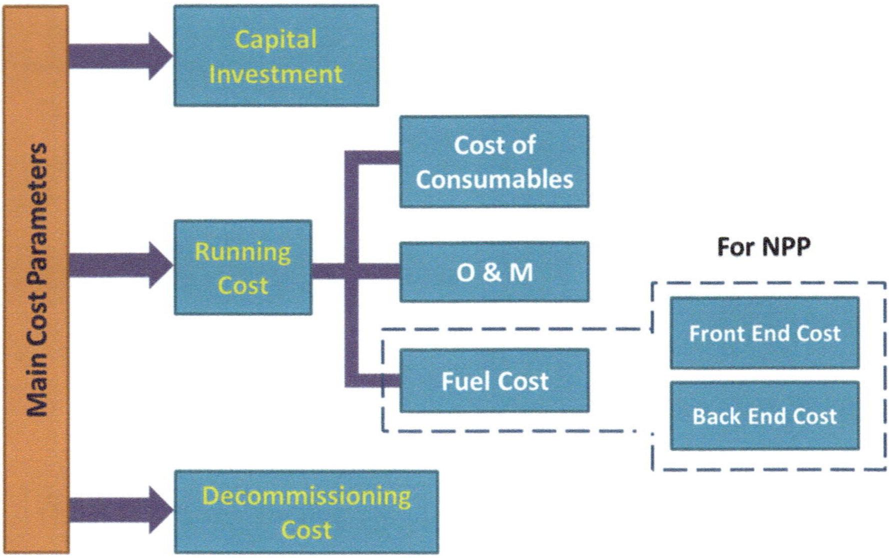

Fig. 44.8 HEEP main categories of cost parameters

the life cycle of the integrated plant for the studied scenario. These events and activities include the following: construction, operation, decommissioning, cooling before decommissioning, refurbishment, waste storage, and used-fuel cooling, where applicable based on the studied plant.

HEEP calculates the levelized cost of hydrogen considering certain cost elements as shown in Fig. 44.8. The capital investment element is the summation of all the design expenses, manufacturing, construction, initial inventory, and annual feed. The second cost element is the running cost which includes operation, maintenance, refurbishment, salaries, and others. The fuel cost is mainly for the nuclear power plant and it has two elements: front end cost including annual feed rate and number of fuel types, and the back-end cost depending on direct disposal or reprocessing of fuel scenarios. The last element is the decommissioning cost, especially the nuclear power plant decommissioning cost which contributes and affects the calculated hydrogen production cost significantly. Other financial parameters including equity to debt ratio, tax rate, interest rate, inflation rate, and discount rate can be entered by the user or used as default values provided by HEEP [15–16].

The calculations are done in two steps, first is calculation of the levelized cost of energy utilized which is delivered by the nuclear power plant and the second step is using the nuclear power plant results as input along with other user specified information to calculate the cost of hydrogen generation. The execution module in HEEP considers the levelized cost of hydrogen as the ratio of sum of present value of production, storage, and transportation to present value of gross hydrogen generated as reported by [7] and as shown in the first equation in Table 44.2.

Table 44.2 Main cost calculation equations considered at HEEP

Cost estimation calculation	
Levelized cost of hydrogen generation	$C_{H_2} = \frac{E_{NPP}(t_0) + E_{H_2GP}(t_0) + E_{H_2T}(t_0)}{G_{H_2}(t_0)}$
Present value of expenditures	$E(t_0) = \sum_{t\_i}^{t\_f} \frac{CI_t(t_0)}{(1+r)^{t-t_0}} + \sum_{t\_i}^{t\_f} \frac{R_t}{(1+r)^{t-t_0}} + \sum_{t\_i}^{t\_f} \frac{DC_t}{(1+r)^{t-t_0}}$
Present value of gross hydrogen generation at time to	$G_{H_2}(t_0) = \sum_{t\_i}^{t\_f} \frac{G_{H_2}(t_0)}{(1+r)^{t-t_0}}$

The subscripts of the parameters in the equations presented in Table 44.2 are NPP, H2GP, and H2T which refer to nuclear power plant, hydrogen generation and storage, and hydrogen transportation, respectively. In the second equation; CI, R and DC are the capital investment expenditures, facility running expenditures, and decommissioning expenditures for the year t, respectively. r is real discount rate. More details of execution module can be found in [7].

44.3 Results and Discussion of Generic Case Studies

In this study, five different cases are considered with HEEP for estimating the hydrogen production cost. Table 44.3 shows a summary of the facilities and plants used in each case. The first three cases use advanced pressurized water reactors (APWR) as nuclear power plant and conventional electrolysis is proposed for hydrogen production. In case 3, two Westinghouse AP1000 reactors operating to deliver all produced electricity to the hydrogen production facility. This corresponds to a production rate of 12.43 kg/s. Cases 1 and 2 consist of smaller reactors and conventional electrolysis plants with correspondingly smaller production rates; 4 and 8 kg/s, respectively. Cases 4 and 5 use next-generation nuclear reactor and hydrogen production technologies. A high-temperature gas-cooled reactors as cooled with helium is considered for the nuclear power plant facility. The reactor in case 4 delivers heat at 900 °C which is utilized to operate HTSE facility for hydrogen generation. In case 5, the reactor delivers process heat at 950 °C, which is used to power S-I thermochemical plant. All the proposed cases use compressed hydrogen gas when storage scenario is considered, and pipelines for transportation. The in-built HEEP algorithms calculate the necessary storage and transportation facilities and costs for a given hydrogen production rate. Table 44.4 shows the assumptions of the economic parameters as input data to HEEP. The main technical and financial parameters for each facility in the proposed cases are listed in Table 44.5 for the nuclear power plant and the hydrogen production facilities.

Table 44.3 The plants considered for the different cases in this study

	Case 1	Case 2	Case 3	Case 4	Case 5
Nuclear power plant, NPP	APWR	APWR	APWR	HTGR	HTGR
Hydrogen generation plant, HG	CE	CE	CE	HTSE	S-I
Storage plant, GS	CG	CG	CG	CG	CG
Transportation	PL	PL	PL	PL	PL
Production rate, kg/s	4	8	12.43	4	4

APWR advanced pressurized water reactor, *HTSE* high-temperature steam electrolysis, *CE* conventional electrolysis, *HTGR* high-temperature gas cooled reactor, *S-I* sulfur Iodine thermochemical cycle, *CG* compressed gas, *PL* pipeline

Table 44.4 Main economic and financial parameters

Economic parameters	
Inflation rate	1.9 %
Total tax rate	18 %
Return on equity	10 %
Equity:debt	85 %:15 %
Plant lifetime	40 years
Property insurance	2 %
Cost of demineralized water	$0.001442/L
Cost of grid electricity	$0.06/kWh

The results in Fig. 44.9 show a comparison of the hydrogen generation cost considering the nuclear power plant and the hydrogen generation plant for the five different cases presented in Table 44.5. There is no storage or transportation integrated for these results. The figure shows that the cost of hydrogen generation varies from 2.90 to 4.39 $/kg of produced hydrogen. The lowest is achieved by case 4 and the highest cost is by case 1. In Fig. 44.10, the contribution of the nuclear power plant and the hydrogen generation plant in the total cost of hydrogen production are presented for the five different cases. The cases in these results are considering the nuclear power plant and hydrogen generation plant only (NPP-HG). It can be seen that for case 1, almost 87 % of the cost is the nuclear power plant share. Case 4 experiences the lowest share from nuclear power plant in the cost with about 35 % of the total cost. This depends on the technology utilized for hydrogen generation and the reactor type, as listed in Table 44.5.

The effect of hydrogen storage and transportation scenarios is studied and the results are presented in Figs. 44.11 and 44.12. In Fig. 44.11, a comparison of the cost of hydrogen production for the five different cases is presented considering production with storage as compressed gas (NPP-HG-CG) and then integrating the transportation option and testing the effect of changing the traveled distance from 300, 600 to 900 km (NPP-HG-GC-P).

Table 44.5 Main technical and cost parameters of plants in the considered cases

		Case 1	Case 2	Case 3	Case 4	Case 5
NPP	NPP capacity	359.5 MWe	719 MWe	1,117 MWe	509.3 MWth	630.7 MWth
	Number of NPP units	2	2	2	2	2
	Capital investment, M$	6,310	9,313	11,928	804.6	1,210
	Annual O&M, M$	194.9	154.8	198.28	46.95	21.97
	Capacity factor, %	93	93	93	90	90
	Construction period, years	5	5	5	3	3
	Annual fuel cost, M$	34.96	51.60	66.09	38.24	69.73
	Decommissioning of NPP	2.8 % CC	2.8 % CC	2.8 % CC	$94.02 M	$101 M
HGP	Capital cost, M$	422.6	845.2	1310	458.5	666.2
	Annual O&M, M$	16.90	33.81	52.52	79.04	44.52
	Water consumption, L/year	1.136×10^9	2.272×10^9	3.530×10^9	1.136×10^9	1.136×10^9
	Decommissioning of HGP	10 %	10 %	10 %	10 %	10 %
	Thermal-H_2 efficiency	26.07 %	26.07 %	26.07 %	39.82 %	35.56 %
	Non-process electricity	–	–	–	–	428 MWe

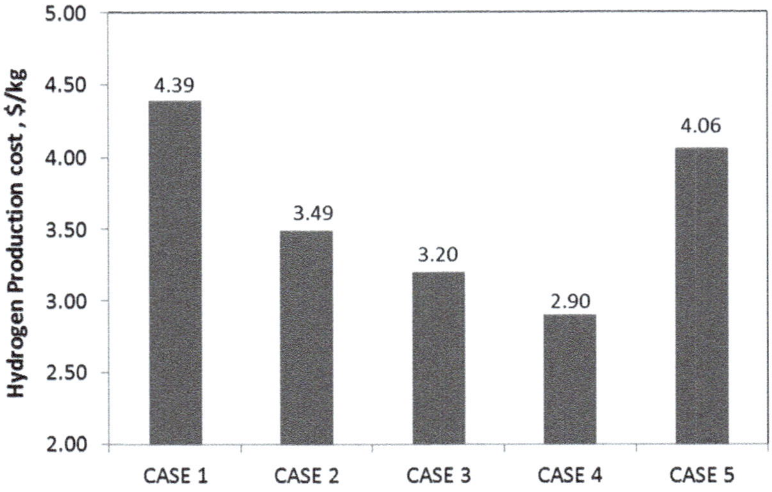

Fig. 44.9 Cost of hydrogen production considering nuclear power and hydrogen generation plants (NPP-HG) for the five presented cases

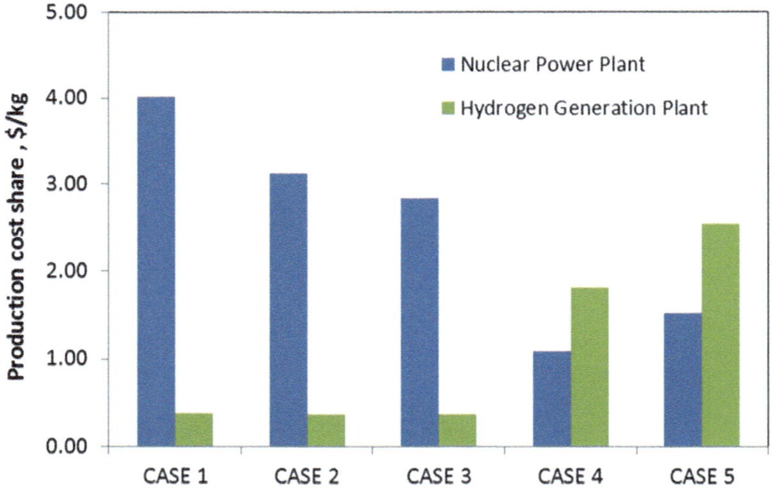

Fig. 44.10 The contribution of each plant in the hydrogen production cost for the five cases for NPP-HG scenario

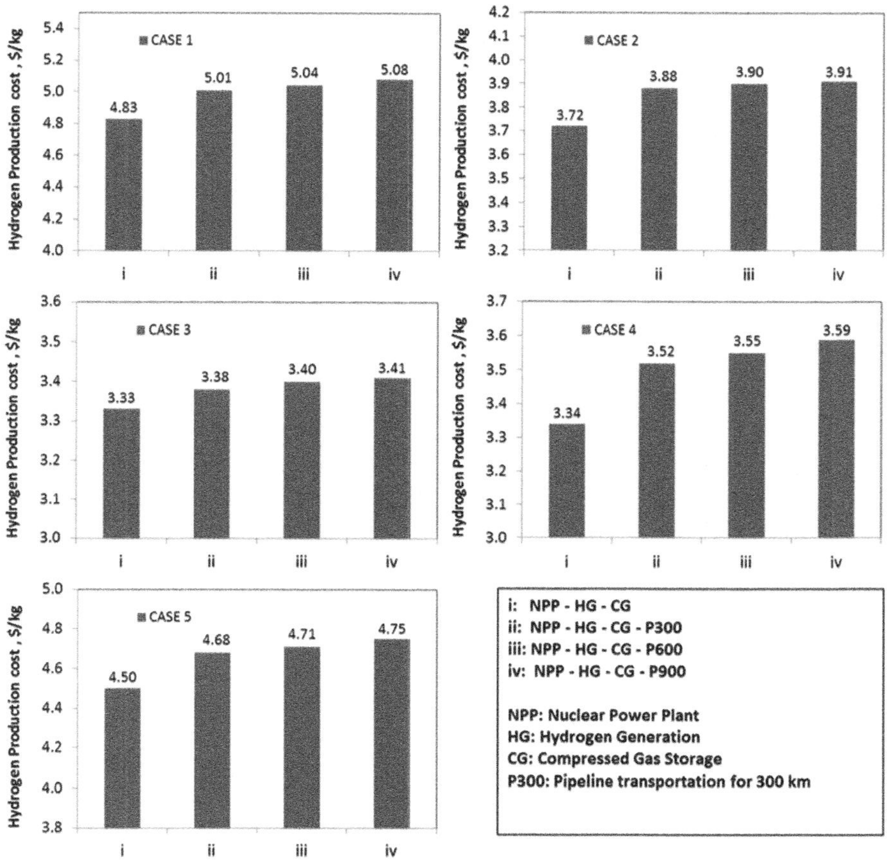

Fig. 44.11 Cost of hydrogen production for different scenarios with storage and transportation for different distances

From Fig. 44.11, for case 1, the hydrogen storage shares with 9.02 % of the production cost considering no transportation. This share decreases to 8.7, 8.64, and 8.58 % when considering transportation for 300, 600, and 900 km through pipelines. These percentages are lower for case 2 where it goes from 6.18 % for no transportation to 5.93, 5.90, and 5.88 % for the three transportation scenarios. These percentages for cases 3, 4, and 5 are as follows: for case 3: 4.32, 4.25, 4.22, 4.21 %, for case 4: 11.06, 10.58, 10.5, 10.4 % and for case 5: 9.8, 9.42, 9.36, and 9.28 %, respectively.

The results in Fig. 44.12 represent the cost of hydrogen production considering transportation through pipelines for different distances without considering the storage plant in the process. It just considers the nuclear power plant, the hydrogen

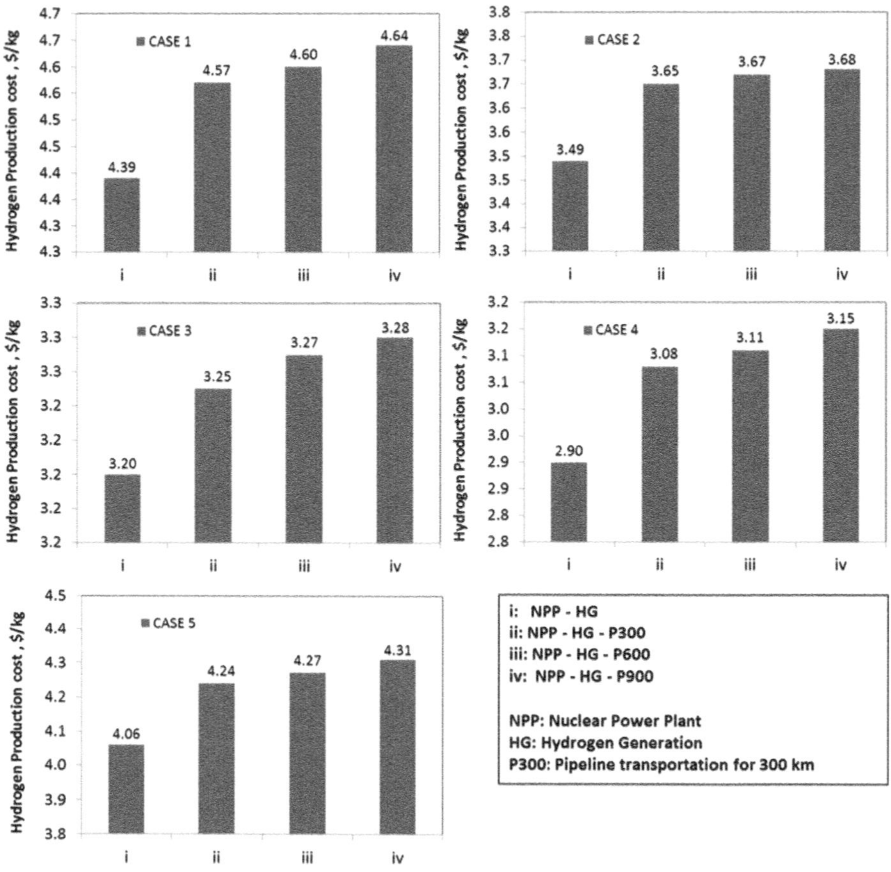

Fig. 44.12 Cost of hydrogen production for different scenarios and the effect of transportation distance on the cost without considering storage

generation plant, and the transportation cost (NPP-HG-P) for 300, 600, and 900 km of traveled distance. From the presented results, the transportation cost varies based on the production capacity for the different cases.

44.4 Conclusions

The HEEP software is utilized in this study to perform a comparative assessment on the hydrogen production cost for five different cases, including different nuclear reactor types and different technologies of hydrogen production. The effects of including a plant for hydrogen storage as compressed gas and transportation through pipelines are also investigated. The results show that case 4 presents the lowest cost with 2.9 $/kg, when the other cases are reported as 4.39, 3.49, 3.2, and

4.06 $/kg for cases 1, 3, 4, and 5, respectively. The integration of hydrogen storage facility causes an increase of the cost of the production to be 4.83, 3.72, 3.33, 3.34, and 4.50 $/kg for cases 1, 2, 3, 4, and 5, respectively. When transportation is considered, it is reported that an amount of 0.18, 0.21, and 0.25$ increase in the cost occurs when changing the traveled distance from 300, 600 to 900 km, when 4 kg of produced hydrogen per second is considered.

References

1. Dincer I (2007) Environmental and sustainability aspects of hydrogen and fuel cell systems. Int J Energ Res 31(1):29–55
2. Dincer I, Zamfirescu C (2012) Sustainable hydrogen production options and the role of IAHE. Int J Hydrog Energ 37(21):16266–16286
3. Ozcan H, Dincer I (2014) Performance investigation of magnesium-chloride hybrid thermochemical cycle for hydrogen production. Int J Hydrog Energ 39:76–85
4. Holladay JD, Hu J, King DL, Wang Y (2009) An overview of hydrogen production technologies. Catal Today 139:244–260
5. Bartels JR, Pate MB, Olson NK (2010) An economic survey of hydrogen production from conventional and alternative energy sources. Int J Hydrog Energ 35:8371–8384
6. Mansilla C, Sigurvinsson J, Bontemps A, Marechal A, Werkoff F (2007) Heat management for hydrogen production by high temperature steam electrolysis. Energy 32:423–430
7. Khamis I (2011) An overview of the IAEA HEEP software and international programmes on hydrogen production using nuclear energy. Int J Hydrog Energ 36:4125–4129
8. Huang CP, Raissi AT (2005) Analysis of sulphur-iodine thermochemical cycle for solar hydrogen production. Sol Energ 78:632–646
9. Keefe DO, Allen C, Besenbruch G, Brown L, Norman J, Sharp R (1982) Preliminary results from bench-scale testing of a sulfur-iodine thermochemical water splitting cycle. Int J Hydrog Energ 7:38–92
10. Gorensek MB, Summers WA (2011) The hybrid sulfur cycle. In: Yan XL, Hino R (eds) Nuclear hydrogen production handbook. CRC Press, Boca Raton, FL, pp 499–545
11. Naterer GF, Suppiah S, Stolberg L, Lewis M, Ferrandon M, Wang Z, Dincer I, Gabriel K, Rosen MA, Secnik E, Easton EB, Trevani L, Pioro I, Tremaine P, Lvov S, Jiang J, Rizvi G, Ikeda BM, Lu L, Kaye M, Smith WR, Mostaghimi J, Spekkens P, Fowler M, Avsec J (2011) Clean hydrogen production with the Cu–Cl cycle—progress of international consortium I: experimental unit operations. Int J Hydrog Energ 36(24):15472–15485
12. Khamis I, Malshe UD (2010) HEEP: a new tool for the economic evaluation of hydrogen economy. Int J Hydrog Energ 35:8398–8406
13. Orhan MF, Dincer I, Rosen MA (2008) Energy and exergy analysis of the hydrogen production step of a copper-chlorine thermochemical water splitting cycle driven by nuclear-based heat. Int J Hydrog Energ 33:6456–6466
14. El-Emam RS, Ozcan H, Dincer I (2014) Cost assessment of nuclear hydrogen production systems using HEEP. International Ege Energy Symposium and Exhibition, Usak, Turkey
15. El-Emam RS, Dincer I (2014) Comparative study on nuclear based hydrogen production: cost assessment. In: International conference on smart energy grid engineering (SEGE'14), Oshawa, ON
16. Ozcan H, El-Emam RS, Dincer I (2014) Comparative assessment of nuclear based hybrid sulfur cycle and high temperature steam electrolysis systems using HEEP. In: Dincer I et al (eds) Progress in sustainable energy technologies: generating renewable energy. Springer, New York, NY, pp 165–180

Chapter 45
Activity of Yttrium Strontium Titanate-Based Catalyst over Steam Reforming of Dodecane

Kais Hbaieb

Abstract Hydrocarbon reforming to produce hydrogen (or syngas) on site or on board a transportation vehicle for fuel cell operation can solve the complicated problem of hydrogen storage and transportation. Using perovskite catalysts can be advantageous due to their high oxygen ion storage and lower cost. In this paper, the use of yttrium strontium titanate (YST)-based catalyst in autothermal reforming of dodecane is studied. The catalyst was prepared by the sol-gel technique and characterized by XRD. The B site was doped by noble metals such as Co, Ru, and Ni and their effect was studied. The operating temperature and pressure are 800 °C and 1 atm, respectively. Steam/carbon ratio was fixed at 3. Each test was conducted for at least 16 h. Hydrogen concentration varied between 65 and 72 % (free of water and nitrogen) with conversion rate as high as 80 to 90 %. The effect of doping at the B site on the catalyst activity is discussed.

Keywords Reforming • Catalysis • Perovskite • Hydrogen

45.1 Introduction

Fuel cell is seen by many as a very prominent alternative to the conventional engine for powering transportation vehicles not only because it is more efficient than the traditional engine but also because of the much less greenhouse emission, thus better compatibility with environment. Possibly the main drawback is the low fuel flexibility as hydrogen is merely the only favorable fuel; however, hydrogen infrastructure, distribution, and storage are a very cumbersome and prohibitively expensive solution. To overcome this difficulty, developing on-board fuel reformer

K. Hbaieb (✉)
Department of Mechanical Engineering, Faculty of Engineering, Taibah University, Al Madinah, Al Munawwara, Saudi Arabia

Strategic Technology Unit, Taibah University, Al Madinah, Al Munawwara, Saudi Arabia
e-mail: hbaiebkais@gmail.com

I. Dincer et al. (eds.), *Progress in Clean Energy, Volume 2*,
DOI 10.1007/978-3-319-17031-2_45

to produce hydrogen from the currently commercial and existing hydrocarbon fuel stands out to be a very attractive alternative.

Producing hydrogen from hydrocarbon fuel requires developing catalyst with high hydrogen yield, selectivity, and reforming efficiency. The most celebrated catalyst is the traditional and very established nickel on alumina support. Farrauto et al. [1] showed that Ni-based catalysts such as Ni/Al_2O_3 or Ni/MgO undergo sintering and coke deposition at high operating temperature. The latter problem may be partially alleviated by feeding oxygen and/or excess water and additional hydrogen [2, 3].

Nobel metals deposited onto an oxide support showed better activity and coking resistance over the traditional Ni catalyst. For example, Pt, Rh, and Ru impregnated onto oxide support exhibited superior catalytic activity and resistance to coke formation than traditional Ni catalyst [4–9]. Moreover, using oxygen-ion conducting catalyst supports ensures better resistance towards carbon formation and sulfur poisoning [10–12] than traditional alumina-based support under methane such as ceria, zirconia, and their mixed composites [13]. With the tendency and desire to direct use of commercial hydrocarbon fuel without deep desulfurization processing stage, sulfur-tolerant catalytic materials have to be sought. Recent progress in developing sulfur-tolerant materials showed that YST is one of the promising materials [14].

The aim of this paper is to explore the use of YST-based material as a catalyst for steam reforming of dodecane. Different doping has been introduced and their effect on catalyst conversion, activity, and selectivity studied. The catalyst will be stand-alone and no support will be used.

45.2 Experimental

YST-based catalysts were prepared using sol-gel (modified Pechini method) technique for the targeted composition of $Y_{0.08}Sr_{0.88}TiO_3$. Titanium isopropoxide was dissolved in ethylene glycol and heated to 70–80 °C before citric acid was added. Metal nitrates/chlorides are the other precursors and are separately dissolved in water. After the isopropoxide solution is stabilized, ethanol and metal nitrate solution were added and pH adjusted to ~6–7. After gel was formed, it was dried overnight. The dried gel was decomposed at ~200 °C and later crushed and transferred to a crucible for calcination. All materials are consistently calcined at 900 °C for 8 h. XRD characterization of the resulting powder revealed the formation of a single phase perovskite, as shown in Fig. 45.1 for the base YST. Ruthenium, nickel, and cobalt were used as active metals doped at the B site to enhance the activity of the catalyst towards hydrogen production. Doping at the B site of YST was conducted by the corresponding metal nitrate. The doping level was 5 % for ruthenium and 10 % for both cobalt and nickel. The designations of the doped catalysts with cobalt, nickel, and ruthenium are YSTC, YSTN, and YSTR, respectively.

Powder of particle size 0.2–0.4 mm was produced by crashing pellets made of calcined powder and sieved through sieves of different sizes. The catalyst was then

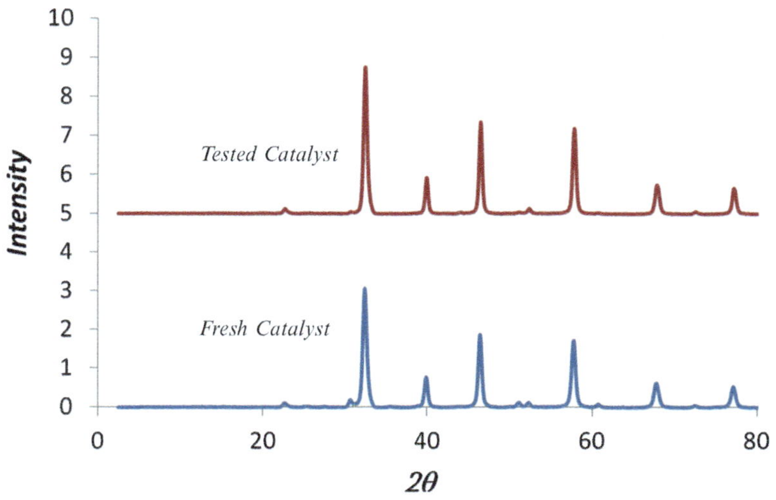

Fig. 45.1 XRD data of yttrium strontium titanate (YST)

loaded in the middle of the reactor tube and held by a rock-wool. During testing, the fuel was mixed with steam and preheated at 400 °C to form input mixtures with a steam-to-carbon ratio of 3. The gas hourly space velocity was between 7,000 and 8,000 h^{-1}. Temperature was maintained at 800 °C. The reformate produced from the catalytic reaction was analyzed by gas chromatography after water and heavy liquids were removed. Typically concentration of hydrogen, methane, and Co_x were measured.

45.3 Results and Discussions

Figure 45.2 shows gas product concentrations of dodecane reforming using the base yttrium strontium titanate (YST) and with a steam to carbon ratio of 3 (after water removal and based on nitrogen free gas mixture). The hydrogen concentration is quite satisfactory. Concentration of carbon monoxide is also considerable, a good indication for high CO selectivity. However, the concentration of methane is as high as 6 % which is unfavorable for hydrogen production.

Doping of YST with cobalt, ruthenium, and nickel showed clear improvement of catalyst activity. Hydrogen concentration increased from 66 % to 71–72 % irrespective of the doping type. Figure 45.3 shows a typical product concentration for dodecane reforming using doped YST catalyst; in this figure, nickel was the active metal doped at the B site of YST.

The figure also shows that concentration of carbon monoxide is nearly 20 % and higher than CO_2 which indicates that even though the temperature is high, water gas shift reaction is operating at a considerable rate. Moreover, doping with active

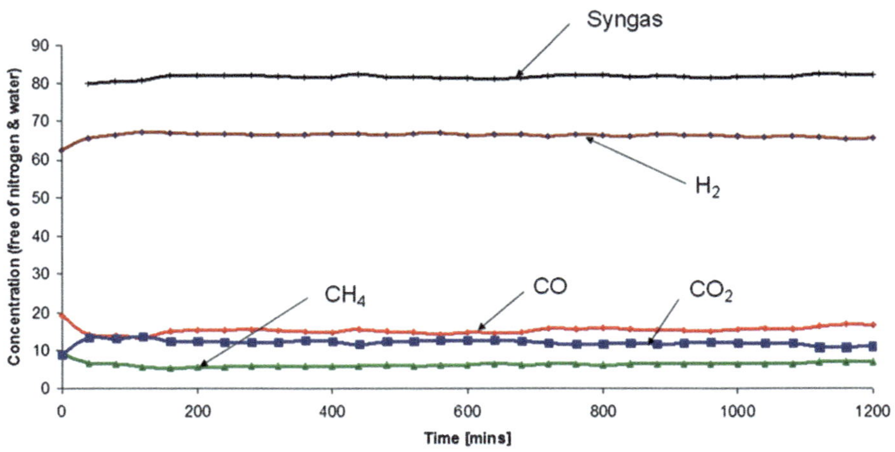

Fig. 45.2 Product concentration for dodecane reforming using yttrium strontium titanate and with a steam to carbon ratio of 3. Testing temperature is 800 °C

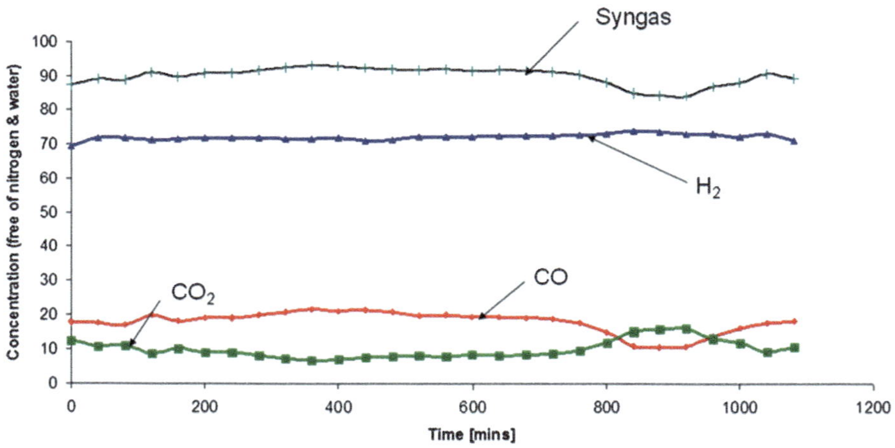

Fig. 45.3 Product concentration for dodecane reforming using nickel-doped YST and with a steam to carbon ratio of 3

metals at the B site of the YST catalyst helped eliminating methane in the product gas mixture and hence enhancing the hydrogen yield.

Table 45.1 shows the product gas distribution for all catalysts. It is noticeable that irrespective of the active metal element used for doping at the B site of the YST catalyst, methane has literally been eliminated, hydrogen concentration has increased by 7 %, and CO concentration has also been improved.

Upon testing completion, XRD characterization has been performed and showed that structure of the catalyst has not been changed (see Fig. 45.1).

Table 45.1 Product gas distributions for base and doped YST catalysts

	CH_4	CO_2	CO	H_2
$YSrTiO_3$	6.3	12.2	15	66.2
$(Y,Sr)(Ti,Ru)O_3$	1.23	9.4	18.7	70.7
$(Y,Sr)(Ti,Ni)O_3$	0	10.14	17.9	72
$(Y,Sr)(Ti,Co)O_3$	0.1	11.4	17.7	70.8

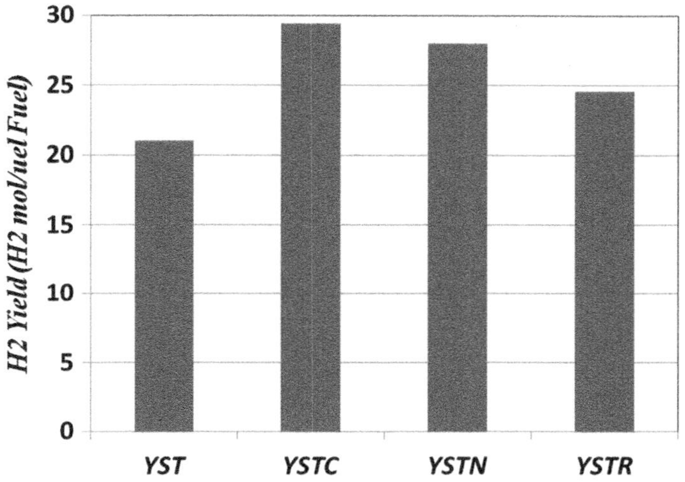

Fig. 45.4 H2 yield for several catalysts under steam reforming

Hydrogen yield calculated as the moles of hydrogen produced per unit mole of input fuel is illustrated in Fig. 45.4 for the different catalysts. Clearly the catalyst activity is boosted upon doping the base YST catalyst with active metal element on the B site of the perovskite structure. As the maximum theoretical yield for steam reforming is 37, the hydrogen yield particularly for YSTC and YSTN is quite appreciable.

45.4 Conclusions

Perovskite-based catalysts have been used for autothermal reforming of dodecane. The performance of the catalyst is promising with a hydrogen concentration at 65 % for the base YST and at 71–72 % for the doped catalysts irrespective of the active metal used for the doping. Carbon monoxide selectivity has also been improved through the doping with noble metals, while the methane concentration has been reduced to nearly zero. This work has demonstrated that doped YST catalysts are promising for reforming of heavy hydrocarbon.

References

1. Farrauto RJ, Batholomew CH (1997) Fundamentals of industrial catalytic process. Chapman and Hall, London
2. Provendier H, Petit C, Kiennemann A (2001) Steam reforming of methane on $LaNi_xFe_{1-x}O_3$ ($0 \leq x \leq 1$) perovskites. Reactivity and characterisation after test. CR ACAD SCI II 4:57–66
3. Dias JAC, Assaf JM (2004) The advantages of air addition on the methane steam reforming over Ni/γ-Al_2O_3. J Power Sources 137:264–268
4. Carpenter I, Hayes J (2004) US patent 2,325,506, 2004
5. Lietz G, Lieske H, Spindler H, Hanke W, Völter J (1983) Reactions of platinum in oxygen- and hydrogen-treated Ptγ-Al2O3 catalysts: II. Ultraviolet-visible studies, sintering of platinum, and soluble platinum. J Catal 81:17–25
6. Cavicchi R, Tarlov M, Semancik S (1991) Reactivity of Pd and Sn adsorbates on plasma and thermally oxidized SnO_2(110). Surface Sci 257:70–78
7. Rousset J, Tardy B, Bertolini C (1994) Pd_1Ni_{99} and Pd_5Ni_{95}: Pd surface segregation and reactivity for the hydrogenation of 13-butadiene. J Catal 149:404–413
8. Ayabe S, Omoto H, Utaka T, Kikuchi R (2003) Catalytic autothermal reforming of methane and propane over supported metal catalysts. Appl Catalysis A 241:261–269
9. Deng Y, Nevell TG, Ewen RJ, Honeybourne CL (1993) Sulfur poisoning, recovery and related phenomena over supported palladium, rhodium and iridium catalysts for methane oxidation. Appl Catalysis A 101:51–62
10. Huang TJ, Lin HJ, Yu TC (2005) A comparison of oxygen-vacancy effect on activity behaviors of carbon dioxide and steam reforming of methane over supported nickel catalysts. Catal Lett 105:239–247
11. Hamakawa S, Yoshino S, Nakamura J, Liu Y, Tsyganok A, Suzuki K, Murata K, Hayakawa T (2001) Role of lattice oxygen migration in Ni-based catalyst for natural gas conversion. Electrochem Solid State Lett 4:D9–D11
12. Zhu T, Flytzani-Stephanopoulos M (2001) Catalytic partial oxidation of methane to synthesis gas over Ni–CeO2. Appl Catalysis A 208:403–417
13. Descorme C, Madier Y, Duprez D, Birchem T (2000) Studies in surface science and catalysis. In: Corma A, Melo FV, Mendioroz S, Fierro JLG (eds) 12th international congress on catalysis, vol 130. Elsevier Science, Granada, Spain, pp 347
14. Kurokawa H, Yang L, Jacobson CP, De Jonghe LC, Visco SJ (2007) J Power Sources 164:510

Part VIII
Power Generation

Chapter 46
Cost Optimization of a Hybrid Off-Grid Power System for Remote Localities: A Statistical Model

Srimanta Ray, Debika Debnath, and Ajoy Kumar Chakraborty

Abstract Renewable energy sources like solar radiation, wind energy, geothermal energy, bioenergy are known to well integrate with off-grid stand-alone power systems. The choice of a renewable energy source depends on the geographical location and climatic condition of the region, available resources, and the economics of the power system. Thus, the evaluation and optimization of hybrid power systems in terms of cost of energy (CoE) with various alternative energy sources considering the demand of the location and available resources is highly pertinent for configuring a hybrid power system for remote localities. The present work describes a case study on development of a response surface (RS) model for predicting the optimum CoE of a hybrid off-grid power system (HOPS). RS model is developed using Box–Behnken experimental design (BBD) technique. A three-factor three-level BBD is used to describe the optimum CoE. The three process variables under consideration in BBD model are size of photovoltaic arrays (PV), diesel generator (DG) capacity, and number of battery (BAT) cells. The analysis of variance (ANOVA) is conducted on the response (CoE) for each factor level settings of a three-factor three-level BBD in order to evaluate a full quadratic factor space. The results of ANOVA established significant linear, quadratic, and interaction terms in the RS model representing the factor space. The coefficients of the RS-model are determined using multiple regression analysis technique at 95 % level of confidence. The RS-model is checked for error in prediction by residual analysis technique and validated against experimental data to confirm the accuracy of the model. The results confirmed that the model has 97.5 % accuracy in

S. Ray, Ph.D. (✉)
Department of Chemical Engineering, National Institute of Technology Agartala, Barjala, Jirania, Tripura (West) 799046, India
e-mail: rays.nita@gmail.com

D. Debnath • A.K. Chakraborty, Ph.D.
Department of Electrical Engineering, National Institute of Technology Agartala, Barjala, Jirania, Tripura (West) 799046, India
e-mail: debikanita@gmail.com; akcalll@yahoo.co.in

I. Dincer et al. (eds.), *Progress in Clean Energy, Volume 2*,
DOI 10.1007/978-3-319-17031-2_46

prediction of CoE for a HOPS. The RS-model is also allowed to identify the optimal configuration of hybrid power system for minimum CoE. A minimum CoE predicted by the model is comparable with that reported in the literature from earlier studies. The model presented in this study can be a useful tool for cost comparison among similar architectures of varying capacities for HOPS.

Nomenclature

IEA	International Energy Agency
PV	Photovoltaic
HPS	Hybrid power system
BAT	Battery
CoE	Cost of energy
HOPS	Hybrid off-grid power system
LOLP	Loss of load probability
LPSP	Loss of power supply probability
FL	Fuzzy logic
NN	Neural network
ML	Machine learning
EC	Evolutionary computation
PSO	Particle swarm optimization
GA	Genetic algorithm
OFAT	One-factor-at-a-time
FFD	Full factorial design
RSM	Response surface methodology
SRSM	Stochastic response surface methodology
RS	Response surface
CCD	Central composite design
BBD	Box–Behnken design
DG	Diesel generator
E	Power kW
S	Surface area m^2
η	Efficiency %
m	Meter
W	Watt
V	Voltage volt
Ah	Ampere hour
kW	Kilowatt
$	Dollar
HOMER	Hybrid Optimization Model for Electric Renewables
kWh	Kilowatt-hour
ANOVA	Analysis of variance
AD	Anderson–Darling

DF	Degree of freedom
Seq	Sequential
SS	Sum of squares
Adj	Adjusted
MS	Mean of square
StDev	Standard deviation
N	Sample size
p	Probability associated
T	Statistical parameter

Greek

ε Error

Subscripts

PV Photovoltaic

46.1 Introduction

God, the Great Giver, can open the whole universe to our gaze in the narrow space of a single land

Rabindranath Tagore

Energy is a crucial factor for the economic development and access to energy source is closely linked with development and economic well-being of a nation [1]. Accordingly, there is a tremendous increase in the demand for electricity across the globe [2, 3]. But in many developing countries the access to electricity is still limited in rural areas and remote localities. As a matter of fact more than half of the population of the developing nations live in the rural areas and suffer from economic impairment due to restricted access to stable electricity source [4]. According to the International Energy Agency (IEA), about 1.4 billion people in the world do not have access to electricity [5]. Thus, a key challenge for developing countries for the last few decades is to improve the living standard of the rural and remotely located communities by providing access to stable electricity source. However, the grid extension is often not considered economically feasible in the remote localities [6, 7]. A large portion of energy requirements of these communities are met through direct or indirect utilization of conventional fossil fuels such as oil, natural gas, and coal. However, fossil fuel sources are limited and fast depleting and are linked to economic instability. In addition, combustion of fossil fuel is associated with various detrimental environmental effects [8]. Accordingly, there is an imperative need to evaluate alternative renewable sources as replacement of fossil fuels and inaccessibility of the grid power for meeting the energy requirement of remote communities, particularly in developing countries.

46.2 Renewable Energy Options

Renewable energy is generally defined as energy that comes from resources which are naturally replenished, such as sunlight, wind, rain, tides, waves, and geothermal heat. Renewable energy replaces conventional fuels in many distinct areas such as electricity generation, hot water, motor fuels, and rural (off-grid) electrification [9].

Solar energy as radiant light and heat from the sun, are harnessed using a wide range of ever-evolving technologies such as solar heating, solar photovoltaic (PV), and solar thermal electricity. Solar power is used by conversion of solar radiation into electricity, either directly using PV or indirectly using concentrated solar power. PV convert light into electric current using the semiconductor based solar cell and photoelectric principle. Thus, solar PV is a significant energy harvesting technique due to abundant availability of solar radiation throughout the globe. However, solar PV is susceptible to the diurnal variation, latitudinal and longitudinal coordinates, and climatic variation [10, 11].

Wind power is derived from the conversion of kinetic energy of the wind into electrical and mechanical energy using conversion devices such as wind turbines to produce electrical power, wind mills or wind pumps to harvest mechanical power. Hundreds of individual wind turbines can form a large wind farms which can be connected to form an alternative electrical power transmission network. However, wind power suffers from geographical and climatic variation. Usually coastal areas are more suited for harvesting wind power. Harvesting energy from onshore wind is involved less installation and maintenance cost about offshore wind is steadier and stronger than onshore wind. But based on the resource availability small onshore wind farms can be a potential energy source for isolated off-grid localities [12].

Biomass is another important renewable energy sources and also important for carbon dioxide sequestration. Biomass is formed through biochemical conversion of carbon dioxide into biological material, collectively referred to as biomass. Biomass for energy sources is mostly obtained as wood, charcoal, rice husk, straw, other agriculture wastes and animal dung. As an energy source, biomass can either be used directly via combustion to produce heat or indirectly after converting to various forms of biofuel [13].

Hydroelectric power is another source of renewable energy that converts the potential energy or kinetic energy of water into mechanical energy and later to electrical energy. Small hydro is the small-scale hydroelectricity facility meant for a small community or an industrial plant. Small hydro plants may be connected to conventional electrical distribution networks as a source of low-cost renewable energy. Alternatively, small hydro projects may be built in isolated areas that would be uneconomic to serve from a network or in areas where there is no national electrical distributed network. Since small hydro projects usually have minimal reservoirs, they are seen as having a relatively low environmental impact compared to large hydroelectric projects [14].

Geothermal energy is generated from heat stored in the earth, or the collection of absorbed heat derived from underground. Geothermal power is cost effective, reliable, sustainable and environmental friendly but has been limited to areas near tectonic plate boundaries. Geothermal wells release greenhouse gases trapped deep within the earth but these emissions are much lower per energy unit than those of fossil fuels. As a result, geothermal power has the potential to help mitigate global warming if widely deployed in place of fossil fuels [15].

Tidal power is a form of hydro power that converts the energy of tides into useful form of power. Among various sources of renewable energy tidal power has traditionally suffered from relatively high cost and limited availability of sites with sufficiently high tidal ranges or flow velocities, thus constricting its total availability [16].

Biogas represents an alternative source of energy, derived mainly from organic wastes such as animal waste. Biodiesel, ethanol, ocean power are the different types of renewable energy sources. Another alternative of electricity generation is fuel cells. Fuel cell is a device that converts the chemical energy from a fuel into electricity through a chemical reaction with oxygen or other oxidizing agent [17, 18].

Renewable energy supplies 18 % of the world's final energy consumption, counting traditional biomass, large hydropower, and "new" renewables (small hydro, modern biomass, wind, solar, geothermal, and biofuels). Traditional biomass represents about 13 %, large hydropower represents 3 % and is growing modestly, primarily in developing countries. New renewables represents 2.4 % and are growing very rapidly in developed countries and in some developing countries [19]. According to the IEA, it has been predicted that solar power would account for 11 % of global electricity production by 2050, and thus, solar energy will play an important role in the future of energy [20].

However, electricity generated from PV power systems is a major renewable energy source which involves zero greenhouse gas emission and no fossil fuel consumption. Thus, hybrid PV systems consisting of solar power and other sources are performing successfully worldwide [21].

46.3 Hybrid Off-Grid Power Systems (HOPS) for Remote Localities

Hybrid power system (HPS) is composed of one renewable and one conventional energy source or more than one renewable with or without conventional energy sources that work in stand-alone (off-grid) or grid connected code [22]. HPS is becoming popular for stand-alone power generation in isolated areas due to the advances in renewable energy technologies and power electronic converters which are used to convert the unregulated power generated from renewable sources into useful power at the load end. However, renewable energy sources are associated with the demerits of power generation instability due to natural variation of the renewable resources and therefore are not reliable to satisfy the power demand of the load at each instant [23]. Accordingly, the choices of a renewable energy sources depends on the geographical location and climatic condition of the region, available resources and the economics of the power system [24, 25]. The major limitation of wind turbines is perennial wind in a specific direction which is very much the case along the coastline [12]. Due to the undulating topography mini hydel is also not very practicable renewable option [25]. The solar PV power systems coupled with energy storage battery units (BAT) to obviate the power fluctuation are best suited for most part of the world. But these PV based stand-alone power systems are highly capital intensive and therefore have high cost of energy (CoE) [26]. Accordingly, PV based stand-alone power systems are often downsized or complemented by additional power sources, leading to hybrid off-grid power system (HOPS), depending on the available solar irradiance, energy demand, and the economics of the power system. As the efficiency and the reliability of power supply increases by the use of different energy sources compared to systems comprising only one renewable energy source, the hybrid power systems consisting of two or more energy sources are considered more appropriate for off-grid power systems particularly in remote localities [27, 28].

46.4 Optimization of HOPS

Technically viable renewable energy sources are available at different levels, in different locations and the optimization of a hybrid system depends on those renewable sources as well as the load demand. The cost of the hybrid system also depends on the abundance of energy source, the quality or nature of source and component prices. Sometimes, involving more sources in the hybrid system makes the system more cost-effective and reliable.

Unit sizing and optimization is basically a method of determining the size of the hybrid system components by minimizing the system cost while maintaining system reliability. Accordingly, the designing or optimization of HOPS has contradictory objectives involving a reasonable and delicate compromise between the sizing objectives. Over sizing the system components will increase the system cost whereas under sizing can lead to failure of power supply. Thus, optimum resource management in a hybrid generation system is crucial to achieve acceptable cost and reliability level.

46.4.1 Optimization of Reliability

The management of HOPS for remote localities is of significant importance in the present scenario with respect to power system stability and reliability. The reliability of power supply is evaluated by various techniques; such as (1) Loss of Load Probability (LOLP) and (2) Loss of Power Supply Probability (LPSP).

The LOLP is a measure of the probability that a system demand will exceed the power supply capacity of the system in a given time period and often expressed as the estimated number of days over a long period [29–31].

LPSP is defined as the probability that an insufficient power supply results when the hybrid system is unable to satisfy the load demand. Also it is a feasible measure of the system performance for an assumed or known load distribution. A LPSP of 0 means the load will always be satisfied; and an LPSP of 1 means that the load will never be satisfied. The LPSP is a statistical parameter; its calculation is not only focused on the abundant or bad resource period. Therefore, in a bad resource year, the system will suffer from a higher chance of losing power.

There are two approaches for the application of LPSP in designing a stand-alone hybrid system. The first one is based on chronological simulation. This approach is computationally heavy and requires the availability of data covering a certain period of time. The second approach uses probabilistic techniques to incorporate the fluctuating nature of the resource and the load, thus eliminating the need for time-series data. For the continuous working of the system more precisely, the chronological method is more useful [29, 32].

However, in general the evaluation of LPSP is an effective and preferred technique for assessing the power supply reliability of a HOPS and often used as a decision criterion for sizing power system components.

46.4.2 Optimization of Cost of Energy (CoE)

The evaluation of HOPS from cost perspective is another important aspect in the optimization of HOPS. Multiple system architectures in a hybrid power system are evaluated by simulating the system performance and the cost heads in order to identify the optimum system configuration. The optimization of HOPS in cost perspective is calculated based on CoE; which is the average cost per kWh of useful energy produced by the system. The average cost includes capital cost, replacement cost, operational and maintenance cost, and fuel cost of the power sources [33–35]. For developing a HOPS in any remote location the important aspect is cost effectiveness of the system to the user. Thus, the evaluation of CoE is an effective technique for assessing the cost of the HOPS and can be used as a decision criterion for sizing power system components.

46.5 Optimization Approaches

There are various optimizational approaches available for optimization of hybrid systems, of which notable are—(1) soft computing approach and (2) statistical approach. The various soft computing techniques include Fuzzy Logic (FL), Neural network (NN), Machine learning (ML), and Evolutionary Computation (EC). Again EC has two subcategories, namely, Evolutionary Algorithm and Swarm Intelligence. EC in general has been proven to be a powerful optimizing technique. Among various types of EC, the particle swarm optimization (PSO) and genetic algorithm (GA) have been widely used by the researchers for optimization of hybrid power system [36, 37]. The various statistical approaches available for optimization include one-factor-at-a-time (OFAT) approach, factorial design, response surface analysis, and stochastic analysis.

The choice of an optimization approach is site specific and based of the availability of weather and load data. Soft computing approaches are well suited for optimization with nonavailability of weather and load data, whereas the statistical approaches are based on available weather and load data. However, for remote inaccessible sites the availability of meteorological data, that is needed for sizing or optimization of renewable resource based HOPS, is often limited. Therefore soft computing techniques are often utilized for optimization of component sizes of a system, but statistical approaches, particularly response surface analysis, are known to perform better in situations where a large number of experiments are not affordable.

Among the various statistical approaches available for optimization of powers source component sizes for HOPS, the one-factor-at-a-time (OFAT) is one such approach wherein one factor is optimized at a time. The OFAT approach is considered good for screening; however; for optimization OFAT results in misinterpretation due to interaction of multiple variables. The factorial design is another statistical method widely used by the researchers to study the effect of multiple variables simultaneously. Among various factorial designs, full factorial design (FFD) is often unpractical due to the requirements for a large number of experiments and fractional factorial design lacks the ability to accurately predict all positions of the factor space that are equidistant from the center (rotatability) [38]. Alternative statistical approaches are response surface methodology (RSM) and stochastic response surface methodology (SRSM). The SRSM is analyzed for uncertainty quantification of an optimized model. The SRSM is adopted to achieve the goal in which the number of model simulations for adequate estimation of uncertainty is substantially reduced as compared to conventional simulation. The stages involved in the uncertainty quantification of a model include, (a) estimation of uncertainties of model inputs, (b) estimation of uncertainty of the model output, and (c) propagation of uncertainty in the model output. Monte-Carlo and fuzzy theory have been widely used for the assessment of the SRSM [39, 40]. While the objective of the RSM is to determine the impact of the variables; and is accurately evaluated by varying all the factors simultaneously in a systematic manner. RSM optimizes multiple variables in a well-designed experiment with a minimum number of experiments. This RSM approach can also be used to establish the relationship between several independent variables and one or more dependent variables [41, 42].

The RSM optimization process follows a sequential approach where the first step is to screen the independent variables (factors) and their levels. The second step is to build the response surface (RS) model using an approximate experimental design method. The third step involves estimating the coefficients of the mathematical model using regression analysis technique and in the final step; the accuracy of the response is assessed using experimental data. Based upon the desirable features of equal predictability throughout the design surface, Central-Composite design (CCD) and Box–Behnken design (BBD) are common choices for RSM optimization. Compared to the CCD, the BBD technique is considered the most suitable for evaluating for response surfaces involving factors having quadratic effect. The BBD technique is a three-level design based upon the combination of two level factorial and incomplete block designs. The BBD method employs a spherical design and requires less experiment than the CCD with the same number of factors. Also the BBD technique is rotatable or nearly rotatable irrespective of the number of factors under consideration, thus helping to optimize the factors by a lesser number of experimental runs [38, 43, 44].

46.6 Factors Affecting the CoE for HOPS

The input information for optimization of HOPS include: load, power sources and their characteristics, costs and sizes of the component power sources and operating constraints. The optimization of a hybrid energy system is site specific and it depends upon the resources available and the load demand for a remotely located site.

46.6.1 System Load

To develop a technically viable system, the system load is one of the most important aspects in the design of the HOPS. Without load the HOPS cannot be developed. Primary load is electrical load that must be met immediately in order to avoid unmet load. Unmet load is electrical load that the power system is unable to serve and it occurs when the electrical demand exceeds the supply. In order to ensure the optimum system, the best possible matching between supply and demand is widely significant. However, the component sizes of the HOPS are directly proportional to the system load [45].

46.6.2 System Architecture

Different hybrid system architectures comprising different power sources in various combinations are evaluated for CoE minimization. As the PV panels are 15–17 % efficient, only that percentage of sunlight hitting a panel gets turned into electricity. Thus, PV alone is not capable to meet the load without storage or complementary power sources, hence configurations with PV either have a battery or a backup (as diesel generator or a biomass generator or a fuel cell) make the system feasible. Wind system is dependent on wind speed. The natural variation of wind speed causes voltage and power fluctuation problems at the load side for wind turbine based HOPS. Thus, PV–Wind hybrid system is not reliable as both of the sources are dependent on climatic conditions. Incorporating backup in hybrid system makes the system realistic in terms of system reliability and stability; however, the CoE of the system is in difficulty [26, 46].

46.6.3 System Component Size

Determining the best geographic location for PV technologies, the system is sized to meet certain demand. By knowing the variability, storage system could be sized so that energy could be provided during cloudy days and nights. BAT sizing was calculated based on the number of days of storage desired. A storage day referring to the amount of capacity a BAT has supply power, without receiving any power input. The effect of PV and BAT size has no significant effect on the CoE of the system depending of system load rather than system reliability. However, diesel generator (DG) size has significant effect on the CoE and similarly on system reliability [47, 48].

46.6.4 System Reliability

To efficiently and economically utilize the renewable energy resources, one optimum sizing method is necessary. The sizing optimization method can help to guarantee the lowest CoE with adequate and full use of the power sources so that the hybrid system can work at optimum conditions in terms of CoE and system power reliability requirement. A reliable HOPS is the one which generates sufficient power to meet the load demand during a certain period. The HOPS is considered unreliable if it fails to satisfy the energy demand. However, CoE is calculated only for the reliable system [49–51].

46.7 Statistical Predictive Model of CoE for HOPS in Remote Localities: Model Development and Validation (A Case Study)

Solar PV technology has received a great deal of attention for efforts to electrify off-grid rural and remote localities because of the well-proven benefits, which include the ability to produce electricity in most environmentally manner with little maintenance hazard and scope for easy integration and future expansion. However, solar PV based power systems are unable to supply constant power to meet energy demand because of the inherent variability of the system. The limitation of solar PV based power system is resolved by adding complementary power source with storage facility. DG has been considered as a preferred complementary nonrenewable energy source with PV based renewable power due to the low capital cost, ease of operation, and simplicity of installation. But hybrid PV-DG system offers the benefits of improved reliability of power supply, crucial criterion for HOPS designed for remote localities. However, optimization of the component sizes is equally critical for ensuring optimum resource utilization, maximum power supply reliability, and minimum energy cost. Thus, the evaluation and optimization of HOPS in terms of CoE with various alternative energy sources considering the demand of the location and available resources is highly pertinent for configuring a hybrid power system for remote localities.

With regard to the above objective a case study conducted in a remotely located academic site in north-eastern India is presented. The study presents a statistical model to optimize the utilization of energy sources of a hybrid stand-alone power system and thereby minimize the CoE. A predictive statistical model to predict the CoE implementing the BBD technique to optimize the power sources component size is discussed. The statistical model proposed in this study is expected to be useful in decision making with regard to the sizing of power source components in a HOPS and determine the impact on the users with respect to CoE.

46.7.1 System Description for Model Development

46.7.1.1 Site Information and Load Data

The HOPS proposed in this study is designed for incessant supply of electricity to a remotely academic institute in north-eastern India considered as a model site. The latitude and longitude of the site are 23.80 N and 91.50 E respectively. The model site has variety of electrical devices commonly encounter in an academic institute

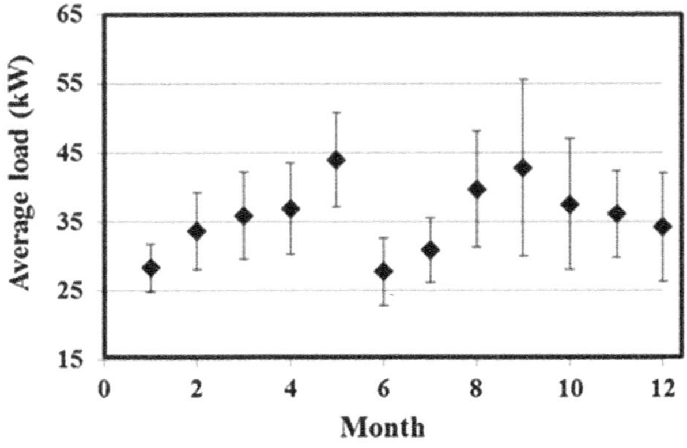

Fig. 46.1A Monthly load variation with standard deviation (average load 35.6 kW)

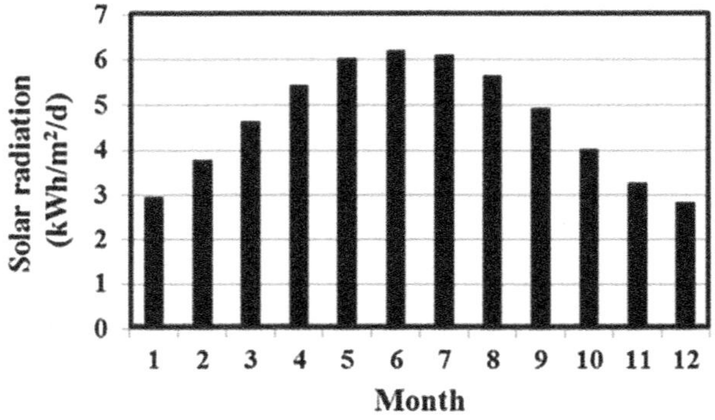

Fig. 46.1B Monthly solar radiation variation of model site

which include 150 units of 40 W fluorescent lights, 100 units of 85 W fans and 140 units of 150 W computers. The average load of the model site is 35.6 kW. The scaled annual average load is 854 kWh/day. The monthly average load profile with standard deviation is shown in Fig. 46.1A.

46.7.1.2 Available Energy Resources

The solar PV is the only viable renewable energy option for the model site due to the geographical location. The solar radiation varies throughout the year from 2.85 to 6.22 kWh/m^2/day with an annual average of 4.7 kWh/m^2/day. The computed month-wise variation of solar radiation is presented in Fig. 46.1B. The power generated in PV source is computed from the solar radiation data (expressed in W/m^2), available PV surface (S_{PV}) and conversion efficiency (η_{PV}) following the relation presented in Eq. (46.1) [52].

$$E_{PV} = \text{Insolation} \times \eta_{PV} \times S_{PV} \tag{46.1}$$

Table 46.1 Unit cost of system components

Components	Units	Capital cost ($)	Replacement cost ($)	O & M cost ($/year)	Lifetime (year)
PV	1 kW	4,200	3,150	10	25
BAT	1 block	3,000	2,550	14	15
CONV	1 kW	750	502.5	8	15
DG	1 kW	500[a]	400	73	150,000 h

[a]Note: ref. [34]

Each PV module with exposed surface area of 3.66 m^2 is computed to have a rated capacity of 205 W PV modules are connected in series in order to generate the desired output. The lifetime of PV array is considered as 25 years.

In order to obviate the natural fluctuation in solar radiation and to include a power storage option, the deep cycle flooded lead-acid BAT are included in the HOPS considered in this study. The BAT are configured as a 40 blocks of 6 V battery delivering 1,156 Ah at a nominal voltage of 240 V with round trip efficiency of 80 %. The power will be stored in the batteries on excess power generation (termed as charging) and will be utilized from stored battery power in case of shortage of power generation (termed as discharging). Depending on the charge or discharge mode of operation, the BAT input power can be positive or negative. The lifetime of BAT is considered 15 years [53].

Nonrenewable DG is added to the HOPS as complementary power source. The cost of a commercially available DG may vary from $250/kW to $500/kW. The larger the DG size, the lower the cost per kW [54]. The nominal oil consumption rate for the considered DG sizes is kept invariant 0.102 L/kWh and the fuel cost for the operation of DG is kept invariant at 0.8 $/L. As the CoE is linked to the duration of operation of DG, hence optimum utilization of DG is a major determinant for the proposed HOPS. The unit cost of various power components of the HOPS are presented in Table 46.1.

46.7.2 Experimental Design for Model Development

46.7.2.1 Determination of Factors and Their Levels

In the considered BBD, each experimental factor is evaluated at three levels, namely, a minimum or low level (denoted as 1), a central or medium level (denoted as 2), and a high or maximum level (denoted as 3). The three factors are PV size, DG size, and BAT cells. The three levels of the PV size are distributed between 100 kW and 300 kW. For the particular system configuration and load profile PV sizes below 100 kW is not sufficient. The utilization of PV size more than 300 kW was ruled out considering the load served and cost implication of a redundant capacity. The DG sizes are varied from 30 kW to 50 kW. Increasing DG capacity (beyond 50 kW) results in higher CoE and lower DG sizes (below 30 kW) are not sufficient for the considered system load. The BAT cells are varied from 600 units

Table 46.2 Factors and associated levels for the study

Levels	Factors PV size (kW)	DG size (kW)	Number of BAT cells (units)
1 (Low)	100	30	600
2 (Medium)	200	40	800
3 (High)	300	50	1,000

Table 46.3 Design matrix for the factor and the response at various factor level settings

Experimental orders	Factors PV size (kW)	DG size (kW)	BAT cell (units)	Response (CoE($/kWh))
1	200	30	600	0.201
2	200	40	800	0.200
3	200	30	1,000	0.202
4	100	30	800	0.214
5	100	40	1,000	0.227
6	100	50	800	0.231
7	200	50	1,000	0.203
8	100	40	600	0.219
9	200	40	800	0.200
10	300	40	600	0.273
11	200	40	800	0.200
12	200	50	600	0.204
13	300	40	1,000	0.280
14	300	30	800	0.275
15	300	50	800	0.276

to 1,000 units. BAT cells below 600 units are incompatible with the considered PV size and BAT cells beyond 1,000 units are excessive for the considered system load. A converter is one of the components for the proposed HOPS. Since varying the converter size has no significant effect on the CoE hence converter has not been considered as a factor and kept invariant at 80 kW. The levels associated with each of the three experimental factors are summarized in Table 46.2.

46.7.2.2 Methodology for Model Development

A full quadratic (second order) model for the response function (CoE ($/kWh)) is evaluated in BBD technique. For the BBD experimental design, the effect of factors other than those evaluated (PV, DG, and BAT) is considered as errors for the experimental design and is accounted as ε (lack-of-fit). The factor space of the BBD is defined from 15 design points marked using the three design factors under evaluation with three associated levels for each of the factors. Out of the 15 design points of BBD, 12 design points are evaluated to define the factor space and 3 design points (central points) are assessed to estimate the error of the model prediction. The factors with their associated levels are used in the BBD without coding (uncoded dimension). The design points of BBD are presented in Table 46.3. For each of the 15 design points, defined by specific size of PV, DG, and BAT in the HOPS, the CoE is computed from the unit cost of the system components summarized in Table 46.1 using Hybrid Optimization Model for Electric Renewables soft computing tool (HOMER, (ver. 2.68; NREL, Golden, CO, USA)). The computed CoE is considered as the response variable for the BBD model.

The quadratic BBD based RS-model (Eq. 46.2) expresses the CoE as a function of the sizes of various power sources (factors).

$$\begin{aligned}\text{CoE (\$/kWh)} &= a_0 + a_1 \times (\text{PV size}) + a_2 \times (\text{DG size}) + a_3 \times (\text{BAT cell}) \\ &+ a_4 \times (\text{PV size})^2 + a_5 \times (\text{DG size})^2 + a_6 \times (\text{BAT cell})^2 + a_7 \times (\text{PV size}) \\ &\times (\text{DG size}) + a_8 \times (\text{PV size}) \times (\text{BAT cell}) + a_9 \times (\text{DG size}) \times (\text{BAT cell}) + \varepsilon\end{aligned} \quad (46.2)$$

In Eq. (46.2), a_0–a_9 are the regression coefficients for the respective model and ε is the error associated with the model. In order to avoid any systematic bias in the outcomes, the experimental order of design points is randomized for the computation of CoE.

46.7.2.3 Coefficient Estimation

An analysis of variance (ANOVA) is performed on the computed CoE (response) to evaluate the full quadratic estimation of the BBD RS-model. The order of the RS-model (linear, square or full quadratic) is determined from the F-values and the associated p-values of the ANOVA results. MINITAB statistical software (version 15, Minitab Inc., State College, PA) is used for all statistical computations presented in this study. A multiple regression analysis (method of least square) was performed to fit the response function (Eq. 46.2) to the experimental data (Table 46.3) and thereby estimate the model coefficients. The significance of the model coefficients are ascertained from the computed t-values and the associated p-values. The model coefficients associated with p values less than 0.05 considered are statistically significant. Accordingly the RS-model is refined by backward elimination method deleting the coefficients that are statistically insignificant ($p > 0.05$) [44, 55, 56].

46.7.2.4 Verification of the Model Accuracy

The difference between the model prediction and the experimental outcome at identical factors level within the design space under consideration is termed as residual. The residuals are expected to follow a normal distribution (occurrences are random) for a well predicted model. Accordingly, an analysis of residuals is performed to verify normal distribution of residuals. The Anderson–Darling (AD) test is a statistical tool used to quantify the deviation of residuals from a normal distribution. The AD test is one sided test where the hypothesis of normal distribution of residuals is rejected if the test statistic is greater than a critical value [44]. An AD test of residuals at 5 % level of significance is used in this study to confirm the model accuracy and assess normal distribution of residuals [55, 56].

46.7.2.5 Optimization for the Minimum CoE

The optimum factor setting for minimum response (CoE) is located within the factor space by the D-optimality based optimization method using the RS-model. The D-optimality procedure is a popular optimization method in which a numerical algorithm is used to calculate the D-optimality value for all possible combinations of the various factor levels. The D-optimality value was varied between zero (worst case) and one (ideal case) for the factor levels under consideration. The factor levels for the minimum response (CoE) are identified at the largest D-optimality value. In the present study, an algorithm from the MINITAB statistical software is used to compute the D-optimality values to identify the minimum response (CoE).

46.7.3 Results and Discussions

46.7.3.1 Effect of Power Sources on System CoE

The effect of three experimental factors, namely, PV size, DG size, and number of BAT cells on the system CoE are evaluated at the different experimental levels (Table 46.2). The effect of varying PV, DG sizes and number of BAT cells for a HOPS are meticulously assessed for minimum CoE. The effect of PV size on system CoE is shown in Fig. 46.2A. Notice initially increasing PV sizes were

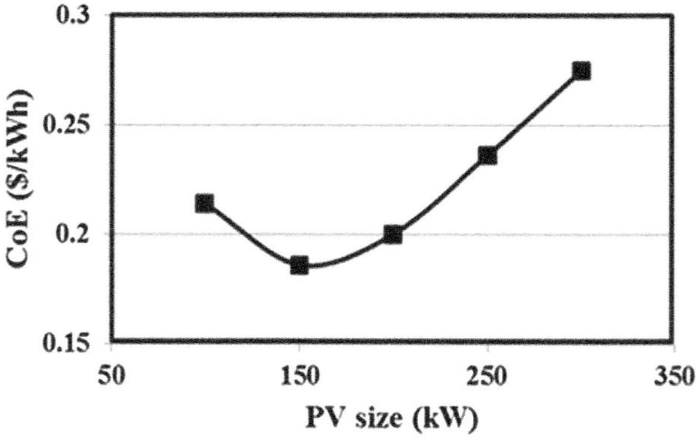

Fig. 46.2A Effect of PV size on system CoE [DG size: 30 kW, BAT cell: 800 units]

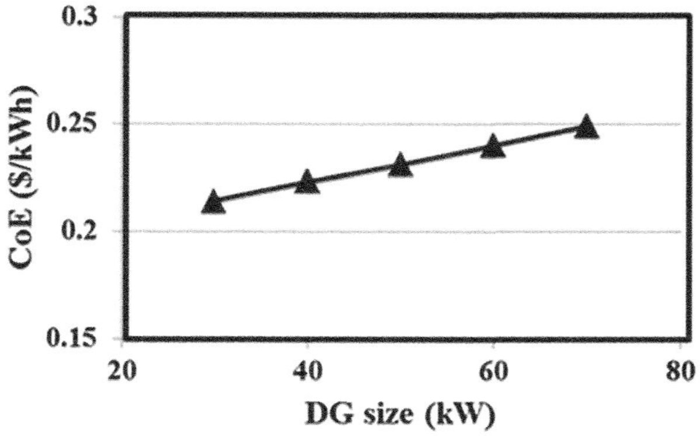

Fig. 46.2B Effect of DG size on system CoE [PV size: 200 kW, BAT cell: 800 units]

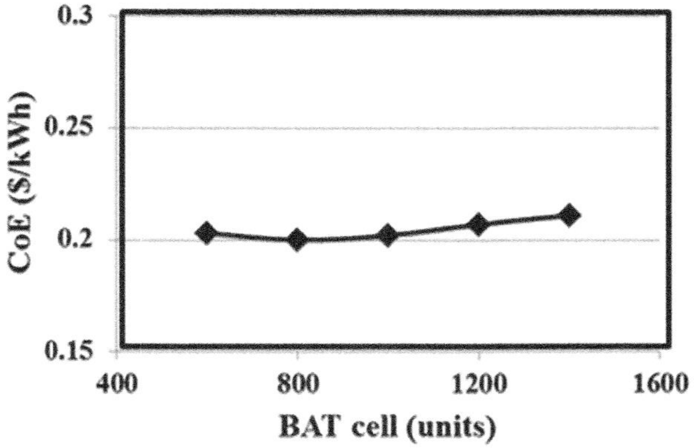

Fig. 46.2C Effect of BAT cell on system CoE [PV size: 200 kW, DG size: 40 kW]

observed with decreasing CoE. The CoE was minimum at 150 kW PV size. However, after 150 kW PV size, the CoE increasing linearly with PV. Initial increments in PV size decrease the operating cost in PV–DG–BAT system and thereby reduce the CoE. However, further increments in PV size beyond 150 kW add to the capital cost and increase the CoE. Figure 46.2B shows the effect of DG size on system CoE. The cost of DG is known to depend on the size, accordingly the CoE increases linearly with DG size [45]. In addition higher DG size adds to increased operating cost. Figure 46.2C reveals that increase in number of BAT cells results in an increase in CoE. Notice that the CoE is nearly invariant for 600–1,000 units of BAT cell.

A plot of two-factor interaction matrix in Fig. 46.3 showed evidence of interaction at all factor-level combination. Significant interactions among factors are noted with regard to the factor–response relationship. Higher CoE is reported in PV–DG–BAT system for PV size of 100 kW compared to that for 200 kW PV due to the PV-DG interactions. Similar interaction is also observed between PV size and BAT cell. A complicated interaction is observed between DG size and number of BAT cells. The strong factor level interactions further emphasize the importance of BBD

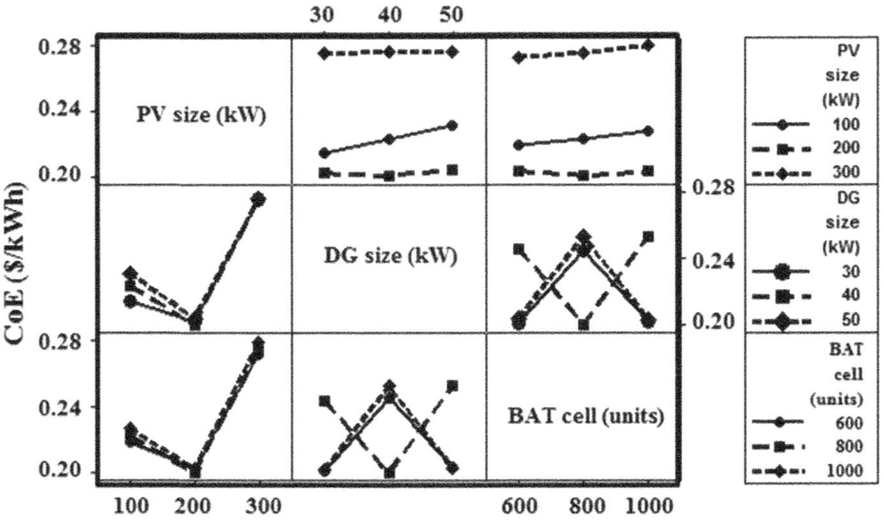

Fig. 46.3 Plot of interaction between experimental factors for system CoE in a three-factor, three-level BBD

Table 46.4 ANOVA results for the experimental response at different factor levels

Source	DF	Seq SS	Adj MS	F	p ($F > F_{0.05}$)
Regression	9	0.0238	0.0026	361.82	0
Linear	3	0.0086	0.0029	389.46	0[a]
Square	3	0.0152	0.0051	693.03	0[a]
Interaction	3	7E−05	2E−05	2.97	0.061
Residual error	17	0.0001	7E−06		
Lack-of-fit	9	8E−05	9E−06	1.6	0.261
Pure error	8	4E−05	6E−06		
Total	26	0.024			

Notes: *DF* degrees of freedom, *Seq SS* sequential sum of square, *Adj MS* adjusted mean of square, *p* a statement describing *F*

[a]Values are statistically significant at 5 % level of significance

based systematic evaluation of factor space for the purpose of optimizing the HOPS for minimum CoE.

46.7.3.2 Data Analysis and Model Development

The CoE for the HOPS involving PV (kW), DG (kW), and BAT cells (units) at various levels are computed for each design points tabulated in Table 46.3 in order to evaluate a full-quadratic BBD RS-model. An ANOVA was performed on the experimental data to evaluate the full quadratic RS-model (Eq. 46.2). The ANOVA

Table 46.5 Regression coefficients for the RS-model

Term	Coefficient	Regression coefficient	T	$P\ (T > T_{0.05})$
Constant	a_0	0.3113	172.464	0
PV	a_1	(−)1.4875	34.04	0
DG	a_2	0.0011	2.668	0.016
BAT	a_3	(−)0.0018	1.601	0.028
PV×PV	a_4	4.8083	43.537	0
DG×DG	a_5	0	0.075	0.941
BAT×BAT	a_6	0.0001	1.434	0.017
PV×DG	a_7	(−)0.004	(−)2.957	0.009
PV×BAT	a_8	(−)0.0005	(−)0.185	0.856
DG×BAT	a_9	0	(−)0.37	0.716

Note: *shaded* values are statistically significant at 5 % level of significance

results presented in Table 46.4 showed that the experimental data bear a strong statistically significant regression with full quadratic approximation of the model at 5 % level of significance. The error associated with the model was evaluated by computing the lack-of-fit. A p-value of 0.261 (associated with the lack-of-fit) suggested that the model error and the lack-of-fit are statistically insignificant at a 5 % level of significance. The statistical significance of linear and quadratic terms in the model is also established from the ANOVA results. The interaction terms were statistically less significant than the linear and quadratic terms.

A multiple regression analysis was performed on the experimental data to estimate the regression coefficient for the model. The computed regression coefficients for the model along with their respective p-values are presented in Table 46.5. A backward elimination method was applied and statistically insignificant terms ($p > 0.05$) were deleted from the full quadratic model to refine the RS-model to the final form (Eq. 46.3). The final RS-model (Eq. 46.3) is thereafter used to obtain the response surface plot and response prediction.

$$\begin{aligned} &\text{CoE (\$/kWh)} = 0.3113 - 1.4875 \times (\text{PV size}) + 0.0011 \times (\text{DG size}) \\ &-0.0018 \times (\text{BAT cell}) + 4.8083 \times (\text{PV size})^2 + 0.0001 \times (\text{BAT cell})^2 \\ &-0.004 \times (\text{PV size}) \times (\text{DG size}) \end{aligned} \tag{46.3}$$

Three dimensional (3D) surface plots of the response variable (CoE ($/kWh)) for the experimental factors (two factor-at-a-time) are shown in Fig. 46.4A, 46.4B, and 46.4C. The surface plot for PV size versus the BAT cells (Fig. 46.4A) show that high value of CoE is associated with high PV size (300 kW) at low BAT (600 units) and high BAT (1,000 units) levels. A region of low CoE is observed for the middle setting of BAT (800 units) for all PV sizes. At high and low BAT levels, the CoE bear a nearly linear relationship to the PV sizes. The lowest CoE is observed at 150 kW PV and 800 units of BAT cell and the maximum CoE is obtained at 300 kW PV and 1,000 units of BAT cell.

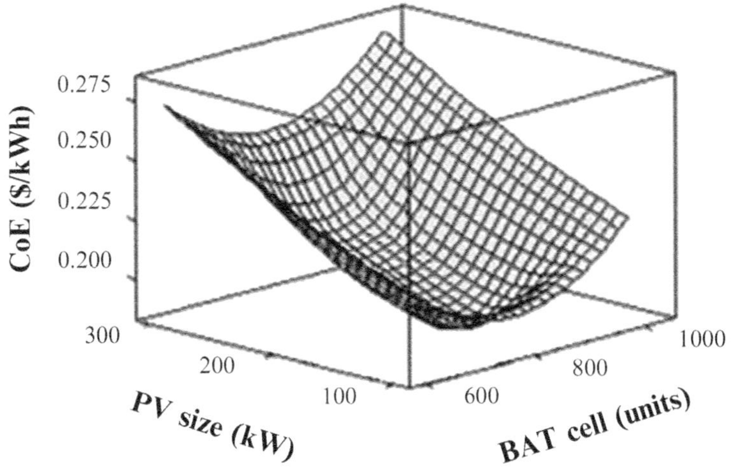

Fig. 46.4A The 3D surface plot of CoE ($/kWh) vs. PV size and BAT cell [DG size = 40 kW]

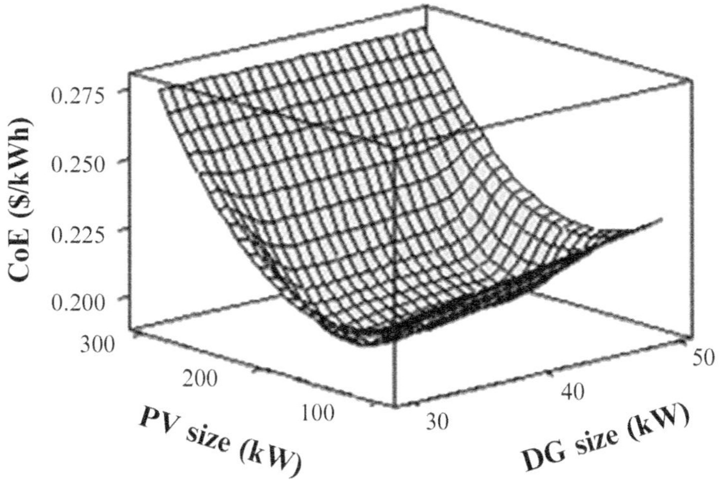

Fig. 46.4B The 3D surface plot of CoE ($/kWh) vs. PV size and DG size [BAT cell = 800 units]

The effect of varying PV and DG sizes on system CoE is shown in Fig. 46.4B. The CoE of the system increases with increasing PV sizes at varying DG sizes. However, at the highest setting of PV (300 kW), the CoE is invariant of DG size. At low PV setting (100 kW), the CoE is linearly associated with the DG sizes. For all DG levels, the low CoE is observed near the middle setting of PV. The surface plot of DG size versus BAT cell is presented in Fig. 46.4C. At high BAT level (1,000 units) the CoE is indifferent to varying DG sizes (similar to Fig. 46.4B). The CoE bears a linear relationship to DG size at low BAT level (600 units). For all DG sizes, the middle setting of BAT is observed to have low CoE. The minimum CoE is

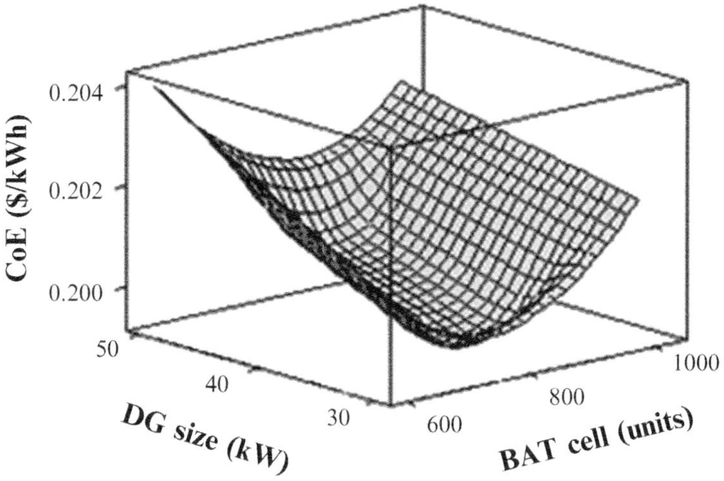

Fig. 46.4C The 3D surface plot of CoE ($/kWh) vs. DG size and BAT cell [PV size = 200 kW]

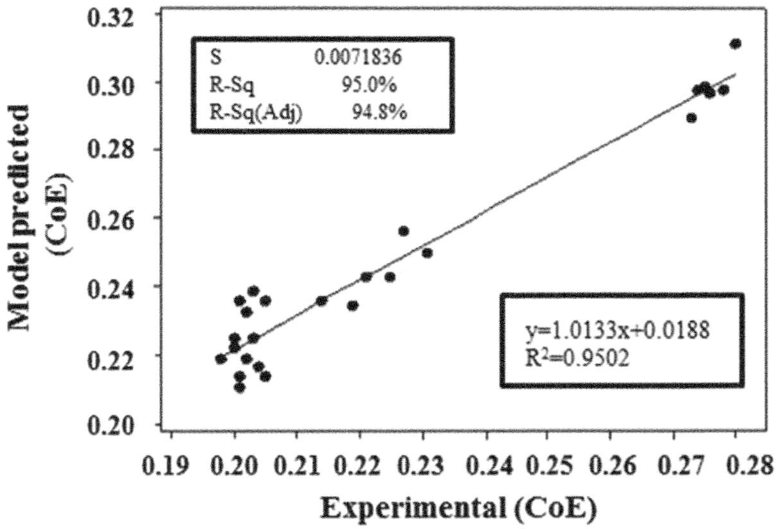

Fig. 46.5A Plot of model predicted CoE against experimental CoE

observed for 800 units BAT cell with 30 kW DG and the maximum value of CoE is obtained at 50 kW DG with 600 units of BAT cell for PV size fixed at 200 kW.

46.7.3.3 Model Verification

The system CoE was computed using the refined model (Eq. 46.3) for the conditions in Table 46.3. The model predicted CoE is correlated with experimentally

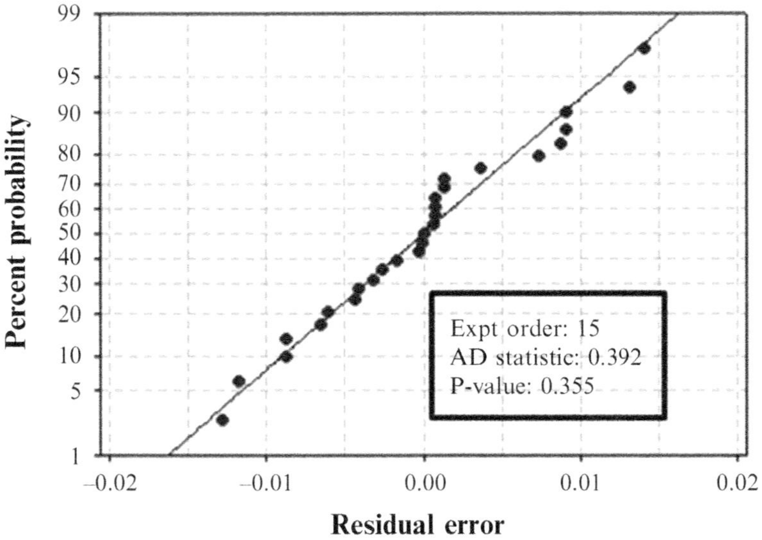

Fig. 46.5B Anderson–Darling normality plot of residuals (model predicted CoE minus experimental CoE). [*Mean* mean value of residual for the CoE (difference between model prediction and experimental result), *StDev* standard deviation of the residuals for 15 experiments, *N* sample size, *AD* Anderson–Darling statistics, *P* level of confidence]

computed CoE to verify the model accuracy. The plot of model predicted CoE versus computed CoE is presented in Fig. 46.5A. The CoE values predicted by the model correlated reasonably well with the computed experimental CoE with a regression coefficient of 0.9748 ($R^2 = 0.9502$) for all the experimental orders. The difference between the model predicted and experimental CoE is computed for each of 15 experimental orders (termed as residual). The residuals are subjected to AD test to assess probability distribution of residuals. A normal distribution of residuals ensures an adequate fit of the model with the experimental data and is useful in judging the adequacy of fitting the model to experimental data.

The probability distribution plot of residuals is presented in Fig. 46.5B. The AD statistic is computed for the model residual. The calculated AD statistic (0.392) was less than the critical value of the statistic (0.752) for a sample size of 15. The computed AD statistic being less compared to the critical value and associated p-value of 0.355 being much greater than $p = 0.05$ confirmed a normal-fit of the probability distribution of residuals. A normal fit of the probability distribution of residuals verifies with 95 % level of confidence that the deviation of the model predictions from the experimental results is purely random [38]. The results confirmed that the model prediction is well correlated with the experimental data over the entire factor-space evaluated in the present study.

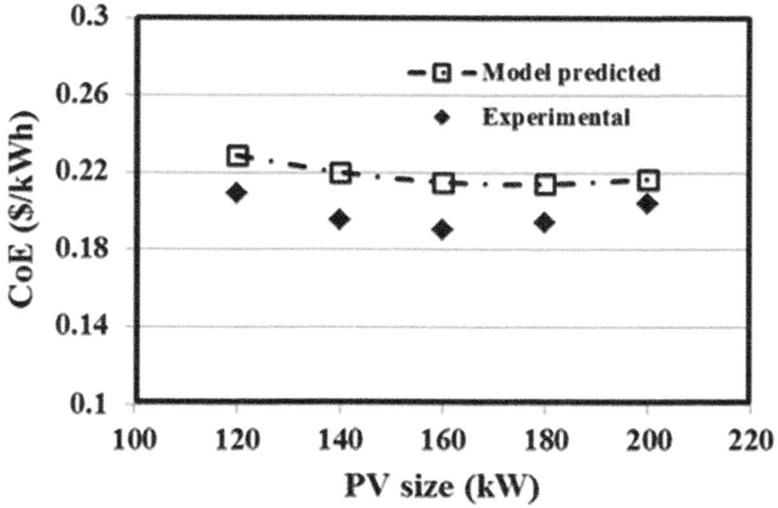

Fig. 46.6A Validation of the model prediction CoE ($/kWh) vs. PV size [DG size: 50 kW; BAT cell: 600 units]

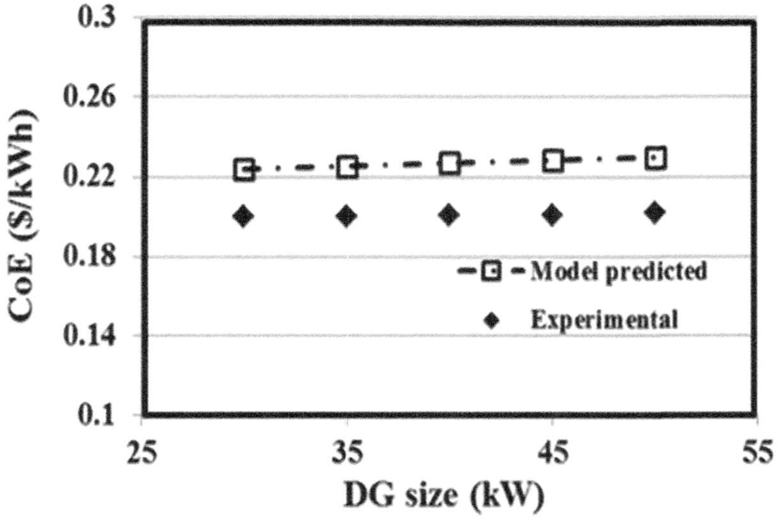

Fig. 46.6B Validation of the model prediction CoE ($/kWh) versus DG size [PV size: 200 kW; BAT cell: 880 units]

46.7.3.4 Model Validation

A validation study is performed by conducting additional experiments at various levels for each of the three factors (PV size, DG size, and BAT cell) under evaluation to confirm the validity of the RS-model over the factor-space. The

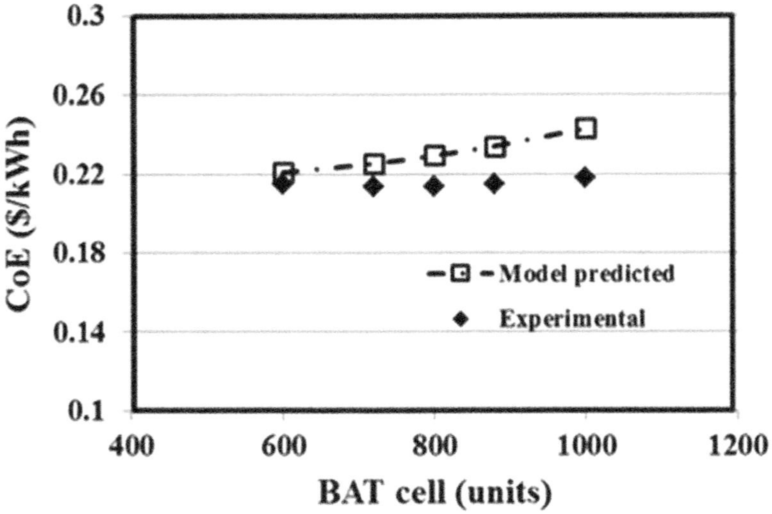

Fig. 46.6C Validation of the model prediction CoE ($/kWh) vs. BAT cell [PV size: 220 kW; DG size: 40 kW]

model predicted CoE values were compared with the experimental CoE for various PV sizes ranging from 100 kW to 220 kW (Fig. 46.6A). The model prediction closely matches the experimental CoE through the entire range of PV sizes. At higher PV sizes (200 kW), the model prediction is better correlated with the experimental CoE; however, for the entire range of PV sizes evaluated the model predicted CoE is overestimated compared to experimental values.

The comparison of the model predicted CoE against computed CoE (experimental) for various DG sizes (Fig. 46.6B) showed well agreement through the entire range with model prediction slightly higher than the computed values. Notice, at lower DG size there is better agreement between the model and the experimental CoE. A plot of model predicted CoE versus experimental CoE against various BAT capacities is presented in Fig. 46.6C. The model predicted CoE is observed to coincide with the experimental CoE for lowest setting of BAT capacity (600 units of BAT cell). The model prediction closely correlates with the experimental value, except the difference between the model and the experimental CoE increases with increase in the BAT capacity. The RS-model proposed in this study has been identified to have the shortcoming of overestimating the system CoE through most of the factor-space under evaluation.

The RS-model developed in this study is used to locate the optimum factor-level settings for minimum response (CoE) within the factor space. The D-optimality based optimization procedure is followed to determine the ideal factor level setting. Accordingly, the D-optimality criterion is computed for all factor levels defined within the factor-space under evaluation (Table 46.2) for the three experimental factors (PV, DG, and BAT). The ideal factor-level settings are identified from the D-optimality criterion closest to 1.00. The largest D-optimality value computed within the factor-space is 0.99 (closest to 1.00). The largest D-optimality value is

Table 46.6 Comparative summary of simulation parameters for various similar architectures from literature

System	PV size (kW)	DG size (kW)	BAT (units)	CoE ($/kWh)	Optimization tool	References
PV + DG + BAT	6	10	10	0.489	HOMER	[45]
PV + DG + BAT	60	50,50	12	0.796	HOMER	[34]
PV + DG + BAT	2,000	2,400	1,300	0.632	HOMER	[33]
PV + DG + BAT	165	45	800	0.188	HOMER	[57]
PV–DG–BAT (present study)	161	30	680	0.194	Statistical model	

associated with a HOPS comprising 161 kW PV, 30 kW DG, and 680 units of BAT cell. The CoE (response) for the optimum HOPS with PV–DG–BAT is computed from the RS-model and is predicted as 0.194 $/kWh.

46.8 Comparison of Optimized CoE with Literature

The optimum size of PV–DG–BAT and the CoE for the HOPS proposed in this study is compared with the CoE reported in earlier studies for similar capacity HOPS. The comparative values are tabulated in Table 46.6.

The tabulated values from literature showed that CoE for PV–DG–BAT based HOPS varied from 0.188 to 0.796 $/kWh. In all the earlier reports HOMER was used as the optimization tool for identifying the ideal component sizing with component size varying from large [33] to small [45]. With exception to a recent report [57], the CoE computed for the optimized HOPS proposed in this study is much lower compared to the other reported CoE values from various system capacities. Hence, the optimization performance of the RS-model (statistical predictive model) proposed in the present study is well comparable with HOMER optimizing tool. Thus, the present study presents an alternative approach of optimizing the sizes of HOPS other than HOMER. Thus, the proposed RS-model can be considered as a useful tool for predicting CoE of similar capacity HOPS and for optimizing component sizes for PV–DG–BAT systems.

46.9 Conclusion

The present study focuses on utilizing electricity demand of a remotely located site using various available off-grid hybrid power sources. This study analyzes the optimization of CoE of a HOPS using statistical technique. A RS-model based on the BBD technique is developed for computing the CoE for PV–DG–BAT based HOPS. The PV size, DG size, and BAT cells are the three experimental factors considered in a three-factor, three-level BBD. Each of the three factors are varied at three well distributed levels to define the intended factor-space for PV–DG–BAT based HOPS sufficient to serve an average energy demand (load) of 35.6 kW. The results of one-factor-at-a-time (OFAT) study showed that increasing DG size and BAT capacities has a nearly identical impact on the CoE and incremental relationship is noted between the factors (DG size and BAT capacity) and the response (CoE). The effect of varying PV sizes on the CoE is rather complicated and improper choice of PV size has unfavorable effect on the CoE. Significant interactions among factors are noted in the factor–response relationship. Accordingly a full quadratic RS-model is evaluated in a three factor BBD with 15 design points. The ANOVA was conducted for each factor level settings of the BBD. The results of ANOVA established statistical significance of the RS-model. The coefficients of the RS-model are computed using multiple regression analysis technique and the model is refined through backward elimination method at 95 % level of confidence. The final RS model is checked for error in prediction by residual analysis using AD test. AD-test concluded normal distribution of residuals. A normal distribution of residuals confirmed that the model prediction correlated reasonably well with the experimental results over the experimental factor space under consideration. The model is further validated against experimental data to confirm the accuracy of the model. The results confirmed that the model has 97.5 % accuracy in prediction of CoE for a HOPS comprising PV and DG as power source and BAT as storage element through the entire design space evaluated in this study. The RS model is used to identify the optimal configuration of the HOPS for minimum CoE. The minimum CoE (response) of 0.194 $/kWh for a HOPS with 161 kW PV, 30 kW DG, and 680 units of BAT cell has been predicted by the model. The CoE of the PV–DG–BAT system presented in this study is comparable with that reported in the literature from earlier studies. The statistical RS-model is developed in this study can be considered as a useful tool for decision making with regard to the sizing of a HOPS and determining the impact on the users with respect to CoE.

Acknowledgement Financial support for this work was provided by the National Institute of Technology, Agartala (NIT Agartala) and All India Council of Technical Education (AICTE).

References

1. Hatti M, Meharrar A, Tioursi M (2011) Power management strategy in the alternative energy photovoltaic/PEM Fuel Cell hybrid system. Renew Sust Energ Rev 15:5104–5110. doi:10.1016/j.rser.2011.07.046
2. Bazmi AA, Zahedia G (2011) Sustainable energy systems: role of optimization modeling techniques in power generation and supply—a review. Renew Sust Energ Rev 15:3480–3500. doi:10.1016/j.rser.2011.05.003
3. Urban F, Benders RMJ, Moll HC (2009) Energy for rural India. Appl Energ 86:S47–S57. doi:10.1016/j.apenergy.2009.02.018

4. Ardehali MM (2006) Rural energy development in Iran: non-renewable and renewable resources. Renew Energ 31:655–662. doi:10.1016/j.renene.2005.08.002
5. Srivastava L, Goswami A, Diljun GM, Chaudhury S (2012) Energy access: revelations from energy consumption patterns in rural India. Energ Policy 47:11–20. doi:10.1016/j.enpol.2012.03.030
6. Asrari A, Ghasemi A, Javidi MH (2012) Economic evaluation of hybrid renewable energy systems for rural electrification in Iran—a case study. Renew Sust Energ Rev 16:3123–3130. doi:10.1016/j.rser.2012.02.052
7. Haidar AMA, John PN, Shawal M (2011) Optimal configuration assessment of renewable energy in Malaysia. Renew Energ 36:881–888. doi:10.1016/j.renene.2010.07.024
8. Erdinc O, Uzunoglu M (2012) Optimum design of hybrid renewable energy systems: overview of different approaches. Renew Sust Energ Rev 16:1412–1425. doi:10.1016/j.rser.2011.11.011
9. Bajpai P, Dash V (2012) Hybrid renewable energy systems for power generation in stand-alone applications: a review. Renew Sust Energ Rev 16:2926–2939. doi:10.1016/j.rser.2012.02.009
10. Singh PP, Singh S (2010) Realistic generation cost of solar photovoltaic electricity. Renew Energ 35:563–569. doi:10.1016/j.renene.2009.07.020
11. Tiwari A, Barnwal P, Sandhu GS, Sodha MS (2009) Energy metrics analysis of hybrid-photovoltaic (PV) modules. Appl Energ 86:2615–2625. doi:10.1016/j.apenergy.2009.04.020
12. Ramachandra TV, Shruthi BV (2005) Wind energy potential mapping in Karnataka, India, using GIS. Energ Convers Manage 46:1561–1578. doi:10.1016/j.enconman.2004.07.009
13. Balamurugan P, Ashok S, Jose TL (2009) Optimal operation of biomass/wind/PV hybrid energy system for rural areas. Int J Green Energ 6:104–116. doi:10.1080/15435070802701892
14. Nfah EM, Ngundam JM (2009) Feasibility of pico-hydro and photovoltaic hybrid power systems for remote villages in Cameroon. Renew Energ 34:1445–1450. doi:10.1016/j.renene.2008.10.019
15. Zarrouka SJ, Moon H (2014) Efficiency of geothermal power plants: a worldwide review. Geothermics 51:142–153. doi:10.1016/j.geothermics.2013.11.001
16. Tang HS, Qu K, Chen GQ, Kraatz S, Aboobaker N, Jiang CB (2014) Potential sites for tidal power generation: a thorough search at coast of New Jersey, USA. Renew Sust Energ Rev 39:412–425. doi:10.1016/j.rser.2014.07.051
17. Gupta A, Saini RP, Sharma MP (2010) Steady-state modelling of hybrid energy system for off grid electrification of cluster of villages. Renew Energ 35:520–535. doi:10.1016/j.renene.2009.06.014
18. Turkay BE, Telli AY (2011) Economic analysis of standalone and grid connected hybrid energy systems. Renew Energ 36:1931–1943. doi:10.1016/j.renene.2010.12.007
19. Kumar A, Kumar K, Kaushik N, Sharma S, Mishra S (2010) Renewable energy in India: current status and future potentials. Renew Sust Energ Rev 14:2434–2442. doi:10.1016/j.rser.2010.04.003
20. Liu SY, Perng YH, Ho YF (2013) The effect of renewable energy application on Taiwan buildings: what are the challenges and strategies for solar energy exploitation? Renew Sust Energ Rev 28:92–106. doi:10.1016/j.rser.2013.07.018
21. Akikur RK, Saidur R, Ping HW, Ullah KR (2013) Comparative study of stand-alone and hybrid solar energy systems suitable for off-grid rural electrification: a review. Renew Sust Energ Rev 27:738–752. doi:10.1016/j.rser.2013.06.043
22. Kaundinya DP, Balachandra P, Ravindranath NH (2009) Grid-connected versus stand-alone energy systems for decentralized power—a review of literature. Renew Sust Energ Rev 13:2041–2050. doi:10.1016/j.rser.2009.02.002
23. Ahmed NA, Miyatake M, Al-Othmana AK (2008) Power fluctuations suppression of stand-alone hybrid generation combining solar photovoltaic/wind turbine and fuel cell systems. Energ Convers Manage 49:2711–2719. doi:10.1016/j.enconman.2008.04.005
24. Ahmed NA, Al-Othman AK, Al-Rashidi MR (2011) Development of an efficient utility interactive combined wind/photovoltaic/fuel cell power system with MPPT and DC bus voltage regulation. Electr Pow Syst Res 81:1096–1106. doi:10.1016/j.epsr.2010.12.015

25. Abbasi SA, Abbasi N (2000) The likely adverse environmental impacts of renewable energy sources. Appl Energ 65:121–144. doi:10.1016/S0306-2619(99)00077-X
26. Rehman S, Alam MM, Meyer JP, Al-Hadhrami LM (2012) Feasibility study of a wind-pv-diesel hybrid power system for a village. Renew Energ 38:258–268. doi:10.1016/j.renene.2011.06.028
27. Yamegueu D, Azoumah Y, Py X, Zongo N (2011) Experimental study of electricity generation by Solar PV/diesel hybrid systems without battery storage for off-grid areas. Renew Energ 36:1780–1787. doi:10.1016/j.renene.2010.11.011
28. Kaviani AK, Riahy GH, Kouhsari SHM (2009) Optimal design of a reliable hydrogen-based stand-alone wind/PV generating system, considering component outages. Renew Energ 34:2380–2390. doi:10.1016/j.renene.2009.03.020
29. Yang H, Zhou W, Lu L, Fang Z (2008) Optimal sizing method for stand-alone hybrid solar-wind system with LPSP technology by using genetic algorithm. Sol Energy 82:354–367. doi:10.1016/j.solener.2007.08.005
30. Zhang P, Li W, Li S, Wang Y, Xiao W (2013) Reliability assessment of photovoltaic power systems: review of current status and future perspectives. Appl Energ 104:822–833. doi:10.1016/j.apenergy.2012.12.010
31. Nelson DB, Nehrir MH, Wang C (2006) Unit sizing and cost analysis of stand-alone hybrid wind/PV/fuel cell power generation systems. Renew Energ 31:1641–1656. doi:10.1016/j.renene.2005.08.031
32. Paliwal P, Patidar NP, Nema RK (2014) Determination of reliability constrained optimal resource mix for an autonomous hybrid power system using Particle Swarm Optimization. Renew Energ 63:194–204. doi:10.1016/j.renene.2013.09.003
33. Demiroren A, Yilmaz U (2010) Analysis of change in electric energy cost with using renewable energy sources in Gokceada, Turkey: an island example. Renew Sust Energ Rev 14:323–333. doi:10.1016/j.rser.2009.06.030
34. Lau KY, Yousof MFM, Arshad SNM, Anwari M, Yatim AHM (2010) Performance analysis of hybrid photovoltaic/diesel energy system under Malaysian conditions. Energy 35:3245–3255. doi:10.1016/j.energy.2010.04.008
35. Merei G, Berger C, Sauer DU (2013) Optimization of an off-grid hybrid PV-Wind-Diesel system with different battery technologies using genetic algorithm. Sol Energy 97:460–473. doi:10.1016/j.solener.2013.08.016
36. Shukla A, Tiwari R, Kala R (2010) Real life applications of soft computing. Taylor and Francis Group, New York, pp 1–18. ISBN: 978-1-4398-2287-6
37. Karray FO, Silva CD (2004) Soft computing and intelligent systems design. Pearson Education Limited, England, pp 38–45. ISBN: 0-321-11617-8
38. Ray S, Reaume SJ, Lalman JA (2010) Developing a statistical model to predict hydrogen production by a mixed anaerobic mesophilic culture. Int J Hydrogen Energ 35:5332–5342. doi:10.1016/j.ijhydene.2010.03.040
39. Booker JM, Ross TJ (2011) An evolution of uncertainty assessment and quantification. Scientia Iranica 18(3):669–676. doi:10.1016/j.scient.2011.04.017
40. Datta D, Kushwaha HS (2011) Uncertainty quantification using stochastic response surface method case study—transport of chemical contaminants through groundwater. Int J Energ Inform Comm 2(3):49–58
41. Singh KP, Singh AK, Singh UV, Verma P (2012) Optimizing removal of ibuprofen from water by magnetic nanocomposite using Box–Behnken design. Environ Sci Pollut Res 19:724–738. doi:10.1007/s11356-011-0611-4
42. Singh KP, Rai P, Pandey P, Sinha S (2012) Modeling and optimization of trihalomethanes formation potential of surface water (a drinking water source) using Box–Behnken design. Environ Sci Pollut Res 19:113–127. doi:10.1007/s11356-011-0544-y
43. Myer RH, Montogomery DC, Anderson-Cook CM (2009) Response surface methodology: process and product optimization using designed experiment, 3rd edn. Wiley, New York, pp 19–26. ISBN: 978-0-470-17446-3

44. Ray S, Lalman JA, Biswas N (2009) Using the Box-Behnken technique to statistically model phenol photocatalytic degradation by titanium dioxide nanoparticles. Chem Eng J 150:15–24. doi:10.1016/j.cej.2008.11.039
45. Mondal AH, Denich M (2010) Hybrid systems for decentralized power generation in Bangladesh. Energy Sustain Dev 14:48–55. doi:10.1016/j.esd.2010.01.001
46. Ngan MS, Tan CW (2012) Assessment of economic viability for PV/wind/diesel hybrid energy system in southern Peninsular Malaysia. Renew Sust Energ Rev 16:634–647. doi:10.1016/j.rser.2011.08.028
47. Luna-Rubio R, Trejo-Perea M, Vargas-Vazquez D, Rios-Moreno GJ (2012) Optimal sizing of renewable hybrids energy systems: a review of methodologies. Sol Energy 86:1077–1088. doi:10.1016/j.solener.2011.10.016
48. Hakimi SM, Moghaddas-Tafreshi SM (2009) Optimal sizing of a stand-alone hybrid power system via particle swarm optimization for Kahnouj area in south-east of Iran. Renew Energ 34:1855–1862. doi:10.1016/j.renene.2008.11.022
49. Phuangpornpitak N, Kumar S (2011) User acceptance of diesel/PV hybrid system in an island community. Renew Energ 36:125–131. doi:10.1016/j.renene.2010.06.007
50. Chen HC (2013) Optimum capacity determination of stand-alone hybrid generation system considering cost and reliability. Appl Energ 103:155–164. doi:10.1016/j.apenergy.2012.09.022
51. Bilal BO, Sambou V, Ndiaye PA, Kebe CMF, Ndongo M (2010) Optimal design of a hybrid solar-wind-battery system using the minimization of the annualized cost system and the minimization of the loss of power supply probability (LPSP). Renew Energ 35:2388–2390. doi:10.1016/j.renene.2010.03.004
52. Touatia S, Belkaid A, Benabid R, Halbaoui K, Chelali M (2012) Pre-feasibility design and simulation of hybrid PV/Fuel Cell Energy system for application to desalination plants loads. Procedia Eng 33:366–376. doi:10.1016/j.proeng.2012.01.1216
53. Lagorse J, Simoes MG, Miraoui A, Costerg P (2008) Energy cost analysis of a solar-hydrogen hybrid energy system for stand-alone applications. Int J Hydrogen Energ 33:2871–2879. doi:10.1016/j.ijhydene.2008.03.054
54. Khan MJ, Iqbal MT (2005) Pre-feasibility study of stand-alone hybrid energy systems for applications in Newfoundland. Renew Energ 30:835–854. doi:10.1016/j.renene.2004.09.001
55. Montgomery DC (2001) Design and analysis of experiments, 5th edn. Wiley, New York, pp 60–88. ISBN: 0-471-31649-0
56. Ray S, Lalman JA (2011) Using the Box–Behnken design (BBD) to minimize the diameter of electrospun titanium dioxide nanofibers. Chem Eng J 169:116–125. doi:10.1016/j.cej.2011.02.061
57. Ray S, Chakraborty AK, Debnath D (2013) Development of a cost-optimized hybrid off-grid power system for a model site in North-Eastern India involving photovoltaic arrays, diesel generators and battery storage. Int J ChemTech Res 5(2):771–779

Chapter 47
Electricity Trade Patterns in a Network

Talat S. Genc, Ege Yazgan, and Pierre-Olivier Pineau

Abstract Using high-frequency trade volume and price data in a transmission network we investigate patterns of trade and its impacts in the market price formation process. In particular, we study the Ontario wholesale electricity market and its trade with multiple interconnected markets, including New York, Michigan, and Minnesota, through 13 interconnections. This research has regulatory implications on integration of electricity markets, and possible investments in transmission and production capacity. The main findings are in order: (a) imports are unambiguously related to prices (significant Granger causality), while exports are not; (b) trade mainly occurs due to the market price differentials between the markets and traders can use past price observation to take trade positions before the markets clear; (c) there is a high degree of integration across the markets in the network, where the speed of convergence of cross prices is almost instantaneous.

Keywords Electricity trade • Transmission network • Electricity prices • Event study • Non-linear Granger causality • Ontario, New York, Michigan, Manitoba, Quebec wholesale electricity markets

47.1 Introduction

Electricity market restructuring throughout the world has led to significant evolutions, especially in the production and transmission sides of the industry. Competitive market designs, efficient auction institutions, and welfare improving transactions have radically changed the wholesale electricity markets. Moreover, open access in transmission has led to more electricity flows across interconnected jurisdictions.

T.S. Genc (✉)
Department of Economics, University of Guelph, Guelph, ON, Canada

Department of Economics, Ipek University, Ankara, Turkey
e-mail: tgenc@uoguelph.ca

E. Yazgan
Department of Economics, Kadir Has University, Istanbul, Turkey

P.-O. Pineau
Decision Sciences, HEC Montreal, Montreal, QC, Canada

I. Dincer et al. (eds.), *Progress in Clean Energy, Volume 2*,
DOI 10.1007/978-3-319-17031-2_47

One of the aims of electricity market reforms is to increase welfare, notably through trade, and reduce price differentials between states/provinces/countries. Exports and imports, in theory, should minimise price differentials in the absence of network externalities and transmission capacity constraints. However, these constraints always exist in electricity networks. Accordingly, we will examine interconnected markets and investigate the role of electricity trade on the price formation process in an electricity market. In particular, we will study the Ontario wholesale electricity market and its trade through 13 interconnections with the neighbouring markets. The current electricity industry reforms also aim to decrease air emissions stemming from power productions. Given that the electricity industry is one of the most polluting sectors in the economy in many parts of the world, the new reforms include retirement of coal plants and institute green energy sources such as wind- and solar-based power generators.

Many studies have analysed several aspects of the restructured electricity markets; however, interregional trades in interconnected electricity markets and their effects on market prices have largely been ignored. This is an important issue, because it can have a significant impact on market price, and therefore on generation and transmission investments, both within a jurisdiction and outside. Furthermore, as electricity market integration between jurisdictions progresses, more trade is expected. Trade effects should be well understood not only for the political economy of the sector but also to foresee investment patterns in production and transmission capacities.

Electricity has been auctioned in wholesale electricity markets in many parts of the world. In electricity auctions (uniform-price or discriminatory) the last accepted energy offer sets the market price (which is paid to all suppliers in uniform-price auctions, and only to the market clearing marginal supplier in discriminatory auctions). As this last accepted offer can come from a local generator or from a wholesaler importing electricity, the market price can be set from outside the home market. For example, in the New York electricity market (run by the New York Independent System Operator, NYISO) and the Midwest electricity market (run by the Midwest Independent System Operator, MISO) exporters and importers can set the market clearing prices in the day-ahead market. The Ontario market (run by the Independent Electricity System Operator, IESO), which does not have a day-ahead market, allows exports and imports to set the pre-dispatch price 1 h before the delivery during the pre-dispatch sequence, and these export and import quantities can be scheduled in the real time. In commodity markets, imports tend to reduce the product price at the home market. A similar argument could be made for electricity markets; that is, imports would lower the wholesale electricity prices by making more energy available, avoiding the need to accept higher energy offers from the suppliers within the home market. Analogously, exports (energy bids from outside buyers) could increase the market prices, as more expensive energy offers have to be accepted to meet the market demand. In this chapter, we test these claims in the Ontario market, and find that they do not hold true in general.

Ontario is the largest province of Canada in terms of population and economy, and its wholesale market has some peculiarities. Despite a competitively set hourly energy market price, some generators are unexposed to market prices because they have the equivalent of contract for difference with the local electricity planning organisation, the Ontario Power Authority. Furthermore, Ontario has important interconnections with large regulated and deregulated neighbouring markets via its transmission grid, leading to important energy transfers. It also has a very volatile market in terms of prices. Ontario has two main physical markets: the real-time energy market and real-time operating market. Contrary to many US electricity markets, it does not have a day-ahead market: market prices are settled every 5 min in real time. Its market price volatility is higher than the ones in neighbouring jurisdictions such as New England (NE), New York (NY), and Pennsylvania-New Jersey-Maryland Interconnection (PJM). These markets have the two-settlement markets (day-ahead and real-time markets) in which most of the real-time demands are cleared in the day-ahead market. The high volatility in the Ontario market is argued to be correlated with the single-settlement nature of the market which is the real-time balancing market.

There are several papers in the literature that examined the Ontario wholesale electricity market. These papers studied production capacity investments (Genc and Sen [1]), identification of variables explaining peak price (Rueda and Marathe [2]), the effects of power outages on prices (Melino and Peerbocus [3]), and the industrial customers' price responsiveness (Choi et al. [4]) in the Ontario market. These studies, among others, have not considered the effects of trade on market prices. An exception is Serletis and Dormaar [5] who examine whether exports and imports are related to changes in Alberta market prices. Also, in a recent paper, Aydemir and Genc [6] examine a Cournot competition model to identify the impact of electricity trade (mainly imports) on market outcomes and the air emissions. They argue that power trade can reduce the levels of air pollutants such as CO_2, NO_x, and SO_2 stemming from power production.

The current work is different than others in several aspects: (1) using an extensive data set (5-min and hourly data from 2002 to 2009 in a context where imports and exports are made with five different jurisdictions—New York, Michigan, Minnesota, Manitoba, and Quebec—interconnected through 13 trading zones) we implement event study analysis and recently developed non-linear causality tests to determine trade effects on Ontario prices; (2) disaggregating the high-frequency price, and export and import data, we study the neighbouring markets and their individual impacts on the Ontario trade. Also, to motivate our empirical methodology we briefly examine two theoretical competition settings and argue why a theoretical framework might offer different predictions (impacts of trade) than the other. Furthermore, our research entails regulatory implications on the integration of electricity markets, and possible investments in transmission capacity.

There has been no electricity trade theory in the literature. We will briefly examine two theoretical models—Cournot and supply function equilibrium (SFE) analyses, which are commonly used in electricity market analyses—and their trade predictions. We will describe their shortcomings in explaining the effects of imports on prices. Virtually all trade theories (Ricardian, Heckscher-Ohlin, etc.)

explain price differentials among autarky markets and trade effects on income distribution and welfare, and predict that goods will flow from low-price markets to high-price markets, and home market prices drop with imports. However, these theories may fall short in explaining the dynamics of trade in electricity markets due to the peculiarities of electricity and the constantly changing supply and demand conditions at each moment. Moreover, trading electricity between jurisdictions is limited by transmission-line capacity and is subject to interventions by system operators. Even exports and imports of electricity can occur simultaneously within a trading period, due to hedging purpose and market rule differences between jurisdictions. In the Ontario market, for example, this type of simultaneous trade (import of energy into Ontario and export of energy from Ontario) is called a wheeling-through transaction.

This chapter contributes to understanding of trade patterns in a transmission network and the relationship between imports/exports and market prices. Our main findings are as follows. First, we find that while Ontario imports can be unambiguously tied to the hourly Ontario energy prices, exports cannot. Utilising linear and non-linear tests, we find Granger causality for all lags in the case of imports. Second, with the disaggregated high-frequency data we find that trade mainly occurs due to the market price differentials between the markets and traders can use past price observation to take their future trade positions. Third, we observe a high degree of integration across the different markets in the network, where the speed of convergence of cross prices is almost instantaneous.

This chapter is structured as follows. Section 47.2 describes the structure of the Ontario market and its interconnections. Section 47.3 briefly examines economic theory predictions of trade behaviour and their shortcomings. Section 47.4 explains the data set, methodology, and some results. In Sect. 47.5, Ontario's trade patterns with neighbouring jurisdictions in the network are analysed. Conclusions are presented in Sect. 47.6.

47.2 The Ontario Electricity Market and Its Interconnections

This section describes the Ontario wholesale electricity market structure and its interties in the power network over which Ontario continuously trades electricity.

The Ontario wholesale electricity market has some interesting features. It has a diversified generation portfolio with all types of production technologies (fossil fuel, nuclear, hydropower, wind, solar, and biomass). It is the most price-volatile market in the region, and among the North American power markets, because only Ontario and Alberta markets in Canada employ one-settlement market mechanism in which the market price is formed in the real time.

The Ontario wholesale electricity market consists of the energy market, operating reserve market, and financial transmission rights market. The IESO publishes dispatch instructions to loads and generators and runs a uniform price auction for every 5 min. The market clearing price (MCP) is set by simply ranking all received energy offers (from generators and wholesalers/importers) in increasing price order, until the forecasted demand is satisfied. The last accepted energy offer sets the market price, which is paid to all suppliers (IESO [7]). The Hourly Ontario Energy Price (HOEP) is the average of the twelve 5-min MCP in a specific hour. HOEP is the price paid by non-dispatchable loads. The IESO governs the wholesale market, ensures the reliability of the integrated power system, and forecasts supply requirements and demand (total Ontario market demand is equal to domestic demand plus export demand). Suppliers submit energy offers (quantity-price pairs) to sell electricity and some wholesale buyers submit energy bids to buy electricity. Although demand is inelastic, some large wholesale customers are able to respond to changes in prices by either shifting some of their demand to off-peak periods or participating in the market and bidding how much electricity they plan to buy at various price levels.[1]

The IESO does not have a day-ahead market. Generation dispatch and market clearing prices are set in the real-time energy market only. However, for reliability purpose the IESO employs a Day-Ahead Commitment Process (DACP), created in 2006, to manage day-ahead available energy units and determine approximate import transactions.[2] On the other hand, US neighbouring jurisdictions (Minnesota, Michigan, and New York) have two-settlement markets, namely day-ahead and real-time energy markets. Day-ahead market has the dominant share of transactions in the US markets. The two-settlement market structure enables that most of the market demand is cleared a day before market opens and generators have enough time to adjust their operations for the instances of unpredictable events in real time.

In 2009 the available generation capacity within Ontario was 35,465 MW (IESO [8]). This capacity has grown at an average rate of 2 % during the period 2002–2009 (Statistics Canada [9]), while the total available energy (generation and imports, minus exports) has remained at the same level from 2002 to 2009 (about 155 TWh per year).

47.2.1 Exports, Imports, and Price Formation in the Ontario Market

The interconnection capacity between Ontario and its neighbours (two regulated Canadian markets: Manitoba and Quebec; and three deregulated US markets: Minnesota, Michigan, and New York) is presented in Table 47.1, in which the export and import capacities are about 6,000 MWh, which would cover almost

---

[1] According to IESO 2010 market's programme (www.ieso.ca), there are 13 facilities operating as "dispatchable load" in the market, offering 700 MW of potential demand response.

[2] In 2008 the IESO Board approved the implementation of an Enhanced Day-Ahead Commitment Process (EDAC) to deliver some minor changes to the existing Day-Ahead Commitment Process.

Table 47.1 Ontario interconnection capacity for exports and imports, in MW, 2009 (IESO [10])

	Exports	Imports
Manitoba	268	336
Minnesota	140	90
Michigan	2,275	1,675
New York	1,925	1,680
Quebec	1,329	2,210
Total	5,937	5,991

one-third of the average load in Ontario if there were no transmission and/or network constraints. The actual total export/import capacity is not equal to the arithmetic sum of individual capacities, as the constraints bring it down to 4,000 MW (IESO [11]). In Table 47.1, exports (imports) column represents the maximum export (import) quantities from (to) Ontario to (from) the interconnection. For instance, in a given hour Ontario can sell up to 1,925 MWh to NY, or can import at most 1,680 MWh from NY. Except for a 625 MW addition to the Quebec interconnection in July 2009, interconnection capacities remained very stable from 2002 to 2009. Detailed individual transmission line information is published several times a year by the IESO (see IESO [10]).

Ontario generators can sell their power at the real-time MCP or sell it in external markets. A home generator can export directly to other markets without participating in the home market. That is, it can submit energy offers in other markets before offering its energy to the Ontario market. Therefore, an Ontario generator can export while having no sales in Ontario. However, the system operator may cancel or curtail already scheduled exports for system adequacy or reliability reasons. Importers of electric power are given a price guarantee for their accepted energy offers. If the MCP is lower than the pre-dispatch price of the importer, the difference is paid to the importer by the operator. To signal market conditions, the IESO releases pre-dispatch schedules with forecasted demand and supply requirements (e.g. generation availability, imports, and exports) along with the price signals (e.g. projected HOEP for the day, and the intertie offer guarantee estimate). Importers use these market signals before placing their bids. Due to unexpected outages, and/or high start-up costs, and/or high ramping rates of power plants, and/or low spinning reserve capacity, the home market price may increase, which can create a trade opportunity for importers. Depending on the price differentials between the home market and interconnected markets, and on the transmission constraints, importers would benefit from arbitrage opportunities.

Currently, the IESO employs "Dispatch Scheduling and Optimization Algorithm" to determine pre-dispatch sequence of prices and demand for the future periods. These are the predicted prices based on demand and supply forecasts. The algorithm is run every hour, and the pre-dispatch prices and quantities calculated for each hour for the future 12–36 h are published at the IESO website.[3] Specifically, market participants

[3] The pre-dispatch prices and quantities are posted at www.theimo.com/imoweb/marketdata/marketToday.asp.

can use the pre-dispatch data to reform their operations planning and participation in the real-time market. For example, to increase price responsiveness the IESO can compensate the market participants who could reduce their demand in real time, if the 3-hour-ahead pre-dispatch price were above $120.

Imports and exports are scheduled 1 h before the delivery hour. Imports and exports are settled in the real time at the MCP plus the congestion price, which is determined during the hour-ahead pre-dispatch sequence. Importers are given a price guarantee: if the MCP is lower than their offer price in the hour-ahead pre-dispatch market then they will be paid at least the average price of their pre-dispatched accepted offers. Therefore, pre-dispatch prices are crucial for payments to importers. Pre-dispatch prices also help determine future import offers.

Based on the operations in the hour-ahead dispatch planning and the market clearing in the real-time market, it is comprehensible that imports and exports can potentially affect the real-time prices. Imports offered below the hour-ahead pre-dispatch price and exports bid above the hour-ahead pre-dispatch price are all scheduled in real-time dispatch with sure probability. Therefore, exports and imports will play an important role in determination of real-time market prices. Although, they do not set the MCP in real time as they cannot be dispatched in every 5 min, imports and exports can set the pre-dispatch price in the hour-ahead market in Ontario. However, the intertie transactions (exports and imports) can set the day-ahead market prices in the US markets.

It is argued by the IESO Market Pricing Working Group that pre-dispatch prices could approach to the real-time prices if the pre-dispatch prices would be determined 5 min ahead of the real-time auction instead of the hour-ahead operations.[4] The IESO is considering a day-ahead market design which will aim to set electricity prices on an hourly basis 1 day ahead of the real time.[5] In such a day-ahead market design, imports and exports would be able to set the day-ahead market prices. The real-time market, however, will remain effective and run the auction in 5-min basis to clear the unmet demand.

47.3 Economic Theory Background

To predict the relationship between electricity trade and market prices, we will briefly overview two microeconomic models—Cournot and SFE settings—that are commonly used to examine the behaviour of power producers and market power issues in the electricity markets.

---

[4] See http://www.ieso.ca/imoweb/pubs/consult/mep/MP_WG_2004Aug20_ISS01_PreDispPrice.pdf and the Issue 30 on forecasting real-time prices.

[5] See http://www.ieso.ca/imoweb/pubs/consult/mep/MP_WG-20060303-Issue7-Imports-Exports-Setting-Price.pdf. The day-ahead market has not been implemented by IESO yet.

First consider the following quantity choice Cournot model in which firms choose production quantities as strategies. Let demand for electricity be price responsive, and assume a constant production cost c. There are n producers and an importer of electricity. For producer i, the profit as a function of output is

$$\pi_i(s_i) = (p(s_i + s_{-i} + I, a) - c)s_i,$$

where $p(Q)$ is the inverse demand with the total market output $Q = s_i + s_{-i} + I$, and I is the quantity imported, and s_i and s_{-i} are the outputs of firm i and rival firms $-i$, respectively. The parameter a denotes the demand intercept. Exports are assumed to be part of total market demand, as in the Ontario market; hence we do not need to explicitly take them into account. Assume a concave and downward sloping inverse demand, that is, $p' < 0$ and $p'' < 0$, and price increases as the market demand increases, that is, $\frac{\partial p}{\partial a} > 0$. Maximisation of the above profit function for firm i leads to the reaction function $s_i = -(p-c)/p'$ where $p' = \partial p/\partial Q$. The strategic firm i adjusts its output with respect to the imports as $\frac{\mathrm{d}s_i}{\mathrm{d}I} = -1 + p''(p-c)/(p')^2$ which is negative since demand is concave. Namely, as imports increase strategic firms tend to decrease their outputs.

If the importer is also strategic it maximises its profit and responds to rival producers as follows. The profit for the importer is

$$\pi(I) = (p(s_i + s_{-i} + I, a) - C)I$$

assuming C as the cost per unit for the importer. The importer's profit maximisation yields, by the first-order necessary conditions, the reaction function $I = -(p - C)/p'$. The importer adjusts its import quantity with respect to the producers' output $s = s_i + s_{-i}$ as $\frac{\mathrm{d}I}{\mathrm{d}s} = -1 + p''(p-C)/(p')^2 < 0$.

The presence of a strategic importer in this static Cournot model will tend to reduce the market price, as it increases the competition. Furthermore, as the (inverse) demand increases, that is, the demand intercept a goes up, the importer imports more: since $p' < 0$ and $\frac{\mathrm{d}p}{\mathrm{d}a} > 0$, we obtain $\frac{\mathrm{d}I}{\mathrm{d}a} = -\frac{\left(\frac{\mathrm{d}p}{\mathrm{d}a}\right)}{p'} > 0$.

On the other hand, if the importer is competitive (it is a price taker and supplies up to its capacity at a price level near its marginal cost), it is easy to verify that import quantity will shift the supply curve downwards and hence causes price reduction. Therefore, competitive imports will reduce market price for a given level of demand.

The above analysis is based on a static model in which firms commit to their production and/or import schedules given the other firms' outputs. An alternative model would be utilising an SFE model (see, e.g., Genc and Reynolds [12]) in which each firm chooses price-quantity pairs as strategies. It is richer and more flexible than a Cournot model in the sense that each firm has the ability to choose a supply schedule as a strategy instead of choosing a fixed quantity (Cournot assumption) or a fixed price (Bertrand assumption). In many electricity markets/auctions, including the Ontario wholesale electricity market, firms compete in supply schedules and intersection of a market demand with aggregate supply function determines the market clearing outcome. How much each firm produces ultimately is directed by its own supply function.

To this end, consider the following simple SFE model in which all firms including importer(s) are strategic and market demand varies over time (say, hourly) and is concave. Each electricity producer chooses a supply function $s_i(p)$, specifying the amount the producer is willing to supply at each possible market price that might occur in a trading period. Denote the aggregate supply function of all producers $S(p) = \sum_i s_i(p)$. As a market rule in electricity auctions, a supply function submitted by any producer must be non-decreasing (see Baldick and Hogan [13], Genc [14]). This is a reasonable market rule as the bids of a supplier are in merit order, and the rule helps equilibrate the market. All suppliers including importer(s) independently and simultaneously choose supply function strategies to maximise their profits. An SFE is a Nash equilibrium in supply function strategies. The supply functions stay constant during the trading period (e.g. a day) in some markets; in other markets suppliers are allowed to submit different supply functions several times a day (e.g. for each hour as in the Ontario market). The intersection of the aggregate supply function and the market demand determines the market price and quantity at the particular time t.

The profit function to be maximised at time t by the importer is

$$\pi_I(t) = (p(t) - C)(Q(p(t)) - S(p(t))),$$

in which the residual demand it faces is $I(p(t)) = Q(p(t)) - S(p(t))$, and the MCP is $p(t)$. Assuming that the residual demand function is differentiable, the necessary condition for optimal price choice $p(t)$ for supplying quantity I is the differential equation

$$I(p) = (p - C)\left(S'(p) - Q'(p)\right).$$

This import function has the characteristic that the importing firm maximises its profit at each point in time (t) during the trading day, given the supply functions chosen by the rival electricity producers.

To see how the imports vary as market prices change, we take the derivative of this expression to obtain

$$\frac{\mathrm{d}I}{\mathrm{d}p} = \left[S'(p) - Q'(p)\right] + (p - C)\left[S''(p) - Q''(p)\right] > 0.$$

This expression is positive because the terms in square brackets are positive as the supply functions are non-decreasing and the demand is concave.

Contrasting both Cournot and the SFE model, it becomes clear that under the SFE model increase in market prices will contribute to higher electricity imports.

In a static Cournot model this result is not immediate: at a given level of demand imports contribute to price reductions.

The above insights are based on stylised economic models, and our assumptions (symmetry, differentiability, concavity, and non-binding constraints) are too simplistic to be true for real electricity markets. For example, due to the market power issues and the capacity constraints of firms supply functions could be discontinuous (see, Genc and Reynolds [12]), and the importer could face a non-differentiable residual demand function, which would create an issue to find out the sign of dI/dp, and to characterise and solve an optimal import strategy $I(t)$. Moreover, we ignored dynamic considerations between importers and producers over time and assumed static models to examine the import behaviour. It is a fact in economic theory that dynamic games in general predict different results than static games do. To the best of our knowledge, there is no dynamic SFE analysis in the literature, nor is any SFE analysis incorporating imports and exports in examining optimum bidding behaviour.

47.4 Econometric Analysis

We show in the theory background section that it could be a daunting task to come up with an electricity trade theory explaining the directions of the imports (and exports) and their influence on the electricity market prices. This is mainly due to the peculiarities associated with the nature of electricity, which makes it different than the other commodities. However, we can apply some time-series econometric methods to investigate the role of imports and exports on market outcomes in the Ontario wholesale electricity market.

Below we explain the data set and the econometric techniques that we employ to study the relationships between trade and market prices. We first begin with an event study analysis, and then employ linear and recently developed non-linear Granger causality tests from export and import volumes to electricity prices.

47.4.1 Data

Our data set includes the Ontario hourly electricity prices HOEP (P, in Can$/MWh), exports ($X$, in MWh) and imports ($M$, in MWh), total market demand (Q, in MWh), and also the 5-min MCP in all neighbouring markets. They span the time period of May 1, 2002, to June 9, 2009, on hourly basis, including all week and weekend days (62,328 data points for each variable).[6] In addition to the HOEP, we also use 1-, 2-, and 3-hour-ahead pre-dispatch prices.[7]

47.4.2 Event Study

To investigate whether imports or exports are able to affect market prices in the Ontario electricity market, we first undertake an event study. Although event study analysis is commonly used in the finance literature, to our knowledge it is the first time that this technique is being applied to trade analysis in electricity markets. The purpose of this event study is to analyse the behaviour of prices before and after the "event". Events, in our case, are defined as "high" export or import values. Consistent with the literature, we define the high levels of exports or imports as a level 4 standard deviation above its mean at a certain point in time. In other words, if the total imports (exports) at time t is 4 standard deviation above its mean we call this as an import (export) event.[8]

By applying this definition to our entire data set we pin down 208 and 541 events for imports and exports, respectively. The dates of these events are illustrated in Fig. 47.1 along with the market prices. As it can be observed in the figure, almost all of the events are concentrated either before/around 2003 (when the Ontario wholesale market collapsed) or after/around 2008 (when the economic crisis in North America started). Most of the import events (blue lines) are observed before 2003 or after 2008, which also coincides with some of the export event dates (red lines). The remaining export events are concentrated near 2003 and 2008 dates in which import events are not shown up.

The next step in our event study is to statistically test whether the electricity prices are affected by these events. To do this we simply test the equality of the mean prices before and after the events by choosing an event window. This window is considered to be the time length in which the effect of the event is supposed to be observed. Then the event window is divided into two halves to test the equality of the mean prices by using the data in the equal parts of the event window.

---

[6] The data is available at www.ieso.ca. We also analyzed this data set in subcategories of peak and off-peak hours and obtained qualitatively similar results. Peak time was defined as hours between 08:00 and 22:00 (including 8.00 and 22:00) during week days and excluding whole weekends (27,825 data points for each variable). Off-peak time data include week day hours between 23:00 and 07:00 and whole weekends (34,503 data points for each variable). These results are available upon request.

[7] The pre-dispatch prices were not published by the IESO for August 14–23, 2003 (the period following the Northeast Blackout of 2003).

[8] We measure the mean and standard deviation by using the data up to point t. We also applied the same event rule by eliminating all prices below the mean; hence we imposed another condition in addition to the above one. This leads to a significant decrease in the number of events; however the results remained qualitatively the same.

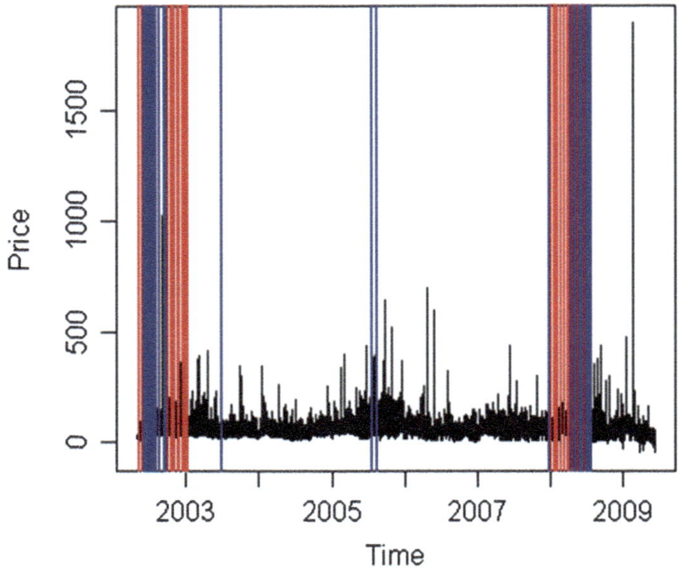

Fig. 47.1 Export and import event dates

Figure 47.2 illustrates the results of t-test applied for the null of equal means versus their inequality by using an event window that is equal to 60 points.[9]

On the y-axis of the figure we report the p-values of the tests corresponding to each import event point represented by an "x". The dashed (red) line represents the 5 % p-value line. The points below this line indicate the events for which the t-test resulted in rejection of the null of equality of mean prices at 5 % significance level. We obtained 85 of such tests, out of 208 events, in which the import events seem to affect price changes. Hence we conclude that we have sufficient evidence on the influence of imports on price behaviour.

In Fig. 47.3 we repeat the same analysis for export events. We find that in 227 tests, out of 541, the mean prices are significantly different between before and after event periods. This analysis also points out that, like imports, exports have an impact on prices.

47.4.3 Testing Granger Causality

During certain export and import events we have found statistically significant evidence of impact of trade activities on the market prices. To gain more insights

[9] Hence we have 30 data points before and after events to carry out the t-test. We have tried different event windows and obtained qualitatively similar results. They are available upon request. Note also that we assume unequal variances in the t-test.

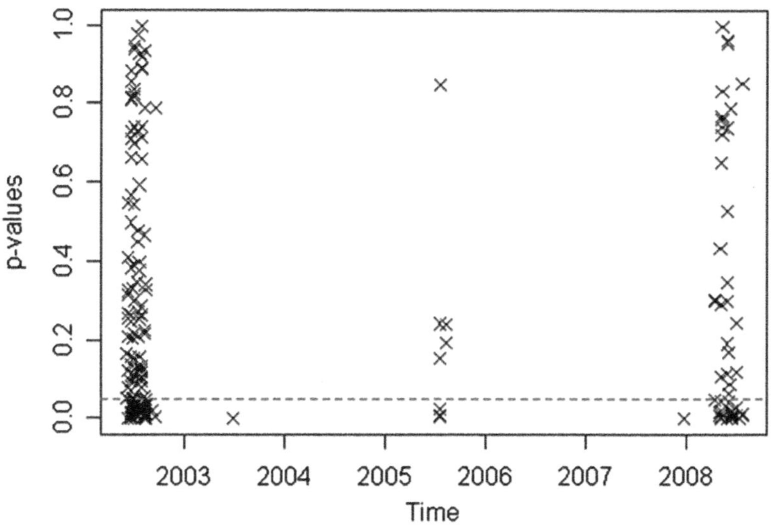

Fig. 47.2 Event study tests (imports)

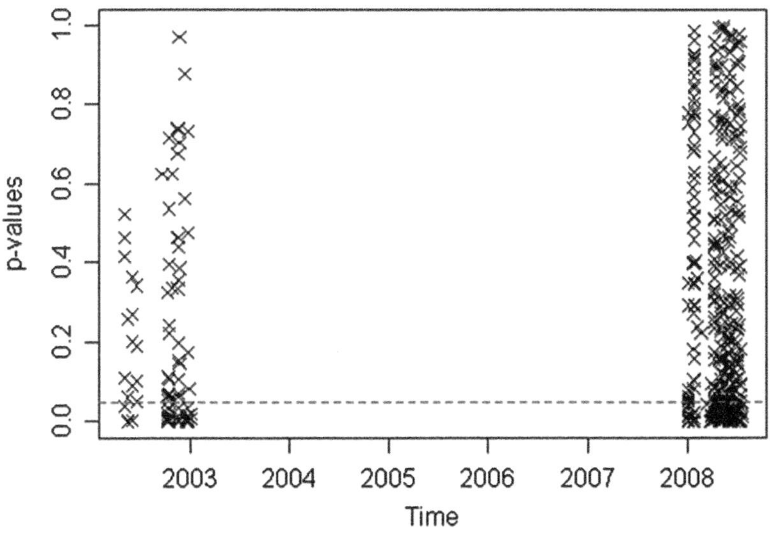

Fig. 47.3 Event study tests (exports)

into the overall trade effects, in particular lingering effects of trade on prices we need to test the Granger causality from export and import volumes to the market prices. The conventional approach of testing Granger causality is to assume a parametric, linear time-series model for the conditional mean and test whether the lags of one variable enter into the equation for another variable. In the linear framework, it is a common practice to test the Granger causality within a vector autoregressive (VAR) model using Wald or augmented Wald test proposed by Toda and Yamamoto [15] and Dolado and Lütkepohl [16].[10] Since the linear test statistics may not be sufficient to detect non-linear effects on the conditional distribution, their sole usage may lead to spurious causality test results. Therefore, we also perform recently developed non-linear tests in addition to the conventional linear Granger causality tests.

Serletis and Dormaar [5] assume four-variable VARs for the Alberta electricity market. Rueda and Marathe [2] use support vector machine-based learning algorithm for sensitivity analysis to detect the main determinants of real-time average peak prices in Ontario. They find that the main explanatory variables of the peak prices are the lagged average peak prices, the actual import peak volumes, the peak Ontario market loads, and the net available supply after accounting for load (excess supply) for the data studied in the period May 2002 to May 2003. Due to these reasons we estimate four-variable VARs for the Ontario market. The variables included in VARs consist of electricity prices, export and import volumes, and total demand (load).[11] We test the null hypothesis of Granger non-causality by restricting the relevant coefficients to zero in the following equation of the VAR(p) model.

$$P_t = \alpha + \beta t + \sum_{i=1}^{p} \phi_{i-1} P_{t-i} + \sum_{i=1}^{p} \varphi_{i-1} X_{t-i} + \sum_{i=1}^{p} \psi_{i-1} M_{t-i} + \sum_{i=1}^{p} \gamma_{i-1} Q_{t-i} + u_t^P, \tag{47.1}$$

where P_t, X_t, M_t, and Q_t stand, respectively, for price, export, import, and total demand at hour t, and u_t^P represents the usual error term of the price equation of VAR. In this equation, the null of "export does not Granger cause to power prices" and the null of "import does not Granger cause to power prices" are tested by using the following restrictions: $H_0 : \varphi_1 = \varphi_2 \ldots = \varphi_p = 0$ and $H_0 : \psi_1 = \psi_2 \ldots. = \psi_p = 0$, respectively.

[10] As is well known, in stationary systems the distribution of Wald test is asymptotically chi-squared. However, the asymptotic theory of Wald tests is typically much more complex in systems that involve variables with stochastic trends and the distribution depends on the number of unit roots and cointegration relations in the system (see Toda and Phillips [17]). The augmented Wald test is indifferent whether the series in VAR are cointegrated or not, or whether they are $I(0)$ or $I(1)$, or mixed. To avoid pretesting biases (either in unit root or cointegration tests) one can directly use this procedure without embarking on problematic unit root or cointegration tests. Also note that we tested time series properties of the above variables with different unit root tests and found that they appear to be stationary.

[11] We also included average temperature in Ontario as an additional variable in the system in (47.1). We find that the results outlined here are not sensitive to the presence of average temperature in the system. The results with temperature data are available upon request:

First, using the Granger causality tests we examine whether Ontario market prices are influenced by the exports to the neighbouring jurisdictions. The lag length p in (47.1) is chosen by Akaike Information Criteria (AIC).[12] Granger causality tests (augmented and non-augmented) indicate that the null of non-Granger causality cannot be rejected with a p-value of 0.206 (augmented) and 0.193 (non-augmented). Second, we investigate the impact of imports on the Ontario prices using the Granger causality tests and find that, contrary to the exports, the null hypothesis of no causality (from imports to power prices) is rejected with p-values equal to 0.000. Consequently, in this linear framework, we obtain conclusive evidence on the causality running from imports to power prices but not for the causality from exports to power prices.

In the parametric linear framework, we also repeat the same analysis by using 1-, 2-, and 3-hour-ahead pre-dispatch prices instead of realised spot prices. We obtain the similar results: there is causality from imports to pre-dispatch prices, but not from exports to the pre-dispatch prices.

47.4.4 A Non-parametric Causality Model

To check the robustness of the results found in the linear model, in this section, we run non-linear causality tests. While the parametric approach, employed in Sect. 47.4.3, is appealing due to its simplicity, the test statistics are only sensitive to causality in conditional mean and may not be sufficient to detect non-linear effects on the conditional distribution. Baek and Brock explain that parametric linear causality tests have low power against certain non-linear alternatives. Non-linear, non-parametric techniques seem to be attractive because they focus on predictions without imposing a certain functional form. Various non-parametric tests have been proposed in the literature. The most commonly used one is developed by Hiemstra and Jones [18] (HJ, henceforth). HJ test is a modified version of Baek and Brock [19] test. However, Diks and Panchenko [20] (DP, hereafter) show that the HJ test is not generally compatible with the Granger causality and leads to the possibility of spurious rejection of the null hypothesis. As an alternative, DP developed a new test statistic that overcomes these limitations. Hence, to test the non-linear causality between P_t and X_t, and P_t and M_t, we use the DP test. The test is applied to the estimated residual series from the VAR model, u_t^p, u_t^x, and u_t^m, where the last two terms refer to the residuals estimated from the export and import equations of the VAR model, similar to the price equation depicted above.

---

[12] To determine the lag length chosen by AIC we initially set the maximum p to be equal to 720 (1 month). To compute the test statistics we run Matlab codes using Matlab 7.9 64-bit version in an Intel(R) Xeon(R) CPU X5570 at 2.93 GHz and 3.14 GHz, 16 GB of RAM machine. Given the huge matrix operations, the codes would not be able to be run in a less qualified machine due to the large-size data set.

By removing linear predictive power, if any, with a linear VAR model, any remaining predictive power of residual series can be considered non-linear predictive power.

By definition, u_t^x (or u_t^m) strictly Granger causes u_t^p if past and current values of u_t^x contain additional information on future values of u_t^x that is not contained in the past and current u_t^p values alone. More formally, let $\mathbf{u}_t^x = \left(u_t^x, \ldots, u_{t-l_x-1}^x\right)$ and $\mathbf{u}_t^p = \left(u_t^p, \ldots, u_{t-l_p-1}^p\right)$, and $\left(l_x, l_p \geq 1\right)$ denote the information sets consisting past observations of u_t^x and u_t^p up to and including time t. Let "$\sim$" denote equivalence in distribution. Then u_t^x does not Granger cause u_t^p if

$$H_0: \ u_{t+1}^p \big| \left(\mathbf{u}_t^x, \mathbf{u}_t^p\right) \sim u_{t+1}^p \big| \mathbf{u}_t^p. \tag{47.2}$$

This is a more general set-up for testing Granger non-causality than the above linear case because it does not involve assumptions on the data generation process and the test of non-causality simply consists of comparing one-step-ahead conditional distribution of u_t^p with and without past and current observed values of u_t^x.

The results of DP tests applied to residuals of the linear VARs chosen by AIC are presented in Table 47.2. As in the previous section the tests are carried out using both actual spot (HOEP) and pre-dispatch prices. The null hypothesis of conditional independence is tested using the lags of VAR residuals in the conditioning set, which is set to 8 as maximum.[13]

The evidence of (non-linear) non-causality from export volumes to power prices, which also indicated non-causality in the linear case, is highly conclusive for

Table 47.2 Non-linear causality tests for different lags with actual market and pre-dispatch prices

	X does not Granger cause *P*		*M* does not Granger cause *P*	
$l_x = l_p$	HOEP	Pre-dispatch	HOEP	Pre-dispatch
1	0.245 (0.403)	−6.362 (1.000)	11.074 (0.000)	3.4915 (0.000)
2	0.399 (0.345)	−8.714 (1.000)	12.451 (0.000)	3.248 (0.000)
3	0.733 (0.232)	−9.579 (1.000)	12.360 (0.000)	2.808 (0.002)
4	0.362 (0.359)	−9.304 (1.000)	10.735 (0.000)	1.879 (0.030)
5	1.073 (0.141)	−9.484 (1.000)	9.601 (0.000)	1.199 (0.115)
6	1.520 (0.064)	−8.983 (1.000)	8.112 (0.000)	1.116 (0.132)
7	1.500 (0.066)	−8.390 (1.000)	7.280 (0.000)	1.103 (0.134)
8	2.455 (0.007)	−8.107 (1.000)	7.010 (0.000)	1.041 (0.148)

Note: T ratios of DP tests are for the bandwidth value of 1.5, the value used by Hiemstra and Jones [18]. p-Values are given in the parentheses. l_x,l_p refer to the lags of the variables in the conditioning set

[13] The C code for computations has been provided by Diks and Panchenko [20].

pre-dispatch prices.[14] This evidence of non-causality from exports to prices is also present for market prices, although not as conclusive as pre-dispatch prices. All non-linear tests cannot reject the null of non-causality at 5 % significance level except the one with $l_x = l_p = 8$. On the other hand, the null hypothesis of non-causality from imports to prices is unambiguously rejected in all tests for actual market prices. However, the evidence of causality from imports to pre-dispatch prices differs across the lag lengths considered for the VAR residuals in the DP test. While the non-causality is rejected, as in the case of actual market prices, up to $l_x = l_p = 5$, it cannot be rejected for the remaining lags. Consequently, our results under non-linear causality tests are in general conformable with the results obtained under the linear counterparts.

The results of both linear and non-linear Granger causality tests indicate that while the evidence for the hypothesis of exports has no causal impact on prices, the evidence for imports is the reverse and points out that imports have an impact on prices. Contrary to the linear case, in the non-linear DP test the latter result on causality seems to be ambiguous for the pre-dispatch prices.

Despite the evidence on the effect of exports on prices in the event study, we are unable to find evidence on causality running from exports to prices. Notice that the causality tests are carried out by using the whole sample. However, as we see in Sect. 47.4.2 most of the export events are concentrated around/after 2008. Therefore, it is a worthwhile exercise to repeat the causality analysis by using this time period of the data rather than the whole sample. Indeed, by using the data from November 1, 2007, to the final date in our sample we obtain strong evidence on causality from exports to prices for this period where the majority of the export events are present. Both Granger causality tests (augmented and non-augmented) indicate that the null of non-Granger causality can be rejected with a p-value of 0.000 for all prices. The results of non-linear causality tests for this period, illustrated in Table 47.3, confirm the results obtained in the linear framework.

Table 47.3 Non-linear causality tests (1 October, 2007, to 6 September, 2009)

$l_x = l_p$	X does not Granger cause to P HOEP
1	4.468 (0.000)
2	4.407 (0.000)
3	4.038 (0.000)
4	3.352 (0.000)
5	2.385 (0.008)
6	2.221 (0.013)
7	2.380 (0.008)
8	2.343 (0.009)

[14] We only report the results of 1-hour-ahead pre-dispatch prices. The results with 2- and 3-hour-ahead pre-dispatch prices are qualitatively similar to the results of 1-hour-ahead pre-dispatch prices.

Although we find the strong evidence on causality from exports to prices for the duration of the financial crisis in the North America, and even the study confirms this evidence, an interesting question arises: Why exports do not (Granger) cause prices for the entire study period of 2002–2009? In the Ontario market exports are scheduled 1 h before the dispatch and performed in the expectation that the market supply is enough to cover Ontario demand. If home market supply security is attained, and the neighbouring markets' expected prices are above the local production cost and/or the home market prices, then the export transactions are carried out. In this case, clearly one does not expect exports to affect the home market prices. On the other hand, the IESO can intervene into export schedules when the home market supply conditions are tight and/or when some home generators fail to deliver the scheduled power. If this case occurs then exports are cancelled out by the system operator, as part of the market rules, to increase the local market supply. Cancelled export schedules become a part of local supply, which should affect the market prices.

In contrast, imports can influence prices in several ways. First, it is clear that the last accepted ask price setting the hour-ahead pre-dispatch price can come from a local generator or a generator from other market via imports. Therefore, the pre-dispatch prices and scheduled imports in the hour-ahead planning can affect the MCP in the real-time uniform-price auction. Second, imports are additional sources of supply, and hence can affect supply schedule and price formation in the market. Therefore, one can expect a significant causal impact (Granger causality) from imports to prices, and the empirical evidence obtained above is in line with this expectation.

In the following section, we examine disaggregated trade data between Ontario and its main trading partners.

47.5 Trade Patterns Between Ontario and the Neighbouring Jurisdictions

Having analysed the impacts of trade on electricity prices we now open the question of why the trade between Ontario and its trading partners takes place in the first place. To address this question we use disaggregated imports and exports data which shows bilateral trade flows between Ontario and others. These data are available for a long horizon in our data set, covering from December 10, 2003, to December 31, 2008, in hourly basis, and we have also a higher frequency data with 5-min intervals (532,512 observations in total). The data consist of imports and exports of Ontario from and to the 13 interconnections, and the actual market prices in these 14 markets. These markets are Ontario (ONT), Manitoba (Man), ManitobaSK (ManSK), Michigan (MICH), Minnesota (MINN), New York (NY), and eight trading zones in Quebec (denoted by PQ.BD, PQ.DZ, PQ.DA, PQ.HZ, PQ.HA, PQ.PC, PQ.QC, PQ.XY).

Table 47.4 The comparison of the prices between Ontario and its trading partners

			Same	
	Same (%)	Different (%)	Trade (%)	No trade (%)
Man-ONT	98.20	1.80	0.30	99.70
ManSK-ONT	89.43	10.57	0.20	99.80
MICH-ONT	85.71	14.29	1.21	98.79
MINN-ONT	87.94	12.06	0.20	99.80
NY-ONT	99.43	0.57	9.45	90.55
PQ.BD-ONT	99.94	0.06	3.72	96.28
PQ.DZ-ONT	99.79	0.21	1.77	98.23
PQ.DA-ONT	99.98	0.02	45.70	54.30
PQ.HZ-ONT	92.96	7.04	1.04	98.96
PQ.HA-ONT	99.84	0.16	27.38	72.62
PQ.PC-ONT	99.97	0.03	10.58	89.42
PQ.QC-ONT	99.97	0.03	0.56	99.44
PQ.XY-ONT	99.93	0.07	50.33	49.67

The comparison of the prices between Ontario and its 13 trading markets is sketched in Table 47.4. Second and third columns of the table illustrate the percentages of the observations for which the prices are the same or different in the bilateral markets out of 532,512 observations. For instance for Ontario's trade with Manitoba, while 98.2 % of the time electricity prices in Manitoba are exactly the same as those of Ontario, only in 1.8 % of the time they are different. In other markets, this ratio of price equivalence is even higher, reaching as high as 99.97 %.

This similar pricing behaviour is certainly an interesting feature in this north-east transmission network connecting Canadian to the US power markets. It reflects the high degree of market integration across the different markets in different locations. It also indicates that the law of one price (LOP) holds in this network and the speed of convergence of cross prices is almost instantaneous.[15] A detailed examination of these interesting findings is beyond the scope of this chapter and could be pursued in a future research.

In the final two columns of Table 47.4 we divide the cases of the same prices with respect to "trade" and "no-trade" situations. Trade refers to the cases where the trade (either export or import) takes place between any two markets when the prices across the markets are equal to each other. As it is clear in the table that in the case of price equality, the dominant case is no trade, i.e., neither import nor export occurs in bilateral markets. For example, when Manitoba and Ontario prices are equal to each other, 99.7 % of the time in the study period no trade is observed between them. However this percentage is smaller in some markets, especially in some Quebec zones where it remains in almost 50 %. In this case an interesting question arises: What would be the incentive for trade when there is no price

[15] As is well known, the speed of price convergence is rather low when measured across aggregate price indices or the same goods of different countries or regions.

differential between any two markets? Clearly the answer is pertaining to the supply, demand, and transmission capacity conditions in the two neighbouring markets, in which the electricity will flow from an excess-production market to a high-demand market obeying the transmission constraints between these markets.

It is clear from Table 47.4 that in most of the cases the prices are equal to each other in the bilateral markets considered above. However, exceptions arise especially in MansSk-ONT, Mich-ONT, Minn-ONT, and PQ.HZ-ONT bilateral markets; the prices are different in significant proportions. Therefore, it is worthwhile to analyse the characteristics of the trade when the prices between the markets are different. Table 47.5 aims to serve this purpose and displays the proportions of the imports and exports in the cases of different prices. This table further divides these cases into two categories: The price of trading market is higher ($P+$) or lower ($P-$) than that of the other market. For example, between the Man-ONT trade $P+$ refers to the situation in which Manitoba prices are higher than the Ontario prices.

When the prices are different, the sign of the price differences ($P+$ or $P-$) in principle should determine the direction of the trade across the bilateral markets. As it can be followed from Table 47.5, for instance, when the Manitoba and Ontario prices are different and they trade with each other, 10 % of the time Manitoba prices happen to be higher than the Ontario prices ($P+$) and 90 % of the time Ontario prices exceed the Manitoba prices ($P-$). In partitioning the case of $P+$, we find that 7.1 % of the 10 % (that is 71 % of the $P+$ cases) Ontario exports electricity to Manitoba; however in the remaining small periods Ontario imports from its counterpart. On the other hand, when the prices of Manitoba are lower than the Ontario prices ($P-$), the dominant trade form is the imports, which constitute 66.8 % of all cases and 74.2 % (=0.668/0.900) of $P-$ cases. Therefore, for the Manitoba-Ontario trade, the direction of exports/imports is predominantly consistent with the sign of the price differences and we can conclude that the electricity trade occurs when regional prices are different; the electricity flows from the high-price market to the low-price market. As it can be followed from the table the same conclusion can be drawn for the remaining markets, with some exceptions in Quebec zones whose sizes are relatively small.

Thus far we have examined trade patterns between the markets using their instantaneous actual prices. Since there is a lag between the decision to trade and the actual market prices, that is to say generators/traders commit to their trade positions before they observe the actual prices, for example importers in the Ontario market must submit their bids (import quantities at certain price levels) to the IESO several hours before auction starts and they are scheduled 1 h before the market clears, a natural question arises: Do we expect to observe similar trade patterns in the case of price lags between any two trading markets? To address this question we repeat the above analysis reported in Tables 47.4 and 47.5 with different lags in the price differences up to 24 lags. We obtain qualitatively the same results as with those in Tables 47.4 and 47.5.[16] This implies that generators/firms/traders do not

[16] We do not report these results here to save some space. They are available upon request.

Table 47.5 The proportions of export and imports when prices are different between any two markets

	Export	Import	Sum	Export	Import	Sum	Export	Import	Sum
	Man-ONT			*ManSK-ONT*			*MICH-ONT*		
$P+$	0.071	0.029	0.100	0.291	0.005	0.297	0.189	0.008	0.197
$P-$	0.231	0.668	0.900	0.041	0.662	0.703	0.296	0.507	0.803
Sum	0.303	0.697	1.000	0.332	0.668	1.000	0.485	0.515	1.000
Regions	*MINN-ONT*			*NY-ONT*			*PQ.BD-ONT*		
$P+$	0.947	0.011	0.958	0.024	0.000	0.024	0.354	0.076	0.430
$P-$	0.011	0.031	0.042	0.129	0.847	0.976	0.231	0.339	0.570
Sum	0.958	0.042	1.000	0.153	0.847	1.000	0.585	0.415	1.000
Regions	*PQ.DZ-ONT*			*PQ.DA-ONT*			*PQ.HZ-ONT*		
$P+$	0.000	0.076	0.076	0.289	0.277	0.566	0.974	0.024	0.998
$P-$	0.022	0.902	0.924	0.145	0.289	0.434	0.002	0.000	0.002
Sum	0.022	0.978	1.000	0.434	0.566	1.000	0.976	0.024	1.000
Regions	*PQ.HA-ONT*			*PQ.PC-ONT*			*PQ.QC-ONT*		
$P+$	0.060	0.000	0.060	0.500	0.000	0.500	0.669	0.000	0.669
$P-$	0.109	0.831	0.940	0.500	0.000	0.500	0.331	0.000	0.331
Sum	0.169	0.831	1.000	1.000	0.000	1.000	1.000	0.000	1.000
Regions	*PQ.XY-ONT*								
$P+$	0.211	0.000	0.211						
$P-$	0.737	0.053	0.789						
Sum	0.947	0.053	1.000						

need to perfectly forecast the future price differences between the markets before they take their trade positions: it is sufficient to look at the past price differentials and their magnitudes to make trade decisions.

47.6 Concluding Remarks

As electricity markets reform and open-access transmission policies increasingly interconnect large territories, it becomes more and more imperative to understand how imports and exports influence local market prices. Due to the characteristics of electricity markets (such as non-storability, continuous match of demand and supply, transmission network constraints, and constantly changing demand and supply conditions), it is a challenging task to develop an electricity trade theory, as we attempted in Sect. 47.3. It can happen that electricity is exported from a high-price market to a low-price market; for instance, during an off-peak time New York exporters may sell electricity to a low-price Quebec market. This may benefit both jurisdictions because New York exporters can recover their export costs and Quebec importers may avoid using power units with high start-up costs or simply save hydro resources for high-priced future time periods. Also, in electricity markets simultaneous exports and imports, called wheeling through transactions, are possible. That is, even though prices are different in both markets a market participant can export electricity to other market and import into the home market at the same time. These factors complicate modelling trade behaviour among electricity markets/jurisdictions and estimating the trade effects on market prices. Even the most commonly used economic theory models, such as Cournot and SFE, may come short in explaining the relationship between trade behaviour and price. Therefore, we empirically analyse the impact of trade in the Ontario wholesale electricity market using time-series techniques. We employ several econometric techniques (event study, linear and non-linear Granger causality tests) to analyse the trade activities between Ontario and its neighbouring jurisdictions in the network, and find that while Ontario exports cannot be unambiguously tied to the hourly Ontario energy prices, imports can. We have shown Granger causality for all lags in the case of imports with linear and non-linear tests.

Our intertie analysis with the disaggregated high-frequency data indicates primarily the price convergence between the trading markets in the network. Whenever prices differ between the markets the electricity mainly flows from the high-price market to the low-price market, although there are some exceptions which can be explained by the regional market conditions. Moreover, electricity traders (merchant firms and generation owners) can use observed price differentials between the markets to place their future trading bids. These findings may imply the strength of markets' integration in the network and suggest that further investments in the interconnection lines would smooth the price spikes and reduce the cross-price differences. All of these would yield more efficient market outcomes facilitated by trade.

To fully grasp the network interactions and trade impacts on market price, more investigation is still required. Using additional empirical data sources (such as local loads in jurisdictions, network constraints data, and possibly other explanatory variables), as well as firm-level data, could be helpful to provide further insights. A trade analysis incorporating such data is a future research direction that one may consider. However, some of the data required, especially firm-level data, is confidential and unavailable to public in some jurisdictions. In such cases, some approximation methods are still available.

Acknowledgements The first author is corresponding author who acknowledges financial support from the Social Sciences and Humanities Research Council of Canada, and can be reached at <tgenc@uoguelph.ca>.

References

1. Genc T, Sen S (2008) An analysis of capacity and price trajectories for the Ontario electricity market using dynamic Nash equilibrium under uncertainty. Energ Econ 30(1):173–191
2. Ruedaa IEA, Marathe A (2005) Important variables in explaining real-time peak price in the independent power market of Ontario. Util Policy 13:27–39
3. Melino A, Peerbocus N (2008) High frequency export and price responses in the Ontario electricity market. Energ J 29(4):35–51
4. Choi WH, Sen A, White A (2011) Response of industrial customers to hourly pricing in Ontario's deregulated electricity market. J Regul Econ 40:303–323
5. Serletis A, Dormaar P (2007) Imports, exports, and prices in Alberta's deregulated power market (Chapter 10). In: Serletis A (ed) Quantitative and empirical analysis of energy markets. World Scientific Publishing, Singapore
6. Aydemir A, Genc TS (2013) Power trade, welfare and air quality. Working paper. University of Guelph, Guelph, ON
7. IESO (2009) Overview of the IESO-administered markets. IESO, Toronto, ON
8. IESO (2009) IESO, 18-month outlook update – tables, released Nov 17. IESO, Toronto, ON
9. Statistics Canada (2009) Electric power generation, transmission and distribution, 57-202-X. Statistics Canada, Ottawa, ON
10. IESO (2009) IESO, Ontario transmission system, IESO_REP_0265v150, Aug 21. IESO, Toronto, ON
11. IESO (2009) IESO, the power system. IESO, Toronto, ON, http://www.ieso.ca/imoweb/siteShared/power_system.asp
12. Genc T, Reynolds S (2011) Supply function equilibria with capacity constraints and pivotal suppliers. Int J Ind Organ 29(4):432–442
13. Baldick R, Hogan W (2002) Stability of supply function equilibria: implications for daily versus hourly bids in a poolco market. Proceedings of the seventh annual POWER research conference. University of California Energy Institute, Berkeley, CA
14. Genc T (2009) Discriminatory versus uniform-price electricity auctions with supply function equilibrium. J Optimiz Theory App 140(1):9–31
15. Toda HY, Yamamoto T (1995) Statistical inference in vector autoregressions with possibly integrated processes. J Economet 66:225–250
16. Dolado J, Lütkepohl H (1996) Making Wald tests work for cointegrated VAR systems. Economet Rev 15(4):369–386
17. Toda HY, Phillips PCB (1993) Vector autoregressions and causality. Econometrica 61(6):1367–1393

18. Hiemstra C, Jones JD (1994) Testing for linear and nonlinear granger causality in the stock price-volume relation. J Fin 49:1639–1664
19. Baek E, Brock W (1992) A general test for granger causality: bivariate model. Technical report. Iowa State University, Ames, IA
20. Diks C, Panchenko V (2005) A new statistic and practical guidelines for nonparametric granger testing. J Econ Dynam Control 1647–1669

Chapter 48
Optimal Sizing of PV Lighting System Using Genetic Algorithms: Application to a Site in South Algeria

S. Makhloufi

Abstract Uncertain renewable energy supplies, load demands, and nonlinear characteristics of some components of photovoltaic (PV) systems make the design problem not easy to solve by classical optimization methods, especially when relevant meteorological data are not available. To overcome this situation, modern methods based on artificial intelligence techniques have been developed for sizing PV systems. In this study, a method for sizing PV lighting systems has been presented; this method is based on genetic algorithms. The method allows determining optimum PV generator size and optimum storage battery capacity that permit minimum system cost with total autonomy of the system. The method has been applied to a PV lighting system with orientation due south and inclination angles from 0° to 90° in Adrar city (south Algeria). Because measured data for the location chosen were not available, a year of synthetic hourly meteorological data of this location, generated by the PVSYST software, have been used in the simulation.

Keywords Genetic algorithms • Lighting system • Photovoltaic • Sizing

Nomenclature

A_{PV}	PV generator area, m^2
A_r	PV generator reference area, m^2
BAT_{LCC}	Battery life cycle cost, €
C_N	Battery nominal capacity, Ah
C_{Nr}	Battery reference nominal capacity, Ah
C_{ul}	Unmet load weighting coefficient
d_1	Empirical diode PV curve fitting factor

S. Makhloufi (✉)
LEESI Laboratory, University of Adrar, Adrar, Algeria
e-mail: makhlofi_s@yahoo.fr

I. Dincer et al. (eds.), *Progress in Clean Energy, Volume 2*,
DOI 10.1007/978-3-319-17031-2_48

I	Total current, A
I_G	Horizontal global irradiance, W/m²
I_L	Photocurrent, A
$I_{L,ref}$	Module photocurrent at reference conditions, A
$I_{mp,ref}$	Reference maximum power current, A
I_o	Reverse saturation current, A
$I_{o,ref}$	Diode reverse saturation current at reference conditions, A
$I_{sc,ref}$	Reference short-circuit current, A
I_T	Incident irradiance, W/m²
$I_{T,ref}$	Reference irradiance, W/m²
$I_{T,\beta}$	Total irradiance received on a tilted surface, W/m²
k_1	Pick power correction coefficient
k_2	Buttery nominal capacity correction coefficient
N_{gen}	Number of generations
n_s	Number of cells in the module connected in series
N_{ul}	Unmet load, hours
$P_{load}(t)$	Charge demand at t time, W
$P_{pv}(t)$	Produced power by the PV generator at t time, W
PV_{LCC}	PV generator life cycle cost,€
R_s	Module series resistance, Ω
SOC_{max}	Nominal capacity of the battery, Ah
S_{pop}	Population size
T_A	Ambient temperature, K
T_C	Cell temperature, K
$T_{c,ref}$	Module reference temperature, K
V	Module voltage, V
$V_{mp,ref}$	Reference maximum power voltage, V
$V_{oc,ref}$	Reference open-circuit voltage, V

Greek Letters

δ	Declination, rad
Δt	Simulation time step, hour
ε	Semiconductor band-gap energy, EV
η_{bat}	Battery efficiency
η_{inv}	DC/AC inverter efficiency
θ	Angle of incidence, rad
θ_z	Zenith angle, rad
$\mu_{I,sc}$	Temperature coefficients of the short-circuit current, A K^{-1}
$\mu_{V,oc}$	Temperature coefficients of the open-circuit voltage, V K^{-1}
ρ	Albedo
τ	Solar time
ϕ	Geographic latitude, rad
ω	Hour angle, rad

48.1 Introduction

Conventional methodologies (empirical, analytical, numerical, hybrid, etc.) for sizing PV systems are generally used when the required weather data (irradiance, temperature, humidity, clearness index, wind speed, etc.) and the information concerning the location of PV system are available. These methods present a good solution for sizing PV systems under the above conditions. However, such techniques cannot be used for sizing PV systems where the required data are not available. Moreover, the majority of the above methods need long-term meteorological data, such as total solar irradiance, air temperature, and wind speed, for their operations. To overcome this situation, newer methods have been developed for sizing the parameters for PV systems based on artificial intelligence techniques [1].

Recently, using PV lighting systems has been considerably increased in Algeria. This is motivated by the enormous potential of PV energy in Algeria, especially in the south. For example, in Adrar city in the south of Algeria (27.51°N, 0.17°W), the annual mean insolation incident on a horizontal surface equals to 5.68 kWh/m^2/day [2]. Consequently developing powerful methods to optimum sizing of these systems becomes very necessary.

This chapter presents a method of optimization for PV lighting systems using genetic algorithm (GA) [3]. Genetic algorithms are an adequate search technique for solving complex problems when other techniques are not able to obtain an acceptable solution. The method allows determining optimum PV generator size and optimum storage battery capacity that permit minimum system cost with total autonomy of the system.

The method has been applied to a PV lighting system with orientation due south and inclination angles between 0° and 90° in Adrar city (south Algeria). Because measured data for the location chosen were not available, a year of synthetic hourly meteorological data, generated by the PVSYST software, have been used in the simulation. The PV lighting system studied is shown in Fig. 48.1.

48.2 Mathematical Modeling

48.2.1 Photovoltaic Array Output Modeling

The "four-parameter" equivalent circuit model that considers a PV cell as an "ideal" irradiance-dependent current source in parallel with a diode was used to model the PV module [4]. The four parameters are module photocurrent at

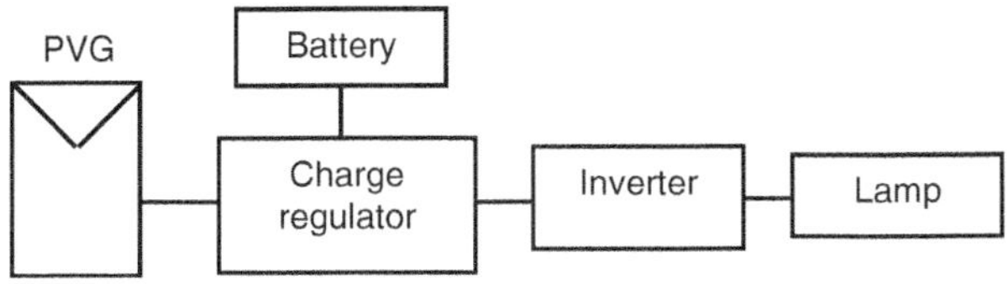

Fig. 48.1 Studied system

reference conditions ($I_{L,ref}$), diode reverse saturation current at reference conditions ($I_{o,ref}$), empirical diode PV curve fitting factor (d_1), and module series resistance (R_s) [5]. The total current (I) is calculated as follows [6]:

$$I = I_L - I_0 \left[\exp\left(\frac{q}{d_1 k T_c} (V + I R_s) \right) - 1 \right] \tag{48.1}$$

The values of parameters d_1 and R_s are fixed for a given PV cell. The photocurrent (I_L) is linearly proportional to the incident irradiance:

$$I_L = I_{L,ref} \frac{I_T}{I_{T,ref}} \tag{48.2}$$

where $I_{L,ref}$ is the photocurrent at the reference conditions and I_T and $I_{T,ref}$ represent incident irradiance at any time and reference insolation, respectively, where the reference insolation is equal to 1,000 W/m^2.

The reverse saturation current (I_o) is expressed in terms of material characteristics and PV module temperature (T_c):

$$I_o = I_{o,ref} \left(\frac{T_c}{T_{c,ref}} \right)^3 \exp\left[\frac{q\varepsilon}{dk} \left(\frac{1}{T_{c,ref}} - \frac{1}{T_c} \right) \right] \tag{48.3}$$

where d is equal to d_1/n_s; n_s is the number of cells in the module connected in series; ε is the semiconductor band-gap energy; and $I_{o,ref}$ and $T_{c,ref}$ are reverse saturation current and module temperature, respectively, at reference conditions.

The parameters $I_{L,ref}$, $I_{o,ref}$, and d_1 can be derived from the general I–V expressions [i.e., (48.1)] at the reference conditions ($I_{T,ref}$ and $T_{c,ref}$). By substituting the current and voltage at the open circuit where current (I) is zero and voltage (V) is $V_{oc,ref}$ (48.4), the short circuit where voltage (V) is zero and current (I) is $I_{sc,ref}$ (48.5), and the maximum power point (MPP) conditions where current (I) is $I_{mp,ref}$ and voltage (V) is $V_{mp,ref}$ (48.6), the following equations are obtained:

$$0 = I_{L,ref} - I_{o,ref} \left[\exp\left(\frac{q V_{oc,ref}}{d_1 k T_{c,ref}} \right) - 1 \right] \tag{48.4}$$

$$I_{sc,ref} = I_{L,ref} - I_{o,ref} \left[\exp\left(\frac{q I_{sc,ref} R_S}{d_1 k T_{c,ref}} \right) - 1 \right] \tag{48.5}$$

$$I_{mp,ref} = I_{L,ref} - I_{o,ref} \left[\exp\left(\frac{q}{d_1 k T_{c,ref}} (V_{mp,ref} + I_{mp,ref} R_S) \right) - 1 \right] \tag{48.6}$$

The magnitude of the reverse saturation current (I_o) is very small, generally in the order of 10^{-5} or 10^{-6}. The term "-1" in (48.4)–(48.6) can be neglected. For the short-circuit condition, the exponential term in (48.5) is very small and can therefore be neglected. These considerations lead to the following three expressions:

$$I_{\mathrm{o,ref}} = \frac{I_{\mathrm{sc,ref}}}{\exp\left(\frac{qV_{\mathrm{oc,ref}}}{d_1 kT_{\mathrm{C,ref}}}\right)} \tag{48.7}$$

$$I_{\mathrm{L,ref}} \approx I_{\mathrm{sc,ref}} \tag{48.8}$$

$$d_1 = \frac{q\left(V_{\mathrm{m\,p,ref}} - V_{\mathrm{oc,ref}} + I_{\mathrm{m\,p,ref}} R_{\mathrm{S}}\right)}{kT_{\mathrm{C,ref}} \ln\left(1 - \frac{I_{\mathrm{m\,p,ref}}}{I_{\mathrm{sc,ref}}}\right)} \tag{48.9}$$

The parameter R_{s} is obtained from the temperature coefficients of the open-circuit voltage and short-circuit current ($\mu_{\mathrm{V,oc}}$ and $\mu_{\mathrm{I,sc}}$).

The values of these temperature coefficients are available from a PV module catalogue. The analytical derivative of voltage with respect to temperature at the reference open-circuit condition is matched to the open-circuit temperature coefficient of the voltage [5]:

$$\frac{\partial V_{\mathrm{oc}}}{\partial T} = \mu_{\mathrm{V,oc}} = \frac{d_1 k}{q}\left[\ln\left(\frac{I_{\mathrm{sc,ref}}}{I_{\mathrm{o,ref}}}\right) + \frac{T_{\mathrm{C}}\mu_{\mathrm{I,sc}}}{I_{\mathrm{sc,ref}}} - \left(3 + \frac{q\varepsilon}{d_1 kT_{\mathrm{C,ref}}}\right)\right] \tag{48.10}$$

The equivalent circuit characteristic parameters were calculated using MATLAB “solve” function to solve these four equations [i.e., (48.7–48.10)]. According to the values shown in Table 48.1 we find

$$\begin{aligned} I_{\mathrm{o,ref}} &= 2.86 \times 10^{-6}\,\mathrm{A} \\ R_{\mathrm{s}} &= 0.2421\ \Omega \\ d_1 &= 120 \end{aligned}$$

The photovoltaic generator (PVG) characteristic parameters used in the study are shown in Table 48.1.

48.2.2 *Storage Modeling*

Several models are proposed in the literature for battery storage modeling. A simple one has been chosen. This model allows calculating state of charge (SOC) of the battery according to the produced power by PV generator and the load. This model does not take temperature effect into consideration.

Table 48.1 PV module characteristic parameters

Parameter	Value
Module short-circuit current at reference conditions	3.45 A
Module open-circuit voltage at reference conditions	43.5 V
Temperature at reference conditions	298 K
Irradiance at reference conditions	1,000 W/m^2
Maximum power point voltage at reference conditions	35.0 V
Maximum power point current at reference conditions	3.15 A
Temperature coefficient of short-circuit current	4.0×10^{-4} A K^{-1}
Temperature coefficient of open-circuit voltage	-3.4×10^{-3} V K^{-1}
Semiconductor band gap	1.12 eV
Number of cells in the module connected in series	72
Peak power	110 Wc
Module area	0.87 m^2

During the charge phase SOC is described by the following equation:

$$SOC(t) = \mathrm{Min}\left(SOC_{max},\ SOC(t - \Delta t) + \left(P_{pv}(t) - \frac{P_{load}(t)}{\eta_{inv}}\right) \times \frac{\eta_{bat}}{U_{bat}} \times \Delta t\right) \tag{48.11}$$

During the discharge phase, battery SOC is described by the following equation:

$$SOC(t) = \mathrm{Max}\left(SOC_{max} \times (1 - DOD),\ SOC(t - \Delta t) + \left(P_{pv}(t) - \frac{P_{load}(t)}{\eta_{inv}}\right) \times \frac{\Delta t}{U_{bat}}\right) \tag{48.12}$$

SOC_{max} is nominal capacity of the battery, Δt is the simulation time step ($\Delta t = 1$ h), $P_{pv}(t)$ is the produced power by the PV generator at time t, $P_{load}(t)$ charge demand at time t, η_{inv} DC/AC inverter efficiency, η_{bat} battery efficiency, and DOD the depth of discharge.

Since the studied system is a lighting system, some particularities must be considered. During the charge phase, i.e., during the daylight, there is no charge demand, so $P_{load}(t)$ is equal to zero. During discharge phase, i.e., during the night,

PV generator does not produce any power, so $P_{pv}(t)$ is equal to zero. Moreover, load is constant because it is a lamp. Therefore, charge and discharge models became

$$\mathrm{SOC}(t) = \mathrm{Min}\left(\mathrm{SOC_{max}}, \mathrm{SOC}(t-\Delta t) + P_{pv}(t) \times \frac{\eta_{bat}}{U_{bat}} \times \Delta t\right) \quad (Charge) \tag{48.13}$$

$$\mathrm{SOC}(t) = \mathrm{Max}\left(\mathrm{SOC_{max}} \times (1-\mathrm{DOD}), \mathrm{SOC}(t-\Delta t) + \frac{P_{load}(t)}{\eta_{inv}} \times \frac{\Delta t}{U_{bat}}\right) \quad (Discharge) \tag{48.14}$$

In this study η_{inv} and η_{bat} have been taken equal to 0.9.

48.3 Meteorological Data Computation

Monthly meteorological data available on the NASA Web site [2] have been used for generating hourly synthetic meteorological data (horizontal global irradiance and ambient temperature) with the aid of PVSYST software.

48.3.1 Module Temperature

To determine module temperature, a simple equation has been developed in [4] using module ambient temperature and incident insolation data. The correlation equation is given as follows:

$$T_C = T_A + 0.031\, I_T \tag{48.15}$$

where T_C is cell temperature, T_A ambient temperature, and I_T incident insolation.

48.3.2 Predicting Hourly Solar Irradiance on Inclined Surface

In many sites, at best, only global irradiances on horizontal planes are available. Because most systems using solar energy are tilted, these data are clearly insufficient. A number of models to estimate global irradiance on an inclined surface from the irradiance on a horizontal surface are available; however, these models require information at the same time on the global and the direct or diffuse irradiance on a horizontal surface. In [7], two models requiring only the global irradiance on horizontal planes as input parameter were developed. This work uses the model given in (48.16), which yields better results:

$$I_{\mathrm{T},\beta} = I_{\mathrm{G}}\left[0.1 + \frac{\rho}{2} + \left(0.1 - \frac{1}{2}\rho\right)\cos\beta + 0.8(\cos\theta/\cos\theta_z)\right] \tag{48.16}$$

$I_{\mathrm{T},\beta}$ is the total irradiance received on a tilted surface; I_{G} is the horizontal global irradiance; and θ_z is the zenith angle calculated by [8]

$$\cos\theta_z = \sin\delta\sin\phi + \cos\delta\cos\phi\cos\omega \tag{48.17}$$

δ is the declination of day D calculated by [9]

$$\delta(D) = 0.4093\sin\left(2\pi\frac{D - 81}{365}\right) \tag{48.18}$$

ρ is the albedo (in this work, the value of albedo has been taken constant and is equal to 0.2), ϕ the geographic latitude, ω the hour angle, and θ the angle of incidence for an arbitrarily inclined surface oriented toward the equator calculated by the following:

$$\cos\theta = \sin\delta\sin(\phi - \beta) + \cos\delta\cos(\phi - \beta)\cos\omega \tag{48.19}$$

In this study, a discrete form of the hour angle ω has been chosen. The action of the sun moving in the sky can be replaced by constant irradiance in short time intervals and sun position by the average position at the same time [9]. If solar time is divided into sufficiently short time intervals, such a description will correspond to the description of a continuous form of the hour angle. Temporary sun positions are considered uniformly distributed in solar time across time intervals; hence, they will be uniform with respect to hour angle. Time Δt can be obtained from the division of 24 h into N equal parts:

$$\Delta t = \frac{24}{N} \tag{48.20}$$

The variable n, which is symmetrical in relation to 12.00, is introduced according to the following formula:

$$\tau = n\Delta t + 12 \tag{48.21}$$

where τ is the solar time. The equation for hour angle determination is

$$\omega(\tau) = 15\tau - 180 \tag{48.22}$$

By substituting (48.21) into (48.22) we obtain

$$\omega(n) = 15n\Delta t \tag{48.23}$$

Taking into consideration (48.20) and (48.23), we obtain

$$\omega(n) = \frac{360n}{N} \tag{48.24}$$

or in radians

$$\omega(n) = \frac{2\pi n}{N} \tag{48.25}$$

In the performed calculations, the following detailed data are assumed: 1-h time intervals and a variability of solar time from 5:00 a.m. until 8:00 p.m. The data given above correspond to the variability of the hour angle from $-105°$ to $+120°$ and the time variable n, assuming integer values from -7 up to $+8$.

48.4 Proposed Method

The flowchart of the proposed method is shown in Fig. 48.2.

The algorithm works with a Boolean vector containing the PVG area correction coefficient (k_1) and buttery nominal capacity correction coefficient (k_2). The algorithm uses S_{pop} of these vectors.

48.4.1 Objective Function

The objective function to be minimized includes the following parameters:

- PV panel life cycle cost (LCC)
- Battery life cycle cost
- Unmet load

Hence the objective function is as follows:

$$ff = N_{ul} \times C_{ul} + \text{PV}_{\text{LCC}} + \text{BAT}_{\text{LCC}} \tag{48.26}$$

PV_{LCC} is the PV generator life cycle cost (€), and BAT_{LCC} life cycle cost (€).

N_{ul} is unmet load (number of hours) and C_{ul} is unmet load weighting coefficient.

To avoid using multi-objective optimization unmet load has been integrated to the objective function using the coefficient C_{ul}, by taking this one sufficiently high N_{ul} can be obtained equal to zero.

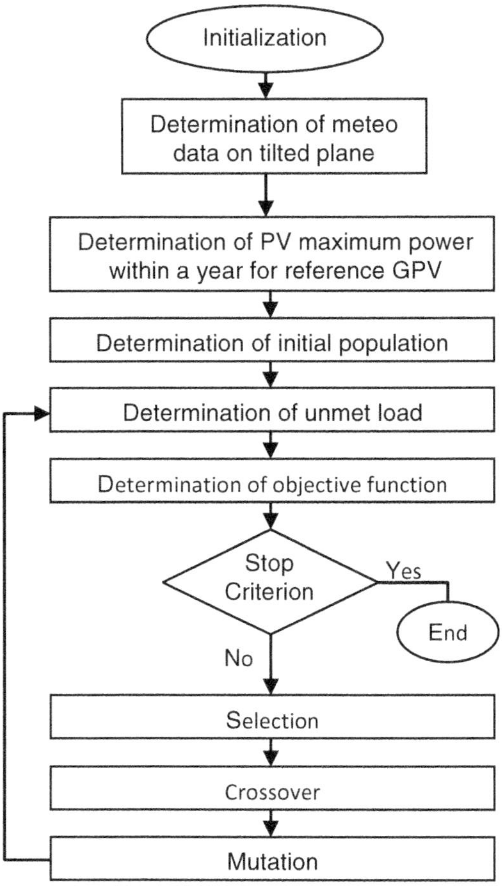

Fig. 48.2 Flowchart of the proposed method

Unmet load has been calculated as follows:

$$\text{if } \mathrm{SOC}(t) < \mathrm{SOC}_{\max} \times (1 - \mathrm{DOD}) \quad \text{then} \quad N_{\mathrm{ul}} = N_{\mathrm{ul}} + 1$$

To calculate $\mathrm{PV}_{\mathrm{LCC}}$ and $\mathrm{BAT}_{\mathrm{LCC}}$ the following models proposed in [10] have been used:

$$\mathrm{PV}_{\mathrm{LCC}}(€) = 650 \times A_{\mathrm{PV}} \tag{48.27}$$

$$\mathrm{BAT}_{\mathrm{LCC}}(€) = 12.411 \times C_{\mathrm{N}} + 69.05 \tag{48.28}$$

A_{PV} is the PV generator area, and C_{N} battery nominal capacity.

These LCC estimations are based on a discount rate of 5 %, an inflation rate of 2 %, and a lifetime of 25 years.

In this study, A_{PV} and C_N are determined as follows:

$$A_{PV} = k_1 \times A_r \tag{48.29}$$

$$C_N = k_2 \times C_{Nr} \tag{48.30}$$

Ar and C_{Nr} are references of PV generator area and battery nominal capacity, respectively. In this study they have been taken equal to 0.87 m^2 and 83.33 Ah, respectively. k_1 and k_2 are the coefficients determined by GA to optimize the cost of the system.

48.4.2 Method Description

The method has been implemented in the following way:

1. The parameters used in the optimization are set (see Table 48.1).
2. The irradiation on a tilted surface is calculated using the model described above. The irradiation is calculated with a step time of 1 h.
3. The irradiation during a year is applied to the model of a 110 W_c PVG (0.87 m^2) to determine the PVG maximum power $(P_{max110}(t))$ produced during every hour of the year. To determine the optimum PV area, the 110 Wc PVG area is used as a reference, and then corrected by the coefficient k_1.
4. S_{pop} vectors are obtained randomly. These vectors have been described above, each one representing a possible configuration of PVG area and battery capacity.
5. For each vector, the maximum power determined in step 3 is applied to the storage model to determine the unmet load parameter N_{ul}. In the model of charge described in (48.13), $P_{pv}(t)$ is substituted by $k_1 \times P_{max110}$. This is a good approximation; this allows avoiding recalculating maximum power for each vector, so the method became faster.
6. The objective function is evaluated for each vector. Best vectors have a greater probability of reproducing themselves, crossing with other vectors. In each cross of two vectors, two new vectors are obtained (descendents).
7. The descendants are evaluated and the best of them replace the worst individuals of the previous generation (iteration).
8. To find the optimal solution and not to stay in local minimal, some solutions randomly change some of their components (mutation). The mutations can change a bit of k_1 or k_2.
9. The individuals (vectors) obtained from reproduction and mutation are evaluated, making the next generation.
10. The process continues (from 5 to 9) until a determined number of generations (N_{gen}) have been evaluated.

48.5 Results

The proposed algorithm has been implemented using MATLAB software; a photovoltaic lighting system located in Adrar (Algeria) has been optimized. Load considered is a simple resistor of 30 W representing the lamp. The parameters used in this case are the following:

The crossover rate is 0.8, the mutation rate is 0.1, population size is 100, and the number of generations is 100.
The DOD has been taken 50 %.
The insolation incident on a horizontal surface (kWh/m^2/day) for Adrar is shown in Table 48.2 [2]. These data have been used for generating hourly synthetic meteorological data (horizontal global irradiance and ambient temperature) with the aid of PVSYST software.

Figures 48.3 and 48.4 show the synthetic meteorological data generated by PVSYS for Adrar located in Southwest Algeria (27.51°N, 0.17°W).

Results obtained using GA for tilted angle from 0° to 90°, with an angle step equal to 5° in the interval 0°–40° and 50°–90°, and with an angle step equal to 1° between 42° and 48°, are given in Fig. 48.5; this figure shows that the lowest LCC is obtained for tilted angle around 45°. System LCC is practically constant between 40° and 46° where the variation of LCC is less than 0.2 %.

According to Fig. 48.5 it can be noted that the best tilted angel is around 45; however changing the tilted angle between 30° and 46° does not affect significantly the LCC.

In Fig. 48.6 it can be seen that between 30° and 46° there is no significant augmentation of the LCC according to the lowest one. However decreasing tilted angle under 30° increases system LCC but less than increasing it above 46°. That means that such a system is more suitable if placed on a building roof than on a wall.

Evolution of battery state of charge during a year for tilted angle of 45° is shown in Fig. 48.7; this figure shows that the battery is charged over 70 % in a large interval of time, so it can be noted that with allowing a minor unmet load, system LCC can be considerably decreased. That can be made with a multi-objective optimization.

Evolution of system LCC during the optimization for tilted angle of 45° is shown in Fig. 48.8.

Table 48.2 Insolation incident on a horizontal surface (kWh/m^2/day) for Adrar

January	February	March	April	May	June	July	August	September	October	November	December	Annual
3.68	4.74	5.90	6.84	7.32	7.70	7.45	6.96	5.86	4.60	3.83	3.32	5.68

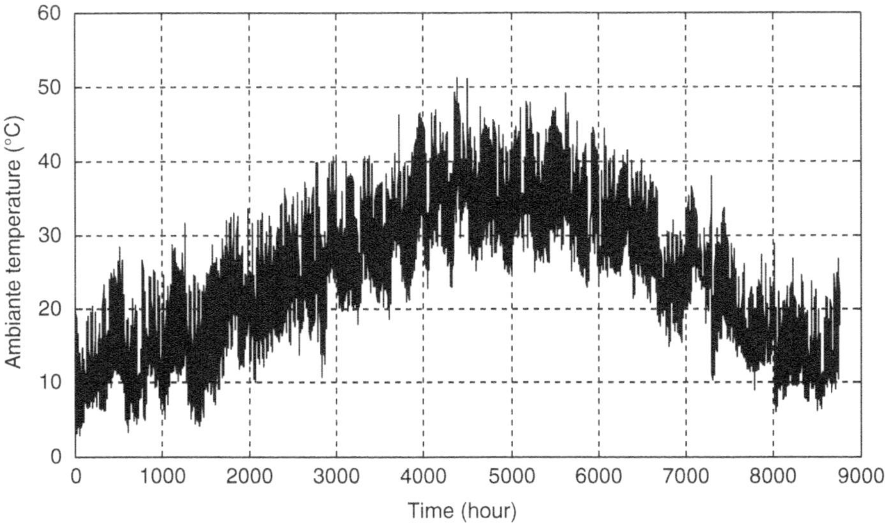

Fig. 48.3 Ambient temperature for Adrar

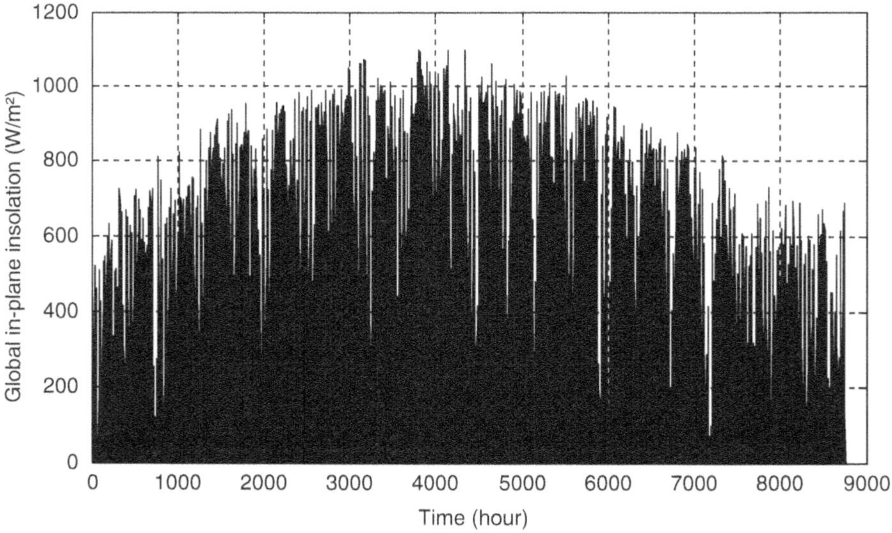

Fig. 48.4 Global in-plane insolation for Adrar

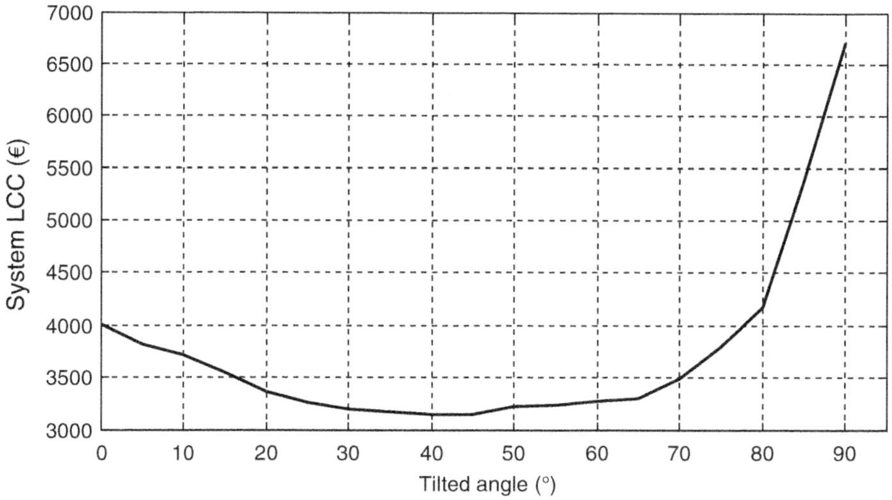

Fig. 48.5 Variation of minimum cost of the system

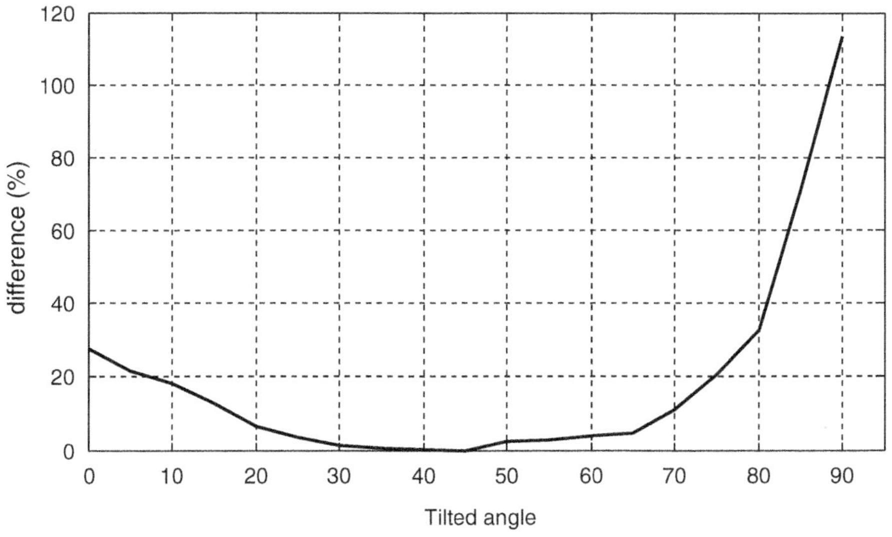

Fig. 48.6 Overcharge according to the lowest LCC

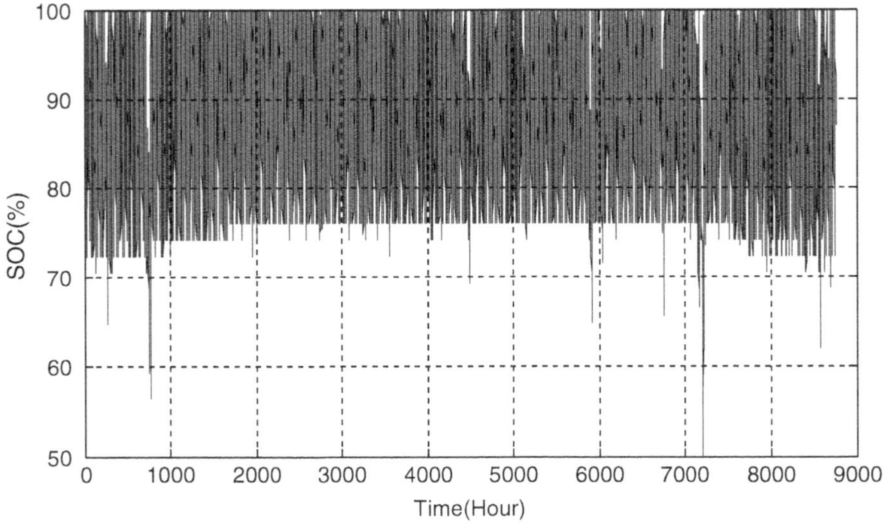

Fig. 48.7 Evolution of battery SOC (45°)

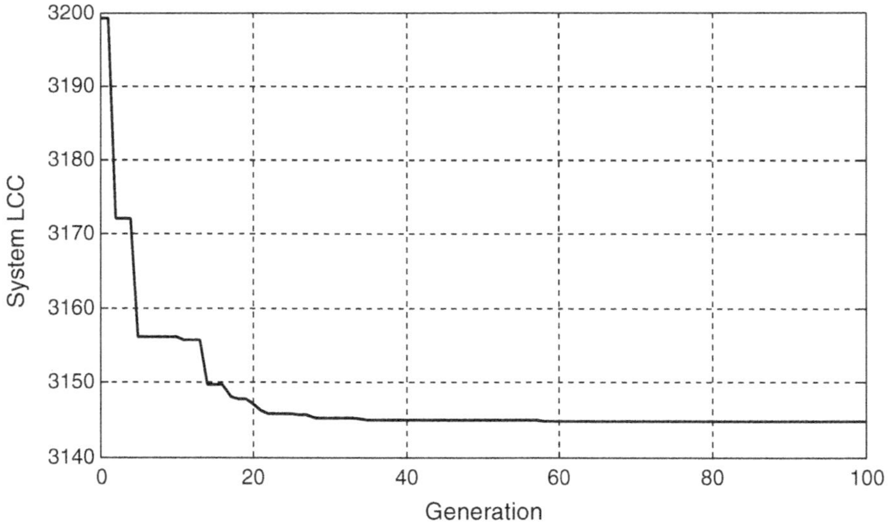

Fig. 48.8 Evolution of system LCC (45°)

48.6 Conclusion

In this study, a developed algorithm using genetic algorithms has been used for optimum sizing of a photovoltaic lighting system located in Adrar (Algeria).

This study shows that the lowest cost (with non-unmet load) has been obtained for a tilted angle around 45°. However changing the tilted angle between 30° and 46° does not affect significantly the LCC. Decreasing tilted angle under 30° increases system LCC but less than increasing it above 46°.

Evolution of battery state of charge during a year shows that with allowing a minor unmet load, system LCC can be considerably decreased. That can be made with a multi-objective optimization.

References

1. Mellit A, Kalogirou SA, Hontoria L, Shaari S (2008) Artificial intelligence techniques for sizing photovoltaic systems: a review. Renew Sust Energ Rev 13(2):406–419
2. http://eosweb.larc.nasa.gov/sse/. Accessed 28 Feb 2014
3. Goldberg DE (1989) Genetic algorithms in search, optimization and machine learning. Addison-Wesley, New York
4. Mondol DJ, Yohanis YG, Norton B (2007) Comparison of measured and predicted long term performance of grid connected photovoltaic system. Energ Convers Manage 48:1065–1080
5. Townsend TU (1989) A method for estimating the long-term performance of direct-coupled photovoltaic. MS thesis, Solar Energy Laboratory, University of Wisconsin, Madison, WI
6. Duffie JA, Beckman WA (1991) Solar engineering of thermal processes, 2nd edn. Wiley, New York
7. Notton G, Poggi P, Cristofari C (2006) Predicting hourly solar irradiations on inclined surfaces based on the horizontal measurements: performances of the association of well-known mathematical models. Energ Convers Manage 47:1816–1829
8. Iqbal M (1983) An introduction to solar radiation. Academic Press, Toronto, ON
9. Owczarek S (1997) Vector model for calculation of solar radiation intensity and sums incident on tilted surfaces. Identification for the three sky condition in Warsaw. Renew Energ 11(1):77–96
10. Abbes D, Martinez A, Champenois G (2013) Life cycle cost, embodied energy and loss of power supply probability for the optimum design of hybrid power system. Math Comput Simul. doi:10.1016/matcom.2013.05.004

Chapter 49
Performance of Induction Motor Under the Effect of Voltage Unbalance with Loading Consideration

Fatma Zohra Dekhandji and Larbi Refoufi

Abstract The performance of induction motors under different voltage unbalance levels and different loading is investigated in this chapter. The stator/rotor currents, torque and additional power losses are determined for each voltage unbalance value. The levels range from low level to severe situations to assess the ability of the induction motor to withstand different stress levels. A MATLAB/SIMULINK implementation is done to produce the currents, torque and power losses.

Keywords Induction motor • Voltage unbalance • Stator/rotor currents • Torque • Power losses

Nomenclature

CVUF	Complex voltage unbalance factor
IEC	International Electromechanical Commission
IEEE	Institute of Electrical Engineering and Electronics
LVUF	Line voltage unbalance factor
NEMA	National Electrical Manufacturers Association
PVU	Percent voltage unbalance
PVUR	Phase voltage unbalance rate
VUF	Voltage unbalance factor

F.Z. Dekhandji (✉) • L. Refoufi
Laboratory Signals and Systems, Institute of Electrical and Electronic Engineering, University M'hamed Bougara of Boumerdes, Avenue de l'Indépedance, Boumerdes, Algeria
e-mail: fzdekhandji@umbb.dz

I. Dincer et al. (eds.), *Progress in Clean Energy, Volume 2*,
DOI 10.1007/978-3-319-17031-2_49

49.1 Introduction

On the industrial or commercial level, assured reliable operation and better performance of various equipment requires understanding of all issues, probable problems, their causes and cost-effective solutions. Power quality problem is any apparent disturbance in voltage, current or frequency deviations which results in failure or mis-operation of customer equipment.

Voltage unbalance is regarded as a power quality problem of significant concern at the electricity distribution level. The nature of the unbalance includes unequal voltage magnitudes at the fundamental system frequency (under-voltages and over-voltages), fundamental phase angle deviation, and unequal levels of harmonic distortion between the phases. An excessive level of voltage unbalance can have serious impacts on mains connected induction motors. The level of current unbalance that is present is several times the level of voltage unbalance. Such an unbalance in the line currents can lead to excessive losses in the stator and rotor that may cause protection systems to operate causing loss of production [1].

The influence of voltage unbalance upon three-phase induction motors has been investigated in literature, as the design, protection, operation, maintenance, and lifetime estimation of the motor are closely related to the level of voltage unbalance to which the motor is to be subjected. The effects of unbalanced voltages on the performance of a three-phase induction motor have been studied widely. The influence of unbalanced on the efficiency [2, 3], power factor [3], derating in the machine [4, 5], temperature rise, and life reduction [6], increase of losses, and negative effects on the insulation [4, 6, 7] are some contributions in the area. Some authors [2–4] have concluded that the efficiency and power factor of the motor depend on the positive sequence voltage, the negative sequence voltage, the voltage magnitude, and the voltage angle. Likewise, these authors show that the derating factor given by the NEMA standard is insufficient to evaluate the effects of unbalance voltage on the motor, because it is based only in NEMA definition. However, these authors do not give a mathematical relation that may be used to evaluate the efficiency and power factor.

There are two widely recognized definitions for voltage unbalance, the IEC (International Electromechanical Commission) definitions [8] and the NEMA (National Electrical Manufacturers Association) definitions [9]. The IEC definition is mathematically more rigorous compared to the NEMA definitions [10].

The method of evaluating the level of voltage unbalance is either the percent voltage unbalance (PVU) defined by the NEMA or the voltage unbalance factor (VUF) defined by the IEC. The PVU is the ratio of the maximum deviation from average voltage, to the average of three voltages, while the VUF is given by the ratio of the negative-to positive-sequence voltage. Both the PVU and the VUF are positive real quantities that reflect the level of voltage unbalance. Generally speaking, the PVU is convenient to field measurements because its calculation involves only the magnitudes of three-phase voltages. On the other hand, the calculation of the VUF requires both the magnitudes and phases of three-phase voltages to be known, which is more difficult in the field. However, the VUF conveys better physical interpretation of the cause of voltage unbalance, and is more useful in prediction and analysis of the effects of voltage unbalance on the motor [10].

An extension of the VUF is the complex voltage unbalance factor (CVUF) that is defined by the ratio of the negative sequence voltage phasor to the positive-sequence voltage phasor. The CVUF is a complex quantity having the magnitude and the angle. Although the CVUF has not yet been widely used by practicing engineers, it has been proposed in some studies [11, 12] due to its richness of information on unbalance. The effect of voltage unbalance on induction motors was studied and formulated using the CVUF by [11]. However, the author focused his analysis on the effect of the magnitude of the CVUF, and neither the physical meaning nor the effect of its angle was discussed. Furthermore, the works used the approximate equivalent circuit of the induction motor. Also, the authors of these works did not investigate the effect of the complex voltage unbalance angle on the induction motor.

In this work, the steady-state performance of an induction motor under voltage unbalance is studied with emphasis on the effect of the magnitude of the CVUF. Specifically, the influence of the variations in the CVUF magnitude on the stator/rotor currents, Torque and stator/rotor copper losses and total losses is dealt with.

49.2 Voltage Unbalance Definitions

The level of voltage unbalance that is present in a system can be specified using two commonly used definitions. The first definition is widely used in European standards and is based on the Theory of Symmetrical Components. The second definition is used in the USA and avoids the use of complex algebra.

49.2.1 IEC (International Electromechanical Commission) Definition

The definition of voltage unbalance used by academic community is the ratio of negative sequence voltage V_n to the positive sequence voltage V_p and is called Negative Sequence Voltage Unbalanced Factor. This definition is adopted by IEC 60034-26 and is also known as the Voltage Unbalance Factor (VUF) or IEC definition.

$$\mathrm{VUF} = \frac{V_n}{V_p} \tag{49.1}$$

For a set of unbalanced voltages V_{ab}, V_{bc},V_{ca}, the positive and negative sequence voltages V_p and V_n are given by:

$$V_p = \frac{V_{ab} + aV_{bc} + a^2V_{ca}}{3} \tag{49.2}$$

$$V_n = \frac{V_{ab} + a^2V_{bc} + aV_{ca}}{3} \tag{49.3}$$

where $a = -0.5 + j0.866$ and $a^2 = -0.5 - j0.866$.

The VUF can also be expressed in a more user-friendly form than given by (49.1), which requires only the three line-line voltage readings V_{ab}, V_{bc},V_{ca}.

$$\text{VUF} = \frac{V_n}{V_p} = \sqrt{\frac{1 - \sqrt{3 - 6\beta}}{1 + \sqrt{3 - 6\beta}}} \tag{49.4}$$

where

$$\beta = \frac{V_{ab}^4 + V_{bc}^4 + V_{ca}^4}{\left(V_{ab}^2 + V_{bc}^2 + V_{ca}^2\right)^2}. \tag{49.5}$$

49.2.2 NEMA (National Electrical Manufacturers Association) Definition

The NEMA Standard MG1.1993 and the IEEE community use the following definition:

$$\text{LVUF} = \frac{\text{maximum voltage deviation from average voltage}}{\text{average voltage}}. \tag{49.6}$$

This definition assumes that the average voltage is always equal to the rated value and avoids the use of complex algebra, and is called Line Voltage Unbalance Factor (LVUF) or NEMA definition.

49.2.3 IEEE (Institute of Electrical Engineering and Electronics) Definition

The IEEE definition of voltage unbalance is known as the Phase Voltage Unbalance Rate (PVUR), is given by

$$\text{PVUR} = \frac{\text{maximum voltage deviation from average phase voltage}}{\text{average phase voltage}}. \tag{49.7}$$

The IEEE definition is the same as the NEMA one; the only difference is that the IEEE uses the phase voltages rather than the line-to-line voltages. Here again the phase angle information is discarded since only the magnitude is considered.

49.3 Simulation Results and Discussions

A three-phase 4-pole, 100 hp, 400-V, 50 Hz, 1,484 rpm squirrel-cage induction motor has been analyzed for the cases of: 0 % (balanced), 2.5 % and the extreme case of 5 % (recommended by NEMA standards) Voltage unbalance factor. The motor parameters are:

$$\mathrm{Rr} = 0.02092\ \Omega;\ \mathrm{Rs} = 0.03552\ \Omega;\ \mathrm{Ls} = 0.000335\ \mathrm{H};\ \mathrm{Lr} = 0.000335\ \mathrm{H};$$
$$\mathrm{Lm} = 0.0151\ \Omega; J = 1.25\ \mathrm{kg/m^2}; B = 0.03914\ \mathrm{Nm/s},$$

where:

Rr: is the rotor resistance referred to stator. Rs: is the stator resistance.
Lr: is the rotor inductance referred to stator. Lm: is the mutual inductance.
Ls: is the stator inductance. *J*: moment of inertia. *B*: friction factor.

The work consists of a supply voltage unbalance investigation on the performance of the induction motor in terms of stator/rotor currents, torque, and additional power losses. The motor loading is taken into account jointly with the voltage unbalance factor. The motor under investigation is loaded on a gradual basis from its full load to the lightly loaded situation. At any value considered, the load is considered to be constant. Figure 49.1 shows an overall view of the MATLAB/SIMULINK implementation used in the investigation.

49.3.1 Case of 0 % Unbalance (Balanced Condition)

This is considered to be a reference case over all the performance criteria. Table 49.1 presents the stator/rotor currents and the torque/speed values over different loading values while Table 49.2 gives the variations in power losses for different loading conditions.

The stator/rotor currents are the same over all the phases regardless of the motor loading. The currents decrease with the decrease in loading which is logic as less demand in current is needed. As for the torque/speed, as the load decreases, the produced torque decreases and the speed increases.

Regarding power losses, these seem to decrease with the load decrease. A similar remark to the stator/rotor currents is noticed regarding the unbalance in power losses. The stator/rotor copper losses are balanced over the three phases which is in agreement with the stator/rotor current balance.

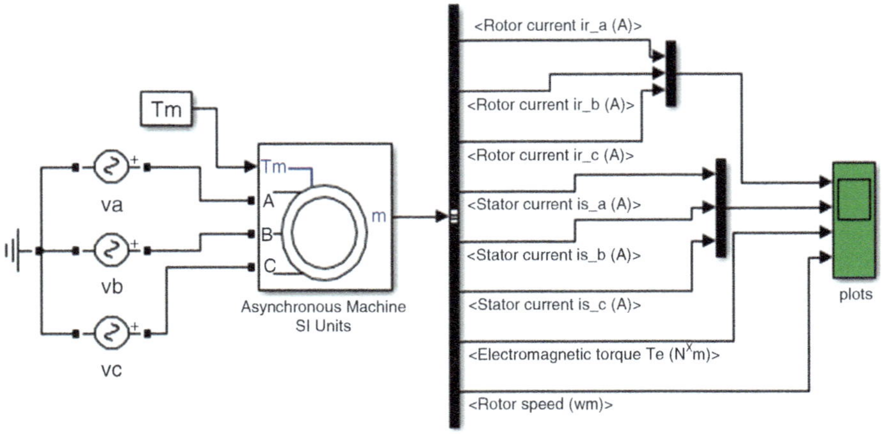

Fig. 49.1 MATLAB/SIMULINK model of the induction motor under voltage unbalance

Table 49.1 Stator/rotor currents and torque/speed variations v.s. for the balanced case

100 hp motor with VUF = 0 %	Full load, 400 Nm	75 %, 300 Nm	50 %, 200 Nm	25 %, 100 Nm
Stator current Isa (A)	109.1	87.58	68.44	53.64
Stator current Isb (A)	109.1	87.57	68.42	53.64
Stator current Isc (A)	109.1	87.57	68.42	53.64
Rotor current Ira (A)	96.22	71.9	47.59	24.84
Rotor current Irb (A)	96.69	71.69	47.61	24.87
Rotor current Irc (A)	96.24	73.73	47.85	24.68
Torque (Nm)	406.1	306.1	206.1	106.1
Speed (rpm)	1,487	1,490	1,494	1,497

49.3.2 Case of 2.5 % Unbalance

This situation has been chosen to be midway between the balanced case and the most severe voltage unbalance situation. The effects of voltage unbalance are summarized in Tables 49.3 and 49.4.

As expected, the voltage unbalance leads to current unbalance. The stator currents are unbalanced regardless of the loading level. The rotor current unbalance turns out, however, to depend on the loading. Indeed, for light loading, no significant current unbalance is noticed while at high loads, a small current unbalance results.

Table 49.2 Stator/rotor power losses in the balanced case

100 hp motor with VUF = 0%	Full load	75 %	50 %	25 %
Stator Copper loss Psa (W)	422.7	272.6	166.5	102.2
Stator Copper loss Psb (W)	422.7	272.6	166.5	102.2
Stator Copper loss Psc (W)	422.7	272.6	166.5	102.2
Stator copper loss (W)	1268.1	817.8	499.5	306.6
Rotor Copper loss Pra (W)	193.7	108.1	47.37	12.9
Rotor Copper loss Prb (W)	195.6	107.5	47.42	12.9
Rotor Copper loss Prc (W)	193.8	111.9	47.91	12.9
Rotor copper loss (W)	583.1	327.5	147.7	38.7

Table 49.3 Stator/rotor currents and torque/speed variations v.s. for the 2.5 % voltage unbalance

100 hp motor with VUF = 2.5 %	Full load	75 %	50 %	25 %
Stator current Isa (A)	113.7	93.87	77.25	65.99
Stator current Isb (A)	94.28	72.32	52.82	38.98
Stator current Isc (A)	121.1	98.71	77.84	59.76
Rotor current Ira (A)	101.3	72.42	50.01	29.39
Rotor current Irb (A)	96.7	72.74	50.28	29.13
Rotor current Irc (A)	94.73	76.45	52.08	29.01
Torque (Nm)	406.1	306.1	206.1	106.1
Speed (rpm)	1,487	1,490	1,494	1,497

Table 49.4 Stator/rotor power losses in the case of 2.5 % voltage unbalance

100 hp motor with VUF = 2.5 %	Full load	75 %	50 %	25 %
Stator Copper loss Psa (W)	459.6	312.9	212.0	154.7
Stator Copper loss Psb (W)	316	185.8	99.09	53.97
Stator Copper loss Psc (W)	521.5	346.1	215.2	128.9
Stator copper loss (W)	1297.1	844.8	526.29	337.57
Rotor Copper loss Pra (W)	190	107.5	51.24	18.08
Rotor Copper loss Prb (W)	192.7	121.5	54.97	17.73
Rotor Copper loss Prc (W)	215.6	113.6	55.73	17.67
Rotor copper loss (W)	598.3	342.6	161.94	53.48
Torque ripple (Nm)	129.6	130.6	131.7	132.33

The torque does not seem to vary with voltage unbalance despite the increase in peak to peak torque ripple.

Once more, power losses seem to decrease with the load decrease. The stator/rotor copper losses go hand in hand with the stator/rotor current unbalance responses. Indeed, the stator losses are unbalanced over the three phases while the rotor losses are fairly balanced for light loading and slightly unbalanced for high loading conditions.

49.3.3 Case of 5 % Unbalance

This voltage unbalance level is the extreme situation according to NEMA and should be mitigated through disconnecting the motor. Tables 49.5 and 49.6 summarize the obtained effects on the motor performance.

Once more, the results reveal clearly that voltage unbalance leads to current unbalance. As it was the situation with the previous voltage unbalance level, the stator currents are unbalanced irrespective of the loading. The rotor current unbalance seems again to depend on the loading. In fact, at small loads, no noticeable current unbalance is seen while at high loads, a relatively remarkable current unbalance results. The torque has not been affected by voltage unbalance despite the further increase in peak to peak torque ripple compared to the previous case.

Power losses turn out once more to be directly proportional to the loading level. The stator/rotor copper losses follow the same behavior as the stator/rotor current unbalance. Explicitly speaking, the stator losses are unbalanced over the three phases, while the rotor losses are almost balanced for light loading and fairly unbalanced for high loading situation.

Table 49.5 Stator/rotor currents and torque/speed variations v.s. for the 5 % voltage unbalance

100 hp motor with VUF = 5%	Full load	75 %	50 %	25 %
Stator current Isa (A)	120	102.1	87.97	79.37
Stator current Isb (A)	79.66	57.24	37.17	25.08
Stator current Isc (A)	133.8	110.9	89.01	68.83
Rotor current Ira (A)	103.6	79.16	56.36	39.64
Rotor current Irb (A)	97.55	80.37	58.39	39.35
Rotor current Irc (A)	102.4	76.13	57.33	39.55
Torque (Nm)	406.1	306.1	206.2	106.2
Speed (rpm)	1,487	1,490	1,494	1,497

Table 49.6 Stator/rotor power losses in the case of 5 % voltage unbalance

100 hp motor with VUF = 5%	Full load	75 %	50 %	25 %
Stator Copper loss Psa (W)	511.7	370.6	274.9	223.8
Stator Copper loss Psb (W)	225.4	116.4	49.08	22.35
Stator Copper loss Psc (W)	636.2	437	281.4	168.3
Stator copper loss (W)	1373.3	924	605.38	414.45
Rotor Copper loss Pra (W)	198.8	137.1	68.19	32.85
Rotor Copper loss Prb (W)	222.3	123.3	71.55	32.38
Rotor Copper loss Prc (W)	221.7	127.1	66.8	32.75
Rotor copper loss (W)	642.8	387.5	206.54	172.23
Torque ripple (Nm)	259.3	261.5	263.29	264.85

49.4 Conclusion

In this chapter, a MATLAB/SIMULINK implementation has been done to investigate the variations in currents, torque and power losses under balanced and unbalanced conditions jointly with different levels of loading. The results show that voltage unbalance is, in general, associated with current unbalance. However, the rotor currents under light loading are not altered by any level of tolerable voltage unbalance while at high loading, they are fairly affected. Power losses are directly proportional to the voltage unbalance and tally with the stator/rotor current response to voltage unbalance. Under unbalanced supply condition, the opposite rotating torque occurs due to the reverse components of currents. Therefore the motor needs to draw more current from the supply to maintain the required mechanical power. As a result, the copper losses heat in the machine causing its lifetime to decrease.

References

1. Refoufi L, Bentarzi H, Dekhandji FZ (2006) New approach of induction motor performance study under different voltage unbalance conditions. WSEAS Trans Power Syst 1(8):1519–1525
2. Faiz J, Ebrahimpour H, Pillay P (2004) Influence of unbalanced voltage on the steady-state performance of a three-phase squirrel-cage induction motor. IEEE Trans Energy Convert 19 (4):657–662
3. Lee CY (1999) Effects of unbalanced voltage on operation performance of a three-phase induction motor. IEEE Trans Energy Convert 14(2):202–208
4. Pillay P (2002) Derating of induction motors operating with a combination of unbalanced voltages and over and under voltages. IEEE Trans Energy Convert 17(4):485–491
5. Kersting WH, Phillips WH (1977) Phase frame analysis effects of voltage unbalance on induction machines. IEEE Trans Indust Appl 33(2):415–420
6. Gafford BN, Duesterhoef WC, Mosher CC (1959) Heating of induction motors on unbalanced voltages. AIEE Trans Power Apparat Syst 78:282–297
7. Quispe E, Gonzalez G, Aguado J (2004) Influence of unbalance and waveform voltage on the performance characteristics of three-phase induction motors. In: Proceedings of international conference on renewable energy and power quality applications, Barcelona, España, ISBN 8460798870
8. IEC (2002) Rotating electrical machines, part 26: effects of unbalanced voltages on the performance of induction motors, IEC Standard 60034-26. IEC, Geneva
9. NEMA (1993) Motors and generators, ANSI/NEMA Standard MG1-1993. NEMA, Rosslyn, VA
10. Wang YJ (2001) Analysis of effects of three-phase voltage unbalance on induction motors with emphasis on the angle of the complex voltage unbalance factor. IEEE Trans Energy Convert 16(3):2001.
11. Wang YJ (2000) An analytical study on steady-state performance of an induction motor connected to unbalanced three-phase voltage. In: IEEE power engineering society winter meeting, Singapore, 23–27 Jan 2000
12. Pierrat L, Meyer JP (1987) Unbalance factor, it is as simple as ABC. Rev Gén Électricité 6:18–26

Chapter 50
A Simple and Accurate Maximum Power Point Tracking Algorithm for Photovoltaic Systems

Aissa Kheldoun, Salim Djeriou, Abdelmalek Kouadri, and Larbi Refoufi

Abstract Many techniques have been proposed to seek the solar photovoltaic system's maximum power point from which one can list the following algorithms Perturb and Observe (P&O) algorithm, the Incremental Conductance (IC) algorithm, among others. Every technique has its relative merits and limitations essentially in terms of stability and ease of implementation. This chapter presents a new algorithm for Maximum Power Point Tracking in PVS. The algorithm uses the Golden Section method to search for the point at which the derivative of the input power versus voltage is zero. The power versus voltage characteristics of the photovoltaic system is a unimodal function making the Golden Section Search technique very suitable. The process of searching uses a variable step to shrink the search space, hence helping to reach the optimum point of the function, the maximum power point, within a small number of iterations. The basic advantage of this search technique is the reduction of tracking convergence time, in comparison with classical techniques with slow convergence time that can lead to instability. The proposed Golden Section Search technique algorithm converges rapidly therefore reducing the number of oscillations due to perturbations and preserving system stability. Matlab/Simulink is used to simulate the Golden search based MPPT.

Keywords Photovoltaic system • Boost DC-DC converter • Maximum power point tracking algorithm • Golden Section Search

---

A. Kheldoun (✉) • S. Djeriou • A. Kouadri • L. Refoufi
Signals and Systems Laboratory, Institute of Electrical and Electronic Engineering, University of Boumerdes, Avenue of the Independence, Boumerdes 35000, Algeria
e-mail: aissa1973@gmail.com

I. Dincer et al. (eds.), *Progress in Clean Energy, Volume 2*,
DOI 10.1007/978-3-319-17031-2_50

Nomenclature

CCM	Continous current mode
GSS	Golden Section Search
IC	Incremental conductance
MPP	Maximum power point
MPPT	Maximum power point tracking
P&O	Perturb and Observe
PV	Photovoltaic
PWM	Pulse width modulation
STC	Standard test conditions

Symbols

ε	Precision
a	Diode ideality constant
E_g	Band gap energy
G	Irradiation level in W/m2
G_1 G_2	Golden sections
G_{STC}	Nominal irradiation level (in 1,000 W/m^2)
h	Search space length
I	PV array current output
I_{MPP}	Current of maximum power point
I_{pv}	Light-generated current
I_{pvn}	Short-circuit current at G_{STC}
I_S	Diode's reverse saturation current
I_{S0}	Reverse current at nominal temperature
K	Boltzmann's constant
K_I	Short circuit current coefficient
K_V	Open circuit voltage coefficient
N_{pp}	Number of parallel-connected modules
N_S	Number of series-connected cells
N_{SS}	Number of series-connected modules
q	Electron's electric charge
R_{sh}	Shunt resistance
R_s	Series resistance
T_{STC}	Nominal temperature (298 K)
V or V_{pv}	PV module's voltage output
V_{MPP}	Voltage of maximum power point
V_{OCN}	PV module's nominal open-circuit voltage
V_t	Thermal voltage of the PV cell

50.1 Introduction

With a strong increase in energy demand and worldwide accelerated depletion of fossil fuel supplies, and increasing costs, there is a continuously growing interest in alternative and renewable energy resources. New technologies are constantly being sought to harness these resources more efficiently, reliably, and cost-effectively. A generating solution receiving a great attention is the use of solar photovoltaic systems that leave no waste behind. In comparison with other green energy sources, the efficiency of solar photovoltaic conversion system is very low. Most of the photovoltaic systems are equipped with maximum power point tracker to extract maximum output power from the PV system, and hence maximum functioning efficiency [1].

The solar photovoltaic system is a set of PV arrays or modules with parallel and/or series connections according to the load's power ratings to which is connected. The solar cell constitutes the basic element of PV modules and arrays. The behavior of this PV cell is very complex as both voltage and current provided by the PV cell are very sensitive to climate conditions, namely: irradiance, temperature, and load [2, 3]. Therefore, when a solar photovoltaic system is directly connected to a resistive load for given climate conditions, the operating point (V, I) will be at the intersection of the I–V characteristics of the PV array and that of the resistive load which is a line. In general this intersection point is not the maximum power point of the PV array. For this reason, PV arrays are connected to loads through DC-to-DC converter. The later allows to PV arrays to be constrained to an appropriate operating voltage (or current) so that maximum power can be delivered. This operation point called Maximum Power Point (MPP) is obtained by tracking algorithms.

Many techniques have been proposed to seek this maximum power point from which one can list the following algorithms Perturb and Observe (P&O) algorithm, the Incremental Conductance (IC) algorithm [2], artificial intelligence based techniques [4, 5], among others [6, 7]. The P&O algorithm is widely used because of its cost effectiveness and ease of implementation [1]. The principle of this technique is based on the perturbation of the operating voltage of the PV array. The main drawback of this algorithm is the difficulty to tune the algorithm in order to achieve fast dynamic response without affecting steady state stability. In this chapter a new tracking technique that is based on Golden Section Search (GSS-MPPT) is proposed and compared with the P&O one.

50.2 The PV System

The photovoltaic system is a set of arrays of PV modules. The PV module comprises a number of PV cells connected in either series or parallel to constitute what is known PV modules and PV arrays. The basic element of PV systems is the PV cell that converts the electromagnetic radiation of solar energy to electricity. PV system is always associated to buck, boost or buck and boost converters. These DC

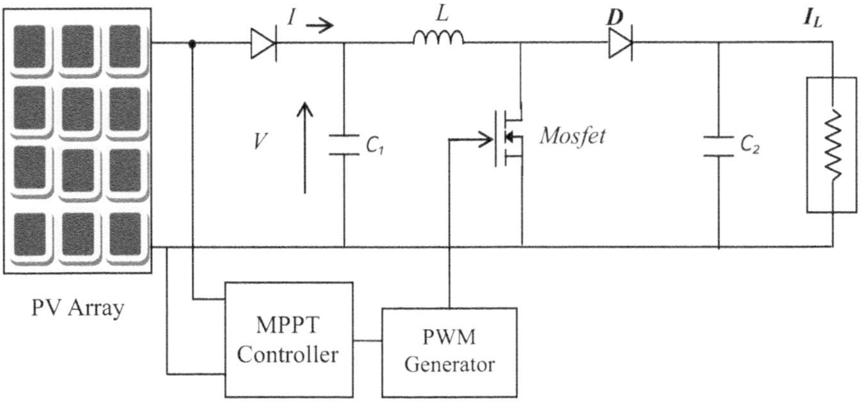

Fig. 50.1 PV system connected to load through a boost converter

to DC converters allow implementing MPPT methods and hence maximum power can be extracted from PV system. In the present work a boost DC-DC converter is used to implement GSS-MPPT and P&O algorithms as shown in Fig. 50.1.

Modeling of DC converters is omitted in this chapter as it can be found in any power electronics book, such as [8]. This section is devoted to PV module modeling which is a matrix of elementary cells that are the heart of PV systems. The modeling of PV systems starts from the model of the elementary PV cell which has the same behavior as the p-n junction.

50.2.1 Photovoltaic Cell

The physical principle governing the behavior of a photovoltaic cell can be explained, based on the behavior of a p-n junction of a semiconductor material being sensitive to sunlight. The ideal solar cell, theoretically, can be modeled as a current source in antiparallel with a diode (Fig. 50.2). The direct current, generated when the cell is exposed to light, varies linearly with solar radiation. An improvement of the model includes effects of a shunt resistance and another in series [9].

Application of KVL to the circuit of Fig. 50.2 results in the characteristic equation of the photovoltaic cell.

$$I = I_{\mathrm{pv}} - I_{\mathrm{s}}\left(\mathrm{e}^{\frac{(V+I \cdot R_{\mathrm{S}})}{aV_{\mathrm{t}}}} - 1\right) - \frac{(V + I \cdot R_{\mathrm{s}})}{R_{\mathrm{sh}}} \tag{50.1}$$

Where I_{pv} is the light generated current, I_{s} is the diode's reverse saturation current, V is the array output voltage, a is the ideality factor, R_{s} and R_{sh} are the series and the shunt resistance, respectively, and V_{t} is the array thermal voltage given by

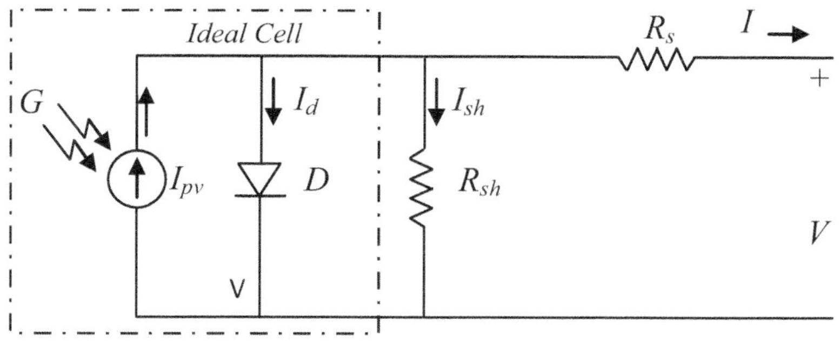

Fig. 50.2 Equivalent circuit for PV cell

$$V_t = \frac{N_s KT}{q}. \tag{50.2}$$

The light generated current of PV cell depends linearly on the irradiance and is also influenced by the temperature

$$I_{pv} = \left(\frac{G}{G_{STC}}\right)(I_{pvn} + K_i(T - T_{STC})). \tag{50.3}$$

I_{pvn} is the short-circuit current provided by the manufacturer at G_{STC}, T_{STC} which refer to the values at nominal or Standard Test conditions (1 kW/m2, 25 °C).

The second term in (50.1) is the diode current which is function of the reverse saturation current or leakage current given by the equation below:

$$I_s = I_{s0}\left(\frac{T}{T_{STC}}\right)^3 e^{\frac{(qE_g)}{aK}} \cdot \left(\frac{1}{T_{STC}} - \frac{1}{T}\right), \tag{50.4}$$

where E_g is the semiconductor band gap energy and I_{s0} is the reverse saturation current at reference or nominal temperature T_{STC}.

50.2.2 PV Array

In a typical PV system, the modules are configured in a series-parallel structure to obtain large power generation or high voltage being suitable for grid connection. Figure 50.3 shows a module of several PV cells which are electrically connected in series and parallel. Equation 50.1 can be modified to (50.5) such as the new PV configuration is taken into account.

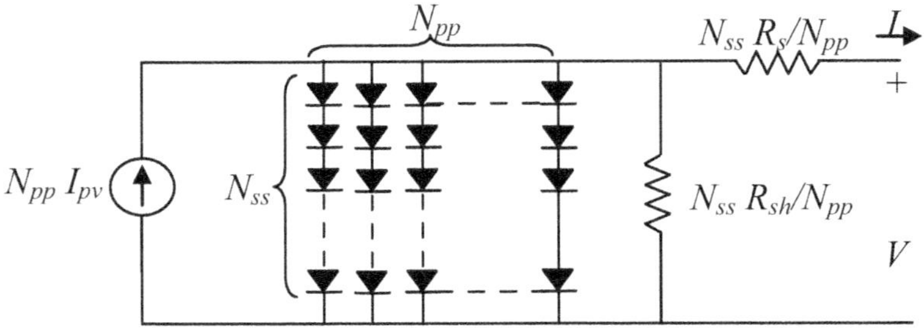

Fig. 50.3 Equivalent circuit of series and parallel connected PV cells

Table 50.1 MSX-60 specifications at STC (air mass = 1.5, $G = 1$ kW/m^2, $T = 25$ °C)

Parameter	Value
Peak power (W), P_{mpp}	60
Peak power voltage (V), V_{mpp}	17.1
Peak power current (A), I_{mpp}	3.5
Open circuit voltage (V), V_{oc}	21.1
Sort-circuit current (A), I_{sc}	3.8
Temperature coefficient of voltage (mV/°C), K_v	−0.08
Temperature coefficient of current (mA/°C), K_I	0.003
Number of series cells, N_s	36

$$I = N_{pp} I_{pv} - I_s \left(e^{\frac{\left(V + I \frac{N_{ss}}{N_{pp}} R_s\right)}{a N_{ss} V_t}} - 1 \right) - \frac{\left(\frac{N_{pp}}{N_{ss}} V + I \cdot R_s\right)}{R_{sh}}. \tag{50.5}$$

As indicated by (50.5), in order to simulate the PV array, five parameters namely I_{pv}, I_s, R_s, R_{sh} and a must be known. Their values are obtained using extraction algorithm which uses the PV data provided by the manufacturer under STC conditions. In the present work the commercial MSX-60 PV module whose specifications are shown in Table 50.1 will be used for simulation. Different algorithms of PV parameters extraction are available and the one presented in [3, 4] can be easily implemented. The results of the algorithm in terms of *P–V* characteristic are shown in Fig. 50.4.

50.3 Conventional P&O Algorithm

The Perturb and Observe (P&O) is one of the most popular algorithms that are being used to track the maximum power point of PV systems. The principle of the algorithm is as explained in the flowchart shown in Fig. 50.5. The algorithm measures the present power and compares it to the initial one and according to the difference the duty cycle is updated. The duty cycle is stepped up or down to

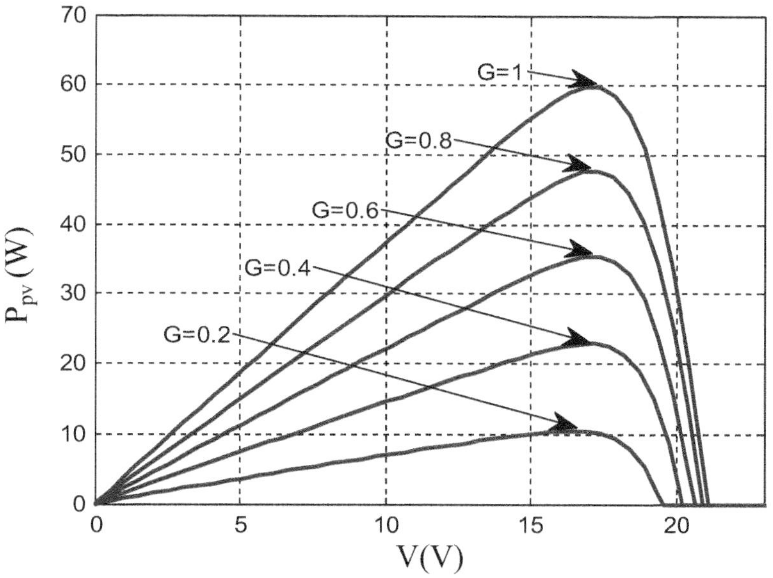

Fig. 50.4 *P–V* characteristics of MSX-60

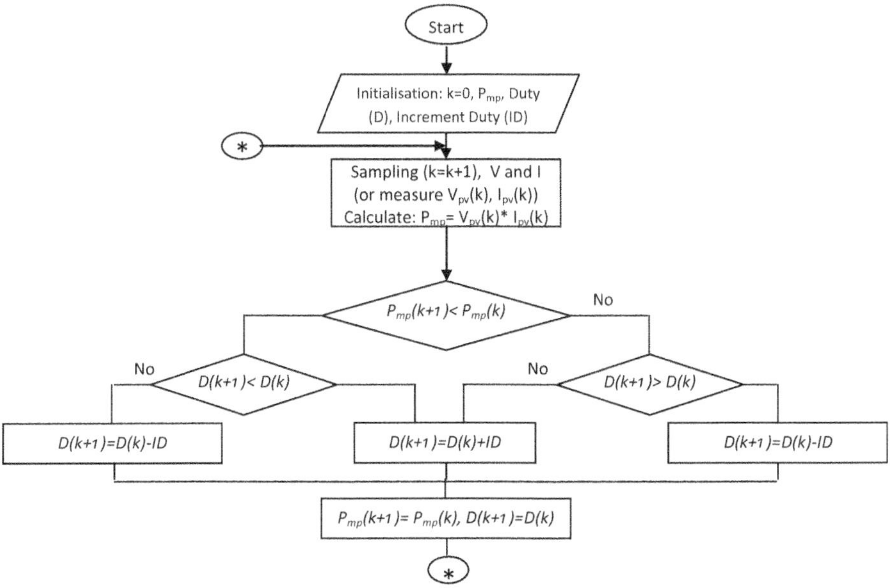

Fig. 50.5 Flowchart of P&O algorithm

move from the present point towards another in the P–V characteristics for which higher extracted power is achieved. The PV system is permanently perturbed so that tracking is ensured.

Other schemes of P&O in which either the current or voltage is perturbed to track the MPP are used also. These schemes require the use a PI controller to have the desired voltage or current at the output of the PV system. From the P–V characteristics of Fig. 50.4, for given climate conditions, there is only one single point of current of voltage at which maximum power is delivered.

50.4 Golden Section Based-MPPT

50.4.1 Overview of Golden Section Search Technique

We recall from calculus that unimodel functions have only one optimum point (maximum or minimum according to the problem under hand). The Golden Section technique can be used to solve the problem as an unconstrained optimization problem if the solution interval $[a, b]$ and the objective function $f(x)$ are both known. The main advantages of the Golden Section Search are the no requirement of both function derivative and the number of iterations. The Golden Section Search technique has a unique way to subdivide the search space (solution interval) to narrowed intervals in which the function will be sampled or evaluated twice. The new interval will be selected according to the comparison of the two samples. The process of dividing the interval will continue till no improvement in terms of function evaluation is obtained or the search space becomes small enough. This technique uses two golden values which are derived from the golden ratio given by

$$G_1 = r = \frac{-1 + \sqrt{5}}{2} \text{ and } G_2 = 1 - r = \frac{3 - \sqrt{5}}{2}.$$

50.4.2 Application of Golden Section Search for MPPT, GSS-MPPT

Agrawal et al. [10] used Golden Section Search technique to develop an MPPT. However, the algorithm was working off-line. This work presents a new architecture of GSS-based MPPT. Figure 50.6 shows the flowchart of the maximum power point tracking algorithm based on the Golden Section Search (GSS-based MPPT). It can be seen that the algorithm starts with some initial values which comprises the initial solution interval, the accuracy epsilon and the golden sections. Two values of power are measured then the process of reducing the interval starts up. Depending on the difference of the measured power values, path 1 or 2 will be chosen to reduce the interval and calculate the new values of voltages. If the difference between the output voltages V_1 and V_2 is small enough, the solution will be the average of them.

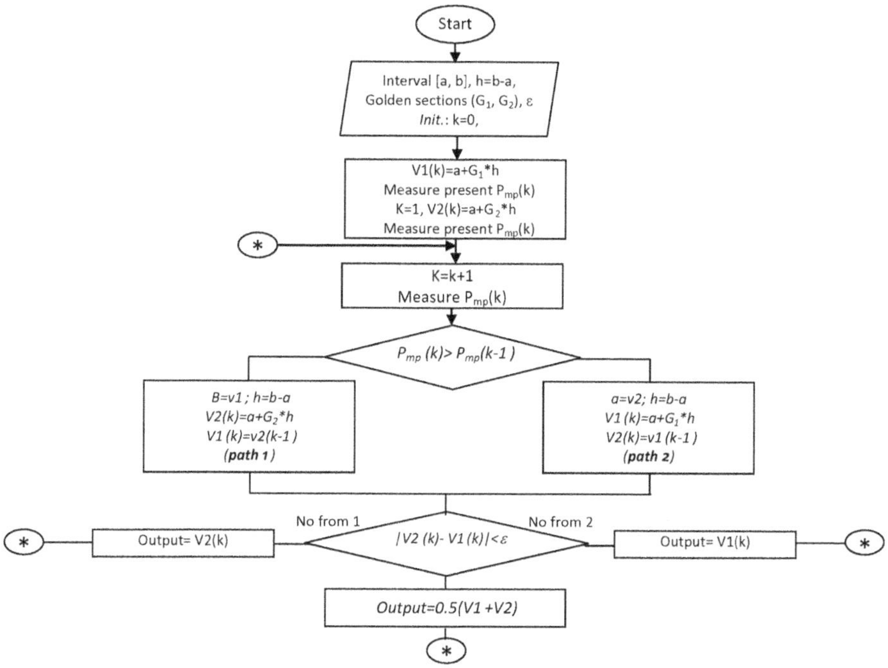

Fig. 50.6 Flowchart of Golden Section based MPPT algorithm

The output voltage is the input of PWM which imposes the voltage through a PI controller. In the same manner, the algorithm can be used to generate the current instead of voltage. In this case the GSS-based algorithm needs the current initial interval of solution.

50.5 Simulation Results and Discussion

In order to verify the effectiveness of the proposed algorithm, the system shown in Fig. 50.1 has been simulated using Matlab software package. The used PV system is the MSX-60 whose parameters are given in Table 50.1. The five parameters of its equivalent circuit have been first identified then introduced in the simulation program. The PV system is associated to DC-to-DC converter allowing to control the output voltage and hence the panel's voltage. In general this converter works as a voltage elevator (boost chopper) which is able to transfer energy even at very low irradiance (PV panel voltage is low). The values of capacitors and inductors are chosen to have CCM and reduced voltage ripple. These values are $L = 350$ μH, $C_1 = C_2 = 560$ μF.

The diode which connects the PV panel to the boost converter is known as the blocking diode. In practice, this diode protects the Battery from discharging through the PV array when there is no sunlight. Regarding the performance of the PV array this diode has no effect, and therefore, it is not included in the simulation framework.

For GSS-based MPPT and P&O MPPT schemes, the irradiance to the PV panel is varied from 0.4 kW/m^2 to 1 kW/m^2 as with a step of 200 W/m^2 at each 1 second as shown in Fig. 50.7.

Figure 50.8 shows the PV system response to successive and sudden variation of the irradiance while the temperature is kept constant, 25 °C. It can be seen from Fig. 50.8 that the PV output power is always around the maximum power point specified by the *P–V* characteristics. It can be noticed also from the graph that GSS-MPPT is fast in tracking the MPP in spite of the abrupt change of the irradiance and the convergence time is less than 0.05 s.

Figure 50.9 shows the output voltage of the PV panel with GSS-based MPPT and for the irradiance variation's profile given in Fig. 50.7. The convergence of the Golden Section Search takes about 6–7 steps to reach the maximum power point voltage.

The output power of the PV panel using P&O MPPT algorithm is shown in Fig. 50.10 with a zoom on the time interval 0–0.5 s. The graph shows that the tie response is about 0.3 s. In comparison with that of the GSS-MPPT, the time response is too much longer and also the power is very noisy due to disturbances of the duty cycle.

It can be easily noticed from Fig. 50.11 that is the PV panel output voltage that the P&O MPPT has a sluggish response. The P&O method has a constant duty cycle step which makes the search process longer than that done by GSS-MPPT. The graph shows also that the voltage is very disturbed as well as the current consequently.

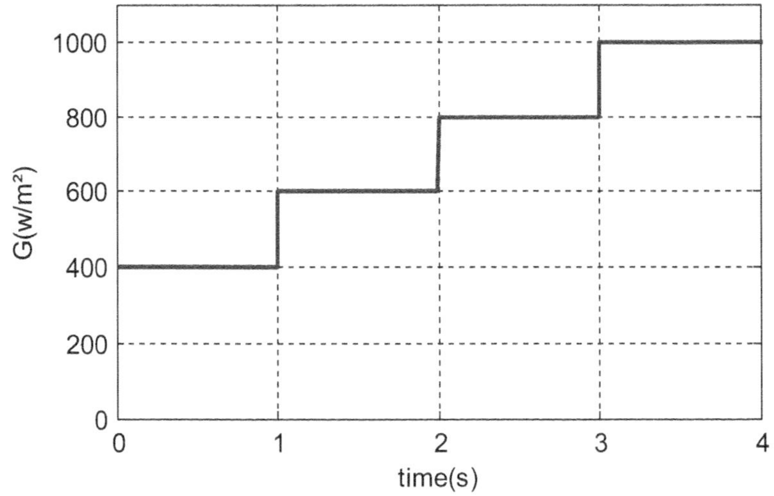

Fig. 50.7 Irradiance variation profile

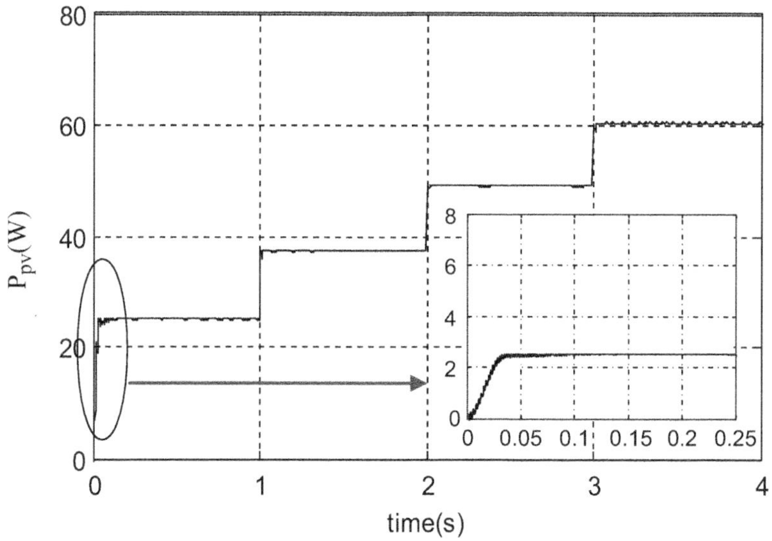

Fig. 50.8 PV output power using GSS-based MPPT

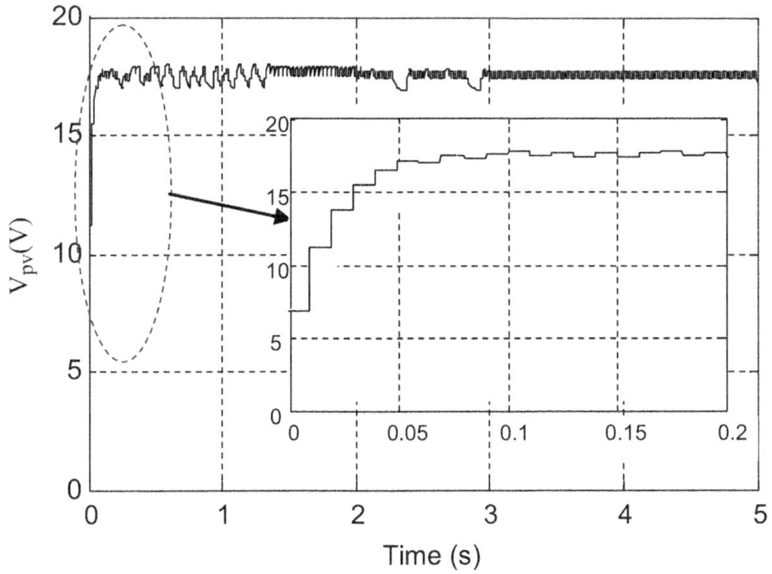

Fig. 50.9 PV output voltage using GSS-based MPPT

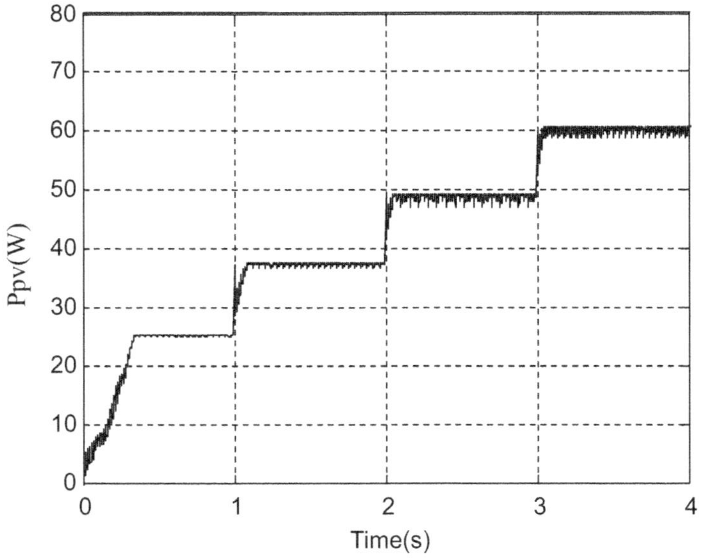

Fig. 50.10 PV output power using P&O MPPT

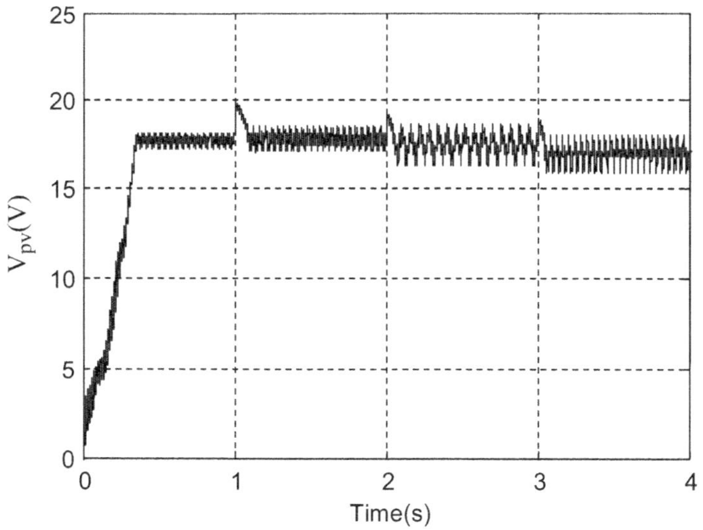

Fig. 50.11 PV output voltage using P&O MPPT

50.6 Conclusion

The chapter has investigated the use of Golden Section Search algorithm to implement the principle of maximum power tracker in photovoltaic systems. First the mathematical models of photovoltaic cell then that of a module have been presented to show the nonlinear behavior of the PV system. This behavior is very sensitive to environment conditions in which the PV panel is used. The main parameters that affect particularly its output power are irradiance and the temperature. Under a given environment condition, the PV output power can vary with the output voltage. The Maximum power tracker algorithm principle is to vary the voltage such that maximum power is reached. The Perturb & Observe algorithm has been implemented also to highlight the advantages of GSS-based MPPT. The comparison of obtained results indicated that GSS-based MPPT outperforms P&O in terms of (1) shorter time convergence and (2) lower voltage and power oscillations. The proposed algorithm uses few arithmetic operations making it simple in terms of implementation. As further work the algorithm can be implemented for grid connected PV systems and tested in case of partial shading.

References

1. Piegari L, Rizzo R (2010) Adaptive perturb and observe algorithm for photovoltaic maximum power point tracking. IET Renew Power Gener 4(04):317–328
2. Esram T, Chapman PL (2007) Comparison of photovoltaic array maximum power point tracking techniques. IEEE Trans Energy Conver 22(2):439–449
3. Villalva MG, Gazoli JR, Filho ER (2009) Comprehensive approach to modeling and simulation of photovoltaic arrays. IEEE Trans Power Electron 24(5):1198–1208
4. Villalva MG, Gazoli JR, Ruppert ER (2009) Modeling and circuit based simulation of photovoltaic arrays. Braz J Power Electron 14(01):35–45
5. Ishaque K, Salem Z, Amjad M, Mekhilef S (2012) An improved particle swarm optimization (PSO)-based MPPT for PV with reduced steady-state oscillation. IEEE Trans Power Electron 27(8):3626–3638
6. Simoes MG, Franceschetti NN (1999) Fuzzy optimization based control of a solar array system. IEEE Electr Power Appl 146(5):552–568
7. Dolara A, Faranda R, Leva S (2009) Energy comparison of seven MPPT techniques for PV systems. J Electromagn Anal Appl 3:152–162
8. Mohan N, Undeland TM, Robbins WP (2003) Power electronics: converters, applications and design. Wiley, Marlton, NJ
9. Ramos JA, Zamora I, Campayo JJ (2010) Modeling of photovoltaic module. In: International conference on renewable energies and power quality (ICREPQ'10), Granada, Spain, pp 23–25
10. Agrawal J, Aware M (2012) Golden Section Search (GSS) algorithm for maximum power point tracking in photovoltaic system. IEEE conference, 2012

Chapter 51
Energetic and Exergetic Performance Evaluations of an Experimental Beta Type Stirling Machine

Houda Hachem, Ramla Gueith, Fethi Aloui, Ibrahim Dincer, and Sassi Ben Nasrallah

Abstract This chapter focuses on a beta type Stirling receptive machine operating under an atmospheric pressure between two heat sources at constant temperatures. Two receiving modes are studied experimentally (refrigerator mode and heat pump mode). Parameters such as hot end temperature (about 110 °C maximum hot temperature output for the heat pump mode), cold end temperature (about −32 °C minimum cold temperature output for refrigerating mode) and coefficient of performance (COP) are studied under different engine speeds. In order to assess the receptive Stirling machine, we carry out energy and exergy performance assessment studies. Energetic and exergetic coefficients of performance and exergy destructions in the heat pump system and the refrigerating machine are quantified at different engine speeds. The results show that optimal operating mode for refrigerating machine is obtained at about 155 rpm. But, around this speed the heat pump mode deliver the worst COP. The influence of heat sources insulation on the exergy

H. Hachem
Laboratoire d'Étude des Systèmes Thermiques et Energétiques,
Université de Monastir, ENIM, Rue Ibn El Jazzar, Monastir 5019, Tunisia

Department of Mechanics, LAMIH UMR CNRS 8201, Université of Valenciennes (UVHC),
Le Mont Houy, Valenciennes Cedex 9 59313, France
e-mail: houdahachem@yahoo.fr

R. Gueith • S.B. Nasrallah
Laboratoire d'Étude des Systèmes Thermiques et Energétiques,
Université de Monastir, ENIM, Rue Ibn El Jazzar, Monastir 5019, Tunisia
e-mail: ramla2gheith@yahoo.fr; sassi.bennasrallah@enim.rnu.tn

F. Aloui (✉)
Department of Mechanics, LAMIH UMR CNRS 8201, Université of Valenciennes (UVHC),
Le Mont Houy, Valenciennes Cedex 9 59313, France
e-mail: Fethi.Aloui@univ-valenciennes.fr

I. Dincer
Faculty of Engineering and Applied Science, University of Ontario Institute
of Technology, 2000 Simcoe Street North, Oshawa, ON, Canada L1H 7K4
e-mail: Ibrahim.Dincer@uoit.ca

I. Dincer et al. (eds.), *Progress in Clean Energy, Volume 2*,
DOI 10.1007/978-3-319-17031-2_51

destruction and coefficient of performance of the system is investigated. Furthermore, the usefulness of good heat insulation is emphasized. In fact, it improves the energetic COP of the heat pump from 2.8 to 3.5 as the exergetic COP from 43 to 75.2 % at the same rotational speed.

Keywords Air cycle • Low source • Hot source • Heat pump • Refrigerating machine • Speed • Insulation • Energetic COP • Exergetic COP

Nomenclature

A	Area, m^2
COP	Coefficient of performance
C_p	Specific heat, J/kg °C
Ex	Exergy, J
ExD	Exergy destruction, kJ kg^{-1} K^{-1}
K	Thermal conductivity, Wm^{-1} K^{-1}
m	Mass, kg
N	Rotation speed, rpm
P	Pressure, Pa
Q	Heat, J
S	Entropy, kJ kg^{-1} K^{-1}
S_{gen}	Entropy generation, kJ kg^{-1} K^{-1}
T	Temperature, K
V	Volume, m^3
W	Work, J

Greek Letters

ρ	Density, kg/m^3
η	Efficiency

Subscripts

C	Carnot
H	High
HP	Heat pump
L	Low
R	Refrigerating machine

51.1 Introduction

The idea of using heat engine in a reverse mode as a heat pump was proposed by Lord Kelvin in the nineteenth century, but it was only in the twentieth century that practical machines came into common use, mainly for refrigeration. Today, a significant portion of global energy consumption is attributable to domestic heating and cooling. Heat pumps are widely used not only for air conditioning and heating but also for cooling, producing hot water and preheating feed water in various types of facilities including office buildings, computer centers, public buildings, restaurants, hotels, district heating or cooling systems and industrial plants. Heat pump is essentially a heat engine operating in reverse and can be defined as a device that moves heat from a region of low temperature to a region of higher temperature. Heat-pump efficiency is determined by comparing the amount of energy delivered by the heat pump to the amount of energy it consumes [1].

The Stirling engine is operating in a closed thermodynamic regenerative cycle in which the working fluid is cyclically compressed and expanded at different temperature levels to achieve net conversion of heat to work or vice-versa. It may be used as refrigerating machines or heat pumps receiving heat at a low temperature and delivering heat at a higher temperature. Such machines require an input of mechanical work to accomplish their operation [2]. The Stirling cycle is fundamentally different from the Rankine cycle used in conventional refrigerators. Unlike the Rankine cycle, the working fluid within the Stirling cycle undergoes no phase change. The Stirling cycle cryocoolers are widely used because of its different advantages: high efficiency, fast cool-down, small size, light weight, low power consumption, high reliability, no pollutant generation like carbon monoxide and so on [3]. Nowadays, the Stirling cycle cryocooler is applied to domestic and commercial refrigerators because it has the advantage of avoiding the use of environmentally harmful refrigerants. Also, the use of this technology gives the ability to get colder temperatures.

The use of air compression-expansion phenomenon to get cooling or heating effects is an interesting field of researches. Many numerical and experimental investigations of Stirling machines can be found in literature. Domestic refrigeration system used Stirling cycle was first studied by Finkelstein and Polonski [4]. A V-type integral Stirling refrigerator (VISR) was developed and tested by Le'an et al. [3]. Parameters such as energy consumption and coefficient of performance (COP) are studied under different speeds and charged pressures. A comparison has been made between nitrogen and helium as a working fluid. It is shown that the optimum rotational speed for the cooling capacity is different between nitrogen and helium. The COP has a peak value around 600 rpm for nitrogen, and for helium around 900 rpm. Numerically, a thermodynamic analysis of a V-type Stirling-cycle refrigerator having air as a working gas is performed by Ataer [5]. It is found that the charge pressure higher than 2 bar will decrease the COP of the Stirling-cycle refrigerator. A refrigeration air cycle was investigated by Giannetti et al. [6] and the impact of regenerator effectiveness was studied. They found that for $\varepsilon r = 0.95$

the COP would increase up to 0.77 for the high and to 0.81 for the low pressure cycle. A beta type Stirling cycle machine of 100 W capacity was designed and tested by Otaka et al. [7]. They studied the effect of many parameters such as dead volume ratio, working fluids, the ratio of the compression volume to the expansion volume, and the phase difference between a power piston and displacer. And they found that the refrigeration produced by nitrogen was 28 % less than that produced by helium. Kaushik et al. [8] studied internal and external irreversibilities inside Stirling and Ericsson heat pumps cycles. External irreversibilities are due to temperature difference between working fluid and external heat sources. Internal irreversibilities are due to generative heat loss and other entropy generation in the cycle. According to their study, the effects of internal irreversibilities appear to be more critical than external ones.

In this paper, a beta type Stirling machine is experimentally used for a refrigerating mode and a heat pump mode. The evolution of different parameters such as pressure, temperature, volume and coefficients of performance are presented and discussed. The effect of the rotation speed is investigated on the coefficient of performance, exergy destruction, maximum heating temperature (when operating as heat pump mode) and minimum cooling temperature (when operating as refrigerator mode). The effect of thermal insulation of the two heat sources on receptive machine's energy and exergy efficiencies are studied.

51.2 Experimental Apparatus and Procedure

This study considers a beta Stirling machine, using air as working fluid. The machine consists of a glass cylinder having 6 cm of diameter and 20 cm of height, inside which slides two pistons: a working piston M and a displacer D. On the displacer there is a copper porous structure called regenerator (Fig. 51.1). The role of the regenerator is to provide a thermal barrier between heat sources and store calories during compression then return them to the working fluid during expansion. The technical and geometrical characteristics of the Stirling engine are summarized in Table 51.1.

The experimental work consists to study the receiving machine mode. An electric engine coupled to the Stirling machine, by a pulley and belt transmission (Fig. 51.2), provided a mechanical work which will be transformed to hot or cold thermal loads, depending on the direction of rotation speed. Stirling machine characteristics: pressure, temperatures and volume are measured and presented during a quasi-steady state. The cooling power (refrigerator) and the heating power (heat pump) are investigated at different speeds of the Stirling machine.

In order to measure temperatures, pressure, volume and speed evolutions during one more cycle, an acquisition system composed of two CASSY cards is used (Fig. 51.3) to convert the analog signal, given by sensors, into a digital signal which will be treated by a processing unit. These cards allow the recording of signals at a

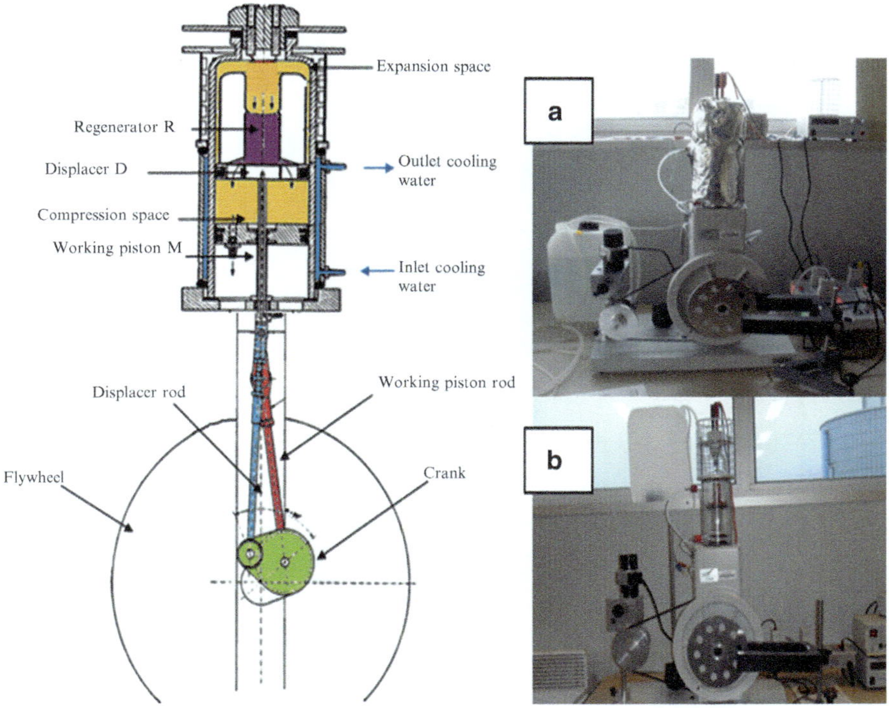

Fig. 51.1 Sectional view of the beta Stirling engine: (**a**) After insulation; (**b**) Before insulation

Table 51.1 Geometrical properties of the experimental engine

Hot space diameter [mm]	60
Hot space height [mm]	49.2
Displacer con rod length [mm]	100
Displacer stroke [mm]	48
Cold space diameter [mm]	60
Cold space height [mm]	35.7
Power piston con rod length [mm]	197
Power piston stroke [mm]	48
Regenerator inner diameter [mm]	22
Regenerator outer diameter [mm]	60
Regenerator height [mm]	59
Regenerator porosity	0.79
Maximum machine volume [cm^3]	300
Minimum machine volume [cm^3]	160
Rotation speed [rpm]	235

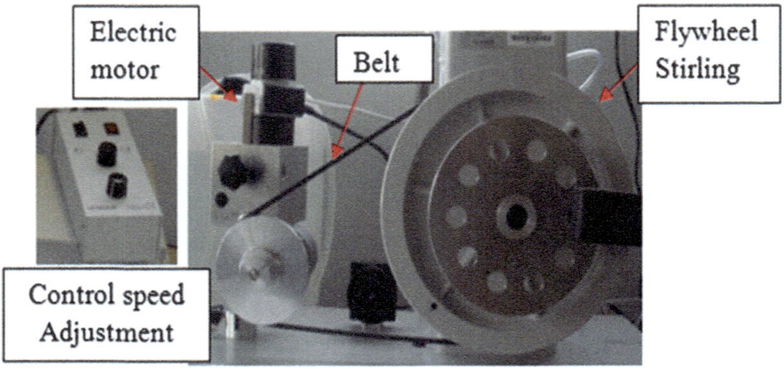

Fig. 51.2 Stirling engine drive system

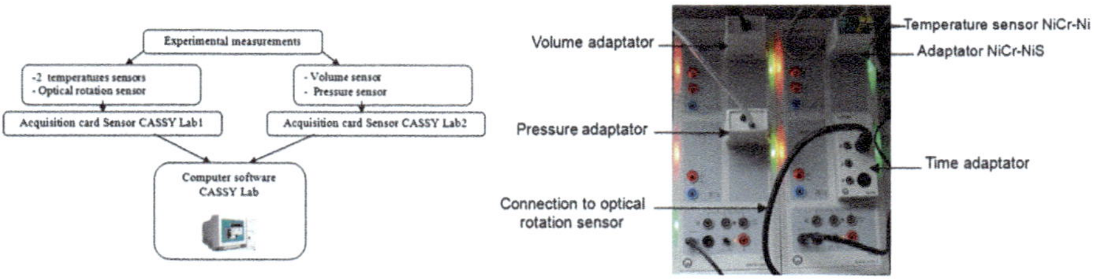

Fig. 51.3 Data acquisition using "Cassy Sensor" system

sampling frequency Fe during a fixed period of measurement. The first card allows pressure and volume signals acquisitions. The second card allows hot and cold spaces temperatures acquisitions and Stirling machine speed acquisition.

51.2.1 Thermodynamic Performance Assessment

An ideal Stirling refrigeration or heat pump cycle executed by a perfect gas in a piston–cylinder assembly is shown in Fig. 51.4. The cycle consists of four processes in series. During the first process ($1 \rightarrow 2$), the gas expands isothermally at TL while receiving energy QL from the cold reservoir by heat transfer. During the second process ($2 \rightarrow 3$), the gas is compressed adiabatically until its temperature is T_H. During the third process ($3 \rightarrow 4$), the gas is compressed isothermally at T_H while it discharged energy Q_H to the hot reservoir by heat transfer. And finally, during the process ($4 \rightarrow 1$), the gas expands adiabatically until its temperature decreases to T_L.

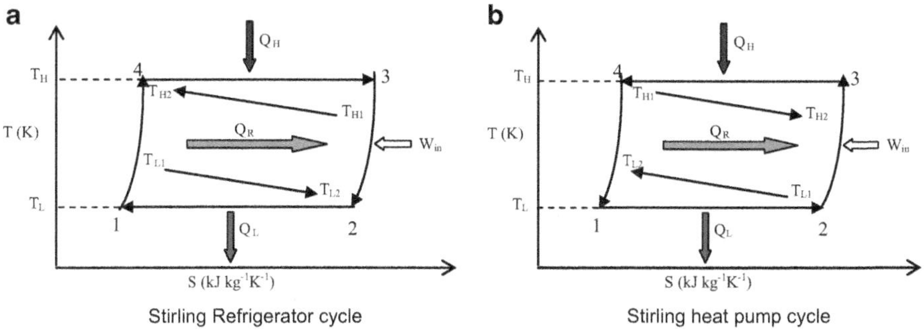

Fig. 51.4 Schematics of reversible Stirling machines cycles for (**a**) refrigeratior (Stirling Refrigerator cycle) and (**b**) heat pump (Stirling heat pump cycle)

For this system, the balance equations for mass, energy, entropy, and exergy are written as follows:
Mass balance equation:

$$m = \text{constant} \tag{51.1}$$

Energy balance equation:

$$Q_L + W = Q_H \tag{51.2}$$

Entropy balance equation:

$$\frac{Q_L}{T_L} + S_{gen} = \frac{Q_H}{T_H} \tag{51.3}$$

Exergy balance equation:

$$\text{Ex}_{Q_L} + W = \text{Ex}_{Q_H} + \text{Ex}^D \tag{51.4}$$

Since refrigeration and heat pump cycles have different objectives, their performance parameters, called coefficients of performance, are defined differently. These energy and exergy coefficients of performance are considered next.

51.2.2 Refrigerating Machine Mode

The Stirling machine works as a refrigerator mode, when its flywheel is driven by an electric engine in the clockwise direction. In the refrigerating mode, heat is transferred from a high temperature medium to a lower temperature medium.

The coefficient of performance, of the refrigerating machine is defined as the ratio of the desired output effect from the refrigerator to the power consumed by it.

$$\mathrm{COP_R} = \frac{\text{Desired output}}{\text{Required input}} = \frac{|Q_L|}{W_{in}} = \frac{|Q_L|}{Q_H - Q_L} = \frac{1}{|Q_H/Q_L| - 1} \tag{51.5}$$

where, T_H is the absolute temperature of the high-temperature reservoir and T_L is the absolute temperature of the low-temperature reservoir.

The COP of Carnot refrigerating cycle is defined as follows:

$$(\mathrm{COP})_C = \left(\frac{Q_L}{Q_H - Q_L}\right)_C = \frac{T_L}{(T_H - T_L)} \tag{51.6}$$

Note that the coefficient of performance of Carnot reversible refrigeration cycle COPc is always great than the coefficient of performance of an irreversible refrigeration cycle when each operates between the same two thermal reservoirs.

The amount of heat rejected to the high-temperature reservoir Q_H during one cycle is calculated as follows:

$$Q_H = \dot{m} C_p \Delta \mathrm{T} dt \tag{51.7}$$

where, $\dot{m}$ is the mass flow of the cooling water, C_p is heat capacity of the cooling water, ΔT is the temperature difference between inlet and outlet cooling water and dt is the cycle time.

The amount of heat received from the low-temperature reservoir Q_L during one cycle is defined as the sum of the inputted work and the amount of heat rejected to the high-temperature:

$$Q_L = W_{in} + Q_H \tag{51.8}$$

The actual work input Win is described as follows:

$$W_{in} = \frac{W_{el}}{\eta_{el}\eta_m\eta_{th}} \tag{51.9}$$

where, W_{el} is the electric power done by the driving electric motor, η_{el} is the electric efficiency of the electric motor, η_{th} is the isothermal efficiency of the machine, and η_m is the mechanical efficiency of the machine.

In order to evaluate the actual inputted work, the area of the PV diagram is calculated referring to experimental data using the trapezoidal method. The actual work input is calculated by integrating the following:

$$W_{in} = \int_{\text{cycle}} -P \cdot dV \tag{51.10}$$

The thermal exergy for the refrigeration mode is defined as follows:

$$\mathrm{Ex}_{\mathrm{QR}} = \left(1 - \frac{T_0}{T_\mathrm{L}}\right)|Q_\mathrm{L}| \tag{51.11}$$

here, T_0 is the reference temperature which is taken as the ambient temperature.

The exergetic coefficient of performance $\mathrm{COP}_{\mathrm{Ex}}$ is calculated as follows:

$$\mathrm{COP}_{\mathrm{ExR}} = \frac{\mathrm{Ex}_{\mathrm{QR}}}{W_\mathrm{in}} \tag{51.12}$$

Referring to the exergy balance:

$$\sum \mathrm{Ex}_\mathrm{in} = \sum \mathrm{Ex}_\mathrm{out} + \sum \mathrm{Ex}^D \tag{51.13}$$

Thus, the amount of exergy destruction ExD for refrigerating machine is defined as:

$$\mathrm{Ex}_{\mathrm{QR}}^{D} = W_\mathrm{in} - \mathrm{Ex}_{\mathrm{QR}} \tag{51.14}$$

The exergy coefficient of performance $\mathrm{COP}_{\mathrm{Ex}}$ frequently gives a finer understanding of performance than the energy coefficient of performance COP. In evaluating COP, the same weight is assigned to energy whether it is shaft work or a stream of low-temperature fluid. The parameter $\mathrm{COP}_{\mathrm{Ex}}$ points out that both external and internal irreversibilities need to be dealt with to improve the machine performance [1].

51.2.3 Heat Pump Mode

When the flywheel is driven counterclockwise, the Stirling machine operates as a heat pump. Heat pump transfers heat from a low temperature medium to a higher temperature medium. The coefficient of performance of the heat pump is defined as the ratio of the desired effect from the heat pump to the power consumed by the heat pump.

$$\mathrm{COP}_{\mathrm{HP}} = \frac{Q_\mathrm{H}}{W_\mathrm{in}} = \frac{Q_\mathrm{H}}{Q_\mathrm{H} - Q_\mathrm{L}} = \frac{1}{1 - (Q_\mathrm{L}/Q_\mathrm{H})} \tag{51.15}$$

The real heat pump coefficient of performance is always less than the reversible Carnot efficiency given as follows:

$$(\mathrm{COP})_C = \left(\frac{Q_\mathrm{H}}{Q_\mathrm{H} - Q_\mathrm{L}}\right)_C = \frac{T_\mathrm{H}}{T_\mathrm{H} - T_\mathrm{L}} \tag{51.16}$$

where, Q_L is the heat removed from the cold reservoir and Q_H is the heat supplied to the hot reservoir.

The thermal exergy for heat pump mode is defined as follows:

$$\mathrm{Ex_{QHP}} = \left(1 - \frac{T_0}{T_H}\right) Q_H \tag{51.17}$$

Then the exergy coefficient of performance for heat pump can be calculated as:

$$\mathrm{COP_{ExHP}} = \frac{\mathrm{Ex_{QHP}}}{W_{in}} \tag{51.18}$$

Finally, energy and exergy formulations even for heat pump and refrigerating machine are applied to the experimental operating data obtained from the Beta Stirling machine.

51.3 Results and Discussion

51.3.1 Refrigerating Machine Performance Results

Internal pressure and volume variations obtained experimentally as a function of time are represented in Fig. 51.5. Their evolutions are perfectly periodic (sinusoidal), as the volume varies from 0 to 140 cm^3 and the pressure from 1 to 1.9 bar. The volume development is always in phase opposition with respect to the pressure.

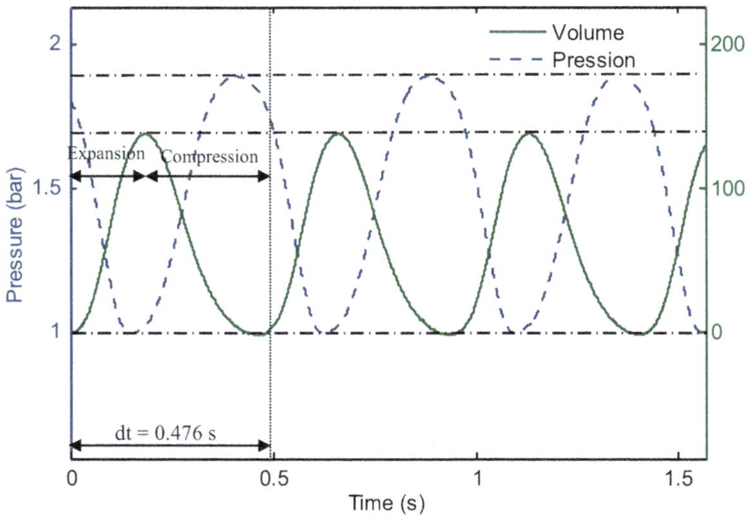

Fig. 51.5 Temporal pressure and volume evolutions at 126 rpm (refrigerating mode)

During a receptive machine cycle ($dt = 0.476$ s), The compression phase lasts longer than the expansion phase. This can be explained by effect of friction losses at reducing net work inputted to the machine.

Referring to pressure and volume experimental data during one cycle, the PV diagram can be represented (Fig. 51.6). The area of this diagram represents the actual work given to the machine. The net work transmitted to the refrigerating machine mode is about 3.513 J at 126 rpm of speed rotation.

In order to evaluate the effect of speed on the output refrigerating performances, the evolution of hot and cold temperatures sources with time are plotted at different speeds of the machine. In view of the fluctuation of experimental data (outcome of thermocouples), fitting curves for temperatures are plotted in Figs. 51.7 and 51.8. According to changing temperatures, hot source quickly reaches the average stable operating temperature about 22 °C. On the other hand, cold source temperature decreases slowly until −32 °C. And it reaches the steady state after about 10 s of operating. The higher the rotation speed of the machine, the faster the steady state would be reached. For different speeds, there is a little bit of difference between the steady state temperatures. Increasing speed decreases the output temperature. A decrease in cold end temperature leads to a decrease in high end temperature and vice versa.

To optimize the refrigerating mode, it would be advantageous to vary the speed. We could then see how changing the exergy flow depending on the speed and find a good compromise between the performances of the machine and low exergy destruction. A summarize of experimental results obtained on the studied refrigerating mode with and without insulation are listed in Table 51.2. It is also observed

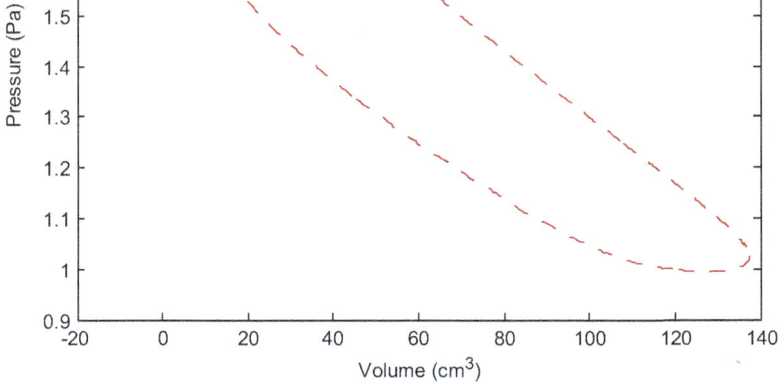

Fig. 51.6 P-V diagram at 126 rpm (refrigerating mode)

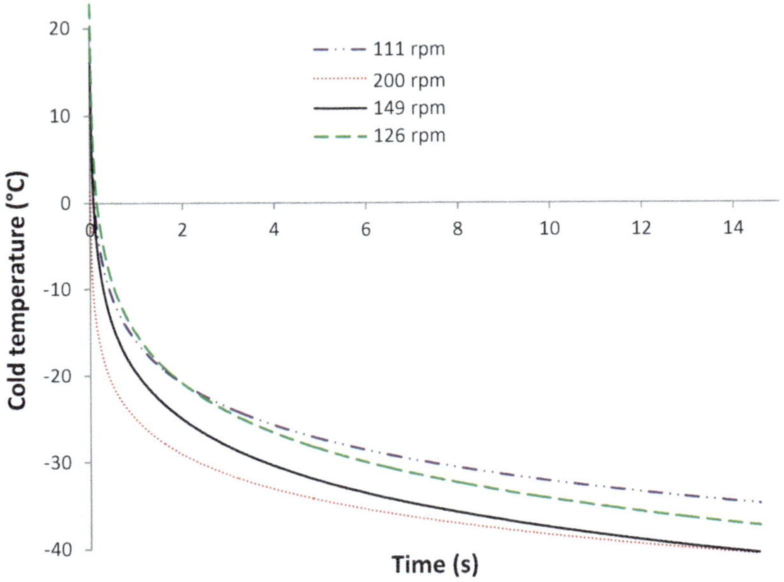

Fig. 51.7 Cold end temperature evolution vs. time at different speeds (refrigerating mode)

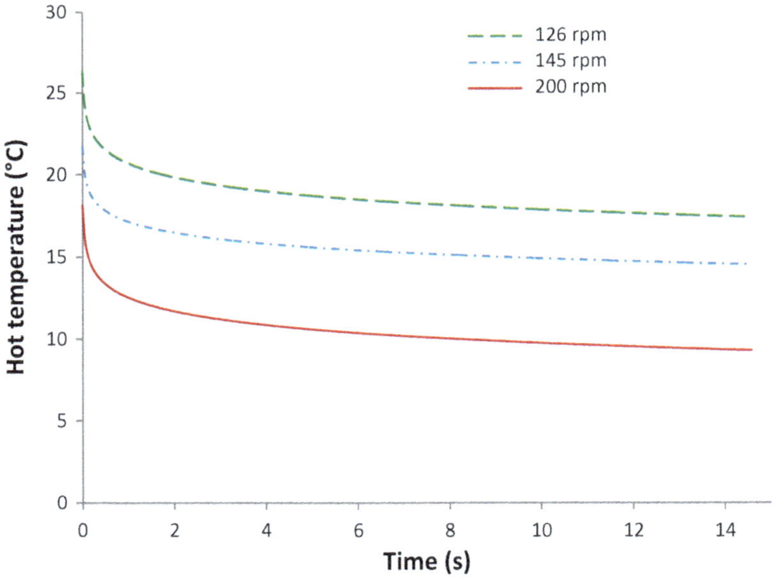

Fig. 51.8 Hot end temperature evolution vs. time at different speeds (refrigerating mode)

Table 51.2 Experimental data for insulated refrigerating mode

Insulated refrigerating mode										
Case	N (rpm)	T_L (K)	T_H (K)	W_{in}	Q_L (J)	Q_H (J)	COP_R	Ex_{QR} (J)	$Ex^D{}_{QR}$ (J)	COP_{ExR}
1	93.74	241	294	3.207	9.609	−6.402	0.6	2.073	1.134	0.647
2	126.089	247	283	3.513	8.749	−5.236	0.626	1.634	1.879	0.465
3	149.153	238	290	4.039	9.108	−5.070	0.642	2.128	1.911	0.527
4	111.082	237	288	6.632	19.274	−12.642	0.604	4.595	2.037	0.693
5	199.58	240	283	4.093	11.309	−7.217	0.610	2.521	1.572	0.616
Non insulated refrigerating mode										
Case	N (rpm)	T_L (K)	T_H (K)	W_{in}	Q_L (J)	Q_H (J)	COP_R	Ex_{QHP} (J)	$Ex^D{}_{QR}$ (J)	COP_{ExR}
6	102.349	243	295	3.345	14.486	−11.141	0.5653	2.995	0.35	0.895
7	152.32	241	289	3.379	10.865	−7.486	0.5921	2.366	1.013	0.7
8	202.49	239	292	2.062	7.693	−5.631	0.5774	1.726	0.336	0.837
9	49.12	247	283	9.325	32.535	−23.21	0.5836	6.075	3.25	0.651
10	111	237	288	6.632	16.897	−10.265	0.622	4.028	2.604	0.607

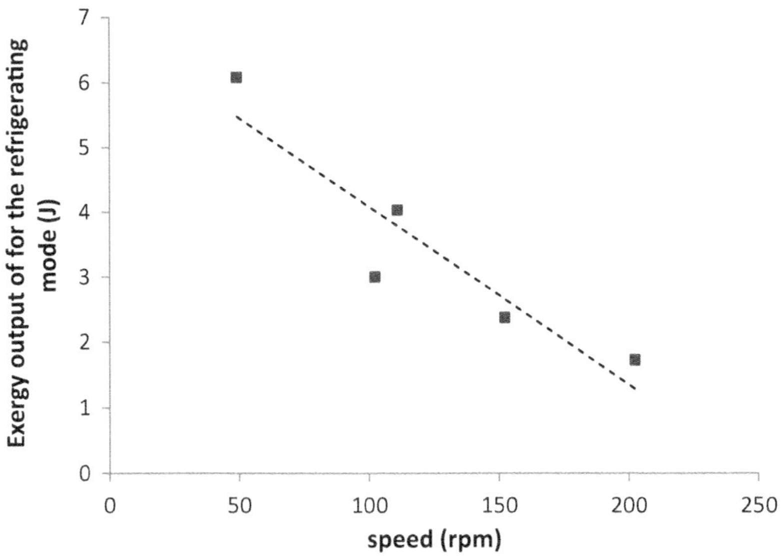

Fig. 51.9 Exergy production of the refrigerating mode vs. speed

that exergy output for the refrigerating mode decreases with increasing speed (Fig. 51.9).

Thermodynamic imperfections can be quantified as exergy destructions, which represent losses in energy quality or usefulness. The quantity of output cooling energy depends on temperature. The higher the temperature, the higher the cooling energy and the associated exergy. This phenomenon explains the reduction of the exergetic refrigerating COP after insulation of the system. The present exergy coefficient of performance for the refrigerating mode varies from 46.5 to 69.3 % for the insulated refrigerating mode and from 98.5 to 60.7 % for the non-insulated mode. Note that at the quasi-steady state, output cooling temperatures for insulated cycle are lower than those corresponding to non insulated cycle. This stresses the usefulness of insulation to improve cooling effects of the Stirling refrigerating machine.

From Fig. 51.9, exergy output of the refrigerating mode decreases with speed increasing. In view of the decrease of the output cooling temperature which leads to the increase of the cooling effects. The energetic COP is greater than the exergetic COP since the definition for the exergetic COP provides a more realistic upper limit for the system performance by considering the external irreversibilities which cannot be eliminated.

The evolution of the energetic COP for the refrigerating mode versus speed is represented in Fig. 51.10. The curve reaches a maximum about 155 rpm. This maximum can be explained as a compromise between losses that disfavor high cooling rates, and losses that disfavor low cooling rates.

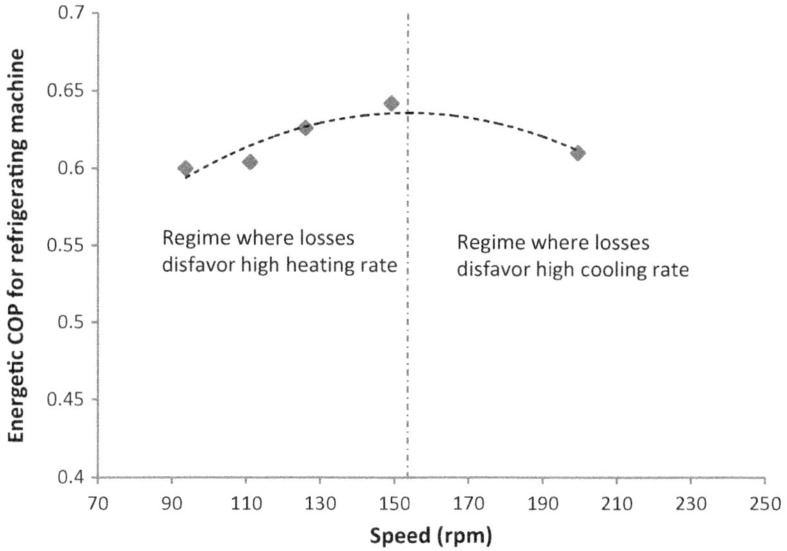

Fig. 51.10 Energetic COPR evolution vs. speed (refrigerating mode)

51.3.2 Heat Pump Performances

As for refrigerating mode, to determine the optimal performances of the heat pump mode, operating parameters are measured at different speeds. Refering to experimental data, temperatures trend curves can be drown for different speeds. Heating temperature evolution versus time is represented in Fig. 51.11. It has a logarithmic evolution. It increases with time until reaching a quasi-steady state on which the maximum heating temperature is about 110 °C for 206 rpm rotation speed. The higher the speed, the higher the output heating temperature and the faster the steady state would be reached. On the other hand cold sink reaches quickly steady temperature about 28 °C at 206 rpm of speed (Fig. 51.12).

Other performance data for the experimental results for insulated and non insulated heat pump mode are given in Table 51.3. Stirling heat pump has COP ranging from 2 to 3.4, implying that they deliver 2–3.4 times more energy than it consume. The best COP is obtained at the less speed inputted in the machine (103 rpm), the heat has an energetic COP of about 3.4 and an exergetic COP of 0.75.

Figure 51.13 presents exergetic COP evolutions for insulated and non insulated heat pump mode versus speed. It is observed that this curve presents a minimum at about 160 rpm. This minimum can devise the operating regime into two parts. When speed is lower than 160 rpm, the exergetic COP of the heat pump decreases. Thus, losses disadvantage the heating effect. By against, when speed is higher than 160 rpm, the exergetic COP of the heat pump increases. In this case, losses promote high heating rate and low cooling rate.

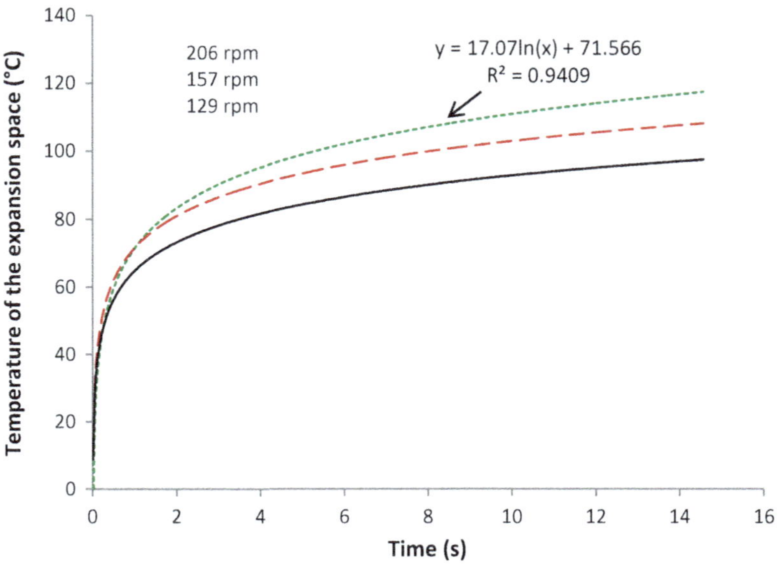

Fig. 51.11 Hot end temperature evolution vs. time at different speeds (heat pump mode)

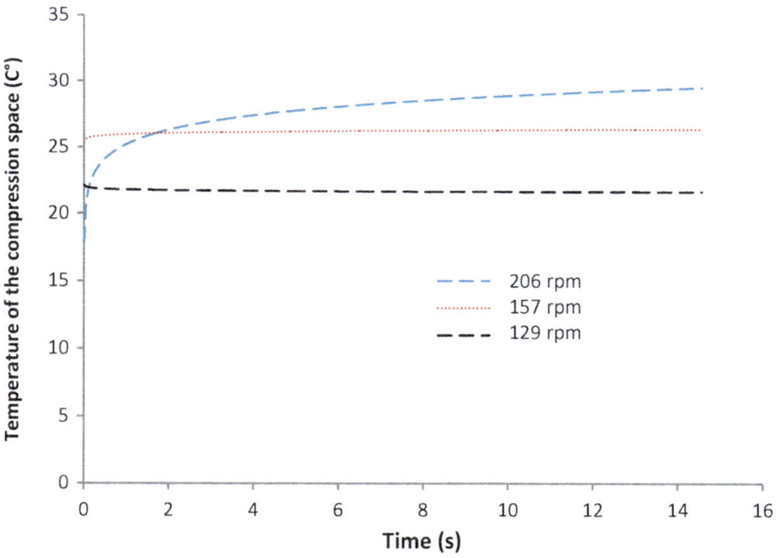

Fig. 51.12 Cold end temperature evolution vs. time at different speeds (heat pump mode)

Figure 51.14 presents the difference between inputted work to the heat pump machine and the output heating exergy. This difference depicts the amount of exergy destruction on each case of experience.

Table 51.3 Experimental data for insulated and non insulated heat pump mode

Case	N (rpm)	T_H (K)	T_L (K)	W_{in} (J)	Q_H (J)	Q_L (J)	COP_{HP}	Ex_{QHP} (J)	Ex^D_{QHP} (J)	COP_{ExHP}
Insulated heat pump mode										
1	206	395	303	3.279	−6.707	3.427	2.045	1.732	1.547	0.528
2	193	386	301	3.346	−6.755	3.409	2.019	1.628	1.718	0.486
3	157	383	298	2.828	−6.802	3.973	2.405	1.598	1.230	0.564
4	129	347	300	3.099	−8.923	5.824	2.879	1.932	1.167	0.624
5	103	374	299	3.152	−10.943	7.790	3.471	2.370	0.782	0.752
Non insulated heat pump mode										
Case	N (rpm)	T_H (K)	T_L (K)	W_{in} (J)	Q_H (J)	Q_L (J)	COP_{HP}	Ex_{QHP} (J)	Ex^D_{QHP} (J)	COP_{ExHP}
6	207.7	383	302	2.030	−4.919	2.889	2.423	1.156	0.874	0.569
7	193.6	386	301	3.346	−6.445	3.099	1.926	1.553	1.790	0.464
8	163.5	367	302	3.111	−6.781	3.670	2.18	1.367	1.744	0.439
9	129.8	374	300	3.099	−7.721	4.622	2.491	1.672	1.427	0.540
10	105	345	303	3.087	−8.799	5.712	2.850	1.326	1.761	0.430

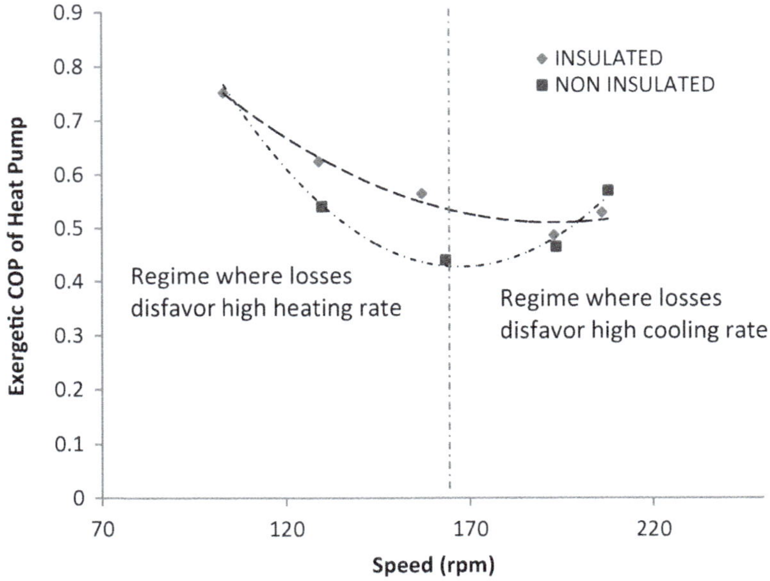

Fig. 51.13 Exergetic heat pump COP evolution vs. speed at insulated and non-insulated cases

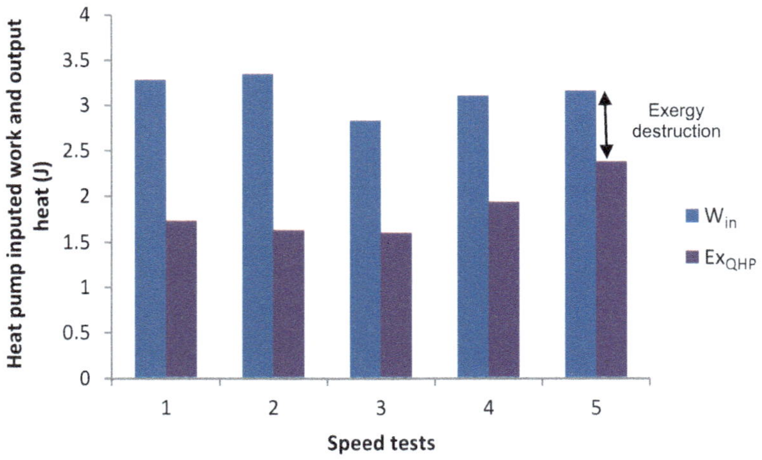

Fig. 51.14 Amount of exergy destruction at different speed tests (heat pump mode)

51.4 Conclusions

Stirling machines are useful for producing heating or cooling effects. The performances of these two modes of operation are investigated. Comprehensive energy and exergy analyses are applied for evaluating heating and cooling systems. Results are shown that the speed increasing of the machine accelerate the steady state attending, increase the output heating temperature for heat pump mode and decrease the output cooling temperature for refrigerating mode. For heat pump mode, when rotation speed is higher than 160 rpm, its COP increases with speed increasing. But when it is less than 160 rpm its COP decreases with speed increasing. For the refrigerating mode, when speed is equal to 155 rpm, the machine produces the maximum useful cooling effect output at less electric power input. As conclusion, the optimal operating speed is about 155 rpm when functioning as a refrigerating machine. On the other hand, when functioning as a heat pump machine, optimal operating speed is obtained out of the interval of speed from to 150–200 rpm. Insulation ameliorate energetic heat pump COPHP respectively from 2.8 to 3.5 as the exergetic coefficient from 43 to 75.2 % at the same rotation speed. A heat pump system can deliver more thermal energy than the electrical energy required to operate it.

References

1. Dincer I, Rosen MA (2013) Exergy: energy, environment and sustainable development. 2nd edition, Elsevier, Oxford
2. Walker G, Fauvel R, Gustafson R, van Bentham J (1982) Stirling engine heat pumps. Int J Refrig 5(2):67
3. Le'an S, Yuanyang Z, Liansheng L, Pengcheng S (2009) Performance of a prototype Stirling domestic refrigerator. Appl Therm Eng 29:210–215
4. Finkelstein T, Polonski C (1959) Development and testing of a Stirling cycle machine with characteristics suitable for domestic refrigerators. Report W/M (3A).U.5. English Electric Company Ltd, Whetstone
5. Ataer OE, Karabulut H (2005) Thermodynamic analysis of the V-type Stirling-cycle refrigerator. Int J Refrig 28:183–189
6. Giannetti N, Milazzo A (2014) Thermodynamic analysis of regenerated air-cycle refrigeration in high and low pressure configuration. Int J Refrig 40:97–110
7. Otaka T, Ota M, Murakami K (2002) Study of performance characteristics of a small Stirling refrigerator, heat transfer. Asian Res 31(5):344–361
8. Kaushik SC, Tyagi SK, Bose SK, Singhal MK (2002) Performance evaluation of irreversible Stirling and Ericsson heat pump cycles. Int J Therm Sci 41:193–200

Chapter 52
Control of Single Stage Grid Connected PV-Inverter Based on Direct Space Vector PWM

F. Akel, T. Ghennam, M. Laour, D. Bendib, E.M. Berkouk, and M. Chikh

Abstract Space vector modulation (SVM) is the preferred modulation technique in three-phase DC/AC converters. The most widely known SVM in the literature is based on generating eight voltage vectors in the Park reference frame. This chapter concentrates on a specific matter of the direct-space-vector modulation analysis for single stage three phase grid connected photovoltaic inverter, and proposes a direct approach selection for the reference vector within the corresponding sector in the space vector area. A particular modeling of this converter is presented and the feasibility of a space vector scheme without using a Park transformation is studied. All voltage vectors are drawn into the line-to-line voltage frame. Then, the switching vectors including the zero one are selected simultaneously and their duty cycles are computed. This method reduces the computation time, increasing the accuracy of positioning the switching instants

---

F. Akel (✉)
Unité de Développement des Equipements Solaires, UDES/Centre de Développement des Energies Renouvelables, CDER, Route Nationale N°11, BP N°386, Bou-Ismail, Tipaza 42415, Algeria

Laboratory of Process Control, Polytechnic National School, Street Hassen Badi, El Harrach, BP N°182, Algiers, Algeria
e-mail: akel.fethielt@gmail.com

T. Ghennam
Unité D'enseignement et de Recherche UER-ELT, Laboratoire d'Electronique de Puissance (LEP), EMP (EX: ENITA), BP 17, Bordj EL Bahri, Alger, Algeria

M. Laour • D. Bendib • M. Chikh
Unité de Développement des Equipements Solaires, UDES/Centre de Développement des Energies Renouvelables, CDER, Route Nationale N°11, BP N°386, Bou-Ismail, Tipaza 42415, Algeria

E.M. Berkouk
Laboratory of Process Control, Polytechnic National School, Street Hassen Badi, El Harrach, BP N°182, Algiers, Algeria

I. Dincer et al. (eds.), *Progress in Clean Energy, Volume 2*,
DOI 10.1007/978-3-319-17031-2_52

for digital implementation purposes. Simulation results are carried out to confirm the validity of the proposed method.

Keywords Space vector modulation • DC/AC converter • Line-to-line voltage • Switching vector

Nomenclature

E_g	Band gap energy of the semiconductor
f_{rc}	Switching function of the row r and the commutation cell c
G	Irradiation on the cell surface (w/m^2)
G_n	Nominal irradiation (w/m^2)
I	PV array output current (A)
$i_{a,b,c}$	Grid currents (A)
i_d, i_q	d- and q-axis grid currents (A)
I_0	Saturation current (A)
I_{ph}	Photocurrent (A)
$I_{sc,n}$	Nominal short-circuit current (A)
K	Boltzmann's constant (J/K)
m_1,m_2	Conversion functions
P	Active power (W)
Q	Reactive power (VAR)
q	Electron charge (C)
R_s	Intrinsic series resistance of the solar cell (Ω)
R_{sh}	Equivalent shunt resistance (Ω)
T	Actual temperature (K)
T_n	Nominal temperatures (K)
t_1, t_2, t_0	Applied vector durations (s)
$\vec{u}_m$	Reference voltage vector (V)
U_{dc}	DC-bus voltage (V)
$\vec{u}_{1n}$, $\vec{u}_{2n}$	Line-to-neutral voltages (V)
$\vec{u}_{m1}$, $\vec{u}_{m2}$	Modulated voltages (V)
V	PV array output voltage (V)
v_d, v_q	d- and q-axis grid voltages (V)
$v_{a,b,c}$	Grid voltages (V)
$V_{oc,n}$	Nominal array open-circuit voltage (V)

Greek Letters

α p-n Junction ideality factor

52.1 Introduction

Photovoltaic applications can be broadly classified into two categories. One is the standalone system and other is the grid connected system [1–3]. The two typical configurations of a grid-connected PV system are single or two stages. In two stages, the first is used to boost the PV array voltage and track the maximum power; the second allows the conversion of this power into high-quality AC voltage. The presence of several power stages undermines the overall efficiency, reliability, and compactness of the system besides increasing the cost [1]. The single stage has numerous advantages, such as simple topology and high efficiency. Nevertheless, the control strategy has to be designed in order to extract the maximum available power and to properly transfer it from the PV array to the grid simultaneously. In this case, an important consideration in the controller design is needed [2, 3].

The voltage source inverter (VSI) as the interface from solar cells to the power grid plays an important role in the development and utilization of solar energy [1, 2]. This converter has attracted and still attracts an impressive number of research works [4–7]. All approaches have advantages and disadvantages, and will compromise various attributes such as harmonic generation, complexity, efficiency, flexibility, reliability, safety, modularity, and cost. At present, Carrier Based Modulation (CBM) and space-vector modulation (SVM) have been considered as the most popular modulation strategies for power converters due to their operation at constant frequency [7, 8]. In case of CBM, the output voltage of each leg of the converter is determined through the comparison of a modulating, high-frequency saw-tooth waveform and the desired sinusoidal voltage. This is done independently for each leg. An alternative method to Carrier Based Modulation, space-vector modulation (SVM) has been proposed in many publications [5, 6]. The SVM method features a higher level of dc-bus voltage utilization compared to the conventional CBM. It also offers flexibility in its digital implementation by providing several optimization parameters [4], such as enabling different approaches to place space-vectors and the number and arrangement of samples in each cycle [7, 8]. The most widely known SVM are based on generating eight voltage vectors in the Park reference frame [4].

In this study, we present the use of specific SVM defined in the real reference space frame based on Line-to-Line voltages representation to control grid connected PV inverter by means of appropriate switch signals to make the output current in phase with the utility voltage for obtaining a unity power factor and ensure a good power quality.

52.2 Photovoltaic Energy Conversion System Description

The structure of the single stage grid-connected system is shown in Fig. 52.1. The system consists of a photovoltaic (PV) array, capacitive dc-link, a three phase voltage-source-inverter (VSI) and an output filter inductor (L). The PV modules

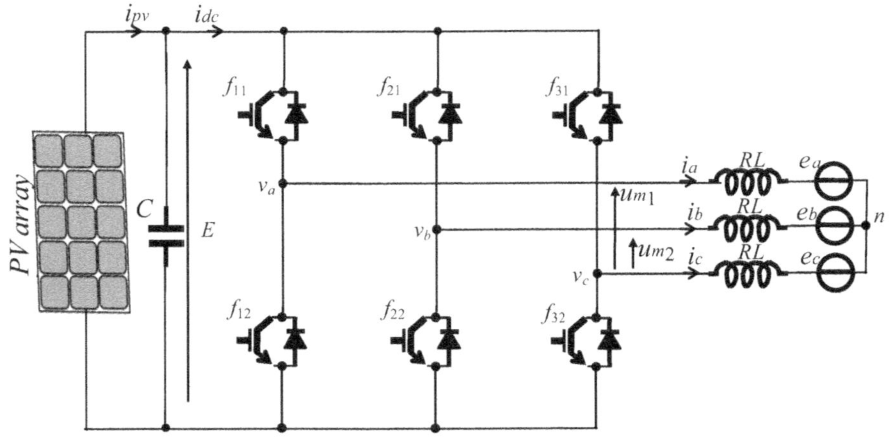

Fig. 52.1 Diagram of single-stage grid connected inverter

are connected in a series-parallel configuration to match the required dc voltage and power rating. The input capacitor supports the solar array voltage for the VSI and aims to reduce the high frequency ripple of the dc voltage in the input side of the inverter. Typically, simple inductors L are used as a filter interfacing inverter and mains, as shown in Fig. 52.1. LCL filter provides advantages in costs and dynamics since smaller inductors can be used. However, in a grid-connected system, an LCL filter may cause resonance, which is a disaster for the system's stability. Hence, control systems are inevitably more complicated.

52.3 Photovoltaic Energy Conversıon System Model

52.3.1 Modeling of the Photovoltaic Generator

Several models have been proposed to describe the behavior of the photovoltaic generator under different weather conditions (irradiation and temperature). In this chapter, we present the model with one exponential (diode) [9, 10]. The equivalent circuit is shown in Fig. 52.2, and the resulting current is expressed as follows:

$$I_{\mathrm{PV}} = I_{\mathrm{ph}} - I_0\left[\exp\left(\frac{q(V + R_{\mathrm{s}}I)}{aKT}\right) - 1\right] - \frac{V + R_{\mathrm{s}}I}{R_{\mathrm{sh}}} \tag{52.1}$$

The generated photocurrent I_{ph} depends linearly on the solar irradiation and is also influenced by the temperature according to the following equation [9–11]:

$$I_{\mathrm{Ph}} = (I_{\mathrm{sc,n}} + K_i(T - T_{\mathrm{n}}))\frac{G}{G_{\mathrm{n}}} \tag{52.2}$$

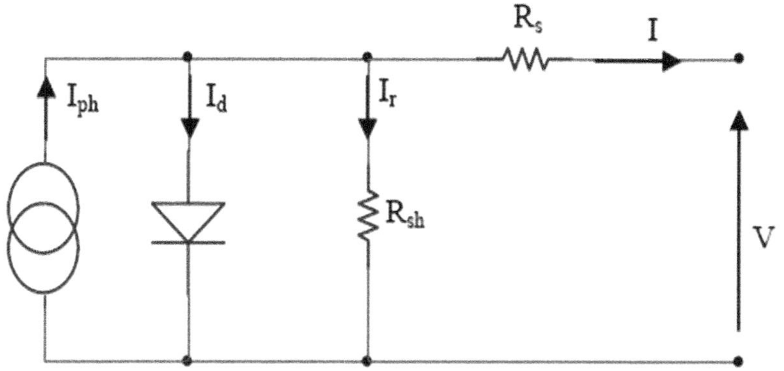

Fig. 52.2 Electrical Scheme of a photovoltaic cell

The diode saturation current I_0 and its dependence on the temperature are expressed as follows [10]:

$$I_0 = I_{0,\text{n}}\left(\frac{T_\text{n}}{T}\right)^3 \exp\left[\frac{qE_\text{g}}{aK}\left(\frac{1}{T_\text{n}} - \frac{1}{T}\right)\right] \tag{52.3}$$

Where the nominal saturation current ($I_{0,\text{n}}$) is given by:

$$I_{0,\text{n}} = \frac{I_{\text{sc},\text{n}} + K_i T}{\exp\left(\frac{q(V_{\text{oc},\text{n}}) + K_v T}{aKT}\right) - 1} \tag{52.4}$$

The parameters of the simulated PV module are summarized in Table 52.1 and the simulation results are presented in Figs. 52.3 and 52.4.

52.3.2 *Three Phase Grid Inverter Model*

The basic structure of the converter is represented in Fig. 52.1, it consists of three commutations cells, which are fed with a capacitor. The converter model used in this study is obtained by considering only two modulated phase-to-phase voltages and replacing all ideal switches with their corresponding switching function (f_{rc}), the index r corresponds to a row and c to a commutation cell. The arrangement of the whole in a matrix containing vertical equivalent commutation circuits gives the depicted Fig. 52.5 [12].

The switching function must satisfy the following condition to avoid the short-circuit of the voltage generator.

Table 52.1 Model parameters

Parameters	Values	Dimensions
I_{scn}	8.21	A
V_{ocn}	32.9	V
I_{mp}	7.61	A
V_{mp}	26.3	V
K_i	3.18e−3	A/K
K_v	−0.123	V/K
N_{cel}	54	
G_n	1,000	W/m^2
T_n	298.15	K

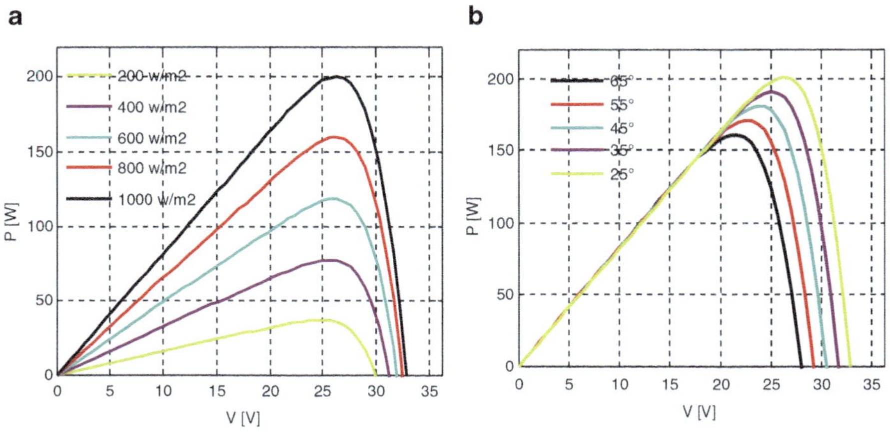

Fig. 52.3 PV DC power output vs. voltage: (**a**) for different temperatures and (**b**) for different solar irradiations

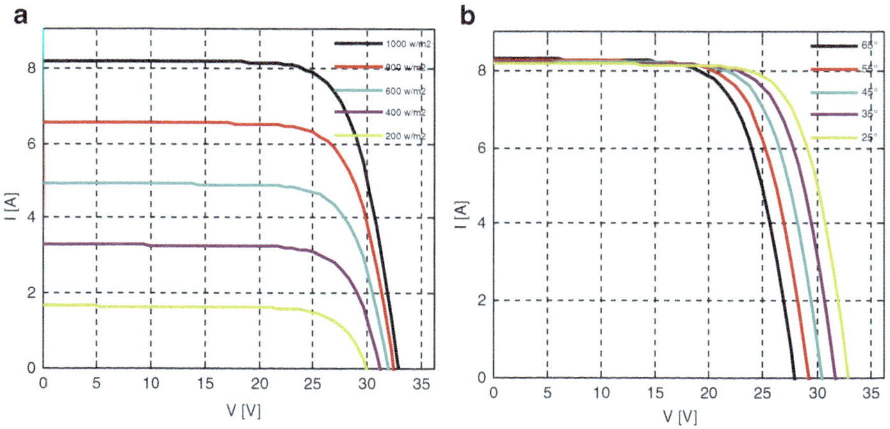

Fig. 52.4 PV output current vs. voltage: (**a**) for different temperatures and (**b**) for different solar irradiations

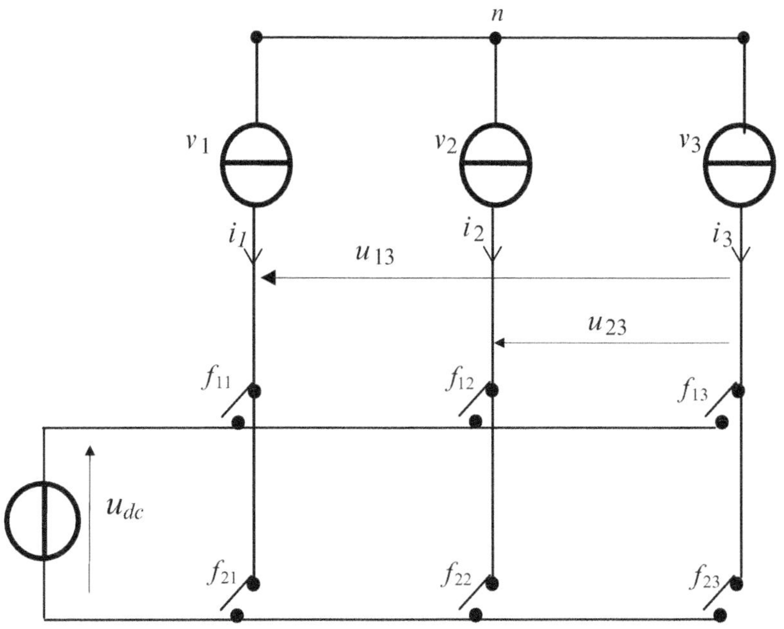

Fig. 52.5 Inverter matrix equivalent model

$$f_{1c} + f_{2c} = 1 \tag{52.5}$$

The modulated voltages are obtained using the DC-bus voltage (u_{dc}) and the switching functions (f_{rc}) as follows:

$$\begin{cases} \vec{u}_{m1} = \vec{u}_{13} = (f_{11} - f_{13})u_{dc} \\ \vec{u}_{m2} = \vec{u}_{23} = (f_{12} - f_{13})u_{dc} \end{cases} \tag{52.6}$$

We define the conversion functions (m_r) as:

$$\begin{cases} m_1 = f_{11} - f_{13} \\ m_2 = f_{12} - f_{13} \end{cases} \tag{52.7}$$

The line-to-neutral voltages are obtained using the modulated voltages u_{m1}, u_{m2} as follows:

$$\begin{cases} \vec{u}_{1n} = \frac{2}{3}\vec{u}_{m1} - \frac{1}{3}\vec{u}_{m2} \\ \vec{u}_{2n} = \frac{2}{3}\vec{u}_{m2} - \frac{1}{3}\vec{u}_{m1} \end{cases} \tag{52.8}$$

Table 52.2 Connection function, conversion function, and corresponding voltage

Connection function			Conversion function		Modulated voltage		Voltage vector
f_{11}	f_{12}	f_{13}	m_1	m_2	U_{m1}	U_{m2}	$\vec{u}$
0	0	0	0	0	0	0	$\vec{u}_0$
1	0	0	1	−1	U_{dc}	0	$\vec{u}_1$
1	1	0	1	0	U_{dc}	U_{dc}	$\vec{u}_2$
0	1	0	−1	1	0	U_{dc}	$\vec{u}_3$
0	1	1	0	0	$-U_{dc}$	0	$\vec{u}_4$
0	0	1	0	1	$-U_{dc}$	$-U_{dc}$	$\vec{u}_5$
1	0	1	−1	0	0	$-U_{dc}$	$\vec{u}_6$
1	1	1	0	−1	0	0	$\vec{u}_7$

The switching configurations and the corresponding line-to-line voltage vectors are reported into Table 52.2. The eight voltage vectors are presented in the reference frame $\left(\vec{u}_{m1}, \vec{u}_{m2}\right)$ by a hexagonal shape as shown in Fig. 52.6.

52.4 Modelling of the Grid and the Connection Interface

According to Fig. 52.1, the delivered currents to the grid across the filter are governed by the following equation:

$$\begin{cases} v_a = Ri_a + L\dfrac{di_a}{dt} + e_a \\ v_b = Ri_b + L\dfrac{di_b}{dt} + e_b \\ v_c = Ri_c + L\dfrac{di_c}{dt} + e_c \\ v_{pv} = C\dfrac{du_{dc}}{dt} + I_{inv} \end{cases} \tag{52.9}$$

It is noted that the model given in (52.9) is time varying and nonlinear. In order to facilitate the control, the model can be transformed into a synchronous Park reference frame (d, q) rotating at the angular frequency of the grid (ω). In this condition the whole dynamic model (52.10) is obtained [1].

$$\begin{cases} v_d = e_d + R.i_d + L\dfrac{di_d}{dt} + \omega.L.i_q \\ v_q = e_q + R.i_q + L\dfrac{di_q}{dt} - \omega.L.i_d \end{cases} \tag{52.10}$$

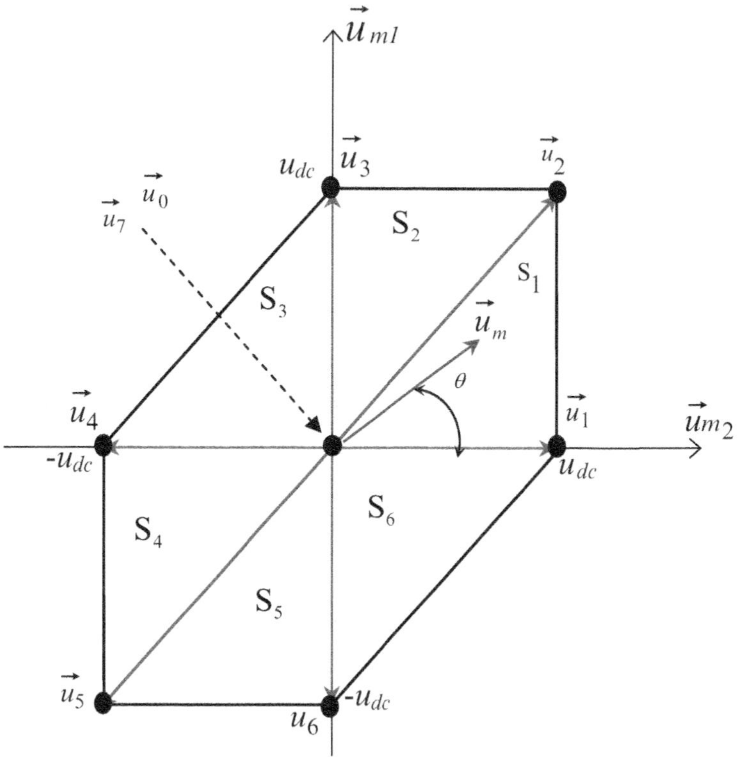

Fig. 52.6 Space vector locality in the $\left(\vec{u}_{m1}, \vec{u}_{m2}\right)$ reference frame

Where:
i_d, i_q: d- and q-axis grid currents, respectively.
v_d, v_q: d- and q-axis grid voltages, respectively.

52.5 Photovoltaic Production System Control

Regarding the control system, this conversion structure has twofold requirements that consists in controlling the current injected to the grid and ensures a DC bus voltage similar to the reference voltage provided by the MPPT algorithm. We opted for a cascade regulation (voltage/current) based on the voltage-oriented control (VOC) method that employs an outer dc link voltage control loop and an inner current control loop to achieve fast dynamic response [2].

The control of the power conversion system is determined from the model given by (52.10), where ω is established by the phase locked loop (PLL) function.

The control system is composed of the following block: MPPT algorithm, DC-bus voltage control, Power control, Current control, Direct Space Vector

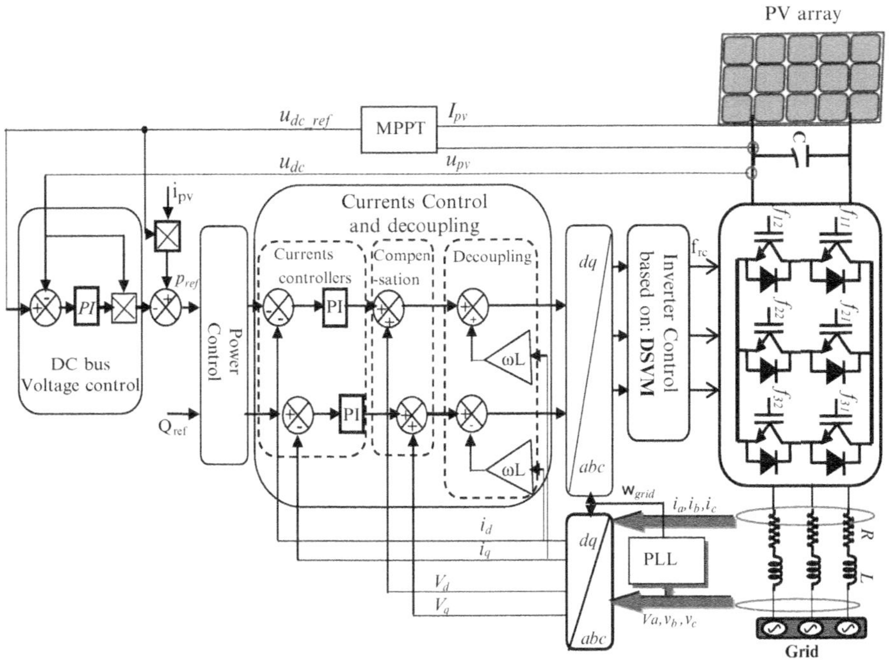

Fig. 52.7 Control structure of single stage grid connected PV inverter

Modulation (DSVM) for switching sequences generation. The whole control scheme is presented in the diagram of Fig. 52.7.

In order to regulate the dc voltage at the DC-reference voltage (U_{dcref}) given by the MPPT algorithm, the error ($\varepsilon = U_{\mathrm{dcref}} - U_{\mathrm{dc}}$) is handed to PI-type controller. The output of this later sets the active power reference P_{cref}, which is necessary to charge the capacitor to the desired value. The reference active power P_{ref} is obtained after being calculated the P_{cref}.

The active and reactive powers (P and Q) can be both expressed using Park grid components (v_{d}, v_{q}) and line current (i_{d}, i_{q}) as follows:

$$\begin{cases} P = \dfrac{3}{2}\left(v_{\mathrm{d}}.i_{\mathrm{d}} + v_{\mathrm{q}}.i_{\mathrm{q}}\right) \\ Q = \dfrac{3}{2}\left(v_{\mathrm{d}}.i_{\mathrm{q}} - v_{\mathrm{q}}.i_{\mathrm{d}}\right) \end{cases} \tag{52.11}$$

The unity power factor is obtained simply by setting the reactive power reference null. From (52.11), It is obviously the presence of cross-coupling between the d and q axes. However, cross-coupling can affect the dynamic performance of the regulator [2]. Therefore, it is very important to decouple the two axes for better performance. This effect can be accomplished with the feed forward decoupling control method as shown in Fig. 52.7.

52.6 Direct Space Vector PWM Strategy

The proposed modulation method generates the switching pattern for the VSI inverter without using a Park transformation.

The reference voltage vector in the space vector $\left(\vec{u}_{\mathrm{m1}}, \vec{u}_{\mathrm{m2}}\right)$ is defined as:

$$\vec{u}_{\mathrm{m}} = \frac{t_1}{T_{\mathrm{m}}}\vec{u}_x + \frac{t_2}{T_{\mathrm{m}}}\vec{u}_y + \frac{t_0}{T_{\mathrm{m}}}\vec{u}_z \tag{52.12}$$

As mentioned previously, the vector can take eight positions according to the states of the switching functions f_{11}, f_{12} and f_{13} (Fig. 52.5). Vectors $\vec{u}_1 - \vec{u}_6$ divide the $\left(\vec{u}_{\mathrm{m1}}, \vec{u}_{\mathrm{m2}}\right)$ plane into six sectors. In each sector, the voltage reference vector is generated by combining the two adjacent vectors $\vec{u}_x$ and $\vec{u}_y$ limiting the sector, in addition to a zero sequence voltage $\vec{u}_z$ ($\vec{u}_0$or $\vec{u}_7$). The DSVM control steps are detailed in the following subsections.

52.6.1 *Detection of Sector*

The sector is defined according to the angle between the $\vec{u}_{\mathrm{m2}}$-axis and the reference voltage vector $\vec{u}_{\mathrm{m}}$.

$$\theta = \mathrm{arctg}\frac{\vec{u}_{\mathrm{m1}}}{\vec{u}_{\mathrm{m2}}} \tag{52.13}$$

The sector number is defined as

$$\begin{cases} \text{if } 0 < \theta < \pi/4 & s = 1 \\ \text{if } \pi/4 < \theta < \pi/2 & s = 2 \\ \text{if } \pi/2 < \theta < \pi & s = 3 \\ \text{if } \pi < \theta < 5\pi/4 & s = 4 \\ \text{if } 5\pi/4 < \theta < 3\pi/2 & s = 5 \\ \text{if } 3\pi/2 < \theta < 2\pi & s = 6 \end{cases} \tag{52.14}$$

52.6.2 *Calculation of the Vector Application Duration*

The application duration of each vector is determined by the obtained projections onto the frame vectors $\left(\vec{u}_{\mathrm{m1}}, \vec{u}_{\mathrm{m2}}\right)$. As example, for a reference vector located in the sector S_1 (Fig. 52.8), the three selected vectors are: $\vec{u}_x = \vec{u}_1$, $\vec{u}_y = \vec{u}_2$ and

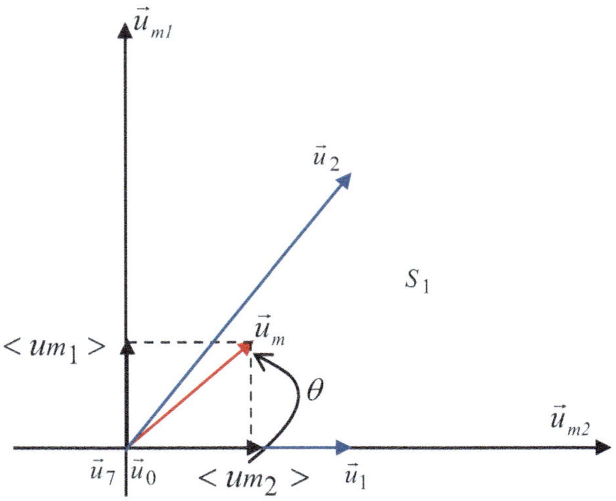

Fig. 52.8 Space vector location into the $\left(\vec{u}_{m1}, \vec{u}_{m2}\right)$ frame

$\vec{u}_z = \vec{u}_{0,7}$. The corresponding durations are t_1, t_2 and t_0, respectively. Thus, the reference vector projection is:

$$\vec{u}_m = \frac{t_1}{T_m}\vec{u}_1 + \frac{t_2}{T_m}\vec{u}_2 + \frac{t_0}{T_m}\vec{u}_0 \tag{52.15}$$

In fact, only two durations must be determined because the modulation frequency is constant and the sum of all durations must be equal to the modulation period:

$$T_m = t_1 + t_2 + t_0 \tag{52.16}$$

The duration of the zero vectors will be always determined according to the others. For our example, the mathematic expansion leads to:

$$\vec{u}_m = \frac{t_1}{T_m}\vec{u}_1 + \frac{t_2}{T_m}\vec{u}_2 + \frac{(T_m - t_1 - t_2)}{T_m}\vec{u}_0 \tag{52.17}$$

$$\vec{u}_m = \frac{t_1}{T_m}\left(\vec{u}_1 - \vec{u}_0\right) + \frac{t_2}{T_m}\left(\vec{u}_2 - \vec{u}_0\right) + \vec{u}_0 \tag{52.18}$$

Using the mentioned projection on $\vec{u}_{m1}$ and $\vec{u}_{m2}$ we get:

$$\text{On } \vec{u}_{m1} : \begin{cases} \vec{u}_1 - \vec{u}_0 = 0 \\ \vec{u}_2 - \vec{u}_0 = u_{dc} \end{cases} \tag{52.19}$$

Table 52.3 Durations calculation in the six sectors

t/S	S_1	S_2	S_3	S_4	S_5	S_6
t_1	$x-y$	x	y	$y-x$	$-x$	$-y$
t_2	y	$y-x$	$-x$	$-y$	$x-y$	x

$$\text{On } \vec{u}_{\mathrm{m2}}:\begin{cases}\vec{u}_1-\vec{u}_0=u_{\mathrm{dc}}\\ \vec{u}_2-\vec{u}_0=u_{\mathrm{dc}}\end{cases} \tag{52.20}$$

After simple manipulation of (52.18), (52.19), and (52.20) the durations are obtained as follows:

$$\begin{cases}t_1=\dfrac{\langle u_{\mathrm{m2}}\rangle-\langle u_{\mathrm{m1}}\rangle}{u_{\mathrm{dc}}}T_{\mathrm{m}}\\ t_2=\dfrac{\langle u_{\mathrm{m1}}\rangle}{u_{\mathrm{dc}}}T_{\mathrm{m}}\end{cases}. \tag{52.21}$$

We assume that:

$$\begin{cases}X=\dfrac{\langle u_{\mathrm{m2}}\rangle}{u_{\mathrm{dc}}}T_{\mathrm{m}}\\ Y=\dfrac{\langle u_{\mathrm{m1}}\rangle}{u_{\mathrm{dc}}}T_{\mathrm{m}}\end{cases} \tag{52.22}$$

Therefore, the durations t_1 and t_2 are given as follows:

$$\begin{cases}t_1=X-Y\\ t_2=Y\end{cases} \tag{52.23}$$

Table 52.3 recapitulates the calculation of the vector's time durations application in the six sectors.

52.7 Results and Discussion

Based on the block diagram of the control scheme illustrated in Fig. 52.7, simulations were done using Matlab/Simulink environment.

The system response to an irradiance step changes between 600 and 1,000 w/m^2 as shown in Fig. 52.9 is investigated.

Figures 52.10 and 52.11 present PV array output voltage behavior towards step change in solar irradiation. It can be seen that the system tracks the new operating point given by the MPPT algorithm very quickly and there are no steady state errors in the DC-voltage compared to its reference.

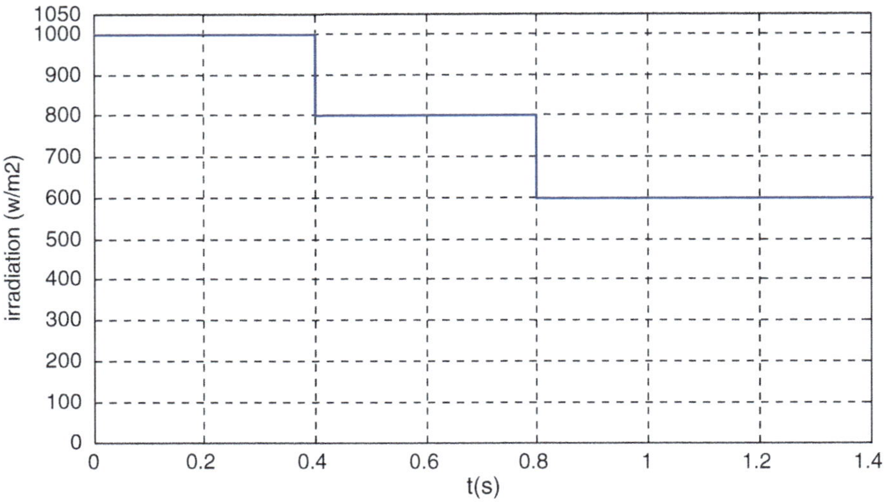

Fig. 52.9 PV output voltage behavior towards step changes in solar irradiation profile

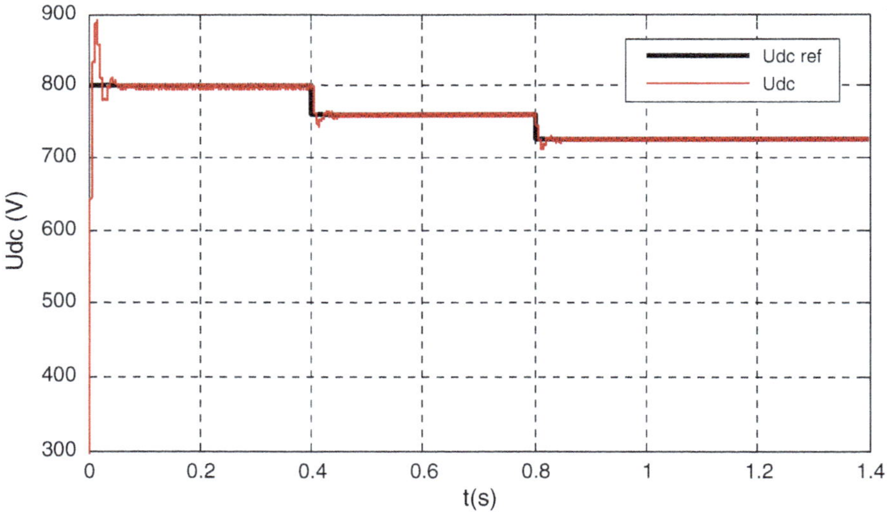

Fig. 52.10 PV output voltage behavior towards step changes in solar irradiation

Figures 52.12 and 52.13 show the injected current into the main utility and the grid side voltage. As it can be noted, the voltage and current are in phase which means that the maximum power extracted from the PV array can pass into the AC-side as the whole system operates at unity power factor with no reactive power exchange.

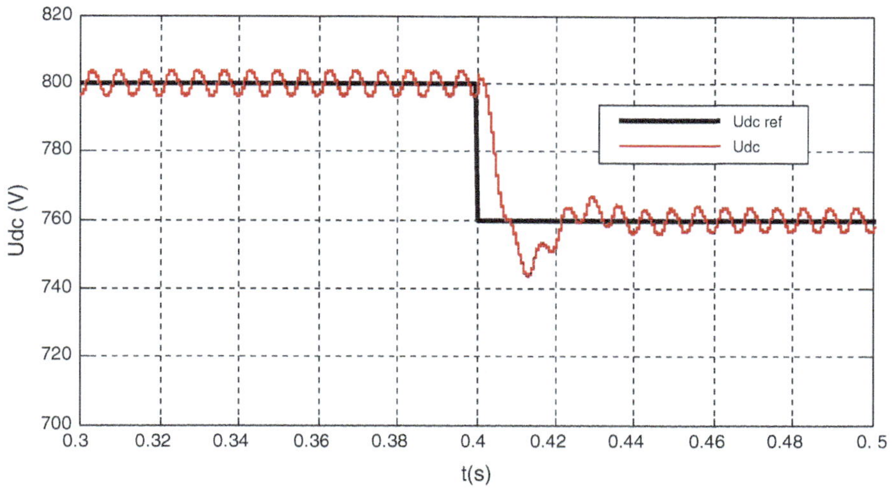

Fig. 52.11 Zoom of PV output voltage

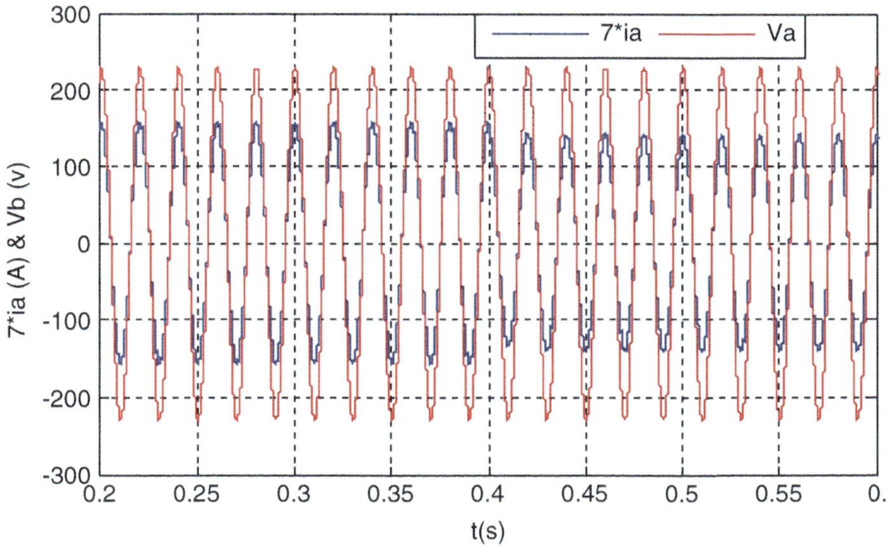

Fig. 52.12 Grid side voltage and injected current ($7 \times I_a$)

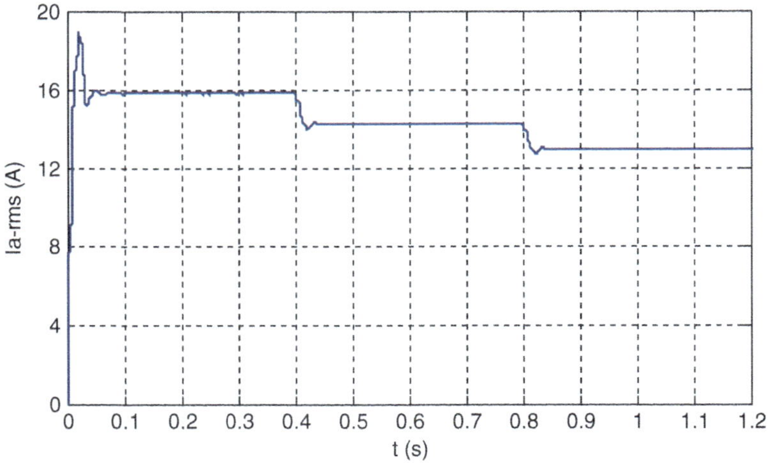

Fig. 52.13 RMS value of the injected current in phase a

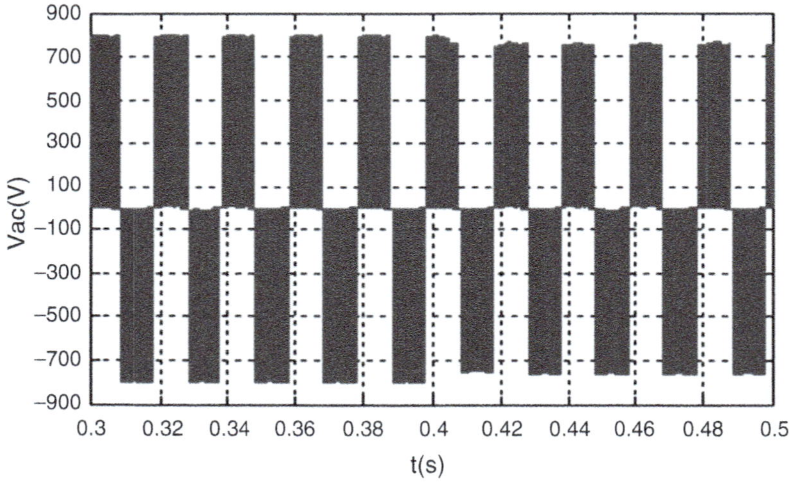

Fig. 52.14 Inverter line-to-line output voltage response

From Figs. 52.14 and 52.15, it is obvious that the control method can ensure constant frequency of the waveforms at the inverter output whatever solar irradiance and temperature conditions. It is clearly observed that the generation of the

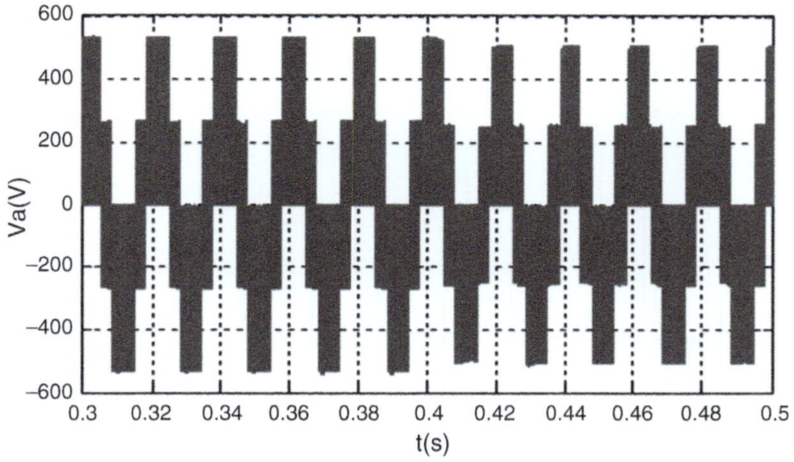

Fig. 52.15 Inverter line-to-neutral output voltage response

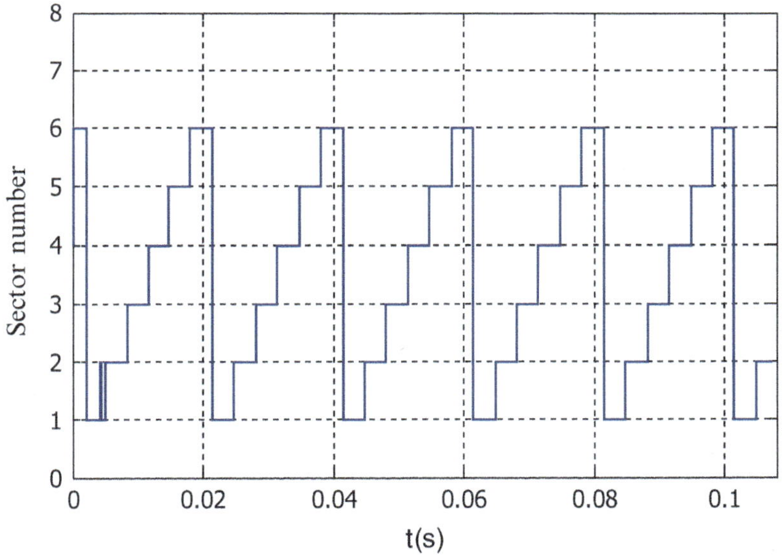

Fig. 52.16 Sector location of the reference voltage vector $\vec{u}_{\mathrm{m}}$

reference voltage vector is guarantee by the application of the adjacent vectors as shown in Figs. 52.16 and 52.17 according to the locality of the reference voltage in the reference frame $\left(\vec{u}_{\mathrm{m1}}, \vec{u}_{\mathrm{m2}}\right)$.

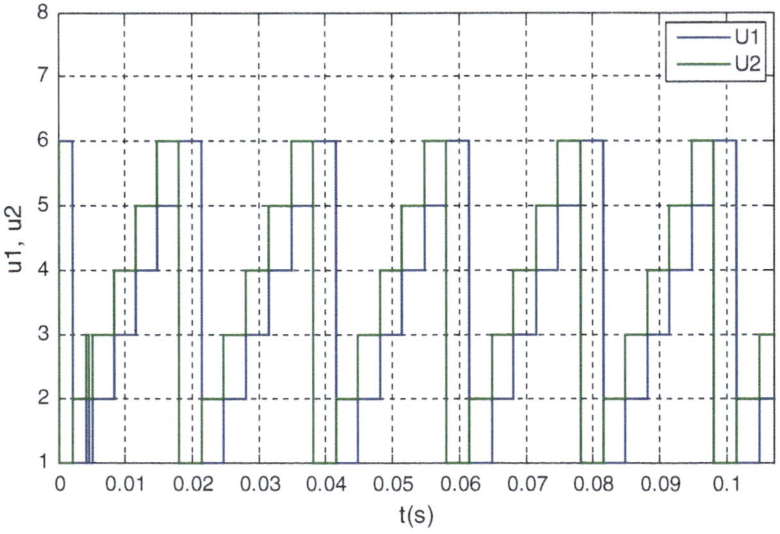

Fig. 52.17 The applied voltage vectors

From Fig. 52.18, It has been demonstrated that the THD of grid current is considerably reduced and the high value of THD (1.47 %) is well below the limit imposed by the standard (5 % in case of IEEE 519-1992).

52.8 Conclusions

In this contribution, vector control of a grid-connected photovoltaic system has been presented and specific modulator based on the line-to-line voltage representation is applied on the inverter. The control method allows us to separately control the power factor of the grid and the dc link voltage. The changes in climatic conditions do not affect the power factor of the grid. It has been demonstrated that the THD of the grid current is considerably reduced which confirms the effectiveness of the control algorithm. The incorporation of the proposed DSVM in the control algorithm would favor simplicity, considerably reduced computational time, and high efficiency.

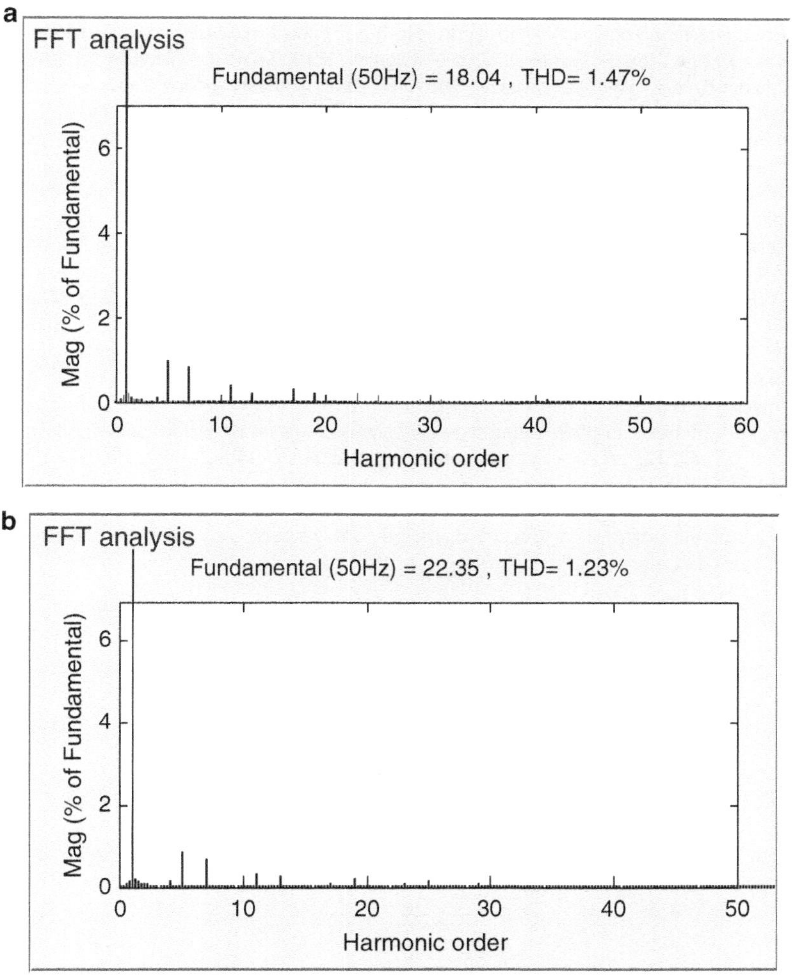

Fig. 52.18 THD of grid current for different irradiation values. (**a**) THD for $G = 1{,}000$ w/m^2. (**b**) THD for $G = 600$ w/m^2

References

1. Lalili D, Mellit A, Lourci N, Medjahed B, Berkouk EM (2011) Input output feedback linearization control and variable step size MPPT algorithm of a grid-connected photovoltaic inverter. Renew Energy 36(12):3282–3291
2. Kadri R, Gaubert JP, Champenois G (2011) An improved maximum power point tracking for photovoltaic grid-connected inverter based on voltage-oriented control. IEEE Trans Ind Electron 58(1):66–75
3. Molina MG, Mercado PE (2008) Modeling and control of grid-connected photovoltaic energy conversion system used as a dispersed generator. In: Proceeding of IEEE transmission and distribution conference and exposition, Latin America, 13–15 Aug 2008

4. Narayanan G, Krishnamurthy HK, Zhao D, Ayyanar R (2006) Advanced bus-clamping PWM techniques based on space vector approach. IEEE Trans Power Electron 21(4):974–984
5. Holtz J (1992) Pulsewidth modulation—a survey. IEEE Trans Ind Electron 39(5):410–420
6. Van der Broeck HW, Skudelny HC, Stanke G (1988) Analysis and realization of a pulse width modulator based on voltage space vectors. IEEE Trans Ind Applicat 24(1):142–150
7. Holmes DG (1996) The significance of zero space vector placement for carrier-based PWM schemes. IEEE Trans Ind Appl 32(5):1122–1129
8. Jobing R, Van der Merwe FS, Kamper MJ (1994) Digital implementation of bus clamped space vector modulation. IEEE Trans Energy Conver 9(2):334–348
9. Narayanan G, Ranganathan VT (2002) Two novel synchronized bus-clamping PWM strategies based on space vector approach for high power drives Two novel synchronized bus-clamping PWM strategies based on space vector approach for high power drives. IEEE Trans Power Electron 17(1):84–93
10. De Soto W, Klein SA, Beckman WA (2006) Improvement and validation of a model for photovoltaic array performance. Sol Energy 80(01):78–88
11. Driesse A, Harrison S, Jain P (2007) Evaluating the effectiveness of maximum power point tracking methods in photovoltaic power systems using array performance models. In: Proceeding of IEEE power electronic specialist conference (PESC), Orlando, FL, 17–21 June 2007
12. Bruno F (1996) Formalisme de modélisation et de synthèse des commandes applique aux convertisseurs statiques a structure matricielle. Ph.D thesis, Université des Sciences et Technologies de Lille

Part IX
Renewable Energy

Chapter 53
Determining the Optimal Capacities of Renewable-Energy-Based Energy Conversion Systems for Meeting the Demands of Low-Energy District Heating, Electricity, and District Cooling: Case Studies in Copenhagen and Toronto

Hakan İbrahim Tol, İbrahim Dinçer, and Svend Svendsen

Abstract This chapter presents a method for determining the optimal capacity of a renewable-energy-based energy conversion system for meeting the energy requirements of a given district as considered on a monthly basis, with use of a low-energy district heating system operating at a low temperature, as low as 55 °C for supply and 25 °C for return, and with additional considerations being directed to supply electricity and cooling. Several optimal solutions with various nominal capacities of the technologies involved were obtained in each of the two case studies, one being for the Greater Copenhagen Area, and the other for the Greater Toronto Area. Various climate conditions of the case areas in question caused different observations of nominal capacities for the energy conversion systems considered with single-production and multi-production based on different renewable energy sources.

Keywords Renewable energy • Low-energy • District heating • Low-temperature • District cooling • Energy conversion system • Municipal decision tool • Exergoeconomy • Lifecycle cost • Optimization

---

H.İ. Tol (✉) • S. Svendsen
Department of Civil Engineering, Technical University of Denmark,
Kongens Lyngby 2800, Denmark
e-mail: hatol@byg.dtu.dk

İ. Dinçer
Faculty of Engineering and Applied Science, University of Ontario
Institute of Technology, 2000 Simcoe Street North, Oshawa, ON, Canada L1H 7K4

I. Dincer et al. (eds.), *Progress in Clean Energy, Volume 2*,
DOI 10.1007/978-3-319-17031-2_53

Nomenclature

Latin Letters

A	The area of the collector
c_1, c_2	First- and second-order heat loss coefficient for flat-plate solar collectors
C	Cost
C_p	The power coefficient for the wind turbines
Ch	Charged heat in the borehole thermal storage system
CF	Capacity factor
d	Depth
dh	Height of the dam
D	The total energy requirement of the district
DisCh	Discharged heat by the borehole thermal storage system
Ex	Exergy value
G	Nominal capacity
I	Intensity
n	Overall amount of the entry type of the indicated in its subscript
p	The end of the period (12 since month is taken as time period in this project)
P	The overall production of the energy conversion systems as a whole
r	The discount rate
Reg_1, Reg_2	Regression parameters
t	Entry type for time
T	Temperature
ν	The wind speed
W	Electrical energy

Greek Symbols

$\forall$	The universal mathematical symbol indicating "for all"
β_0, β_1, and β_2	The parameters used in calculating investment cost of hydroelectric plant
η	Energy efficiency
ξ	The factor representing the building facilities required for the hydroelectric plant
ρ	Air density
τ	Exergetic temperature factor
Ψ	Exergetic efficiency

Subscripts

b	Brine present in the geothermal reservoir
BT	Biomass-sourced trigeneration
c	Cooling energy
Cell	The cell of PV panels
C	Cooling energy
CC	Compressor chiller
DH	District heating
DF	Double-flashing geothermal plant
DS	Dry steam geothermal power plant
DU	Direct utilization of geothermal source (i.e., for heat production)
DSP	Dry steam plants
e	Energy type
fuel	Fuel input to the cogeneration plant
F	Fuel cost
FC	Free cooling
Gain	Energy gain involved for heat recovery units
h	Heat energy
hydro	Hydroelectric plant
H	Heat energy
HHP	Hybrid heat pump
HR	Heat recovery unit
i	The index for the energy conversion system (used for the optimization)
in, out, /s	Input and output of the exergy for the boundary conditions of the energy conversion plant (/s refers to plurality of the inputs and outputs)
I/P	Specific investment cost
LCC,*i*	Lifecycle cost
m	Month studied
Max	Maximum amount of the entry type
MSW	Municipal solid waste
O	Outdoor temperature
O&M	The operating and management cost
PV	Photovoltaic unit
PV/T	Hybrid photovoltaic and thermal unit
R	The return temperature
subS	Geothermal subsurface practices
surf	Geothermal surface plant
S	Single production
S,p	Salvage cost (at the period p)
Sol	Solar
SF	Single-flashing geothermal plant
SR	Solar radiation

STC	Solar thermal collector
tdc	The temperature degression coefficient of the PV cells
TF	Triple-flash geothermal plant

53.1 Introduction

Most countries aim, in their long-term energy plans, to integrate their renewable energy (RE) with the energy conversion systems at their disposal. Denmark, for example, aims at becoming totally fossil-fuel-free by the 2050s [1, 2]. Ontario in Canada plans to have clean, reliable, and cost-effective electrical energy production systems based on renewable energy sources by the 2030s [2]. The aim of lowering the supply temperature in district heating (DH) systems has been achieved in successive generations of DH technology. The first generation involved use of steam, the second generation of superheated (pressurized) water at around 120 °C, and the third generation the use of hot water at around 90 °C [3]. Foged [4] found there to be an 18.5 % reduction in the heat loss of a Danish DH network when the supply temperature was reduced from 85 °C down to 70 °C. The fourth generation of DH technology involved use of very low temperatures, such as 55 °C, for supply and 25 °C for return, allowing for the use of low-grade (low-quality) heat sources and enabling the efficiency of heat production at plant sites to be improved [3, 5–7]. Low-energy DH systems have been shown to be successful in demonstration projects carried out in Lystrup, Denmark [8–10]; in the SSE Greenwatt Way development project in Slough, UK [3]; in the Drake Landing Solar Community in Okotoks in Alberta, Canada [3, 11]; and in Kırşehir, Turkey [12], and in various research studies in this area [7, 13–19]. Because of their very large energy-saving potential and the ease of integrating them with virtually any type of heat source, low-energy DH systems are regarded as being the heating infrastructure needed to meet the requirements of the national energy plans of most countries.

53.1.1 Aim and Objectives

This study is aimed in particular at determining the optimal capacities for a mixture of energy conversion systems making use of RE sources, which are to supply heat to low-energy DH networks. Hence, the residential heat demands of both space heating and domestic hot water consumption are considered to be supplied with low-temperature supply of 55 °C. In view of the improvements in efficiency that are possible with use of multigeneration systems such as cogeneration plants [20] consideration was also given to other forms of residential energy requirements that need to be satisfied for a given district, such as those of electricity and cooling. The objective of this study, therefore, focuses on developing a method based on reducing the overall lifecycle costs of a variety of energy conversion systems through determining the optimal capacity of each with an integrated approach of satisfying various forms of energy demand together. Since there is a month-to-month variation in the energy production of various energy conversion systems in the energy demand of various energy types that are dissimilar with respect to each other and to the energy production, the analysis given in this chapter was carried out on a monthly basis for a period of a year. The method developed was later employed in two case studies, the one being for the Greater Copenhagen Area in Denmark and the other being for the Greater Toronto Area, ON, in Canada.

The method provided in this chapter is novel in comparison to the studies described in Sect. 53.1.2. Its novelty is based on various factors such as (1) the optimization algorithm that determines the optimal capacities to meet three different energy requirements simultaneously while taking into consideration of the monthly variation of the energy generation and of the energy requirement in each energy type considered, (2) detailed focus on lifecycle cost calculations and also the technical considerations, (3) waste heat that was involved with the renewable energy sources, and (4) comparison of the energy conversion systems in the point of exergoeconomic values and the evaluation of the environmental protection compared to the fossil-fuel-based energy conversion systems involved in this study.

53.1.2 Background

Various studies in this area have concerned the integration of RE sources with DH systems and/or with energy-supply decision tools. Lund et al. [21] discussed possibilities of establishing a DH system based solely on the use of RE sources, comparing these with (1) the use of existing natural gas supplies and (2) such alternative individual heating options as micro-CHP systems and individual heat pumps. They emphasized the need of expanding DH networks in stages in the urban areas and use of heat pumps in the rural areas. Sperling and Möller [22] pointed out that expanding DH networks together with the energy savings at residential sites provides results in reduced fuel consumption at heat production sites. Østergaard and Lund [23] described the transformation of the energy infrastructure in the city of Frederikshavn as being based on the use of %100 RE for such differing forms of energy requirements as those of electricity, of the heat and fuel demand of industry, and of transportation. Mathiesen [24] analyzed the effects of the use of DH systems on biomass consumption, its being found that the reduced demand for biomass in the heating sector led to an increase in the potential of biomass for other sectors. Weber and Shah [25] developed the tool DESDOP, which assesses the capacity of each energy conversion system to meet the energy requirements in terms of electricity and heating for a given district, such as a town. Niemi [26] discussed a method based on a multi-energy approach to meeting the energy requirements for electricity and heat, one providing a distributed energy solution for each energy form. Ozlu et al. [27] compared various RE-based energy conversion systems with thermodynamic assessment in terms of their ability to satisfy electricity and heat requirements. An extensive review of the literature in this field can be found in [28].

53.2 System Description

In addition to the matters just referred to, the study was concerned with developing an energy system planning tool being able to determine the optimal capacities of RE-based energy conversion systems (also called as "plant" in this chapter manuscript) based on different primary RE sources available locally, involving different types of production, such as of the single-generation and the multigeneration (poly-generation) type. (The terms involved here, and in connection with this in the remainder of the chapter, are based on two reports [28, 29].) Since the residential energy requirements and the energy production of each energy conversion system vary over the course of a year, the analyses were carried out for successive monthly periods. More specifically, the method involved consideration of the degree to which, month by month, the energy demands for electricity, heating (both space heating and production of hot water for domestic use), and cooling were adequately met. A black-box representation of each RE source and of the energy conversion systems involved is shown in Fig. 53.1 as used in [28].

53.2.1 Optimization Algorithm

An aim pursued here was to minimize the sum of the lifecycle costs of the various energy conversion systems involved, concerned with production of the various final forms of energy as whole that were needed for satisfaction of the residential energy requirements of different forms selected for study here, those of electricity, heating, and cooling. The decision variable used in the optimization algorithm employed here was defined as the nominal (installed) capacity of each energy conversion system, as shown in (53.1), as being employed in [30]:

$$\begin{array}{ll} \text{Minimize} \sum_{i=1}^{n_i} C_{\text{LCC},i} \text{ with respect to } G_i & \text{(as the fitness function)} \\ \text{Subject to the demand constraints} & \\ P_{\text{e,m}} \geq D_{\text{e,m}} \quad \forall e = 1, \ldots, n_{\text{e}} \Lambda \forall m = 1, \ldots, n_{\text{m}} & \text{(as the constraint function)} \\ \text{Subject to the resource availability constraints} & \\ G_i \leq G_{\text{Max},i} \quad \forall i = 1, \ldots, n_i & \text{(as the upper-bound limits)} \end{array} \tag{53.1}$$

where $C_{\text{LCC,i}}$ is the lifecycle cost and G is the nominal capacity of the energy conversion system I where P and D refer, respectively, to the overall production of the systems as a whole and to the total energy requirement of the district, the subscripts e and m referring to the energy type and the month studied, respectively. The clarifications provided in parenthesis indicate the terms they refer to, as being used in the optimization solver, which involves use of the genetic algorithm (GA) with the commercial software Matlab [31].

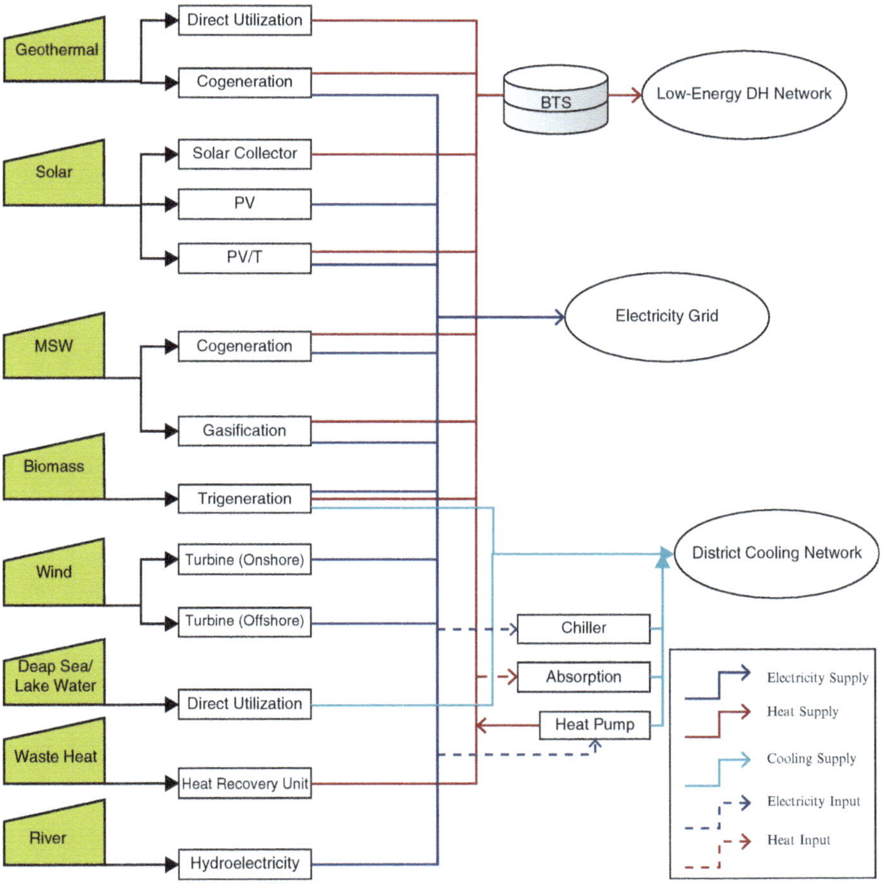

Fig. 53.1 A *black-box* representation of the RE sources and the energy conversion systems involved

53.2.2 Economic Calculations

The economic calculations performed were based on a lifecycle cost analysis carried out, concerned with investment costs; net present value (NPV) of future annual costs, such as operating and maintenance (O&M) costs, and of fuel costs; and salvage costs, which represent the values of the plants at the end of their lifetime, its expression being shown in (53.2) [32, 33], which excludes any investment grants, tax exemptions, or financial incentives provided:

$$C_{\mathrm{LCC},i} = C_{\mathrm{I}} + \sum_{t=1}^{p} \frac{C_{\mathrm{O\&M}} + C_{\mathrm{F}}}{(1+r)^{t}} - \frac{C_{\mathrm{S,\,p}}}{(1+r)^{p}} \tag{53.2}$$

where C is the general symbol indicating the costs; the subscripts I, O&M, F, and S,p refer, respectively, to the investment, operating and management, fuel, and salvage cost at the end of the period p; and r refer to the discount rate used in the NPV calculations.

A base period was taken into account in the NPV calculations and in the salvage cost calculations so as to permit a comparison between different types of energy conversion systems to be made. The reason behind this was the considerable diversity found in the economic data pertaining to the different energy conversion systems. The smallest lifetime length was chosen as the base period amongst the range of various lifetimes of the energy conversion systems.

The energy conversion system for which the possible smallest size was regarded amongst the range of lifetimes varying in length was considered to be the base in defining the length of the base period involved. For any of the energy conversion systems that had longer lifetime than the base period as thus defined, the salvage value at the end of the base period was calculated by assuming a linear reduction in the expected value of the plant in question [34]. One should note that an energy conversion system that has a zero salvage value at the end of its lifetime cannot be assumed to have a linear reduction in its salvage value, since a salvaging of it during any period shorter than its lifetime still accounts for its having a zero salvage value.

The specific investment cost (unit cost per nominal capacity) of any energy conversion system becomes smaller as the nominal capacity increases. This is the basic idea behind the economy of scale [35, 36], being included in determining the investment costs for all of the energy conversion systems. The specific investment costs for all of the plants were formulated in this manner, its expression being given in (53.3):

$$C_I/_{P_i} = \mathrm{Reg1} \times G^{\mathrm{Reg2}} \tag{53.3}$$

where the subscript I/P refers to the specific investment costs and G the nominal capacity of the plant as established, the units employed depending upon the energy conversion system employed, where, for example, the units are in [m^2] for solar systems and in [m^3] for borehole thermal storage. Reg1 and Reg2 are the regression parameters used to calculate the specific investment costs involved with use of an economy of scale.

Particular attention was directed at estimates of the costs of the hydroelectric plant, since a different model applies to it, one shown in (53.4) [37]:

$$C_{\frac{I}{P_{\mathrm{hydro}}}} = (1+\xi)\left[\beta_0 \Big/ \left(G_{\mathrm{hydro}}{}^{\beta_1}\,\mathrm{dh}^{\beta_2}\right)\right] \tag{53.4}$$

where ξ is the factor representing the building facilities required for the hydroelectric plant, and dh and G_{hydro} refer to the height of the dam in [m] and the nominal capacity of the hydroelectric plant in [kW].

An extensive literature review was performed for deciding upon the economic data to be used in the study. The cost data employed was obtained from earlier studies involving values from past years, i.e., earlier than the reference year of the study, for which the year 2012 was decided upon, the data being converted to prices which applied that year, use being made here of cost-index factors as given in [38].

53.2.3 Energy and Exergy Analyses

The energy output, appropriate for a particular energy conversion system, was defined in terms of some performance measure, such as the energy conversion efficiency (in accordance with the first law of thermodynamics) and a capacity factor, these being obtained for all systems and electricity-to-heat ratios (commonly expressed as power-to-heat ratio) in a cogeneration plant [39]. Along with the energy output obtained in connection with the energy conversion efficiency of a system, the capacity factor represents the utilization rate of the nominal capacity, this being affected by the maintenance requirements of the energy conversion system, temporary shortage of the primary source, and over-dimensioning of the nominal capacity [35, 40]. The capacity factors used in this study were derived on the basis of only the first two factors referred to above, however. The latter of these two is valid for RE sources such as wind and solar energy, which thus differs in terms of the capacity factor rates involved, in accordance with the location of the energy conversion system which has been installed [40, 41]. The annual production of an energy conversion system can be calculated by use of the general expression presented in (53.5). For cogeneration systems, the electricity-to-heat ratio was employed for determining the shares of each output from the Pe that was obtained, where a 1,500 kW energy output together with an electricity-to-heat ratio of 1:2 refers to an electricity output of 500 kW and 1,000 kW, as described in [39]:

$$P_{\mathrm{e}} = \mathrm{CF}_i \times \eta_i \times G_i \times 8,765 \tag{53.5}$$

where η and CF refer to the energy efficiency and the capacity factor, respectively, of the energy conversion system i. Here, the number 8,765 refers to the hours of operation over a period of a year.

An exergy analysis was performed in the study for evaluating the performance of the energy conversion systems, which consisted of various modes of energy production, such as single- and multi-output generation. This is because of the exergy measure providing the common quality level for comparing and uniting various forms of energy at different levels of quality [42]. The performance measures for the technologies involved in the study were compared by means of exergoeconomic analysis concerned with the costs of the exergy produced.

The general equations used in the exergy calculations are given below. Details, if called for, concerning calculations of the exergy values pertaining to each of the energy conversion systems involved are given under the heading of each (see Sect. 53.2.4). Since electricity is a pure form of energy, the energy value of it was taken as its exergy value. However the exergy value of the heat energy involved was calculated by use of an exergetic temperature factor, as shown in (53.6) [7, 43]:

$$\mathrm{Ex_h} = \tau_\mathrm{h} \times Q_\mathrm{h} \tag{53.6}$$

where $\mathrm{Ex_h}$ refers to the exergy value of the heat energy, which is represented by h. τ_h is the exergetic temperature factor, its expression in the case of heat energy being given in (53.7):

$$\tau_\mathrm{h} \equiv 1 - {}^{T_\mathrm{O}}/_{T_\mathrm{h}} \tag{53.7}$$

where T refers to the temperature and the subscript O the outdoor temperature, presented as a reference.

The expression for converting the cooling energy to its exergy value is presented in (53.8):

$$\mathrm{Ex_c} = \tau_\mathrm{c} \times Q_\mathrm{c} \tag{53.8}$$

where $\mathrm{Ex_c}$ refers to the exergy value of the energy from cooling, denoted as c, and τ_c is the exergetic temperature factor, its expression in the case of cooling being given in (53.9):

$$\tau_\mathrm{c} \equiv \left| 1 - (T_\mathrm{O}/T_\mathrm{c}) \right| \tag{53.9}$$

The exergetic efficiency can be calculated for any type of energy conversion system through dividing the overall exergy output by the exergy input, its general expression being shown in detail in (53.10). In this study the exergetic efficiency was calculated on a monthly basis due to the varying outdoor temperature. The seasonal efficiency and/or the monthly variations in the capacity factor were involved in the energy conversion systems being based on the varying availability of the RE source, i.e., the solar system, wind turbines, and hydroelectricity plants:

$$\Psi = \mathrm{Ex_{out/s}}/\mathrm{Ex_{in/s}} \tag{53.10}$$

where Ex refers to the exergetic value of the energy, and the subscripts in, out, and /s represent the input, output, and plurality of the energy forms, respectively.

The exergoeconomic calculations were based on the idea of the cost of exergy produced, the expression for it being given in (53.11) [44]:

$$C_\mathrm{Ex} = C_{\mathrm{LCC},i}/\mathrm{Ex_{out/s}} \tag{53.11}$$

Table 53.1 The overall efficiencies and emission factors considered for the fossil-fuel-based energy conversion systems

Energy conversion systems	Fuel type	Overall efficiency [−] (values taken from [47]) (%)	Emission factors [kg/MJ] (values taken from [48])
Back-pressure steam turbine	Coal	88	9.25E−02
Extraction-condensing steam turbine	Coal	70	9.25E−02
Gas turbine	Natural Gas	80	5.03E−02
Reciprocating engine	Propane	80	5.97E−02

53.2.4 Environmental Assessment

Despite the well-known idea of RE-based technologies being totally clean, they nevertheless involve the risk of harming the environment if appropriate precautions are not taken. For example, manufacturing PV cells involves an environmental threat in the form of the fossil fuels and the hazardous materials required for their manufacture. Another concern has to do with the exploitation of the geothermal energy sources due to the possible release of underground gases in the drilling of wells and due to the metals, gases, and minerals contained within the geothermal medium if reinjection of the waste geothermal medium is not employed after its heat content has been transferred to the DH network or to the plants producing electricity [45]. This study, however, is comparative in this respect, involving an investigation of the environmental impact of fossil fuels and the greenhouse gas emissions they can release [46]. Cogeneration was considered as a possibility for the energy conversion systems based on fossil-fuel sources, their electricity-to-heat ratios being assumed to correspond to the electricity-to-heat profiles of the case areas of concern in the study, and monthly assessments of energy demands regarding production of electricity and heat, ranging from 0.15 to 2.55 (Table 53.1).

53.3 Energy Conversion Technologies

In this section, the large-scale energy conversion technologies involved in the study are described briefly and the exergetic calculations specific to each technology involved are presented. No comprehensive overview of the renewable energy sources available or of the relevant energy conversion technologies is undertaken, but an introduction to what is to be taken up in this respect is provided in order to specify clearly the energy conversion systems the study takes up. A review of energy conversion technologies of relevance here is to be found in [20, 49].

53.3.1 *Solar Energy*

The utilization of solar energy today used by humankind already for many millenniums evolved through its applications in passive heating, burning mirrors, and bathing, for example [50], and evolved to its use in recent technologies such as photovoltaic (PV) cells, solar thermal collectors, and hybrid PV/T collectors [51, 52]. Solar energy from the sun shows a variation in terms of obtainability and supply rate, both on a diurnal and on a yearly basis. The specific energy outputs of PV cells, solar thermal collectors, and hybrid PV/T collectors can be determined by means of the commercial software "Polysun," which is concerned with the differing of solar radiation in accordance with the specific location on the earth involved, its also taking account of the local weather conditions (e.g., the frequency of cloudy days and the amount of daylight), inclination of solar light, and thickness of the atmospheric layer [53].

The exergy input gained by solar energy $\mathrm{Ex}_{\mathrm{Sol}}$ was defined using the Petela theorem, its expression being given in (53.12) [54]:

$$\mathrm{Ex}_{\mathrm{Sol}} = A_{\mathrm{SC}} \times I_{\mathrm{SR}} \times \left[1 - \frac{4}{3}\frac{T_{\mathrm{O}} + 273}{T_{\mathrm{Sol}}} + \frac{1}{3}\left(\frac{T_{\mathrm{O}} + 273}{T_{\mathrm{Sol}}}\right)^4\right] \quad (53.12)$$

where A_{SC} refers to the area of the solar (thermal) collector or of the solar (PV) cell [m^2] (according to where the equation was employed), I_{SR} indicates the overall intensity of solar radiation for a given period (in this study evaluation was monthly) [Wh/m^2], and Tsol is the temperature of the sun, defined as being 6,000 K [49, 52].

53.3.1.1 PV Cells

PV cells involve the technology of converting solar photons to electricity energy. Those absorbed photons not utilized by the PV cell cause an increment in the temperature of the PV cells and as a consequence a reduction in the energy conversion efficiency [55]. The expression for calculating the electrical output of the PV system is shown in (53.13) as a function of the PV cell area:

$$W_{\mathrm{PV}} = A_{\mathrm{SC}} \times \eta_{\mathrm{PV}} \times I_{\mathrm{SR}} \times \mathrm{CF}_{\mathrm{PV}} \quad (53.13)$$

where W_{PV} is the electrical energy produced by the PV cells [Wh], and η_{PV} is the electrical efficiency of the PV cells [–], their performance as a function of cell temperature T_{Cell} [°C] being given in (53.14) [56, 57]:

$$\eta_{\mathrm{PV}} = \eta_0[1 - \mathrm{tdc} \times (T_{\mathrm{Cell}} - 25)] \quad (53.14)$$

where η_0 is the electrical efficiency of the PV cells under standard test conditions [–] and tdc is the temperature degression coefficient of the PV cells [1/°C] [58].

53.3.1.2 Solar Thermal Collector

The expression for calculating the thermal output of a solar thermal collector is given in (53.15) as a function of the solar collector area and collector efficiency [59]:

$$Q_{\mathrm{STC_h}} = A_{\mathrm{SC}} \times \eta_{\mathrm{STC}} \times I_{\mathrm{SR}} \times \mathrm{CF_{STC}} \tag{53.15}$$

where Q_{STC} is the heat accumulated by means of the solar thermal collector [Wh] and η_{STC} is the collector efficiency [−], its expression being given in (53.16):

$$\eta_{\mathrm{STC}} = \eta_{\mathrm{O}} - c_1 \times x - c_2 \times {}^{x^2}\!\big/_{I_{\mathrm{SC}}} \tag{53.16}$$

where η_{O} is the optical efficiency and x is the characteristic variable [m^{2}K/W], the definition of it being given in (53.17), c_1 and c_2, respectively, being the first- and second-order heat loss coefficients, defined as being 3.05 [W/m^{2}K] and 0.0051 [W/m^{2}K^2], respectively, for flat-plate solar collectors, as being used in studies [60, 61]:

$$x = {}^{(T_{\mathrm{R}} - T_0)}\!\big/_{I_{\mathrm{SR}}} \tag{53.17}$$

The exergetic efficiency of the solar thermal collector can then be calculated by means of (53.10).

53.3.1.3 PV/T

The output of a PV system is directly proportional to the solar radiation, although the efficiency decreases slightly with increasing temperature [36]. The defect caused by the effect of the increasing temperature of the cells on the efficiency of electricity output can be avoided by use of hybrid PV/T cells in which the heat accumulated by the cells is swept away by the air or water circulating through the cells, this producing heat as a useful energy output [62–64].

The detailed application of (53.10) is shown in (53.18), which serves as an example of how cogeneration systems of other forms function:

$$\psi_{\frac{\mathrm{PV}}{T}} = {}^{\left(W_{\frac{\mathrm{PV}}{T}} + \tau_{\mathrm{h}} \times Q_{\frac{\mathrm{PV}}{T_{\mathrm{h}}}}\right)}\!\Big/_{\mathrm{Ex_{Sol}}} \tag{53.18}$$

53.3.2 *Wind Energy*

Earlier applications of harvesting the kinetic energy of wind involved use of windmills, which led then to the wind turbine technology of today in which large-scale use of wind farms produces electricity energy as a usable energy output [65]. Wind turbine technology is based on the principle of converting the kinetic energy of wind into rotation energy, which in turn is converted into electrical energy. In addition to the wind speed, which is the major factor affecting the efficiency of the system, there are numerous other factors that need to be taken into consideration, such as the direction of the wind, permanence of the wind speed, how plane the site area is, and air properties of the medium [66, 67]. These factors need to be considered in defining the location and properties of a wind farm at a given establishment site.

The theoretical conversion efficiency of any wind turbine cannot exceed the Betz limit, which is a power coefficient with a magnitude of 0.593 [68]. The expression for calculating the electricity output of a wind turbine on this basis is given in (53.19):

$$W_{\mathrm{Wind}} = \tfrac{1}{2} \times \rho \times A_{\mathrm{S}} \times \nu^{3} \times C_{\mathrm{p}} \tag{53.19}$$

where ρ is the air density, A_{S} is the area being swept, ν is the wind speed, and C_{p} is the power coefficient.

Wind farms can be located either on land (onshore) or out at sea or in freshwater (offshore), both of these having various constraints that are of major concern for designers. Environmental defects can be encountered in onshore applications in the form of noise pollution or deaths of birds or bats during operation. Also, it is difficult to apply an economy of scale to onshore wind turbine systems due to possibly differing costs of land because of there being a variety of landlords to deal with, the costs of land often representing a sizeable share of the overall costs. The frequent deficiencies of onshore establishments can be avoided by use of offshore installations, for which the capacity factors are often larger, the wind speed more stable, and the disadvantages for humans less than for onshore installations [69].

The exergy value of the wind energy serving as input to a wind turbine is equal to its kinetic energy prior to its passing through the turbine-swept area, the expression for it being given in (53.20). Further information regarding this can be found in the studies [44, 70]:

$$\mathrm{Ex}_{\mathrm{Wind}} = \tfrac{1}{2} \times \rho \times A_{\mathrm{S}} \times \nu^{3} \tag{53.20}$$

53.3.3 Geothermal Energy

Geothermal energy, which is the domestic source of thermal energy in the form of hot water or steam found in the crust of the earth, is widely utilized, for two types of applications in particular: (1) production of electricity and (2) heat production. The temperature gradient of the geothermal energy source varies according to the location at which it is extracted, its ranging from 30 to 100 °C/km in size [71].

The quality of a geothermal resource, in the wellhead site, can best be described by use of the specific exergy index (SExI), defined in study [72].

53.3.3.1 Direct Utilization with Use of DH Systems

Low-quality SExI sources can be exploited by means of DH systems. Note here that the exergy level is always evaluated with respect to a reference environment (i.e., dead state). The exergetic efficiency varies with changes in the temperature of the reference (dead) state adopted for use in the calculations. Since in study [73], however, the exergetic efficiency was found to range from 45.5 to 47.5 % when the temperature of the reference state ranged from 0 to 25 °C, the reference-state temperature effect was not taken account of in this study.

53.3.3.2 Indirect Utilization with Geothermal Power Plants

There are three types of plants commonly used in electricity production sourced by geothermal energy, the type employed depending upon the type of content of the brine present in the reservoir, as shown in Table 53.2.

1. Dry steam plants are erected in fields that contain a reservoir, pure steam from it being used directly in a turbine [77]. The regression equation for the dependence of the exergetic efficiency on the brine inlet temperature is given in (53.21), which is based on results of study [78]:

$$\psi_{\mathrm{DSP}}(T_{\mathrm{W}}) = -3.175 \times 10^{-6} \times T_{\mathrm{b}}^{2} + 1.789 \times 10^{-3} \times T_{\mathrm{b}} + 0.347 \tag{53.21}$$

2. Flash plants are used in cases in which the geothermal source is provided with a mixture of steam and hot water of high enthalpy. In plants of this type the steam is extracted from the mixture during a flashing process. The maximum capacity of a plant of this type is one of up to 45 $\mathrm{MW_e}$ [75]. The exergetic efficiencies of plants of this type with single flash ranges of between 20.5 and 43.35 % were determined in studies [74, 79]. The regression equation of the exergetic efficiency as dependent upon the reservoir temperature is given in (53.22) for

Table 53.2 Detailed description of the geothermal power plant technologies employed [74–76]

Power plant technology	Geothermal source content	The base of the technology
Dry steam	Vapor dominated	Direct utilization of the steam from the source
Flash	Two-phase mixture of liquid and vapor	Deriving the steam by means of a steam separator
Binary	Liquid dominated	Production of steam by means of a heat exchanger

plants operating on a single-flash basis, in (53.23) for plants operating on a double-flash basis, and in (53.24) for plants operating on a triple-flash basis, the equations being derived from the data obtained in the study [78]:

$$\psi_{SF} = -1.050 \times 10^{-6} \times T_b^2 + 1.305 \times 10^{-3} \times T_b + 5.335 \times 10^{-2} \qquad (53.22)$$

$$\psi_{DF} = -2.945 \times 10^{-6} \times T_b^2 + 2.337 \times 10^{-3} \times T_b - 1.053 \times 10^{-2} \qquad (53.23)$$

$$\psi_{TF} = -3.956 \times 10^{-6} \times T_b^2 + 3.016 \times 10^{-3} \times T_b - 7.664 \times 10^{-2} \qquad (53.24)$$

3. Binary plants are used around the world for fields having liquid-dominated brine in low and medium enthalpy, which can easily be reached in shallow depths. They operate with use of a binary working fluid with a low boiling point, such as R123a, R600a, and R152a [80]. The exergetic efficiency of binary plants was found to range in level between 16.3 and 29.6 % according to the studies [49, 81–83]. The exergetic efficiency was found to be improved to a level ranging between 20 and 40 % with use of advanced improvements in the energy conversion cycle in studies [84, 85]. In this study a regression equation for the exergetic efficiency of a binary power plant dependent on the brine temperature was derived from the data given in study [86], as can be seen in (53.25):

$$\psi_{B} = -1.339 \times 10^{-6} \times T_b^2 + 2.673 \times 10^{-3} \times T_b - 9.271 \times 10^{-3} \qquad (53.25)$$

The exergetic efficiency of each plant type is shown in Fig. 53.2, the data regarding the binary types being taken from [86] and regarding the flash and dry steam types from [78].

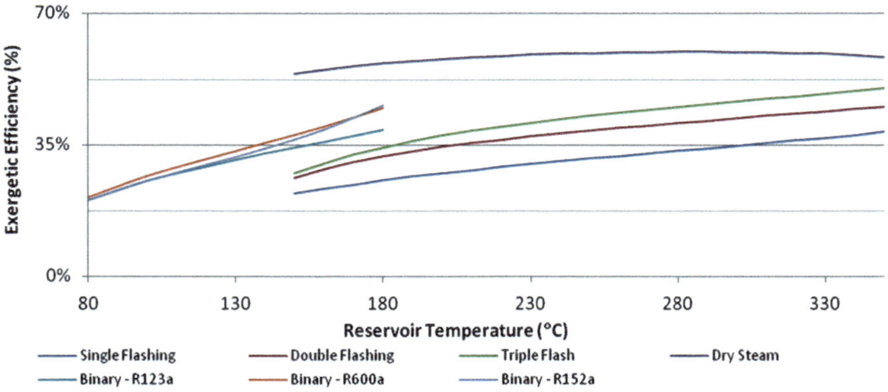

Fig. 53.2 Illustration of the dependency of the exergetic efficiency on the reservoir temperature, given for each plant type (data taken from [78, 86])

53.3.3.3 Indirect Utilization (Cogeneration Plants)

The efficiency of utilizing the source involved can be improved by means of multigeneration, which has been investigated thus far in cogeneration plants and shown to be the case in recent studies carried out at tri-generation plants [80, 87]. In a geothermal power plant, brine reinjection results in the major part of the overall exergy loss at a reasonable temperature that is suitable to be utilized in DH systems and can be further utilized in heating systems such as in greenhouse heating and drying processes [80, 88]. Accordingly, in this study a cogeneration plant based on the use of geothermal sources was considered, one involving the extraction of heat from the brine just prior to its reinjection back into the ground [89].

The exergetic efficiency of a geothermal cogeneration plant producing electricity and heat for DH was found to be 44.1 % for a binary power plant that was defined as producing 10 MW_e and 13.5 MW_{th} through use of a geothermal source at a temperature of 165 °C [90]. An increase in the exergetic efficiency of the cogeneration plant from 38 to 60 % coupled with an increase in the inlet temperature of the brine from 140 to 170 °C was observed. Heberle and Brüggemann [77] studied the change in the overall exergetic efficiency of a geothermal cogeneration plant with a heat production unit coupled in series and in parallel to an organic Rankine cycle of the plant that utilized a geothermal source with a capacity of 4,208 kW_{Ex}. The electricity-to-heat ratio in terms of the exergy connected with the geothermal cogeneration plant was given as 887.70 kW_e/968.92 kW_{Ex} in the case of parallel coupling and as 1,219.99 kW_e/968.92 kW_{Ex} in the case of series coupling, the net electricity production being given as 1,446.40 kW_e with use of the same configuration but this time in the geothermal power plant, the rate of exergy input from the geothermal source being the same in each case. They investigated the overall exergetic efficiency of the plant by changing the inlet temperature of the brine within the range of between 80 and 180 °C. The overall exergetic efficiency of the geothermal cogeneration plant with use of isopentane as the working fluid was found to change between 50.7 and 42.1 % and between 55.5 and 52.1 % in the case of the parallel and series circuit, respectively.

53.3.3.4 Borehole Thermal Energy Storage

The excessive production of heat during the summer months stemming from the solar thermal collector systems and from the waste heat captured from supermarket chiller systems results in the need of storing the excessive heat which is generated during the summer months and using this stored heat during the winter months to supply the low-energy DH network. Borehole thermal energy storage systems consist of aggregated circulation lines of boreholes underground in which the hot water produced during the summer (for example by use of a solar collector field) is supplied to the circuit of the boreholes to store the heat as radial thermal stratification in the ground stratum [91, 92].

The exergetic efficiency of a borehole thermal energy storage system under set boundary conditions, quite apart from the heat production facility involved, was defined as shown in (53.26):

$$\psi_{BHTS} = {}^{Ex_{DisCh}}\!/_{Ex_{Ch}} \tag{53.26}$$

where the subscripts Ch and DisCh refer, respectively, to the charged heat in the borehole thermal storage system and the heat discharged from it during the winter period.

53.3.4 *Biomass*

Biomass, with its history in humankind dating back to the times when the fire was first discovered, still has its role through its properties of being renewable and easy to reach, and its ability to take the place of coal in current fossil-fuel-based technologies [93]. Along with its advantages, there are certain disadvantages of it connected with the increasing demand for biomass within the DH sector as a fuel source for heat production as well as in other sectors, such as the chemical sector and additional industrial sectors [24, 94], loss of soil nutrients, and soil erosion occurring due to rainfall after the harvesting of biomass from the field [95]. Municipal solid waste (also called as MSW) can be regarded as totally renewable after replacement of the fossil-fuel-based plastics it contains by bio-based polymers [96]. Biomass energy conversion technologies are considered in this study, together with three different technologies, those of cogeneration via the incineration and gasification of municipal solid wastes and the trigeneration sourced by the biomass. More regarding energy conversion systems in this area can be found in [97].

53.3.4.1 Municipal Solid Waste Incineration

MSW, amounting to 3.5 million tons per annum (tpa) in Denmark, has a share of 20 % in the overall DH heat supply and of 4 % in the overall electricity supply [98]. Despite their being capital intensive, MSW incineration plants (also called as waste-to-heat plants) are an efficient and environment-friendly way of managing this waste, as compared with the alternative waste management method of landfilling, when special attention is directed at removing from waste, prior to its incineration, such substances as heavy metals that can cause toxic air emissions [99].

In this study, MSW incineration with the cogeneration production of electricity and heat with energy conversion efficiencies of 15 % and 43 %, respectively, was examined [100].

Computing the exergetic efficiency of the incineration was based on the use of (53.10), here as the ratio of the sum of the electricity production and of the exergetic value of the heat energy to the exergy rate of the fuel employed, the expression for this being given in (53.27):

$$\psi_{\mathrm{MSW}} = \left(W_{\mathrm{MSW}} + \tau_{\mathrm{h}} \times Q_{\mathrm{MSW_h}}\right) \Big/ \mathrm{Ex_{fuel}} \tag{53.27}$$

where $\mathrm{Ex_{fuel}}$ is calculated by multiplying the lower heating value of any biomass product by the exergy coefficient factor for it, as described in detail in [39, 101].

53.3.4.2 Municipal Solid Waste Gasification

The gasification of MSW into syngas (or some other form of gas) for use as an energy source in a combined-cycle gas turbine has the advantage of providing greater efficiency in the production of electricity than incineration while having lesser efficiency in the production of heat [100, 102]. In this study the use of MSW gasification in the cogeneration production of electricity and heat was found to have energy conversion efficiencies of 27 % and 24 %, respectively [100]. The exergetic evaluation of an energy conversion system of this type can be carried out with use of the same equation as that for MSW incineration plants, namely (53.27).

53.3.4.3 Biomass-Sourced Trigeneration

This energy conversion system was considered with trigeneration production type, involving the simultaneous production of three forms of energy, that of electricity, heating, and cooling. The trigeneration configuration is based on utilizing an organic Rankine cycle for production of electricity during which the waste heat produced is being recovered by use of a heat exchanger employed for heat production and/or being utilized by an absorption chiller for the production of cooling medium, as described in [103]. Maraver et al. [104] studied trigeneration configurations having different layouts in energetic and exergetic terms. The expression for calculating the exergetic efficiency of a trigeneration based on the use of biomass is shown in (53.28):

$$\psi_{\mathrm{BT}} = \left(W_{\mathrm{BT}} + \tau_{\mathrm{h}} \times Q_{\mathrm{BT_h}} + \tau_{\mathrm{c}} \times Q_{\mathrm{BT_c}}\right) \Big/ \mathrm{Ex_{fuel}} \tag{53.28}$$

53.3.5 Waste Heat

Industrial waste heat (also termed surplus heat) occurs in part due to the inefficiencies of the equipment employed, and in part to the limitations of the thermodynamic cycle, heat of this type possessing strong potential for use either in increasing the efficiency of the industrial process or in producing usable energy forms that can be utilized in industrial plants, for example for capturing the heat from the exhaust gas of a boiler by means of a heat recovery exchanger unit in order to supply heat to the DH network. It is suggested, however that, reutilizing the recovered high-quality waste heat in the same industrial cycle is suggested first before looking for heat recovery for producing usable energy forms for other applications [105–107]. The quality of the waste heat varies in terms of the industrial process and the chiller equipment involved, thus determining the type of the heat recovery equipment to be employed and the usable energy form produced [107, 108].

53.3.5.1 Heat Recovery

Low-quality forms of waste heat can be connected to a local DH network, either directly or indirectly, and be boosted to a high level of quality [109]. A successful example of utilizing waste heat from industry is that of a project in the refining industry in Sweden which has been in operation since the 1970s. The waste heat of 543.8 GWh_{th}/year recovered from two refineries with production capacities of 11 Mtons/year and 6 Mtons/year, respectively, was utilized in the DH systems in Göteborg and in Lysekil, this providing savings in emissions corresponding to 152,400 tons of CO_2 [110]. Werner [99] described the Swedish heat recovery factors that can be used for determining the waste heat potential of different large-scale industrial sectors (see [111] for further details). In this study attention in regard to industrial waste heat was directed at recovering the condensing heat of supermarket refrigeration, which is considered of being boosted with the aid of a heat pump, details of this and success stories regarding its use being given in [112–114]. Since waste heat recovery was considered to be beneficial due to the saving of the energy that would otherwise be lost, the cost of the waste heat that regard to the cost of the input energy consumed in the refrigeration cycle was not considered in the analysis carried out in this study. The exergy value of the heat recovered can be calculated with use of the expression given in (53.29):

$$\mathrm{Ex}_{\mathrm{Gain}} = {(\dot{Q}_{\mathrm{DH}} \times \tau)} \Big/ {\eta_{\mathrm{HR}}} \tag{53.29}$$

where $\mathrm{Ex}_{\mathrm{Gain}}$ refers to the exergy gain [kW] achieved by the implementation of the heat recovery unit, which has a recovery efficiency of $\eta_{\mathrm{HR}}[-]$, assumed to be 0.92 based on the studies of [106, 115].

53.3.5.2 Hybrid Heat Pumps

Another energy conversion system, one involving use of large-scale hybrid heat pump technology and able to produce heating and cooling simultaneously, was considered; its successful application can be seen in detail in the studies [91, 116]. The exergetic efficiency of such systems can be given in terms of either of the two different modes of operation, the one being production of heating alone (single production), as shown in (53.30), and the other being the cogeneration of heating and cooling, as shown in (53.31). In this study the cogeneration operation was considered for the summer period and single production for the winter period:

$$\psi_{\mathrm{HHP_s}} = \left(\tau_{\mathrm{h}} \times Q_{\mathrm{BT_h}}\right) \Big/ W_{\mathrm{HHP_s}} \tag{53.30}$$

$$\psi_{\mathrm{HHP_c}} = \left(\tau_{\mathrm{h}} \times Q_{\mathrm{BT_h}} + \tau_{\mathrm{c}} \times Q_{\mathrm{BT_c}}\right) \Big/ W_{\mathrm{HHP_c}} \tag{53.31}$$

53.3.6 Hydroelectric Plants

Use of hydroelectric plants is an efficient way of producing electricity, permitting rapid response to sudden variations in amounts of electricity required, through taking advantage of the potential energy of a water body gathered by storing of river water behind a dam [117]. A minimum rate of flow on the river in question may be required for aesthetic reasons because of the attractiveness of the region for tourists and/or due to the additional requirements for agricultural fields and/or for urban clean water supply [117, 118]. The exergetic efficiency of a hydroelectric plant is given as 85–95 %, which is the same as the energetic efficiency, as shown in studies [118, 119]. The monthly capacity factor for a small-scale hydroelectricity plant also varies, as can be seen in Fig. 53.3 [47] (valid for the northern hemisphere). In this study, small-scale use of hydroelectric plants was considered, in view of the fact that their large-scale use is already working in a highly satisfactory way in the Niagara Falls [118] as well as in other powerful rivers around the world.

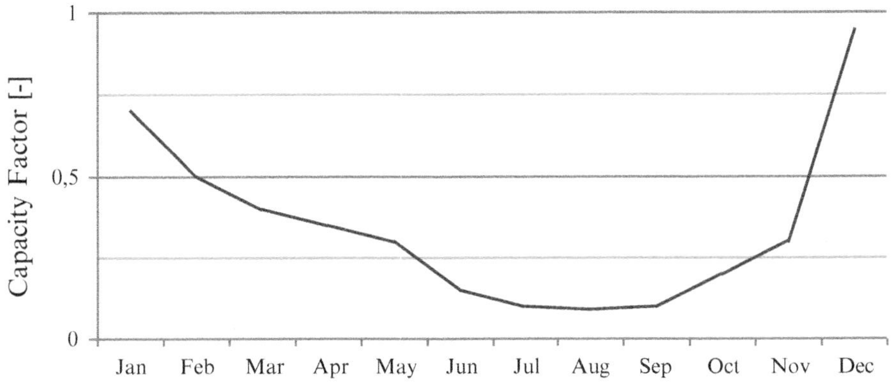

Fig. 53.3 An illustration of the variation in capacity factor for the hydroelectricity production [37]

53.3.7 Cooling

Districts exposed both to very cold winter periods and hot summer periods need to be considered, the district energy systems involved consisting of both district heating and district cooling distribution lines, as can be seen in the case of Toronto, Canada [120]. Compared with that, regions having short summers and lesser requirements regarding district cooling can be considered of having district cooling network for the commercial buildings, as can be seen in Copenhagen, Denmark [121, 122]. For both types of climatic conditions, the cooling energy requirements were considered to be supplied by use of a district cooling network, which operates according to the same principles as a district heating network, but involves chilled water instead of hot water being circulated through the network as the energy-carrying medium [123]. This study deals with three different energy conversion technologies: that for the generation of cooling energy alone, such as in the case of compressor-driven chillers, that of absorption chillers, and that involving free cooling, the exergetic evaluations of these being given in detail below.

53.3.7.1 Free Cooling

Free cooling is an economical way of producing chilled water, involving exploiting of naturally chilled reserves that can be found in the depth of such bodies of water as seas, lakes, and rivers [123, 124]. Also, replacing conventional individual cooling systems with a district cooling system, its base load provided mainly by free cooling, provides a great saving in the peak electricity consumption [125].

For a free cooling system, as being similar in principle to the direct utilization of geothermal energy on the basis of its supplying of heat to a DH system, the expression of exergetic efficiency can be rearranged to assume the simple form shown in (53.32):

$$\psi_{FC} = {}^{\tau_c} \times Q_{FC}/_{W_{FC}} \tag{53.32}$$

53.3.7.2 Compressor Chillers

The variation of water temperature in the case of utilizing free cooling or peak cooling requirements can be compensated by means of large-scale applications in order to improve the security of the supply in district cooling systems [125]. Large-sale compressor chillers are considered in this study with having coefficient of performance (COP) ranging from 4 to 5 [123]:

$$\psi_{CC} = {}^{\tau_c} \times Q_{CC}/_{W_{CC}} \tag{53.33}$$

53.3.7.3 Absorption Chillers

Another type of equipment to be used for increasing the security of supply in district cooling systems can be referred to as absorption chillers. They are driven by the heat source to produce the cooling energy needed (directly fired options of this sort are not considered in this study). The heat source for absorption chillers can be the waste heat recovered from the industrial processes taking place or from the cogeneration facilities present that have low-cost surplus heat available. Although the COP of an absorption chiller is smaller than that of a compressor chiller, its consumption of electricity is significantly lower for absorption technology than for compressor chillers [123].

53.4 Economic Data

Indicative economic data was collected for the purpose of forming a rational basis for comparing different energy conversion technologies considered in the study, their values being given in Table 53.3 with reference to their being used to derive the formulation parameters employed and to define the economic data that is reasonable to use for comparison purposes to be used by the optimization algorithm. Some assumptions were included in defining the cost data which, together with the assumptions, in a base manner depends on the references indicated in Table 53.3 to form functions for economic calculation in the optimization as a rational basis for comparing different energy conversion technologies. The values are given in Table 53.3 together with the other economic data used in the calculations, such as the maximum capacity permitted for use of (53.3) and the ratio of the value of O&M to the salvage value, both of them given as ratio to the investment costs.

Special focus was given to an expression for the investment cost of geothermal systems which was derived by means of cost breakdown method, i.e., surface cost as a function of the nominal capacity and subsurface cost as a function of the depth of the geothermal pipeline that is dependent on the temperature gradient of the area in focus.

Evaluation of the costs connected with geothermal utilization can be undertaken, involving cost breakdowns into such as (1) subsurface costs, including exploration and drilling (C_{subS}), and (2) surface costs (C_{surf}), which vary in the case of different power plants and cogeneration plants, and direct utilization, all of them using the same (53.34) [80, 150]. A regression equation for calculating the subsurface costs (well costs) was derived by use of data involved in cost index conversion, a factor of 1.8 being applied for the period from 2004 to 2012, the data being taken from the study [151] and the cost data being based on [38, 152]:

Table 53.3 The economic data used in the lifecycle cost analysis, the values taken from references being given in the last column

Energy conversion technologies	Reg1	Reg2	Output unit	Maximum capacity limit[a]	Unit of capacity	Ratio ($C_{O\&M}$ to C_I) (%)	Ratio (C_S to C_I) (%)	Lifetime	References
PV	8.635E+04	−2.951E−01	€/m²	2.0E+08	m²	0.6	10	20	[64, 126–128]
PVT	1.243E+05	−2.951E−01	€/m²	2.0E+08	m²	1.9	10	30	[62, 126, 127]
SC	8.557E+03	−2.010E−01	€/m²	2.0E+08	m²	4.0	7	20	[126, 129, 130]
Wind (onshore)	2.656E+03	−1.091E−01	€/kW[b]	7.8E+02	MW	8.3	20	20	[36, 126, 131–133]
Wind (offshore)	2.257E+03	−5.053E−02	€/kW[b]	1.3E+03	MW	5.1	20	20	[36, 126, 131, 132]
BTS	8.376E+03	−5.474E−01	€/m³	2.8E+04	m³	7.0	0	50	[36, 134, 135]
Geothermal—direct	3.399E+03	−9.428E−01	€/kW[b]	5.0E+01	MW	2.5	57	25	[80, 135–138]
MSW—incineration	4.864E+04	−3.716E−01	€/tpa	6.0E+05	tpa	7.0	10	30	[36, 100, 126]
MSW—gasification	3.847E+04	−3.716E−01	€/tpa	6.0E+05	tpa	11.5	10	25	[100, 126]
Biomass	3.119E+04	−3.411E−01	€/tpa	6.6E+04	tpa	6.0	25	30	[103, 139]
Hydroelectric	$\beta_0 = 3{,}300$, $\beta_1 = -1.22E-01$, and $\beta_2 = -1.07E-01$[c]		€/kW	2.0E+03	kW	5.0	0	20	[140–142]
Chiller	4.967E+02	−2.384E−02	€/kW	5.0E+04	kW	10.9	5	23	[143–145]
Absorption	5.549E+02	−2.384E−02	€/kW	5.0E+04	kW	16.8	5	23	[143–145]
Free cooling	1.651E+03	−9.428E−01	€/kW[b]	2.0E+02	MW	3.6	57	25	[125, 144, 146]
Heat pump	6.354E+02	−2.384E−02	€/kW	5.0E+04	kW	0.8	10	25	[36, 116, 147, 148]
Heat recovery	5.777E+02	−2.384E−02	€/kW	7.5E+02	kW	8.0	10	25	[112, 149]

[a]The maximum capacity limits were defined on the basis of the maximum capacity available as found in the literature and/or of the application limits of the economy-of-scale expressions

[b]Unit conversions must be included together with the multiplication of specific cost and an installed capacity, i.e., [MW] × 1,000 × [€/kW]

[c]The equations connected with these parameters are given in Sect. 53.2.2

$$C_{\text{Geo}} = C_{\text{surf}} + C_{\text{subS}} \tag{53.34}$$

$$C_{\text{subS}} = 1.934 \times 10^{-7} \times d^2 + 1.664 \times 10^{-3} \times d + 0.380 \tag{53.35}$$

C_{subS} [M\$'12] is the drilling costs as a function of the depth d in [m]. In the appendix here, \$ refers to Canadian dollars (CAD).

Surface costs are defined in accordance with the power plant technology adopted, which is defined in accordance with the properties of the brine found in the area in question. Regression equations as a function of the brine inlet temperature were derived for calculating the capital costs of geothermal power plants involving different technologies, with use of data provided in the studies [75, 153–157]:

$$C_{\text{C-B}} = 1.115 \times 10^{-1} \times T_{\text{bi}}^2 - 4.031 \times 10^{+1} \times T_{\text{bi}} + 5.104 \times 10^{+3} \tag{53.36}$$

where $C_{\text{C-B}}$ refers to the specific capital costs of the binary geothermal power plant [\$/kW$_e$] as a function of the brine inlet temperature T_{bi} [°C].

$$C_{\text{C-SF}} = -3.704 \times T_{\text{bi}} + 2.338 \times 10^{+3} \tag{53.37}$$

where $C_{\text{C-SF}}$ refers to the specific capital costs of a single-flash geothermal power plant [\$/kW$_e$].

$$C_{\text{C-DF}} = -3.974 \times T_{\text{bi}} + 2.336 \times 10^{+3} \tag{53.38}$$

where $C_{\text{C-DF}}$ refers to the specific capital costs of a double-flash geothermal power plant [\$/kW$_e$].

$$C_{\text{C-DS}} = -2.091 \times 10^{+1} \times T_{\text{bi}} + 7.788 \times 10^{+3} \tag{53.39}$$

where $C_{\text{C-DS}}$ refers to the specific capital costs of a dry steam geothermal power plant [\$/kW$_e$].

$$C_{\text{C-DU}} = 1.007 \times 10^5 \times \dot{Q}^{0.4225} \tag{53.40}$$

where $C_{\text{C-DU}}$ refers to the specific capital costs of a heat production plant for the direct utilization of geothermal sources [\$/MW$_{\text{th}}$], the regression being derived from data provided in [136], and $\dot{Q}$ is the nominal capacity in terms of the heat demand of the district.

53.5 Case Studies

The method described as defining the most economically optimal solution for supplying the mixture of renewable energy sources to satisfy the energy requirements of a particular district, as referred to in Sect. 53.2.1, was carried out in two case areas of differing climatic conditions, the one being the Greater Copenhagen Area (GCA), and the other the Greater Toronto Area (GTA). The heating- and cooling-degree-days [158] for these two areas are shown in Tables 53.4 and 53.5, respectively, to indicate the climatic conditions.

In this study we took into consideration the avoiding of individual heating or cooling systems that run on a source of energy of some other form, such as an individual heat pump system (air conditioner) that produces cooling energy with use of electricity. The indicative data representing the energy requirements considered in the study, such as heating, electricity, and cooling, are shown in Tables 53.6 and 53.7, for GCA and GTA, respectively.

The parameters used in calculating the monthly energy production and exergy considerations are shown in Tables 53.8 and 53.9, for GCA and GTA, respectively. The long-term interest rates were found to be 2.73 % for Denmark and 2.80 % for Canada, the data being taken from [164].

Although the fuel costs were not determined for most of the energy conversion technologies involved in the study, the prices of biomass fuel do differ in line with the availability of the fuel locally (excluding the MSW). On this basis, the cost of biomass shows variations, such as of the market prices at seaports, excluding the inland transportation costs, given as around 9.8–11.1 \$/GJ for Europe and 3.7 \$/GJ for the USA in the report [169], presented here so as to show the differences. Thus, in this study the cost of biomass was assumed to be about 166 €/ton for district heating plants in Denmark, as taken from study [170], which was based on statistics of the Danish District Heating Association, and as 130 €/ton for wood pellets in Ontario, Canada, as derived from the report [171], taking account of a lower heating value of 18 MJ/kg. Also, it should be noted that in Copenhagen hydrothermal resources do not exist [1]. Since, the geothermal temperature gradient was reported to be 30 °C/km for both cases [172–174], only the direct utilization of geothermal energy (details regarding it are given in Sect. 53.3.3.1) was taken account of in the study. However, for cases of high levels of the geothermal temperature gradient, the other geothermal based energy conversion technologies should be taken into consideration, details regarding them being given in Sect. 53.3.3 and the cost considerations involved in Sect. 53.4. In the case of GTA, the dam of the hydroelectricity plant was considered as 50 m in height [118].

Table 53.4 Heating-degree-days with respect to the base temperature of 15 °C [158]

	January	February	March	April	May	June	July	August	September	October	November	December
GTA	404	481	414	254	214	32	7	0	1	29	136	310
GCA	320	389	439	280	254	104	55	9	7	52	178	252

Table 53.5 Cooling-degree-days with respect to the base temperature of 20 °C [158]

	January	February	March	April	May	June	July	August	September	October	November	December
GTA	0	0	0	2	1	31	75	145	89	27	0	0
GCA	0	0	0	0	0	6	1	10	12	1	0	0

Table 53.6 The residential energy requirements for GCA [121, 159–161]

	January	February	March	April	May	June	July	August	September	October	November	December
Heat	1,392	1,308	1,274	771	536	386	352	268	402	754	1,107	1,358
Electricity	850	767	787	661	680	636	630	685	689	726	751	786
Cooling[a]	0.77	0.72	0.80	0.99	0.99	1.27	1.65	1.65	1.10	0.69	0.69	0.69

[a]The residential cooling for GCA was found to be excessively small but the commercial requirement for cooling was considered here to benefit the efficiency improvements with the cogeneration facilities considered

Table 53.7 The residential energy requirements for GTA [2, 162, 163]

	January	February	March	April	May	June	July	August	September	October	November	December
Heat	3,652	3,369	1,872	1,419	961	869	861	883	861	2,158	3,072	3,502
Electricity	553	502	502	470	486	482	573	541	478	478	474	514
Cooling[a]	214	199	222	275	343	252	458	281	306	191	191	191

[a]The residential cooling requirements were calculated for the period of May–September alone. Additional consideration was given to the commercial cooling requirements, these being associated with the data of the case GCA for the periods of January–May and September–December

Table 53.8 Parameters considered for the case of GCA [165–167]

Parameter	Units	January	February	March	April	May	June	July	August	September	October	November	December
Outdoor temperature	°C	0.5	1.1	2.4	6.6	11.1	14.3	16.8	17.6	13.6	9.3	5	1.7
Solar irradiation	MWh	19.4	28.5	47.9	67.3	79.9	78.0	79.4	74.5	55.2	38.6	19.7	13.4
Rel. humidity	%	86	84	82	76	72	72	73	75	78	83	84	85
Maximum atmospheric pressure	hPa	1,047	1,048	1,044	1,035	1,041	1,037	1,031	1,032	1,039	1,043	1,048	1,052
Wind velocity	m/s	7.72	7.72	6.69	6.17	6.17	6.17	5.66	5.66	6.69	7.20	7.72	7.20

Table 53.9 Parameters considered for the case of GTA [166–168]

Parameter	Units	January	February	March	April	May	June	July	August	September	October	November	December
Outdoor temperature	°C	−5.2	−2.9	0.5	6.6	12.9	18.8	21.8	21.4	17.5	10.5	4.6	−0.7
Solar irradiation	MWh	45.5	56.9	70.8	73.3	80.6	80.3	87.0	82.7	72.5	60.1	33.9	31.2
Rel. humidity	%	72	68	62	55	55	55	54	57	59	62	67	72
Max. atmospheric pressure	hPa	980	980	980	980	980	980	980	980	980	980	980	980
Wind velocity	m/s	9.26	8.75	8.75	8.75	7.20	6.69	6.17	5.66	6.17	6.69	8.23	8.23

53.6 Results

The study provides a novel method of finding the minimal capacities of RE-based energy conversion systems, their production being considered on a monthly basis and satisfaction of the various types of energy requirements of relevance, such as for the heat involved in space heating and for producing hot water for domestic use, for electricity, and for cooling purposes. The energy requirements were considered to be provided through an integrated distribution infrastructure consisting of a low-energy district heating system, an electricity grid, and a district cooling system. The optimal solutions obtained for the case areas GCA and GTA, differing in the climatic conditions involved, were given in this section.

53.6.1 Case Study of GCA

Optimization that was carried out showed various optimal solutions involving different combinations of the optimal capacities for the energy conversion systems, each satisfying the energy requirements decided upon for the study, the results being presented in Table 53.10 and the share on the overall lifecycle costs in Fig. 53.4.

The net overproduction observed (after taking account of the capacity factors in question) for the optimal solutions obtained is shown in Figs. 53.5, 53.6, and 53.7, respectively, as regards heating, electricity, and cooling.

A detailed presentation of the energy outputs for an optimal solution is shown in Figs. 53.8, 53.9, and 53.10 representing electricity, heat, and cooling, respectively.

A detailed presentation of the proportion of each energy conversion system involved in the annual production of energy is shown in Figs. 53.11, 53.12, and 53.13.

53.6.2 Case Study of GTA

The optimal solution obtained for the GTA with use of the nominal capacities for the energy conversion systems is shown in Table 53.11, their share of the overall costs being given in Fig. 53.14. The monthly overproduction of energy in terms of electricity, heat, and cooling is given for the optimal solution in the case of GTA in Fig. 53.15.

A detailed presentation of the proportion of each energy conversion system in the annual energy production is shown in Figs. 53.19, 53.20, and 53.21.

Table 53.10 Various optimal solutions obtained in the case of GCA for different installed capacities of the energy conversion systems, these involving different lifecycle costs and exergetic costs

Energy conversion systems	Capacity units	Optimal solutions (OS)				
		OS 1	OS 2	OS 3	OS 4	OS 5
PV	m^2	6.00E+06	–	–	3.00E+05	2.16E +03
PVT	m^2	1.50E+04	–	4.00E+06	3.00E+05	2.16E +03
SC	m^2	1.00E+05	–	5.00E+06	5.00E+04	1.91E +03
Wind (onshore)	MW	3.00E+03	–	1.00E+03	4.00E+03	3.90E +03
Wind (offshore)	MW	4.02E+03	7.56E+03	1.00E+03	4.00E+03	4.34E +03
BTS	m^3	4.00E+05	–	4.90E+05	2.50E+05	3.00E +05
Geothermal—direct	MW	2.00E+03	2.15E+03	6.00E+02	1.90E+03	1.80E +03
MSW—incineration	tpa	1.50E+05	–	3.00E+06	6.00E+04	2.16E +03
MSW—gasification	tpa	5.75E+03	–	3.00E+06	6.62E+03	2.15E +03
Trigeneration—biomass	tpa	1.42E+04	–	5.50E+05	1.02E+04	2.18E +03
Chiller	kW	5.78E+03	–	1.00E+04	6.76E+03	1.88E +03
Absorption	kW	5.99E+03	–	4.00E+04	6.93E+03	1.94E +03
Free cooling	MW	2.31E+02	2.51E+02	1.50E+02	2.32E+02	2.47E +02
Heat pump (hybrid)	kW	1.89E+04	–	1.00E+03	1.28E+04	2.04E +03
Heat recovery (chillers)	kW	1.00E+05	–	2.50E+04	1.10E+01	1.63E +04
Hydroelectric	MW	–	–	–	–	–
Overall lifecycle cost		M€ 25,111	M€ 19,963	M€ 19,307	M€ 23,658	M€ 22,530
Overall exergetic cost		€/MWh 1,514	€/MWh 1,142	€/MWh 1,497	€/MWh 1,406	€/MWh 1,309

53.6.3 Environmental Assessment

The amounts of carbon dioxide emissions avoided by the utilization of RE sources as compared with use of various fossil-fuel-based energy conversion technologies are shown in Table 53.12.

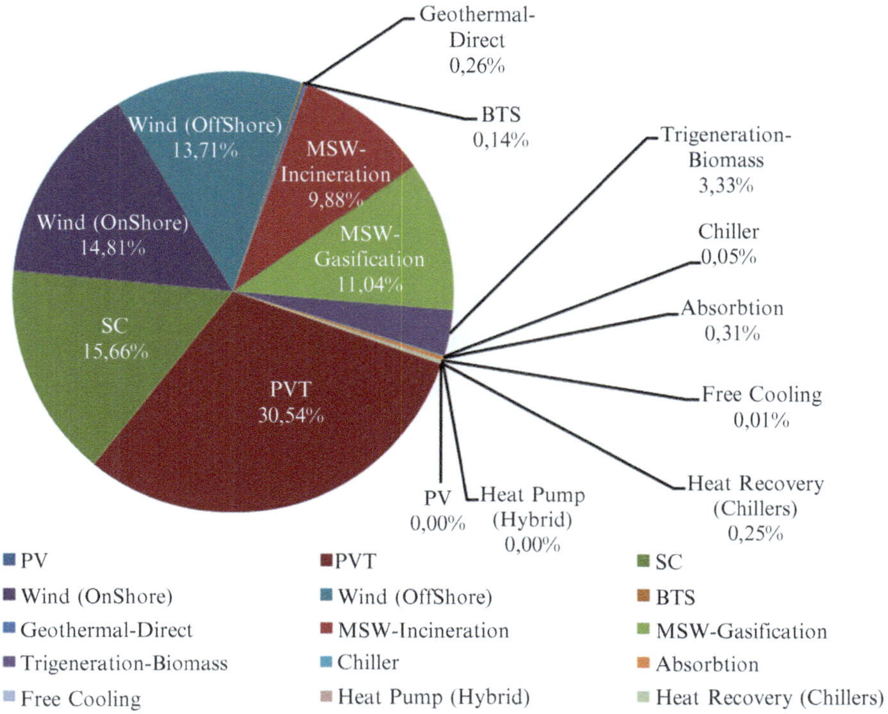

Fig. 53.4 The share on the overall costs of the energy conversion systems in the case of GCA

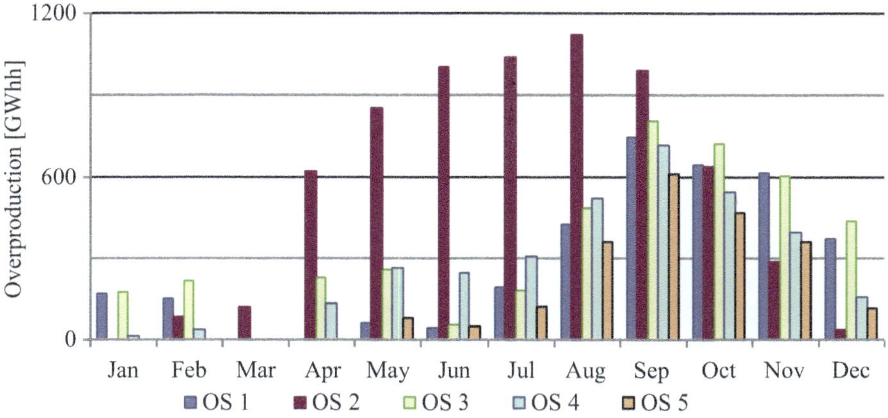

Fig. 53.5 The monthly overproduction in terms of heating energy observed for the optimal solutions obtained in the case of GCA

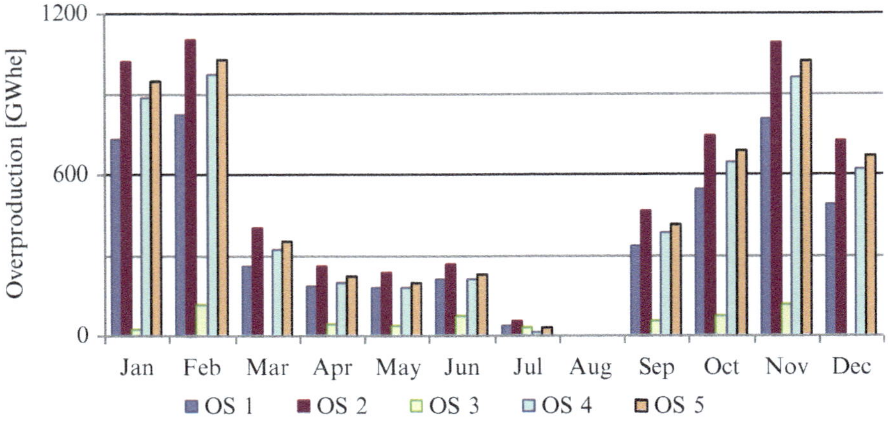

Fig. 53.6 The monthly overproduction in terms of electricity energy observed for the optimal solutions obtained in the case of GCA

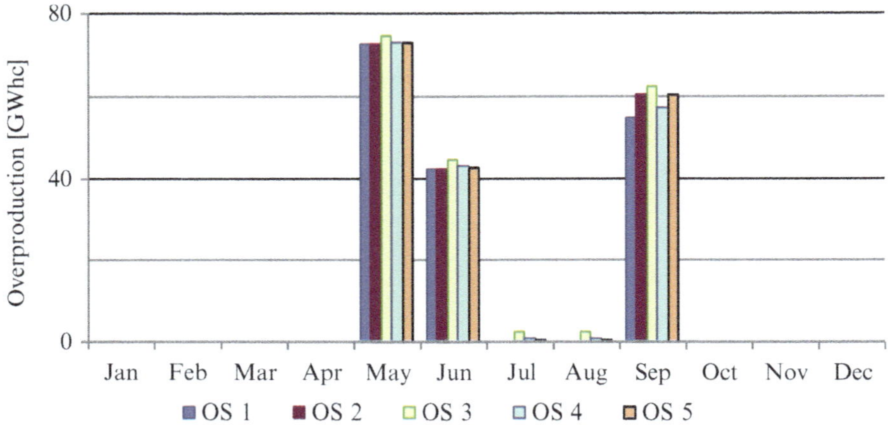

Fig. 53.7 The monthly overproduction in terms of cooling energy observed for the optimal solutions obtained in the case of GCA

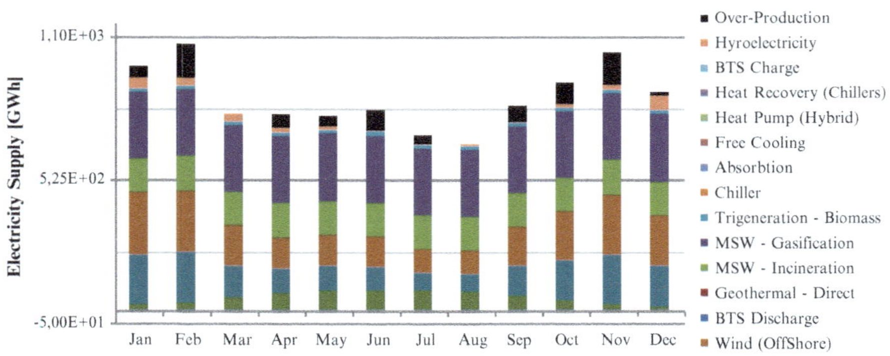

Fig. 53.8 The electricity production for the period of a year as obtained for OS 3 in the case of GCA (the chiller, free cooling system, and heat pump being considered here with their consumptions of electricity) (also published in [175] by the same authors)

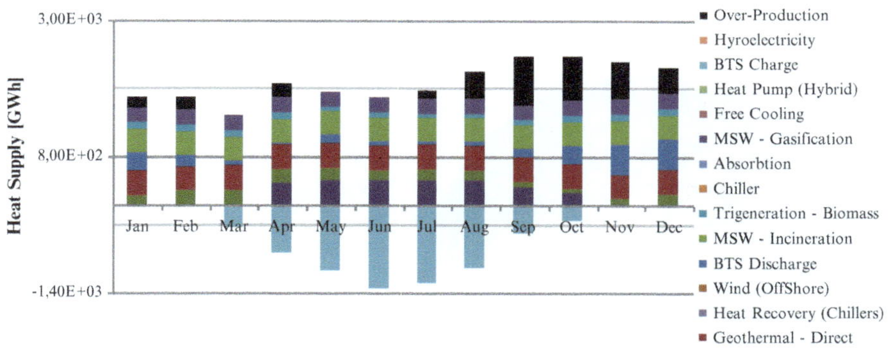

Fig. 53.9 The heat production for the period of a year as obtained for OS 3 in the case of GCA (the absorption system with its consumption of heat being considered here) (also published in [175] by the same authors)

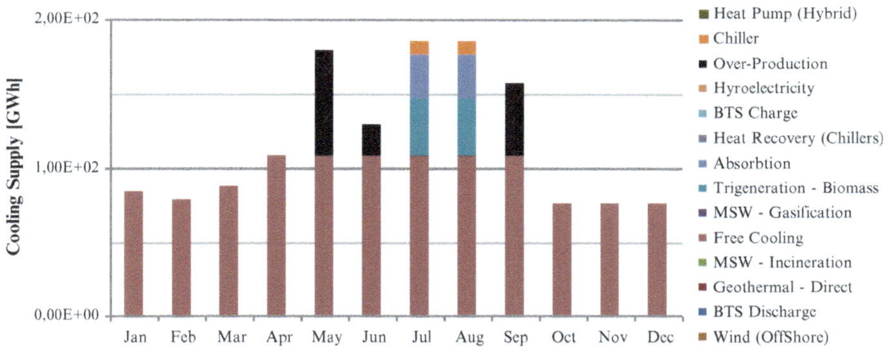

Fig. 53.10 The cooling production for the period of a year as obtained for OS 3 in the case of GCA (also published in [175] by the same authors)

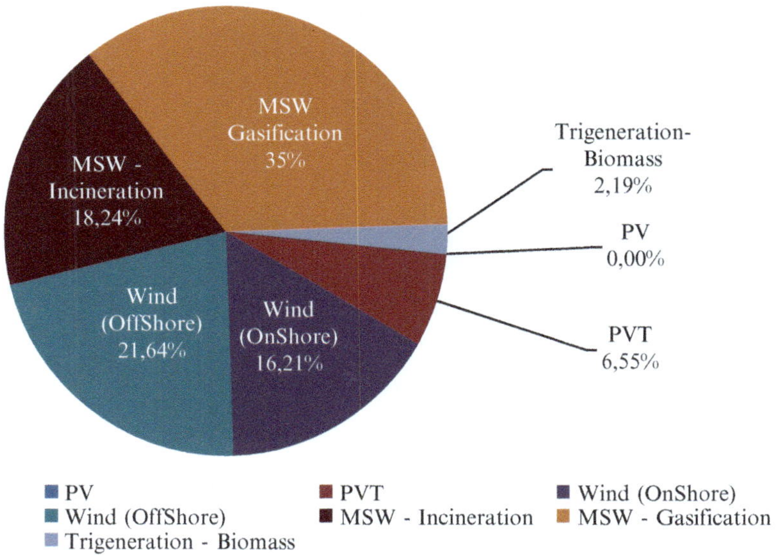

Fig. 53.11 The proportion of the different energy conversion systems involved in the overall annual production of electricity in the case of GCA

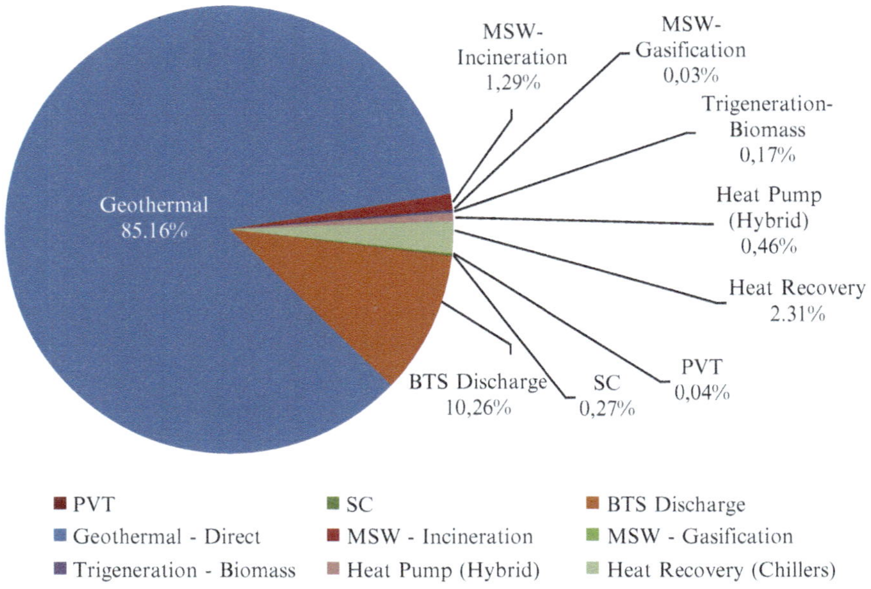

Fig. 53.12 The proportion of the different energy conversion systems involved in the overall annual production of heat in the case of GCA

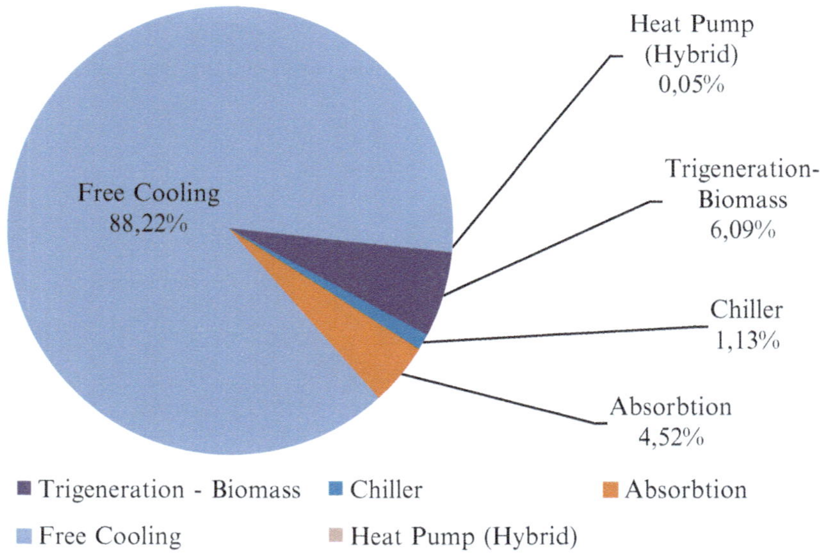

Fig. 53.13 The proportion of the different energy conversion systems on the overall annual cooling production in the case of GCA

Table 53.11 The optimal solution being obtained for the case of GTA

Energy conversion systems	Capacity units	OS
PV	m^2	2.05E+07
PVT	m^2	5.00E+04
SC	m^2	0.00E+00
Wind (onshore)	MW	8.50E+02
Wind (offshore)	MW	8.00E+02
BTS	m^3	1.10E+06
Geothermal—direct	MW	4.30E+03
MSW—incineration	tpa	5.32E+03
MSW—gasification	tpa	1.00E+06
Trigeneration—biomass	tpa	9.45E+05
Chiller	kW	0.00E+00
Absorption	kW	8.00E+04
Free cooling	MW	5.40E+02
Heat pump (hybrid)	kW	1.00E+02
Heat recovery (chillers)	kW	8.00E+05
Hydroelectricity	kW	1.00E+05
Lifecycle cost		M€ 23,751
Overall exergetic cost		€/MWh 1,576

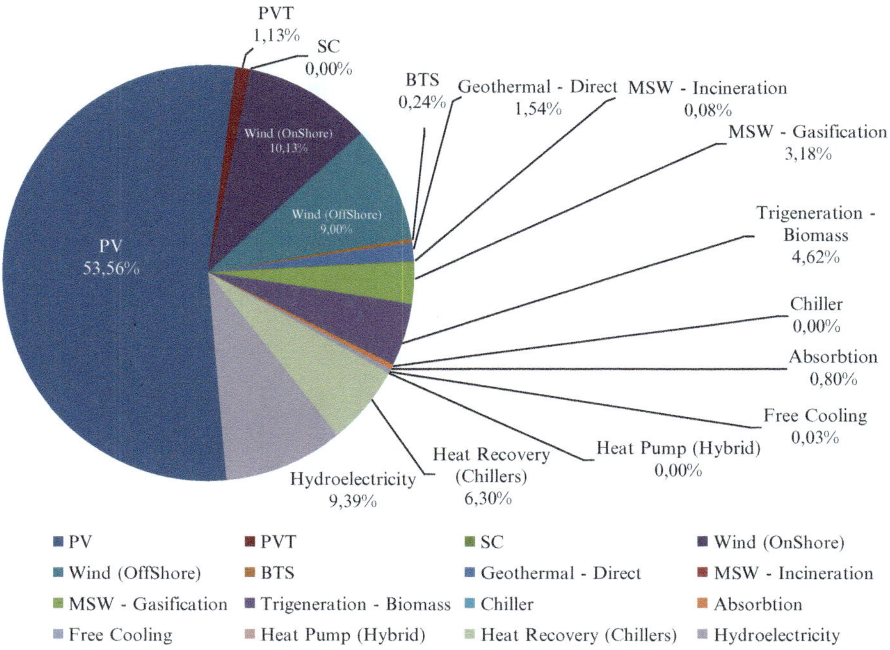

Fig. 53.14 The share on the overall costs of the different energy conversion systems in the case of GTA

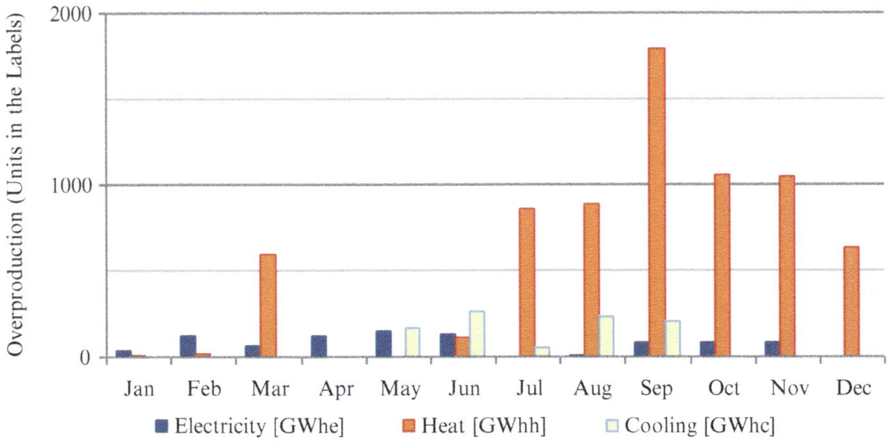

Fig. 53.15 The monthly overproduction of energy in terms of electricity, heat, and cooling observed for the optimal solution in the case of GTA

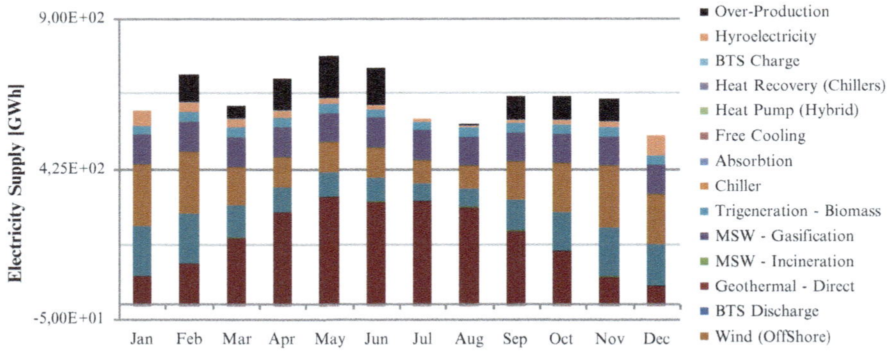

Fig. 53.16 The electricity production for the period of a year as obtained for OS in the case of GTA (also published in [175] by the same authors)

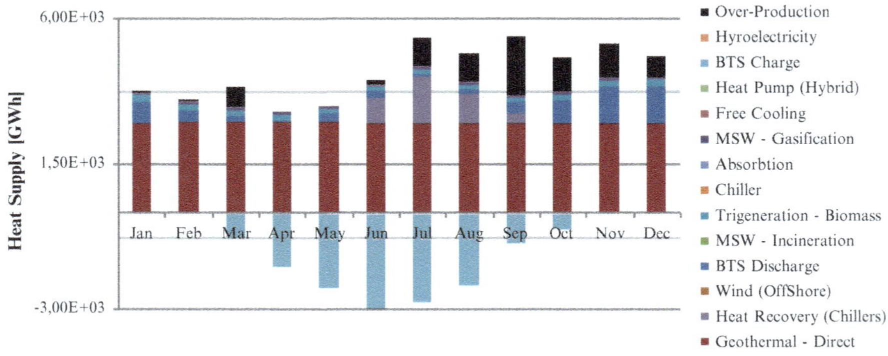

Fig. 53.17 The heat production for the period of a year as obtained for OS in the case of GTA (also published in [175] by the same authors)

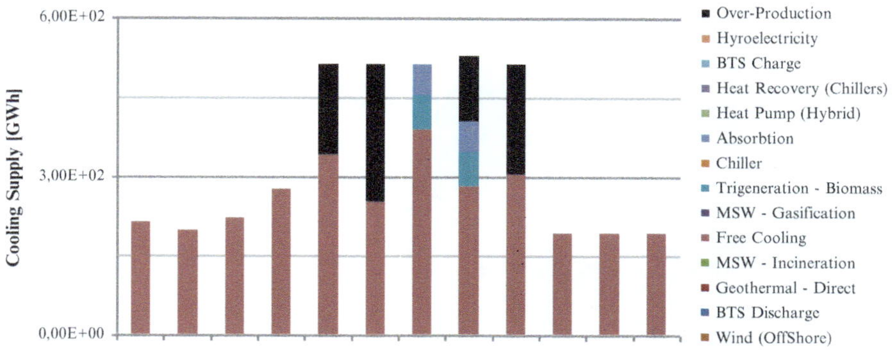

Fig. 53.18 The cooling production for the period of a year period obtained for OS in the case of GTA (also published in [175] by the same authors)

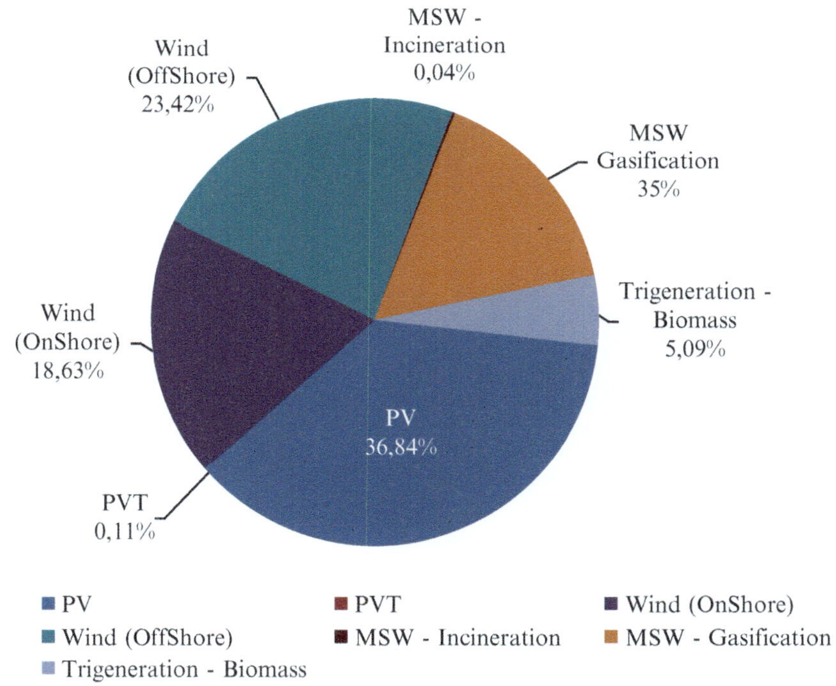

Fig. 53.19 The proportion of the different energy conversion systems on the overall annual production of electricity in the case of GTA

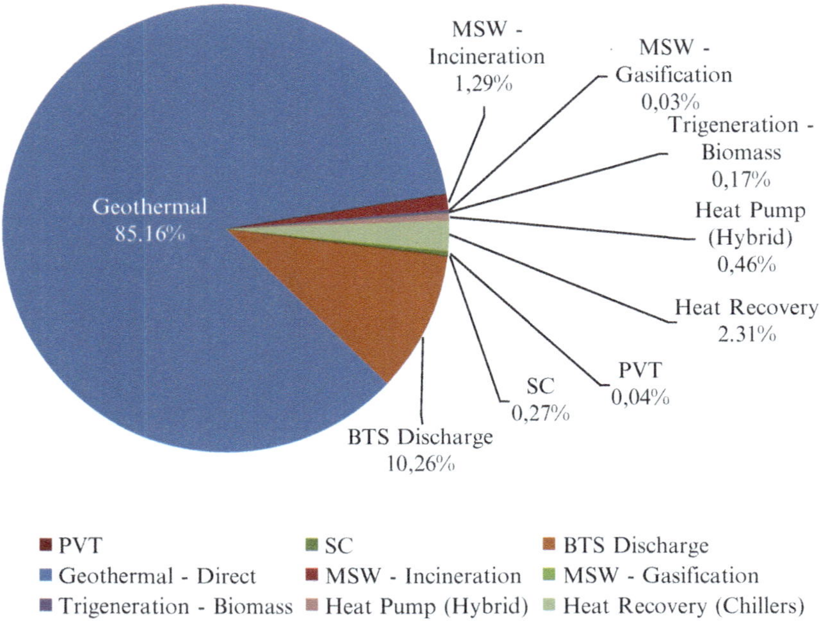

Fig. 53.20 The proportion of the different energy conversion systems on the overall annual heat production in the case of GTA

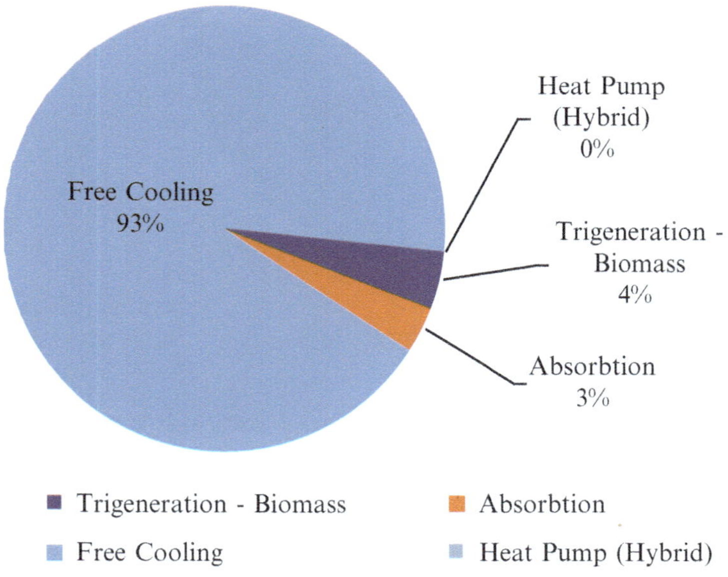

Fig. 53.21 The proportion of the different energy conversion systems on the overall annual cooling production in the case of GTA

Table 53.12 The amounts of carbon dioxide emissions avoided by use of RE sources of a cogeneration type as compared with use of fossil-fuel-based energy conversion technologies

	Emissions (CO_2) [Mton]	
Fossil-fuel-based energy conversion technologies	GCA	GTA
Coal-fired back-pressure steam turbine	703	1,118
Coal-fired extraction-condensing steam turbine	883	1,406
Natural gas-fired gas turbine	420	668
Propane-fired reciprocating engine	499	794

53.7 Discussion

Despite the genetic algorithm being powerful for handling optimization models of virtually any type, the optimal solutions obtained were found on certain rather infrequent trials to show considerable error in their satisfaction of the constraints involved, especially when evaluating the heat supply-demand match. The thermal storage that could take place in the boreholes was considered to be a major reason for this, due to the marked differences that existed in the manner in which the heat energy that was produced was utilized, through the excessive heat produced during the summer being stored and being discharged during the winter. Thus, manual arrangement of the variable data (in part the nominal capacities of the different systems involved but mainly of the borehole storage system) was employed in order to adjust the variation in the energy supply to match the variation in demand.

Several OSs of differing capacity and differing overall lifecycle costs were obtained in both of the case studies. Since different aspects of OSs can affect decision making differently, the evaluation of the OSs was carried out with use of different parameters, such as lifecycle costs, exergetic costs, and overproduction assessed on a monthly basis. In the case of GCA, one can note in Table 53.10 that the energy supply configuration given for OS 2 as based on certain of the energy conversions systems, namely those of offshore wind turbines, of the direct utilization of geothermal data, and of free cooling, resulted in lower overall costs than those based on most of the other OSs. However the difference between production and demand (in the sense of overproduction) was found to be significantly greater than for the other solutions, as can be seen in Figs. 53.5 and 53.6. The other solution, that of OS 3, was taken as the best solution for the case study in question, due to its having the lowest cost and the least degree of overproduction and due also to the diversity of the RE sources involved when note is taken of the security of supply, a matter not included in the algorithm of the optimization.

The exergetic costs of each energy conversion system are dependent upon several factors, such as the efficiency of the energy conversion system, together with the capacity factor, and the economic performance of the technology employed, such as the rate of gain in the economy of scale for investment costs, the superiority of the other economic values described in Sect. 53.2.2, and the efficiency improvements that could be made through multigeneration, if the equipment available were adequate for this. One should take account also of the different exergetic costs, each of which was observed as being unique to the technology employed, and of the economy of scale being uniquely dependent upon the nominal capacity of the energy conversion system. For example, the borehole thermal storage systems, direct utilization of geothermal sources, and free-cooling technologies employed resulted in a significant reduction in exergetic costs in all OSs. This is due to the very low specific costs assigned to the systems in question. Regarding the solar-based energy conversion systems, for example, the order of size of the lifecycle cost observations obtained (i.e., of PV, PV/T, and the solar collectors) does not correlate appreciably with their exergetic cost values, where the order of size of the lifecycle cost is that of $LCC_{PV/T} > LCC_{PV} > LCC_{SC}$, whereas the order of size of the exergetic costs is that of $C_{Ex,SC} > C_{Ex,PV} > C_{Ex,PV/T}$, despite the same nominal capacity being employed in both cases. A further and final point to be discussed is that of the disharmony to be noted in comparing the lifecycle costs with the exergetic costs, and also with the exergetic improvement (i.e., reduction in exergetic costs) that occurred through the hybrid production of heat and electricity with use of the PV/T system that occurred. However, simply comparing the PV cells with the solar collectors does not indicate clearly how matters lie, because of the different forms of the final useable energy that they provide, the one supplying electricity and the other heat energy. This was one of the main reasons for the carrying out of a lifecycle cost analysis, which can be regarded as embodying the major aim in the optimization model, rather than carrying out an exergetic evaluation, meanwhile considering evaluating the satisfaction level of energy requirements in question. Still another reason for carrying out such an analysis is that in economic terms renewable energy sources (with the exception of biomass-based systems) are free.

There was a high degree of overproduction of heat energy in the summer months due in part to the increased efficiency and/or increased capacity of some energy conversion systems, such as solar systems, due in part to the recovery of heat from the waste heat of supermarket cooling plants, assumed to only be possible during the summer months, and due in addition to the reduced demand for heat then because of the lack of need for space heating and the lesser need for domestic hot water production. The borehole storage systems present appear to be the best way of overcoming the inequities between summer and winter through their enabling the excessive amounts of heat generated during the summer months to be carried over to the cold winter months, allowing for a reduction in the nominal capacities required to satisfy demands for heat. The same sort of variation can be observed for the electricity and cooling supply. Matters of the storage of cooling were not considered in the study and there is no available technology for storing electricity for long periods of time. However satisfying the minimum demands for electricity in August was found to result in a high degree of overproduction of electricity during the rest of the year, as can be seen in Fig. 53.6, this being due to lower exergetic costs of a wind turbine system than of PV cells (both of them producing electricity alone), this despite the ability of the electricity demands during the summer months to be readily met, due to the increased solar insolation.

The annual energy requirements in terms of the ratio of heat to electricity was found to be higher for GTA than for GCA, the respective values for it obtained being 3.9 and 1.1, respectively, whereas the heat-to-cooling ratio values of the two were quite similar. The considerably larger gap between the peak and the minimum of the heat energy requirements observed for GTA can be seen as reflecting the greater use of a borehole storage system than in the case of GCA. The nominal capacity of the biomass trigeneration system of GTA was found to be double the capacity of it for GCA. The contribution of the hydroelectric capacity results in a slight reduction in the capacity of wind farms that are established offshore. The high rate of solar insolation, found to be higher in GTA than in GCA, showed a tendency to increase the capacity of solar energy systems, except for the capacity of the solar collector systems of GTA being reduced in regard to OS. The high level of the cooling demands in connection with GTA was found to result in a slight increase in the free cooling capacity together and in a higher degree of trigeneration based on biomass, with use of an absorption system which installed capacity being doubled.

53.8 Conclusions

The major aim of the study was to investigate a novel method for determining the optimal capacities of RE-based energy conversion systems that can handle different climatic conditions readily and can be used to evaluate the degree to which the energy requirements of different types for different periods of the year can be satisfied. An optimization model was formulated aimed at minimizing the overall lifecycle costs of the energy conversion system as a whole. The constraints of this optimization model, designed to maximize the satisfaction of monthly energy requirements in terms of providing the heat energy needed for space heating and for domestic hot water production in a low-energy district heating system having a temperature scheme of 55/25 °C for supply/return, electricity energy, and cooling, were described. The lifecycle cost analysis carried out was found to provide a reasonable basis for finding an optimal solution, rather than using an exergetic cost model taking account only of RE sources that are economically free to make use of. It can be noted that large energy savings together with low lifecycle costs without any appreciable emission release are possible through optimizing the RE sources available locally and integrating them with citywide distribution systems and utilizing the low-temperature operations available in a DH network. The assessment of locally available RE sources then together with the optimization of their nominal capacities which is required to meet the demands of the community can be seen as a must for any district. Borehole storage systems can be the saving element here in matching the varying heat production taking place with the varying heat requirements during the course of the year. Wind farms should be considered promising as a source of electricity due to their low exergetic costs for districts that have plain land areas or available neighboring offshore areas. The same applies to two additional technologies that are readily available, the one being the direct utilization of geothermal energy and the other being the free cooling, to meet energy requirements in terms of heating and cooling, respectively.

Various shortcomings of the method as described above should be taken account of in future work. The ideal energy-supply decision tool should take into consideration such matters as the following: (1) the existence of such multi-input multi-output integrated energy conversion systems as those considered in studies [29, 176]; (2) there being several useable outputs or inputs (not necessarily in the form of energy) pertaining to the geographical area involved that should be considered, such as the availability of clean water, hydrogen, and biogas, together with storage facilities for them [177]; (3) possibilities for hourly evaluation of the demand satisfaction in connection with the requirements taken up in [1]; and (4) site-specific considerations concerning the location of local RE sources, land area requirements of each energy technology, and geographical information regarding the area [25]. One should also note that there is a great variety of different RE sources other than what is considered in this study, such as tidal energy, wave energy, and several types of energy conversion systems, such as concentrated solar electricity-production plants, that could be of relevance [20].

Acknowledgements The authors gratefully acknowledge the support of the COWIfonden, Marie & MB Richters Fond, and Oticon Fonden, which provided the financial support for the external study carried out by the PhD Candidate Hakan İbrahim Tol at the UOIT, Oshawa, ON, Canada. We would also like to thank Henrik Lorentsen Bøgeskov for his guidance in accessing the data of the Danish district cooling network.

References

1. Lund H, Mathiesen BV (2009) Energy system analysis of 100 % renewable energy systems—the case of Denmark in years 2030 and 2050. Energy 34:524–531
2. Duguid B (2010) Ontario's long-term energy plan – building our clean energy future. Queen's Printer for Ontario. http://www.energy.gov.on.ca/docs/en/MEI_LTEP_en.pdf
3. Wiltshire R (2011) Low temperature district energy systems In: Proceedings of the urban energy conference, Debrecen, Hungary, 14–15 Oct, 2011
4. Foged M (1999) Low temperature operation, DBDH. http://www.dbdh.dk. Accessed 10 Jan 2011
5. Olsen PK, Lambertsen H, Hummelshøj R, Bøhm B, Christiansen CH, Svendsen S et al (2008) A new low-temperature district heating system for low-energy buildings. In: The 11th international symposium on district heating and cooling, Reykjavik, Iceland, 31 Aug to 2 Sept 2008
6. Paulsen O, Fan J, Furbo S, Thorsen JE (2008) Consumer unit for low energy district heating net. In: The 11th international symposium on district heating and cooling, Reykjavik, Iceland, 31 August to 2 Sept 2008
7. Torio H, Schmidt D (2011) Low exergy systems for high-performance buildings and communities. ECBCS – Annex 49

8. Christiansen CH, Worm J, Jørgensen H, Thorsen JE, Bennetsen J, Larsen CT et al (2011) Demonstration of low energy district heating system for low energy building in ringgårdens Afd. 34 in Lystrup (Demonstration af lavenergifjernvarme til lavenergibyggeri i boligforeningen ringgårdens afd. 34 i Lystrup). DEA, Århus, Denmark
9. Thorsen JE, Christiansen CH, Brand M, Olesen PK, Larsen CT (2011) Experiences on low-temperature district heating in Lystrup – Denmark. In: Proceedings of the international conference on district energy international conference on district heating and cooling, Portorož, Slovenia, 20–22 Mar 2011
10. Christiansen CH, Paulsen O, Bøhm B, Thorsen JE, Ting Larsen C, Jepsen BK et al (2009) Technical report for project development and demonstration of low-energy district heating for low energy buildings (Udvikling og demonstration af lavenergi-fjernvarme til lavenergibyggeri. Hovedrapport og bilag). Teknologisk Institut. Copenhagen, Denmark
11. Sibbitt B, Onno T, McClenahan D, Thronton J, Brunger A, Kokko J et al (2007) The Drake Landing Solar Community project – early results. In: Canadian solar buildings conference, Canmet Energy Technology Centre, Calgary, 10–14 June 2007
12. Mertoglu O, Bakir N, Kaya T (2003) Geothermal applications in Turkey. Geologija 32:419–428
13. Olsen PK, Christiansen CH, Hofmeister M, Svendsen S, Thorsen JE, Gudmundsson O et al (2014) Guidelines for low-temperature district heating. EUDP – DEA. Denmark
14. Schmidt D, Ala-Juusela M (2004) Low exergy systems for heating and cooling of buildings. In: PLEA, Eindhoven, The Netherlands, 23 Sept 2004
15. Hasan A, Kurnitski J, Jokiranta K (2009) A combined low temperature water heating system consisting of radiators and floor heating. Energ Buildings 41:470–479
16. Brand M, Thorsen JE, Svendsen S (2012) Numerical modelling and experimental measurements for a low-temperature district heating substation for instantaneous preparation of DHW with respect to service pipes. Energy 41:392–400
17. Tol Hİ, Svendsen S (2012) Improving the dimensioning of piping networks and network layouts in low-energy district heating systems connected to low-energy buildings: a case study in Roskilde, Denmark. Energy 38:276–290
18. Tol Hİ, Svendsen S (2012) A comparative study on substation types and network layouts in connection with low-energy district heating systems. Energ Conver Manage 64:551–561
19. Tol Hİ, Svendsen S (2013) The exergetic, environmental, and economic effect of the hydrostatic design static pressure level on the pipe dimensions of low-energy district heating network. Challenges 4:1–16
20. Dincer I, Zamfirescu C (2012) Renewable-energy-based multigeneration systems. Int J Energ Res 36:1403–1415
21. Lund H, Möller B, Mathiesen BV, Dyrelund A (2010) The role of district heating in future renewable energy systems. Energy 35:1381–1390
22. Sperling K, Möller B (2012) End-use energy savings and district heating expansion in a local renewable energy system – a short-term perspective. Appl Energ 92:831–842
23. Østergaard PA, Lund H (2011) A renewable energy system in Frederikshavn using low-temperature geothermal energy for district heating. Appl Energ 88:479–487
24. Mathiesen BV, Lund H, Connolly D (2012) Limiting biomass consumption for heating in 100 % renewable energy systems. Energy 48:160–168
25. Weber C, Shah N (2011) Optimisation based design of a district energy system for an eco-town in the United Kingdom. Energy 36:1292–1308
26. Niemi R, Mikkola J, Lund PD (2012) Urban energy systems with smart multi-carrier energy networks and renewable energy generation. Renew Energy 48:524–536
27. Ozlu S, Dincer I, Naterer GF (2012) Comparative assessment of residential energy options in Ontario, Canada. Energ Buildings 55:674–684
28. Chicco G, Mancarella P (2009) Distributed multi-generation: a comprehensive view. Renew Sustain Energy Rev 13:535–551

29. Dincer I, Zamfirescu C (2004) Integrated multigeneration energy systems. In: Dincer I, Zamfirescu C (eds) Sustainable energy systems and applications. Springer, New York
30. Kiani B, Hamamoto Y, Akisawa A, Kashiwagi T (2004) CO_2 mitigating effects by waste heat utilization from industry sector to metropolitan areas. Energy 29:2061–2075
31. The Mathworks Inc (2010) Matlab help manual optimization toolbox [disk]. Version 7.11.0.584 (R2010b). Matworks Inc, Natick, MA
32. Bejan A, Tsatsaronis G, Moran M (1996) Thermoeconomic optimization. In: Bejan A, Tsatsaronis G, Moran M (eds) Thermal design & optimization. Wiley, New York
33. Bøhm B (1988) Energy-economy of Danish district heating systems: a technical and economic analysis. Laboratory of Heating and Air Conditioning. Technical University of Denmark, Kongens Lyngby
34. Henning D (1997) MODEST—an energy-system optimisation model applicable to local utilities and countries. doi:10.1016/S0360-5442(97)00052-2
35. Werner S (2004) District heating and cooling. In: Cleveland CJ (ed) Encyclopedia of energy. Elsevier, New York
36. The Danish Energy Agency, Energinet.dk (2012) Technology data for energy plants – generation of electricity and district heating, energy storage and energy carrier generation and conversion. http://www.energinet.dk/SiteCollectionDocuments/Danske%20dokumenter/Forskning/Technology_data_for_energy_plants.pdf. Accessed 18 June 2012
37. Kaldellis JK (2007) The contribution of small hydro power stations to the electricity generation in Greece: technical and economic considerations. Energy Policy 35:2187–2196
38. Downey K (2011) Power plant construction costs: cost pressures returning. Englewood, CO. http://press.ihs.com/press-release/energy-power/power-plant-construction-costs-cost-pressures-returning. Accessed 23 July 2012
39. Dincer I, Rosen MA (2013) Exergy analysis of cogeneration and district energy systems. In: Dincer I, Rosen MA (eds) Exergy, 2nd edn. Elsevier, Amsterdam
40. Yang M, Patiño-Echeverri D, Yang F (2012) Wind power generation in China: understanding the mismatch between capacity and generation. Renew Energy 41:145–151
41. Croucher M (2011) Capacity factors and solar job creation. Energy Policy 39:6914–6915
42. Rosen MA, Le MN, Dincer I (2005) Efficiency analysis of a cogeneration and district energy system. Appl Therm Eng 25:147–159
43. Dincer I, Rosen MA (2013) Exergy and energy analyses. In: Dincer I, Rosen MA (eds) Exergy, 2nd edn. Elsevier, Amsterdam
44. Hepbasli A, Alsuhaibani Z (2011) Exergetic and exergoeconomic aspects of wind energy systems in achieving sustainable development. Renew Sustain Energy Rev 15:2810–2825
45. Brower M (1992) Cool energy: renewable solutions to environmental problems. MIT Press, Cambridge, MA
46. Rosen M (2009) Energy and environmental advantages of cogeneration with nuclear and coal electrical utilities. In: Proceedings of 4th IASME/WSEAS international conference on energy & environment, Cambridge, UK, 24–26 Feb 2009
47. CEN/CENELEC (2004) Manual for determination of combined heat and power (CHP). CWA 45547, Sept 2004
48. EIA (N/A) Emission factors and global warming potentials. http://www.eia.gov/survey/form/eia_1605/emission_factors.html. Accessed 18 June 2012
49. Hepbasli A (2008) A key review on exergetic analysis and assessment of renewable energy resources for a sustainable future. Renew Sustain Energy Rev 12:593–661
50. Butti K, Perlin J (1980) A golden thread: 2500 years of solar architecture and technology. Cheshire Books, Palo Alto, CA
51. Bosanac M, Sørensen B, Katic I, Sørensen H, Nielsen B, Badran J (2003) Photovoltaic/thermal solar collectors and their potential in Denmark. Final report. Danish Technological Institute, Solar Energy Centre, Novator Advanced Technology Consulting, Consulting Engineers Ltd, Copenhagen, Report no.: EFP project 1713/00-0014, p 114

52. Kalogirou SA (2004) Solar thermal collectors and applications. Prog Energ Combust Sci 30:231–295
53. Vela Solaris AG (2010) Polysun simulation software – user manual [disk]. Version 5.5.10.14394:123. Vela Solaris AG, Solartechnik Prüfung Forschung, Vinterthur
54. Petela R (2003) Exergy of undiluted thermal radiation. Sol Energ 74:469–488
55. Tripanagnostopoulos Y, Souliotis M, Battisti R, Corrado A (2005) Energy, cost and LCA results of PV and hybrid PV/T solar systems. Prog Photovolt Res Appl 13:235–250
56. Joshi AS, Dincer I, Reddy BV (2009) Thermodynamic assessment of photovoltaic systems. Sol Energ 83:1139–1149
57. Akyuz E, Coskun C, Oktay Z, Dincer I (2012) A novel approach for estimation of photovoltaic exergy efficiency. Energy 44:1059–1066
58. Masters GM (2004) Renewable and efficient electric power systems. Wiley, Hoboken, NJ
59. Gunerhan H, Hepbasli A (2007) Exergetic modeling and performance evaluation of solar water heating systems for building applications. Energ Buildings 39:509–516
60. Fan J, Chen Z, Furbo S, Perers B, Karlsson B (2009) Efficiency and lifetime of solar collectors for solar heating plants. In: ISES solar world congress, Johannesburg, South Africa, 11–14 Oct 2009
61. Dinçer İ, Zamfirescu C (2011) Sustainable energy systems and applications. Springer, New York
62. Dubey S, Tiwari GN (2009) Analysis of PV/T flat plate water collectors connected in series. Sol Energ 83:1485–1498
63. Charalambous PG, Maidment GG, Kalogirou SA, Yiakoumetti K (2007) Photovoltaic thermal (PV/T) collectors: a review. Appl Therm Eng 27:275–286
64. Tselepis S, Tripanagnostopoulos Y (2002) Economic analysis of hybrid photovoltaic/thermal solar systems and comparison with standard PV modules. In: PV in Europe – from PV technology to energy solutions, Rome, Italy, 7–11 Oct 2002
65. Burton T, Jenkins N, Sharpe D, Bossanyi E (2011) Wind turbine installations and wind farms. In: Jenkins N, Sharpe D, Bossanyi E, Burton T (eds) Wind energy handbook. Wiley, New York
66. Østergaard PA, Mathiesen BV, Möller B, Lund H (2010) A renewable energy scenario for Aalborg Municipality based on low-temperature geothermal heat, wind power and biomass. Energy 35:4892–4901
67. Gökçek M, Genç MS (2009) Evaluation of electricity generation and energy cost of wind energy conversion systems (WECSs) in Central Turkey. Appl Energ 86:2731–2739
68. Betz A (1926) Wind-energie und ihre ausnutzung durch windmühlen. Ökobuch Verlag, Göttingen
69. Uslu A (2009) Europe's onshore and offshore wind energy potential – an assessment of environmental and economic constraints. Final report. European Environment Agency (EEA), Luxembourg, Report no.: 6/2009, p 90
70. Ozgener O, Ozgener L (2007) Exergy and reliability analysis of wind turbine systems: a case study. Renew Sustain Energy Rev 11:1811–1826
71. Dincer I, Hepbasli A, Ozgener L (2007) Geothermal energy resources. In: Capehart BL (ed) Encyclopedia of energy engineering and technology – 3 volume Set (Print Version). Taylor and Francis, London
72. Lee KC (2001) Classification of geothermal resources by exergy. Geologija 30:431–442
73. Ozgener L, Hepbasli A, Dincer I (2006) Effect of reference state on the performance of energy and exergy evaluation of geothermal district heating systems: Balcova example. Build Environ 41:699–709
74. Jalilinasrabady S, Itoi R, Valdimarsson P, Saevarsdottir G, Fujii H (2012) Flash cycle optimization of Sabalan geothermal power plant employing exergy concept. Geologija 43:75–82
75. Lako P (2010) Geothermal heat and power. Final report. International Energy Agency, Paris, Energy Technology Systems Analysis Programme, 2010 May. Report no.: E07, p 6

76. Kanoğlu M, Çengel Y, Dinçer İ (2012) Efficiency evaluation of energy systems. Springer, New York
77. Heberle F, Brüggemann D (2010) Exergy based fluid selection for a geothermal Organic Rankine Cycle for combined heat and power generation. Appl Therm Eng 30:1326–1332
78. Chamorro CR, Mondéjar ME, Ramos R, Segovia JJ, Martín MC, Villamañán MA (2012) World geothermal power production status: energy, environmental and economic study of high enthalpy technologies. Energy 42:10–18
79. Yari M (2010) Exergetic analysis of various types of geothermal power plants. Renew Energy 35:112–121
80. Kanoğlu M, Çengel YA (1999) Economic evaluation of geothermal power generation, heating, and cooling. Energy 24:501–509
81. DiPippo R (2004) Second law assessment of binary plants generating power from low-temperature geothermal fluids. Geologija 33:565–586
82. Coskun C, Oktay Z, Dincer I (2011) Modified exergoeconomic modeling of geothermal power plants. Energy 36:6358–6366
83. Kanoglu M, Bolatturk A (2008) Performance and parametric investigation of a binary geothermal power plant by exergy. Renew Energy 33:2366–2374
84. Kanoglu M, Dincer I, Rosen MA (2007) Understanding energy and exergy efficiencies for improved energy management in power plants. Energy Policy 35:3967–3978
85. Franco A, Villani M (2009) Optimal design of binary cycle power plants for water-dominated, medium-temperature geothermal fields. Geologija 38:379–391
86. Luo C, Huang L, Gong Y, Ma W (2012) Thermodynamic comparison of different types of geothermal power plant systems and case studies in China. Renew Energy 48:155–160
87. DiPippo R (2008) Advanced geothermal energy conversion systems. In: DiPippo R (ed) Geothermal power plants – principles, applications, case studies and environmental impact, 2nd edn. Butterworth-Heinemann, Oxford
88. Kaya E, Zarrouk SJ, O'Sullivan MJ (2011) Reinjection in geothermal fields: a review of worldwide experience. Renew Sustain Energy Rev 15:47–68
89. Mago PJ, Fumo N, Chamra LM (2009) Performance analysis of CCHP and CHP systems operating following the thermal and electric load. Int J Energ Res 33:852–864
90. Kanoglu M, Dincer I (2009) Performance assessment of cogeneration plants. Energ Conver Manage 50:76–81
91. Kizilkan O, Dincer I (2012) Exergy analysis of borehole thermal energy storage system for building cooling applications. Energ Buildings 49:568–574
92. Chapuis S, Bernier M (2009) Seasonal storage of solar energy in borehole heat exchangers. In: 11th International IBPSA conference, Glasgow, UK, 27–30 July 2009
93. Faaij A (2006) Modern biomass conversion technologies. Mitig Adapt Strat Global Change 11:335–367
94. RISI (2013) European biomass review – new comprehensive analysis and forecast of the European bioenergy sector. http://www.risiinfo.com/risi-store/do/product/detail/European-Biomass-Review.html?source=D16J. Accessed 3 June 2012
95. Patzek TW, Pimentel D (2005) Thermodynamics of energy production from biomass. Crit Rev Plant Sci 24:327–364
96. Wolf O, Crank M, Patel M, Marscheider-Weidemann F, Schleich J, Hüsing B et al (2005) Techno-economic feasibility of large-scale production of bio-based polymers in Europe. Final report. Seville, Spain: European Commission – Joint Research Centre; 2005 May; Report no.: EUR 22103 EN, p 260
97. Münster M (2009) Energy systems analysis of waste to energy technologies by use of EnergyPLAN. Risø DTU – National Laboratory for Sustainable Energy, Risø, Denmark. Report no.: Risø-R-1667(EN)
98. Fruergaard T, Christensen TH, Astrup T (2010) Energy recovery from waste incineration: assessing the importance of district heating networks. Waste Manag 30:1264–1272

99. Werner S (2006) Possibilities with more district heating in Europe. Ecoheatcool. Report no.: Work Package 4. Brussels, Belgium
100. Murphy JD, McKeogh E (2004) Technical, economic and environmental analysis of energy production from municipal solid waste. Renew Energy 29:1043–1057
101. Saidur R, BoroumandJazi G, Mekhilef S, Mohammed H (2012) A review on exergy analysis of biomass based fuels. Renew Sustain Energy Rev 16:1217–1222
102. Young GC (2010) Municipal solid waste to energy conversion processes: economic, technical, and renewable comparisons. Wiley, Hoboken, NJ
103. Huang Y, Wang Y, Rezvani S, McIlveen-Wright D, Anderson M, Mondol J et al (2013) A techno-economic assessment of biomass fuelled trigeneration system integrated with organic Rankine cycle. Appl Therm Eng 53:325–331
104. Maraver D, Rezeau A, Sebastian F, Royo J (2009) Thermodynamic optimization of a trigeneration system based on biomass combustion. In: Proceedings of the 17th European biomass conference. Hamburg, Germany, 29 June to 3 July 2009
105. Bédard S (2009) Waste heat to power – a process integration perspective. In: ENERGY 2009 Lean and Green. Toronto, ON, 24–25 Nov 2009
106. Pulat E, Etemoglu AB, Can M (2009) Waste-heat recovery potential in Turkish textile industry: case study for city of Bursa. Renew Sustain Energy Rev 13:663–672
107. Johnson I, Choate WT (2008) Waste heat recovery: technology and opportunities in U.S. industry. Report no.: (N/A). BCS, Incorporated, Laurel, MD
108. Punte S, Repinski P (2006) Thermal energy equipment: waste heat recovery. In: Töpfer K (ed) Energy efficiency guide for industry in Asia. United Nations Environment Programme, Bangkok
109. Gabrielii C (2002) The potential for heat pumps utilizing industrial waste heat (Potentialen för värmepumpar som utnyttjar industriell spillvärme). Report no.: N/A. CPL, Chalmers Publication Library, Göteborg
110. Samuelsson H (2009) SETatWork Goog practice – waste heat from industry. SETatWork report no.: N/A
111. Land AL, Feldhusen H, Tvärne A, Cronholm L, Sundlöf C (2002) Industrial waste heat – processes and potentials (Industriell spillvärme – processor och potentialer). The Swedish District Heating Association. Stockholm, Sweden. Report no.: FVF 02 11 49
112. Baxter VD (2003) Advanced supermarket refrigeration/heat recovery systems. Report no: IEA Annex 26. Oak Ridge National Laboratory, Oak Ridge, TN
113. EUREKA Wärmerückgewinnung und Kühltechnik GmbH & Co. KG (2002) Heat recovery http://www.eureka-emsdetten.de/2-1-heat-recovery.html. Accessed 16 Sept 2012
114. York by Johnson Controls (2010) Optimize your facility's energy utilization with free heat. http://www.johnsoncontrols.com/content/dam/WWW/jci/be/integrated_hvac_systems/hvac_equipment/chiller_products/centrifugal-packaged/PUBL-6317_(510).pdf. Accessed 16 Sept 2012
115. Nieuwlaar E, Dijk D (1993) Exergy evaluation of space-heating options. Energy 18:779–790
116. Bailer P, Pietrucha U (2004) District heating and district cooling with large centrifugal chiller – heat pumps. In: The 9th international symposium on district heating and cooling, Espoo, Finland, 30–31 Aug 2004
117. Chatterjee B, Howitt RE, Sexton RJ (1998) The optimal joint provision of water for irrigation and hydropower. J Environ Econ Manag 36:295–313
118. Rosen MA, Scott DS (2003) Entropy production and exergy destruction: part I—hierarchy of earth's major constituencies. Int J Hydrogen Energy 28:1307–1313
119. Ertesvåg IS, Mielnik M (2000) Exergy analysis of the Norwegian society. Energy 25:957–973
120. Rogner H (1993) Clean energy services without pain: district energy systems. Energ Stud Rev 5:114–120
121. Københavns Energi A/S (2011) KE Fjernkøling A/S http://www.ke.dk/portal/page/portal/Fjernkoeling/ke_fjernkoeling_as?page=970. Accessed 19 Sept 2012

122. Ritola J, Ala-Juusela M (2004) Ground source cooling. Report no.: 61-02. VTT Building and Transport, Espoo
123. Skagestad B, Mildenstein P (2002) District heating and cooling connection handbook. Report No.: Annex VI. IEA, Paris
124. Looney CM, Oney SK (2007) Seawater district cooling and lake source district cooling. Energ Eng 104:34–45
125. Dalin P, Ivancic A, Tvärne A, Penthor A, Martin B, Ricci F et al (2006) District cooling – cooling more with less. Report no.: N/A. Euroheat & Power, Brussels
126. de Jager D, Klessmann C, Stricker E, Winkel T, de Visser E, Koper M et al (2011) Financing renewable energy in the European energy market. Report no: PECPNL084659. ECOFYS, Amsterdam
127. Hearps P, McConnell D (2011) Renewable energy cost review Report no: N/A. Melbourne Energy Institute, Melbourne, VIC
128. Sherwani AF, Usmani JA, Varun S (2010) Life cycle assessment of solar PV based electricity generation systems: a review. Renew Sustain Energy Rev 14:540–544
129. Fisch MN, Guigas M, Dalenbäck JO (1998) A review of large-scale solar heating systems in Europe. Sol Energ 63:355–366
130. Zhivov A, Woody A, Walker A, Croy R, Richter S, Meißner R et al (2011) Central solar hot water systems design guide. Report no: N/A. WBDG, Washington, DC
131. Wiser R, Bolinger M (2010) 2009 wind technologies market report. Report no: N/A. US Department of Energy – Energy efficiency & Renewable Energy, Washington, DC
132. Ortegon K, Nies LF, Sutherland JW (2013) Preparing for end of service life of wind turbines. J Clean Prod 39:191–199
133. Ertürk M (2012) The evaluation of feed-in tariff regulation of Turkey for onshore wind energy based on the economic analysis. Energy Policy 45:359–367
134. Ucar A, Inalli M (2008) Thermal and economic comparisons of solar heating systems with seasonal storage used in building heating. Renew Energy 33:2532–2539
135. Golabi K, Scherer CR, Tsang CF, Mozumder S (1981) Optimal energy extraction from a hot water geothermal reservoir. Water Resour Res 17:1–10
136. Rafferty K (1996) Selected cost considerations for geothermal district heating in existing single-family residential areas. GeoHeat tp 93:1–28
137. Erdogmus B, Toksoy M, Ozerdem B, Aksoy N (2006) Economic assessment of geothermal district heating systems: a case study of Balcova–Narlidere, Turkey. Energ Buildings 38:1053–1059
138. Wang W (2000) The opportunity and risk of geothermal heating investment in China. In: Proceedings world geothermal congress, Beppu, Morioka, Japan, 28 May to 10 June 2000
139. Ozdenefe M, Atikol U (2010) Applicability of trigeneration systems in the hotel buildings: the North Cyprus case. WSEAS Trans Power Syst 5:98–107
140. Pöttler R (N/A) Energy from hydropower – Innovative and regenerative. ILF Beratende Ingenieure ZT Gesellschaft mbH, Innsbruck, Austria. Report no.: N/A
141. Mittal G (2004) Real options approach to capacity planning under uncertainty. Dissertation, Massachusetts Institute of Technology
142. Kaldellis JK, Vlachou DS, Korbakis G (2005) Techno-economic evaluation of small hydro power plants in Greece: a complete sensitivity analysis. Energy Policy 33:1969–1985
143. Udomsri S, Martin AR, Martin V (2011) Thermally driven cooling coupled with municipal solid waste-fired power plant: application of combined heat, cooling and power in tropical urban areas. Appl Energ 88:1532–1542
144. Svensson I, Moshfegh B (2011) System analysis in a European perspective of new industrial cooling supply in a CHP system. Appl Energ 88:5164–5172
145. Murugavel V, Saravanan R (2010) Life cycle cost analysis of waste heat operated absorption cooling systems for building HVAC applications. In: Proceedings of the 10th international conference enhanced building operations, Kuwait, 26–28 Oct 2010

146. Energy and Waste Management Office (2012) Metro Hall – deep lake water cooling. Toronto, ON. Report no.: (N/A)
147. Lund H, Morten BB (2007) Large-scale heat pumps in sustainable energy systems: system and project perspectives. Therm Sci 11:143–152
148. Rivas F, Khire RA, Messac A, Van Dessel S (2006) Life cycle cost based economic assessment of active building envelope (ABE) systems. In: 2nd AIAA multidisciplinary design optimization specialist conference, Newport, Rhode Island, 1–4 May 2006
149. NRCAN (2002) Heat recovery from kitchen refrigeration Report no.: M92-242/2002-6E. NRCAN, Ottawa, ON
150. Sanyal SK (2004) Cost of geothermal power and factors that affect it. In: Twenty-ninth workshop on geothermal reservoir engineering, Stanford, CA, 26–28 Jan 2004
151. Tester JW, Anderson BJ, Batchelor AS, Blackwell DD, DiPippo R, Drake EM et al (2007) Impact of enhanced geothermal systems on US energy supply in the twenty-first century. Phil Trans R Soc A 365:1057–1094
152. European Geothermal Energy Council (EGEC) (N/A) The future of geothermal development – presentation. http://www.iea-gia.org/documents/Antics-EGEC-THEFUTUREOFGEOTHERMALDEVELOPMENT.pdf. Accessed 12 July 2012
153. INL (2006) The future of geothermal energy. Report no: INL/EXT-06-11746. INL, Idaho Falls, ID
154. Entingh DJ, Easwaran E, McLarty L (1994) Small geothermal electric systems for remote powering. The Geothermal Division – US Department of Energy, US. Report no.: GTSMAL-3.GTD
155. Brugman J, Hattar M, Nichols K, Esaki Y (1994) Next generation geothermal power plants. Report no: EPRI RP 3657-01. EPRI, Pasadena, CA
156. Lund JW, Lienau PJ (2009) Geothermal district heating. In: International geothermal days, conference & summer school, Slovakia, 26–29 May 2009
157. Piatti A, Piemonte C, Szegö E (1992) Planning of geothermal district heating systems. Kluwer Academic Publisher, Dordrecht
158. Durmayaz A (2000) An application of the degree-hours method to estimate the residential heating energy requirement and fuel consumption in Istanbul. Energy 25:1245–1256
159. Elsman P (2009) Copenhagen district heating system. Report no: N/A. Københavns E, Copenhagen
160. ENeRGI STYRELSEN (2010) Energy statistic 2010 (Energistatistik 2010). Report no.: (N/A). ENeRGI STYRELSEN, Copenhagen
161. CTR, Københanvs E, and VEKS (2009) Data for teknologier til produktion af varme (data of technologies for the production of heat). Copenhagen. Report no.: N/A
162. NRCAN (2011) Monthly market update. http://www.nrcan.gc.ca/energy/sources/natural-gas/monthly-market-update/2258. Accessed 9 Nov 2012
163. OECD (2012) Economic outlook. http://stats.oecd.org/. Accessed 15 June 2012
164. Energy Efficiency Office (2007) Toronto's sustainable energy plan. Report no.: N/A. Energy Efficiency Office, Toronto, ON
165. Laursen EV, Thomsen RS, Cappelen J (1999) Observed air temperature, humidity, pressure, cloud cover and weather in Denmark – with climatological standard normals, 1961–90. Report no.: 99-5. Danish Meteorological Institute, Copenhagen
166. NREL (2012) International wind resource maps [map]. NREL, Washington, DC
167. NREL (2011) System advisor model – version 2.3.1 [disk]. NREL, Washington, DC
168. The Weather Network (2012) Statistics: Toronto Lester B. Pearson Int'l, Mississauga, ON
169. The International Renewable Energy Agency (IRENA) (2012) Renewable energy technologies: cost analysis series. Report no: issue 5/5. The International Renewable Energy Agency, Abu Dhabi
170. Nikolaisen L (2012) Country report 2011 for Denmark. Report no.: IEA bioenergy task 40. Danish Technological Institute

171. Oo A, Kelly J, Lalonde C (2012) Assessment of business case for purpose-grown biomass in Ontario. Report no: (N/A). The Western University Research Park, London
172. Mahler A, Magtengaard J (2005) Geothermal development in Denmark, Country update WGC 2005. In: Proceedings world geothermal congress, Antalya, Turkey, 24–29 April 2005
173. Magtengaard J, Mahler A (2010) Geothermal reserves and sustainability in the Greater Copenhagen Area. In: World geothermal congress, Bali, Indonesia, 25–30 April 2010
174. Bristow D (2009) The business case for alternative energy technologies in Ontario. Dissertation, University of Toronto
175. Tol Hİ, Dincer I, Svendsen S (2013) Regional energy planning tool for renewable integrated low-energy district heating systems: environmental assessment. In: Dincer I, Colpan CO, Kadioglu F (eds) Causes, impacts and solutions to global warming, 1st edn. Springer, New York
176. Ahmadi P, Rosen MA, Dincer I (2012) Multi-objective exergy-based optimization of a polygeneration energy system using an evolutionary algorithm. Energy 46:21–31
177. Ozbilen A, Dincer I, Naterer GF, Aydin M (2012) Role of hydrogen storage in renewable energy management for Ontario. Int J Hydrogen Energy 37:7343–7354

Chapter 54
Thermal Modeling of a Modified Basin-Type Solar Still

Ayman G.M. Ibrahim, Elsayed E. Allam, and Salman E. Elshamarka

Abstract The mathematical model of a modified solar still that operates under vacuum condition and is equipped with an air-cooled condenser is developed. Simulation results for the behavior of such a system under the meteorological conditions of Cairo, Egypt, are presented. In addition, the proposed system is designed and constructed. The experimental validation of simulation results is carried out. Satisfactory matching between simulation results and experimental measurements is obtained. Performance investigation of the proposed system during a whole year is presented and conclusions are drawn.

Keywords Desalination • Solar energy • Solar still • Productivity • Efficiency

Nomenclature

A	Area (m^2)
C	Solution concentration (%); specific heat (J/kg °C)
cl	Constant
D	Diameter (m)
d_w	Depth of saline water (m)
g	Gravitational acceleration (9.81 m/s^2)
h	Convection heat transfer coefficient (W/m^2 K)
H	Length of fin (m)
$\overline{h}_c$	Average film heat transfer coefficient (W/m^2 K)
h_{fg}	Latent heat of vaporization (J/kg)
I	Incident solar radiation (W/m^2)
I_o	Solar constant (1,353 W/m^2)
k	Thermal conductivity (W/m K)
l	Length (m)

A.G.M. Ibrahim (✉) • E.E. Allam • S.E. Elshamarka
Department of Mechanical Power and Energy, Military Technical College, Cairo, Egypt
e-mail: agmibrahim@hotmail.com

I. Dincer et al. (eds.), *Progress in Clean Energy, Volume 2*,
DOI 10.1007/978-3-319-17031-2_54

m	Mass (kg)
n	Day of the year (1–365)
N	Number of fins
Nu	Nusselt number (dimensionless)
P	Vapor pressure (N/m^2)
Pr	Prandtl number (dimensionless)
$\dot{Q}$	Heat transfer rate (W)
R	Universal gas constant (8.3145 kJ/kmol K)
r	Radius (m)
Ra	Rayleigh number (dimensionless)
Re	Reynolds number (dimensionless)
S	Distance between two successive fins (m)
t	Thickness (m); time (s)
T	Temperature (°C)
w	Average wind speed (m/s)

Greek Letters

α	Absorptivity (dimensionless); thermal diffusivity (m^2/s)
α_m	An experimental coefficient (10^{-7}–10^{-6} kg/m^2 Pa s K$^{-0.5}$)
β	Slope (°) inclined surface; thermal expansion coefficient (K^{-1})
γ	Surface azimuth angle (°); kinematic viscosity (m^2/s)
δ	Solar declination angle (°)
ε	Emissivity (dimensionless)
η	Efficiency (%)
θ	Solar incidence angle (°)
ϑ	Half angle between the center of the tube and the edge of the liquid at the bottom (°)
λ	Constant
$\dot{\nu}_e$	Evaporation rate per unit area (m^3/s m^2)
ξ	Ratio of diameters (dimensionless)
ρ	Reflectivity (dimensionless); density (kg/m^3)
σ	Stefan-Boltzmann constant (5.67×10^{-8} W/m^2 K^4)
τ	Transmissivity (dimensionless)
φ	Latitude (°)
ω	Hour angle (°)

Subscripts

a	Ambient; air
b	Direct beam solar radiation; base, bare
c	Condenser

c_i	Condenser inner surface
c_o	Condenser outer surface
d	Diffuse solar radiation
e	Evaporation
eff	Efficiency; effective
exp	Experimental
f	Freshwater
g	Glass cover
h	Hour
i	Inside
i_f	Interface of film condensation
in	Input
l	Liquid
o	Reference state; outside
r	Reflected solar radiation
s	Saline water; side
sat	Saturated conditions
sb	Surface of the bottom wall of solar still
sim	Simulation
ss	Surface of the side wall of solar still
t	Total
v	Vapor
w	Water

54.1 Introduction

There is an increasing demand of freshwater due to population growth and rapid increase of agricultural and industrial activities. One way of meeting this demand is desalination of the saline water. There are several desalination methods like reverse osmosis, electrodialysis, vapor compression, multistage flash distillation, and multiple-effect distillation. However, these processes usually consume significant amount of energy. Many countries in the world, particularly those suffering from severe water shortages, cannot afford the energy required for desalination. Fortunately, many of those countries lie in areas with high solar insolation rates. Solar desalination can be a suitable choice to utilize the solar energy in a cost-effective way.

The most widely used solar desalination system is a simple solar still, where the heat collection and distillation processes take place in the same equipment. Conventional solar stills use a relatively simple technique to produce a small amount of freshwater based on the greenhouse effect using solar energy. They are easy to build and simple to operate and maintain. However, the main disadvantage of the solar still is its low efficiency and daily freshwater production [1]. Many research investigations explored new construction materials or new design and operational concepts that are extensively reviewed elsewhere [2–4].

Evaporation at a low temperature under vacuum conditions leads to a good improvement in the system efficiency as energy losses are reduced when water is evaporated at lower temperature and with less energy requirements. In the presence of vacuum the effect of noncondensable gases which reduce the rate of condensation is also avoided. Productivity from solar desalination systems operating under subatmospheric pressure was investigated theoretically by Al-Hussaini and Smith [5]. The theoretical results indicated that an increase of productivity could be 100 %. Kabeel et al. [6] carried out an experimental modification into the conventional solar still using a vacuum fan operated by the photovoltaic solar panels. The obtained results showed a considerable increase of distillate water productivity.

Desalination system operating at reduced pressure using natural forces of gravity and atmospheric pressure was first proposed by Al-Kharabsheh et al. [7, 8]. This new concept has the advantages of vacuum distillation without requiring additional energy to create the vacuum. An electric water heater was used to supply the hot water at 60 °C, assuming that in real life the unit will be supplied with its energy requirements from a solar collector. The experimental results showed that the system efficiency may reach 80 %.

Gude et al. [9] used a 12-V DC heater powered by grid power to drive desalination system operating at low evaporation pressure. Results of this study showed that efficiency of 69 % can be sustained at evaporation temperatures of 40 °C. Further, they investigated a modified system driven solely by solar energy without any reliance on grid power [10, 11]. The average cumulative production over 24 h of 4.95 L/day/m^2 was obtained during tests. However, the plotted results for daylight operation indicated a distillate production of 1 L/day/m^2. With the addition of a photovoltaic panel as a source to drive the proposed desalination system, efficiencies range from 65 to 90 %.

Natural vacuum desalination (NVD) system powered by renewable energy was studied by Ayhan et al. [12]. A pilot unit of the NVD system was built to demonstrate the working principle. System tests were performed to estimate the production rate of distilled water on the basis of the climatic data of Bahrain.

In Egypt, there is great demand for potable water due to population and improvement in standard of living. Egypt has already exhausted her fixed Nile water share, and the groundwater requires expensive energy to abstract more and costs to transport. Horizontal expansion, especially in remote areas for either agriculture or other purposes, is a must for many social, economic, environmental, strategic, and other reasons. Desalination might be the proper way of water supply in arid area expansions. However, desalination uses significant amount of energy, today mostly from fossil fuels. It is, therefore, reasonable to rely on renewable energy sources such as solar energy. Egypt is one of the Sunbelt countries. In 1991, a solar atlas for Egypt was issued, concluding that the yearly direct solar radiation intensity ranges between 1,900 kW h/m^2 in the north and 2,800 kW h/m^2 in the south and the sunshine duration ranges between 9 and 11 h/day, with very few cloudy days [13]. Under these situations, solar stills are highlighted for low-capacity freshwater demand, less than 200 m^3/day, since other desalination plants are uneconomical [14, 15].

This study investigates solar desalination under subatmospheric conditions, where a vacuum is created using vacuum pump. Modification to the conventional solar still is proposed by adding a separate air-cooled condenser, whereby solar energy is used directly to produce distillate in the solar basin. The thermal model of this system is developed along with experimental validation under the meteorological conditions of Cairo, Egypt. Further, the monthly variation of system performance is predicted.

54.2 System Description

The proposed system consists of an evaporation chamber with batch-wise operation mode which receives solar energy directly from the sun transmitted through glass cover. The evaporation chamber is equipped with purge connection to a vacuum pump. An air-cooled condenser is connected to the evaporation chamber. The freshwater tank collects the condensate from both the condensate receiver at the end of the condenser and the glass cover of the evaporator chamber. A schematic arrangement of the proposed system is shown in Fig. 54.1.

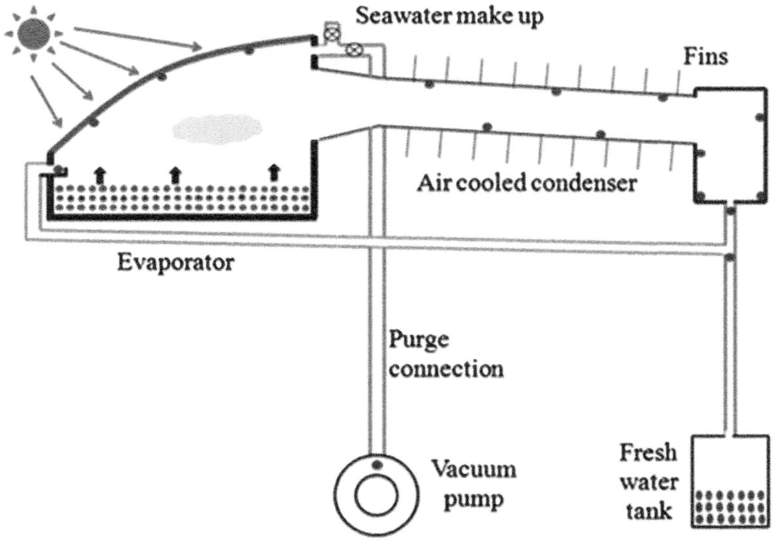

Fig. 54.1 Schematic arrangement of the proposed system

54.3 Model Development

A complete mathematical description of the system requires continuity and energy equation balances. In applying those balances, these assumptions are used:

- Hourly steady-state conditions are prevailed.
- Uniform temperature is prevailed throughout the elements of the solar still.
- Heat capacities of evaporator and condenser are neglected.
- Glass cover curvature has no effect on solar still inclination angle.
- Heat transfer through the piping system is neglected.
- Input energy required to drive the vacuum pump is neglected.

Under these assumptions the system of equations describing the physical model of desalination system is described below.

54.3.1 Calculation of Solar Radiation

To start desalination, a source of heat is required, which goes to preheat the feed, evaporate the water, and compensate for the heat losses. This heat is supplied by solar energy which is the sum of hourly terrestrial direct normal and diffuse solar radiation on inclined glass surface.

The total solar irradiance incident on a tilted surface with angle β is calculated as

$$I_{\mathrm{T}} = I_{\mathrm{b}}R_{\mathrm{b}} + I_{\mathrm{d}}R_{\mathrm{d}} + R_{\mathrm{r}}(I_{\mathrm{b}} + I_{\mathrm{d}}) \tag{54.1}$$

where $R_{\mathrm{b}} = \cos\theta/\cos\theta_{\mathrm{z}}$, $R_{\mathrm{d}} = (1 + \cos\beta)/2$, and $R_{\mathrm{r}} = \rho(1 - \cos\beta)/2$

The angle of incidence of beam radiation, θ, is related to other angles by relation

$$\begin{aligned}\cos\ \theta = {} & \sin\varphi(\sin\delta\cos\beta + \cos\delta\cos\gamma\cos\omega\sin\beta)\\ & + \cos\varphi(\cos\delta\cos\omega\cos\beta - \sin\delta\sin\beta\cos\gamma)\\ & + \cos\delta\sin\beta\sin\gamma\sin\omega\end{aligned} \tag{54.2}$$

The zenith angle, θ_{z}, can be calculated by

$$\cos\theta_{\mathrm{z}} = \sin\varphi\sin\delta + \cos\varphi\cos\delta\cos\omega \tag{54.3}$$

The hourly extraterrestrial solar radiation incident on a horizontal plane is given by [16]

$$\begin{aligned}I_{\mathrm{o}} = {} & \frac{24 \times 3,600}{2\pi} I_{\mathrm{sc}}\left(1 + 0.033\cos\frac{360\ \mathrm{n}}{365}\right)\\ & \left[\frac{2\pi(\omega_2 - \omega_1)}{360}\sin\delta\sin\varphi + \cos\delta\cos\varphi(\sin\omega_2 - \sin\omega_1)\right]\end{aligned} \tag{54.4}$$

where

I_{sc} solar constant is the heat energy received by a surface perpendicular to the sun's rays outside the atmosphere (1,353 W/m^2), n Julian day is the day of the year with range from (1 to 365), and ω_1 and ω_2 are any two successive hour angles.

The direct solar radiation on horizontal surface, I_b, is given by

$$I_b = I_o \tau_b \quad (54.5)$$

where the atmospheric transmittance for direct radiation, τ_b, is calculated according to Hottel [17].

The diffuse solar radiation on horizontal surface is

$$I_d = I_b \tau_d \quad (54.6)$$

Liu and Jordan [18] developed an empirical relationship between the transmission coefficient for direct and diffuse radiation for clear days as

$$\tau_d = 0.2710 - 0.2939\ \tau_b \quad (54.7)$$

54.3.2 Evaporation Chamber

The net rate of mass transfer at the phase interface which is the sum of the rates of vaporization and condensation can be calculated based on Bemporad correlation [19], for evaporation rate between two chambers, first containing seawater and the other containing freshwater, which are connected via a vacuum chamber. The evaporation rate per unit area, $\dot{\nu}_{e\text{-}c}$, from seawater chamber (evaporator) to freshwater chamber (condenser) is given as (assuming that evaporation is not impeded by foreign gas molecules)

$$\dot{\nu}_{e\text{-}c} \left(\frac{\alpha_m}{\rho_f}\right) \left(f(c)\frac{P_{T_w}}{(T_w + 273)^{0.5}} - \frac{P_{T_f}}{(T_f + 273)^{0.5}}\right) \quad (54.8)$$

where $f(c)$ is a correction factor to account for the presence of a solute concentration and is given as

$$f(c) = 1 - 0.0054 \times c \quad (54.9)$$

The evaporation rate per unit area, $\dot{\nu}_{e-g}$, on the glass cover is given by

$$\dot{\nu}_{e-g} = \left(\frac{\alpha_m}{\rho_f}\right) \left(f(c)\frac{P_{T_w}}{(T_w + 273)^{0.5}} - \frac{P_{T_{gi}}}{(T_{gi} + 273)^{0.5}}\right) \quad (54.10)$$

The heat losses from the bottom and side walls of the evaporator are calculated by

$$\dot{Q}_b = h_b A_b (T_{sb} - T_a) \tag{54.11}$$

$$\dot{Q}_s = h_s A_s (T_{ss} - T_a) \tag{54.12}$$

where the convective heat transfer coefficients at the bottom h_b and side walls h_s are calculated using the following [20]:

$$\mathrm{Nu}_b = 0.27\ \mathrm{Ra}_b{}^{0.25} \tag{54.13}$$

$$\mathrm{Nu}_s = 0.68 + \left(0.67\ \mathrm{Ra}_s{}^{0.25}\right) / \left(1 + (0.492/\mathrm{Pr}_s)^{9/16}\right)^{4/9} \tag{54.14}$$

The heat interaction at the glass cover includes the reflected fraction of solar radiation and heat transfers by convection and radiation from glass cover to ambient. The total heat dissipation from the glass cover is estimated as follows:

$$\dot{Q}_g = \left\{\varepsilon_g \sigma \left((T_{go} + 273)^4 - (T_{sky} + 273)^4\right) + h_o (T_{go} - T_a) + \rho_g I_T \right\} A_g \tag{54.15}$$

where

The outside convection heat transfer coefficient is calculated as [16]

$$h_o = 2.8 + 3\,w \tag{54.16}$$

and the sky temperature is calculated as [21]

$$T_{sky} = \left(0.0559 \times (T_a + 273)^{1.5}\right) - 273 \tag{54.17}$$

Eventually, the energy balance equations within the evaporation chamber are as follows:

- For glass cover:

$$\alpha_g A_g I_T + \dot{m}_{fg}\left(h_{fg} + C_{wg}(T_w - T_{gi})\right) + \varepsilon_{eff}\,\sigma A_b \left((T_w + 273)^4 - (T_{gi} + 273)^4\right) = \varepsilon_g \sigma A_g \left((T_{go} + 273)^4 - (T_{sky} + 273)^4\right) + h_o A_g (T_{go} - T_a) + m_g C_g \Delta T_g / \Delta t \tag{54.18}$$

where

$$\frac{k_g A_g (T_{gi} - T_{go})}{t_g} = \left\{ \left\{ \varepsilon_g \sigma \left((T_{go} + 273)^4 - (T_{sky} + 273)^4 \right) + h_o (T_{go} - T_a) \right\} A_g \right. \tag{54.19}$$

- For saline water:

$$h_w A_b (T_b - T_w) + \frac{k_w A_b (T_b - T_w)}{d_w} + \varepsilon_{eff} \sigma A_b \left((T_b + 273)^4 - (T_w + 273)^4 \right) = m_w C_w \Delta T_w / \Delta t + \dot{m}_{ft} h_{fg} + \varepsilon_{eff} \sigma A_b \left((T_w + 273)^4 - (T_{gi} + 273)^4 \right) \tag{54.20}$$

and convective heat transfer coefficient of saline water, h_w, is calculated using [16]

$$\mathrm{Nu}_w = 0.54 \mathrm{Ra}_w^{0.25} \tag{54.21}$$

- For black base:

$$\alpha_b \tau_g \tau_w A_b I_T = Q_b + h_w A_b (T_b - T_w) + \frac{k_w A_b (T_b - T_w)}{d_w} + \varepsilon_{eff} \sigma A_b \left((T_b + 273)^4 - (T_w + 273)^4 \right) \tag{54.22}$$

54.3.3 Condenser

Evaporation of water from saline water chamber tends to cool it, while condensation of vapor tends to heat freshwater. For the process of distillation to be continuous, heat is to be added continuously to the evaporator and rejected from freshwater chamber. The condensation heat transfer will mainly be dissipated to the environment via the condenser. So, the condenser should be able to dissipate the amount of energy given by [20]

$$\dot{Q}_C = \dot{m}_{fc} \left(h_{fg} + 0.68 C_w (T_w - T_f) \right) \tag{54.23}$$

where the second term in right-hand side accounts for the sensible heat transfer from the condensate. This amount of heat is transferred through the condensate film, then conducted through the condenser wall, and eventually transferred to the environment by convection only. The heat transfer by radiation is neglected due to small temperature difference.

The condensation heat transfer coefficient is expected to be very large compared to the free convection heat transfer coefficient from the outside surface of the condenser. Since the velocity of vapor is small, and the condensation rate is low, the condensate will flow as a thin annular film inside the tube. Then it flows in a longitudinal direction along the bottom side of the tube. For flow with (Re < 30,000) the average film heat transfer coefficient is given by [22]

$$\bar{h}_c = 0.728\left[\frac{g h_{fg}\, k_c^3\, (\rho_l - \rho_v)}{\gamma_l\, D_{ci}(T_w - T_{if})}\right]^{1/4} \tag{54.24}$$

Film properties involved are to be evaluated at an intermediate temperature between the interface, T_i, and the inside surface of the condenser, T_{ci}.

For heat conduction through the condenser wall:

$$\dot{Q}_C = \frac{2\pi l_c k_c (T_{ci} - T_{co})}{ln(r_{co}/r_{ci})} \tag{54.25}$$

The condenser is assumed to be a horizontal tube with squared fins. The heat transfer from the cylinder and lateral fin surfaces is given by [22]

$$\mathrm{Nu} = \frac{\mathrm{Ra_s}}{12\pi}\left\{2 - exp\left[-\left(\frac{\mathrm{cl}}{\mathrm{Ra}_s}\right)^{3/4}\right] - exp\left[-\lambda\left(\frac{\mathrm{cl}}{\mathrm{Ra_s}}\right)^{3/4}\right]\right\} \tag{54.26}$$

where $\mathrm{cl} = \left[23.7 - 1.1\frac{\left(1+152\xi^2\right)^{1/2}}{1+\lambda}\right]^{4/3}$, $\lambda = 0.17\,\xi + exp(-4.8\,\xi)$, $\xi = \frac{D_{co}}{D_{fin}}$ and $D_{fin} = 1.23\,H$.

Rayleigh number is given by

$$\mathrm{Ra_s} = \frac{g\beta(T_{co} - T_a)S^3}{\alpha\gamma}\frac{S}{D_{fin}} \tag{54.27}$$

The rate of heat transferred from the condenser can be calculated as

$$\dot{Q}_c = [h_{co} N A_{fin,\,sides}\eta_{fin} + h_{co}A_b](T_{co} - T_a) \tag{54.28}$$

where η_{fin} is calculated as reported in [23].

54.3.4 Performance Indices

The performance of the desalination system will be assessed based on the total productivity of distillate water and the thermal efficiency which are defined as

$$\dot{m}_{ft} = \dot{m}_{fc} + \dot{m}_{fg} \tag{54.29}$$

$$\eta = \frac{\dot{m}_{\text{ft}}\, h_{\text{fg}}}{I_T\, A_{\text{g}}} \tag{54.30}$$

54.4 Experimental Setup

For conducting the experiment, an experimental setup is designed and constructed. Figure 54.2 shows the schematic diagram of the experimental setup. Two experiments are carried out during August 2011 and April 2012. The main components of experimental setup are explained in the following.

54.4.1 Evaporation Chamber

The evaporator is a square in shape of 0.25 m^2 cross-sectional area and 0.4 m height at one side and is sloped with angle 30° to be 0.1 m height at the other side. At the side of 0.1 m height, a 0.01 m lip is fixed at a height of 0.08 m from the bottom of the evaporator to collect any condensate, which results from vapor condensation on the glass cover before reaching the condenser. The lip has a slop sufficient for the condensate to flow to a common point, which connected to the condensate receiver tank. The evaporator internal side walls and base are painted black. The evaporator has provision for supply and removal of saline water and brine through 0.0127 m

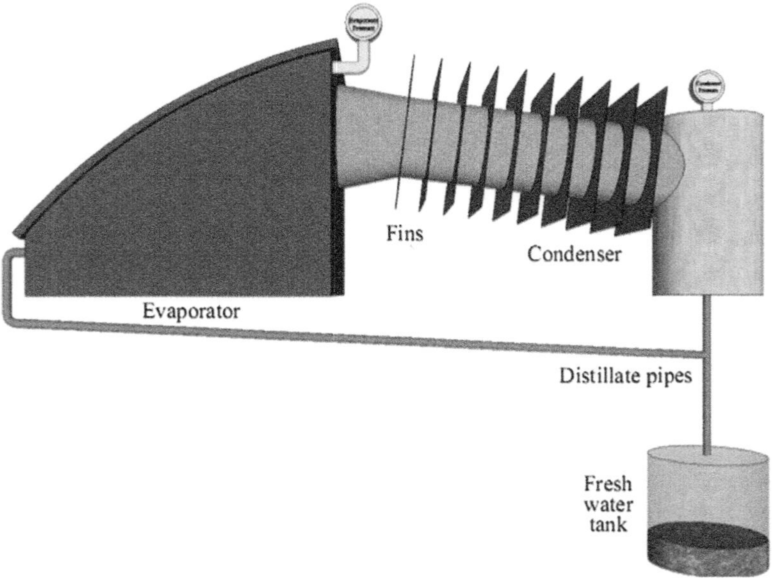

Fig. 54.2 Schematic diagram of the experimental setup

in diameter polypropylene pipe equipped with isolating valve. The evaporator is 0.003 m thickness carbon steel.

The evaporator's top surface is a 0.015 m thickness security glass sheet to allow solar energy transmission to the base of the evaporator. This glass cover has an average transmittance 0.65 and is bent to withstand extra stresses due to pressure difference. Meanwhile, a glass cover is sealed with silicon rubber to ensure strong contact between glass and the frame.

54.4.2 Separate Condenser

Condensation process on glass cover of conventional solar still encountered fluctuation of the glass cover transmittance. This led to a cyclic process of clear and misty glass. Therefore, a separate air-cooled condenser is added. The condenser is made of a copper tube of 0.1 m diameter, 0.6 m length, and 0.001 m thickness. On its lateral surface, 10 squared fins of 0.25 m length and thickness of 0.001 m were soldered 0.04 m apart. The condenser was connected to the evaporator by welding. At the other end, the condenser was connected to a condensate receiver via 0.0127 m polypropylene pipe. The condenser receiver is cylindrical in shape made of 0.003 m thickness galvanized iron with 0.12 m^3 in volume.

54.4.3 Freshwater Receiver and Piping

The freshwater receiver is cylinder of 0.15 m^3 volume and made from polycarbonate with 0.01 m thickness. The freshwater receiver is scaled and connected to both the evaporator and condensate receiver using 0.0127 m polypropylene pipes.

54.4.4 Instrumentation

Temperatures are measured using six DST-30 solar energy digital thermocouples k-type with accuracy ± 0.5 °C, all over the system. Two thermocouples are used to measure black base temperature and saline water temperature in the evaporator. Inner glass temperature and outer glass temperature are measured using other two thermocouples. Another thermocouple is used to measure temperature of water condensate in the condenser. The last thermocouple is used to measure ambient air temperature.

Solar radiation drives the heat and mass transfer processes in a solar still. So, it is necessary to measure this environmental factor with a high degree of accuracy. To achieve this, HANEEI® digital solar radiation gauge Model SOLAR 119 is used to measure global solar irradiance. The pyranometer is fixed to be at the same slope and the same direction of the glass cover.

The pressures are measured using SOLO® analog-type vacuum pressure gauge. One vacuum pressure gauge is fixed in the evaporator to measure water vapor pressure. The other is fixed in the condenser to measure condensation pressure.

The cumulative volume of distilled water is measured by collecting the productivity in a scaled tank of a least count 1 ml. The distillate water is collected from two passes (water condensates in the condenser and water condensates in glass sheet cover).

A two-stage vacuum pump (Edwards® Model E2M18) is utilized to create vacuum through a purge connection that is equipped with isolated valve.

54.5 Results and Discussion

The equations of mathematical model presented in the above section are employed in MATLAB code and solved numerically. The simulations were carried out for the climatic conditions of Cairo, Egypt (latitude 30.05°N, longitude 31.25°E). The inlet temperature of the saline water was assumed as the ambient. The initial saline water concentration was taken as 3.5 %. The properties of air and water were calculated from correlations given by [23, 24], respectively.

The simulation results for the aforementioned system specifications are obtained and compared with experimental measurements to validate the accuracy of simulation program. The simulation results are compared with experimental measurements of the constructed system for 2 typical days in August and November 2011.

Figure 54.3 demonstrates comparison of simulation results and experimental data that are recorded in August 2011. It can be seen from Fig. 54.3a that hourly experimental and simulation values are rather close together. The maximum global solar radiation on tilted glass cover is 1,018 W/m^2 at 12 pm.

Experimental and simulation values of base temperature are shown in Fig. 54.3b. As shown in the figure, they have the same trend. The maximum base temperature was 70.9 °C and occurred at 1 pm. There are differences in results from the thermocouple which used to measure the base temperature. The reason could be that the thermocouple was not stuck properly to the base.

Figure 54.3c presents the hourly variation of saline water temperature inside the evaporation chamber. The maximum temperature for the experiment was 60 °C between 1 and 2 p.m. The predicted maximum temperature was 64 °C and occurred at 2 p.m.

Predicted and experimental glass bottom and upper temperatures are shown in Fig. 54.3d, e, respectively. As shown in the figure, there is a satisfactory matching of predicted values and experimental data. Due to relatively large thickness of glass cover, 15 mm, the temperature difference between bottom and upper surface could reach 8 °C. However, the peak values occurred at 1 p.m.

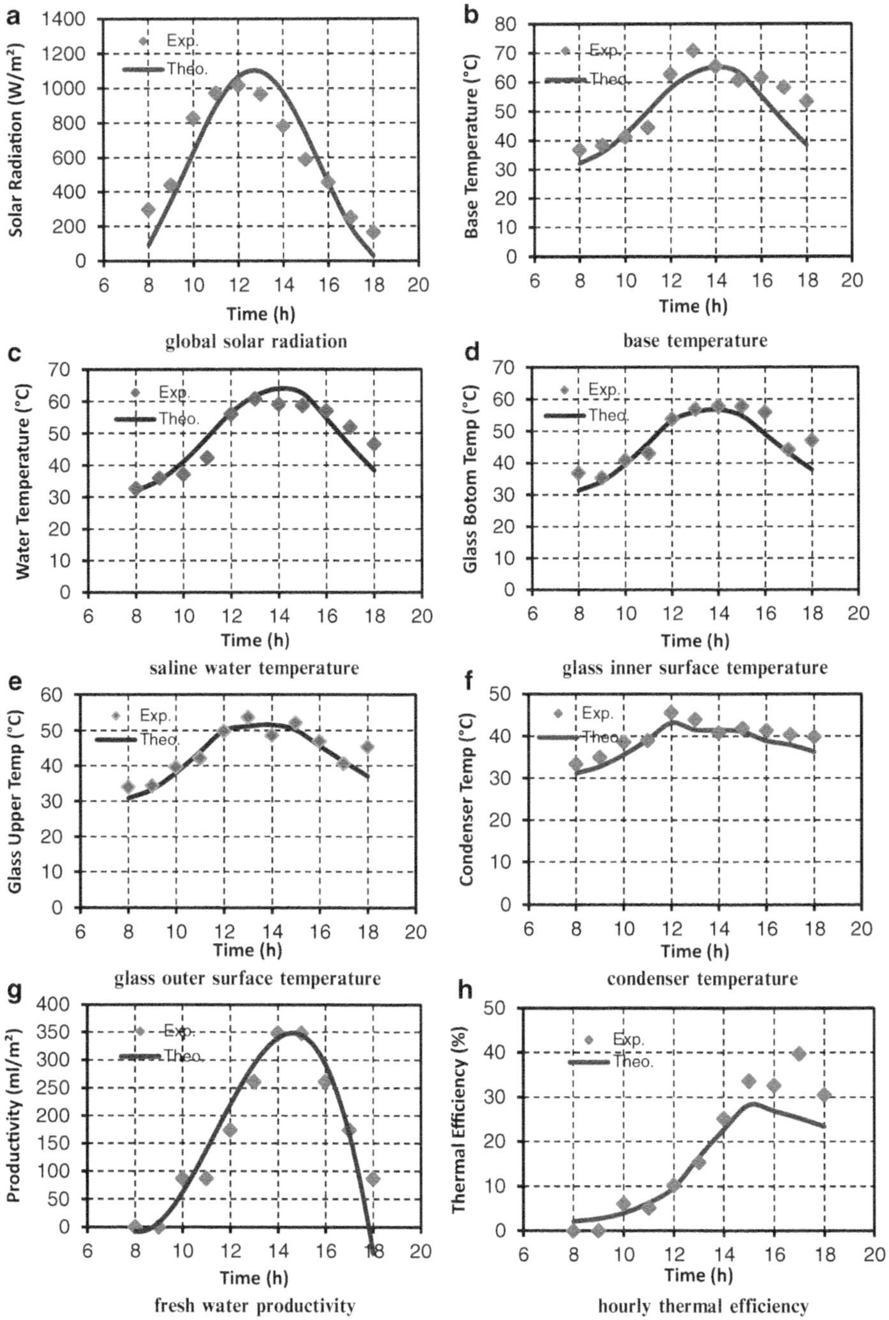

Fig. 54.3 Simulation results vs. experimental measurements in typical day of August. (**a**) Global solar radiation. (**b**) Base temperature. (**c**) Saline water temperature. (**d**) Glass inner surface temperature. (**e**) Glass outer surface temperature. (**f**) Condenser temperature. (**g**) Freshwater productivity. (**h**) Hourly thermal efficiency

Figure 54.3f shows the hourly condenser temperature. As shown in the figure, there is a good matching of simulation results and experimental data. It is noted that the peak condenser temperature is 45.6 °C and occurred at 12 p.m.

A satisfactory matching between the simulation results and measured freshwater productivity is shown in Fig. 54.3g. The production rate starts very slowly due to warming of the unit and the somewhat low solar energy during the morning hours. A peak productivity is obtained at about 16.00, and then it starts to decrease. The freshwater production is somewhat small due to bad transmittance of the glass cover. The thick glass cover has major drawbacks on system performance. The thick glass cover is characterized with low transmittance that has major drawbacks on system performance as follows. The solar radiation reaches the saline water is decreased which in turn reduces the rate of saline water evaporation drastically. Moreover, a substantial part of solar radiation is absorbed by the glass cover which results in the increase of glass cover temperature and consequent decrease of freshwater codensation on its inner surface.

The comparison of simulated and measured thermal efficiency of the system is shown in Fig. 54.3f. As shown in the figure, the simulation model predicts well the system performance. The peak thermal efficiency is 39.7 % at 3 p.m. Irrespective of potential improvements of the proposed system, its thermal efficiency is comparable to conventional solar still which is 30 % (maximum) and daily freshwater production is about 2 L/m^2, under climatic conditions of Cairo, Egypt [25].

In the second part of simulation program validation, the experiment was repeated in November 2011. Figure 54.4 demonstrates comparison of simulation results and experimental data.

Hourly variation of global solar radiation and different temperatures of the components of the proposed system are shown in Fig. 54.4a–f. As shown in the figure, the agreement between measured and predicted values for different parameters is satisfactory. The base temperature was always the highest between all the temperatures since it absorbs the solar radiation. The maximum base temperature is 44.3 °C and occurs at 2 p.m.

The simulation results of hourly productivity and thermal efficiency are drawn with the experimental data and are shown in Fig. 54.4g, h. No productivity has been noticed from experimental system at early hours of system operation because minimum value of freshwater that can be seen is 0.001 m height in distillate tank, which is noticed at 10 a.m.; however efficiency and productivity from simulation had no zero values.

Generally, the difference between the simulation results and experimental measurements is very close. This is due to uncertainties in correlations used for calculations of different heat transfer coefficients and solar radiation incident on the glass cover.

The simulation results of proposed desalination unit during a whole year are presented in Figs. 54.5, 54.6, 54.7, and 54.8. The monthly variation is calculated for average day of each month [26]. Figure 54.5 shows the variation of average monthly extraterrestrial solar radiation I_o, average monthly daily solar radiation on a horizontal surface I_h, and the average monthly of solar radiation on a tilted

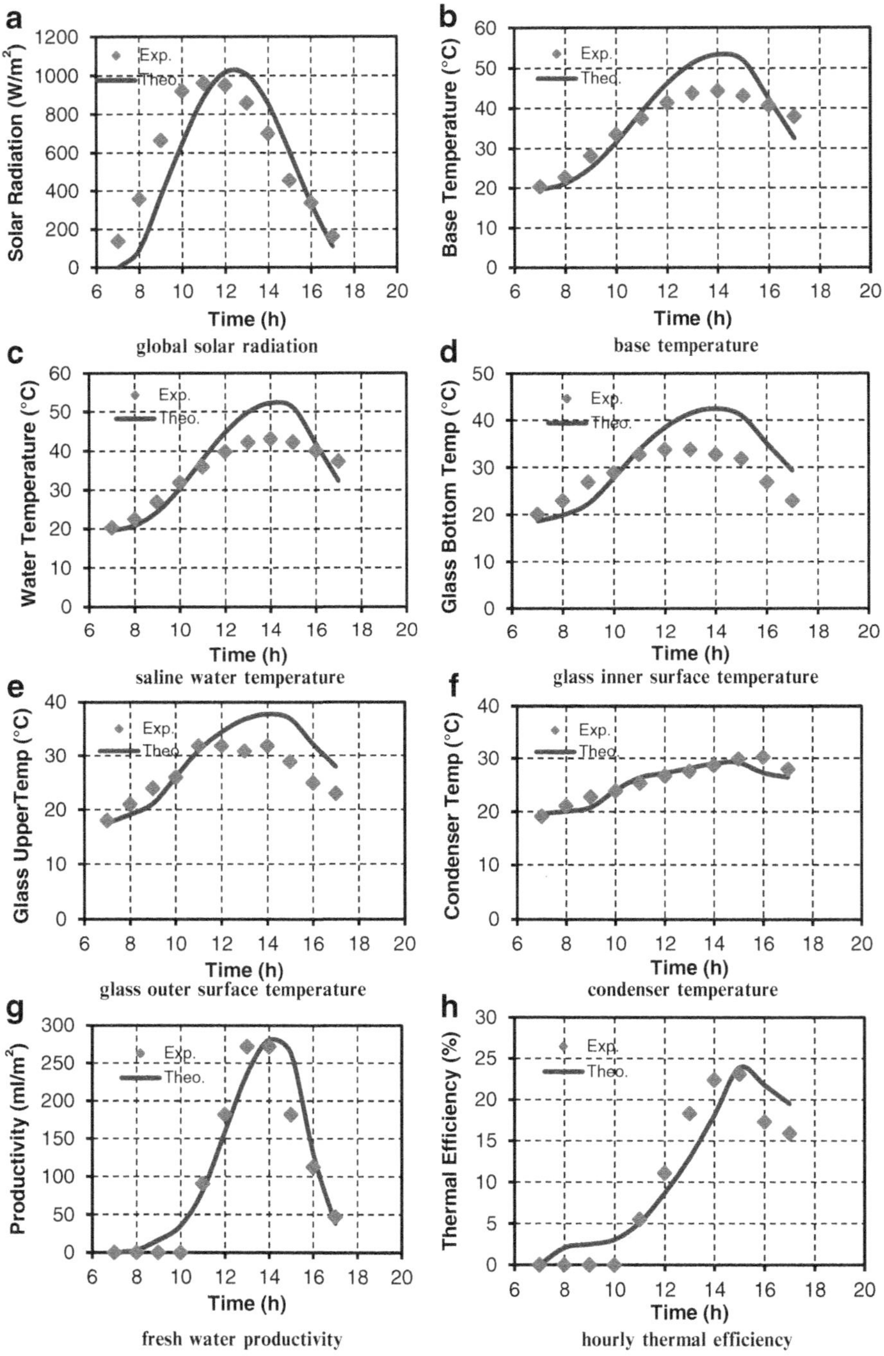

Fig. 54.4 Simulation results vs. experimental measurements in typical day of November. (**a**) Global solar radiation. (**b**) Base temperature. (**c**) Saline water temperature. (**d**) Glass inner surface temperature. (**e**) Glass outer surface temperature. (**f**) Condenser temperature. (**g**) Freshwater productivity. (**h**) Hourly thermal efficiency

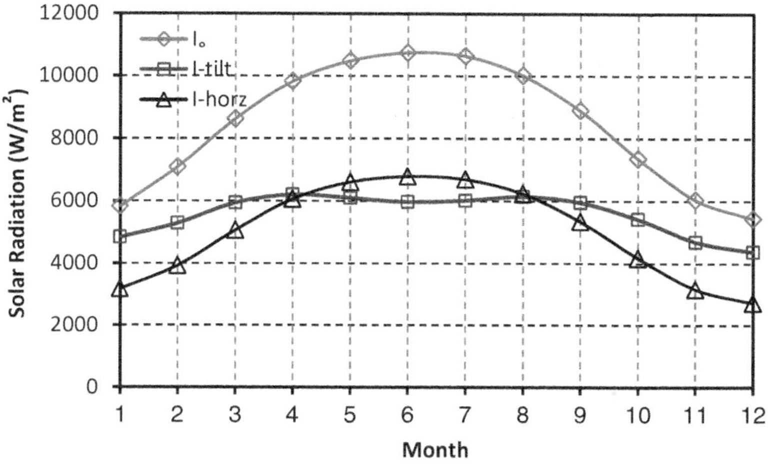

Fig. 54.5 Monthly variation of daily solar radiation for average day

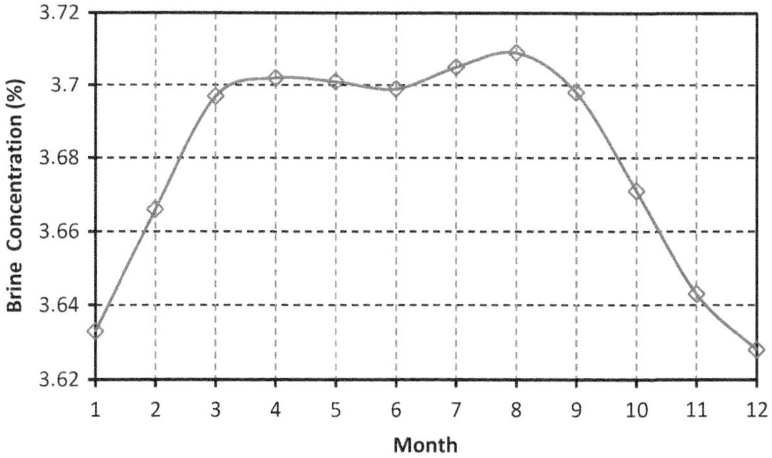

Fig. 54.6 Monthly variation of brine concentration for average day

surface I_T facing south during a whole year in Cairo where β equals the latitude of Cairo. It is shown that the maximum values are obtained in June. The minimum value occurred in December. It is noted that I_T exceeds I_h except during the summer months.

The monthly variation of brine concentration is shown in Fig. 54.6. Its trend follows the monthly variation of productivity as shown in Fig. 54.7. The maximum value, 3.71 %, is obtained in August. The minimum productivity occurs in December. As shown in Fig. 54.8 the maximum and minimum of daily thermal efficiencies occur in August (21 %) and January (17.8 %), respectively.

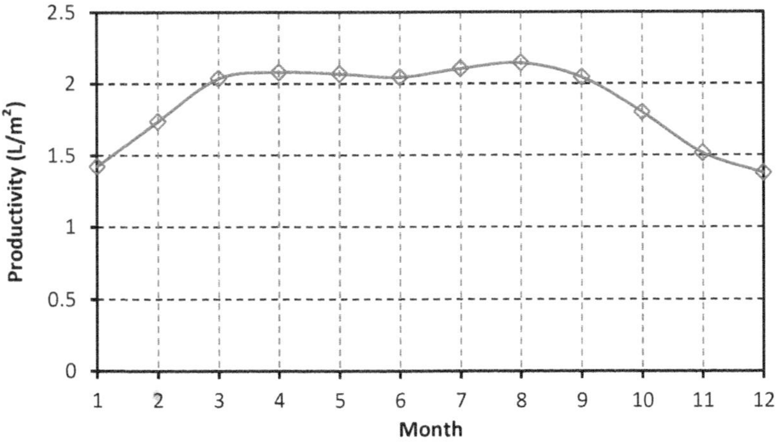

Fig. 54.7 Monthly variation of daylight productivity for average day

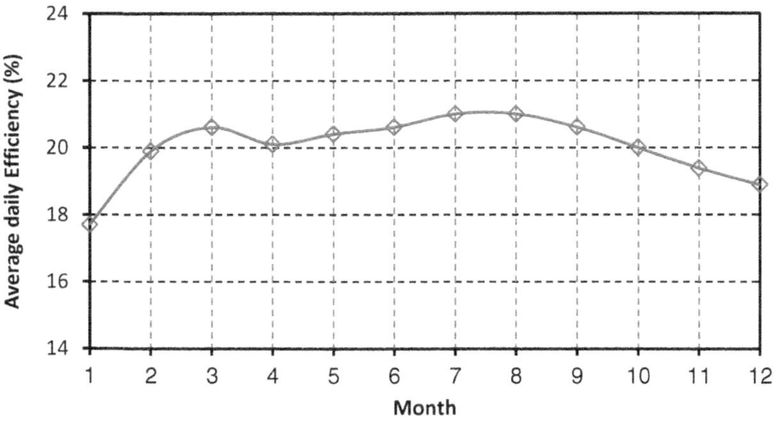

Fig. 54.8 Monthly variation of daily thermal efficiency for average day

It is noted that the predicted performance of the proposed desalination system is inferior to the expected values. This is mainly due to the transmissivity of the thick glass cover, 0.65, which is significantly lower than commonly used glass cover in conventional solar stills. Secondly, the high concentration of water on the inner surface of glass cover due to its reduced temperature compared with one of conventional solar stills [25]. These two effects caused a significant reduction of glass cover optical transmissivity bringing it down to 0.5.

54.6 Conclusions

In this study mathematical modeling of a solar still operating under vacuum conditions was conducted. The simulation was run for the climatic conditions of Cairo, Egypt. The results of hourly water productivity, thermal efficiency, and temperature variations for two typical days in August and April are obtained. An experimental test rig for the proposed system is constructed, as well. Investigation of the validity of thermal model is performed on the basis of experimental data. The results of investigation proved its validity. It can be concluded that the model is capable to predict in a satisfactory way the behavior of the proposed system. Hence, it will be a valuable tool for sensitivity analysis in the stage of evaluation of existing system. On the other hand, in the stage of system design, it could be used to explore the potential to produce higher quantity of freshwater for the same input and evaporator area.

The year-round performance of proposed system is presented, where the pest performance is obtained in August. It is concluded that the influence of the optical properties of glass cover is very important, since it determines the exploitation of falling solar radiation by the desalination system.

References

1. Kalogirou S (2014) Solar energy engineering: processes and systems, 2nd edn. Academic, San Diego, CA
2. Murugavel K, Chockalingam K, Srithar K (2008) Progresses in improving the effectiveness of the single basin passive solar still. Desalination 220:677–686
3. Kabeel AE, El-Agouz SA (2011) Review of researches and developments on solar stills. Desalination 276:1–12
4. Velmurugan V, Srithar K (2011) Performance analysis of solar stills based on various factors affecting the productivity—a review. Renew Sust Energ Rev 15:1294–1304
5. Al-Hussaini H, Smith IK (1995) Enhancing of solar still productivity using vacuum technology. Energ Convers Manage 36:1047–1051
6. Kabeel AE, Omara ZM, Essa FA (2014) Improving the performance of solar still by using nanofluids and providing vacuum. Energ Convers Manage 86:268–274
7. Al-Kharabsheh S, Goswami DY (2003) Analysis of an innovative water desalination system using low-grade solar heat. Desalination 156:323–332
8. Al-Kharabsheh S, Goswami DY (2003) Experimental study of an innovative solar water desalination system utilizing a passive vacuum technique. Sol Energ 75:395–401
9. Gude VG, Khandan NN (2009) Desalination at low temperatures and low pressures. Desalination 244:239–247
10. Gude VG, Khandan NN (2010) Sustainable desalination using solar energy. Energ Convers Manage 51:2245–2251
11. Gude VG, Khandan NN, Deng S (2011) Desalination using solar energy: towards sustainability. Energy 36:78–85
12. Ayhan T, Al MH (2010) Feasibility study of renewable energy powered seawater desalination technology using natural vacuum technique. Renew Energ 35:506–514
13. Shaltout M (1998) Egyptian solar radiation atlas. Geography Institute, Cairo

14. Tiwari GN, Singh HN, Tripathi R (2003) Present status of solar distillation. Sol Energ 75:367–373
15. Kalogirou SA (2005) Seawater desalination using renewable energy sources. Prog Energ Combust Sci 31:242–281
16. Duffie JA, Beckman WA (1980) Solar engineering of thermal processes, 2nd edn. Wiley, New York
17. Hottel HC (1976) A simple model for estimating the transmittance of direct solar radiation through clear atmospheres. Sol Energ 18:129–134
18. Liu BYH, Jordan RC (1960) The interrelationship and characteristic distribution of direct, diffuse and total solar radiation. Sol Energ 4:1–19
19. Bemporad GA (1995) Basic hydrodynamic aspects of a solar energy based desalination process. Sol Energ 54:125–134
20. Incropera FP, DeWitt DP, Bergman TL, Lavine AS (2007) Fundamentals of heat and mass transfer, 6th edn. Wiley, New York
21. Elsayed MM, Taha IS, Sabbagh JA (1994) Design of solar thermal system. Scientific Publishing Center, King Abdulaziz University, Jeddah
22. Rohsenow WM, Hartnett JP, Cho YI (1998) Handbook of heat transfer, 3rd edn. McGraw-Hill Book Company, New York
23. Al-Kharabsheh S, Goswami DY (2004) Theoretical analysis of a water desalination system using low-grade solar heat. ASME J Sol Energ Eng 126:774–780
24. El-Dessouky H, Ettouney H (2002) Fundamentals of salt water desalination. Elsevier, New York
25. Zeinab SA, Ashraf L (2007) Experimental and theoretical study of a solar desalination system located in Cairo, Egypt. Desalination 217:52–64
26. Klein SA (1977) Calculation of monthly average insolation on tilted surfaces. Sol Energ 4:325–329

Chapter 55
Potentials and Prospects of Renewables in Libya

Satya P. Bindra, Farag Soul, S.D. Jabu, A. Allawafi, A.M. Belashher, H. Reani, S. Abulifa, and Khalifa Hammuda

Abstract As elsewhere, Libya too is affected by human-induced climate change and the volatility of world oil markets. This chapter is designed to present the potential and prospects for replacing nonrenewable fossil fuels (i.e., hydrocarbon) as a backbone of the economy by a range of renewables including wind, solar-PV, hydropower, geothermal, and biomass to deliver energy. The focus is on land opportunities and constraints and on production costs as a function of resource availability and depletion and of innovation dynamics. After outlining a range of renewable production technologies it demonstrates how to try to spread and instill the culture within Libyan society of the idea of renewable energy (RE). It uses best

S.P. Bindra (✉)
Principal Resource Person, UNCSD Rio + 20, Libya National Report Focal Point, Tripoli, Libya
e-mail: s.p.bindra@gmail.com

F. Soul
Faculty of Engineering, Al Garabouli Branch, Al-Mergeb University, Al-Khums, Libya
e-mail: faragsoul@yahoo.com

S.D. Jabu
Colleges Directorate, Libya
e-mail: sami110@wp.pl

A. Allawafi
Technical College of Civil Aviation and Metrology Esphea, Libya
e-mail: Alwafiali111@gmail.com

A.M. Belashher
Ministry of Agriculture, Libya

H. Reani
Ministry of Water Resources, Libya
e-mail: h_s_reani@yahoo.com

S. Abulifa
Free Excellency Company for Oil and Environmental Solution Services, Libya
e-mail: Mscgm@yahoo.com

K. Hammuda
Ministry of Labour, Tripoli, Libya
e-mail: khalifa3030@yahoo.com

I. Dincer et al. (eds.), *Progress in Clean Energy, Volume 2*,
DOI 10.1007/978-3-319-17031-2_55

practices and lessons learned to start this culture at an early stage via universities, Libyan academies, high institutes and technical colleges in Libya, and even lower schools. Finally lessons learned are described to highlight a way forward towards a low carbon economy in the country to combat the dangers of global warming through the use of abundantly available renewable potential sources of clean energy.

Abbreviations

AC	Assessment center
APCI	African productive capacity initiative
GDP	Gross domestic product
GECOL	General Electric Company of Libya
GSMI	Global Strategic Management Institute
EU	European Union
ESCAP	Economic and Social Commission for Asia and Pacific
ICT	Information and communication technology
IEA	International Energy Agency
MENA	Middle East and North Africa
MITA	Malta Information and Telecommunication Authority
NEPAD	New Economic Partnership for Africa
REAoL	Renewable Energy Authority of Libya
UNDP	United Nations Development Program
UNCSD	United Nations Commission for Sustainable Development
UNIDO	United Nations Industrial Development Organization
WEF	World Economic Forum

55.1 Introduction

Libya is a *rentier* economy where oil and gas account for 98 % of GDP. It is the second biggest North African country spreading over an area of 1.76 million sq km, located between Algeria and Tunisia in the west and Egypt in the east, bordering the Mediterranean Sea in the north, and (from west to east) Niger, Chad, and the Sudan in the south. Almost all its land territory is land area apart from access to the Mediterranean Sea (about 1,770 km of coastline) [1, 2]. The country is looking forward to economic diversification of its GDP away from hydrocarbon by a diversification deemed imperative by the World Bank and IMF to create jobs and move beyond the export of hydrocarbons. This is a rich country in renewable energy (RE) resources. It has the potential to produce the equivalent of almost seven million barrels of crude oil per day in energy. It could generate enough renewable power to meet its own demand and a significant part of the world energy demand by exporting electricity. As per an estimate Libya has an average daily solar radiation rate of about 7.1 kW h per square meter per day (kWh/m^2/day) on a flat plane on the coast and 8.1 kWh/m^2/day in the southern region. The country uses only 0.1 % of its estimated 88 % desert territory for solar power.

This chapter demonstrates how the UNCSD Rio + 20 Focal Point responsible for sustainable livelihood in Libya is devising a clear strategy to demonstrate that within a defined timetable renewable energy technology using solar, wind, and biomass has the potential by taking it forward. This is being planned by assisting the country in developing its manpower as human resources capable of installing and maintaining renewable energy systems. The strategy considers untapped the country's dry, hot, and prolonged gusts as having great potential for wind power. All it needs is to harness only a tiny fraction of the renewable energy resources integration it has available in the form of solar, wind, and biomass. Thus it would not only meet its own demands for energy, but also a significant part of the world's demands by exporting electricity as a gateway to MENA, the European Union, and Africa.

55.2 Appraisal of Potentials and Prospects of Energy Situation in Libya

Libya has many advantages including its geographic location—as a tourist destination, a gateway to Africa, and proximity to Europe. Libya consists of 94.5 % desert and semi-desert and has virtually no accessible water resources on the earth's surface. The Libyan climate ranges from Mediterranean along the coastline to extremely dry in the interior south. The barren, flat, and undulating plains have barely 1.03 % arable land. The country has many natural resources including gold, iron, copper, and even uranium. Mining offers a huge potential for growth. Out of petroleum, natural gas, and gypsum as the most prominent natural resources, the first two are the main driving factor for the Libyan economy. Hydrocarbons contribute about 95 % of export earnings, 65 % of the GDP, and about 80 % of government revenue [2]. It is also due to fossil resources that Libya has always enjoyed fairly high international interest and relevance. Although the February 17, 2011 revolution has impeded its economic activities, it is still widely considered to be a land of unique opportunity for entrepreneurial activity and thus economic growth. Ethnic Arabs, Arabized Berbers, Tuareg, and Tubu populations who form around 97% of the population follow Islam (Sunni Muslim approximately). The official national language is Arabic.

The oil sector dominates 95 % of Libyan export revenues [3]. In addition to the petroleum and petrochemicals industry, the country is also active in the aluminum, iron and steel, and cement industries. In 2012, crude oil, refined petroleum products, natural gas, and chemicals were exported to Italy (23.5%), Germany (12.5%), China (11.3%), France (9.7%), Spain (7.6%), the United Kingdom (4.7%), and the United States (4.5%). Libya mainly imports machinery, semi-finished goods, food, transport equipment, and consumer products. The main import trading partners in 2012 were China (13.7%), Turkey (12.3%), Italy (8.7%), Tunisia (7.3%), South Korea (6.2%), Greece (5.4%), and Germany (4.9%) [4, 8, 9]. As per estimates in

Table 55.1 Total energy production (2011)

Energy source	In ktoe	In %
Coal and peat	0	
Crude oil	24,371	78.7
Oil products	0	
Natural gas	6,419	20.7
Nuclear	0	
Hydro	0	
Geothermal, solar, etc.	0	
Biofuels and waste	172	0.55
Electricity	0	
Heat	0	
Total[a]	30,962	100.00

[a]Due to rounding, the percentages may not add up to a hundred.

2011, Libya has produced about 30,962 ktoe of energy, which is about 360,088.19 GWh. Crude oil makes up almost 79% of energy production. Renewable energies have been neglected and have only risen to about 0.06% [5, 9]. IEA estimates during the year 2011 are given in Table 55.1.

Results of a recent estimate [11] indicate a potential for liquid biofuels on the order of 75–300 EJ $year^{-1}$ and for electricity from biomass options at production costs below 10¢ kWh^{-1} of 200–300 PWh $year^{-1}$. Theoretically, future electricity demand can be amply met even from biomass sources in most regions by 2050 below 10¢ kWh^{-1}, but major uncertainties are the degree to which land is actually available and the rate and extent to which specific investment costs can be reduced.

55.2.1 Social Acceptance of Renewables

The special issue of social acceptance of renewable energy innovation in Libya as elsewhere [14] is increasingly recognized as a constraining factor. This is especially the case with respect to wind energy, due to its visual impact on landscapes. Three dimensions of social acceptance, that is, sociopolitical, community, and market acceptance, are the subject of present debate. The factors influencing sociopolitical and community acceptance are increasingly recognized as being important for understanding the apparent contradictions between general public support for renewable energy innovation and the difficult realization of specific projects. The third dimension, market acceptance, has thus far received less attention and requires further research.

55.2.2 Electricity Demand Supply and Consumption in Libya

World Bank estimates show that 99.8 % of the Libyan people have access to electricity, which is the highest rate among African countries [6–9]. Estimates

Table 55.2 Sector-specific Libyan energy consumption

Sector	Consumption in GWh	In %
Industry	4,864	22
Transport	0	0
Residential	5,365	24
Commercial and public services	7,915	36
Agriculture/forestry	2,292	10
Fishing	0	0
Other nonspecified	1,599	7
TFC	22,035	100

show that electricity generation has more than doubled from 2000 to 2010. The country in 2010 had a total electricity installed capacity of 6.8 GW, which is generated by power plants either fueled by oil or natural gas. Out of the total final consumption of 22,035 GWh in 2011, commercial and public services accounted for 36 % whereas the residential sector amounted to 24 % and industry to 22 %. Electricity consumption in Libya during 2011 was on the order of total 32.96 TWh, meaning 3.73 MWh per capita.

The national electric grid consists of a high-voltage network of about 12,000 km, a medium voltage network of about 12,500, and 7,000 km of low-voltage network. Some villages and remote areas that are located far away from these networks cannot be connected to the grid for economic reasons. Those locations with a small population and a small amount of energy demand use diesel generators as a power supply, requiring regular maintenance and supply of fuel [10, 12].

There is an operating grid interconnection to Egypt with a capacity of 240 MW which is 180 km long. Regarding grid access of renewable energy projects, there is neither a priority access granted to RE by law nor has a grid code been developed yet. Additionally, a detailed map for potential RE sites is missing (Table 55.2).

55.3 UNCSD Rio + 20 Focal Point Program for Renewables in Libya

Libya is developing local Libyan resources to meet the anticipated future renewable energy needs in cooperation with global, regional, and UN partners who are already engaged to review, amend, modify, and add value to our renewable business plans. With the state-of-the-art input from world-class partners (UNCSD Rio + 20 Libya National Focal Point, US GSMI, WEF Switzerland, MITA Malta, UNDP ICT Practice area, UNIDO/NEPAD APCI, EU 2014 ERASMUS Plus, etc.), Libya aims to develop the renewables industry by matching available resources to upcoming projects. This is in line with the national economic development strategy in New Libya using the Assessment Centre Approach for tapping underutilized and critical links in the organizational value chain. Using reliable (i.e., consistent in measurement) and valid (i.e., measuring what they purport to measure) figures, the value added by Assessment Centres (ACs) in development and selection practices are based on validated design, compared to conventional standardized psychometric measures. Against this background, the Libyan Assessment Centre uses these sophisticated tools, especially in light of a hibernating global economy for rebuilding Libya. The Libyan Assessment Centre methodology is designed so as to demonstrate the (1) strong return on investment; (2) precision and accuracy in measurement; (3) integration of technology; (4) support better talent management, selection, and development decisions; (5) reduction of cultural bias; (6) clear linkage to key organizational outcome (i.e., higher performance, lower turnover, less absenteeism); (7) justification of further investments in HR initiatives to senior management; and (8) increasing the perceived value of HR to the organization.

Thus the main purpose of Assessment Centres in the Libyan energy organizational environment is to help successfully and accurately measure job-related traits of individuals to help predict their future performance and assess their suitability for roles. Effective human resource strategies being employed include: selection and development decisions, workforce capability, employee engagement, talent management, and succession planning that would be built on the foundation of accurate and reliable information. It uses proven expertise to help in providing rich layers of information for managerial decision making. Human-related diagnostic information is intended in all likelihood to result only in valued organizational outcomes when assessment practices are clearly linked to long-term strategy.

Numerous methods employed by UNCSD Rio + 20 Libya Focal Point to estimate the potential of renewables in Libya shows that because energy prices are heavily subsidized in all economic sectors in Libya, it is difficult to estimate the potential of renewables and energy efficiency on a cost-effective basis. The present status is that renewables are not utilized in significant amounts. Barely 5 MW of solar energy are separated into several small PV projects and have been installed yet. The country fails to meet its international obligations requirements envisaged under various UN conventions on climate change.

55.3.1 Solar Power Potential

Recognizing the sun as a major natural resource with which Libya is blessed in abundant measure it is believed that solar energy is a valuable and renewable energy source that should be fully exploited for the benefit of the country.

A general solar map developed using satellite data shows that Libya has great potential for solar energy. However, it requires developing a detailed solar atlas [11, 12]. Estimates in the coastal regions show that the daily average of solar radiation on a horizontal plane amounts to 7.1 kWh/m^2/day. However, the radiation is 8.1 kWh/m^2/day in the southern region. The average sun duration of more than 3,500 h per year means it is equivalent to a layer of 25 cm of crude oil per year on the land surface.

The total installed capacity of solar energy which was barely 5 MW in 2012 has been achieved from small PV projects since 1976. In the first instance the solar system was used to supply cathodic protection for the oil pipeline connecting the Dahra oil field with the Sedra port. In1980, the first PV system was used in the communications sector to supply energy to the microwave repeater station near Zella. In all up to 2006, 80 stations running by PV in the field of communications had been established.

As per estimates in 2005, the total installed photovoltaic peak power was around 420 KWp. In 2012, it exceeded 950 kWp. At El-Agailat, a PV pumping system was installed at the beginning of the 1980s. Up to 2006, 35 PV water pumping projects had been installed (~110 KWp). The total capacity of the PV water pumping system was 120 kWp in 2012. The use of PV systems for rural electrification was only starting in 2003. By 2006, the total number of remote systems installed by General Electric Company of Libya (GECOL) was 340. It had a total capacity of 220 kWp. The Center of Solar Energy Studies (CSES) and the Saharan Center also installed 150 with a total power of 125 KWp. In 2012, the rural electrification PV systems had a capacity of 725 kWp. Currently, there are three PVs projected in the pipeline: a 14-MW power station in Houn, a 40-MW project in Sabha, and a 15-MW power station in Ghat.

The estimates on the technical potential with respect to concentrated solar power (CSP) shows that the country can generate around 140,000 TWh/year which is equivalent to 27,000 GW of capacity at 60 % load factor [13]. An overview further shows that in 1983, ten solar heater systems were installed. Up to 2006, an additional 2,000 had been deployed.

55.3.2 Biomass Potential

An estimate concerning Libya's biomass potential shows that presently biomass energy sources are small and can only be used on an individual level as an energy source. It is not suitable to produce energy at present. However, as stated elsewhere, biomass sources in Libya too, as in most regions, by 2050 can help produce electricity below 10 ¢ kWh^{-1} [11].

55.3.3 Wind Power Potential

Although there is no detailed wind atlas yet, the general wind map based on satellite data shows that the wind potential is good. The average wind speed at a 40-m height is between 6 and 7.5 m/s. The most attractive locations along the Libyan coast are at Dernah. It shows that the average wind speed is around 7.5 m per second. In many desert oases in the south of the country wind energy has been used to pump water since 1940. However, because the windmills need to be maintained regularly, this way of producing energy has not been developed on a large scale. Using the manufacturers' premass production unit capital costs ($/kW) to estimate the cost of electricity produced, in cents/kWh, it is concluded that further reduction in the manufacturers' unit capital cost is still required to enable wind energy to compete with other conventional energy sources in Libya.

55.3.4 Hybrid System

The variations in resource availability and end-use suitability in Libya tend to limit any particular single renewable technology to specific locations and uses. The solution that is increasingly being favored as the best means of providing decentralized power with high reliability is the hybrid system. Hybrid systems do not rely on a single energy source, but two, three, or even four potential sources. For economic reasons there are typically only two, but all hybrid systems will normally include a diesel generator set. The required electrical energy for desert/rural electrification can be met using one or a combination of options that include wind, solar, or hybrid (solar and wind) conversion systems in addition to the diesel and grid extension. The selection of one or more of these options depends mainly on the available energy, the performance of the conversion system, and the characteristics of the electrical load. It also depends on the capital and operational costs of each option. The maintenance requirement and the availability of the required support are also important parameters to be considered in the selection of suitable options. Very few studies have been conducted on the feasibility of utilizing hybrid systems in meeting power demand in Libya. This is a potential area that is receiving more attention and needs to be explored more.

55.4 Discussion of Some Salient Aspects of the Energy Sector in Libya

An in-depth overview of the Libyan situation prepared and presented by UNCSD Rio + 20 Focal Point reveals that the country has a fairly one-sided economy that heavily relied (and still relies) on the occurrence of fossil fuels, that is, oil and gas. Due to inherent overdependence and ease of accessibility to a huge quantity of fossil energy sources, renewable energy sources were and are still now considered to be of secondary relevance. Thus the efforts for pursuit of a diversified and sustainable energy sector has been and still are limited by the old regime mentality that had strongly subsidized energy coming from domestic fossil sources. It developed and maintained the hydrocarbon sector with little or no economic incentive to shift to a more sustainable energy mix. The current status and the future potentials of renewable energy applications in Libya show that the power in the earth's wind and in solar radiation, which reaches the earth, is more than sufficient to make significant as well as strategic contributions to the country's energy supply. Applications of solar energy in the country have not grown since 1958 after the discovery of oil. Thus effective utilization of solar energy in Libya has not yet made reasonable progress mainly due to several obstacles. But, valuable lessons have been learned and a wealth of experience has been gained from the limited experience mainly from neighboring countries in the region. The technical and economic feasibility of wind energy utilization in the country has not yet been fully explored.

The postrevolutionary establishment of a Ministry of Electricity and Renewable Energy is an encouraging step towards the integration of the subjects of renewable energy and energy efficiency into the national agenda. The Renewable Energy Authority of Libya or REAoL soon after the February 17, 2011 revolution has established a target of 10 % renewables by 2025, which would account for a total capacity of 2,219 MW. Intermediate targets are 389 MW by 2015, and 1,069 MW by 2020.

An overview of the laws and regulations shows that there is no legislation covering financial support for RE or which addresses the issue of financing additional costs of RE projects. In addition much in demand pressure for the need for clear legislation for the participation of private capital in the power sector is missing.

There is an absence of policies for RE in Libya. There is no public competitive bidding for large-scale private RE projects. In addition, there is no obligation to conclude long-term power purchase agreements with RE producers. There are also no feed-in tariffs and no net-metering policy for small-scale RE projects.

With respect to finance and investments, Libya does not have an RE fund for financing RE projects. Further review of the present situation shows that currently planned projects are intended to be financed through the government budget. The power sector is still closed to private investors. There are no financial guarantees to private investors to ensure payment under power purchase agreements by the Libyan government. Internal tax privileges are not provided for RE projects. Libya needs a sea change. The keys to success are:

1. Accelerate technology transfer to Libya and neighboring Arab, MENA, Mediterranean, Euro-Med, and African countries, through Libya (the gateway to both the European Union and Africa).
2. Promote renewable culture within our region and establish related industries.
3. Strengthen strategic partnerships with first-tier renewable technology providers.
4. Create successful business projects through technology investments.

The Libyan government needs to follow good practices that give multiple benefits to renewable energy projects such as no need for industrial clearance, availability of loans, excise duty exemptions, customs duty concessions, financial support to renewable energy's R&D projects, income tax holidays, accelerated depreciation, preferential tariffs, interest and capital subsidies, energy buy-back, and third party sale and trading. At the same time, the Libyan government must reduce the capital cost of renewable energy projects so that more and more companies can invest in the sector. To fulfill this aim, the government must adopt the latest and most suitable technologies in the sector and promote healthy competition. REAoL employees need to be leaders in their field capable of delivering valued organizational outcomes where assessment practices are clearly linked to long-term strategy.

55.5 Concluding Remarks

Clean and renewable energy is the key for future life, not only for Libya but for the rest of the world. Renewables are deemed the future of energy. This chapter demonstrates that UNCSD Rio + 20 Focal Point Libya-led various policy think tanks and working groups are in line with the Renewable Energy Authority of Libya (REAoL) to translate theory into practice. It has the potential and prospects of a turn of the tides if the ambitious goals are included in legal frameworks and official goals to develop and apply technologies that will enable them to use renewable energy sources in the most efficient ways. Libya's geographical location has several advantages for extensive use of most of these sources. This renewable energy source potential and their present use are evaluated here based on the available data.

Libya needs to grasp cash offers from donors such as German GIZ that has rightly and timely initiated a 1,000 Roofs Program, which seeks to install 1,000 PV roof-top systems with a total capacity of 3 MW. These installations can be on-grid and off-grid, and have the potential to offer a basis to establish further support instruments such as feed-in tariffs. These international joint programs have the potential to help the establishment of a series of independent R&D projects throughout the country. The country needs to learn from a wealth of experience that has been gained in the assessment, instrumentation, calibration, data collection, monitoring, and analysis of solar energy projects in neighboring countries such as Tunisia. Low and medium solar thermal energy applications are technically and economically feasible and should be encouraged and supported by the government. The feasibility of wind energy utilization in the country has not yet been fully explored. More feasibility studies have to be conducted in the field of hybrid systems. Efforts should be directed at finding applications of those renewable systems that have already been developed in industrialized nations. Effective utilization of renewable energy systems requires government subsidies. Interaction between regional renewable research centers and local research centers and industries must be promoted. Public awareness about the use and importance of utilization of renewable energy has to be increased. Renewable energy capacity development through education and training programs must be incorporated as part of educational programs.

References

1. Auswärtiges Amt (German Ministry of Foreign Affairs), Last Updated: Jan, 2012
2. BBC, Libya profile. Accessed 15 Dec, 2013
3. Bindra SP, Hasan Lahmer (2002) Change and development of education & training in petroleum industry in Libya. Mediterranean petroleum conference, Jan 2002, Tripoli, Libya
4. Bindra SP (1998) A state of technology transfer and vocational training. Sharif University of Technology Publication Office, Tehran, Iran
5. Bindra SP (1998) Science and technology parks in Mediterranean region. Proc. workshop on insular coastal area risk management in the Mediterranean. University of Malta, Malta
6. Bindra SP (1998) Change and development in technological education for commercialisation of knowledge .Proc. XV IASP World conference on science and technology parks, Perth, Western Australia, pp 180–199
7. Bindra SP, et al. (1995) Total technological education in Iran. World Conference, Sydney, Australia, Sept
8. Central Intelligence Agency (CIA). World Fact book, Last Updated: July 10, 2013 and Feb 23, 2012
9. RCREEE Country Profile (2012) Renewable energy in Libya
10. International Energy Agency (2009) Energy balance for Libyan Arab Jamahiriya. last updated: 2012
11. Bert JM de Vries, Detlef P. van Vuuren, Monique M Hoogwijk (2007) Renewable energy sources: their global potential for the first-half of the 21st century at a global level: an integrated approach. Energy Policy, Elsevier Science Direct
12. Saleh, Ibrahim M (2006) Prospects of renewable energy in Libya. International symposium on solar physics and solar eclipses (SPSE)
13. RCREEE Provision of technical support/services for an economical, technological and environmental impact assessment of national regulations and incentives renewable energy and energy efficiency. Country report: Libya
14. Ushiyama O (1998) Wind energy activities in Japan. World renewable energy congress V, 20–25 Sept, Florence, Italy, pp 811–816

Chapter 56
Synthesis of Titanium-Decorated Graphene for Renewable Energy Applications

Zahra Gohari Bajestani and Yuda Yürüm

Abstract Reduced graphene oxide (RGO) was prepared from natural graphite by Hummers' Method. Few-layer graphene was decorated with titanium by an incipient wetness impregnation method. The pristine graphene shows hydrogen storage capacity equal to 1.3 wt%, while graphene decorated by titanium (RGO–Ti) enhanced hydrogen storage capacity to 1.4 wt%. We showed that titanium addition improved hydrogen storage capacity by chemical interactions. These interactions can be used for fabrication of different graphene-based materials as potential candidates for developing new absorbents for energy application.

Keywords Graphene • Titanium • Hydrogen storage • Adsorption isotherm

56.1 Introduction

Hydrogen is the most abundant element in the universe and it can be generated from renewable energy sources. It has great potential to be used as an energy source, but its storage is one of the most challenging barriers in hydrogen system applications. Many experimental and theoretical methods were used to determine safe and effective hydrogen storage materials [1]. Various properties such as large gravimetric and volumetric storage capacity and low cost are necessary for practical storage of hydrogen [2]. High surface area, low density, and large microporosity make carbon-based materials good options to use as adsorbent in hydrogen storage systems. These materials including nanotubes, nanofibers, and fullerene have attracted much attention for efficient and safe storage [3].

Recently, graphene, a single atomic layer sheet of sp^2-bonded carbon, has gained tremendous attention because of its unique properties such as high surface area [4, 5]. It was proposed that graphene has the potential to become an ideal substrate for hydrogen storage application. Srinivas et al. [6] measured the capacity of

Z.G. Bajestani • Y. Yürüm (✉)
Faculty of Engineering and Natural Sciences, Sabanci University,
Orhanli, Tuzla, Istanbul 34956, Turkey
e-mail: yyurum@sabanciuniv.edu

I. Dincer et al. (eds.), *Progress in Clean Energy, Volume 2*,
DOI 10.1007/978-3-319-17031-2_56

graphene for hydrogen storage at different temperatures that is 1.2 and 0.1 wt% at 77 and 298 K, respectively.

Doping graphene with appropriate chemical groups is a method to change the properties of graphene. Investigations show that dispersion of transition metals such as Pd on carbon materials improves hydrogen storage capacity by chemical adsorption [3]. Lebon et al. [7] studied the hydrogen storage of graphene doped with 3d transition metals by using density functional calculations. They found that the titanium-doped graphene sheets are capable of adsorbing hydrogen molecules and can meet the hydrogen storage target of the US Department of Energy. Rojas and Leiva [8] showed that each titanium adatom can adsorb four hydrogen molecules with an average binding energy up to −0.42 eV/H_2 molecule.

In this work, we prepared graphene by thermal exfoliation of graphite oxide obtained by Hummers' Method. We dispersed titanium on graphene sheets by incipient wetness impregnation method and measured hydrogen storage capacity by hydrogen adsorption isotherm. The effect of titanium addition on graphene sheets was studied by different methods.

56.2 Experiment

To form graphite oxide (GO), graphite flakes were first mixed with sulfuric acid (H_2SO_4) and sodium nitrate, ($NaNO_3$) and then potassium permanganate ($KMnO_4$) was added slowly during 2 h. After completion of the oxidation stage, mixture was added to water and filtered. Then, it is rinsed and redispersed in 5 % HCl solution. Finally, the mixture was washed with water till reaching neutral pH and dried at 60 °C under vacuum. RGO was obtained by thermal exfoliation/reduction of GO at 1,000 °C under controlled atmosphere for 12 min.

The decoration of graphene with titanium was performed with titanium chloride as the precursor. GO was dispersed in water and sonicated for 2 h and then mixed for 24 h with titanium chloride under vigorous stirring to form a stable suspension. RGO–Ti was obtained by the same thermal reduction method as the RGO sample.

As-prepared samples were characterized by X-ray diffraction (XRD), Raman spectroscopic analysis, X-ray photoelectron spectroscopy (XPS), and scanning electron microscopy (SEM) equipped with an energy-dispersive X-ray (EDX) analysis. The specific surface area of samples was determined by analyzing the standard nitrogen adsorption isotherms (BET) at 77 K. To measure the hydrogen storage capacity, samples were degassed for 12 h under high vacuum at 100 °C. The hydrogen adsorption isotherms were measured at room temperature up to 10 bar.

56.3 Results and Discussion

It has been reported that the intensity of (002) diffraction line at $2\theta = 26°$ is inversely proportional to oxidation degree of the graphite oxide, whereas the absence of this (002) line implies a fully oxidized graphite. On the other hand, the change of crystal structure from raw graphite to oxidized state results in an appearance of (001) peak at $2\theta = 10.5°$. Figure 56.1 shows XRD diffractograms of the graphite, GO, RGO, and RGO–Ti. Intercalation of oxygen-containing groups between graphene layers changed the interlayer spacing (d-spacing) from 3.3 Å in graphite to 8.37 Å in GO. Observation of (002) peak and absence of (001) peak in the XRD pattern of RGO suggest that this RGO sample is reduced completely and composed of free graphene-like sheets with poorly ordered structure along the stacking direction. The absence of titanium-characteristic peaks in RGO–Ti is an indication of too low metal content and/or extremely small metal particle size [9, 10].

Raman spectroscopy was used to find the structural information required to define the density of defects in graphene sheets [11]. Raman spectra (Fig. 56.2) of all samples display two characteristic bands at 1,607 cm^{-1} (G band) and 1,373 cm^{-1} (D band) related to the vibration of the sp^2-bonded carbon atoms in a two-dimensional hexagonal lattice and defects in the hexagonal graphitic layers respectively [11–13].

Thermal exfoliation and the presence of various functional groups develop different degrees of disorder in carbon-based material. Many researchers have used the intensity ratio of the D and G bands (D/G ratio) to compare the degree

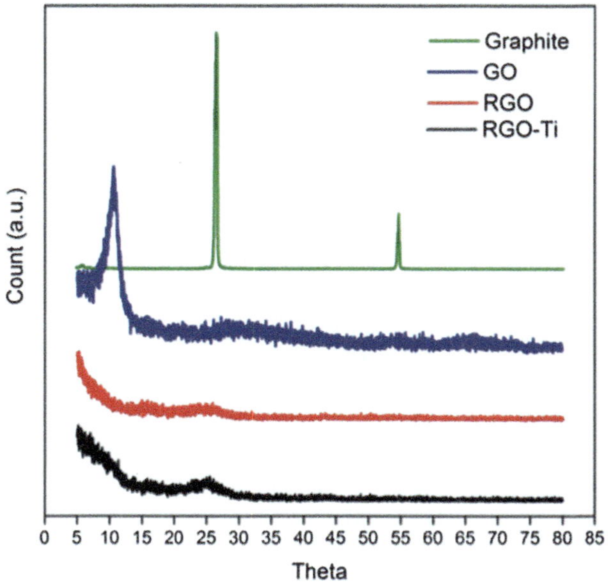

Fig. 56.1 XRD pattern of graphite (*green*), GO (*blue*), RGO (*red*), and RGO–Ti (*black*)

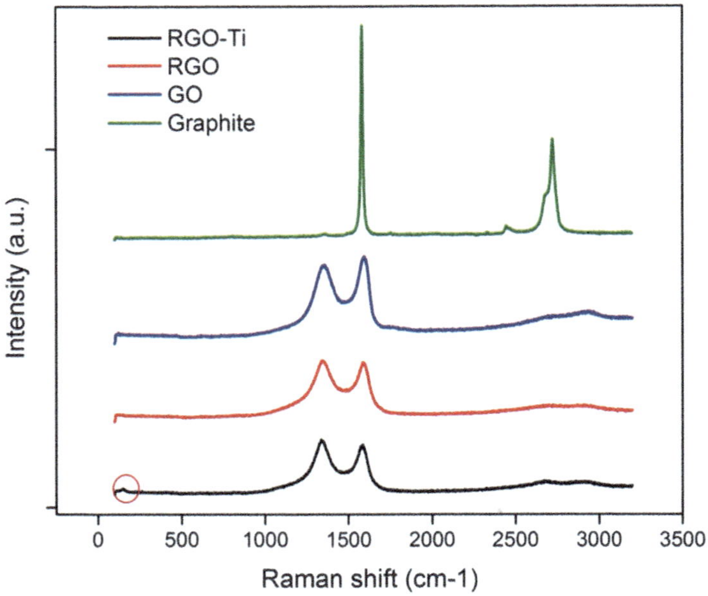

Fig. 56.2 Raman spectra of graphite (*green*), GO (*blue*), RGO (*red*), and RGO–Ti (*black*)

of disorder of graphene sheets qualitatively [11, 12, 14]. Multilayer graphene structures such as graphite show a peak at ~2,720 cm^{-1} (2D band). The intensity of 2D peak in GO, RGO, and RGO–Ti declined compared to that of graphite, indicating the presence of structural defects. The intensity of D band in graphite is the lowest one among the D bands of all samples. After oxidation a high D/G = 0.88 was observed in the Raman spectrum of GO due to the existence of a large amount of oxygen-containing functional groups. Reduced samples, RGO and RGO–Ti, exhibited higher D/G ratios of 1.02 and 1.1, respectively, showing that thermal reduction increased the density of defects in these samples. These results are consistent with the previous data published in literature [14].

In addition to G and D bands, we observed an extra peak at 148 cm^{-1} in RGO–Ti. It has been reported that anatase TiO_2 exhibits Raman band at ~144 cm^{-1} (E_g). The coexistence of E_g band from anatase and D and G bands from RGO confirms the presence of TiO_2 on graphene sheets [15–19].

We characterized the chemical structure of GO, RGO, and RGO–Ti by using XPS. We observed four peaks (Fig. 56.3a) at 284.56, 285.02, 286.8, and 288.8 eV in the C 1 s XPS spectrum of GO that are related to the C–C (aromatic) [20], C–OH [21, 22], C (epoxy/alkoxy)/C=O [23], and O=C–O (carboxylic) [20] groups, respectively. The presence of these peaks implies the occurrence of a high degree of oxidation during the oxidation process. Figure 56.3b shows the C 1 s XPS of graphite oxide after thermal reduction. The absence of C–OH peak and decrease of intensity in oxygen-containing group components demonstrated that most of the functional groups were removed from graphene sheets [20, 22]. The survey scan

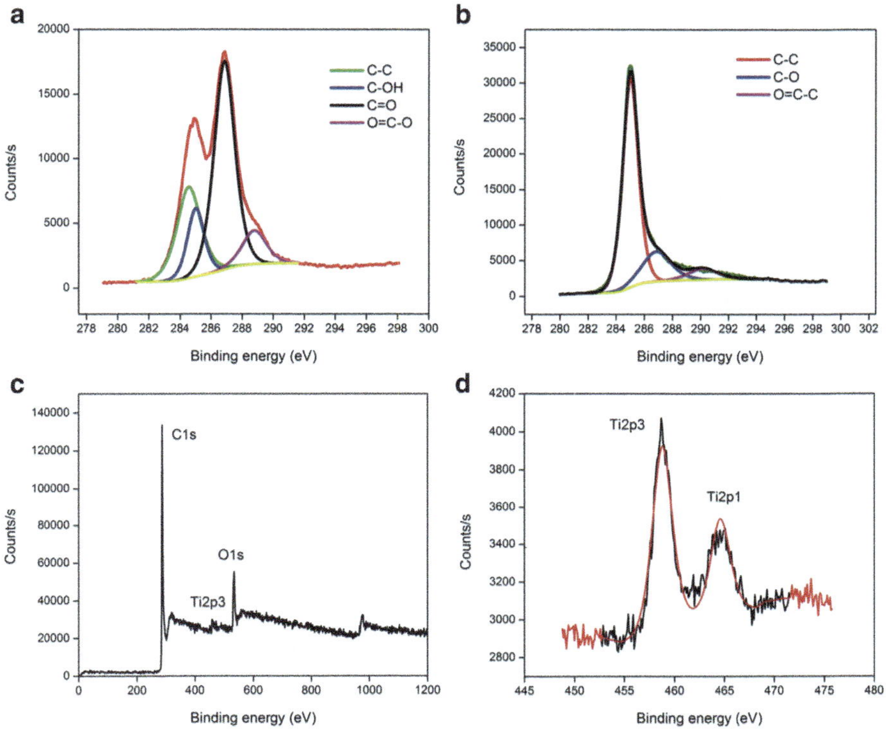

Fig. 56.3 XPS spectrum of C 1 s of GO (**a**) and C 1 s of RGO (**b**); survey scan (**c**) and Ti2p (**d**) of the RGO–Ti

result of the RGO–Ti (Fig. 56.3c) indicates a sharp peak at 285 eV and two low-intensity peaks at 458 and 530 eV that are in agreement with the C1s, Ti2p3, and O1s spectra of RGO–Ti. Figure 56.3d displays the Ti2p XPS spectrum of RGO–Ti in which the peaks located in 464.58 and 458.75 eV correspond to Ti 2p1/2 and Ti 2p3/2, respectively. It has been reported that titanium atoms react with any available oxygen to form TiO_2 with the bonding energy similar to what we observed. Many researchers [24–26] suggested that the abovementioned bonding energies belong to titanium in the fully oxidized state (Ti^{+4}).

We compared the BET surface area of reduced samples by using nitrogen adsorption–desorption isotherms at 77 K (Fig. 56.4). The nitrogen isotherms on RGO and RGO–Ti show that the graphene adsorbs more N_2 than the graphene-containing titanium. The reasons behind this observation are higher surface area and lower density of RGO compared to RGO–Ti [27]. The surface area and total pore volume of RGO were measured to be 655.799 m^2/g and 2.41 cc/g compared to 371.144 m^2/g and 1.25 cc/g for those of RGO–Ti.

We monitored the morphological structure of graphene sheets in different specimens by using SEM. Figure 56.5 exhibits exfoliated graphene sheets with layered and transparent structure. A higher degree of exfoliation was detected in RGO

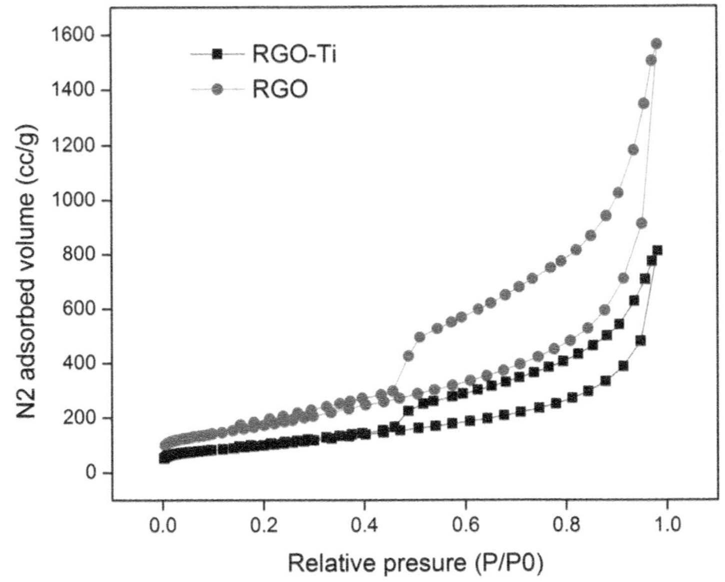

Fig. 56.4 Nitrogen isotherms measured for RGO and RGO–Ti

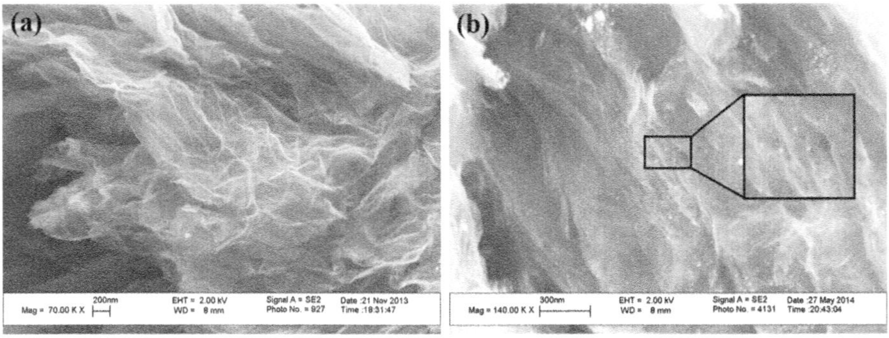

Fig. 56.5 SEM image of (**a**) RGO and (**b**) RGO–Ti

compared to RGO–Ti that can be ascribed by the restacking of graphene sheets in the mechanical mixing of graphene with titanium [28]. This observation is in agreement with the decrease of BET surface area that was extracted from N_2 adsorption–desorption measurement. Figure 56.5b shows a SEM micrograph of RGO–Ti sample. Since the results of XPS and Raman spectroscopy confirmed the existence of TiO_2 in this sample, we conclude that bright points with diameter of less than 25 nm (shown with the box in Fig. 56.5b) illustrate titanium-containing particles.

An elemental analysis of the RGO–Ti sample was shown in Fig. 56.6. The appearance of a peak at 4.5 keV indicates the presence of titanium in the sample [29].

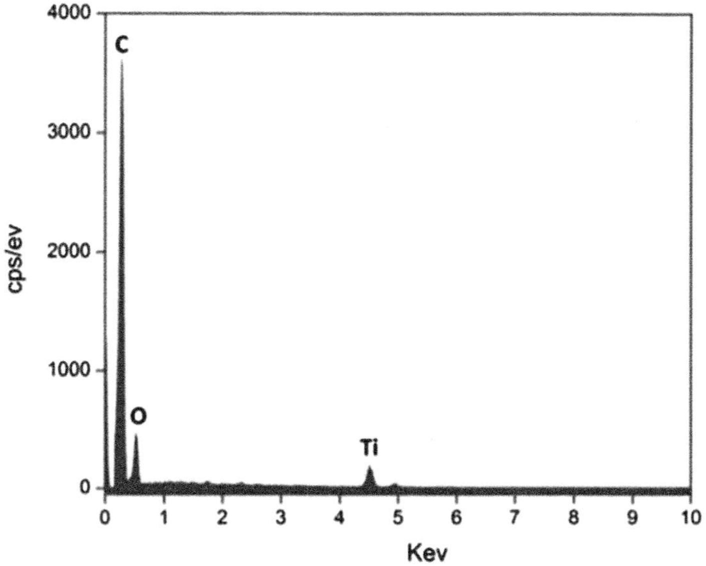

Fig. 56.6 EDX spectra of RGO–Ti

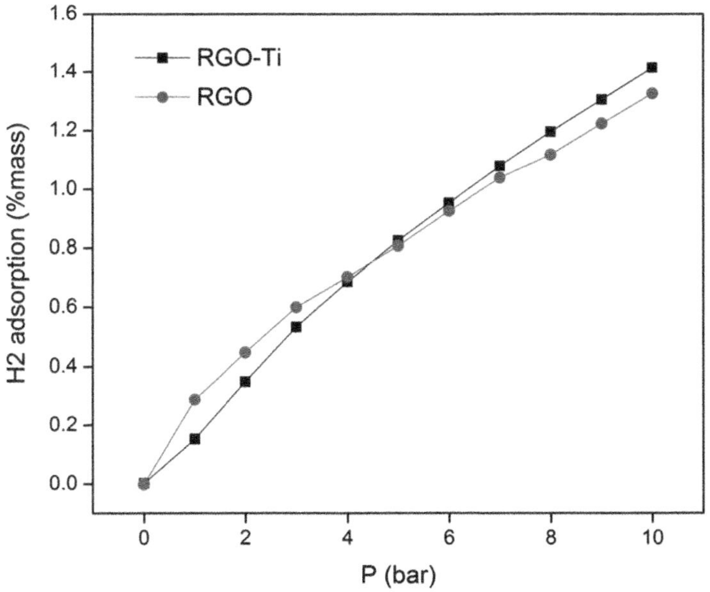

Fig. 56.7 H_2 adsorption of RGO and RGO–Ti

Figure 56.7 presents the hydrogen adsorption isotherms at room temperature for RGO and RGO–Ti up to 10 bar. Increasing the pressure of hydrogen improved the hydrogen uptake of samples. The pristine graphene shows hydrogen storage capacity equal to 1.3 wt%, while decoration of graphene with titanium enhanced the hydrogen storage capacity to 1.4 wt%.

According to BET results, decrease of surface area and total pore volume, we can conclude that hydrogen uptake via the physisorption mechanism is lowered in RGO–Ti, but based on Wang et al.'s [27] work on Ni–B nanoalloy-doped graphene, we suggest that decoration of graphene with titanium improves hydrogen storage capacity by chemisorption of hydrogen.

56.4 Conclusion

Graphene and graphene supported by titanium were prepared by chemical method. XPS results showed that titanium reacted with any available oxygen atoms at first to form TiO_2. Analysis of nitrogen isotherm indicated a decrease of surface area and total pore volume in the titanium-decorated sample; however, we found that decoration of graphene with titanium increased the hydrogen storage capacity by chemisorption mechanism.

References

1. Zhang HP, Luo XG, Lin XY, Lu X, Leng L (2013) Density functional theory calculations of hydrogen adsorption on Ti-, Zn-, Zr-, Al-, and N-doped and intrinsic graphene sheets. Int J Hydrogen Energy 38:14269–14275
2. Vinayan BP, Sethupathi K, Ramaprabhu S (2013) Facile synthesis of triangular shaped palladium nanoparticles decorated nitrogen doped graphene and their catalytic study for renewable energy applications. Int J Hydrogen Energy 38:2240–2250
3. Chen CH, Chung TY, Shen CC, Yu MS, Tsao CS, Shi GN, Huang CC, Ger MG, Lee WL (2013) Hydrogen storage performance in palladium-doped graphene/carbon composites. Int J Hydrogen Energy 38:3681–3688
4. Spyrou K, Gournis D, Rudolfa P (2013) Hydrogen storage in graphene-based materials: efforts towards enhanced hydrogen absorption. ECS J Solid State Sci Technol 2(10):M3160–M3169
5. Hong W, Kim BH, Lee SM, Yu HY, Yun YJ, Jun Y, Lee JB, Kim HJ (2012) Agent-free synthesis of graphene oxide/transition metal oxide composites and its application for hydrogen storage. Int J Hydrogen Energy 37:7594–7599
6. Srinivas G, Zhu Y, Piner R, Skipper N, Ellerby M, Ruoff R (2010) Synthesis of graphene-like nanosheets and their hydrogen adsorption capacity. Carbon 48:630–635
7. Lebon A, Carrete J, Longo RC, Vega A, Gallego LJ (2013) Molecular hydrogen uptake by zigzag graphene nanoribbons doped with early 3d transition-metal Atoms. Int J Hydrogen Energy 38:8872–8880
8. Rojas MI, Leiva EPM (2007) Density functional theory study of a graphene sheet modified with titanium in contact with different adsorbates. Phys Rev B 76:155415
9. Huang CC, Pu NW, Wang CA, Huang JC, Sung Y, Ger MD (2011) Hydrogen storage in graphene decorated with Pd and Pt nano-particles using an electroless deposition technique. Sep Purif Technol 82:210–215
10. Saner B, Okyay F, Yürüm Y (2010) Utilization of multiple graphene layers in fuel cells. 1. An improved technique for the exfoliation of graphene-based nanosheets from graphite. Fuel 89:1903–1910

11. Poh H, Sanek F, Ambrosi A, Zhao G, Soferb Z, Pumera M (2012) Graphenes prepared by Staudenmaier, Hofmann and Hummers methods with consequent thermal exfoliation exhibit very different electrochemical properties. Nanoscale 4:3515–3522
12. Giovanni M, Poh HL, Ambrosi A, Zhao G, Sofer Z, Sanek F, Khezri B, Webstera R, Pumera M (2012) Noble metal (Pd, Ru, Rh, Pt, Au, Ag) doped graphene hybrids for Electrocatalysis. Nanoscale 4:5002–5008
13. Wang X, Zhang X (2013) Electrochemical co-reduction synthesis of graphene/nano-gold composites and its application to electrochemical glucose biosensor. Electrochim Acta 112:774–782
14. Ambrosi A, Bonanni A, Sofer Z, Cross J, Pumera M (2011) Electrochemistry at chemically modified graphenes. Chem Eur J 17:10763–10770
15. Choi HC, Jung YM, Kim SB (2005) Size effects in the Raman spectra of TiO2 nanoparticles. Vib Spectrosc 37:33–38
16. Yoshitake H, Abe D (2009) Raman spectroscopic study of the framework structure of amorphous mesoporous titania. Micropor Mesopor Mat 119:267–275
17. Kadam RM, Rajeswari B, Sengupta A, Achary SN, Kshirsagar RJ, Natarajan V (2014) Structural characterization of titania by X-ray diffraction, photo acoustic, Raman Spectroscopy and electron paramagnetic resonance spectroscopy. Spectrochim Acta A S1386–1425
18. Liu H, Duan C, Su X, Dong X, Huang Z, Shen W, Zhu Z (2014) A hemoglobin encapsulated titania nanosheet modified reduced graphene oxide nanocomposite as a mediator-free biosensor. Sensors Actuators B 203:303–310
19. Li K, Chen T, Yan L, Dai Y, Huang Z, Xiong J, Song D, Lv Y, Zeng Z (2013) Design of graphene and silica co-doped titania composites with ordered mesostructure and their simulated sunlight photocatalytic performance towards atrazine degradation. Colloids Surf A 422:90–99
20. Liu YW, Guan MX, Feng L, Deng SL, Bao JF, Xie SY, Chen Z, Huang RB, Zhen LS (2013) Facile and straightforward synthesis of superparamagnetic reduced graphene oxide–Fe3O4 hybrid composite by a solvothermal reaction. Nanotechnology 24:025604
21. Compagnini G, Russo P, Tomarchio F, Puglisi O, D'Urso L, Scalese S (2012) Laser assisted green synthesis of free standing reduced graphene oxides at the water–air interface. Nanotechnology 23:505601
22. Kim JM, Hong GW, Lee SM, Chang SJ, Jun Y, Kim BH, Kim HJ (2014) Energy storage of thermally reduced graphene oxide. Int J Hydrogen Energy 39:3799–3804
23. Wang P, Zhai Y, Wang D, Dong S (2011) Synthesis of reduced graphene oxide-anatase TiO2 nanocomposite and its improved photo-induced charge transfer properties. Nanoscale 3:1640–1645
24. Felten A, Suarez-Martinez I, Ke X, Tendeloo G, Ghijsen J, Pireaux JJ, Drube W, Bittencourt CP, Ewels C (2009) The interface between titanium and carbon nanotubes. ChemPhysChem 10:1799–1804
25. Pisarek M, Roguska A, Marcon L, Andrzejczuk M (2012) Biomimetic and electrodeposited calcium-phosphates coatings on Ti – formation, surface characterization, biological response. Biomed Eng, Chapter one, InTech
26. Biesingera MC, Lau LWM, Gerson AR, St C Smart R (2010) Resolving surface chemical states in XPS analysis of first row transition metals, oxides and hydroxides: Sc, Ti, V, Cu and Zn. Appl Surf Sci 257:887–898
27. Wang Y, Liu J, Wang K, Chen T, Tan X, Li CM (2011) Hydrogen storage in Ni-B nanoalloy-doped 2D graphene. Int J Hydrogen Energy 36:12950–12954
28. Parambhath VB, Nagar R, Ramaprabhu S (2012) Effect of nitrogen doping on hydrogen storage capacity of palladium decorated graphene. Langmuir 28:7826–7833
29. Ullah K, Zhu L, Meng ZD, Ye S, Sun Q, Oh WC (2013) A facile and fast synthesis of novel composite Pt–graphene/TiO2 with enhanced photocatalytic activity under UV/visible light. Chem Eng J 231:76–83

Chapter 57
Maximisation and Optimisation of the Total Solar Radiation Reaching the Solar Collector Surfaces

Kacem Gairaa, Abdallah Khellaf, Farouk Chellali, Said Benkaciali, Yahia Bakelli, and Salah Bezari

Abstract The performance of solar conversion systems is affected by its orientation and tilt angle with the horizontal plane. This is because both of these parameters change the amount of solar energy received by the collector surfaces.

In this study, the optimum tilt angle of solar systems is determined in order to increase the solar radiation amount reaching the collector plane. Mathematical models have been used for estimating the total (global) solar radiation on a sloped surface and to determine the optimum tilt and surface azimuth angles in Ghardaïa area, southern Algeria, on a daily basis, as well as for a specific period of the year. Then, these models are compared with experimental data for choosing the best one, through the statistical test calculation.

For this purpose, a database for one complete year of solar radiation components (global, diffuse and direct) has been used.

The results obtained show that the optimum angle for each month allows us to collect the maximum solar radiation; the annual tilt angle is approximately equal to the latitude of the site and the surface azimuth is due south. The loss in the amount of energy between a fixed plane of 32.6° and another one adjustable according to the monthly optimum tilt angle is around 10 %.

Keywords Solar collector • Inclined surface • Solar radiation • Optimum angle

---

K. Gairaa (✉)
Unité de Recherche Appliquée en Energies Renouvelables, URAER, Centre de Développement des Energies Renouvelables, CDER, 47133 Ghardaïa, Algeria

Department of Electrical Engineering, Faculty of Applied Science, Ibn Khaldoun University, BP P 78, Zaaroura 14000, Tiaret, Algeria
e-mail: giso147@gmail.com

A. Khellaf
Centre de Développement des Energies Renouvelables, CDER, BP 62 Route de l'Observatoire, Bouzaréah 16340, Algiers, Algeria

F. Chellali • S. Benkaciali • Y. Bakelli • S. Bezari
Unité de Recherche Appliquée en Energies Renouvelables, URAER, Centre de Développement des Energies Renouvelables, CDER, 47133 Ghardaïa, Algeria

I. Dincer et al. (eds.), *Progress in Clean Energy, Volume 2*,
DOI 10.1007/978-3-319-17031-2_57

Nomenclature

A_I	Anisotropy index in HDKR model
B_n	Beam solar irradiance W/m^2
D_h	Diffuse solar irradiance on horizontal surface W/m^2
f'	Modulating factor in HDKR model
F	Modulating function in Klucher model
f'_1	Circumsolar brightening coefficient in Perez model
f'_2	Horizon brightening coefficient in Perez model
G_h	Global solar irradiance on horizontal surface W/m^2
G_β	Global solar irradiance on inclined surface W/m^2
$G_{\beta,e}$	Estimated global irradiation on inclined surface $kWh/m^2/day$
$G_{\beta,m}$	Measured global irradiation on inclined surface $kWh/m^2/day$
I_{on}	Extraterrestrial solar irradiance W/m^2
MBE	Mean bias error
MPE	Mean relative percentage error
R_b	Ratio of the beam radiation on the tilted surface to that on horizontal surface
R_d	Ratio of the diffuse radiation on the tilted surface to that on horizontal surface
RMSE	Root mean square error

Greek Letters

A	Solar elevation angle °
ρ	Ground albedo
β	Tilt angle °
θ	Incidence angle °
θ_z	Zenith angle °
δ	Sun declination angle °
Ø	Latitude °
γ	Azimuth angle °
ω	Hour angle °

57.1 Introduction

In the recent years, the world is interested by the renewable energies more than before to reduce the greenhouse effect, caused by the absurd exploitation of conventional sources. This interest is resulted in the launch of several ambitious programs in order to develop cleaner energy sources and energy efficiency.

Measurements of global solar radiation reaching the surface of the earth and its two components direct and diffuse are essential in the most research fields of solar energy. The daily values as well as the monthly ones are needed to evaluate the performance of existing solar devices, and to estimate the efficiency of the future installations [1].

The optimum design of photovoltaic or thermal solar systems for any application is important to determine their performance at the site of installation. The amount of power produced by a solar energy conversion system depends upon the amount of sunlight reaching the collector. So, more light means more power; the performance of a solar collector is highly influenced by its orientation and its angle of tilt with the horizontal. This is due to the fact that both the orientation and tilt angle change the solar radiation reaching the collector surfaces. The best way to collect maximum daily energy is the use of tracking systems, but in most practical cases, the tracker cannot be used due to its expensive cost and for the complication of some requested PV or thermal configuration systems.

In the literature, many researchers in different sites of the world have determined the optimum tilt angle and orientation of solar collectors. In general, the obtained results suggested that the yearly optimum tilt angle is found to be around the latitude of the site, while the optimum azimuth is full south in the hemisphere north (due north in the hemisphere south). A brief state of the art of some works regarding optimum tilt angle is cited below:

Benghanem [2] has optimised the tilt angle for solar panel in Madinah (Saudi Arabia); he has used measurements of daily global and diffuse solar radiation on a horizontal surface; the results gotten showed that the optimum tilt angle for each month allows to collect the maximum solar energy. He found too the loss in the amount of collecting energy when using the yearly average fixed angle of around 8 % compared with the monthly optimum tilt angle.

Ghosh et al. [3] have used isotropic and anisotropic models for estimating the global solar radiation on an inclined plane. They found that the anisotropic model of Perez gives the best result based on a statistical test with an RMSE of 0.09. Thereafter, with this model they determined the optimum tilt and orientation angles for Dhaka (Bangladesh) site. Koray et al. [4] performed the optimum tilt angle in Izmir (Turkey), in order to maximise the solar radiation reaching solar collectors. He has employed the isotropic model of Liu and Jordan to calculate the total solar radiation on inclined surface. His obtained results indicated that the seasonal tilt angle should be 55.7° in winter, 18.3° in spring, 4.3° in summer and 43° in autumn. Murat et al. [5] have examined the optimum tilt angle and orientation of PV panels in Sanliurfa (Turkey); they found that the monthly optimum tilt angles change throughout the year with the minimum values obtained in June and the maximum ones achieved in December. Mohd et al. [6] have determined the collector optimum tilt angle and orientation in Brunei Darussalam; they revealed that changing tilt angle 12 times in a year maintains approximately the total solar radiation near the maximum value that is found by changing the tilt angle daily to its optimum value. The yearly gain is 5 % more than the case of collector fixed on a horizontal surface. Morcos [7] studied the optimum tilt angle and orientation in Assiut (Egypt); he noted that changing the tilt angle eight times in a year is necessary to maintain the global solar radiation near its maximum values. He has also found that the gain in global solar radiation is 6.85 % more than the case of collector fixed at the latitude of the site.

The goal of this study is the determination of the optimum tilt angle and orientation in order to increase the amount of solar radiation reaching the collector surfaces on a daily and on a monthly basis as well as for a specific period of the year. Four mathematical models have been tested to calculate the global solar radiation on inclined surface. A case study of a Ghardaïa site (south Algeria) has been taken as an example of application.

57.2 Description of the Site and Data Source

Ghardaïa site is considered an arid and dry area, located in the south of Algeria, about 600 Km south of the capital city (Fig. 57.1); it is framed by the following geographical coordinates: latitude of 32°36′ N, longitude of 3°48′ E and altitude of 450 m above mean sea level [8].

The measured data on solar radiation used in this study were collected by a radiometric station presented in Fig. 57.2 with high precision, installed on the roof of solar radiation laboratory. The station has two parts:

- A fixed part consists of two *EKO MS-64* pyranometers for the measurement of global solar radiation on a horizontal surface (its short-wave sensitivity is 7.0 (mV/kW/m^2)) and on inclined surface at the latitude of the site.
- A moving part, which is able to track the path of the sun from sunrise to sunset.

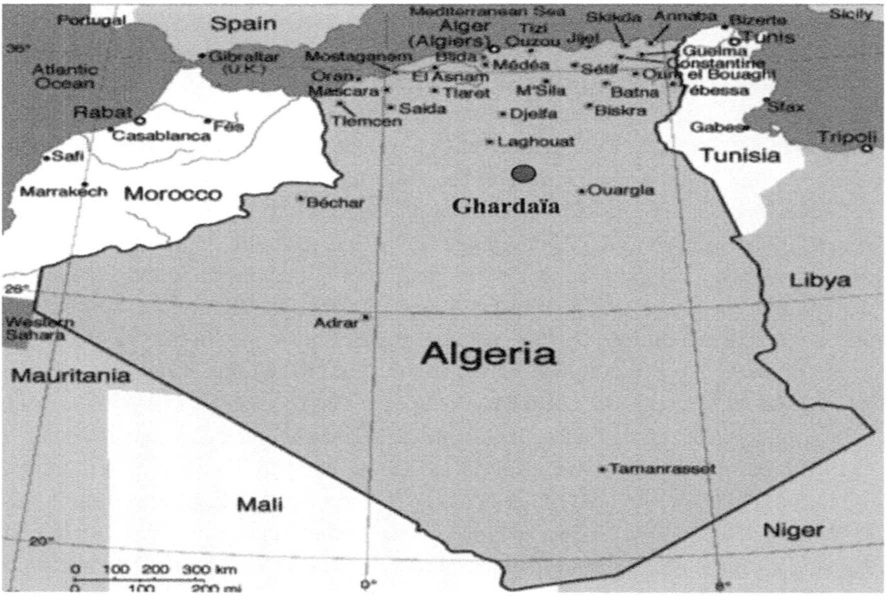

Fig. 57.1 Ghardaïa site location

Fig. 57.2 Radiometric station

The last one consists of an *EKO MS-101D* pyrheliometer with short-wave sensitivity of 6.71 (mV/kW/m^2), which is pointed at the sun disc for measuring the DNI component. Another *EKO MS-64* pyranometer with short-wave sensitivity of 7.0 (mV/kW/m^2) for the measurement of diffuse solar radiation on the horizontal plane is equipped with a shadow band for hiding the radiant flux coming directly from the sun. All solar components are made with an interval of five minutes for each one. Figure 57.3 illustrates the monthly average daily solar radiation components measured in Ghardaïa area.

57.3 Global Solar Radiation on Inclined Surface (G_β)

Several models have been proposed by various researchers to calculate global solar radiation on tilt surfaces from the available data on a horizontal surface. The only difference among the models appears in the assessment of sky diffuse components; the global solar radiation on sloped surface is the sum of three components, direct, diffuse and reflected radiation (Fig. 57.4), and can be expressed as

$$G_\beta = (G_h - D_h)R_b + D_h R_d + G_h R_r \quad (57.1)$$

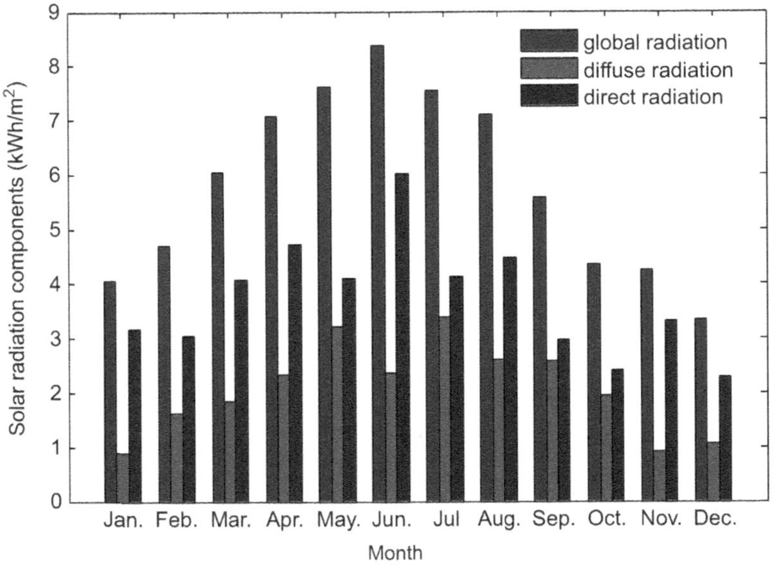

Fig. 57.3 Monthly average global, diffuse and direct solar radiation falling on horizontal surface measured in Ghardaïa site

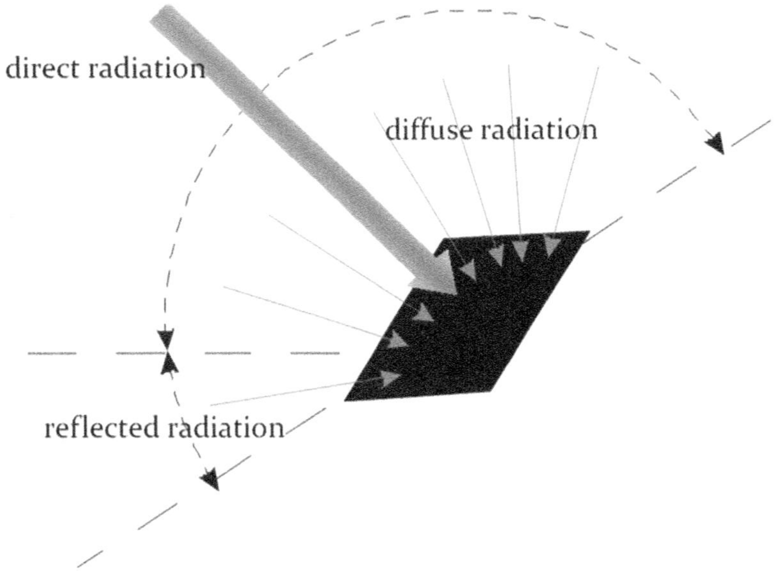

Fig. 57.4 Composition of the global solar radiation on a tilted surface

where R_r is ground reflection term

$$R_r = \rho\left(\frac{1 - \cos\beta}{2}\right) \quad (57.2)$$

R_b is the ratio of the beam radiation on the tilted surface to that on horizontal surface calculated as

$$R_b = \frac{\cos\theta}{\cos\theta_z} \quad (57.3)$$

θ and θ_z are calculated by the following equations:

$$\cos\theta = A + B \quad (57.4)$$

$$A = \sin\delta\sin\emptyset\cos\beta - \sin\delta\cos\emptyset\sin\beta\cos\gamma + \cos\delta\cos\emptyset\cos\beta\cos\omega \quad (57.5)$$

$$B = \cos\delta\sin\emptyset\sin\beta\cos\omega\cos\gamma + \cos\delta\sin\gamma\sin\beta\sin\omega \quad (57.6)$$

$$\cos\theta_z = \sin\delta\sin\emptyset + \cos\emptyset\cos\delta\cos\omega \quad (57.7)$$

Based on the suppositions made by Iqbal [9] and Duffie et al. [12], the estimation models can be classified into isotropic or anisotropic models.

57.3.1 *Isotropic model*

57.3.1.1 Liu and Jordan model [9]

This model assumes that the diffuse radiation is uniformly distributed over the sky dome; the radiation tilt factor (ratio of the diffuse radiation on inclined surface to that on horizontal surface) is given by

$$R_d = (1 + \cos\beta)/2 \quad (57.8)$$

57.3.2 *Anisotropic Model*

These models assume that the anisotropy of the diffuse radiation is in the circumsolar region resulting from forward scattering of solar radiation and concentrated in the part of the sky, while it is isotropically distributed in the rest of the sky dome.

Figure 57.5 shows the schematic distribution of the diffuse solar radiation for both isotropic and anisotropic models.

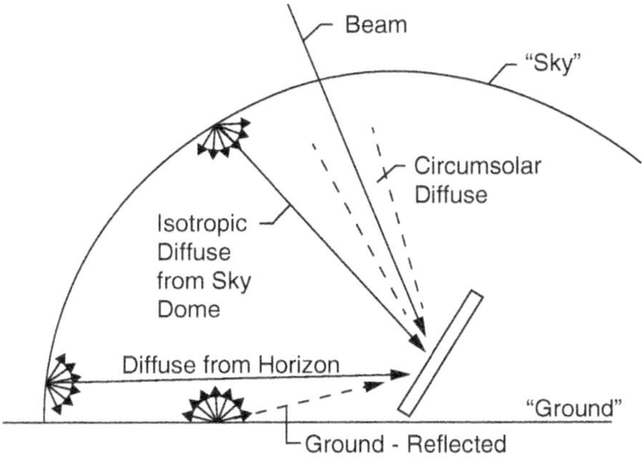

Fig. 57.5 Isotropic and anisotropic diffuse solar radiation [9, 12]

57.3.2.1 Klucher Model [10]

In this model, the condition of cloudiness of the sky has been taken into account; the diffuse radiation tilt factor is obtained from the relation

$$R_{\mathrm{d}} = \frac{1}{2}(1 + \cos\beta)\left[1 + F\sin^{3}(\beta/2)\right]\left(1 + F\cos^{2}\theta_{z}\sin^{3}\left(\frac{\pi}{2} - \alpha\right)\right) \quad (57.9)$$

while

$$F = 1 - \left(\frac{D_{\mathrm{h}}}{G_{\mathrm{h}}}\right)^{2} \quad (57.10)$$

57.3.2.2 Perez Model [11]

The diffuse tilt factor in this model is described as

$$R_{d} = \frac{1}{2}(1 + \cos\beta)\left(1 - F_{1}^{'}\right) + F_{1}^{'}{}^{\left(a/b\right)+F_{2}^{'}\sin\beta} \quad (57.11)$$

where $F_{1}^{'}$ and $F_{2}^{'}$ are circumsolar and horizon brightness coefficients and the values are presented in a table given by Perez using data for nine stations having various climatic conditions. a and b are terms that account for the angles of incidence of the cone of circumsolar radiation on the tilted and horizontal surfaces ($a = \max(0, \cos\theta)$ and $b = \max(0.087, \cos\theta_{z})$).

57.3.2.3 HDKR Model [12]

The diffuse tilt factor in the HDKR model is given as follows:

$$R_{\mathrm{d}} = (1 - A_l)\left(\frac{1 + \cos\beta}{2}\right)\left(1 + f^{'}\sin^{3}\left(\frac{\beta}{2}\right)\right) + A_l R_{\mathrm{b}} \tag{57.12}$$

The modulating factor f' accounts for the effect of the cloudiness, while the coefficient A_l is the anisotropy index and it defines the portion of horizontal diffuse radiation to be treated as circumsolar:

$$A_l = \frac{B_{\mathrm{n}}}{I_{\mathrm{on}}} \tag{57.13}$$

$$f^{'} = \sqrt{\frac{B_{\mathrm{h}}}{G_{\mathrm{h}}}} \tag{57.14}$$

57.4 Statistical Test

The accuracy of the estimated models will be judged by the statistical indicators, such as the mean bias error (MBE), root mean square error (RMSE) and mean relative percentage error (MPE). These indicators are usually applied in the comparison of solar radiation models. The mean square error provides information about the performance of correlations, which enable comparison of the real differences between the estimated values and the measured ones; a low RMSE is desirable. The MBE provides the long-term performance of the model; in general, the positive MBE shows overestimation while the negative MBE indicates underestimation. The expression of each statistical indicator is given below:

$$RMSE = \left[\frac{1}{n}\sum\nolimits_{i=1}^{n}\left(\frac{G_{\beta,\mathrm{e}} - G_{\beta,\mathrm{m}}}{G_{\beta,\mathrm{m}}}\right)^{2}\right]^{1/2} \tag{57.15}$$

$$MBE = \frac{1}{n}\left(\sum\nolimits_{i=1}^{n}\frac{G_{\beta,\mathrm{e}} - G_{\beta,\mathrm{m}}}{G_{\beta,\mathrm{m}}}\right) \tag{57.16}$$

$$MPE = \left[\left(\frac{G_{\beta,\mathrm{m}} - G_{\beta,\mathrm{e}}}{G_{\beta,\mathrm{m}}}\right)\right]100 \tag{57.17}$$

57.5 Results and Discussion

Using the equations described above, the global solar radiation falling on an inclined surface has been calculated for different angles of inclination. A program written in Matlab is used for the calculation of G_β for the isotropic and anisotropic models as well as the statistical errors. Measurements of global and diffuse solar radiation on a horizontal surface and the direct normal components with an interval of 5 min for the whole year are used as input parameters. Table 57.1 summarises the results of the statistical comparison obtained from the four tested models. It is clear that the anisotropic *HDKR* model gives least errors monthly and annually. Hence, this model is used therefore for determining the different tilt angles.

In order to determine the yearly optimum tilt angle in Ghardaïa site, the tilt angle has been varied from 0° to 90° in steps of 1°, when the azimuth angle has been varied from −90 to +90 covering all orientations. The optimum tilt angle is corresponding to the maximum of energy; the obtained results show that the yearly angle is around the latitude of the site ($\beta = 32.6°$) and the azimuth angle is due full south ($\gamma = 0°$); this result indicates an optimum fixed tilt throughout a year. Figure 57.6 illustrates the contour of global solar radiation on a sloped surface for

Table 57.1 Statistical comparison for the four studied models

Model	Statistical indicators		
	RMSE	MBE	MPE
Liu and Jordan	0.021	-0.0189	2
HDKR	*0.0216*	*0.0013*	*0.212*
Klucher	0.0344	0.0284	2.69
Perez	0.109	0.0518	5.54

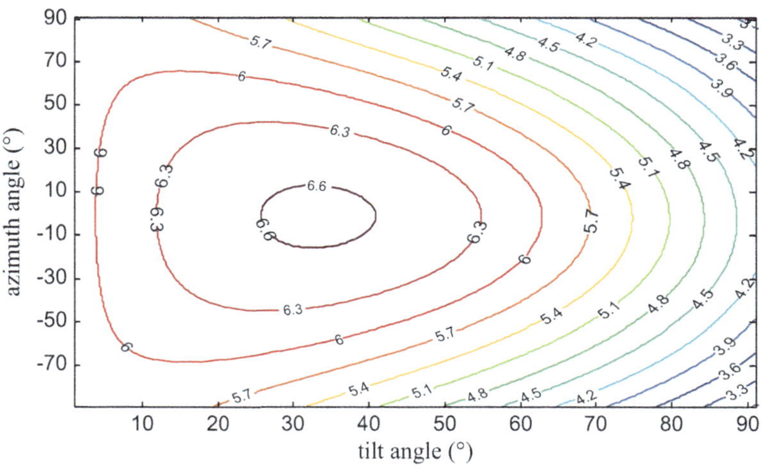

Fig. 57.6 Contour of global solar radiation (kWh/m^2) falling on inclined surface

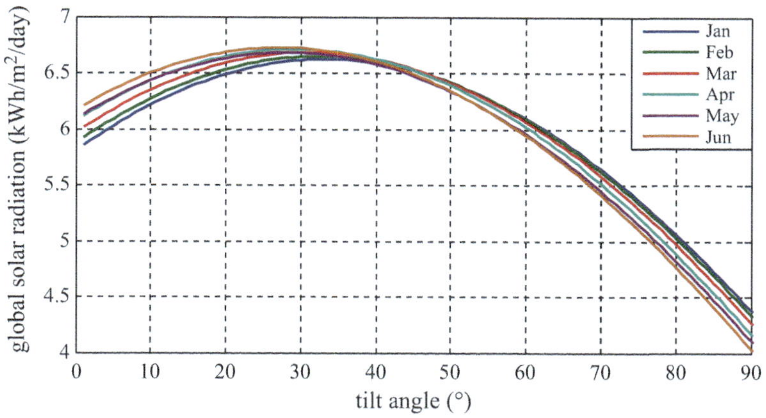

Fig. 57.7 Monthly average daily global solar irradiations for different tilt angles (January–June)

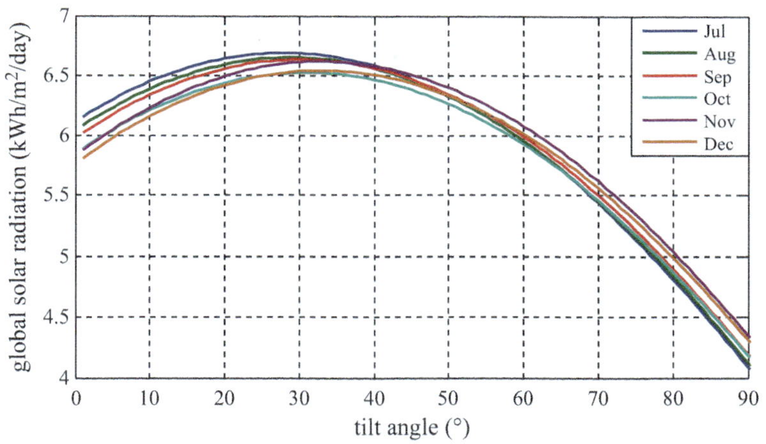

Fig. 57.8 Monthly average daily global solar irradiations for different tilt angles (July–December)

different tilt and azimuth angles; the area of energy peak is found around the latitude tilt angle and the maximum yearly energy in this tilt angle is 6.6 (kWh/m^2/day).

Figures 57.7 and 57.8 present the monthly average daily global solar irradiation on a south-facing surface. It is clear from the figures that a unique optimum tilt angle exists for each month of the year that corresponds to the maximum point of each curve. Optimum tilt angle and monthly average daily total solar radiation on an optimum tilted surface are given in Table 57.2. The optimum tilt angle was changing between a minimum of 1° in June and a maximum of 61° in December throughout a year. The loss in the amount of energy between a fixed collector tilted at the latitude of the site and another one adjustable according to the monthly

Table 57.2 Optimum tilt angle and daily average total solar radiation for each month of the year in Ghardaïa site

Month	β (°)	G_β ($kWh/m^2/day$)	G_h ($kWh/m^2/day$)
January	60	7.27	4.07
February	52	7.24	4.69
March	38	7.16	6.04
April	20	7.4	7.08
May	6	7.88	7.61
June	1	8.31	8.38
July	4	7.7	7.68
August	14	6.7	7.17
September	28	6.34	6.06
October	43	5.39	4.36
November	58	7.08	4.24
December	61	5.93	3.34

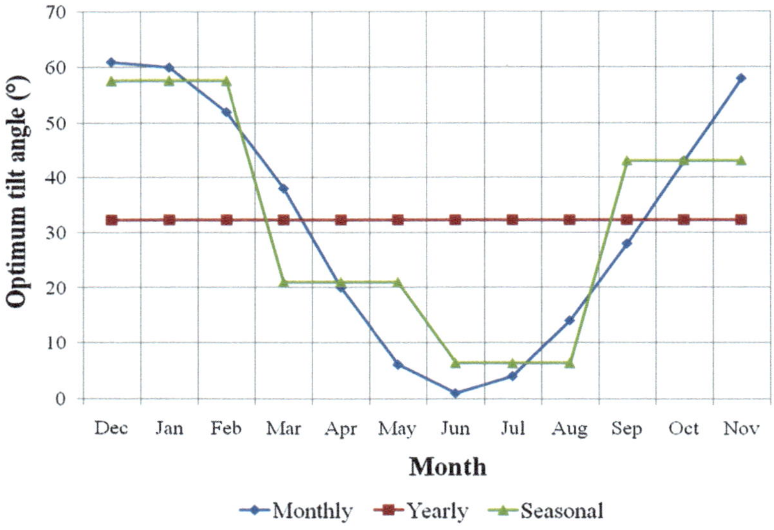

Fig. 57.9 Variation of optimum tilt angle for Ghardaïa site

optimum tilt angle is around 10 %, while the loss is 37 % if the fixed collector tilted at the latitude of the site is compared with another one placed on a horizontal surface, mainly in the winter months.

The seasonal optimal tilt angle was calculated by the same method as the monthly tilt, by looking for the maximum energy for the four seasons of the year. This requires that the collector angle must be changed four times a year, as shown in Fig. 57.9. In winter the optimum tilt angle should be 57.6°, in spring 21°, in summer 43° and in Autumn 6.3°. Figure 57.10 gives the average solar energy collected when the angle of the tilt is optimum, and when the seasonal and yearly average angles are used.

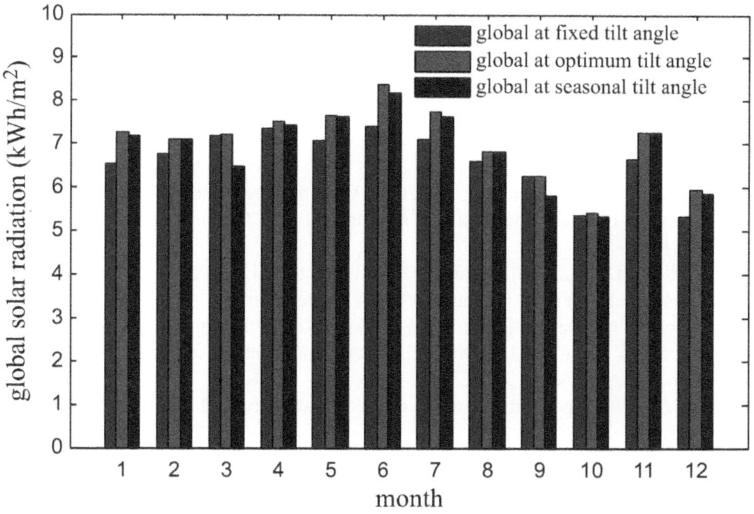

Fig. 57.10 Monthly average daily global solar radiation for optimum, seasonal and yearly tilt angles

57.6 Conclusions

In this study, the optimum tilt angle has been determined in order to increase the amount of total solar radiation reaching solar collectors. Four models have been tested for calculating the global solar radiation on inclined surfaces. The statistical test allowed identifying the HDKR model for the calculation of global solar radiation on inclined surface with the least errors monthly and annually.

The yearly optimum tilt angle is found to be 32.6° and the optimum azimuth is full south which permit to collect the yearly solar energy of 2,424.5 (kWh/m^2/day). The monthly optimum tilt angle has a minimum value of 1° in June and the monthly average total solar radiation at this angle is 8.34 (kWh/m^2/day). Then, the optimum tilt angle reaches its maximum in winter months and has a value of 61° in December which collects 5.93 (kWh/m^2/day) of solar energy monthly; the average optimum tilt angle for the summer months is 6.3° and for the winter months 57.6°. With the seasonally adjusted tilt angles, the yearly collected solar energy was 2,539.69 (kWh/m^2/day).

In the light of the obtained results, users are advised to set their collectors optimally to improve efficiency and get better output of solar collectors.

References

1. Gairaa K, Benkaciali S (2011) Analysis of solar radiation measurements in Ghardaïa area, south Algeria. Energy Procedia 6:122–129
2. Benghanem M (2011) Optimization of tilt angle for solar panel: case study for Madinah, Saudi Arabia. Appl Energy 88:1427–1433
3. Ghosh HR, Bhowmik NC, Hussain M (2010) Determining seasonal optimum tilt angles, solar radiation on variously oriented, single and double axis tracking surfaces at Dhaka. Renew Energy 35:1292–1297
4. Koray U (2006) Optimum tilt angle for solar collectors. Energ Source A 28:1171–1180
5. Murat K, Mehmet S, Yunus B, Sedat D (2004) Determining optimum tilt angles and orientations of photovoltaic panels in Sanliurfa, Turkey. Renew Energy 29:1265–1275
6. Mohd AH, Malik AQ (2001) Optimum tilt angle and orientation foe solar collector in Brunei Darussalam. Renew Energy 24:223–234
7. Morcos VH (1994) Optimum tilt angle and orientation for solar collectors in Assiut, Egypt. Renew Energy 4:291–298
8. Gairaa K, Bakelli Y (2013) A comparative study of some regression models to estimate the global solar radiation on a horizontal surface from sunshine duration and meteorological parameters for Ghardaïa site, south Algeria. ISRN Renew Energ: Article ID 754956, 11 pages
9. Iqbal M (1983) An introduction to solar radiation. Academic, Toronto, Canada
10. Klucher TM (1979) Evaluation of models to predict insolation on tilted surfaces. Sol Energy 23:111–114
11. Perez R, Steward R, Arbogast C, Seals R, Scott J (1986) An anisotropic hourly diffuse radiation model for sloping surfaces: description, performance, validation, site dependency evaluation. Sol Energy 36:481–497
12. Duffie JA, Beckman WA (2006) Solar engineering of thermal process, 3rd edn. Wiley, New York, NY

Chapter 58
Technical Study of a Hybrid Solar–Geothermal Power Plant and Its Application to a Thermal Design Course

Derek K. Baker, C. Cihan Özalevli, and S. Kazım Sömek

Abstract An energetic model for a hybrid solar–geothermal electric power plant (HSGEPP) is developed to explore the extent to which solar thermal resources can extend and enhance marginal and declining geothermal fields. The model is developed and presented to allow replication in a 4th-year solar engineering thermal design course. The model is applied to a HSGEPP being developed in Turkey, and simulations are run using a typical meteorological year formatted data set. The solar fraction (f_s) of the HSGEPP is equal to the fractional decrease in the geothermal resource usage. The increase in annual f_s with the collector field's solar multiple (M_s) is linear up to approximately $f_s = 0.25$ for $M_s = 1.25$, after which the rate of increase in f_s begins to decay and f_s approaches 0.37 for $M_s = 5$. For $M_s = 1.25$, the monthly solar fraction ranges from 0.05 in December to 0.43 in July.

Keywords Concentrating solar thermal • Geothermal • Hybrid • Undergraduate engineering education • Thermal design course

---

D.K. Baker (✉)
Centre for Solar Energy Research and Applications (GÜNAM), Middle East Technical University (METU), 06800 Ankara, Turkey
e-mail: dbaker@metu.edu.tr

C.C. Özalevli • S.K. Sömek
Department of Mechanical Engineering, Middle East Technical University (METU), 06800 Ankara, Turkey
e-mail: cihan.ozalevli@ygaventures.org; e174426@metu.edu.tr

I. Dincer et al. (eds.), *Progress in Clean Energy, Volume 2*,
DOI 10.1007/978-3-319-17031-2_58

Nomenclature (The constants and variables for the HSGEPP model are listed in Table 58.1 and are not repeated here for brevity)

Abbreviations

2A	Two-axis tracking
CST	Concentrating solar thermal
DNI	Direct normal insolation or irradiance
EW	One-axis tracking in east–west direction
GEPP	Geothermal electric power plant
HCE	Heat collecting element
HSGEPP	Hybrid solar–geothermal electric power plant
HTF	Heat transfer fluid
IAM	Incidence angle modifier
NS	One-axis tracking in north–south direction
PTC	Parabolic trough collector
STE	Solar thermal electric
TMY2	Typical meteorological year 2

Variables

$A_{coll,o}$	Nominal collector area m^2
a_o a_1, a_2	Constants for IAM, units vary and are as shown in Table 1
c_{geo}	Specific heat capacity for geothermal brine
c_o c_1, c_2	Constants for L_t, units vary and are as shown in Table 1
f	Fraction
$G_{o,n}$	Extraterrestrial normal irradiance W m^{-2}
h_{fg}	Enthalpy of evaporation kJ kg^{-1}
I	Insolation Wh m^{-2}
L	Loss (units vary)
M_s	Solar multiple
n	Day number
q	Heat transfer (Wh) or rate of heat transfer (W)
T	Temperature °C
t	Time hour with minutes expressed as a fraction
Δt	Time period hour with minutes expressed as a fraction

Greek Letters

ε_{orc}	ORC exergetic efficiency
ϕ	Latitude °

γ	Azimuth °
η	Efficiency fraction or %
θ	Angle of incidence °
θ_z	Solar zenith angle °
ζ	Error units vary

Subscripts

b	Beam
coll	Collector
d	Diffuse
db	Dry bulb
dct	Dry cooling tower
ew	East–West tracking surface
geo	Geothermal
inj	Injection
j	Type of tracking: 2A EW or NS
h	ORC hot side
n	Normal
o	Maximum or nominal
opt	Optical
orc	Organic Rankine cycle
s	Solar
t	Thermal
wb	Wet bulb
wct	Wet cooling tower

Superscripts

n	Collector thermal loss exponent
*	Assumes no night

58.1 Introduction

Due to the intermittent nature of solar resources and the usage of a heat engine in solar thermal electric (STE) power plants, interest exists in hybridizing concentrating solar thermal (CST) technologies with other thermal energy sources to achieve economically viable power plants with high capacity factors, high dispatchability, and superior environmental performance. The possibility of hybridizing CST technologies with geothermal resources in a hybrid solar–geothermal electric power plant (HSGEPP) is particularly attractive as, if the geothermal resources are managed properly, both thermal resources can be considered as sustainable.

Due to their low driving temperatures, the power output of geothermal electric power plants (GEPPs) tends to be very sensitive to the ambient temperature. Greenhut [1] predicts an ~40 % decrease in power output for a reference GEPP in June relative to January. Conversely, solar resources tend to peak in the summer. Several theoretical studies have been conducted to increase the power output of GEPPs in the summer using CST technologies. The most detailed studies are by Lentz and Almanza [2, 3], Greenhut [1], Greenhut et al. [4], Astolfi et al. [5], and Kuyumcu et al. [6]. The Gümüşköy GEPP in Turkey's southwest province of Aydin is the focus of Kuyumcu et al.'s study. In 2014, a pilot row of parabolic trough collectors (PTCs) was installed at this GEPP, which made it one of the first, if not the first, demonstration HSGEPPs in the world. One of the primary objectives for the Gümüşköy HSGEPP is to develop and then transfer the necessary know-how and technologies to commercialize other HSGEPPs. A secondary objective is to support engineering education. The research presented herein supports both these objectives. During a visit to the Gümüşköy GEPP by the present authors, the engineers noted that in addition to reduced power output in the summer, GEPPs can also suffer from declining geothermal resources, which if sufficiently severe can render the GEPP inoperable. The decline of geothermal resources can be slowed and possibly reversed if the rate of extraction is reduced, such as through by CST energy for geothermal energy. Therefore, in addition to increasing power output in the summer, a second objective for HSGEPPs can be to enhance and extend the life of marginal geothermal fields. However, all the research on HSGEPPs found in the literature focused on increasing the summer power output of GEPPs, and no research was found exploring the feasibility of using CST to improve geothermal resource management.

The primary objective for the current work is to explore the extent to which CST energy can substitute for geothermal energy to enhance marginal and extend declining geothermal fields. A secondary objective is to structure this work such that it can enhance engineering education. Within this context, a general model of a HSGEPP is developed and presented in a manner that can be replicated in a fourth-year engineering thermal design course. This general model is then applied to the Gümüşköy HSGEPP. Parametric studies based on annual simulations using an hourly time step are performed to understand how the performance of the HSGEPP varies with changes in key design and operating parameters. Results are presented and discussed, and conclusions drawn.

58.2 Methodology

The goal of the methodology is to develop and present a system-level model of a CST system that can be used for feasibility analyses and design purposes. The scope of the model is limited to that which can be replicated in a typical one-semester three-credit 4th-year thermal design course typical for an ABET accreditation curriculum. The system-level model is broken down into discrete sub-models, and these sub-models are developed and linked through a series of four projects. The order in which the sub-models are developed and presented parallels the flow of energy and starts with a maximum (extraterrestrial) solar resource model and ends with a heat engine model. For brevity, references rather than equations are given herein for classical models that can easily be found in solar engineering textbooks. Programming logic and equations not easily found in solar engineering textbooks are provided to ease replication within a class. The class is typically structured such that the early projects are very prescribed (closed ended) to develop basic skills related to computer programming and scientific communication, while in later projects, students apply these skills to solve design (open-ended) problems. For the scientific communication skills, a heavy emphasis is placed on having students present their results in professional tables and figures appropriate for a scientific journal article and discussing their results in terms of the underlying theory and their significance for design. A two-column report format is used so that most tables and figures must fit in a single column, which forces students to be careful about how they format their tables and figures and results in a more professional tables and figures. The emphasis on having students discuss their unique results is important as finding relevant content on the web to copy and paste is almost impossible (which makes plagiarism more difficult), and while almost all 4th-year engineering students can produce the required numeric results, many are weak in interpreting and communicating the significance of these results, especially in terms of the underlying theory that leads to these results and what they mean in terms of design. Since each project builds on the previous project and meaningful interpretation of results is only possible if the model is producing correct results, having students develop working models for each project is critical. Therefore, for each project, benchmark data are provided to allow model verification. Because the benchmark data and the results required for the report use different sets of constants and input data, this verification process teaches students to structure their program such that the values for constants and inputs can easily be changed (specifically, students learn to use variables and not numbers in equations), the necessity for verifying computer models, and how to systematically debug a program. Typically, students work until their program output matches the benchmark data, which means that most students develop working models. Projects 1–3 are usually individual, while Project 4 is group. Projects 1 and 2 require written reports focusing on presenting and interpreting the results. Project 3 requires an oral presentation in which students demonstrate that their model can work for some arbitrary set of constants and inputs. These individual projects are important for assuring that all students demonstrate basic computer programming and written and oral scientific communication skills. The final project is a group design project and contains both a written report and oral presentation. As an outcome of these linked projects, a general system-level model of a CST system is developed that can be applied/adapted to analyze a range of CST applications. Herein, this general CST model is used to model the Gümüşköy HSGEPP.

The following four subsections present the necessary information to develop this system-level HSGEPP model through a series of 4 projects, with each subsection corresponding to one project. These subsections not only present the theoretical and mathematical underpinnings of these sub-models but also discuss the important educational outcomes for each project and how the project is structured to achieve these outcomes. For convenience, all the constants and inputs for the system-level model and their nominal values for the Gümüşköy HSGEPP case study are summarized immediately in Table 58.1 except for the hourly meteorological model (termed TMY2 inputs herein).

Table 58.1 Constants and nominal inputs for the HSGEPP model

Variable	Value	Units	Description	Source/ comment
Collector				
Collector aperture area				
$A_{coll,o}$	55,500	m^2	Nominal collector aperture area	Variable input
M_s	1.25	–	Solar multiple	Variable input
Optical efficiency				
$\eta_{opt,n}$	0.773	–	Normal optical efficiency	[7]
a_o	1.00	–	0th IAM coefficient	[8]
a_1	8.840×10^{-04}	deg^{-1}	1st IAM coefficient	[8]
a_2	-5.369×10^{-05}	deg^{-2}	2nd IAM coefficient	[8]
Thermal losses (L_t)				
T_{coll}	200	°C	Characteristic collector temperature	Variable input
T_{lab}	25	°C	Lab ambient temperature	[7]
c_1	6.240×10^{-04}	$(Wh\ °C^{-1}\ m^{-2})^n$	1st thermal loss coefficient	[7]
c_2	-3.020×10^{-05}	$(Wh\ °C^{-1}\ m^{-2})^{2n}$	2nd thermal loss coefficient	[7]
c_3	1.260×10^{-06}	$(Wh\ °C^{-1}\ m^{-2})^{3n}$	3rd thermal loss coefficient	[7]
n	0.33	–	Thermal loss exponent	[7]
Geothermal resources				
T_{geo}	150	°C	Wellhead temperature	[5]
T_{inj}	70	°C	Reinjection temperature	[5]
c_{geo}	4.2	$kJ\ kg^{-1}\ K^{-1}$	Specific heat capacity	[9]
ORC				
q_{orc}	40	MW_t	Nominal ORC thermal input	Adapted from [5] and [6]
ε_{orc}	0.67	–	Cycle second law efficiency	Adapted from [5]

(continued)

Table 58.1 (continued)

Variable	Value	Units	Description	Source/ comment
$T_{c,min}$	15	°C	Minimum condenser temperature	Personal conversation w/operator
ΔT_h	5	°C	Expander inlet approach temperature: $\Delta T_H = T_{exp,} - T_{geo}$	[5]
ΔT_{dct}	12	°C	Cond. approach temperature with dry cooling tower: $T_{cond,e} - T_{db}$	[5]
ΔT_{wct}	5	°C	Cond. approach temperature with wet cooling tower: $T_{cond,e} - T_{wb}$	Assumed as equal to ΔT_h
Wet cooling tower				
h_{fg}	2454	kJ kg^{-1}	Enthalpy of evaporation for water	[9]

TMY2 inputs not included

58.2.1 *Project 1: Maximum Solar Resource Model for Tracking Surfaces*

For the first project, students develop and apply geometric and solar resource models to predict the maximum solar resources on three characteristic tracking surfaces: 2-axis (2A) tracking, East–West (EW) tracking about a horizontal North–South axis, and North–South (NS) tracking about a horizontal East–West axis.

58.2.1.1 Geometric Model

A Geometric Model is developed with inputs of location on earth, day of the year, and time of the day as described by the latitude (ϕ), day number (n), and solar time (t_s), respectively. For these inputs, this Geometric Model predicts the location of the sun in terms of solar zenith (θ_z) and solar azimuth (γ_s) angles, the orientation for tracking surface j in terms of the tilt (β_j) and surface azimuth (γ_j) angles, and the angle of incidence for tracking surface j (θ_j). The methodology and related equations for this model are very classical, can be found in any good solar engineering textbook, and are not repeated here for brevity. For the present work, the model by Duffie and Beckman [9] is used. This solar geometry model is developed for an hourly time step, and over each hour, the position of the sun is assumed fixed and at the position described by ½ past the hour.

58.2.1.2 Maximum Solar Resource Model

The maximum solar resource model defines the limits to solar resources imposed by the distance of the earth from the sun for the condition of no atmosphere (or equivalently for extraterrestrial conditions). The earth's orbit about the sun is not perfectly concentric, and Duffie and Beckman [9] provide a model for the variation in extraterrestrial irradiance on a normal surface with day number, $G_{o,n} = G_{o,n}(n)$. The maximum hourly insolation on a normal surface assuming no night is $I^*_{o,n} = G_{o,n}(n) \times 1$ h. Day and night are modeled based on θ_z at half past the hour using Eq. (58.1):

$$I_{o,j} \begin{cases} = I^*_{o,n} \cos(\theta_j) & \text{if } \theta_z \leq 90^\circ \\ = 0 & \text{if } \theta_z > 90^\circ \end{cases} \quad j = \text{2A, EW, or NS} \tag{58.1}$$

58.2.1.3 Outcomes

Students with little programming and scientific writing experience typically spend considerable time developing this computer model and the tables and plots. In response, the first report is kept relatively brief, and results are only requested for the location of interest. To visualize the data, graphs are required for irradiance (line graphs) or hourly insolation (bar graphs) versus hour for the summer and winter solstices and spring and fall equinoxes. To check the accuracy of the model, hourly insolation for 1 day is presented in a table. Additionally, students are encouraged to explore other locations (latitudes) to better understand how maximum solar resources vary with location, season, and hour. Typical conceptual outcomes from this project are as follows. 2A tracking maximizes the solar resources. EW tracking has maximum resources in the morning and evening and approaches the performance of 2A tracking throughout the entire day as the zenith angles at solar noon approaches 0° (e.g., equinoxes at the equator and summer solstice for $\phi = 35°$ N, which is the latitude at the Gümüşköy GEPP); the solar resources reach a daytime minimum at solar noon, and this minimum becomes smaller as the solar noon zenith angle increases (e.g., at $\phi = 35°$ N, solar noon resources are smaller in the winter than the summer). For all locations and days, NS tracking has maximum solar resources at solar noon which equal the solar noon resources for 2A tracking. For $\phi = 35°$ N and 1-axis tracking, EW and NS tracking maximize daily resources on the summer and winter solstice, respectively.

58.2.2 Project 2: Actual Solar Resource Model for Tracking Surfaces

While maximum solar resources are completely deterministic, actual solar resources also contain a strong stochastic component due to variations in meteorological conditions. Project 2 builds on Project 1 through the addition of a location-specific hourly meteorological model such as measured meteorological data or typical meteorological year (TMY) data. This discussion is based on using TMY2-formatted data. Depending on the level of the students, the TMY2 data can either be given in raw form, in which case the students must parse the data, or in parsed form, in which case the students only use the data. The Project 1 model is revised and extended as follows.

58.2.2.1 Geometric Model

TMY2-formatted data sets are tabulated based on hour ending standard (local) time (t_{std}) for every hour of the year. The Project 1 Geometric Model uses solar time as the input. The model to convert standard to solar time from Duffie and Beckman [9] is used, which requires inputs of the location's longitude (L_W) and standard longitude (L_{std}), both in degrees West. Additionally, the Project 1 Geometric Model requires latitude in degrees North (ϕ_N). TMY2-formatted data sets provide latitude as degrees North or South (ϕ_{TMY2}), longitude as degrees East or West (λ_{TMY2}), and time zones as a positive (East of Greenwich mean time, GMT) or negative (West of GMT) as an integer. The TMY2 location data are converted to inputs for the Geometric Model as follows:

$$\phi_N \begin{cases} = \phi & \text{if } \phi \text{ in } ^\circ\text{N} \\ = -\phi & \text{if } \phi \text{ in } ^\circ\text{S} \end{cases} \tag{58.2}$$

$$L_W \begin{cases} = L_{tmy} & \text{if } L_{tmy} \text{ in } ^\circ\text{W} \\ = 360^\circ - L_{tmy} & \text{if } L_{tmy} \text{ in } ^\circ\text{E} \end{cases} \tag{58.3}$$

$$L_{std} \begin{cases} = 360^\circ - (TZ \times 15^\circ) & \text{if } TZ > 0 \\ = -TZ \times 15^\circ & \text{if } TZ \le 0 \end{cases} \tag{58.4}$$

The calculation for θ_j at half past the hour for each hour interval proceeds as for Project 1.

58.2.2.2 Actual Solar Resource Model

In TMY2-formatted data sets, the hourly extraterrestrial normal insolation ($I_{o,n}$) is given explicitly, and $I_{o,n} = 0$ for night hours, $0 < I_{o,n} < I^*_{o,n}$ for sunrise and sunset hours, and $I_{o,n} = I^*_{o,n}$ for daytime hours. To allow night losses (defined below) to be calculated, $I^*_{o,n}$ is modeled using TMY2 data by taking the maximum value for $I_{o,n}$ over a 24-h period:

$$I^*_{o,n} = \max_{24h}[I_{o,n}] \tag{58.5}$$

Relative to the Project 1 model, this more accurately models the maximum solar resources for sunrise and sunset hour intervals, but Eq. (58.5) leads to errors for days when the sun is never above the horizon for an entire hour interval, e.g., for all locations above the Arctic circle on the winter solstice. Hourly terrestrial direct (beam) normal insolation ($I_{b,n}$) and horizontal diffuse insolation (I_d) are also given in TMY2-formatted data sets. Assuming an isotropic diffuse sky model (Duffie and Beckman [9]) and neglecting terrestrial reflection, the actual hourly beam ($I_{b,j}$), diffuse ($I_{d,j}$), and total (I_j) insolation on tracking surface j are modeled as Eqs. (58.6–58.8):

$$I_{b,j} = I_{b,n} \cos(\theta_j) \tag{58.6}$$

$$I_{d,j} = I_d \left(\frac{1 + \cos(\beta_j)}{2} \right) \tag{58.7}$$

$$I_j = I_{b,j} + I_{d,j} \tag{58.8}$$

Quantifying the magnitudes of beam and diffuse actual resources and the losses that cause these resources to be less than the maximum resources in the report helps students conceptually frame, interpret, and discuss the results. For non-concentrating surfaces, the solar resources are the total insolation given by Eq. (58.8). For concentrating surfaces, the solar resources are only the beam insolation given by Eq. (58.6), while the diffuse resources given by Eq. (58.7) are losses. For each hour for CST technologies, night (L_n), atmospheric (L_{atm}), orientation (also called cosine) (L_j), and diffuse (L_d) losses are defined:

$$L_n = I^*_{o,n} - I_{o,n} \tag{58.9}$$

$$L_{atm} = I_{o,n} - I_n \tag{58.10}$$

$$L_j = I_n - I_j \tag{58.11}$$

$$L_d = I_{d,j} \tag{58.12}$$

Losses can be partitioned into those that the engineer can control (termed *design* losses herein) and cannot control (termed *intrinsic* losses herein). Assuming a limited ability to change the location of the system, the night and atmospheric losses are intrinsic losses. The orientation losses are design losses that change with the type of tracking. For concentrating systems, the diffuse losses are design losses if the engineer has the option of switching to a non-concentrating technology but are intrinsic losses if this option is not available. To develop good programming skills and help debugging, an internal consistency error ζ_j is defined:

$$\xi_j = \sum_{\text{All}\,t} \left[I^*_{o,n} - \left(I_{b,j} + L_n + L_{atm} + L_j + L_d \right) \right] \tag{58.13}$$

If $\zeta_j \neq 0$, the student knows that their model has a programming error.

58.2.2.3 Outcomes

The Project 2 model predicts the actual solar resources for tracking surface j for all 8,760 h of the year. For many students, the programming challenge for Project 2 is dealing with these large data sets and post-processing the results so they can be visualized, interpreted, and communicated in a meaningful way. Students present resources and losses as daily average values for each month and for the year in both tabular and graphical forms. Note presenting monthly results in terms of average daily rather than average monthly values removes the impact of varying number of days between months and therefore provides a better basis for comparison, which is emphasized in the lecture. The variability in solar resources can be visualized for example by plotting hourly data for representative days or series of days for different seasons or through scatter plots of daily solar resources versus day number for all days in the year. Among 1-axis tracking options, students typically find EW tracking maximizes summer and annual resources, while NS tracking maximizes winter resources.

58.2.3 Project 3: Collector Model

58.2.3.1 Collector Model

Projects 1 and 2 develop very broad and fundamental solar resource skills applicable to a wide range of solar conversion technologies. Beginning with Project 3, the scope of the projects begins to narrow. SkyTrough parabolic trough collectors (PTCs) (www.skyfuel.com) were installed at the Gümüşköy GEPP. The objective for Project 3 is to develop and apply an energetic model for these collectors. Based on the Project 2 results, to maximize summer and annual resources, EW tracking is assumed. The useful thermal energy supplied per hour by the collector (q_{coll}) is modeled using Eq. (58.14) where η_{coll} is the PTC's energetic efficiency, $A_{\mathrm{coll,o}}$ is the nominal collector aperture area which is held constant during parametric studies, and M_{s} is a solar multiple used to vary the collector area for parametric studies:

$$q_{\mathrm{coll}} = \eta_{\mathrm{coll}} M_{\mathrm{s}} A_{\mathrm{coll,o}} I_{\mathrm{b,ew}} \tag{58.14}$$

The collector efficiency is generally modeled as the difference between an optical efficiency (η_{opt}) and a dimensionless thermal loss term (L_{t}). Based on experimental testing, Kutscher et al. [7] suggest an empirical collector efficiency model for a SkyTrough PTC with a constant η_{opt} and variable L_{t}. Although not explicitly stated, based on the presentation, it appears that the model is for zero angle of incidence. Therefore, a modified version of Kutscher et al.'s model appropriate for nonzero angle is developed and used herein. The optical efficiency is modeled as the normal optical efficiency suggested by Kutscher et al. multiplied by an incidence angle modifier (IAM) function:

$$\eta_{\text{opt}} = \eta_{\text{opt,n}}\text{IAM}(\theta_{\text{ew}}) \tag{58.15}$$

An IAM function specific to a SkyTrough collector was not found in the open literature. Therefore, an IAM model originally proposed by Dudley et al [8] and adapted to a form appropriate for Eq. (58.15) by Patnode [10] is used. This model gives negative IAM values for incidence angles greater than 77°, which is not physically meaningful. These negative IAM values are eliminated by modeling IAM as Eq. (58.16), where θ_j is in degrees and not radians:

$$\text{IAM} = \max\left[0, a_o + a_1\frac{\theta_{\text{ew}}}{\cos\left(\theta_{\text{ew}}\right)} + a_2\frac{\theta_{\text{ew}}^2}{\cos\left(\theta_{\text{ew}}\right)}\right] \tag{58.16}$$

Dudley et al. determined values for the empirical constants a_1, a_2, and a_3 for the Luz PTCs used at the SEGS plant in California, and as a first approximation, these values are used herein. Kutscher et al. [7] model the thermal loss term using Eq. (58.17):

$$L_{\text{t}} = c_1\tau + c_2\tau^2 + c_3\tau^3 \tag{58.17}$$

Here c_1, c_2, and c_3 are collector-specific empirically determined constants. Kutscher et al. model τ as $(T_{\text{htf}} - T_{\text{lab}})/G_{\text{b,n}}^{\text{n}}$ where T_{htf} is the temperature of the heat transfer fluid (HTF) flowing through the heat collecting element (HCE). To determine the empirical constants c_1, c_2, and c_3, and the empirical exponent n, Kutscher et al. fit the model to experimental data measured under laboratory conditions with ambient temperature T_{lab}. They benchmarked the resultant model against field data and found excellent agreement. Through a sensitivity analysis, they found that over the range of expected ambient conditions, replacing T_{lab} with the ambient temperature resulted in negligible changes in predicted output and therefore suggests treating T_{lab} as a constant. Applying Kutscher et al.'s model to night conditions ($G_{\text{b,n}} = 0$) results in infinite losses. Therefore, for programming purposes, a modified version of Kutscher et al.'s model based on hourly insolation and a characteristic collector temperature (T_{coll}) is used herein as given by Eq. (58.18):

$$\tau \begin{cases} = \dfrac{T_{\text{coll}} - T_{\text{lab}}}{I_{\text{b,ew}}^n} & \text{if } I_{\text{b,ew}} > 0 \\ = 0 & \text{if } I_{\text{b,ew}} = 0 \end{cases} \tag{58.18}$$

The relative contributions of normal optical efficiency, IAM function, and thermal losses to the deviation of the collector efficiency from 100 % can be quantified by defining normal optical ($L_{\text{opt,n}}$) and IAM (L_{IAM}) losses in addition to the existing definition for L_{t}:

$$L_{\text{opt,n}} = 1 - \eta_{\text{opt,n}} \tag{58.19}$$

$$L_{\text{IAM}} = \eta_{\text{opt,n}}\left[1 - \text{IAM}\left(\theta_{\text{pt,ew}}\right)\right] \tag{58.20}$$

Using these losses, students can check the internal consistency of their model as follows:

$$\xi_{\text{coll}} = \sum_{\text{all t}} \left[1 - \left(\eta_{\text{coll}} + L_{\text{opt,n}} + L_{\text{IAM}} + L_{\text{t}}\right)\right] \tag{58.21}$$

As before, if $\zeta_{\text{coll}} \neq 0$, the student knows that their model has a programming error. Importantly, the optical, IAM, and thermal losses are design losses if one is designing a collector but are intrinsic losses for a specific collector. The difference $\eta_{\text{opt}} - L_{\text{t}}$ will be positive at night and can be negative for hours with low solar resources, both of which are not meaningful. Therefore, the instantaneous collector efficiency is modeled using Eq. (58.22):

$$\eta_{\text{coll}} \begin{cases} = \max\left[\eta_{\text{opt}} - L_t, 0\right] & \text{if } I_{\text{ew}} > 0 \\ = 0 & \text{if } I_{\text{ew}} = 0 \end{cases} \tag{58.22}$$

The nominal values for all inputs in this collector model are presented in Table 58.1. For presenting collector efficiencies over some time period Δt, the efficiency at each time step is weighted by the beam insolation:

$$\eta_{\text{coll},\Delta\text{t}} = \frac{\sum_{\Delta\text{t}} \eta_{\text{coll}} I_{\text{b, j}}}{\sum_{\Delta\text{t}} I_{\text{b, j}}} \tag{58.23}$$

Equations (58.22) and (58.23) are the collector energetic efficiencies relative to beam (direct) solar resources and have an upper limit of 100 %. A collector energetic efficiency can also be defined relative to the total solar resources as in Eq. (58.24):

$$\eta_{\text{coll},\Delta\text{t},\text{b + d}} = \frac{\sum_{\Delta\text{t}} \eta_{\text{coll}} I_{\text{b, j}}}{\sum_{\Delta\text{t}} \left(I_{\text{b, j}} + I_{\text{d, j}}\right)} \tag{58.24}$$

Note $\eta_{\text{coll},\Delta\text{t,b+d}}$ is always less than $\eta_{\text{coll},\Delta\text{t}}$ such that $0 \leq \eta_{\text{coll},\Delta\text{t,b+d}} < \eta_{\text{coll},\Delta\text{t}} \leq 100$ %. Equation (58.24) is useful for comparing the energetic efficiencies for concentrating and non-concentrating technologies, since non-concentrating technologies utilize both beam and diffuse resources, and therefore their efficiencies are defined relative to beam and diffuse resources.

58.2.3.2 Outcomes

By this third project, students tend to be proficient at developing, programming, and running mathematical models and at post-processing, presenting, and interpreting results in writing. Therefore, for this third project, students are expected to add value by exploring this problem beyond the minimum requirements and present their model and results in an 8-min presentation. The 8-min presentation time limit is based on the time limit for funding presentations imposed by one venture capitalist group. The importance in being able to quickly and effectively communicate one's ideas for a new business to potential investors is emphasized in the lecture, which provides further context and motivation for this individual presentation. For this presentation, students first demonstrate the robustness of their model by quickly inputting (e.g., copy and pasting) a new set of inputs and constants that they have never seen before and showing that their model gives the correct results. The rest of the presentation is completely open-ended, and students are merely asked to communicate how they added value by exploring the problem beyond the minimum requirements. Students have explored the problem in a wide variety of ways, including improving the mathematical model based on a literature review, exploring new conditions and locations, and creating a beta version of a commercial software package by adding a user friendly graphical user interface (GUI).

58.2.4 Project 4: ORC and Complete HSGEPP Models

58.2.4.1 ORC and Complete HSGEPP Models

The content for this fourth project varies with semester depending on the specific CST technology being studied. Based on the objectives and scope of the present work, and considering the time limitations and appropriate content for a 1-semester solar engineering course, a simple model of an ORC is developed, which is then integrated into a complete model of a HSGEPP. Both Greenhut [1] and Astolfi et al. [5] develop a numeric model of an ORC by modeling each component separately and then combining these individual models in a system-level model. Both Greenhut and Astolfi et al. found for a fixed maximum temperature that the ORC's power output increases linearly with decreasing ambient temperature. Furthermore, Greenhut bounded the maximum power output due to practical limitations to the minimum condenser pressure. Greenhut validated his model with actual operating data and achieved good results. For a fixed maximum temperature, a heat engine's power output increasing linearly with decreasing ambient temperature is consistent with trends in the Carnot efficiency. Assuming a maximum temperature fixed by the geothermal resources, the thermal efficiency of the ORC (η_{orc}) can be modeled as the product of a constant exergetic (second law) efficiency (ε_{orc}) and Carnot efficiency, $\eta_{\text{orc}} = \varepsilon_{\text{orc}}(1 - T_c/T_h)$. From a pedagogical standpoint, this thermal efficiency model is significant as it reinforces basic thermodynamic concepts. However, rather than using the Carnot efficiency, which is applicable to a heat engine operating between 2-temperature reservoirs, herein, the thermal efficiency for the reversible Lorenz cycle is used, which is based on a driving heat transfer from a hot stream being cooling from $T_{\text{h,in}}$ to $T_{\text{h,out}}$ and heat rejection to a cold temperature reservoir at T_c. The ORC efficiency is then modeled as the product of a constant exergetic efficiency and the Lorenz efficiency:

$$\eta_{\text{orc}} = \varepsilon_{\text{orc}}\left(1 - T_{\text{c}}/\overline{T}_{\text{h}}\right) \tag{58.25}$$

Here $\overline{T}_{\text{h}}$ is a hot entropic mean temperature which assuming a constant specific heat for the hot stream is defined by Eq. (58.26) [11]:

$$\overline{T}_{\text{h}} = \frac{T_{\text{h,in}} - T_{\text{h,out}}}{\ln(T_{\text{h,in}}/T_{\text{h,out}})} \tag{58.26}$$

Astolfi et al. [5] define $T_{\text{hot,in}}$ and $T_{\text{hot,out}}$ as the temperatures at the production (T_{geo}) and injection (T_{inj}) wellheads, and $T_{\text{c}} = T_{\text{db}}$ (ambient dry-bulb temperature since a wet cooling tower is used). They report an exergetic efficiency of 55.5 %. However, this definition of the relevant temperatures does not differentiate between the heat transfer irreversibilities in the heat exchangers from the irreversibilities internal to the ORCs. Therefore, herein $T_{\text{h,in}}$ and $T_{\text{h,out}}$ are modeled as follows:

$$T_{\text{h,in}} = T_{\text{geo}} - \Delta T_{\text{h}} \tag{58.27}$$

$$T_{\text{h,out}} = T_{\text{inj}} - \Delta T_{\text{h}} \tag{58.28}$$

Here ΔT_{h} is the temperature difference between the collector's heat transfers fluid and the ORC's working fluid. As a first approximation, ΔT_{h} is assumed the same at the inlet and outlet of the heat exchanger, and the value reported by Astolfi et al. is used [5]. At the Gümüşköy GEPP, variable speed fans are used for the dry cooling towers, and the operators decrease the fan speed as the condenser temperature approaches 15 °C to prevent the turbine outlet pressure from going below a minimum value. Therefore, 15 °C is taken as the minimum condenser temperature ($T_{\text{c,min}}$), and T_{c} is modeled using Eq. (58.29):

$$T_{\text{c}} \begin{cases} = \max\left[T_{\text{c,min}}, T_{\text{db}} + \Delta T_{\text{dct}}\right] & \text{for dry cooling tower} \\ = \max\left[T_{\text{c,min}}, T_{\text{wb}} + \Delta T_{\text{wct}}\right] & \text{for wet cooling tower} \end{cases} \tag{58.29}$$

The subscripts db and wb denote dry- and wet-bulb temperatures, and the terms ΔT_{dct} and ΔT_{wct} are the temperature differences between the ORC's working fluid in the condenser and the air's dry- and wet-bulb temperatures, respectively. Astolfi et al. [5] report a value for ΔT_{dct}, and this value is used herein. Although a definitive value for ΔT_{wct} was not found from a reliable source, various sources indicate minimum values on the order of 3 °C. The wet cooling tower has heat transfer from a condensing vapor to an evaporating liquid and therefore should have low thermal resistances. As a first approximation, $\Delta T_{\text{wct}} = \Delta T_{\text{h}}$ is assumed. For the $\Delta T_{\text{h}} = 5$ °C, $\Delta T_{\text{dct}} = 12$ °C, and $T_{\text{db}} = 15$ °C reported in Astolfi et al. [5], $\varepsilon = 0.67$ is required to yield the same cycle thermal efficiency and is assumed herein. In TMY2-formatted data sets, T_{db}, dew-point temperature (T_{dp}) and relative humidity (rh) are given, but T_{wb} is not given. Two approaches have been taken for having students calculate T_{wb}. The first is to calculate T_{wb} based on thermodynamic first principles, which requires a relatively complex iterative solution and strengthens both thermodynamic psychrometric and computer programming skills. The second is to have students find an empirical equation for T_{wb} given T_{db}, T_{dp}, and rh in the open literature, which strengthens students' research skills.

Astolfi et al. [5] assume values for T_{geo}, T_{inj}, production mass flow rate (m_{geo}), and rated gross electrical power output of 150 °C, 70 °C, 100 kg/s, and 4.6 MW_e, respectively. For the Gümüşköy GEPP, Kuyumcu et al. [6] report for these same parameters 165 °C, 80 °C, 120 kg/s, and 6.6 MW_e, respectively. However, in informal discussions with the plant's operators, the ORC size was described as 5.5 MW. The working fluid for both Kuyumcu et al. and Astolfi et al. is R134a. Note that the Carnot efficiencies defined by these production and injection well temperatures are practically identical, and scaling the 4.6 MW_e capacity for Astolfi et al. by 1.2 based on the larger mass flow rate for Kuyumcu et al. yields an estimated capacity of 5.5 MW_e for Gümüşköy, which is consistent with the informal discussions but not Kuyumcu et al. It may be that 6.6 MW_e is the turbine power, and 5.5 MW_e is the net power output, which would be consistent with the 5.7 and 4.6 MW turbine and net power given by Astolfi et al. Astolfi et al. provide greater detail about their ORC than Kuyumcu et al., and therefore herein the intensive characteristics of the ORC described by Astolfi et al. are assumed, while the extensive characteristics from Astolfi et al. are scaled by a factor of 1.2 to model the Gümüşköy power plant.

As stated in Sect. 58.1, in this work, the scientific objective for the present work is to explore the feasibility of conserving geothermal resources by using solar thermal resources. Based on discussions with the plant's operators, the mass flow rate of the geothermal brine can easily be controlled. Therefore, in this model, the thermal power supplied to the ORC (q_{orc}) is assumed constant, the output from the PTCs (q_{coll}) is always maximized, and any difference between q_{orc} and q_{coll} is supplied by the geothermal resources (q_{geo}) by varying the mass flow rate of the geothermal brine (m_{geo}). If the collector field is oversized, q_{coll} can exceed q_{orc}, and, assuming no thermal storage, some of the collector's thermal power must be dumped (q_{dumped}) by defocusing some of the collectors. The model is then as follows:

$$q_{\text{dumped}} = \max[0, q_{coll}(t) - q_{\text{orc}}] \tag{58.30}$$

$$q_{\text{geo}} = \max[0, q_{\text{orc}} - q_{coll}(t)] \tag{58.31}$$

$$m_{\text{geo}} = \frac{q_{geo}(t)}{c_{\text{geo}}\left(T_{\text{geo}} - T_{\text{inj}}\right)} \tag{58.32}$$

Astolfi et al. report nominal q_{orc} values of approximately 33 MW_t and production and injection temperatures of 150 and 70 °C. Assuming $c_{geo} = 4.2$ kJ kg^{-1} K^{-1} and $q_{geo} = 33$ MW, Eq. (58.32) predicts $m_{geo} = 98$ kg/s, which is very close to the value reported by Astolfi et al. of 100 kg/s.

The nominal collector area is calculated based on the maximum hourly insolation for the year:

$$A_{\text{coll,o}} = q_{\text{orc}}/\max[I_{\text{b,ew}}] \tag{58.33}$$

This nominal collector area is important as it maximizes annual q_{coll} and minimizes annual q_{geo} under the constraint $q_{dumped} = 0$. Noting that the actual solar collector area is $A_{coll,o}\ M_s$ (see Eq. (58.14)), $M_s > 1.0$ results in annual $q_{dump} > 0$, while $M_s < 1.0$ results in $q_{geo} > 0$ for every hour of the year. Alternately, assuming that the cost of the collectors scales linearly with collector area, the cost of solar thermal energy (e.g., USD/kWh) is constant for $0 \leq M_s \leq 1.0$ but increases with M_s for $M_s \geq 1.0$. For any time interval, collector (f_{coll}), geothermal (f_{geo}), solar (f_s), and dump solar (f_{ds}) fractions are defined in Eqs. (58.34–58.37):

$$f_{coll} = \frac{\sum_{\Delta t} q_{coll}}{\sum_{\Delta t} q_{orc}} \tag{58.34}$$

$$f_{geo} = \frac{\sum_{\Delta t} q_{geo}}{\sum_{\Delta t} q_{orc}} \tag{58.35}$$

$$f_s = \frac{\sum_{\Delta t} \min[q_{orc}, q_{coll}]}{\sum_{\Delta t} q_{orc}} = 1 - f_{geo} \tag{58.36}$$

$$f_{ds} = \frac{\sum_{\Delta t} q_{dumped}}{\sum_{\Delta t} q_{orc}} \tag{58.37}$$

For $0 \leq M_s \leq 1.0$, $f_s = f_{coll}$, while for $1 < M_s$, $f_s < f_{coll}$, due to $q_{dump} > 0$. Note the fractional decrease in geothermal resource usage is equal to f_s.

While wet cooling towers decrease T_c and therefore increase η_{orc}, they also consume water. As a first approximation, the water consumption is proportional to the waste heat rejected (q_c) and electricity produced (w_{orc}):

$$q_c = q_{orc}(1 - \eta_{orc}) = w_{orc}\left(\frac{1}{\eta_{orc}} - 1\right) \tag{58.38}$$

Therefore, water consumption increases as η_{orc} decreases. Often locations with large solar resources do not have large water resources, and water consumption by wet cooling towers becomes a large concern for STE systems. Water consumption problems become exacerbated for ORCs since they are characterized by relatively low T_h and therefore small η_{orc}, large q_c, and large water consumption per unit electricity production. Since wet cooling towers work through evaporative cooling, the simplest model for volumetric water consumption (V_{h2o}) based on thermodynamic first principles is based on the mass specific volume of liquid water (υ_f) and enthalpy of evaporation (h_{fg}):

$$V_{h2o} = q_c v_f / h_{fg} \tag{58.39}$$

In the SkyTrough PTC product literature (www.skyfuel.com), SkyFuel reports 3.2 liters/kWh$_e$ wet cooling tower water consumption for a gross steam cycle thermal efficiency of 37.5 %. Combining Eqs. (58.38) and (58.39) and rearranging predict water consumption of approximately 2.7 l/kWh$_e$ for this same thermal efficiency. The difference between the value reported by SkyFuel and that predicted herein can be attributed to neglecting blowdown, drift, and other non-evaporative water losses in Eq. (58.39). For predicting orders of magnitudes and trends in water consumption, Eq. (58.39) is considered sufficient and appropriate. Water consumption per unit electricity produced over some time period Δt is given by Eq. (58.40):

$$\frac{V_{h2o}}{w_{orc}} = \frac{\sum_{\Delta t} V_{h2o}}{\sum_{\Delta t} w_{orc}} \tag{58.40}$$

58.2.4.2 Outcomes

This last project is given as a very open-ended group design problem. Students are expected to showcase the skills they have developed over the semester through both a written report and oral presentation.

58.3 Results and Discussion

The objective for the methodology section is to develop a fairly general model of an STE system in a manner that can be replicated in a fourth engineering thermal design class. The resulting model can be directly applied or modified for design and feasibility studies for a wide range of STE systems. In this Results and Discussion section, this general model is specifically applied to the Gümüşköy HSGEPP. A TMY2-formatted data set for Mugla, Turkey, is used as the meteorological model for all analyses, which is approximately 100 km south of the Gümüşköy GEPP and in the same climatic region (Aegean coast of Turkey). Nominal values for constants and inputs for all analyses are reported in Table 58.1. While the general trends and relative magnitudes for these results are expected to be meaningful and therefore appropriate for an initial feasibility study, sufficient uncertainty exists in many of the input values that strong conclusions should not be drawn from the exact values.

58.3.1 Performance of a Stand-Alone Geothermal Power Plant Using Dry and Wet Cooling Towers

Power output and ratio of actual to nominal power output for a stand-alone geothermal power plant coupled to dry and wet cooling towers are presented as monthly average trends and average annual values in Fig. 58.1a, b assuming ΔT_{wct}

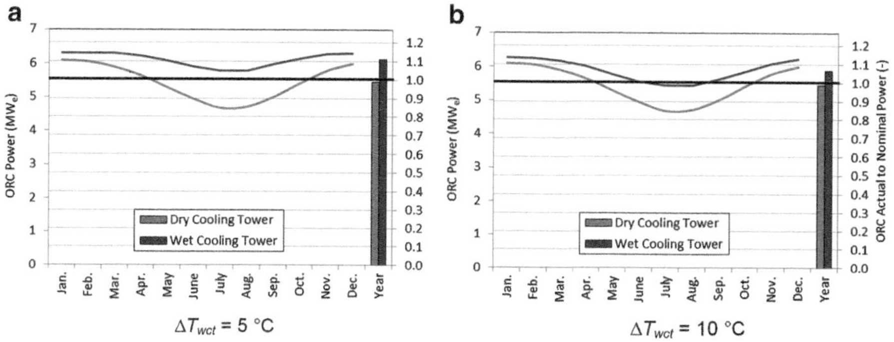

Fig. 58.1 Trends in and annual average of ORC power and ratio of actual to nominal power. (All inputs at nominal values given in Table 58.1 except ΔT_{wct})

is 5 (Fig. 58.1a) and 10 °C (Fig. 58.1b). The maximum dry-bulb temperature according to the meteorological model is 37.8 °C. An approximately 25 % decrease in power output is predicted for the dry cooling tower case in the summer relative to winter. As expected, the wet cooling tower results in significantly higher and more consistent power output throughout the year, due to both $T_{\text{wb}} < T_{\text{db}}$ if rh < 100 % and temporal variations for T_{wb} being smaller than those for T_{db}. However, the annual average water consumption for the wet cooling is estimated to be approximately 8 l/kWh, which is a high value and does not include water losses due to blowdown, drift, etc. The flat power output for the winter for the wet cooling tower case is due to the condenser operating at its minimum allowable temperature. The performance of the ORC during the summer for the wet cooling tower case is fairly sensitive to the assumed ΔT_{wct}, indicating that this is both an important design parameter and must be modeled correctly to achieve meaningful results. All these results are consistent with the literature reviewed in Sect. 58.1.

58.3.2 Actual Solar Resources

Solar resources for EW tracking are presented in Fig. 58.2a, b. The annual direct (beam) normal insolation (DNI) is estimated as 1,940 kWh m^{-2}, which is a relatively large value and appropriate for CST technologies. Figure 58.2a presents daily beam insolation versus day number and visually communicates seasonal and daily variations in beam resources. Specifically, clear summer days have approximate 2.5x the resources as clear winter days; all days from May through September have at least some beam resources, and November through April have many days with no significant beam resources. In Fig. 58.2b, daily average insolation and losses for each month and the year are presented. The total height of each bar represents the maximum (extraterrestrial) solar resources if there is no night. The dip in the total bar height in the summer is due to the earth being farther from the sun, while the increase in the beam and diffuse resources in the summer is due to longer and clearer days (or equivalently smaller night losses). The atmospheric

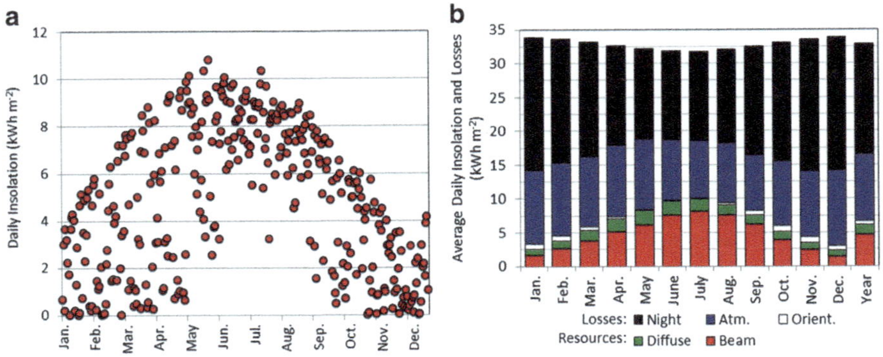

Fig. 58.2 Resources and losses for an EW tracking surface with a horizontal North–South axis. (**a**) Daily beam resources. (**b**) Daily average resources and losses

losses are relatively constant. The daily atmospheric losses increase with cloudiness, distance the solar radiation passes through the atmosphere (which increases with solar zenith angle), and day length. The relatively constant atmospheric losses can be understood as relative to winter; the decreases in cloudiness and solar zenith angle in the summer are offset by longer days. The diffuse insolation is a loss for concentrating technologies but is a resource for non-concentrating technologies. The orientation losses are relatively small, which suggests that alternate tracking options could not increase the resources significantly. Additionally, these orientation losses approach zero in the summer due to 1-axis EW tracking closely approximating 2-axis tracking in the summer at this latitude. The atmospheric losses can only be changed by changing the location. The distribution of night losses can be changed by changing the latitude, but the total annual night losses are constant due to 50 % of a year being night at all locations.

58.3.3 Collector Performance

SkyFuel reports a relatively constant η_{coll} over the temperature range 200–400 °C at the design conditions. However, the collectors will rarely operate at these conditions, and under actual conditions characterized by lower beam resources, η_{coll} may be more sensitive to changes in the collector temperature. Therefore, annual simulations were performed where all constants and inputs are kept at their default values given in Table 58.1 except $A_{\mathrm{coll}} = 1\ \mathrm{m}^2$ is assumed and T_{coll} is varied from 200 to 350 °C in 50 °C increments. Trends in monthly average and the annual average collector efficiency are presented in Fig. 58.3 relative to both beam (Eq. (58.23), Fig. 58.3a) and total (Eq. (58.24), Fig. 58.3b) resources for a range of characteristic collector temperatures.

In the summer, the collector has minimal IAM and thermal losses since the collector efficiency is relatively close to the normal optical efficiency, i.e., the best

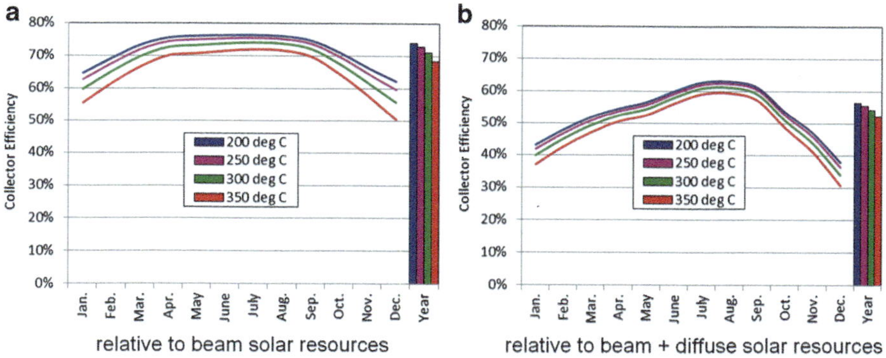

Fig. 58.3 Trends in and annual average of collector efficiency relative to aperture normal (**a**) beam and (**b**) beam + diffuse resources. (All inputs at nominal values given in Table 58.1 except T_{coll})

way to improve summer performance is to improve the collector's normal optical efficiency. The decrease in the collector's efficiency in the winter relative to the summer is due to two effects. The first is an increase in the IAM losses due to increases in θ_{ew} in the winter months. The second is an increase in L_t due to decreases in aperture normal beam resources caused by both decreases in beam normal resources and increases in orientation (cosine) losses associated with larger θ_{ew}. The decreases in the collector's efficiency in winter months due to increases in IAM losses are independent of collector temperature. For characteristic collector temperatures of 200 and 350 °C and comparing winter to summer, increases in IAM losses cause ~12 % absolute decrease in collector efficiency for both temperatures, while increases in thermal losses cause ~2 and 12 % absolute decrease in collector efficiency, respectively. In conclusion, seasonal variations in collector efficiency are primarily due to variations in IAM losses at 200 °C and are due approximately equally to variations in IAM and thermal losses at 350 °C. While higher collector temperatures decrease the collector's performance, they also allow smaller heat exchangers to be used.

58.3.4 Collector Sizing and System Performance

For the conditions in Table 58.1, the variation in the annual collector, solar, and geothermal fractions with M_s is shown in Fig. 58.4a, where from Eq. (58.35) $f_{geo} = 1 - f_s$. By definition of M_s and as noted above, $f_s = f_{coll}$ for $0 \leq M_s \leq 1.0$ and $f_s < f_{coll}$ for $M_s > 1.0$. Practically, the deviation between f_s and f_{coll} only becomes significant for $M_s \geq 1.25$. Therefore, assuming the cost of collectors scales linearly with collector area, the unit cost of usable solar thermal energy is relatively independent of size up to $M_s = 1.25$ and then begins to increase dramatically. Additionally, without storage, the limit to the solar fraction is on the order of 0.35–0.40,

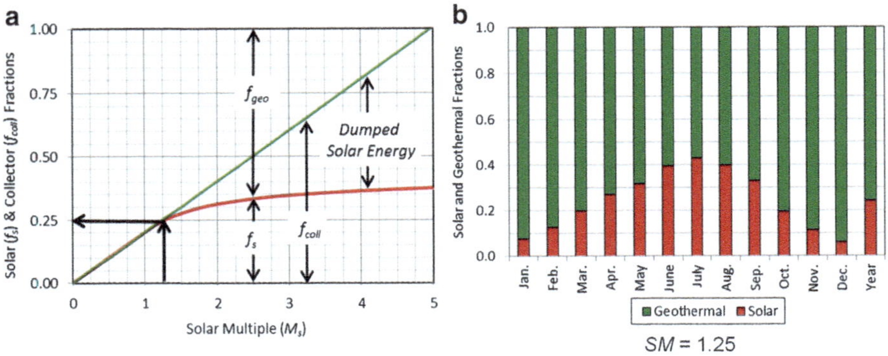

Fig. 58.4 Solar (f_s) and geothermal ($f_{geo} = 1 - f_s$) fractions. (**a**) Versus M_s. (**b**) Monthly values for $M_s = 1.25$. All other inputs as per Table 58.1

which makes sense considering that 50 % of the time in a year is night. Alternately, if the objective is to maximum the conservation of geothermal resources while supplying solar thermal power at the minimum unit cost, the solar field should be sized with $M_s \approx 1.25$. Note this results in consistent with the typical recommendation of M_s on the order of 1.25 for STE power plants with no storage.

The remaining results assume $M_s = 1.25$. In Fig. 58.4b, the solar and geothermal fractions are shown on a monthly and annual basis. The monthly solar fractions are consistent with the solar resources and collector efficiency results and show large seasonal variations.

In Fig. 58.5, hourly solar, geothermal, and solar dump fractions are shown for four 14-day periods centered around the (a) spring equinox, (b) summer solstice, (c) fall equinox, and (d) winter solstice. The plots show that hourly variations in solar and geothermal fractions tend to be periodic in the summer but exhibit strong variations between days in the winter, including having some days where no solar thermal energy is supplied. The solar collector field is slightly oversized around the summer solstice and well matched for the spring and fall equinoxes in terms of the amount of excess solar energy dumped. The solar field is severely undersized in the winter. While increasing M_s beyond 1.25 would in general increase the winter solar fractions, it would not increase solar fractions on days with no beam resources nor decrease the variations in solar fractions between days.

58.4 Conclusions

In Sect. 58.2, the structure for a 4th-year solar engineering thermal design course is presented. The broad structure of the class strengthens core engineering skills related to mathematical modeling, computer programming, parametric and simulation studies, post-processing and interpreting results, and written and oral scientific communication and the application of these skills to solve open-ended design problems. These core engineering skills are important not only for solar engineering

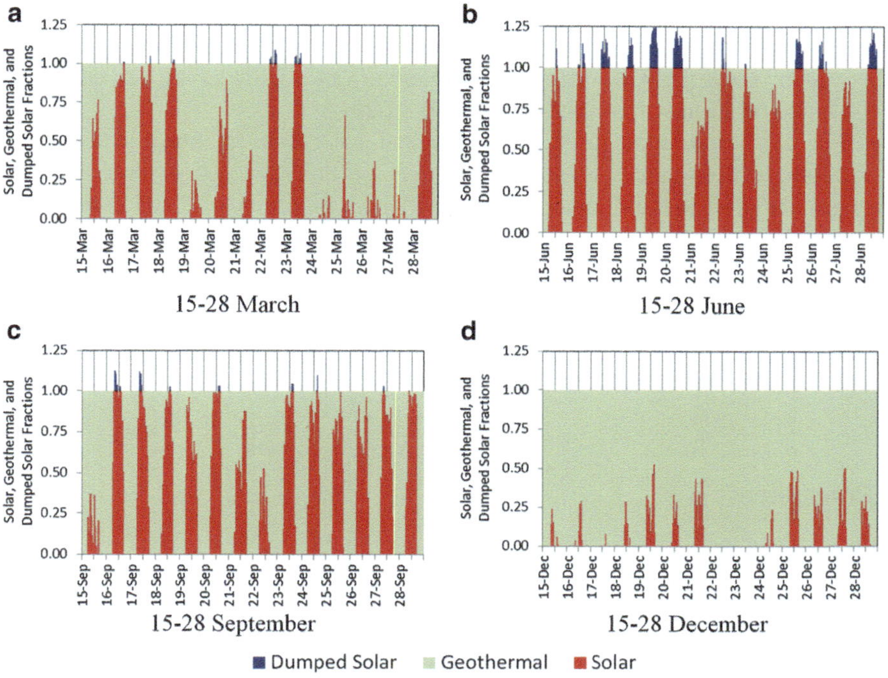

Fig. 58.5 Solar, geothermal, and dumped solar fractions for 4-characteristic 2-week periods. All inputs as listed in Table 58.1

but for all areas of engineering. This pedagogical framework to strengthen core engineering skills is filled in with solar engineering skills to model solar resources, collector performance, and energy conversion technologies. Through a series of linked projects, students develop a system-level model of a CST system that can be applied or adapted to analyze a wide range of different systems. The results and discussions for the current paper are based on the application of this general model to analyze the Gümüşköy HSGEPP in Turkey's Aegean region.

Significant conclusions are as follows:

- Relative to winter, the elevated ambient temperatures in the summer are predicted to decrease the power output of the Gümüşköy GEPP by ~10–25 % depending on the type and design of the cooling tower.
- When analyzing HSGEPP, several previous authors (e.g., [1, 6] assumed a constant collector efficiency and solar resources equal to the DNI. For EW tracking, these assumptions seem appropriate as a first approximation, especially if a detailed performance model of the collectors is not available. However, in reality at the Gümüşköy GEPP, while the summer solar resources are close to the DNI for EW tracking, the winter solar resources are less than the DNI due to orientation (cosine) losses. Additionally, in the winter, the efficiency will be lower than in the summer due to larger incident angle modifier and thermal losses.

- Assuming that the collector costs scale linearly with collector area, if the goal is to maximize the conservation of geothermal resources while keeping the unit cost of solar thermal energy at a minimum (e.g., USD/kWh), the solar field should be sized with a solar multiple of ~1.25.
- For a solar multiple of 1.25, the annual solar and fractional reduction in geothermal resources is ~0.25. The average daily values of these fractions peak at ~0.45 in the summer months and have a minimum of ~0.05 in the winter months.

Acknowledgments This work is supported with grants from TUBITAK (7120763) and EU-SFERA (228296). The educational component of this work supports the outreach and dissemination of activities of the EU-Solaris Project (FP7 312833). The authors would like to thank the support of BM Holding, TYT, and YGA.

References

1. Greenhut AD (2010) Modeling and analysis of hybrid geothermal-solar thermal energy conversion systems. MSc Thesis, Massachusetts Institute of Technology
2. Lentz Á, Almanza R (2006) Solar–geothermal hybrid system. Appl Therm Eng 26 (14–15):1537–1544
3. Lentz Á, Almanza R (2006) Parabolic troughs to increase the geothermal wells flow enthalpy. Sol Energy 80(10):1290–1295
4. Greenhut AD, Tester JW, DiPippo R, Field R, Love C, Nichols K, Augustine C, Batini F, Price B, Gigliucci G, et al. (2010) Solar–geothermal hybrid cycle analysis for low enthalpy solar and geothermal resources. Proceedings world geothermal congress, 2010
5. Astolfi M, Xodo L, Romano MC, Macchi E (2011) Technical and economical analysis of a solar–geothermal hybrid plant based on an Organic Rankine Cycle. Geothermics 40(1):58–68
6. Kuyumcu OC, Solaroglu ZDU, Akar S, Serin O (2013) Hybrid geothermal and solar thermal power plant case study: Gumuskoy GEPP. Geo-Heat Center Bulletin 31:19–24
7. Kutscher C, Burkholder F, Kathleen Stynes J (2011) Generation of a parabolic trough collector efficiency curve from separate measurements of outdoor optical efficiency and indoor receiver heat loss. J Sol Energy Eng 134(1):011012–011012
8. Dudley VE, Kolb GJ, Mahoney AR, Mancini TR, Matthews CW, Sloan M, Kearney D (1994) Test results: SEGS LS-2 solar collector. NASA STIRecon Tech Rep N 96:11437
9. Duffie JA, Beckman WA (2006) Solar engineering of thermal processes, 3rd edn. John Wiley, New York, NY
10. Patnode AM (2006) Simulation and performance evaluation of parabolic trough solar power plants. MSc Thesis, University of Wisconsin-Madison
11. Vijayaraghavan S (2003) Thermodynamic studies on alternate binary working fluid combinations and configurations for a combined power and cooling cycle. MSc Thesis, University of Florida

Chapter 59
Technical Analysis of a Hybrid Desiccant Cooling in a Mediterranean Climate

Arash Karshenass, Derek Baker, Cemil Yamali, and Rahul Singh

Abstract This chapter presents a performance analysis of a hybrid liquid desiccant cooling system (DCS) with lithium chloride as the desiccant material in a Mediterranean climate. Mathematical models of desiccant contactors are adopted from the literature. The complete system is modeled in the TRNSYS platform and is simulated using Typical Meteorological Year data. The building model was developed in accordance with the building's construction material and usage style. Simulations are performed over the summer period of a typical year and the results of the system are compared with the results of a conventional vapor compression cooling system (VCCS) from viewpoint of system characteristics and energy savings. One of the important contributions is the conclusion that a transient cycle analysis is necessary to understand the DCS cycle's performance. The results also indicate that the consumed energy in both systems is approximately equal in magnitudes but different in type; DCS shifts the energy required for the operation of the system from electrical energy to thermal energy. It is also observed that for large supply-airflow rate applications, DCS would be more beneficial than VCCS.

Keywords HVAC • Desiccant cooling • Energy saving • Solar cooling

A. Karshenass • C. Yamali • R. Singh
Centre for Solar Energy Research and Applications (GUNAM), ODTÜ, 06800 Ankara, Turkey

Mechanical Engineering, Middle East Technical University, 06800 Ankara, Turkey
e-mail: arash.karshenass@metu.edu.tr; cemily@gmail.com; rahulsingh284.nitjsr@gmail.com

D. Baker (✉)
Centre for Solar Energy Research and Applications (GUNAM), Middle East Technical University (METU), 06800 Ankara, Turkey

Mechanical Engineering, Middle East Technical University, Northern Cyprus Campus, Kalkanlı, Guzelyurt District, Cyprus
e-mail: dbaker@gmail.com

I. Dincer et al. (eds.), *Progress in Clean Energy, Volume 2*,
DOI 10.1007/978-3-319-17031-2_59

Nomenclature

A Air superficial flow rate (kg m^{-2} h^{-1})
a Specific area (m^2 m^{-3})
C_p Solution-specific heat at constant pressure (kJ kg^{-1} K^{-1})
g Humidity ratio (kg_{water} kg_{air}^{-1})
h Specific enthalpy (kJ kg^{-1})
L Solution superficial flow rate (kg m^{-2} h^{-1})
$\dot{m}$ Mass flow rate (kg h^{-1})
q Heat input (kJ h^{-1})
T Temperature (°C)
X Solution concentration (–)
Z Dehumidifier height (m)

Greek Letters

ε Effectiveness (–)
β Regenerator temperature difference ratio (–)
γ Solution surface tension (N m^{-1})

Subscripts

A Air
cond Condensation
deh Dehumidifier
evap Evaporation
g Dehumidification
H Enthalpy
I Inlet
L Solution
O Outlet
reg Regeneration solution
ru Reusing solution

59.1 Introduction

Energy saving and indoor environment quality are two major concerns in building HVAC systems. Generally, buildings in the USA and the UK consume 38.9 % and 47 %, respectively, of total primary energy [1, 2] and 97.5 % of the energy consumed by buildings in the USA is obtained from fossil fuels [3]. Interest in thermally operated cooling systems has been increasing in recent years because of the potential for lower effects on global warming and ozone layer depletion and the capability to drive these systems with clean thermal resources such as solar energy.

Desiccant cooling systems (DCSs) are considered as one of the powerful alternatives to current vapor-compression cooling systems (VCCS) particularly in hot and humid climates, due to the fact that they provide better air quality while they have the potential to save energy.

Desiccants are one subset of sorbents which attract and hold moisture because of the lower water vapor pressure on the desiccant surface compared to the surroundings. The desiccants are available in solid and liquid phases. When air is exposed to a dry desiccant surface (in the contactor), moisture is removed from the air until the vapor pressures of the vapor in the air and on the desiccant surface reach the same value, which is the equilibrium condition [4]. In DCS, dehumidification (latent cooling) is achieved by using a desiccant material, and sensible cooling is achieved by using cooling coils; hence independent control of humidity and temperature becomes possible. Desiccant solution vapor pressure depends on the aqueous solution's concentration and temperature. Higher concentrations and lower temperatures result in lower solution vapor pressures. During dehumidification, the humidity of the air is transferred to the solution which is followed by dilution of the solution and release of thermal energy, which results in an increase in the solution temperature and consequently vapor pressure. To be able to use the warm, diluted solution (weak solution) again, it has to be regenerated and cooled in order to have a low vapor pressure. Regeneration is done by heating diluted solution resulting in a high vapor pressure and subsequently water transfer from solution to the exhaust air in a process that is exactly the opposite of dehumidification. Also it should be mentioned that the mass flow rate of the solution in the dehumidifier is usually high enough so that the concentration does not drop significantly but the temperature increment is the determining parameter for further moisture attraction which has to be eliminated. Hence, the outlet solution from the dehumidifier can be reused several times just by cooling without any need for regeneration.

In the open literature, several experimental investigations on the dehumidifier and regenerator are presented with lithium chloride (LiCl) as the desiccant substance. A counterflow random-packed bed dehumidifier is investigated by Chung et al. [5] with 40 % LiCl as the desiccant solution and ratio of solution to air flow rate of about 9 which is high in magnitude but is required for full wetting of the random contactor. Chung [6] investigated a counterflow packed bed dehumidifier with 40 % LiCl as the solution but for a much smaller solution to airflow rate of about 0.6. Chung and Ghosh [7] investigated two different counterflow structured packed bed dehumidifiers using LiCl solution with concentrations in the range of 30–38 % and solution to airflow rate in the range of 6.5–15. Fumo and Goswami [8] used LiCl with a concentration of 34 % in a structured packed bed dehumidifier and regenerator with solution to airflow rate in the range of 4.5–7.7 and detailed results are presented.

In addition, many theoretical investigations are presented using three main methods: (1) computational fluid dynamics (CFD); (2) number of transfer units (NTU); and (3) fitting algebraic equations. Kronaki et al. [9] provide a finite difference model for air dehumidification in a structured packed bed tower and use the model to compare the three most commonly used liquid desiccant solutions: LiCl; lithium bromide (LiBr); and calcium chloride ($CaCl_2$). According to the results, LiCl operates better than the others due to its lower surface vapor pressure.

According to the study, inlet air temperature increment has a negligible effect on condensation rate and dehumidification efficiency but increasing inlet air humidity results in increasing condensation rate and efficiency, meaning that the system has better performance in humid climates. While the desiccant flow rate does not affect condensation significantly, it must be high enough to ensure that full wetting of the packing material occurs, meaning a need for a minimum desiccant flow rate. The results indicate that increasing solution concentration results in more water condensation (lower outlet air humidity) but the efficiency remains almost constant. The same trends exist for decreasing solution temperature. Fumo and Goswami [8] performed a similar study for both the regenerator and dehumidifier for the same system. From a regeneration point of view, water evaporation increases with increasing air and desiccant flow rate and temperature. Also, as the concentration increases, evaporation decreases which means that regeneration is easier at lower concentrations. Liu and Jiang [10] focused on the simultaneous heat and mass transfer interactions in the packed bed contactor. Their results indicate that the variance of the desiccant concentration within the contactor may be neglected in most conditions but the variance of the desiccant temperature is the main factor that affects the desiccant absorption ability.

Analysis of a finite difference model is quite complicated and iterative numerical solutions are required. To overcome this problem, algebraic methods can be derived which are simple equations used for quick prediction of effectiveness (ε) with the aim of determining air outlet properties using Eqs. (59.1) and (59.2). Note that Eqs. (59.1) and (59.2) explicitly show that the contactor effectiveness is a function of the humidity ratio (g) and enthalpy (h) of the inlet air (subscript A) and solution (subscript L) streams:

$$\varepsilon_g = \frac{g_{A,i} - g_{A,o}}{g_{A,i} - g_{L,i}} \tag{59.1}$$

$$\varepsilon_H = \frac{h_{A,i} - h_{A,o}}{h_{A,i} - h_{L,i}} \tag{59.2}$$

Using this set of equations, air outlet humidity ratio and enthalpy are obtained if dehumidification effectiveness and enthalpy effectiveness of dehumidifier are known and, subsequently, the air outlet temperature. While the accuracy of the method is slightly lower compared to finite difference, it is still acceptable for some applications such as air-conditioning and in fact simplicity is the main advantage of this method as it enables hourly system analysis in a reasonable amount of time. Algebraic equations are obtained through curve fitting of the available input and output data of dehumidifiers and regenerators, either using experimental tests or finite difference results. Unfortunately, this type of information is rare and only a handful of such studies are available in the open literature. Martin and Goswami [11] developed correlations for dehumidification effectiveness and enthalpy effectiveness of a counterflow random-packed bed dehumidifier for LiCl and triethylene glycol (TEG). Gandhidasan [12] developed two equations for dehumidification based on several assumptions, and these equations are used for obtaining condensation rate and fluid outlet temperature for a fixed air outlet temperature. Since outlet air temperature has to be known in this model, it has to be used in accordance with experimental data. Similar to that, Gandhidasan [13] also developed two equations for regeneration.

In addition to dehumidifier/regenerator investigations, several authors investigated the whole cooling system rather than just a specific component. Dhar et al. [14], Katejanekarn and Kumar [15], and Heidarinejad and Pasdarshahri [16] have studied stand-alone desiccant cooling systems while Dai et al. [17] and Kinsara et al. [18, 19] have studied hybrid desiccant cooling systems.

Nearly all the studies in the open literature have investigated steady-state systems subjected to fixed weather condition. Differences in specific inlet conditions have led to different and often contradictory results. To check the feasibility of a system in a particular climate, hourly investigations are required because ambient conditions (temperature, humidity, and solar irradiation) and cooling load vary continuously.

In summary, experimental and numerical studies have been performed on solid and liquid desiccants, with stand-alone and hybrid DCS. However, these studies are based on a single or limited set of environmental conditions which do not fully reflect the true variation in environmental conditions. Therefore the objective for this research is to develop a mathematical model for a hybrid DCS appropriate for performing annual simulations with an hourly time step in order to predict the cycle's performance over a typical year as ambient conditions and operating parameters are varied, and compare this system's performance to a VCCS. To fulfill this objective, two models are developed in TRNSYS [20], one for a hybrid DCS and the other for a conventional VCCS. Annual simulations are performed using Typical Meteorological Year-formatted data files created by using Meteonorm [21] for Middle East Technical University Northern Cyprus Campus (METU NCC), and finally results are compared with each other and discussed, and conclusions drawn.

59.2 Model

Two buildings are modeled that are identical except that one is equipped with an all-air VCCS and the other with an all-air DCS. Both systems are designed with the aim of maintaining desired indoor humidity and temperature in addition to providing the minimum required fresh air.

59.2.1 Building Model

The modeled building is a one-floor sports center, with a length of 33 m, width of 24 m, and height of 3.5 m. All surfaces are exposed to ambient conditions except the eastern wall. It is assumed that the building is occupied by 80 people [22] from 9 O'clock to 20 O'clock with lightning of 13 W m^{-2} [23] from 17 O'clock to 20 O'clock. The ventilation system operates continuously during the time of use and has the task of keeping indoor air temperature and relative humidity at 24 °C and 50 %, respectively [22]. Due to a positive inside pressure during operating hours, infiltration is considered to be zero. The supply air includes the minimum required fresh air [24] while the remaining is from recirculation air.

59.2.2 Vapor Compression Cooling Cycle

A schematic diagram of the VCCS modeled is shown in Fig. 59.1. In this constant air volume (CAV) system, fresh air is combined with recirculation air coming from the conditioned zone to be used as supply air. The supply air is first dehumidified to the required humidity ratio using a cooling coil. Because the temperature of air at the outlet of dehumidification coil is too low to be supplied directly to the building, the supply air is reheated again to a suitable temperature to meet the sensible load. Finally, the supply air with desired humidity and temperature is sent into the space. The required supply airflow rate is determined such that the humidity ratio does not fall below 0.007, because with a contact factor of 0.93 at the cooling coil, the required chilled water temperature from the chiller would drop below 5 °C which is difficult to obtain in ordinary HVAC chillers due to low COPs. Hence, the minimum value of supply airflow rate is calculated as 4.004 $m^3\ s^{-1}$. Because the enthalpy of the fresh air is significantly higher compared to that for the air at room conditions, and hence requires more dehumidification and cooling, the fresh airflow rate is set to the minimum level required to meet the fresh air requirements of the conditioned zone. According to ASHRAE [24], this minimum flow rate is 1.37 $m^3\ s^{-1}$ for this case and consequently 2.634 $m^3\ s^{-1}$ of recirculation air is desired.

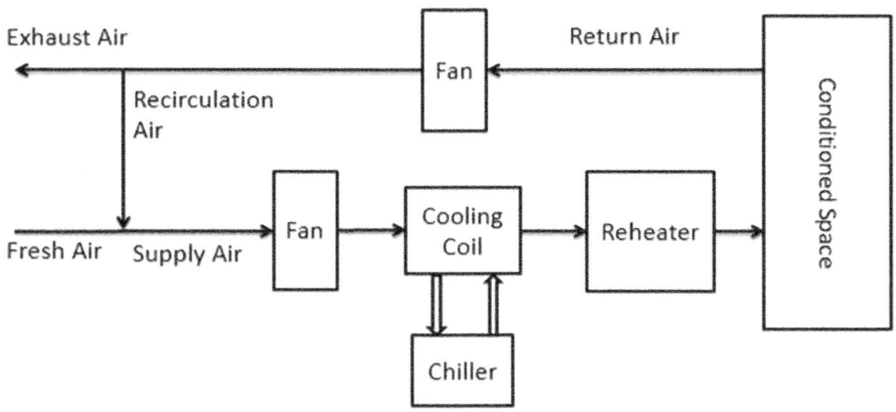

Fig. 59.1 Schematic diagram of VCCS

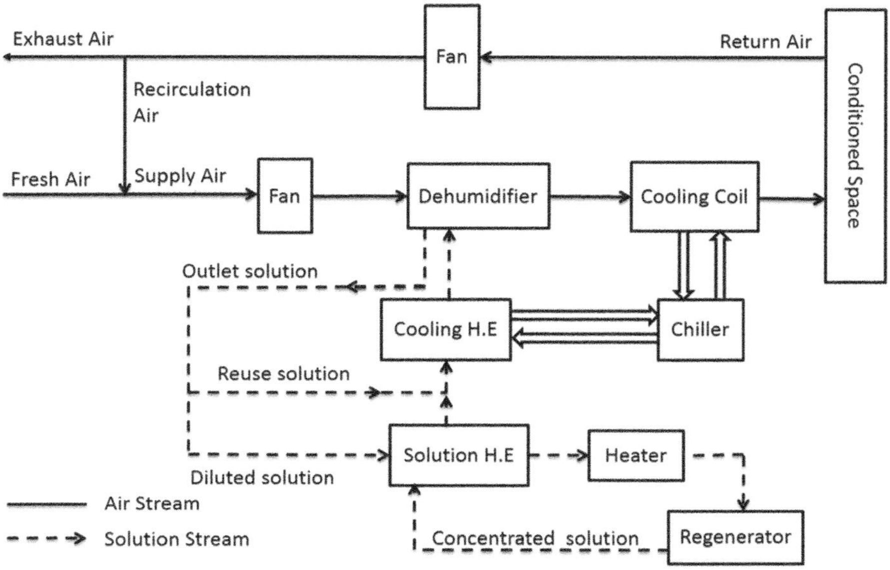

Fig. 59.2 Schematic diagram of DCS

59.2.3 Desiccant Cooling Cycle

A schematic diagram of the DCS modeled is shown in Fig. 59.2. Similar to the VCCS model, the minimum fresh air is combined with recirculation air but instead of using cooling coils for mechanical dehumidification, a packed bed dehumidifier with aqueous LiCl as a desiccant is adopted for dehumidification of the air to the desired value and following that, a cooling coil is only used to cool the air (sensible load). Since the air exits the DCS at a temperature above the supply temperature, there is no need for a reheat. To compare both systems, investigations are done for the same airflow rates. The model of the dehumidifier is adopted from the work of Martin and Goswami [11], Eqs. (59.3) and (59.4), which is based on experiments by Chung et al. [5] while the solution thermal properties are adopted from Conde [25] and Chaudhari and Patil [26]:

$$\varepsilon_g = 1 - 48.3\left(\frac{L}{A}\right)^{0.396\frac{\gamma_L}{0.029} - 1.57}\left(\frac{h_{A,i}}{h_{L,i}}\right)^{-0.751}(a_t Z)^{0.0331\frac{\gamma_L}{0.029} - 0.906} \tag{59.3}$$

$$\varepsilon_H = 1 - 3.77\left(\frac{L}{A}\right)^{0.289\frac{\gamma_L}{0.029} - 1.12}\left(\frac{h_{A,i}}{h_{L,i}}\right)^{-0.528}(a_t Z)^{-0.0044\frac{\gamma_L}{0.029} - 0.365} \tag{59.4}$$

For Chung et al.'s experiments, the air and solution inlet flow rates are 0.028 $m^3\ s^{-1}$ and 0.00019 $m^3\ s^{-1}$, respectively, with a 0.01826 m^2 dehumidifier cross-sectional area, resulting in 0.767 $m^3\ m^{-2}\ s^{-1}$ and 0.010405 $m^3\ m^{-2}\ s^{-1}$ of air and solution superficial flow rates, respectively. Since our supply airflow rate is 4.004 $m^3 s^{-1}$, the required dehumidifier cross-sectional area is 2.58 m^2 in order to have the same air superficial flow rate and consequently the solution flow rate is 0.02687 $m^3 s^{-1}$.

According to the previous work of Ge at al. [27], the dehumidification rate is easier to control by controlling the solution temperature than the concentration. Hence, liquid desiccant enters the dehumidifier with a constant flow rate and concentration of 35.2 % and at the temperature required to absorb the desired amount of moisture, as given by Eq. (59.5), and produces dry air and a dilute and warm desiccant solution:

$$\dot{m}_{cond} = \dot{m}_{A,i}\left(\frac{g_{A,i} - g_{A,o}}{1 - g_{A,o}}\right) \tag{59.5}$$

Significantly, Chung's experiments are done for desiccant solution inlet concentrations of 30 and 40 %, meaning that Eq. (59.5) is validated for these two concentrations. On the other hand, available data that are used for regeneration [8] are for regeneration of solutions in the range of 32–35.5 % which is not appropriate for the dehumidifier. But as mentioned above [9], while the solution inlet concentration affects the amount of condensed moisture in the dehumidifier, it does not have an effect on the moisture effectiveness and the reason for that is when the inlet solution concentration increases, its humidity ratio decreases and consequently the air outlet concentration decreases. As a result and according to Eq. (59.1), the moisture effectiveness remains approximately constant. While enthalpy effectiveness is not investigated in the paper by Koronaki et al. [9], similar conclusions can be drawn. Specifically, as $X_{L,i}$ increases, $g_{A,o}$ decreases, and lower $h_{A,o}$s will be obtained. From the solution point of view, higher $X_{L,i}$s result in lower $h_{L,i}$s because the enthalpy of pure water is much higher than that for pure LiCl. Finally, according to Eq. (59.2), enthalpy effectiveness can remain approximately constant and in fact, this is an assumption in this study. In conclusion, when calculating the enthalpy and moisture effectiveness using Eqs. (59.3) and (59.4), the inlet concentration can be assumed as 30 or 40 %. But when calculating the air outlet properties by Eqs. (59.1) and (59.2), the exact solution concentration has to be applied. After obtaining the air outlet humidity, enthalpy, and following them temperature, other outputs can be obtained through mass and energy balance equations at the dehumidifier using Eqs. (59.6)–(59.9).

For air:

$$\dot{m}_{A,o} = \frac{\dot{m}_{A,i}\left(1 - g_{A,i}\right)}{1 - g_{A,o}} \tag{59.6}$$

And for the solution:

$$\dot{m}_{L,o} = \dot{m}_{L,i} + \dot{m}_{cond} \tag{59.7}$$

$$X_{deh,o} = \frac{\dot{m}_{L,i} X_{deh,i}}{\dot{m}_{L,o}} \quad (59.8)$$

$$h_{L,o} = \frac{\dot{m}_{L,i} h_{L,i} + \dot{m}_{A,i} h_{A,i} - \dot{m}_{A,o} h_{A,o}}{\dot{m}_{L,o}} \quad (59.9)$$

Also, the ratio of solution to airflow rate must be very high to achieve perfect wetting of the contactor. As a result, a very small change in solution concentration occurs enabling the outlet solution to be used again if it is only cooled and without any regeneration. To model this, the outlet solution flow rate is split into two flows. One flow is reused again only by cooling it from the dehumidifier outlet temperature to the dehumidifier inlet temperature at constant concentration, 35.2 %, and the other flow is sent to the regenerator at the dehumidifier outlet temperature and regenerator inlet concentration of 34.5 % and are calculated by Eqs. (59.10) and (59.11):

$$\dot{m}_{ru} = \frac{\dot{m}_{L,o}\left(X_{deh,o} - X_{reg,i}\right)}{X_{Deh,i} - X_{reg,i}} \quad (59.10)$$

$$\dot{m}_{reg} = \frac{\dot{m}_{L,o}\left(X_{deh,i} - X_{deh,o}\right)}{X_{Deh,i} - X_{reg,i}} \quad (59.11)$$

The part of the solution which goes to the regenerator passes through a solution heat exchanger for preheating, and then through a heater where it is heated to the desired temperature, and is then brought into contact with the regeneration air in the regenerator to be reconcentrated to 35.2 % to be used in the dehumidifier again. The concentrated solution passes through the solution heat exchanger which results in the preheating of the diluted solution and precooling of the concentrated solutions. Then the concentrated solution is mixed with the reused part of the solution coming from the dehumidifier. The combined solutions then pass through a heat exchanger and are cooled to the temperature required to obtain the demanded condensation in the dehumidifier. The TRNSYS model is shown in Fig. 59.3 and flowchart of the processes in Fig. 59.4.

The regenerator is modeled according to another theoretical study [13] through Eqs. (59.12) and (59.13), and regeneration experimental tests [8] summarized in Table 59.1, because in the theoretical model [13] the outlet air temperature has to be known, and hence experimental data suitable to the inlet conditions of the solution are required:

$$\beta = \frac{T_{A,o} - T_{A,i}}{T_{L,i} - T_{A,i}} \quad (59.12)$$

$$\dot{m}_{evaporation} = \frac{1}{2346}\left[q - \dot{m}_A C_{P_A} \beta (T_{L,i} - T_{A,i})\right] \quad (59.13)$$

Similar to the dehumidifier, data are available as a superficial flow rate; thus it is possible to change the regenerator cross-sectional area in accordance with our solution flow rate to have the same superficial flow rate.

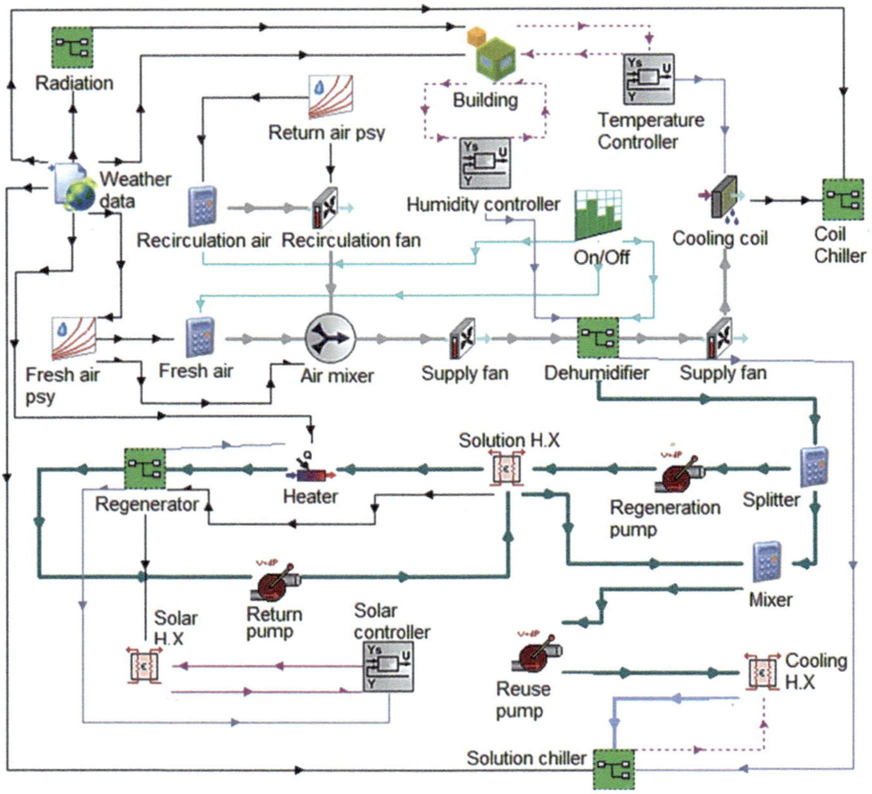

Fig. 59.3 TRNSYS diagram of DCS

59.3 Results

Analyses are presented for both DCS and VCCS systems under the same operating and ambient conditions. The supply airflow rate is adjusted so that the chilled water temperature requirement is not less than 5 °C in VCCS, which results in 4 $m^3 s^{-1}$ of supply air. Ambient temperature and humidity ratio, and required temperature and humidity ratio of supply air for July 21st, are plotted in Fig. 59.5. The required temperature of supply air is higher in the morning than evening because the eastern wall is shaded by a building next to it. Also there exists a sudden decrease in supply air temperature at 18:00 since the lights are turned on at that time. On the other hand, the required humidity ratio is constant, meaning that the latent load does not vary with time. The reason for this is zero infiltration and the only source of moisture generation being occupancy which is considered constant. Figure 59.6 exhibits temperatures and concentrations at the inlet and outlet of the random-packed bed dehumidifier also for July 21st. Since inlet solution concentration and required humidity ratio of supply air are constant, as the ambient humidity ratio increases, the solution temperature has to be decreased to absorb more humidity;

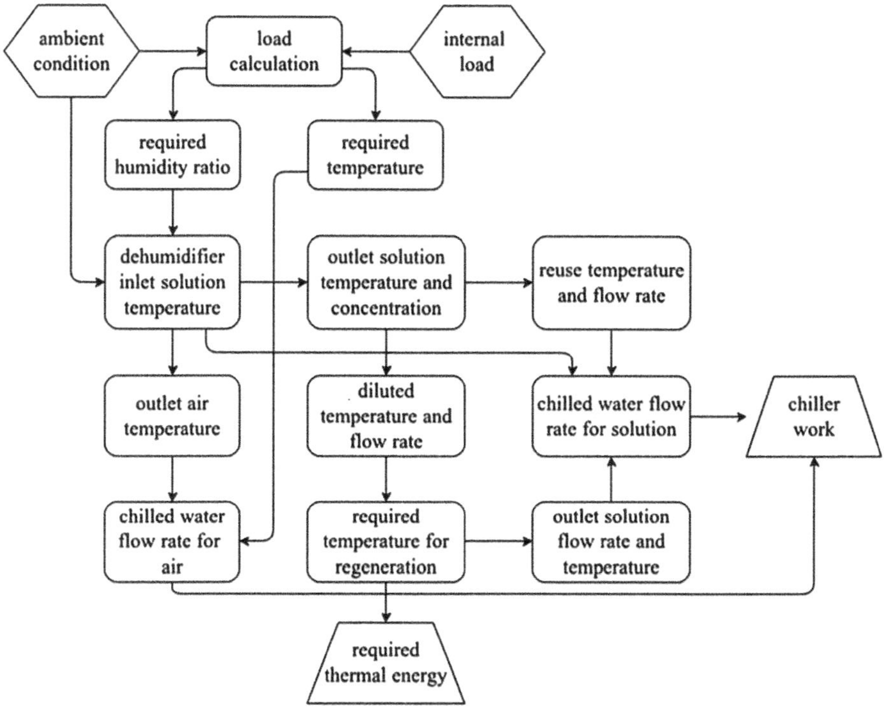

Fig. 59.4 Flowchart of DCS modelling in TRNSYS

thus its temperature increases at midday when the surrounding humidity ratio is lower compared to morning times. In the evening, a significant decrease in temperature occurs due to the increasing ambient humidity ratio. Following these fluctuations in desiccant solution inlet temperature, the desiccant solution and air outlet temperatures also fluctuate since their temperatures strongly depend on the solution inlet temperature and this characteristic is also observed in previous studies [5, 8]. From a concentration point of view, since the inlet concentration is constant, as ambient absolute humidity ratio increases, the desiccant solution becomes more diluted, followed by a higher mass flow rate of solution to the regenerator and lower mass flow rate for reuse.

From an energy point of view, the DCS power requirement and chiller COP on July 21st are displayed in Fig. 59.7. As ambient temperature increases, the COP of the chiller experiences a large decrement but since the evaporator temperature is sufficiently high and is set at 15 °C, its magnitude is still high. While the chiller brings the solution temperature to a lower magnitude in the morning compared to midday, its required power is lower thanks to a higher COP in the morning which results from lower ambient temperatures. After midday, the power consumption increases at a higher rate because of a decreasing COP and the need to provide a lower temperature solution compared to midday. The chiller duty for the cooling supply air is very low compared to that of the solution cooling. Since the air mass

Table 59.1 Experimental data [8]

Inlet						Outlet				
A_i (kg m^{-2} s^{-1})	$T_{A,i}$ (°C)	$g_{A,i}$ (kg kg^{-1})	L_i (kg m^{-2} s^{-1})	$T_{L,i}$ (°C)	X_i (%)	$T_{A,o}$ (°C)	$g_{A,o}$ (kg kg^{-1})	$T_{L,o}$ (°C)	X_o (%)	$\dot{m}_{evap}$ (g m^{-2} s^{-1})
1.438	29.8	0.0177	6.479	65.1	34.5	57.5	0.0488	56.6	35.2	2.10

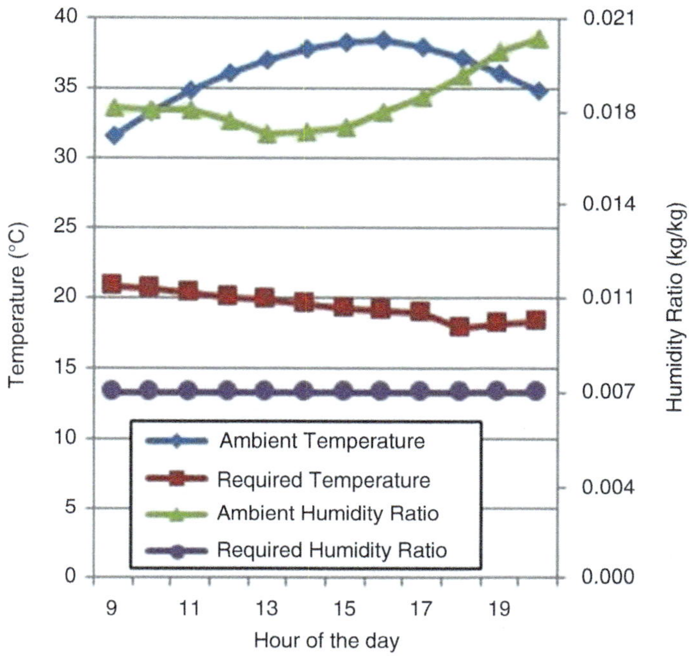

Fig. 59.5 Ambient condition and supply air required state

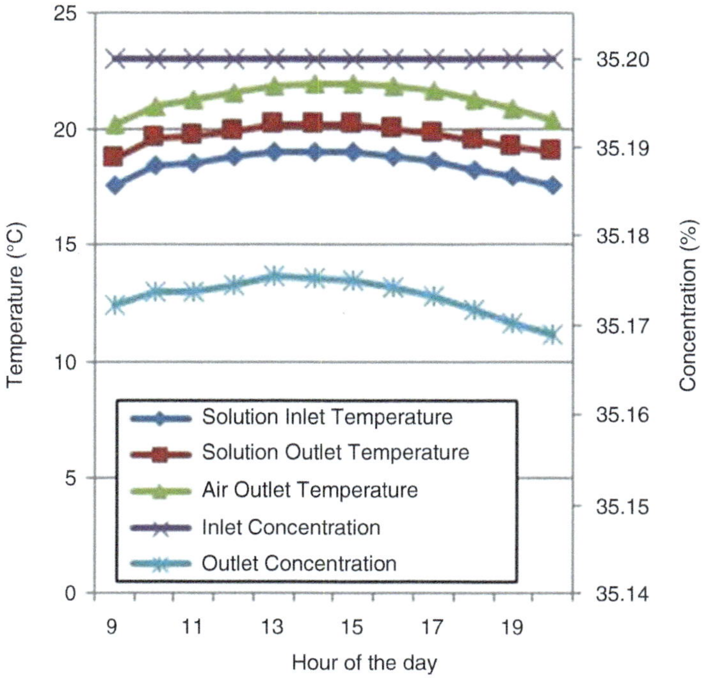

Fig. 59.6 Dehumidifier in/out temperature and concentration

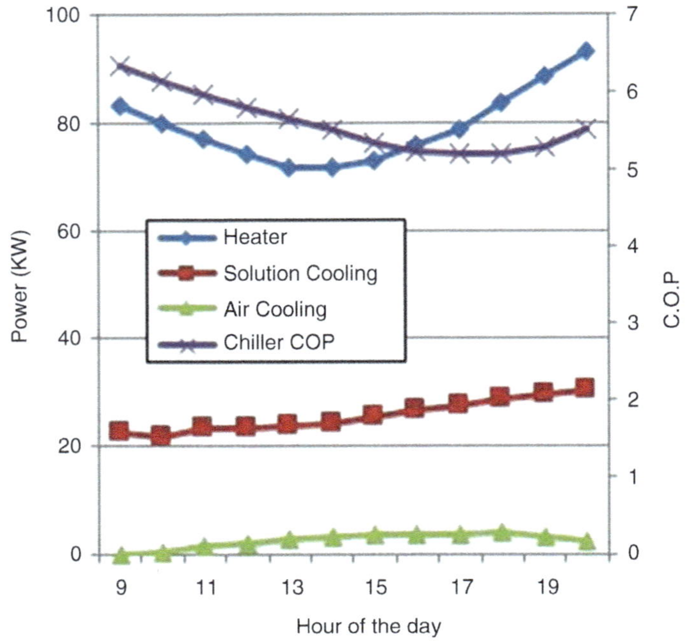

Fig. 59.7 Power of chiller and heater and chiller COP of DCS

flow rate passing over the cooling coil is constant, the temperature difference is the varying parameter in the cooling coil, which is the difference between the required temperature and the dehumidifier outlet air temperature. The chiller duty increases slightly due to decreases in both the COP and required temperature of supply air until hour 18 when it reaches its peak. In addition, the increasing humidity ratio of the surroundings increases the thermal energy required for regeneration due to two facts. First, the inlet solution temperature at the dehumidifier decreases and consequently, that of the outlet solution decreases. Thus more thermal energy is required to bring the solution to the temperature required for regeneration. Second, as per Eq. (59.11), the ratio of the solution mass flow rate to be regenerated increases as the outlet concentration decreases. These two facts result in a significant increase in the thermal energy required for regeneration. The required thermal energy is minimum at noon and is maximum in the morning and late evening, which means that if solar-thermal is adopted for solution heating purposes, thermal storage is required. According to Fig. 59.7, most of the required energy is consumed for regeneration of the diluted desiccant and it is sensitive to the ambient humidity ratio when compared to Fig. 59.5. The sensitivity of performance to ambient humidity ratio indicates that DCS characteristics have to be investigated in transient conditions rather than at specific outside conditions. In this regard, it is important to note that in nearly all the existing literature a constant ambient humidity ratio is investigated while in fact the absolute humidity ratio of the ambient air varies hourly and even a small change in the humidity ratio results in significant variations

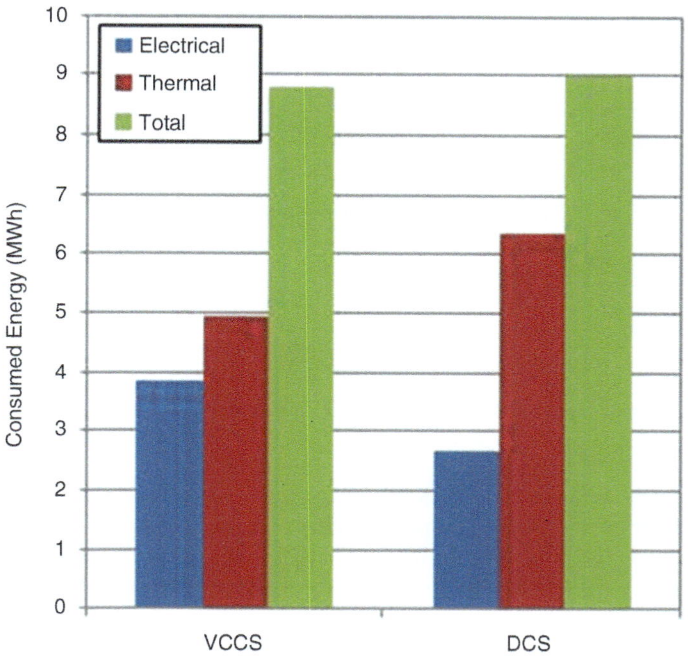

Fig. 59.8 Energy demand of the two systems

in the system behavior. Hence, both the VCCS and DCS systems are analyzed from an energy point of view for 3 months, namely June, July, and August, and the results are shown in Fig. 59.8. Energy demand is divided into two major groups, thermal and electrical energy. Thermal energy is used for heating in DCS and reheating in VCCS while electrical energy is consumed by all other components in both systems, including fans, chillers, and pumps. According to Fig. 59.8, the total energy consumption of the two systems is approximately equal in magnitude but it should be considered that electricity is of a higher quality. As a first approximation, the electricity consumption can be multiplied by three when compared to thermal energy because electricity is usually generated using fossil fuels and the efficiency of the process is about 33 % due to plant inefficiencies and grid losses. Therefore for equal total energy consumption, electricity requires three times more fossil fuel consumption than thermal energy, and consequently has three times higher CO_2 emissions. Thus the DCS is preferable from a pollution point of view and according to the current model, VCCS results in 23 % more CO_2 emissions if the thermal energy for both systems is provided by the same type of fossil fuel. Also, it is possible to provide the thermal energy of the DCS from flat-plate solar-thermal collectors because the regeneration temperature of the solution is only 65 °C. Considering a collector working temperature of 85 °C and heat exchanger effectiveness of 0.7, the required hot water flow rate on July 21st from solar thermal collectors varies between 1,800 and 2,500 kg h^{-1} but according to the

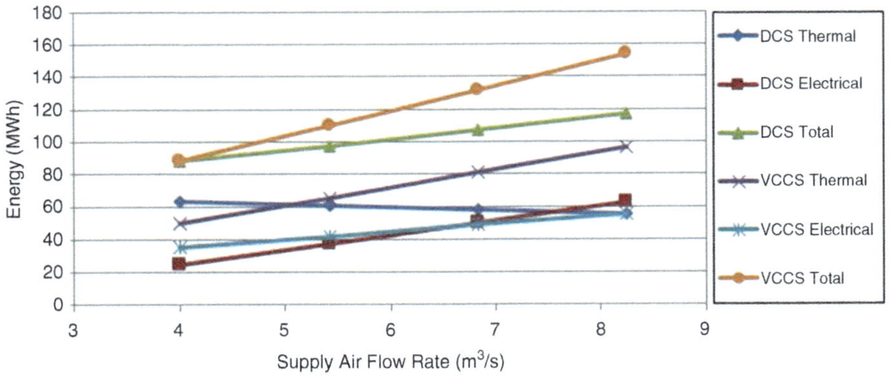

Fig. 59.9 Effect of supply air flow rate on energy consumption

meteorological data set this day is one of the most humid days during the year and for an ordinary summer day, humidity ratio values would be lower. Starting from about 2,000 kg h^{-1} in the morning, the required hot water follows a gradual decrease up to hour 13 where a minimum of about 1,800 kg h^{-1} occurs and then it undergoes a significant increment to hour 20 when it reaches a maximum of 2,500 kg h^{-1}. Finally, the sum of the consumed energy by the two systems in June, July, and August is compared at different supply airflow rates in Fig. 59.9. Minimum supply airflow rate is about 4 m^3 s^{-1} because a lower magnitude requires chilled water temperature below 5 °C in the VCCS.

In DCS, as airflow rate increases, the required temperature of the dehumidifier inlet solution rises and hence the chiller can work at higher temperatures which results in higher COP. But higher flow rates also mean increasing work of hydronic units, particularly fans. In an opposite manner, airflow increment results in higher temperature of the dehumidifier outlet solution and consequently a lower temperature increment for regeneration because the regeneration temperature is constant, and the reason for this is the constant outlet regenerator concentration. However, the energy consumption trend for VCCS is entirely different because higher airflow rate results in both higher electrical energy and thermal requirements due to higher fan loads and higher mass of air to be reheated. In conclusion, the increase in total energy consumption with supply air mass flow rate is smaller for DCS than VCCS.

59.4 Conclusion

A hybrid desiccant cooling system is investigated using TRNSYS and is compared to a conventional all-air VCCS from both operating characteristic and energy consumption point of views using meteorological data for the METU Northern Cyprus Campus. According to the results, the ambient temperature effect on DCS is very small compared to ambient humidity ratio. The former only affects the COP of the chiller while the latter results in significant changes in both thermal and electrical energy consumption. Thus, to understand the system's performance for a specific location the system should be investigated over a range of operating conditions rather than for specific conditions because the outside humidity ratio, the main parameter, varies hourly during a day and its impact on the system's performance is significant. Herein both systems are investigated over three summer months assuming similar airflow rates. The results show that the total energy demand of VCCS and DCS are approximately the same but differences exist in the type of energy demand. While VCCS requires more electrical energy, the DCS requires more thermal energy which could prove beneficial from pollution and economical points of view. Specifically, for the same required energy, electricity is more expensive than heat and can produce approximately three times more greenhouse gas emission due to inefficiencies in electricity production and losses in the grid. Also it is possible to provide some or all of the required heating energy for DCS using solar energy because the required temperature from the heating source can be as low as 80 °C. Hence glazed flat-plate collectors could be a suitable choice, since they have a lower initial cost than higher temperature solar-thermal collectors. If a solar-thermal system is adopted, the best results would be obtained when ambient humidity and solar resources are in the same phase temporally. However in the studied case, they are out of phase diurnally, meaning that when irradiation peaks at midday, the outside humidity ratio has its minimum magnitude and conversely in the morning and evening when solar resources are low, humidity reaches its peak, and thus thermal storage may be required. Finally, the operating characteristics of both systems are investigated for different airflow rates. Results show that as airflow rate increases, the energy demand of VCCS increases significantly and more than that for the DCS. Thus, DCS would save more energy when it is necessary to have an air-conditioning system with high flow rates such as for hospitals or industrial applications. Also, because limitations of the VCCS's chiller temperature are not present for the DCS, running the DCS at lower flow rates than for the VCCS may be possible, and may be appropriate for some applications such as residential applications. Lower airflow rates mean lower electricity demand and higher heat demand. Thus it is better to run the system at as low an airflow as possible to shift more electricity to heat, resulting in lower operating costs and CO_2 emissions.

References

1. Halliday S, Beggs C, Sleigh P (2002) The use of solar desiccant cooling in the UK: a feasibility study. Appl Therm Eng 22(12):1327–1338
2. US EPA (2013) US Environmental Protection Agency. http://www.epa.gov/. Accessed: 2 Jan 2013
3. US Energy Information Administration "EIA" (2012) US Energy Information Administration. www.eia.gov. Accessed: 2 Jan 2014

4. Mei L, Dai YJ (2008) A technical review on use of liquid-desiccant dehumidification for air-conditioning application. Renew Sustain Energy Rev 12(3):662–689
5. Chung T-W, Ghosh TK, Hines AL (1993) Dehumidification of air by aqueous lithium chloride in a packed column. Sep Sci Technol 28(1–3):533–550
6. Chung T-W (1997) Principles and applications of gas separation in nonfrosting technology. Sep Sci Technol 32(8):1389–1402
7. Chung T, Ghosh TK (1996) Comparison between random and structured packings for dehumidification of air by lithium chloride solutions in packed column and their heat and mass transfer correlations. Ind Eng Chem Res 35(1):192–198
8. Fumo N, Goswami DY (2002) Study of an aqueous lithium chloride desiccant system: air dehumidification and desiccant regeneration. Sol Energy 72(4):351–361
9. Koronaki IP, Christodoulaki RI, Papaefthimiou VD, Rogdakis ED (2012) Thermodynamic analysis of a counter flow adiabatic dehumidifier with different liquid desiccant materials. Appl Therm Eng 50(1):361–373
10. Liu XH, Jiang Y (2008) Coupled heat and mass transfer characteristic in packed bed dehumidifier/regenerator using liquid desiccant. Energy Convers Manag 49(6):1357–1366
11. Martin V, Goswami DY (2000) Effectiveness of heat and mass transfer processes in a packed bed liquid desiccant dehumidifier/regenerator. HVAC&R Res 6(1):21–39
12. Gandhidasan P (2004) A simplified model for air dehumidification with liquid desiccant. Sol Energy 76(4):409–416
13. Gandhidasan P (2005) Quick performance prediction of liquid desiccant regeneration in a packed bed. Sol Energy 79(1):47–55
14. Dhar P, Kaushik S, Jain S, Pahwa D, Kumar R (1995) Thermodynamic analysis of desiccant-augmented evaporative cooling cycles for Indian conditions. Trans Am Soc Heat Refrig Airconditioning Engin 101(1):735–752
15. Katejanekarn T, Kumar S (2008) Performance of a solar-regenerated liquid desiccant ventilation pre-conditioning system. Energ Build 40(7):1252–1267
16. Heidarinejad G, Pasdarshahri H (2011) Potential of a desiccant-evaporative cooling system performance in a multi-climate country. Int J Refrig 34(5):1251–1261
17. Dai Y, Wang R, Zhang H, Yu J (2001) Use of liquid desiccant cooling to improve the performance of vapor compression air conditioning. Appl Therm Eng 21(12):1185–1202
18. Kinsara AA, Al-Rabghi OM, Elsayed MM (1998) Parametric study of an energy efficient air conditioning system using liquid desiccant. Appl Therm Eng 18(5):327–335
19. Kinsara AA, Elsayed MM, Al-Rabghi OM (1996) Proposed energy-efficient air-conditioning system using liquid desiccant. Appl Therm Eng 16(10):791–806
20. University of Wisconsin–Madison. TRNSYS: transient system simulation tool. www.trnsys.com/. Accessed 2 Jan 2014
21. Meteotest Company (2013) Meteonorm: irradiation data for every place on Earth. www.meteonorm.com/. Accessed 2 Jan 2014
22. Bell AA (2000) HVAC equations, data, and rules of thumb. McGraw-Hill, New York, NY
23. American society of heating, refrigerating and air-conditioning engineers (2009) Handbook, ASHRAE fundamentals. ASHRAE, Atlanta, GA
24. American society of heating, refrigerating and air-conditioning engineers (2007) Handbook, ASHRAE HVAC application. ASHRAE, Atlanta, GA
25. Conde MR (2004) Properties of aqueous solutions of lithium and calcium chlorides: formulations for use in air conditioning equipment design. Int J Therm Sci 43(4):367–382
26. Chaudhari SK, Patil KR (2002) Thermodynamic properties of aqueous solutions of lithium chloride. Phys Chem Liq 40(3):317–325
27. Ge G, Xiao F, Niu X (2011) Control strategies for a liquid desiccant air-conditioning system. Energ Build 43(6):1499–1507

Chapter 60
Certification Issue of Solar Photovoltaic Modules for Local Production, Problems, and Solutions in Turkey

Yusuf Bicer, Cevat Ozarpa, and Y. Erhan Boke

Abstract In parallel to Turkey's developing technology and growing economy, to meet the increasing energy needs, Turkey should focus on domestic and renewable energy resources. Any growth in energy supply must be achieved in a low carbon way. In this process, the production of electricity from solar energy, namely photovoltaic technology, will have a large share. Hence determining the quality of photovoltaic modules has great importance in both financial terms and manufacturing.

Determination of quality and reliable photovoltaic modules is only possible by the product certification and inspection services. Thus, the production stage of the module and installation process can be demonstrated by checking all of the quality system standards. In order to ensure the most accurate and independent service, international accreditation is crucial.

There are three steps in front of Turkey during the adaptation of solar energy. First of them is quality production of photovoltaic modules, second is testing and certification, and lastly proper installation of photovoltaic modules onto the power plants. This research finds solutions for the problems faced in testing and certification process of solar photovoltaic modules for local producers by local certification laboratories and certification bodies.

As there are different elements which affect the quality of photovoltaic modules, this research focuses on production of solar photovoltaic modules in Turkey, certification and accreditation for photovoltaic modules, regulations in Turkey, and encountered problems and solutions.

Y. Bicer (✉)
Energy Science and Technology, Energy Institute, Istanbul Technical University, Maslak, Istanbul 34469, Turkey
e-mail: bicery@itu.edu.tr

C. Ozarpa
İGDAŞ (Istanbul Gas Distribution Corporation) Maintanance & Metering Management, Sultangazi, Istanbul 34265, Turkey

Y.E. Boke
Faculty of Mechanical Engineering, Istanbul Technical University, Gümüşsuyu, Istanbul 34437, Turkey

I. Dincer et al. (eds.), *Progress in Clean Energy, Volume 2*,
DOI 10.1007/978-3-319-17031-2_60

Keywords Solar energy • Photovoltaic module • Product certification • Domestic production • Accreditation

60.1 Introduction

In order to have a reliable photovoltaic power plant, each unit of the power plant should work in harmony. If a photovoltaic module is not sufficient to work in harmony, in that case there is no meaning of group work. Hence, photovoltaic module's quality and reliability have the priority.

Turkey is the world number two after China in the addition of solar collectors in year 2012 as shown in Fig. 60.1 [1]. However for photovoltaic technology, Turkey is recently trying to progress together with the last announcement of 600 MW licensed solar photovoltaic power plants. In June 2013, approximately 9,000 MW applications were made by 496 different applicants. Conformably, as Fig. 60.2 shows, there is a high interest for installing unlicensed solar photovoltaic power plants. Approximately 75 % of total unlicensed applications are for solar energy which corresponds to around 675 MW installed power [2]. Most of them are in the process of project evaluation. However there was no licensed solar power plant installed till 2015. The aim of the Ministry of Energy and Natural Resources is to reach 30 % renewable energy share by 2023. Hence solar energy plays an important role in this stage [3].

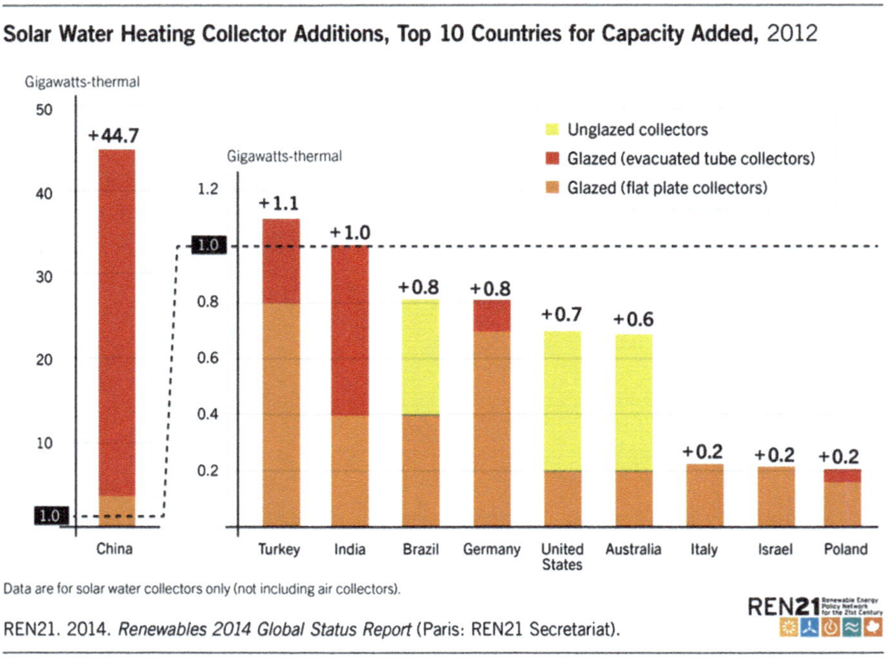

Fig. 60.1 Solar water heating collector additions, 2012

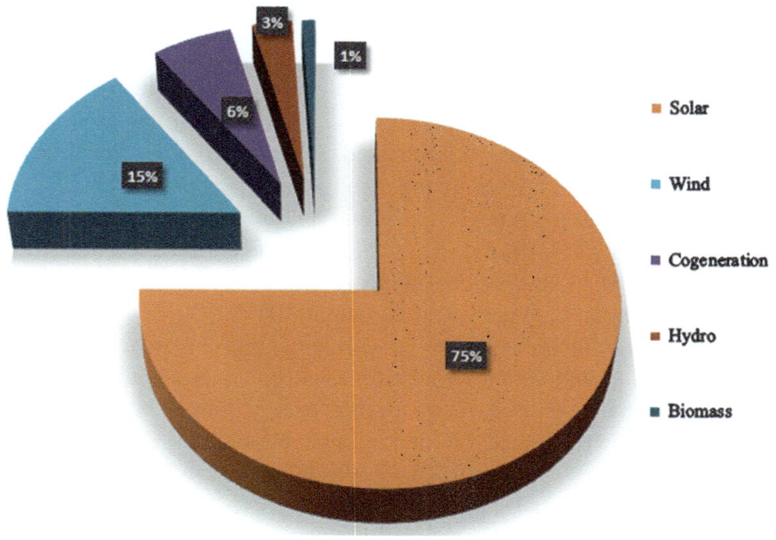

Fig. 60.2 Unlicensed electricity generation applications till August 2014

60.2 Local Production in Turkey

Solar panel costs are probably partly down due to technological and manufacturing advancements. Production of any machine or material brings the research and development activities together with some innovations. Therefore, manufacturing of solar photovoltaic modules in Turkey is important for the Turkish market for the coming years. There are many Chinese manufacturers which produce photovoltaic modules not suitable according to international standards. However, the quality of photovoltaic modules produced in Turkey will certainly catch European standards because of Turkey's European Union integration.

Turkey has around 15 photovoltaic module manufacturers, some of which having partially large capacities around 100 MW. Although half of them have really quality production line, other producers are still investing more to increase the quality level. These producers need to test their products before marketing and certificate them according to international standards. Since there are currently no test laboratories and certification institutions for photovoltaic modules in Turkey, the producers have to send their modules to outside especially to accredited European photovoltaic certification bodies to get the required certificates.

60.3 Necessity of Accreditation and Certification

In the scope of accredited certification, schemes for certification of photovoltaic modules which include conditions, factory production of photovoltaic modules, and inspection of on-site modules are prepared. The designed certification system contains the conditions to be met by the manufacturer such as tests during production line, installation requirements, and important points to be evaluated. Turkish Accreditation Agency (TÜRKAK) has carried out the certification audit for certification body in terms of international standard TS EN 45011 General requirements for bodies operating product certification system [4].

On the other hand, according to the Turkish regulations, in order to certificate the domestic production and benefit from additional financial incentives, equipment and components that are used in renewable energy power plants are required to carry conformity certificate of international or national standards given by organizations of TS EN 45011 or TS EN ISO/IEC 17065 Product Certification Bodies [4–6].

One of the important points in certification process is testing laboratories. They are eligible when they bear the standard of "TS EN ISO/IEC 17025 General requirements for the competence of testing and calibration laboratories" [7].

Product certification satisfies product-specific standards and other standard documents to provide assurance for compliance. Product certification systems include initial experimental evaluation of the supplier's quality system, subsequent plant periodic inspection, and testing of sample products of the suppliers taken from free market. TS EN 45011 product certification system assures that a photovoltaic module is manufactured according to the standards, stages, and processes of international standards IEC 61215 for crystalline silicon photovoltaic modules, IEC 61646 for thin-film photovoltaic modules, and IEC 61730 for safety requirements [8–10].

60.4 Legislation Related to Certification in Turkey

In the Official Gazette Nr. 27969 dated 19 June 2011, a new regulation called "Local manufacturing of the equipments used in power plants producing electrical energy from renewable energy sources" was published [6].

Turkish Government guarantees to buy solar electricity from investors for 10-year duration with the amount of 13.3 cents/kWh. Additionally, according to previously published regulations, renewable energy power plants which use domestically manufactured equipment are provided by the state up to 6.7 cents/kWh additional incentives for solar photovoltaic plants. This incentive was for the plants to be commissioned by 2015 and valid for 5 years. By the recent legislation, it has been extended for the plants which will be commissioned by 2020. When renewable energy market's high costs are considered, any additional incentives provided by the government will facilitate the issue for investors. In this context, investors who plan to invest in this sector are willing to benefit from additional incentives by using domestic products as much as they can.

Nevertheless, Turkey's renewable energy technologies and production have great significance for domestic production. Developed countries all over the world in their respective industries have production-oriented work and have been pioneers in the market. Therefore it has great importance for Turkey in the coming years to encourage domestic production and its use, constituting itself as having high solar energy potential in order to be a pioneer in renewable energy.

Hence, it was a key point for the legislation to emphasize domestic production by stating that TS EN 45011 or TS EN ISO/IEC 17065 Product Certification is mandatory in accordance with the standard of product [4, 5].

There are three documents to be prepared and presented to the ministry for benefiting additional incentives. These are:

- Domestic Manufacturing Status Certificate received from Chamber of Commerce and Industry.
- Type Certificate or Product Certificate or Certificate of Authentication Unit issued by "TS EN 45011 General requirements for bodies operating product certification system" or "TS EN ISO/IEC 17065 Conformity assessment—Requirements for bodies certifying products, processes and services" national accreditation bodies of the International Accreditation Forum (IAF) and the mutual recognition agreement [4, 5].
- In case type certification is declared, documents showing holder of the certificate's legal person domestic producer information and details of the assembly according to the type certificate.

With this regulation, presence of mandatory product certification will be held in the country of manufacture and assemblies of photovoltaic modules comply with relevant standards. Thus, the relevant domestic component or system manufacturers will not only focus on developing their own technologies, but they will also have a chance to follow the most current national and international standards at the same time and fulfill the requirements of this standard. This will undoubtedly bring quality and reliable products together for the complex assembly structures in renewable energy.

60.5 Problems and Solutions in Turkey's Market

When it comes to components of photovoltaic systems, the most common problem encountered during certification process is the reliable testing laboratory and certification institution. Secondly, the distance of the laboratory plays an important role for the producers. In case there will be local laboratories around the production area, they will send their products for testing more often and analyze the results. In that case, they will be able to improve their modules which lead research and development activities for the producers. If there is a local laboratory and certification body, the prices of services and modules will be more convenient.

Transportation issue plays an important role during testing process. In case there will be some failures during transportation from factory to testing laboratory, it will affect the results of the tests and may cause negative consequences.

Nearby laboratories will shorten the testing time and certification period which is an important advantage. Especially for situations of material changes, some additional tests are required and producers are not willing to send their modules for every change. Therefore, they do not change their raw material although they are suspicious about the quality of material.

Local certification bodies are in a critical position for the investors of module production line. Because in case they will receive support and advise from local test laboratories and certification bodies, the production line can be designed according to the experiences and results of the tests.

International standards for photovoltaic modules are designed basically for one type of moderate climate. However, climatic conditions of Turkey are more different than European regions. This affects the module efficiencies. In Turkey, average temperatures are higher, some regions are dusty, and there are coasts which are salty and humid [11]. In this case, modules may need to have additional tests and certification procedures such as "Salt mist corrosion testing of photovoltaic modules" and "Environmental testing—Test of Dust and Sand." These tests are important and reliable quality indicators for the modules used in Turkey.

Additionally, in the certification process factory inspection is crucial. If the certification body is close to the manufacturer, they will always follow the qualification procedure without any problem. Hence some type of tests might be repeated in a defined period [12, 13].

A few producers in the world have their own testing mechanism in their production facilities for some safety tests in IEC 61730. In the proposed certification scheme, it is stated that producers need to complete two tests, reverse current overload test and ground continuity test, during yearly factory inspection process. Additionally, in the production line, electroluminescence (EL) tester or laser tester is important. Each PV module needs to be tested in those devices after or before lamination. The criteria for the evaluation of EL test should be determined by producer and approved by certification body.

60.6 Results and Conclusion

For a growing photovoltaic market in Turkey, regulations and standards need to be stable. Nongovernmental organizations in Turkey such as GÜNDER-ISES (*International Solar Energy Society—Turkey Section*) and GENSED (*Solar Energy Industrialists and Industry Association*) have critical status in this process. There should be fixed rules for the acceptation of the solar energy projects especially for unlicensed applications.

Local certification and testing institutions are a background for high-quality modules. Although domestic production of solar modules is an advanced sector for the countries such as Germany and Far Eastern countries, it is a recent topic for Turkey. Many countries have experience in how to determine the quality of photovoltaic modules and how to make investments accordingly.

Since Turkey is a developing country for solar energy sector, investors need to be convinced to enter the market. In this context, they need some additional supports from the government. Turkish Government is trying to regulate the market by publishing such type of legislations. The additional incentives are really an important factor for the investors. Therefore the process of benefiting from these incentives needs to be realized as soon as possible. This research focuses on how to realize the domestic production incentives in terms of product certification system which is a mandatory topic as stated in the legislation.

Each region has different climatic conditions which brings varied testing techniques. For Turkish region, a detailed and more composite certification process is necessary including additional tests and factory inspection criteria. By realizing the local certification body and test laboratory, it can be proved that additional incentives for local product users are utilizable in terms of product certification system.

References

1. Renewables 2014 Global Status Report. Renewable energy policy network for the 21st century (REN21), Paris, France, 2014
2. Unlicensed Electricity Production Applications (2014) www.lisanssizelektrik.org. Accessed 14 Aug 2014
3. 2010-2014 Strategic Plan. Ministry of Energy and Natural Resources, Ankara, Turkey, 2010
4. General requirements for bodies operating product certification system, TS EN 45011
5. Conformity assessment—requirements for bodies certifying products, processes and services, TS EN ISO/IEC 17065
6. Official Gazette Nr. 27969 (June 19, 2011) Local manufacturing of the equipments used in power plants producing electrical energy from renewable energy sources, Ankara, Turkey
7. General requirements for the competence of testing and calibration laboratories, TS EN ISO/IEC 17025
8. Crystalline silicon terrestrial photovoltaic (PV) modules, IEC 61215
9. Thin-film terrestrial photovoltaic (PV) modules—design qualification and type approval, IEC 61646
10. Photovoltaic (PV) module safety qualification, IEC 61730
11. Biçer Y (2012) Control of a dual axis solar panel. B.Sc. thesis, Control Engineering, Istanbul Technical University, Istanbul, Turkey
12. TamizhMani G, Kuitche J (2013) Accelerated lifetime testing of photovoltaic modules. Arizona State University, Arizona, USA
13. Bicer Y, Özarpa C, Böke E (2014) Product certification of solar photovoltaic modules for local production, problems and solutions in Turkey. Paper presented at 7th International PV Power Plants Conference PV Power Plants—Turkey, Solarpraxis, Istanbul, 9–10 April 2014

Chapter 61
Statistical Analysis of Short-Term Solar Radiation Data over Aligarh (India)

Basharat Jamil and Naiem Akhtar

Abstract This study aims at exploring solar radiation in the Aligarh region of India (latitude, 27.88°N; longitude, 78.08°E), due to the fact that measured solar radiation data for this region is scarce. A new set of regression coefficients for the Angström-type correlation for estimation of monthly average daily solar radiation on horizontal surfaces in Aligarh (India) has been presented. The regression coefficients were applied to the data of bright sunshine hours obtained from the Solar Energy Laboratory, Department of Mechanical Engineering, Aligarh Muslim University, during the period of September 2013 to February 2014. The results were compared with the correlations of other researchers available in literature and solar radiation data obtained from measurements. Correlation for the estimation of monthly average diffuse solar radiation is also developed in terms of the ratio of diffuse to global radiation. The performances of the proposed correlations were analyzed by mean bias error and root mean square error. A good agreement was found between the estimated and measured values following the application of the new proposed correlations. Thus, monthly average global, diffuse, and direct solar radiation on a horizontal surface in the Aligarh region of India has been obtained.

Keywords Aligarh (India) • Experimental measurement • Horizontal surface • Regression coefficients • Solar radiation

Nomenclature

$\overline{H}$	Monthly average global solar radiation (MJ/m^2-day)
$\overline{H}_0$	Monthly average extraterrestrial solar radiation (MJ/m^2-day)
$\overline{H}_d$	Monthly average diffuse solar radiation (MJ/m^2-day)
$\overline{H}_b$	Monthly average beam solar radiation (MJ/m^2-day)
H_{sc}	Solar constant

B. Jamil (✉) • N. Akhtar
Department of Mechanical Engineering, Zakir Husain College of Engineering and Technology, Aligarh Muslim University, Aligarh, Uttar Pradesh 202002, India
e-mail: basharat1986@gmail.com; akhtar_n@rediffmail.com

I. Dincer et al. (eds.), *Progress in Clean Energy, Volume 2*,
DOI 10.1007/978-3-319-17031-2_61

n	Day of the year
$\overline{S}$	Monthly average duration of bright sunshine (h)
$\overline{S}_0$	Monthly average maximum possible sunshine duration (h)
$\overline{H}_{i,\mathrm{e}}, \overline{H}_{i,\mathrm{m}}$	ith calculated and measured values (MJ/m^2-day)
K_T	Monthly average of daily clearness index
K_D	Monthly average of daily diffuse index
$a, b, c,$ and d	Empirical constants
$\varnothing$	Latitude (degree)
δ	Solar declination angle (degree)
ω_s	Sunset hour angle (degree)
θ	Incidence angle (degree)

61.1 Introduction

The increase in energy demands over the globe because of the dramatic rise in human population and consequent increase in fossil fuel prices have led developing countries like India to shift their focus on renewable energy sources. Additionally, severe environmental problems (like global warming) caused by the unattended use of fossil fuels release a large amount of greenhouse gases (GHGs). Worldwide concern over the issue of rapid fossil fuel depletion and environmental degradation is critical.

Solar energy is a readily available source of alternative energy and is free of cost. It is therefore among one of the top contenders of alternative energy. Since solar energy is available in abundance, it can be utilized in varied applications. For this reason the focus on solar energy research has increased tremendously in recent years. Typical applications in different areas include solar water heating and distillation, solar crop drying, photovoltaics, building energy estimation, daylighting, energy storage, desalting, meteorological forecasting, etc., to name a few. Knowledge of solar energy is fundamental for analysis of such solar energy applications.

Global solar radiation is fundamentally composed of two components, namely, direct (or beam) radiation and diffuse radiation [1]. Solar energy applications require a complete detailed analysis of these components so that the potential of the site for solar energy utilization can be evaluated.

61.2 Solar Radiation Measurements

For the purpose of analysis and further utilization in solar energy conversion systems, measurement of solar radiation at the ground level is an important issue. The radiation information at the ground level can be gathered by pyranometers because it's the best way available for the price measurement of solar radiation and its components. Also, their operational cost is quite low with easy day-to-day maintenance of the apparatus.

Solar radiation data (global and diffuse) in this study were obtained from the Solar Energy Laboratory at the Department of Mechanical Engineering, Aligarh Muslim University, Aligarh (India). The data was collected over a period of 6 months from September 2013 to February 2014. Global solar radiation was measured using Kipp & Zonen CM-11 pyranometer, and diffuse horizontal solar radiation was measured using Kipp & Zonen CM-11 pyranometer with a shading ring (CM-121B). These measurements were continuously recorded for the sunshine duration and stored in Kipp & Zonen Datalogger LogBox. The difference of the global and diffuse solar radiation data provides the beam (or direct) horizontal solar radiation.

The hourly average daily values of global, diffuse, and direct solar radiations on a horizontal surface have been obtained during sunshine hours for the 6 months and are represented in Fig. 61.1.

The daily values of solar radiation are obtained by integrating the hourly distribution. All the data used in this study are recorded live for the past 6 months which are used for the ongoing research on solar energy. The daily average values of solar radiation and its component are represented in Fig. 61.2.

It shows the monthly average daily global, direct, and diffuse solar radiation data measured for the period of study. The monthly average daily global solar radiation is observed to be the highest for the month of September of the order of 21.7 MJ/m^2-day, while it is observed to be the lowest in January (15.1 MJ/m^2-day). The maximum and minimum values of monthly average daily diffuse radiation are 8.2 MJ/m^2-day in October and 5.4 MJ/m^2-day in December, respectively. The monthly average direct solar radiation values are 14.2 MJ/m^2-day in September to 8.1 MJ/m^2-day in January. The monthly average hours of bright sunshine for the 6 months under study are also shown in Fig. 61.3.

61.3 Materials and Methods

61.3.1 Correlations for Estimation of Solar Radiation Data

Solar radiation data are commonly available in the form of monthly average daily global radiation on a horizontal surface $(\overline{H})$ and monthly average daily diffuse radiation $(\overline{H}_\mathrm{d})$.

For the estimation of monthly average global solar radiation on a horizontal surface, the modified Angström-type regression equation is the most commonly used, which is of the form

$$\frac{\overline{H}}{\overline{H}_0} = a + b\frac{\overline{S}}{\overline{S}_0} \tag{61.1}$$

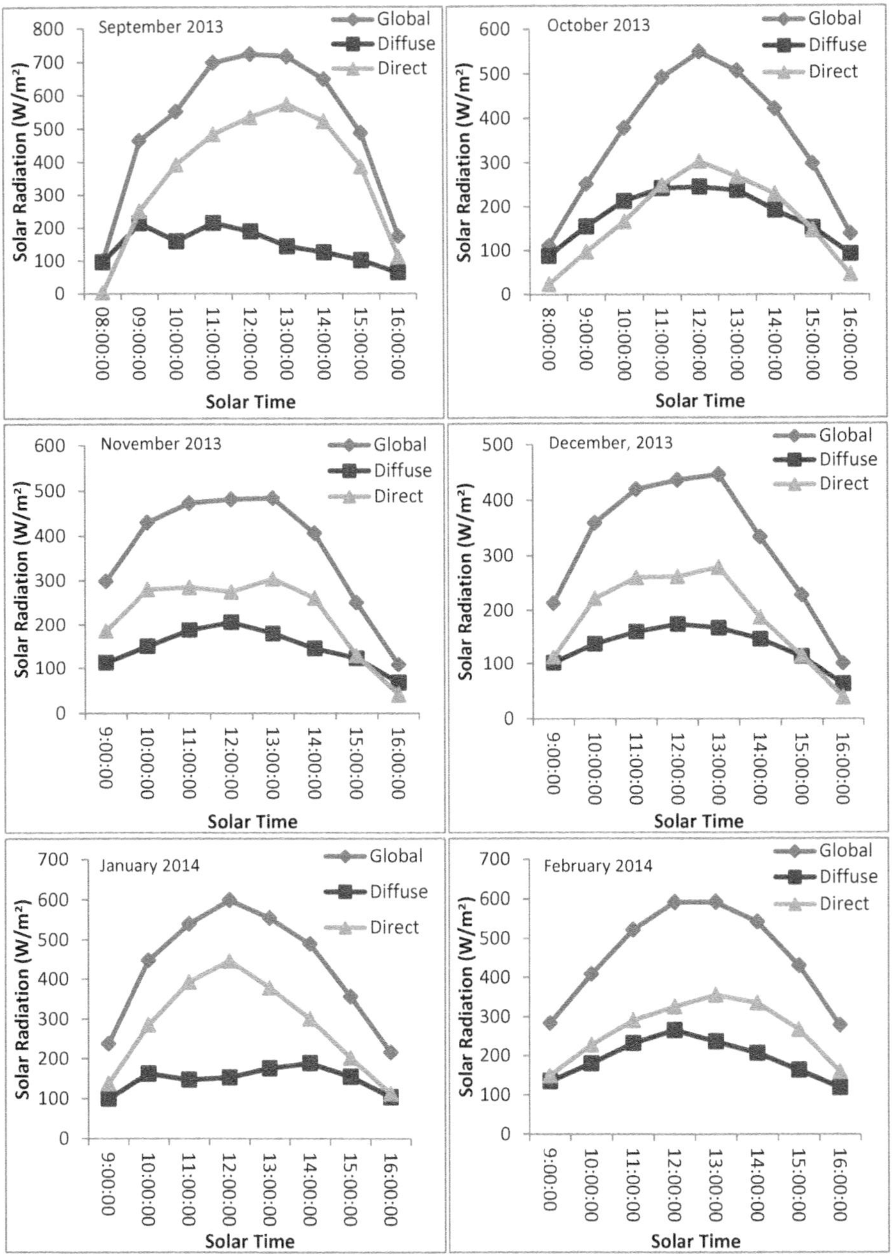

Fig. 61.1 Daily solar radiation for 6 months for the sunshine duration

where $\overline{H}$ is the monthly average global solar radiation on a horizontal surface, $\overline{H}_0$ is the monthly average extraterrestrial solar radiation, $\overline{S}$ is the monthly average duration of bright sunshine, $\overline{S}_0$ is the monthly average maximum possible sunshine duration, and a and b are the empirical coefficients [2–4].

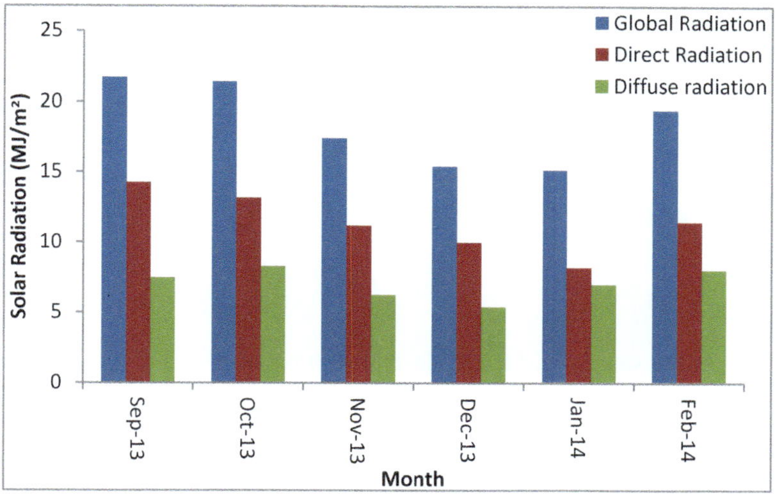

Fig. 61.2 Monthly average daily solar radiation on a horizontal surface in Aligarh

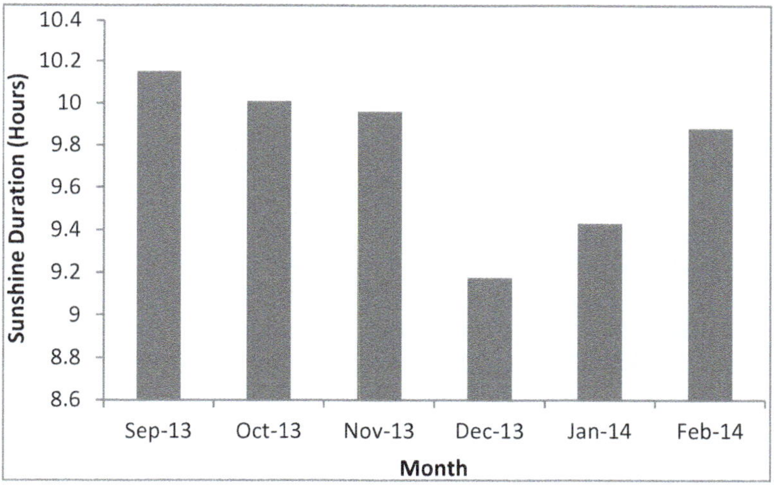

Fig. 61.3 Monthly average daily hours of sunshine in Aligarh

The monthly average extraterrestrial solar radiation $\left(\overline{H}_0\right)$ is computed from the following equation

$$\overline{H}_0 = \frac{24}{\pi} H_{sc}\left(1 + 0.033\cos\left(\frac{360}{365}n\right)\right)\left(\cos\varnothing\cos\delta\sin\omega_s + \frac{\pi}{180}\omega_s\sin\varnothing\sin\delta\right) \tag{61.2}$$

where H_{sc} is the solar constant, n is day of the year for each month, $\varnothing$ is the latitude, and δ is the solar declination, which can be expressed as:

$$\delta = 23.45^{\circ} \sin\left[\frac{360(284+n)}{365}\right] \tag{61.3}$$

The maximum possible sunshine duration is given by

$$\overline{S}_0 = \frac{2}{15}\omega_s \tag{61.4}$$

where ω_s is the sunset hour angle, described in terms of inclination angle (θ) and declination angle (δ) given as:

$$\cos\omega_s = -\tan\theta\tan\delta \tag{61.5}$$

The calculation of monthly average diffuse solar radiation takes on a similar form as Eq. (61.1), with the ratio of monthly average diffuse solar radiation to global solar radiation described as a function of the ratio of monthly average global to extraterrestrial solar radiation

$$\frac{\overline{H}_d}{\overline{H}} = c + d\frac{\overline{H}}{\overline{H}_0} \tag{61.6}$$

where $\left(\overline{H}_d\right)$ is the monthly average daily diffuse radiation on a horizontal surface and c and d are the empirical coefficients.

Klein [5] has simplified the calculation of $\left(\overline{H}_0\right)$, by determining the particular day of each month for which the extraterrestrial radiation is nearly equal to the monthly mean value. These are as follows: January 17, February 16, March 16, April 15, May 15, June 11, July 17, August 16, September 15, October 15, November 14, and December 10.

The regression coefficients a and b in Eq. (61.1) have been estimated from $\overline{H}/\overline{H}_0$ and $\overline{S}/\overline{S}_0$ by fitting the data with a linear curve. Similarly, c and d in Eq. (61.6) are obtained from the linear fitting of $\overline{H}_d/\overline{H}$ and $\overline{H}/\overline{H}_0$ data points.

61.3.2 Performance of Correlations

The most commonly used statistical measures to analyze the performance of a correlation in estimating a value are *mean bias error* and *root mean square error*. These are described below.

61.3.2.1 Mean Bias Error

The long-term performance of a correlation for estimating a value is provided by the mean bias error (MBE). It allows the comparison of actual deviation between the estimated and measured value for each term. A smaller value of MBE is preferred and ideally it should be zero. Mathematically, it is defined as:

$$\mathrm{MBE} = \frac{1}{n}\sum_{i=1}^{n}\left(\overline{H}_{i,\mathrm{e}} - \overline{H}_{i,\mathrm{m}}\right) \tag{61.7}$$

A positive value gives the average amount of overestimation in the calculated value and vice versa. One drawback of this test is that overestimation of an individual observation will cancel underestimation in a separate observation.

61.3.2.2 Root Mean Square Error

The root mean square error (RMSE) provides information on the short-term performance of the correlation. The value of RMSE is always positive and ideally it should be zero. It is mathematically represented as:

$$\mathrm{RMSE} = \left[\frac{1}{n}\sum_{i=1}^{n}\left(\overline{H}_{i,\mathrm{e}} - \overline{H}_{i,\mathrm{m}}\right)^2\right]^{1/2} \tag{61.8}$$

61.4 Results and Discussion

In the present study, observations of bright sunshine hours and monthly average daily solar radiations values are utilized to compute the regression coefficients a and b. The monthly average values of $\overline{H}/\overline{H}_0, \overline{H}_\mathrm{d}/\overline{H}$, and $\overline{S}/\overline{S}_0$ during the period of 6 months from September 2013 to February 2014 are provided in Table 61.1.

Table 61.1 Monthly average values of $\overline{S}/\overline{S}_0$ and K_T and K_D in Aligarh

Month	$\frac{\overline{S}}{\overline{S}_0}$	$K_\mathrm{T} = \frac{\overline{H}}{\overline{H}_0}$	$K_\mathrm{D} = \frac{\overline{H}_\mathrm{d}}{\overline{H}}$
September 2013	0.835973	0.641421184	0.342954825
October 2013	0.886182	0.755465365	0.385027375
November 2013	0.940192	0.743277021	0.358245175
December 2013	0.894261	0.727941221	0.351644088
January 2014	0.902465	0.671231903	0.45988537
February 2014	0.894948	0.72442449	0.411624032

where K_T and K_D are the monthly average sky clearness index and monthly average diffusion index, respectively.

61.4.1 Correlations for Estimation of Monthly Average Global Solar Radiation

Many researchers have modified the Angström-type model for the conditions of solar radiation over a variety of locations all over the world and have presented the values of coefficients a and b.

Rietveld [6] proposed a model for monthly average global solar radiation that is applicable worldwide. It is given as follows:

$$\frac{\overline{H}}{\overline{H}_0} = 0.18 + 0.62\frac{\overline{S}}{\overline{S}_0} \tag{61.9}$$

Garg and Garg [7] obtained the following equation from the experimental data of 11 stations in India:

$$\frac{\overline{H}}{\overline{H}_0} = 0.3156 + 0.4520\frac{\overline{S}}{\overline{S}_0} \tag{61.10}$$

For the present case, the regression coefficients a and b of the Angström-type correlation for the monthly average daily values of solar radiation was determined and is given in Table 61.2.

To evaluate the suitability of the present correlation, the mean bias error (MBE) and root mean square error (RMSE) are evaluated and the values are shown in the table below.

The measured and estimated values of monthly average global solar radiation using the described correlations are displayed in Fig. 61.4.

It can be seen from Fig. 61.4 that a good agreement is found between estimations and actual measurement using the new correlation. This is also justified by the values of MBE and RMSE in Table 61.2.

61.4.2 Correlations for Estimation of Monthly Average Diffuse Solar Radiation

As stated earlier, for the estimation of monthly average daily diffuse radiations, the ratio of diffuse to global radiation could be correlated against the ratio of global to extraterrestrial solar radiation as given by Eq. (61.6).

Table 61.2 Performance of correlations in estimating monthly average global solar radiation

	Coefficient			
Correlation	a	b	MBE	RMSE
Rietveld	0.1800	0.6200	0.60	0.34
Garg and Garg	0.3156	0.4520	0.24	0.31
Present case	0.0609	0.8646	−0.01	0.25

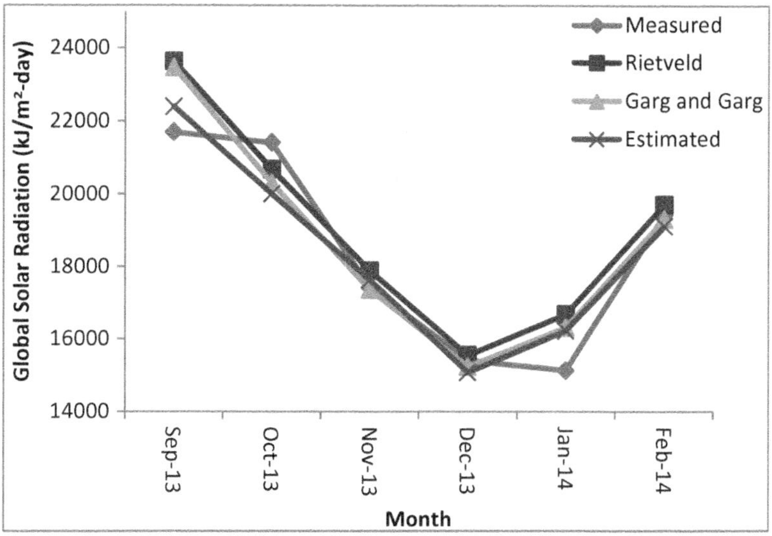

Fig. 61.4 Comparison of the measured and estimated values of monthly average global solar radiation

Table 61.3 Performance of correlations in estimating monthly average diffuse solar radiation

Correlation	Coefficient		MBE	RMSE
	c	*d*		
Modi and Sukhatme	1.411	−1.696	−1.56	0.29
Gupta et al.	1.354	−1.570	−0.73	0.26
Present Case	0.465	−0.113	0.03	0.07

Modi and Sukhatme [8] proposed a model for the monthly average diffuse solar radiation for Indian data. It is given as follows:

$$\frac{\overline{H}_{\mathrm{d}}}{\overline{H}} = 1.411 - 1.696\frac{\overline{H}}{\overline{H}_0} \tag{61.11}$$

Gupta et al. [9] obtained the following equation:

$$\frac{\overline{H}_{\mathrm{d}}}{\overline{H}} = 1.354 - 1.570\frac{\overline{H}}{\overline{H}_0} \tag{61.12}$$

Based on the measured data, a new correlation has been proposed for the estimation of monthly average diffuse radiation for the period of study. This case has been compared against the work of previous researchers described above. The results are exhibited in Table 61.3.

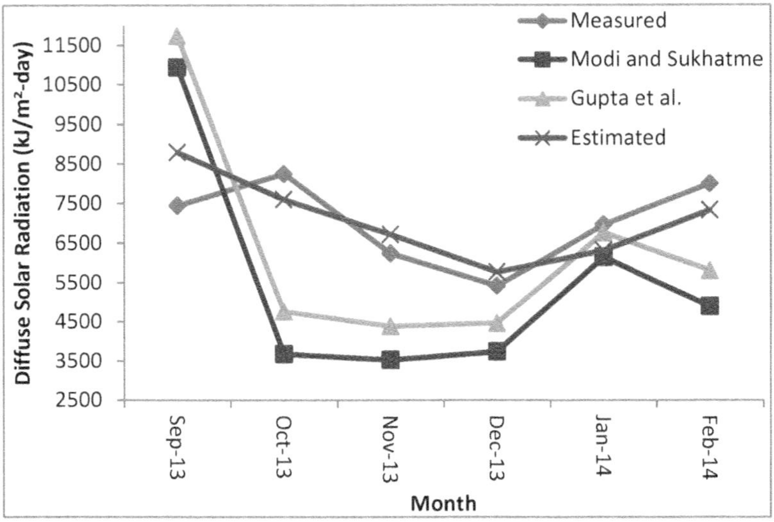

Fig. 61.5 Comparison of the measured and estimated values of monthly average diffuse solar radiation

It can be observed that the new proposed correlation is the most accurate. This statement is well supported by the fact that the values of MBE and RMSE are the least for new correlation (Fig. 61.5).

61.4.3 Estimation of Monthly Average Direct Solar Radiation

For the estimation of monthly average direct solar radiation, the difference of the monthly average global solar radiation and monthly average diffuse solar radiation (obtained from the new set of correlations) is done:

$$\overline{H}_{\mathrm{b}} = \overline{H} - \overline{H}_{\mathrm{d}} \tag{61.13}$$

The results are shown in Fig. 61.6. It can be observed that the estimated values of direct solar radiation closely follow the measured data.

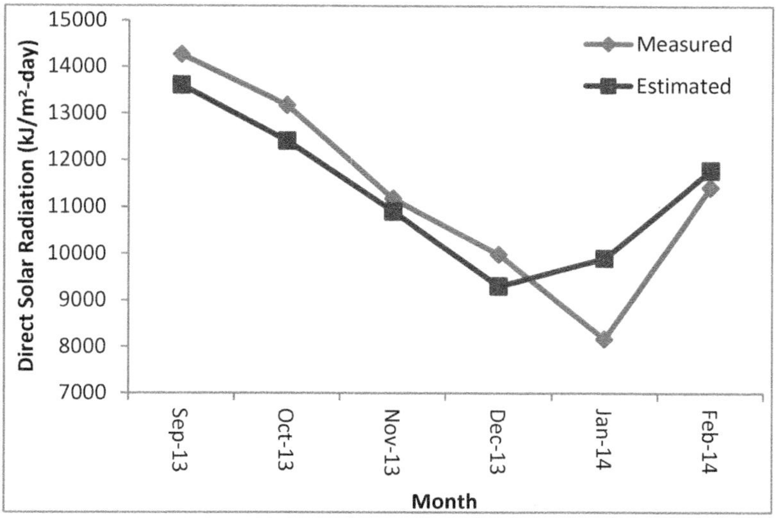

Fig. 61.6 Comparison of the measured and estimated values of monthly average direct solar radiation

61.5 Conclusions

Correlations for estimation of monthly average global solar radiation incident on a horizontal surface were applied to the sunshine hour data and compared statistically in terms of mean bias error and root mean square error. The new model correlation showed the highest level of accuracy when compared with the existing models. Also, the correlation for monthly average diffuse solar radiation on a horizontal surface was proposed. Again, the new correlation was found to be the best compared to other models. Thus, it can be concluded that the proposed correlation models are recommended for the estimation of solar radiation for the Aligarh region of India.

References

1. Oğulata RT, Oğulata SN (2002) Solar radiation on Adana, Turkey. Appl Energy 71(4):351–358. doi:10.1016/S0306-2619(01)00050-2
2. Angström A (1924) Solar and terrestrial radiation. Report to the international commission for solar research on actinometric investigations of solar and atmospheric radiation. Q J Roy Meteor Soc 50(110):121–126. doi:10.1002/qj.49705021008
3. Page JK (1961) The estimation of monthly mean values of daily total short wave radiation on vertical and inclined surface from sunshine records for latitudes 40N-40S. Proc UN Conf New Sources Energy 4(S98):378–390
4. Duffie JA, Beckman WA (1991) Solar engineering of thermal processes. Wiley, New York

5. Klein SA (1977) Calculation of monthly average insolation on tilted surfaces. Solar Energy 19:325–329. doi:10.1016/0038-092X(77)90001-9
6. Rietveld MR (1987) A new method for estimating regression coefficients in the formula relating solar radiation to sunshine. Agr Meteorol 19:243–252. doi:10.1016/0002-1571(78)90014-6
7. Garg HP, Garg SN (1985) Correlation of monthly average daily global, diffuse and beam radiation with bright sunshine hours. Energy Convers Manage 25:409–417. doi:10.1016/0196-8904(85)90004-4
8. Modi V, Sukhatme SP (1966) Estimation of daily total and diffuse insolation in India from weather data. Solar Energy 22(5):407–5411. doi:10.1016/0038-092X(79)90169-5
9. Gupta CL, Rao KU, Reddy TA (1979) Radiation design data for solar energy applications. Energy Manage 3(4):299–313

Chapter 62
Characterization and Simulation of Silicon Oxynitride Films Deposited by ECR-PECVD: For Solar Applications

Lilia Zighed and Abdelhakim Mahdjoub

Abstract Graded refractive index silicon oxynitrides are deposited by electron cyclotron resonance plasma-enhanced chemical vapor deposition (ECR-PECVD). The analysis of engraving speeds and infrared absorption spectra taken on samples informed us about their chemical compositions. This analysis shows the presence of refractive index gradient. To benefit at the same time from antireflective and passive properties of oxynitrides, we proposed graded index layer having Fermi profile, which we optimized by simulation. This coating would replace the classic double layer antireflective coating (ARC) made in two different technological steps. Calculation predicts an enhancement of photogenerated current exceeding 45 % and a weighted reflectance (between 300 and 1,100 nm) around 5.6 %.

Keywords Silicon oxynitride • Graded refractive index • ARC • Solar cells

62.1 Introduction

Silicon oxynitride (SiO_xN_y) thin films have attracted the attention of scientists for many years. They are used in integrated circuit technology [1] in microelectronics and sensors [2–4] and optoelectronics industry [5, 6]. Furthermore, graded refractive index SiO_xN_y layers were used for optical applications, in particular in rugate filters [7–9]. In our study we investigate their use as antireflective coatings (ARC) for solar cells.

In this work, in Sect. 62.1 we briefly describe the experimental setup used for fabrication of silicon oxynitride layers. In order to characterize them, thickness

L. Zighed (✉)
Laboratoire LGCES, Faculty of Technology, Université 20 Août 1955, Skikda 21000, Algeria
e-mail: lzighed@gmail.com

A. Mahdjoub
Laboratoire LMSSEF, Faculty of Sciences, Université Larbi Benmhidi, O. E. Bouaghi 04000, Algeria

I. Dincer et al. (eds.), *Progress in Clean Energy, Volume 2*,
DOI 10.1007/978-3-319-17031-2_62

measurements and IR spectra were acquired by Fourier transform infrared spectroscopy (FTIR) at different stages of pickling.

In Sect. 62.2 we propose a model which allows the calculation of optical properties of graded refractive index layers. We also describe an ARC in which refractive index varies according to Fermi profile. The optimization of the physical parameters of this last, namely thickness and index profile allows to minimize surface reflection and improve photovoltaic cell performance.

62.2 Experimental Details

62.2.1 Deposit of Homogeneous Layers

SiO_xN_y films were deposited by ECR-PECVD. The plasma consists of a mixture of nitrogen and oxygen. Silane (SiH_4) was used as silicon precursor. Since oxygen is highly reactive [6], it is easy to make an oxynitride with a high oxygen concentration. By cons, producing silicone oxynitride rich in nitrogen is more difficult because it requires working with low ratios of O_2/N_2 gas. Thus, the silicon oxynitride layers may be deposited under the following experimental conditions: the flow of silane and nitrogen is maintained at 20 sccm and 4 sccm, respectively, for O_2/N_2 gas ratios ranging from 0.025 to 0.25 (or more). The total pressure in the deposition chamber was about 1.5 mTorr. Silicon substrates (100) were heated to 200 °C and the distance from the ECR source was 15 cm. The deposit was controlled by monochromatic ellipsometry.

62.2.2 Deposit of Graded Index Layers

Graded index layers were deposited in the same conditions as homogeneous SiO_xN_y. However, to obtain graded index films, we have to change the composition of SiO_xN_y by varying O_2/N_2 ratios during film growth. As oxygen is much reactive than nitrogen, a small variation of O_2 flow rate results in sizeable index changes. For this reason, it is more convenient to deposit graded index films keeping a constant nitrogen flow, while the flow of oxygen varies (typically between 0.5 and 8 sccm). SiH_4 and N_2 flow rates were set at 4 sccm and 20 sccm, respectively.

62.2.3 Characterization

We have made a comparative study between a homogeneous oxynitride film and a graded index layer. A qualitative analysis of the deposited layers is made by FTIR. To show that the composition of the graded index samples varies in depth, the samples were etched in hydrofluoric acid (HF) diluted to 1 %. After each etching (for 30 s), a thickness measurement is made by profilometry and an FTIR spectrum is raised. The FTIR measurements are made by a SHIMADZU 8300 spectrometer operating in horizontal attenuated total reflection (HATR) using a ZnSe crystal. The absorption spectra are acquired at room temperature in the range 400–4,000 cm^{-1} with 8 cm^{-1} resolution. Thickness measurements are made through a profilometer DEKTAK 3. These measurements are performed in the Laboratoire de Recherche sur la Physico-Chimie des Surfaces et Interfaces (LRPCSI), University of Skikda.

62.3 Results and Discussion

Our comparative study is made on two samples deposited on Si (100). The OX35 sample deposited at constant flow O_2/N_2 and sample Ox30 deposited with variable flow.

62.3.1 Homogeneous Layer

Measurement for the remaining thickness of SiO_xN_y after each etching for 30 s of the sample OX35 gave profile of Fig. 62.1.

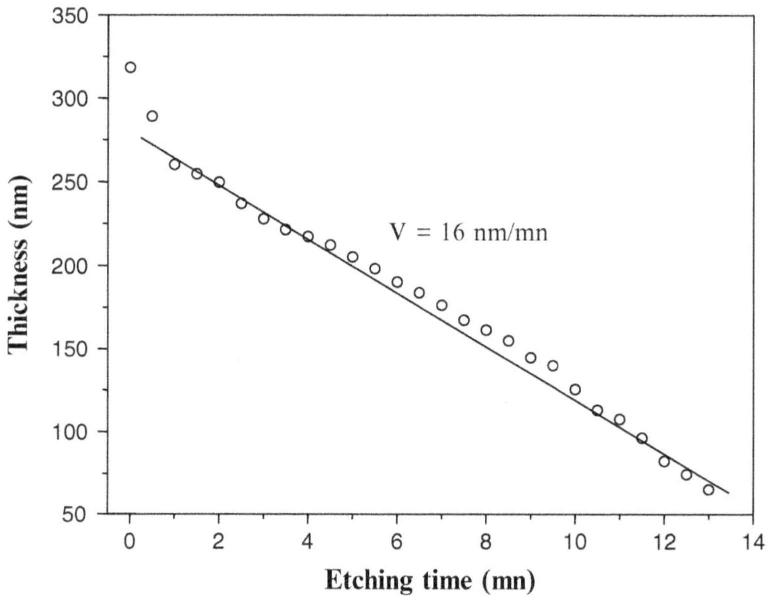

Fig. 62.1 Remaining thickness of SiO_xN_y as function of etching time in sample OX35

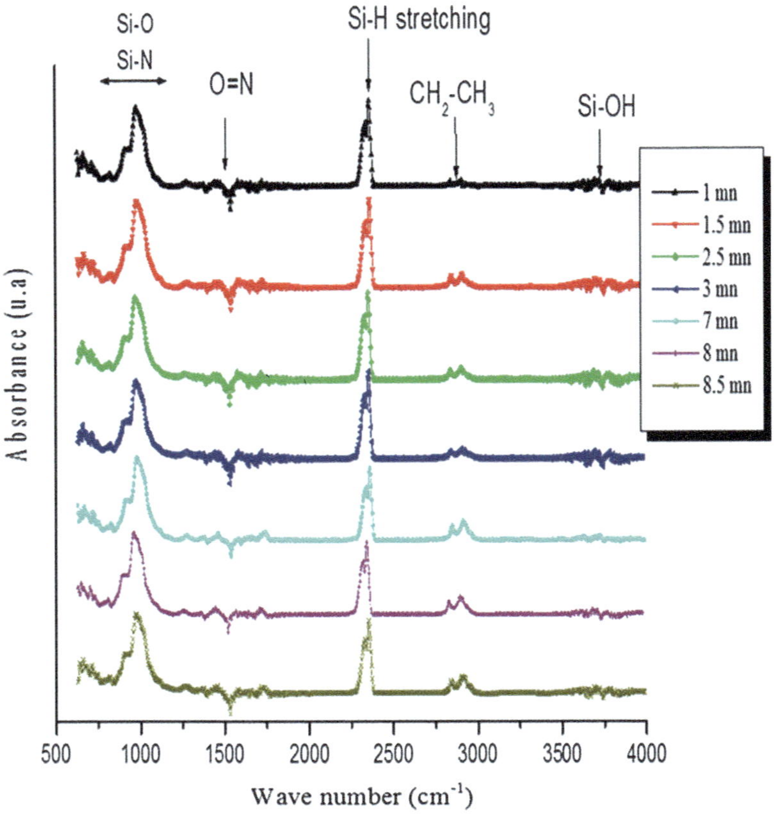

Fig. 62.2 FTIR spectra measured on the sample OX35 after each etching

This figure shows a significant speed engraving during the first minute of etching. This is due to the oxidation and hydration of the surface layer in contact with the ambient atmosphere. This has less dense layer and therefore quickly dissolved.

After 1 min of etching, the etching becomes less fast and uniform for the rest of the deposited layer. The etching rate is about 16 nm/min. This leads us to believe that the sample OX35 is a homogeneous silicon oxynitride therefore with constant refractive index. The FTIR measurements on the same sample after each etching step are shown in Fig. 62.2.

Figure 62.2 shows the obtained spectra have the same shape. We note in all spectra the presence of noise around 1,550 cm^{-1} associated with the O=N bond [6]. Broadband close to 3,600 cm^{-1} is attributed to Si–OH bond [10, 11]. The peak at 2,340 cm^{-1} is due to the presence of CO_2 in the measurement chamber [10], and vibration to 2,900 cm^{-1} is due to stretching modes of CH_2–CH_3 bonds. The peak centered at 1,000 cm^{-1} contains information about the different Si–O–Si and Si–N bonds. Therefore we analyzed in detail the peak centered at 1,000 cm^{-1}. We had broken it into three Gaussians as shown in Fig. 62.3. The first peak located at 900 cm^{-1} is related to the Si–N bond, the second peak centered at 980 cm^{-1} is

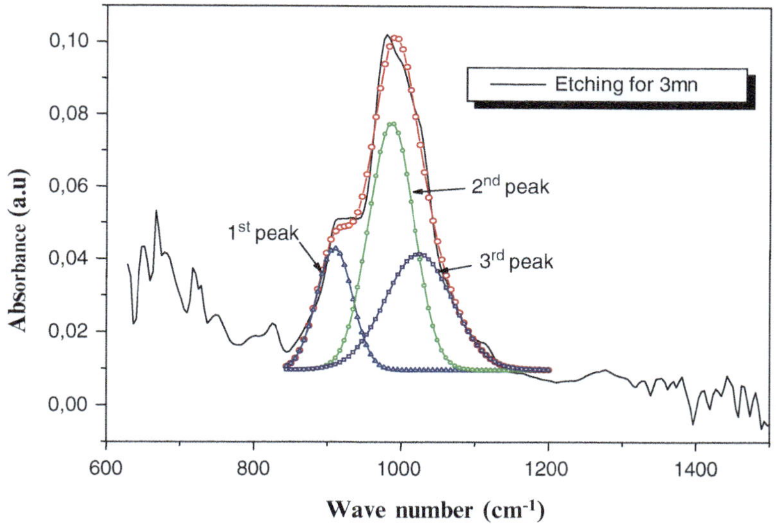

Fig. 62.3 Deconvolution of the peak centered at 1,000 cm^{-1} in three Gaussians

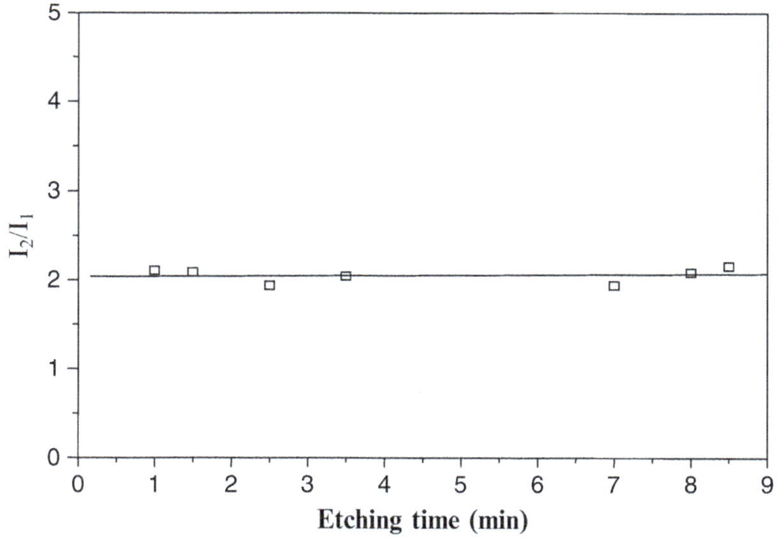

Fig. 62.4 Intensity ratio of the second to the first peak in depth of sample OX35

associated with the Si–O bond and the third peak at around 1,060 cm^{-1} is due to the stretching mode (TO) of the Si–O–Si bond [10].

Figure 62.4 shows that the intensity ratio I_2/I_1 is almost constant in depth of the SiO_xN_y film which confirms our initial conclusion on the homogeneity of the chemical composition of the sample OX35.

62.3.2 Graded Index Layer

In the analysis of the sample Ox30, we proceeded in a similar manner to that of the sample OX35. Successive engravings in the same solution of diluted HF are followed each time by a thickness and FTIR measurements.

Figure 62.5 shows a nonlinear variation of the thickness according to etching time. That is to say, in this case the etch rate varies according to the thickness of the layer. This rate passes from 33 nm/min at the beginning of the etching to 6.6 nm/min at the end of pickling. This allows us to say first that the SiO_xN_y deposited layer is an inhomogeneous layer whose composition varies according to thickness.

The FTIR spectra measured after each engraving of the sample Ox30 are illustrated in Fig. 62.6.

In Fig. 62.6, the peak centered at 1,000 cm^{-1} contains information about the layer composition. We note that the amplitude of this peak decreases in function of time (remaining thickness of the layer). For more information about the layer composition variation in depth, we have followed the same steps for the sample OX35: we decomposed the peak centered at 1,000 cm^{-1} in three Gaussians (see Fig. 62.3). The deconvolution of this peak allows us to have an idea about the variation of the Si–O and Si–N bond in depth and thus the variation of the composition of the layer.

Figure 62.7 shows the variation of the ratio of the second to first peak as a function of etching time, that is to say in depth of the layer.

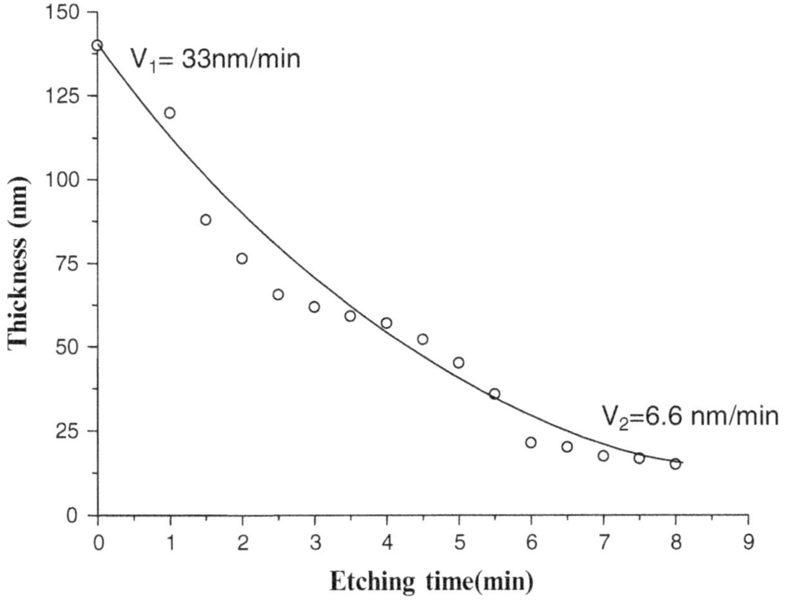

Fig. 62.5 Remaining thickness of SiO_xN_y as function of etching time in sample OX30

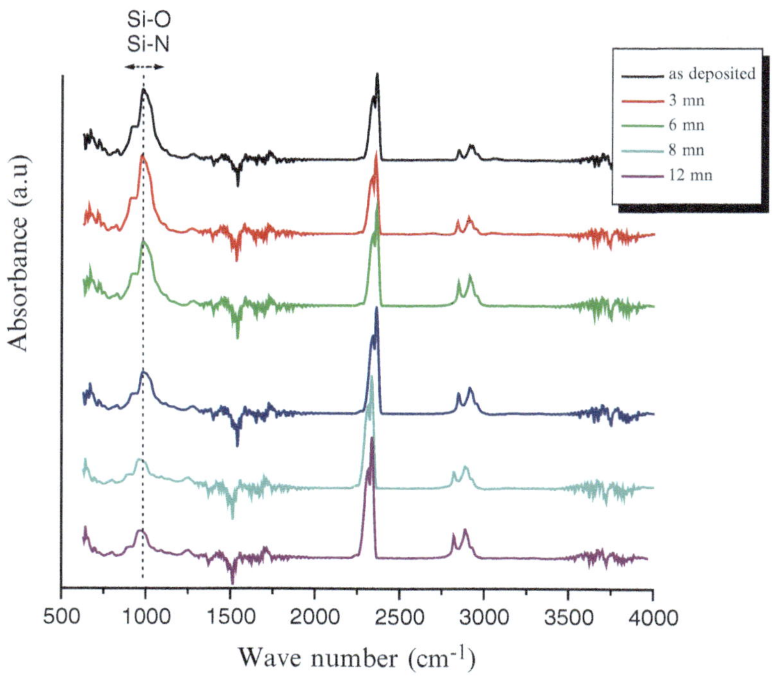

Fig. 62.6 FTIR spectra measured on the sample OX30 after each etching

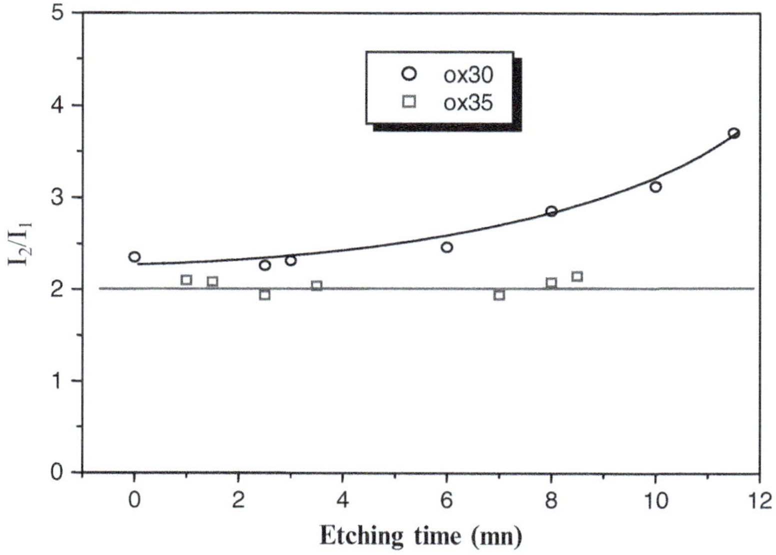

Fig. 62.7 Intensity ratio of the second to the first peak in depth of SiO_xN_y layer

We see that this ratio varies in the sample Ox30 and it remains practically constant in the sample OX35. This confirms our first hypothesis: the sample OX35 is formed by a homogeneous composition of SiO_xN_y; therefore a constant index and the sample Ox30 is a SiOxNy whose composition varies in depth of the layer, which resulted in a graded refractive index. We arrive thus effectively to realize films with graded refractive index by varying the oxygen flow during deposition.

62.4 Modeling and Optimization

We saw in the first part of this work that it is possible to fabricate SiO_xN_y graded index layers by ECR-PECVD. In what follows, we model and optimize SiO_xN_y antireflective coatings used for silicon solar cells.

62.4.1 *Modeling*

The calculation of reflectivity of graded index dielectric systems is a classic problem treated by various approaches [12, 13]. The optical stratified medium theory by its matrix representation has the advantage of a simple formalism and flexibility of use [14, 15]. The inhomogeneous film, in our case the ARC based on SiO_xN_y, is subdivided into homogeneous strata of equal thickness, with index N_{j} variable from strata to strata. The Bruggeman effective medium approximation (BEMA) [14–16] permits the determination of the index of each stratum from the volume fractions of f_{SiO2} and f_{Si3N4} as well as the refractive indices of $\widetilde{N}_{\mathrm{SiO_2}}$, $\widetilde{N}_{\mathrm{Si_3N_4}}$, respectively, of silica (SiO_2) and nitride silicon (Si_3N_4):

$$f_{\mathrm{SiO_2}}\frac{\widetilde{N}^2_{\mathrm{SiO_2}}-\widetilde{N}^2_{\mathrm{j}}}{\widetilde{N}^2_{\mathrm{SiO_2}}+2\widetilde{N}^2_{\mathrm{j}}}+f_{\mathrm{Si_3N_4}}\frac{\widetilde{N}^2_{\mathrm{Si_3N_4}}-\widetilde{N}^2_{\mathrm{j}}}{\widetilde{N}^2_{\mathrm{Si_3N_4}}+2\widetilde{N}^2_{\mathrm{j}}}=0 \tag{62.1}$$

For silicon bulk, silica, and silicon nitride, we used refractive indices published by Palik [17].

To obtain graded index layer, we have to change volume fractions f_{SiO2} and f_{Si3N4} following a variation law versus depth according to the desired profile. The method is described in detail in reference [18].

62.4.2 Quality Criteria of ARC

To take into account the spectral aspect of sunlight, and the spectral response of the cell, we used to optimize the performance of an ARC weighted average reflectivity [19–22]. It is set between λ_1 and λ_2 by

$$R_W = \frac{\int_{\lambda_1}^{\lambda_2} R(\lambda)\Phi(\lambda)S(\lambda)\mathrm{d}\lambda}{\int_{\lambda_1}^{\lambda_2} \Phi(\lambda)S(\lambda)\mathrm{d}\lambda} \tag{62.2}$$

where $R(\lambda)$ is the total reflectance, $\Phi(\lambda)$ is the incident photon flux spectrum of AM1.5 [23] normalized to a power density of 100 mW/cm^2, and $S(\lambda)$ is the spectral sensitivity of the cell.

In our study, the values of $S(\lambda)$ are calculated for a single crystal silicon solar cell characterized by the following parameters: a cell having a thickness of 300 μm, n-type doped emitter 10^{19} cm^{-3}, the base that is of type p doped 10^{16} cm^{-3}, and the diffusion length of electrons in this base which is 150 μm. The junction is at 0.3 μm into depth. Several authors [15, 24, 25] consider that the photocurrent generated by the solar cell is the best criterion for judging the quality and design of an ARC.

It is defined by

$$J_{SC} = \int_{\lambda_1}^{\lambda_2} [1 - R(\lambda)]\Phi(\lambda)S(\lambda)\mathrm{d}\lambda \tag{62.3}$$

The photocurrent is maximum when the weighted average reflectivity R_W is minimum.

The gain in photocurrent due to antireflective coating will be

$$G_P = \frac{\Delta J_{SC}}{J_{SC}} = \frac{J_{SC}(\text{with ARC}) - J_{SC}(\text{without ARC})}{J_{SC}(\text{without ARC})} \tag{62.4}$$

We have therefore use in this work R_W and G_p for optimizing the ARC parameters.

62.4.3 Optimization

The model we used to describe the profile of the gradient films had already been validated in a previous study [18]. In what follows we will optimize the parameters of the graded index ARC to have the minimum weighted average reflectance and maximum photocurrent gain. To describe the profile of the refractive index, we have used a function describing the variation of silica volume fraction through the silicon oxynitride layer. Our optimization was done on an ARC in which silica volume fraction variation is described by the following expression:

$$f_{SiO_2}(x_d) = \left(1 + e^{a_f\,(x_d - b_f)}\right)^{-1} \tag{62.5}$$

where x_d is the number of the normalized sublayer, a_f is a form factor that determines the stiffness of the profile, and b_f is its inflection point (see Fig. 62.8).

This profile "Fermi profile" becomes equivalent to a conventional double layer ARC when the value of a_f is great. The optimization will focus on two parameters: the thickness and the inflection point of the profile (b_f). a_f is set at 20.

Figure 62.9 shows the variation of the gain in photocurrent and the average weighted reflectivity of an ARC having Fermi profile. The maximum gain in photocurrent obtained is equal to 45.06 % and corresponds to the minimum weighted average reflectivity equal to 6.31 %. These values are obtained with a thickness of 120 nm and $b_f = 0.48$. This is a considerable improvement compared to the performance of a monolayer classical ARC made of Si_3N_4 that to say, respectively photocurrent gain and a weighted average reflectivity of 42.72 % and 9.78 %.

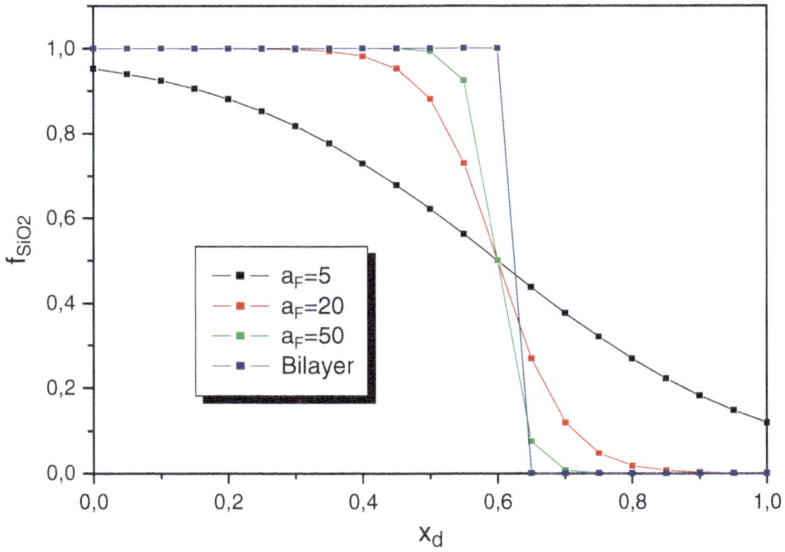

Fig. 62.8 Variation of silica volume fraction according to Fermi profile

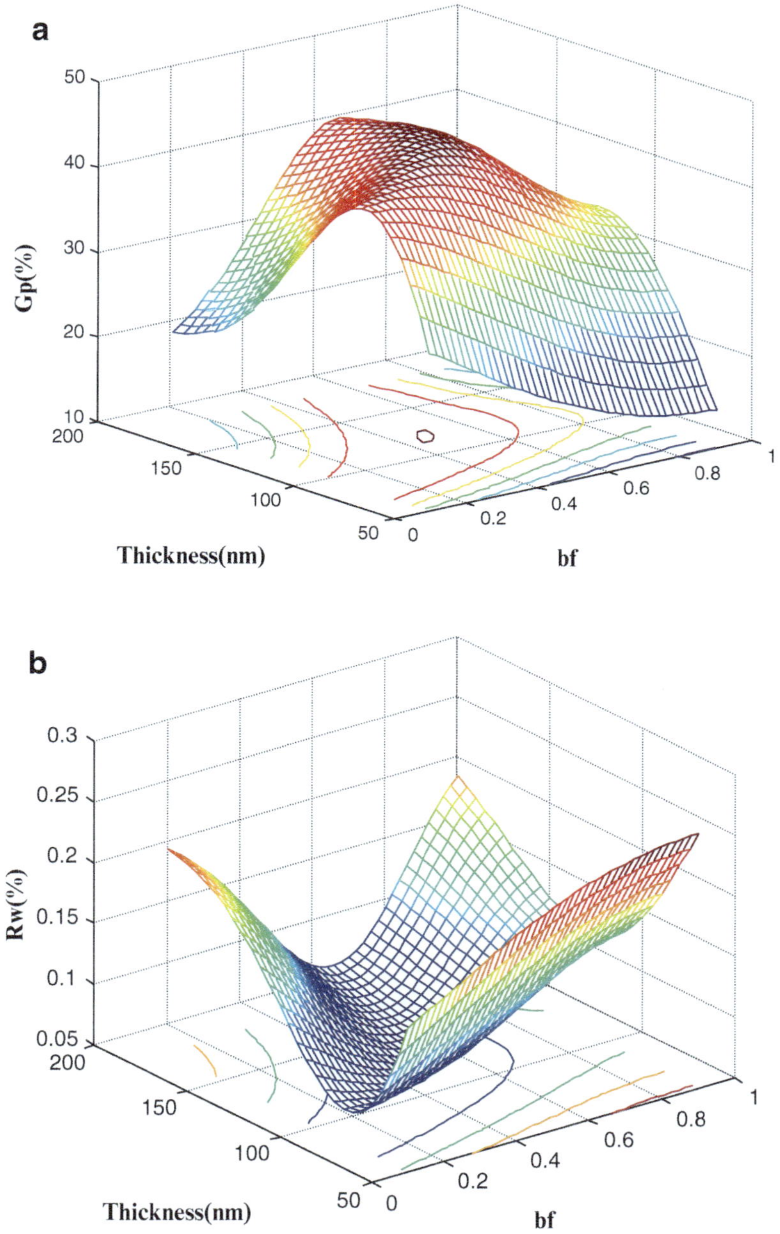

Fig. 62.9 Optimization of graded index ARC. (**a**) Photocurrent gain. (**b**) Weighted average reflectivity

62.5 Conclusion

In this work, we have shown through a characterization by profilometry and FTIR spectroscopy that graded refractive index of silicon oxynitride layers can be fabricated by ECR-PECVD. The optimization of layer parameters with a carefully chosen profile would reduce the weighted average reflectivity to 6.3 % and improve photocurrent to 45 %.

References

1. Mohite KC, Khollam YB, Mandale AB, Patil KR, Takwale MG (2003) Characterization of silicon oxynitride thin films deposited by electron beam physical vapor deposition technique. Mater Lett 57(26–27):4170–4175
2. Aspnes DE, Theeten JB (1979) Dielectric function of Si-SiO_2 and Si-Si_3N_4 mixtures. J Appl Phys 50(7):4928–4935
3. Brunet-Bruneau A, Vuye G, Frigerio M, Abelès F, Rivory J, Berger M, Chaton P (1996) Infrared ellipsometry investigation of SiO_xN_y thin films on silicon. Appl Optics 35(25):4998–5004
4. Szekeres A, Nikolova T, Simeonov S, Gushterov A, Hamelmann F, Heinzmann U (2006) Plasma-assisted chemical vapor deposited silicon oxynitride as an alternative material for gate dielectric in MOS devices. Microelectron J 37(1):64–70
5. Diatezua MD, Thiry PA, Lambin P, Caudano R (1993) Infrared response of silicon oxynitrides investigated by high-resolution electron-energy-loss spectroscopy. Phys Rev B 48(12):8701–8708
6. Pandey RK, Patil LS, Bange JP, Patil DR, Mahajan AM (2004) Growth and characterization of SiON thin films by using thermal-CVD machine. Opt Mater 25(1):1–7
7. Snyder PG, Xiong Y, Woollam A, Al-Jumaily GA, Gagliardi FJ (1992) Graded refractive index silicon oxynitride thin film characterized by spectroscopic ellipsometry. J Vac Sci Tech A 10(4):1462–1466
8. Callard S, Gagnaire A, Joseph J (1997) Fabrication and characterisation of graded refractive index silicon oxynitride thin films. J Vac Sci Tech A 15(4):2088–2094
9. Bulkin PV, Swart PL, Lacquet BM (1995) Properties and applications of electron cyclotron plasma deposited SiO_xN_y films with graded refractive index profiles. J Non-Cryst Sol 187:484–488
10. Pfuch A, Heft A, Weidl R, Lang K (2006) Characterization of SiO_2 thin films prepared by plasma-activated chemical vapour deposition. Surf Coat Tech 201(1–2):189–196
11. Vassallo E, Cremona A, Laguardia L, Mesto E (2006) Preparation of plasma-polymerized SiO_x-like thin films from a mixture of hexamethyldisiloxane and oxygen to improve the corrosion behaviour. Surf Coat Tech 200(10):3035–3040
12. Rivory J (1998) Characterization of inhomogeneous dielectric films by spectroscopic ellipsometry. Thin Solid Films 313–314:333–340
13. Kildemo M (1998) Real-time monitoring and growth control of Si-gradient-index structures by multiwavelength ellipsometry. Appl Optics 37(1):113–123
14. Orgeret M (1985) Les Piles solaires: Le composant et ses applications. Masson, Paris, p 138
15. Nubile P (1999) Analytical design for antireflection coatings for silicon photovoltaic devices. Thin Solid Films 342(1–2):257–261
16. Gao L, Gu JZ (2002) Effective dielectric constant of two component material with shape distribution. J Phys D Appl Phys 35(3–7):267–271

17. Palik ED (1985) Handbook of optical constants of solids, Academic Press Handbook series. Academic, Orlando, FL
18. Mahdjoub A, Zighed L (2005) New designs for graded refractive index antireflection coatings. Thin Solid Films 478(1–2):299–304
19. Lipiński M, Panek P, Beltowska E, Czternastek H (2003) Reduction of surface reflectivity by using double porous silicon layers. Mater Sci Eng B 101(1–3):297–299
20. Strehlke S, Bastide S, Lévy-Clément C (1999) Optimization of porous silicon reflectance for silicon photovoltaic cells. Sol Energy Mater Sol Cells 58(4):399–409
21. Duerincks F, Szlufcik J (2002) Defect passivation of industrial multicrystalline solar cells based on PECVD silicon nitride. Sol Energy Mater Sol Cells 72(1–4):231–246
22. Born M, Wolf E (1970) Principles of optics. Pergamon Press, London, p 228
23. Ricaud A (1997) Photopile solaire, 1st edn. Presses polytechniques et universitaires Romandes, Lausanne, Switzerland, p 39
24. Zhao J, Green MA (1991) Optimized antireflection coatings for high-efficiency solar cells. IEEE Trans Electron Devices 38(8):1925–1934
25. Aiken DJ (2000) Antireflection coating design for series interconnected multi-junction solar cells. Prog Photovolt Res Appl 8(6):563–570

Chapter 63
Effective Manufacture of Free-Standing TiO_2 Nanotube Arrays Without Bottom Barrier Layer for Dye-Sensitized Solar Cell

Wen-Kai Tu, Jia-Shiang Chen, and Shu-Hua Chien

Abstract Different from other procedures, we introduce a simple procedure for preparing large-area, free-standing, non-curling, opened-end TiO_2 nanotube array (TiNT) films by selected annealing temperature and oxalic acid treatment. The selective dissolution properties of oxalic acid are used to easily separate TiNT off Ti substrates and dissolve the bottom caps in this one-pot reaction. These TiNT films were attached on fluorine-doped tin oxide (FTO) glasses and used in the front-side-illuminated dye-sensitized solar cell (DSSC). As compared to the DSSC made of closed-end TiNT, the DSSC made of opened-end TiNT exhibited an enhancement in efficiency from 4.65 to 7.82 %, corresponding to 70 % improvement.

Keywords Titanium oxide • TiO_2 nanotube arrays • Dye-sensitized solar cell

Nomenclature

1-D	One-dimensional
DSSC	Dye-sensitized solar cell
D_{eff}	Effective diffusion coefficient ($cm^2\ s^{-1}$)
EIS	Electrochemical impedance spectroscopy
FTO	Fluorine-doped tin oxide
FE-SEM	Field emission scanning electron microscopy
FF	Fill factor
IPCE	Incident photon-to-current conversion efficiency

W.-K. Tu • J.-S. Chen
Institute of Chemistry, Academia Sinica, Taipei 11529, Taiwan

S.-H. Chien (✉)
Institute of Chemistry, Academia Sinica, Taipei 11529, Taiwan

Department of Chemistry, National Taiwan University, Taipei 10617, Taiwan
e-mail: chiensh@gate.sinica.edu.tw

I. Dincer et al. (eds.), *Progress in Clean Energy, Volume 2*,
DOI 10.1007/978-3-319-17031-2_63

J	Current density (mA cm^{-2})
J_{SC}	Short-circuit current density (mA cm^{-2})
k_{eff}	First-order reaction rate constant for the loss of electrons (s^{-1})
L_F	Film thickness (μm)
L_n	Diffusion length (μm)
N719	$(Bu_4N)_2Ru(dcbpyH)_2(NCS)_2$
R_k	Dark reaction impedance (Ω)
R_w	Diffusion reaction impedance (Ω)
TiNT	TiO_2 nanotube arrays
TNP	Titania nanoparticles
V	Voltage (V)
V_{OC}	Open-circuit potential (V)
XRD	X-ray diffraction
η	Energy conversion efficiency (%)
τ	Lifetime of an electron (ms)

63.1 Introduction

In order to directly utilize solar energy as an alternative energy resource, low-cost solar cells have been developed. One of the most promising solar cells would be the DSSC, which can be produced in a simple process. In 1991, DSSCs were developed by O'Regan and Grätzel [1] based on transparent film of titania nanoparticles (TNP) and Ru-based dye. The overall conversion efficiency was reported over 7 %. DSSCs were gradually developed and possessing high conversion efficiency (>10 %) [2–5]. Recently, Yella et al. designed a redox mediator containing cobalt complexes leading to a superior conversion efficiency of 12.3 % [6]. The porous thin film composed of TNP is used as an anode material because of the high specific surface area for dye adsorption. However, the electron transport through the random network of porous thin film results in the electron loss occurring in grain boundaries at the contacts between nanoparticles. Therefore, the use of one-dimensional (1-D) titania nanostructure is expected to enhance the charge transport and reduce the recombination between photoelectron and the oxidized dye and/or the electron acceptors in the electrolyte [7].

1-D nanostructure such as nanowires, nanorods, and nanotubes can be prepared by hydrothermal treatment [8–11], sol–gel and template synthesis [12,13], and chemical vapor deposition method [14]. 1-D TiO_2 nanotube arrays can be fabricated by anodization of Ti foil in acidic electrolytes with fluoride ions. The morphology of the TiNT is influenced by different parameters such as anodization time, anodization temperature, applied voltage, and electrolyte composition. The anodization process is a relatively simple approach for altering the diameter, wall thickness, and length of the nanotube by the controlling parameters [15–21]. Furthermore, Zhu et al. [22] showed that dye molecules covered both the interior and exterior walls of the TiNT. Consequently, TiNT films and TNP films have comparable dye coverage. Nevertheless, the application of TiNT attached to an opaque Ti foil for DSSC requires back-side illumination, which leads to an inferior conversion efficiency because the partial incident light is absorbed by the Pt electrode and the electrolyte [15,16,20]. TiNT and Ti foil may greatly obstruct the transfer of electrons [23].

The transparent TiNT-based photoanodes on conductive glass substrates were carried out by Grimes's group [17–19]. A thin titanium film was sputtered on the FTO glass, and the resulting film was then anodized to form a 3.6-μm-thick nanotube film on the FTO glass. Though the advantage of the front-side illumination led to a conversion efficiency of 4.7 % [18], the limited TiNT thickness and the poor adhesion had to be improved. In 2009, Grimes's group further improved the efficiency. A 20-μm-thick titanium film was sputtered on the FTO glass, and the resulting film was then anodized in fluorine-containing dimethyl sulfoxide electrolyte to form nanotube films up to 33 μm in length. This procedure yielded the conversion efficiency of 6.9 % [19]. However, the procedure was relatively expensive and time-consuming. On the other hand, the free-standing TiNT was fabricated and then attached to the FTO glass by Park and coworkers [24]. However, the detached TiNT normally curled and collapsed during annealing process. In 2009, Chen and Xu [23] developed two-step anodization procedure for obtaining large-area non-curling crystallized free-standing TiNT films. The TiNT films mentioned above were closed-end structures which might cause near-UV absorption and front surface light reflection as well as prevent the diffusion of the electrolyte. Recently, we developed opened-end TiNT films, which were attached on the transparent conductive glasses by sol–gel-synthesized TNP and applied to photoelectrocatalysis [21] and DSSC [25]. Though the opened-end TiNT films were successfully obtained, the producing process was complicated and time-consuming. The TiNT film was first detached by dipping in H_2O_2 solution and then protected by hot-melt adhesive layer. The bottom caps were removed by soaking in oxalic acid, and the hot-melt adhesive layer was subsequently removed by immersing in acetone solution.

Herein, we introduce a simple and handy procedure for large-area opened-end TiNT without cracks, referring to the two-step anodization procedure [23]. The TiNT films were firstly fabricated and then annealed at certain temperature to get crystallized TiNT films. Next, a thin underlayer of amorphous TiNT layer was grown by a second anodization. After dipping in oxalic acid solution, amorphous underlayer was selectively dissolved, and free-standing crystalline TiNT membranes were detached from the Ti substrate. At the same time, the bottom caps were removed, and the opened-end TiNT films were obtained without any cracks. The front-side-illuminated DSSC composed of the as-prepared TiNT films was shown to possess the superior photoelectronic properties.

63.2 Experimental

63.2.1 Preparation of Free-Standing and Opened-End TiNT Film

Figure 63.1 shows the scheme of preparing the free-standing opened-end TiNT film and fabricating the TiNT/FTO electrode. The growth of vertically oriented TiNT films was carried out by anodization of Ti foil at 80 V, 15 °C in an ethylene glycol electrolyte containing 0.5 wt.% NH_4F and 3 wt.% H_2O_2 (35 wt.%) for X h ($X = 2, 4, 6$, and 8). The anodization was completed by using a platinum plate as the cathode and the Ti foil as the anode. After first anodization, the different TiNT/Ti samples were annealed at Y °C ($Y = 200$, 300, and 400) in air for 1 h. Each as-annealed TiNT/Ti was anodized again in the same stock electrolyte at 40 V for 1 h, forming an underlayer amorphous TiO_2. Subsequently, the TiNT/Ti was immersed in 5 wt.% oxalic acid solution at 80 °C for Z h ($Z = 6$, 10, 14, and 16). In this one-pot reaction, the TiNT of the first layer was separated from the substrate, and the caps of the tube bottoms were removed. The large-area, non-curling, free-standing, and opened-end TiNT film was easily obtained. The prepared TiNT film was denoted as TiNT-X-Y-Z. For comparison, a free-standing and closed-end TiNT film was also prepared. The Ti foil was anodized at 80 V in the same stock electrolyte mentioned above for 6 h and then annealed at 300 °C for 1 h. The TiNT/Ti film was anodized again at 40 V for 1 h. The last step was slightly different from preparing opened-end TiNT. In the last step, the TiNT/Ti film was immersed in oxalic acid only for 20 s to lift the TiNT film off the Ti foil and keep the TiNT bottoms closed.

63.2.2 Fabrication of DSSC

TNP paste was prepared by mixing sol–gel-processed TiO_2 nanoparticles with water and polyethylene glycol. The preparation of sol–gel-processed TiO_2 nanoparticles was referred to Grätzel's group [26]. FTO glasses were immersed

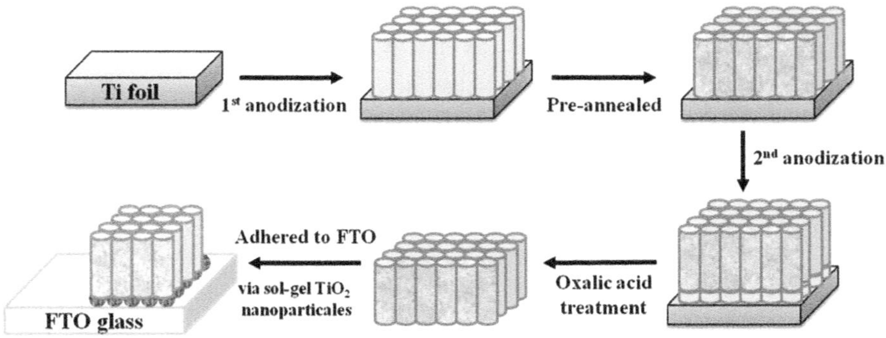

Fig. 63.1 Scheme for fabricating a TiNT/FTO electrode with opened-end TiNT film

in 0.2 M $TiCl_4$ solution at 70 °C for 30 min and then annealed at 450 °C for 30 min [27, 28]. The as-prepared TiNT film was adhered onto FTO glass with TNP paste, which was first printed onto FTO glass by doctor blade technique. The TiNT/FTO was dried at room temperature, followed by annealing at 450 °C for 30 min. After immersing in 0.2 M $TiCl_4$ solution and annealing [29,30], the excess TNP portion uncovered by TiNT film was shaved off, and the active area was 0.25 cm^2. The working electrode was then immersed in a 0.3 mM $(Bu_4N)_2Ru(dcbpyH)_2(NCS)_2$ (N719 dye, Solaronix) ethanol solution for 24 h. The counter electrode was a platinum-coated FTO substrate which is prepared by sputtering. The solar cell device was prepared by infiltrating between TiNT/FTO photoanode and Pt counter electrode with iodide-based electrolyte solution containing 0.5 M LiI, 0.05 M I_2, and 0.5 M *tert*-butylpyridine in anhydrous acetonitrile.

63.2.3 Characterization

The morphology and crystallinity of TiNT films and TiNT/FTO photoanodes were examined by field emission scanning electron microscopy (FE-SEM, JEOL JSM-6500 F) and X-ray diffraction (XRD, Philips X'pert Pro diffractometer with Cu Kα radiation of 0.15405 nm). The N719 dye adsorbed on the TiNT/FTO photoanode was dissolved completely in 0.2 M sodium hydroxide solution. The amount of adsorbed N719 dye was determined by UV–Vis spectroscopy (Hitachi U-3410).

63.2.4 Photovoltaic Measurements

Photoelectrochemical experiments were performed using an electrochemical analyzer (650B, CH Instruments), and devices were illuminated with a calibrated AM 1.5 solar light simulator (Oriel Class A Solar Simulators, Model 91195A) operating at an intensity of 100 mW cm^{-2}. The light intensity was adjusted with a monocrystalline Si reference cell equipped with an IR cutoff filter (KG-5). The current density–voltage measurements (*J*–*V*) were conducted through a voltage range from 0 to 1 V with illumination. The open-circuit potential (V_{OC}), short-circuit current density (J_{SC}), fill factor (FF), and energy conversion efficiency (η) were obtained. The incident photon-to-current conversion efficiency (IPCE) experiments were carried out using the system which contained a Xe lamp (300 W) with a monochromator (Oriel 74100). The light intensity was measured with an optical power meter (Ophir Optronics 70310) equipped with a calibrated thermopile head (Ophir Optronics 71964). The electrochemical impedance spectroscopy (EIS) measurement was performed by applying a frequency range from 0.05 to 105 Hz with AC amplitude of 10 mV under AM 1.5 illumination at an applied bias of V_{OC}.

63.3 Results and Discussion

63.3.1 Characterization of Free-Standing and Opened-End TiNT Film

We discussed the influences of anodization time, annealing temperature, and oxalic acid treating time on the characteristics of TiNT bottom layer. Figure 63.2 indicates the relation between the first anodization time and the TiNT film thickness. When the Ti foils were anodized at 80 V for 1, 2, 4, 6, and 8 h, the TiNT film thicknesses (L_F) were 13, 20, 30, 35, and 40 μm. The inner diameters of those TiNT films were around 130 nm, and the wall thickness was around 25 nm. In Fig. 63.3, XRD patterns of TiNT/Ti film anodized for 2 h and then annealed at 200, 300, and 400 °C are shown. The TiNT/Ti film after annealing at 200 °C was amorphous titania, and only diffraction peaks of Ti (JCPDS # 89-5009) could be found. The diffraction peaks of TiNT/Ti film after annealing at 300 and 400 °C were observed and referred to Ti and anatase titania (JCPDS # 84-1286). However, the intensity and sharpness of the anatase peaks grew with raising the annealing temperature, indicating the increase in the crystallinity of the anatase form.

The effect of annealing temperature was studied. After first anodization of 2 h, the TiNT films were individually annealed at 200, 300, and 400 °C and then immersed in oxalic acid solution for 8 h. After the oxalic acid treatment, the TiNT-2-200-8 was totally dissolved by oxalic acid, and the Ti foil was left in the solution. On the other hand, both TiNT-2-300-8 and TiNT-2-400-8 were flaked off from the Ti foil. The FE-SEM bottom view images were further examined and displayed in Fig. 63.4. In Fig. 63.4a, the bottom caps of TiNT-2-300-8 are almost removed, while in Fig. 63.4b, the bottom caps of TiNT-2-400-8 remain closed end. Consequently, the annealing condition at 300 °C was preferred because the free-standing and opened-end TiNT films were easily achieved in this simple one-pot reaction.

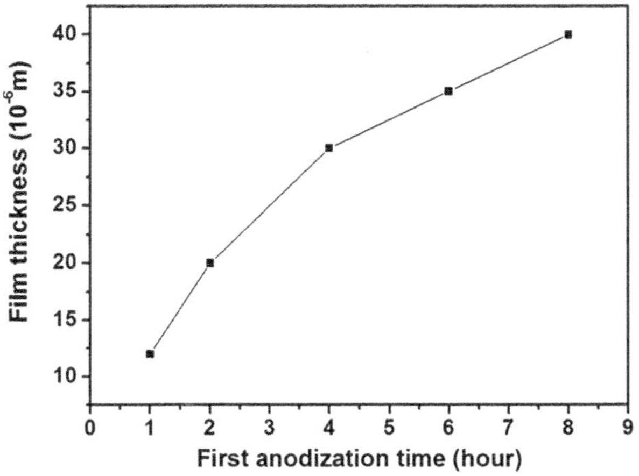

Fig. 63.2 Variation of TiNT film thickness as a function of first anodization time at 80 V

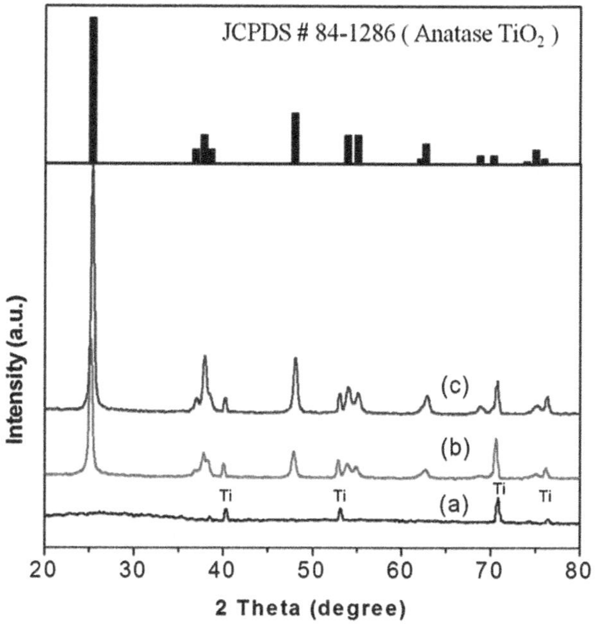

Fig. 63.3 XRD patterns of TiNT/Ti film annealed at (*a*) 200 °C, (*b*) 300 °C, and (*c*) 400 °C

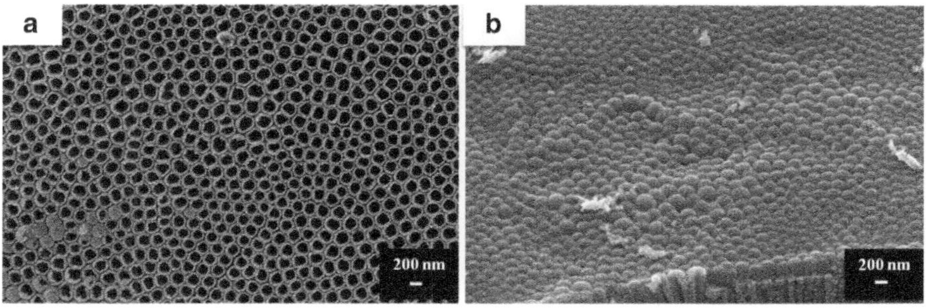

Fig. 63.4 FE-SEM bottom view images of TiNT films anodized for 2 h, annealed at different temperatures, and immersed in oxalic acid solution for 8 h. (**a**) 300 °C and (**b**) 400 °C

The effect of anodization time was studied. Figure 63.5 shows FE-SEM bottom view images of TiNT films anodized for different times, annealing at 300 °C and immersing in oxalic acid solution for 8 h. In the oxalic acid treatment, titanium oxide between the TiNT and the bottom caps was etched by oxalic acid, forming yellow titanium oxalate complex in the solution. In Fig. 63.5a, most of the bottom caps of TiNT-2-300-8 (20 μm thick) were removed by oxalic acid solution, while in Fig. 63.5c, most of the bottom caps of TiNT-6-300-8 (35 μm thick) remained. In addition, the TiNT-8-300-8 (40 μm thick) remained closed end after the oxalic acid treatment. It indicates that the thicker TiNT film may limit the diffusion of oxalic acid through the TiNT and thus retard the progress of removing bottom caps.

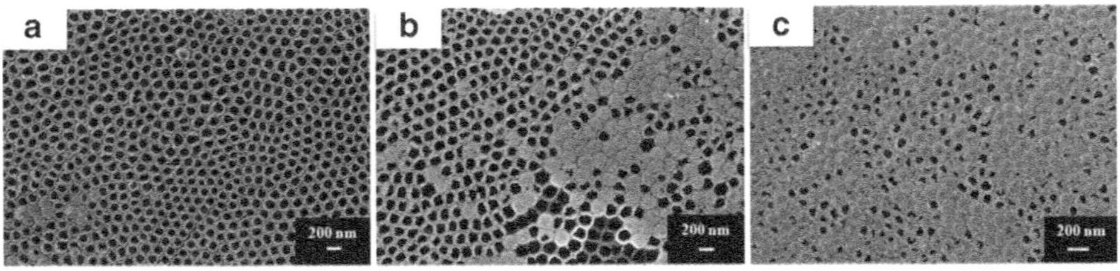

Fig. 63.5 FE-SEM bottom view images of TiNT films anodized for different times, annealed at 300 °C, and immersed in oxalic acid solution for 8 h. (**a**) 2 h, (**b**) 4 h, and (**c**) 6 h

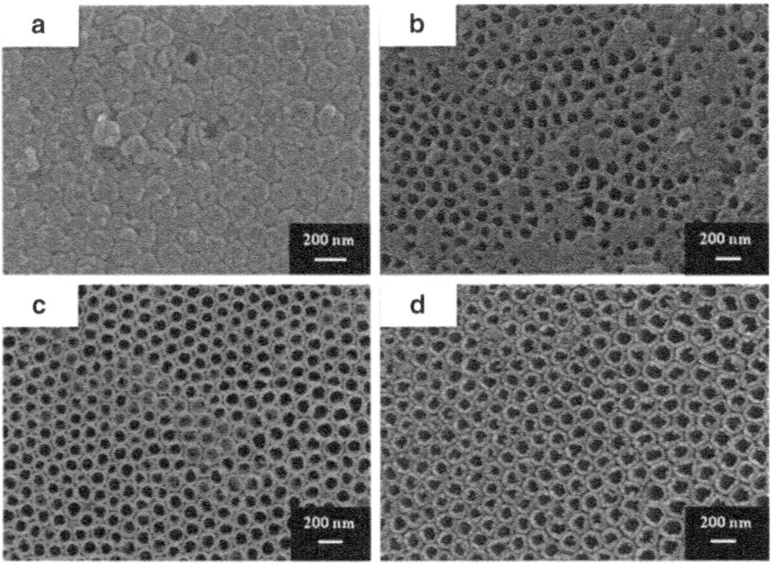

Fig. 63.6 FE-SEM bottom view images of TiNT films anodized for 6 h, annealed at 300 °C, and immersed in oxalic acid solution for different times. (**a**) 6 h, (**b**) 10 h, (**c**) 14 h, and (**d**) 16 h

The effect of oxalic acid treating time was studied. Figure 63.6 shows FE-SEM bottom view images of the evolution of the TiNT films anodized for 6 h, annealing at 300 °C and immersing in oxalic acid solution for different times. As shown in Fig. 63.6a, the bottom caps still exist after immersing in oxalic acid solution for 6 h. However, the TiNT bottom caps are gradually removed with immersing time, and the bottoms are nearly fully opened after immersing in oxalic acid for 16 h. Therefore, the most efficient preparation of opened-end and free-standing TiNT films was determined that the TiNT films were annealed at 300 °C and treated in oxalic acid solution for 16 h. Furthermore, Fig. 63.7 shows the FE-SEM side view image of TiNT anodized for 6 h and removed from the Ti foil. In Fig. 63.7a, the TiNT film was about 35 μm. As Fig. 63.7b shown, the 35-μm-thick TiNT under this preparation possesses excellent mechanical strength and is easily handled with

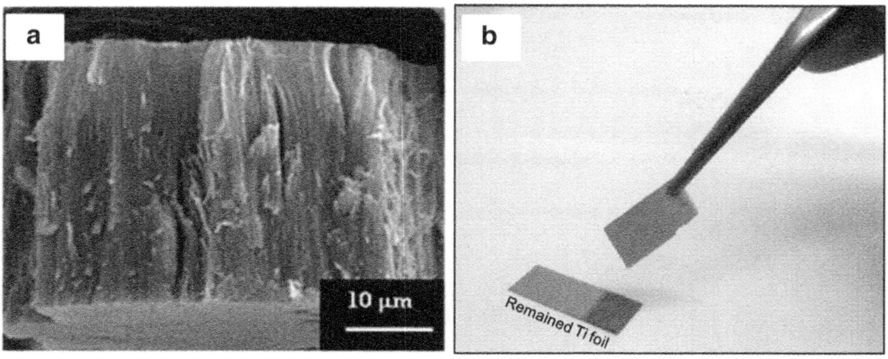

Fig. 63.7 (**a**) FE-SEM side view image of TiNT anodized for 6 h. (**b**) Photograph of 35-µm-thick TiNT film

tweezers. In order to prepare the TiNT films with good quality in an efficient way, all of the following samples for analysis and examination were annealed at 300 °C before the second anodization and then treated in oxalic acid solution for 16 h.

63.3.2 Photovoltaic Measurements

For the fabrication of DSSC photoanode, the annealed TiNT film was adhered onto FTO glass using sol–gel-processed TNP paste. The 3-µm-thick TNP layer formed between the FTO glass and the TiNT film. DSSCs made of free-standing and opened-end TiNT with different thickness were investigated below. The DSSCs using 20-, 30-, 35-, and 40-µm-thick TiNT/FTO as an anode were labeled as TiNT-2-300-16, TiNT-4-300-16, TiNT-6-300-16, and TiNT-8-300-16. The *J–V* curves for the DSSCs made of various TiNT films under AM 1.5 illumination are presented in Fig. 63.8. The measured photovoltaic performances are presented in Table 63.1. With an increase of TiNT film thickness from 20 to 35 µm, J_{SC} increased from 10.8 to 15.2 mA cm^{-2} and η also increased from 5.49 to 7.82 %. The main enhancement of J_{SC} and η was attributed to the increasing amount of adsorbed dye molecules. The amount of adsorbed dye molecules increased from 0.11 to 0.22 µmol cm^{-2} and led to enhanced photoelectron production. When the TiNT film thickness further reached 40 µm, the decreases in J_{SC} and η were found with a slight increase of the dye loading. The increases in surface area and adsorbed dye molecules would produce more recombination centers, resulting in a lowered J_{SC}. The J_{SC} of opened-end TiNT had an upper limit at 35 µm. Besides, V_{OC} decreased with the increasing L_F because of the charge recombination and mass transport limitations in the thicker film [31]. Figure 63.9 displays the IPCE of DSSCs made of various TiNT films with different thickness. TiNT-6-300-16 possessed the highest IPCE values over a wide range from 400 to 750 nm, while TiNT-2-300-16 possessed the lowest IPCE values. The trend of IPCE values matched the trend of J_{SC}.

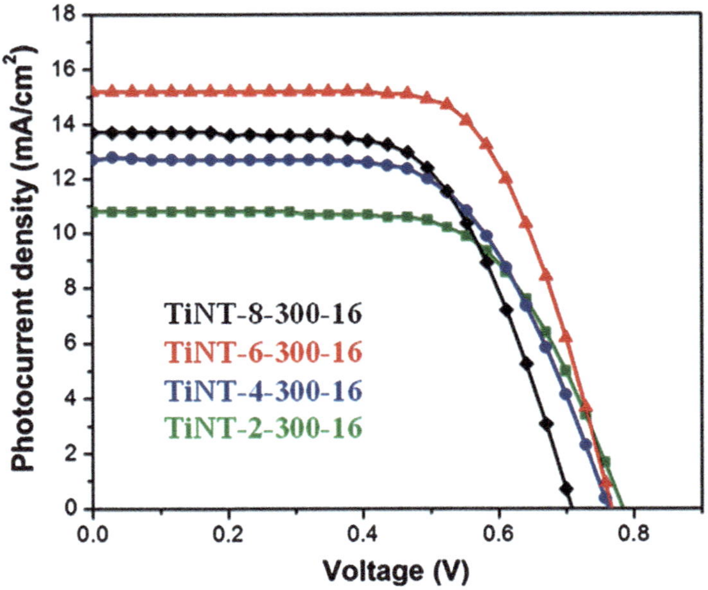

Fig. 63.8 *J–V* curves of DSSCs made of various TiNT films

Table 63.1 Photovoltaic performances of DSSCs made of various TiNT films

Sample	L_F (μm)	Dye adsorption (μmol cm^{-2})	V_{OC} (V)	J_{SC} (mA cm^{-2})	FF	η (%)
TiNT-2-300-16	20	0.11	0.78	10.8	0.65	5.49
TiNT-4-300-16	30	0.15	0.77	12.7	0.62	6.05
TiNT-6-300-16	35	0.22	0.76	15.2	0.67	7.82
TiNT-8-300-16	40	0.24	0.70	13.7	0.64	6.14

For comparison, the *J–V* characteristics of DSSCs made of opened-end TiNT-6-300-16 film and closed-end TiNT-6-300-16 film under AM 1.5 illumination are showed in Fig. 63.10. The measured photovoltaic performances are presented in Table 63.2. The DSSC made of closed-end TiNT film exhibited the J_{SC} of 10.4 mA cm^{-2}, the V_{OC} of 0.71 V, the FF of 0.63, and the η of 4.65 %. The J_{SC} of opened-end TiNT film dramatically increased up to 15.2 mA cm^{-2}, with V_{OC} of 0.76 V, FF of 0.67, and η of 7.82 %. The higher V_{OC} was most likely attributed to an increase in electron density in the opened-end TiNT and thus the negative shift of the quasi-Fermi level [25]. The dye loading of opened-end TiNT film and closed-end TiNT film was almost the same, indicating that dye molecules could diffuse into the TNP underlayer through the interpore region during the immersion in dye.

The IPCE of opened-end and closed-end TiNT films are shown in Fig. 63.11. Though the dye loading and film thickness of both two DSSCs were similar, the DSSC made of opened-end TiNT film showed the superior electron-collecting

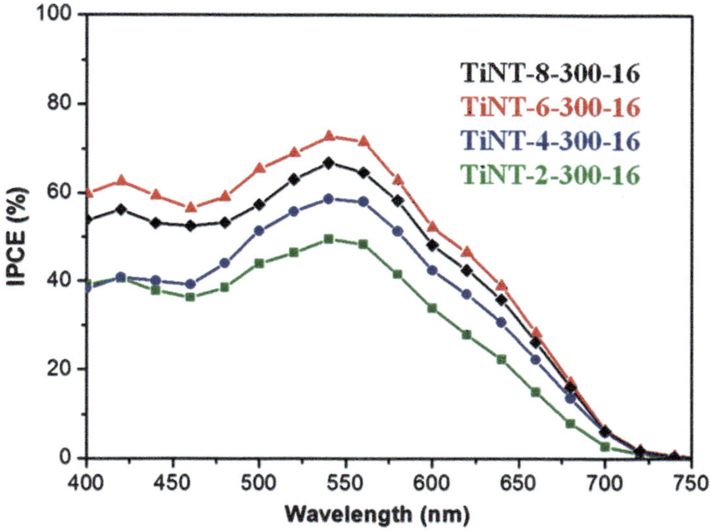

Fig. 63.9 Dependence of the IPCE on wavelength for DSSCs made of various TiNT films

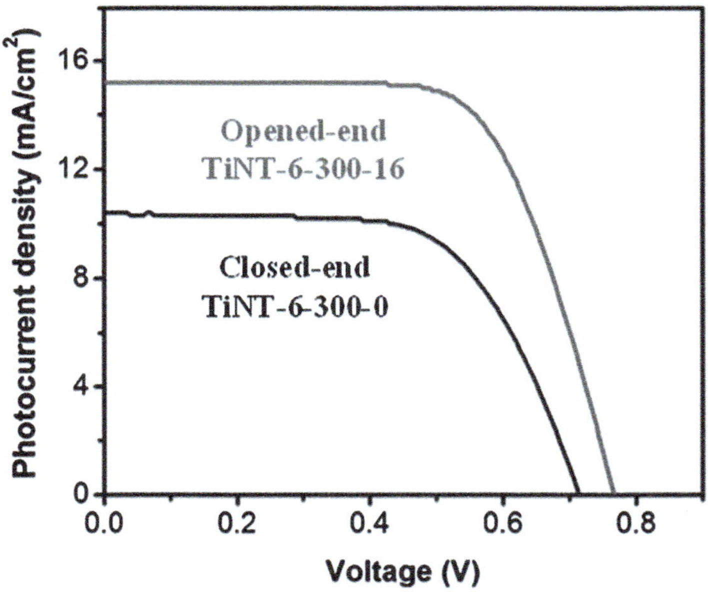

Fig. 63.10 *J–V* curves of DSSCs made of opened-end TiNT film and closed-end TiNT film

efficiency by the higher IPCE values across the entire wavelength range. As reported in the previous study [21,25], the lower amount of photoelectrons generated in closed-end TiNT film may be ascribed to the partial filtration of the incident light due to the presence of the bottom caps. Besides, the bottom caps of closed-end

Table 63.2 Photovoltaic performances of DSSCs made of opened-end TiNT film and closed-end TiNT film

Sample	L_F (μm)	Dye adsorption (μmol cm^{-2})	V_{OC} (V)	J_{SC} (mA cm^{-2})	FF	η (%)
Opened-end TiNT-6-300-16	35	0.22	0.76	15.2	0.67	7.82
Closed-end TiNT-6-300-16	35	0.21	0.71	10.4	0.63	4.65

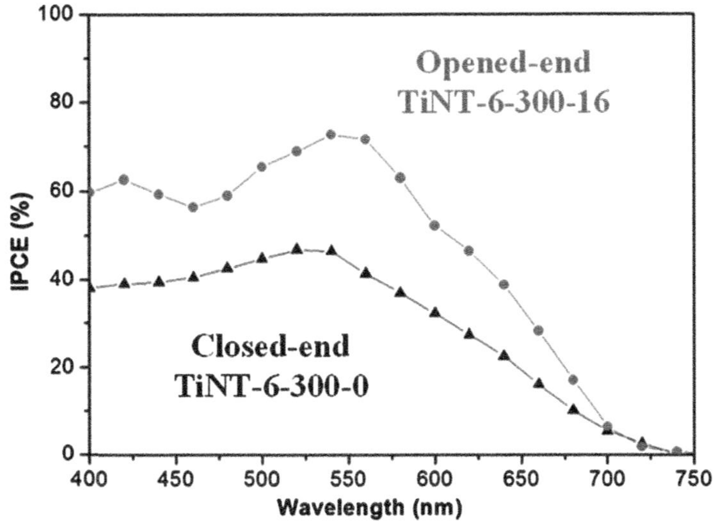

Fig. 63.11 Dependence of the IPCE on wavelength for DSSCs made of opened-end TiNT film and closed-end TiNT film

TiNT film would obstruct I^-/I_3^- electrolyte diffusing into the TNP underlayer and limit the dye regeneration.

The electron transport properties of DSSCs made of opened-end and closed-end TiNT films were investigated by EIS measurement under AM 1.5 illumination. Figure 63.12 shows the Nyquist plots of the obtained impedance data. The data were fitted in terms of the appropriate equivalent circuit using the ZView software as described previously [32]. The derived parameters, including the first-order reaction rate constant for the loss of electrons (k_{eff}), the lifetime of an electron ($\tau = 1/k_{eff}$), the dark reaction impedance (R_k), and the diffusion reaction impedance (R_w) in a TiO_2 film with thickness (L_F), are summarized in Table 63.3. The effective diffusion coefficient (D_{eff}) and diffusion length (L_n) were calculated using Eqs. (63.1) [33] and (63.2) [34], respectively.

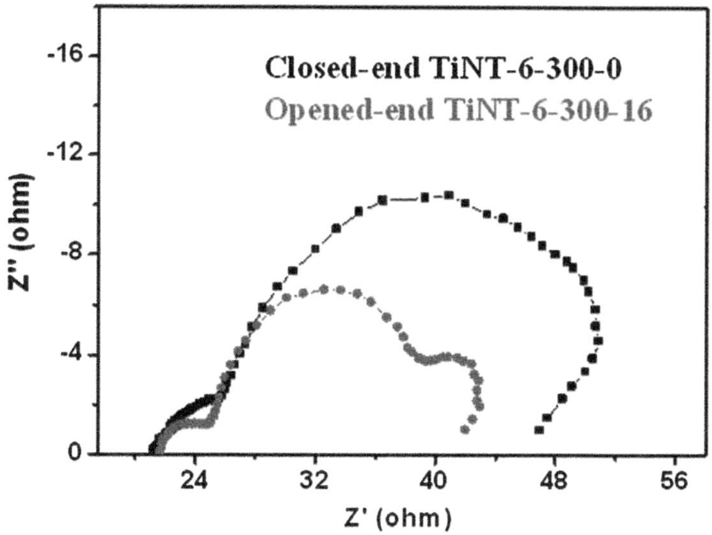

Fig. 63.12 Nyquist plots of the impedance data of the DSSCs made of opened-end TiNT film and closed-end TiNT film

Table 63.3 Photovoltaic performances of DSSCs made of opened-end TiNT film and closed-end TiNT film

Sample	L_F (μm)	k_{eff} (s^{-1})	τ (ms)	R_k (Ω)	R_w (Ω)	D_{eff} ($cm^2\ s^{-1}$)	L_n (μm)
Opened-end TiNT-6-300-16	35	3.09	51.5	19.58	1.26	3.69×10^{-3}	138
Closed-end TiNT-6-300-16	35	5.49	29.0	23.93	4.90	2.06×10^{-3}	77

$$D_{eff} = (R_k/R_w) \times (L_{F^2}/\tau) \tag{63.1}$$

$$L_n = (\tau \times D_{eff})^{1/2} \tag{63.2}$$

As shown in Table 63.3, the τ of the opened-end TiNT-6-300-16 film is 51.5 ms and of the closed-end TiNT-6-300-16 film is 29.0 ms, while the D_{eff} of the opened-end TiNT film is 3.69×10^{-3} cm^2 s^{-1} and of the closed-end TiNT film is 2.06×10^{-3} cm^2 s^{-1}. Both τ and D_{eff} of the opened-end TiNT film were much higher than that of the closed-end TiNT film, attributing to the electron recombination suppression and the enhancement of electron collection. Moreover, the L_n in the opened-end TiNT film was almost twice longer than that in the closed-end TiNT film, implying that opened-end TiNT film possessed superior electron transport ability. The values of L_n in the opened-end and closed-end TiNT films both exceeded than film thickness. Therefore, all photogenerated electrons could be collected efficiently.

63.4 Conclusion

In conclusion, we have developed a simple method to prepare large-scale, anatase, free-standing, and opened-end TiNT film with two-step anodization, selected annealing temperature, and oxalic acid treatment. Annealing the TiNT film at 300 °C is helpful to form the anatase phase and to remove the bottom caps. Furthermore, the efficient oxalic acid treatment can not only separate the TiNT film from Ti foil but also remove the bottom caps of the TiNT film in the one-pot reaction. In DSSC, the TiNT-6-300-16 of 35 μm achieves the highest η of 7.82 %. Furthermore, the η of opened-end TiNT film improves ~70 % as compared with the closed-end TiNT film and indicates that opened-end TiNT film enhances both the photoelectron-collecting efficiency and the utilization of injected light.

Acknowledgments We acknowledge the financial support from Academia Sinica and Ministry of Science and Technology of Taiwan, R.O.C.

References

1. Oregan B, Grätzel M (1991) A low-cost, high-efficiency solar cell based on dye-sensitized colloidal TiO_2 films. Nature 353:737–740
2. Wang ZS, Kawauchi H, Kashima T, Arakawa H (2004) Significant influence of TiO_2 photoelectrode morphology on the energy conversion efficiency of N719 dye-sensitized solar cell. Coord Chem Rev 248:1381–1389
3. Nazeeruddin MK, De Angelis F, Fantacci S, Selloni A, Viscardi G, Liska P, Ito S, Bessho T, Grätzel M (2005) Combined experimental and DFT-TDDFT computational study of photoelectrochemical cell ruthenium sensitizers. J Am Chem Soc 127:16835–16847
4. Ito S, Murakami TN, Comte P, Liska P, Grätzel C, Nazeeruddin MK, Grätzel M (2008) Fabrication of thin film dye sensitized solar cells with solar to electric power conversion efficiency over 10 %. Thin Solid Films 516:4613–4619
5. Cao YM, Bai Y, Yu QJ, Cheng YM, Liu S, Shi D, Gao FF, Wang P (2009) Dye-sensitized solar cells with a high absorptivity ruthenium sensitizer featuring a 2-(hexylthio)thiophene conjugated bipyridine. J Phys Chem C 113:6290–6297
6. Yella A, Lee HW, Tsao HN, Yi C, Chandiran AK, Nazeeruddin MK, Diau EWG, Yeh CY, Zakeeruddin SM, Grätzel M (2011) Porphyrin-sensitized solar cells with cobalt (II/III)-based redox electrolyte exceed 12 percent efficiency. Science 334:629–634
7. Adachi M, Murata Y, Takao J, Jiu JT, Sakamoto M, Wang FM (2004) Highly efficient dye-sensitized solar cells with a titania thin-film electrode composed of a network structure of single-crystal-like TiO_2 nanowires made by the "oriented attachment" mechanism. J Am Chem Soc 126:14943–14949
8. Nian JN, Teng HS (2006) Hydrothermal synthesis of single-crystalline anatase TiO_2 nanorods with nanotubes as the precursor. J Phys Chem B 110:4193–4198
9. Armstrong, G, Armstrong AR, Canales J, Bruce PG (2005) Nanotubes with the TiO2-B structure. Chem Commun:2454–2456
10. Liu B, Boercker JE, Aydil ES (2008) Oriented single crystalline titanium dioxide nanowires. Nanotechnology 19:50

11. Liu B, Aydil ES (2009) Growth of oriented single-crystalline rutile TiO_2 nanorods on transparent conducting substrates for dye-sensitized solar cells. J Am Chem Soc 131:3985–3990
12. Lin Y, Wu GS, Yuan XY, Xie T, Zhang LD (2003) Fabrication and optical properties of TiO_2 nanowire arrays made by sol-gel electrophoresis deposition into anodic alumina membranes. J Phys Cond Matter 15:2917–2922
13. Liu SM, Gan LM, Liu HL, Zhang WD, Zeng HC (2002) Synthesis of single-crystalline TiO2 nanotubes. Chem Mater 14:1391–1397
14. Pradhan SK, Reucroft PJ, Yang FQ, Dozier A (2003) Growth of TiO_2 nanorods by metalorganic chemical vapor deposition. J Cryst Growth 256:83–88
15. Macak JM, Tsuchiya H, Ghicov A, Schmuki P (2005) Dye-sensitized anodic TiO_2 nanotubes. Electrochem Commun 7:1133–1137
16. Paulose M, Shankar K, Varghese OK, Mor GK, Hardin B, Grimes CA (2006) Backside illuminated dye-sensitized solar cells based on titania nanotube array electrodes. Nanotechnology 17:1446–1448
17. Mor GK, Shankar K, Paulose M, Varghese OK, Grimes CA (2006) Use of highly-ordered TiO_2 nanotube arrays in dye-sensitized solar cells. Nano Lett 6:215–218
18. Grimes CA (2007) Synthesis and application of highly ordered arrays of TiO_2 nanotubes. J Mater Chem 17:1451–1457
19. Varghese OK, Paulose M, Grimes CA (2009) Long vertically aligned titania nanotubes on transparent conducting oxide for highly efficient solar cells. Nat Nanotechnol 4:592–597
20. Lin CJ, Yu WY, Chien SH (2008) Rough conical-shaped TiO_2-nanotube arrays for flexible backilluminated dye-sensitized solar cells. Appl Phys Lett 93:133107
21. Lin CJ, Yu WY, Lu YT, Chien SH (2008) Fabrication of open-ended high aspect-ratio anodic TiO2 nanotube films for photocatalytic and photoelectrocatalytic applications. Chem Commun:6031–6033
22. Zhu K, Neale NR, Miedaner A, Frank AJ (2007) Enhanced charge-collection efficiencies and light scattering in dye-sensitized solar cells using oriented TiO_2 nanotubes arrays. Nano Lett 7:69–74
23. Chen QW, Xu DS (2009) Large-scale, noncurling, and free-standing crystallized TiO_2 nanotube arrays for dye-sensitized solar cells. J Phys Chem C 113:6310–6314
24. Park JH, Lee TW, Kang MG (2008) Growth, detachment and transfer of highly-ordered TiO2 nanotube arrays: use in dye-sensitized solar cells. Chem Commun:2867–2869
25. Lin CJ, Yu WY, Chien SH (2010) Transparent electrodes of ordered opened-end TiO_2-nanotube arrays for highly efficient dye-sensitized solar cells. J Mater Chem 20:1073–1077
26. Barbe CJ, Arendse F, Comte P, Jirousek M, Lenzmann F, Shklover V, Grätzel M (1997) Nanocrystalline titanium oxide electrodes for photovoltaic applications. J Am Ceram Soc 80:3157–3171
27. Wang P, Zakeeruddin SM, Comte P, Charvet R, Humphry-Baker R, Grätzel M (2003) Enhance the performance of dye-sensitized solar cells by Co-grafting amphiphilic sensitizer and hexadecylmalonic acid on TiO_2 nanocrystals. J Phys Chem B 107:14336–14341
28. Ito S, Liska P, Comte P, Charvet RL, Pechy P, Bach U, Schmidt-Mende L, Zakeeruddin SM, Kay A, Nazeeruddin MK, Grätzel M (2005) Control of dark current in photoelectrochemical (TiO2/I–I3-) and dye-sensitized solar cells. Chem Commun:4351–4353
29. Nazeeruddin MK, Kay A, Rodicio I, Humphrybaker R, Muller E, Liska P, Vlachopoulos N, Grätzel M (1993) Conversion of light to electricity by Cis-X_2Bis(2,2′-bipyridyl-4,4′-dicarboxylate) ruthenium(II) charge-transfer sensitizers ($X = Cl^-$, Br^-, I^-, CN^-, and SCN^-) on nanocrystalline TIO_2 electrodes. J Am Chem Soc 115:6382–6390
30. Charoensirithavorn P, Ogomi Y, Sagawa T, Hayase S, Yoshikawaa S (2010) Improvement of dye-sensitized solar cell through TiCl4-treated TiO_2 nanotube arrays. J Electrochem Soc 157: B354–B356

31. Ohsaki Y, Masaki N, Kitamura T, Wada Y, Okamoto T, Sekino T, Niihara K, Yanagida S (2005) Dye-sensitized TiO_2 nanotube solar cells: fabrication and electronic characterization. Phys Chem Chem Phys 7:4157–4163
32. Wu JJ, Chen GR, Yang HH, Ku CH, Lai JY (2007) Effects of dye adsorption on the electron transport properties in ZnO-nanowire dye-sensitized solar cells. Appl Phys Lett 90:213109
33. Adachi M, Sakamoto M, Jiu JT, Ogata Y, Isoda S (2006) Determination of parameters of electron transport in dye-sensitized solar cells using electrochemical impedance spectroscopy. J Phys Chem B 110:13872–13880
34. Nakade S, Matsuda M, Kambe S, Saito Y, Kitamura T, Sakata T, Wada Y, Mori H, Yanagida S (2002) Dependence of TiO_2 nanoparticle preparation methods and annealing temperature on the efficiency of dye-sensitized solar cells. J Phys Chem B 106:10004–10010

Chapter 64
Effect of Different Parameters on the Solar Drying of Henna; Experimental Investigation in the Region of Biskra (Algeria)

Adnane Labed, Noureddine Moummi, Miloud Zellouf, Kamel Aoues, and Amar Rouag

Abstract This paper presents the results of an experimental investigation on the thermal performances of two models of solar air flat-plate collectors (FPCs) for drying applications (indirect dryers). We carried out studies in order to determine the best-performing model and to achieve a clean product in the least possible time by using a sustainable energy. Thus, we have proceeded to the application of the best-performing dryer for the drying of the henna (*Lawsonia inermis*); we have sought to determine the influence of some parameters on the drying kinetic of the product for the forced convection hot-air drying and their temperature dependence. The drying air was heated by solar energy under variable outdoor conditions. Both collectors were designed, constructed, and tested in the University of Biskra (Algeria) on a stand facing south at an inclination angle equal to the local latitude.

Keywords Solar energy • Flat-plate collector • Efficiency factor • Drying • Henna • *Lawsonia inermis*

Nomenclature

A_C	Collector surface area (m^2)
c_p	Specific heat of air at constant pressure (kJ/kg K)
db	Dry basis
DR	Drying rate (kg/kg/h)

A. Labed (✉) • N. Moummi
Laboratoire de Génie Mécanique (LGM), Université de Biskra,
B.P. 145 R.P., 07000 Biskra, Algeria
e-mail: adnanelabed@yahoo.fr; a.labed@univ-biskra.dz

M. Zellouf • K. Aoues • A. Rouag
Laboratoire de Génie Energétique et Matériaux (LGEM), Université de Biskra,
B.P. 145 R.P., 07000 Biskra, Algeria

I. Dincer et al. (eds.), *Progress in Clean Energy, Volume 2*,
DOI 10.1007/978-3-319-17031-2_64

F_R	Collector heat removal factor depending on air inlet temperature (dimensionless)
HR	Relative humidity of ambient air (%)
I_G	Global irradiance incident on solar air heater collector (W/m^2)
L_c	Length of the flat-plate collector (m)
l_c	Width of the flat-plate collector (m)
M_{db}	Moisture content (dry basis) (kg/kg)
$\dot{m}$	Air mass flow rate (kg/s)
Q_U	Useful energy gain of the collector (W/m^2)
T_a	Ambient temperature (°C)
T_{fi}	Inlet air temperature of the collector (°C)
T_{fo}	Outlet fluid temperature of the collector (°C)
U_L	Collector overall heat loss coefficient (W/m^2 °C)
ΔM	Loss of mass of the drying product (%)
ΔP	Pressure loss [Pa]

Greek Letters

α_{abs}	Absorptance (dimensionless)
β	Collector tilt (degrees)
τ_v	Transparent cover transmittance (dimensionless)
η	Thermal efficiency (dimensionless)

64.1 Introduction

Henna is an odoriferous plant very widespread in Algeria, because it is used for the tattooing of the hand, feet, and the hair, but it can be used for various reasons all the time: drug, dye, etc.

Women constitute the major part of the users of henna, but considering its religious dimension and its various benefits, the men also use it to protect themselves from the malevolent eye or simply to dye the hair and the beard.

All fresh and dry leaves and seeds of the henna demonstrated antibacterial activity. Henna dry leaves demonstrated the best in vitro antimicrobial activity and in particular against *Shigella sonnei* [1].

Traditional drying of henna can take two days under direct sunlight with free air and three days in the shade. Leaves are dried in the shade to retain the green color with two drying periods a year to facilitate postharvest leaf drying. This form of drying has many drawbacks such as degradation by windblown debris, rain, insect infestation, and human and animal interference that will result in contamination of the product. Drying rate will reduce due to intermittent sunshine, interruption, and wetting by rain [2].

Solar drying technology offers an alternative that can process the plants, vegetables, and fruits in clean, hygienic, and sanitary conditions to national and international standard.

Many types of solar air heaters and dryers have been developed in the past for the efficient utilization of solar energy [3]. Among stationary solar heaters, air flat-plate collectors, FPCs, have been widely used for energy conservation and management in an increasing number of installations. They are quite attractive for low-temperature solar energy technology which requires air temperatures below 100 °C. In fact, solar air FPCs are extensively used over the years, because they are relatively simple with a minimal use of materials, are easy to operate, and have low capital costs [4–6]. Solar air FPCs are used for space heating, dehydration of industrial products, and drying of agricultural crops such as vegetables, fruits, grains, spices, medicinal plants, lumber, tobacco, and fish [7]. Furthermore, it is established that the introduction of different geometries of artificial roughness, in the dynamic air vein of the FPC, increases the transfer rate and favors the creation of turbulences near the absorber plate. These techniques can improve the thermal performance either by increasing the heat transfer area with the use of FPC having aluminum cans and corrugated/finned absorber surfaces or by enhancing the absorber to air convective heat transfer coefficient with the use of roughened absorber surfaces. Indeed, the use of artificial roughness in different forms and shapes is known us the most effective and economic way of improving the performance of a solar air heater [8, 9].

Many studies on the drying of agricultural products have also shown that the use of the obstacles in the collector duct improves the performance of the drying unit [10, 11].

In our laboratories, several theoretical and experimental investigations were carried out on the enhancement of thermal performances of solar air FPCs designed for drying applications [12–18]. In all these studies, the authors used different forms of obstacles mounted under the absorber plate on the air channel duct.

In the present study, we have conducted a comparative study between two types of solar air FPCs with artificial roughness in air channel duct; model (I) is a simple-pass solar FPC having trapezoidal obstacles, and model (II) is a double-pass FPC having trapezoidal obstacles, and the end of the flow channel was brought together by a metal transition, which turned the airflow 180 over the back plate, and the gape under the back plate is reserved to the crop drying. Thus, we have compared these two models, in order to determine the best-performing model for the drying uses. As application, we have used these two dryers in the drying of henna and compared with the traditional open sun drying of this product. We have also used the best dryer (model II) for the drying of the henna with different airflow rates.

64.2 Experimental

64.2.1 Experimental Setup

All the solar air FPCs were tested in the University of Biskra under similar environmental conditions in the spring of 2010. Biskra is located in the east of Algeria with a latitude of 34°48′N, longitude of 5°44′E, and altitude of 85 m.

The collectors were placed in a stand facing south at an inclination angle equal to the local latitude.

In all steadied collectors, the air outlet and inlet cross sections are equipped by divergent channel duct; the test facility permits to vary the mass flow rate of the air. In summary, the above experimental setup was instrumented for the measurement of the solar radiation, wind velocity, pressure drop, temperature of the atmosphere air, inlet and outlet air temperatures, surface temperature of absorber plate, and the air mass flow rate.

To carry out these experiments, 12 insulated thermocouples have been used for the mean temperature measurement. Five thermocouples were provided in the air channel along the direction of flow. Inlet and outlet air temperatures were measured by two thermocouples, and the ambient temperature was measured by a mercury thermometer. Thus, five thermocouples were positioned on the top surface of the absorber plates, at identical positions along the axial center line, to obtain the average temperature of the absorber.

The test data were measured at an average interval of 10 min, so the temperatures, solar radiation intensity, pressure drop, the airflow rate and wind velocity, relative humidity, and weight of drying product are respectively measured with the K-type thermocouples with an accuracy of 0.01 °C; Kipp & Zonen pyranometer CM 11 with 1 % accuracy, pressure transducer accuracy (Kimo CP301) with ±1 Pa and 0.5 % of reading, a Kimo-type anemometer with hot wire (VT300) with ±3 % of reading and ±10 m^3 for the flow rate measurement and ±3 % of reading and ±0.1 m/s accuracy, Q.C.58 hygrometer with error ranges of 3 % for RH (relative humidity of ambient air), and laboratory balance with 1 % accuracy and maximum weight of 200 g were used in most experiments.

Both collector components have the same size: thickness of the single cover glass (5 mm), height of the air gap between the cover and the absorber plate (25 mm), height of the air duct (25 mm), dimensions of the absorber (1.96 m × 0.9 m with the thickness of 0.4 mm), and thickness of the rear insulation (40 mm). Therefore, the materials used in the fabrication of both FPC components are the same. The absorbers were made of galvanized steel with black coating. The heated air flows between the inner surface of the absorber plate and the back plate, with or without obstacles. The rear insulation is provided by a polystyrene sheet (30 mm of thickness), which is sandwiched between two plywood sheets.

Figure 64.1 shows the experimental device of the first dryer (model I). We show that the absorber plate was placed behind the transparent glass cover with a layer of static air separating it from the cover of the collector.

In the second FPC, we have a trapezoidal offset plate fin attached to the back plate of the collector. The end of the flow channel was brought together by a metal transition, which turned the airflow 180°. This was made so as to conduct tests in the double-pass mode in the first part of this study and to dispose the drying product in the second part. The materials used in the fabrication of these FPC components are the same as in the other FPCs employed in the previous section. An exception is that the height of the gape under the back plate is different ($H = 12$ cm), and it is equipped with wire mesh trays to help in crop drying (Fig. 64.2).

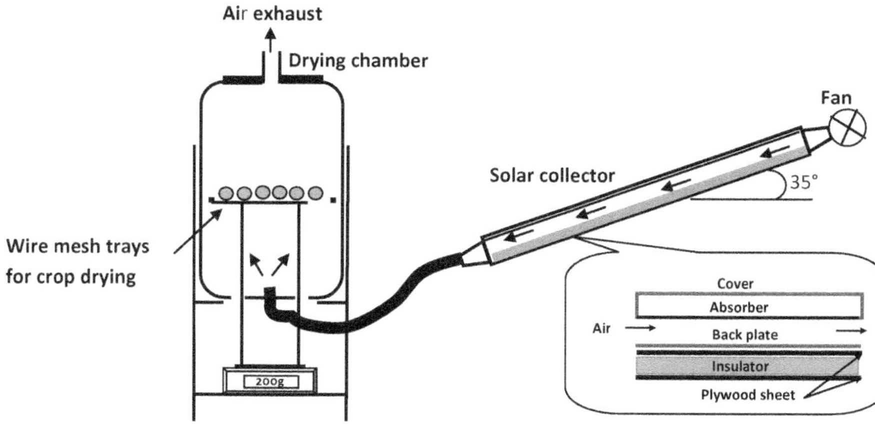

Fig. 64.1 Experimental device I (model I)

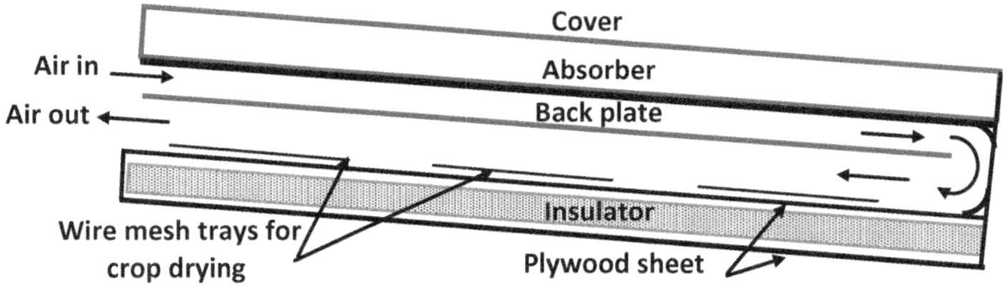

Fig. 64.2 Experimental device II (model II)

In Fig. 64.3, we show a schematic view of the trapezoidal fin obstacles used in the dynamic air vein of these two FPC models. The fin obstacles are oriented parallel to the fluid flow and are soldered to the upper side of the back plate (Fig. 64.3).

64.2.2 Experimental Analysis

Before the presentation of these six configurations in the previous section, we present the expressions used for the calculation of global heat loss, useful energy, and efficiency of the solar collectors:

The useful energy gain had been established by Duffie as below [4]:

$$Q_{\mathrm{u}} = \dot{m}\, c_P (T_{\mathrm{fo}} - T_{\mathrm{fi}}). \tag{64.1}$$

A measure of collector performance is the collector efficiency, defined as the ratio of useful heat gain over any time period to the incident solar radiation over the same period; we can, thus, define efficiency as

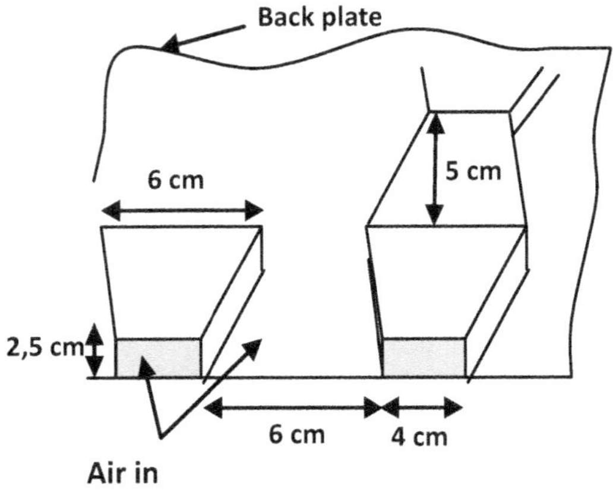

Fig. 64.3 Schematic view of fin obstacles in dynamic air vein

$$\eta = \frac{Q_U}{I_G A_C}. \tag{64.2}$$

Solving Eqs. (64.1) and (64.2), it can be obtained as

$$\eta = \frac{\dot{m}\, c_P (T_{fo} - T_{fi})}{I_G A_C}. \tag{64.3}$$

The drying characteristics of the henna such as moisture content and drying rate were determined by using Eqs. (64.4) and (64.5), respectively.

The moisture content, M_{db}, on dry basis was calculated by using Eq. (64.4):

$$M_{db} = \frac{(M_t - M_d)}{M_d} \times 100\,\%. \tag{64.4}$$

The drying rate, DR, should be proportional to the difference in moisture content between the material to be dried and the equilibrium moisture content [19]. The concept of thin-layer drying was assumed for the experiments as reported by Eq. (64.5):

$$\mathrm{DR} = \frac{dM}{dt} = -k(M_t - M_e). \tag{64.5}$$

Based on the analysis of the errors in the experimental measurements through the used instruments, the uncertainties in experimental measurement and results are often used to refer to possible values that may include errors. According to [20], the result R of an experiment is assumed to be calculated from a set of measurements; it is given as a function of the independent variables $X_1, X_2, \ldots, X_n$:

$$R = R(X_1, X_2, X_3, \ldots, X_N) \tag{64.6}$$

where in X_1, X_2, and X_3 are measured variables.

Let δR be the uncertainty in the result and $\delta X_1, \delta X_2, \ldots, \delta X_n$ be the uncertainties in the independent variables. If the uncertainties in the independent variables are all given with same odds, then uncertainty in the result having these odds is calculated by [4, 20]

$$\delta R = \left[\left(\frac{\partial R}{\partial X_1}\delta X_1\right)^2 + \left(\frac{\partial R}{\partial X_2}\delta X_2\right)^2 + \ldots + \left(\frac{\partial R}{\partial X_n}\delta X_n\right)^2\right]^{1/2}. \tag{64.7}$$

The independent parameters measured in the experiments reported here are collector inlet temperature T_{fi}, collector outlet temperature T_{fo}, ambient temperature T_{a}, mass flow rate, and solar irradiation.

If A_{c} and C_p are considered constants in the Eq. (64.4), it can be written as

$$\eta = f(T_{\text{fo}}, T_{\text{fi}}, I_{\text{G}}, \dot{m}). \tag{64.8}$$

The total uncertainty equation for collector efficiency η, moisture content M_{db}, and drying rate DR can be written as

$$\delta\eta = \left[\left(\frac{\partial\eta}{\partial\dot{m}}\delta\dot{m}\right)^2 + \left(\frac{\partial\eta}{\partial T_{\text{fo}}}\delta T_{\text{fo}}\right)^2 + \left(\frac{\partial\eta}{\partial T_{\text{fi}}}\delta T_{\text{fi}}\right)^2 + \left(\frac{\partial\eta}{\partial I_{\text{G}}}\delta I_{\text{G}}\right)^2\right]^{1/2} \tag{64.9}$$

$$\delta M_{\text{db}} = \left[\left(\frac{\partial M_{\text{db}}}{\partial M_t}\delta M_{\text{t}}\right)^2 + \left(\frac{\partial M_{\text{db}}}{\partial M_{\text{d}}}\delta M_{\text{d}}\right)^2\right]^{1/2} \tag{64.10}$$

$$\delta \text{DR} = \left[\left(\frac{\partial \text{DR}}{\partial M_{\text{t}}}\delta M_{\text{t}}\right)^2 + \left(\frac{\partial \text{DR}}{\partial M_{\text{d}}}\delta M_{\text{d}}\right)^2 + \left(\frac{\partial \text{DR}}{\partial t}\delta t\right)^2\right]^{1/2}. \tag{64.11}$$

Calculations show that the total uncertainty in calculating efficiency, η, moisture content M_{db}, and drying rate DR is almost in the order of 1, 2.8, and 3 %, respectively.

64.3 Results and Discussions

64.3.1 FPC Performances

The basic method of measuring collector performance is to expose operating collector to solar radiation and measure the fluid inlet and outlet temperature, and the fluid flow rate addition, radiation on the collector, ambient temperature, and wind speed are also recorded.

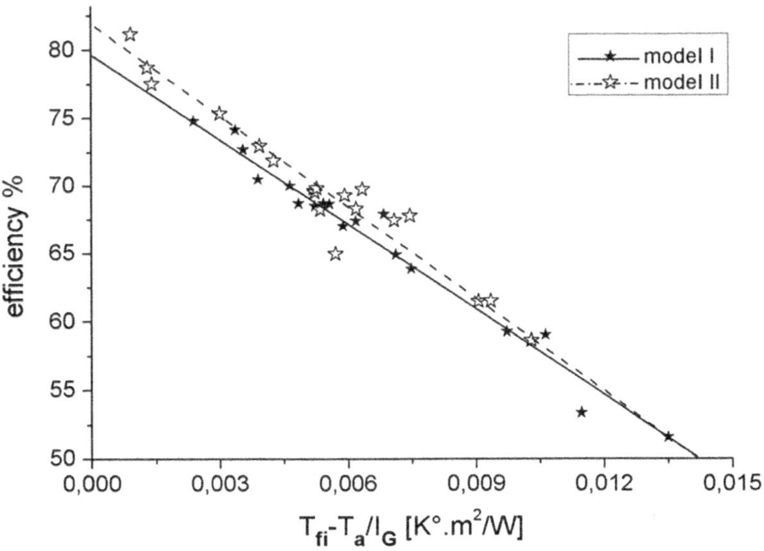

Fig. 64.4 Efficiencies versus $(T_{fi} - T_a)/I_G$ for different collector models

ASHRAE standard requires that, for the collector efficiency test, the solar insulation must be above 630 W/m^2 [9]. Instantaneous efficiencies are determined from $\eta = \frac{\dot{m}\ c_p(T_{fo} - T_{fi})}{I_G A_C}$ for the averaged pairs and are plotted as a function of $(T_{fi} - T_a)/I_G$. The thermal efficiency curves for both collectors are plotted in Fig. 64.4. The scatter of the data around the straight line is mainly attributed to the pump temperature, angle of incidence variations, wind speed, and the dependence of U_L on the plate temperature. Also, the variations of the relative proportions of beam diffuse and ground reflective components of solar radiation are participating in the data scattering. Thus, scatters in the data are to be expected [21]. However, it is acceptable that collectors are characterized by the line intercept $F_R(\tau_v \alpha_{abs})$ and the slope $F_R U_L$. From Fig. 64.4, it is evident that the efficiency decreases as the temperature parameter increases. That means, at higher reduced temperature parameters, the overall loss is lower.

It is also found that model II has the highest efficiency; the maximum efficiencies for these two FPCs (model I and model II) are determined as 77.8 % and 83.7 %, respectively, at $\dot{m} = 0.028\ \text{kg/s}$.

Figure 64.5 presents the variation of the pressure drop as a function of variation of the airflow rate for the two solar FPC models (Fig. 64.5).

The air in the double-pass FPC (model II) has a long trajectory, and substantial heat transfer enhancement is obtained. There is, of course, an associated increase in pressure drop due to increased friction with the lower airflow channel duct. It attains the values 8, 20, and 31.5 Pa for airflow rates of 0.011, 0.022, and 0.028 kg/s, respectively. The simple-pass FPC (model I) presents the low pressure drops. They attain the value 7, 17.5, and 25 Pa for airflow rates of 0.011, 0.022, and 0.028 kg/s, respectively.

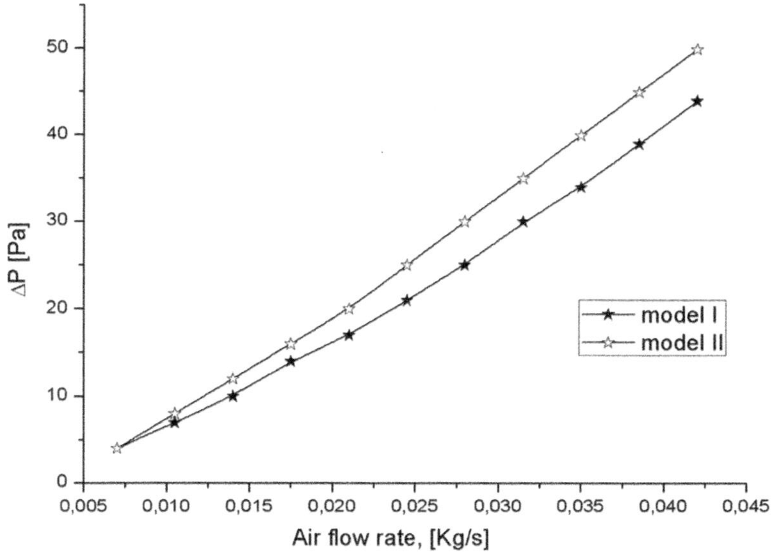

Fig. 64.5 Pressure drop (Pa) vs. the air mass flow rate (kg/s) for different studied FPCs

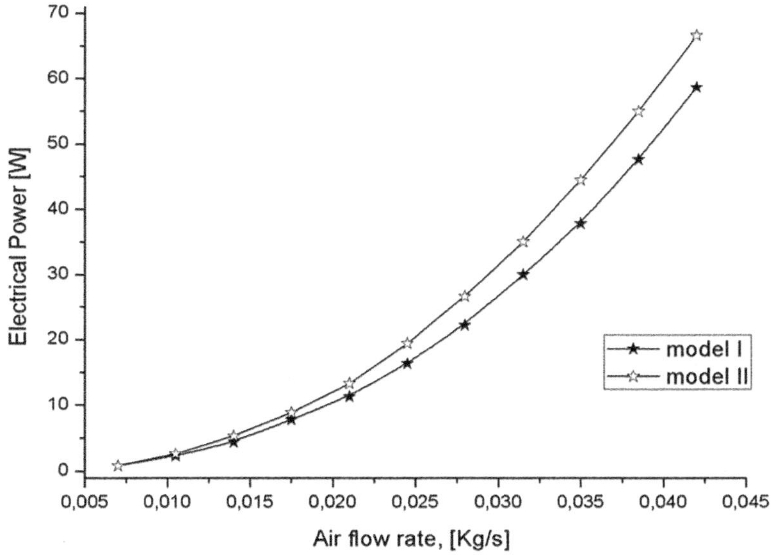

Fig. 64.6 Electrical power consumption (W) vs. the air mass flow rate (kg/s) for different studied FPCs

A series of experiments were conducted to establish the fan power consumption curves for the two studied systems. Electrical power consumption of the fan represented in Fig. 64.6 increases with increasing total pressure drop. This means that the highest values of power consumption are attained with the double-pass solar FPC (model II), while the simple-pass FPC (model I) presents the lower ones. These low values of the fan electrical power consumption are the results of the lower pressure drops.

64.3.2 *Drying Time Improvement*

64.3.2.1 Influence of the FPC Performances

Improvement in drying time allows a reduction in the operating costs. For the drying applications, we have selected two systems with forced convection. The first one (models I) is composed of an enclosure with one rack, a solar air flat-plate collector, and a fan. In both FPCs models (N and B-1) the drying chamber is a hard plastic drum of 8 mm thickness, 50 cm in diameter, and 80 cm height, positioned meadows of the collectors and supported by a metal frame.

The second system (model II) is a single frame dryer, where the end of the airflow channel of the FPC was brought together by a metal transition, which turned the airflow 180°. The materials used in the fabrication of these FPC components are the same as in the first FPC employed in the previous section. An exception is that the height of the gape under the back plate is different ($H = 12$ cm), and it is equipped with wire mesh trays for the crop depositing (Fig. 64.2). This was made so as to conduct tests of the FPC in the double-pass mode in the previous part of this study and to dispose the drying product in this part [13].

Field tests of the dryer for henna drying also were carried out in the spring of 2010; the typical results are shown in Figs. 64.7, 64.8, and 64.9. In all experiments, we have considered the initial moisture content (dry basis) of the product to be 4.6 kg_{water}/kg, and the final moisture content must be equal to 0.05 kg_{water}/kg (db).

Figure 64.7 shows a typical experimental evolution of the henna moisture content (dry basis) as a function of the drying time, in the cases of models I and II with $\dot{m} = 0.024$ kg/s compared to the drying in the open air under direct sun rays. Drying is started at 09:00 with an initial moisture content of 4.6 kg_{water}/kg (dry basis) and continued until 18:00 (9 h 00 of drying time/day). The final moisture content of the samples is obtained after 7 h 45 min and 4 h 15 min in the cases of model I and II, respectively, and it takes 15 h under direct sun.

64.3.2.2 Effect of the Airflow Rate on the Drying Time

The drying kinetic of the product can be represented on the quantity of moisture present in the sample or expressed on drying rate. To study the influence of the airflow rate on the drying time, we have used model II with three airflow rates 0.036, 0.024, and 0.012 kg/s. Figure 64.8 illustrates how the moisture content (wet basis) variation is affected during the drying time. As expected, there is an inverse relationship between the moisture content and drying time. Drying was started with

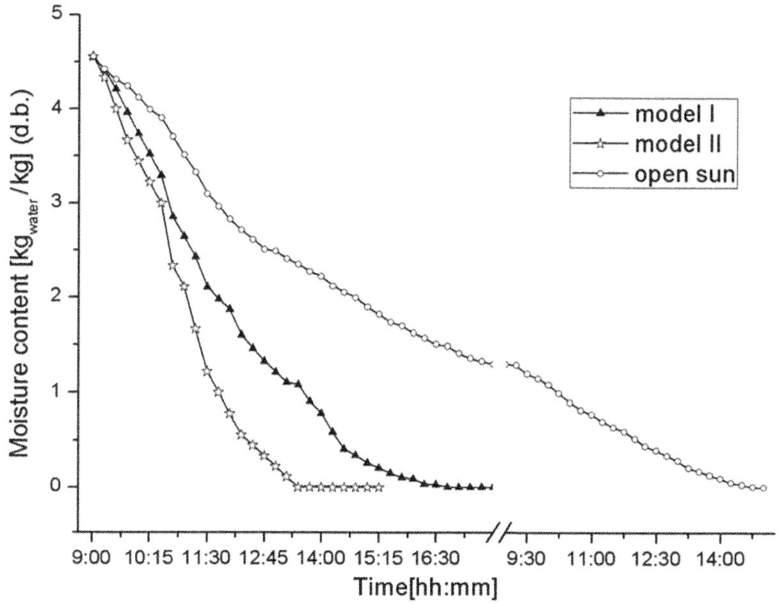

Fig. 64.7 Evolution of the henna moisture content (wet basis) as a function of the drying time for different dryer models ($\dot{m} = 0.024$ s), compared to the drying under open sun

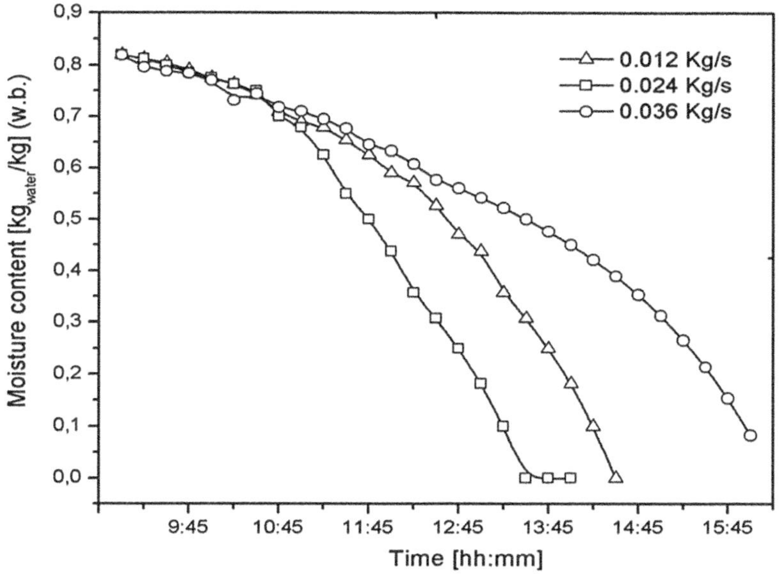

Fig. 64.8 Evolution of the henna moisture content (db) vs. drying time, for different airflow rates: 0.012, 0.024, and 0.036 kg/s, respectively, in the case of model II

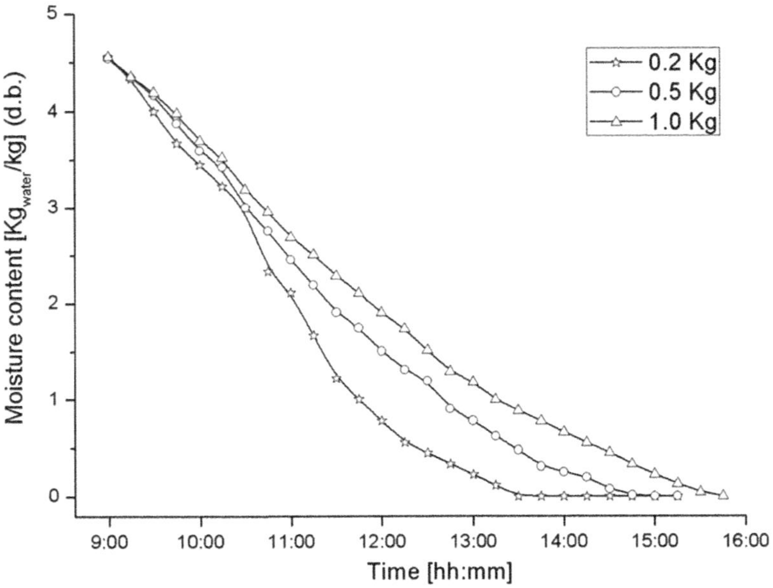

Fig. 64.9 Evolution of the henna moisture content (db) vs. drying time, for different product weights: 0.2, 0.5, and 1 kg, respectively, in the case of model II

initial moisture content of 0.82 kg_{water}/kg (wb) and continued, until a final moisture content of the samples that ranged from 0.00 to 0.083 kg_{water}/kg (wb) is attained after 7 h.

The time taken to reach a particular moisture content from the initial moisture content increases with an increase in the mass flow rate of the drying air. To lead to moisture content value of 0.083 kg_{water}/kg(wb), the drying time with the first airflow rate (0.036 kg/s) is longer (7 h 00 min), but is equal to 4 h 15 min and 5 h 15 min with the second and the third flow rates (0.024 and 0.012 kg/s), which correspond to a relative reduction in drying time of 39.29 % and 25 %, respectively (Fig. 64.8).

Figure 64.9 illustrates the influence of the henna quantity existing in the drying chamber on its moisture content (db) evolution as a function of the drying time in the case of model II. There is a relationship between the initial product weight and its moisture content during drying. After 6 h 45 min of drying of 1 kg of henna, the resulting moisture content in the product is equal to 0.05 [kg_{water}/kg] (db), while it takes 5 h 45 min and 4 h 30 min [kg_{water}/kg h] (db) to get a similar value of moisture content when the weight is 0.5 and 0.2 kg, corresponding to a relative reduction of 15 and 33 %, respectively.

64.4 Conclusion

An experimental study was conducted to evaluate the thermal performances of two types of solar air FPCs for drying applications: (1) simple-pass collector with trapezoidal obstacles (model I) and (2) double-pass collector with trapezoidal obstacles in the airflow duct (model II).

This study has allowed us to show that, at different air mass flow rates, the highest efficiencies were obtained from the double-pass solar air FPC with trapezoidal obstacles. In addition, the use of the best-performing FPC (model II) reduces the product drying time in comparison with model I and the traditional open sun drying.

The effect of the air mass flow rate was highlighted; it is clear that drying time with 0.024 kg/s was shorter than those with 0.012 and 0.036 kg/s.

However, the influence of some parameters specific to the product such as the quantity of the product on the drying time was investigated in outdoor conditions. It was found that there is an inverse relationship between the initial product weight and its drying time.

References

1. Habbal OA, Al-Jabri AA, El-hag AH, Al-Mahrooqi ZH, Al-Hashmi NA (2005) In vitro antimicrobial activity of *Lawsonia inermis* Linn (henna). A pilot study on the Omani henna. Saudi Med J 26(1):69–72
2. Mohanraj M, Chandrasekar P (2009) Performance of a forced convection solar drier integrated with gravel as heat storage material for chili drying. J Eng Sci Technol 4(3):305–314
3. Sami S, Rahimi A, Etesami N (2011) Dynamic modeling and a parametric study of an indirect solar cabinet dryer. Dry Technol 29(7):825–835
4. Duffie JA, Beckman WA (2013) Solar engineering of thermal processes. Wiley, New York
5. Kalogirou SA (2004) Solar thermal collectors and applications. Prog Energy Combust Sci 30 (3):231–295
6. Moummi N, Youcef-Ali S, Moummi A, Desmons JY (2004) Energy analysis of a solar air collector with rows of fins. Renew Energy 29(13):2053–2064
7. Fudholi A, Sopian K, Yazdi MH, Ruslan MH, Gabbasa M, Kazem HA (2014) Performance analysis of solar drying system for red chili. Sol Energy 99:47–54
8. Ozgen F, Esen M, Esen H (2009) Experimental investigation of thermal performance of a double-flow solar air heater having aluminium cans. Renew Energy 34(11):2391–2398
9. Karim MA, Hawlader MNA (2006) Performance investigation of flat plate, v-corrugated and finned air collectors. Energy 31(4):452–470
10. Abene A, Dubois V, Le Ray M, Ouagued A (2004) Study of a solar air flat plate collector: use of obstacles and application for the drying of grape. J Food Eng 65(1):15–22
11. Ahmed-Zaid A, Messaoudi H, Abenne A, Le Ray M, Desmons JY, Abed B (1999) Experimental study of thermal performance improvement of a solar air flat plate collector through the use of obstacles: application for the drying of 'Yellow Onion'. Int J Energy Res 23 (12):1083–1099
12. Labed A, Moummi N, Aoues K, Zellouf M, Moummi A (2009) Etude théorique et expérimentale des performances d'un capteur solaire plan à air muni d'une nouvelle forme de rugosité artificielle. Rev Energ Renouv 12(4):551–561

13. Labed A, Moummi N, Benchabane A, Aoues K, Moummi A (2012) Performance investigation of single-and double-pass solar air heaters through the use of various fin geometries. Int J Sustain Energy 31(6):423–434
14. Aoues K, Moummi N, Zellouf M, Benchabane A (2011) Thermal performance improvement of solar air flat plate collector: a theoretical analysis and an experimental study in Biskra, Algeria. Int J Ambient Energy 32(2):95–102
15. Labed A, Moummi N, Benchabane A (2012) Experimental investigation of various designs of solar flat plate collectors: application for the drying of green chili. J Renew Sustain Energy 4(4):043116. doi:10.1063/1.4742337
16. Labed A, Moummi N, Benchabane A (2012c) Experimental investigation of various designs of solar dryer: application for the drying of hot chili. Sixth international conference on thermal engineering: theory and applications, 29–1st Jun 2012, Istanbul, Turkey
17. Mahboub C, Moummi N (2012) Calculation of the glass cover temperature and the top heat loss coefficient for 60° vee corrugated solar collectors with single glazing. Sol Energy 86 (2):804–808
18. Moummi N, Moummi A, Aoues K, Mahboub C, Youcef Ali S (2010) Systematic forecasts of solar collector's performance in various sites of different climates in Algeria. Int J Sustain Energy 29(3):142–150
19. Ekechukwu OV (1999) Review of solar-energy drying systems I: an overview of drying principles and theory. Energy Convers Manag 40(6):593–613
20. Esen H (2008) Experimental energy and exergy analysis of a double-flow solar air heater having different obstacles on absorber plates. Build Environ 43(6):1046–1054
21. Karsli S (2007) Performance analysis of new-design solar air collectors for drying applications. Renew Energy 32(10):1645–1660

Chapter 65
Parametric Study of Tower Power Plant Performances for Its Implementation in Algeria

Noureddine Yamani, Abdallah Khellaf, and Kamel Mohammedi

Abstract There is a great interest in the development of the concentration power technologies to produce electricity. Among these technologies, the power tower technology plays an essential role. Beside the heliostat field, the receiver is an important part in this technology and its performance highly affects the efficiency of the whole system.

In the present work, the performance of a solar power tower receiver is studied. First, the case of the conventional water/steam receiver is considered, and then the open volumetric receiver is investigated. The study is carried out for the site of Hassi R'mel in Algeria.

An analysis of the results has indicated that the open volumetric receiver is more suitable for the Algerian climate than the conventional water/steam receiver.

Keywords Solar tower power • TRNSYS • Simulation • Solar energy • STEC

65.1 Introduction

From all the concentrating solar power (CSP) technology available today, the solar power tower is moving to the forefront and it is attracting a lot of interests every day. This type of CSP plant is expected to have the lowest solar levelized electricity cost (LEC) in the near future [1, 2]. A typical solar power tower plant consists of a heliostat filed, a solar receiver, and a thermodynamic power cycle system. The receiver is the key element of the power plant: it absorbs the solar radiation collected by the heliostats, converts it to heat, and then by means of the heat transfer fluid (HTF) transferred it to the thermodynamic cycle bloc. There are

N. Yamani (✉) • K. Mohammedi
L.E.M.I Laboratory, University of M'hamed Bougara, Boumerdes, Algeria
e-mail: n_yamani@hotmail.fr

A. Khellaf
Center for renewable energies development, CDER, Bouzareah, Algeria

I. Dincer et al. (eds.), *Progress in Clean Energy, Volume 2*,
DOI 10.1007/978-3-319-17031-2_65

various types of solar receiver depending on the geometrical configuration, the used absorber materials, and the heat transfer fluid. The well-known receivers are water/steam receiver and air volumetric receiver. The former was initially applied in solar towers such as PS10 and PS20, whereas the latter has been tested and developed at Jülich demonstrative solar power plant.

Considering the volumetric receiver, K. Hennecke et al. [3] have reported the technical data about the engineering and the development of the Jülich solar power tower. M. Schmitz et al. [4] have suggested the use of pressurized air solar receiver with secondary concentrator and an elliptic heliostat field design for powering a Brayton power conversion system. G. B Coelho et al.[5] have used Ebsilon Professional software to design and optimize the power conversion system, while the solar subsystems, i.e., heliostat field and volumetric receiver, have been designed and optimized using HFLCAL software. Barigozzi et al. [6] have used TRNSYS-STEC and Thermoflex software to simulate the off-design performance of a Brayton cycle.

With the focus on the water/steam receiver, R. J. Zoschak and S. F. Wu [7] have examined seven configurations that integrate solar tower power system into hybrid steam cycle of 80 MW, including feedwater heating, water evaporation, steam superheating, combined evaporation and superheating, steam reheating, air preheating, and combined air preheating and feedwater heating. Q. Yu et al. [8] have combined the collector and receiver models using STAR-90 software to simulate the steady and dynamic performance of Dahan solar tower power. Ershu Xu [9] has applied a modular modeling method to develop the dynamic simulation models of Dahan plant, based on STAR-90 simulation software. Zhihao Yao et al. [10] have designed the heliostat field using HFLD code while of whole power plant have been simulated using TRNSYS software. Z. Yao et al. [11] have presented a simulation model to be used for predicting the performance of the Dahan solar plant that uses the Rankine cycle to convert solar energy into electricity. Myung-Hoon Moon et al. [12] have developed a program using Visual Basic 6.0 to predict the performance of Dahan solar thermal power plant. Nils Ahlbrink et al. [13] have coupled three simulation environments, i.e., STRAL, LabView, and Dymola software, to investigate transient performance of solar tower power plants. Pierre Garcia et al. [14] have compared six simulation environments for solar tower power investigation, namely, UHC, DELSOL, HFLCAL, MIRVAL, FIAT LUX, and SolTrace; Peter Schwarzbozl et al. [15] have used the HFLCAL software to evaluate the solar subsystem of a power tower plant, while the TRNSYS-STEC has been applied to predict the thermal performance of the power block, and the field efficiency matrix tool is used to link the software. O. Behar et al. [16] have developed a mathematical model to evaluate the performance of commercial Brayton cycle with volumetric air receiver, under Algerian desert climate.

In this study, the performances of solar power tower coupled to a Rankine cycle under Algerian climate have been originally investigated using the powerful software, TRNSYS-STEC. More in depth, we have compared the state-of-the-art central receiver technologies to select the best suitable one for Algeria.

65.2 Algeria Energy Status

Algeria is located in Northwest Africa. It lies at latitudes from 18° to 38° north and at longitude from 9° west to 12° east. It covers a total area of 2,381,741 km^2. Its coast on the Mediterranean Sea extends over 1,200 km. It stretches southward on 1,800 km as far as the Tropic of Cancer.

According to its geographical situation, Algeria holds one of the highest solar reservoirs in the world.

The insolation time over almost the totality of the national territory exceeds 2,000 h annually and may reach 3,900 h (high plains and Sahara). The daily obtained energy on a horizontal surface of 1 m^2 is of 5 kWh over the major part of the national territory, or about 1,700 kWh/m^2/year for the north and 2,263 kWh/m^2/year for the south of the country (Fig. 65.1).

The promotion of renewable energy is a major thrust of the Algerian energy and environmental policy, and the benefits of these energies have been recognized for the economic development of the country. Each of the channels of renewable energies in Algeria was examined, taking into account its potential for development, given the nature of the needs to be met and the resources (oil, sunlight, wind).

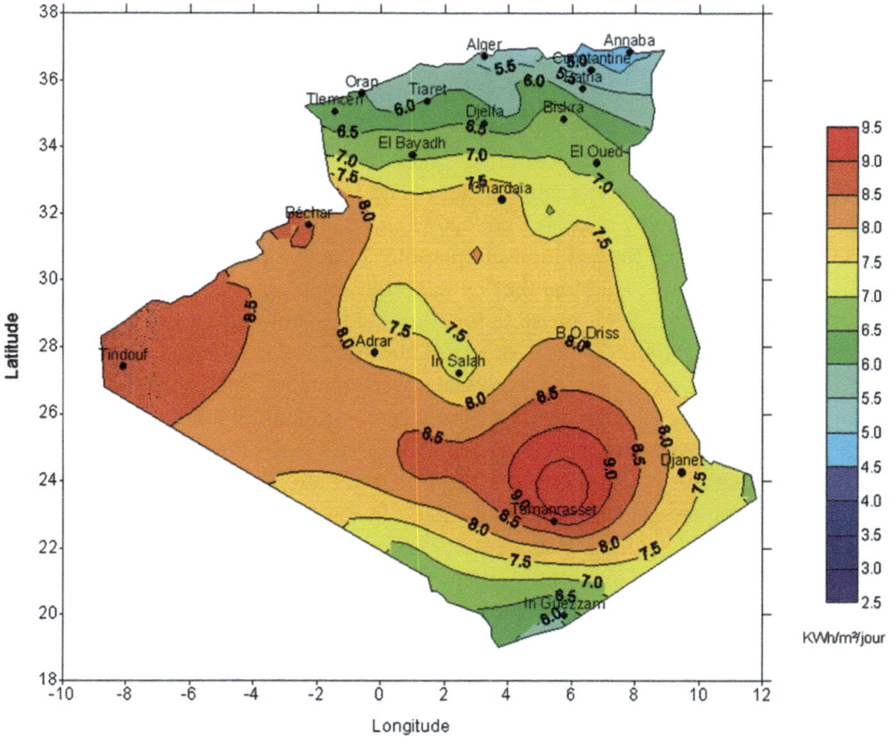

Fig. 65.1 Daily direct irradiation for the month of April

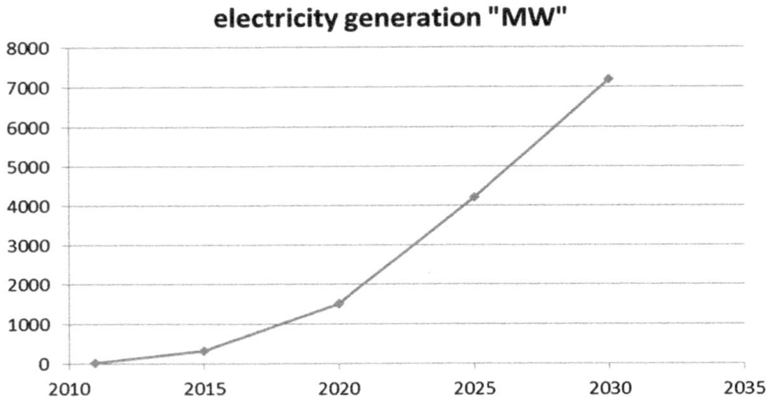

Fig. 65.2 Evolution of energy generation in Algeria [16]

In the context of the implementation solar energy program adopted by Algeria, the evolution of the electrical energy production form CSP is reported in Fig. 65.2.

This produced electrical power is meant not only for local consumption but also to export to Europe.

65.3 Concentration Principle

A typical CSP plant consists of three main subsystems: solar collector field, solar receiver, and a power conversion system. The solar receiver absorbs the concentrated solar radiation by means of collectors and transfers it to the heat transfer fluid (HTF) which is used to feed high-temperature heat to a power conversion system. The subsystems are linked together by heat transfer.

A diversity of options is possible for solar field layout, tracking system, receiver type, heat transfer fluid (HTF), storage technology, and power conversion system.

Many configurations exist for power tower concept. These configurations, for some of them under design, test, or improvement, include the volumetric receiver, the particle receiver, and the cavity receiver [17]. Concerning heat transfer fluids (HTF), molten salt is widely used as HTF in commercial plants. Synthetic oil and saturated steam are also currently used as HTFs in commercial plants. Superheated steam has been recently introduced as HTF. Pressurized air and other gases, in particular CO_2 and N_2, nanofluids, concrete and fluidized bed particles, and circulating particles, are under development for both trough and tower, while helium or hydrogen is used in dish Stirling (Fig. 65.3).

The solar receiver is the heat exchanger where the solar radiation is absorbed and transformed into thermal energy. There are different classification criteria for solar receivers, depending on the geometrical configuration and the absorber materials used to transfer the energy to the working fluid.

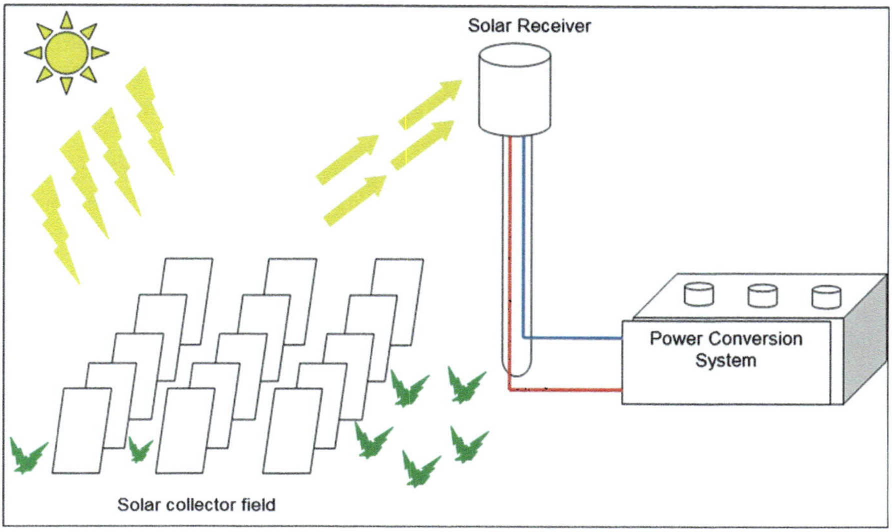

Fig. 65.3 Principle of concentration of a typical solar power tower

Table 65.1 Central receiver solar power tower

Name	Date of creation	Power	Receiver type, HTF
EURELIOS plant	1981 (Italy)	1 MW	Steam receiver
SUNSHINE	1981 (Japan)	1 MW	Steam receiver
Thémis	1982 (France)	2–2.5 MW	450 °C molten salt receiver
Solar One	1982 (USA)	10 MW	516 °C steam receiver
CESA-1	1983 (Spain)	1 MW	525 °C steam receiver
SPP-5	1986 (Ukraine)	5 MW	250 °C steam receiver
Planta Solar 10	2007 (Spain)	11 MW	257 °C steam receiver
Jülich Solar Tower	2008 (Germany)	1.5 MW	700 °C air receiver
Planta Solar 20	2009 (Spain)	20 MW	257 °C steam receiver
Luz Power Tower	2009 (Israel)	4–6 MW	550 °C steam receiver
Sierra SunTower	2009 (USA)	5 MW	440 °C steam receiver

A summary of the main previous work results dealing with receivers design, experiments, and improvements is presented in Table 65.1.

65.4 Modeling of the Power Plant

65.4.1 Power Plant Design

As mentioned above, the power plant consists of three main subsystems, the heliostat field, the receiver, and the power conversion subsystem.

The heliostat field of the presently selected power plant is assumed to be made of 100 mirrors with each one having an area of 100 m^2. The power conversion bloc is a Rankine type. It is mainly composed of a heat recovery steam generator (HRSG), a steam turbine, and a condenser. A deaerator is also included to improve steam cycle performance.

For the case of the receiver, we have selected two types, i.e., volumetric air and water/steam receivers. More details about this subsystem will be presented in the sections below.

65.4.2 TRNSYS Modeling

The modeling of the system is carried out using TRNSYS software. A model of the present system is represented in Fig. 65.4 for the case of volumetric receiver and Fig. 65.5 for the case of water/steam receiver.

TRNSYS is a software platform that enables the user to model different transient systems using modular components. A component reads in a text-based input file and provides output through the solution of algebraic or differential equations. Components include weather data, solar thermal collectors, heat exchangers, power conversion cycles, etc.

As it is illustrated in Figs. 65.4 and 65.5, the weather data is linked to heliostat field by radiation transfer. The field in its turn linked to solar receiver by fluid transfer. The weather data reading and processing by using the component "weather data," the heliostat field efficiency model "FEffMatx," and the central receiver "air

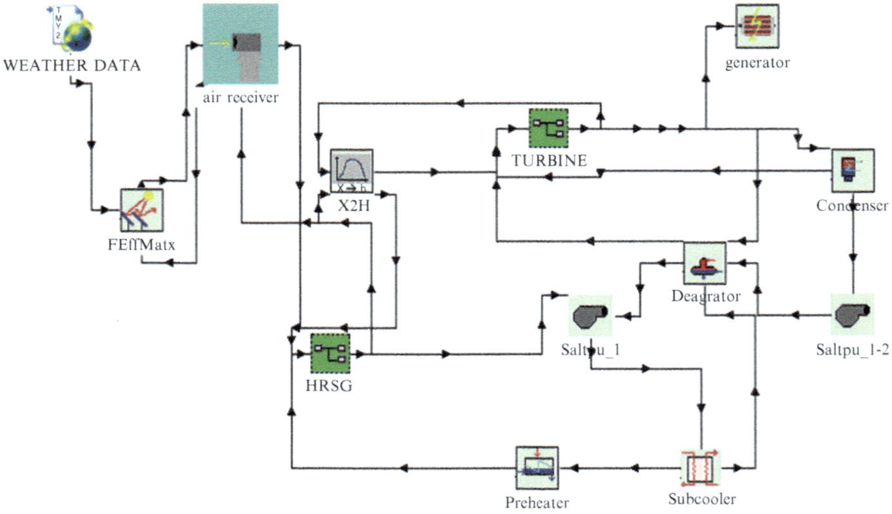

Fig. 65.4 TRNSYS modeling of the solar tower with a volumetric receiver

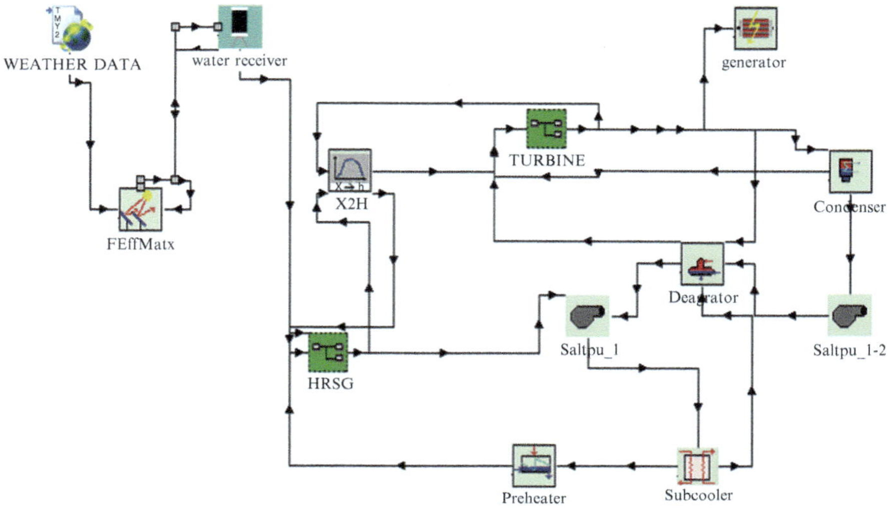

Fig. 65.5 TRNSYS modeling of the solar tower with a water/steam receiver

Table 65.2 Heliostat field input parameters

Input data	Value
Mirror number	100
Mirror area	100 m^2
Tower height	80 m
Reflectivity	0.85

receiver in the first case and water receiver the second case." The links between these components are the direct normal radiation from the data reader to the field efficiency model, the solar zenith and azimuth angle from the radiation processor to the field efficiency model, and the power from the field efficiency model to the central receiver. The radiation processor additionally requires radiation data to evaluate azimuth and elevation angle.

The heat recovery steam generator (HRSG) is linked to the receiver. The HRSG section consists of the economizer heat exchanger, the evaporator heat exchanger, and the superheater heat exchanger. The link between the components is the heat transfer fluid (HTF) which is circulated through the receiver to the steam generator.

In the turbine model, steam quality and enthalpy are used instead of temperature. These quantities are also connected in the flow direction. A steam turbine with two extraction lines is selected. The final turbine stage is connected to the condenser. The two extraction splitters are linked to the deaerator and the feedwater heater. A condensing preheater and a subcooler and a feedwater pump are part of the system.

The system has been set to work under the conditions reported in Tables 65.2 and 65.3.

Table 65.3 Receiver input parameters

Input data	Value
Optical efficiency	0.9
Emissivity of absorber	0.8
Receiver aperture	10 m^2
Design inlet pressure	15 Bar
Design inlet temperature	450 °C

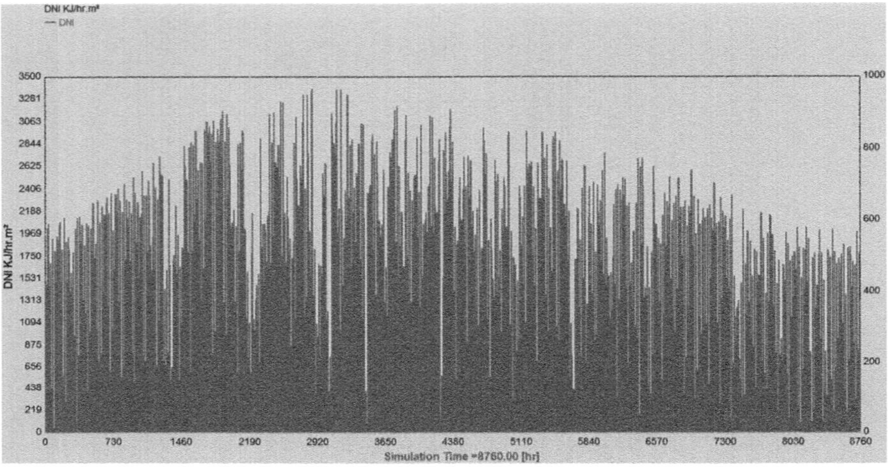

Fig. 65.6 Hourly direct solar irradiation at the selected site, Tamanrasset

65.5 Simulation and Results

65.5.1 *The Direct Normal Irradiation (DNI)*

From TRNSYS layout, the hourly DNI reaches the heliostats at the chosen site over a period of 1 year.

Figure 65.6 illustrated the hourly DNI at the selected site. It is observed that the average solar radiation that can be absorbed by the receiver is about 2,500 W/m^2. It can reach the values of 3,300–3,400 W/m^2 in summer (2,300–3,000 h). The variation of solar radiation at the selected location is not very large; thus, it allows stable operation of the power conversion system.

65.5.2 *Comparative Performance of Volumetric Air and Water/Steam Technologies*

From the thermodynamic point of view, the higher the operating temperature, the higher the performance of the power plant. In Fig. 65.7, it is represented the outlet

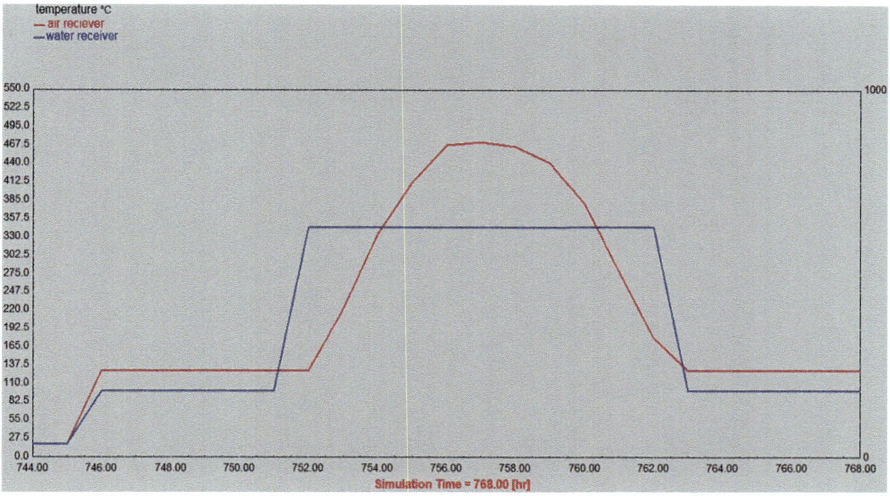

Fig. 65.7 Receiver outlet temperatures during the selected day

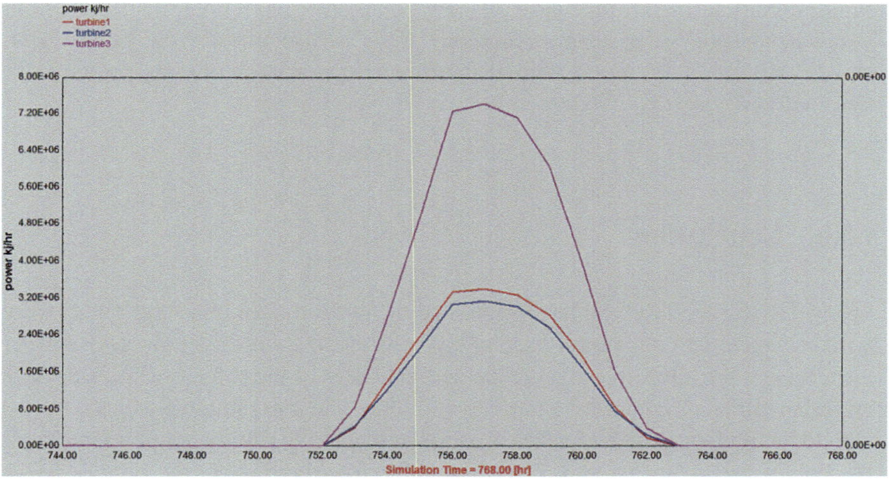

Fig. 65.8 Graphic variation in power turbines for the first day of February, "air case"

temperature of the receiver is reported. The blue color represents the operating temperature of the water/steam receiver, while the red color illustrates the volumetric air receiver outlet temperature.

From sunrise to sunset of the selected day, air receiver offers higher temperatures and therefore better performance for the power conversion system. At midday, the volumetric air can reach up to 470 °C while the water/steam receiver provides steam around 370 °C.

The effect of receiver outlet temperature on the performance of the power conversion system (steam turbine) is illustrated in Fig. 65.8. The output of the

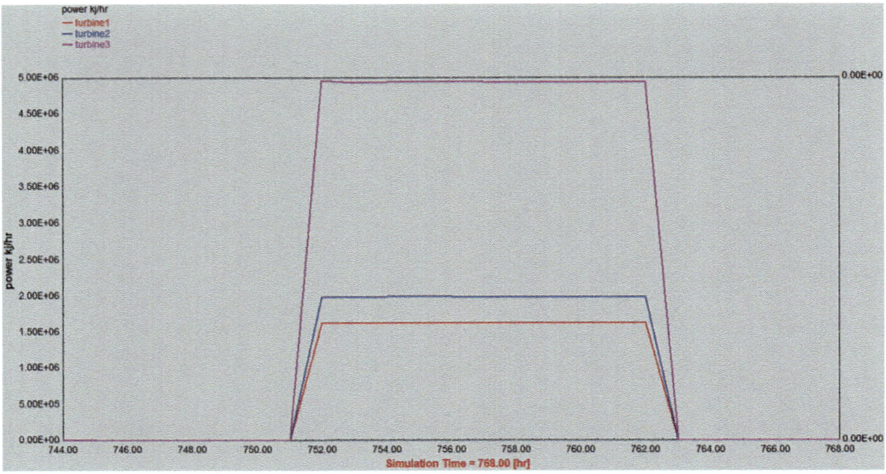

Fig. 65.9 Graphic variation in power turbines for the first day of February, "water case"

high-pressure turbine for the case of using volumetric air technology is higher than that when using water/steam technology. This is also true for the other turbine stages. Moreover, the more normal the direct irradiation, the more preferred the volumetric receiver (Fig. 65.9).

65.6 Conclusion

Using TRNSYS-STEC, we have simulated the performance of a solar power tower with two receiver configurations, namely, a water/steam receiver and a volumetric air receiver. We have investigated the performances of the solar power tower under Algerian climate in order to determine and select the most suitable technology that should be implanted in the future solar power plants. The study has pointed out that the volumetric air technology is more suitable for Algeria than the state-of-the-art water/steam technology when it is coupled to a Rankine cycle. The simulation of annual performance has shown that the overall efficiency and the solar electricity ratio could reach 32 % and 85 %, respectively. Moreover, such a hybrid concept is capable of saving about 2,000 T/year of fossil fuel.

Algeria is making concerted efforts to harness its renewable energy potential despite being a hydrocarbon-rich nation; Algeria's renewable energy program is one of the most comprehensive in the MENA region, and the concerned authorities are determined to secure investments and reliable technology partners for the ongoing and the upcoming projects.

References

1. Pitz-Paal R, Dersch J, Milow B (2005) European Concentrated Solar Thermal Road-Mapping (ECOSTAR): roadmap document. SES6-CT-2003-502578. http://www.promes.cnrs.fr/uploads/pdfs/ecostar/ECOSTAR.Summary.pdf
2. Kalogirou SA (2013) Solar thermoelectric power generation in Cyprus: selection of the best system. Renew Energy 49:278–281
3. Hennecke K, Schwarzbözl P, Hoffschmidt B, Göttsche J, Koll G, Beuter M, Hartz T (2007) The solar power tower jülich1a solar thermal power plant for test and demonstration of air receiver technology. In: Proceedings of ISES Solar World Congress: solar energy and human settlement, vol 5, pp 1749–1953
4. Schmitz M, Schwarzbözl P, Buck R, Pitz-Paal R (2006) Assessment of the potential improvement due to multiple apertures in central receiver systems with secondary concentrators. Sol Energy 80:111–120
5. Coelho B, Schwarzbozl P, Oliveira A, Mendes A (2012) Biomass and central receiver system (CRS) hybridization: volumetric air CRS and integration of a biomass waste direct burning boiler on steam cycle. Sol Energy 86:2912–2922
6. Barigozzi G, Bonetti G, Franchini G, Perdichizzi A, Ravelli S (2012) Thermal performance prediction of a solar hybrid gas turbine. Sol Energy 86:2116–2127
7. Zoschak RJ, Wu SF (1975) Studies of the direct input of solar energy to a fossil-fueled central station steam power plant. Sol Energy 17:297–305
8. Qiang Y, Wang Z, Ershu X, Li X, Guo M (2012) Modeling and dynamic simulation of the collector and receiver system of 1 MWe DAHAN solar thermal power tower plant. Renew Energy 43:18–29
9. Ershu X, Qiang Y, Wang Z, Yang C (2001) Modeling and simulation of 1 MW DAHAN solar thermal power tower plant. Renew Energy 136:848–857
10. Yao Z, Wang Z, Zhenwu L, Wei X (2009) Modeling and simulation of the pioneer 1 MW solar thermal central receiver system in China. Renew Energy 34:2437–2446
11. Yao Z, Wang Z, Kang Y-H, Kim J-K, Wei X (2007) Theoretical simulation investigation of the pioneer 1 MW solar power tower system in china. In: Proceedings of ISES Solar World Congress: solar energy and human settlement, vol 5, pp 1733–1937
12. Moon M-H, Kim Y, Kang K-M, Ko JH, Seo TB (2007) System performance estimation for a solar tower power plant. In: Proceedings of ISES Solar World Congress 2007: solar energy and human settlement, vol 5, pp 1899–1903
13. Ahlbrink N, Belhomme B, Pitz-Paal R (2009) Modeling and simulation of a solar tower power plant with open volumetric air receiver. In: Proceedings 7th Modelica Conference 20–22 Sep 2009
14. Garcia P, Ferriere A, Bezian J-J (2008) Codes for solar flux calculation dedicated to central receiver system applications: a comparative review. Sol Energy 82:189–197
15. Schwarzbozl P, Buck R, Sugarmen C, Ring A, Crespo JM, Altwegg P, Enrile J (2006) Solar gas turbine systems: design, cost and perspectives. Sol Energy 80:1231–1240
16. Behar O, Khellaf A, Mohammedi K (2013) A review of studies on central receiver solar thermal power plants. Renew Sustain Energy Rev 23:12–39
17. Boudries R, Dizène R (2008) Potential of hydrogen production in Algeria. Int J Hydrog Energy 33:4476–4487

Chapter 66
Experimental and Theoretical Modeling of 5 MW Offshore Wind Turbine with TLP Platform

M.M. Ettefagh, Mobin Alipour, Yousef Golizadeh Akhlaghi, and Ebrahim Akbari

Abstract The importance of substituting the conventional fossil fuels by clean energies exhorts the scientists and specialists of this major to present new and effective methods for this substitution. These new methods must have enough capability for utilizing the clean energies more than the previous operated methods. The main goal of this research is based on experiencing the experimental ways of floating the tension leg platform (TLP) and analyzing the dynamic responses of this structure. For realization of this goal, 1/100 scaled model of the real structure was designed and during the test process several probable defects were assumed. Consequently, results of each step were obtained and recorded. In addition, for analyzing the responses, diagrams of power spectral density (PSD) of each phase of the test were derived. Furthermore, the 2-D mathematical model of this structure was considered and the results were used for validating the experimental responses.

Keywords Wind energy • Offshore wind turbines • Experimental modeling

Nomenclature

A	Area of cross section of TL
C_D	Drag coefficient
C_L	Lift coefficient
E	Young's modulus of TL
$F_{1,3,5}$	Wave force
F_B	Buoyancy force

M.M. Ettefagh • M. Alipour (✉) • Y.G. Akhlaghi • E. Akbari
Department of Mechanical Engineering, Faculty of Mechanical Engineering, University of Tabriz, 29 Bahman Blvd., Tabriz, Iran
e-mail: ettefagh@tabriz.ac.ir; alipour.mobin@metu.edu.tr; golizadehakhla1@itu.edu.tr; ebrahim.akbari91@gmail.com

I. Dincer et al. (eds.), *Progress in Clean Energy, Volume 2*,
DOI 10.1007/978-3-319-17031-2_66

F_w	Aerodynamic rotor trust
L_0	Initial length
L_t	Initial lengths of tension legs
ΔL_i	Change in lengths
M	Mass of the platform
T_0	Total initial balance
T_i	Instantaneous tension
ΔT_i	Change in tension
V	Speed
W	Total gravitational force of the system
X_B	Center of buoyancy
X_g	Center of gravity

Greek Letters

ρ	Density kg/m^3
$\eta_{1,3,5}$	Surge heave and pitch displacement
σ'	Local solidity

Abbreviations

GWEC	Global Wind Energy Council
NREL/MIT	National Renewable Energy Laboratory/Massachusetts Institute of Technology
TLP	Tension leg platform

66.1 Introduction

By increasing the worldwide demand of energy and electricity, utilization of the operation methods of fossil fuels has been increased since past decades. By considering the point that the fossil fuels will be finished in the near future and may have irreparable damages for our environment, operating and substituting the renewable form of energy sources is the crucial issue. Operating wind energy by wind turbines is one of the effective methods in this substitution process. According to the statistics that is published by the GWEC up to 2013, 282,482 MW wind energy has been operated in the world and the share of Turkey in this operation is 2,312 MW. The noise pollutions that could be produced by the wind turbines and harmful effects on the weather conditions may occur after long times of operation, and motivate the scientist to analyze the other possible environmental areas for operation. Because of these cited problems, global tendency to utilize the

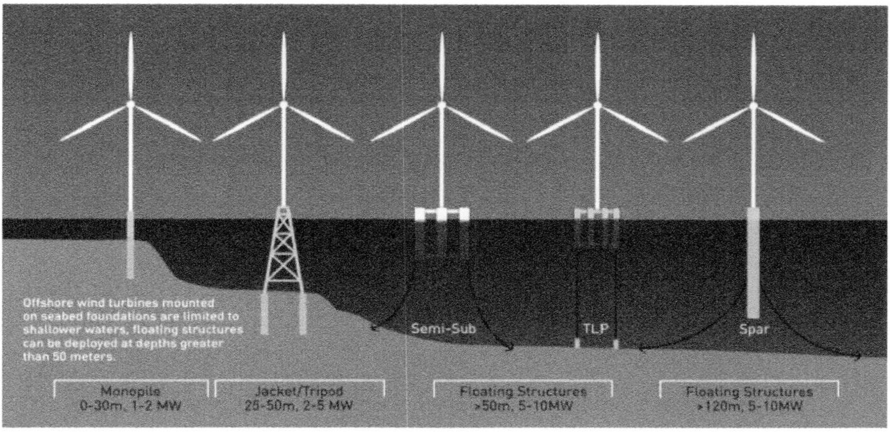

Fig. 66.1 Different types of floating wind turbines [1]

wind energies in the offshore areas has been increased since the last decade. Besides eliminating the onshore wind farm's listed problems, the capacity of the wind energy in the offshore areas is more than the onshore areas. For the depth of less than 60 m, specially for the depth of 20–50 m, Monopile and Jacket/Tripod platforms can be used, but by considering the point that the large capacity of the wind energy of offshore areas is in the depth of above the 60 m, the necessity of designing the floating platforms is significant. In Fig. 66.1 you can see the various types of fixed and floating platforms [1].

Designing and producing the experimental scaled model of the giant structures is one of the effective methods for gaining and analyzing the similar response with the responses of the real structure. Up to now a few similar works for modeling of the offshore wind turbines have been done. Some of these tests were supported by the industrial companies which finally have led to real structures. In 2010 the researchers of the Marin University designed the 1/50 scaled model for each three types of the floating platforms [2]. In 2011 Nicolas Psichogios [3] designed the 1/200 scaled model of the 5 MW NRELL/MIT wind turbine at the DTU University. In 2012 Seyed Kazim Knagvi [4] designed the 1/100 scaled model with the similar method in the WPI University. Also the 1/67 experimental scaled model of the offshore wind turbine, which had been designed and produced by Principle Power Company, was modeled by Roddier [5] in the California Berkeley. As it is mentioned in abstract, the purpose of this research is studying the dynamical behavior of the offshore wind turbine and the process of experimental modeling of it. Also, in this chapter, the responses and behaviors of the structure with and without the probable defects are shown. By applying a poor displacement in the x and y directions separately and finally by applying the wind power which speed is 2.8 m/s in the water depth of 65 cm, the response has been recorded and test has been done. In addition for analyzing the responses, the 2-D mathematical model's responses will be used to validate the test.

66.2 Theoretical and Numerical Modeling

66.2.1 Mathematical Modeling

In this research we used the 2-D mathematical model for analyzing the responses of structure. The model has been used by Ramachandran [6] previously and we have adopted the test situation like water depth, constant, or coefficient of used materials in model and scaled structure's dimensions to mathematical model. Finally results were derived by using MATLAB software. By considering approximately 10 cm as the initial displacement in x direction the obtained frequency will be 0.525 Hz based on this code for our scaled model:

$$m_1\ddot{\eta}_{1\mathrm{f}} = F_1 + k_{\mathrm{t}}\eta_{1\mathrm{t}} + c_{\mathrm{t}}\dot{\eta}_{1\mathrm{t}} - T_1\left(\frac{\eta_1 + \tilde{d}(1-\cos\eta_5)}{L_1}\right) - T_2\left(\frac{\eta_1 + \tilde{d}(1-\cos\eta_5)}{L_2}\right) \tag{66.1}$$

$$m_{\mathrm{t}}\ddot{\eta}_{1\mathrm{t}} = F_{\mathrm{w}} - k_{\mathrm{t}}\eta_{1\mathrm{t}} - c_{\mathrm{t}}\dot{\eta}_{1\mathrm{t}} - m_{\mathrm{t}}\ddot{\eta}_{1\mathrm{f}} \tag{66.2}$$

$$m_3\ddot{\eta}_3 = F_3 + T_0 - \rho g\frac{\pi}{4}D_{\mathrm{t}}^2\eta_3 - T_1\left(\frac{L_0+\eta_3+\tilde{d}\sin\eta_5}{L_1}\right) - T_2\left(\frac{L_0+\eta_3-\tilde{d}\sin\eta_5}{L_2}\right) \tag{66.3}$$

$$\begin{aligned}\ddot{\eta}_5 = {} & F_5 + Wx_{\mathrm{g}} - F_{\mathrm{b}}x_{\mathrm{b}} + F_{\mathrm{w}}(H_{\mathrm{t}}+h)\cos\eta_5 + T_2\left(\frac{L_0+\eta_3-\tilde{d}\sin\eta_5}{L_2}\right)\tilde{d}\cos\eta_5 \\ & + T_2\left(\frac{\eta_1-\tilde{d}(1-\cos\eta_5)}{L_2}\right)\tilde{d}\sin\eta_5 - T_1\left(\frac{L_0+\eta_3+\tilde{d}\sin\eta_5}{L_1}\right)\tilde{d}\cos\eta_5 \\ & - T_1\left(\frac{\eta_1-\tilde{d}(1-\cos\eta_5)}{L_1}\right)\tilde{d}\sin\eta_5\end{aligned} \tag{66.4}$$

$$\mathrm{d}F_1 = C_{\mathrm{m}}\rho A(\dot{u} - \ddot{\eta}_1)\mathrm{d}z + \rho A\dot{u}\mathrm{d}z + C_{\mathrm{D}}\frac{1}{2}\rho s|u - \dot{\eta}_1|(u - \dot{\eta}_1)\mathrm{d}z \tag{66.5}$$

$$F_3 = \frac{\pi}{4}s^2(p_{\mathrm{b}} - p_{\mathrm{t}}) + \frac{\pi}{8}s^2\rho C_{\mathrm{DZ}}|w - \dot{\eta}_3|(w - \dot{\eta}_3) + \frac{4}{3}\pi\left(\frac{s}{2}\right)^3\rho C_{\mathrm{mz}}(\dot{w} - \ddot{\eta}_3) \tag{66.6}$$

$$\mathrm{d}F_5 = C_{\mathrm{m}}\rho A(\dot{u} - \ddot{\eta}_1)z\mathrm{d}z + \rho A\dot{u}z\mathrm{d}z + C_{\mathrm{D}}\frac{1}{2}\rho s|u - \dot{\eta}_1|(u - \dot{\eta}_1)z\mathrm{d}z \tag{66.7}$$

It must be mentioned that $\widetilde{d} = (s + d/2)$ and F_1, F_3, and F_5 are wave forces in three possible directions (surge, heave, and pitch). Also F_W is wind force and η is displacement in each direction. Based on our test situation and our facilities we could not produce wave for our test, so the wave forces will be considered as zero. Figure 66.2 shows the schematic figure of the structure and applied forces to the TLP configuration [6] (Fig. 66.3).

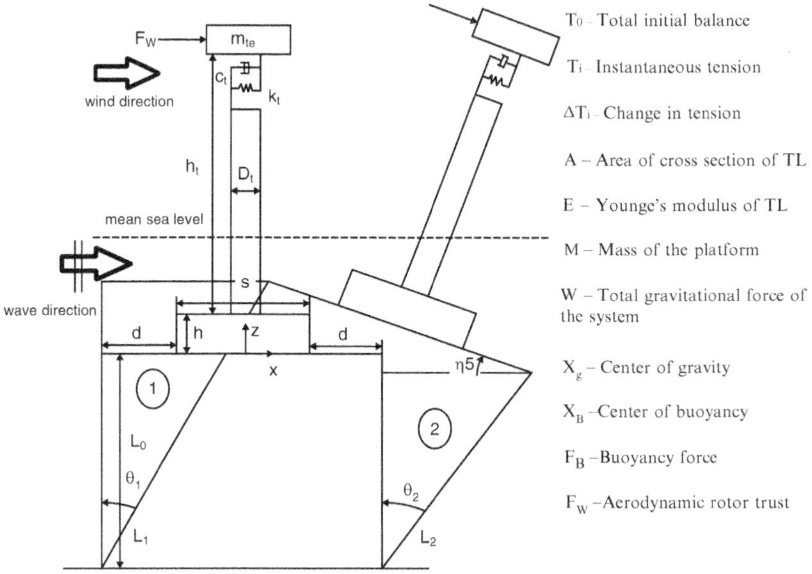

Fig. 66.2 TLP configuration (schematic) [6]

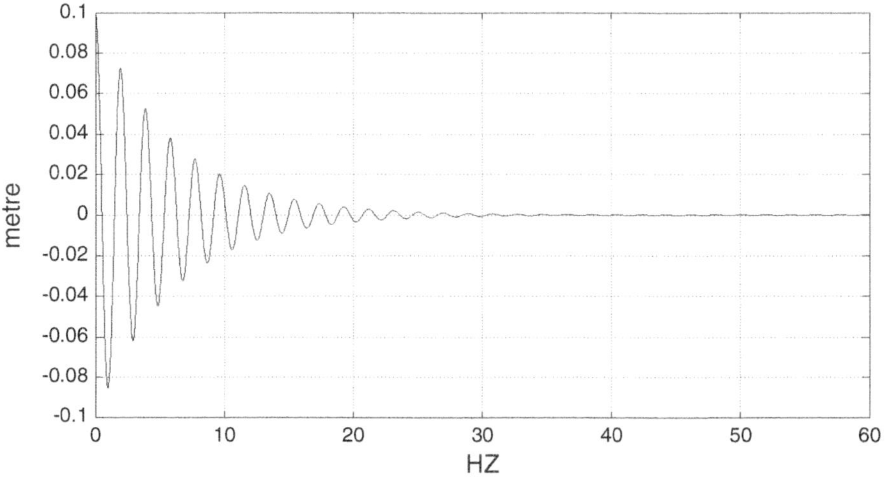

Fig. 66.3 The response of mathematical model by considering the experimental model's dimension and test conditions

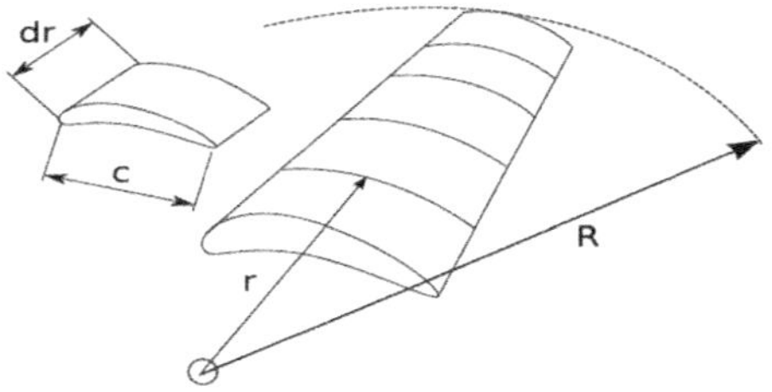

Fig. 66.4 The blade element model [7]

66.2.2 BEM Method

For calculating the wind trust force, one of the famous methods is blade element momentum (BEM). In this method blade is divided into some elements and each element experiences the different flow due to different rotational speed as shown in (Fig. 66.4). By using 10–20 elements and integrating in the blade length direction, the total thrust force can be calculated [7, 8]:

$$\mathrm{d}Fx = \sigma' \pi P = \frac{V^2(1-a)^2}{\cos^2\beta}(C_\mathrm{L}\sin\beta + C_\mathrm{D}\cos\beta)r\mathrm{d}r \tag{66.8}$$

$$\mathrm{d}T = \sigma' \pi p \frac{V^2(1-a)^2}{\cos^2\beta}(C_\mathrm{L}\cos\beta - C_\mathrm{D}\sin\beta)r^2\mathrm{d}r \tag{66.9}$$

where

$$a = \frac{V_1 - V_2}{V_1}$$

$$w = \Omega r \qquad C = \text{chord}$$

$$a' = \frac{w}{2\Omega} \qquad B = \text{number of blades}$$

$$\tan\beta = \frac{\Omega r(1+a')}{V(1-a)}$$

For using the unknown parameters related to the blade specifications, reference of [9] was being used during the code operation. In addition, for calculating the C_d and C_l of blades, related drag and lift coefficient diagrams of cited reference were used. By using the mentioned code which was written based on BEM theory in the FORTRAN software, thrust force of 5 MW NREL prototype size wind turbine was calculated to be 745,291 N. All of the numbers and coefficients are used according to Table 66.1 and with several trial and errors. In this calculation the average speed of wind is assumed to be 11.2 m/s by which the highest efficiency of the turbine could happen.

Table 66.1 Properties of turbine blades' elements

Rows	Radius r (m)	Chord C (m)	Aerofoil inlet angle $\gamma(°)$	Incidence $i(°)$	Relative flow angle onto blades $\beta(°)$	Axial induction factor a	Angular induction factor $a' \times 10^{-2}$	Local solidity σ'	Lift coefficient C_L	Drag coefficient C_D
1	2.867	3.542	72.69	–	42.00	0	–	0.5900	0	0.50000
2	5.600	3.854	76.69	–	51.61	0	–	0.3280	0	0.50000
3	8.333	4.167	76.69	–	58.92	0	–	0.2390	0	0.35000
4	11.75	3.542	76.69	16.78	59.91	0.1711	7.4392	0.1439	1.666	0.24670
5	15.85	4.652	78.52	9.68	68.84	0.2648	5.7018	0.1401	1.436	0.01500
6	19.95	4.458	79.84	7.65	72.19	0.2457	3.4894	0.1067	1.200	0.01180
7	24.05	4.249	80.99	6.12	74.87	0.2449	2.4277	0.0843	1.085	0.01010
8	28.15	4.007	82.21	4.97	77.24	0.2643	1.8754	0.0679	1.058	0.00790
9	32.25	3.748	83.46	4.54	78.92	0.2727	1.4610	0.0555	1.018	0.00760
10	36.35	3.502	84.64	4.16	80.47	0.2927	1.1914	0.0460	0.999	0.00720
11	40.45	3.256	85.81	4.11	81.70	0.3129	0.9951	0.0384	0.998	0.00715
12	44.55	3.010	86.87	4.67	82.21	0.2986	0.8091	0.0323	0.980	0.00570
13	48.65	2.764	87.68	4.60	83.07	0.3091	0.6801	0.0271	0.965	0.00560
14	52.75	2.518	88.47	4.90	83.58	0.3128	0.5887	0.0228	1.006	0.00575
15	56.17	2.313	89.14	5.13	84.01	0.3147	0.5187	0.0196	1.024	0.00630
16	58.90	2.086	89.63	5.41	84.22	0.3049	0.4619	0.0169	1.057	0.00710
17	61.63	1.419	89.89	6.11	83.78	0.2069	0.3226	0.0109	1.120	0.00960

66.3 Experimental Modeling

For experimental modeling and calculating the proportional dimensions, the Froude method, which is one of the common methods among the dimensional modeling methods, was selected. According to Table 66.2 all the dimensions and other dynamic and kinematic factors can be modeled by this method. The TLP kind of offshore wind turbine that has been designed by MIT/NRELL [10] in 2006 was chosen as reference for scaling process and by using Froude method [11, 12], 1/100 scaled model dimensions were calculated.

66.3.1 *Design and Construction of the Components of Experimental Model*

One of the most important factors in the modeling process is selecting the most suitable material for model's components. The materials of the components should have the enough durability like water persistence and also have the proper density for adaptation of the center of mass in both theory and experimental models. For achieving this adaptation between theory and experimental models' CM, other factors like added masses and thicknesses of the components are important. Finally the ABS plastic was chosen as the suitable material. Designed model has been made by including nine main parts: (1) TLP tank, (2) tower, (3) nacelle, (4) thrust disk, (5) cover of tank, (6) rods of tension cables, (7) spokes, (8) box of accelerometer,

Table 66.2 Froude scaling method

Parameter	Unit(s)	Scale factor
Length (e.g. displacement, wave height and length)	L	λ
Area	L^2	λ^2
Volume	L^3	λ^3
Density	M/L^3	1
Mass	M	λ^3
Time (e.g. wave period)	T	$\lambda^{0.5}$
Frequency (e.g. rotor rotational speed, structural)	T^{-1}	$\lambda^{-0.5}$
Velocity (e.g. wind speed, wave celerity)	LT^{-1}	$\lambda^{0.5}$
Acceleration	LT^{-2}	1
Force (e.g. wind, wave, structural)	MLT^{-2}	λ^3
Moment (e.g. structural, rotor torque)	ML^2T^{-2}	λ^4
Power	ML^2T^{-3}	$\lambda^{3.5}$
Stress	$ML^{-1}T^{-2}$	λ
Mass moment of inertia	ML^2	λ^5
Area moment of inertia	L^4	λ^4

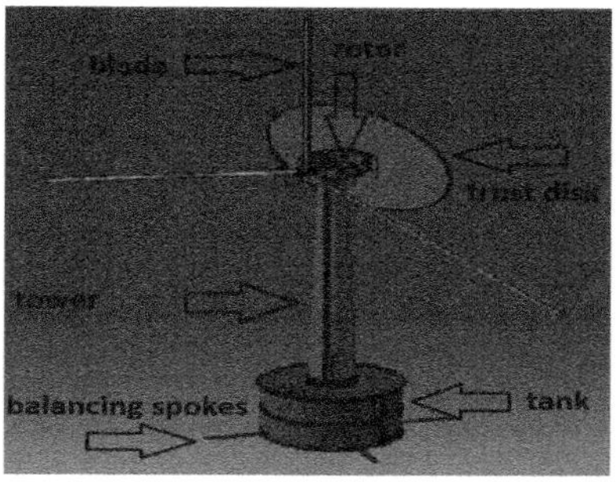

Fig. 66.5 Designed model of TLP and other components in CATIA

and (9) blades and armature. All the components were drawn in the CATIA (Fig. 66.5) and designed according to the dimension of reference [10] as shown in Tables 66.3 and 66.4.

66.3.2 Designing Method of the Thrust Disk

One of the difficulties of modeling process is how to apply the thrust force of the wind to the modeled structure. For applying this force, there are two solutions. First one is modeling the new profile for the blades of 1/100 modeled structure. In this method, for applying the proportional force which must be applied depending on the scaling ratio, the new blade angles must be considered. The second method is applying the force by adding the thrust disk to the modeled structure. In this research the second method was used and the diameter of the disk was calculated by using Eq. (66.8). One of the most important factors in this phase is the amount of the thrust force of the real structure. Kazim Nagvi [3] used 1.7 N which is scaled down amount of the 1.7 MN thrust force of prototype turbine which is reasonable by using the Betz method but in this research the thrust force was calculated by BEM method which is mentioned in the mathematical modeling section of this chapter. The thrust force was applied as 0.745 N for 1/100 model. By assuming the speed 2.8 m/s in the laboratory and the cited thrust force, the diameter of trust disk calculated is 0.4 m.

66.3.3 Choosing the Armature and Blade

By assuming all the degrees of freedoms of tower, blades, and tank, offshore wind turbine could have 24 degrees of freedom (DOF). By considering the components to be rigid except the floating platform, this structure has six DOF. According to the gyroscope effect, having the pitch because of the torque of thrust force and rotating blades, a rotation around the *Z*-axis that is called yaw must be created. In addition because of the created yaw and rotating blades, one rotation around the *x*-axis that is

Table 66.3 Properties of the MIT/NREL TLP [10]

System properties		
Radius	11.00	m
Cylinder height	21.5	m
Concrete ballast height	4.5	m
Steel thickness	0.01	m
Steel mass	176	Metric ton
Concrete mass	4375	Metric ton
Turbine mass	698	Metric ton
System in towing conditions		
Water ballast height	6.55	m
Water mass	2548	Metric ton
Total mass	7798	Metric ton
Towing draft	20.01	
Deck clearance	1.49	
Center of gravity	−11.07	
Center of buoyancy	−10.01	
C55, towing	1.97E+08	N m
System in installed, operating conditions		
Number of tethers	4	
FTethers, total	2.50E+07	N
FTethers, each	6.25E+06	N
FTethers, max	2.38E+07	N
ΔF	3.33E+06	N
FT,3	9.58E+06	N
FT,1	2.92E+06	N
Installed draft	20.01	m
Deck clearance	1.49	m
Total mass	5,249	Metric ton
Buoyant mass	7,797	Metric ton
Reserve buoyancy	48.54 %	
Center of gravity	−9.40	m
Center of buoyancy	−10.01	m

Table 66.4 NREL 5-MW offshore wind turbine

Rotor orientation	Upwind
Control	Variable speed, collective pitch
Rotor diameter/hub diameter	126 m/3 m
Hub height	90 m
Max rotor/generator speed	12.1 rpm/1,173.7 rpm
Maximum tip speed	80 m/s
Overhang/shaft tilt/precone	5 m/5°/−2.5°
Rotor mass	110,000 kg
Nacelle mass	240,000 kg
Tower mass	347,460 kg
Overall c.g. location	$(x,y,z)t = (-0.2, 0, 64)$m

called roll could be created similarly. By using the thrust disk instead of the real blades for creating the mentioned yaw and roll we had to use the gearbox and three blades in the inverse side of the disk. Blades should be selected from the balsa material for having the proportional moment of inertia with the modeling scale ratio. The calculated moment of inertia was $I_m = 0.0048$ kg m^2. It is 10^{-10} times of real moment of inertia in prototype size and it has 0.00015 differences with the inertia of the three blades with balsa material. Also the rotating speed of blades should be 120 rpm which is proportional with the speed of 12 rpm in the real structure.

66.3.4 Test, DATA Acquisition, and Adding Fault

By using two p&k accelerometer sensors located in the center of mass and nacelle, the response of the model in different conditions was recorded. According to Fig. 66.6 that shows the laboratory, the behavior of the model under the effect of aerodynamic force has been studied. Because of the several added mass like armature belonging, wires of accelerometers and glue of the components, achieving the estimated scaled center of mass was difficult. By adding balancer that had 4 kg weight, the center of mass was located in the coordinate of (5, 0, −2), respectively. Free vibrations of model structure were obtained by applying a poor displacement in the x and y directions. Also by applying the wind with 2.8 m/s speed and by adding the one logical defect on the x-direction cable, the behavior of the structure was recorded separately.

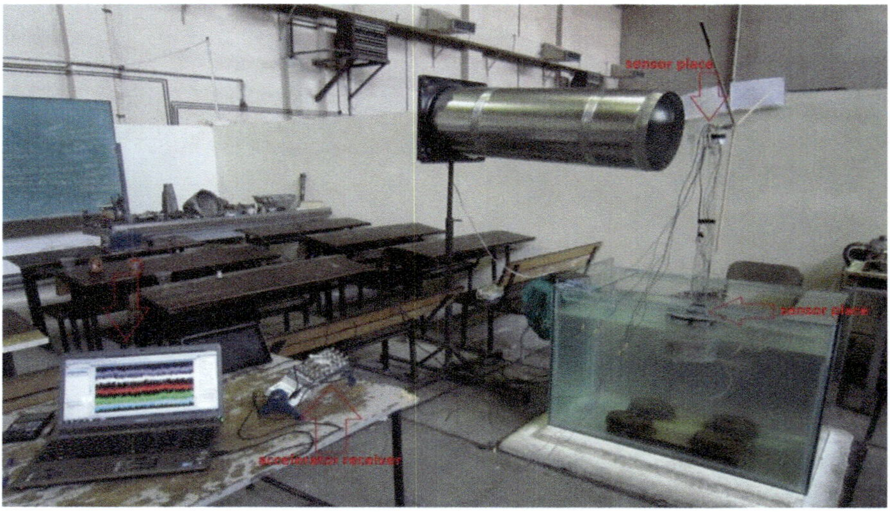

Fig. 66.6 The instruments used in the test, modeled structure, and location of sensors

66.4 Analyzing DATA

66.4.1 Analyzing the Obtained DATA

Because of having noise, the recorded data should be filtered before analyzing. So by using the CHEBYSHEV filter in the low pass the data was filtered. Then the diagrams of power spectral density (PSD) were drawn and the phases of this analysis were shown in the figures. Finally the dominant frequencies of all the power spectral densities of the acceleration diagrams have shown in Table. 66.3 and Figs. 66.7 and 66.8.

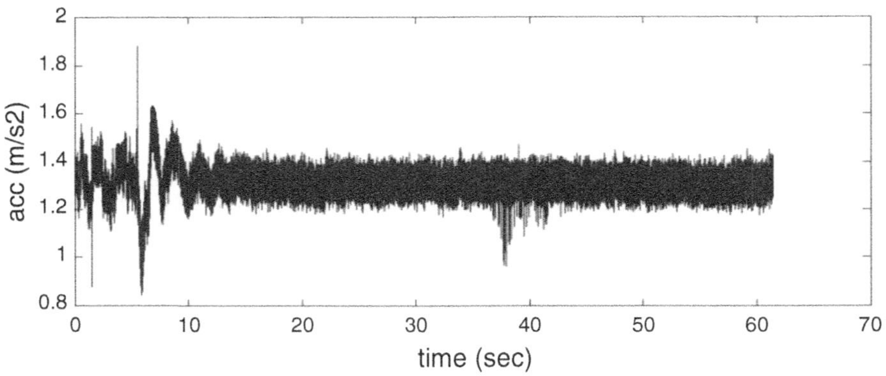

Fig. 66.7 Acceleration-time diagram for floater sensor in x direction before filtering with initial displacement in x direction

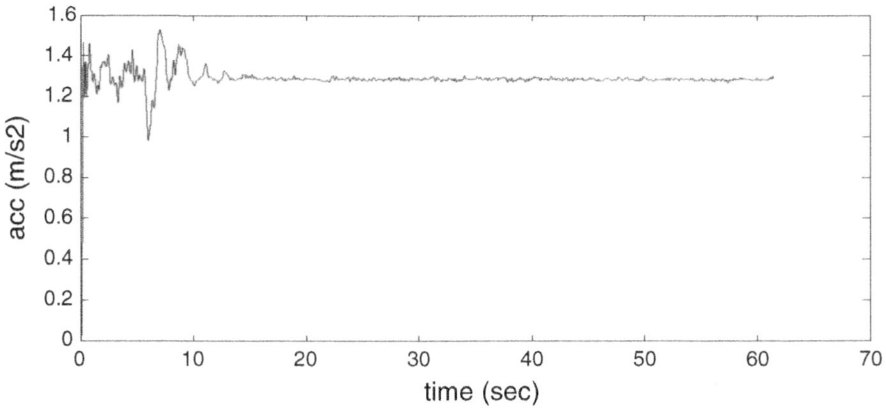

Fig. 66.8 Acceleration-time diagram for floater sensor in x direction after filtering with initial displacement in x direction

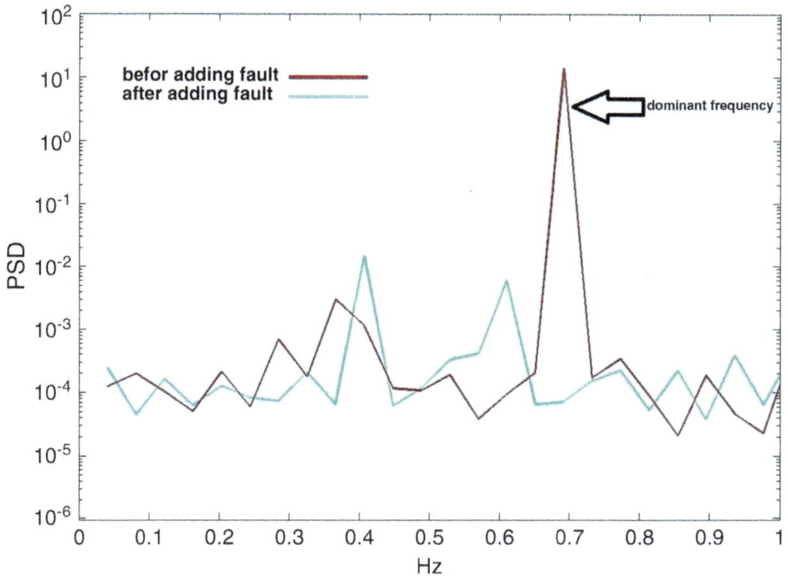

Fig. 66.9 PSD diagram of the x-direction frequency of the model while it was accelerating because of 0.1 m initial displacement in x direction

66.4.2 *Analyzing the Data of Initial Displacement*

By deriving the PSD figure of x direction when the model was being moved with 0.1 initial displacement in x direction, it is obvious that the 0.69 Hz is the main or dominant frequency which has a good validation with mathematical model's result that is 0.525 Hz. After adding a possible fault on the spokes of TLP, the frequency of system and power of that frequency in that direction usually should be reduced. The power of frequencies decreased and the previous main frequency approximately reduced to 0.40 Hz too, which is applicable for us (Fig. 66.9).

66.4.3 *Analyzing the Data of Wind Loading*

The dominant frequency of this model in wind loading condition is approximately 2 Hz. Also adding fault makes the system to have a reduction in its dominant frequency based on Fig. 66.10. The acceleration-time diagram of model under the loading of wind in x direction is shown in Figs. 66.11 and 66.12.

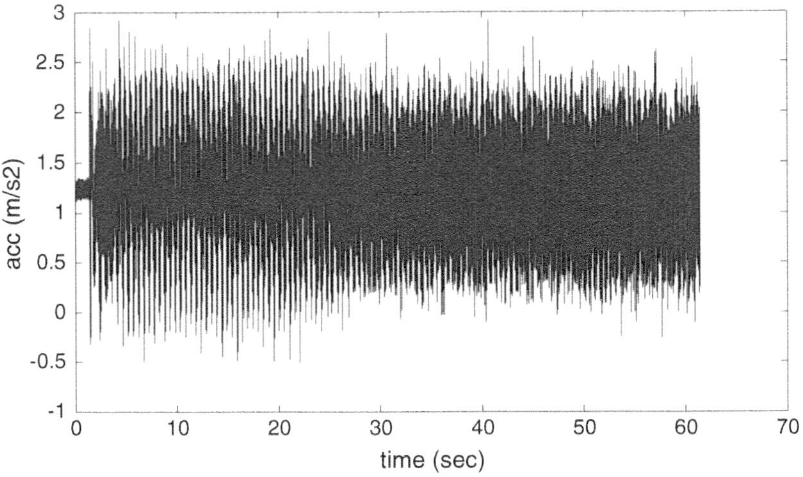

Fig. 66.10 Acceleration-time diagram for floater sensor in x direction before filtering with wind loading in x direction

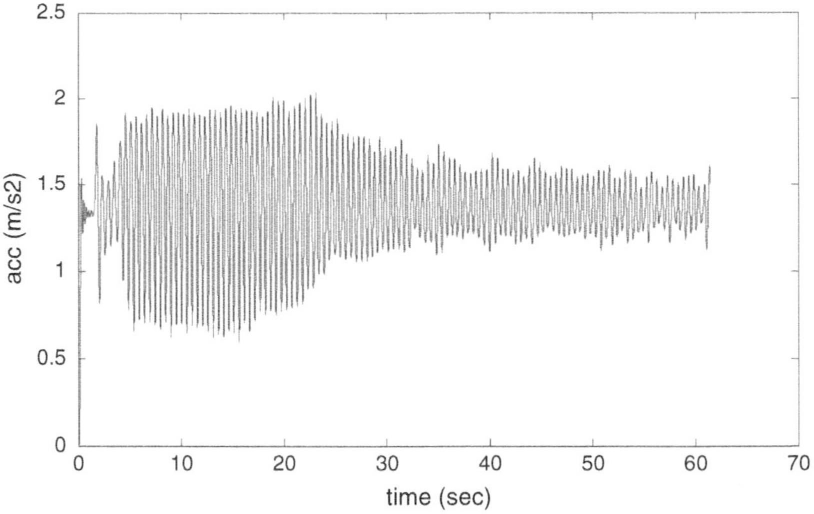

Fig. 66.11 Acceleration-time diagram for floater sensor in x direction after filtering with wind loading in x direction

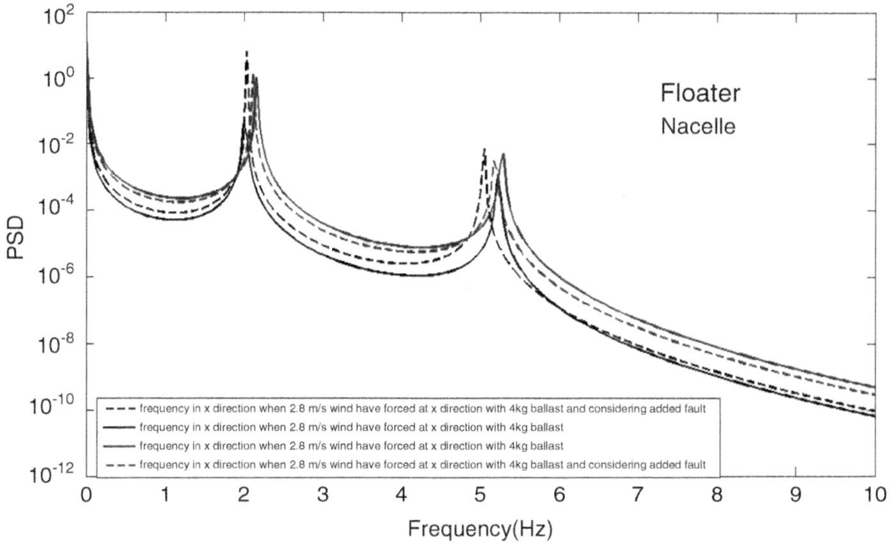

Fig. 66.12 Power spectrum density (PSD) diagram of model in x direction with wind loading

66.5 Results, Discussion, and Conclusion

According to the visual data and results by increasing the balancer weight, the system becomes more stable and has less vibration but floating process of the model with heavy balancer and lower draft is harder than light balancers. Based on the visual data and sensorial data it could be concluded that the sensor which is located in the nacelle has sensed more vibration than floater but after adding the fault this mentioned result is not always correct (based on the results in Table 66.3). Although all the frequencies have shown the reduction in their power, but not in all cases the frequency amount has shown the estimated reduction because of adding fault. In some cases they have been increased which is not logical. This result indicates that using this method may not have a sufficient accuracy and other methods of signal processing like time-frequency and time-series should be used for analyzing this data in further works (Table 66.5).

Table 66.5 The dominant frequency of the structure with 4 kg ballast weight

	Nacelle sensor in the direction of			Floater sensor in the direction of		
Attitude	x	y	z	x	y	z
Obtained data from model						
Free frequency with initial displacement in x direction	0.71	0.24	0.25	0.69	0.37	0.24
Free frequency with initial displacement in y direction	1.46	0.20	0.37	0.69	0.65	0.28
Frequency in the presence of a wind with the speed of 2 m/s	2.20	2.00	2.50	1.98	2.10	2.50
Obtained data from model by considering fault						
Free frequency with initial displacement in x direction	0.31	0.45	0.23	0.41	0.12	0.89
Free frequency with initial displacement in y direction	1.34	0.43	0.94	1.54	0.37	1.38
Frequency in the presence of a wind with the speed of 2 m/s	2.15	2.00	2.55	2.02	5.05	2.70

References

1. Roddier D, Weinstein J (2010) Floating wind turbines, ASME, Mechanical engineering, Focus on offshore technologies. ASME_100401
2. Psichogios N (2011) Design of a hydro-aeroelastic mode scale TLP wind turbine. DTU, Kongens Lyngby
3. Naqvi SK (2012) Scale model experiments on floating offshore wind turbines. WPI
4. Roddier D, Cermelli C, Aubault A, Weinstein (2010) A wind float: a floating foundation for offshore wind turbines. J Renew Sustain Energy 2,033104
5. Ramachandran GKV, Bredmose H, Sørensen JN, Jensen JJ (2011) Response of a TLP floating wind turbine subjected to combined wind and wave loading. Department of Mechanical Engineering, Technical University of Denmark
6. Cermelli C, Aubault A, McCoy T (2010) Qualification of a semi-submersible floating foundation for multi megawatt wind turbines. This paper was prepared for presentation at the Offshore Technology Conference held in Houston, TX, USA, 3–6 May 2010
7. GrantIngram (2011) Wind turbine blade analysis using the blade element momentum method
8. Andersen B (2010) A comparison of two and three bladed floating wind turbine. The University of Toledo
9. Jonkman J, Butterfield S, Musial W, Scott G (2009) Definition of a 5-MW reference wind turbine for offshore system development. Technical report, NREL/TP-500-38060
10. Wayman EN, Sclavounos PD, Butterfield S, Jonkman J, Musial W (2006) Coupled dynamic modeling of floating wind turbine systems. Offshore technology conference, Houston, TX, NREL/CP-500-39481. doi:10.1002/we.347
11. Martin HR (2011) Development of a scale model wind turbine for testing of offshore floating wind turbine systems. The University of Maine
12. Chakrabarti S (2005) Handbook of offshore engineering. Elsevier Ltd, Amsterdam

Chapter 67
Wind Resource Assessment and Micrositting: A Case Study for Turkey

Mustafa Süel and Barış Özerdem

Abstract Due to earth's limited fossil energy resources along with their environmental and health hazards, the usage of renewable energy resources is very crucial. Among all the renewable energy resources, wind is one of the most promising. Wind resource assessment which estimates the strength of wind resources at a planned site and micrositting which determines the highest annual energy production for turbines' layout are very important in installing wind farm. The output of wind resource assessment is wind conditions and annual energy production at a site. The output of micrositting is the optimal layout for the highest annual energy production through the turbines. In this paper, wind resource assessment of a site located in Seferihisar-İzmir, Turkey, is accomplished by using wind data collected between September 2008 and June 2010. The micrositting study is done for five different wind turbines. The commercially available softwares WindPRO and WAsP are used for these analyses. As a result, the best alternate is obtained and presented.

Keywords Wind energy • Wind resource assessment • Wind farm • Micrositting

67.1 Introduction

Energy has always been an enormously important parameter that symbolizes the level of economic and social development. Since the beginning of the industrialization era, the main energy source is fossil fuels like coal, natural gas, and petroleum. But the earth's fossil energy resources are very limited. Therefore, fossil fuels will not be able to solely meet the energy demand of the world. Furthermore,

---

M. Süel
M.Sc. Program of Energy and Environmental Management, Bahcesehir University, Çırağan Cad., 34353 Beşiktaş-İstanbul, Turkey
e-mail: mustafa.suel@std.bahcesehir.edu.tr

B. Özerdem (✉)
Department of Energy Systems Engineering, Bahcesehir University, Çırağan Cad., 34353 Beşiktaş-İstanbul, Turkey
e-mail: mehmetbaris.ozerdem@bahcesehir.edu.tr

I. Dincer et al. (eds.), *Progress in Clean Energy, Volume 2*,
DOI 10.1007/978-3-319-17031-2_67

environmental and health hazards of fossil fuel consumption are becoming more apparent, and concerns on this matter are increasing. They are responsible for 50–60 % greenhouse gases released into the atmosphere. This emission is the main cause of global warming and climate change. Hence, a shifting focus toward renewable energy resources occurred in recent years. Many countries have been working on sufficient utilization of renewable energy resources, in addition to efficient utilization of conventional energy resources. European Union's targets for renewable energy in total electricity consumption are 20 % and 30 % in 2020 and 2030, respectively. It is obvious that wind energy has the main portion to achieve this goal. But, on the other hand, currently available renewable energy technologies are far away from replacing fossil fuels' technologies, yet.

Among all the renewable energy resources, wind is one of the world's fastest growing resources. Human efforts to harness wind for energy go back to the ancient times. But those efforts were for sailing ships and boats, grinding grains, and pumping water until the end of the nineteenth century. The first modern wind turbine, specifically designed for electricity generation, was constructed in 1890. The rapid growth in wind power which started after the mid-twentieth century in Europe and the USA is a result of improvements accomplished in power electronics and material science technologies. The world cumulative wind power installed is 318,000 MW at the end of 2013 which corresponds to 12.4 % increase with respect to previous year. Installations completed in 2013 are 35,500 MW. The main contributors to this figure are China and Canada [1].

There is a growing awareness for renewable energy resources such as wind, solar, and geothermal in Turkey as a result of rapidly increasing population and industrial development, as well. Especially wind energy seems to be the most promising one for electricity production among those mentioned above. Turkey's installed wind farm capacity is 2,958.45 MW as end of the year 2013 [2]. It must be noted that, despite Turkey's capacity, the usage of wind energy is still very limited in Turkey. Therefore, the target mentioned in strategical governmental documents for wind energy is 20,000 MW installed capacity until the year 2023 [3].

Wind resource assessment at a planned site is very crucial in terms of installation of an efficiently operating wind farm. Since using meteorological data directly will underestimate the true wind energy potential on site, a properly erected mast should be located on site where characteristic wind parameters of the location can be measured as a representative of the whole site. Main measuring parameters are wind speed, wind direction, temperature, relative moisture, and barometric pressure. After collecting the data, the processing takes place by using proper softwares. These milestones are very important to determine the viability of the project.

In this paper, wind resource assessment of a site located in Seferihisar-İzmir, Turkey, is analyzed by using wind data collected between September 2008 and June 2010. The micrositting study which determines annual energy production for turbines' layout is done for five different wind turbines. The commercially available softwares called WindPRO and WAsP are used for these analyses.

67.2 Literature Review

Many studies have been conducted to estimate the wind resource assessment in different parts of Turkey, in the last two decades. The western coast of Anatolia studies have been done by Incecik and Erdogmus [4], Sen and Sahin [5], Ozerdem and Turkeli [6, 7], and Oner et al. [8]. An island on Aegean Sea called Bozcaada has been investigated by Turksoy [9], Dundar, and Inan [10] for its wind potential. Gokceada which is another island on Aegean Sea has also been studied for the same purpose by Tolun et al. [11] and Eskin et al. [12]. Some other inner locations of Anatolia such as Bursa, Kutahya, and Kayseri have been investigated by Ucar and Balo [13], Ozgur et al. [14], and Genc and Gokcek [15], respectively. Bilgili and Sahin have assessed the southern and southwestern regions of Anatolia in terms of wind energy density [16]. Sahin and Bilgili have also studied southeastern coast of Anatolia in terms of wind characteristics and wind energy potential [17]. Some areas in the eastern Anatolia such as Elazığ have been investigated by Akpınar and Akpınar [18]. As a result of these studies, the western coast line of Anatolia has been found very promising in terms of wind power potential.

67.3 Material and Method

67.3.1 The Site

The site which is located in Seferihisar county is on the southwestern area of İzmir province. The county has 49 km costal line of Aegean Sea. It includes several hills covered with typically Mediterranean bush. There were no obstacles detected around the measurement area. Surface roughness was low due to low plant heights, which is important in wind shear. Figure 67.1 shows the site which has 40 km distance from İzmir province.

67.3.2 The Measuring Station

Measuring the wind on site is the most reliable method, and the energy yield of a wind turbine can only be predicted with statistically validated values of mean wind speed, wind speed distribution, and the vertical wind profile. Statistically reliable values require long-term measurements. Additionally, if the wind speed is plotted over time, indications as to the turbulence intensity can be derived from the wind variations.

Usable data can only be obtained by means of measurements over a certain period and recording of the measured values. This requires a stationary measuring system on a mast with a logging device for the measured data. The sensors of a wind

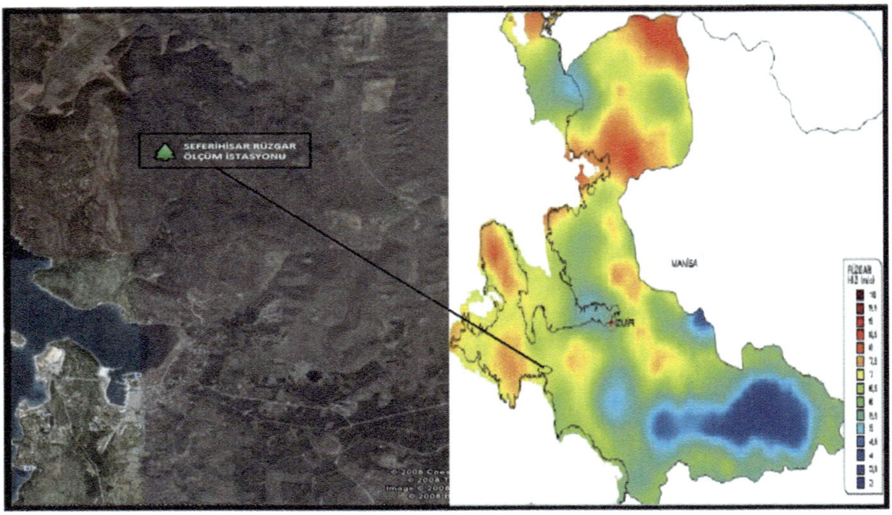

Fig. 67.1 The site

measuring system suitably consist of a combination of an anemometer and a wind vane. For this mast, four anemometers at 50, 65, 78.5, and 80 m heights, two wind vanes at 64.5 and 78 m heights, a temperature-relative humidity sensor at 10 m height, and a barometer at 3 m height are used (Fig. 67.2). The sensors are mounted on top of or on the side arms of a mast or a tower. The data logger mounted on the mast was capable of collecting mean, maximum, and minimum wind speeds, prevailing wind direction, mean temperature, relative humidity, and barometric pressure. The mast is also equipped with a GSM system to make remote communication possible with the station by a computer or mobile phone. A backup power supply is included to minimize chances of data loss caused by power failure, as well. This is accomplished by a solar panel and a battery.

67.3.3 Data Processing

Some softwares, namely, Witerm, WAsP, and WindPRO, are used in this study. First input is the topographic map with 1:25,000 scale. The data for counters are taken from Radar Topography Mission (SRTM) database obtained from the US Geographical Survey. Witerm is used to collect the daily data taken from the measurement station. WAsP is used to prepare wind resource map as an interval module runs under WindPRO. WindPRO which composes many modules is used for energy calculations and micrositting. Main models processed during this study are Basic, Meteo, Model, Optimize, and Park.

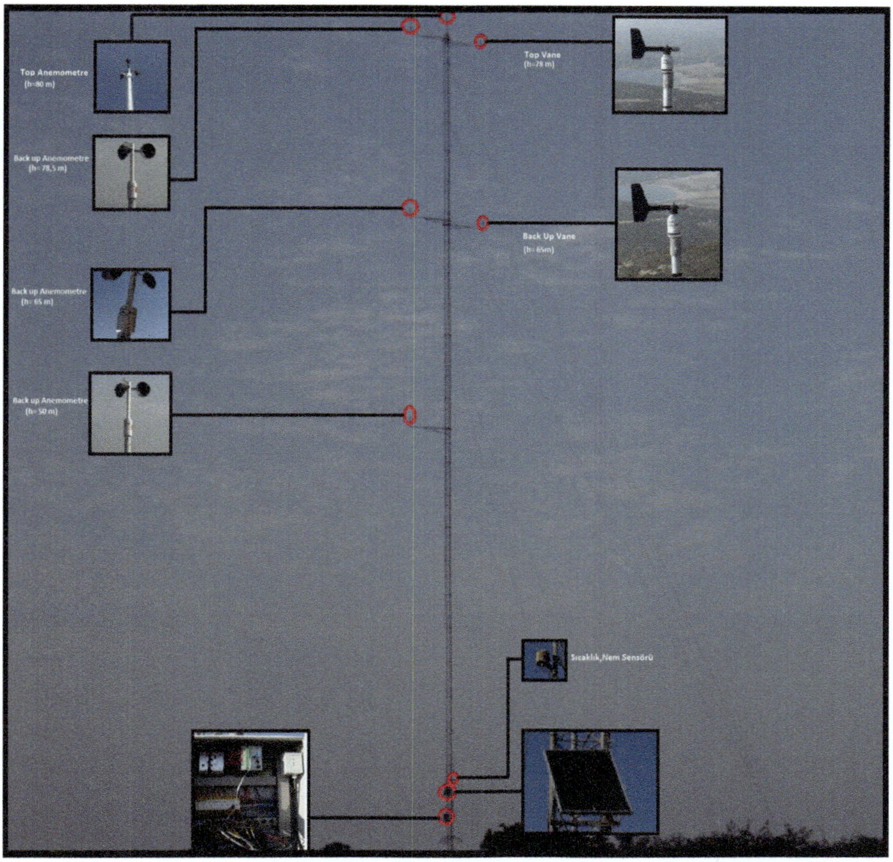

Fig. 67.2 The measurement station

67.4 Results and Discussion

The parameters were measured for 21 months between September 2008 and June 2010. Mean wind speed values were 6.90, 6.88, 6.66, and 6.30 m/s for 80, 78.5, 65, and 50 m heights. Months having the highest values were February in winter and August in summer. Prevailing wind direction was north (N) with 34.86 %. The second wind direction was north–northeast (NNE) with 18.46 %. Average temperature, relative humidity, and barometric pressure values were 15.87 °C, 67.17 %, and 999 hPa, respectively. Figure 67.3 shows the prevailing directions at two different heights. Wind speed and wind power intensity maps were obtained for 80 m height as shown in Figs. 67.4 and 67.5, respectively.

Micrositting is done for each turbine type by using wind resource maps. Distances between the turbines are kept 7 and 3 times of rotor diameters parallel and normal to prevailing wind direction, respectively. Gross energy production

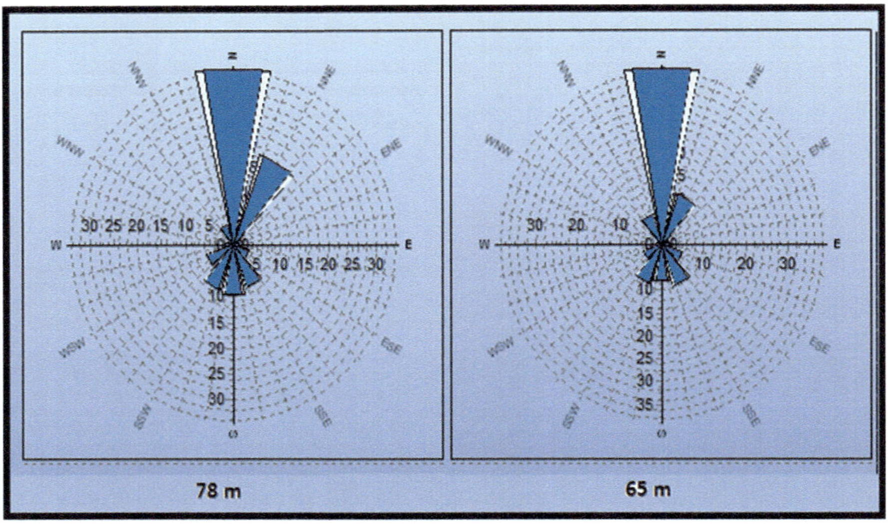

Fig. 67.3 Prevailing wind directions

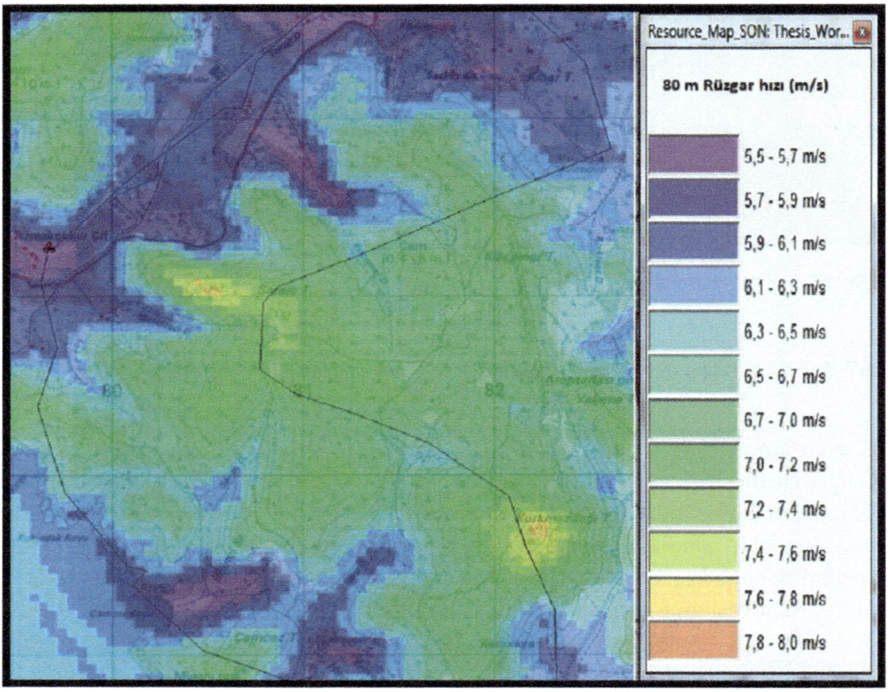

Fig. 67.4 Wind speed map at 80 m height

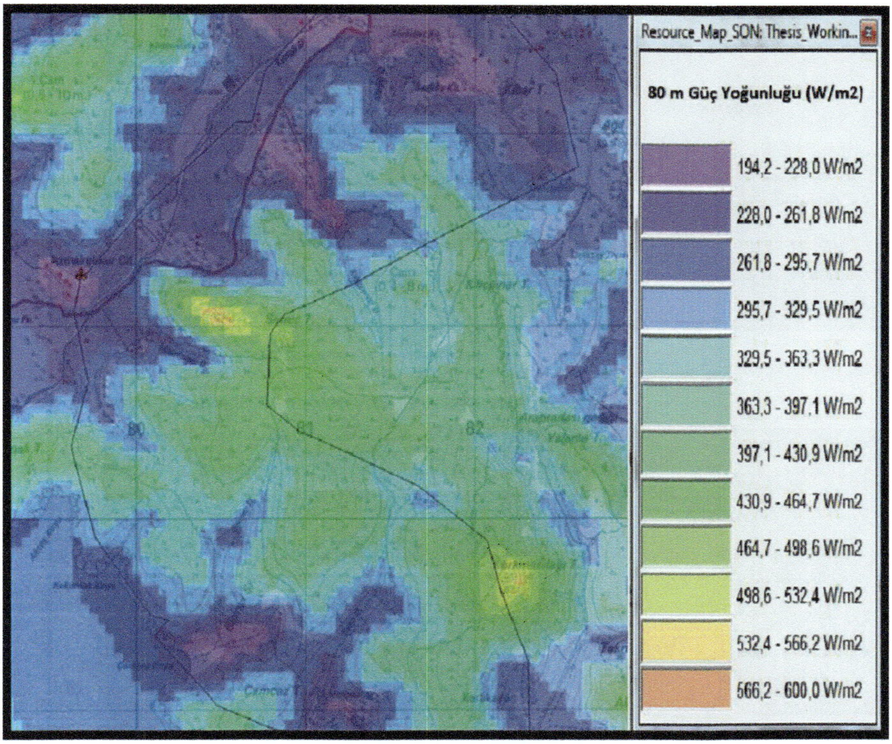

Fig. 67.5 Wind power intensity at 80 m height

predictions and capacity factors are determined by using wake effect model. Uncertainties and other losses are not considered during this study. Park calculations are conducted for each turbine. 2,000 kW turbine with 97 m rotor diameter (IEC Class IIIA), 1,600 kW turbine with 100 m rotor diameter (IEC Class III), 2,400 kW turbine with 117 m rotor diameter (IEC Class III), 2,300 kW turbine with 113 m rotor diameter (IEC Class III), and 1,800 kW turbine with 100 m rotor diameter (IEC Class III) are used for scenarios A, B, C, D, and E, respectively.

Figure 67.6 represents scenario A which corresponds to a wind farm with 17.6 MW rated power having eleven wind turbines with 1,600 MW each. The gross energy production and capacity factor found were 71,594 MWh/year and 46.4 %, respectively. Although the mean wind speed was 7.0 m/s, park efficiency found was 97.2 %.

Figure 67.7 represents scenario B which corresponds to a wind farm with 18 MW rated power having ten wind turbines with 1,800 MW each. The gross energy production and capacity factor found were 68,306 MWh/year and 43.3 %, respectively. Although the mean wind speed was 7.1 m/s, park efficiency found was 97.4 %.

Figure 67.8 represents scenario C which corresponds to a wind farm with 18 MW rated power having nine wind turbines with 2,000 MW each. The gross energy production and capacity factor found were 64,283.4 MWh/year and 40.7 %,

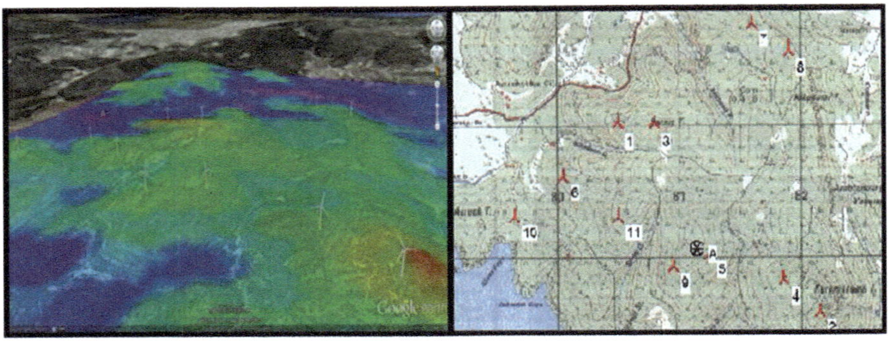

Fig. 67.6 Wind farm layout for scenario A

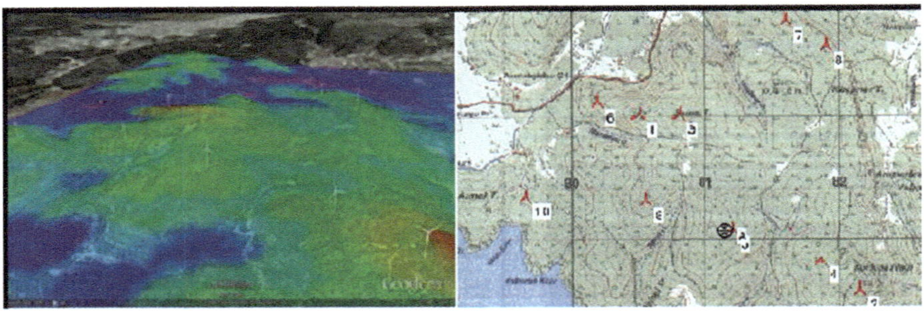

Fig. 67.7 Wind farm layout for scenario B

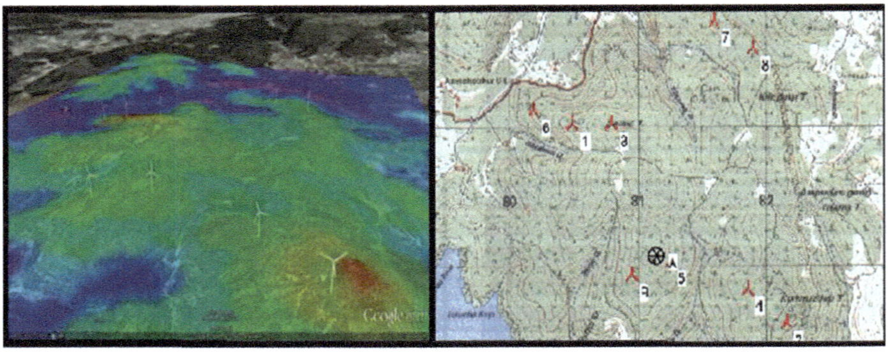

Fig. 67.8 Wind farm layout for scenario C

respectively. Although the mean wind speed was 7.1 m/s, park efficiency found was 97.7 %.

Figure 67.9 represents scenario D which corresponds to a wind farm with 16.1 MW rated power having seven wind turbines with 2,300 MW each. The gross energy production and capacity factor found were 65,673.5 MWh/year and 45.7 %,

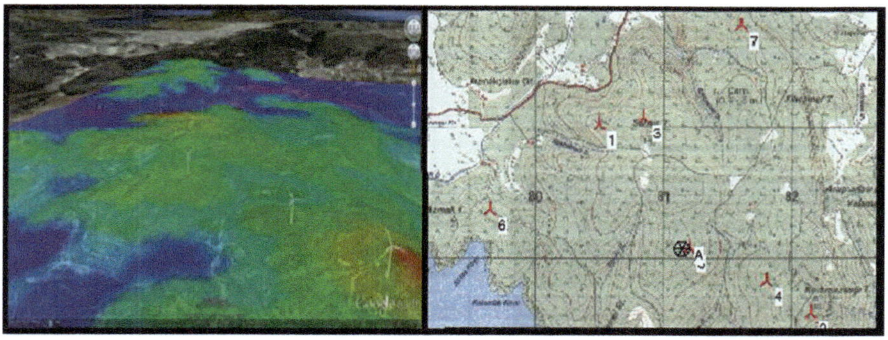

Fig. 67.9 Wind farm layout for scenario D

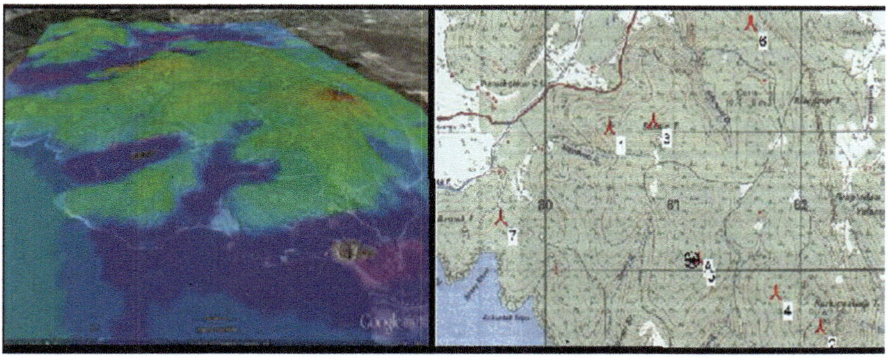

Fig. 67.10 Wind farm layout for scenario E

respectively. Although the mean wind speed was 7.1 m/s, park efficiency found was 98.3 %.

Figure 67.10 represents scenario E which corresponds to a wind farm with 16.8 MW rated power having seven wind turbines with 2,400 MW each. The gross energy production and capacity factor found were 65,673.5 MWh/year and 44.6 %, respectively. Although the mean wind speed was 7.1 m/s, park efficiency found was 98.3 %, as previous scenario.

67.5 Conclusions

The wind analysis done by the data obtained via measurement station looks similar to the data given by Turkey's wind atlas for the same region. Mean wind speed, c, and k parameters found were 6.92 m/s, 7.81 m/s, and 1.997 for the site's Weibull distribution, respectively. According to the park energy calculations, the highest electricity production and capacity factor have occurred for scenario B. When energy consumption per capita in Turkey is taken as 3,199 kWh for 2012, this scenario could have met the energy consumption of 22,380 people, if the energy is produced permanently. This figure corresponds to more than 2/3 of the population of Seferihisar county of Izmir.

Acknowledgment Special thanks to the Ağaoğlu Energy Group for giving permission to use the data in this study.

References

1. Global Wind Energy Council (2012) Global wind report. Global Wind Energy Council, Brussels, Belgium
2. Turkish Wind Energy Association (2014) Statistical report on wind energy of Turkey. Turkish Wind Energy Association, Ankara, Turkey (in Turkish)
3. State Planning Organization (2009) Electrical energy market and supply safety strategical document. State Planning Organization, Ankara, Turkey (in Turkish)
4. Incecik S, Erdogmus F (1995) An investigation of the wind power potential of the western coast of Anatolia. Renew Energy 6:863–865
5. Sen Z, Sahin AD (1997) Regional assessment of wind power in western Turkey by the cumulative semivariogram method. Renew Energy 12:169–177
6. Ozerdem B, Turkeli MH (2003) An investigation of wind characteristics on the campus of Izmir Institute of Technology, Turkey. Renew Energy 28:1013–1027
7. Ozerdem B, Turkeli MH (2005) Wind energy potential and micrositting on Izmir Institute of Technology campus, Turkey. Renew Energy 30:1623–1633
8. Oner Y et al (2009) Prediction wind energy potential using by wind data analysis in Bababurnu-Turkey. Proceedings of international conference on clean electrical power, 9–11 Jun, Capri, Italy, pp 232–235
9. Türksoy F (1995) Investigation of wind power potential at Bozcaada, Turkey. Renew Energy 6:917–923
10. Dundar C, Inan D (1996) Investigation of wind energy application possibilities for a specific island (Bozcaada) in Turkey. Proceedings of world renewable energy conference, 15–21 Jun, Colorado, USA, pp 822–826
11. Tolun S et al (1995) The wind energy potential of Gökçeada in the northern Aegean Sea. Renew Energy 6:679–685
12. Eskin E et al (2008) Wind energy potential of Gokceada Island in Turkey. Renew Sustain Energy Rev 12:839–851
13. Ucar A, Balo F (2009) Investigation of wind characteristics and assessment of wind-generation potentiality in Uludag-Bursa, Turkey. Appl Energy 86:333–339
14. Ozgur MA et al (2009) Statistical evaluation of wind characteristics in Kutahya, Turkey. Energy Sourc Part A 31:1450–1463
15. Genc MS, Gokcek M (2009) Evaluation of wind characteristics and energy potential in Kayseri, Turkey. J Energy Eng 135:33–43
16. Bilgili M, Sahin B (2009) Investigation of wind energy density in the southern and southwestern region of Turkey. J Energy Eng 135:12–20
17. Sahin B, Bilgili M (2009) Wind characteristics and energy potential in Belen-Hatay, Turkey. Int J Green Energy 6:157–172
18. Akpinar EK, Akpinar S (2004) Determination of wind energy potential for Maden-Elazig, Turkey. Energy Convers Manag 45:2901–2914

Chapter 68
Investigation of the Long-Term Resource Variation in Western Coast (Aegean Sea) of Turkey Through Use of MERRA Reanalysis Data

Konstantinos C. Gkarakis

Abstract The onsite measurement periods, typically ranging from 1 to 5 years, are short when compared with the standard period for average climate definition of 20–30 years and the lifetime of a wind farm. The annual variability of the wind regime, if not accounted for, adds to the uncertainty of the site's resource assessment and can lead to serious misevaluations. The use of long-term wind data is an old issue in wind resource assessment. The typical methods of eliminating the annual variations of the wind regime from the average are correlations like the measure–correlate–predict (MCP) technique. However, the availability of such long periods of wind data is not as frequent as desired. At the end of 2009 a new reanalysis dataset named MERRA (Modern Era Retrospective analysis for Research and Applications) was published by NASA's Global Modeling and Assimilation Office/Goddard Space Flight Center. The MERRA analysis is being conducted with the GEOS-5 Atmospheric Data Assimilation System (ADAS). The model grid is 0.5° latitude and 2/3° longitude and temporal resolution of 1 h. The MERRA data considers the orography and roughness of the site, and has a high availability.

In this work, the author analyzes the MERRA data for the level of 50 m above ground level (agl), period 1979–2013, in the wider area of Western Coast (Aegean Sea) of Turkey (17 grid points from the area of Enez to Antalya cities, 26.0°E–30.67°E and 40.50°N–36.0°N) and studies the yearly variability. Also, a correlation study (using MCP techniques) is done between the different grid points of MERRA reanalysis project and their combinations should also be evaluated. Possible relationships between terrain characteristics, local wind systems, and correlation values should also be investigated. The establishment of the existence of a predominant periodicity in the long-term wind variation could provide valuable information on the behavior of the wind climate for the upcoming years.

K.C. Gkarakis (✉)
RES Lab, Department of Energy Technology Engineering, Technological Educational Institute (TEI Athens), 12210 Egaleo, Attica, Greece
e-mail: ape@teiath.gr

I. Dincer et al. (eds.), *Progress in Clean Energy, Volume 2*,
DOI 10.1007/978-3-319-17031-2_68

Keywords Long-term resource variation • Turkey • MERRA • Aegean Sea

68.1 Introduction

The onsite measurement periods, typically ranging from 1 to 5 years, are short when compared with the standard period for average climate definition of 20–30 years and the lifetime of a wind farm. The annual variability of the wind regime, if not accounted for, adds to the uncertainty of the site's resource assessment and can lead to serious misevaluations. The use of long-term wind data is an old issue in wind resource assessment. The typical methods of eliminating the annual variations of the wind regime from the average are correlations like the measure–correlate–predict (MCP) technique. However, the availability of such long periods of wind data is not as frequent as desired.

So, it is common practice to find a suitable source of long-term data in the vicinity of the wind farm site. This will allow a correlation analysis to be undertaken; it is likely to result in an analysis with significantly less uncertainty than that resulting from use of the site data alone. It is important that the used dataset of long-term reference station can be used in an analysis and to have the necessary quality and validity. The use of long-term data is to correct the mean long-term wind speed of the measuring mast (of the short-term period) and to have a prediction of the energy production of the proposed wind farm based on the long-term wind regime at the site.

Mainly for the scope of long-term correction of measured wind speeds and consequently the estimated energy production, the MERRA (Modern Era Retrospective analysis for Research and Applications) reanalysis data is used.

68.2 MERRA Reanalysis Wind Data

At the end of 2009 a new reanalysis dataset named MERRA was published by NASA's Global Modeling and Assimilation Office/Goddard Space Flight Center [1].

The MERRA analysis is being conducted with the GEOS-5 Atmospheric Data Assimilation System (ADAS). The model grid is 0.5° latitude and 2/3° longitude and has a temporal resolution of 1 h. It covers a period from 1979 until present. Coverage is global for land areas and coastal regions. Offshore coverage is only to be expected up to approximately 50 km from the coastline.

The MERRA data considers the orography and roughness of the site, and has a high availability. As well as other reanalysis databases have different pressure levels; however only data at 50 m height above ground has been used in this study.

Data from the MERRA reanalysis project are currently available to users of WindPRO as part of the online services from the EMD server [2].

Table 68.1 Overview of MERRA dataset parameters

Parameter	Description
Time	UTC time stamp
Pressure	Sea-level pressure
Temperature	Temperature at 2 m above the displacement height
wdirection	Wind direction at 50 m above surface
wspeed	Wind speed at 50 m above surface
dispHeight	Displacement height

MERRA data is commonly available within WindPRO with a time lag of 1–2 months. This is due to the processing time of the data at NASA and the subset processing at EMD. When the data have been processed at EMD, then the following parameters are available from within the online data services within WindPRO. In Table 68.1 is presented the overview of MERRA dataset parameters.

A complete description of MERRA is found in [3].

68.3 MCP Overview

MCP is the abbreviation for measure–correlate–predict technique, which is widely in use for establishing a long-term wind statistic using limited wind data from a site and long-term data from a more or less nearby site.

The goal of any estimation of long-term wind statistic is to establish a transfer model between the available short-term or long-term wind data and the long-term statistic on the site.

In our analysis we use the wind index MCP technique through EMD WindPRO software.

The index correlation method is a method creating the MCP analysis by means of monthly averages of the energy yield, thus disregarding the directional distribution of the winds. Even though this method may seem rather crude and primitive when comparing to other more advanced MCP methods, which take the wind veer into account, this method has its advantages in stability and performance as it may even succeed in the cases where other MCP methods seem to fail. This is due to the fact that the wind indexes are related directly to WTG energy yield and that the method allows the production calculation to be completed using actual measured data before applying the correction [4].

68.4 Methodology

The methodology of the work is quite simple. A correlation study has been done between the different grid points of MERRA reanalysis project and their combinations should also be evaluated. Possible relationships between terrain characteristics, local wind systems, and correlation values should be investigated.

Table 68.2 Coordinates and area separation of used MERRA grid points

No	Lon (deg.) E	Lat (deg.) N	Study area
1	26.0	40.50	Northern Aegean Sea
2	26.0	40.00	Northern Aegean Sea
3	26.0	39.50	Northern Aegean Sea
4	26.67	39.50	Northern Aegean Sea
5	26.67	39.00	Northern Aegean Sea
6	26.67	38.50	Central Aegean Sea
7	26.67	38.00	Central Aegean Sea
8	26.67	37.50	Central Aegean Sea
9	27.34	37.50	Central Aegean Sea
10	27.34	37.00	Southern Aegean Sea
11	27.34	36.50	Southern Aegean Sea
12	28.00	37.00	Southern Aegean Sea
13	28.00	36.50	Southern Aegean Sea
14	28.67	36.50	Eastern Mediterranean Sea
15	29.34	36.00	Eastern Mediterranean Sea
16	30.00	36.00	Eastern Mediterranean Sea
17	30.67	36.50	Eastern Mediterranean Sea

The main data for the correlation study was the following:

- Analysis of the MERRA reanalysis data for the level of 50 m above ground level (agl), period 1979–2013, in the wider area of Western Coast (Aegean Sea) of Turkey (17 grid points from the area of Enez to Antalya cities, 26.0°E–30.67°E and 40.50°N–36.0°N) and the study of the yearly variability: In Table 68.2 are displayed the coordinates (WGS'84) of each grid point and the areas which are separated. In Fig. 68.1 are displayed the 17 nodes of MERRA database which are used in this work.
- Grid point coordinates (WGS'84): 26.0–30.67 E and 40.50–36.50 N.
- Time period 1979–2013 (1/1/1979 to 1/12/2013) at 50 m agl for standard atmosphere conditions, instantaneous values at the reference time. Grid of 0.5° latitude × 0.67° longitude globally.
- Correlation study using EMD WindPRO MCP Module [5].
- Use of the 1-h data. Extract of correlation factor Rw (weighted mean of sectorwise correlations). Calculation for each 30° direction bin weighted with the frequency of that particular sector at the reference. Choice of the nearest grid points.
- Also, use of the wind index MCP technique through EMD WindPRO and the extract of correlation factor R. This correlation method is typically making the MCP analysis by using the monthly averages of the energy yield, thus disregarding the directional distribution of the winds.

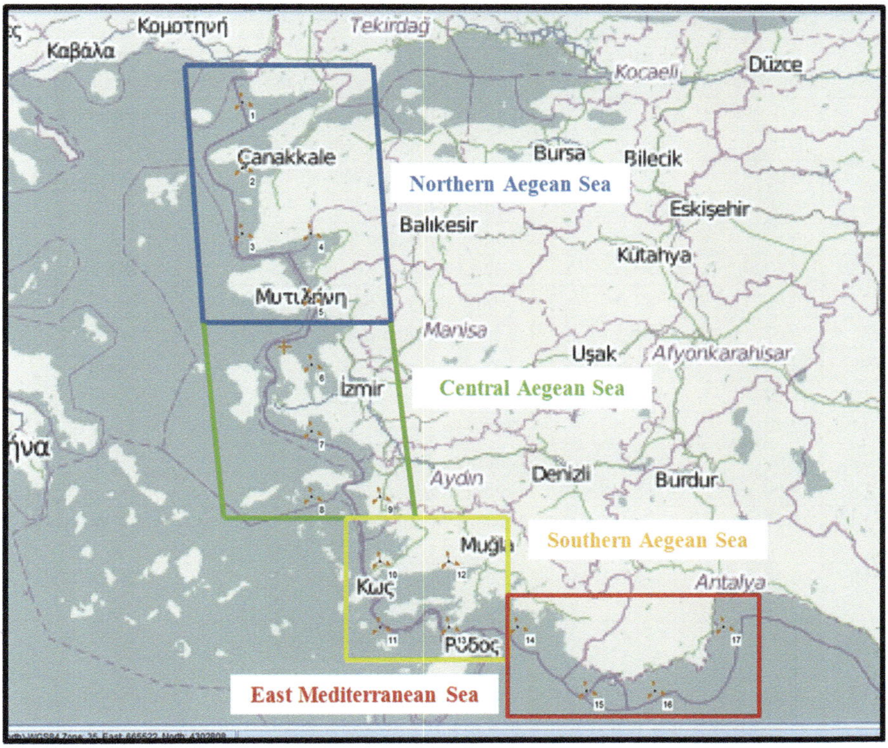

Fig. 68.1 Map with the selected grid points of MERRA and their zones from North to South

- Presentation of long-term resource variation in four study areas (minimum-maximum), year-to-year variability, wind direction (energy based) change, and relationship between grid points.

68.5 Results and Discussion

The wind speed and direction for the time period of 1979–2013 for each under study grid point are presented in Table 68.3.

The area of Aegean Sea has an attractive wind potential which is quite stable. The mean wind speed (at 50 m agl) of the period of 1979–2013 is 6.4 m/s for the Northern Aegean Sea, 6.2 m/s for the Central Aegean Sea, 6.1 m/s for the Southern Aegean Sea, and 5.1 m/s for the Eastern Mediterranean Sea. In Northern Aegean Sea the wind direction changes from ENE to NNE, and in Central Aegean Sea starts from NNE and continues to N-NNW. Also, in Southern Aegean Sea the energy-based wind direction is NNW and N. In the end, in East Mediterranean Sea there is a reduction of mean wind speed and in the grid points of 14–16 the directions are

Table 68.3 Wind speed and direction of MERRA grid points at 50 m agl (1979–2013)

No	Lon (deg.) E	Lat (deg.) N	Study area	*V* mean at 50 m (m/s)	Wind directions (energy based)
1	26.0	40.50	Northern Aegean Sea	6.32	ENE–NNE
2	26.0	40.00	Northern Aegean Sea	6.52	ENE–NNE
3	26.0	39.50	Northern Aegean Sea	6.60	NNE–ENE
4	26.67	39.50	Northern Aegean Sea	6.39	NNE–ENE
5	26.67	39.00	Northern Aegean Sea	6.12	NNE
6	26.67	38.50	Central Aegean Sea	5.88	NNE–N
7	26.67	38.00	Central Aegean Sea	6.46	N–NNE
8	26.67	37.50	Central Aegean Sea	6.62	N–NNW
9	27.34	37.50	Central Aegean Sea	5.79	N–NNW
10	27.34	37.00	Southern Aegean Sea	6.32	NNW–N
11	27.34	36.50	Southern Aegean Sea	6.72	NNW–N
12	28.00	37.00	Southern Aegean Sea	5.41	N–NNW
13	28.00	36.50	Southern Aegean Sea	5.98	NNW–WNW
14	28.67	36.50	Eastern Mediterranean Sea	5.40	WNW–NNW
15	29.34	36.00	Eastern Mediterranean Sea	5.45	WNW–W
16	30.00	36.00	Eastern Mediterranean Sea	5.30	W–WNW
17	30.67	36.50	Eastern Mediterranean Sea	4.35	N–NNE

WNW and W. In the last grid point due to its special position, the mean wind speed is quite reduced and the wind direction is completely different (N-NNE).

Despite the complexity of the terrain, seasonal airflow follows in most cases larger regional synoptic patterns over the Eastern Mediterranean. The Western Coast of Turkey and the neighboring islands of Aegean Sea have been influenced by a combination of synoptic wind systems (larger and regional-local). The study area is influenced by northern winter wind and in the period of May to October from the regional Etesian Winds ("meltemi"), a northerly airflow resulting from the large pressure contrast between the Azores High in the western Mediterranean and the thermal depression over the Persian Gulf. After labilization in the late morning, such continental air masses may reach stormy wind speeds over the Aegean Sea. The intensity and frequency of appearance are high in the sea and are reduced, as we pass to the coast and the mainland.

In winter, this regional pressure contrast is replaced by the establishment of either East European Anticyclones or cyclonic depressions originating in the Tyrrhenian Sea or further East as Cyprus Lows. The resulting airflow will depend on the position and track of the individual pressure center; in most cases it will be from southerly and westerly directions.

It is worthy to be mentioned that the coast has been influenced by the local daily wind systems of sea/land breeze.

Moreover, in the North Aegean Sea a strong NE wind invades from the Black Sea through low orography of East Thrace under conditions of favorable prevalence isobaric condition.

The change in the wind direction is mainly due to the arrangement of isobars and the influence of the local orography.

The percentage wind speed variation from 1979 to 2013 based on MERRA reanalysis data at 50 m agl is displayed in Figs. 68.2, 68.3, 68.4, and 68.5.

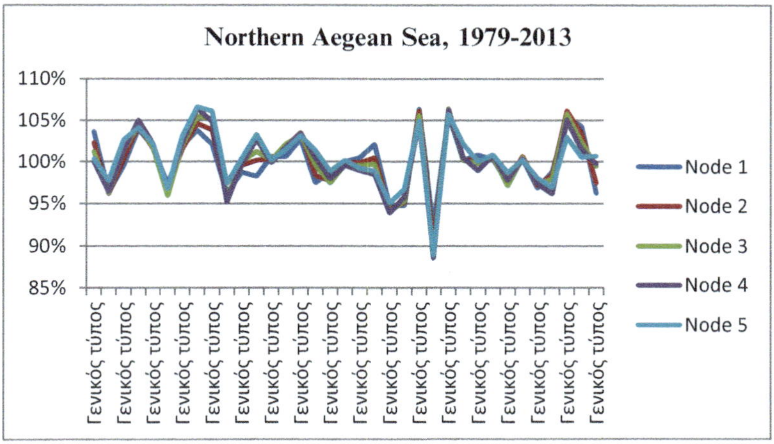

Fig. 68.2 Percentage wind speed variation from 1979 to 2013 based on MERRA reanalysis data at 50 m agl in the area study "Northern Aegean Sea"

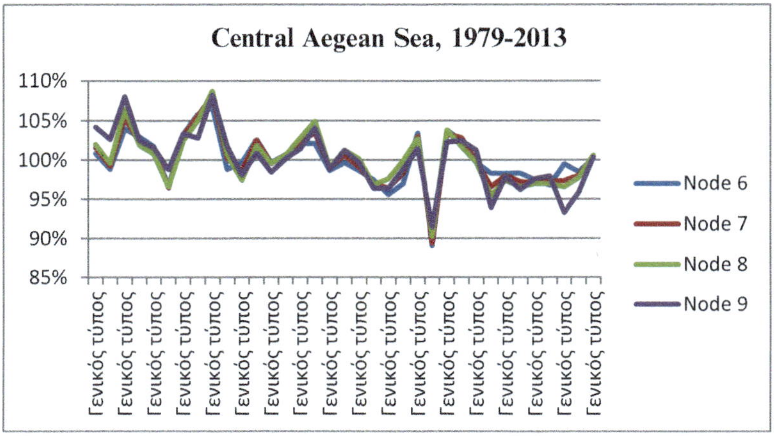

Fig. 68.3 Percentage wind speed variation from 1979 to 2013 based on MERRA reanalysis data at 50 m agl in the area study "Central Aegean Sea"

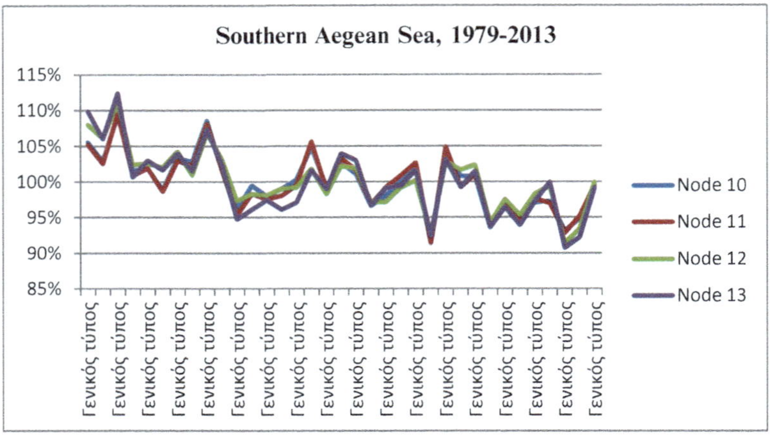

Fig. 68.4 Percentage wind speed variation from 1979 to 2013 based on MERRA reanalysis data at 50 m agl in the area study "Southern Aegean Sea"

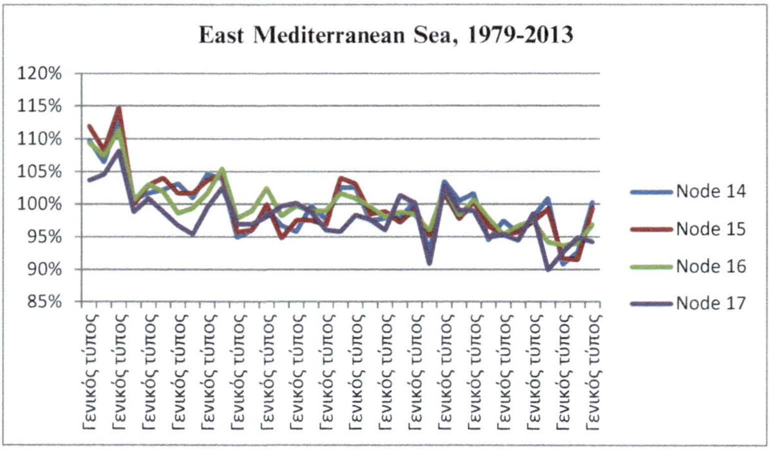

Fig. 68.5 Percentage wind speed variation from 1979 to 2013 based on MERRA reanalysis data at 50 m agl in the area study "Eastern Mediterranean Sea"

In Table 68.4 are displayed the percentage minimum and maximum deviations of the annual wind speed of the 17 MERRA grid points.

In the first study area, the five grid points show the same trend with minimum of 11 % (2002) and maximum of 7 % (2003 and 1986 in no. five grid point).

In the second study area, the four grid points show the same trend with minimum of 11 % (2002) and maximum of 9 % (1987). The grid point no. 9 has divergence from the others due to the fact that it is in mainland.

In the third study area, the four grid points show the same trend with minimum of 9 % (2002 and 2011 at no. 12, 13) and maximum of 12 % (1981).

Table 68.4 Variation of annual wind speed of MERRA grid points at 50 m agl and the corresponding years (1979–2013)

Node	Min (%)	Max (%)	Year of min.	Year of max.
1	−8	6	2002	2003
2	−10	6	2002	2003
3	−11	6	2002	2003
4	−11	6	2002	2003
5	−11	7	2002	2003, 1986
6	−11	7	2002	1987
7	−11	8	2002	1987
8	−10	9	2002	1987
9	−9	8	2002	1987
10	−9	9	2002	1981
11	−9	10	2002	1981
12	−9	11	2011	1981
13	−9	12	2011	1981
14	−9	13	2011	1981
15	−8	15	2011	1981
16	−6	11	2011	1981
17	−10	8	2010	1981

Table 68.5 The year-to-year variability of MERRA grid points at 50 m agl (1979–2013)

Grid point	%	Grid point	%	Grid point	%	Grid point	%
1	3.4	**6**	3.2	**11**	4.1	**16**	3.9
2	3.6	**7**	3.4	**12**	4.2	**17**	3.7
3	3.8	**8**	3.5	**13**	4.8		
4	3.8	**9**	3.6	**14**	4.6		
5	3.5	**10**	4.0	**15**	4.9		

In the fourth study area, the nos. 14 and 15 grid points show a similar trend, no. 16 has a small difference. and no. 17 appears totally different variation. In nos. 14–16 the years with the lower wind speed are 2011 and 2010 (no. 17). The year of maximum annual wind speed is 2010. The intense differentiation of the no. 17 grid point is due to its position and the limited influence from the larger regional synoptic patterns over the Aegean Sea and the proximity to the high mountains of the area (Tahtali mountain) which creates different regional wind circulations.

In Table 68.5 is displayed the year-to-year variability of wind speed of the 17 MERRA grid points.

The annual variability for the period of 1979–2013 is quite low from 3.2 to 4.9 %, evidence of the wind potential stability and quality of the study area.

In Table 68.6 are displayed the correlation coefficient (Rw) table MERRA grid points at 50 m agl–1-h data (Rw: the weighted mean of sectorwise correlations) and in Table 68.7 are presented the correlation coefficient (R) monthly averages (using wind index). The included table shows a qualitative rating of the reference as a function of correlation value. The red-colored numbers are presented as the poor correlation coefficients.

Table 68.6 Correlation coefficient table MERRA grid points at 50 m agl–1-h data (wind and direction data)

Rw	1	2	3	4	5	6	7	8	9	10	11	12	13	14	15	16	17
1		**0.93**	**0.82**	**0.75**	**0.64**								Correlation Coefficient		Quality of reference		
2			**0.95**	**0.87**	**0.77**												
3				**0.95**	**0.88**	**0.75**							0.5–0.6		Very poor		
4					**0.95**	**0.82**							0.6–0.7		Poor		
5						**0.94**	**0.83**						0.7–0.8		Moderate		
6							**0.96**	**0.88**	**0.75**				0.8–0.9		Good		
7								**0.97**	**0.86**	**0.80**			0.9–1.0		Very good		
8									**0.91**	**0.89**	**0.85**						
9										**0.96**	**0.89**	**0.85**					
10											**0.97**	**0.88**	**0.87**				
11												**0.84**	**0.89**	**0.70**			
12													**0.94**	**0.85**	**0.67**		
13														**0.74**	**0.73**		
14															**0.88**	**0.69**	
15																**0.87**	**0.49**
16																	**0.72**
17																	

Table 68.7 Correlation coefficient table MERRA grid points at 50 m agl–monthly data (only wind data)

Wind index	1	2	3	4	5	6	7	8	9	10	11	12	13	14	15	16	17
1		**0.93**	**0.91**	**0.76**	**0.70**								Correlation Coefficient		Quality of reference		
2			**0.97**	**0.87**	**0.81**												
3				**0.95**	**0.92**	**0.80**							0.5–0.6		Very poor		
4					**0.98**	**0.90**							0.6–0.7		Poor		
5						**0.96**	**0.89**						0.7–0.8		Moderate		
6							**0.98**	**0.94**	**0.83**				0.8–0.9		Good		
7								**0.99**	**0.91**	**0.86**			0.9–1.0		Very good		
8									**0.96**	**0.93**	**0.90**						
9										**0.98**	**0.95**	**0.91**					
10											**0.99**	**0.94**	**0.92**				
11												**0.94**	**0.93**	**0.85**			
12													**0.98**	**0.94**	**0.80**		
13														**0.86**	**0.86**		
14															**0.93**	**0.85**	
15																**0.94**	**0.61**
16																	**0.75**
17																	

The correlation coefficients between the closest grid points are good and very good, so there are solid indications of good reference quality. It is obvious that where there is good correlation, these grid points are influenced by the same wind synoptic system.

Moderate-quality references appear between grid points due to their position (mainland/sea) or orographic obstacles (mountains, etc.) which change the air circulation. As it is expected the correlation coefficients are reduced due to the distance between the grid points.

68.6 Conclusions

1. The long-term resource variation in the Western Coast of Turkey follows the 17 above-mentioned MERRA grid points with minimum of 11 % and maximum of 15 %.
2. The year-to-year wind speed variability is quite low, evidence of the wind potential stability and quality of the study area.
3. In Northern Aegean Sea, in the period 2000–2013 (where there is important activity in installation of wind farms), 2002 was the worst year and the best years were 2001, 2003, and 2011 in annual wind speed.
4. In Central Aegean Sea, in the period 2000–2013, 2002 was the worst year and the best year was 2004 (only 2 % increase) in annual wind speed. The time period from 2006 to 2013 presents lower wind potential than the average of the total period (1979–2013).
5. In Southern Aegean Sea, in the period 2000–2013, 2002 (nos. 10, 11 grid points) and 2011 (12, 13) were the worst years. The time period from 2006 to 2013 presents lower wind potential than the average of the total period (1979–2013).
6. In Eastern Mediterranean Sea, in the period 2000–2013, 2010 and 2011 were the worst years. The time period from 2006 to 2013 presents lower wind potential than the average of the total period (1979–2013) except grid point no. 14 (1 % increase in 2010).
7. Summarizing, there appeared to be good and very good correlation between grid points in areas with similar geographical characteristics, sea or common atmospheric circulation systems. There was existence of subsets of nodes with similar trends in total study area (except no. 17).
8. The correlation coefficients between grid points are improved when we use monthly data only of wind speed (wind index MCP method).
9. MERRA data is a useful information for the study of the long-term wind regime and its yearly wind resource variation.
10. The finest spatial and temporal resolutions of MERRA data contribute to considerable improvements in the degree of correlation with local onsite measurements and allow extracting long-term corrected energy production results for wind farms.

References

1. Lileo S, Petrik O (2012) Investigation on the use of NCEP/NCAR, MERRA and NCEP/CFSR reanalysis data in the wind resource analysis. Proceedings EWEA 2012, Copenhagen
2. Thøgersen ML et al (2010) WindPRO/ONLINE – Remote sending data and other data for download in WindPro, August 2010, EMD A/S: 5.1–5.4
3. Rienecker MM, Suarez MJ, Gelaro R, Todling R, Bacmeister J, Liu E, Bosilovich MG, Schubert SD, Takacs L, Kim GK, Bloom S, Chen J, Collins D, Conaty A, da Silva A et al (2011) MERRA - NASA's modern-era retrospective analysis for research and applications. J Clim 24(14):3624–3648. doi:10.1175/JCLI-D-11-00015.1
4. Thøgersen ML et al (2010) WindPRO/MCP, June 2010, EMD A/S: 1.4, 6.1–6.2
5. Thøgersen ML et al (2007) Measure-Correlate-Predict Methods: case studies and software implementation. Proceedings EWEC 2007, Milan

Chapter 69
Using Remote Sensing Technologies for Wind Turbine/Farm Health Monitoring

Muammer Ozbek and Daniel J. Rixen

Abstract Efficient use of contemporary measurement systems (accelerometers, piezoelectric or fiber-optic strain gauges) in structural health monitoring of wind turbines is mainly limited due to high sensor installation costs, practical limitations in placing these sensors on existing structures, low spatial resolution, and similar disadvantages caused by the complicated nature of wind loading and the turbine structure. The factors affecting the performance of these sensors such as sensitivity to lightning, electromagnetic fields, humidity and temperature variations, and the corresponding error compensation methods are still being investigated. Similarly, additional long-term durability tests are required to determine whether the bonding between the sensor and composite blade material deteriorates over time due to repetitive loading and severe environmental factors or not.

In this work, two optical measurement techniques (photogrammetry and laser interferometry), which do not require any sensors to be installed on the structure, are introduced as promising and versatile alternatives for measuring the vibration response of wind turbines. By using LDV (laser Doppler vibrometer), the dynamic behavior of the turbine at parked condition can be measured with a very high accuracy (in micron level). Similarly, photogrammetry enables the in-operation vibration response of the turbine to be measured with an average accuracy of ± 25 mm from a measurement distance of 220 m. Considering the fact that during rotation peak-to-peak blade deformations can be as high as 1,000 mm, this accuracy can be considered as quite high and still be improved further.

Keywords Remote Sensing Technologies • Laser interferometry • Photogrammetry • Wind turbine • Structural health monitoring

---

M. Ozbek (✉)
Civil Engineering Department, Istanbul Bilgi University, Santral Campus, Eyup, 34060 Istanbul, Turkey
e-mail: muammer.ozbek@bilgi.edu.tr

D.J. Rixen
Institute of Applied Mechanics, Technische Universität München, Boltzmannstr. 15, 85748 Garching, Germany

I. Dincer et al. (eds.), *Progress in Clean Energy, Volume 2*,
DOI 10.1007/978-3-319-17031-2_69

69.1 Introduction

State-of-the-art monitoring systems require different types of sensors to be installed in the turbine. Accelerometers, piezoelectric film sensors, and fiber-optic strain gauges are some of the most important sensors which are widely used for condition-based monitoring applications. However, these sensors are sensitive to lightning and electromagnetic fields (e.g., the magnetic field caused by the generator). Besides, some extra installations inside the blades such as placement of cables for power supply and data transfer are required for these applications. The signals from rotating sensors on the blades are transferred to a stationary computer via slip rings or by radio/wireless transmission. For large commercial turbines, the required installations and preparations (sensor calibration) may be very expensive and time consuming [1, 2]. Accelerometers and strain gauges can be applied on an existing turbine, but due to high sensor installation costs and technical limitations in placing the sensors, in practice, these sensors are only placed at some limited number of locations (tower or root region of the blades) which can be accessed easily.

In addition to the aforementioned practical limitations, the complicated nature of wind load makes the efficient use of accelerometers on condition monitoring of wind turbines very difficult. The deflections under the action of wind loading can be considered as the sum of a static component due to average wind speed and a dynamic component due to turbulence [3]. Accelerometers cannot provide very accurate information about the static component. Therefore, several researchers suggest that in wind response measurements, accelerometers should be used together with other systems such as GPS (Global Positioning System) which are able to detect the static deformations accurately [4–6]. Although GPS-accelerometer combination is widely used to monitor the response of several structures such as high-rise buildings and bridges, it cannot easily be applied to wind turbines because of the technical difficulties in placing GPS sensors in the blade and reduced measurement accuracy due to secondary reflections caused by the rotational effects.

Fiber-optic strain gauges are proposed to be a promising alternative to accelerometers and conventional strain gauges since optical sensors are not prone to electromagnetic fields or lightning. However, it is reported that some feasibility tests are still needed to ensure the effective and cost-efficient use of this measurement system. The factors affecting the performance of fiber-optic sensors such as sensitivity to humidity and temperature variations and the corresponding error compensation methods should also be investigated further [7, 8]. Similarly, additional long-term durability tests are required to determine whether the bonding between fiber-optic and composite blade material deteriorates over time due to repetitive loading and severe environmental factors or not.

Fiber-optic strain gauges are expected to provide a high spatial resolution, but installation costs significantly increase depending on the number of sensors. Besides, high-capacity decoders are needed to be able to acquire data from many sensors simultaneously resulting in a further increase in the hardware costs. Fiber-optic sensors can be applied throughout the blade only if the installation is performed during the manufacturing stage in the factory. The system cannot be easily applied to existing turbines.

69.2 Optical Measurement Techniques

In this work, two noncontact optical measurement systems (photogrammetry and laser interferometry) are proposed to be promising and cost-efficient alternatives for condition-based health monitoring of wind turbines. Unlike conventional measurement systems (accelerometers, piezoelectric or fiber-optic strain gauges), optical measurement techniques do not require any sensors to be placed on the turbine. However, some reflective markers should be placed (or painted) on the structure. These markers are made up of a retroreflective material, which is 1,000 times more reflective than the background blade material. Since the markers are in the form of very thin stickers, they do not have any effect on aerodynamic performance of the blades.

The markers which are used as displacement sensors can easily be placed on an existing turbine. No extra cable installations for data transfer and power supply are required inside the turbine. Therefore, compared to the conventional sensors (accelerometers, piezoelectric or fiber-optic strain gauges), marker installations are very cost efficient and can be completed within very short periods of time. Retroreflective paints applied on the turbine components during manufacturing stage in the factory can substitute these markers which may result in a further decrease in the installation costs.

During the tests, a total of 55 markers were placed on the turbine (11 markers for each blade and 22 markers on the tower). Placement of the markers on the blade and their final distribution throughout the structure can be seen in Fig. 69.1a, b, respectively. Although the pictures shown in Fig. 69.1 were captured by a handheld digital camera using its flash light only, the markers can be seen easily. It should be noted that it only took two professional people 6 h to place 55 markers on the turbine. Considering the fact that each marker acts as an independent sensor, it can be concluded that it is almost impossible to reach such a high sensor installation speed by using conventional sensor technologies.

Fig. 69.1 (**a** and **b**) Placement of markers and final layout

These markers are essential for both photogrammetry and laser interferometry, but they are used for different purposes in each method. Photogrammetry is a proven measurement technique based on the determination of the 3D coordinates of the points on an object by using two or more images taken from different orientations and positions [9]. Although each picture provides 2D information only, very accurate 3D information related to the coordinates and/or displacements of the object can be obtained by simultaneous processing of these images. In photogrammetry, markers are used as the targets to be tracked by the camera systems, and all the targets can be tracked simultaneously.

In laser interferometry, a laser vibrometer continuously sends a laser beam to the target and receives the beam reflected from its surface. If the object is moving, this causes a frequency change and phase shift between the sent and the reflected beams. By detecting this frequency change (using Doppler principle), the velocity of the moving object can be found. If the object itself has a reflective surface, no extra retroreflective markers are needed. However, because the blade material was not reflective enough and the distance between the laser source and the turbine was very long, high-quality laser signals could only be acquired if the laser was targeted to the markers. Figure 69.2 shows the reflection of the laser beam from the marker on the blade.

Different from photogrammetry, laser vibrometer can only measure the motion of a single point at a time. However, it is still possible to successively measure all the markers distributed throughout the blade. During the tests, the laser interferometry measurements were taken by using a Polytec OFV 505 laser head and OFV 5000 controller with VD06 velocity decoder. These systems were located in the field at a distance of 200 m from the turbine. An SLR (superlong-range) lens which enables an increased measurement range up to 300 m was also required to take measurements from this distance. It should be noted that because it is very difficult

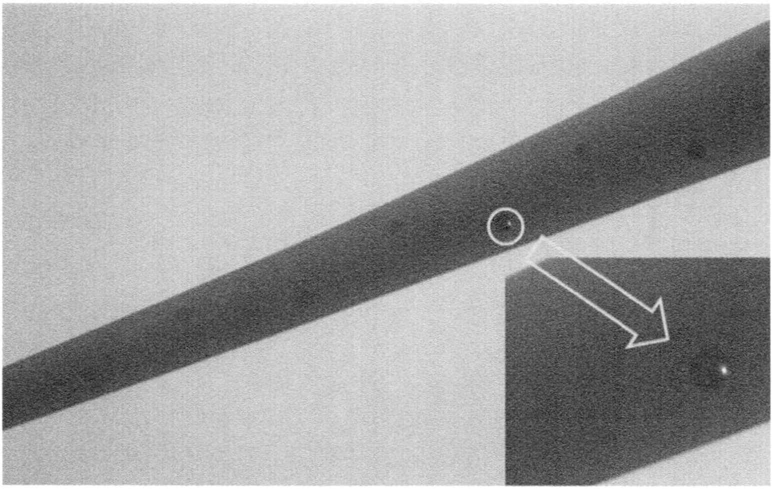

Fig. 69.2 Reflection of the laser beam from the marker on the blade

to keep the laser on the same marker while the turbine is rotating, LDV (laser Doppler vibrometer) was only used for the measurements taken on the parked turbine. Similarly, photogrammetric measurements could not be conducted when the turbine was at parked condition because the low wind speeds could not excite the structure sufficiently resulting in high noise to signal ratios.

A third system, which has already been installed in the turbine as a part of a long-term wind load monitoring campaign, consists of six strain gauges placed at the root region of the three blades (two strain gauges per blade) and two strain gauges located at the tower base.

These strain gauges are used to measure the flapwise and edgewise vibrations of the blades and the fore-aft and side to side vibration of the tower at a sampling frequency of 32 Hz. All the data recorded by the three different systems were then synchronized using a GPS clock whose absolute time accuracy is approximately 10 ms. Considering the fact that the frequencies that are expected to dominate the response of the wind turbine are mostly in the low frequency range [0–5 Hz], this accuracy can be considered as sufficient.

69.3 Test Turbine

Our tests were conducted on a pitch-controlled, variable speed Nordex N80 wind turbine with a rated power of 2.5 MW. The turbine has a rotor diameter and tower height of 80 m. The photogrammetric measurements were performed by GOM mbH [10] (GOM Optical Measuring Techniques) at the ECN (Energy Research Center of the Netherlands) wind turbine test site located in Wieringermeer, the Netherlands. More detailed information about the facilities of the test site can be found through the related website [11].

69.4 Measurements on the Parked Turbine (Laser Interferometry)

In order to verify that laser optical sensors (vibrometers) can effectively be used in measuring the vibration response of the structure, it is required to demonstrate that all the vibration modes, which are identified from conventional sensor (strain gauge) data, can also be extracted from laser measurements.

Strain gauges installed in the turbine are placed at some specific locations (rotor, tower, and nacelle) and orientations (flapwise, edgewise) to ensure that all the modes can be observed. Indeed, gauges placed on the rotor may not detect the frequencies related to the tower modes. Similarly, blade sensors oriented in edge-wise direction may not provide accurate information about vibration in flapwise direction. Complete description of the dynamic characteristics of the structure can only be obtained by combining the information coming from different sensors. Since LDV can measure the vibration of a single point at a time, this can only be provided by taking measurements at different locations on the turbine.

Table 69.1 summarizes the modal parameters (frequencies) calculated by using strain gauge and LDV measurements. During the measurements at standstill, the turbine was kept at a fixed orientation, and yawing motion was prevented by application of the yaw brakes. Blade pitch angles were fixed at zero degree where flapwise blade vibration exactly corresponds to motion out of the rotor plane. This is the same as the angle of the blade during rotation below rated wind speeds (<15 m/s for the test turbine). Similarly, the brakes were applied to prevent the movement of the rotor.

Frequencies shown in Table 69.1 were extracted by using the NExT (natural excitation technique) together with the LSCE (least-squares complex exponential) time domain identification method. When the turbine starts rotating, the name of the mode changes to the one indicated in parentheses. The abbreviations FW and BW stand for forward and backward whirling, respectively. These modal parameters are important for tuning and validation of numerical models and for the verification of prototype designs. They can also be used for health monitoring applications.

The signals shown in Table 69.1 were investigated in four groups, namely:

- Edgewise direction (in-plane direction) blade strain measurements shown as S_E
- Flapwise direction (out-of-plane direction) blade strain measurements shown as S_F
- Tower strain measurements shown as S_T
- Laser measurements on the blade shown as S_L

In the table, the symbol "X" is used to show the frequencies that can be extracted from a certain signal combination. Considering the strain gauge signals only, it can be easily seen that some frequencies can be identified either from in-plane strains or from out-of-plane strains only but not from both. It can also be seen that all the frequencies can be identified by analyzing LDV measurements. However, these frequencies cannot be detected from a single data block only. LDV measurement

Table 69.1 Modal parameters calculated for the parked turbine

Mode	Frequency (Hz)	S_F	S_E	S_T	S_L
1st longitudinal tower	0.345	X		X	X
1st lateral tower	0.347		X	X	X
1st yaw (BW flapwise)	0.902	X		X	X
1st tilt (FW flapwise)	0.974	X		X	X
1st symmetric flapwise	1.077	X		X	X
1st vertical edgewise (BW)	1.834		X	X	X
1st horizontal edgewise (FW)	1.855		X	X	X
2nd tilt (FW flapwise)	2.311	X	X	X	X
2nd yaw (BW flapwise)	2.430	X	X	X	X
2nd symmetric flapwise	3.000	X		X	X
2nd edgewise	6.360	X	X		X
Tower torsion mode	6.154	X			X

contains a single channel data recorded on a specific marker at a time. The targeted marker may not be at a suitable location to detect some of the modes (or frequencies). Therefore, it is required to try several locations (markers) and time series to identify all the modes. However, the frequencies identified by using two different systems are always in good coherence.

69.5 Measurements on the Rotating Turbine (Photogrammetry)

During in-field tests, dynamic behavior of the turbine was monitored from a measurement distance of 220 m by using four CCD cameras. The vibration characteristics of the rotor were measured at 33 different points simultaneously. Although photogrammetry is efficiently used in smaller scales by a wide variety of disciplines, this method was applied for the very first time to a MW scale wind turbine within the scope of this research project.

Analysis results showed that the deformations of the turbine can be measured with an average accuracy of ±25 mm from a measurement distance of 220 m [12, 13]. Considering the fact that for a rotating turbine, deformations measured in flapwise direction can be as high as ±1,000 mm, this accuracy can be considered as high. The measurement error is observed to be mainly caused by calibration problems due to the rotation of the turbine. Therefore, their magnitudes are frequency dependent.

The frequency domain analysis of the calculated measurement errors provides important information to be used in determining the source of the error. Figure 69.3 aims at showing the important frequencies identified in the corresponding

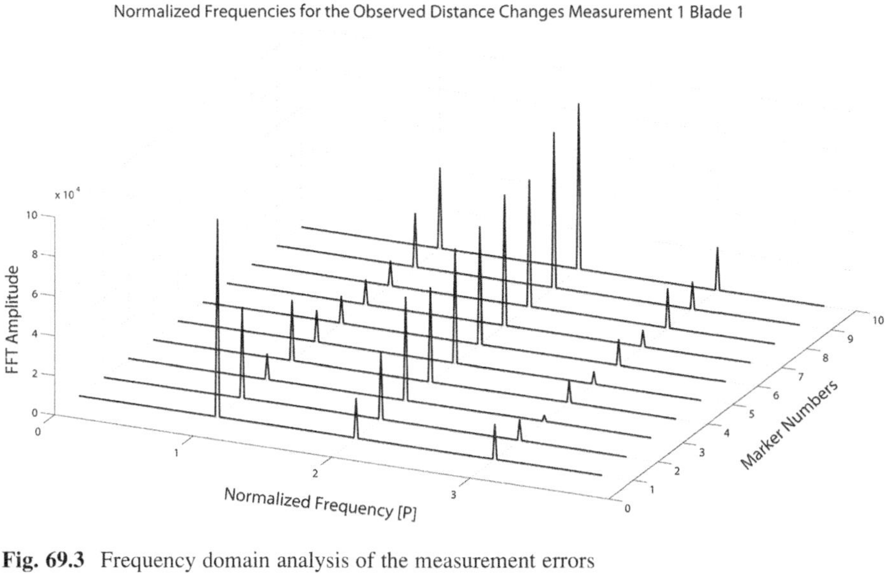

Fig. 69.3 Frequency domain analysis of the measurement errors

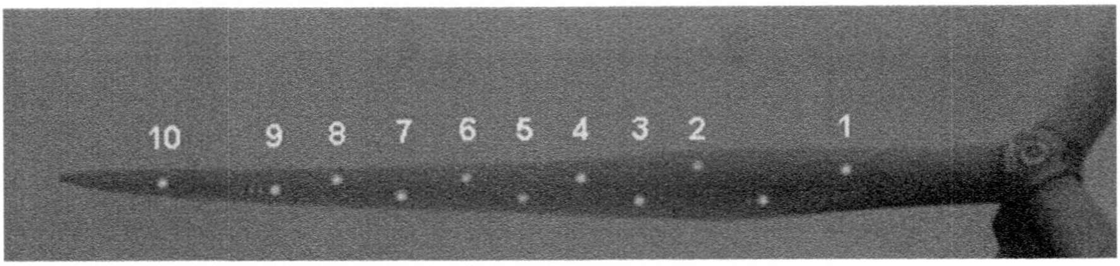

Fig. 69.4 Numbering of the markers on the blade

coordinate measurement error. The frequencies, normalized with respect to the rotational frequency (also called as P), are displayed along the *X*-axis. The *Y*-axis represents the marker number. The numbering of the markers is depicted in Fig. 69.4. The *Z*-axis represents the amplitude of the Fourier transform.

As can be seen from the figure, the 3D displacement measurement error calculated for marker 10 has a strong 2P and a relatively weaker 1P component, whereas the one calculated for marker 1 has a very dominant 1P and a much weaker 2P component. 1P and 2P components always exist in the Fourier transforms of the data, but their relative amplitudes differ depending on the location of the marker.

Markers 1 and 2 are placed on curved surfaces close to the root of the blade. The measurement errors calculated for these markers are expected to be mainly related to the reflection quality and marker orientation varying during the rotation. Although 1P component of the error can be seen in all the ten series, its contribution in the overall error is dominant for markers 1 and 2 only. As the distance from the center of rotation increases, the contribution of 2P component gradually becomes more important, probably because those markers are closer to the boundaries of the field of view where calibration errors and artifacts due to projection on the plane of rotation become more dominant. Let us observe in Fig. 69.3 that the contribution of order of 3P (as well as all higher orders) is significantly lower than the components of order 1P (0.28 Hz) and 2P. For higher frequencies, measurement accuracy is much higher and in the range of ± 5 mm [13, 14].

A typical displacement time history measured in flapwise direction for the tip markers of three blades can be seen in Fig. 69.5. In the figure, horizontal axis represents the number of rotation cycles in the recorded data. It can be seen that the tip of the blade can experience a relative displacement up to 102.4 cm during rotation. Frequency domain analyses of these deformation time histories provide very useful information about the vibration characteristics of the turbine.

Figures 69.6 and 69.7 show the PSD (power spectral density) graphs of edgewise and flapwise direction blade vibration photogrammetric data, respectively. These figures are presented to provide a 3D frequency distribution that also includes information related to the measurement location. The *X*-axis represents the frequencies normalized with respect to rotational frequency (P-0.28 Hz); therefore, it is dimensionless. The *Y*-axis corresponds to the marker number. Marker 1 is placed at the blade root, whereas marker 10 is located at the tip of the blade. The *Z*-axis represents the computed PSD amplitude. 1P and 2P components and the first edgewise mode can be recognized from Fig. 69.6.

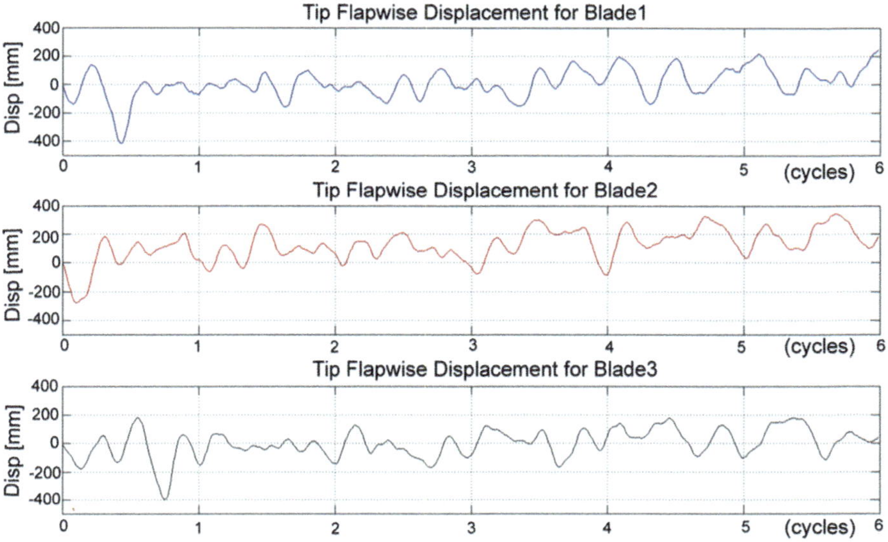

Fig. 69.5 Time normalized tip flapwise displacement records

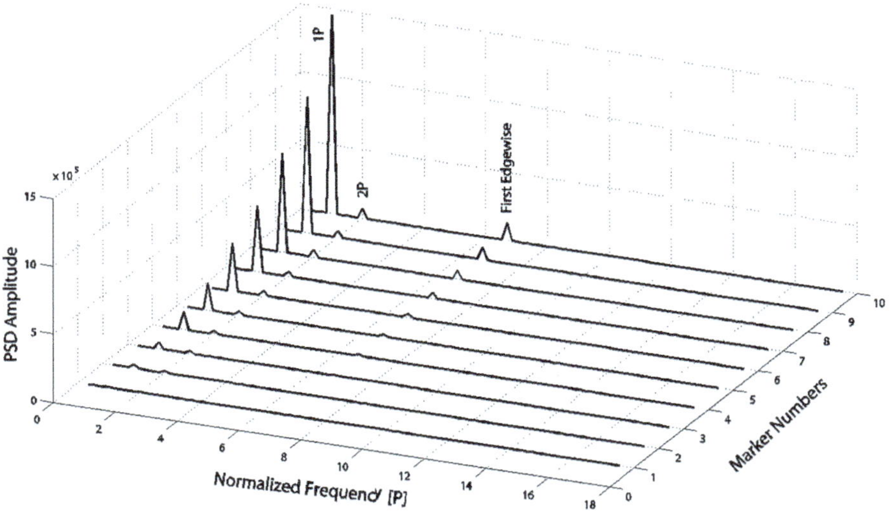

Fig. 69.6 Normalized PSD of edgewise blade vibration for measurement 1 blade 2

As can be seen in Fig. 69.7, flapwise vibration data enables more frequencies to be identified. Integer multiples of rotational frequency up to 4P can be detected from the corresponding PSD graph. Besides these P components, the first flapwise mode can also be seen in Fig. 69.7. Since the response is mainly dominated by P harmonics, other turbine modes, which have relatively weaker modal participations in the response, cannot be identified easily from PSD plots only.

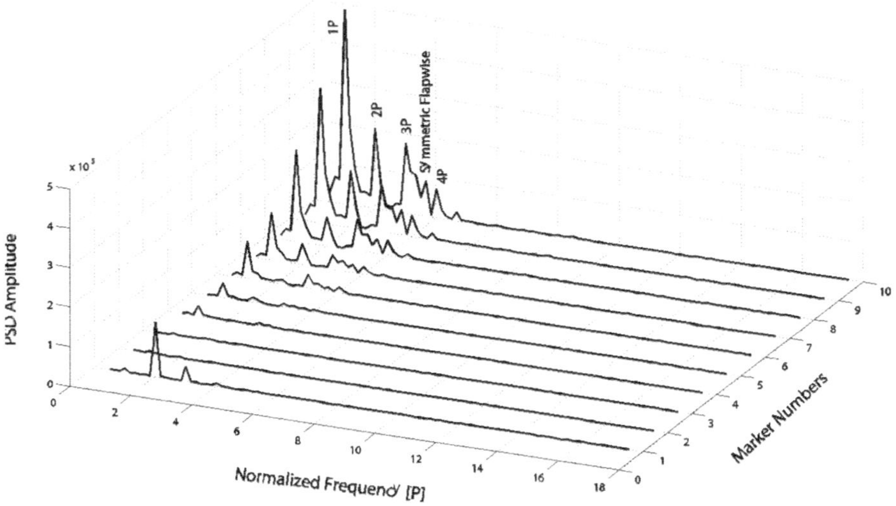

Fig. 69.7 Normalized PSD of flapwise blade vibration for measurement 3 blade 3

69.6 Conclusions

The proposed optical measurement techniques, photogrammetry and laser interferometry, enable the deformations on the turbine to be measured with an average accuracy of ±25 mm from a measurement distance of 220 m. The error calculated is composed of several components such as random and systematic errors. The random component of the coordinate measurement error is in the range of ±5 mm. This amount is consistent with the random measurement errors reported in literature. If compared to the amplitudes of the deformations (±1,000 mm) that are expected, this accuracy can be considered to be high and even be improved further by using higher-resolution cameras. The systematic component is frequency dependent and found to be maximum for the outermost markers. However, using more sophisticated camera calibration, data processing, and comparison techniques can reduce the systematic error further. 1P component of the systematic error can easily be eliminated by spatial filtering; therefore, it is ignored.

The data obtained from photogrammetry appeared to be suitable to identify the 1P–4P harmonics, as well as some of the lower eigenfrequencies of the blades in operation. First edgewise and first flapwise modes can easily be identified from Figs. 69.6 and 69.7, respectively.

Some other turbine modes having relatively weaker participations can also be identified by using more sophisticated identification algorithms which were not explained in detail in this article. A more comprehensive discussion about the identification of these modes can be found in [14–17].

Developments over the last decade have resulted in cheaper, higher-resolution, and more-sensitive cameras and in efficient software for photogrammetry, so that we believe that photogrammetry can be a versatile and cost-effective technique for health monitoring and dynamic validation of wind turbines.

The markers which are used as displacement sensors can easily be placed on an existing turbine. No extra cable installations for data transfer and power supply are required inside the structure. Therefore, compared to the conventional sensors (accelerometers, piezoelectric or fiber-optic strain gauges), marker installations are very cost efficient and can be completed within very short periods of time. Retroreflective paints applied on the turbine components during manufacturing stage in the factory can substitute these markers which may result in a further decrease in the installation costs.

A photogrammetric measurement system which consists of several CCD cameras, flashes, and a central PC can be reused for monitoring several turbines. If continuous monitoring is not required, all the turbines in a wind farm can be observed by using a single system. The measured data can be stored and used to build a condition monitoring archive. Since all the measurement systems are located on the ground, a possible technical problem can be detected and solved easily.

Wind turbines have very specific characteristics and challenging operating conditions. Although the photogrammetric measurement systems (including both the hardware and image processing software), calibration methods, and utilized operational modal analysis techniques were not specifically designed and optimized to be used for monitoring large wind turbines, the accuracy reached in this feasibility study is very promising. It is believed that this accuracy can easily be increased further by utilizing more specialized hardware and data processing methods.

Laser optical devices were also observed to provide very useful information about the dynamic characteristics of the turbine. The system parameters obtained by using LDV (laser Doppler vibrometer) measurements were always consistent with those obtained by using eight strain gauges installed on the structure.

All the frequencies that can be identified by using the eight strain gauges can also be identified by using LDV measurements only. However, since one single point might not be sufficient to detect all the frequencies, different locations on the blade should be measured which results in longer measurement periods. Provided that the quality of the laser reflecting from the blade is sufficient, laser vibrometers can reach to very high accuracies (even in micron scale). The tests performed in this study show the high-quality signals can easily be obtained from a distance of 200 m by using some special lenses. The markers required for photogrammetry can also be used for laser measurements; therefore, no additional preparations on the turbine are needed.

Laser vibrometers can be used as a separate measurement system or together with photogrammetry. Developing hybrid systems aiming at combining the advantages of the both techniques and obtaining very accurate 3D coordinate measurements are at the center of our ongoing research programs.

Acknowledgments This research project was partly funded by the We@Sea research program, financed by the Dutch Ministry of Economical Affairs. The authors would like to thank the ECN (Energy Research Center of the Netherlands) for providing the test turbine and the other technical equipment. The authors also acknowledge the extensive contribution of Pieter Schuer (GOM mbH), Wim Cuypers (GOM mbH), Theo W. Verbruggen (ECN), and Hans J. P. Verhoef (ECN) in organizing and performing the field tests.

References

1. Corten GP, Sabel JC (1995) Optical motion analysis of wind turbines. Technical Report, Delft University of Technology, SV Research Group. ISBN 90-75638-01-9
2. Corten GP (1996) Optical motion analysis of wind turbines. In: European Union Wind Energy Conference, Bedford, UK
3. Tamura Y, Matsui M, Pagnini LC, Ishibashi R, Yoshida A (2002) Measurement of wind-induced response of buildings using RTK-GPS. J Wind Eng Ind Aerodyn 90:1783–1793
4. Nakamura S (2000) GPS measurement of wind-induced suspension bridge girder displacements. ASCE J Struct Eng 126(12):1413–1419
5. Breuer P, Chmielewski T, Gorski P, Konopka E (2002) Application of GPS technology to measurements of displacements of high-rise structures due to weak winds. J Wind Eng Ind Aerodyn 90(3):223–230
6. Nickitopoulou A, Protopsalti K, Stiros S (2006) Monitoring dynamic and quasi-static deformations of large flexible engineering structures with GPS: accuracy, limitations and promises. Eng Struct 28:1471–1482
7. Rademakers LWMM, Verbruggen TW, van der Werff PA, Korterink H, Richon D, Rey P, Lancon F (2004) Fiber optic blade monitoring. In: European Wind Energy Conference, London
8. Schroeder K, Ecke W, Apitz J, Lembke E, Lenschow G (2006) A fiber Bragg grating sensor system monitors operational load in a wind turbine rotor blade. Meas Sci Technol 17 (1167):1172
9. Mikhail EM, Bethel JS, McGlone JC (2001) Introduction to modern photogrammetry. Wiley, New York
10. GOM Optical Measuring Techniques. http://www.gom.com. Accessed Apr 2014
11. ECN Energy Research Center of the Netherlands. https://www.ecn.nl/expertises/wind-energy/tests-measurements/. Accessed Apr 2014
12. Ozbek M, Rixen DJ, Erne O, Sanow G (2010) Feasibility of monitoring large wind turbines using photogrammetry. Energy 35:4802–4811
13. Ozbek M, Rixen DJ (2013) Operational modal analysis of a 2.5 MW wind turbine using optical measurement techniques and strain gauges. Wind Energy 16:367–381
14. Ozbek M, Meng F, Rixen DJ, van Tooren MJL (2010) Identification of the dynamics of large wind turbines by using photogrammetry. Proceedings of the IMAC International Modal Analysis Conference, Jacksonville, FL
15. Ozbek M, Rixen DJ (2011) Optical measurements and operational modal analysis on a large wind turbine: lessons learned. Proceedings of the IMAC International Modal Analysis Conference, Jacksonville, FL
16. Ozbek M, Rixen DJ, Verbruggen TW (2009) Remote monitoring of wind turbine dynamics by laser interferometry: phase 1. Proceedings of the IMAC International Modal Analysis Conference, Orlando, FL
17. Ozbek M, Meng F, Rixen DJ (2013) Challenges in testing and monitoring the in-operation vibration characteristics of wind turbines. Mech Syst Signal Process 41:649–666

Chapter 70
Dynamic Stability Analysis of Wind Turbines Through In-Field Vibration Tests

Muammer Ozbek and Daniel J. Rixen

Abstract Depending on their types and sizes, MW-scale wind turbines are usually designed to be operational for wind speeds between 4 and 25 m/s. In order to reach this goal, most of the turbines utilize active pitch control mechanisms where the angle of the blade (pitch angle) is changed as a function of wind speed. Similarly, the whole rotor is rotated toward the effective wind direction by using the yaw mechanism.

The ability of the turbine to adapt to the changes in operating conditions plays a crucial role in ensuring maximum energy production and the safety of the structure during extreme wind loads. This on the other hand makes it more difficult to investigate the system from the dynamic analysis point of view. Unlike ordinary engineering structures, the modal damping ratios identified for wind turbines are not constant; they change depending on wind speed, rotor speed, and blade pitch angle. Unexpected resonance problems due to dynamic interactions among the aeroelastic modes and/or excitation forces can always be encountered. Therefore, within the design wind speed interval, for each velocity increment, it has to be proven that there are no risks of possible resonance problems and that the structure is dynamically stable.

This work presents the results of in-field vibration tests and the corresponding data analysis performed on a 2.5 MW, 80 m diameter wind turbine. Within the scope of the research, 12 different modes were identified for the turbine at parked conditions. Similarly, seven different aeroelastic modes were extracted for the rotating turbine. These results were then qualitatively compared with a reference study in literature which includes in-field vibration tests and aeroelastic stability analysis performed on a similar size and capacity wind turbine.

Keywords Aeroelastic Stability Analysis • Wind turbine • Modal analysis • In-field vibration test • Aeroelastic damping

---

M. Ozbek (✉)
Civil Engineering Department, Istanbul Bilgi University, Santral Campus, E2-209, Eyup, 34060 Istanbul, Turkey
e-mail: muammer.ozbek@bilgi.edu.tr; ozbek.muammer@yahoo.com

D.J. Rixen
Institute of Applied Mechanics, Technische Universität München, Boltzmannstr. 15, 85748 Garching, Germany
e-mail: rixen@tum.de

I. Dincer et al. (eds.), *Progress in Clean Energy, Volume 2*,
DOI 10.1007/978-3-319-17031-2_70

70.1 Introduction

Identification of the dynamic properties and the corresponding structural response of wind turbines is essential for optimizing the energy produced, ensuring safe and reliable operation and increasing the lifetime of the system. As the sizes of modern wind turbines increase, their dynamic behaviors get more complicated, and it becomes more important to predict the response characteristics of new designs through simulations.

Modern computation and simulation tools provide designers with great opportunities to detect and solve most of the possible problems at very early stages and to improve their designs. Indeed, several important system properties such as eigenfrequencies and mode shapes, which govern the dynamic response of the turbine, can be estimated very accurately by using structural analysis programs. However, some important dynamic parameters (e.g., damping) cannot be modeled precisely without supplementary information obtained from in-field tests and measurements.

Damping, together with the abovementioned system characteristics, plays a crucial role in predicting maximum dynamic loads and fatigue stresses acting on the structure. The overall damping has a very complex mechanism which requires considering several physical factors simultaneously. Thus, it is very difficult to be modeled without supplementary information obtained from in-field tests and measurements. Some of its components like material damping can nowadays be reasonably well taken into account. However, modeling the damping occurring in bearings, joints, and gearing or the damping due to ground (soil)-structure interaction is still not possible. Similarly, identifying the aeroelastic component of damping which is due to the combined effect of structural deformations and aerodynamic forces is a very challenging task and always needs experimental verification.

Considering the fact that only the models based on real response measurements are able to represent the complicated interactions among different parts of the structure, several tests have been applied on both parked and rotating turbines [1–4]. Although there are numerous studies conducted on wind turbines at parked condition, the information related to dynamic testing and modal analysis of MW-scale large wind turbines during operation is quite limited. This work aims at making a contribution to this challenging field of experimental and operational modal analysis by presenting the results of the in-field vibration tests performed on a 2.5 MW, 80 m diameter wind turbine and the corresponding data analyses.

For this purpose, the dynamic response of the test turbine was monitored by using three different measurement systems, namely, conventional strain gauges, photogrammetry, and laser interferometry, while the turbine was both at parked condition and rotating. The recorded data was analyzed by using OMA (operational modal analysis) methods, and eigenfrequencies and damping ratios were extracted. The obtained system parameters were then qualitatively compared with the results presented in a study from literature [3], which includes both aeroelastic simulations and in-field measurements performed on a similar size and capacity wind turbine.

70.2 Test Turbine

Our tests were conducted on a pitch-controlled, variable speed Nordex N80 wind turbine with a rated power of 2.5 MW. The turbine has a rotor diameter and tower height of 80 m. The photogrammetric measurements were performed by GOM mbH [5] (GOM Optical Measuring Techniques) at the ECN (Energy Research Center of the Netherlands) wind turbine test site located in Wieringermeer, the Netherlands. More detailed information about the facilities of the test site can be found through the related website [6].

The reference turbine used for qualitative comparison, General Electric NM80, is also a pitch-regulated, variable speed wind turbine with a rotor diameter of 80 m. This turbine has a rated power of 2.75 MW and is used as a test case for validation of new aeroelastic stability tools developed within the scope of the European Commission-supported STABCON project [3].

70.3 Measurement Systems

One of the main objectives of the research project is to investigate how optical measurement systems (photogrammetry and laser interferometry) can be used to measure the dynamic response of large wind turbines. Unlike conventional measurement systems (accelerometers, piezoelectric or fiber-optic strain gauges), optical measurement techniques do not require any sensors to be placed on the turbine. Therefore, no additional preparations such as cable installations for power or data transfer are needed inside the blade or the tower. However, some reflective markers should be placed (or painted) on the structure. These markers are made up of a retroreflective material, which is 1,000 times more reflective than the background blade material. Since the markers are in the form of very thin stickers, they do not have any effect on the aerodynamic performance of the blades.

These markers are essential for both photogrammetry and laser interferometry, but they are used for different purposes in each method. Photogrammetry is a proven measurement technique based on the determination of 3D coordinates of the points on an object by using two or more images taken from different orientations and positions. Although each picture provides 2D information only, very accurate 3D information related to the coordinates and/or displacements of the object can be obtained by simultaneous processing of these images. In photogrammetry, markers are used as targets to be tracked by the camera systems, and all the targets can be tracked simultaneously. The layout of the markers throughout the turbine can be seen in Fig. 70.1.

Although photogrammetry is efficiently used in smaller scales by a wide variety of disciplines, this method was applied for the very first time to a MW-scale wind

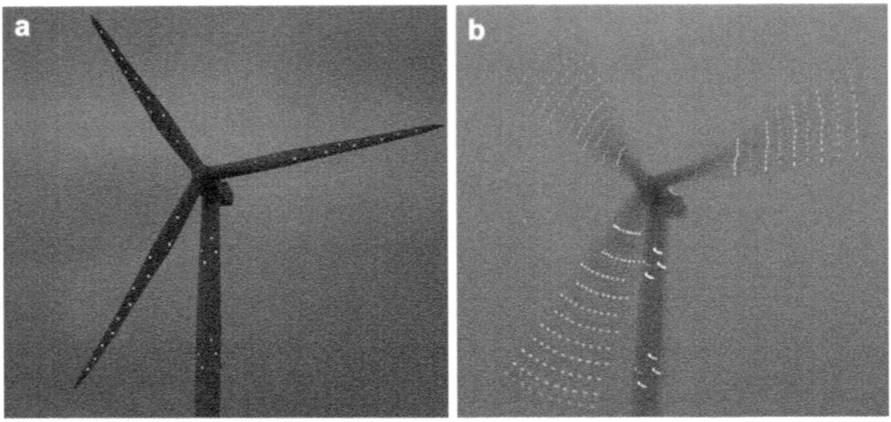

Fig. 70.1 (**a** and **b**) The layout of markers on the turbine

turbine within the scope of this research project. The analysis results showed that deformations of the turbine can be measured with an average accuracy of ±25 mm from a measurement distance of 220 m [7, 8]. Considering the fact that for a rotating turbine deformations measured in flapwise direction can be as high as ±1,000 mm, this accuracy can be considered as high.

In laser interferometry, a laser vibrometer continuously sends a laser beam to the target and receives the beam reflected from its surface. If the object is moving, this causes a frequency change and phase shift between the sent and reflected beams. By detecting this frequency change (Doppler principle), the velocity of the moving object can be found. If the object itself has a reflective surface, no extra retroreflective markers are needed. However, since the blade material was not reflective enough and the distance between the laser source and the turbine was very long, high-quality laser signals could only be acquired if the laser was targeted to the markers. Once the quality of the reflected laser beam is assured, laser vibrometer can measure the vibration of the blade with very high accuracy (in micron level) [9].

A third system, which has already been installed in the turbine as a part of a long-term wind load monitoring campaign conducted by ECN, consists of six strain gauges placed at the root region of the three blades (two strain gauges per blade) and two strain gauges located at the tower base. These strain gauges are used to measure the flapwise and edgewise vibration of the blades and fore and aft and side to side vibration of the tower at a sampling frequency of 32 Hz.

All the data recorded by three different systems were then synchronized by using a GPS clock whose absolute time accuracy is approximately 10 m. Considering the fact that frequencies that are expected to dominate the response of the wind turbine are mostly in the low-frequency range [0–5 Hz], this accuracy can be considered as sufficient. Table 70.1 summarizes the main features of measurements taken by these three different measurement systems.

Table 70.1 The main features of measurements

Measurement type	Strain gauge	Photogrammetry	Laser interferometry
Measurement block (s)	Continuous	21	294
Measurement accuracy	–	±25 mm	In micron scale
Measurement distance	On the turbine	220 m	200 m (max 300 m)
Sampling frequency (Hz)	32	28	1,280
Number of sensors	8	33	1
Parked turbine tests	X	–	X
Rotating turbine tests	X	X	–

70.4 Analysis Results and Identified System Parameters

Researchers [3, 9, 10] agree on the fact that performing modal analysis on a rotating turbine is much more challenging than performing the same analysis on a parked turbine due to the facts that:

- For a rotating wind turbine, some of the important turbine modes have very high aeroelastic damping ratios ranging between 10 and 30 % (in terms of critical damping ratio) which makes them very difficult to be detected by most of the identification algorithms that are currently in use. Aeroelastic damping is a combination of both structural and aerodynamic dampings but mostly dominated by the aerodynamic component caused by rotation of the blades. However, for a parked turbine, the aerodynamic component is small (at low wind speeds); therefore, the identified damping is generally considered to be composed of only structural damping which is usually less than 1 %. On the other hand, some exceptions to this are also described in this section.
- For a rotating turbine, integer multiples of rotational frequency (also called P harmonics where P denotes the rotational frequency) always dominate the response of the structure. These frequencies can be effective up to 24P and sometimes coincide with the true eigenfrequencies of the system [11–13].
- Besides, for rotating turbines, these P harmonics cause violation of the steady-state random excitation assumption which is one of the most important requirements of OMA algorithms.
- Another important assumption, time-invariant system requirement, is also difficult to accomplish for rotating wind turbines because of the rotation of the blades and yawing, pitching motion of the turbine. However, for parked turbines, all these motions of the different components are prevented which makes the time-invariant system assumption much easier to fulfill.

70.5 Tests on the Parked Turbine: Strain and LDV

This section summarizes the results of the analyses of strain gauge and laser interferometry measurements taken on the parked turbine. During the measurements, the turbine was kept at a fixed orientation and yawing motion was prevented by application of the yaw brakes. Blade pitch angles were fixed at zero degree where flapwise blade vibration exactly corresponds to the motion out of the rotor plane. This is the same as the angle of the blade during rotation below rated wind speeds (<15 m/s for the test turbine). Similarly, the brakes were applied to prevent the movement of the rotor.

Table 70.2 summarizes the modal parameters (frequencies and damping ratios) calculated by using strain gauge and LDV measurements. LDV measurements were taken on the markers that were close to the tip of the blades. Frequencies and damping ratios were extracted by using the NExT (Natural Excitation Technique) approach [1, 2] together with the LSCE (Least Square Complex Exponential) time domain identification method. When the turbine starts rotating, the name of the mode changes to the one indicated in parentheses. The abbreviations FW and BW stand for forward and backward whirling, respectively. Damping ratios are given in terms of critical damping ratio.

These modal parameters are important for the tuning and validation of numerical models and for the verification of prototype designs. They can also be used for health monitoring applications. As can be seen in the table, frequency values are relatively stable and do not change depending on the measurement block analyzed. However, damping values may differ slightly. The damping scatter encountered in the first flapwise and side to side tower modes is mostly related to the aerodynamic drag phenomenon [4]. Since the turbine is kept at a fixed orientation during the tests, the relative angle between the effective wind direction and the normal of the rotation plane continuously changes depending on the instantaneous wind

Table 70.2 Modal parameters calculated for the parked turbine

Mode	Frequency (Hz)	Damping
1st fore and aft tower	0.345	0.003
1st side to side tower	0.347	0.003–0.009
1st yaw (BW flapwise)	0.902	0.010–0.020
1st tilt (FW flapwise)	0.974	0.011–0.020
1st symmetric flapwise	1.077	0.010–0.020
1st vertical edgewise (BW)	1.834	0.004
1st horizontal edgewise (FW)	1.855	0.004
2nd tilt (FW flapwise)	2.311	0.005
2nd yaw (BW flapwise)	2.430	0.004
2nd symmetric flapwise	3.00	0.005
2nd edgewise	6.36	0.005
Tower torsion mode (needs further verification)	6.154	0.005

direction, resulting in a different aerodynamic coupling for each measurement. This also shows that it is not possible to completely eliminate the aerodynamic component of damping even for low wind speeds.

70.6 Tests on Rotating Turbine: Strain and Photogrammetry

This section summarizes the results of the analyses of strain gauge and photogrammetry measurements taken on the rotating turbine. During the test period, the response of the turbine was continuously measured by strain gauges. Therefore, modal parameters could be extracted for various operating conditions and wind speeds. Calculated modal parameters were then compared with the results presented by [3]. The work mentioned includes the results of both aeroelastic simulations performed by the stability tool HAWCStab [14, 15] and the real measurements taken on a wind turbine which has a similar size and capacity as the test turbine in our work. Therefore, some of the graphs presented below include the parameters extracted from our study and two additional graphs taken from the simulations and the measurements presented in the reference study [3].

Figure 70.2 shows the aeroelastic frequencies we identified for different wind speeds. As can be seen in the figure, some of the modes extracted for the parked turbine (shown in Table 70.2) could not be detected for the rotating turbine. The first tilt (FW flapwise), first yaw (BW flapwise), and first symmetric flapwise modes could not be identified due to their very high damping ratios. Hansen et al. [3]

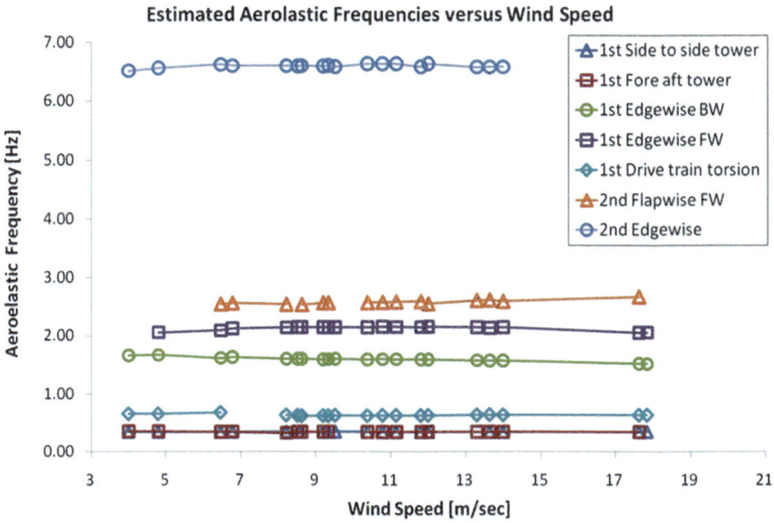

Fig. 70.2 Extracted eigenfrequencies

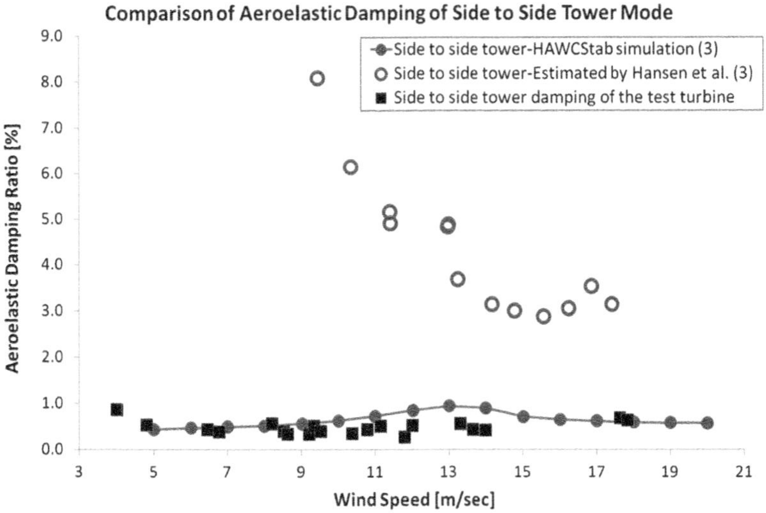

Fig. 70.3 Side to side tower mode damping comparison

experienced the same problem and reported that these three flapwise modes have too high aerodynamic dampings for identification in response to the excitation by turbulence. Similarly, the second BW flapwise mode, which has a relatively lower damping, could not be observed in the rotation data due to its weak modal participation in the overall motion.

Figure 70.3 shows the change in aeroelastic damping ratio calculated for the side to side tower mode as a function of wind speed. Identified values are in a very good agreement with the HAWCStab simulation results both in terms of trend and magnitude. The values found are less than 1 % through different operating conditions and wind speeds.

The same comparison is made for the fore and aft tower mode and the results are shown in Fig. 70.4. Although the two tower modes have almost the same frequencies, aeroelastic damping calculated for the fore and aft mode is greater due to the motion of the tower in the direction perpendicular to the rotor plane.

Comparison of aeroelastic damping ratios found for the first BW edgewise mode is shown in Fig. 70.5. The extracted damping ratios are slightly higher than the HAWCStab results, but are very close to the estimations [3] made by using in-field vibration data. Edgewise modes are very straightforward to identify because they have very high modal participation in the overall response of the turbine and low aeroelastic damping.

Similarly, Fig. 70.6 displays the same damping comparison for the first edgewise FW mode. Acquired damping ratios are again very close to both simulations and estimations given in [3].

Figure 70.7 shows aeroelastic damping ratios identified for the drivetrain torsion mode. This mode could not be detected by most of the sensors placed on the rotor and the tower. Edgewise direction strain measurements taken on the blades were the

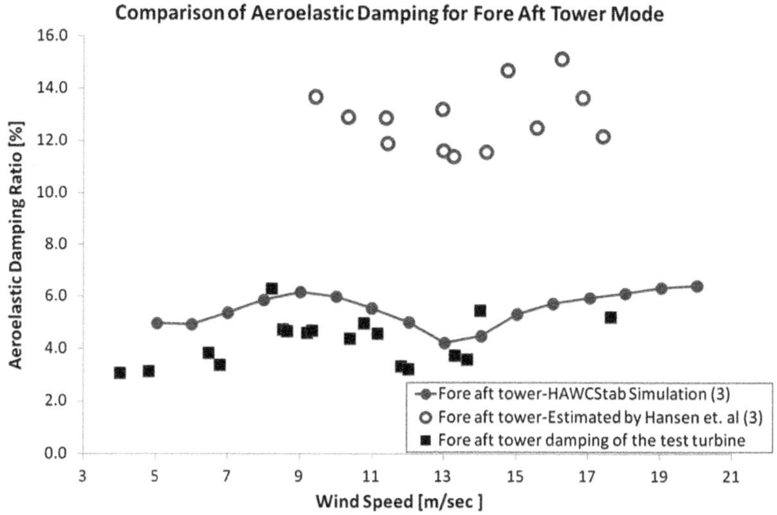

Fig. 70.4 Fore and aft tower mode damping comparison

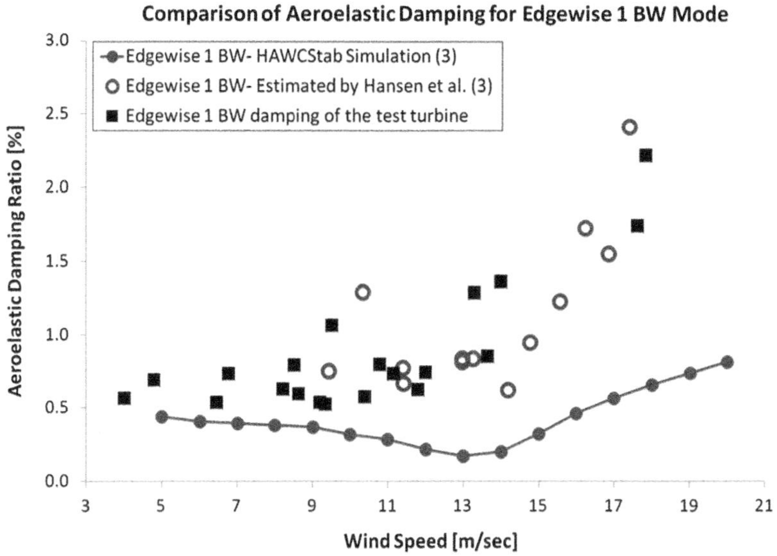

Fig. 70.5 The first edgewise BW mode damping comparison

only signals that could be used to analyze this mode. Obtained results are similar to those predicted by HAWCStab simulations. Since any damping estimation for this mode was not given in the reference study, the identified damping ratios could only be compared with the simulation results.

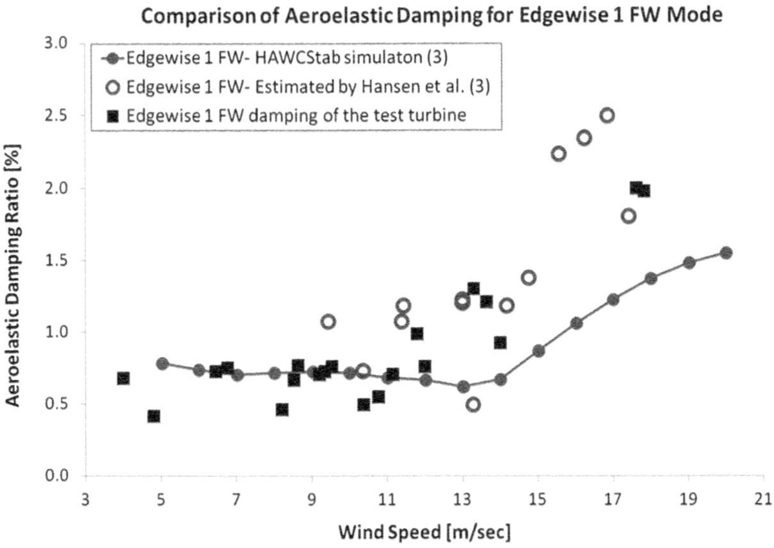

Fig. 70.6 The first edgewise FW mode damping comparison

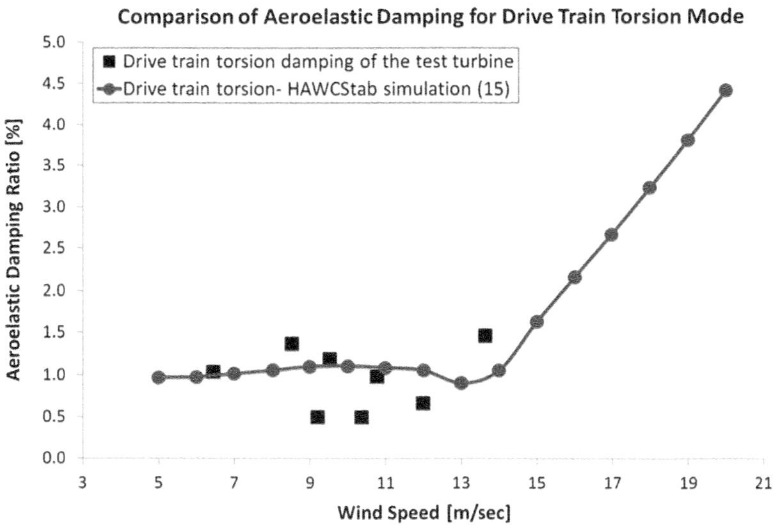

Fig. 70.7 The first drivetrain torsion mode damping comparison

Figure 70.8 shows aeroelastic damping ratios calculated for the second FW flapwise and second edgewise modes. Because Hansen et al. did not include the damping estimations or simulation results for these modes in their work [3], the identified damping ratios could not be compared with any reference values.

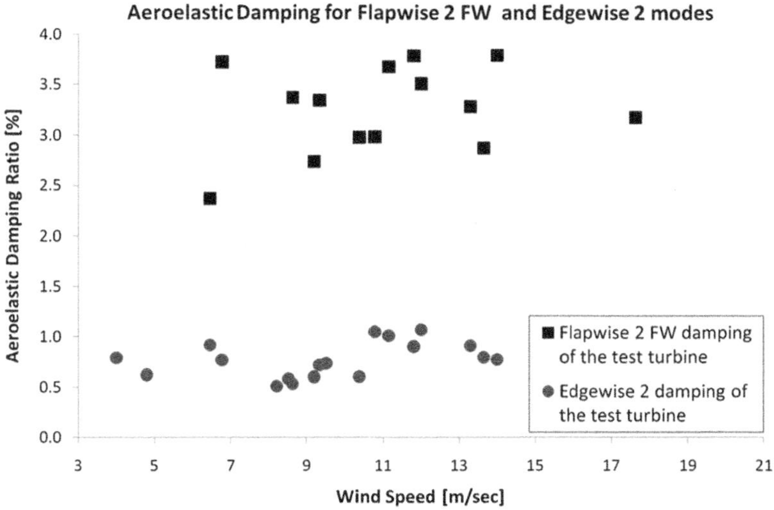

Fig. 70.8 Damping ratios extracted for the second FW flapwise and second edgewise modes

70.7 Conclusions

Identification of the modal parameters of wind turbines are very important for optimizing the energy produced, ensuring safe and reliable operation, and increasing the lifetime of the system. However, conventional dynamic testing techniques based on exciting the structure from several locations with sufficient force amplitudes cannot be easily applied to these challenging structures due to their size and the technical difficulties in providing very large forces that are required to reach sufficient excitation levels.

OMA (operational modal analysis) tools, namely, the analysis methods that do not require the forces acting on the system to be measured, can be a solution to some important problems encountered in dynamic testing of wind turbines. Since estimation of the modal parameters is solely based on the use of measured response, these methods can easily be used to extract the dynamic properties of these large structures excited by natural environmental inputs (winds).

Analyses performed by using OMA methods seem very promising in extracting the modal parameters. Within the scope of the research, 12 different turbine modes were successfully calculated from the measurements taken on the parked turbine using strain gauges and LDV.

Similarly, several turbine modes could be identified from in-operation measurements using strain gauges and photogrammetry. Obtained results are in good coherence with those presented in similar studies in literature.

Performing modal analysis on a rotating turbine is much more challenging than performing the same analysis on a parked turbine due to the high aeroelastic damping of some important modes, rotational P harmonics that dominate the dynamic response, and the difficulties in fulfilling some important system identification assumptions such as time-invariant system and steady-state random excitation.

During the analyses, it was observed that frequency values are more easily identified and the calculated values are mostly stable. However, some scatter can be encountered in estimated damping ratios. This scatter can be caused by physical factors such as the change in operating conditions or mathematical uncertainty related to the applied algorithms.

Acknowledgments This research project was partly funded by the We@Sea research program, financed by the Dutch Ministry of Economic Affairs. The authors would like to thank ECN (Energy Research Center of the Netherlands) for providing the test turbine and the other technical equipment. The authors also acknowledge the extensive contribution of Pieter Schuer (GOM mbH), Wim Cuypers (GOM mbH), Theo W. Verbruggen (ECN), and Hans J. P. Verhoef (ECN) in organizing and performing the field tests.

References

1. James GH, Carne TG, Lauffer JP (1995) The natural excitation technique (NExT) for modal parameter extraction from operating structures. J Anal Exp Modal Anal 10(4):260–277
2. James GH, Carne TG, Veers P (1996) Damping measurements using operational data. ASME J Solar Energy Eng 118:190–193
3. Hansen MH, Thomsen K, Fuglsang P (2006) Two methods for estimating aeroelastic damping of operational wind turbine modes from experiments. Wind Energy 9:179–191
4. Carne TG, James GH (2010) The inception of OMA in the development of modal testing technology for wind turbines. Mech Syst Signal Process 24:1213–1226
5. GOM Optical Measuring Techniques. http://www.gom.com. Accessed Apr 2014
6. ECN Energy Research Center of the Netherlands. https://www.ecn.nl/expertises/wind-energy/tests-measurements/. Accessed Apr 2014
7. Ozbek M, Rixen DJ, Erne O, Sanow G (2010) Feasibility of monitoring large wind turbines using photogrammetry. Energy 35:4802–4811
8. Ozbek M, Meng F, Rixen DJ, van Tooren MJL (2010) Identification of the dynamics of large wind turbines by using photogrammetry. Proceedings of the IMAC international modal analysis conference, Jacksonville, FL
9. Ozbek M, Rixen DJ, Verbruggen TW (2009) Remote monitoring of wind turbine dynamics by laser interferometry: phase 1. Proceedings of the IMAC international modal analysis conference, Orlando, FL
10. Tcherniak D, Chauhan S, Hansen MH (2010) Applicability limits of operational modal analysis to operational wind turbines. Proceedings of the 28th international modal analysis conference, Jacksonville, FL
11. Ozbek M, Rixen DJ (2011) Optical measurements and operational modal analysis on a large wind turbine: lessons learned. Proceedings of the IMAC international modal analysis conference, Jacksonville, FL
12. Ozbek M, Rixen DJ (2013) Operational modal analysis of a 2.5 MW wind turbine using optical measurement techniques and strain gauges. Wind Energy 16:367–381
13. Ozbek M, Meng F, Rixen DJ (2013) Challenges in testing and monitoring the in-operation vibration characteristics of wind turbines. Mech Syst Signal Process 41:649–666
14. Hansen MH (2004) Aeroelastic stability analysis of wind turbines using an eigenvalue approach. Wind Energy 7:133–143
15. Riziotis VA, Voutsinas SG, Politis ES, Chaviaropoulos PK (2004) Aeroelastic stability of wind turbines: the problem, the methods and the issues. Wind Energy 7:373–392

Chapter 71
PV/Wind Installation: Powered Rural Zone Family House in Bechar (Algeria)

B. Bouchiba, M. Dahbi, F. Benhamida, A. Hazzab, and M. Habbab

Abstract In the development of energy sources in rural regions in Algeria, it is necessary to view the use of solar and wind energy in all applications as one of the most promising new and renewable energy sources. This paper presents the methodology of the feasibility study using hybrid (wind-solar) energy conversion system for providing the electrical loads in a family house according to their energy requirements. A long-term data of wind speed and solar radiation for every hour of the day were used. These data were used to calculate the average power generated by a PV/wind installation for every hour of a typical day in a month. A load of a typical family house in a rural zone in Sahara of Algeria (desert area) was used as a load demand of the system. The hybrid (wind-solar) system considered in the present analysis consists of one 1 kW wind energy conversion system (WECS), 125 W of (10) BP photovoltaic panels together with battery storage system. A computer program has been developed to achieve this and to determine the specification of hybrid system component. The study was performed using a graphical user interface programmed in MATLAB/Simulink environment.

Keywords Renewable energy • Solar and wind energy • Battery storage system • Load demand

Nomenclature

PV	Photovoltaic generator
WG	Wind generator
WECS	Wind energy conversion system

B. Bouchiba (✉) • M. Dahbi • A. Hazzab • M. Habbab
CAOSEE Laboratory, Faculty of sciences and technology, University of Bechar, Bechar, Algeria
e-mail: bouchiba_bousmaha@yahoo.fr; m.dahbi@yahoo.fr; a_hazzab@yahoo.fr; m_habbab@yahoo.fr

F. Benhamida
IRECOM Laboratory, Department of Electrical Engineering, University of Djillali Liabes, Sidi Bel-Abbes, Algeria
e-mail: farid.benhamida@yahoo.fr

I. Dincer et al. (eds.), *Progress in Clean Energy, Volume 2*,
DOI 10.1007/978-3-319-17031-2_71

C_p	Power coefficient
λ	Tip-speed ratio
ρ	Air density
A_w	Swept area of the rotor (m^2)
V	Wind speed (m/s)
η_{ad}	Efficiency of the AC/DC converter
Ins	Insulation data
A	Area of the single PV panel
η_{PV}	Overall efficiency of the PV panel and the DC/DC converter
η_{pc}	Power conditioning efficiency
Pf	Packing factor
N_w	Number of wind turbines
N_{PV}	Number of PV panels
P_{wind}	Wind power
P_{PV}	Solar PV power
P_T	Equivalent to the energy generated at a particular hour
$E_B(t)$	Charge quantities of battery bank
σ	Hourly self-discharge rate
W_L	Load demand
η_{inv}	Efficiency of inverter
η_B	Efficiency of battery bank

71.1 Introduction

Renewable energy is expected to play a very important role in meeting energy demands in the near future. Since it is a clean type of energy with a diversity of applications, decentralized nature, and availability, the hybrid wind-solar energy will represent a suitable solution for energy requirements especially in rural areas. It is important to state that the use of hybrid wind-solar energy in rural regions will protect these areas from pollution, since the use of hybrid wind-solar energy home systems avoids large amounts of CO_2 emissions [1].

In Algerian context, a company of electricity, Sonelgaz, has used hybrid system-based solar and wind energy to power the isolated villages and remote houses of south Algeria [3]. These locations include Behar Adrar, Timimoun (Tala Hamou-Moussa), and Tindouf (Gara Djebilet, Hassi Mounir, Draa el Khadra).

Bechar (southwest Algeria) in new valley was selected as the site under consideration in this work since it is the second largest province in Algeria. This work presents a study and design of a complete hybrid system (1 wind generator, 10 PV panels, and 12 batteries for 2-day autonomy) for providing the electrical loads in a load of a typical family house in a rural zone in Sahara of Algeria (desert area) was used as a load demand of the system. For a given load and a mixed multiple-criteria integer programming problem, the types and sizes of PV generator, wind turbine generators (WTG), and storage system were calculated. A computer program is developed to achieve this and to determine the specification of hybrid system component. The paper is organized as follows. In Sect. 71.2, the environmental data is presented. The system modeling (wind system, PV system, and battery system) is outlined in Sect. 71.3, and energy requirements of a typical house in rural zones are given in Sect. 71.4. The system design is presented in Sect. 71.5. Simulation results and discussion are shows in Sect. 71.6. Conclusions are drawn in Sect. 71.7.

71.2 Environmental Data

Climatic conditions determine the availability and magnitude of wind and solar energy at a site. Bechar is a province of southwestern Algeria; it is based around an oasis of the Sahara Desert. Two distinct seasons are noticed in this region: a very hot season (April to October) and a cold season (November to March); monthly mean temperature reaches close to 35 °C for hot months, and in cooler months, the mean temperatures drop by about 15 °C as compared to the hot months [4]. For the site under consideration, 22-year long-term wind speed data and solar irradiation data have been used [5]. Figure 71.1 shows the daily average wind speed and daily average solar insulation profile from the data recorded for the site (Bechar). It is clear from Fig. 71.1 that wind speed is generally higher in spring and summer months (March to August), where it exceeds 6 m/s. The minimum of wind speed during the year is 5.38 m/s in October, and the maximum is 6.22 m/s in July.

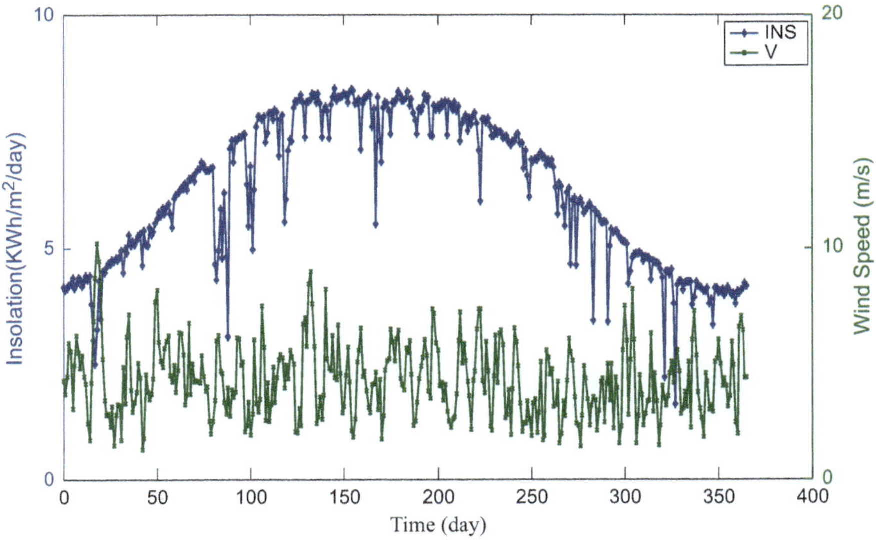

Fig. 71.1 Daily average wind speed (m/s) and daily average solar insolation of Bechar site (kWh/m^2/day)

This clearly reflects that a WECS would produce appreciably more energy during the year. The wind energy calculations are made by matching the power-wind speed characteristics of commercial wind machines with the long-term hourly wind speed data.

The monthly average values of solar energy incident on the horizontal in the Bechar. It is clear from the result that solar energy incident in this region is very high, especially during summer months, where it exceeds 7 kWh/m^2/day as compared to other months. The average solar radiation is 5.84 kWh/m^2/day. As it can be seen, the variation between the years is minimal.

71.3 The System Modeling

To design a hybrid wind photovoltaic system (PV/WG) for a family house in rural zones, Fig. 71.2 shows the component of the system.

71.3.1 Modeling of Wind System

The wind turbine is characterized by the nondimensional curve of coefficient of performance C_p as a function of tip-speed ratio. λ is the ratio of linear speed of the tip of blades to the rotational speed of wind turbine. It can be expressed as follows [6, 7]:

$$\lambda = \omega_W \times r/V \tag{71.1}$$

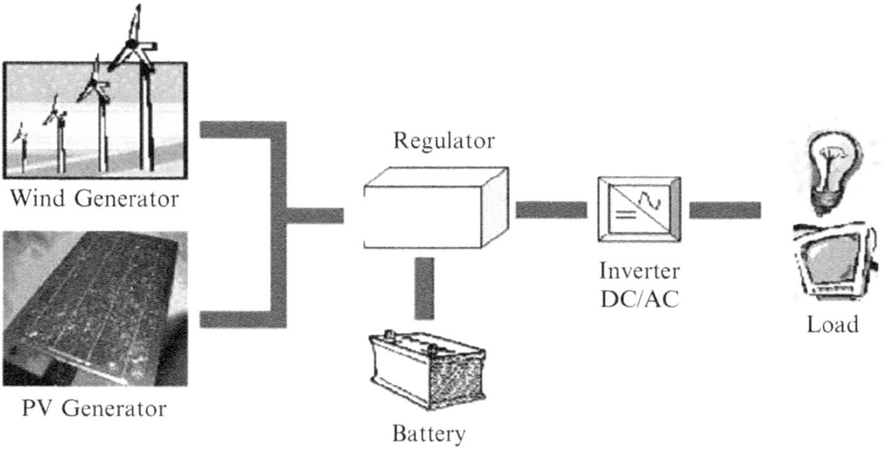

Fig. 71.2 Hybrid system component

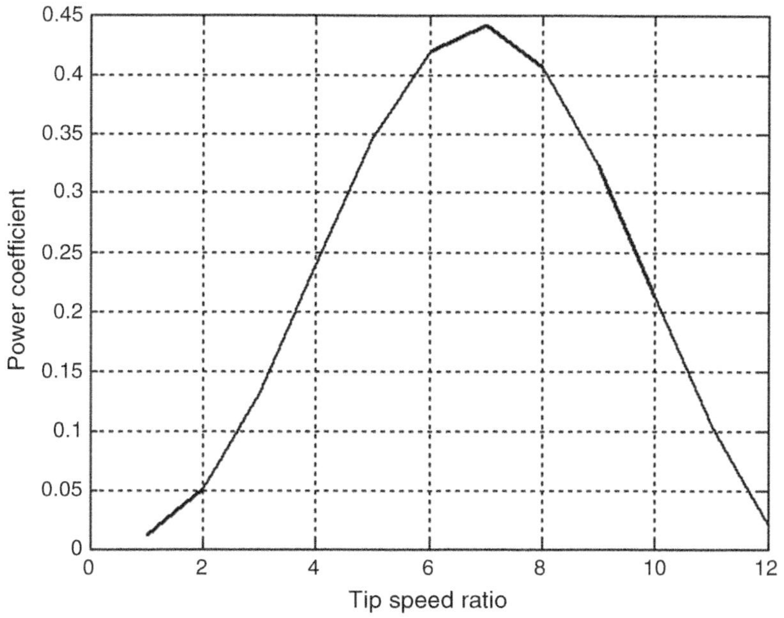

Fig. 71.3 Power coefficient C_p versus tip-speed ratio λ

where r is the radius of the rotor, ω_w is mechanical angular velocity of the rotor, and V is speed of wind.

The coefficient of performance is also known as power coefficient. The power coefficient C_p versus tip-speed ratio λ is given in Fig. 71.3. For the wind turbine used in the study (BWC XL.1 1 kW), the C_p-opt = 0.44 as of tip-speed ratio λopt = 6.9.

Average hourly wind speed data is evaluated and converted to wind turbine power. If the wind speed is between the cut-in and the rated speed of the wind turbine, then the power output is defined as [8]

$$P_{\mathrm{Wind}}(t) = \frac{1}{2}\rho A_w V(t)^3 C_P \eta_{ad} \tag{71.2}$$

where ρ is the air density (kg/m^3), A_w is the swept area of the rotor (m^2), V is the wind speed (m/s), C_p is the efficiency of the wind turbine (0.442), and η_{ad} is the efficiency of the AC/DC converter (assumed to be 95 % in this study).

If wind speed is between the rated wind speed and the furling speed of the wind turbine, the power output will be equal to the rated power of the wind turbine. Finally, if the wind speed is less than the cut-in speed or greater than the furling speed, there will be no output power from the turbine.

For example, the relation between the electrical power output of a BWC XL.1 1 kW generator and mechanical angular velocity of the rotor ωw, for different wind

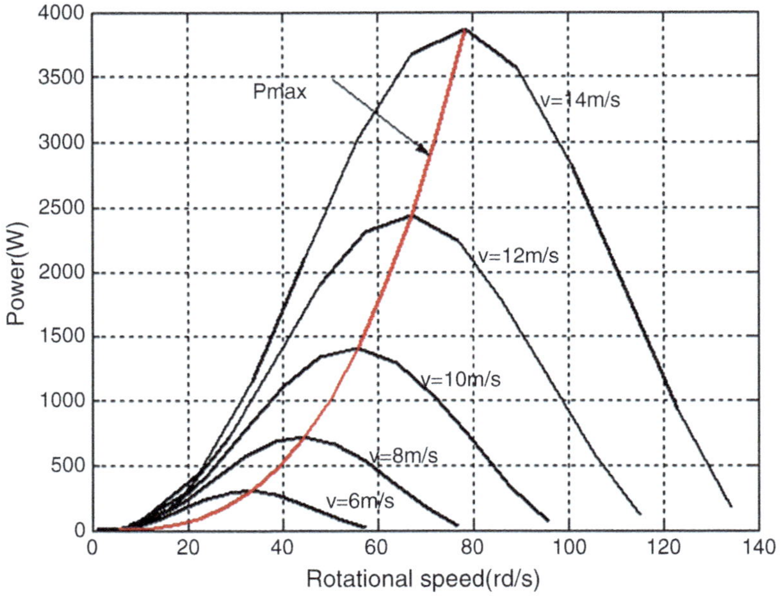

Fig. 71.4 Wind turbine power characteristics

Table 71.1 Wind turbine parameters

Rated output (kW)	1
Cut-in speed (m/s)	2.5
Rated speed (m/s)	11
Cut-out speed (m/s)	13
Rotor diameter (m)	2.5
Air density (kg/m^2)	1.225
Efficiency of wind turbine (%)	0.442

speed, is shown in Fig. 71.4. This illustrates the fact that for any given wind speed, there is a rotation speed ωw, which generates the maximum power.

The performance data of the wind turbine is provided in Table 71.1. A Bergey BWC XL.1 1 kW wind turbine is selected for the study.

71.3.2 Modeling of PV System

Insulation data is converted into power output from the PV array using the following equation [2]:

$$P_{\mathrm{PV}}(t) = \mathrm{Ins}(t) A \eta_{\mathrm{PV}} \eta_{\mathrm{ad}} P_{\mathrm{f}} \tag{71.3}$$

where Ins(t) is the insulation data at time t (kWh/m^2), A is the area of the single PV panel (m^2), η_{PV} is the overall efficiency of the PV panel and the DC/DC converter, η_{pc} is the power conditioning efficiency (0.86), and P_{f} is the packing factor (0.9).

Figure 71.5 gives the result for different PV solar radiation levels with constant PV temperature (25 °C). The current and the power of a PV module change significantly with the change in PV solar radiation.

The total wind and PV-generated power during each hour are first computed as follows [4]:

$$P_{\mathrm{T}}(t) = N_{\mathrm{w}} P_{\mathrm{wind}}(t) + N_{\mathrm{PV}} P_{\mathrm{PV}}(t) \tag{71.4}$$

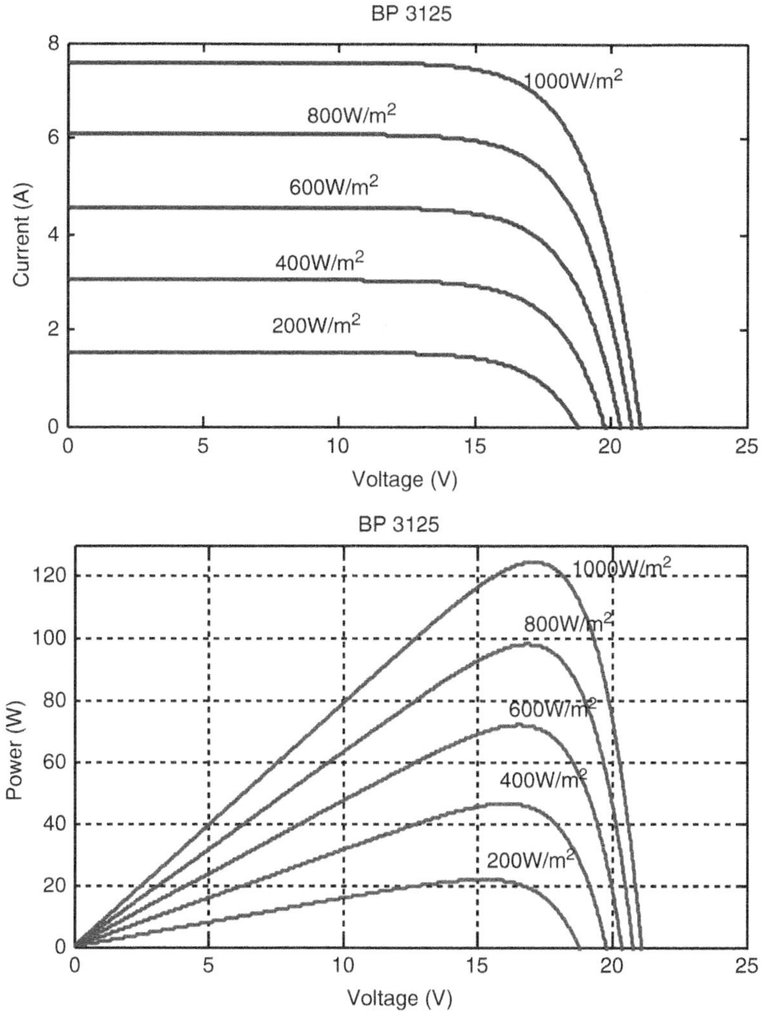

Fig. 71.5 The IV and PV characteristic of PV module

where N_w and N_{PV} are the number of wind turbines and PV panels and $P_{wind}(t)$ and $P_{PV}(t)$ are the power from wind and solar PV at time t, respectively. $P_T(t)$ is equivalent to the energy generated at a particular hour (since there is a 1 h time step).

When the total power (PV/wind) is greater than the power demand, the battery bank is in a charging state. The charge quantity of the battery bank at the time t can be described by [9]

$$E_B(t) = E_B(t-1)(1-\sigma) + (P_T(t) - W_L(t)/\eta_{inv})\eta_B. \tag{71.5}$$

On the other hand, the battery bank is in a discharging state. In this paper, the discharging efficiency of the battery bank is assumed to be 1. Therefore, the discharge quantity of the battery bank at the time t can be expressed as

$$E_B(t) = E_B(t-1)(1-\sigma) + (W_L(t)/\eta_{inv} - P_T(t)) \tag{71.6}$$

where $E_B(t)$ and $E_B(t-1)$ are the charge quantities of the battery bank at the time t and $t-1$, σ is the hourly self-discharge rate, $P_T(t)$ is the total power generated by renewable energy source after energy loss in the controller, $W_L(t)$ is load demand at time t, and η_{inv} and η_B are the efficiency of inverter and charge efficiency of battery bank. The performance data of PV panels is provided in Table 71.2; BP SX-12S (120 W) solar panel is selected for the study.

71.4 Energy Requirements of a Typical House in Rural Zones

The family houses in the rural zones of the Bechar in Algeria are expected to be very simple and do not need large quantities of electrical energy for lighting or operating electric appliances. It is expected that most of these houses will have electrical loads as shown in Table 71.3.

In Fig. 71.6, a suggested load curve for a complete typical day's energy consumption is shown.

Table 71.2 PV panel parameters

Maximum power (W)	125
Efficiency (%)	12
Area (m^2)	1.01
VOC (V)	22.1
ISC (A)	7.54
Vn (V)	24

Table 71.3 Daily energy requirement for a typical house in rural zones

Load type	No. of unit	Load power (W)	Operating period (h/day)
Lamps (light)	5	5 × 40	From 20 to 24 h
Refrigerator	1	100	24 h/day
TV	1	80	From 17 to 24 h
Motor pump	1	120	From 10 to 14 h
Washing machine	1	250	From 12 to 16 h
Electric fan	1	100	From 11 to 16 h
Total energy (Wh/day)			5,740

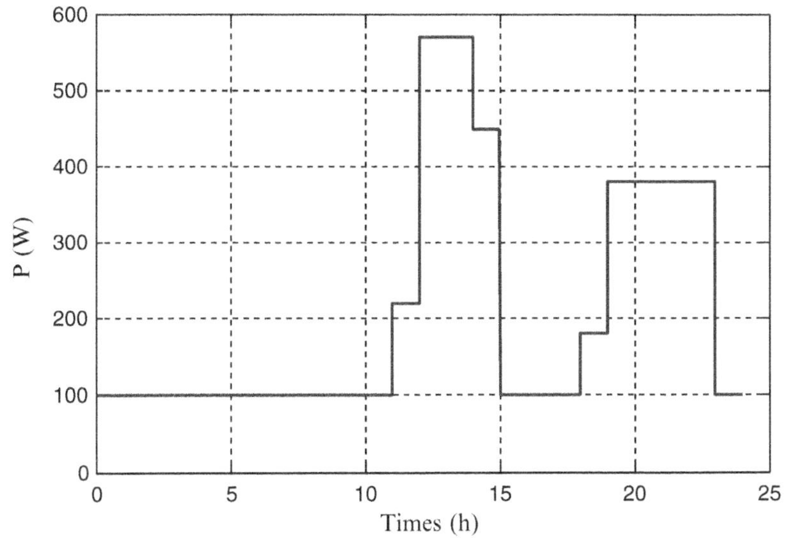

Fig. 71.6 The load curve of a typical complete day's consumption in the spring

Then according to this, the total required daily demand is [10]

$$W_{\mathrm{T}} = \frac{W}{\eta_{\mathrm{Total}}} = \frac{5,740}{0.46} = 12,478\,\mathrm{Wh/day}. \tag{71.7}$$

71.5 The System Design

71.5.1 Wind Generator

The acquired power from the wind turbine (BWC XL.1 1 kW) considering 16 h is [11]

$$W_{\mathrm{wind}} = W \times N_{\mathrm{wind}} \times H_{\mathrm{wind,day}} = 370 \times 1 \times 16 = 5,920\,\mathrm{Wh}. \tag{71.8}$$

71.5.2 PV Array Sizing

The required power from the PV panels will be

$$W_{\mathrm{PV}} = W_{\mathrm{T}} - W_{\mathrm{wind}} = 12{,}478 - 5{,}920 = 6{,}558\,\mathrm{Wh}. \tag{71.9}$$

For the hybrid system considered in the present paper, the PV module type BP3125 solar module was used [17]:

PV panel efficiency $\eta_{\mathrm{PV}} = 12\ \%$
Nominal capacity of PV panels: 125WP
Voc = 22.1 V; Isc = 7.54 A
Max power = 125 W
Voltage selection for the system: VDC = 24 V
Then, the required daily [Ah] is

$$E(\mathrm{Ah}) = \frac{W_{\mathrm{PV}}}{V_{\mathrm{DC}}} = \frac{6{,}558}{24} = 271.21\,\mathrm{Ah/day}. \tag{71.10}$$

Based on the average solar intensity per month where the number of daily hour solar intensity is 7.78 h [11], the total current required for the system is

$$I_{\mathrm{Total}} = \frac{E(\mathrm{Ah})}{\mathrm{Hours}} = \frac{271.21}{6} = 34.85A. \tag{71.11}$$

To calculate the required number of PV panels that would be connected in series and in parallel and based on the system voltage (i.e., 24 V), then

$$N_{\mathrm{SPV}} = \frac{V_{\mathrm{DC}}}{V_{\mathrm{CC}}} = \frac{24}{22.1} = 1.08 \approx 2 \tag{71.12}$$

$$N_{\mathrm{PPV}} = \frac{I_{\mathrm{Total}}}{I_{\mathrm{SC}}} = \frac{34.85}{7.54} = 4.62 \approx 5. \tag{71.13}$$

The total required and applied number of PV panels is

$$N_{\mathrm{PV}} = N_{\mathrm{SPV}} \times N_{\mathrm{PPV}} = 2 \times 5 = 10. \tag{71.14}$$

71.5.3 Design of the Storage System

The battery capacity is obtained by the following equation [12]:

$$C_{\mathrm{bat}}(\mathrm{Wh}) = \frac{W_{\mathrm{T}} \times \mathrm{autonomy(day)}}{D \times \eta_{\mathrm{B}}} \tag{71.15}$$

$$C_{\text{bat}}(\text{Ah}) = \frac{C_{\text{bat}}(\text{Wh})}{V_{\text{n}}} \tag{71.16}$$

where W_{T} (Wh) is the total daily load demand and D (%) and η_{B} are the degree of discharge of the batteries and charge efficiency of battery bank. Two (2) days of autonomy were fixed for this study:

$$C_{\text{bat}}(\text{Wh}) = \frac{12,478 \times 2}{0.5 \times 0.8} = 62,391\,\text{Wh} \tag{71.17}$$

$$C_{\text{bat}}(\text{Ah}) = \frac{62,391}{24} = 2,599\,\text{Ah} \tag{71.18}$$

To meet the load profile, the battery capacity needed is 385 Ah (for 12 V), and thus number of batteries should be six. The performance data of battery storage used in this evaluation is provided in Table 71.4.

71.6 Results and Discussion

The hybrid system simulated in the present investigation consists of 1 kW wind generator, ten panels of photovoltaic solar array (120 W), and six batteries.

Figures 71.6 and 71.7 show the daily curve and output power from the hybrid PV/wind system and charge/discharge of battery storage.

The daily energy production and consumption situations of a typical hybrid system are given in Fig. 71.7. The maximum daily energy consumption is 570 W between 12 and 16 h, and the minimum value is 100 Wh during the morning. The fluctuation of the total outputs is all fairly violent. Furthermore, the output of wind generator changes more fiercely than that of the PV array, which is determined by the inhomogeneity nature of wind power and solar energy distribution.

The production exceeds the loads between 8 h and 17 h; the battery is also charged. It is found that the low PV power during this period can be observed. Moreover, the wind production exceeds the load demand. From the results obtained from the simulation of typical days, it is clear,that the advantage of the complementarily of wind power and PV power and while minimizing the use of battery storage. The surplus of power is stored in batteries in order to be used during the insufficient production of PV/WG system (the low insolation and low wind speed period) between 17 and 20 PM (Fig. 71.7).

Table 71.4 Parameters of battery

Manufacturer	TOYO
Type	GFM-100
Rated voltage	12 V
Rated capacity	385 Ah
Hourly self-discharge rate	0.0001
Efficiency (%)	80

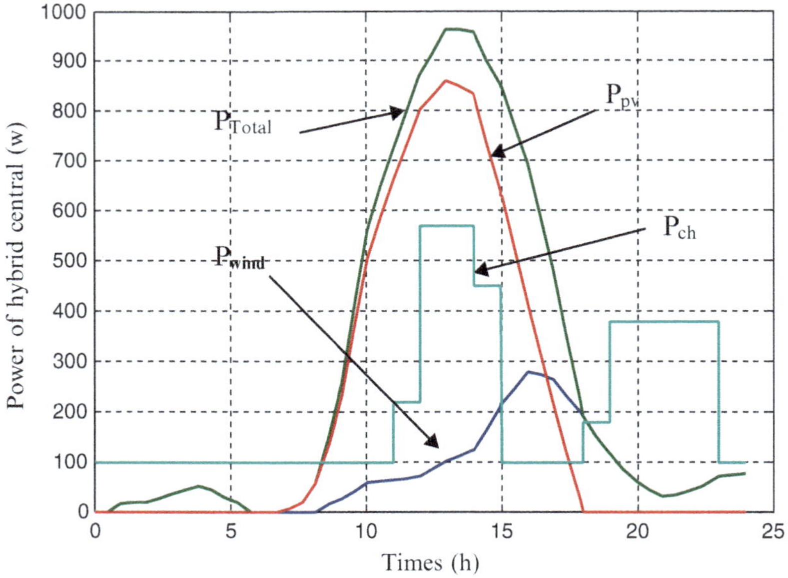

Fig. 71.7 The daily energy production and consumption situations of typical 1 hybrid system (Wh)

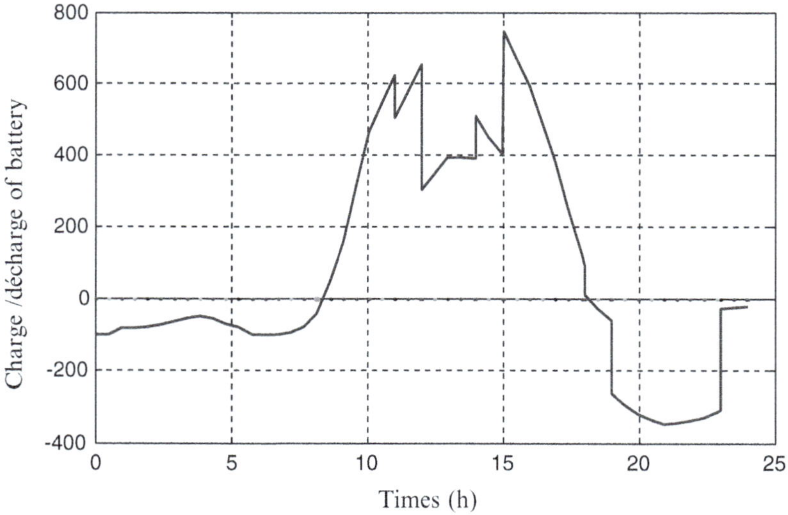

Fig. 71.8 The hourly charge/discharge quantity of battery (Wh)

The batteries are necessary for a backup during a period of insufficient energy of both two energy sources (WG system and PV system) due to weak wind speed and solar radiation. The hourly charge/discharge quantity of battery is given in Fig. 71.8. It is clear from the results obtained from the simulation of the typical days that the maximum charge quantity of battery E_{Bmax} is 750 W of 15 AM and the maximum discharge quantity is 350 W of 21 PM. The situation of hourly discharge quantity of battery losses is also comparatively severe in 19 h which are caused by the poorest wind power in this hour of day.

In summary, the operating situation of the system is determined by various factors such as the load demand, distribution of the natural resources, and the capacities of chosen components. Also it can be noticed that with the advantage of the complementariness of either wind power or solar energy, the power failure time is dramatically decreased.

It is noted that surplus energy is generated in each hour, which results from the inhomogeneity of the distributions of natural resources.

71.7 Conclusions

This paper presents the feasibility of using autonomous hybrid photovoltaic-wind generator with battery storage (PV/WG) in the site of Bechar (Algeria). The monthly average wind speeds for Bechar range from 5.38 to 6.22 m/s. The monthly average daily value of solar radiation range from 3.21 to 7.49 kWh/m^2. It indicates that this site consists of an appropriate candidate for deployment of hybrid PV/WG systems. The batteries are necessary for a backup during a period of insufficient energy of both two energy sources (WG system and PV system) due to weak wind speed and solar radiation. The complementary feature of PV and wind makes the system more reliable. Finally, it can be concluded that the use of renewable energy source as a reliable option for rural site in southwest of Algeria is justified. In addition, the impact of this system in providing electricity will create many new opportunities and avenues leading to national development through rural development.

References

1. Hongxing Y, Wei Z, Chengzhi L (2009) Optimal design and techno-economic analysis of a hybrid solar–wind power generation system. Appl Energy 86:163–169
2. Celik AN (2003) A simplified model for estimating the monthly performance of autonomous wind energy system with battery storage. Renew Energy 28(4):561–572
3. Dahbi M, Benatiallah A, Benslimane T (2009) Modeling and simulation of hybrid (wind + photovoltaic + battery) power system in south west Algerian site (Bechar). Int Rev Model Simul 2(2):178–183
4. Office National de la Météorologie: Atlas climatologique national, Recueil de données, période (1975–1984)
5. Browyy BS, Salameh MZ (1997) Dynamic response of a stand-alone wind energy conversion system with battery energy storage to wind gust. IEEE Trans Energy Convers 12(1):73–78
6. Johnson GL (1985) Wind energy systems. Prentice-Hall, New York

7. El-Shatter TF, Eskander MN, El-Hagry MT (2006) Energy flow and management of a hybrid wind/PV/fuel cell generation system. Energy Convers Manag 47:1264–1280
8. Deshmukh MK, Deshmukh SS (2008) Modeling of hybrid renewable energy systems. Renew Sustainable Energy Rev 12(1):235–249
9. Essalaimeh S, Al-Salaymeh A, Abdullat Y (2013) Electrical production for domestic and industrial applications using hybrid PV-wind system. Energy Convers Manag 65:736–743
10. Dahbi M, Benatiallah A, Sallem M (2013) The analysis of wind power potential in Sahara site of Algeria-an estimation using the 'Weibull' density function. TERRAGREEN13 international conference advancements in renewable energy and clean environment (ARECE), 15–17 Feb 2013, Beirut, Lebanon
11. Nfah EM, Ngundam JM, Tchinda R (2007) Modelling of solar/diesel/battery hybrid power systems for far-north Cameroon. Renew Energy 32:832–844
12. Capderou M (1987) Atlas Solaire de l'Algérie, Tome 01, vol 2, OPU Alger

Chapter 72
Prospects of a Settlement in Southern Georgia Based on Green Energy

Alexander G. Tvalchrelidze

Abstract Georgia is an ideal country for the project of a settlement entirely based on renewable energy: its renewable energy resources are huge, diversified, and entirely meeting the demands of the project. As a result of a thorough investigation, the Zveli Site in Southern Georgia was identified. The site represents a land plot of about 25 km^2 situated on the second plane above the flood of the Mtkvari (Kura) River. Hydropower resources of the Mtkvari River and its inflows near the Zveli Site are about 400 MW, 40 times exceeding the preliminary assessed needs of the site. Geothermal resources of Samtskhe-Javakheti are assessed to be 13.87 MW including 2.16 MW in the Aspindza deposit situated in 5 km from the site. Photovoltaic and wind resources are also huge. The settlement will consist of residential + administrative and industrial zones. The settlement and its industries are suggested to use energy-saving technologies. Wind, photovoltaic, biomass, hydropower, and geothermal sources of energy will be used. The main fields of business activities would be agriculture and food product processing and marketing at the background of 100 % employment. The settlement will have a kindergarten, a school, a police post, a fire station, municipality services, supermarkets, and cafés, as well as a park zone.

Keywords Sustainable development • Project • Green energy • Zveli Site • Georgia

72.1 Introduction

According to the classical definition [1], sustainable development is a pattern of resource use that aims to meet human needs while preserving the environment so that these needs can be met not only in the present but also for future generations. In other words, sustainable development consists in the synergy of three constituent parts—(1) environmental development, (2) social development, and (3) economic development (Fig. 72.1).

A.G. Tvalchrelidze (✉)
Georgian Academy of Natural Sciences, 3, Bazalety St., 0179 Tbilisi, Georgia
e-mail: sandrotval@yahoo.com

I. Dincer et al. (eds.), *Progress in Clean Energy, Volume 2*,
DOI 10.1007/978-3-319-17031-2_72

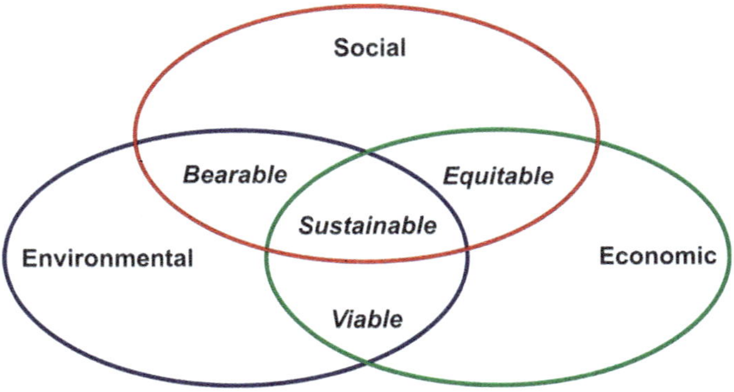

Fig. 72.1 Model of sustainable development

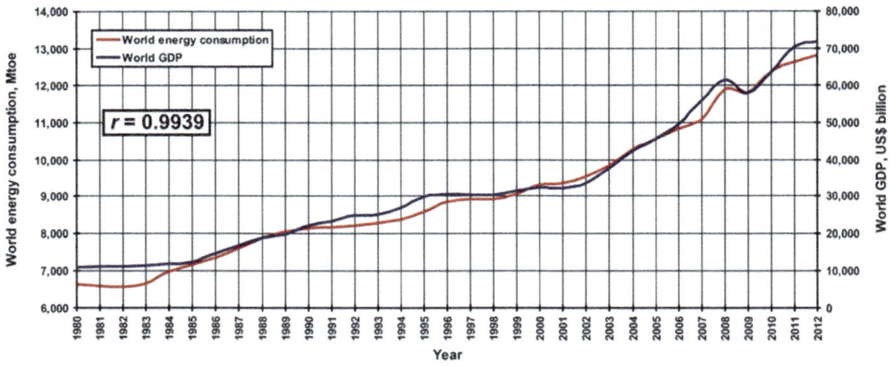

Fig. 72.2 Interdependence between world energy consumption and world GDP in 1980–2012

Sustainable development is a complex issue, with a variety of features, though its gist is obvious: environmental security, renewable energy systems, and development of relevant technologies. And yet, the idea of sustainable development primarily implies the use of renewable energy (that of water, sun, wind, geothermal sources, biomass) and the introduction of relevant renewable technologies. Hence, sustainable development of the globalized world is essentially determined by energy supply. Figure 72.2 demonstrates the correlation between the world energy consumption (in million tons of oil equivalent), as reported by the US Energy Information Administration [2], and the world gross domestic product (GDP), as released by the World Bank Group [3].

In the structure of world final energy consumption, the share of renewable energy sources is as large as 19 %; however, among renewable energy sources, only 20.6 % is used for power generation [4]. According to the appraisal by the US Energy Information Administration [2], world energy consumption by 2030 would rise to 678 quadrillion Btu (British thermal unit) or to 642 quadrillion kJ but the share of renewable energy sources would not change.

On the other hand, the Kyoto Protocol [5], which entered into force on February 16, 2005, and today is ratified by 191 states and 1 regional economic integration

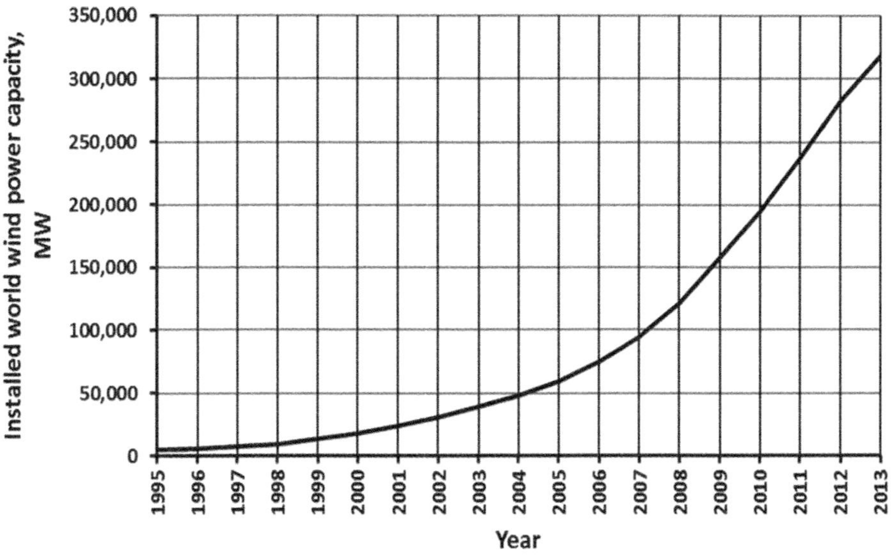

Fig. 72.3 Installed world wind power capacity in 1995–2014

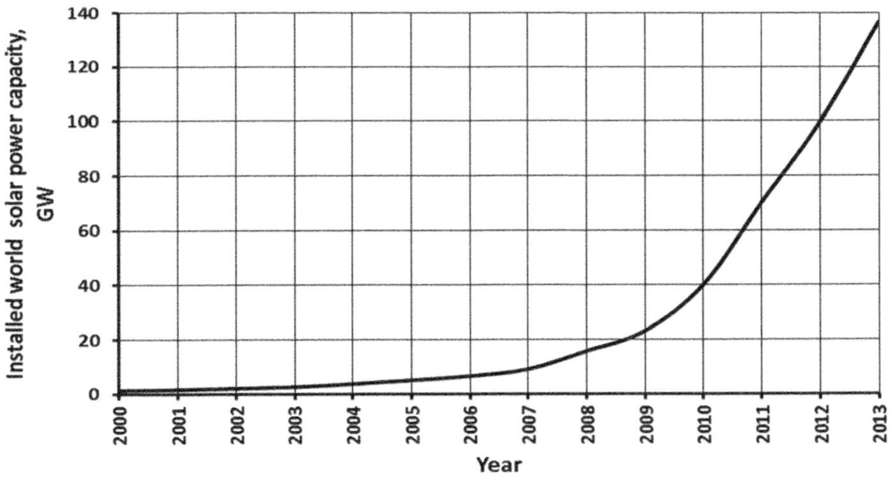

Fig. 72.4 Installed world photovoltaic capacity in 2000–2014

organization [6], imposes limitations of greenhouse emissions into the atmosphere. Correspondingly, the twenty-first century is considered to represent the beginning of the green energy era [7]. Today, the world wind power capacity is 318.5 GW [8], having increased roughly 66 times since 1995 (Fig. 72.3), and the global solar photovoltaic power capacity is 136.6 GW [9], which increased 107.2 times since 2000 (Fig. 72.4). Hence, the European Commission calls for the development of green energy and on launching settlements entirely provided with renewable energy [10].

As a response to this appeal, we are proposing a pilot project of a settlement where the energy for households and economic activities would entirely be provided from renewable sources. At our opinion, this project is extremely important as far as it provides a guideline to sustainable economic and social development based on renewable energy supply.

72.2 Conceptual Layout of the Project

The basic goal of the project is to elaborate a model of a settlement, which, on one hand, would be ensured by autonomous sources of renewable energy and, on the other hand, would provide the population with the multiple possibilities of employment and self-employment: unemployment incidence within the settlement should be zero. Of course:

1. The settlement shall be ultimately self-financed and profitable.
2. In addition to employment in the energy and social spheres (managerial bodies of the settlement, municipal services, health-care institutions, a kindergarten, a school, sport, private businesses of recreation, etc.), only businesses related to the processing of local resources shall be developed.
3. The acquisition of dwellings and industrial assets shall be performed under the condition of cheap long-term loans.
4. Correspondingly, the centralized management of the settlement and its infrastructure step by step, pro rata reimbursement of the loan, shall be replaced by an elected local government and elected management of businesses under, for instance, the French model of the collective farm.

As far as this pilot project shall demonstrate the new philosophy of sustainable development and environmentally friendly lifestyle, the construction of a settlement for 2,500 inhabitants would be enough.

Georgia is an ideal country for the project of a settlement entirely based on renewable energy: its renewable energy resources are huge, diversified, and entirely meeting the demands of the project [11–13]. Figure 72.5 demonstrates the distribution of these resources at the territory of the country, whereas hydroenergy resources are characterized in Table 72.1. According to available data [14], the geothermal resources of Georgia are equal to 250 MW.

Identification of the site for the settlement met several extremely important pre-requirements:

1. Availability of different renewable energy resources in the needed amount
2. Developed infrastructure
3. Availability of free land
4. Availability of good-quality arable land and pasture
5. Availability of potable water
6. Good climatic conditions

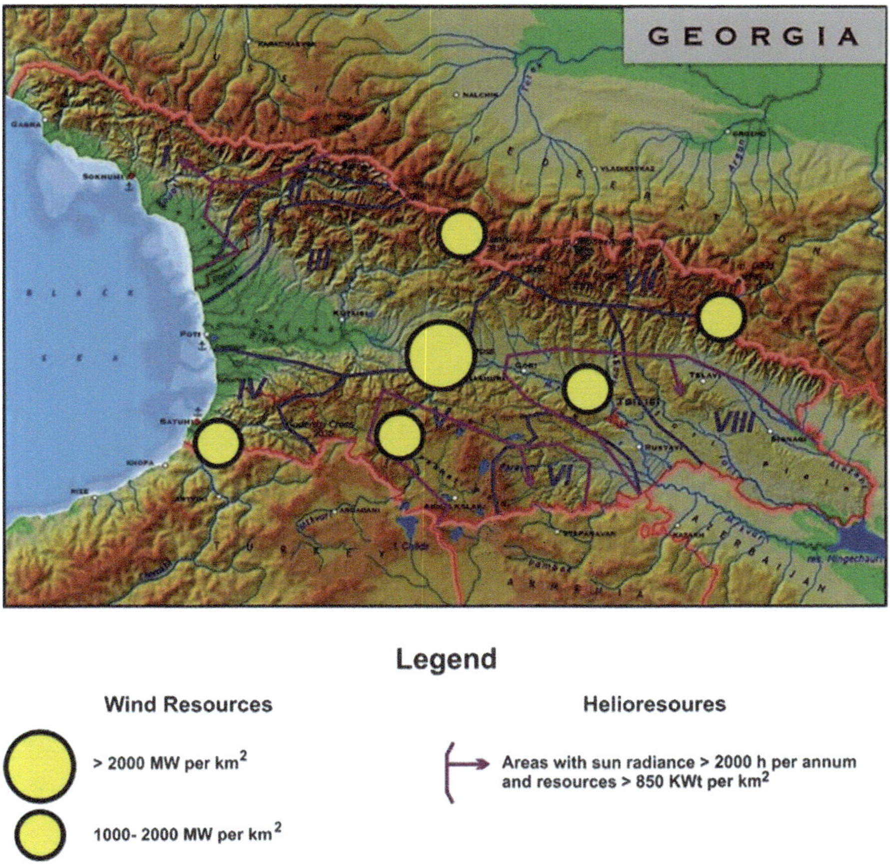

Fig. 72.5 Renewable energy resources of Georgia

Table 72.1 Hydroenergy resources of Georgia

No. on Fig. 72.3	River basin	Resources (GW)	
		Potential	Technical
I	Northern Black Sea coastal rivers	3,558	1,766
II	Enguri	2,402	1,280
III	Rioni and Khobi	3,772	1,787
IV	Southern Black Sea coastal rivers	1,259	616
V	Mtkvari (Kura)	2,572	1,248
VI	Khrami and Algeti	427	340
VII	Greater Caucasus northern slope rivers	483	243
VIII	Alazani and Iori	1,044	525
Total		15,518	7,803

Fig. 72.6 Geographic setting of the Zveli Site

7. Corresponding relief
8. Vicinity of railway station
9. Comparative independence of the site providing the possibility of self-governing community development at the background of good access

As a result of thorough investigations, the Zveli Site in Southern Georgia was identified, the geographic setting of which is shown in Fig. 72.6.

The Zveli Site coincides with an abandoned village situated in Southern Georgia, in 15 km from Akhaltsikhe, the center of the Samtskhe-Javakheti Province, and in 6 km from Aspindza Village, along a good-quality asphalt motor road from Akhaltsikhe to Akhalkalaki, at the altitude of 1,800 m. The site represents a land plot of about 25 km^2 situated on the second plane above the flood of the Mtkvari River. The land plot is covered with grass and rare bushes and trees and is crossed with three narrow river valleys with available potable water. Relief is gentle and entirely available for settlement and infrastructure construction.

The Zveli Site is characterized by a mountainous continental climate with an average air temperature in January of −2.2 °C and in August +20.8 °C. The absolute temperature minimum equals −19 °C and absolute maximum +39 °C. Average annual temperature is +9.4 °C. The average annual amount of atmospheric precipitations, mainly in the form of rain, is as low as 500–600 mm. Maximum precipitations fall in May (15 %) and minimum in February (6 %). The average thickness of snow cover is 18 cm and duration is 63 days. The site is characterized by northwestern (48 %) and southern and southeastern (31 %) winds. Winds are

Table 72.2 Reserves of fresh mineral waters near the site

Source	Debit (m^3/day)
Daba	151,500
Albari	79,800
Bejano	33,700
Ikhtila	38,000
Total	303,000

Table 72.3 Geothermal resources of Samtskhe-Javakheti

Geothermal field	No. of wells	Temperature (°C)	Debit (m^3/day)
Abastumani	3	48	1,040
Vardzia	3	45–58	1,330
Tmogvi	1	62	520
Nakalakevi	3	34–58	795
Aspindza	1	42	864
Tsikhisjvari	1	32	1,000
Borjomi	25	30–41	537
Akhaldaba	4	33–42	500
Total	41		

also typical along river valleys. The angle of incidence for sunbeam is 25–27° in winter and 70–72° in summer. Correspondingly, day length varies from 9 to 15 h. The radiation balance is 1.8 million kJ/m^2 and annual sun radiance equals 5.4–5.8 million kJ/m^2. The annual duration of sun shining is 1,951 h [15].

The Zveli Site is characterized by the availability of different renewable resources. The preliminary assessed maximum daily demand on tap water is about 500,000 l or only 500 m^3. This amount may easily be supplied from either ecologically pure inflows of the Mtkvari River or sources of underground fresh mineral water of excellent quality situated in the vicinity of the site [16] (Table 72.2). Hydropower resources of the Mtkvari River and its inflows near the Zveli Site are about 400 MW [17], and this amount 40 times exceeds the preliminary assessed needs of the site. Geothermal resources of Samtskhe-Javakheti are evaluated to be 13.87 MW (Table 72.3) including 2.16 MW in the Aspindza spring situated in 5 km from the site [13, 14].

72.3 Landscape and Architectural Design

Table 72.4 and Fig. 72.7 describe the landscape zoning of the Zveli Site. It may be seen that the land plot is naturally divided into two parts: (1) the settlement as such and (2) industrial and farming plot. The settlement consists of the administrative and the residential zones crossed with access roads. The administrative zone is composed of administrative, health-care, and educational buildings as well as of cafés, restaurants, and other recreation businesses.

Table 72.4 Landscape zoning of the Zveli Site

#	Item	Area		Percent of total
		m^2	ha	
1	Land plot total	2,155,000	215.5	100.00
2	Administrative zone	90,000	9.0	4.18
3	Individual lend plots and housing	650,000	65.0	30.16
4	Green zone	120,000	12.0	5.57
5	Residential zone total	770,000	77.0	35.73
6	Roads and parking	170,000	17.0	7.89
7	Industrial zone	589,000	58.9	27.33
8	Farming zone	536,000	53.6	24.87
9	Industrial and farming zone total	1,125,000	112.5	52.20

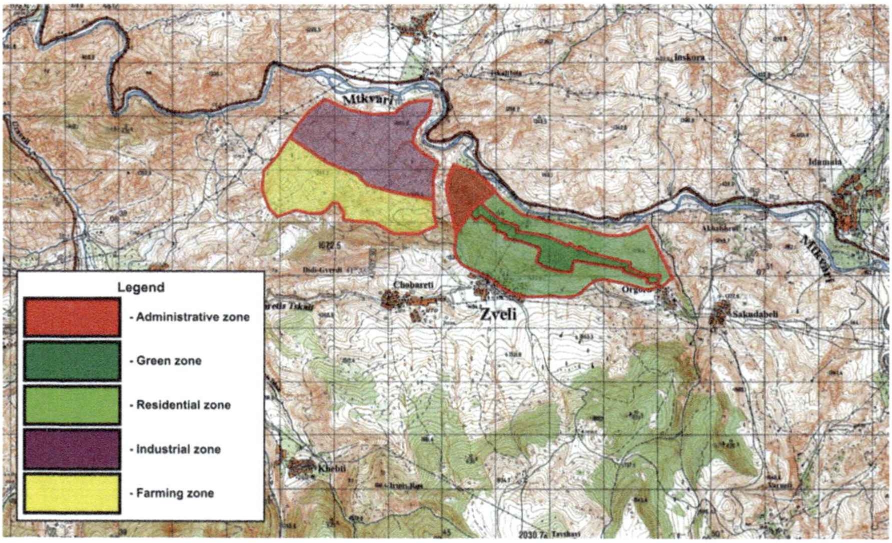

Fig. 72.7 Map of landscape zones of the Zveli Site

The landscape design of the settlement is shown in Fig. 72.8. The total length of the settlement is 2.5 km including 1.8 km of the residential zone. The green zone follows the residential zone in a sub-latitudinal direction and separates it into two parts. The green zone hosts a stadium, a public pool, tennis courts, a gymnasium, as well as a set of shops and cafés. The residential zone consists of 500 land plots with an area varying from 800 to 1,200 m^2.

The settlement would consist of typical houses (Fig. 72.9). Each house would have a ground and a first (mansard) floor and a veranda and a greenhouse. The roof of the house would be equipped with sun batteries with the possibility to change their inclination angle. Houses would be oriented in such a manner that sun batteries would be attacked by direct sunrays from southwest. Other energies would be

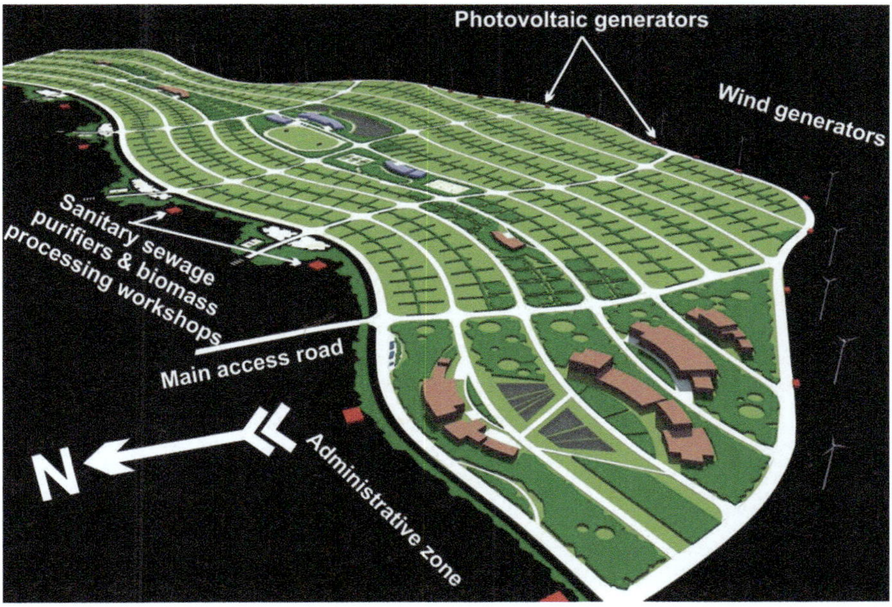

Fig. 72.8 Landscape design of the settlement

Fig. 72.9 Typical house of the settlement

supplied from the central energy distribution station. A cluster of six houses would be equipped by united water supply and canalization collector. Three types of houses are planned with, correspondingly, four, three, and two bedrooms. Figure 72.10 provides the design of ground floors of these dwellings, whereas Fig. 72.11 is that of the first floors.

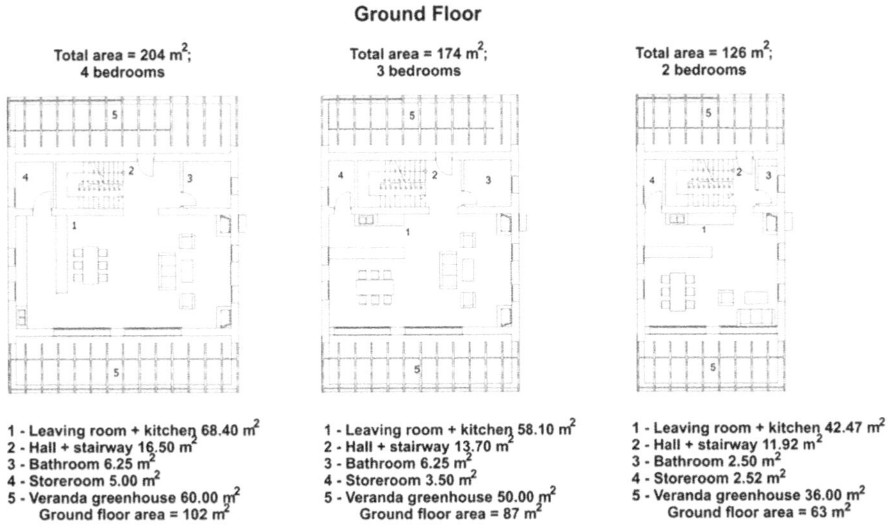

Fig. 72.10 Architectural design of the ground floors

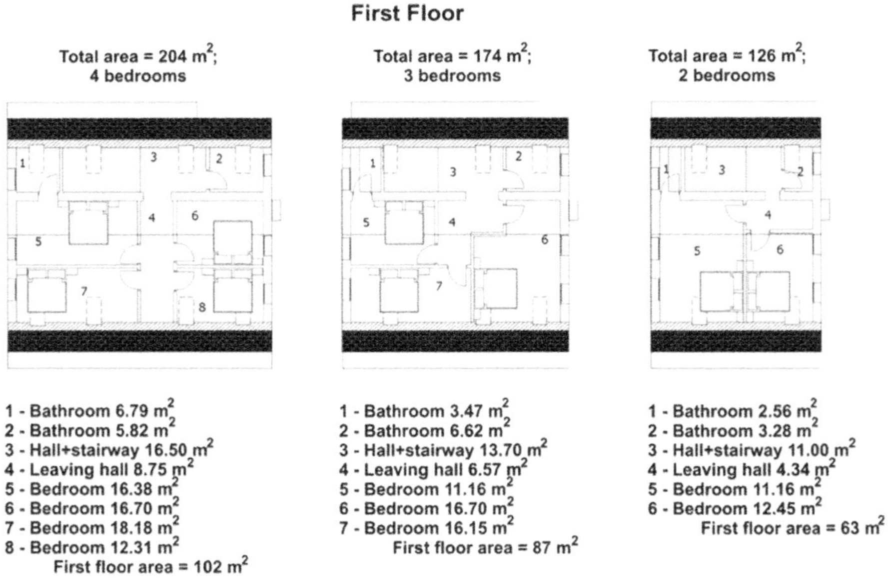

Fig. 72.11 Architectural design of the first floors

Industrial water in any quantity may be supplied directly from the Mtkvari River. As far as the needed quantity of tap water is not very high (about 500 m^3/day or four times over the EU norms), there are different alternatives for tap water supply:

1. Drilling of 20 boreholes for water supply
2. Intake of tap water from nearest inflows of Mtkvari River

3. Construction of 20 km water pipeline to Ikhtila source of underground mineral water (resources 38,800 m^3/day)
4. Construction of a 25 km water pipeline to the Tabatskuri Lake

The third alternative is the most efficient and the fourth one is the cheapest.

The settlement and its industries are suggested to use energy-saving technologies: hot water and heating supply would be mainly ensured by solar batteries at the roof of dwellings. The same may be said for industrial greenhouses with planted vegetables. So, the energy of 20 kWh per household (or 10 MWh for the entire settlement) would be absolutely enough plus 10 MWh of energy for the industrial and farming businesses.

Thus, the land plot is suggested to be equipped with:

1. Fifteen wind power machines produced by the Enercon Company, Germany, with a capacity of 1 MW each
2. Five photovoltaic batteries of Eurosolar, Germany, with a capacity of 1 MW each
3. Two small hydropower plants with a capacity of 5 MW each produced by Alstom Power, Switzerland

Thus, the maximum power generation would be 30 MWh and the routine generation 25 MWh. The total amount of generated electricity is assessed to be 219 million kWh and the demand would be 175.2 million kWh. In other words, the minimum annual amount of generated energy to be sold is suggested to be 43.8 million kWh.

The settlement energy system could be linked to the country electricity distribution grid to sell excessive energy or to buy it in force majeure circumstances.

At the given stage, assessment of the biomass impact is unfeasible.

72.4 Settlement Population and Businesses

According to the official statistical data [18], in Georgia unemployment incidence is 15.0 % of the economically active population or embraces 305.1 thousand persons. Unemployment covers all social and age groups of population including white collars, engineers, agricultural staff, teachers, workers, etc. Unemployment is extremely high in cities and towns, namely, in Tbilisi, the capital of Georgia. Unemployed people consist of an internally displaced population as a result of conflicts in Abkhazia and South Ossetia as well as of people fired in the course of the economic crisis. Among the unemployed population, there are people with an extremely high level of education and professional skills.

The main idea of the project hereto is to provide employment possibilities to the different social groups of the population according to the tender procedure. The designed social indices of the settlement are given in Table 72.5.

Table 72.5 Basic social indices of the settlement

Index	Persons
Population	2,500
Labor force	1,200
Children under 17 years	1,000
Housekeepers and pensioners	300
Pupils of the school	400
Pupils of the kindergarten	250
Civil servants	10
Administration	35
Health-care employees	35
Teachers	55
Fire squad	15
Law enforcement	6
Community servants	20
Cattlepersons	45
Industrial employees	700
Farm employees	279
Employees total	1,200

It may be seen that the project provides different employment possibilities to the eventual population of the settlement, namely:

1. Civil, administrative, and community servants
2. Health-care and education employees
3. Fire squad and law enforcement
4. Industrial employees
5. Farm employees
6. Cattlepersons

In addition to the services needed for good management of the settlement such as education, health care, civil order, local government, etc., the settlement business structure shall be oriented on local resources. From this point of view, the main fields of business activities would be agriculture and food product processing and marketing. Agriculture will be related with cattle and sheep husbandry; fodder production for them; potato, milk, and meat production; and tomato, cucumber, and fruit growing in greenhouses. Industrial business will be related to butter and cheese production from milk and packaging; meat packaging and storage; and tomato, cucumber, and fruit packaging and storage as well as the distribution of these products directly to hyper- and supermarkets throughout Georgia. Optional is the export of ecologically clean agricultural production abroad.

Taking into account the average crop capacity in Samtskhe-Javakheti [18] and capacity of pastures leased from the state, Table 72.6 describes the assessed minimum basic indices of economic activity.

Table 72.7 introduces the minimum wholesale prices of agricultural production in Georgia. With respect to these data, Table 72.8 shows that the assessed volume of sales would be more than US$ 55 million.

Table 72.6 Main eventual indices of business activity

Index	Unit of measure	Value
Cattle	No	500
Sheep	No	500
Milk production	t	12,500
Cheese production	t	25,000
Meat production	t	75
Potato production	t	5,000
Tomato production	t	250
Cucumber production	t	200
Fruit production	t	100
Fodder production	t	1,500

Table 72.7 Minimum wholesale prices of agricultural products

Product	Unit measure	Price (US$)
Milk	l	0.3
Cheese	kg	2.0
Meat	kg	3.0
Potato	kg	0.2
Tomato	kg	0.5
Cucumber	kg	0.3
Fruits	kg	0.5

Table 72.8 Annual assessed sales of agricultural products

Product	Sales (US$)
Milk	3,750,000
Cheese	50,000,000
Meat	225,000
Potato	1,000,000
Tomato	125,000
Cucumber	60,000
Fruit	60,000
Total	55,220,000

These data allows performing simple annual cash-flow analysis, which proves the entire profitability and self-sufficiency of the settlement. Moreover, as it is shown in Table 72.9, the settlement not only would be able to reimburse US$ 5 million per annum but also to found an investment and reserve fund in an annual amount of also US$ 5 million for the development and renovation of businesses.

At the background of total employment, the average monthly salary would equal US$ 1,110 or three times more than average in the country [18].

The average assessed cost of the house in the settlement switched to the infrastructure is US$ 50,000. As the annual earning of the average household would be US$ 26,650, a 5-year term loan with the interest rate of 4 % would be absolutely acceptable for the population.

Table 72.9 Assessed average annual economic indices

Index	Value (US$)
Annual production sales	55,220,000
Annual energy sales	4,400,000
Total annual sales	59,620,000
Production cost	21,125,000
Income	38,495,000
Taxes	5,774,250
Net income	32,720,750
Managerial cost	3,000,000
Earning	29,720,750
Reimbursement of loan	5,000,000
Investment fund	5,000,000
Pensions	200,000
Average monthly salary	1,110
Average annual household income	26,650

Table 72.10 Assessment of the needed investment

Investment article	Value (US$)
Houses	2,500,000
Roads and parking	5,500,000
Administrative buildings	3,300,000
School and kindergarten	1,500,000
Water pipeline	1,200,000
Canalization and purifiers	2,500,000
Small hydropower plants	3,000,000
Wind generators	9,000,000
Photovoltaic batteries	20,000,000
Farm	1,000,000
Agricultural plant	6,000,000
Lorry fleet	1,500,000
Subtotal	57,000,000
Feasibility study	2,500,000
Municipal vehicles	500,000
Turnover assets	2,000,000
Total investment	59,500,000

It should be noted that this analysis outlines only the minimum possible cash flow as far as it does not explore auxiliary businesses and services like insurance, turnover of cafés and recreation facilities, etc. Even the pessimistic assessment demonstrates that income from these activities in any case would be over US$ 10 million.

Now, Table 72.10 demonstrates that in totality investment of about US$ 60 million is needed. This sum with the acceptable interest rate may easily be reimbursed within 15 years.

72.5 Basic Conclusions

The idea of a settlement entirely based on renewable energy has never been embodied as it was suggested not to be economically profitable. However, the project hereto clearly outlines the extreme profitability of such a settlement, which may be attractive to investors. Moreover, the proposed managerial chart will insure the wealth of the population much over the mean life conditions in the country.

The network of such settlements is the challenge of the twenty-first century.

Acknowledgments The author is sincerely grateful to the Shota Rustaveli National Science Foundation and to the Open Society Georgia Foundation for the financial support of this investigation.

References

1. Adams WM (2006) The future of sustainability: re-thinking environment and development in the twenty-first century. In: Report of the IUCN Renowned Thinkers Meeting 29–31 January 2006, Washington DC, pp 1–19
2. (2013) International energy outlook. U.S. Energy Information Administration, Washington DC
3. The World Bank Group. Database. http://data.worldbank.org/indicator/. Accessed 22 Apr 2014
4. Renewables (2010) Global status report. Renewable energy policy network for the 21st century. Deutsche Gesellschaft für Technische Zusammenarbeit (GTZ) GmbH, Eschborn
5. (1998) Kyoto Protocol to the United Nations Framework Convention on Climate Change. United Nations, New York
6. Status of Ratification of the Kyoto Protocol. UNFCCC. http://unfccc.int/kyoto_protocol/status_of_ratification/. Accessed 5 May 2010
7. El-Asry M (ed) (2013) Renewables. Global futures report 2013. REN21, Paris
8. The windpower. http://www.thewindpower.net/statistics_world_en.php. Accessed 15 Apr 2014
9. World Solar Power Capacity. http://cleantechnica.com/2014/04/13/world-solar-power-capacity-increased-35-2013-charts/. Accessed 15 Apr 2014
10. (2006) Environment fact sheet: energy for sustainable development. The European Commission, Brussels
11. Svanidze GG, Gagua VP, Sukhishvili EP (1987) Renewable energy resources of Georgia. Hydrometeo Press, Leningrad (in Russian)
12. Chkhaidze B (1998) Energy resources of Georgia. Bull Center Strateg Stud Dev 7:2–17 (in Georgian)
13. (2006) Renewable resources of Georgia. EBRD, London
14. Tsertsvadze N, Vardigoreli O (2007) Utilization of geothermal energy in Georgia. Techinformi Press, Tbilisi
15. Tivadze E, Tsitlanadze T (2008). Climate of Aspindza. http://lemill.net/content/webpages/10d010e110de10d810dc10eb10d810e1-10f010d010d510d0. Accessed 15 Mar 2010
16. Tvalchrelidze A (2006) Mineral resources and the mineral resource base of Georgia. Rudy I Metally Press, Moscow (in Russian)
17. Tvalchrelidze A, Silagadze A, Keshelashvili G, Gegia D (2011) Georgia's social & economic development program. Nekeri Press, Tbilisi (in Georgian). English version is available under http://www.ifsdeurope.com
18. (2013) Statistical yearbook of Georgia. National Statistics Office of Georgia, Tbilisi

Chapter 73
Modeling and Transient Simulations of 30 MW Solar Thermal Electric Power Plants in the Northeast Mediterranean Region

Serhat Bilyaz, Rahul Singh, Arash Karshenass, and Derek Baker

Abstract The annual performance of a parabolic trough collector (PTC) concentrating solar power (CSP) system without storage or backup is predicted for Antalya (Turkey) and Güzelyurt and Larnaca on the island of Cyprus by modeling the system with TRNSYS software. The model is established to be similar to the 30 MW_e CSP plant (SEGS VI) in Kramer Junction, California. The model is benchmarked with published results for Kramer Junction and Antalya, and good agreement is found. Based on irradiation maps, Güzelyurt is expected to be one of the sunniest areas on Cyprus, but hourly meteorological data are not supplied with TRNSYS for this location. Meteorological data in TMY2 format are available in TRNSYS for Larnaca. In this work, TMY2 data for Güzelyurt is created using Meteonorm software. The investigation of the performance of a CSP plant for these two locations on the island of Cyprus is one of the main objectives of this paper. The annual performance of these systems is predicted by performing simulations using an hourly time step. Results for typical clear and cloudy days are compared for the three locations. The highest monthly average net power outputs are observed for Güzelyurt. Finally, the capacity factor of the CSP plant is investigated for different solar multiples for a solar-only PTC plant without hybridization or thermal storage for all three locations. Results show that maximum power outputs are obtained for Güzelyurt. For our solar-only system, solar multiples of 1.1–1.3 have good results, but for solar multiples greater than 1.3, an excess amount of solar resources starts to be wasted, so that storage is recommended especially for Güzelyurt.

S. Bilyaz (✉) • R. Singh • A. Karshenass
Mechanical Engineering, Centre for Solar Energy Research and Applications (GUNAM), Middle East Technical University (METU), 06800 Ankara, Turkey
e-mail: bilyaz@metu.edu.tr; rahulsingh284.nitjsr@gmail.com; arash.karshenass98@gmail.com

D. Baker
Mechanical Engineering, Centre for Solar Energy Research and Applications (GUNAM), Middle East Technical University (METU), Güzelyurt, Turkey
e-mail: dbaker@metu.edu.tr

I. Dincer et al. (eds.), *Progress in Clean Energy, Volume 2*,
DOI 10.1007/978-3-319-17031-2_73

Keywords CSP • Solar multiple • Capacity factor • Solar energy

Nomenclature

$Q_{\text{solarinput}}$	Net heat input at the collector
m_{HTF}	Mass flow rate of heat transfer fluid
$C_{\text{p,HTF}}$	Constant pressure specific heat of heat transfer fluid
T_{output}	Outlet temperature of collector
T_{input}	Inlet temperature of collector
$\text{SM}_{\text{designpoint}}$	Solar multiple at design point
CF	Capacity factor
θ	Angle of incidence
K	Constant relating solar multiple to collector field area

Subscript

e	Electricity

73.1 Introduction

Energy consumption of humans throughout the world increases year by year. In recent decades, an energy crisis is prominent due to high magnitudes of consumption which is mostly harvested from fossil fuels, followed by two major problems, resource depletion and environmental pollution such as global warming. Solar radiation is the mostly widely available renewable energy resource, hence could be considered as an alternative to hydrocarbon-based fossil fuels without any limitations or pollution. There are several types of solar-driven power systems, but among them, concentrating solar power (CSP) is the most suitable for electricity generation at large scales. CSP refers to those plants which concentrate incoming solar radiation to obtain high temperatures with the aim of electricity production. The system consists of four main sections, namely, (1) concentrator, (2) receiver, (3) heat transfer fluid (HTF), and (4) power block which is usually a Rankine cycle. Power tower, parabolic dish, and parabolic trough collector (PTC) are three designs of CSP plants that work at different temperatures. The most mature design among these three is PTC. PTCs consist of rows of trough-shaped mirrors which concentrate solar insolation to a receiver tube placed along the focal axis of each trough where the HTF flows. Radiation focused by the trough-shaped concentrators on the receiver pipe results in heating of the HTF, and the obtained thermal energy is transferred to an energy conversion system by the use of HTF which flows inside the receiver.

There are many theoretical and experimental studies examining existing CSP systems throughout the world. Solar electric generating systems (SEGS) III to VI are four PTC plants located in Kramer Junction, California USA. SEGS IV is a 30 MW_e CSP plant built in 1989 which has been studied in several papers. Lippke [1] studied part-load behavior of the plant by developing a solar field model based on LS-2 collector parameters and using EASY code [2] and found that the net output would be maximized when the superheating temperature is decreased as much as possible. Jones et al. [3] developed a transient model of the entire plant but without any backup in the TRNSYS software [4], adopting the STEC library [5] and including startup, shutdown, and cloud responses. Results were in good agreement with measured plant data on both sunny and cloudy days. Since the HTF circulation is controlled manually in SEGS plants with the aim of obtaining a specified temperature outlet, an algorithm is developed [6] for automatic control of HTF flow rate. Another study has been done by Patnode [7] on the SEGS IV PTC plant for detailed investigation of operation parameters' effect on system outputs. In the study, a solar field model was developed in TRNSYS, while the power block was modeled by engineering equation solver (EES) [8] and imported to TRNSYS. The entire model is validated with SEGS IV experimental data collected over 1996–2005. As results, daily and annual power outputs are presented, optimization of solar field flow rate is done, and alternative condensers are analyzed. Usta et al. [9] apply a similar model in TRNSYS and got results for Antalya, Turkey. Results of California have also been checked for verification and compared to the results of Antalya. It is concluded that monthly averaged power output of Antalya is approximately 30 % less than of California.

Today, electricity and fuel are very expensive in Cyprus where most of the electricity is generated using oil fired power plants. The objective of this paper is to examine the performance of solar-only CSP power plants in the Northeast Mediterranean region, especially for the island of Cyprus which has substantial solar resources and has not been investigated for CSP power plants before. In the present work, hourly net power outputs of Antalya, Turkey, and Güzelyurt and Larnaca on Cyprus are obtained for one year. Daily averaged power outputs for months are presented. For Antalya, clear and winter week results are shown for benchmarking with previous work [9, 10]. For Güzelyurt and Larnaca, results for clear and cloudy summer and winter days are investigated and discussed. The effect of solar field size is studied for the power block with nominal electrical power of 30 MW_e for three locations. The study is based on determination of solar multiple, which is the ratio between the thermal power produced by the solar field at the design point and the thermal power required by the power block at nominal conditions [11]. The variation in the capacity factor with respect to solar multiple for solar-only parabolic trough plant, that is, no electricity production from fossil fuel, has been studied in this paper. Generally, solar multiples are greater than one, and the higher value allows the power plant to operate at peak power for longer time periods, but the oversized field leads to higher kWh_e cost. A parametric analysis is performed by examining change of capacity factor with increasing solar multiple. After discussing the results, uncertainties in the results are discussed, and future work is suggested.

73.2 Model

73.2.1 Cycle

A PTC power plant model is developed in TRNSYS platform based on the work of Usta [9, 10]. The model consists of 100 rows of north-south axis tracking PTCs with 50 loops of HTF flow with a total area of 1.82×10^5 m^2. Collectors are of LS-2 type with concentration ratio of 71:1, and Therminol VP-1 is the HTF. Tracking collectors focus solar irradiation on the linear receiver where HTF is heated up to 390 °C as it circulates through each loop. The receiver is made of an inner stainless steel absorber with high absorptivity at short length waves, an outer glass tube with high transparency for these short waves, and a vacuum between the two to minimize heat losses by removing conduction and convection heat transfer. HTF flow is controlled in a way to reach 390 °C before leaving the solar field. Heated fluid from all 50 loops is collected in headers and moves to heat exchangers to produce steam in a Rankine cycle with one high and one low pressure turbine. After leaving the field and passing through an expansion vessel, the HTF flow is split into two flows. One is pumped to produce superheated steam through a counterflow system of heat exchangers consisting of a super heater, steam generator, and preheater respectively, and the other flow is pumped to a reheat heat exchanger, located between the two power turbines. Following that, the two HTF flows are mixed again and sent back to the solar field. The return HTF temperature is assumed to be fixed at 297.8 °C [1]. In the steam generator heat exchanger series, HTF with temperature of 390 °C firstly enters the superheater which results in producing superheated steam. Then the lower temperature HTF enters the evaporator to change the phase of saturated liquid water into saturated vapor and, finally, by passing through the preheater increases the feedwater temperature to the saturation point. On the other hand, water moves exactly in the opposite direction to the HTF, meaning it passes through the preheater, evaporator, and superheater, respectively, as shown in Fig. 73.1. The generated superheated steam goes to the two-stage, high pressure turbine (HPT) and leaves with lower temperature and pressure. Also, a part of the steam in the HP turbine is extracted through two extractors with the aim of feedwater preheating in two high pressure feedwater heaters (HPFWHs). Then the remaining steam is brought into the reheater, as mentioned above, to increase the cycle efficiency before it enters the low pressure (LP) turbine. Next, steam enters the five-stage LP turbine, and similar to the HP turbine, some of the steam is extracted through four extractors to preheat feedwater using three LP feedwater heaters (LPFWH) and a deaerator. After leaving the LP turbine, the remaining steam is condensed in a condenser. By the help of two pumps, liquid water

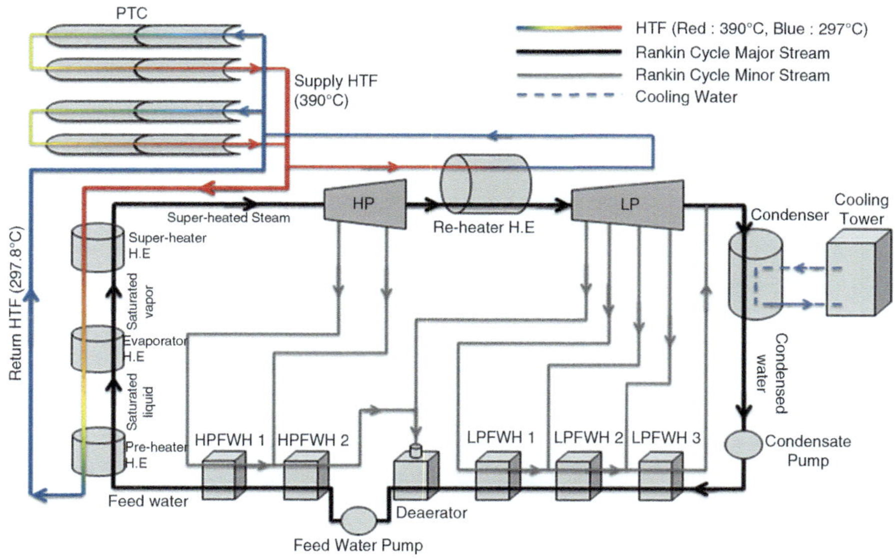

Fig. 73.1 Schematic diagram of the cycle

(feedwater) passes through the LP and HP feedwater heaters and deaerator and, then finally, reaches the preheater in the cycle. The TRNSYS diagram of the cycle and system parameters are shown in Fig. 73.2 and Table 73.1, respectively.

73.2.2 Trough Control Mechanism

In the model, the mass flow rate of the HTF is adjusted in response to transient conditions. In the TRNSYS trough model (type 369), inlet and outlet temperatures of the heat transfer fluid are specified. When the solar heat input is changed hourly, the mass flow rate of the HTF passing through the collector is adjusted so that it can reach the specified outlet temperature. Assuming constant specific heat through the collector, the heat transfer to the working fluid can be defined as Eq. (73.1).

$$Q_{\text{solarinput}} = m_{\text{HTF}} C_{\text{p,HTF}} \left(T_{\text{output}} - T_{\text{input}} \right) \tag{73.1}$$

Since input and output temperatures are defined in the power cycle model and both the temperature difference and specific heat are constant, the mass flow rate of the HTF is proportional to the solar heat input. Sometimes, heat transfer oil becomes very viscous during the night, and in order to avoid freezing, the HTF is circulating in the system in the night. Nevertheless, such kind of a practice has not been considered in this analysis. In the present model, when the direct normal irradiation (DNI) is zero (e.g., during night), the system should be shut down, and the mass

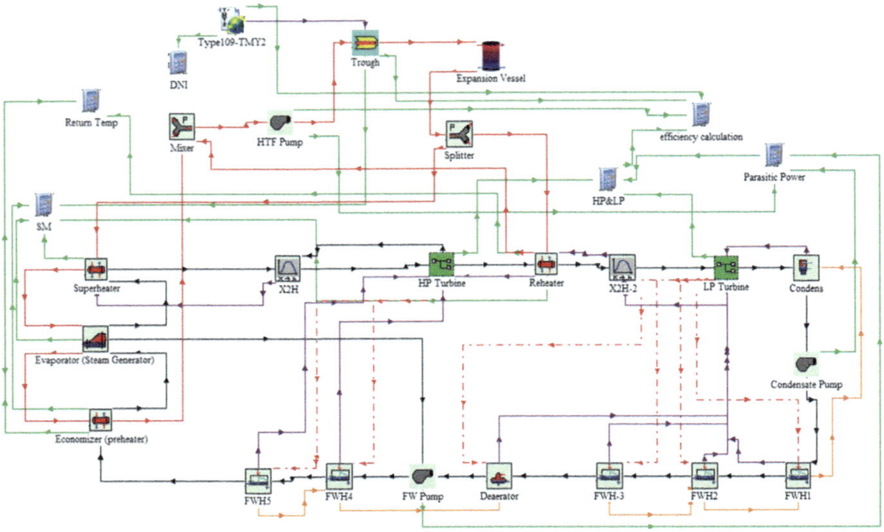

Fig. 73.2 TRNSYS diagram of system model

Table 73.1 TRNSYS model parameters

Parameters	Values	Units
Tolerance convergence	0.05	–
Simulation time step	1	h
Tracking	N-S axis tracking	–
Collector field area	182,000	m^2
Inlet temperature solar field	297.78	°C
Outlet temperature solar field	390.56	°C
Pump max flow rate	396.4	kg s^{-1}
Specific heat of HTF	2.59	kJ kg^{-1} K^{-1}

flow rate should be zero. However, the collector inlet temperature of the working fluid is constant in the component model (type 369), which cannot be reached when the system is completely shut down. Therefore, there is a parameter called a night down ratio in the component which defines a minimum mass flow rate of the working fluid during night conditions. In other words, the system is operating at a very small mass flow rate. This situation causes negative network output results due to parasitic expenditures. These parasitic negative outputs are set to zero in excel, which provides zero work output at night conditions. Sometimes, heat transfer oil becomes very viscous during the night, and in order to avoid freezing, the HTF is circulating in the system in the night. Nevertheless, such kind of a practice has not been considered in this analysis.

73.2.3 Solar Multiple and Capacity Factor

Solar multiple and capacity factor are calculated using Eqs. 73.2 [11] and 73.3, respectively, as shown below:

$$\mathrm{SM}_{\mathrm{design\_point}} = \left.\frac{\dot{Q}_{\mathrm{th,\,solar\_field}}}{\dot{Q}_{\mathrm{th,\,power\_block}}}\right|_{\mathrm{design\_point}} \tag{73.2}$$

$$\mathrm{CF} = \frac{\text{Actual Power Output for given time period}}{\text{Maximum Power Output for given time period}} \tag{73.3}$$

The thermal power input by the solar field depends on the DNI and plant location, while the thermal power demand by the power block at design point remains constant as 30 $\mathrm{MW_e}$ for all selected plant configurations. The heat engine thermal requirement includes steam generators (preheater, boiler, and superheater) and reheater. The thermal power supplied to the Rankine cycle at nominal conditions is shown in Table 73.2. The capacity factor is calculated for each solar field size for a period of one month for which net electricity output is maximum, which corresponds to June, July, and September for Antalya, Güzelyurt, and Larnaca, respectively.

Turbine parameters and stages are fixed at design conditions, and the only parameter that affects the output is the mass flow rate of the heat transfer fluid. Mass flow rate is proportional with solar heat input, and this is also proportional with the collector area. It is expected that there is a linear relation between total power output and collector area; this linear dependence is obtained within the simulations, and a parametric analysis is performed for net power output dependence versus collector area. Starting from $1.82 \times 10^5\ \mathrm{m}^2$ collector area, the collector field area is increased in 3,640 m^2 increments, and simulations are performed for the month that the location has the largest average power output. Daily averaged net power outputs are calculated until the simulation gives an error due to reaching the maximum capacity of power cycle at an hour. Daily averaged power output is plotted with respect to collector area, and a linear increase is observed due to the reasons explained above. A line is fit to graph, and the slope value is used for extrapolations. A new line equation is developed for the next step using the slope obtained from the last simulation, and new hourly results are calculated for larger collector areas. Power outputs exceeding 30 $\mathrm{MW_e}$ are set to 30 $\mathrm{MW_e}$ to simulate defocusing the collectors, which occurs when the collectors produce excess thermal

Table 73.2 Power block thermal demand at design point

Component	Thermal power demand ($\mathrm{MW_{th}}$)
Preheater	3.31
Boiler	58.89
Superheater	10.75
Reheater	12.17
Total	85.12

power. Daily averaged results are obtained accordingly. Capacity factor is calculated from daily averaged results. After solar multiples are calculated as explained above, capacity factors are plotted with respect to solar multiple.

73.3 Results and Discussion

73.3.1 Clear and Cloudy Days

In this section, clear and cloudy day's results are presented. Collectors are aligned though north-south axis and tracking sun from east to west. In Fig. 73.3a, net power outputs for Antalya for a clear summer week are presented. The clear week is July 4–10 which is same week used by Usta et al. [9, 10]. Therefore, the graph can be used to verify the current work. Good agreement is achieved. Twenty-five to thirty MW_e peak power output is obtained due to available DNI in July. There are no fluctuations of power output during the week which shows there are no clouds during the week. At solar noon time, the power output decreases slightly, and this "sag" is characteristic of tracking in the east-west direction at this latitude. The reason for this is explained previously by Patnode [7]. Specifically, at solar noon, the collector is set to be vertical, but a large angle of incidence (θ), which is the angle between the sun-earth line and a line normal to the collector surface, decreases the product DNI cos(θ). In Fig. 73.3b, the net power output over a cloudy winter week is presented. Date is selected as January 21–28 which is the same dates that Usta et al. [9, 10] used. First, it can be easily seen that for the last two days of the week, there is no power output due to insufficient DNI resources. Since the desired outlet temperature could not be achieved, the system is shut down for these two days. For the week, the peak power output is ~16 MW_e which is half of the values reached in a clear summer day. The total power output over the week is even less relative to a summer week since the 16 MW_e is only reached on one day.

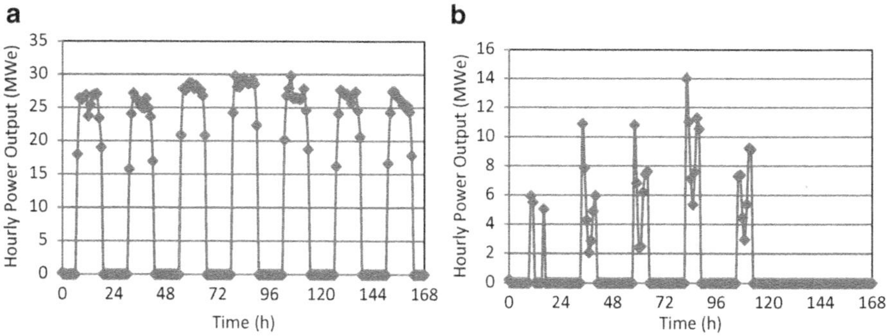

Fig. 73.3 Hourly power output for Antalya (**a**) for a clear summer week (July 4–10) and (**b**) for a cloudy winter week (October 18–24)

Additionally, the peak output is typically obtained about 10 am during the day. For the first day of the week, the system shuts down for 4 h around noon which can be due to clouds or rainy weather. It can be seen that the characteristic "sags" in output at solar noon are also obtained in the winter week, but these "sags" are deeper since the zenith angle, and therefore the angle of incidence, is greater at solar noon in the winter than in the summer.

Net electricity generation for Güzelyurt for a clear summer day (July 3rd) and a partially cloudy summer day (August 7th) are shown in Fig. 73.4a, b, respectively. The power output is nearly 30 MW_e throughout the sunshine hours except for the slight "sag" at solar noon. This "sag" is more prominent on a slightly overcast day (Fig. 73.4b) due to the larger angle of incidence (θ) as well as high cloud cover during 11:00 to 13:00. Maximum power production for a partially cloudy summer day is 22.7 MW_e at 15:00. Figure 73.4c shows net power output for Güzelyurt for a clear winter day (December 1st). Peak electricity generation is only 14.6 MW_e because of lower DNI. The "sag" at solar noon during winter is sharper as compared to the summer day due to a larger incidence angle (θ). Power generation is more erratic for a cloudy fall day which can be seen in Fig. 73.4d. In the case of Larnaca, net power generation for a clear summer day (July 10th) is shown in Fig. 73.5a and partially cloudy summer day (July 6th) in Fig. 73.5b. The average power output during sunshine hours is 21.2 MW_e, which is less than the nominal 30 MW_e due to deficient DNI values. The "sag" at solar noon is observed for Larnaca also, as in the

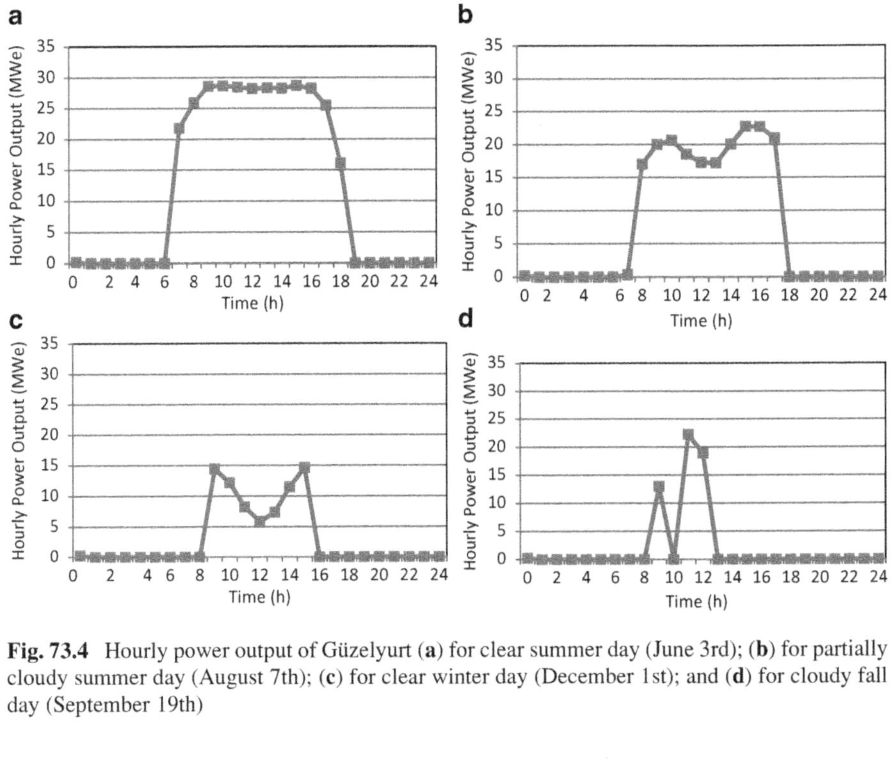

Fig. 73.4 Hourly power output of Güzelyurt (**a**) for clear summer day (June 3rd); (**b**) for partially cloudy summer day (August 7th); (**c**) for clear winter day (December 1st); and (**d**) for cloudy fall day (September 19th)

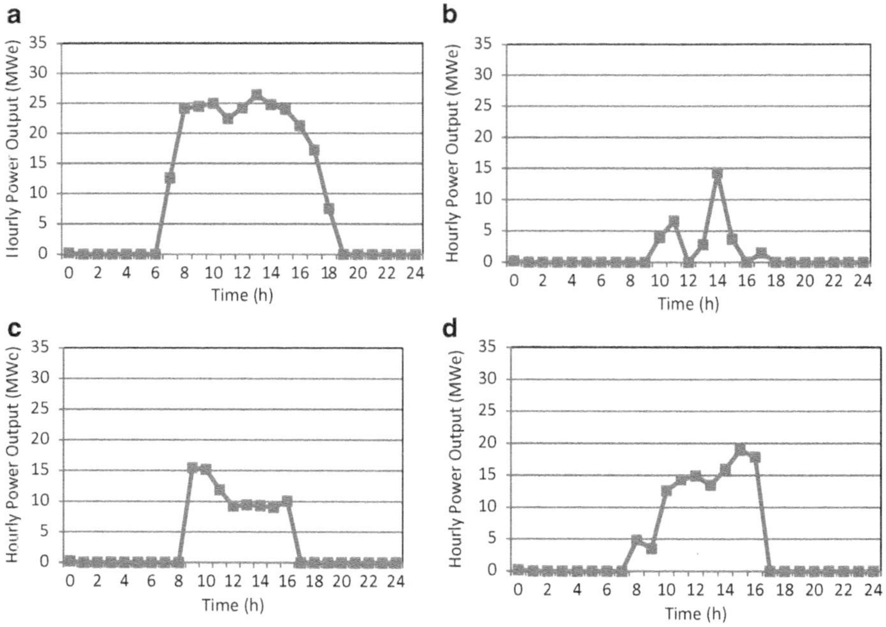

Fig. 73.5 Hourly power output of Larnaca (**a**) for clear summer day (July 10th); (**b**) for cloudy summer day (July 6th); (**c**) for clear winter day (January 23rd); and (**d**) for slightly overcast spring day (March 25th)

cases of both Antalya and Güzelyurt. Figure 73.5b shows the electricity generation for cloudy summer day (July 6th) for Larnaca, and significant fluctuations in power output are seen owing to intermittent cloud coverage. Net power generation for Larnaca for a clear winter day (January 23rd) is shown in Fig. 73.5c. The peak power output is 15.5 MW_e, and average electricity output during daytime is 11.22 MW_e. Figure 73.5d shows a partially cloudy spring day (March 25th) power output for Larnaca in which electricity production is higher toward the end of the daytime as compared to power generation starting time (7:00 to 10:00) due to higher density of clouds during the first half of the day.

73.3.2 Daily Average Gross Power Outputs for Three Locations

Daily average power outputs for three sites are compared in Fig. 73.6. The maximum daily power generations for Antalya, Güzelyurt, and Larnaca are 245 MWh_e, 312 MWh_e, and 187 MWh_e, respectively. Longer periods of clear sky and substantially larger DNI values account for Güzelyurt's higher electricity output in contrast to the other two locations. Larnaca's performance is better in fall-winter than that of

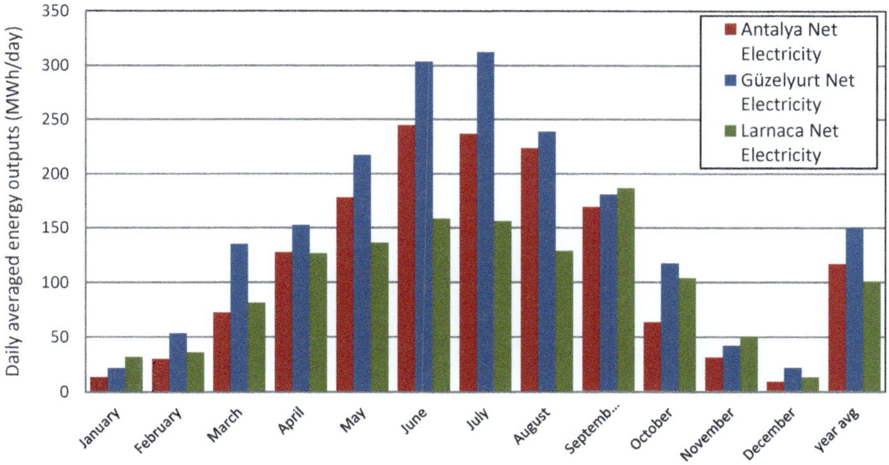

Fig. 73.6 Monthly performance of CSP power plant for Antalya, Güzelyurt, and Larnaca

Antalya with total power production of 421.8 MW_e during the six months of September to January, while Antalya generates 316.34 MWh_e for the same time period. However, the yearly average output of Antalya is higher than Larnaca.

73.3.3 Parametric Study: Solar Multiple and Capacity Factor

Simulations are performed for each location, that is, Antalya, Larnaca, and Güzelyurt, and results are tabulated in Table 73.3. All the parameters are fixed at the design point except for the solar field area so solar multiple and collector field size have a linear relationship, and the ratio of solar multiple to area, shown in Eq. 73.4, is found constant (K).

$$K = \frac{\text{Solar Mutiple}}{\text{Area}} \tag{73.4}$$

The plot of variation of monthly capacity factor with respect to solar multiple for the three locations is shown in Fig. 73.7 for June for Antalya, July for Güzelyurt, and September for Larnaca, which represents the months with the maximum output for each location. Capacity factor represents how long a PTC plant develops peak power, that is, 30 MW_e for the Rankine cycle considered in this paper. Simulation results (Fig. 73.7) show zenithal value of capacity factor and least area requirement for Güzelyurt compared to the other two locations owing to its high DNI value. Initially, capacity factor increases with solar multiple for each location, but the percentage gain in capacity factor decreases for higher values of solar multiples. In the case of Antalya, from solar multiple 1 to 1.2, the capacity factor increases by

Table 73.3 Simulation results for PTC plant considered

Antalya			Larnaca			Güzelyurt		
$K = 6.1 \times 10^{-6}$			$K = 5.18 \times 10^{-6}$			$K = 6.45 \times 10^{-6}$		
Area (m^2)	Solar multiple	Capacity factor	Area (m^2)	Solar multiple	Capacity factor	Area (m^2)	Solar multiple	Capacity factor
163,800	1.00	0.277	192,920	1.00	0.266	156,520	1.01	0.405
196,560	1.20	0.329	232,960	1.21	0.305	185,640	1.20	0.452
214,760	1.31	0.356	251,160	1.30	0.314	203,840	1.31	0.465
229,320	1.40	0.368	269,360	1.40	0.323	218,400	1.41	0.472
247,520	1.51	0.378	291,200	1.51	0.334	232,960	1.50	0.477

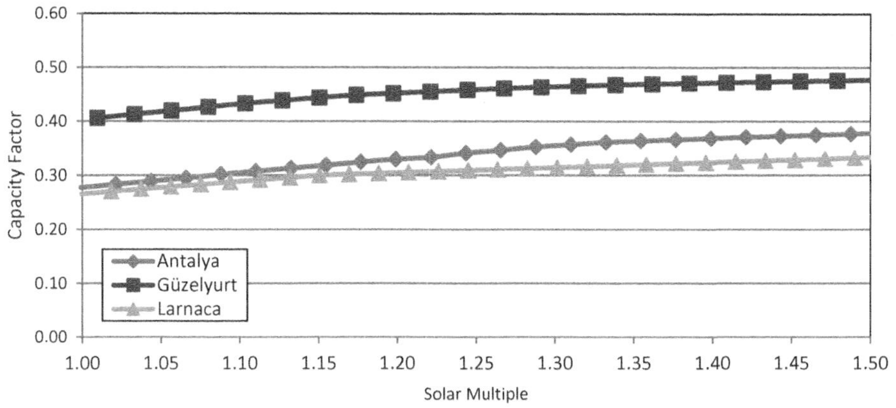

Fig. 73.7 Solar multiple versus capacity factor considered for select months. Antalya: June; Güzelyurt: July; Larnaca: September

18 % and from 1.3 to 1.5, the increment in capacity factor is just 6.1 %. Similar trend is observed in the case of both Larnaca and Güzelyurt. Considering Larnaca, capacity factor augmentation is 14.67 % for solar multiple increases from 1 to 1.2, on the other hand, capacity factor increases by 6.37 % for increases in solar multiple values from 1.3 to 1.5. Furthermore, for Güzelyurt, capacity factor increment is 11.6 % from an increase in solar multiple from 1 to 1.2 and only 2.6 % from 1.3 to 1.5.

Although an increment in the solar field size allows the plant to operate at nominal conditions for longer durations, it also leads to production of excessive thermal power at times of peak solar resources which is usually shed by defocusing some of the PTC modules. So the percentage rise in capacity factor is smaller for higher values of solar multiple due to increases in shedding. The optimal range of solar multiple for solar-only PTC power plant presented in this paper is from 1.1 to 1.3.

73.4 Uncertainties in Results

In this section, uncertainties in the results are discussed. First, typical meteorological data of Güzelyurt and Antalya are generated using Meteonorm software [12], which means real data are not available, but data is created theoretically by considering the nearest real data and geographical and site conditions. It is obvious that using theoretically generated data would not be the same as using real data.

As is discussed in the trough control section, one of the weaknesses in the system model is that the system cannot be shut down completely during the night; in other words, there is a very small minimum mass flow rate in order to maintain numerical calculations within the trough component which causes parasitic power losses (pumps are working). Inlet and outlet temperatures of the collector are assumed constant, so that when the system is just opened in the early morning, heating up the system after night conditions is not considered.

In the parametric analysis of solar multiple and capacity factor, solar multiple is increased by increasing the area and handled in Excel with linear extrapolation. Applying such linear extrapolation for higher collector areas is a rough assumption, and although it gives an approximation about high solar multiple conditions, a significant amount of irradiation is shed by defocusing, and therefore, storage could be appropriate for high solar multiples.

73.5 Conclusion and Future Work

The annual performance of a solar-only PTC plant is predicted using TRNSYS for Antalya, Güzelyurt, and Larnaca. Simulation results show the highest monthly average net power output for Güzelyurt among the three locations. Daily average electricity generation for the peak month is 245 MWh_e for Antalya, 312 MWh_e for Güzelyurt, and 187 MWh_e for Larnaca. Between Larnaca and Antalya, Larnaca performs better during the months of fall-winter, but Antalya's yearly performance is higher than that of Larnaca. During clear summer days with a solar field of 1.82×10^5 m^2, both Güzelyurt and Antalya generate nominal power while Larnaca's power output is slightly less due to smaller solar resources. However, peak power generation is both low (14 MW_e–22 MW_e) and erratic for cloudy days and winter months for all the three locations. During winter and cloudy days, PTC plant works on part-load condition for all three locations. So the addition of a backup boiler or an integrated solar combined cycle (ISCC) to the Rankine cycle is required to provide steady power delivery throughout winter and cloudy time period. Apart from removing fluctuations in power generation, a standby boiler or ISCC will lead to augmentation in full load hours as well.

The monthly capacity factor and solar multiple curve show maximum capacity factor among three locations for Güzelyurt which is 0.48 for July. Despite a high monthly capacity factor value for Güzelyurt, the percentage increment in capacity factor decreases notably with increasing solar field size. For instance, for a 15 % increase in solar multiple starting from SM = 1.3 (i.e., field size), only a 2.6 % increase in capacity factor is achieved. Due to the larger levels of electricity generation all around the year, Güzelyurt seems to be the best choice for a PTC power plant among the three locations. However, excessive thermal power production during times of large solar resources is a major concern for solar-only PTC plants with no storage and high solar multiples. So in order to ensure efficient energy utilization, an energy storage system will not only provide energy during night time or low sunshine hours but also increase overall plant efficiency along with a significant increase in capacity factor. For this solar-only system, solar multiples of 1.1–1.3 appear to be good since little excess solar thermal power is shed. However, for the solar multiple greater than 1.3, thermal storage is suggested in order to be able to use these excess solar resources especially in Güzelyurt.

As a future work, the biggest weakness in the current model is the assumption of constant collector inlet and outlet temperature and the method of controlling the trough mass flow rate. These weaknesses are related to the trough component, and some modifications are required for the solution algorithm. A new component can be modeled in FORTRAN and used to replace the existing model. Transition from night conditions to day conditions can be modeled more accurately. Real meteorological data will increase the validity of the results, but it requires experimental setup and long-term data. Since islands like Cyprus have limited fresh water sources, water consumption for wet cooling towers may be problematic. Wet cooling towers can be replaced with dry cooling towers. Finally, in order to improve the system's energetic performance, thermal storage and hybridization with other energy sources (fossil fuel, wind energy) can be added. Different heat transfer fluids can be explored, and new more efficient collectors can be adapted. Considering different solar resources and available collector field areas, new power cycles can be designed for different locations to obtain power output greater than or less than 30 MW_e.

References

1. Lippke F (1995) Simulation of the part load behavior of a 30 MWe SEGS Plant. Prepared for Sandia National Laboratories, Albuquerque, NM, SAND95-1293
2. Wahl P (1992) Object-oriented design and implementation of a software system for calculating the part-load conditions of solar electric power plants (in German). ZSW report, SOT-WS 12/92
3. Jones SA, Pitz-Paal R, Schwarzboezl P, Blair N, Cable R (2001) TRNSYS modeling of the SEGS VI parabolic trough solar electric generating system. Proceedings of solar forum, Washington, DC, USA
4. University of Wisconsin-Madison. Solar Energy Laboratory, Klein SA (1979) TRNSYS, a transient system simulation program. Solar Energy Laboratory, University of Wisconsin-Madison
5. Pitz-Paal R, Jones SA (1998) A TRNSYS model library for solar thermal electric components (STEC), a reference manual, Release 1.0. IEA-Solar Power and Chemical Energy Systems, Task III: Solar Technologies and Applications
6. Stuetzle T, Blair N, Mitchell JW, Beckman WA (2004) Automatic control of a 30 MWe SEGS VI parabolic trough plant. Sol Energy 76:187–193
7. Patnode MA (2006) Simulation and performance evaluation of parabolic trough solar power plants. Thesis, University of Wisconsin, Madison, USA
8. Klein SA, Alvardo FL (2002) Engineering equation solver. F-Chart Software. Madison, WI
9. Usta Y, Baker D, Kaftanoğlu B (2011) Modeling and simulations of a 30 Mwe solar electric generating system using parabolic trough collectors in Turkey. Proceedings of international green energy conference VI (IGEC VI), Eskisehir, Turkey
10. Usta Y (2010) Simulations of a large scale solar thermal power plant in Turkey using concentrating parabolic trough collectors. Thesis, Middle East Technical University, Turkey

11. Montes MJ, Abánades A, Martinez-Val JM, Valdés M (2009) Solar multiple optimization for a solar-only thermal power plant, using oil as heat transfer fluid in the parabolic trough collectors. Sol Energy 83:2165–2176
12. Remund J, Kunz S (1997) METEONORM: global meteorological database for solar energy and applied climatology. Meteotest

Chapter 74
An Experimental Study on the Drying of Peanuts Using Indirect Solar Dryer

D. Mennouche, B. Bouchekima, S. Zighmi, A. Boubekri, S. Boughali, and A. Matallah

Abstract An indirect type natural convection solar dryer is designed, constructed, and investigated experimentally to study the drying of peanut (*Arachis hypogaea* L.) under the weather conditions of Ouargla, Algeria.

During the experiments, the samples of peanut were dried to the final moisture content of 8.31 from 34.15 % (w.b) in three days of drying. Experimental drying curves showed only a falling drying rate period. Measured air-drying temperatures, relative humidity, and air velocity in the dryer ranged between 20 and 46.06 °C, 8.03 and 35.82 %, and 0 and 1.68 m/s, respectively.

D. Mennouche (✉)
Laboratoire de Développement des Energies Nouvelles et Renouvelables dans les Zones Arides et Sahariennes, Université Kasdi Merbah Ouargla, Ouargla 30000, Algeria

Univ. Ouargla, Fac. Sciences Appliquées, Département de Génie des Procédés, Université Kasdi Merbah, Ouargla 30000, Algeria

Faculté de Technologie, Département de Génie des Procédés, Université Abderrahmane Mira, Bejaia 06000, Algeria
e-mail: mennouche.dj@univ-ouargla.dz

B. Bouchekima
Laboratoire de Développement des Energies Nouvelles et Renouvelables dans les Zones Arides et Sahariennes, Université Kasdi Merbah Ouargla, Ouargla 30000, Algeria

Univ. Ouargla, Fac. des Hydrocarbures, des Energies Renouvelables, des Sciences de la Terre et de l'Univers, Département des Energies Renouvelables, Université Kasdi Merbah, Ouargla 30000, Algeria

S. Zighmi
Univ. Ouargla, Fac. Sciences Appliquées, Département de Génie des Procédés, Université Kasdi Merbah, Ouargla 30000, Algeria

Univ. Ouargla, Fac. Sciences de la nature et de la vie, Lab. Laboratoire Génie de l'Eau et de l'Environnement en Milieu Saharien, Université Kasdi Merbah, Ouargla 30000, Algeria

A. Boubekri
Laboratoire de Développement des Energies Nouvelles et Renouvelables dans les Zones Arides et Sahariennes, Université Kasdi Merbah Ouargla, Ouargla 30000, Algeria

Univ. Ouargla, Fac. Sciences Appliquées, Département de Génie Mécanique, Université Kasdi Merbah, Ouargla 30000, Algeria

I. Dincer et al. (eds.), *Progress in Clean Energy, Volume 2*,
DOI 10.1007/978-3-319-17031-2_74

The data of sample weight, relative humidity, and temperatures were recorded continuously from morning to evening for each test.

The peanut oil extraction (solid-liquid extraction by using Soxhlet extractor) shows that the yield of oil in the dried product increased by 16.26 % compared to the fresh product.

Peanut dried in the indirect solar dryer (type natural convection) was completely protected from insects, rain, and dusts, and the dried samples were of high yield in terms of oil. This system can be used for drying various agricultural products. Also, it is simple in construction and can be constructed at a low cost with locally obtainable materials.

Keywords Solar drying • Natural convection • Peanut • Extraction • Moisture content

Nomenclature

$T_{\text{bef-tr}}$	Temperature before tray (°C)
$T_{\text{aft-tr}}$	Temperature after tray (°C)
T_{out}	Outside temperature (°C)
Rh_{out}	Relative humidity outside the dryer (%)
$\text{Rh}_{\text{bef-tr}}$	Relative humidity before tray (%)

Notations

w.b Wet basis

74.1 Introduction

Solar energy is an important alternative source of energy. It is free, endless, clean, and available abundantly almost the world over [1–5].

Peanut (*Arachis hypogaea* L.) is an annual crop grown predominantly in the Mediterranean region [6]. It is widely cultivated in warm climates and has

S. Boughali
Laboratoire de Développement des Energies Nouvelles et Renouvelables dans les Zones Arides et Sahariennes, Université Kasdi Merbah Ouargla, Ouargla 30000, Algeria

Univ. Ouargla, Fac. des Mathématiques et des Sciences de la Matière, Département de Physique, Université Kasdi Merbah, Ouargla 30000, Algeria

A. Matallah
Laboratoire de Développement des Energies Nouvelles et Renouvelables dans les Zones Arides et Sahariennes, Université Kasdi Merbah Ouargla, Ouargla 30000, Algeria

short-lived yellow flowers. Most peanuts grown in the world are used for oil production, peanut butter, confections, and snack products [7]. Peanut oil is pale yellow oil with a distinctive nutty taste [8].

Traditionally, peanut samples were brought back to normal moisture by direct exposure to the sun for a few days. In direct sun-drying process, products are subjected to rain, wind, and dust. They are exposed to birds, insects, and rodents. Substantial losses take place during this drying process and the food quality may be seriously tarnished [9–16]. Some of these problems have been overcome using an indirect solar dryer.

Indirect solar dryer is simple equipment that can be built by village artisans, using locally existing materials and allows a hygienic drying. Indirect solar dryers are also easy to manipulate and maintain. Many researches and performance studies have been reported on indirect solar dryers [16–21].

The present work is aimed to study the drying of peanut using an indirect solar dryer (type natural convection) realized at LENREZA, University of Ouargla (UKMO). In order to know the effect of drying process on the peanut oil, the extraction of this oil is performed at the end of drying operation.

74.2 Materials and Methods

74.2.1 Sample Preparation

Peanut samples were harvested in the Oasis of Ghardaia (Region of Metlili), southern of Algeria, during the spring season of 2013. At the laboratory, the samples were cleaned manually to remove all foreign matter such as dust, dirt, stones, and chaff as well as immature, broken seeds. Only the clear samples were maintained, sealed in plastic bags, and stored in a cold room (+4 °C).

74.2.2 Drying Experiments

Drying experiments were performed on an indirect solar dryer (Fig. 74.1) assembled at LENREZA (Laboratory of Development of new and Renewable Energy in Arid and Saharan zones), University of Ouargla, Algeria. It is composed of an air plane solar collector $(2.0 \times 1.0 \times 0.13)$ m with an inclined angle of 31° (latitude of Ouargla city) regarding the horizontal and directed to south. The drying room is made up of coated sheet of $(1.0 \times 0.8 \times 0.8)$ m in dimensions thermally isolated with polystyrene on all the external walls and provided with a chimney in galvanized sheets $(1.0 \times 0.02 \times 0.02)$ m.

The total incident solar radiation was measured throughout the day by a solar integrator (Kipp & Zonen). A K-type thermocouple (chromel-alumel) connected to Cobra apparatus was used for temperature measurements at different points in the

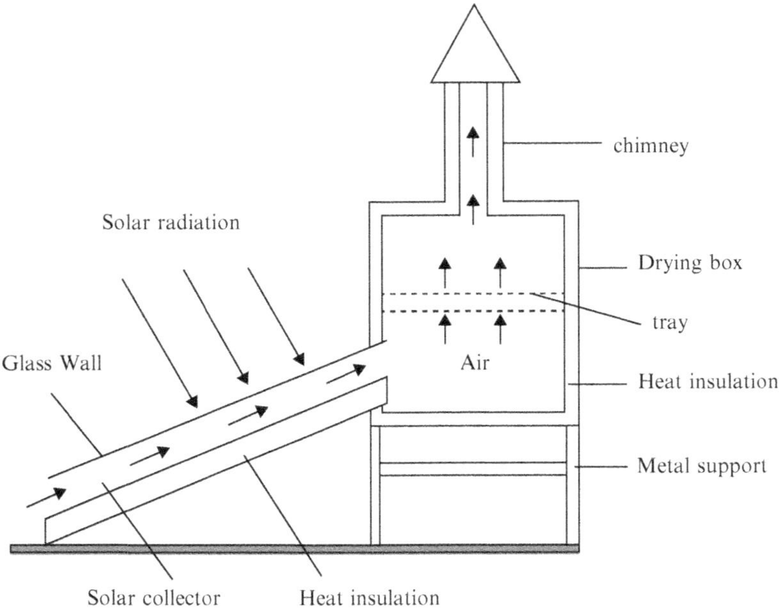

Fig. 74.1 Solar dryer (LENREZA)

solar dryer (outside, after, and before tray). Relative humidity, before and after the tray of drying, was measurement with two hygrometers. Air velocity in the dryer was measured using a hot wire anemometer (Chauvin Arnoux CA 1051).

74.2.3 Moisture Measurements (Water Content)

Water content was determined using a moisture analyzer (Sartorius MA 45, accuracy ± 0.01) [22] operating at the following conditions: sample weight (3 g) and temperature analysis (105 °C). The device stops automatically once the weight of ground peanut samples becomes constant. Moisture content is expressed on a wet basis.

74.2.4 Extraction Process of Peanut Oil

Extraction of the oil content of the peanut seed was carried out in Soxhlet extractor by standard methods using 150 mL n-hexane and 58 g peanut seeds. The yield was calculated as follows:

$$\text{Yield}\ (\%) = \frac{\text{Extracted peanut oil weight}}{\text{initial peanut weight}}$$

74.3 Results and Discussion

74.3.1 *Temperature: Solar Radiation*

Figure 74.2 represents the temperature distribution ($T_{bef\text{-}tr}$, $T_{aft\text{-}tr}$, T_{out}) in the solar dryer and the outside medium. It is observed that the outside temperature (T_{out}) varies between 19 and 35 °C during the daily drying period. The temperature increases once the air arrives at the tray supporting the dried product ($T_{bef\text{-}tr}$). Temperature ranges between 20 and 46.06 °C for all the daily drying period.

On the same figure, we can also note that once the air passes through the surface of the product, the temperature ($T_{aft\text{-}tr}$) decreases and varies from 22 to 45 °C. Such an observation could be related to the energy consumption while the product water vapor is diffused in the drying air.

Figure 74.3 displays the daily solar radiation variation versus local time during the drying period. It should be noted that the radiation curve presents a Gaussian form and varies from 12 to 731 W/m^2. It is also important to note that solar radiation intensity affects directly the drying air temperature within the chamber (Fig. 74.2). This result reveals that the instantaneous drying air temperature depends upon the solar radiation evolution.

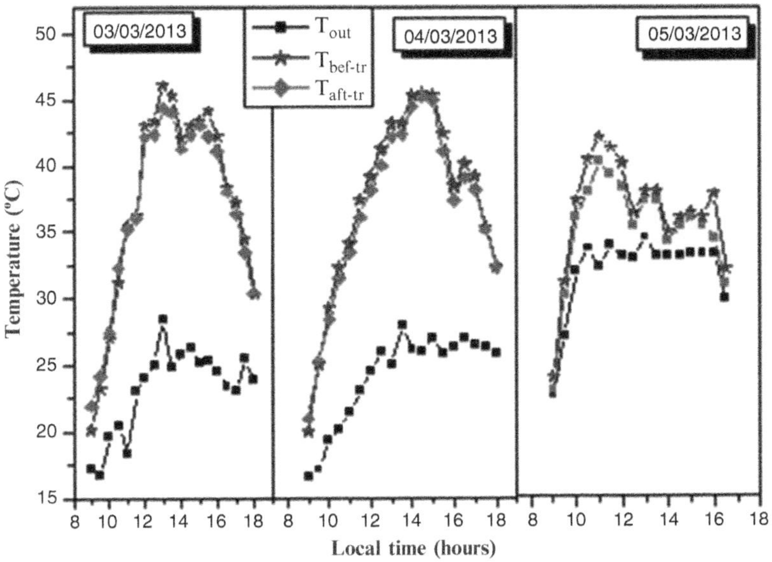

Fig. 74.2 Daily measured temperature changes

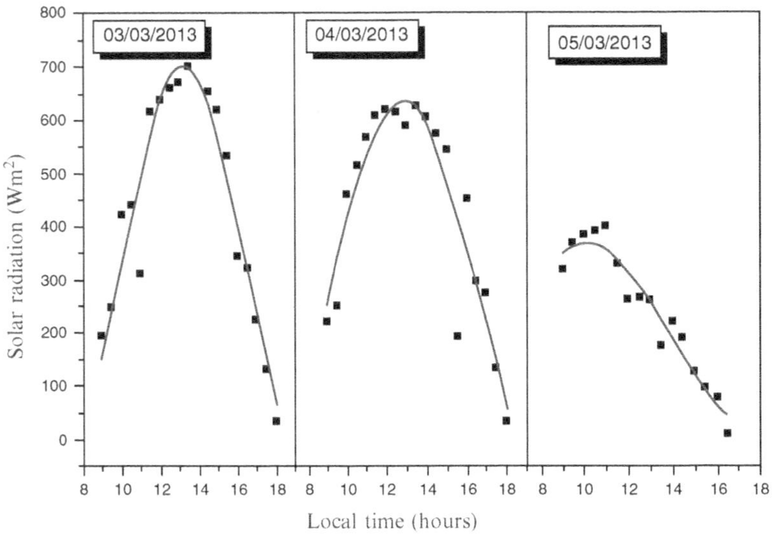

Fig. 74.3 Daily solar radiation changes

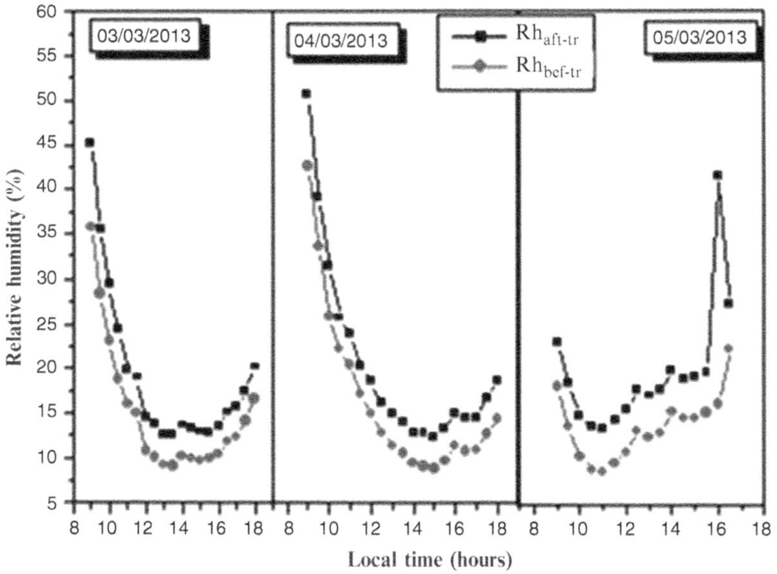

Fig. 74.4 Daily measured air relative humidity changes

74.3.2 Relative Humidity of the Drying Air

Figure 74.4 shows the drying air relative humidity changes, before and after the drying tray with respect to local time. It was observed that the Rh_{bef-tr} varied between 8.03 and 35.82 % during the drying period.

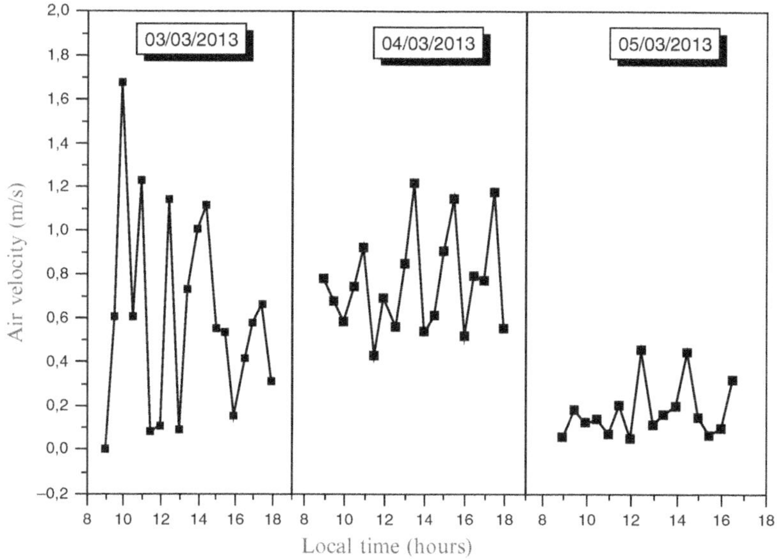

Fig. 74.5 Daily air velocity changes

It can be seen also in Fig. 74.4 that the relative humidity of the air passing through the product surface ($Rh_{aft\text{-}tr}$) increases (compared to $Rh_{bef\text{-}tr}$) and reaches a minimal value of 11.01 % in the second day of drying. This increase is due to water evaporation taking place in the boundary layer upon the surface of the product.

74.3.3 Air Velocity Evolution

Figure 74.5 illustrates the air velocity variation inside the dryer with time. It is worth noticing that the velocity of the air was somewhat low and varied between 0 and 1.68 m/s.

74.3.4 Drying Curve

Figure 74.6 demonstrates the variation of the moisture content with time during the drying process. The drying period necessary to reach the final water content 8.31 % (w.b) was found to be 15 h of drying (3 days). The high rate of water loss was observed within the first and the second days of drying period where the decrease was from 34.15 to 10.33 % (w.b). In the third day, the water loss became less important, and 8.31 % was observed at the end of the drying process (17 h).

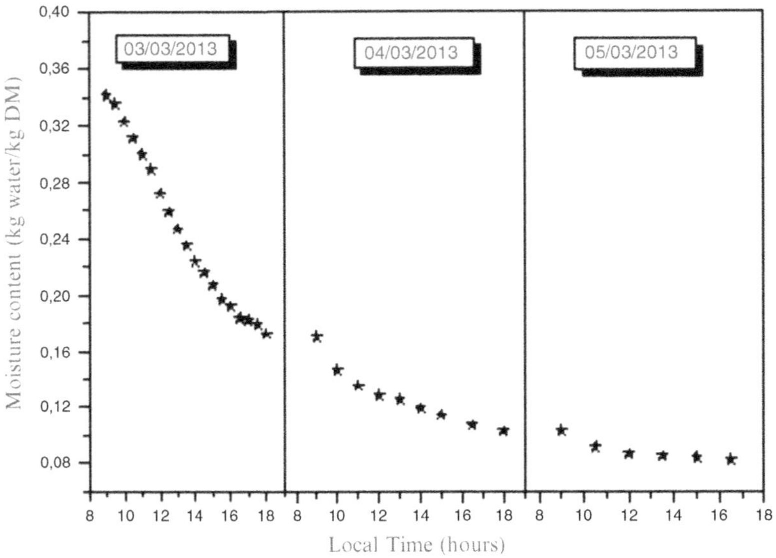

Fig. 74.6 Moisture content changes

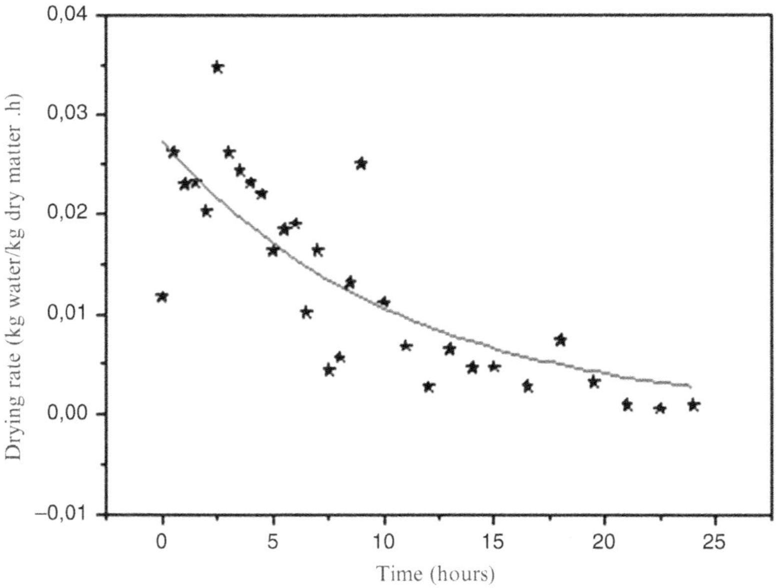

Fig. 74.7 Variation of drying rate as a function of drying time

Figure 74.7 shows the variation in drying rate as a function of the drying time. This curve presents no constant drying rate period to be considered, which shows that the important part of drying process has been occurred in the falling rate period. This implies that a film of water reached the surface from within the body of the dried sample, almost immediately.

Table 74.1 Extraction of peanut oil

	Fresh product	Dried product
Initial peanut weight (g)	259.21	58
Extracted oil weight (g)	22.29	14.42
Yield (%)	8.60	24.86

74.3.5 Extraction of Peanut Oil

Table 74.1 shows the results of the extraction of peanut oil; it is noted that the yield of dried peanut (24.86 %) is higher than the fresh product (8.60 %). This result leads to conclude that the oil in the dried product increased by 16.26 % compared to the fresh product.

74.4 Conclusion

In the present work, the drying of peanut in indirect solar dryer has been studied experimentally under Ouargla city prevailing conditions. From the experimental drying curves obtained, it is showed that the duration of the drying operation to achieve moisture content of 8.31 % (w.b) was three days of drying. Also, it is noted that only the falling drying rate period exists.

It was observed that the measured air-drying temperatures, relative humidity, and air velocity in the dryer ranged between 20 and 46.06 °C, 8.03 and 35.82 %, and 0 and 1.68 m/s, respectively.

The extraction of the peanut oil shows that the yield of oil in the dried product increased by 16.26 % compared to the fresh product.

Acknowledgments The authors would like to thank Pr. Aomar Boukraa, director of LENREZA, University Kasdi Merbah Ouargla, for his help and assistance to this project.

References

1. Panna LS (2011) Silk cocoon drying in forced convection type solar dryer. Appl Energy 88:1720–1726
2. Sreekumar A, Manikantan PE, Vijayakumar KP (2008) Performance of indirect solar cabinet dryer. Energy Convers Manag 49:1388–1395
3. Şevik S (2013) Design experimental investigation and analysis of a solar drying system. Energy Convers Manag 68:227–234
4. Reyes A, Mahn A, Cubillos F, Huenulaf P (2013) Mushroom dehydration in a hybrid-solar dryer. Energy Convers Manag 70:31–39
5. Mohajer A, Nematollahi O, Joybari MM, Hashemi SA, Assari MR (2013) Experimental investigation of a hybrid solar drier and water heater system. Energy Convers Manag 76:935–944

6. Aydin C (2007) Some engineering properties of peanut and kernel. J Food Eng 79:810–816
7. Admedna M, Yu J, Goktepe I (2007) Peanut protein concentrate: production and functional properties as affected by processing. Food Chem 103(1):121–129
8. Oyinlola AOA, Adekoya LO (2004) Development of a laboratory model screw press for peanut oil expression. J Food Eng 64:221–227
9. Amer BMA, Hossain MA, Gottschalk K (2010) Design and performance evaluation of a new hybrid solar dryer for banana. Energy Convers Manag 51:20–813
10. Lahsasni S, Kouhila M, Mahrouz M, Idlimam A, Jamali A (2004) Thin layer convective solar drying and mathematical modeling of prickly peel (*Opuntia ficus indica*). Energy 29:211–224
11. Prasad J, Vijay VK, Tiwari GN, Sorayan VPS (2006) Study on performance evaluation of hybrid drier for turmeric (*Curcuma longa* L.) drying at village scale. J Food Eng 75:497–502
12. Lalit MB, Sanatoch S, Naik SN (2010) Solar dryer with thermal energy storage systems for drying agricultural food products: a review. Renew Sust Energ Rev 14:2298–2314
13. Aghbashlo M, Kianmehr MH, Arabhosseini A (2009) Performance analysis of drying of carrot slices in a semi-industrial continuous band dryer. J Food Eng 91:99–108
14. Goyal RK, Tiwari GN (1997) Parametric study of a reverse flat plate absorber cabinet dryer: a new concept. Sol Energy 60:41–48
15. Ertekin C, Yaldiz O (2004) Drying of eggplant and selection of a suitable thin layer drying model. J Food Eng 63:349–359
16. Boubekri A, Benmoussa H, Mennouche D (2009) Solar drying kinetics of date palm fruits assuming a step-wise air temperature change. J Eng Sci Technol 4(3):308–320
17. Sharma A, Chen CR, Vu Lan N (2009) Solar-energy drying systems: a review. Renew Sust Energ Rev 13:1185–1210
18. Jairaj KS, Singh SP, Srikant K (2009) A review of solar dryers developed for grape drying. Sol Energy 83:1698–1712
19. Ekechukwu OV, Norton B (1999) Review of solar-energy drying systems II: an overview of solar drying technology. Energy Convers Manag 40:615–655
20. VijayaVenkataRamana S, Iniyanb S, Ranko G (2012) A review of solar drying technologies. Renew Sust Energ Rev 16:2652–2670
21. Sodha MS, Chandra R (1994) Solar drying systems and their testing procedures: a review. Energy Convers Manag 35:219–267
22. Boughali S, Benmoussa H, Bouchekima B, Mennouche D, Bouguettaia H, Bechki D (2009) Crop drying by indirect active hybrid solar—electrical dryer in the eastern Algerian Septentrional Sahara. Sol Energy 83(12):2223–2232

Chapter 75
Comparative Study of Conventional and Renewable Energy Sources for HVAC Systems

M. Malik, I. Dincer, and M.A. Rosen

Abstract In this chapter, the life cycle aspects and environmental impact of substituting various renewable and non-renewable energy sources are analysed for several heating, ventilation and air conditioning (HVAC) systems, and several case studies are considered. For the case of conventional energy sources, it is found that substitution of natural gas for other conventional energy sources can significantly reduce CO_2 emissions. For the case of biomass, biogas is found to be the most effective type of biomass for reducing CO_2 emissions. The case studies demonstrate that building design and insulation significantly affect HVAC energy consumption. The results are significant because HVAC is the major source of energy consumption in most residential and other buildings.

Keywords Energy • Renewable energy • Heating • Cooling • Residential building • Environmental impact

75.1 Introduction

Efforts are ongoing in most countries to reduce consumption of energy in general and conventional fossil fuels in particular, and to increase use of renewable energy. Heating ventilation and air conditioning (HVAC) is widely utilized in most countries in buildings in all sectors, e.g. residential, commercial, institutional and industry, the latter of which encompasses various types of industries (e.g. cement, computer chip manufacturing). The energy consumption by HVAC systems in buildings varies from 16 to 50 % of total energy consumption, depending on the country and its sectorial energy use patterns [1]. Breakdowns of the residential sector contribution to the total energy use for various countries are shown in Fig. 75.1.

M. Malik (✉) • I. Dincer • M.A. Rosen
Clean Energy Research Laboratory (CERL), Faculty of Engineering and Applied Science, University of Ontario Institute of Technology, 2000 Simcoe Street North, Oshawa, ON, Canada, L1H 7K4
e-mail: monu.malik@uoit.ca; ibrahim.dincer@uoit.ca; marc.rosen@uoit.ca

I. Dincer et al. (eds.), *Progress in Clean Energy, Volume 2*,
DOI 10.1007/978-3-319-17031-2_75

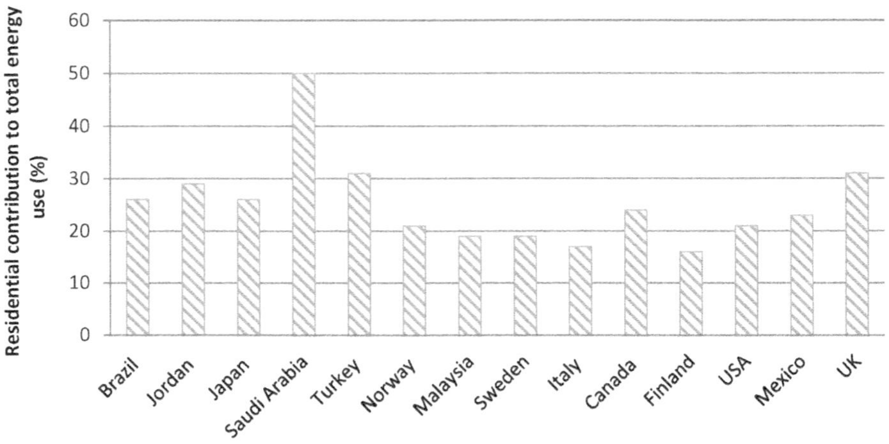

Fig. 75.1 Contribution of residential energy consumption to national energy consumption for selected countries. (Data from [1])

Numerous studies have been reported on various aspects of HVAC systems. Approximately 22 % of energy consumption in the USA occurs in the residential sector [2], of which 48 % is used for heating and cooling, depending on the region [3]. Annual carbon dioxide (CO_2) emissions due to energy consumption in the residential sector in 2010 were 313.4 million tonnes in the USA [4]. Heating and cooling have significant potential for improved efficiency [5], although some barriers to improving efficiency of residential buildings exist, like longer life times of existing building stock and insufficient private and public awareness and support [6]. Although significant energy research is ongoing in the residential sector on technological aspects of buildings, little attention is given to resident/public behaviour towards energy savings, e.g. encouraging people to reduce energy consumption in heating/cooling systems by using night-time temperature setback thermostats [7].

HVAC systems can use conventional and/or renewable energy resources. Each has advantages and disadvantages, in terms of technical performance, economics, environmental impact and other factors, and many investigations have been reported on these. For instance, exergy and energy analyses have been reported using four heating options (heat pump, condensing boiler, conventional boiler and solar collector) for a building using fossil fuel and renewable energy sources [8]. Through a case study, the environmental impacts of HVAC systems were shown to depend mainly on type of energy resource used, type of appliance, distribution method and material used in the HVAC system [9]. Similar results were reported in another study [10], in which environmental impact was seen to depend greatly on the form and source of energy and the type of heating and cooling systems. Geothermal heating systems have been shown to have less environmental impacts than solar thermal systems [11], and it has been demonstrated that the installation stage of geothermal systems exhibits the highest environmental impact. For solar energy systems, solar cooling was shown to have the highest environmental impact due to the use of high quantities of plastic and steel in coil units and the cooling tower [12], which ultimately increase quantities of harmful pollutant emissions. Nonetheless, this impact only occurs during the manufacturing process and there is negligible environmental impact during operation. Solar thermal storage systems have been shown to be capable of improving the capacity, availability, reliability and overall performance of heating and cooling [13].

In the present study, conventional and renewable energy options for HVAC in residential buildings are identified and analysed. Through several case studies, the environmental effects of energy source substitution are examined in terms of CO_2 emissions. Further three cases are compared to determine the effect of substitution of energy resources on environmental impact in terms of CO_2 emissions. The objective of this study is to determine the environmental impacts of using various energy sources for HVAC systems and to provide enhanced knowledge that can assist in identifying the most advantageous energy sources and designs for HVAC system in residential buildings.

75.2 Energy Sources for HVAC

Energy consumption patterns in heating, cooling and ventilation mainly depend on the building location and time of year, as well as the type of HVAC appliances used. Numerous sources of energy are available for HVAC systems, and substitution of one energy resource with another can increase or decrease the associated environmental impact. The energy resources available for HVAC systems can be grouped into two main types, as shown in Fig. 75.2:

1. Non-renewable energy sources (including most conventional energy sources).
2. Renewable energy sources.

It is difficult to include and classify all the renewable and non-renewable energy sources, so only major sources are included and discussed in the present study. Some energy resources can be classified as both renewable and non-renewable energy sources, like energy from waste, as shown in Fig. 75.2. Each energy

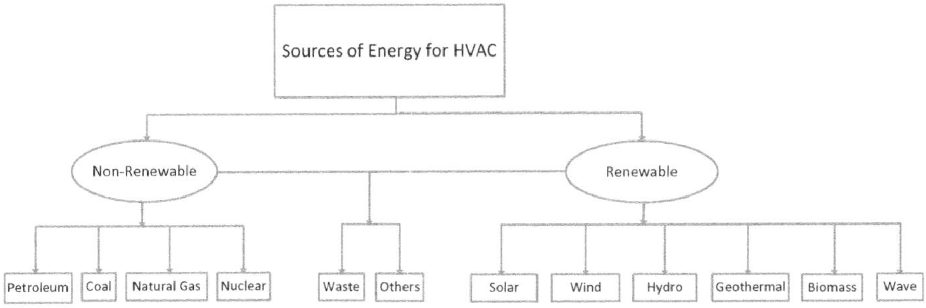

Fig. 75.2 Classification of major renewable and non-renewable energy sources for HVAC

resource has advantages and disadvantages in terms of utilization efficiency, economics, environmental impact and other factors. Usually, the conversion of these resources to electricity for heating, cooling and ventilation causes impacts on the environment due to the low conversion efficiency of fuel to electricity.

75.2.1 Non-renewable Sources

Non-renewable energy resources (e.g. fossil fuels) are those that are not replenished at the rate at which they are used. As previously discussed, three main non-renewable energy sources, i.e. petroleum, coal natural gas and nuclear, can drive HVAC systems. A breakdown of these sources is shown in Fig. 75.3.

Each energy resource in Fig. 75.3 has its own characteristics, e.g. composition and energy content, and produces CO_2 when combusted. It is useful to compare non-renewable energy resources based on their energy (or heat) content and CO_2 emissions, as one input for determining the fuel best suited for a given HVAC system. The heat content, carbon content and total CO_2 emissions for selected non-renewable energy resources are given in Table 75.1 [14]. The CO_2 emissions are dependent on the carbon content of the fuel.

Figure 75.4 compares the specific heat contents of several fuels and shows that the specific heat content is highest for kerosene (14.40 kWh/kg) and lowest for lignite (4.59 kWh/kg). Note that the heat content of a fuel is based on its higher heating value (HHV). The most widely used energy sources for HVAC systems are

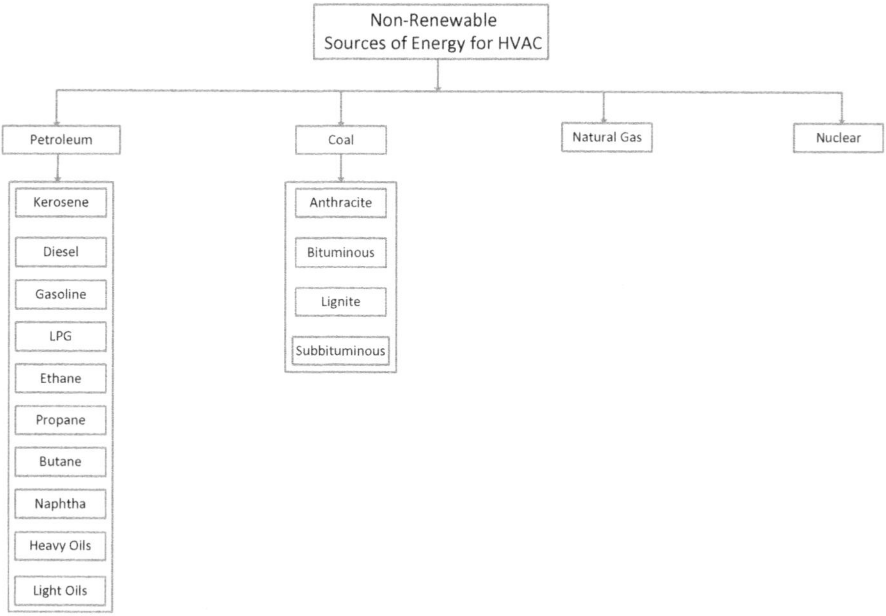

Fig. 75.3 Classification of main non-renewable energy sources for HVAC

Table 75.1 Characteristics of several fuel types

Source	Fuel type	Heat content (kWh/kg)	Carbon content (per unit energy) (g C/kWh)	CO_2 emission (per unit energy) (kg CO_2/kWh)	CO_2 emission (per unit mass) (kg CO_2/kg)
Coal and coke	Anthracite	8.11	96.36	0.353	2.86
	Bituminous	8.05	86.91	0.319	2.57
	Subbituminous	5.57	90.29	0.331	1.84
	Lignite	4.59	89.67	0.329	1.51
	Coke	8.01	94.96	0.348	2.79
Natural gas	Natural gas	12.65	49.34	0.181	2.29
Petroleum products	Still gas	13.50	62.10	0.228	3.47
	Kerosene	14.40	69.98	0.257	3.27
	LPG	13.26	58.62	0.215	2.85
	Propane (liquid)	14.26	57.19	0.210	2.99
	Ethane	9.37	58.28	0.214	2.00
	Ethylene	6.56	62.75	0.230	1.51
	Isobutane	12.65	60.39	0.221	2.80
	Butane	13.01	60.63	0.222	2.89
	Natural gasoline	11.98	62.20	0.228	2.73
	Special naphtha	12.62	67.32	0.247	3.11
	Petroleum coke	4.96	95.30	0.349	1.73
	Heavy gas oils	12.32	69.71	0.256	3.15
	Light oil	11.70	59.83	0.260	2.60

Source: [14]
LPG liquid petroleum gas

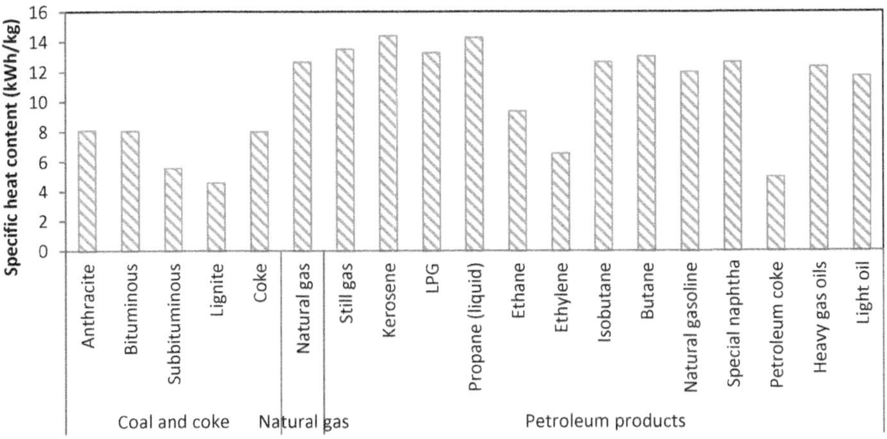

Fig. 75.4 Comparison of specific heat contents of selected non-renewable energy sources

oil and natural gas. Natural gas is mostly used for direct heating while oil is usually indirectly used for heating and cooling by converting it into electricity [15, 16].

The carbon content per unit energy is shown in Fig. 75.5 for various fuels. The carbon content is observed to be highest for anthracite (96.36 g C/kWh) and lowest for natural gas (49.34 g C/kWh). Much CO_2 is produced when burning a non-renewable fuel in the presence of air for space heating, electricity generation and other purposes, as shown in Fig. 75.6 on an energy basis and in Fig. 75.7 on a mass basis. The CO_2 emission per unit energy production is observed in Fig. 75.6 to be highest for petroleum coke (0.349 kg CO_2/kWh) and lowest for natural gas

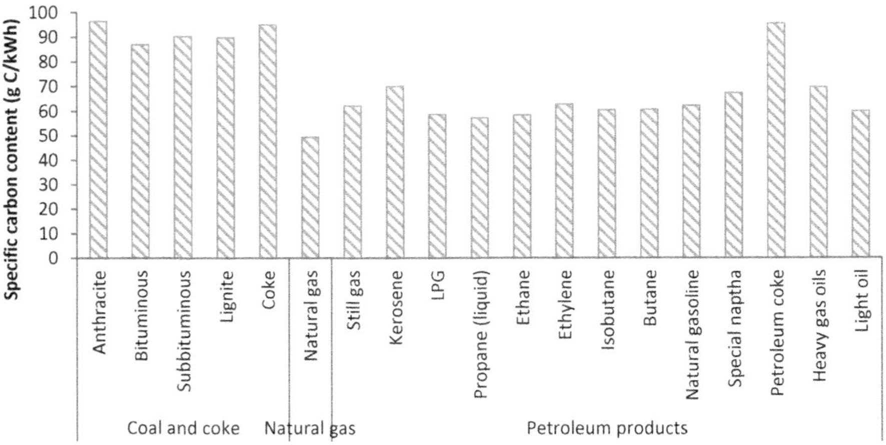

Fig. 75.5 Comparison of specific carbon content for selected non-renewable energy sources

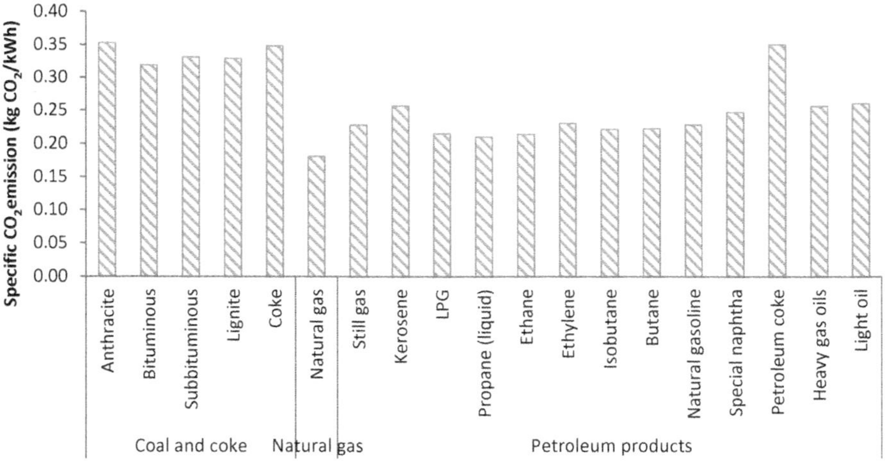

Fig. 75.6 Comparison of CO_2 produced from combustion per unit energy content for selected non-renewable energy sources

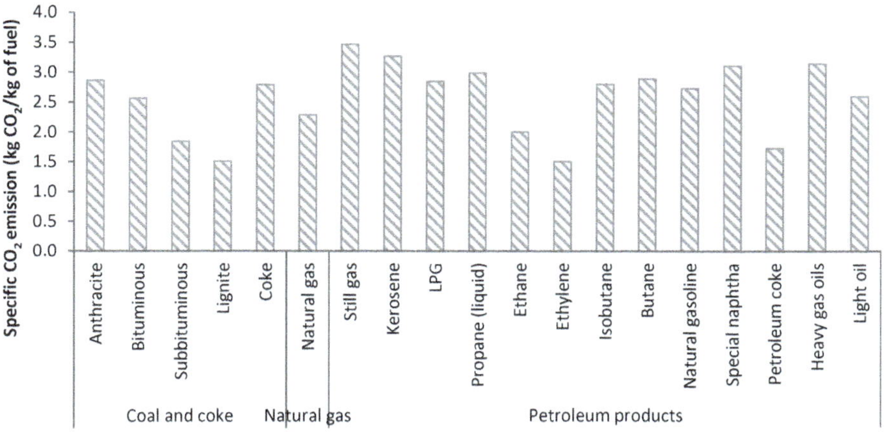

Fig. 75.7 Comparison of CO_2 produced from combustion per unit fuel consumption for selected non-renewable energy sources

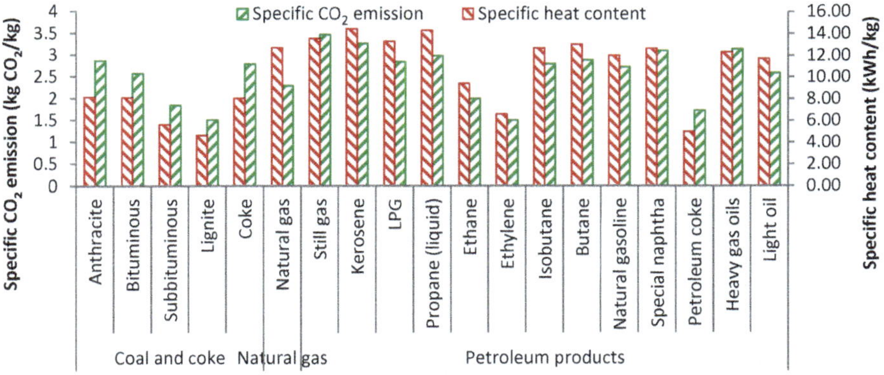

Fig. 75.8 Comparison of specific heat content and specific CO_2 emission from combustion, on mass basis, for selected non-renewable energy sources

(0.181 kg CO_2/kWh), while the CO_2 emission per unit mass of fuel is seen in Fig. 75.7 to be highest for still gas (3.47 kg CO_2/kg) and lowest for lignite and ethylene (1.51 kg CO_2/kg).

Figure 75.8 compares the heat content and CO_2 emissions from combustion for various fuels, on a unit mass basis. The ratio of CO_2 emission per unit energy is observed to be highest for natural gas and lowest for anthracite.

75.2.1.1 Case Study for Non-renewable Energy Sources

To demonstrate the effects of fuel substitution in an HVAC system on CO_2 emissions, a case study involving a hypothetical house is analysed considering

Table 75.2 Annual fuel requirement, rate of CO_2 emission and CO_2 emission reduction through different options in a typical house

Source	Fuel type	CO_2 emission (kg CO_2/kWh)	Fuel required (kg/year)	Annual CO_2 emissions to heat a typical house having heat load 30,000 kWh/year		
				Option 1: CO_2 emission while producing electricity (kg/year)	Option 2: CO_2 emission reduced in direct heating (kg/year)	Option 3: CO_2 emission reduced by using natural gas in direct heating (kg/year)
Coal and coke	Anthracite	0.353	3,700	27,900	16,100	5,700
	Bituminous	0.319	3,730	25,200	14,500	4,600
	Subbituminous	0.331	5,400	26,100	15,100	5,000
	Lignite	0.329	6,540	25,960	15,000	4,900
	Coke	0.348	3,740	27,500	15,900	5,600
Natural gas	Natural gas	0.181	2,370	14,300	8,300	0
Petroleum products	Still gas	0.228	2,220	18,000	10,400	1,600
	Kerosene	0.257	2,080	20,300	11,700	2,500
	LPG	0.215	2,260	17,000	9,800	1,100
	Propane (liquid)	0.21	2,100	16,600	9,600	960
	Ethane	0.214	3,200	16,900	9,800	1,100
	Ethylene	0.23	4,570	18,100	10,500	1,600
	Isobutane	0.221	2,370	17,500	10,100	1,300
	Butane	0.222	2,300	17,600	10,100	1,400
	Natural gasoline	0.228	2,500	18,000	10,400	1,600
	Special naphtha	0.247	2,380	19,500	11,300	2,200
	Petroleum coke	0.349	6,040	27,600	15,900	5,600
	Heavy gas oils	0.256	2,430	20,200	11,700	2,500
	Light oil	0.26	2,560	20,500	11,900	2,600

two energy inputs. The annual heating requirement of the house is taken to be 30,000 kWh (see Table 75.2).

The following assumptions are invoked to simplify the analyses [17]:

1. The overall fuel-to-electricity conversion efficiency for all types of fuel is 38 %. This efficiency actually depends on the type of fuel and the system used, e.g. a conventional coal fired power plant typically has a much lower efficiency than a gas turbine plant. A common conversion efficiency, roughly based on the average of efficiencies for different systems, is used for all fuels here to simplify the analyses.

2. The overall fuel-to-heat efficiency for the direct heating process for all types of fuel is 90 %. This efficiency also depends on type of fuel and conversion system in practice, but a constant value is used, again to simplify the analyses.
3. The efficiency of an electric resistance heating is 100 %, which is reasonable since the conversion of electricity to heat incurs only very minor losses.

Three options are considered for the energy input to the house.

1. Annual rate of CO_2 emissions if fuel is used for electricity generation with an overall fuel-to-electricity conversion efficiency of 38 % and then the electricity is supplied to an electric heater for space heating with an efficiency of 100 %.
2. Reduction in annual rate of CO_2 emissions, relative to option 1, if fuel is used directly for space heating with a 90 % efficiency.
3. Reduction in annual rate of CO_2 emissions, relative to option 1, if the various conventional fuels which are currently used for HVAC are substituted with natural gas and natural gas is used directly for space heating with a 90 % efficiency.

In option 1, fuel is converted to electricity, which is used for space heating. In option 2, fuel is used directly for space heating and the reduction in annual CO_2 emissions relative to option 1 is calculated. In option 3, all fuels are substituted with natural gas, which is used directly for space heating, and the reductions in annual CO_2 emissions relative to option 2 are calculated. The annual CO_2 emissions are compared in Table 75.2 for the three options.

The quantity of each type of fuel required per year, the CO_2 emissions per year while producing electricity for HVAC systems, and the CO_2 emissions reduced per year while using fuel directly for heating are determined (see Table 75.2). It can be seen in Table 75.2 that the greatest quantity of fuel required per year is for lignite (6,540 kg/year) due to its low heat content, and the lowest quantity is for kerosene (2,100 kg/year) due to its high heat content. The reduction in annual CO_2 emissions when using fuel directly for heating is also calculated for various fuels (see Fig. 75.9). The result shows that CO_2 emissions per year can be greatly reduced by using fuel for direct heating. The largest reduction in CO_2 emissions per year occurs for anthracite (16,100 kg/year) when it is used directly for heating in place of electric heating.

The reductions in CO_2 emissions per year are also determined from substituting various fuels with natural gas in direct heating (see Fig. 75.10). The result shows that annual CO_2 emissions can be greatly reduced by replacing coal, coke and petroleum coke with natural gas and somewhat reduced by replacing other fossil fuels with natural gas.

75.2.2 Types of Renewable Energy Sources

Renewable energy sources are those that are replenished at the same rate at which they are used. The main renewable energy sources which can be used to drive

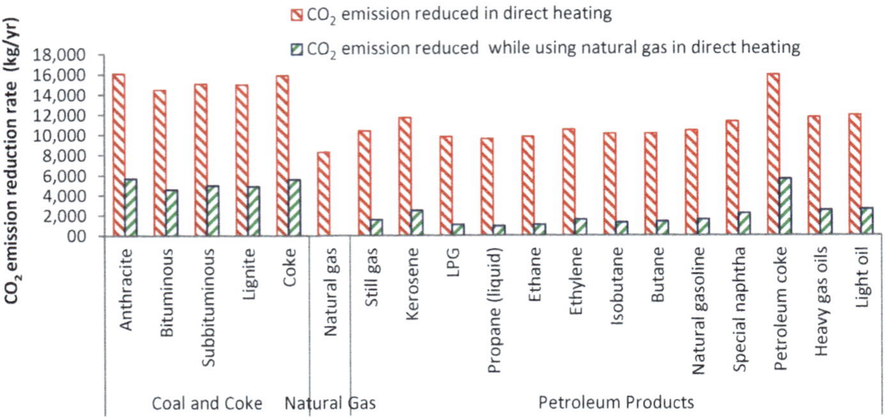

Fig. 75.9 Annual CO_2 emission reduction with direct heating and reduction by replacing with natural gas in direct heating, for various fuels

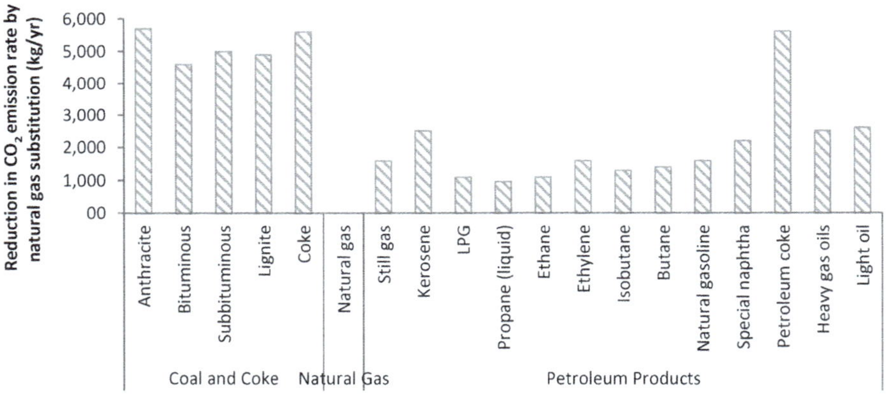

Fig. 75.10 Annual reduction in CO_2 emissions from substituting natural gas with other non-renewable fuels, in direct heating

HVAC systems are solar, wind, hydro, geothermal, wave and biomass. Further details of these energy sources are shown in Fig. 75.11.

75.2.2.1 Major Renewable Sources

The main renewable energy sources for HVAC system include solar, wind, hydro, geothermal, ocean, biomass and wave energy. Renewable energy can also be exploited to offset HVAC needs through building design, e.g. making use of natural heating, cooling and ventilation where possible. Electricity generation energy efficiencies and life cycle CO_2 emissions are given in Table 75.3 for various

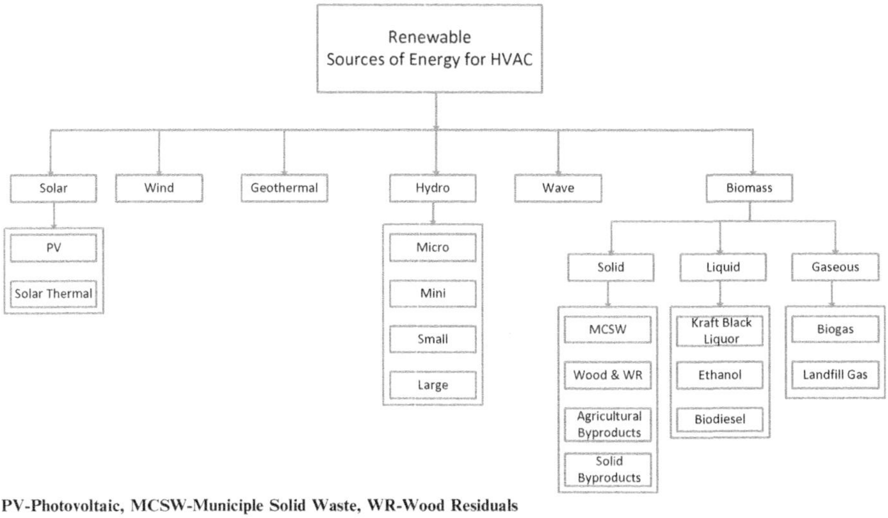

Fig. 75.11 Breakdown of non-renewable energy sources usable for HVAC. *PV* photovoltaic, *MCSW* municipal solid waste, *WR* wood residuals

Table 75.3 Life cycle CO_2 emissions from selected renewable energy sources and their overall energy efficiencies

Energy source	Overall energy efficiency (%)	Life cycle CO_2 emission per unit of electricity produced		
		Minimum ($g\ CO_2/kWh$)	50th percentile ($g\ CO_2/kWh$)	Maximum ($g\ CO_2/kWh$)
Wind	30–45	2	12	81
Solar thermal	15–25	5	22	217
Solar PV	12–22	7	46	89
Hydroelectric	85–95	0	4	43
Geothermal	20–35	6	45	79
Ocean energy	2–4	2	8	23

Source: [18]

types of renewable energy. The values in Table 75.3 are based on reports from various studies and authors.

The variation in Table 75.3 in the overall energy efficiencies is due to changes in the environmental conditions and the type of conversion system used, e.g. in the case of hydro power, the overall efficiency of the power plant depends on the type of turbine used and the mass flow rate of water, while in case of geothermal energy, overall efficiency depends on the temperature of water.

The specific CO_2 emissions from the renewable sources are negligible compared to the specific CO_2 emissions from non-renewable resources (i.e. fossil fuels). The minimum, maximum and 50th percentile life cycle specific CO_2 emissions from several renewable energy sources are adapted from a study [18] (see Table 75.3).

There is a large difference between the minimum and maximum values of the life cycle CO_2 emission for renewable sources, because in the study a number of references (based on a literature review of LCAs of GHG emissions from electricity generation) are considered, and these often involve different parameter values. The 50th percentile life cycle specific CO_2 emission is highest for solar photovoltaics (46 g CO_2/kWh) and lowest for hydroelectric power (4 g CO_2/kWh). Thus, from the perspective of environmental impact due to carbon dioxide emissions, it may be advantageous to utilize hydroelectric power to drive HVAC systems.

It is difficult to discuss all renewable sources of energy for HVAC in detail. So, only one renewable source of energy (biomass) is selected and discussed in detail here.

75.2.2.2 Biomass

Biomass is biological material derived from living or recently living organisms. It often can be used directly as energy source or converted to other fuels like biogas or biodiesel. Biomass can be separated into three categories: solid, liquid and gaseous (see Fig. 75.11). The specific heat content, carbon content and unit CO_2 emission for various biomass types are given in Table 75.4. Biomass can be used for applications like heating, cooling and electricity generation [19], especially because of its availability and low operating cost. Biogas and biodiesel are the best suited types of biomass for driving HVAC systems due to their high energy contents and low environmental impacts.

Figure 75.12 compares the specific heat contents of various biomass fuels, in terms of higher heating value (HHV). The specific heat content is seen to be greatest for biogas (12.80 kWh/kg) and lowest for peat (2.58 kWh/kg).

Carbon dioxide is produced when biomass is combusted, and Fig. 75.13 compares the carbon content per unit energy for various types of biomass. The carbon content per unit energy is seen to be greatest for agricultural by-products (109.97 g C/kWh) and lowest for biogas (captured methane) (48.45 g C/kWh). The quantities of specific CO_2 emissions from combustion of various types of biomass are shown in Fig. 75.14 on an energy basis and in Fig. 75.15 on a mass basis. The CO_2 emission from combustion per unit energy production is observed in Fig. 75.14 to be greatest for agricultural by-products (0.403 kg CO_2/kWh) and lowest for biogas (0.176 kg CO_2/kWh). The CO_2 emission from combustion per unit mass is observed in Fig. 75.15 to be greatest for petroleum coke (solid) (3.39 kg CO_2/kg) and lowest for municipal solid waste (0.99 kg CO_2/kg).

Figure 75.16 compares unit specific heat contents and specific CO_2 emissions for biomass combustion, on mass bases. The ratio of CO_2 emission per unit energy is seen to be greatest for biogas and lowest for solid by-products.

Table 75.4 Biomass fuels types, specific heat contents and unit carbon emissions

Form	Biomass fuel type	Heat content (kWh/kg)	Carbon content (per unit energy) (g C/kWh)	CO_2 emission (per unit energy) (kg CO_2/kWh)	CO_2 emission (per unit mass) (kg CO_2/kg)
Solid	Municipal solid waste	3.21	84.42	0.309	0.99
	Tyres	8.68	80.01	0.293	2.55
	Plastics	12.28	69.78	0.256	3.14
	Petroleum coke (solid)	9.69	95.30	0.349	3.39
	Wood & wood residuals (12 % MC)	4.97	87.28	0.320	1.59
	Agricultural by-products	2.67	109.97	0.403	1.07
	Peat	2.58	104.07	0.382	0.99
	Solid by-products	8.34	98.20	0.360	3.00
Liquid	Kraft black liquor (hardwood)	3.87	87.86	0.322	1.25
	Kraft black liquor (softwood)	3.95	88.55	0.325	1.28
	Ethanol (100 %)	8.24	63.70	0.234	1.93
	Biodiesel (100 %)	11.26	68.72	0.252	2.84
Gas	Biogas (captured methane)	12.80	48.45	0.178	2.27
	Landfill gas (50 % CH_4/ 50 % CO_2)	7.65	48.45	0.178	1.36

MC moisture content

75.2.2.3 Case Study for Renewable Energy Sources

The effect of substituting various fuels for HVAC systems on environmental impact in terms of CO_2 emissions per year is investigated. A similar approach is used as in the case study of non-renewable sources. A hypothetical house with an assumed annual heating requirement of 30,000 kWh is considered (see Table 75.5), and the same assumptions regarding device efficiencies are invoked as in the case of case study for non-renewable sources.

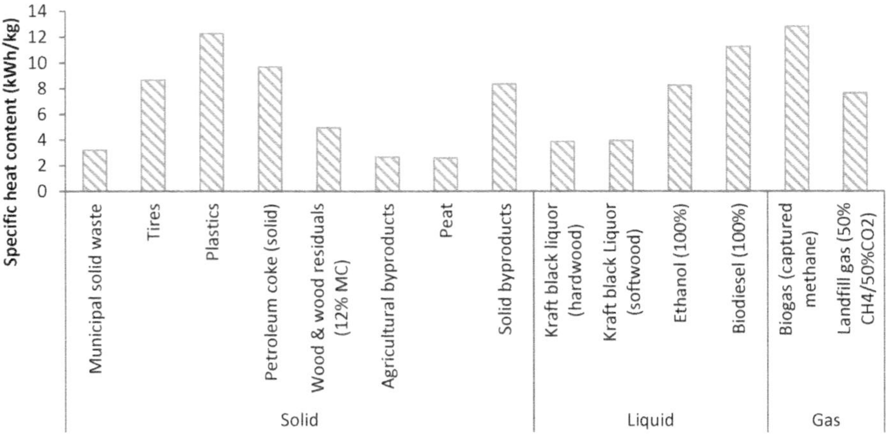

Fig. 75.12 Specific heat contents of various biomass fuels

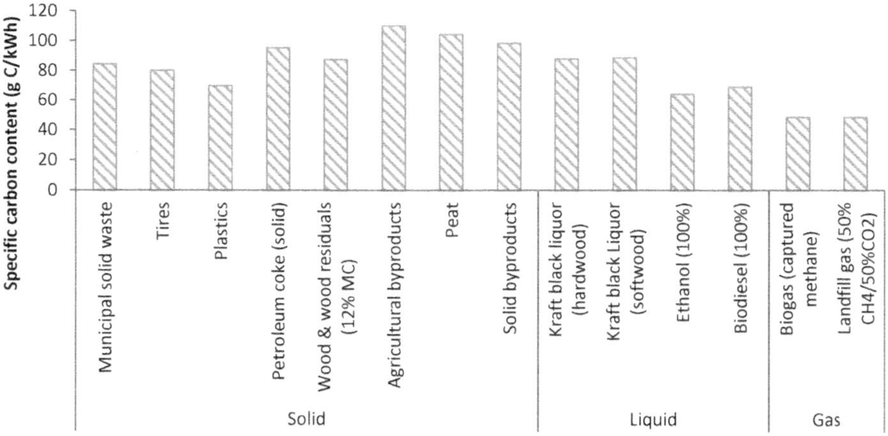

Fig. 75.13 Carbon content per unit energy for various types of biomass

Four options, all involving biomass fuels, are considered for the energy input to the house and comparisons are made in terms of CO_2 emission per year in each case:

1. Biomass is used for electricity generation, with an overall fuel-to-electricity energy conversion efficiency of 38 %, and the electricity is supplied to an electric resistance space heater having an energy efficiency of 100 %.
2. Biomass is used directly for space heating, with a 90 % overall fuel-to-heat energy efficiency.
3. Biomass is replaced with natural gas, which is directly used for space heating with a 90 % overall fuel-to-heat energy efficiency.

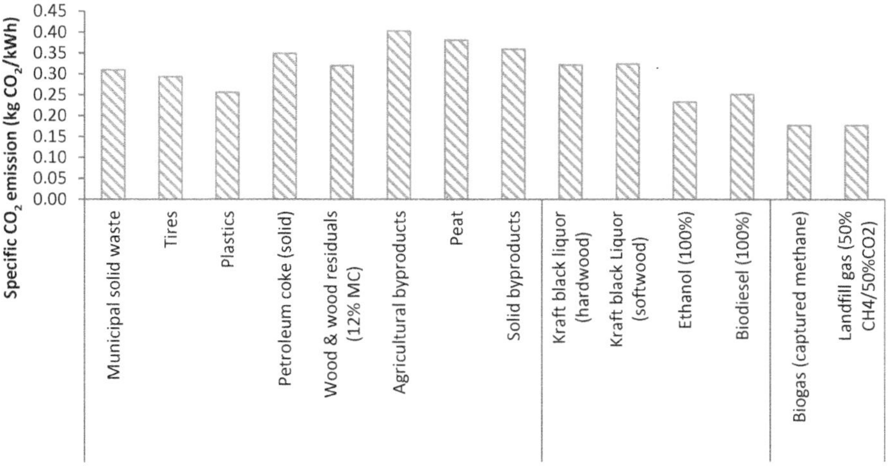

Fig. 75.14 CO_2 emission per unit energy from combustion for various types of biomass

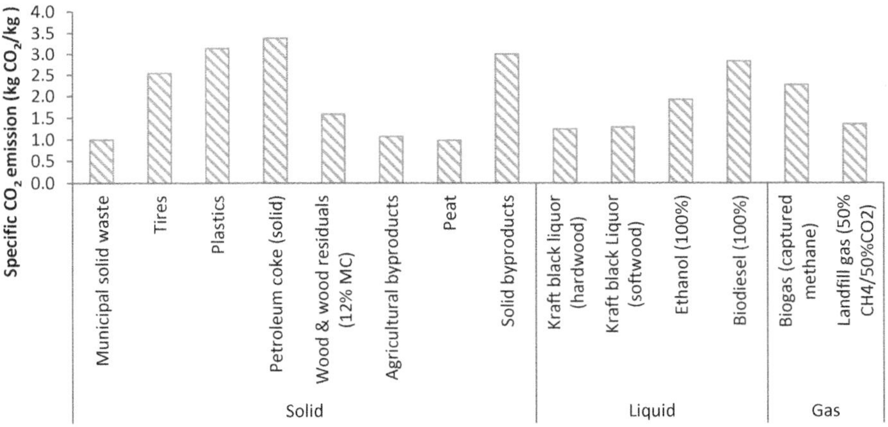

Fig. 75.15 CO_2 emission per unit mass for various types of biomass

4. Biomass is replaced with biogas prepared from biomass, which is directly used for space heating with a 90 % overall fuel-to-heat energy efficiency.

In option 1, biomass is converted to electricity, which is used for heating, and the total CO_2 emission per year is calculated. In option 2, biomass is directly used for heating and the CO_2 emission reduction per year compared to option 1 is calculated. In option 3, all the biomass is replaced with natural gas, which is used for direct heating, in order to observe the effect on the CO_2 emission reduction per year. The annual CO_2 emission reduction compared to option 2 by substituting natural gas for biomass is calculated. In option 4, biomass is replaced with biogas (prepared from

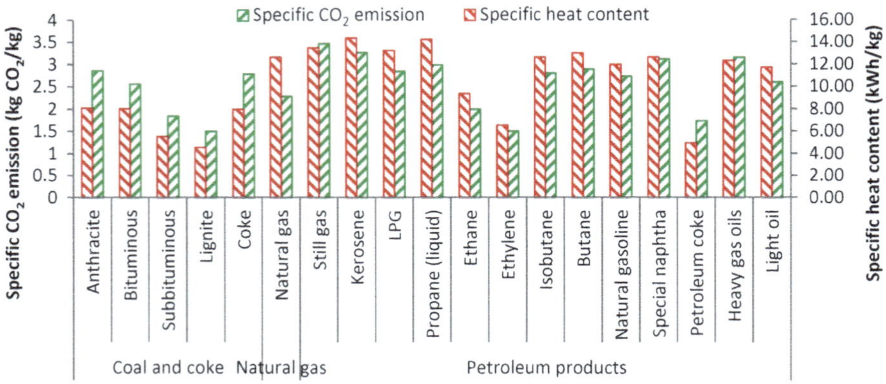

Fig. 75.16 Comparison of specific carbon contents and specific CO_2 emissions for various types of biomass, on mass bases

biomass) and the biogas is used for direct heating to assess the effect on CO_2 emission reduction per year. The four options are compared in terms of quantity of CO_2 emission per year (see Table 75.5). We determine the quantity of each type of biomass required per year, the CO_2 emissions per year from producing electricity for HVAC systems, the CO_2 emissions per year while using biomass directly for heating, and the reduction in CO_2 emissions per year from using biomass directly for heating.

It can be seen in Table 75.5 that the greatest quantity of biomass required is for agricultural by-product (11,300 kg/year) due to its low specific heat content, and the lowest quantity for biogas (2,340 kg/year) due to its high specific heat content. The reduction in CO_2 emissions per year when using biomass directly for heating is also calculated for various biomass types (see Fig. 75.17). The results show that CO_2 emissions per year can be greatly reduced by using biomass for direct heating. The largest reduction in annual CO_2 emissions occurs for agricultural by-product (18,400 kg/year) when it is used directly for heating in place of electric heating.

The reductions in CO_2 emissions per year are also determined from substituting various biomass types with natural gas and biogas in direct heating (see Fig. 75.18). The results show that CO_2 emissions per year can be greatly reduced by replacing agricultural by-product, peat and solid by-product with natural gas and biogas, and somewhat reduced by replacing other biomass types with natural gas and biogas. Natural gas is non-renewable energy source but the results nonetheless suggest that natural gas is more suitable than most biomass types for direct heating in terms of quantity of CO_2 emission. But in terms of cost and renewability, biomass is more suitable.

Table 75.5 Annual biomass requirement and CO_2 emission and emission reduction for various biomass options in a typical house

Source	Biomass	CO_2 emission (kg CO_2/kWh)	Biomass required (kg/year)	Annual CO_2 emissions to heat a typical house having heat load of 30,000 kWh/year			
				Option 1: CO_2 emission while producing electricity (kg/year)	Option 2: CO_2 emission reduced in direct heating (kg/year)	Option 3: CO_2 emission reduced by using natural gas in direct heating (kg/year)	Option 4: CO_2 emission reduced by using biogas in direct heating (kg/year)
Solid	Municipal solid waste	3.21	9,330	24,400	14,100	4,280	4,390
	Tyres	8.68	3,460	23,200	13,400	3,750	3,850
	Plastics	12.28	2,440	20,200	11,700	2,500	2,600
	Petroleum coke (solid)	9.69	3,100	27,600	15,900	5,620	5,730
	Wood and wood residuals (12 % MC)	4.97	6,040	25,300	14,600	4,640	4,750
	Agricultural by-products	2.67	11,300	31,800	18,400	7,410	7,520
	Peat	2.58	11,600	30,100	17,400	6,690	6,800
	Solid by-products	8.34	3,600	28,400	16,400	5,970	6,100
Liquid	Kraft black liquor (hardwood)	3.87	7,750	25,400	14,700	4,700	4,820
	Kraft black liquor (softwood)	3.95	7,600	25,600	14,800	4,790	4,900
	Ethanol (100 %)	8.24	3,640	18,400	10,700	1,750	1,860
	Biodiesel (100 %)	11.26	2,660	19,900	11,500	2,370	2,500

(continued)

Table 75.5 (continued)

Source	Biomass	CO_2 emission (kg CO_2/kWh)	Biomass required (kg/year)	Annual CO_2 emissions to heat a typical house having heat load of 30,000 kWh/year			
				Option 1: CO_2 emission while producing electricity (kg/year)	Option 2: CO_2 emission reduced in direct heating (kg/year)	Option 3: CO_2 emission reduced by using natural gas in direct heating (kg/year)	Option 4: CO_2 emission reduced by using biogas in direct heating (kg/year)
Gas	Biogas (captured methane)	12.80	2,340	14,000	8,100	−110	0.0
	Landfill gas (50 % CH_4/50 % CO_2)	7.65	3,920	1,4026.6	8,100	−110	0.0

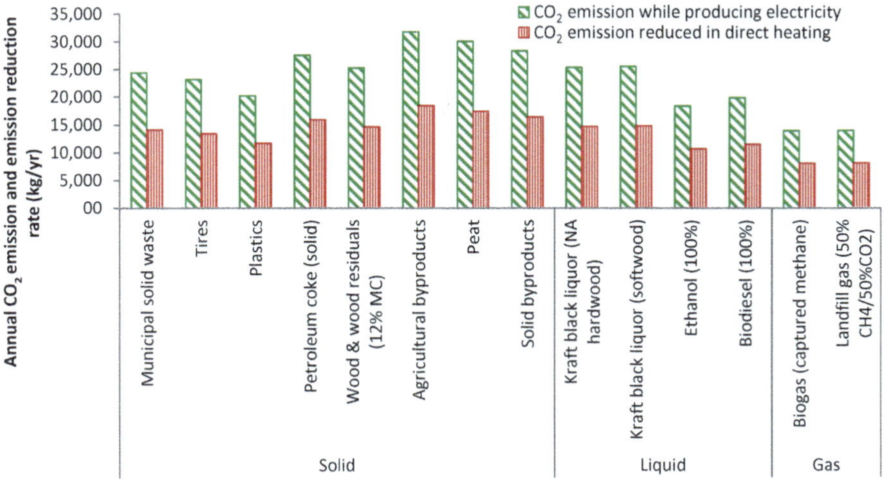

Fig. 75.17 Annual CO_2 emission from electricity production and CO_2 emission reduction in direct heating

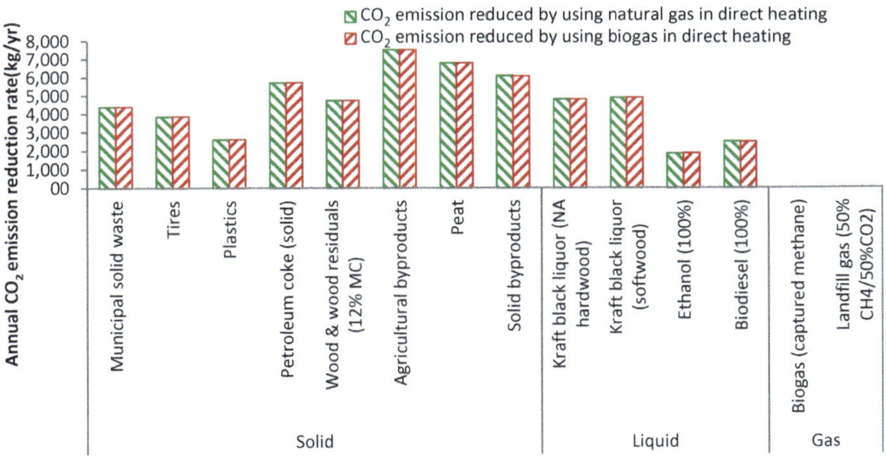

Fig. 75.18 Annual CO_2 emission reduction by substituting natural gas and biogas in place of different biomass sources, in direct heating

75.3 Comparative Study of Energy Consumption in Three Residential Buildings

The effects of changing the energy source for HVAC systems, and of building design, are investigated through a case study. In the case study, three different hypothetical buildings and several energy sources for them are considered.

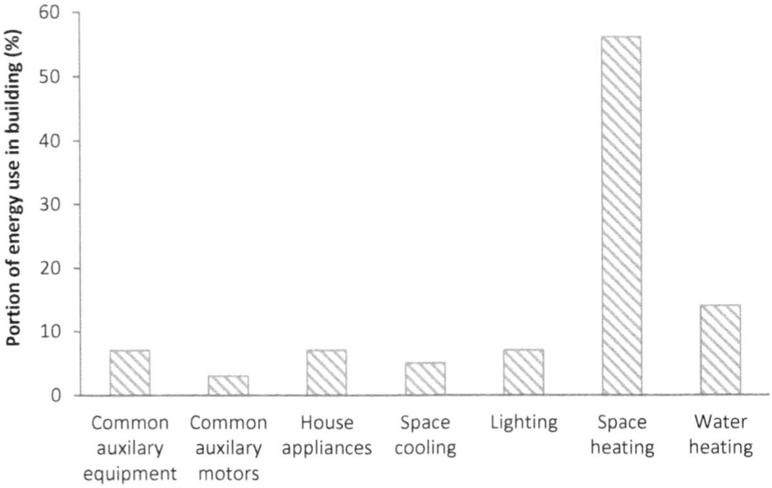

Fig. 75.19 Breakdown of energy use in an average residential building in the USA [Data from 20]

The effects of various combinations of building design and energy source on environmental impact measures are analysed.

For accurate and meaningful analyses, it is necessary to quantify for the buildings the energy requirements and their breakdown by application. Although the energy needs of a building vary spatially and with building characteristics, three typical buildings in USA are selected for the current study.

The electricity demand of each building is assumed to be constant (20 kW). A breakdown of the energy use in the building is shown in Fig. 75.19. For the present case study only HVAC loads of the buildings, i.e. space heating, space cooling and ventilation loads, are considered. These are 56, 5 and 9 % of the total load, respectively.

75.3.1 Description of Residential Buildings

Building 1 is taken to be an old building located where weak environmental standards are applied during design and construction. Poor insulation is utilized and the HVAC technology is old and inefficient. A coal fired thermal power plant is taken to be the electricity source for building 1 and the building design allows for no natural space heating and cooling. A heat pump with a low coefficient of performance (COP) is used for space heating and cooling.

Building 2 is taken to be a new building for which modern standards are followed during design and construction. Reflective materials are used with standard insulation levels to reduce radiative heat transfer losses. The building has a high efficiency HVAC system using modern technology. As for building 1, a coal fired thermal power plant is taken to be the electricity source and the building design allows for no natural heating and cooling. A heat pump with a high COP is used for heating and cooling.

Building 3 is taken to be a new building, for which modern standards are followed during design and construction. Reflective materials are used with standard insulation levels to reduce radiative heat transfer losses. The building has a high efficiency HVAC system using modern technology. In this case, the building design allows some natural heating and cooling (20 % heating, 15 % cooling) and full natural ventilation. The remaining heating and cooling load is supplied through a solar photovoltaic system, as shown in Fig. 75.20. A heat pump with a high COP is used for the remaining heating and cooling.

The parameter values used in the case studies for all three buildings are given in Table 75.6. A heat pump is used for both heating and cooling in all three building. The COP of the heat pump is considered for building 1 to be 2.5 in heating mode and 1.5 in cooling mode, and for building 2 and building 3 to be 5.0 in heating mode and 4.0 in cooling mode. The overall energy efficiency of the coal fired thermal power plant is taken to be 34 % and of the solar photovoltaic system is taken to be 22 %. The energy efficiency of the mechanical ventilation fan system is assumed to be 85 % in building 1 and 95 % in building 2. There is no mechanical ventilation fan system in building 3, as there is 100 % natural ventilation in the building. For CO_2, SO_2 and NO_x emissions, unit values per kW of electricity production reported in the literature for a coal fired power plant [21] are used.

The results of the examination of energy demand in the residential buildings (see Fig. 75.21) show that building 1 has the highest rate of electricity consumption

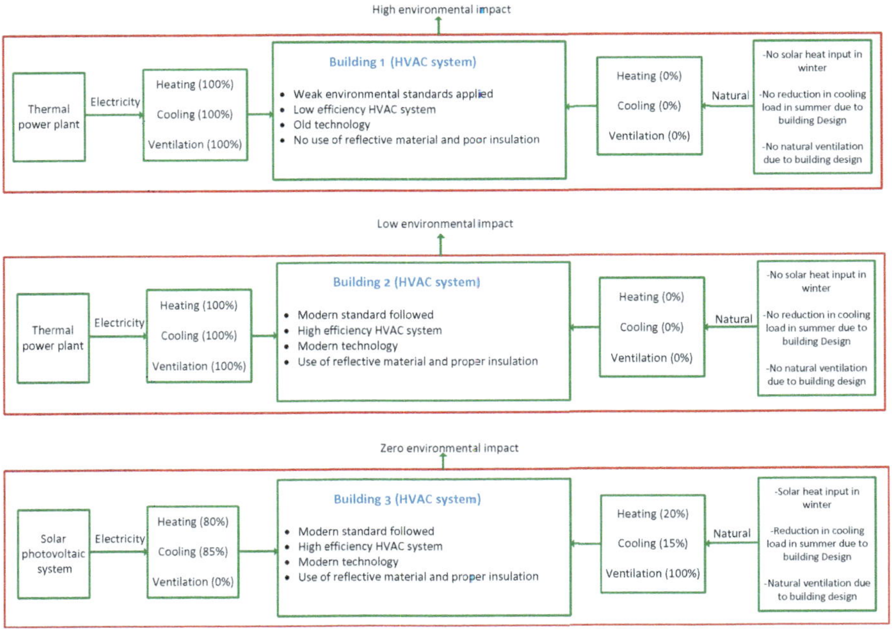

Fig. 75.20 Energy sources for HVAC systems for three residential buildings in the USA and their environmental impacts

Table 75.6 Total load requirements and other design parameters of three building case studies

Parameter	Building 1	Building 2	Building 3
Input parameters			
Electric load of building (kW)	20	20	20
Space heating load (kW)	11.20	11.20	11.20
Space cooling load (kW)	1.0	1.0	1.0
Ventilation load (kW)	1.80	1.80	1.80
COP of heat pump in heating mode	2.50	5.0	5.0
COP of heat pump in cooling mode	1.5	4.0	4.0
Energy efficiency of mechanical ventilation fan system (%)	85	95	N/A
Overall energy efficiency of thermal power plant (%)	34	34	N/A
Energy efficiency of solar photovoltaic power plant (%)	N/A	N/A	22
Output parameters			
Space heating load reduced through building design (kW)	0.0	0.0	2.20
Space cooling load reduced through building design (kW)	0.0	0.0	0.20
Ventilation load reduced through building design (kW)	0.0	0.0	1.80
Electricity demand of building HVAC (kW)	7.30	4.40	2.0
Actual energy required to produce electricity for HVAC (kW)	18.20	13.0	9.10
CO_2 emission rate (g/s)	1.90	1.10	0.0
SO_2 emission rate (g/s)	1.60	0.9	0.0
NO_x emission rate (g/s)	0.008	0.006	0.0

Noted that only operational CO_2, SO_2 and NO_x emissions rate are considered here, not full life cycle emissions
N/A not applicable

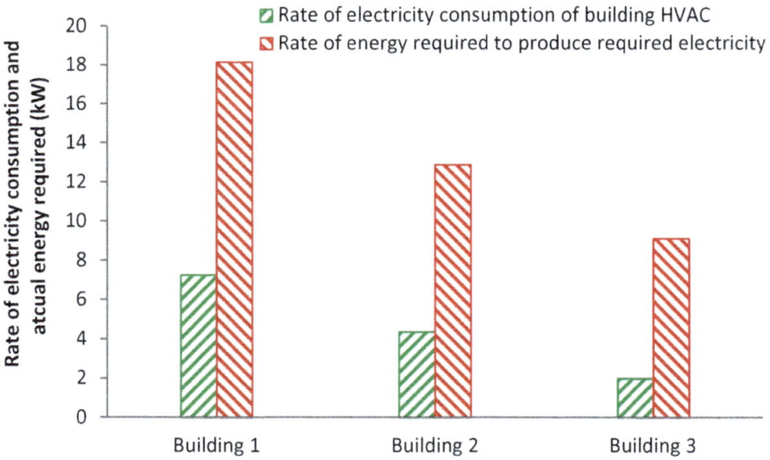

Fig. 75.21 Rate of electricity consumption and actual energy requirement in three buildings

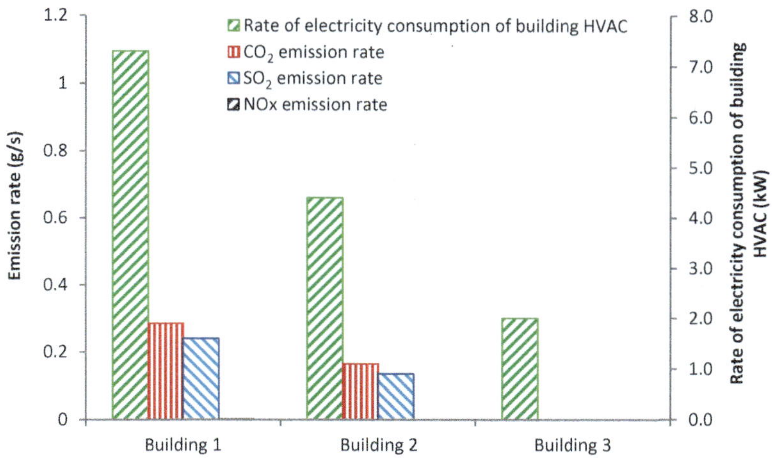

Fig. 75.22 Rate of electricity consumption of HVAC system of buildings and respective CO_2, SO_2 and NO_x emissions rates for the three buildings

(7.30 kW) whereas building 3 has the lowest (2.0 kW). The high rate of electricity use exhibited by building 1 mainly occurs because its heat pump has a low COP, while the rate of electricity consumption for building 3 is low because a high efficiency heat pump is used and natural heating, cooling and ventilation are incorporated. The differences in the overall energy efficiencies of the systems cause their actual energy demands to vary. Figure 75.21 shows that the energy rate needed to produce the required electricity is highest for building 1 (18.20 kW), based on the 38 % energy efficiency of the coal fired power steam power plant, and lowest for building 3 (9.10 kW). Note that this value is still a high proportion of the rate of electricity consumption in building 3. This is mainly due to the low overall energy efficiency of the solar photovoltaic system, which is taken to be 22 %.

Figure 75.22 shows for all three buildings considered the rate of electricity consumption and the CO_2, SO_2 and NO_x emissions rate. The rate of CO_2 emission is observed in Fig. 75.22 to be 1.90 g/s for building 1, 1.10 g/s for building 2 and 0.0 g/s for building 3. Similarly, the SO_2 emission rate is observed to be 1.6 g/s for building 1, 0.9 g/s for building 2 and 0.0 g/s for building 3. The emissions rates are much lower for NO_x than CO_2 and SO_2 for building 1 and 2, and zero for building 3. Figure 75.22 shows that building 3 has a zero emission rate for CO_2, SO_2 and NO_x, and this is because all the electricity demand in building 3 is fulfilled through the solar photovoltaic system and the building design.

75.4 Conclusions

Various non-renewable and renewable energy sources available for HVAC systems are investigated. Heat content, carbon content and CO_2 emissions for selected fuels are examined. CO_2 emissions are calculated considering both conventional and renewable sources of energy. Also, CO_2 emissions are compared in electricity conversion and in direct heating. The following conclusions are drawn from the results:

- Considering non-renewable energy sources, the maximum CO_2 emission rate while producing electricity is associated with anthracite (27,900 kg/year) and minimum with natural gas (14,300 kg/year).
- CO_2 emissions can be greatly reduced by replacing coal, coke and petroleum coke with natural gas.
- Considering biomass energy sources, the maximum CO_2 emission rates while producing electricity are associated with agricultural by-products (31,800 kg/year) and minimum with biogas (14,000 kg/year).
- CO_2 emissions can be greatly reduced by replacing agricultural by-product, peat and solid by-product with natural gas and biogas, because natural gas and biogas have a higher specific heat content than other biomass. A larger reduction in CO_2 emissions is found for the case in which biogas is substituted for natural gas.
- Environmental impact in terms of CO_2, SO_2 and NO_x emissions can be significantly reduced by using high efficiency HVAC systems and modern technology having lower energy consumption.
- Environmental impact can be reduced to zero, for the operational period of energy system, by using natural heating, cooling and ventilation and renewable energy sources.

Based on the case study the authors suggest use of highly efficient HVAC system with modern technology. Building design can permit reductions in HVAC electricity consumption through incorporation of natural heating, cooling and ventilation. At the same time economic analysis of such highly efficient system, modern technology and building design is required.

References

1. Saidur R, Masjuki HH, Jamaluddin MY (2007) An application of energy and exergy analysis in residential sector of Malaysia. Energy Policy 35(2):1050–1063
2. U.S. Energy Information Administration (EIA) (2012) Annual energy review 2011. DOE/EIA-0384. http://www.eia.gov/totalenergy/data/annual/pdf/aer.pdf. Accessed 11 June 2014
3. U.S. Energy Information Administration (EIA) (2012) Residential energy consumption survey 2009. http://www.eia.gov/consumption/residential/data/2009/. Accessed 11 June 2014
4. U.S. Energy Information Administration (EIA) (2011) State energy data 2011: consumption, DOE/EIA-0214. http://www.eia.gov/state/seds/sep_use/notes/use_print.pdf. Accessed 10 June 2014
5. Mardookhy M (2013) Energy efficiency in residential buildings in Knoxville. In: University of Tennessee at Knoxville (ed) Industrial engineering. University of Tennessee at Knoxville, Knoxville, TN
6. Painuly JP, Park H, Lee MK, Noh J (2003) Promoting energy efficiency financing and ESCOs in developing countries: mechanisms and barriers. J Clean Prod 11(6):659–665
7. Mardookhy M, Sawhney M, Ji R, Zhu X, Zhou W (2014) A study of energy efficiency in residential buildings in Knoxville, Tennessee. J Clean Prod 85:241–249

8. Balta MT, Dincer I, Hepbasli A (2010) Performance and sustainability assessment of energy options for building HVAC applications. Energ Buildings 42(8):1320–1328
9. Prek M (2004) Environmental impact and life cycle assessment of heating and air conditioning systems, a simplified case study. Energ Buildings 36(10):1021–1027
10. Heikkilä KN (2004) Environmental impact assessment using a weighting method for alternative air-conditioning systems. Build Environ 39(10):1133–1140
11. Chiavetta C, Tinti F, Bonoli A (2011) Comparative life cycle assessment of renewable energy systems for heating and cooling. Proceed Eng 21:591–597
12. Koroneos C, Tsarouhis M (2012) Exergy analysis and life cycle assessment of solar heating and cooling systems in the building environment. J Clean Prod 32:52–60
13. Chidambaram LA, Ramana AS, Kamaraj G, Velraj R (2011) Review of solar cooling methods and thermal storage options. Renew Sustain Energy Rev 15(6):3220–3228
14. The Climate Registry: U.S. (2013) 2013 Climate registry default emission factors, table 12.1 U.S. Default factors for calculating CO_2 emissions from fossil fuel and biomass combustion. U.S.
15. International Energy Agency (2010) Energy technology perspectives, scenarios & strategies to 2050. International Energy Agency, Paris
16. International Energy Agency (2013) 2013 Key world energy statistics. International Energy Agency, Paris
17. Rosen MA, Sy E, Gharghouri P (1996) Substituting natural gas heating for electric heating: assessment of the energy and environmental effects in Ontario. Energ Stud Rev 8(2):143–154
18. IPCC (2011) Special report on renewable energy sources and climate change mitigation. Cambridge University Press, New York, NY
19. Anselmo FP, Badr O (2004) Biomass resources for energy in North-Eastern Brazil. Appl Energ 77(1):51–67
20. U.S. Energy Information Administration (EIA) (2001) Residential energy consumption survey. http://www.eia.gov/consumption/residential/data/2001/. Accessed 11 June 2014
21. Mittal ML, Sharma C, Singh R (2012) Estimates of emissions from coal fired thermal power plants in India. In: 20th Emission inventory conference 2012, Tampa, FL

Index

I. Dincer et al. (eds.), *Progress in Clean Energy, Volume 2*,
DOI 10.1007/978-3-319-17031-2

H

R

GPSR Compliance

The European Union's (EU) General Product Safety Regulation (GPSR) is a set of rules that requires consumer products to be safe and our obligations to ensure this.

If you have any concerns about our products, you can contact us on ProductSafety@springernature.com

In case Publisher is established outside the EU, the EU authorized representative is:

Springer Nature Customer Service Center GmbH
Europaplatz 3
69115 Heidelberg, Germany

Batch number: 10761746

Printed by Printforce, the Netherlands